Phytopathogenic Bacteria

The field of Plant Bacteriology/Phytobacteriology is of great importance as it is rapidly changing and advancing due to recent advances in genomics and molecular plant bacteriology, as well as the global spread of bacterial plant diseases and the emergence of new bacterial plant pathogens. So, there is a need to integrate the understanding of bacterial taxonomy, genomics and basic plant bacteriology that reflects state-of-the-art knowledge about plant-disease mechanisms. In the present book, the first eight chapters contain general information about bacteria, the classification of plant pathogenic bacteria and the molecular mechanisms of virulence and pathogenesis of phytopathogenic bacteria. The ninth chapter of the book contains a description of 62 specific bacterial diseases of plants. Some of the described diseases, such as bacterial blight of rice, wilt of solanaceous plants, soft rot of fleshy fruits and vegetables, citrus canker, citrus huanglongbing etc., are of international importance and cause huge economic losses to mankind.

This volume

- Contains the latest, updated information on plant pathogenic bacteria as of December 2025.
- Presents the classification of the bacteria and associated nomenclature based on *Bergey's Manual of Systematic Bacteriology*, 2nd edition and the *International Journal of Systematic and Evolutionary Microbiology*. Seventy-three genera of bacteria containing plant pathogenic species are described, and their plant pathogenic species are listed.
- Discusses practical and thoroughly tested disease management strategies that would help control the enormous losses caused by these diseases.
- Reviews the role of Type I-VI secretions and peptide- or protein-containing toxins produced by bacterial plant pathogens.
- Briefs about plants and plant products that act as carriers of human enteric bacterial pathogens, emphasizing, for example, the role of seed sprouts as a common vehicle in causing food-borne illness.

Dr B. S. Thind was Ex-Professor-cum-Head, Department of Plant Pathology, Punjab Agricultural University, Ludhiana. He has 34 years of experience in teaching, research and transfer of technology in the subdiscipline of plant bacteriology. He has conducted research investigations on bacterial blight of rice, bacterial stalk rot of maize, bacterial blight of cowpea, bacterial leaf spot of green gram, bacterial spot of chillies and bacterial soft rot of potatoes. He also acted as Principal Investigator of two ICAR (Indian Council of Agricultural Research)-funded research schemes entitled, "Detection and control of phytopathogenic bacteria from cowpea and mungbean seeds" from 1981 to 1986 and "Perpetuation, variability and control of Xanthomonas oryzae pv. oryzae, the causal agent of bacterial blight of rice" from 1989 to 1993, and also a DST-funded research scheme entitled "Biological control of bacterial blight, sheath blight, sheath rot and brown leaf spot of rice" from 1999 to 2002. He has also authored a manual entitled *Plant Bacteriology* and a textbook entitled *Phytopathogenic Procaryotes and Plant Diseases*, published by Scientific Publishers (India) in 2012 and another book, entitled *Phytopathogenic Bacteria and Plant Diseases*, published by CRC Press, Taylor & Francis Group in 2019. He is a life member of the Indian Phytopathological Society, the Indian Society of Mycology and Plant Pathology, the Indian Science Congress and the Indian Society of Plant Pathologists.

Phytopathogenic Bacteria
An Imminent Threat to Sustainable Food Production

B. S. Thind

CRC Press
Taylor & Francis Group
Boca Raton London New York

CRC Press is an imprint of the
Taylor & Francis Group, an **informa** business

Designed cover image: Cover image designed by Dr. Parminder Singh Tak, Department of Plant Pathology, Punjab Agricultural University, Ludhiana

First edition published 2027
by CRC Press
2385 NW Executive Center Drive, Suite 320, Boca Raton FL 33431

and by CRC Press
4 Park Square, Milton Park, Abingdon, Oxon, OX14 4RN

CRC Press is an imprint of Taylor & Francis Group, LLC

Library of Congress Cataloging-in-Publication Data
Names: Thind, B. S. author
Title: Phytopathogenic bacteria : an imminent threat to sustainable food production / B.S. Thind.
Description: First edition. | Boca Raton, FL : CRC Press, 2026. | Includes bibliographical references and index.
Identifiers: LCCN 2025048247 (print) | LCCN 2025048248 (ebook) | ISBN 9781041159148 hardback | ISBN 9781041159209 paperback | ISBN 9781003681878 ebook
Subjects: LCSH: Phytopathogenic bacteria | Bacterial diseases of plants
Classification: LCC QR100 .T485 2026 (print) | LCC QR100 (ebook)
LC record available at https://lccn.loc.gov/2025048247
LC ebook record available at https://lccn.loc.gov/2025048248

ISBN: 978-1-041-15914-8 (hbk)
ISBN: 978-1-041-15920-9 (pbk)
ISBN: 978-1-003-68187-8 (ebk)

DOI: 10.1201/9781003681878

Typeset in Times
by SPi Technologies India Pvt Ltd (Straive)

Dedication

For Mehar, my granddaughter-
the most precious flower
in the garden of our lives.

Contents

SECTION I General Aspects

SECTION II Specific Plant Diseases

About the Author

The author has 55 years' experience into the research, teaching and transfer of technology in plant pathology. Retired as Professor and Head of the Department of Plant Pathology, Punjab Agricultural University, Ludhiana, India in 2003 and then served there as Emeritus Scientist (ICAR, New Delhi, India) from July 2003 to July 2005. Thind has taught various courses in plant pathology and plant bacteriology disciplines, including advanced courses in plant bacteriology. To date, he has guided 10 PhD and 13 MSc. students in the field of plant bacteriology. He has published more than 125 research papers in reputed national and international research journals related to plant pathology and plant bacteriology.

Thind is a life member of five professional societies and has held different positions in the executive committees of these societies. He attended the VIIth International Conference on Plant Pathogenic Bacteria held in Budapest, Hungary in 1989 and the VIIIth International Conference on Plant Pathogenic Bacteria held in Versailles, France in 1992. As a Member of the National Organising Committee, he helped to organise the IXth International Conference on Plant Pathogenic Bacteria held in Chennai, India in 1996, at which he also chaired one session in this conference. Dr Thind has chaired/co-chaired various sessions of different professional societies, the Jeersannidhi Lecture Award and was a member of the panel of judges for the Prof. M. J. Narasimhan Academic Award Contest. Over the course of his long career, Thind has attended various national/international symposia, conferences, meetings and workshops of different professional societies.

He has acted as a paper setter and external examiner of four universities and also for the Agricultural Scientists' Recruitment Board, New Delhi. He has also acted as an expert of selection committees for different posts and as an external examiner of 17 Ph. D. and 10 M. Sc. students for different universities. Acted as reviewer/referee of various national and international research journals.

Dr B. S. Thind

Abbreviations of Bacterial Genera Used in the Book

Genus	Abbreviation	Genus	Abbreviation
Acetobacter	*Ace.*	*Listeria*	*Lis.*
Achromobacter	*Achr.*	*Lonsdalea*	*Lon.*
Acidovorax	*Aci.*	*Methanothermus*	*Met.*
Acinetobacter	*Acin.*	*Musicola*	*Mus.*
Affinibrenneria	*Aff.*	*Neisseria*	*Nes.*
Agrobacterium	*Agr.*	*Nocardia*	*Noc.*
Allorhizobium	*All.*	*Pantoea*	*Pan.*
Arsenophonus	*Ars.*	*Paenibacillus*	*Paen.*
Arthrobacter	*Art.*	*Paraburkholderia*	*Parab.*
Bacillus	*Bac.*	*Paracidovorax*	*Parac.*
Bacterium	*Bact.*	*Pectobacterium*	*Pec.*
Bradyrhizobium	*Bra.*	*Phaseolibacter*	*Phas.*
Brenneria	*Bre.*	*Phytomonas*	*Pht.*
Burkholderia	*Bur.*	*Providencia*	*Pro.*
Campylobacter	*Cam.*	*Pseudoalteromonas*	*Psa.*
'*Candidatus* Liberibacter'	'*Can.* Lib.'	*Pseudomonas*	*Psm.*
'*Candidatus* Phlomobacter'	'*Can.* Phl.'	*Ralstonia*	*Ral.*
'*Candidatus* Phytoplasma'	'*Can.* Phy.'	*Rathayibacter*	*Rat.*
Citrobacter	*Cit.*	*Raoultella*	*Rao.*
Clavibacter	*Cla.*	*Rhizobacter*	*Rhib.*
Clostridium	*Clo.*	*Rhizobium*	*Rhiz.*
Corynebacterium	*Cor.*	*Rhizorhapis*	*Rhir.*
Curtobacterium	*Cur.*	*Rhodococcus*	*Rho.*
Dickeya	*Dic.*	*Robbsia*	*Rob.*
Enterobacter	*Entb.*	*Salmonella*	*Sal.*
Enterococcus	*Entc.*	*Samsonia*	*Sam.*
Erwinia	*Erw.*	*Sanguibacter*	*San.*
Escherichia	*Esc.*	*Serratia*	*Ser.*
Ewingella	*Ewi.*	*Sphingobacterium*	*Sphb.*
Gibbsiella	*Gib.*	*Sphingomonas*	*Sph.*
Gluconobacter	*Glu.*	*Spiroplasma*	*Spr.*
Herbaspirillum	*Her.*	*Staphylococcus*	*Stp.*
Janibacter	*Jani.*	*Stenotrophomonas*	*Stn.*
Janthinobacterium	*Jant.*	*Streptococcus*	*Stc.*
Klebsiella	*Kle.*	*Streptomyces*	*Str.*
Kosakonia	*Kos.*	*Tatumela*	*Tat.*
Lactobacillus	*Lactob.*	*Thermus*	*Thm.*
Lactococcus	*Lactoc.*	*Xanthomonas*	*Xan.*
Leifsonia	*Lef.*	*Xylella*	*Xyl.*
Leucobacter	*Leb.*	*Xylophilus*	*Xylo.*
Leuconostoc	*Len.*	*Yersinia*	*Yer.*
Liberibacter	*Lib.*		

Preface

The study of phytopathogenic bacteria, and the plant diseases caused by them, is of the utmost importance to mankind, as these diseases cause huge economic losses. Some of these diseases, such as bacterial wilt of solanaceous plants, soft rot of fleshy vegetables and fruits, crown gall of plants, citrus canker, citrus huanglongbing, fire blight of pome fruit trees and bacterial blight of rice, are internationally significant. Plant diseases caused by phytopathogenic bacteria are also disastrous because many of these diseases cannot be effectively controlled due to the lack of effective chemicals, in contrast to the diseases caused by fungi. Moreover, the secondary spread of bacterial diseases is very fast compared to the diseases caused by fungi, resulting in great economic damage.

The classification and nomenclature of bacteria has undergone a sea change over the past four decades. The sequence analysis of highly conserved regions of the bacterial genome, such as the small subunit rRNA gene, has provided very valuable information concerning numerous taxonomic changes and the realignment of different taxonomic groups. The sequencing of genomes of many phytopathogenic bacteria has certainly stabilised their ever-fluctuating nomenclature and taxonomy. Since the publication of the 1st edition of *Bergey's Manual of Systematic Bacteriology* in 1984, the number of published bacterial species has more than tripled, i.e., more than 390 new genera and 2200 new species have been described. Many new genera containing plant pathogenic bacteria have been recognised, and the rearrangement of the species among the genera has also occurred. During the 1970s, there were only 6 genera of bacteria that contained plant pathogenic bacteria; this number has now risen to 71, including 3 candidatus categories.

In contrast to direct monetary losses and other harmful effects of phytopathogenic bacteria, they have a great deal of value. They are used for the commercial production of many industrial products, as well as tools for various scientific studies. Their use in industrial fermentation, and the production of antibiotics, xanthan gums and polysaccharides such as curdlans are all important applications. The production of pectin-degrading enzymes produced by soft-rotting erwinias and restriction nucleases produced by *Agrobacterium* and *Xanthomonas* species are other notable examples of their commercial use. *Gluconobacter* species, similarly, have a range of biotechnological applications. Their biotransformed strains are used for the production of rare and special sugars like l-ribose and miglitol, both very promising pharmaceutical lead molecules. *Agr. tumefaciens* (the causal agent of crown gall of many plants), apart being a plant pathogen, is one of the best-known vehicles for gene transfer and a most useful genetic engineering tool for developing transgenics.

Our knowledge of phytopathogenic bacteria, the plant diseases caused by them, host-pathogen interactions, molecular mechanisms involved in virulence and pathogenesis and disease management strategies developed, has now reached a stage where it would be useful to integrate the concepts and ideas into a book. Hence the production of this volume: *Phytopathogenic Bacteria: An Imminent Threat to Sustainable Food Production.*

This volume contains nine chapters. Chapter 1 focuses on the economic importance of phytopathogenic bacteria and a historical review of phytobacteriology. Chapter 2 contains basic information about the bacteria and the bacterial cell. Chapter 3 deals with variations in bacteria. Chapter 4 contains information on the evolution of phytopathogenic bacteria. Chapter 5 deals with the diagnosis of bacterial diseases of plants, and the up-to-date classification of plant pathogenic bacteria is given in Chapter 6. The bacterial names included are based on information published in the *International Journal of Systematic Bacteriology* (*IJSB*), the *International Journal of Systematic and Evolutionary Microbiology* (*IJSEM*), and related journals up to, and including, the June issues of 2025 and supplemented with the information given in *Names of Plant Pathogenic Bacteria, 1864 - 2004* prepared by Young *et al.* (2004), *Comprehensive List of Names of Plant Pathogenic Bacteria, 1980 - 2007* prepared by Bull *et al.* (2010), *List of New Names of Plant Pathogenic Bacteria (2008 - 2010)* prepared by Bull *et al.* (2012) and *List of New Names of Plant Pathogenic Bacteria (2011 - 2012)* prepared by Bull *et al.* (2014). Chapter 7 deals with information on the molecular mechanisms of phytopathogenic bacteria involved in virulence and pathogenesis. Chapter 8 then deals with a new topic, i.e., plants acting as carriers of human enteric bacterial pathogens. It cautions consumers to decontaminate plant and plant products before human consumption.

The core of the book, Chapter 9, contains descriptions of specific bacterial plant diseases. Although it was not possible to include all the bacterial plant diseases, the chapter contains descriptions of 62 specific diseases, including economically important diseases of international importance and representative diseases of different crops and regions. To familiarise the readers with the symptoms of different diseases, coloured photographs of the symptoms of these diseases are given. Information published in plant pathological and related biological journals up to June issues of 2025 have been included in the book.

The abbreviations of names of bacterial genera, included in the book, are given on pages xiii. As many of the genera included in this book begin with the same letter, a deviation has been made from the tradition of abbreviating the genus name by using only the first letter of the genus. In

Chapter 9, the names of causal agents of different diseases are given along with author citations. In author citations, when there are more than two authors, the name of the first author is given followed by "*et al.*" One or two, and in a few cases, up to three recent synonyms of each causal organism are also given. With regard to the citation of references, the names of all the authors are given when there are five authors in a reference. When there are more than five authors in a reference, the names of the first five authors are given followed by "*et al.*" The names of all the authors/editors are given for book references.

At the undergraduate level, students have limited exposure to phytobacteriology. However, at the postgraduate level, students are taught phytobacteriology in detail. State agricultural universities and the universities teaching biological sciences have courses dealing exclusively with phytobacteriology. PhD students specialising in phytobacteriology are taught advanced courses. This book will serve as a textbook for advanced undergraduate and postgraduate students and also as a reference book for research workers engaged in the discipline of plant pathology and phytobacteriology.

I am indebted to numerous colleagues and friends for suggestions made during the preparation of the manuscript. Also, in the preparation of this manuscript, I have referred to a number of publications, and I would like to acknowledge all the authors of these excellent and valuable publications/articles.

There might be some errors, mistakes and shortcomings in this publication. I will greatly appreciate healthy criticism and suggestions for the improvement of this publication, to be published in future.

B. S. Thind

Acknowledgements

It is my pleasure to acknowledge the help of all those who contributed and helped me in the preparation of this publication.

I am greatly indebted to Dr Krishan Kumar Jindal, formerly Associate Professor of Plant Pathology at Dr Y. S. Parmar University of Horticulture and Forestry, Nauni, Solan, Himachal Pradesh, who introduced me to the profession of book writing. The omission of his acknowledgement in my first book, *Phytopathogenic Procaryotes and Plant Diseases* (2012), is sincerely regretted. The frequent counselling and help rendered by Dr G. S. Miglani, formerly Professor of Genetics, Punjab Agricultural University (PAU), Ludhiana, is gratefully acknowledged. I am also indebted to the late Dr J. M. Young, Landcare Research, Auckland, New Zealand and Dr N. W. Schaad, ARS-USDA, Foreign Disease-Weed Science Research Unit, Ft. Detrick, MD, USA for their expert advice and frequent counselling. My sincere thanks are also due to Professor Ian K. Toth, The James Hutton Institute Invergowrie, Dundee UK. I am also grateful to my friend and ex-colleague, Dr R. S. Kahlon, Ex-Prof-cum Head, Department of Microbiology, PAU Ludhiana, for his guidance, suggestions and encouragement.

I am greatly indebted to Dr Mandeep S. Hunjan, Principal Scientist, Plant Bacteriology, Department of Plant Pathology, PAU, Ludhiana, for his day-to-day consultations and guidance in the finalisation of this book. The generous help rendered by my ex-colleagues and friends, namely Dr T. S. Thind, ex-Senior Plant Pathologist-cum-Head, Department of Plant Pathology, PAU, Ludhiana, Dr Bipen Kumar, ex-Senior Plant Pathologist, and Dr S. K. Thind, ex-Professor of Plant Protection, members of the Plant Pathology Faculty, PAU, Ludhiana; Dr R. P. Maharshi, Ex-Head, Department of Plant Pathology, SKN College of Agriculture, Jobner, Rajasthan, Dr Kuldip Singh and the late Dr Ram Kishun in lending me the photographs of the symptoms of different plant diseases is sincerely acknowledged. I owe my special thanks to Dr Parminder Singh Tak, Plant Pathologist, Department of Plant Pathology, PAU, who helped me immensely in downloading a large number of pictures and photographs from the internet and also guided me in the editing and arranging of these photographs. I owe a special gratitude to my wife, Mrs. Ravinder Kaur Thind, for her continuous and untiring encouragement throughout this entire period. My sincere thanks are also to those many others who have helped me, directly or indirectly, in the preparation of this manuscript, but whose names have not been mentioned here.

B. S. Thind

Section I

General Aspects

1 Introduction

1.1 PHYTOBACTERIOLOGY

Phytobacteriology, or plant bacteriology, is a subdiscipline of plant pathology, which deals with plant-associated bacteria and their interactions with each other and with their hosts (Dumenyo *et al.*, 2001). This subdiscipline covers all aspects of plant-associated bacteria, including both their basic and applied nature. The basic aspects comprise the morphology, physiology, taxonomy, genetics, and serology of the bacteria. Applied aspects of this subdiscipline include host-pathogen interactions, including the pathogenesis and infection process, epiphytology, the effect of environment on disease development and disease management. In the past, the emphasis has been mainly on the investigation of plant diseases caused by bacteria on account of the heavy losses these cause. Lately, however, the field of phytobacteriology has expanded considerably to include: beneficial bacterial-plant interactions such as nodulation and nitrogen fixation; the promotion of plant growth by phyllosphere, rhizosphere and soil-inhabiting bacteria; and the control of ice-nucleation active bacteria by phyllosphere bacteria. Phytobacteriology also includes the management and control of bacterial plant diseases using various methods such as cultural practices, resistant cultivars, chemical and biological control and molecular genetics. A basic knowledge of various branches of science such as microbiology, genetics, chemistry, biochemistry, molecular biology, plant physiology, agronomy and soil science is essential for a better understanding of plant diseases caused by microorganisms. Molecular biology is now becoming increasingly important in every biological science, including phytobacteriology.

To be pathogenic in a host, the bacteria must sense the presence of a host and activate virulence mechanisms while outcompeting host microbiota and evading host immunity. Bacteria likely sense multiple inputs, ranging from changes in nutrient availability to host immune signals. At present, there is little understanding of how bacteria sense a host, and how they integrate multiple aspects of a host environment and adapt to promote their own survival. Hamilton and Haney (2023) investigated how two-component systems integrate environmental signals additively or synergistically to aid the transition to a pathogenic lifestyle. They used the genetically pliable model opportunistic pathogen, *Pseudomonas aeruginosa*. *Psm. aeruginosa* causes a broad range of opportunistic diseases, but, more commonly, it lives a commensal lifestyle on plants and animals. They used a Tn7 *Psm. aeruginosa* library to screen for the loss of pathogenicity on an *Arabidopsis* host and identified two-component systems involved in shifting from a commensal to a pathogenic lifestyle. Among others, they found ColR/S to be a confirmation that their screen-identified gene related to host association. The authors are currently characterising the host signals responsible for regulating the identified two-component systems. They believe that investigation of *Psm. aeruginosa*'s host responses will identify conserved virulence determinants and symbiotic mechanisms. This will contribute to our understanding of why otherwise commensal bacteria become opportunistic pathogens. In addition, the researchers can use this knowledge to interfere with opportunistic pathogens' plastic responses to control their virulence in a sustainable manner.

1.2 PROCARYOTES

On the basis of cellular organisation, the living organisms are divided into two groups, namely, procaryotes and eucaryotes. The procaryotes have a specific set of characteristics, which differentiate them from eucaryotes. The main characteristic of the procaryotes is that, with a few exceptions, nucleoplasm or genophore (formerly known as the nucleoid) is not separated from the cytoplasm by a unit-membrane system, i.e., the nuclear membrane is absent. Other characteristic features are the presence of 70S ribosomes (with the exception of archaea, whose ribosomes have a slightly higher S value) dispersed in the cytoplasm, no cytoplasmic streaming and an absence of endoplasmic reticulum with attached ribosomes. They take the nutrients in liquid form. They possess a rigid cell wall, with the exception of one group, i.e., **mollicutes**. These occur either as single cells or as a simple association of similar cells. The cells may be motile or nonmotile; in the case of motile, the motility is mediated by bacterial flagella or gliding motility on solid surface.

The procaryotes inhabit moist environments and predominantly occur as unicellular microorganisms. However, these also occur in other forms, such as filaments, mycelia, or colonies. In some cases, differentiating structures such as holdfasts and resting cells are formed. The gene transfer and recombinations occur but in these processes gametogenesis and zygote formation never occurs.

The above given characteristics of procaryotes are sufficient to distinguish them from eucaryotes, but they do share some common features. For example, the hyphae formed in actinomycetes appear similar to those of moulds; the ability of spirochetes to twist and contort their shape matches the flexibility shown by certain protozoa. Some large bacteria may approach the size of eucaryotic cells while some eucaryotic cells are as small as bacteria. However, procaryotic and eucaryotic cells can be easily differentiated on the basis of fluorescent-labelled genes.

DOI: 10.1201/9781003681878-2

The procaryotes comprise two different groups, namely, the **bacteria** and the **archaea**, which have been recognised as different domains for the purpose of classification (Krieg, 2001).

1.2.1 Bacteria

On the basis of Gram-stain-reaction and the presence or absence of a cell wall, the bacteria have been divided into the following three phenotypic subgroups:

1. Gram-stain-negative bacteria having cell walls
2. Gram-stain-positive bacteria having cell walls
3. Bacteria lacking cell walls

1.2.1.1 Gram-stain-negative Bacteria Having Cell Walls

These bacteria possess a Gram-stain-negative type of cell wall, which is composed of an outer membrane, and an inner relatively thin peptidoglycan layer containing muramic acid; the latter may be absent in a few organisms. The reproduction is usually by binary fission; some groups show budding and a very few divide by multiple fission. The cells may occur in various shapes, i.e., spheres, ovals, straight or curved rods, helicals, or filaments. The cells may be non-motile or motile showing swimming or gliding motility. With regard to oxygen requirement, the members may be either aerobic, anaerobic, facultatively anaerobic, or microaerophilic.

1.2.1.2 Gram-stain-positive Bacteria Having Cell Walls

These bacteria have a Gram-stain-positive type of cell wall, in which there is no outer membrane and the peptidoglycan layer is relatively thicker than that of the Gram-stain-negative type of cell wall. The cell shapes include spheres, rods or filaments; the rods and filaments may show true branching. The reproduction is mainly by binary fission, but actinomycetes and their relatives produce sporogenous hyphae. Some members produce spores, also called **endospores**, which help to tide over the adverse conditions but are not a mean of multiplication. The oxygen requirement of the members is the same as that of Gram-stain-negative bacteria.

1.2.1.3 Bacteria Lacking Cell Walls

These are commonly called **"mycoplasmas"** (and also known as mollicutes or soft-skin microorganisms). These lack a cell wall, but are enclosed by a unit membrane, the plasma membrane. As they have no cell wall (peptidoglycan layer), these are not sensitive to β-lactam antibiotics or other antibiotics, which inhibit cell wall synthesis. The members are highly pleomorphic, and have different cell forms which vary in size; some of these forms are filterable. Cells stain Gram-stain-negative, no resting forms are produced, and usually, these are non-motile; however, some members show gliding motility. Different modes of reproduction occur, such as, budding, fragmentation and/or binary fission. Most members require complex media and all require cholesterol or related sterols for growth. The typical colony under adequate growth conditions is biphasic with a **fried-egg appearance**, i.e., it consists of an opaque, granular central area which grows down into the medium as well as a flat, translucent peripheral zone. Mol% G + C contents of r RNA (43-48) and of DNA (23-46) are lower than that of walled bacteria.

1.2.2 Archaea

Archaea, predominantly inhabitants of terrestrial and aquatic habitats, are found in anaerobic, hypersaline or hydrothermally and geothermally heated environments. Some members also live symbiotically in the digestive tracts of animals. Archaea grow as mesophiles and thermophiles, but some species grow at very high temperatures, i.e., up to 110 °C. These grow as aerobes, anaerobes and facultative anaerobes. Archaea are not sensitive to β-lactam antibiotics as their cell walls do not contain murein, which is peptidoglycan containing muramic acid. All archaea contain glycerol isopranyl ether lipids, which is their unique biochemical property.

Gram-stain reaction varies depending upon the composition of the cell envelopes. This may be either Gram-stain-positive or -negative within the same group. The cells have different shapes, including spherical, spiral, plate or rod. Unicellular or multicellular forms also occur in filaments or aggregates. The modes of reproduction include binary fission, budding, fragmentation, constriction or mechanisms which are not yet known. The cell masses exhibit a variety of colours, including pink, red, purple, green, yellow, grey, white etc. Five major groups recognised in archaea are given below.

1. Methanogenic archaea
2. Sulphate-reducing archaea
3. Extremely halophilic archaea
4. Cell wall-lacking archaea
5. Extremely thermophilic S°-metabolising archaea

1.3 PHYTOPATHOGENIC BACTERIA

Phytopathogenic or plant pathogenic bacteria include cell-walled bacteria, including both Gram-stain-negative and Gram-stain-positive, and cell wall-lacking bacteria (mollicutes). The general characteristics of these phytopathogens are that the majority of them are non-spore-forming, rod-shaped organisms. Only *Streptomyces* species produce spores and are filamentous. Most plant pathogenic bacteria are nutritionally non-fastidious. Their optimal growth temperature is around 23-28 °C. However, a few of them, such as *Ralstonia solanacearum* and *Burkholderia cepacia*, can grow at 42 °C. Plant pathogenic bacteria can be either strictly aerobic or facultatively anaerobic.

Phytopathogenic and photosymbiotic procaryotes have one unique feature, i.e., their ability to multiply in the plants that separates these from all other procaryotes. Within a plant host, plant pathogens can increase from an initially low number of cells to a higher number in a short period of time. No other feature clearly distinguishes them from the myriad of saprophytes that occupy every other plant-associated habitat (Kennedy and Lacy, 1982). These have evolved to grow well in their plant hosts, often in specific hosts as well as in a specific plant part, such as blossoms, leaves, roots, tubers or corms. Inside the plant hosts, phytopathogenic bacteria multiply in the intercellular spaces and the penetration of the cells generally occurs after the disruption and disorganisation of the cell walls. Genetic studies have shown the involvement of numerous genes in virulence and pathogenesis. A number of plant pathogenic bacteria harbour virulence and pathogenicity genes on plasmid (an extrachromosomal element), whereas other pathogens bear virulence and pathogenicity genes in their chromosomes. The conservation of virulence and pathogenicity genes in pathogenicity islands strongly suggests the interchangeable movement of these genes across species boundaries via lateral gene transfer flow (Hacker and Kaper, 2000; Syvanen and Kado, 2002). Many of these genes provide an inherent ability to the pathogen to grow in its host and hence, are essential for successful colonisation of the host.

Hemibiotrophs share characteristics with both biotrophs and necrotrophs, initially invading cells that are maintained alive, prior to a transition to a necrotrophic lifestyle in which nutrients are obtained by killing host cells. All phytopathogenic bacteria possess weapons to suppress or avoid the premature elicitation of plant defences, indicating that they initially develop on living hosts, and later on use weapons involved in cell death induction, probably resulting in nutrient acquisition. Therefore, phytopathogenic bacteria should always be considered as **hemibiotrophs** (Kraepiel and Barny, 2016).

Phytopathogenic procaryotes cause plant diseases on a wide range of host plants. Some of these diseases are highly devastating and cause serious economic losses. Bacterial plant pathogens develop new survival strategies to thwart efforts to control and eliminate the plant diseases which they cause. Kado (2009) reported that the acquisition of novel genes through lateral gene transfer is a powerful mean for pathogen development. The generation, through selective mutations, of novel and functional genes whose products enhance successful colonisation can also facilitate pathogen fitness. Irrespective of the mechanism, genetic changes that are beneficial to pathogen survival occur due to selection pressure put on the pathogen by various control measures adopted in farming practices.

As the procaryotes include both bacteria and archaea, but only bacteria are pathogenic to plants, the terms "**phytopathogenic procaryotes**" and "**phytopathogenic bacteria**" have been used interchangeably in this book.

Phytopathogenic procaryotes differ from other procaryotes in that they have the ability to multiply inside the plants, thereby causing plant diseases. Phytopathogenic procaryotes cause plant diseases on a wide range of host plants. Some of these diseases are extremely devastating and cause serious economic losses.

1.3.1 Economic Importance of Phytopathogenic Bacteria

There is insufficient information available with relation to losses due to plant diseases caused by phytopathogenic procaryotes, in spite of the fact that some of these diseases are of international importance and can be responsible for huge economic losses. The loss caused by a plant disease varies from region to region due to variations in environmental factors, which influence the development of diseases, and also the susceptibility of the varieties under cultivation. For example, the bacterial blight of rice is severe, occurring mainly in Southeast Asia. Moreover, the importance of a crop also varies in a particular region or country.

The chemical control of bacterial plant diseases is less effective than that of diseases caused by fungi because of the availability of a smaller number of effective chemicals and the relatively low efficacy of the available chemicals. Moreover, many antibiotics that are effective against many plant pathogenic bacteria are also used for the control of human and animal bacterial pathogens. As a result, their use in plant disease control in many countries is not permitted due to the risk of transfer of resistance to human and animal pathogens. The crop losses due to plant pathogenic procaryotes in the United States in 1976 are given in Table 1.1 (Kennedy and Alcorn, 1980).

Bacterial wilt of solanaceous plants is considered the world's single most destructive bacterial plant disease. Brown rot bacterium is responsible for an annual estimated global loss of one billion US dollars, mostly impacting poor small-scale farmers in tropical highlands (Champoiseau and Allen, 2009). Bacterial wilt of potato is also considered the second most important disease of potato worldwide, after late blight.

Bove (2006) has rightly stated that citrus huanglongbing (greening) is the most important, serious, severe, destructive and devastating global disease affecting citrus species. The reduction in yield can range from 30 to 100% depending on the proportion of the affected canopy. The affected orchards become economically unviable some 7-10 years after planting. About 100 million trees have been destroyed, in many countries across South and Southeast Asia, India, Indonesia, the Philippines, Arabian Peninsula and South Africa. Since 2004, more than 500,000 trees have been officially destroyed in Brazil due to this disease, with around 300,000-400,000 trees unofficially destroyed by commercial citrus growers (Gottwald *et al.*, 2007). Catastrophic

TABLE 1.1
Estimates of US Crop Losses Caused by Phytopathogenic Bacteria in 1976[a]

Pathogen[b]	Disease	Loss (millions of US dollars)
Ralstonia solanacearum	Bacterial wilt of tomato and tobacco	9.4
Agrobacterium tumefaciens	Crown gall of fruits and nuts	23.0
Erwinia amylovora	Fire blight of pear	4.7
Erw. Tracheiphila	Cucumber wilt	0.9
Pantoea stewartii subsp. *stewartii*	Stewart's wilt of corn	0.2
Pectobacterium atrosepticum and/or *Pec. carotovorum* subsp. *carotovorum*	Blackleg and/or soft rot of potato	14.0
Clavibacter insidiosus	Bacterial wilt of alfalfa	17.0
Pseudomonas savastanoi pv. *glycinea*	Bacterial blight of soybean	65.0
Psm. syringae pv. *syringae*	Bacterial leaf blight of wheat	18.0
Xanthomonas citri subsp. *malvacearum*	Angular leaf spot of cotton	15.0
Xan. axonopodis pv. *phaseoli*	Common bacterial blight of bean	5.0
Leifsonia xyli subsp. *xyli*	Ratoon stunting disease of sugarcane	10.0
Xylella fastidiosa subsp. *fastidiosa*	Pierce's disease of grapevine	3.0
Spiroplasma citri	Citrus stubborn disease	1.0
'*Candidatus* Phytoplasma spp.'	Pear decline	1.6
	Lethal yellowing of coconut	3.0

[a] Based on data summarised from Kennedy and Alcorn (1980).
[b] Recent names of the pathogenic bacteria are given.

losses due to the disease from India during 1960s were reported by Fraser *et al.* (1966).

The incomplete eradication of citrus canker from the United States, particularly from Florida state, took a very heavy toll on citrus trees. After its first introduction in Florida in 1912, it took about 20 years to eradicate the disease from Florida by destroying and burning more than three million nursery and a quarter-million fruit-bearing trees. The second introduction of the disease in Florida in 1986 resulted in the destruction of more than 20 million nursery and orchard trees and the eradication programme continued till 1992. The third introduction of the disease in Florida in 1995 and subsequent years led to the destruction of 1.56 million commercial trees and 600,000 dooryard trees, leaving the eradication target incomplete. Successful eradication of the disease in Australia, South Africa and New Zealand, and partially successful eradication (incidence below 0.22% in Sao Paulo) in Brazil, was achieved after the destruction of citrus trees on a large scale.

Pierce's disease of grapevine caused by *Xyl. fastidiosa* subsp. *fastidiosa* was noticed in the US state of California in the 1880s, but the lack of an appropriate vector of causal agent and other non-favouring factors kept the disease under check. The introduction of the glassy-winged sharpshooter (*Homalodisca coagulate*), a highly efficient vector of the bacterium, in southern California changed the status of the disease from minor to severe, resulting in an enormous increase in crop losses. It has threatened the cultivation of not only the European grape (*Vitis vinifera*), but also of the many ornamental crops in the southeastern United States.

In 1918, 75% of the bean fields in New York state of the USA were affected with common blight (*Xan. axonopodis* pv. *phaseoli*), and the disease caused heavy losses. In the subsequent years, losses ranged from 20-50%. In 1953, the disease was widespread in Nebraska (USA) and it caused an estimated loss of over 1 million US dollars. In 1976, in the USA, common blight was economically the most important disease among the bacterial diseases of beans and it caused an estimated loss of 4 million US dollars. According to Gilbertson and Maxwell (1992), in 1967 the disease caused yield losses up to 38% in dry bean in Ontario (Canada) and 10-20% in navy bean in Michigan (USA). Yield losses ranging from 17-22% have also been reported from Colombia. In addition to the loss in yield, the disease causes reduction in quality of seeds through the discolouration of infected seeds.

Although the bacterial blight of rice, caused by *Xan. oryzae* pv. *oryzae*, occurs in many countries, it is most destructive in Southeast Asia. The losses are generally high in an early-infected crop and the occurrence of the kresek phase may lead to the total failure of the crop. In Japan, prior to introduction of resistant varieties, losses generally ranged from 20-30%; in some cases, however, they went as high as 50% (Ou, 1972). The reports from the Philippines, India and Indonesia estimate that losses due to the kresek phase have reached 60-75%, depending on weather conditions, location and rice variety (Reddy *et al.*, 1979; Ou, 1985). In addition to reduction in yield, the disease may also affect grain quality by interfering with maturation.

In another example, bacterial canker (black spot) of mango causes heavy losses because most of the commercial varieties of mango are highly susceptible to the pathogen, and the infection results in drastic yield losses. The fruit infection ranging from 50-80% is common on highly susceptible varieties. In 1966 and 1967, severe epiphytotics of the disease caused almost 100% fruit loss on most susceptible cultivars in the principal mango-growing areas of South Africa. Fruit loss was mainly due to fruit drop and non-marketability of the harvested fruit. The estimated loss in 1966 alone was approximately one million US dollars (Gagnevin and Pruvost, 2001). Kishun (1981) observed 20-80, 10-70, 10-40 and 10-55% fruit drop in Alphonso, Pairi, Totapuri and local cultivars, respectively, grown in India; the maximum fruit drop in all the cultivars occurred 30-60 days after fruit set. The author further observed that the rot in storage in these varieties was 5-80, 11-67, 10-100 and 5-80%, respectively.

Lethal yellowing of coconut palm trees has killed several palm trees in Florida state of US and some other countries. First observed in the state in 1971, it killed 40,000 trees by 1974; by 1975, 75% of palm trees in Dade county of the state were either dead or dying as a result of this disease. In Jamaica, 5.4 million trees were killed during a span of 20 years, starting from 1961. In Mexico and Tanzania, thousands of hectares of palm trees, and in Ghana more than a million coconut palm trees were killed within 30 years. The disease also killed more than 60,000 palm trees in Togo up to 1964 (Agrios, 2004).

Common scab of potato is one of the most economically important diseases of the crop. It has been estimated to cost the Tasmanian potato industry in excess of 3.6 million Australian dollars per annum (Wilson, 2004).

An estimated loss of 4.7 million US dollars occurred due to fire blight in California in 1976, while in Southwest Michigan alone, the loss in 1991 was 3.8 million US dollars. Again, in 2000, a severe outbreak of the disease occurred on apples in Michigan, resulting in the removal and destruction of 35,000-45,000 apple trees from approximately 1550-2300 acres, amounting to a loss of 42 million US dollars. In 1998, apple and pear growers in Washington and northern Oregon suffered a loss of 68 million US dollars due to fire blight damage.

Stewart's wilt of corn, caused by *Pan. stewartii* subsp. *stewartii*, is another good example of a disease that continues to have enormous implications for trade and commerce. This disease has been a major threat to corn production in the eastern corn belt of the United States for more than a century. Severe outbreaks of the disease occurred in Nebraska in 1999 and 2000. In 1999, the disease was found to occur in 762 out of 1317 fields in Iowa (58% prevalence). Due to the impact of the disease, the seed produced from 58% of the seed corn fields in Iowa in 1999, could not be exported to the countries due to phytosanitary (quarantine) restrictions for this pathogen.

In addition to causing direct loss in terms of the quality and yield of crops, phytopathogenic bacteria cause other harmful effects to mankind. *Rathayibacter toxicus* causes gumming disease in ryegrass and other fodder grasses and produces toxins, which when consumed, often result in the fatal poisoning of livestock. Various species of *Anguina* (nematode) act as a vector for this bacterium. Glycolipid toxins, known as corynetoxins, are produced by the bacterium as the infected grass matures and becomes senescent. In Australia, *Rat. toxicus* is found most commonly in *Lolium rigidum* (annual rye grass) with *A. funesta* as a vector and in *Polypogon monspeliensis* (rabbit-foot grass) and *Lachnagrostis filiformis* (syn. *Agrotis avenacea*, annual blown grass) with an undescribed *Anguina* vector. In Australia, more than 100,000 sheep and thousands of cattle have died from this disease in certain years (Edgar, 2004). In addition to affecting farmers or growers, bacterial diseases also affect transport and other industries dependent on the affected crops

In contrast to direct monetary losses and other harmful effects caused by phytopathogenic bacteria, they have a great deal of commercial value (Starr, 1984). Studies of these bacteria have yielded very useful products and tools. For example, several phytopathogenic bacteria are used in industrial fermentation processes. The most prominent case is the large-scale use of *Xan. campestris* to make xanthan gums, extracellular polysaccharides which are of great economic importance. Xanthan gums are used for various purposes, including oil-well drilling muds, tertiary recovery of petroleum from spent wells, ceramic glazes, antidrip agents in paints and thickening agents in foods. Another class of economically important polysaccharides produced by phytopathogenic bacteria are curdlans, which are produced by various species of *Agrobacterium* and other bacteria. On heating, curdlans form very elastic and resilient gums, which have potential use in various food products as noncaloric gelling, thickening, binding and stabilising additives. Similarly, a pseudomonad, isolated from an aquatic plant, produces an agar-like polysaccharide that has the potential to be used as a substitute for agar in microbiological media.

In addition to vinegar production, *Gluconobacter* species have many other biotechnological applications. In recent decades, many bioconversion routes for rare and special sugars involving *Gluconobacter* spp. have been developed. Among the more recent uses, are the biotransformations involved in the production of L-ribose and miglitol, both very promising pharmaceutical lead molecules. Most of these processes make use of *Gluconobacter*'s membrane-bound polyol dehydrogenases.

Various enzymes produced by phytopathogenic bacteria have different applications and are produced on commercial scale. Pectin-degrading enzymes have found an important application in the food industry, where these are used in the clarification of fruit juices. For instance, an enzyme produced by an *Agrobacterium* strain is used for converting super-coiled, covalently closed, circular DNA into a simple, covalently closed, circular form. Various *Xanthomonas* species and other phytopathogenic bacteria produce restriction endonucleases, which have a range of uses in genetic engineering and basic research. An L-asparaginase (protein synthesis inhibitor drug) produced by soft rot bacterium,

Pec. chrysanthemi has great potential for use as an antileukemic agent.

The potential of Ti-plasmid of *Agr. tumefaciens* to move genetic material from bacteria to plants have made it an important tool in plant genetic engineering, for which it is widely used. This technology is widely responsible for the development of the plant biotechnology industry, and is revolutionising agriculture and health care. The production of agrocin by an avirulent bacterium, *Agr. radiobacter* (strains K 84 and K 1026), and its role in controlling crown gall disease on a commercial scale are well documented. Presently, there are more than 16 commercial biocontrol products based on different bacteria, including *Agr. radiobacter*, which are used to control plant diseases caused by different pathogens. The detailed studies of the nodulation process by *Rhizobium* and *Bradyrhizobium* species have led to the preparations of bacterial inoculum for their application to legume crops (Dumenyo *et al.*, 2001).

1.4 HISTORICAL REVIEW OF PHYTOBACTERIOLOGY

In 1676, Antony van Leeuwenhoek discovered the bacteria for the first time and called them animalcules, i.e., small animals. Some two centuries later, in 1876, Robert Koch showed that anthrax in animals was caused by a bacterium, *Bacillus anthracis*. Robert Koch also devised the pour plate method for the purification of bacteria, and outlined Koch's postulates to prove the pathogenicity of causal agents of pathogenic diseases.

In 1866, Woronin, a Russian scientist, was the first person to show the association and multiplication of bacteria in the root nodules of leguminous plants. However, the multiplication of bacteria in root nodules is generally considered to be a symbiotic relationship rather than a strictly pathogenic action. Professor T. J. Burrill of the University of Illinois was the first person to show a bacterium as a causal agent of a plant disease. In a series of experiments conducted between 1878 and 1884, he showed that so-called fire blight in pears was caused by a bacterium, which he identified as *Micrococcus amylovorus*.

As the pure culture technique was in its early stages of development in the laboratory of Robert Koch in Germany, Burrill was unaware of this modern microbiological technique. Therefore, he did not make use of a pure culture of the bacterium in his work. Later on, J. C. Arthur working at New York Experiment Station confirmed Burrill's work using pure culture of the causal bacterium. Around the same time, in 1883, J. H. Wakker from Holland showed that yellow-slime disease in hyacinths was caused by a bacterium. A few years later, in 1887, Luigi Savastano, an Italian, demonstrated that a bacterium was responsible for knot disease in olive trees. In 1893, L. H. Pammel found that the bacteriosis of cabbage is caused by a yellow bacterium, *Bacillus campestris*, a name that was later changed to *Xan. campestris* pv. *campestris*.

Frediano Cavara (1857–1929), an Italian botanist who was the director of the Royal Botanical Gardens at the University of Naples, conducted a detailed study of galls on grapevines. He successfully isolated a bacterium from the galls and demonstrated that it caused an identical gall disease on healthy grapevines (Cavara, 1897). Later on, Erwin F. Smith identified this bacterium, naming it *Bacterium tumefaciens* (Smith and Townsend, 1907), which was changed to *Phytomonas tumefaciens* and then again to *Agrobacterium tumefaciens*. In reports by Cavara (1897), it is clear that Smith, who visited Cavara in 1904, was shown the bacterial cause of the crown gall disease. Although Erwin F. Smith and C. O. Townsend (Smith and Townsend, 1907) are cited as the discoverers of the bacterial nature of crown gall, credit should also be given to Cavara for his initial discovery and the demonstration of the bacterial etiology of crown gall. In fact, Smith and Townsend's major contribution was to confirm the earlier findings of Cavara. Smith became intrigued with this disease, using it as a paradigm of cancer in animals and humans (Smith, 1911). In 1901, L. R. Jones showed that soft rot of vegetables is caused by *Bacillus carotovorus*.

Soon after 1890, Erwin F. Smith began to work on bacterial diseases; the first disease which he investigated was bacterial wilt of cucurbits. Later on, he worked on many bacterial diseases of plants, including bacterial wilt of solanaceous plants and black rot of crucifers. Erwin F. Smith's monumental contributions are included in his monograph, which was titled *Bacteria in Relation to Plant Diseases*, and published in three volumes between 1905 and 1914. In 1896, Smith postulated that there are, in all probability, as many bacterial diseases of plants as there are of animals (Smith, 1896). In 1920, Smith predicted that there will be bacterial diseases, found on plants of every plant family.

A number of plant pathologists and authors have given credit to T. J. Burrill for providing the first credible proof that bacteria could cause plant diseases. However, Kennedy *et al.* (1979), in their critical appraisal of research work on bacterial diseases of plants carried out by T. J. Burrill and others, expressed the opinion that the credit for this discovery should be shared by E. Mitscherlich, a German chemist, T. J. Burrill and J. C. Arthur, rather than to T. J. Burrill alone. They have further stated that Burrill's place in history with respect to these studies is perhaps attributable to the fact that he was proven by others to have been right, and that he had a strong advocate in Erwin F. Smith.

Most of the research work on bacterial diseases in plants carried out between 1878 and 1900, and even later, was carried out in the United States. This is because there was a scepticism among European botanists about the role of bacteria in causing plant diseases. This was mainly attributable to the influence of De Bary, who had ascribed minor importance to bacteria as pathogens of plants. When he published his *Comparative Morphology and Biology of Fungi, Mycetozoa and Bacteria* in 1884, De Bary was aware of Wakker's work on the hyacinth disease but not of Burrill's work on fire blight. In this publication, in the section of bacteria, he devoted just one short paragraph to those parasitic on plants and also pointed out that there had been very few studies of this phenomenon. Similarly, in his *Lectures on Bacteria* published in 1887, De Bary devoted only 2 out of 146 pages to bacteria parasitic to plants, stating that saprophytic bacteria may also, under special conditions, attack the tissues of living plants as facultative parasites, producing disease in them and destroying them.

In 1897, Alfred Fischer, a professor of botany at Leipzig, Germany, who had been a student of De Bary back in 1881, challenged the work of E. F. Smith and others concerning the role of bacteria in causing plant diseases. In his book entitled *Lectures on Bacteria*, Fischer refused to accept the findings of Smith and others, describing them as unreliable. Smith replied to Fischer in 1899, pointing to the latter's almost complete ignorance of the subject. Fischer wrote back in the same year, and Smith replied in two rebuttal papers, which seemed to bring an end to the controversy, with Smith emerging as the clear winner of the debate. This debate has been duly recorded in one of the Phytopathological Classics published by APS Press (Campbell, 1981). Walker (1969) has rightly commented on the controversy created by Alfred Fischer, stating that while this debate is now chiefly of historical importance, it is surprising that more than 20 years after the work of both Robert Koch and Professor T. J. Burrill; such a prominent botanist was still challenging the existence of bacterial plant pathogens. By 1900, the concept that bacteria could cause plant diseases had become solidly established (Starr, 1984), leading to the establishment of **phytobacteriology** as a new subdiscipline of plant pathology. Subsequently, Erwin F. Smith's contributions to the field of plant bacteriology, and his efforts to put it on a sound footing, earned him the title of "**father of plant bacteriology**". Focusing on developments in another part of the globe, Goto (1992) has given a brief account of research work done on bacterial diseases of plants in Japan from the late 1880s to the 1970s.

Although E. F. Smith made highly significant contributions to the field of plant bacteriology, he did not establish a dynasty or a school within which young scientists could nurture their interest and enthusiasm in the newly discovered subdiscipline of phytobacteriology, as observed by Sequeira (2000) in his review article. Therefore, after his death in 1927, phytobacteriology began to wane. Around the middle of the 20th century, plant pathology was dominated by subdisciplines of mycology and virology.

In the initial stage of development of the subdiscipline of plant bacteriology, the emphasis was on reporting new bacterial diseases in plants and providing a description and identification of their causal bacteria. Then the emphasis shifted to the investigations on the physiology of diseased plants and, during the period from the 1940s to the 1960s, the significant contributions were made in this field, along with the ecology of plant pathogenic bacteria. The work of R. M. Klein (1954) and A. C. Braun and Ross Pringle (1959) provided the evidence that the crown gall bacterium, *Agr. tumefaciens* caused the permanent transformation of host cells, leading to their autonomous and rapid growth in culture. This work, along with the pioneering work of A. C. Braun and his colleagues on the structure and mode of action of wildfire toxin produced by *Psm. syringae* pv. *tabaci*, also had a great impact on many plant bacteriologists in terms of determining the pathogenicity determinants in other diseases (Sequeira, 2000).

The discovery of mycoplasma-like organisms by Doi *et al.* (1967) outlined the correct etiology of yellow-type diseases, which had previously been thought to be caused by viruses. The discovery of mycoplasma-like organisms also resulted in the subsequent discovery of spiroplasmas and fastidious procaryotes. The mycoplasma-like organisms are now called phytoplasmas. The demonstration of the conjugative transfer of bacterial genes in *Erwinia* by Chatterjee and Starr (1972) shifted the emphasis of research to the molecular genetics of bacterial plant pathogens. Another notable contribution during the 1970s was the success achieved in the biological control of crown gall with *Agr. radiobacter* strain K 84 (New and Kerr, 1972). Subsequently, an effective and viable biological control of fire blight was also obtained. The introduction of the pathovar system by Young *et al.* (1978) has substantially streamlined the taxonomy of plant pathogenic bacteria. Rapid advances in DNA sequencing technology and the sequence analysis of highly conserved regions of the bacterial genome, such as the small subunit rRNA gene, have led to the natural classification of procaryotes which reflects the evolutionary history of bacteria and archaea. The widespread application of new methods for the classification of procaryotes has led to an explosive growth in the number of validly published species and higher taxa. Since the completion of the first edition of *Bergey's Manual of Systematic Bacteriology* in 1989, the number of published species has more than tripled; this has been accompanied with numerous taxonomic rearrangements and changes in nomenclature. The number of bacterial genera containing phytopathogenic bacteria, which stood at only six during 1970s, now stands at 73, including three '*Candidatus* genera'.

The first avirulence gene (*avrA*) characterised was cloned from a race 6 strain of the soybean pathogen, *Psm. syringae* pv. *glycinea* (recent name, *Psm. savastanoi* pv. *glycinea*) by Staskawicz *et al.* (1984). It elicits a resistant reaction on soybean cultivars which have the resistant gene *Rpg2*. Avirulence genes determine the race specificity of a pathogen in a compatible host pathogen interaction. The products of *avr* genes apparently function as recognition factors in resistant hosts (i.e., those carrying the *R* gene) and virulence determinants in susceptible hosts (i.e., those carrying the *r* allele). In most cases, bacterial avirulence gene function is dependent on interactions with hypersensitive reaction and pathogenicity (*hrp*) genes. The *hrp* genes which control the ability of phytopathogenic bacteria to cause disease on susceptible plants and elicit hypersensitive reaction on resistant or non-host plants were first discovered for *Psm. syringae* pv. *phaseolicola* by Lindgren and associates in 1986. The role of avirulence genes and of *hrp* genes in host parasite interaction have been reviewed by Leach and White (1996) and Lindgren (1997), respectively.

Investigations carried on *Agr. tumefaceins*, the causal agent of crown gall, since it was first reported by Smith and Townsend (1907), have caused this bacterium to be transformed from being merely a plant pathogen to being a household name in plant genetic engineering and plant molecular genetics (Nester, 2008). The basic research carried out with this bacterium has provided very useful insights into host–parasite interactions and the mechanism of gene transfer from a procaryote to a eucaryote. The main problem of using Ti-plasmid as a gene vector due to its larger size has been solved by separating it into two parts: the 25 bp borders flanking the T-DNA between which the genes of interest are inserted; and the vir region, which is necessary for the processing and transfer of the T-DNA. This binary vector system is used in most laboratories. *Agrobacterium* is truly a **natural genetic engineer**.

Lately, atmospheric pressure plasmas have received considerable attention as a promising physical tool for biological decontamination and sterilisation. Several types of plasma devices have been described for the sterilisation of animal and human tissues, and the system can also be used for reducing the population of various plant microbes that cause infectious diseases. Mraz *et al.* (2014) used *Cla. michiganensis* subsp. *michiganensis*, *Erw. amylovora* and *Escherichia coli* (as a control) to study the bacterial growth mechanism after exposure to the atmospheric pressure GlidArc plasma. It was found that low-temperature plasma treatment affected all the three bacterial species by slowing down their growth and their reproduction rate. After plasma treatment, both *Esc. coli* and *Erw. amylovora* reached their maximum growth sooner than the *Cla. michiganensis* subsp. *michiganensis*. This is, however, the result of a longer culturing period for members of *Clavibacter* genus. The application of GlidArc plasma is a suitable treatment that can be used in agriculture for plant protection. Further studies of the effect of low-temperature atmospheric pressure plasmas on bacteria could be focused on its practical use in seed treatment of various crops.

A brief account of research in phytobacteriology in India is given below. In India, Cappel (1892) reported bangle blight (or *bungdi* disease) of potato from Pune, Maharashtra. Butler in 1903 also reported this disease, stating that it was similar to brown rot of potato caused by *Psm. solanacearum*. Coleman in 1909 proved the bacterial nature of this disease. Later, Dr M. K. Patel started work on the investigation of bacterial diseases in 1948. He, along with his associates, worked for nearly 15 years and reported nearly 40 bacterial diseases, most of which were new records. Based on his contribution to the field of plant bacteriology in India, Dr M. K. Patel is regarded as the "**father of plant bacteriology**" in India. A review of bacterial plant disease investigation published by Patel and Kulkarni (1953) included 27 species of bacterial pathogens reported from India. The first book on bacterial plant diseases in the country was written by G. Rangaswami in 1962, in which he listed 73 bacterial plant pathogens. In 1965, Mathur and associates published a list of bacterial plant pathogens found in India. Phytopathogenic bacteria of India and bibliography published by Chakravarti *et al.* (1973) contains a list of 483 references on plant pathogenic bacteria and 148 bacterial plant pathogens belonging to six genera, namely, *Agrobacterium, Bacillus, Corynebacterium, Erwinia, Pseudomonas* and *Xanthomonas*. A summer institute on Plant Bacteriology under the directorship of Dr P. N. Patel was held in the Division of Mycology and Plant Pathology (now, Division of Plant Pathology), Indian Agricultural Research Institute, New Delhi in 1972 and its proceedings were published in three volumes edited by Dr P. N. Patel. First National Symposium on Plant Bacterial Diseases was held at Sri Venkateshwara University, Triputi in 1980 under the directorship of Dr M. V. Nayudu and several important recommendations were made. The Ninth International Conference on Plant Pathogenic Bacteria was held in 1996 at the University of Madras, Chennai under the chairmanship of Dr A. Mahadevan.

At present, phytobacteriology is one of the leading areas in plant pathology, and has attracted a large number of scientists from areas other than plant pathology discipline. This is evident from the large number of papers on phytobacteriology published in scientific journals, namely, *Molecular Plant-Microbe Interactions, The Plant Cell* and *Applied and Environmental Microbiology.*

REFERENCES

Agrios, G. N. (2004). *Plant Pathology*. 5th Ed. Elsevier Academic Press, San Diego, CA, 922 pp.

Bove, J. M. (2006). Huanglongbing: A destructive, newly emerging century-old disease of citrus. *J. Plant Pathol.* 88: 7–37.

Braun, A. C. and Pringle, R. B. (1959). Pathogen factors in the physiology of disease-toxins and other metabolites. In: S. Holton, G. W. Fischer, R. W. Fulton, H. Hart and S. E. A. McCallan (eds.). *Plant Pathology: Problems and Progress*. Univ. Wisconsin, Madison, pp. 88–99.

Campbell, C. L. (ed.) (1981). *The Fischer–Smith Controversy: Are There Bacterial Diseases of Plants?* Phytopathol. Classics No. 13, APS Press, St. Paul Minn. XVIII + 65 pp.

Cappel, E. L. (1892). A note on a potato disease prevalent in Poona district and elsewhere. Bombay Directorate of Department of Land Records and Agriculture (Abstract by Butler, 1903).

Cavara, F. (1897). Eziologia di alcune malattie de piante coltivate. *Staz. Sper. Agr. Ital.* 30: 482–509.

Chakravarti, B. P., Hegde, S. V., Ahmed, S. R., Gupta, D. K. and Rangarajan, M. (1973). Phytopathogenic bacteria of India and bibliography (1892–1972). Bulletin, Department of Plant Pathology, University of Udaipur, Udaipur.

Champoiseau, P. G. and Allen, C. (2009). *Ralstonia solanacearum race 3 biovar 2: Tropical losses, temperate anxieties*. APSnet Feature Story, [http://www.apsnet.org/online/feature/ralstonia/].

Chatterjee, A. K. and Starr, M. P. (1972). Genetic transfer of episomic elements among *Erwinia* species and other Enterobacteria: F'lac+. *J. Bacteriol.* 111: 169–76.

Doi, Y. M., Teranaka, M., Yora, K. and Asuyama, H. (1967). Mycoplasma- or PLT group-like microorganisms found in the phloem elements of plants infected with mulberry dwarf, potato witches' broom, aster yellows, or paulownia witches' broom. *Ann. Phytopathol. Soc. Jpn.* 33: 259–66.

Dumenyo, C. K., Chatterjee, A. and Chatterjee, A. K. (2001). Phytobacteriology. In: O. C. Maloy and T. D. Murray (eds.). *Encyclopedia of Plant Pathology*. Vol. II. John Wiley and Sons, Inc., New York, pp. 762–69.

Edgar, J. (2004). Future impact of food safety issues on animal production and trade: Implication for research. *Aust. J. Exp. Agri.* 44: 1073–78.

Fraser, L. R., Singh, D., Capoor, S. P. and Nariani, T. K. (1966). Greening virus, the likely cause of citrus dieback in India. *FAO Plant Prot. Bull.* 14: 127–30.

Gagnevin, L. and Pruvost, O. (2001). Epidemiology and control of mango bacterial black spot. *Plant Dis.* 85: 928–35.

Gilbertson, R. L. and Maxwell, D. P. (1992). Common blight of beans. In: H. S. Chaube, U. S. Singh, A. N. Mukhopadhyay and J. Kumar (eds.). *Plant Diseases of International Importance*. Vol. II, Diseases of Vegetables and Oilseed Crops. Prentice Hall, Englewood Cliffs, NJ, pp. 18–39.

Goto, M. (1992). *Fundamentals of Bacterial Plant Pathology*. Academic Press, Inc., San Diego, CA, 342 pp.

Gottwald, T. R., da Graca, J. V. and Bassanezi, R. B. (2007). Citrus huanglongbing: The pathogen and its impact. APSnet Feature Story, [http://www.apsent.org/online/feature/huanglongbing/].

Hacker, J. and Kaper, J. B. (2000). Pathogenicity islands and evolution of microbes. *Annu. Rev. Microbiol.* 54: 641–79.

Hamilton, C. and Haney, C. (2023). How bacteria integrate multiple inputs to establish commensal or pathogenic lifestyles. *Phytopathology*, 113: S3.76.

Kado, C. I. (2009). Horizontal gene transfer: Sustaining pathogenicity and optimizing host–pathogen interactions. *Mol. Plant Pathol.* 10: 143–50.

Kennedy, B. W. and Alcorn, S. M. (1980). Estimates of U.S. crop losses to procaryote plant pathogens. *Plant Dis.* 64: 674–76.

Kennedy, B. W., Widin, K. D. and Baker, I. S. F. (1979). Bacteria as the cause of disease in plants: A historical perspective. *ASM News*, 45: 1–5.

Kennedy, R. W. and Lacy, G. H. (1982). Phytopathogenic procaryotes: An overview. In: M. S. Mount and G. H. Lacy (eds.). *Phytopathogenic Procaryotes*. Vol. 1. Academic Press, New York, pp. 3–17.

Kishun, R. (1981). Loss in mango fruit due to bacterial canker *Xanthomonas mangiferaeindicae*. *Proc. 5th Int. Conf. Plant Path. Bact. Cali, Colombia*, pp. 181–84.

Klein, R. M. (1954). Mechanisms of crown-gall induction. In: *Abnormal and Pathological Plant Growth. Brookhaven Symposia in Biology*. No. 6, pp. 97–114, Brookhaven National Laboratory.

Kraepiel, Y. and Barny, M.-A. (2016). Gram-negative phytopathogenic bacteria, all hemibiotrophs after all? *Mol. Plant Pathol.* 17: 313–16.

Krieg, N. R. (2001). Procaryotic domains. In: D. R. Boone and R. W. Castenholz (eds.). *Bergey's Manual of Systematic Bacteriology*. Vol. I – The Archaea and the Deeply Branching and Phototrophic Bacteria. 2nd Ed. Springer Verlag, Inc., New York, pp. 21–25.

Leach, J. E. and White, F. F. (1996). Bacterial avirulence genes. *Annu. Rev. Phytopathol.* 34: 153–79.

Lindgren, P. B. (1997). The role of *hrp* genes during plant bacterial interactions. *Annu. Rev. Phytopathol.* 35: 129–52.

Mraz, I., Beran, P., Šera, B., Gavril, B. and Hnatiuc, E. (2014). Effect of low-temperature plasma treatment on the growth and reproduction rate of some plant pathogenic bacteria. *J. Plant Pathol.* 96: 63–67.

Nester, E. (2008). *Agrobacterium*: The natural genetic engineer: 100 years later. APSnet Feature Story, [http://www.apsnet.org/online/feature/Agrobacter-ium/].

New, P. B. and Kerr, A. (1972). Biological control of crown gall: Field observations and glasshouse experiments. *J. Appl. Bacteriol.* 35: 279–87.

Ou, S. H. (1972). *Rice Diseases*. 1st Ed. Commonwealth Mycological Institute, Kew, Surrey, 368 pp.

Ou, S. H. (1985). *Rice Diseases*. 2nd Ed. C. A. B. International Slough, Oxfordshire, 380 pp.

Patel, M. K. and Kulkarni, Y. S. (1953). A review of bacterial plant disease investigation in India. *Indian Phytopathol.* 5: 131–40.

Reddy, A. P. K., MacKenzie, D. R., Rouse, D. I. and Rao, A. V. (1979). Relationship of bacterial leaf blight severity to grain yield of rice. *Phytopathology*, 69: 967–69.

Sequeira, L. (2000). Legacy for the millennium: A century of progress in plant pathology. *Annu. Rev. Phytopathol.* 38: 1–17.

Smith, E. F. (1896). The bacterial diseases of plants: A critical review of the present state of our knowledge. *Am. Nat.* 30: 797–804.

Smith, E. F. (1911). Crown-gall and sarcoma. *U.S. Dep. Agric. Bur. Plant Ind. Circ.* 85: 1–4.

Smith, E. F. and Townsend, C. O. (1907). A plant-tumor of bacterial origin. *Science*, 25: 671–73.

Starr, M. P. (1984). Landmarks in the development of phytobacteriology. *Annu. Rev. Phytopathol.* 22: 169–88.

Staskawicz, B. J., Dahlbeck, D. and Keen, N. T. (1984). Cloned avirulence gene of *Pseudomonas syringae* pv. *glycinea* determines race-specific incompatibility on *Glycine max* (L.) Merr. *Proc. Natl. Acad. Sci. USA*, 81: 6024–28.

Syvanen, M. and Kado, C. I. (eds.) (2002). *Horizontal Gene Transfer*. 2nd Ed. Academic Press, San Diego, CA.

Walker, J. C. (1969). *Plant Pathology*. 3rd Ed. McGraw-Hill Book Co., New York, 819 pp.

Wilson, C. R. (2004). A summary of common scab disease of potato research from Australia. In: S. Naito, N. Kondo, S. Akino, *et al.* (eds.). *Proc. Int. Potato Scab Symp. 2004*, Hokkaido Univeristy, Sapporo, pp. 198–214.

Young, J. M., Dye, D. W., Bradbury, J. F. Panagopoulos, C. G. and Robs, C. F. (1978). A proposed nomenclature and classification for plant pathogenic bacteria. *NZ J. Agric. Res.* 21: 153–77.

2 Bacterial Cell

On the basis of cellular organisation, the living organisms are divided into two groups, namely, procaryotes and eucaryotes. The members of the procaryotic world comprise a vast heterogeneous group of very small, principally unicellular organisms.

Stanier and van Niel (1962) proposed that it is their procaryotic cell structure that distinguishes bacteria from other forms of life, including viruses, protists, fungi, algae, plants and animals. They defined the procaryotic cell according to three principal criteria: the absence of internal membranes that compartmentalise the nuclear material and the enzymatic machinery for respiration and photosynthesis; nuclear division that occurs by fission and not mitosis; and the presence of peptidoglycan in the cell wall.

Over the past 50 years, a wealth of new information has been generated that greatly enriches our understanding of procaryotes. These organisms have proven to be of enormous abundance and diversity, the product of complex evolutionary processes over billions of years. These dominated life on Earth prior to the appearance of eucaryotes. Their distant past implies that many of the most salient features of modern life evolved in an entirely procaryotic world, including: most of the organising principles of the cell; the basic mechanisms of replication, transcription and translation; the major catabolic and anabolic pathways; and the biogeochemical cycles that maintain the biosphere (Whitman, 2009). Although Stanier and van Niel (1962) provided a very precise definition of the term procaryote, it has occasionally been misused, being used as synonymous with bacteria.

The procaryotes include two types of organisms, namely, bacteria and archaea. The majority of procaryotes, including the photosynthetic cyanobacteria, are included in the bacteria. Bacteria appear quite simple, when we look at their size, shape and arrangement of their cells through an ordinary light microscope. However, modern electron microscopy has revealed an amazing complexity and details of external and internal cell structures.

2.1 SIZE

Given their microscopic nature, bacteria are measured in micrometres (μm), which are equivalent to 1/1000 mm (10^{-3} mm). The cell size of bacteria varies with the species, but most cells are approximately 0.5–1.0 μm in diameter or width and 2.0–5.0 μm in length. The cells of *Staphylococcus* and *Streptococcus* species typically show a slightly larger diameter, measuring 0.75–1.25 μm, and some filamentous forms may be as long as 100 μm.

A few cases of extremely large bacteria have come to the knowledge of bacteriologists. *Epulopiscium fishelsoni*, found in the gut of surgeon fish, measures 50–100 μm in width and is 0.5 mm in length. Another bacterium, *Thiomargarita nomibiensis*, measures 100–300 μm in diameter, making it apparently the largest bacteria discovered to date. The eucaryotic cells on an average measure from 2–200 μm in diameters.

The size of the bacterial cells is influenced by age and cultural conditions. Actively growing cells are usually larger than the cells in their stationary phase. In stained preparations, the cell size is also reduced due to the shrinkage of cells.

The extent of small size of a bacterium is evident from the fact that approximately 1 trillion (1,000,000,000,000 or 10^{12}) bacterial cells weigh a mere 1 gram.

The extremely minute size of a bacterial cell means that it has a very high ratio of its surface area to its volume. This highlights the fact that they have a relatively large surface through which nutrients can enter for a relatively small volume of cell substance to be nourished. This is partly responsible for the high rate of metabolism and growth of bacteria. Due to their rapid growth and multiplication, the bacteria are frequently used in biological research.

2.2 SHAPE

Every bacterial species is associated with a constant shape. There are three basic shapes of bacteria: spherical, cylindrical and spiral. The most commonly found bacteria are either cocci or bacilli.

2.2.1 Spherical

Spherical cells are called **cocci** (singular, **coccus**). They are usually round, but they can also be ovoid or ellipsoidal (Figure 2.1a). Their size is measured in terms of their diameter.

2.2.2 Cylindrical

These are rod-shaped and have both length and breadth. They are called **bacilli** (singular, **bacillus**) and the ends of their cells may be flat, rounded, pointed or tapered (Figure 2.1b). Their size is measured in length and breadth. There is little difference in the width of different species, but the length varies considerably.

2.2.3 Spiral

The cells are helical in shape and are called **spirilla** (singular, **spirillum**). Spiral bacteria have one or more twists and they are never straight (Figure 2.1c). Certain spiral bacteria

DOI: 10.1201/9781003681878-3

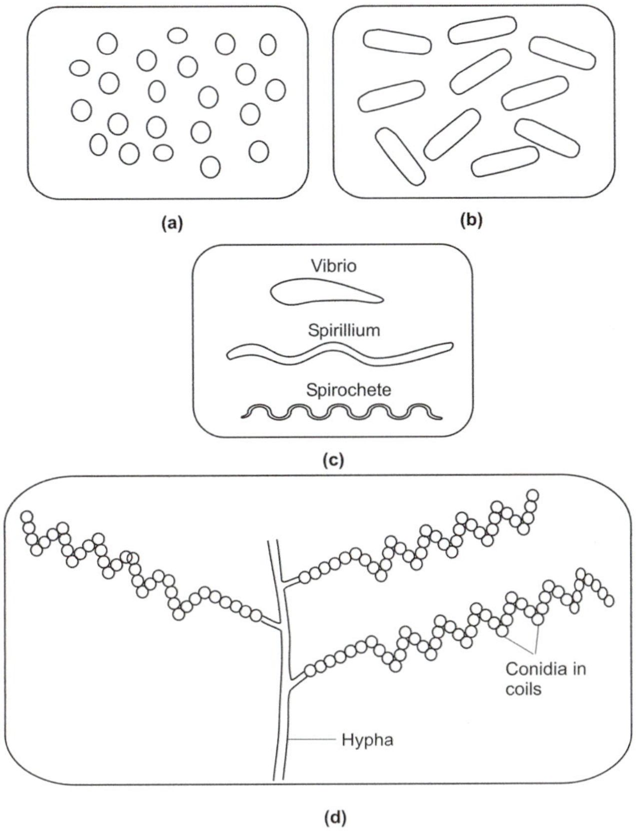

FIGURE 2.1 Shapes of bacterial cells: (a) cocci (b) bacilli (c) spirals (d) filamentous hyphae.

are curved short rods and resemble distinctive commas. These are called **vibrios**; *Vibrio cholerae*, the causal agent of cholera, is a typical example. The other spiral bacteria are long rods twisted into spirals; if the rods are rigid having thick cell walls, they are called **spirilla**; by contrast, **spirochetes** have thin flexible cell walls (Tortora *et al.*, 2007). *Treponema pallidum*, the causal agent of syphilis, is a typical example of a spirochete.

There are many modifications of these three basic shapes. The cells of *Pasteuria* are pear-shaped while those of *Caryophanon* are disc-shaped. Square and star-shaped forms also occur. The species of *Caulobacter* genus contain appendaged bacteria.

The members of Actinomycetes produce long filamentous hyphae that may branch to produce a mass of hyphae called mycelium (Figure 2.1d). They morphologically appear like fungi, but the cells are typically procaryotic, and hence they are classified with eubacteria.

Although the shape of the cells of most bacterial species is constant under a definite set of environmental conditions, a few species have a variety of cell shapes; these are called pleomorphic. Mycoplasmas are one example of just such a cell.

2.3 ARRANGEMENT

The majority of bacterial species exist in unicellular forms; however, some species show arrangements or attachment of cells. These arrangements are usually found in spherical and cylindrical forms. In spherical cells, the following five types of arrangement of cells are found.

2.3.1 Diplococci

A coccus cell divides in one plane forming a pair of two cells, e.g., some spp. of *Neisseria* (Figure 2.2a).

2.3.2 Streptococci

The cells divide in one plane forming a chain of cells, e.g., *Streptococcus* and *Lactococcus* spp. (Figure 2.2b).

2.3.3 Tetrads

The cocci divide at a right angle to the first plane of division, forming a group of four cells in the shape of a square, e.g., *Pediococcus* and *Micrococcus* spp. (Figure 2.2c).

2.3.4 Sarcinae

In this case, the third division occurs in a plane different to the first two divisions, resulting in a cuboidal or packet arrangement of eight cells, e.g., *Sarcina* spp. (Figure 2.2d).

2.3.5 Staphylococci

When the divisions in the three planes occur in an irregular pattern, the arrangement of cells looks like a bunch of grapes, e.g., *Staphylococcus* spp. (Figure 2.2e).

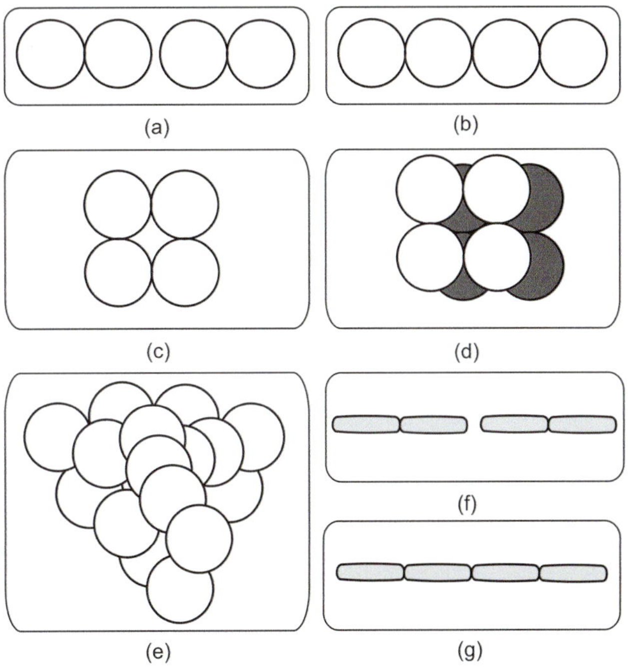

FIGURE 2.2 Arrangement of cells: (a) diplococci (b) streptococci (c) tetrads (d) sarcina (e) staphylococci (f) diplobacilli (g) streptobacilli.

In cylindrical bacteria, only the following two types of cell arrangement are found.

2.3.6 Diplobacilli

Two cells are arranged from end to end (Figure 2.2f).

2.3.7 Streptobacilli

Three or more cells are arranged from end to end, forming a chain, e.g., *Bacillus* spp. (Figure 2.2g).

Each of these arrangements is typical of a particular species and can prove helpful in its identification. However, only rarely are all the cells of a given species arranged exactly in the same pattern. It is, therefore, the predominant arrangement that is taken account of when studying bacteria. The size, shape and arrangement of bacterial cells are important parameters of their gross morphology and are used in their identification/classification.

2.4 CELL STRUCTURE

The cell structures/parts described below are not common to all bacterial species. Some of these structures, such as cytoplasmic membrane, cytoplasm, ribosomes, nucleoid and region containing all or most of the genophore, are found in all the bacteria, whereas the others are associated with certain particular species. Some of these structures are present inside the cell while others are present outside the cell wall (Figure 2.3).

2.4.1 Flagella and Pili

Flagella: Many bacteria are motile; this motility is due to the presence of flagella (singular, flagellum). Many species of bacilli and spirilla possess flagella while cocci rarely have these organelles. Flagella are thin, hair-like filaments with a helical shape. These originate from the cytoplasmic membrane and extend to the surface after penetrating the cell wall. Flagella are many times longer than the cells, measuring 15–20 μm in length. However, the diameter of a flagellum is only a fraction of the cell's width, i.e., 12–20 nanometres (nm). Being too thin in width, unstained flagella cannot be seen with an ordinary light microscope. The staining procedures using a mordant make flagellum thicker due to the deposition of dye and thus making it visible by light microscopy.

The number of flagella, and the pattern of flagellar attachment on a bacterial cell, are used to classify bacteria. Four types of flagellar attachment found in bacteria are given below.

2.4.1.1 Monotrichous

One flagellum is present at one end of the cell, e.g., *Xanthomonas* spp. (Figure 2.4a).

2.4.1.2 Lophotrichous

A cluster of flagella present at one end of the cell, e.g., *Pseudomonas* spp. (Figure 2.4b).

2.4.1.3 Amphitrichous

One flagellum or cluster of flagella present at both ends of the cell, e.g., *Spirillum* spp. (Figure 2.4c).

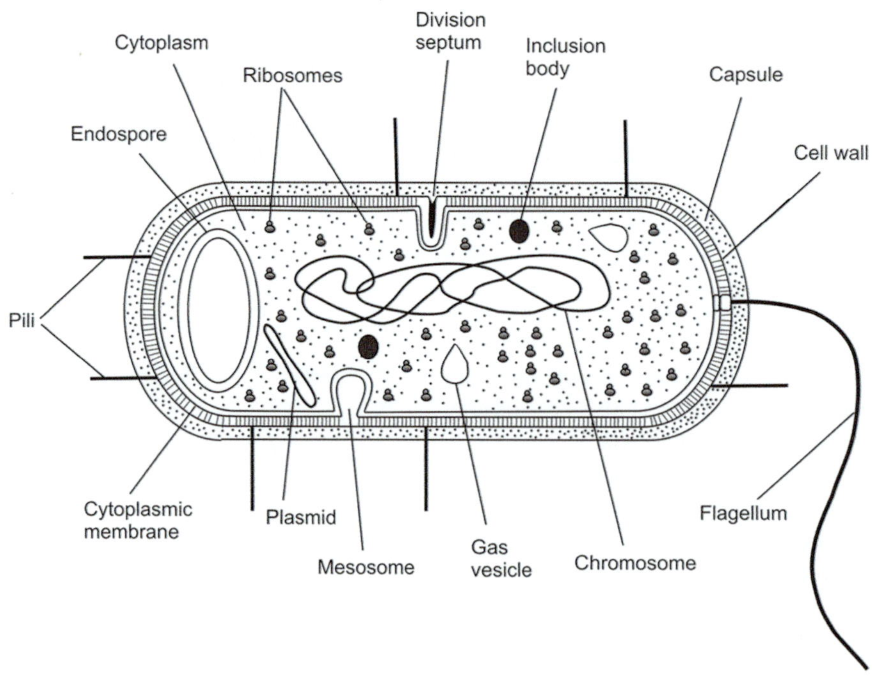

FIGURE 2.3 Bacterial cell.

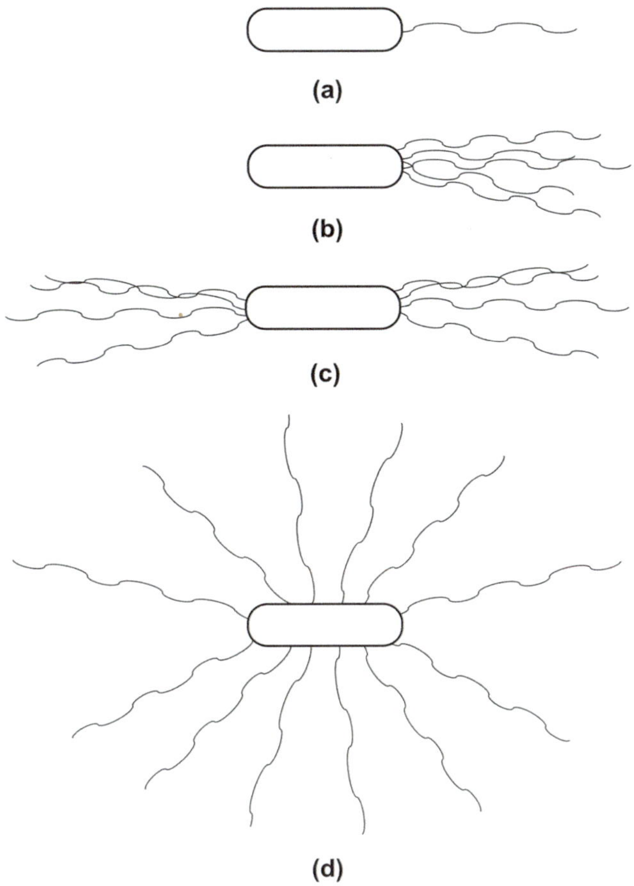

FIGURE 2.4 Arrangement of flagella: (a) monotrichous (b) lophotrichous (c) amphitrichous (d) peritrichous.

2.4.1.4 Peritrichous

The flagella scattered over the entire surface of the cell, e.g., *Erwinia* and *Escherichia* spp. (Figure 2.4d).

Some bacteria, e.g. spirochetes, have specialised flagella called **periplasmic flagella** (also known as **axial filaments**). The helical filaments of these flagella are not free and loose, in contrast to those of normal flagella. These arise at the poles and twine around the protoplasmic cylinder beneath the outer membrane of the cell wall. The spirochetes move in a corkscrew-like manner with the help of these flagella.

A flagellum contains three parts: the basal body, a hook and a long filament. The basal body is embedded in the cell and consists of a central rod surrounded by rings. Gram-stain-positive bacteria have only one pair of rings surrounding the central rod, one ring present in each of cytoplasmic membrane and cell wall. Gram-stain-negative bacteria contain two pairs of rings, with outer rings attached to the cell wall and inner rings lying in the cytoplasmic membrane. The helical filament is made of protein molecules called flagellin. These molecules are made within the cell and then transported along the hollow core of the flagellum for adding to the distal end, thereby confining the growth of the flagellum to its distal end (Pelczar *et al.*, 2004).

Flagella function by rotating in a corkscrew-like fashion, which propels the bacteria through liquid. The rings of the basal body rotate the flagella through chemical reaction. Those bacteria with polar flagella show a more rapid movement than those with peritrichous flagella. The environment in which they are present affects the motility of the bacteria, i.e., they move towards a favourable environment or away from a harmful environment. Swimming bacteria also exhibit chemotaxis, which is movement in response to chemicals in the environment. Bacteria usually swim towards increasing levels of nutrients and away from increasing levels of harmful substances.

Much is known about the flagellum, i.e., its basic structure, the function of its individual motor components and the regulation of its synthesis. However, this is only the beginning of an attempt to identify the dynamics of flagellar proteins and to understand how the motor structurally adapts to environmental stimuli. Baker and O'Toole (2017), in their review article, discuss both the external and the cellular factors that influence the dynamics of stator complexes (the ion-conducting channels of the flagellar motor). There is also a focus on recent discoveries, suggesting that stator dynamics are a means for controlling flagellar function in response to different environments.

The flagella, being the organs of locomotion, play a role in the virulence of some phytopathogenic bacteria. The bacteria swim towards their host plants, often after sensing a chemical signal from the host. In *Agrobacterium tumefaciens*, aflagellated mutants show approximately a 38% reduction in virulence and produce smaller and fewer tumours on the host in comparison to their wild–type parent. In addition, Rvi$^-$ (reduced in virulence) mutants of *Pectobacterium atrosepticum* that are deficient in flagella biosynthesis and motility, but produce extracellular enzymes levels similar to the wild-type, are less virulent than the parents.

Pili: The pili (singular, pilus), also called **fimbriae**, are short, straight, non-helical structures present on the surface of bacterial cells. These are thinner (3–10 nm in diam) than flagella and can be seen only by using an electron microscope. The pili are more common in Gram-stain-negative bacteria and are not associated with the motility of the cells. These are hollow like flagella and are made of protein subunits called pilin, which are arranged in a spiral manner around the central core. These originate beneath the cell wall but there are no anchoring structures, analogous to the basal bodies of flagella. These are present all along the cell and their number may range from one to several hundred per cell.

In *Esc. coli*, there are 100–300 pili arranged peritrichously around a cell, each pilus measuring 2–20 nm in length and being 7 nm in diameter. Different morphological types of pili occur, each associated with a different function, but attachment to the substrate or host cell is the primary function. A special type of pilus, called **F** or **sex pilus**, is associated with the bacterial mating. Bacteria that have this pilus are called F$^+$ (donor cells) and those without it are called F$^-$ (recipient cells). In bacteria pathogenic to animals, pili help bacteria to attach to host cells lining the respiratory and intestinal tracts. *Nes. gonorrhoeae*, the causal agent of gonorrhea, bears pili that recognise and adhere to receptor on human cells. Pili also provide the site of attachment to some bacteriophages.

As a result of their adhesive power, fimbriae are a major virulence factor for many pathogenic Gram-stain-negative bacteria. They are also potential targets for antibodies. Fimbriae are commonly required to initiate the colonisation, which leads to disease, and their success as adhesion organelles lies in their ability to both initiate and sustain bacterial attachment to epithelial cells. The ability of fimbriae to unwind and rewind their helical filaments presumably reduces their detachment from tissue surfaces with the shear forces that accompany significant fluid flow. Hence, the disruption of functional fimbriae by inhibiting this resilience should have high potential for use as a vaccine to prevent disease.

Singh *et al.* (2017) showed that two characteristic biomechanical features of fimbrial resilience, namely, the extension force and the extension length, were significantly altered by the binding of antibodies to fimbriae. The fimbriae that were studied are normally expressed on enterotoxigenic *Esc. coli*, which are a major cause of diarrheal disease. This alteration in biomechanical properties was observed with bivalent polyclonal antifimbrial antibodies that recognise major pilin subunits, but not with the antigen-binding fragments (Fab) of these antibodies. The authors proposed that the mechanism by which bound antibodies disrupt the uncoiling of natural fimbria under force is clamping together layers of the helical filament, thereby increasing their stiffness and reducing their resilience during fluid flow. Another proposal was that antibodies tangle fimbriae via bivalent binding, i.e., by binding to two individual fimbriae and linking them together. The use of antibodies to disrupt physical properties of fimbriae may be generally applicable to the large number of Gram-stain-negative bacteria that rely on these surface-adhesion molecules as an essential virulence factor.

2.4.2 Surface Layers (Capsule and Slime Layer)

Two principal types of surface layers, namely capsule and slime layers, are found in bacteria. If the layer is definite in outline and shape and is attached firmly to the cell wall, it is called a **capsule**. A capsule may be found in many bacteria, but this is not a universal occurrence. It is generally made of polysaccharides; in some cases, however, a capsule may consist of polypeptides. It can be best seen by negative staining or through electron microscopy. The capsule is a protective layer and often plays a major role in the pathogenicity of pathogenic bacteria. The capsules of *Stc. pneumoniae* protect the bacterial cells from being engulfed by the white blood cells (phagocytosis) of eucaryotes, thereby increasing the chances of infection. In many pathogenic species, the loss of capsule may result in the loss of pathogenicity but not the viability.

When the surface of the external layer is not definite in shape, is disorganised and is loosely attached to the cell wall, it is called a **slime layer** (e.g., *Rhizobium trifolii*). The slime layer being soluble in water creates many industrial problems in the industry. The slime-producing bacteria clog filters in sugar manufacture by increasing the viscosity of the fluids and coat pipes and other equipment.

The slime layer and capsule (also termed the **glycocalyx**) also perform other functions, depending on the bacterial species involved. Among these, the adherence of cells to rocks in fast-moving water, plant roots, teeth etc. is a major function. A capsule may prevent the attachment and lysis of cells by bacteriophages. These may also act as a reservoir of stored food and protect against temporary drying of cells by binding water molecules.

2.4.3 Cell Wall

Barring a few groups of bacteria, such as mycoplasmas, the cell wall is the universal outermost layer of bacterial cells. Being a rigid structure, it is responsible for the shape of the cell, and it also accounts for 20–30% of the dry weight of the cell. Besides giving a characteristic shape to the cell, the cell wall serves as a selective barrier, allowing the influx of desired nutrients and liquids and preventing the escape of certain enzymes. It also prevents the movement of dyes, bile salts, heavy metals, antibiotics and degradative enzymes into the cell.

In bacteria, peptidoglycan (also called **murein** or **mucopeptide**) is mainly responsible for determining the shape of the cell wall. The cell wall of bacteria is also unique in the sense as the peptidoglycan is found neither in archaea nor in eucaryotes. The basic structure of peptidoglycan in all the bacterial species consists of the following three building blocks: (1) N-acetyl glucosamine (NAG), (2) N-acetylmuramic acid (NAM), and (3) a peptide made of four amino acids, or tetrapeptide.

Alternating units of NAG and NAM linked together by glycosidic bonds form the backbone and give strength to the cell wall. The small peptides attached to NAM give an additional rigidity to the cell wall. In Gram-stain-positive bacteria and actinomycetes, these peptides consist of four amino acids: L-alanine, D-glutamic acid, L-lysine and D-alanine. Diaminopimelic acid (DAP) replaces L-lysine in Gram-stain-negative bacteria, cyanobacteria, myxobacteria and some Gram-stain-positive bacteria. D-configuration of glutamic acid and alanine makes the peptide bond resistant to pepsin and trypsin enzymes. The peptide chains are cross-linked with each other either through a direct bond (Gram-stain-negative bacteria) or through a pentaglycine bridge that is formed between terminal D-alanine of one peptide chain and the third amino acid (lysine or DAP) of the other.

More than five decades of research work supports the idea that cell envelope synthesis, including the inward growth of cell division, is tightly coordinated with DNA replication and protein synthesis through central metabolism. Remarkably, no unifying model exists to account for how these fundamentally disparate processes are functionally coupled. Recent studies show that proteins involved in carbohydrate and nitrogen metabolism can moonlight as direct regulators of cell division, coordinate cell division and DNA replication, and even suppress defects in DNA replication. Sperber and Herman (2017), in their mini-review, focus on studies illustrating the intimate link between metabolism and the regulation of peptidoglycan synthesis during growth and cell division. They identify the following three recurring themes:

1. Nutrient availability, rather than growth rate, is the primary determinant of cell size.
2. As the degree of gluconeogenic flux is likely to have a profound impact on the metabolites available for cell envelope synthesis, the selection of a growth medium is a critical consideration when designing and interpreting experiments related to morphogenesis.
3. Perturbations in pathways relying on commonly shared and limiting metabolites, like undecaprenyl phosphate, can lead to pleotropic phenotypes in unrelated pathways.

Enzyme lysozyme can completely destroy the cell wall of Gram-stain-positive bacteria, leaving a spherical cell called a **protoplast**. However, the cell walls of Gram-stain-negative bacteria are relatively resistant to this enzyme and are not completely destroyed; the **spheroplast** produced in this manner is equally susceptible to osmotic lysis as the protoplast.

Penicillin is a potent inhibitor of cell wall synthesis of Gram-stain-positive bacteria. It acts by preventing the formation of cross-linkages. Penicillin does not affect the pre-existing cross-linkages, but it does prevent the formation of new cross-linkages. Therefore, the non-growing population of bacteria is not inhibited. Mycoplasmas are similarly unaffected by penicillin because they lack the cell wall.

The cell walls of bacteria and archaea differ in terms of both structure and chemical function. The cell walls of archaea, as already stated, lack peptidoglycan but do contain proteins, glycoproteins or complex polysaccharides. Some species contain pseudomurein, which consists of *N*-acetyltalosaminuronic acid in place of N-acetylmuramic acid and lacks D-amino acid.

In spite of the fact that peptidoglycan is common to the cell walls of Gram-stain-positive and Gram-stain-negative bacteria, the cell walls of both the groups differ in thickness and chemical composition. These differences help in the classification of some bacteria and also explain the basis of differential Gram's staining reaction. The cell walls of Gram-stain-positive species are generally thicker (20–25 nm), consisting of several layers of peptidoglycan and show a high degree of cross-linking up to 90%; by contrast, those of Gram-stain-negative species (10–15 nm thick) show a low degree of cross-linking, up to only about 10%. The similar situation occurs with respect to Gram-stain-positive and Gram-stain-negative archaea.

2.4.3.1 Cell Walls of Gram-stain-positive Bacteria

Their cell walls have much greater amounts of peptidoglycan and hence are much thicker than those of Gram-stain-negative bacteria. In some species, the weight of this polymer may be up to 50% or more of the dry weight of the cell wall, compared with 10% in Gram-stain-negative species. Teichoic acids (acidic polysaccharides) are also present in the walls of many species. Teichoic acids, which are polymers of a carbohydrate like glucose, an alcohol (glycerol or ribitol) and phosphate, are attached to the peptidoglycan and to the cytoplasmic membrane. These may help in the storage of phosphorus and, being negatively charged, also help in the transport of positive ions into and out of the cell. Gram-stain-positive bacteria have neither the outer membrane nor the periplasmic space found in Gram-stain-negative species.

2.4.3.2 Cell Walls of Gram-stain-negative Bacteria

These are more complex, but thinner than the cell walls of Gram-stain-positive bacteria. The cell walls of these bacteria are bilayered structures and consist of an outer membrane and a peptidoglycan layer; the latter is present in the periplasmic space (the space between the outer membrane and the cytoplasmic membrane). The peptidoglycan layer of Gram-stain-negative bacteria constitutes only 5–10% of the dry weight of the cell wall. The outer membrane is responsible for the distinguishing character of these bacteria.

The outer membrane also acts as a selective barrier for the movement of some substances into and out of the cell. This selective permeability is attributed to the presence of proteins called porins, which form diffusion channels for the passage of molecules. Permeability is also related to the electric charge and size of molecules. Both specific and non-specific porins are present. The outer membrane is attached to the underlying peptidoglycan layer by a lipoprotein.

The outer membrane, a bilayered membrane, contains phospholipids, lipopolysaccharides (LPSs), integral membrane proteins and lipoproteins. These components are synthesised in the cytoplasm or at the inner leaflet of the inner membrane, and have to be transported across the inner membrane and through the periplasm to assemble eventually in the correct membrane (Bos *et al.*, 2007). In the outer membrane, lipopolysaccharides are present exclusively in the upper layer and phospholipids almost entirely in the lower layer.

Lipopolysaccharides found only in Gram-stain-negative bacteria consist of lipid A, core polysaccharide and O antigen. The lipid part of a lipopolysaccharide, called an endotoxin, is poisonous in nature. It can cause diarrhoea, fever and the destruction of red blood cells in the host.

The O antigens present in the outer membrane are responsible for many serological reactions of these bacteria. These also act as sites for the attachment of bacteriophages.

Some outer member proteins also act as receptor sites for the attachment of bacteriocins and bacteriophages.

2.4.3.3 Gram Staining/Reaction

The differences in the chemical composition and structures of cell walls of Gram-stain-positive and Gram-stain-negative bacteria are the bases for differential Gram-stain-reactions. The treatment of bacterial cells with crystal violet followed by Lugol's iodine solution results in the formation of crystal violet-iodine (CVI) complex within the cells. The subsequent washing of Gram-stain-negative bacterial cells with ethanol dissolves the lipids and removes them. This disrupts the outer membrane, increases its permeability and results in the washing away of the dye complex. Hence, decolourised cells are stained pink with the counter stain, safranin. On the other hand, in the Gram-stain-positive bacteria, ethanol makes the pores in peptidoglycan layer to shrink, thus trapping the CVI dye complex inside the cells.

2.4.4 Cytoplasmic Membrane

The cytoplasmic membrane present below the cell wall is the site of many enzymatic activities. It also acts as a selective barrier for the transport of water-soluble molecules into and out of the cell and is much more selective than the cell wall. It contains various enzymes, some of which are involved in energy production and cell wall synthesis. Permeases proteins present in the membrane transport small molecules into the cell.

Cytoplasmic membrane is 7.5 nm thick and, when seen by an electron microscope, it looks like a bilayered membrane. It is principally composed of phospholipids and proteins. The phospholipids form a bilayer in which most of the proteins are partially or totally buried (intrinsic proteins). Some of the proteins, called extrinisic proteins, are also present at the surface of the phospholipid bilayer. The fluidity nature of the membrane, allowing both proteins and lipids to move within phospholipids matrix, is called the **fluid mosaic model**.

Cytoplasmic membranes of most procaryotes do not contain sterols, such as cholesterol, and hence they are less rigid than those of eucaryotes. However, among the procaryotes, the cytoplasmic membranes of the mycoplasmas contain the sterols, which impart rigidity to these membranes. In the mycoplasmas, cytoplasmic membrane is the outer most layer of the cells.

The electron micrographs of thin sections of bacterial cells reveal extensive invaginations of the cytoplasmic membranes, which are called **mesosomes**. These are more prominent in Gram-stain-positive bacteria. Mesosomes may be present either near the cytoplasmic membrane or deep inside the cytoplasm. Earlier, these were thought to be associated with various functions such as cell division, metabolic process, enzyme secretion etc. Presently, however, these are considered to be more of an artifact than the real structure.

In the absence of membrane-bounded organelles in bacterial cells, the cytoplasmic membranes act as multiple structures. The bacteria in which the metabolism is based on the use of light energy or change of gases, their cytoplasmic membranes show extensive and elaborate intracellular extensions. Such types of extensions increase the surface area available for these activities. In phototrophic bacteria, the photosynthetic pigments are attached to these membrane systems. The nature of these membrane systems varies with different groups of phototrophic bacteria. In purple bacteria, these modifications are quite elaborate and often occupy a sizable internal cell area. In *Ectothiorhodospira*, the systems consist of flat sheets called **lamellae**, while in *Chromatium* these occur as round tubes or vesicles. In *Pelodictyon*, a green bacterium, chlorosomes constitute the photosynthetic apparatus which are cylindrical structures attached to the lower side of the cytoplasmic membrane. The lamellar membrane systems of oxygenic cyanobacteria are often complex and consist of multilayered structures called **thylakoids**.

The cytoplasmic membrane plays an important role in regulating the transport/movement of water-soluble molecules. It involves simple diffusion or facilitated diffusion down a concentration gradient or energy-based active transport.

2.4.5 Cytoplasm

Below the cytoplasmic membrane is the cytoplasmic area. The cytoplasmic area contains the cytoplasm, ribosomes, enzymes, genetic material (chromosomes and plasmids), granules, vacuoles etc. The cytoplasm is a matrix of gel-like consistency consisting of water (about 80%), nucleic acids, carbohydrates, proteins, lipids, inorganic ions etc. Many chemical reactions, involving the synthesis of cell components from nutrients, occur in the cytoplasm. Contrary to eucaryotic cells, there is no cytoplasmic streaming in procaryotic cells, i.e., cytoplasm does not flow around within the cell.

The **cytoskeleton** is a structure that helps cells maintain their shape and internal organisation; it also provides mechanical support which enables cells to carry out essential functions such as division and movement. It is a network of protein fibres supporting the cell shape and anchoring organelles within the cell. Three main structural components of the cytoskeleton are microtubules (formed by tubulins), microfilaments (formed by actins) and intermediate filaments. Only in recent decades, it has been widely accepted that bacteria are internally organised and possess a complex multifaceted cytoskeleton similar to that observed in eucaryotic cells. Proteins that make up the bacterial cytoskeleton can largely be grouped into families based on homology to the major types of eucaryotic cytoskeletal filaments, i.e., tubulin-like, actin-like and intermediate filament-like. Additionally, bacteria are host to some filament systems that lack recognisable homology to eucaryotic cytoskeletons (Barry and Gitai, 2011).

Tubulin-related bacterial protein FtsZ and actin-related bacterial proteins MreB/Mbl have been described as the constituents of bacterial cytoskeletons. Genes coding for MreB/Mbl could only be found in elongated bacteria, but not in coccoid forms. It was speculated that constituents of today's eucaryotic cytoskeleton (tubulin, actin) may have evolved from procaryotic precursor proteins closely related to today's bacterial proteins FtsZ and MreB/Mbl. Prior to the description of proteins MreB/Mbl, evidence was obtained for the existence of a shape-preserving cytoskeleton ubiquitously present in all bacteria (Mayer, 2003).

The presence of actin, tubulin and intermediate filament homologs in bacteria suggests that primitive cytoskeletons first arose in bacteria. Busiek and Margolin (2015) have reviewed the function and direct interactions between these actin and tubulin homologs. Actin and tubulin homologs are prominent in the bacterial cytoskeleton, but other proteins, such as RodZ, DivIVA and intermediate filament homologs, are seen as important supporting players. Though we know many of these players and how some of them interact, the mechanisms by which cytoskeletal components dictate cell shape have yet to be discovered. As it is generally true that cell architecture is inherited, it is also true that rod-shaped cells can emerge from round cells or from round dormant spores, which must depend on *de novo* assembly of cytoskeletal structures (Margolin, 2009).

2.4.6 Ribosomes

The ribosomes, the sites of protein synthesis, are densely packed throughout the cytoplasm. During protein synthesis, many ribosomes may be held together to form polyribosomes or polymers. In procaryotic cells, the ribosomes are 70S (Svedberg unit) type, (compared to 80S of eucaryotic cells), each having two subunits: the smaller 30S, squash-shaped, and the larger 50S, somewhat spherical. The Svedberg unit is a coefficient of sedimentation in the ultracentrifugation. High levels of Mg^{2+} ions are required to keep both the subunits of ribosomes associated. Antibiotics like streptomycin, neomycin, tetracycline etc. that inhibit protein synthesis cause a malfunctioning of the ribosomes. In *Esc. coli*, the ribosomes are made up of 66% RNA, known as ribosomal RNA (r RNA) and 34% proteins.

2.4.7 Inclusions and Storage Products

Different kinds of chemical substances often accumulate, forming insoluble deposits in the cytoplasm. Although they differ in terms of their chemical composition and other aspects, these function as the reservoir of energy and other structural building blocks. Most of them are bounded by a non-unit, thin, lipid-based membrane and are called inclusion bodies. These can be easily seen with light microscope in suitably stained preparations.

Some phototrophic (purple and green) bacteria store large amounts of elemental sulphur in the form of granules by oxidising sulphur compounds such as hydrogen sulphide, thiosulphate etc. These granules may be present inside the cells (e.g., *Chromatium* spp.) or on the cell surface (e.g., *Chlorobium* spp.). They may serve as stored energy for the bacteria. Another commonly stored product in bacteria is a lipid material called poly-β-hydroxybutyrate (PHB). Poly-β-hydroxyalkanoate (PHA), a derivative of PHB, also accumulates as granules in bacteria. In recent years, the plastic-like consistencies of many of these PHAs, and the biodegradable nature of their plastics, have generated a considerable amount of interest. These granules can be stained with fat-soluble dyes like Sudan black and Nile blue. Glycogen, a polymer of glucose subunits, is another stored product found in bacterial cells. Being polysaccharide, glycogen granules stain brown when iodine solution is applied. Both PHB and glycogen granules serve as a reservoir for carbon and energy.

Volutin granules or metachromatic granules, formed of polyphosphates, are also present in some bacterial species. In electron microscopy, these appear as round, dark areas and stain intense reddish purple with dilute methylene blue dye.

Intracellular crystals of iron oxide, magnetite, and Fe_3O_4 are found in some aquatic bacteria. These crystals are bounded by a membrane and are called **magnetosomes**. These vary in shape from square to rectangular, or, alternatively, look like a spike. As such, bacteria possess the ability to orient and migrate along geomagnetic or exhibit magnetotaxis; they are called **magnetotactic bacteria**.

2.4.8 Gas Vesicles

The bacteria living in water bodies possess gas vesicles. With the use of these, the bacteria can move up and down in response to different environmental conditions. The large blooms produced by some cyanobacteria are endowed with these gas vesicles. These are hollow but rigid, spindle-shaped structures and are proteinaceous in nature. These vary in length and width and also in number, from a few to several hundred per cell. The vesicle membrane is permeable to gasses but does not allow the water and solutes to enter the vesicle.

2.4.9 Nucleoid (Genophore)

The genetic material in bacteria, comprising of a single haploid chromosome and extrachromosomal genetic elements, is called the **nucleoid** or **genophore**. The earlier concept that the nuclear material of bacteria is dispersed throughout the cell was due to the inadequacy of the staining techniques. With the advance of knowledge, the exact nature of bacterial hereditary material can be easily demonstrated. It is a fairly compact clump, organised into a definite body and occupies roughly one-third of the volume of the cell. Unlike the nucleoplasm of eukaryotes, the nucleoid is **not surrounded by a nuclear membrane**. In contrast, the nuclear material in eucaryotes or higher organisms is surrounded by a well-defined nuclear membrane.

Fluorescent microscopy has been found to be very helpful for delineating the gross boundaries of the nuclear region. The fluorescent probes, such as 4,6-diamidino-2-phenylindole dihydrochloride (DAPI), capable of entering the cell and binding to DNA at a non-toxic concentration, greatly help in visualisation of nucleoid boundaries. The electron micrograph of *Esc. coli* revealed that its nucleoid has a highly clefted appearance and is confined to a ribosome-free zone in the cell. The nucleoid is largely composed of DNA (60%), with small amounts of RNA and proteins. Nucleoid proteins or nucleoid-associated proteins are distinct from histones of eucaryotic nuclei and do not form nucleosomes. Common nucleoid associated proteins are HU, histone-like nucleoid-structuring protein (H-NS), Fis, curved DNA-binding protein (cbpa) and DNA-binding proteins from starved cells (DPS) that help in the compaction of DNA by bending, bridging and aggregation. Proteins like HU and Fis probably play a role in transcription.

Xu and Wang (2019) have summarised the application of genomics in different aspects of plant bacterial pathosystems. Genomics has been used in studying the mechanisms of plant–bacteria interactions and host specificity. It also helps in taxonomy, the study of non-cultured bacteria, the identification of a causal agent, single cell sequencing, population genetics and meta-transcriptomic. Overall, genomics has significantly improved our understanding of plant–microbe interaction.

2.4.9.1 Chromosome

The bacterial genome primarily comprises of single, haploid, circular chromosome and, in addition, there may also be present some extrachromosomal genetic elements, referred to as plasmids, potentially accounting for about 1% of the total genome. However, some bacteria, such as *Vibrio cholerae* and *Rohodobacter sphaeroides*, have two chromosomes. The chromosome of *Esc. coli* is a closed circular DNA molecule approximately 1 mm in length. Taking into account the 1–2 μm length of this bacterial cell, the chromosome must have condensed about 500–1000 times to fit into the cell. The condensation of the chromosome occurs by organising the DNA into a series of loops and super-coiling of these loops in an underwound fashion. The transcription, recombination and other functions are not hindered by the folding pattern (Srivastava and Srivastava, 2003). In most bacteria, the chromosome is circularly closed, but in a few cases linear chromosomes have also been reported, e.g., *Streptomyces lividans* and *Rhodococcus fascians*. Among the procaryotes, *Mycoplasma genitalium* possesses the smallest genome, containing only 470 genes (~10–15% of *Esc. coli*).

2.4.10 Plasmids

The plasmids are extra chromosomal genetic material found in most bacteria; they play a significant role in their adaptation and evolution. Most plasmids are covalently closed circular structures consisting of double-stranded DNA and the DNA is negatively supercoiled. However, some bacteria such as *Borrelia burgderferi* and some species of genus *Streptomyces* possess linear plasmids. These plasmids have the ability to replicate independently. The number of copies of a plasmid per cell varies with different plasmids. F-plasmid or P1 prophage-plasmid exists as a single copy while some 40–300 copies of plasmid PIJ 101 are found in *Str. coelicolor.* Many bacteria may contain more than one type of plasmids. Initially, the plasmids were named on the basis of functions which they performed, e.g., the F-plasmid mediates fertility and the Ti-plasmid of *Agr. tumefaciens* induces tumour formation in some vascular plants. Now a more standardised method to name plasmids has been worked out. In this method, the letter 'p' is placed before the capital letters that either describe the plasmid or denote the initials of the person or persons who isolated or constructed the plasmid, e.g., $pPATH_{Pag}$, pathogenicity plasmid of *Pantoea agglomerans* pv. *gypsophilae.* The methods for the detection and isolation of plasmids have also been devised.

Plasmids are a common occurrence in plant pathogenic bacteria and contribute significantly to host evolution in a multifaceted manner. Plasmids tend to encode determinants of virulence and ecological fitness that can enhance adaptation to a specific niche or can influence niche expansion. Many of these determinants appear to have been acquired from other bacteria via horizontal transfer, showing an important function of plasmids in the acquisition of sequences that enable rapid evolution. Finally, these genes are transferred to the host chromosomes through plasmid integration events, thereby resulting in the stabilisation of important acquired determinants within the genome. Most plasmids characterised in phytopathogenic bacteria are self-transmissible and possess suites of genes encoding type IV secretion systems. Sundin (2007) has given an excellent account of genomic insights into the contribution of phytopathogenic bacterial plasmids to the evolution of their hosts. The role of plasmids in the evolution of phytopathogenic bacteria has also been reviewed earlier by Coplin (1989).

Most genera of plant pathogenic bacteria harbour one or more plasmids containing genes that encode traits of importance to ecological fitness and/or virulence. By 2007, complete genome sequence information for 22 medium- to large-sized plasmids belonging to six plant pathogenic species was available (Sundin, 2007).

Bacterial plasmids are well-known carriers of antibiotic and heavy metal resistance genes in all ecosystems. This fact is also true for plasmids of plant pathogenic bacteria. Plasmids-encoded resistance determinants to copper and streptomycin in *Psm. syringae* pv. *tomato* and *Xan. vesicatoria* were among the first plasmid-encoded determinants to be cloned and characterised in these bacteria. With few exceptions, all strains of *Erw. amylovora* isolated from diseased tissues contain a nonconjugative plasmid called the ubiquitous plasmid pEA29. This plasmid can be cured in the laboratory, and the cured strains are reduced in virulence and show thiamine auxotrophy. The presence of pEA29 influences the expression of the chromosomal amylovoran biosynthetic operon during infection. Exopolysaccharide amylovoran is a pathogenicity determinant of the fire blight pathogen and thiamine biosynthesis is required for the high-level expression of the amylovoran biosynthetic operon during infection. Among the other plasmids found in *Erw. amylovora*, the plasmid pEL60 is of interest because it is an IncL/M plasmid with a high degree of similarity to the backbone of pCTX-M3, a multiple antibiotic resistance plasmid identified in the opportunistic human pathogen, *Citrobactor freundii*. pEL60 is a cryptic plasmid, encoding only genes involved in replication, maintenance and conjugative transfer, while pCTX-M3 contains two additional insertions totalling 29.6 kb that encode a number of antibiotic resistance genes.

The acquisition and expression of novel sequences may confer pathogenicity on a saprophytic strain. A classic example is the transformation of *Pan. agglomerans*, a widespread epiphytic and commensal bacterium into a host-specific tumorigenic pathogen by acquiring a plasmid called pPATH (plasmid governing pathogenicity of *Pan. agglomerans*) containing a pathogenicity island (PAI). The PAI was evolved on an interon plasmid of the IncN family, which is distributed among genetically diverse populations of *Pan. agglomerans*. The horizontal gene transfer was the major contributing force in the creation of PAI, although a pathoadaptive mechanism might also be involved.

Vivian *et al.* (2001) found that most strains of *Psm. syringae* analysed, regardless of pathovar, contained at least one indigenous plasmid. Genetic studies of plasmids from different pathovars reveal that a large majority of plasmids in *Psm. syringae* share a major replication gene *repA*. The sequencing and functional characterisation of *repA* from 100 kb plasmid pPT23A of *Psm. syringae* pv. *tomato* PT 23 and the analysis of the distribution of *repA* in diverse pathovars of *Psm. syringae* led to the recognition of the pPT23A plasmid family as a group, called pPT23A family plasmids (PFPs). Most plasmids of this family are in the range 35–100 kb and are believed to have originated from a common ancestor because they share the *repA* gene that encodes an essential replication protein. The pPT23A plasmid family is considered to be universally distributed among *Psm. syringae* pathovars. The *repA* replication gene is the only gene conserved in all PFPs studied by Zhao *et al.* (2005). Genes encoding virulence factors such as type III effectors, coronatine toxin biosynthesis, plant hormone production and the determinants important for epiphytic fitness, including UV radiation tolerance, are commonly found on PFPs (Vivian *et al.*, 2001). Genes of potential ecological importance or virulence genes of general importance to *Psm. syringae* are more widely distributed among pathovars. The loss of virulence in *Psm. syringae* pv. *phaseolicola* to bean is associated with the loss of plasmid pAV511.

The plasmids act as carriers of virulence/avirulence genes in *Xanthomonas* species. Additional avirulence/effector genes, such as those of *avrBs3/pthA* family and *avrBs4*, are plasmid encoded in *Xanthomonas* spp., and in some cases, these genes are located on conjugative plasmids. *avrBs3/pthA* factors play a role in pathogen aggressiveness and virulence. Fujikawa *et al.* (2006) have shown that *avrBs3/pthA* effectors function as suppressors of host defence responses.

The integration of plasmids into the host chromosomes occurs, and it represents one delivery method for horizontally acquired sequences. The integration of plasmid pMC7105 from *Psm. syringae* pv. *phaseolicola* was first demonstrated in the mid-1980s. It was shown to replicate autonomously or integrate into the host chromosome. Integration was achieved via recombination between copies of the insertion sequence IS*801* on the plasmid and chromosome. An extensive sequence homology found between pMC7105 and plasmid DNA from *Psm. savastanoi* (=*syringae*) pv. *glycinea* indicates that the integration of plasmids may also occur in other *Psm. syringae* pathovars. The plasmid pFKN, a PFP from *Psm. syringae* pv. *maculicola*, has also been shown to integrate into the chromosome through recombination between alleles of the effector gene *avrPphE*. This integration was considered to comprise an evolutionary event resulting in the formation of a new pathogenicity island in the *Psm. syringae* pv. *maculicola* chromosome. The integration of plasmids into chromosomes can be viewed as the first step in the generation of new island sequence.

Insertion sequence (IS) elements also play a role in the evolution of bacteria in different ways. These sequences move within plasmids, between plasmids and between plasmid and chromosome. IS elements can facilitate recombination between plasmids and chromosomes. Plasmid-encoded IS elements effect pathogen virulence and evolution. The element IS*801* appears prominently in the evolution of certain *Psm. syringae* pathovars but is absent in other pathovars. IS*801*-like sequences occur close to effector genes in *Psm. savastoni* pvs. *glycinea* and *phaseolicola*, and downstream of plant hormone biosynthetic genes in *Psm. savastanoi* pv. *savastanoi*. Due to this close proximity, IS*801* has been considered to be involved in the mobility of virulence genes. The IS*3* family is one of the largest families of the IS element. IS*476* encoded on plasmids of *Xan. campestris* pv. *campestris* and *Xan. vesicatoria*, IS*1389* of *Xan. campestris* pv. *amaranthicola*, IS *Xac3* of *Xan. citri* subsp. *citri*, IS *Psy26* of *Psm. syringae* pv. *tomato* and an unnamed element of pEA29 of *Erw. amylovora* are members of the IS*3* family. IS elements like IS*801* serve as sites for recombination, facilitating integration of plasmids into chromosomes, and these elements can also mobilise adjacent sequences during the transposition process.

The plasmids also play a role in other plant pathogenic bacteria. *Psm. syringae* pv. *syringae*, incitant of bacterial canker of stone fruits contains plasmid-mediated genes for copper resistance. Patro and Jindal (2000) showed the existence of plasmids in *Xan. axonopodis* pv. *vignaeradiatae*, incitant of bacterial leaf spot of green gram, and their correlation with virulence. Highly virulent isolates contained the maximum number of plasmids. The plasmid-borne genes were shown to govern virulence, exopolysaccharide production, antibiotic resistance and pigmentation.

The plasmids have played a very important role as vectors in genetic engineering. These have been used to transfer required genes into the strains of the same species, into different species or even into organisms of different kingdoms. Their small size, easy isolation and introduction, a greater number of copies, possession of an easily selectable marker, the presence of sites for restriction endonucleases and amenability to the desired modifications have greatly enhanced their utility in genetic engineering. A tumour-inducing (Ti) plasmid of *Agr. tumefaciens* is commonly used for transferring genetic material into different bacterial species, or even into plants. *Agr. tumefaciens* is an excellent genetic engineering tool. The DNA transformation capabilities of this bacterium have been extensively exploited in biotechnology as a means of inserting foreign genes into plants. Therefore, Nester (2008) has appropriately called it a **natural genetic engineer** and also a **household name** in plant genetic engineering and plant molecular genetics.

2.4.11 Dormant Forms

Some bacteria produce dormant forms or resting stages called spores and cysts. These dormant forms help the bacteria to tide over, or withstand adverse climatic conditions such as heat or drying. They are metabolically inactive, i.e., not growing but under favourable conditions, they can germinate and grow into metabolically active vegetative cells.

2.4.11.1 Spores

The spores are thick-walled, dormant and highly refractile structures. As they are formed inside the cells, they are also called **endospores**. Spores are highly resistant to heat, drying and chemicals toxic to the vegetative cells. The endospores of *Clostridium botulinum*, the cause of botulism (food poisoning), can resist boiling for several hours while the vegetative cells of most bacteria are killed above 70 °C. As the spores are resistant to adverse environmental conditions, they help the bacteria to tide over the adverse/harsh environmental conditions.

Among the rod-shaped bacteria, the members of genera *Bacillus* and *Clostridium* are exclusively spore-formers. Among the spherical bacteria, it is the members of genus *Sporosarcina* that form the spores. Generally, one spore is produced per vegetative cell and these are usually produced in old bacterial cultures. The spores vary in shape (being either spherical or ovoid) and location within the cell. The location in the cell can be central, sub-terminal or terminal (Figure 2.5a–c, respectively). In some cases, terminal spores are formed in swollen sporangia (Figure 2.5d). Spore position in the mother cell or sporangium frequently differs

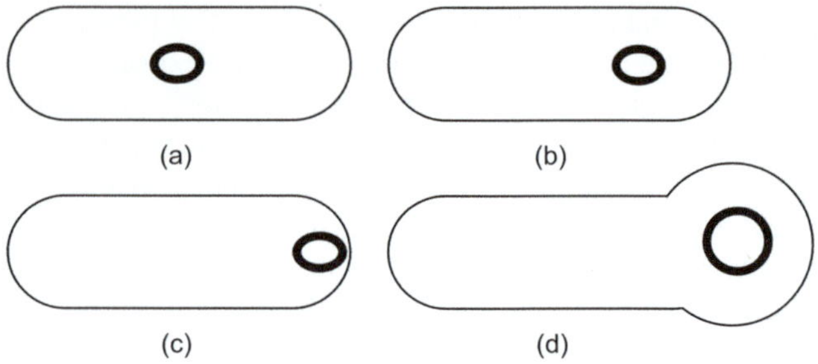

FIGURE 2.5 Location of spores in the cell: (a) central (b) sub-terminal (c) terminal (d) terminal spore with swollen sporangium.

among species and is helpful in the identification of bacteria. It is essential to apply heat to the endospores when staining for light microscopy. In bacteria, endospores are not the means of multiplication; rather, they are means to tide over the adverse conditions, as one vegetative cell produces one spore and, on germination, one spore produces one cell.

During sporulation, a dehydration process occurs which expels most of the water from the spores. All endospores contain large amounts of dipicolinic acid (DPA), a compound not found in vegetative cells. DPA accounts for 5–10% of the endospore's dry weight and occurs in combination with large amounts of calcium, both accounting for 15% of the spore's dry weight. Earlier, it was thought that DPA was directly responsible for the resistance of spores to heat. However, the occurrence of heat-resistant mutants lacking DPA has seriously weakened this viewpoint. Calcium does help in resistance to wet heat, oxidising agents and sometimes dry heat. It may be that calcium-dipicolinate often stabilises spore nucleic acids. The calcium-dipicolinate complex, in combination with the acid-soluble protein stabilisation of DNA, protoplast dehydration, DNA repair, the spore coat and the greater stability of cell proteins in bacteria adapted to growth at high temperatures, is assumed to be responsible for imparting resistance to spores against heat and other lethal agents (Prescott *et al.*, 2005).

The outermost exosporium layer of *Clo. difficile* spores is likely to play a major role in spore–host interactions during recurrent infections, contributing to the persistence of the spores in the host. Pizarro-Guajardo *et al.* (2016a) demonstrated, by transmission electron microscopy, the presence of two ultra-structural morphotypes of the exosporium layer in spores formed from the same sporulating culture. However, it was not clear whether these distinct morphotypes appeared due to purification techniques or during biofilm development. Pizarro-Guajardo *et al.* (2016b) demonstrated, through transmission electron microscopy, that these two exosporium morphotypes are formed under sporulation conditions and are also present in spores formed during biofilm development. This investigation provides definite evidence that, in a population of sporulating cells, spores with a thick outermost exosporium layer and spores with a thin outermost exosporium layer are formed.

Another type of spores produced by bacteria is **conidia** (singular, conidium), produced by actinomycetes. These spores have little resistance to heat, but they are resistant to drying. As one thallus of actinomycetes produces many conidia, these are the means of reproduction, but not the means to tide over the adverse conditions.

2.4.11.2 Cysts

Cysts are also dormant, thick-walled structures that resist drying. However, in contrast to endospores, they are not highly resistant to heat. Vegetative cells give rise to cysts; under favourable conditions, cysts germinate to give rise to vegetative cells. The vegetative cells of several bacterial genera have been reported to produce cysts, but those produced by the members of *Azotobacter* are considered the typical bacterial cysts.

REFERENCES

Baker, A. E. and O'Toole, G. A. (2017). Bacteria, rev your engines: Stator dynamics regulate flagellar motility. *J. Bacteriol.* 199: e00088–117.

Barry, R. and Gitai, Z. (2011). Self-assembling enzymes and the origins of the cytoskeleton. *Curr. Opin. Mirobiol.* 14: 704–11.

Bos, M. P., Robert, V. and Tommassen, J. (2007). Biogenesis of the Gram-negative bacterial outer membrane. *Ann. Rev. Phytopathol.* 61: 191–214.

Busiek, K. K. and Margolin, W. (2015). Bacterial actin and tubulin homologs in cell growth and division. *Curr. Biol.* 25: R243–54.

Coplin, D. L. (1989). Plasmids and their role in the evolution of phytopathogenic bacteria. *Ann. Rev. Phytopathol.* 27: 187–212.

Fujikawa, T., Ishihara, H., Leach, J. E. and Tsuyumu, S. (2006). Suppression of defence response in plants by the *avrBs3/pthA* gene family of *Xanthomonas* spp. *Mol. Plant–Microbe Interact.* 19: 342–47.

Margolin, W. (2009). Sculpting the bacterial cell. *Curr. Biol.* 19: R812–22.

Mayer, F. (2003). Cytoskeletons in prokaryotes. *Cell Biol. Int.* 27: 429–38.

Nester, E. (2008). *Agrobacterium*: The Natural genetic engineer: 100 years later. APSnet Feature Story, [http://www.apsnet.org/online/feature/Agrobact-erium/].

Patro, T. S. S. K. and Jindal, J. K. (2000). Expression of plasmid-borne characters in *Xanthomonas axonopodis* pv. *vignaeradiatae*. *Trop. Agri. Res.* 12: 86–94.

Pelczar, M. J. Jr., Chan, E. C. S. and Krieg, N. R. (2004). *Microbiology: Concepts and Applications*. 5th Ed. Tata McGraw-Hill, New Delhi, 918 pp.

Pizarro-Guajardo, M., Calderon-Romero, P., Castro-Cordova, P., Mora-Uribe, P. and Paredes-Sabja, D. (2016a). Ultrastructural variability of the exosporium layer of *Clostridium difficile* spores. *Appl. Environ. Microbiol.* 82: 2202–09.

Pizarro-Guajardo, M., Calderon-Romero, P. and Paredes-Sabja, D. (2016b). Ultrastructure variability of the exosporium layer of *Clostridium difficile* spores from sporulating cultures and biofilms. *Appl. Environ. Microbiol.* 82: 5892–98.

Prescott, L. M., Harley, J. P. and Klein, D. A. (2005). *Microbiology*. 6th Ed. McGraw-Hill Companies, New York, 1130 pp.

Singh, B., Mortezaei, N., Savarino, S. J., Uhlin, B. E., Bullitt, E. *et al.* (2017). Antibodies damage the resilience of fimbriae, causing them to be stiff and tangled. *J. Bacterol.* 199: e00665–16. https://doi.org/10.1128/JB.00665-16

Sperber, A. M. and Herman, J. K. (2017). Metabolism shapes the cell. *J. Bacteriol.* 199: e00039–e00117.

Srivastava, S. and Srivastava, P. S. (2003). *Understanding Bacteria*. Kluwer Academic Publishers, Dordrecht, 469 pp.

Stanier, R. Y. and van Niel, C. B. (1962). The concept of a bacterium. *Arch. Microbiol.* 42: 17–35.

Sundin, G. W. (2007). Genomic insights into the contribution of phytopathogenic bacterial plasmids to the evolutionary history of their hosts. *Ann. Rev. Phytopathol.* 45: 129–51.

Tortora, G. J., Funke, B. R. and Case, C. L. (2007). *Microbiology: An Introduction*. 9th Ed. Benjamin Cummings, San Francisco, CA, 958 pp.

Vivian, A., Murillo, J. and Jackson, R. W. (2001). The roles of plasmids in phytopathogenic bacteria: Mobile arsenals? *Microbiology* 147: 763–80.

Whitman, W. B. (2009). The modern concept of the procaryote. *J. Bacteriol.* 191: 2000–05.

Xu, J. and Wang, N. (2019). Where are we going with genomics in plant pathogenic bacteria? *Genomics* 111: 729–36. https://doi.org/10.1016/j.ygeno.2018.04.011

Zhao, Y., Ma, Z. and Sundin, G. W. (2005). Comparative genomic analysis of the pPT23A plasmid family of Pseudomonas syringae. *J. Bacteriol.* 187: 2113–26.

3 Variation in Bacteria

3.1 MUTATION

A sudden heritable change in the genetic material of an organism is called a **mutation**. This term refers both to the change in the genetic material and to the process by which the change occurs. An organism in which this change has occurred is called a **mutant**. A mutant results from a change in the DNA base sequence, but it may or may not give rise to a changed phenotype. Sometimes the phenotypic changes are so minor that these can be detected only through the use of special genetic or biochemical techniques.

The mutation may be either spontaneous or induced; the latter is caused by treatment with mutagens. As spontaneous mutations are very rare, mutagens are used to create mutants needed in various types of studies. The commonly used mutagens include physical agents (UV light, X-ray etc.) and chemicals such as nitrous oxide, acridine dyes etc. The mutations are also classified on the basis of nature of change, i.e., addition/deletion/substitution occurred in the DNA. The mutations that result from the change of one base in DNA are called **point mutations**, while those resulting from the removal of a part or whole of gene are known as **deletion mutations**. **Insertion mutations** result from the addition of a chunk of DNA into the base sequence of a gene.

Mutants have played an important role in helping to understand variation in bacteria. As the main source of allelic variation, they have made enormous contributions toward biological variations. Various biochemical processes and pathways have been elucidated employing different mutants. These have also contributed to an understanding of the functioning of various genes. The haploid nature of bacteria greatly facilitates the isolation and characterisation of mutants, which can be detected through either selection or screening.

3.1.1 Selection

The selection is based on the principle that the mutant is able to grow in the presence of a specific chemical while the parent cells are unable to grow. An antibiotic-resistant mutant can be isolated by growing it on a medium containing that particular antibiotic, since only the mutant cells will grow.

3.1.2 Screening

In this method, both mutants and wild-type strains are able to grow on the routine media but they are differentiated by growing on some specific media or on minimal media. MacConkey agar is used to isolate mutants unable to use lactose (Lac^-). Both Lac^+ and Lac^- bacteria can grow on this medium, but the utilisation of lactose by Lac^+ cells leads to acid production and consequently, a drop in local pH, turning the colonies pink. On the other hand, Lac^- mutant cells, being unable to utilise lactose, use amino acids as a carbon source and produce ammonia. This increases the local pH, resulting in the decolourisation of the dye due to which colonies appear white.

The detection of the mutants unable to grow on a minimal medium is generally carried out by two methods: (1) **patch replication** and (2) **replica-plating**. In patch replication, the mutagenised colonies are first grown on the master plate on the complete medium. After the growth, the colonies on the master plate are numbered on the backside of the plate and then transferred individually to a minimal medium. The mutants do not grow on a minimal medium. The cells of a parent colony, which do not grow on the minimal media, are the mutant cells. The replica-plating method developed by Lederberg and Lederberg (1952) proves more accurate and efficient than patch replication in detecting the mutants. It is also more helpful in screening different types of mutants like auxotrophic and temperature-sensitive conditional mutants. The bacterial culture exposed to the mutagen is first grown on the master plate containing the complete medium. Following the appearance of colonies, the plate is inverted and pressed against sterile velvet mounted on a wooden block. Some of the cells from each colony stick to the velvet. A sterile plate of a minimal medium is pressed onto the velvet and incubated. The failure of the imprinted colony to grow on the minimal medium indicates the presence of mutant cells in the corresponding colony on the master plate (Figure 3.1a). To select an antibiotic-resistant mutant, the medium containing that particular antibiotic is used in place of the minimal medium (Figure 3.1b).

Most point mutations revert, restoring the lost function and giving rise to a wild-type allele. This process of restoring the original phenotype is called **reversion** and the restored organism is described as **revertant**. It should be noted that reversion can be either spontaneous or induced by mutagens. The reversion rate for single point mutations is comparatively higher than those for double point mutations. By contrast, a deletion mutation does not revert at all.

3.2 RECOMBINATION

In recombination, the genetic material from two related organisms is physically exchanged. As the recombination generally occurs between two related or homologous DNAs, it is also called a **homologous recombination**. This mechanism

DOI: 10.1201/9781003681878-4

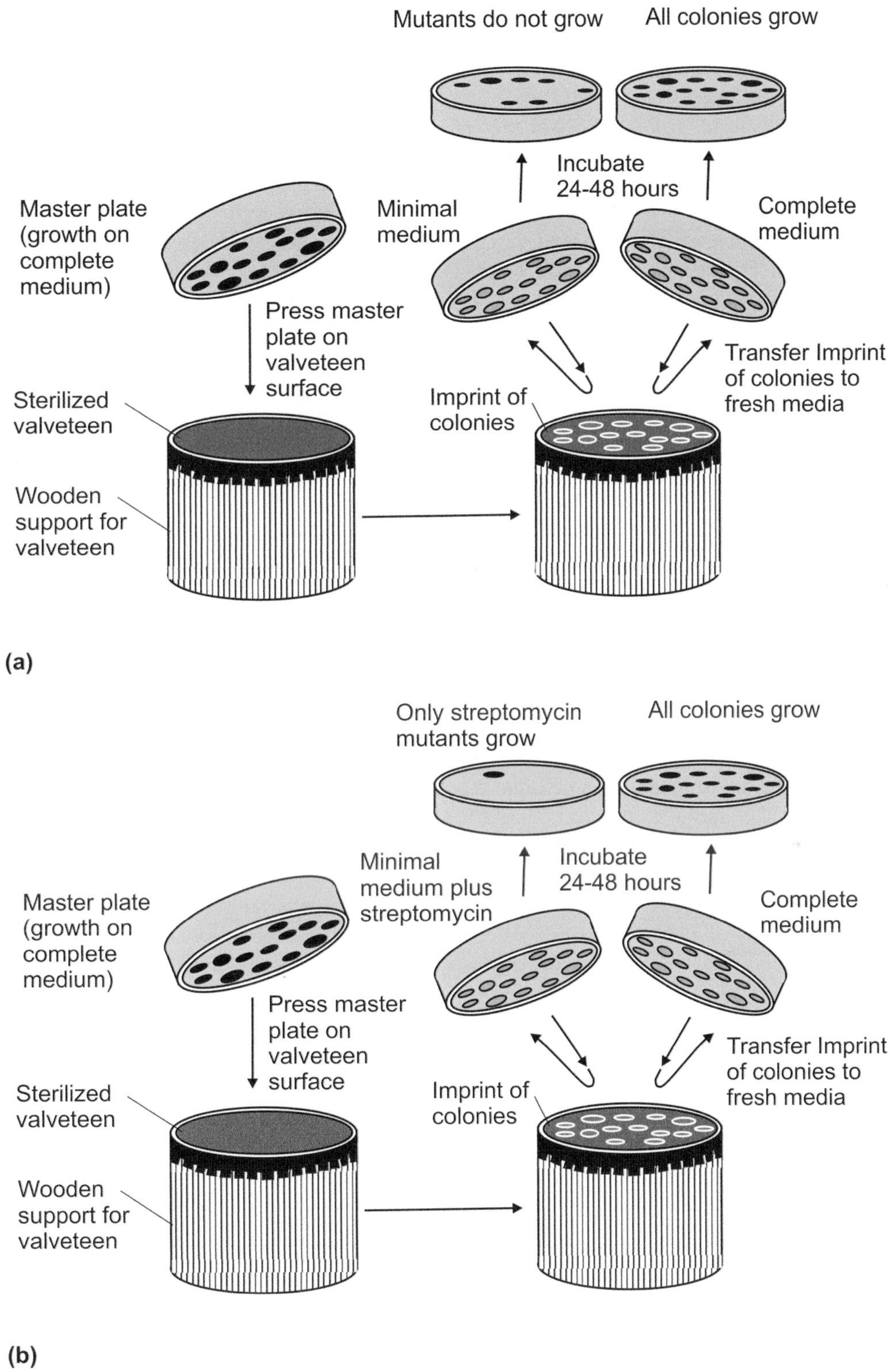

FIGURE 3.1 Replica-plating: (a) selection of auxotrophic mutants (b) selection of drug resistant mutants.

plays an important role in generating variation in bacteria. Recombination, along with mutation, has made an immense contribution to the process of evolution. Non-homologous recombinations also occur in bacteria in which the combining molecules are not similar in their DNA sequences. The integration of the phage genome into a bacterial genome, which have only very limited homology, is one typical example of this kind of recombination. Recombination in bacteria occurs through three mechanisms, namely, **transformation, transduction** and **conjugation**. The major difference between these three processes is related to their respective modes of transfer of DNA from one cell to another.

The first method, transformation, involves the uptake of DNA molecules from one bacterium by another bacterium. In transduction, the transfer of bacterial genes from a donor cell to a recipient cell is mediated by a phage. Finally, in conjugation, the genetic material from the donor cell is transferred to the recipient female cell through a **conjugation tube** or **sex pilus**. These three mechanisms can be distinguished according to two criteria, i.e., (1) sensitivity to the presence of deoxyribonuclease (DNase) and (2) the occurrence of cell contact. The occurrence of the recombination in the absence of DNase but not in its presence indicates that the recombination process is transformation. Transduction and conjugation processes are differentiated from this because the bacteria are grown in a Davis U-tube (Figure 3.2). Donor and recipient cells are placed in separate arms of this tube, and the two arms of the tube are separated by a sintered glass filter. The bacterial cells cannot pass through this filter, but DNA molecules and viruses (including bacteriophages) can. The liquid in one arm of the tube is sucked in and pushed back with force to mix the contents of both the arms. If the recombination occurs in the Davis U-tube, i.e., in the absence of cell contact, it is transduction.

3.2.1 Transformation

This can be defined as an uptake of a naked DNA molecule or fragment from the environment by a cell and the incorporation of this molecule into the chromosome of the recipient cell in a heritable form. This phenomenon was first discovered by Frederick Griffith (1928) during his work on *Streptococcus pneumoniae*. After years of research, Avery *et al.* (1944) showed that the component responsible for transformation was DNA. Their work also provided one of the conclusive indications that DNA was indeed the carrier of genetic information.

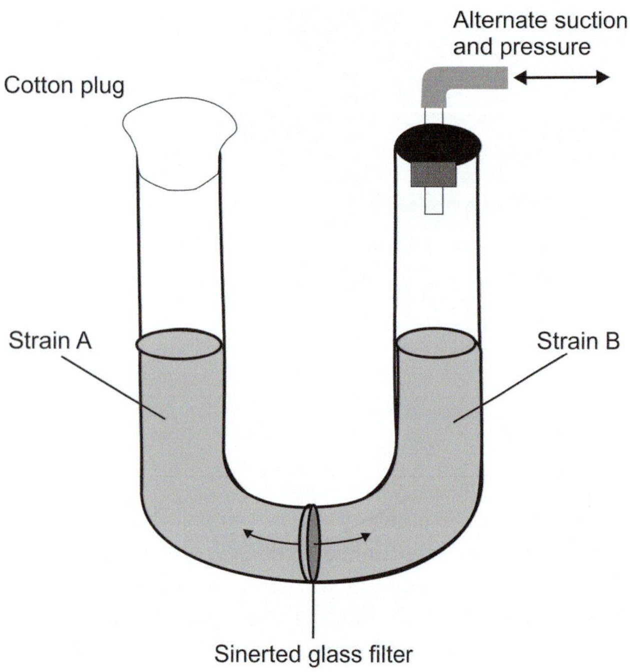

FIGURE 3.2 Davis U-tube

After lysis, bacteria release considerable DNA into the surrounding environment. Gene transfer through this process occurs in soil and marine environments and may be an important method of genetic exchange in nature. In nature, transformation does not occur in all species of bacteria. It occurs only in those species which have a competency factor. The transformation frequency of very competent cells is around 10^{-3} for most genera when sufficient DNA is available. Natural transformation has been reported in only a few genera, namely, *Bacillus, Pseudomonas, Azotobacter, Helicobacter, Acinetobacter, Haemophilus, Thermoactinomyces, Neisseria* etc. Most of the work on transformation has been carried out with *Stc. pneumoniae, Bac. subtilis* and *Hae. influenzae*, but the mechanism of transformation has been intensively studied in *Stc. pneumoniae*.

The number of bacterial cells in a culture that are in a competent state varies according to the growth conditions and the stage of the growth curve. The cells of *Stc. pneumoniae* become competent during the exponential phase and secrete a small protein called the competence factor, which stimulates the production of 8–10 new proteins needed for transformation. A component cell binds a double-stranded DNA fragment. This process is random; DNA fragments compete with each other for binding. After binding, endonucleases cleave the DNA into double-stranded fragments, ranging in size from approximately 5 to 15 kilo bases. During the uptake of the DNA, envelope-associated exonuclease hydrolyses one strand while the other strand associates with small proteins and moves through the plasma membrane. The single-stranded fragment then aligns with a homologous region of the chromosome of the recipient cell and becomes integrated with the chromosome after replacing a part of it. The replaced segment of the host DNA degrades. The integration of segment of heterologous DNA also occurs, but at frequencies very much lower than the frequencies observed using homologous DNA. The mechanism of transformation in *Hae. influenzae* differs from that of *Stc. pneumoniae* in several respects. *Hae. influenzae* does not produce any competency factor; secondly, the absorption of DNA occurs only from the closely related species. The basis for the specificity of absorbed DNA is that *Hae. influenzae* has a special 11-base pair sequence (5'AAGTGCGGTCA3') that is repeated over 1400 times in its DNA (Prescott *et al.*, 2005).

The cell walls of Gram-positive and Gram-negative bacteria are thick and complex. Hence an efficient and elaborate machinery or enzyme system is required to transport the exogenous DNA bound to a protein complex across the cell walls and membranes. In competent cells, the correct genes are expressed for the proper functioning of the transformation machinery. The expression of these genes is controlled by growth conditions. Generally, very small fragments of DNA are taken up by competent cells, but a minimum length of about 500 nucleotide-pairs seems to be needed for the integration to occur. In most cases, donor DNA fragments are about 20,000 nucleotide-pairs (or about 1/200 of the total chromosome length).

In addition to natural transformation, artificial or induced transformation can also be achieved in the laboratory through the use of different techniques. The chilling of cells in the presence of divalent cations such as Ca^{2+} makes the cell walls permeable to plasmid DNA. Cells suspended in the calcium chloride solution containing DNA are incubated on ice and then briefly heat-shocked at 42 °C for 30–120 seconds, which causes the DNA to enter the cells. This method works very well for circular plasmid DNAs. This approach also succeeds with species such as *Escherichia coli*, which are not naturally competent. To increase the transformation frequency, relatively higher concentrations of DNA, than are present in the nature, are used. An excellent preparation of competent cells gives ~10^8 transformants per microgram of plasmid, whereas a good non-commercial preparation gives 10^5 to 10^6 transformants per microgram of plasmid. This method usually does not work well with linear DNA, because exonuclease in the cell degrades the linear DNA. In contrast, naturally competent cells are more efficiently transformed with linear DNA than with plasmid. When linear DNA fragments are used for the transformation of *Esc. coli*, the cells are rendered deficient in one or more exonuclease activities to protect the DNA fragments. As plasmids are not as easily degraded as linear fragments, it is more appropriate to transform bacteria with plasmid DNA. The discovery of the calcium chloride treatment has enabled researchers to use *Esc. coli* as an efficient tool for genetic engineering. Moreover, DNA from any source can be incorporated into bacteria by splicing it into a plasmid before transformation (Prescott *et al.*, 2005).

Another method employed for the entry of plasmid DNA into the bacterial cells is electroporation, in which cells are briefly shocked in an electric field of 10–20 kV/cm. This creates holes in the cells through which the plasmid DNA, even of large plasmids, can enter. After the shock, natural membrane-repair mechanisms close these holes rapidly.

Transformation usually produces a mixture of a very few transformed cells and an abundant number of non-transformed cells. Methods are available to ascertain whether or not the transformation has occurred. The plasmids used in transformation usually contain genes which impart resistance to antibiotics. The incorporation of the antibiotic into the medium and the ability of the cells to grow on this medium confirms that transformation has taken place. Another marker used for identifying *Esc. coli* cells that have acquired recombinant plasmids is the *lacZ* gene, which codes for β-galactosidase.

As β-galactosidase is a homo-tetramer, with each monomer made up of one *lacZ-∝* and one *lacZ-ω* protein, no functional enzyme will be formed if only one of the two requisite proteins is expressed in the resulting cell. Thus, if a strain of *Esc. coli* without *lacZ-∝* in its genome is transformed using a plasmid containing the missing gene fragment, the transformed cells will produce β-galactosidase, whereas untransformed cells will not, as they are only able to produce the omega half of the monomer. In this type of transformation, the polylinker region of the plasmid lies in the *lacZ-∝* gene fragment, meaning that successfully produced recombinant plasmids will have the desired genes inserted somewhere within *lacZ-∝*. The expression of this disrupted gene fragment by *Esc. coli* does not result into the production of any useable *lacZ-∝* protein; hence, no usable β-galactosidase is formed. When grown on media containing the colourless modified galactose sugar X-gal, colonies that are able to metabolise the substrate and have been transformed but not by recombinant plasmids will appear blue, while those colonies that are not able to metabolise the substrate but have been transformed by recombinant plasmids will appear white (Wikipedia, 2008).

The introduction of DNA into animal cells is usually called **transfection**.

The evolution or emergence of new bacterial plant pathogens has long been a subject of great interest to plant pathologists or plant bacteriologists. In their review article, Barash and Manulis-Sasson (2009) describe how *Pantoea agglomerans*, a widespread epiphyte and commensal bacterium, has evolved or transformed into an hrp-dependent and host-specific tumorigenic pathogen by the acquisition of a plasmid containing pathogenicity island (PAI). The PAI was evolved on an interon plasmid of the IncN family, which is distributed among genetically diverse populations of *Pan. agglomerans*. The horizontal gene transfer was the major contributing force to the creation of PAI, although a patho-adaptive mechanism might also be involved. Plant pathogenic strains of *Pan. agglomerans* isolated from gypsophila contain a pathogenicity plasmid designated as pPATH. The pathogenicity plasmid of *Pan. agglomerans* pv. *gypsophilae* (*Pag*), designated as $pPATH_{Pag}$, is approximately 135 kb and it contains a pathogenicity island of ~75 kb. The $pPATH_{Pag}$ harbours an *hrp*/*hrc* gene cluster, type III effectors, a gene cluster encoding for biosynthesis of indole acetic acid (IAA) and cytokinins, and many diverse insertion sequence elements and partial sequences from known genes of other bacteria. Mutations in genes encoding components of the type III secretion system or the *hrp*/*hrc* regulatory gene cascade completely abolished gall formation while mutations in genes encoding type III effectors or the biosynthesis of IAA and cytokinins reduced gall development, either substantially or almost totally. These findings suggest that the pathogenicity of *Pan. agglomerans* pv. *gypsophilae* and *Pan. agglomerans* pv. *betae* is hrp-dependent, and that the acquisition of the plasmid-borne PAI was a crucial step in the transformation of *Pan. agglomerans* into a gall-forming and host-specific plant pathogen.

The pathogenicity plasmid of *Pan. agglomerans* pv. *gypsophilae* is nonconjugative. The nonconjugative nature of this plasmid was also supported by the lack of putative genes that showed significant homology to *tra* genes. Plant-synthesised IAA and cytokinins are absolutely necessary for gall initiation, whereas the pathogen-secreted phytohormones are not needed. However, pathogen-secreted IAA and cytokinins do have a significant effect on gall size and might be important for the expression of the *hrp* regulatory genes as signalling molecules.

3.2.2 Conjugation

Conjugation, also known as **sexual reproduction**, is the transfer of genetic material from one bacterium to another bacterium by direct cell-to-cell contact. The transfer of genetic material is unidirectional, i.e., only from donor to a recipient cell. The donor cells possess F factor, also called **fertility** or **sex factor**, and are designated as F^+. The recipient cells do not contain F factor and are designated as F^-.

The initial evidence for conjugation was provided by Lederberg and Tatum (1946) while working with the K-12 strain of *Esc. coli*. They used two auxotrophic strains of *Esc. coli* K-12, one strain required biotin (Bio^-), phenylalanine (Phe^-) and cysteine (Cys^-) for growth while the other needed threonine (Thr^-), leucine (Leu^-) and thiamine (Thi^-). Both the strains were inoculated together, incubated for several hours and then plated on the minimal medium. Recombinant phototrophic colonies appeared on the minimal medium after incubation, indicating that the chromosomes of both the auxotrophs came into contact and underwent recombination. They purposely used double or triple auxotrophs, assuming that two or three reversions would not often occur simultaneously. However, Lederberg and Tatum (1946) did not prove that the bacterial cells came into direct contact with each other for the transfer of genetic material. This evidence was later provided by Davis (1950), who used a U- tube (Figure 3.2) fitted with a sintered glass filter in the centre. The filter allows the passage of medium and very small particles, but not bacterial cells. Each side of the tube filled with nutrient medium was inoculated with a different auxotroph of *Esc. coli*. During the incubation, the medium was sucked in and pushed back with force to ensure an exchange of medium between the two halves. After incubation, the cultures from both the sides were plated on the minimal medium. The non-appearance of colonies indicated the lack of gene transfer because the bacterial cells did not come into contact with each other. Conjugation is mainly of the following three kinds.

3.2.2.1 $F^+ \times F-$ Mating

This is a basic type of conjugation. The F^+ strains contain an extra chromosomal F factor carrying the genes for pili formation and plasmid transfer. It is a circular DNA molecule, contains 19 genes and replicates independently of bacterial chromosome. The donor (F^+) and recipient (F^-) cells of *Esc. coli* do not differ morphologically, except that the donor or male cells contain F pilus or sex pilus. Each male cell may carry 1–5 pili, which are 1–2 µm long and formed at a right angle to the cell surface. These are hollow, made up of a protein called pilin, with an external diameter of 85 A° and an axial hole of 20–25 A° in diameter. The sex pili unite the donor and recipient cells together and also act as a conduit for the transfer of the genetic material. At the time of transfer, one strand of F factor is nicked; it starts to unwind and separates from the other strand. The nicked strand begins to move to the F^- cell through the pilus; during this process, the entering strand is replicated to produce double-stranded DNA. The circular strand left behind in the F^+ cell is also replicated. Eventually, the nicked strand completely passes through to the recipient cell and is completely replicated. The pilus is broken, separating the cells. As a result of this process, F factor is introduced into a recipient cell, making it F^+ (Figure 3.3a). This new F^+ cell can then act as a donor and can initiate conjugation with another F^- cell. As both the cells contain an F factor, both cells are F^+. As the bacterial chromosome genes are rarely transferred with the independent F factor, the level of recombination frequency is low. This is due to the fact that the F factor is not homologous to the DNA of a bacterial chromosome.

Most of the work on plasmids and conjugation has been carried out with *Esc. coli* and other Gram-negative bacteria. However, some Gram-positive bacterial genera, namely, *Enterococcus, Streptococcus, Bacillus, Staphylococcus* and *Streptomyces*, also contain self-transmissible plasmids. Very little is known about the transfer of genetic material in such cases. The recipient cells of *Esc. faecalis* release short peptide chemical signals, which activate transfer genes in donor cells having the required plasmid. Activated donor cells release special plasmid-encoded proteins, which allow donor and recipient cells to adhere. Thereafter, the plasmid transfer occurs.

3.2.2.2 Hfr Conjugation

This is also known as Hfr × F^- mating. Hfr donor strains give a 10^3 times higher frequency recombination than F^+ strains and hence are called **high-frequency recombinants** (Hfr). In F^+ cells, the F factor/plasmid occurs in an autonomous state, independent of bacterial chromosome. Occasionally, the F factor becomes integrated into bacterial chromosome at a random location by recombination between homologous insertion sequences (IS) present on the plasmid and the host chromosome. In *Esc. coli*, the sites of integration are IS with F plasmid carrying two copies of IS 3 and one copy each of IS 2 and y δ (TN 1000). The bacterial chromosome carries several copies of these homologous elements, which can be used by F plasmid for integration. The nature and orientation of chromosome transfer of an Hfr is governed by the location and orientation of these elements. The transfer of chromosome by Hfr is initiated by nicking the chromosome at the origin of transfer (*OriT*) of the integrated F plasmid. As it is replicated, the chromosome moves into the recipient cell, the leading part being origin of transfer along with a part of F plasmid followed by the chromosome. As the main body of the F plasmid is at the tail end of the chromosome being transferred, it rarely gets transferred. Hence, the recipient cell is rarely converted into a F^+ strain (Figure 3.3b). The DNA received from the Hfr donor can undergo recombination with the host chromosome at a high frequency, as the DNA sequences are homologous. This recombination can

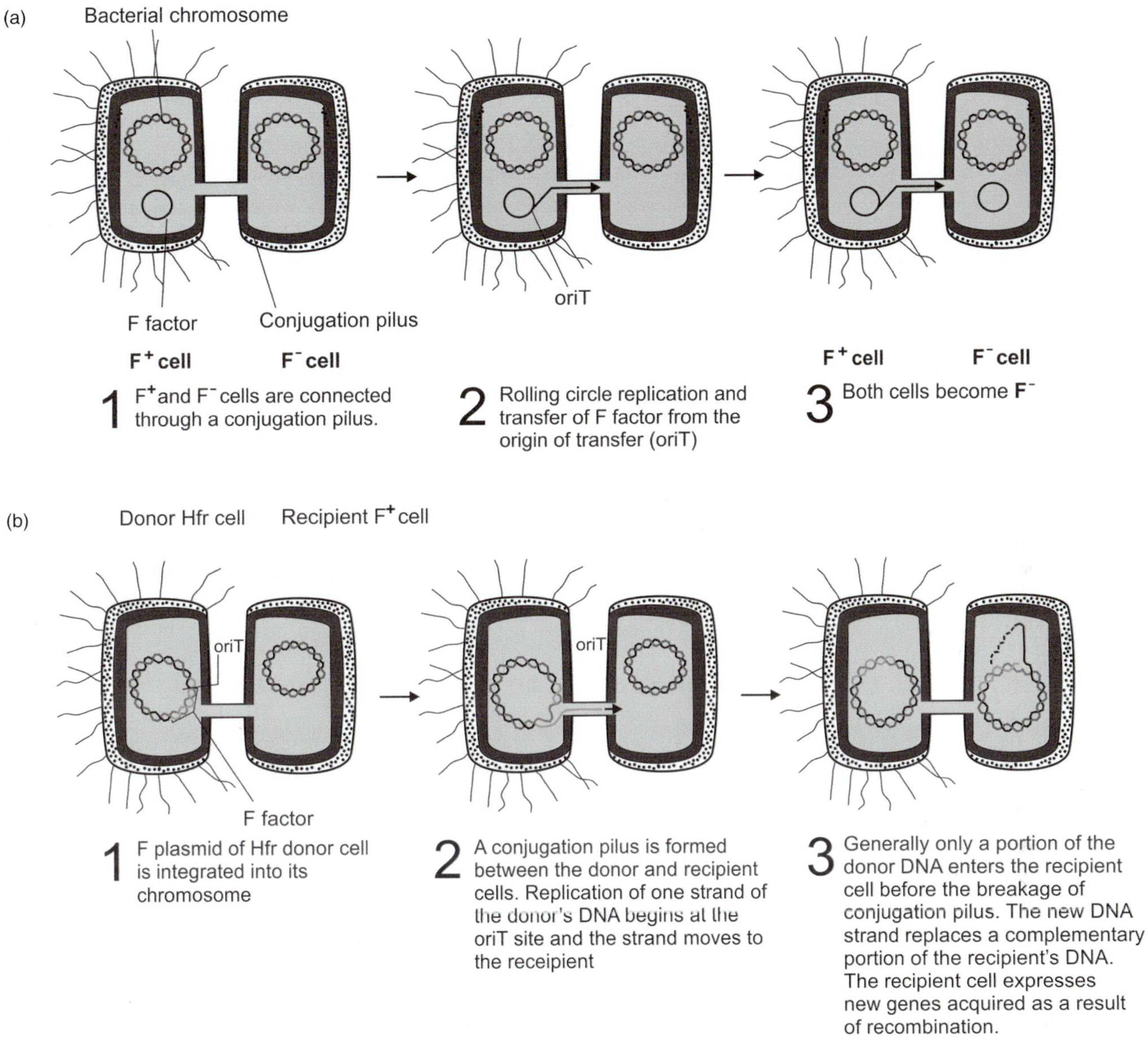

FIGURE 3.3 (a) Conjugation between an F$^+$ bacterial cell and an F$^-$ bacterial cell (b) conjugation between an Hfr bacterial cell and an F$^-$ bacterial cell.

also result in gene conversion, if the transferred DNA and the corresponding region of the host DNA contain different alleles of the same gene.

The finer details of chromosome transfer by Hfr donor were revealed by Jacob and Wollman (1955) using the **"interrupted-mating"** procedure. They disrupted the mating at different time intervals by vortexing the mating population. The cells were then plated to determine their genotypes. In general, the transfer starts from a specified point, i.e., origin of transfer or *OriT*, and proceeds linearly while the two cells remain in contact. There is a clear relationship between the time at which a given marker is transferred from Hfr to F$^-$ and its map location on the chromosome. After ascertaining the time of entry of several other donor markers in the subsequent studies, a temporal map of *Esc. coli* was constructed indicating the location of genes on a time scale (in minutes). These findings also revealed that chromosomal transfer by Hfr is never complete, leading to only a part of the chromosome being transferred. The markers which are close to the origin of transfer are transferred efficiently and those which are distant do not get a chance to be transferred and hence do not appear in the recombinants. Therefore, it leads to the formation of a **merozygote**. The map of *Esc. coli* has been standardised for 100 temporal map units. Time taken for complete transfer of Hfr chromosome to F$^-$ is 100 minutes (Srivastava and Srivastava, 2003).

3.2.2.3 F′ × F– Mating

This is also called F′ conjugation or F-mediated sexduction. As F plasmid is an episome, it can leave the bacterial chromosome. During this process, if only the plasmid portion is excised from the chromosome, the Hfr cell becomes an F^+ cell again. However, sometimes the excision is not proper and the F factor takes a segment of the bacterial chromosome along with it. When a part of the chromosomal DNA becomes associated with the episome, this is called an F′ or F-prime. The F′ factors range in size from those carrying a single bacterial marker to those carrying up to half of the bacterial chromosome. F′ plasmid is transferred to the F^- cell and it converts the recipient cell into an F′ cell. Adelberg and Burns (1959) were the first to identify the F′ factor in *Esc. coli*.

The mechanism of F′ factor transfer to the recipient F^- cell is the same as that of F factor transfer in $F^+ \times F^-$ mating and chromosome transfer in Hfr × F^- mating. The recipient cell, having its own complete chromosome, becomes diploid for the segment of the chromosomal DNA on the episome. Such a partial diploid cell is called a **merozygote**. The chromosomal DNA on the episome can undergo recombination at high frequency with its homologous sequence on the chromosome. F′ conjugation is very important to microbial geneticists because F′ factors play important roles in the genetic analysis of bacteria. Partial diploidy provides an important method for determining a dominance relationship between alleles and defining genes by complementation tests in bacteria. As F′ factors are available carrying almost any part of *Esc. coli* chromosome, sexduction analysis can be made for almost any mutation of interest. The creation of F′ plasmid is also useful in mapping the chromosome.

3.2.3 Transduction

Zinder and Lederberg (1952) discovered another mechanism of gene transfer in bacteria while working with *Salmonella typhimurium*. They mixed two multiple auxotrophic strains, one requiring phenylalanine, tyrosine and tryptophan (phe⁻ tyr⁻ trp⁻) and the other requiring methionine and histidine (met⁻ his⁻). After incubation, when this mixture was plated on an amino acid-free minimal medium, prototrophic colonies appeared at a frequency of ~10^{-5}. The appearance of the prototrophic strain even after placing the auxotrophic strains in two arms of a Davis U-tube showed that the cell-to-cell contact was not required and hence this process is different from conjugation. The transfer of the genetic material was due to a filterable agent, which was identified as phage P 22 of *Sal. typhimurium*. This process of gene transfer mediated by a phage was referred to as **transduction**.

In 1956, Lederberg and associates discovered the transfer of *gal* marker in *Esc. coli* through Lambda (λ) phage. However, there are remarkable differences in terms of the gene transfer carried out by these two phages, namely, P 22 and Lambda. The former can transfer virtually any gene, leading the process to be called **generalised transduction**. By contrast, phage Lambda transfers only *gal* or *bio* or, very rarely, both the genes, earning the name **restricted** or **specialised transduction**. These two transduction pathways are related in a way that both employ a phage for the transfer of genetic material, but differ in many other respects. The recombinant cells formed through either of these pathways are called **transductants**.

3.2.3.1 Generalised Transduction

Phage P 22 of *Sal. typhimurium* and phage P 1 of *Esc. coli* are well-known examples which bring about generalised transduction, although many other phages are also capable of achieving this. During the lytic cycle of the phage, viral DNA and proteins are synthesised for the assembly of phage particles. At the same time, the bacterial genome is degraded into random DNA fragments. By mistake, in some of the protein capsids, DNA fragments of bacterial genome are packed instead of viral DNA; these particles act as transducing phages (Figure 3.4). The evidence for the presence of bacterial DNA, but not a mixture of bacterial and viral DNA, was provided by Ikeda and Tomizawa in 1965 based on the density labelling of the DNA of donor cells with 5-bromouracil before infection and replacing 5-bromouracil with thymidine at the time of phage infection. The quantity of the bacterial DNA carried by transducing phages is governed principally by the size of the capsid. Phage P 22 carries about 1% of the bacterial genome while in phage P 1 of *Esc. coli* and other Gram-negative bacteria, it ranges from 2.0–2.5% of the genome.

Transducing phages are capable of injecting the DNA into other bacterial cells; this does not cause lysis of the cells, however, as they do not contain viral DNA. They simply act as a carrier of genetic information from the donor bacterium to the recipient bacterium. As the DNA fragment enclosed in the capsid can be any segment of the bacterial genome, it will transfer any gene(s); hence it is called **generalised transduction**. After the transducted DNA is delivered to the recipient cell, three possible fates await this DNA: (a) degradation by nucleases, (b) recombination with homologous recipient DNA, and (c) occurrence in the non-replicating stable form. About 70–90% of the transferred DNA is not integrated, but often it is able to survive and express. The bacteria which contain this non-integrated DNA are called **abortive transductants** and are **partial diploids**. The process of occurrence of donor DNA in a recipient cell without recombination and replication is known as **abortive transduction**. The non-integrated DNA exists in the cells in a circular super-coiled form and the circularity is brought about by the association of ends with a protein. Its absence in infectious phage particles is explained as being cellular rather than a phage-encoded protein. Circularised DNA is protected from degradation by the enzymes. Once destined to abort, this DNA does not form a complete transductant. Complete transductants are formed when transduced DNA integrates with the genome of the recipient cell (Srivastava and Srivastava, 2003).

3.2.3.2 Specialised or Restricted Transduction

This involves the transfer of either one or a few specified genes from the donor to the recipient cell. It is caused by

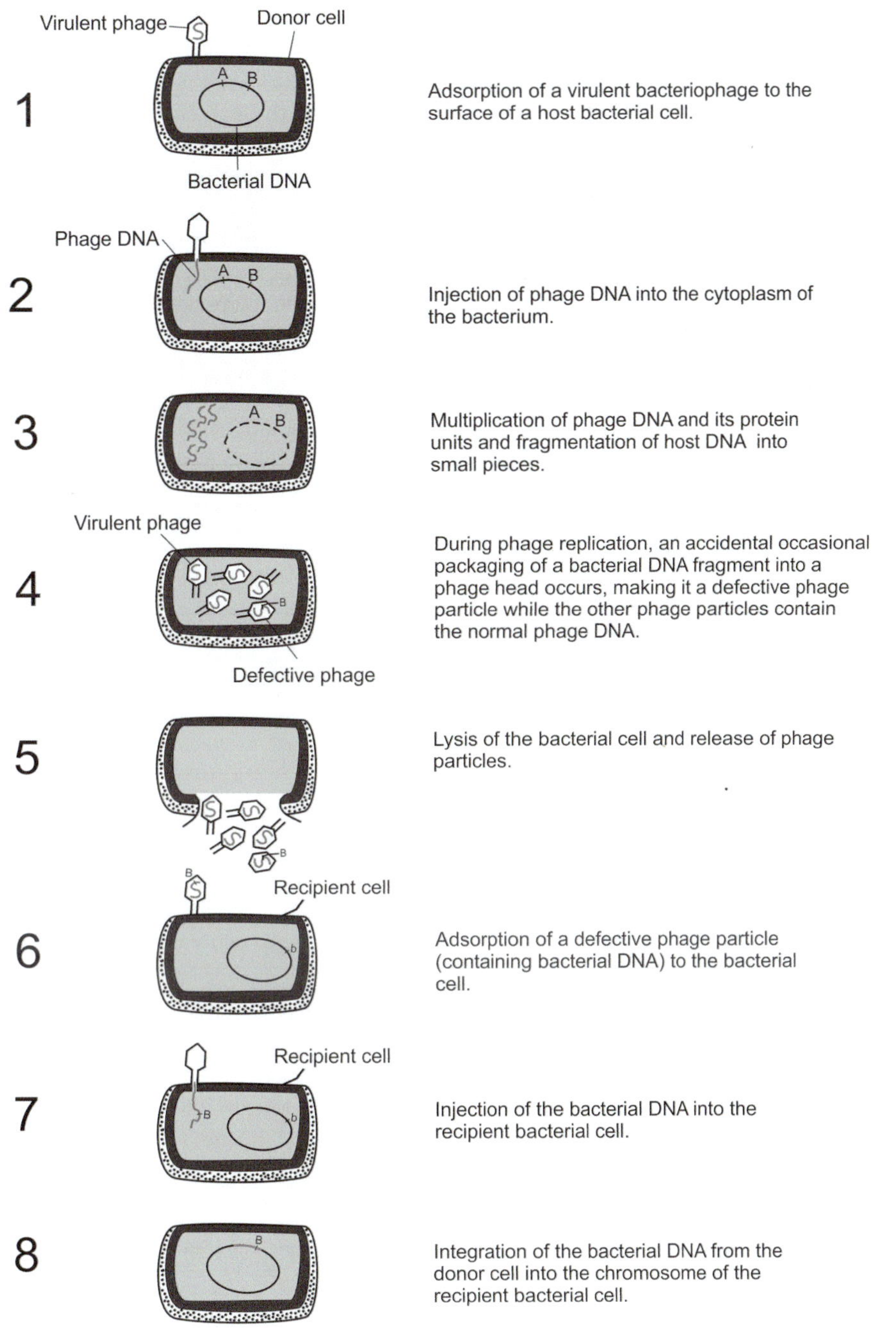

FIGURE 3.4 Generalised transduction.

the temperate phages, which attach their genomes at one or a few specified sites of the bacterial chromosome. Phage λ is the best-studied example for this purpose. As mentioned earlier, this phenomenon was discovered by Lederberg and his associates in 1956 while they were trying to discover whether phage λ could transfer genes of *Esc. coli* from a donor to a recipient cell. They lysogenised prototrophic wild-type *Esc. coli* K-12 strain with phage λ and treated this bacterium with UV irradiation to produce phage progeny. Different kinds of non-lysogenic mutants (deficient in different characters) of *Esc. coli* were infected with these phages and then grown on different selective media to know the gene transfer. The results were negative in all except one case; in the exception about 10^{-6} *gal*$^-$ (deficient for galactose utilisation) cells were converted into *gal*$^+$ after phage infection. Thus, it became evident that phage λ can also transfer the genes from one cell to another, but the transfer is limited to only a few gene(s). Later on, it was found that phage λ can also transfer *bio* (for biotin utilisation) gene and rarely both *gal* and *bio* genes (Figure 3.5).

The life cycle of λ phage and the generation of transducing phage in *Esc. coli* is given below. After adsorption to the host cells, λ phage injects its DNA into the host cell. The DNA, which is originally double-stranded, circularises and finally gets converted into a covalently closed circular DNA. Then the phage genome is attached to the bacterial chromosome at a site called *att* λ situated between gal and bio genes. After integration, it multiplies as prophage with the multiplication of bacterial cells and is passed on to all the progeny cells. It can also be transferred through conjugation to the recipient cells. Temperate phages undergo spontaneous transition from the lysogenic state to the lytic state but it is very rare, i.e., occurring in about one in 10^5 cell divisions. The change from the lysogenic state to the lytic state can also be induced artificially by irradiation with ultraviolet light. During this change, prophage is freed from the host chromosome. The excision of prophage from the host cell is site-specific, as with the integration process. Usually, the excision process is very precise and the phage genome is cut out exactly in the form in which it existed prior to integration. However, on rare occasions the error occurs in the excision of phage genome; in such cases, a segment of the bacterial chromosome is excised with the phage genome leaving behind a segment of the phage genome. Only the host genes situated close to attachment sites of the prophage are excised with the phage genome and packaged in the phage particles. As the genome of λ phage is attached between *bio* and *gal* genes of *Esc. coli* chromosome, it can take along with it either the *gal* gene or the *bio* gene, or, very rarely, both. This event is similar to the excision of the F plasmid and the creation of F′s. Another specialised transducing phage 80 integrates near *trp* (for the synthesis of tryptophan) genes of *Esc. coli* chromosome and transduces *trp* marker.

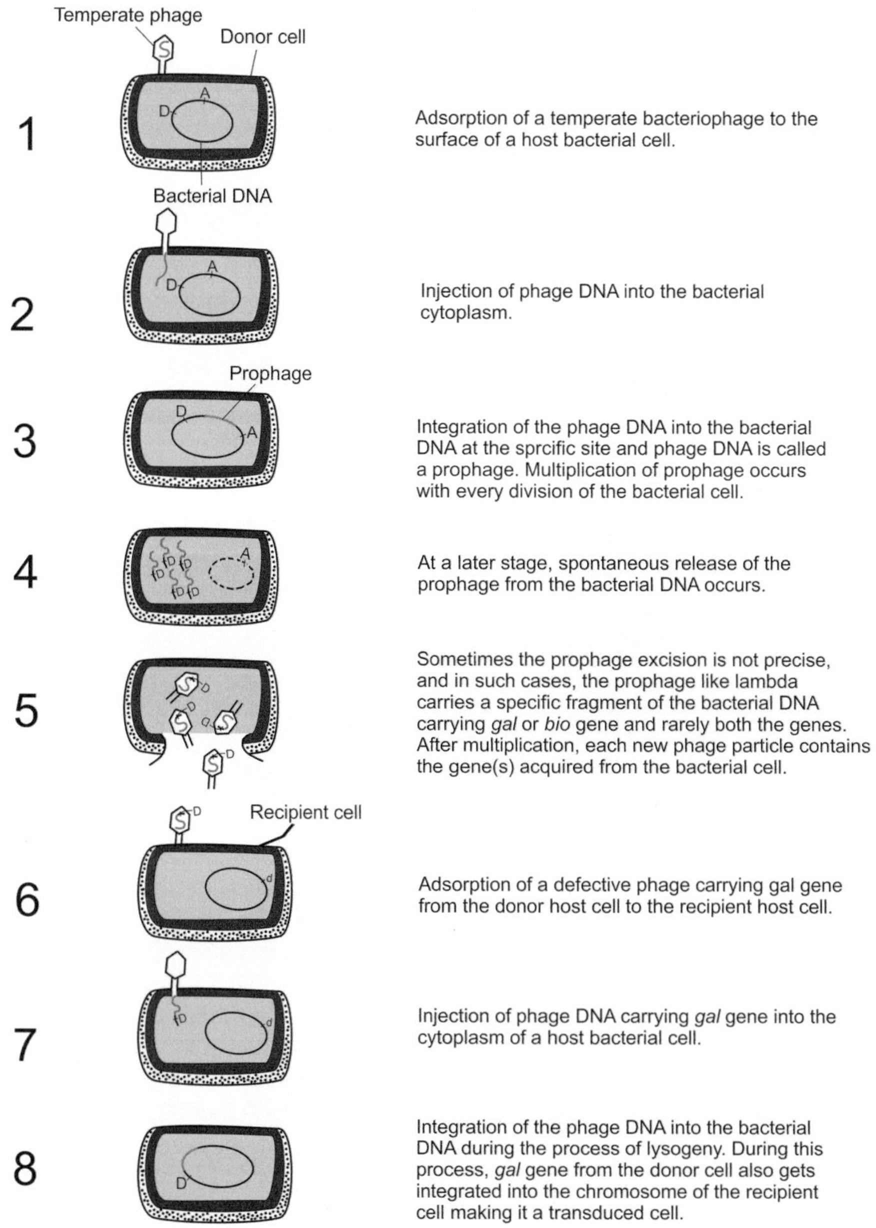

FIGURE 3.5 Specialised transduction.

The induction of *Esc. coli* cells lysogenised with λ phage produces a mixture of abundant normal phage particles and very few, i.e., about 10^{-6} defective transducing particles. As the lysate contains very few transducing particles, it is often called low-frequency transduction lysate (LFT lysate). The transducing λ *gal* phage, formed due to the integration of *gal* gene into λ genome, is defective and is called λ *dgal* (lambda defective *gal*). It has been rendered defective due to the replacement of phage genes, required for growth and maturation under lytic conditions, with bacterial DNA. Since it is defective, it cannot multiply without the help of another phage called a helper phage. A wild-type λ phage, or any other appropriate phage that can provide the missing function of the defective phage, can act as a helper phage. In contrast, the creation of λ *bio* phage does not result in the loss of essential genes, generating a normal phage called λ *pbio*. Of these two phages, only λ *pbio* can multiply at a low multiplicity of infection, i.e., only one phage particle infects a single host cell. λ *dgal* reproduces at a high multiplicity of infection, meaning that a host cell is infected with one particle of λ *dgal* and one or more particles of wild-type λ phage.

λ phage can transfer bacterial genes in two ways. In the first method, two crossovers replace the *gal*$^-$ gene of the recipient's chromosome with donor *gal*$^+$ gene. The transductants produced by this method are haploid, stable and non-lysogenic. In the second method, there is a single crossover between the recipient and phage-carried alleles. This results in the addition of the phage genome and donor genes to the genome of the recipient cell without any replacement of donor DNA. The transductants produced as a result of single crossover are partially diploids *gal*$^+$/*gal*$^-$ heterogenotes, unstable and lysogenic. As the *gal*$^+$/*gal*$^-$ heterogenotes are unstable, the segregation of *gal*$^-$ cells occurs at a frequency of about one in 1000 cell divisions. Since λ *dgal* phage particles cannot reproduce without the help of a helper phage, λ *dgal* chromosome is excised and may simply be lost or diluted out during cell divisions. Recombination may also occur between *gal*$^+$ exogenote (donor DNA fragment) and a *gal*$^-$ endogenote (recipient chromosome) resulting in the transfer of gal$^+$ marker to the endogenote and the production of stable *gal*$^+$ transductants (Gardner *et al.*, 1991). The host cells infected by both the phages, i.e., λ *dgal* and a normal λ (λ$^+$) phage, are called double lysogens. This results from the insertion of λ$^+$ genome at attachment site *attB* and that of λ *dgal* on the left attachment site. The inducement of these heterozygous lysogens with UV irradiation produces 50% λ *dgal* particles and 50% λ$^+$ particles in the lysate. Such lysates are called high-frequency transduction (Hft) lysates, because in these lysates 50% of the phage particles contain transducing *gal* genes.

Transduction is an important tool for transferring genes in bacteria. It is also used in genetic mapping, although it is not an efficient and reliable tool for this purpose. Hft lysates greatly facilitate the genetic analysis by increasing the frequency of transduction events. As the amount of bacterial DNA carried by a phage is limited, the transduction is suitable for transferring a small, but specified portion of a bacterial genome. Many specialised phages or their regions have been used to develop cloning vectors.

Li *et al.* (2009) reviewed the current bacterial genotyping methods and classified them into three main categories: (1) DNA banding pattern-based methods, which classify bacteria according to the size of fragments generated by amplification and/or enzymatic digestion of genomic DNA, (2) DNA sequencing-based methods, which study the polymorphism of DNA sequences, and (3) DNA hybridisation-based methods using nucleotidic probes. They described and compared the applications of genotyping methods to study the bacterial strain diversity. The authors also discussed the selection of appropriate genotyping methods and the challenges of bacterial strain typing, described the current trends of genotyping methods and investigated the progresses allowed by the availability of genomic sequences.

Whole-genome sequencing, in particular, is becoming an increasingly feasible and cost-effective approach for near real-time high-resolution strain typing, genomic characterisation and comparative analyses. In his review article, Maccannell (2013) has discussed the current state of the art in bacterial strain typing for outbreak investigation and infectious disease surveillance, and the impact of emerging genomic technologies in the field of public health microbiology.

Sindt *et al.* (2018) presented a novel workflow that applies the statistical method of partial least squares discriminant analysis (PLS-DA) to MALDI-TOF-MS protein fingerprint data of *Xan. axonopodis*. Mass spectra of 32 *Xan. axonopodis* strains was used to create a mass spectral library and PLS-DA was employed to model the closely related strains. A robust workflow was designed to optimise the PLS-DA model by assessing the model performance over a range of signal-to-noise ratios (s/n) and mass filter (MF) thresholds. The optimised parameters were observed to be s/n = 3 and MF = 0.7. The model correctly classified 83% of spectra withheld from the model as a test set. A new decision rule was developed, termed the rolled-up Maximum Decision Rule; this method improved identification rates to 92%. These results demonstrate that MALDI-TOF-MS protein fingerprints of bacterial isolates can be utilised to enable identification at the strain level. Furthermore, the open-source framework of this workflow allows for broad implementation across various instrument platforms as well as integration with alternative modelling and classification algorithms.

With the availability of genome sequence information for four strains of *Xyl. fastidiosa*, Doddapaneni *et al.* (2006) performed genome-wide analyses to identify and categorise DNA variations and to understand their influence on strain functional divergence. They identified INDELs and strain-specific genes as the main source of variations among strains, with individual strains showing different rates of genome evolution. Based on these genome comparisons, it appears that the Pierce's disease strain Temecula-1 genome represents the ancestral genome of *Xyl. fastidiosa*. Results of this analysis are publicly available in the form of a web database.

Straub *et al.* (2021) have provided an overview of key methods in bacterial population genomics and described recent work focusing on three topics of critical importance to plant pathology: (i) resolving pathogen origins and transmission pathways during outbreak events, (ii) identifying the genetic basis of host specificity and virulence, and (iii) understanding how pathogens evolve in response to changing agricultural practices.

3.2.4 Insertion Sequences and Transposons in Phytopathogenic Bacteria

The simplest form of a transposon is an insertion sequence (IS), also known as IS element. It is a short DNA sequence that functions as a basic transposable element. Insertion sequences are characterised by two main features: (i) they are relatively small compared to other transposable elements, typically ranging from 700 to 2,500 base pairs (bp) in length, and (ii) they only code for proteins involved in the transposition process. This distinguishes them from transposons, which also carry accessory genes, such as those conferring antibiotic resistance. The proteins encoded by insertion sequences usually include a transposase, which catalyses the enzymatic reaction that enables the IS to move, and a regulatory protein that either stimulates or inhibits transposition activity. The coding region of an insertion sequence is typically flanked by inverted repeats. For instance, the well-known IS*911* (1,250 bp) is bordered by two 36 bp inverted repeats, and its coding region contains two partially overlapping genes, *orfA* and *orfAB*, which encode the transposase (*OrfAB*) and a regulatory protein (*OrfA*). Insertion sequences and transposons are often named using the format IS*N* or Tn*N*, where 'N' is a number (e.g., Tn*6768*, Tn*6769*, Tn*6770*, IS*1*, IS2, IS*1112*, etc.). In addition to occurring autonomously, insertion sequences may also occur as parts of composite transposons; in a composite transposon (also known as a "complex transposon"), two insertion sequences flank one or more accessory genes, such as an antibiotic resistance gene (e.g. Tn*10*, Tn*5*) or a group of other genes (ISXo5).

All sequenced *Xanthomonas* genomes contain numerous IS elements, but the rice pathogen *Xan. oryzae* pv. *oryzae* (*Xoo*) genomes contain the most diverse pool (Monteiro-Vitorello *et al.*, 2005). Of the 19 known families of IS elements, eight families composed of 28 distinct elements appear in well-documented genomes of *Xoo*. MAFF311018 and KACC10331 have nearly identical numbers of IS elements, while PXO99^A contains fewer elements overall, but more copies of ISXo*8*, IS*1114* and ISXo2. In *Xoo* strain MAFF311018, 611 insertion sequences (ISs) of 25 types make up about 10% of the genome, a significantly higher percentage than in other plant pathogenic bacteria. These ISs are scattered across the genome, often forming tandem repeats and IS "islands" of up to 30 kb. Of these, 386 are full-length ISs, while 225 are truncated. Other *Xoo* genomes have between 251 and 267 complete IS elements. In the PXO99A strain, at least ten major genomic rearrangements relative to MAFF have produced 29 distinct syntenic blocks. These rearrangements are largely mediated by transposable elements like ISXo5, ISXo8 and IS1389/ISXoo3. For example, ISXo5 is located near the ends of a 212,087 bp tandem repeat, which includes a 116,872 bp inversion flanked by inverted ISXo5 copies. The *Xoo* strain KACC10331 contains five key IS elements and has more than twice the number of transposable elements compared to genomes of other *Xanthomonas* species. Among the 478 protein-coding sequences in these IS elements, 271 are similar to transposases, indicating their role in horizontal gene transfer and genomic rearrangements. Additionally, 207 genes in this strain are associated with mobile genetic elements, including 37 prophage-related genes.

Ochiai *et al.* (2005) presented the complete genome sequence of *Xoo* strain MAFF 311018. The size of the genome was 4,940,217 bp, in a single circular chromosome. The genome structure of *Xoo* MAFF 311018 was characterised by large numbers of effector (avr) genes of the avrBs3/pth family and insertion sequences (ISs). RFLP analysis of diverse strains using ISXol as a probe suggests that the prevalence of mobile elements in this species, which can bring about genome inversions and rearrangement, may have played a major role in generating the high degree of genetic diversity and race differentiation characteristic of this pathogen. The *Xoo* MAFF 311018 sequence was also highly similar to those of *Xan. axonopodis* pv. *citri* and *Xan. campestris* pv. *campestris*, with the exception of the large number of effectors and IS elements, and numerous inversions and rearrangements.

3.2.4.1 How Do Transposons Affect Diversity in Phytopathogenic Bacteria?

Transposons or insertion sequences (IS elements) can function as promoters or terminators due to their strong polarity on downstream genes, as they contain both RNA and protein termination signals. In some cases, specific insertion mutations, such as those involving IS*1*, IS*2*, IS*3* and IS*5*, result in significant expression of downstream genes, likely because the transposon provides a new promoter for those genes. Conversely, IS elements can cause insertional mutagenesis by inserting into coding or regulatory regions of the genome, disrupting gene function or altering gene expression. This process can create genetic diversity within a population, generating variants with different pathogenic traits. For instance, the loss of extracellular polysaccharide production in *Xoo* was observed due to IS-mediated inactivation of the *gumM* gene, part of the *gum* cluster responsible for extracellular polysaccharide biosynthesis (Rajeshwari and Sonti, 2000). Similarly, IS elements disrupted the *egl* (endoglucanase) gene in *Ralstonia solanacearum*, reducing virulence by impairing the bacterium's ability to degrade plant cell walls.

In *Ral. solanacearum* strain GMI1000, the active transposition of IS elements led to the disruption of genes involved in stress response and metabolism, suggesting an ongoing role of IS elements in fine-tuning the genome for optimal fitness in varying environments. Transposons can also facilitate horizontal gene transfer (HGT) by moving between different bacterial strains or species, carrying virulence-associated genes such as those encoding type III secretion system effectors, which enhance virulence or resistance. This process increases the genetic diversity of the *Ral. solanacearum* species complex, contributing to the emergence of more virulent strains. Additionally, complex transposons can increase a bacterium's total genome content through gene duplication events. For example, in the PXO99^{A} strain of *Xoo*, a near-perfect tandem duplication of 212,087 bp is flanked by the ISXo5 element, suggesting this IS element may have caused the duplication event (Salzberg *et al*., 2008).

It was previously believed that the introduction of resistance genes in rice agro-ecosystems was the primary factor driving pathogenic and genetic diversity in phytopathogenic bacteria. However, race diversity in these bacterial pathogens can now be partially attributed to the role of insertion sequences (ISs). The *Xoo* genome contains the highest number of IS families and copy numbers, particularly flanking avirulence (*avr*) gene clusters or pathogenicity islands (Salzberg *et al*., 2008). The ability of ISs to cause deletions, inversions and other genetic rearrangements plays a crucial role in the bacterium's genetic diversity. The presence of diverse IS families in specific *Xoo* races also provides insight into the pathogen's rapid evolution. Additionally, more than half of the protein-coding sequences within the identified IS elements of the *Xoo* genome show significant similarity to transposases, suggesting that ISs have played a vital evolutionary role in facilitating horizontal gene transfer and internal genomic rearrangements. Transposons play a critical role in increasing genetic variability in plant pathogenic bacteria, driving their evolution and adaptability by inducing mutations, facilitating gene transfer and promoting genomic rearrangements. This genetic variability is crucial for the survival and success of these pathogens in the face of environmental challenges and host defences.

3.3 PLASMIDS

Plasmid is an extra chromosomal unit of genetic material that replicates independently of a chromosome and is stably inherited to the offspring. Most, but not all, plasmids are dispensable, because they are not essential for the survival of cells in which they reside. However, in some cases they are required under certain environmental conditions. The presence of R-plasmid is essential in a bacterium in enabling it to grow on a medium containing that particular antibiotic to which it imparts resistance. Based on their ability to mediate conjugative self-transfer, they are divided into two groups, namely, conjugative and non-conjugative plasmids. Conjugative or transmissible plasmids mediate transfer of DNA by conjugation. All F and F′ plasmids, many R plasmids and some col plasmids belong to this group. By contrast, non-conjugative or non-transmissible plasmids do not mediate transfer of DNA by conjugation. Many R and col plasmids are non-conjugative.

The ability of conjugative plasmids to transfer the genetic material has played a major role in bacterial evolution. This ability has also been used to affect horizontal gene transfer not only between the strains of the same species but also across the species, and even in organisms belonging to different kingdoms. The most important achievement made through the use of plasmids is their role as vectors in genetic engineering experiments. Due to their some desirable properties, they have been used extensively, as such or after required modifications, as cloning vectors in molecular biology.

The spread of multiple antibiotic and drug resistance in bacteria, including pathogens and instability of industrially important bacteria used in fermentation, have resulted in serious consequences. The evolution of compound R plasmids, i.e., carrying two or more R- determinants (the segment carrying the resistance gene or genes), has aggravated the situation further. The occurrence of bacteria containing compound R plasmids has resulted in their ability to acquire resistance to commonly used antibiotics like penicillin, streptomycin, tetracycline etc. A study in Japan revealed that the frequency of cells carrying R plasmid-mediated antibiotic resistance in the natural population of sewers and in polluted lakes and streams, increased from less than 1% to 50–80% in less than 10 years. Moreover, R plasmids are not rapidly dispersed only within a bacterial species but also transmitted among different species and even genera. The dispersal of *Esc. coli* R plasmids to different genera, namely, *Salmonella, Pasteurella, Shigella, Proteus* and *Haemophilus*, that also include pathogenic species, is well known (Gardner *et al*., 1991). The dispersal of R plasmids is not as rapid as those of F plasmids because only about 0.02% of cells containing R plasmids are competent donors. However, if, after the transfer of R plasmid, a large number of R^{-} recipient cells are added, R plasmid quickly spreads in the recipient cells. The recipient cell, which receives R plasmid, becomes a donor capable of subsequent transfer.

Stc. lactis and *Stc. cremoris* are commonly used as starter cultures in the dairy fermentation process for cheese making. Both of these bacteria contain multiple plasmids and these plasmids carry genes that encode enzymes, which play essential roles in the fermentation processes. Two important key factors for proper cheese ripening are lactose metabolism and proteinase activity. The genes carried by the plasmids encode for three enzymes involved in lactose metabolism and proteinase activity. As most plasmids are unstable by nature, the presence of key genes on these plasmids indicates the unpredictable functioning of these fermentation starter cultures.

REFERENCES

Adelberg, E. A. and Burns, S. N. (1959). A variant sex factor in *Escherichia coli*. *Genetics* 44: 497.

Avery, O. T., Macleod, C. M. and McCarty, M. (1944). Studies on the chemical nature of the substance inducing transformation of pneumococcal types: Induction of transformation by a deoxyribonucleic acid fraction isolated from pneumococcus type III. *J. Exp. Med.* 79: 137–58.

Barash, I. and Manulis-Sasson, S. (2009). Recent evolution of bacterial pathogens: The gall-forming *Pantoea agglomerans* case. *Ann. Rev. Phytopathol.* 47: 133–52.

Davis, B. D. (1950). Nonfiltrability of the agents of genetic recombination in *Escherichia coli*. *J. Bacteriol.* 60: 507.

Doddapaneni, H., Yao, J., Lin, H., Walker, M. A. and Civerolo, E. L. (2006). Analysis of the genome-wide variations among multiple strains of the plant pathogenic bacterium *Xylella fastidiosa*. *BMC Genom.* 7: 225. https://doi.org/10.1186/1471-2164-7-225

Gardner, E. J., Simmons, M. J. and Snustad, D. P. (1991) *Principles of Genetics*. 8th Ed. John Wiley and Sons Inc., New York, 811 pp.

Griffith, F. (1928). The significance of pneumococcal types. *J. Hyg.* (Lond.). Jan. 27(2): 113–59. https://doi.org/10.1017/s0022172400031879

Jacob, F. and Wollman, E. L. (1955). Etude génétique d'un bactériophage tempéré *d'Escherichia coli*. III. Effet du rayonnement ultraviolet sur la recombinaison génétique. *Ann. Inst. Pasteur Paris* 88: 724–49.

Lederberg, J. and Lederberg, E. M. (1952). Replica plating and indirect selection of bacterial mutants. *J. Bacteriol.* 63: 399–406.

Lederberg, J. and Tatum, E. L. (1946). Novel genotypes in mixed cultures of biochemical mutants of bacteria. *Cold Spring Harbor Symposia Quant. Biol.* 11: 113.

Li, W., Raoult, D. and Fournier, P.-E. (2009). Bacterial strain typing in the genomic era. *FEMS Microbiol. Rev.* 33: 892–916.

Maccannell, D. (2013). Bacterial strain typing. *Clin. Labor. Med.* 33: 629–50.

Monteiro-Vitorello, C. B., de Oliveira, M. C., Zerillo, M. M., Varani, A. M., Civerolo, E. *et al.* (2005). Xylella and *Xanthomonas* Mobil'omics. *Omics* 9: 146–59.

Ochiai, T., Inoue, Y., Takeya, M., Sasaki, A. and Kaku, H. (2005). Genome sequence of *Xanthomonas oryzae* pv. *oryzae* suggests contribution of large numbers of effector genes and insertion sequences to its race diversity. *JARQ* 39: 275–87.

Prescott, L. M., Harley, J. P. and Klein, D. A. (2005). *Microbiology*. 6th Ed. McGraw-Hill Companies, Inc., New York, 1130 pp.

Rajeshwari, R. and Sonti, R. V. (2000). Stationary-phase variation due to transposition of novel insertion elements in *Xanthomonas oryzae* pv. *oryzae*. *J. Bacteriol.* 182: 4797–4802. https://doi.org/10.1128/JB.182.17.4797-4802.2000

Salzberg, S. S., Sommer, D. D., Schatz, M. C., Phillippy, A. M., Rabinowicz, P. D. *et al.* (2008). Genome sequence and rapid evolution of the rice pathogen *Xanthomonas oryzae* pv. *oryzae* PXO99^{A}. *BMC Genom.* 9: 534.

Sindt, N. M., Robison, F., Brick, M. A., Schwartz, H. F., Heuberger, A. L. *et al.* (2018). MALDI-TOF-MS with PLS modeling enables strain typing of the bacterial plant pathogen *Xanthomonas axonopodis*. *J. Am. Soc. Mass Spectrom.* 29: 413–21.

Srivastava, S. and Srivastava, P. S. (2003). *Understanding Bacteria*. Kluwer Academic Publishers, Dordrecht, 469 pp.

Straub, C., Colombi, E. and McCann, H. C. (2021). Population genomics of bacterial plant pathogens. *Phytopathology* 111: 23–31. https://doi.org/10.1094/PHYTO-09-20-0412-RVW

Wikipedia (2008). Transformation (genetics). [http://en.wikipedia.org/wiki/Transformation_(genetics)].

Zinder, N. D. and Lederberg, J. (1952). Genetic exchange in *Salmonella*. *J. Bacteriol.* 64: 679–99.

4 Evolution of Phytopathogenic Bacteria

The association of bacteria with plants is a long-standing one. The initial intimate association probably began with the appearance of photosynthetic cyanobacteria as endophytes of a heterotrophic eukaryote that evolved into the first primitive plants (McFadden, 1999; Moreira *et al.*, 2000). Microbial and ecological diversity then led to microbe–plant interactions between free-living bacteria and plants. These bacteria were capable of utilising different type of substrates shed by the plants. Due to their nutritional diversity, these bacteria, particularly the chemolithotrophs, had no specific ecological habitat. Such bacterial types are generally classified as saprophytic bacteria, due to their ability to consume a wide range of inorganic and non-living organic substrates.

Through evolutionary steps, the involvement of chance associations and the acquisition of metabolic networks through horizontal gene transfer events in response to changing environments (Pal *et al.*, 2005), some saprophytic bacteria gradually developed a preference for plants as a niche to escape competitors and predators. Initially, these bacteria evolved into epiphytic types, choosing the plant environment as a more favourable niche rather than those in water and soil, where the microbial competition for space and food remained extremely high. The abundance of fissures and crevices of the cuticular layer, composed of cutin bounded by a layer of wax on its surface and a layer of pectin bound to the cellulose of the cell wall (Eglinton and Hamilton, 1967), provided excellent sites for bacterial colonisation. As epiphytic bacteria became increasingly specialised, their habitat, in turn, became considerably narrow and yet provided a niche sufficient for competitive advantages. In the modern era, epiphytes are specialised bacteria whose sustained presence on the host plant is not harmful to either party.

Long-term and constant associations of epiphytic bacteria and plants have led some of them to become mutually beneficial by exchanging requirements for growth and maintenance to ensure the health of both partners. This type of mutualism is exemplified by *Janthinobacterium lividum*, whose niche is the *Ardisia* plant (Rodrigues-Pareira *et al.*, 1972). These purple-pigmented bacteria are ectosymbionts, whose beneficial contributions are in the form of phytohormones (cytokinins) that are elaborated by these bacteria to stimulate and promote plant growth as well as seed germination (Rodrigues-Pereira *et al.* 1972, Tsavkelova *et al.*, 2007). In turn, there is sufficient substrate on the plant surface to provide the food base for the ectosymbiont. These ectosymbionts have retained the purple insoluble pigmentation, presumably as protection against harmful ultraviolet light.

Endosymbionts are highly evolved to promote the growth of their plant hosts and simultaneously to create a niche that prohibits the habitation of foreign bacteria. These intimate associations require numerous genes to fix nitrogen and reduce it to ammonia ions, neutralise toxic plant phenols and produce structures such as infection threads culminating in nodule or gall formation, which serve as unique homes for these bacteria. These genes are frequently harboured on large plasmids (e.g., Sym plasmids), suggesting that these genes were obtained and sequestered in more recent evolutionary times by horizontal gene transfer.

Environmental shifts can perturb the existing endosymbiotic association to gradually change it into a pathogenic association with the plant. In this case, the plant no longer benefits from the association, and bacteria remain as long as the pathogenic association continues to benefit the survival and perpetuation of the pathogen. Here, the host provides a unique niche to the pathogen that allows it to circumvent competing microorganisms and to replicate to sufficient numbers to be transmitted to a new susceptible host. The pathogen has also evolved to circumvent host defence barriers.

In nature, there is usually some base equilibrium in pathogenic associations with the host plant so that the host is not killed outright. Disruption in this equilibrium results in severe reactions by the host to the pathogen. One classic example in which the base equilibrium has been drastically changed is the generation of crop plants from wild-type plants through plant breeding and domestication. The wild-type plant is highly tolerant to the pathogen, whereas the corresponding crop plant is highly susceptible. Thus, the series of pandemics that have occurred throughout history simply reflect the vulnerability of the crop plant to a pathogen that has long existed in equilibrium with the wild-type plant host.

The highly developed interrelationship of a pathogen and a host plant is exemplified by crown gall and hairy root. *Agrobacterium tumefaciens* causes crown gall and *Agr. rhizogenes* causes hairy root. *Agr. tumefaciens* and *Agr. rhizogenes* have created highly specific niches for themselves by converting normal plant tissue into crown galls and hairy roots, respectively, by the transfer of specific genes into the host cells and host chromosomes. These specific genes encode the biosynthesis of auxins, cytokinins and a histone-binding protein for abnormal appendage formation and encode enzymes to synthesise a specific food base composed of arginine or other basic amino acids conjugated with a keto organic acid to form unusual compounds called opines. Opines are not usually catabolised by bacteria other than *Agr. tumefaciens* or *Agr. rhizogenes*.

DOI: 10.1201/9781003681878-5

This type of substrate specifically provides a highly competitive advantage over other competing bacteria that are unable to utilise opines. Since *Agr. tumefaciens* and *Agr. rhizogenes* cleverly genetically engineer the host plant to provide a specific compound for nutritional utilisation, and a niche in the form of galls or hairy galls, this plant–microbe association epitomises one of the highest levels of pathogen–host interactions.

The plant–microbe interactions are dynamic associations. Any imbalance in the environmental conditions that would perturb the association might favour one partner of the association over the other. Hence, the environmental changes can disturb the nature of association between the plant and its associated microbial symbiont. When this occurs, either the habitat becomes unfit for the microbe or the plant harbouring the microbe becomes susceptible to the advances of the microbe with the creation of a pathogenic relationship. Plant pathologists have routinely used cultural practices as a means of controlling the advance of the pathogens. Such practices are used to favour the crop plant over the pathogen by shifting the environmental conditions, in one manner or another, to make the crop unfit for habitation by the pathogen. Cultural adjustments in watering and fertilisation schedules, calcium (gypsum) applications and soil pH, as well as aeration and pruning, are among the methods used for this purpose. For example, gypsum is routinely applied to young pineapple plants to reduce the incidence of pink disease (Anthony Hepton, Dole Foods, Inc., personal communication). The interaction between the pathogen and its host plant remains dynamic since there is enormous evolutionary pressure to remain successful in a given niche. Among microbes, the transition from the saprophytic way of life to a highly specialised host-specific relationship is a sophisticated means of escaping the constant competition among microbes in the soil and on plant surfaces. Kado (2010) has given an excellent account of evolution of bacterial plant pathogens.

4.1 BACTERIAL GENOME

The genomes of bacteria contain deoxyribose nucleic acid (DNA) that is composed of four nucleic acid bases, namely, adenine (A), guanine (G), thymine (T) and cytosine (C). It is the specific number and order of these bases that gives an organism its unique identity. Bacterial genomes are made up of large DNA molecules arranged generally in a single chromosome. In general, the chromosomes are circular, but they may be linear in some cases. Besides chromosomes, bacterial genomes often contain extra chromosomal DNA molecules, called plasmids. These plasmids may vary in length from a few kilo bases to over 400 kilo bases (kb).

The genome size of plant pathogenic bacteria varies according to different genera and species. Even different strains of the same species show a high degree of variability (Vinatzer and Yan, 2008). The smaller genomes are estimated to be around 2.5 million base pairs (mbp), the larger ones can be greater than 5 mbp (Pierson and Ishimaru, 2000). Among the phytopathogenic bacteria, phytoplasmas have smaller genomes, in the range of 530–1350 kilo base pairs (kbp), than the other bacteria. '*Candidatus* Phytoplasma cynodontis', the causal agent of Bermuda grass white leaf disease, has the smallest genome size (530 kbp) and may represent the smallest chromosome known for any living organism. An exact proportion and distribution of each of the four bases in the genome is obtained by sequencing the genome completely. The relative amount of each base, reported as the G + C content, is a characteristic of a genus. Some genera, like *Erwinia*, have a low G + C content, ranging from 49.8–54.1 mol%, while others, such as *Clavibacter* (70 ± 5 mol%), *Curtobacterium* (66.0–71.0 mol%) and *Streptomyces* (69.0–78.0 mol%), have higher G + C contents.

4.2 PATHOGENICITY ISLANDS

Plant pathogenic bacteria have large segments of DNA associated with pathogenicity; these are known as '**pathogenicity islands**'. These pathogenicity islands (PAIs) can be located using genetic tools like transposon mutagenesis and complementation. The G + C contents of some of these PAIs are distinct from the overall G + C content of the organism. From this and other relevant information, it is postulated that the presence of PAIs can indicate how some bacteria may have evolved to become pathogens. PAIs are clusters of virulence genes that have been acquired through transfer and recombinant events, and contribute to a pathogenic phenotype. As these genes often come from phylogenetically diverse genomes, G + C content varies among genes in a PAI. PAIs have a mosaic structure and are subjected to a continuous process of gene acquisition, gene loss and gene decay (Loria *et al.*, 2006).

The emergence of new plant pathogens, or an increase in the virulence of the existing pathogens, has always been a matter of great concern to plant pathologists. An understanding of the evolutionary mechanisms, leading to the emergence of new pathogens or of aggressive virulent strains of the existing ones, is essential to reducing or minimising the losses caused by plant pathogens. The development of analytical tools and the availability of genomes' sequenced data have greatly increased our understanding of the mechanisms affecting the evolution of plant pathogens. In contrast to eucaryotes, which evolved mostly through modifications of existing genetic information, and in which horizontal gene transfer played a minor role, bacteria have acquired a significant proportion of their genetic diversity through the acquisition of DNA sequences from distantly related organisms, including eucaryotes.

4.3 HORIZONTAL GENE TRANSFER

Horizontal gene transfer (HGT), also called **lateral gene transfer**, is defined to be the movement of genetic material between bacteria other than by descent, in which the information travels through the generations as the cell divides. By contrast, vertical transfer occurs when an organism receives genetic material from its ancestor, i.e., its parent or the species from which it is evolved. HGT is most often thought of as a sexual process that requires a mechanism for the mobilisation of a chromosomal DNA among bacterial cells. As the bacteria do not reproduce sexually, they have acquired several mechanisms, namely, transformation, conjugation and transduction, for the exchange of genetic material. Natural genetic transformation is believed to be the essential mechanism for the attainment of genetic plasticity in many species of bacteria. During bacterial evolution, the ability of bacteria and the archaea to adapt to new environments most often results from the acquisition of new genes through horizontal transfer rather than by the alteration of gene functions through numerous point mutations.

HGT is influenced more by physical than by phylogenetic proximity, and genes carried by HGT are often associated with pathogenicity and adaptation to a changing environment or to new environmental niches. Thus, coding sequences that are retained following HGT are more likely to be derived from bacteria sharing a particular niche, and thus coding, for functions related to survival in that niche.

HGT is very important in the sense that it may cause the almost instant emergence of new pathogens, in comparison to the hundreds or even thousands of years needed for the emergence of new pathogens by other mechanisms. HGT is undoubtedly an important driving force for the evolution of bacterial genome and for the rapid adaptation to changing environments; pathoadaptive mutations, which increase virulence by means of minor genetic alterations such as point mutation, and gene rearrangement or deletion, also contribute to the evolution of virulence.

The evolution of bacteria in different niches is mainly driven by two factors: vertical gene transfer (VGT) and horizontal gene transfer (HGT), which occur through transduction, transformation, or conjugation events (Lawrence, 2005). The impact of mutations on the evolution of microorganisms is well known, but mobile genetic elements (MGEs) are also important drivers of bacterial evolution and may be the primary evolutionary source in phytopathogenic bacteria. Transposons, a key component of MGEs, are sequences of DNA that can move to different positions within the genome of a single cell through a process called **transposition**. During this process, they can cause mutations and alter the amount of DNA in the genome. There are various types of transposons, which can be categorised based on their mechanisms of transposition. Class I transposons, or retrotransposons, copy themselves by first being transcribed into RNA, then reverse-transcribed back into DNA by reverse transcriptase, and finally inserted at another position in the genome. Class II transposons move directly from one position to another, using a transposase to "cut and paste" them within the genome.

Earlier, all bacteria were assumed to be clonal, i.e., the variation between strains of their species is caused more by mutation than by recombination. Smith *et al.* (1993) revolutionised the field of microbial evolution by developing molecular evolutionary tests that showed that the recombination occurs frequently in some species. However, with more publications coming in favour of recombination, it is clearly established that recombination is an important evolutionary mechanism in numerous bacterial species. A notable example is *Escherichia coli*, which was previously considered to be clonal. Extensive sequence analyses of many isolates of *Esc. coli* has now confirmed the earlier evidence of recombination in some *Esc. coli* clades (Wirth *et al.*, 2006).

Recombinations occur frequently only when the DNA sequences of the donor and the recipient are very similar. However, these can be detected only when DNA sequences of the donor and recipient are sufficiently different from each other for recombination events to be inferred from sequence data with statistical significance. Multilocus sequence typing (MLST) allows for the determination of the contribution of homologous recombination resulting from conjugation, transduction, or transformation, to variation between strains of a species. Recombination in many bacterial species was overlooked before the advent of MLST. Sarkar and Guttman (2004) found little recombination in *Pseudomonas syringae* on the basis of analysis of distantly and closely related isolates and mutation contributed approximately four times more than the recombination to variation. The basis for the relatively small contribution of recombination to variation between the analysed isolates was probably due to the low percentage (91%) of DNA sequence identity between distantly related isolates. Additionally, the adaptation of *Psm. syringae* pathovars to different plant species probably makes physical contact between distantly related isolates more difficult. Yan *et al.* (2008) reported that recombination contributed 5.8 times more than mutation to variation between isolates when closely related isolates with a DNA sequence identity of approximately 99% were analysed. Many isolates studied had a partially overlapping host range, thus giving an ample opportunity for recombination. Even in cases where a common host between isolates was not identified, isolates may have evolved only recently from ancestors with overlapping host ranges. They were also able to confirm high rates of recombination between closely related *Psm. syringae* isolates by identifying recombination breakpoints within several genes and by applying various recombination tests. However, for some gene fragments, the DNA identity between isolates was more than 99.3% and it was too high to allow detection of recombination.

The analysis of a mix of closely and distantly related isolates of *Ralstonia solanacearum* (Castillo and Greenberg, 2007) and *Xylella fastidiosa* (Scally *et al.*, 2005) revealed that both of these pathogens evolved mainly by mutation rather than recombination. However, the analysis of a large number of closely related isolates of *Ral. solanacearum* (Bertolla *et al.*, 1999) and *Xanthomonas axonopodis* pv. *vesicatoria* (Basim *et al.*, 1999) showed that *Ral. solanacearum* was found to be naturally competent during infection, potentially allowing a very high rate of recombination; and recombination rates were also high enough in *Xan. axonopodis* pv. *vesicatoria* to be detected in controlled coinfections. Thieme *et al.* (2005) sequenced the complete genome of *Xan. campestris* pv. *vesicatoria* and found novel insights into the genome plasticity and pathogenicity of the bacterium. They further found that several regions of atypical G + C content were associated with both insertion sequence elements and tRNA genes, and contained putative pathogenicity determinants which might have been acquired by horizontal gene transfer. Goss *et al.* (2005) also found high recombination rates in *Psm. viridiflava* after analysing a larger number of closely related isolates. The results of majority of the above-mentioned studies clearly show that recombination probably plays an important role in the evolution of many bacterial plant pathogenic species, but it is easily overlooked when an insufficient number of closely related isolates is sampled.

Toth *et al.* (2006), in their review article, compared the genome of *Erw. carotovora* subsp. *atroseptica* (recent name, *Pectobacterium atrosepticum*) with those of human and animal pathogenic enterobacteria, namely, *Esc. coli, Salmonella enterica, Yersinia pestis, Yer. pseudotuberculosis and Shigella flexneri*, and suggested HGT in *Erw. carotovora* subsp. *atroseptica* may directly have led to the acquisition of number of determinants that mediate its interactions, pathogenic or otherwise, with plants, offering a glimpse into its evolutionary divergence from animal pathogenic enterobacteria. *Erw. carotovora* subsp. *atroseptica* genome contains 17 major putative horizontally acquired islands and several minor horizontally acquired islands. In both of the cases, the genes involved in pathogenesis and/or ecological adaptation are found. Two of the horizontally acquired islands represent complete prophages, whereas the others show evidence of phage remnants and transposase, and integrase genes that are often associated with HGT.

Immediate public access to the data of complete genome sequences opens up a new biological age. It is considered to be the birth of a new science, **genome-based biology**. In addition to the implications for medicine, the knowledge of microbial complete genome sequence also provides a wealth of information for tracing evolutionary networks. Hundreds of bacterial genomes, including the genomes of dozens of plant pathogenic bacteria, have been sequenced. The genome-sequenced data have provided valuable information pertaining to sources of genomic variation and the evolution of new strains. Some of the phytopathogenic bacteria, whose genomes have been sequenced, are detailed below.

1. *Xyl. fastidiosa* clone 9a5C
2. *Ral. solanacearum* strains GMI 1000, MOIK 2 and 1609
3. *Psm. syringae* pv. *tomato* strains DC 3000 and T 1
4. *Xan. axonopodis* pv. *citri* (recent name, *Xan. citri* subsp. *citri*)
5. *Xan. campestris* pv. *campestris* strains 85-10, ATCC 33913 (CFBP 5241) and 8004 (CFBP 6650)
6. *Xan. campestris* pv. *vesicatoria* (recent name, *Xan. vesicatoria*)
7. *Xan. oryzae* pv. *oryzae* strains MAFF 311018, KACC 10331 and PXO 99^A
8. *Liberibacter asiaticus*
9. *Erw. carotovora* subsp. *atroseptica* (recent name, *Pectobacterium atrosepticum*) strain 1043

In their review article, investigations by Timilsina *et al.* (2024), of whole-genome sequencing, comparative genomics and population genomics provide unprecedented insights into past and ongoing pathogen evolution. These include new understandings of the adaptive evolution of secretion systems and their effectors. The focus on *Xanthomonas* pathosystems has seen an intensive study and improved our understanding of pathogen emergence and evolution, particularly in the context of host specialisation related to citrus canker, bacterial blight of rice and bacterial spot of tomato and pepper. Across pathosystems, pathogens appear to follow a pattern of bursts of evolution and diversification that impact host adaptation. There remains a need for studies on the mechanisms of host range evolution and genetic exchange among closely related, but differentially host-specialised species and to start to move beyond the study of specific strain and host cultivar pairwise interactions to thinking about these pathosystems in a community context.

Plant pathogenic bacteria, amplified by their continuous and rapid evolution and resulting in the emergence of virulent, competitive and elusive plant pathogens, are among the major challenges in crop production. To further understand these bacteria, comprehensive methods are necessary to know their interactions with host plants and their responses to disease management. In recent years, plant bacteriologists have used population and comparative studies to quantify the diversity in plant pathogenic bacteria and their virulence mechanisms, including the effectors secreted from different protein secretion systems, phytohormones, or toxins that mediate their interactions with the hosts and/or competing microbes in the phyllosphere. Studies comparing strains over time, within and among geographic regions, and across phylogenetic trees has provided insights into the scales of variation and the changes in pathogens over time and in response to selection pressures. In this series of articles, population and comparative genomics were used to investigate plant pathogens representing diverse bacterial genera, including Gram-negative bacteria composed of facultative anaerobes (i.e., *Dickeya, Pantoea* and *Pectobacterium*) as well as obligate aerobes (i.e., *Pseudomonas, Xanthomonas* and *Xylella*) and also the Gram-stain-positive genus, *Clavibacter* (Timilsina *et al.*, 2022).

Outbreaks of bacterial plant diseases can occur through the rapid evolution of a pathogen by overcoming host defences. Over the past decade or so, the advent of genome sequencing, especially next-generation technologies, has seen a revolution in the study of plant pathogen evolution. The review article by Arnold and Jackson (2011) highlights the developments in understanding bacterial plant pathogen evolution, enabled by genomics and specifically focusing on type III protein effectors. The genotypic changes and mechanisms involved in pathogen evolution are now much better understood. However, there is still need to learn about the drivers of pathogen evolution, in terms of both plant resistance and bacterial lifestyle. Limited numbers of studies show pathogen evolution in the field.

To investigate the evolution of the host specificity of bacterial plant pathogens, Cai *et al.* (2011) reported a molecular evolutionary analysis of the model plant pathogen, *Psm. syringae* pv. *tomato* DC3000 and of close relatives that are pathogens of a diverse set of crop plants. Extensive host range tests on five different plant species were performed. Combining phylogenetic data with host range data, a reconstruction of the host range of all putative ancestors was performed. In particular, the hypothesis that highly virulent narrow host range pathogens of today's crops grown in monoculture evolved from ancestors with wider host range that were adapted to natural mixed-plant communities of pre-agricultural times, was tested. They found support for this hypothesis in individual clades. However, reconstruction of host range of the most recent common ancestor of all analysed strains was not conclusive. On the basis of the results obtained, they stressed the importance of including pathogens from wild plants when reconstructing the evolution of plant pathogenic bacteria.

Monteil *et al.* (2013) analysed phylogeny, virulence genes, host range and aggressiveness of *Psm. syringae* strains closely related to the tomato pathogen *Psm. syringae* pv. *tomato* (*Pst*), including strains isolated from snowpack and streams. The population of *Pst* relatives in non-agricultural environments was estimated to be large and its diversity was found to be higher than that of the population of *Pst* and its relatives on crops. Ancestors of environmental strains, *Pst* and other genetically monomorphic crop pathogens were inferred to have recombined frequently, suggesting an epiphytotic population structure for *Psm. syringae.* Some environmental strains having repertoires of type III-secreted effectors very similar to *Pst*, are almost as aggressive on tomato as *Pst*, but have a wider host range than typical *Pst* strains. They concluded that crop pathogens may have evolved through a small number of evolutionary events from a population of less aggressive ancestors with a wider host range present in non-agricultural environments.

Goss *et al.* (2013) have given new insights into the evolution of plant pathogenic bacteria. The conventional wisdom regarding bacterial plant pathogens is that strains found causing disease on plants are specialised and distinct from saprophytic strains found in the environment (soil, water, decaying plant material) that have no apparent role in disease.

Analogous population structures have been demonstrated for environmental human pathogens, including *Psm. aeruginosa* (Pirnay *et al.*, 2002). An extended life cycle of *Psm. syringae*, which incorporates movement within the freshwater cycle as well as enrichment and diversification in non-agricultural ecosystems, was proposed by Morris *et al.* (2008). *Psm. viridiflava*, a close relative of *Psm. syringae*, also has diverse and recombinogenic populations in weed communities (Goss *et al.*, 2005). These environmental populations may be key to understanding the emergence of pathogens, because there is increasing evidence that the ecology of plant and animal pathogens, across variable environments, has shaped their evolution as pathogens (Pallen and Wren, 2007; Morris *et al.*, 2009).

The findings of Lu *et al.* (2008) provide insights into the ancestral *Xanthomonas* genome and indicate that differentiation with respect to host- and tissue-specificity involved not major modifications or wholesale exchange of clusters, but subtle changes in a small number of genes or in non-coding sequences, and/or differences outside the clusters, potentially among regulatory targets or secretory substrates.

Deciphering mechanisms shaping bacterial diversity should help to build tools to predict the emergence of infectious diseases. Xanthomonads are plant pathogenic bacteria and are encountered all over the world. *Xanthomonas axonopodis* is a genetically heterogeneous species, clustering into six groups, strains that are collectively pathogenic on a large number of host plants. However, each strain displays a narrow host range. Mhedbi-Hajri *et al.* (2013) have addressed the question of the nature of the evolutionary processes, geographical and ecological speciation, that shaped this diversity. They assembled a large collection of *Xan. axonopodis* strains that were isolated over a long period, over continents, and from various hosts. Based on the sequence analysis of seven housekeeping genes, they found that recombination occurred as frequently as point mutation in the evolutionary history of *Xan. axonopodis*. However, the impact of recombination was about three times greater than the impact of mutation on the diversity observed in the whole dataset. The authors then reconstructed the clonal genealogy of the strains using coalescent and genealogy approaches and studied the diversification of the pathogen using a model of divergence with migration. The suggested scenario involves a first step of generalist diversification that spanned the past 25 000 years. A second step of ecology-driven specialisation has occurred over the past two centuries. Eventually, secondary contacts between host-specialised strains probably occurred as a result of agricultural development and intensification, allowing genetic exchanges of virulence-associated genes. These transfers may have favoured the emergence of novel pathotypes. The final argument of the authors is that the largest ecological entity within *Xan. axonopodis* is the pathovar. *Arsenophonus* bacteria are among the most biologically diverse and widespread endosymbionts of arthropods. Notably, two species, *Phlomobacter fragariae* and *Arsenophonus phytopathogenicus*, have been characterised as phloem-restricted plant pathogens that are obligatorily transmitted by and hosted in planthoppers of the family

Cixiidae (*Hemiptera: Auchenorrhyncha*). Bressan (2014) has reviewed the current understanding on the lifestyle transition, evolution, host interaction and infection cycles of these emerging plant pathogens.

Savory *et al.* (2017) coupled ecological sampling and genome sequencing to characterise the population genetic history of *Rhodococcus* and the distribution patterns of virulence plasmids in isolates from nurseries. An analysis of chromosome sequences showed that plants host multiple lineages of *Rhodococcus*, and suggested that these bacteria are transmitted due to independent introductions, reservoir populations and point source outbreaks. They demonstrated that isolates lacking virulence genes promote beneficial plant growth, and that the acquisition of a virulence plasmid is sufficient to transition beneficial symbionts to phytopathogens. This evolutionary transition, along with the distribution patterns of plasmids, reveals the impact of horizontal gene transfer in rapidly generating new pathogenic lineages and provides an alternative explanation for pathogen transmission patterns. Results also uncovered a misdiagnosed epiphytotic that implicated beneficial *Rhodococcus* bacteria as pathogens of pistachio. The misdiagnosis perpetuated the unnecessary removal of trees and exacerbated economic losses.

4.4 GALL-FORMING *PANTOEA AGGLOMERANS*: A CASE STUDY

The most classical example of the evolution of a new bacterial plant pathogen is *Pan. agglomerans*. This bacterium, which is widespread in nature as a commensal and epiphytic bacterium associated with many plants, was transformed into a host-specific gall-forming pathogen on gypsophila and beet. In their review article, Barash and Manulis-Sasson (2009) have given an excellent account of transformation of *Pan. agglomerans* into an hrp-dependent and host-specific pathogen. This transformation was achieved by the evolution of a pathogenicity plasmid that harbours most, if not all, of the genes required for gall formation and host specificity. The pathogenicity plasmid designated as pPATH is unique in structure and contains a pathogenicity island (PAI). The PAI was evolved on an interon plasmid of the IncN family, which is distributed among genetically diverse populations of *Pan. agglomerans*. The horizontal gene transfer was the major contributing force in the creation of PAI, although a pathoadaptive mechanism might also be involved. Plant pathogenic strains of *Pan. agglomerans* isolated from gypsophila contain a pPATH. The presence of six insertion sequence (IS) elements belonging to five families (i.e., IS*1*, IS*3*, IS*5*, IS*6* and IS*630*) along the PAI is strong evidence that an intensive HGT may have taken place during the evolution of the pPATH. The IS*3* family is one of the largest families of IS elements, with members found in diverse bacterial groups. Several IS elements found in phytopathogenic bacterial plasmids are members of the IS*3* family. These elements were identified in *Xan. campestris, Psm. syringae, Erw. amylovora* (Sundin, 2007) and *Agrobacterium tumefaciens.*

The pathogenicity plasmid of *Pan. agglomerans* pv. *gypsophilae* (*Pag*), designated as $pPATH_{Pag}$, is approximately 135 kb and it contains a pathogenicity island of ~75 kb. The pPATH harbours an *hrp/hrc* gene cluster, type III effectors, a gene cluster encoding for the biosynthesis of indole acetic acid (IAA) and cytokinins, and many diverse insertion sequence elements and partial sequences from genes of other bacteria. Mutations in genes encoding components of the type III secretion system or the *hrp/hrc* regulatory gene cascade completely abolished gall formation while mutations in genes encoding type III effectors or biosynthesis of IAA and cytokinins reduced gall development substantially, or almost totally. These findings suggest that pathogenicity of *Pan. agglomerans* pv. *gypsophilae* and *Pan. agglomerans* pv. *betae* is hrp-dependent, and that the acquisition of the plasmid-borne PAI was a crucial step in the transformation of *Pan. agglomerans* into a gall-forming and host-specific plant pathogen.

The pathogenicity plasmid of *Pan. agglomerans* pv. *gypsophilae* is nonconjugative, which is supported by the lack of putative genes that show significant homology to *tra* genes. Plant-synthesised IAA and cytokinins are absolutely required for gall initiation, whereas the pathogen-secreted phytohormones are not needed. However, pathogen-secreted IAA and cytokinins do have a significant effect on gall size and might be important for expression of the *hrp* regulatory genes as signalling molecules.

Plasmids are common residents of phytopathogenic bacteria and contribute significantly to host evolution in a multifaceted manner. Plasmids tend to encode determinants of virulence and ecological fitness that can enhance adaptation to a specific niche or can influence niche expansion. Many of these determinants appear to have been acquired from other bacteria via horizontal transfer, illustrating an important function of these plasmids in the acquisition of sequences that enable rapid evolution. These genes can ultimately be delivered to the host chromosome through plasmid integration events, thus stabilising important acquired determinants within the genome. Sundin (2007) has given an excellent account of contribution of phytopathogenic bacterial plasmids to the evolutionary history of their hosts.

As plasmids confer a wide variety of traits on the bacteria that harbour them and are able to promote exchange of genetic information across taxonomic boundaries, they have undoubtedly figured heavily in bacterial evolution. The review by Coplin (1989) focuses on the distribution, phenotypes and possible evolutionary roles of plasmids in plant pathogenic bacteria.

4.5 EVOLUTION IN *STREPTOMYCES* SPECIES

Streptomyces are multicellular, Gram-positive, high G+C content bacteria, with a complex development cycle, large linear chromosomes and a high capacity for catabolism and secondary metabolism. In addition to linear chromosomes, *Streptomyces* species contain both circular and linear plasmids that add to adaptability and enhance horizontal gene transfer. Minorities of *Streptomyces* species are plant pathogens and the emergence of new pathogenic species has occurred many times in agricultural systems. Plant pathogenic species contain a large PAI that contains the thaxtomin biosynthetic genes as well as other virulence determinants.

The emergence of new plant pathogenic species of *Streptomyces* suggests that the *Streptomyces* PAI is mobilisable, at least in some strains. The appearance of *Str. turgidiscabies* in Japan is the classical example of a newly emerged species. Due to its recent emergence, *Str. turgidiscabies* was used to test the hypothesis that the PAI could be transferred to other *Streptomyces* species via conjugation. *Str. coelicolor* and *Str. diastatochromogenes* were used as recipients. An analysis of transconjugates indicated that all known pathogenicity loci were mobilised during the conjugation events. Additional analysis indicated that a DNA element of approximately 660 kb integrated into the *Str. coelicolor* chromosome in most of the transconjugates. Mobilisation of the 660 kb version of the PAI to *Str. coelicolor* did not confer a pathogenicity phenotype, while its mobilisation to *Str. diastatochromogenes* enabled the transconjugates to produce thaxtomin (toxin involved in pathogenesis) and colonise excised tuber tissue (Kers *et al.*, 2005). It is evident from these results that PAI is not the sole pathogenicity determinant.

The unhindered transfer of *Str. turgidiscabies* PAI to saprophytic *Streptomyces* species *in vitro* suggests that the mobilisation of pathogenicity determinants among *Streptomyces* species occurs frequently in nature. However, the occurrence of a limited number of pathogenic *Streptomyces* species in contrast to a large number of saprophytic *Streptomyces* species in nature suggests that there may be a genetic barrier in most *Streptomyces* species that results in the instability of PAI or prevents the expression of the pathogenicity genes in the recipient strains. It is further confirmed by the fact that the mobilisation of PAI into *Str. coelicolor* did not confer pathogenic ability to this well-known saprophytic species.

4.6 *XANTHOMONAS ORYZAE* PV. *ORYZAE*: A CASE OF RAPID EVOLUTION

Complete genome sequences have been published for three strains of *Xan. oryzae* pv. *oryzae* (*Xoo*), namely, MAFF311018 (MAFF), a Japanese race 1 strain, also referred to as T7174 (Ochiai *et al.*, 2005), KACC10331 (KACC), a Korean race 1 isolate, also known as KXO85 (Lee *et al.*, 2005), and $PXO99^A$, a strain of Philippines race 6 (Salzberg *et al.*, 2008). The genomes of MAFF and KACC overall are highly similar to each other in terms of gene content and organisation. $PXO99^A$ is relatively more distant from either MAFF or KACC. In contrast to MAFF and KACC, $PXO99^A$ is virulent on a large number of rice varieties representing diverse genetic sources of resistance, including the broad-spectrum recessive resistance gene *xa5*. Relatively few resistance genes, including *xa13*, are effective against $PXO99^A$ while *xa13* is ineffective against MAFF and KACC. As MAFF and KACC are highly similar with regard to genome content and organisation, Salzberg *et al.* (2008) have compared $PXO99^A$ and MAFF and have drawn the following conclusions.

The *Xoo* genome has a remarkable plasticity, which is most strikingly evident in the large number of major rearrangements and indels between these strains. The differences suggest that the *Xoo* genome evolves rapidly and this conclusion is perhaps best supported by the 212 kb sequence duplication in $PXO99^A$. The duplication represents a remarkable recent event, with only a single nucleotide difference differentiating between two copies in $PXO99^A$. Gene duplication contributes to gene diversification, permitting for the unconstrained evolution of otherwise-indispensable sequences. The abundant duplications in $PXO99^A$ can act as an important source of genomic variation in *Xoo*. It is clear from the analysis of 212 repeat kb that IS elements play an important role in generating duplications. A comparison of MAFF and $PXO99^A$ provided clear evidence of the involvement of homologous recombination among transcription activator like (TAL) effector genes in generating differences in their structure and copy number at genomic locations that were otherwise conserved, indicating that the sequences themselves play a major role in generating that diversity.

The analysis of sequenced genomes of *Xanthomonas* species reveals that the genus as a whole shows a high degree of genomic variation. Notably, *Xoo* strains contain the greatest number and diversity of IS elements of all the sequenced xanthomonads. The genome sequence of $PXO99^A$, and its comparison with those of MAFF and KACC, have clearly shown that the *Xoo* genome is highly plastic and rapidly evolving. These findings help to explain the extraordinary diversity of *Xoo* genotypes and development of a large number of races isolated from different parts of the world.

The development of microbial genomics poses a new challenge for plant pathologists, who seek to discover how to make the best use of this available information. They should apply this information in the most meticulous manner to reduce the threats and losses from plant pathogenic bacteria and also improve the viable use of beneficial bacteria. The limited number of effective chemicals and the sources of disease resistance compel us to think up and evolve other strategies to control bacterial diseases of plants. Plant virologists have already utilised their knowledge of complete plant viral genomes to develop transgenic virus-resistant plants. The efforts should also be made to do the same with bacterial genomics. The possibilities for developing methods of disease control in a manner analogous to pharmaceutical approaches for veterinary and clinical medicine should also be explored. The knowledge of complete genome sequences of many plant pathogenic bacteria may allow us to identify the Achilles' heel of a plant pathogenic bacterium.

REFERENCES

Arnold, D. L. and Jackson, R. W. (2011). Bacterial genomes: Evolution of pathogenicity. *Curr. Opin. Pl. Biol.* 14: 385–91. https://doi.org/10.1016/j.pbi.2011.03.001

Barash, I. and Manulis-Sasson, S. (2009). Recent evolution of bacterial pathogens: The gall-forming *Pantoea agglomerans* case. *Annu. Rev. Phytopathol.* 47: 133–52.

Basim, H., Stall, R. E., Minsavage, G. V. and Jones, J. B. (1999). Chromosomal gene transfer by conjugation in the plant pathogen *Xanthomonas axonopodis* pv. *vesicatoria*. *Phytopathology* 89: 1044–49.

Bertolla, F., Frostegard, A., Brito, B., Nesme, X. and Simonet, P. (1999). During infection of its host, the plant pathogen *Ralstonia solanacearum* naturally develops a state of competence and exchanges genetic material. *Mol. Plant–Microbe Interact.* 12: 467–72.

Bressan, A. (2014). Emergence and evolution of *Arsenophonus* bacteria as insect-vectored plant pathogens. *Infect. Genet. Evol.* 22: 81–90. https://doi.org/10.1016/j.meegid.2014.01.004

Cai, R., Yan, S., Liu, H., Leman, S. and Vinatzer, B. A. (2011). Reconstructing host range evolution of bacterial plant pathogens using *Pseudomonas syringae* pv. *tomato* and its close relatives as a model. *Infect. Genet. Evol.* 11: 1738–51. https://doi.org/10.1016/j.meegid.2011.07.012

Castillo, J. A. and Greenberg, J. T. (2007). Evolutionary dynamics of *Ralstonia solanacearum*. *Appl. Environ. Microbiol.* 73: 1225–38.

Coplin, D. L. (1989). Plasmids and their role in the evolution of plant pathogenic bacteria. *Annu. Rev. Phytopathol.* 27: 187–212.

Eglinton, G. and Hamilton, R. J. (1967). Leaf epicuticular waxes. *Science* 156: 1322–35.

Goss, E. M., Kreitman M. and Bergelson, J. (2005). Genetic diversity, recombination and cryptic clades in *Pseudomonas viridiflava* infecting natural populations of *Arabidopsis thaliana*. *Genetics* 169: 21–35.

Goss, E. M., Potnis, N. and Jones, J. B. (2013). Grudgingly sharing their secrets: New insight into the evolution of plant pathogenic bacteria. *New Phytol.* 109: 630–32. https://doi.org/10.1111/nph.12397

Kado, C. I. (2010). Chapter 2: Evolution of bacterial pathogens. In: *Plant Bacteriology*. The American Phytopathological Society, St. Paul, MN, pp. 2–19.

Kers, J. A., Cameron, K. D., Joshi, M. V., Bukhalid, R. A., Morello, J. E. *et al.* (2005). A large, mobile pathogenicity island confers plant pathogenicity on *Streptomyces* species. *Mol. Microbiol.* 55: 1025–33.

Lawrence, J. G. (2005). Horizontal and vertical gene transfer: The life history of pathogens. In: W. Russell and H. Herwald (eds.). *Concepts in Bacterial Virulence*. KARGER, Basel, pp. 255–71.

Lee, B.-M., Park, Y.-J., Park, D.-S., Kang, H.-W., Kim, J.-G. *et al.* (2005). The genome sequence of *Xanthomonas oryzae* pathovar *oryzae* KACC10331, the bacterial blight pathogen of rice. *Nucleic Acids Res.* 33: 577–86.

Loria, R., Kers, J. and Joshi, M. V. (2006). Evolution of plant pathogenicity in *Streptomyces*. *Annu. Rev. Phytopathol.* 44: 469–87.

Lu, H., Patil, P., Van Sluys, M.-A., White, F. F., Ryan, R. P. *et al.* (2008). Acquisition and evolution of plant pathogenesis-associated gene clusters and candidate determinants of tissue-specificity in *Xanthomonas*. *PLoS ONE* 3: e3828. https://doi.org/10.1371/journal.pone.0003828

McFadden, G. I. (1999). Endosymbiosis and evolution of the plant cell. *Curr. Opin. Pl. Biol.* 2: 513–19.

Mhedbi-Hajri, N., Hajri, A., Boureau, T., Darrasse, A., Durand, K. *et al.* (2013). Evolutionary history of the plant pathogenic bacterium *Xanthomonas axonopodis*. *PLoS ONE* 8(3): e58474. https://doi.org/10.1371/journal.pone.0058474

Monteil, C. L., Cai, R., Liu, H., Llontop, M. E. M., Leman, S. *et al.* (2013). Nonagricultural reservoirs contribute to emergence and evolution of *Pseudomonas syringae* crop pathogens. *New Phytol.* 199: 800–11. https://doi.org/10.1111/nph.12316

Moreira, D., Le Guyader, H. and Philippe, H. (2000). The origin of red algae and the evolution of chloroplasts. *Nature* 405: 69–72.

Morris, C. E., Bardin, M., Kinkel, L. L., Moury, B., Nicot, P. C. *et al.* (2009). Expanding the paradigms of plant pathogen life history and evolution of parasitic fitness beyond agricultural boundaries. *PLoS Pathog.* 5: e1000693.

Morris, C. E., Sands, D. C., Vinatzer, B. A., Glaux, C., Guilbaud, C. *et al.* (2008). The life history of the plant pathogen *Pseudomonas syringae* is linked to the water cycle. *ISME J.* 2: 321–34.

Ochiai, H., Inoue, Y., Takeya, M., Sasaki, A. and Kaku, H. (2005). Genome sequence of *Xanthomonas oryzae* pv. *oryzae* suggests contribution of large numbers of effectors genes and insertion sequences to its race diversity. *Jpn. Agric. Res. Q.* 39: 275–87.

Pal, C., Papp, B. and Lercher, M. J. (2005). Adaptive evolution of bacterial metabolic networks by horizontal gene transfer. *Nat. Genet.* 37: 1372–75.

Pallen, M. J. and Wren, B. W. (2007). Bacterial pathogenomics. *Nature* 449: 835–42.

Pierson III, L. S. and Ishimaru, C. A. (2000). Genomics of plant-associated bacteria: A glimpse of the future that has become reality. APSnet Feature Story, http://www.apsnet.org/online/feature/Genomics/.

Pirnay, J.-P., De Vos, D., Cochez, C., Bilocq, F., Vanderkelen, A. *et al.* (2002). *Pseudomonas aeruginosa* displays an epidemic population structure. *Environ. Microbiol.* 4: 898–911.

Rodrigues-Pereira, A. S., Houwen, P. J. W., Deurenberg-Vos, H. W. J. and Pey, E. B. F. (1972). Cytokinins and the bacterial symbiosis of *Ardisia* species. *Z. Pflanzenphysiol.* 68: 170–77.

Salzberg, S. L., Sommer, D. D., Schatz, M. C., Phillippy, A. M., Rabinowicz, P. D. *et al.* (2008). Genome sequence and rapid evolution of the rice pathogen *Xanthomonas oryzae* pv. *oryzae* PXO99^A. *BMC Genom.* 9: 204.

Sarkar, S. F. and Guttman, D. S. (2004). Evolution of the core genome of *Pseudomonas syringae*, a highly clonal, endemic plant pathogen. *Appl. Environ. Microbiol.* 70: 1999–2012.

Savory, E. A., Fuller, S. L., Weisberg, A. J., Thomas, W. J., Gordon, M. I. *et al.* (2017). Evolutionary transitions between beneficial and phytopathogenic *Rhodococcus* challenge disease management. *eLife*. https://doi.org/10.7554/eLife.30925

Scally, M., Schuenzel, E. L., Stouthamer, R. and Nunney, L. (2005). Multilocus sequence type system for the plant pathogen *Xylella fastidiosa* and relative contributions of recombination and point mutation to clonal diversity. *Appl. Environ. Microbiol.* 71: 8491–99.

Smith, J. M., Smith, N. H., O'Rourke, M. and Spratt, B. G. (1993). How clonal are bacteria? *Proc. Natl. Acad. Sci.* USA 90: 4384–88.

Sundin, G. W. (2007). Genomic insights into the contribution of phytopathogenic bacterial plasmids to the evolutionary history of their hosts. *Annu. Rev. Phytopathol.* 45: 129–51.

Thieme, F., Koebnik, R. and Bekel, T. (2005). Insights into genome plasticity and pathogenicity of the plant pathogenic bacterium *Xanthomonas campestris* pv. *vesicatoria* revealed by the complete genome sequence. *J. Bacteriol.* 187: 7254–66.

Timilsina, S., Goss, E. M., Koebnik, R., Potnis, N. and Jones, J. B. (2022). Editorial: Population and comparative genomics of plant pathogenic bacteria. *Front. Microbiol.* 13: 1012034. https://doi.org/10.3389/fmicb.2022.1012034

Timilsina, S., Kaur, A., Sharma, A., Ramamoorthy, S., Vallad, G. E. *et al.* (2024). *Xanthomonas* as a model system for studying pathogen emergence and evolution. *Phytopathology* 114: 1433–46. https://doi.org/10.1094/PHYTO-03-24-0084-RVW

Toth, I. K., Pritchard, L. and Birch, P. R. J. (2006). Comparative genomics reveals what makes an enterobacterial plant pathogen. *Annu. Rev. Phytopathol.* 44: 305–36.

Tsavkelova, E. A., Cherdyntseva, T. A., Klimova, S. Y., Shestakov, A., Botina, S. G. *et al.* (2007). Orchid-associated bacteria produce indole-3-acetic acid, promote seed germination and increase their microbial yield response to exogenous auxin. *Arch. Microbiol.* 188: 1405–10.

Vinatzer, B. A. and Yan, S. (2008). Mining the genomes of plant pathogenic bacteria. How not to down into gigabases of sequence. *Mol. Pl. Pathol.* 9: 105–18.

Wirth, T., Falush, D., Lan, R., Colles, F., Mensa, P. *et al.* (2006). Sex and virulence in *Escherichia coli*: An evolutionary perspective. *Mol. Microbiol.* 60: 1136–51.

Yan, S., Liu, H., Mohr, T. J., Jenrette, J., Chiodini, R. *et al.* (2008). Role of recombination in the evolution of model plant pathogen *Pseudomonas syringae* pv. *tomato* DC3000, a very atypical tomato strain. *Appl. Environ. Microbiol.* 74: 3171–81.

5 Diagnosis of Bacterial Diseases of Plants

5.1 DIAGNOSIS

For an effective bacterial plant disease management, the correct diagnosis of a disease based on the identification of its causal agent is particularly essential. A satisfactory diagnosis of some diseases can be made on the basis of symptoms alone, but the identification of the causal agent is necessary for confirmation. In order to diagnose a plant disease correctly, a comprehensive host list of the pathogen under investigation is required, including its typical symptoms and also known potential pathogens for that specific host. For this purpose, the compendia of different crop diseases published by the American Phytopathological Society are very valuable publications. A list of plant diseases found on the American Phytopathological Society website (http://www.apsnet.org/online/common/top.asp) is another source of valuable information. *Westcott's Plant Disease Handbook* is also useful because it contains a breakdown of the specific symptoms associated with each disease (Horst, 2001). Photographs of symptoms, especially those in colour, can also help in the identification of plant diseases. A list of the known pathogens for a given crop generally reduces the choices to one or two suspected genera for the given symptoms.

5.2 ASSESSMENT OF SYMPTOMS

The assessment of symptoms and the diagnosis of the disease based on the symptoms in a laboratory depend upon the sample sent to the laboratory. It must consist of plant material showing different stages/phases of the disease, i.e., whether the infection is young or old. If the disease affects different plant parts, i.e., leaves, stems, shoots, bulbs, roots, etc., the sample should contain all of those plant parts that exhibit symptoms. In a completely diseased tissue, it is difficult to find the primary cause of the disease since it is very often replaced by secondary pathogens or saprophytes. Therefore, the symptoms in the sample must be on the **borderline between healthy and diseased plant tissue**. These samples should be wrapped in paper and put in a plastic bag, leaving some space for air, to prevent rotting and decay by secondary organisms during transit. Information, including the name of cultivar, date of sowing, date of sampling, place, weather conditions, soil type, etc. should accompany the sample. Finally, the sample should be statistically representative.

The symptoms of plant diseases caused by different groups of pathogens, namely, bacteria, fungi, viruses, phytoplasmas and nematodes, have many similarities, and there is often an overlapping of the symptoms. However, two signs, i.e., **water-soaking** and **bacterial exudation**, are associated with most, but not all of the bacterial plant diseases. The amount of bacterial ooze varies with the nature of disease and the environmental conditions. The ooze is generally abundant and more forceful in vascular diseases and in freshly caused lesions. The intensity of ooze is also greater under humid conditions. In a dry atmosphere, by contrast, the bacterial ooze is converted into scales, flakes or beads, often becoming detectable only with the help of a magnifying glass, especially when it is sparse. Riley *et al.* (2006) have also stressed the **importance of signs** by stating that signs are much more specific to disease-causing agents than the symptoms, and are extremely helpful in the diagnosis of a disease and identification of the agent causing the disease.

In the case of bacterial wilt of solanaceous plants, the surface of the transverse section is moister in appearance than is the case with fungal wilt. White droplets of bacterial exudates generally appear on the cut surface and the ooze can be intensified by squeezing the infected plant tissue. One diagnostic feature of bacterial wilt of cucurbits is the emergence of the bacterial ooze from the cut ends of the infected stems. This sticky bacterial mass can be drawn into fine threads by touching the cut end of the stem with a finger or a knife blade and gently pulling the finger or a knife blade away from the cut end. The fine threads of bacterial ooze can also be seen by pulling away the cut ends of an infected stem.

The microscopic examination of diseased plants is extremely helpful in establishing the bacterial nature of plant diseases. Bacterial streaming is found in almost all the bacterial diseases, with the exception of the crown gall. A portion of the lesion, along with some adjoining healthy tissue is cut, mounted in a drop of water and placed under the microscope, without covering it with a cover slip. The clouds of bacteria streaming out of the cut end of the mounted tissue are now visible. Ooze is more forceful in vascular diseases such as bacterial blight of rice. Phase contrast microscopy is more suitable for this purpose, because starch or latex particles can be distinguished from the bacterial cells. A bacterial ooze test can also be performed in the field by cutting an infected leaf (bacterial blight of rice) or stem (bacterial wilt of cucurbits) and placing it in a beaker containing clean water. After a few minutes, fine threads of bacterial ooze start to come out of the cut end and the water turns milky. In diseases caused by fastidious procaryotes such as spiroplasmas, phytoplasmas and *Xylella fastidiosa*, the vascular bundles of the sectioned tissues are

DOI: 10.1201/9781003681878-6

examined with an electron microscope to detect the presence of these pathogens.

5.3 KOCH'S POSTULATES

To prove that a particular pathogen/organism is the causal agent of the disease, it is essential to prove the four postulates given below, which were first outlined by Robert Koch, a renowned German bacteriologist:

1. An organism/pathogen is always associated with a disease.
2. Isolation of the pathogen or causal agent and its identification.
3. Inoculation of the pathogen on the host and production of symptoms identical to the symptoms observed in nature.
4. Reisolation of the pathogen from inoculated plants; it should be identical in all respects to the inoculated pathogen.

5.3.1 Isolation of Bacteria

The isolation method described below is applicable only to the culturable plant pathogenic bacteria. It is not applicable to the spiroplasmas, phytoplasmas and fastidious xylem- or phloem-limited bacteria as they are non-culturable and the different methods are used to study them. In diseases caused by fastidious bacteria such as spiroplasmas, phytoplasmas and *Xylella fastidiosa*, the vascular bundles of the sectioned tissues are examined with an electron microscope to confirm the presence of these pathogens.

Take a fresh disease sample and wash it thoroughly with clean water. Cut small pieces/bits of diseased tissue bordering the healthy tissue and place them in sodium hypochlorite solution or alcohol for 1–2 minutes for surface sterilisation. These bits can also be disinfected with 70% alcohol. After surface sterilisation, the bits are washed four times with sterile water to remove the residue of chlorine/alcohol. These bits are then placed in a tube containing sterile water or phosphate-buffered saline (PBS, pH 7.2) and left for 30 minutes to allow the bacteria to ooze out from the diseased tissue. Alternatively, a few pieces of sterilised diseased tissue can be crushed in a sterile small dish containing phosphate-buffered saline or sterile water to make the cell suspension. For the isolation of *Xanthomonas citri* subsp. *malvacearum*, leaf tissue showing vein infection should be preferred over the other symptomatic tissue.

Isolation of bacteria is more difficult when dealing with seed samples, since host tissue has a very low number of bacterial cells or plant samples having latent infection or epiphytic populations. In traditional isolation, plant tissue is washed and macerated, often filtrated and centrifuged; then a small aliquot of the suspension is streaked on a suitable non-selective or semi-selective medium. It is also advisable to inoculate a part of this suspension into a test plant, which may act as a selective medium. However, these methods are labour-intensive and time-consuming. Selective media have a fairly high level of detection ($10^{3\text{-}5}$ cells mL^{-1}).

Generally, the isolation of the bacterium should be made on a general, non-selective medium instead of a semi-selective medium. A general purpose, non-selective but differential medium having tetrazolium chloride is useful for distinguishing a potential pathogen from saprophytes during the initial isolation, as long as the concentration of the tetrazolium chloride is reduced to 0.001% to avoid inhibitory effects on the growth of members of *Xanthomonas* genus. However, semi-selective media prove useful in confirming the presence of suspected pathogens. Semi-selective media are more helpful for isolations from soil, but generally not from plant tissues because the latter are generally surface-sterilised before isolation. A few of the bacterial plant pathogens, such as phytoplasmas and papaya bunchy top pathogen, have not been cultured on artificial media. The presence of such pathogens in the host plants should be confirmed by other methods, such as electron microscopy.

The following two methods are used for plating the cell suspension on the solid medium:

1. Streak method
2. Pour plate method

5.3.1.1 Streak Method

In the streak method, the dilute cell suspension is streaked on the solidified culture medium poured in the plates. An aliquot of 20–100 µL of the cell suspension is placed on the solid media in a plate and spread with a sterile spreader. The same spreader is then used to further spread bacteria on two more plates. In this way, a dilution of approximately 1:10 on the second plate and 1:100 on the third plate is obtained, yielding separate colonies that can be examined for the desired colony type. Cell suspension can also be streaked on the solid media using an inoculating needle. The loop of an inoculating needle, charged with cell suspension, is dragged from one end to the other end of the plate; brought back to the first end and again moved to the other end. In this way, the whole of the plate is streaked in a zigzag fashion. The same loop, without charging with cell suspension, is streaked on two or three more plates. After streaking, the plates are incubated, in an inverted position, at the required temperature. An advantage of streaking with inoculating needle is that colonies growing on the streaks are supposed to have originated from the streaked cell suspension, and hence are picked up.

In the case of most phytopathogenic bacteria, a general, non-selective media are generally used. These media contain usually water and non-defined nutrients (e.g., peptones, yeast extract and beef extract) or defined inorganic salts and organic nutrients, making them suitable for growth of most bacterial species. A general purpose, non-selective, but differential medium containing tetrazolium chloride is useful for distinguishing a potential pathogen from saprophytes during the initial isolation. For this purpose the concentration of the tetrazolium chloride should be reduced to 0.001%

to avoid inhibitory effects on the growth of members of *Xanthomonas* genus. Semi-selective or selective media are used for specific bacterial pathogens. Semi-selective media contain substances to enhance the production of pigments and/or substances which can only be used by certain bacteria and/or inhibitory substances like antibiotics for non-desired bacteria. Selective media contain nutrients and inhibitors, which, by their quantity and quality, allow only a specific bacterium to form colonies in/on the medium in an ideal situation. To enhance the isolation of target bacteria and to exclude non-pathogenic bacteria, semi-selective or selective isolation media are often used. However, many of the non-desirable bacteria can also form colonies that may be very similar to those of the target bacterium. Therefore, the isolated bacterium must be compared with the pathogenic bacterium. Culture media are solidified by adding agar-agar, a polysaccharide, i.e., sulphuric acid ester of linear galactan, obtained from red algae of the genera *Gracilaria*, *Gelidium*, *Pterocladium* and *Acanthopeltis*. Media are sterilised in the autoclave at 115 °C or 121 °C for 20 minutes.

5.3.1.2 Pour Plate Method

In this method, 20–40 µL of cell suspension is poured into each plate; then a molten, but cool solid medium is poured onto these plates. The plates are rotated clock- and anti-clock-wise to mix the cell suspension with the medium. After the solidification of the medium, the plates are incubated, in an inverted position, at the required temperature.

Most plant pathogenic bacteria grow well in an oxygen-rich atmosphere at 20–28 °C. In most cases, the colonies appear on the isolation plates after around 2–5 days of incubation. Colony morphology is very important for the identification of a bacterium. Based on the colony characteristics of the bacterium under investigation, isolated single colonies of the desired bacterium are transferred on medium slants in the tubes and incubated for growth. The characteristics of the colonies in the plates and of mass growth on the slants are recorded. Many plant pathogenic bacteria produce pigmentation in the culture media that helps greatly in their identification. *Xanthomonas* species produce yellow, water-insoluble colour (xanthomonadins). Some bacteria produce water-soluble pigments, and these diffuse into the medium, e.g., the brown melanoid pigment of *Ralstonia solanacearum* and *Streptomyces scabiei* and the fluorescent pigments (pyoverdins) of many *Pseudomonas* species.

The pure culture of the bacterium is obtained using the single colony culture method. It can be obtained by both streak and pour plate methods. In the streak method, a loop-full of a very dilute cell suspension of the isolated bacterium is streaked continuously (without recharging) on 5–6 plates. If the isolated colonies appearing in the last streaked plates appear similar in morphology, including shape, size, colour, etc., this means that the colonies formed are descendants of a single colony and the culture is pure. To make it certain, this process is repeated 3–4 times, and the single colony culture obtained from the last repeat is labelled as pure culture and used for further studies. A pure culture forms the basis for all further studies. In the pour plate technique, the serial dilutions of bacterial cell suspension are mixed (1 mL per plate) with the molten, but cool solid medium and plates are incubated. Isolated colonies are supposed to appear in the highest dilutions. The remaining procedure is same as in the streak method.

5.3.2 Identification of Bacteria

Identification of the bacteria is very time-consuming as it involves the conducting of a large number of tests, and some of the tests take a relatively long time. The following morphological, physiological and biochemical characteristics are determined using conventional methods of identification.

5.3.2.1 Morphological Characteristics

These include Gram-stain reaction, cell shape and size, spore formation, and the presence or absence of flagella and capsule. Most plant pathogenic bacteria are Gram-stain-positive or -negative rods, with the exception of *Streptomyces* spp., which produce filamentous hyphae. The colony morphology is also very important; the form, colour and smell of a colony may be an indication of a bacterial pathogen, but many other non-pathogenic bacteria may also form similar types of colonies. Diffusible pigment production also plays an important role in the identification of some bacteria.

5.3.2.2 Physiological Characteristics

These include the temperature range for growth, thermal death point (temperature at which the bacterial cells in dilute cell suspension die, when kept for 10 minutes at that temperature) and tolerance to NaCl concentration. The temperature range for the growth of plant pathogenic bacteria ranges approximately from 4 to 37 °C; the thermal death point usually varies from 50–55 °C. The determination of antibiotic resistance may be useful because the antibiotics are generally used to control bacterial pathogens in the field (e.g. streptomycin against *Erwinia amylovora*). Furthermore, screening for toxin production and ice-nucleation activity will yield additional characteristics which are useful in the identification of plant pathogenic bacteria, especially those belonging to the *Psm. syringae* group and *Clavibacter* spp.

5.3.2.3 Biochemical Characteristics

Determination of biochemical characteristics of plant pathogenic bacteria is important as the reactions of these tests are based on the diverse enzyme systems of the bacteria, and many of these enzymes are involved in plant pathogenic action. These tests include the ability of the bacteria to utilise certain carbon sources like sugars, alcohols, organic acids, glycosides, and/or nitrogen sources, e.g., amino acids. The determination of oxidative and fermentative metabolism of glucose is important for the differentiation of *Pseudomonas* and *Xanthomonas* from *Erwinia*. The former two genera are oxidative positive and fermentative negative, while the latter is positive for both tests. Some biochemical tests give the information regarding the formation of certain end products by the bacteria, e.g., formation of H_2S from cysteine, indole from tryptophan, and NO_2 or N_2 from nitrate. The action of enzymes, namely, pectinases, gelatinase, amylases and casease, is judged by the hydrolysis of pectin, gelatin, starch and casein, respectively. Cellulases and lipases (esterases) cause hydrolysis of cellulose and fats/fatty acids, respectively.

The diagnosis of a new or an undescribed disease is certainly more difficult and requires more skill and expertise than confirming the presence of an already known disease. Therefore, the entire set of diagnostic principles should be used to narrow down the choices, as no single tests are suitable. More emphasis should be placed on microscopic examination and the initial association of the pathogen with the disease syndrome. Some basic information about the pathogen should be obtained to decide which tests are to be used for further identification before spending lot of time on numerous phenotypic and genotypic tests for pathogen identification. Tests for basic information include Gram-staining reaction, oxidative versus fermentative metabolism and the presence or absence of spores and flagella. To further characterise the bacterium, the relatively inexpensive tests include metabolic tests (analytical profile index, API, strip tests, bioMerieux, Inc.), the analysis of fatty methyl esters (MIDI, Newark, DE), metabolic substrate analysis (MicroLog™, Biolog, Inc., Hayward, CA), or 16S ribosomal (r)DNA sequence analysis. Extreme caution must be exercised in interpreting results of any one test, particularly for the identification of an unknown disease caused by an unidentified bacterial pathogen. The past experience emphasises the importance of performing basic bacteriological tests before accepting a name, which comes from a database, regardless of the refined analytical methods used to generate a similarity index.

In addition, routine morphological, biochemical and physiological tests, serological and molecular techniques have been developed, which are more specific and sensitive. These techniques are routinely used in various laboratories throughout the world for the identification of bacteria and diagnosis of bacterial plant diseases.

A polyphasic analysis still appears to be the most reliable approach for the identification of new pathogens. After the pathogenicity has been proved and the genus determined with relatively simple basic bacteriological tests, the final identification of the bacterium is interpreted from the results of various genotypic methods. Hu *et al.* (2001) reported that comparison of 16S rDNA sequence data with phenotypic data for type strains assisted in selection of determinative tests that may discriminate distinct taxa for simplified laboratory analyses.

The ultimate confirmation of a plant pathogen should be done by proving its pathogenicity. However, this requires time, the availability of host plants and optimum environmental conditions for disease development. Moreover, in some bacterial pathogens, the time required to produce the disease symptoms is very long. For example, some strains of *Clavibacter michiganensis* require 21–28 days to produce the symptoms. On the contrary, the confirmation by pathogenicity tests is often quick for leaf-spotting pathogens as they take only 3–4 days to produce the symptoms under optimum conditions. For some bacterial pathogens, as an alternative to the pathogenicity test, the hypersensitivity test on tobacco is quite useful, but this technique does not work in the case of others. Lelliot and Stead (1987) have given a detailed account for the diagnosis of bacterial diseases of plants by giving their detailed symptoms, diagnostic procedures, host inoculation tests, the composition of culture media and procedures for staining techniques and biochemical tests.

5.3.2.4 Serological Techniques

Serology is an important tool for providing additional information for the identification of bacteria, but a bacterium cannot be identified solely on the basis of its serological reactions. Many times, antisera are not specific enough, and they may cross-react with non-pathogenic bacteria possessing some of the antigenic determinants of the target bacterium in common, especially at lower dilutions of the serum. In serology, specific antisera are usually preferred to react solely with the target bacterium or its products. However, sometimes antisera are too specific and may miss certain strains of the same pathogen.

The agglutination test or latex agglutination test (antibodies are coated with sensitised latex beads, as a result of which the reaction is more readily visible and more sensitive) and the precipitation test are not highly sensitive and are also liable to disturbing cross-reactions. However, the immunofluorescence (IF) test is a very robust and sensitive serological test (it can detect 10^{3-4} cells mL^{-1} in plant extract), and the primary reaction of antigen and antibody is clearly visible. Moreover, binding reactions can be observed at very high titres. In the IF test, the antibodies are marked with a chemical dye, usually fluorescein isothiocyanate (FITC), which fluoresces in blue light. What is required is a light microscope fitted with an epi-fluorescent light, having suitable excitation and barrier filters for FITC. Indirect IF is slightly more sensitive and less specific than direct IF (Anonymous, 1998). Immunofluorescence is widely used in Europe to detect bacterial pathogens in seed and propagating materials. Van der Wolf and Schoen (2004) reported its use to screen 60,000 potato seed pieces annually to determine the presence of *Ral. solanacearum* in the Netherlands. It is also used to detect *Cla. sepedonicus* in potato seed pieces; in France, it is used to screen tomato seed lots for the presence of *Cla. michiganensis*.

5.3.2.4.1 Monoclonal Antibodies

The development of hybridoma technology has led to considerable improvements in immunodiagnostic techniques. Monoclonal antibodies (MAbs) produced by this method have a defined specificity to a single epitope. The reactivity of monoclonal antisera is usually lower than that of polyclonal antisera, because only one epitope of the bacterium will show the binding reaction. To enhance the reactivity of monoclonal antisera, mixtures of monoclonals can be made, if available (De Boer, 1987).

Hybridomas produce consistent antibodies/antisera, which give consistent results in different laboratories. Several MAbs produced by plant pathogenic bacteria and used in their epidemiological studies are listed in Table 5.1. Many of these antibodies have been characterised by testing their specificities with a large number of strains of the target pathogens. Some bacterial taxa are relatively homozygous in the sense that they possess common antigenic determinants and one antibody usually reacts with all, or nearly all, strains of the taxon. *Cla. michiganensis* and *Xan. axonopodis* pv. *pelargonii* fall in this category. On the other hand, many plant pathogenic bacteria, such as *Pectobacterium carotovorum* subsp. *carotovorum* and *Pec. atrosepticum*, are serologically heterogeneous; therefore, not all members of the population react with a polyclonal antiserum or a taxon-specific MAb. Pathogen-specific MAbs that react with heat stable lipopolysaccharides enable the development of robust detection kits for use in rapid laboratory and

TABLE 5.1
Taxon-specific monoclonal antibodies produced for plant pathogenic bacteria[a]

Genus, species/ subspecies or pathovar	MAb designation	No. of target strains tested	Reference
Acidovorax avenae subsp. *citrulli*	3D1F3	26	M. Bandla (personal communication)
Agrobacterium tumefaciens biovar 3	A6F21-1D3G7C8		Bishop *et al.* (1989)
Clavibacter michiganensis	Cmm1	88	Alvarez *et al.* (1993b)
Cla. sepedonicus	McAb1 to McAb-5	19	De Boer and Wieczorek (1984)
Erwinia carotovora subsp. *atroseptica*	14/18.6, 14/2, 14/8.6	3	Vernon-Shirley and Burns (1992)
Erw. chrysanthemi	6A6	36	Singh *et al.* (1999)
Erw. stewartii	C/G7/B2	43	Lamka *et al.* (1991)
Pseudomonas avenae	Pa1 to Pa-5	20	Alvarez et *al.* (1993a)
Psm. syringae pv.	Six MAbs[b]	223	Ovod *et al.* (1995)
phaseolicola	AG-1, AG-2	9	Wong, W.C. (1990)
Ralstonia solanacearum species-specific strain-specific	Rs1, Rs1a Mab	75	Alvarez et *al.* (1992) Griep *et al.* (1998)
Xanthomonas albilineans	12 MAbs	1	Tsai *et al.* (1990)
	7 MAbs	38	Alvarez *et al.* (1996)
Xan. axonopodis pv. *begoniae*	Xcb-1	26	Benedict *et al.* (1990)
Xan. axonopodis pv.	X-4600	30	Permar and Gottwald (1989)
citri	Xcd 1, Xcd 3, Xcd7	329	Lipp *et al.* (1992)
Xan. axonopodis pv. *dieffenbachiae*	Xcd 108	302	Norman and Alvarez (1994)
Xan. campestris pv.	X9, X13, X17	200	Alvarez et *al.* (1985)
campestris	10C5, 20H6, 16B5, 17C12, 10H12, 11B6	37	Franken, A. A. J. M. (1990)
Xan. hortorum pv.	Xcp-1	76	Benedict *et al.* (1990)
pelargonii	XP2	18	Wong, W. C. (1990)
Xan. hortorum pv. *phaseoli*			
Xan. oryzae pv. *oryzae*	Xco1, Xco2, Xco5	178	Benedict *et al.* (1989)
Xan. oryzae pv. *oryzicola*	Xco1a	8	Benedict *et al.* (1989)
Xan. translucens pv. *undulosa*	AB3-B6	44	Bragard *et al.* (1993)
Xan. campestris pv. *mangiferaeindicae*	XCM-1-XCM-6	4	Sanders *et al.* (1994)
Xylophilus ampelinus		63	Gorris *et al.* (1989)

[a] Data summarised from Alvarez (2004).
[b] Two MAbs (Ps core-1 and Ps core-2) specific to core lipopolysaccharide and four O-chain-specific MAbs (Ps-O:2–1, Ps-O:2–2, Ps-O:2–3 and Ps-O:3–1) were used to classify 223 strains belonging to 19 pathovars of *Psm. syringae*.

field diagnosis. MAbs directed towards capsule and/or extracellular polysaccharides in pathogens, such as *Cla. michiganensis* and *Ral. solanacearum*, are useful in a number of immunodiagnostic formats.

Most of the MAbs produced for different bacterial plant pathogens have been tested using an enzyme-linked immunosorbent essay (ELISA). The ability of ELISA assays to detect $10^{5\text{-}6}$ cfu mL^{-1} is adequate for the identification of bacterial pathogens from symptomatic plants and colonies on the selective media. Moreover, the sensitivity of ELISA can be enhanced tenfold by using an extraction buffer containing ethylenediamine tetra acetic acid and lysozyme, which releases lipopolysaccharides into solution, thereby enhancing the antibody–antigen reaction without increasing background readings. ELISA techniques using both monoclonal and polyclonal antibodies are available for several taxa of phytopathogenic bacteria, and rapid detection kits are also commercially available. Multitarget ELISA, which can detect more than one species in the same ELISA plate wells using different enzyme labels, have been developed by Agdia, Inc. A multiple ELISA for detection of *Cla. michiganensis* and *Xan. axonopodis* pv. *vesicatoria* is available (Alvarez, 2004). A brief description of ELISA technique is given below.

5.3.2.4.2 Enzyme-linked Immunosorbent Assay

In this test, the marker is an enzyme responsible for the reaction. The antibodies are first adsorbed (coated to) in the wells of a plastic ELISA plate. Subsequently, bacterial antigens in buffered sample solution are pipetted in the wells, which are trapped by the coated antibodies. After incubation and washing, an enzyme-labelled antiserum (conjugate) is added. Only the wells, where antigens react with the enzyme, change in colour after incubation with a suitable enzyme substrate. The enzyme used is generally alkaline phosphatase and substrate *p*-nitrophenyl phosphate. One major drawback of ELISA is that, in many cases, bacterial products and not the whole cells are detected, as the cells are washed out due to insufficient binding capacity of coating antibodies. Moreover, there is no information about the viability of cells and the dead cells may give false reading.

5.3.2.4.3 Flow Cytometry

The development of flow cytometry has enhanced the immunodiagnostic detection of bacteria. In this technique, as the bacterial cells or other particles pass individually through a sensor in a liquid stream, they are rapidly identified and quantified. Cells are identified by the conjugation of fluorescent dyes to specific antibodies and multiple cellular parameters are determined simultaneously based on a cell's fluorescence and its ability to scatter light. Cells may be sorted electronically, permitting purification and/or culture of subpopulations of selected cells for further confirmatory tests (Alvarez and Adams, 1999; Alvarez, 2001). Flow cytometry has been used for the detection of *Cla. michiganensis* in tomato seed extracts, of *Xan. axonopodis* pv. *dieffenbachiae* in anthurium (Alvarez and Adams, 1999), of *Xan. campestris* pv. *campestris* in seed extracts of *Brassica* sp. (Chitarra *et al.*, 2002) and to determine the viability of *Ral. solanacearum* in potato seed tubers (Van der Wolf *et al.*, 2004).

Other techniques, such as immunofluorescence, immunomagnetic separation and lateral flow devices, are also used for the detection of plant pathogenic bacteria. Immunofluorescence is widely used in Europe to detect bacterial pathogens in seed and propagating materials. In the Netherlands, it was used to screen 60,000 potato seed pieces annually to determine the presence of *Ral. solanacearum* (Van der Wolf and Schoen, 2004). It is also used to detect *Cla. michiganensis* in potato seed pieces, and in France it is used to screen tomato seed lots for the presence of *Cla. michiganensis.*

Lateral flow devices use the principles, primarily those of ELISA, but various types of filters used act as solid support for the initial binding reaction (Danks and Barker, 2000). A lateral flow device test kit, developed by Central Science Laboratory, UK, detects *Ral. solanacearum* in a 3-minute single step process. Rapid ImmunoStrip® kits for the detection of *Ral. solanacearum, Cla. michiganensis* and *Xan. hortorum* pv. *pelargonii* are sold by Agdia, Inc. In immunomagnetic separations, target cells are isolated from a mixed solution using paramagnetic polystyrene beads coated with specific antibodies. After washing, bound cells can be used for polymerase chain reaction or, as they are viable, they can be grown on semi-selective media for further confirmation.

It is emphasised here that commercially available modern serological kits can only be used for presumptive identification in a presumptive diagnosis. As these systems use only one characteristic or one set of similar characteristics, no authentic identification can be made. There are so many risks of errors resulting from cross-reactions, background errors, etc. To date, only the fulfilling of **Koch's postulates** has led to satisfactory identifications and a correct diagnosis.

5.3.2.5 Genomic Techniques

During the last three decades, many genomic techniques have been developed for the detection of known pathogens in plant samples. The brief description of four techniques is given below.

5.3.2.5.1 Polymerase Chain Reaction

At the start of polymerase chain reaction (PCR), the DNA preparation from which the desired segment is to be amplified, an excess of two primer molecules, the four deoxyriboside triphosphates and the DNA polymerase are mixed together in a reaction mixture. The multiplication of nucleic acid is achieved by repeated cycles of the following steps performed sequentially.

1. *Denaturation* (*melting*): The reaction mixture is heated usually to 95 °C, ensuring denaturation.
2. *Annealing*: The mixture is now cooled to 68 °C (depending on the primers used), which permits

the annealing of the primer to the complementary sequences in the DNA; these sequences are located at the 3′ ends of the two strands of the desired segment.

3. *Extension of nucleic acid strands*: The temperature is now so adjusted (72 °C, if Taq polymerase is used) that the DNA polymerase synthesises the complementary strands by utilising 3′ OH of the primers. The primers are extended towards each other so that the DNA segment lying between two primers is copied; this is ensured by employing primers complementary to the 3′ ends of the segment to be amplified. After multiplication, the PCR products can be visualised on an ethidium bromide-stained agarose gel via electrophoresis.

Taq polymerase (originally isolated from the hot-spring thermophilic bacterium *Thermus aquaticus*, but now also available as an artificially synthesised product) is generally used and the temperature is adjusted to 72 °C. At this temperature, the base pairing among about 20 bases long primers and the DNA is much more specific than at 37 °C, the optimal temperature for *Escherichia coli* DNA polymerase. This minimises the chances of annealing of primers to imperfectly matched sequences and, thereby, the amplification of unwanted DNA. The sensitivity of PCR is very high, and, theoretically, it can detect even one copy of the target DNA in a sample. However, in practice, usually a lower sensitivity is achieved due to the presence of inhibitory substances in the plant extract. Hybridisation with probes via blotting is often used to verify the identity of PCR products. If specific restriction sites are present in the product, restriction enzyme analysis can also be performed. If there is enough information about the specificity of the primers used, the method can also be used simultaneously for detection and identification. The negative PCR results can be the fallout of the following factors.

1. Absence of the target or target sequence in the sample; Rico *et al.* (2003) reported that when non-toxigenic strains are important under field infections and the PCR uses a sequence of DNA coding for the toxin, the pathogen will not be detected.
2. Degradation of the target DNA in the sample.
3. Failure of reaction due to wrong experimental conditions or presence of inhibitory compounds.

For the successful completion of PCR, an internal control such as primers against plant DNA (which is always present in the plant extract) that will multiply this DNA, should also be included in the same PCR experiment. Unfortunately, false-positive reactions in a bacteriological context are still possible due to the homology of the non-target organism DNA with the primers thought to be specific for the target. In order to increase sensitivity and/or specificity of PCR, it can also be performed in combination with pre-enrichment in a selective liquid medium, immuno-magnetic capture and fluorescent labels. Lopez *et al.* (2003) reported an achievement of an ELISA plate set up in some tests.

5.3.2.5.2 Real-time (TaqMan®) PCR

The development of detectors (sequence detection systems) that can measure fluorescence which is emitted during the PCR cycle has led to the invention of a quantitative PCR called real-time (TaqMan®) PCR. The method is based on the 5′→3′ exonuclease activity of the *Taq* DNA polymerase, which results in cleavage of fluorescent dye-labelled TaqMan® probes during PCR. In this technique, as the multiplication of DNA during early stages of reaction can be made visible; the positive samples can be detected before the inhibitory compounds in the extract block the reaction. Moreover, the detection is rapid compared to conventional PCR; it is more accurate and it has many other advantages.

5.3.2.5.3 DNA/RNA (Dot/Slot-Blot) Hybridisation

In dot-blot hybridisation, genetic material (DNA, RNA) of the bacterium (under investigation) is used to bind with the target bacterium. It is a very specific reaction and, due to its high specificity, it can be used simultaneously for detection and identification. The diseased plant sample is treated to remove all possible non-DNA material. Thereafter, DNA cleavage is allowed to take place under low pH conditions. After fixing a single-stranded DNA of the target bacterium on a nitrocellulose filter, a DNA probe is added. This probe contains a large number of specific nucleic acid molecules of the target bacterium, which have been multiplied before in another bacterium, mostly using a virus as a vector. The nucleic acid probe binds to the complementary parts, if present, on the single strand fixed on the filter, under a process called hybridisation. After washing, the hybridised DNA or RNA or RNA:DNA complexes can be visualised, using either a radioactive or a nonradioactive-labelled nucleic acid probe. The detection level of this technique is high (ca. 10^5 cells mL^{-1}) and can be further enhanced by the application of enrichment culture.

Alvarez *et al.* (2015) devised a portable field pathogen detection system by developing primers for the loop-mediated isothermal amplification (LAMP) kit from the 16S rDNA gene because it is a tandem region and it is polymorphic for different phytoplasma groups. For LAMP reactions, fluorochrome hydroxynaphthol blue, which need not to be observed under UV light, was used. In the case of reactions in the field, a prototype incubator was developed, involving a 5-W solar panel, 12 V 12 Ah battery, a voltameter and a mini dry bath with a 35-sample capacity. Kit's sensitivity is equal to that of the quantitative PCR technique and 1000 times more than that of the nested PCR technique. This LAMP kit can process 35 samples in 90 min without requiring electrical connections or any necessity to leave the field.

Getaz *et al.* (2017) designed a reliable and sensitive LAMP assay using a unique marker, providing a highly specific and rapid detection technique, convenient for on-site detection. The specificity of the designed assay was tested on 37 strains from a culture collection of *Xan. fragariae*, 82 strains of other *Xanthomonas* species and pathovars and 11 strains of other bacterial genera isolated from strawberry leaves. A detection limit of 10^2 fg was achieved, approximating to 20 genome copies per reaction. A consistent lower detection efficiency of 10^2 cfu mL^{-1} was achieved when the analysis was performed with crude plant material. The LAMP assay designed in this study was adapted to work on crude plant material, without any prior extensive extraction steps or incubation period. Additionally, it requires no advanced analytical knowledge or a fully equipped laboratory. Results were achieved within 7–20 minutes, depending on the pathogen concentration, thus providing a high-throughput and user-friendly method for the detection and screening of plant material for quarantine regulations.

Fuller *et al.* (2017) designed oligonucleotide primers and probes to target *virD2* (virulence gene D2 of *Agrobacterium tumefaciens*) for use in a molecular diagnostic tool, which relies on isothermal amplification and lateral-flow-based detection. The oligonucleotide tools were tested in the assay and evaluated for their detection limit and specificity in detecting alleles of *virD2*. One set of primers that successfully amplified *virD2* when used with an isothermal recombinase was selected. Both the tested probes had detection limits in picogram amounts of DNA. Probe 1 could detect all the tested pathogenic isolates that represented the maximum diversity of *virD2*. Finally, the coupling of lateral-flow detection to the use of these oligonucleotide primers in isothermal amplification made the process simple and easy, and alleviated reliance on the specialised tools necessary for molecular diagnostics. The assay is an advancement for the rapid molecular detection of pathogenic *Agrobacterium* spp.

5.3.2.5.4 Fluorescent In Situ Hybridisation (Fish) using rRNA (rDNA) Oligonucleotide Probes

In this method, short i.e. 20–30 mers oligonucleotide probes, are used against 16S or 23S (ribosomal) RNA/DNA. These oligonucleotide probes are used for *in situ* hybridisation as they are able to diffuse through the cell walls of microorganisms present in thin tissue sections or in plant or soil extracts fixed on a microscopic slide. For Gram-stain-positive bacteria, sometimes degradation of peptidoglycan layer of the cell wall with lysozyme is necessary to enhance the penetration of the probe into the cells. Sensitivity depends also on the metabolic activity of the cells, as the dead cells show a positive reaction for a considerable period of time. After labelling the probes with a fluorescent dye or a gold label, the microorganisms can be visualised by incident light (fluorescence) microscopy. The specificity of the method is relatively high, but false-positive reactions occur due to the homology of RNA of non-target organisms with the specific probe sequence, as reported by Moter and Gobel (2000) and Wullings *et al.* (1998).

The below given techniques involving the separation of bacterial components, namely, proteins and nucleic acids in an electric field, are also used for the identification of plant pathogenic bacteria.

1. Polyacrylamide gel electrophoresis (PAGE), one- or two-dimensional plasmid electrophoresis.
2. Immuno-electrophoresis.
3. Native protein electrophoresis (enzyme detection).
4. Restriction fragment length polymorphism (RFLP) analysis.
5. Random amplified polymorphism DNA (RAPD) analysis.
6. Repetitive sequence-based (REP) fingerprinting PCR (REP, ERIC or BOX).
7. Amplified fragment length polymorphisms (AFLP) analysis.

Several workers have reported the development and/or application of pathogen-specific primers for the detection of bacterial plant pathogens in heterogeneous mixtures. Most of these involve amplification by PCR coupled with one or more other techniques. Another PCR technique, real-time PCR, is rapid and more accurate for the detection of bacterial pathogens and has several other advantages. Other genomic methods include DNA-hybridisation, dot-blots and nucleic acid sequence-based amplifications. The main emphasis has been on the development of specific probes for the detection of target pathogens. Manulis *et al.* (1994) developed pathovar-specific probes and PCR tests for *Xan. campestris* pv. *pelargonii*, which could detect 10–50 cfu of the bacterium per sample. The purpose of developing such probes was to also detect the pathogens in asymptomatic plant tissue.

The primers and a probe were also developed for the detection of *Psm. syringae* pv. *actinidiae* in asymptomatic kiwi fruits (Koh and Nou, 2002). Cubero and Graham (2002) developed primer sets for *Xan. axonopodis* pv. *citri* that could distinguish pathotype A from *Xan. aurantifolii* pathotypes B and C. The primers were based on sequence differences in the internal transcribed spacer (ITS) region and *pthA* gene. Primer sets based on ribosomal sequences had a high level of specificity for *Xan. axonopodis* pv. *citri* A (of citrus canker type A), whereas those based on *pthA* were universal for pathogens of all types of citrus bacterial cankers. Multiplex PCR is used to identify several pathogens simultaneously (Bertolini *et al.* 2003). Caruso *et al.* (2003) developed cooperational PCR (Co-PCR) to detect *Ral. solanacearum* in water.

5.3.2.5.5 Enrichment Techniques

Enrichment techniques have frequently been used to increase the sensitivity of immunofluorescence (IF), ELISA, and/or PCR. An enrichment-ELISA assay using pathogen-specific MAbs was used to detect *Xan. axonopodis* pv. *dieffenbachiae* in leaf samples, *Pec. atrosepticum* or *Ral. solanacearum* in potato seed pieces. Alvarez *et al.* (1997) reported that even a limited culturing step, which is still insufficient for colony identification, increases the sensitivity of ELISA about 10,000-fold. Enrichment assays have the added advantage that only the viable cells multiply to give a positive signal. The immunofluorescence colony-staining (IFC) technique combines an enrichment step in pour plates for colony detection with fluorochrome-labelled antibodies and also increases the sensitivity of IF approximately 1,00,000 times (Veena and Van Vuurde, 2002). The microcolonies formed due to the multiplication of viable cells can be seen with an epifluorescence stereomicroscope at 40–60X whereas individual cells (dead) are not visible at this low magnification. IFC is used mainly for epidemiological studies and for the confirmation of results achieved with other detection methods, which do not distinguish between living and dead cells. The optimisation of buffers and conjugates also improves test efficacy. Antibodies of subclass IgG are highly suitable for IFC formats because their molecules are smaller than those of IgAs (dimers) and IgMs (pentamers) and they diffuse readily through agarose gel.

The enrichment of target bacteria on semi-selective media also improves the sensitivity of PCR reaction. Such tests are often termed "BIO-PCR" because the living cells of colonies harvested from culture plates are used in the PCR reaction. The major bands result from amplifications of DNA from living cells, although small amplicons may also be formed from dead bacterial cells. Sakthivel *et al.* (2001) were able to detect 55 fg of DNA per reaction tube, which was equivalent to 70 colony-forming units of *Xan. oryzae* pv. *oryzae* mL^{-1} of original sample and about 7 cells per reaction tube. To detect *Burkholderia cepacia* using a "BIO-PCR" assay, contaminated samples were incubated only for 24 h in broth; DNA was extracted and added to PCR beads with specific primers. The entire process was completed in 27 h in comparison to 5–6 days required for isolation and identification of the target bacterium.

Certain compounds that interfere with DNA amplification have limited the use of direct PCR for the detection of target bacteria from natural samples. In a number of such cases, immunocapture or immunomagnetic separation has been employed to solve the problem. Optimum time for the immunocapture of *Psm. syringae* pv. *syringae* with Advanced Magnetics™ beads coated with a polyclonal antiserum was only one hour of incubation. Detection limits were significantly improved when immunocapture was followed by a specific PCR test. Walcott and Gitaitis (2000) reported that immunomagnetic separation plus PCR resulted in a 100-fold increase in the detection of *Aci. avenae* subsp. *citrulli* compared to direct PCR. Van Overbeek *et al.* (2002) also supported the polyphasic approach for studying the interaction between *Ral. solanacearum* and its biocontrol agent, *Psm. corrugata*.

Schaad *et al.* (2003) have reviewed the advances made in molecular-based diagnostic techniques used for the detection of plant pathogens, including bacteria. They have compared the efficiency and sensitivity of conventional PCR, real-time PCR, BIO-PCR and immunocapture PCR. Real-time BIO-PCR is much more sensitive than conventional real-time PCR and it can detect 2 cfu mL^{-1} of *Psm. syringae* pv. *phaseolicola* in seed extracts and of *Cla. michiganensis* pv. *sepedonicus* in potato tuber extracts. Multiplex PCR can be used to detect more than one species of bacteria in the same reaction tube using probes labelled with different fluorescent dyes. To avoid cross-absorption, the wavelength of each dye must be well separated.

Newly developed serological and molecular techniques have played an important role in the diagnosis of bacterial plant diseases and the identification of phytopathogenic bacteria. However, after going through the literature carefully, it is apparent that conventional methods are equally valid, important and still used by workers. Many merits and plus points of molecular methods have been advocated and those that have proven to be often true, are given below:

1. Rapid, sensitive and cost-effective.
2. Have been integrated into certification/inspection schemes.
3. Standardised test kits are commercially available for use.
4. Can be applied to non-culturable microorganisms such as phytoplasmas/spiroplasmas.
5. Genetically modified organisms can be traced in the environment more easily.
6. Less sensitive to mutation or variation.
7. Differentiation at low taxonomic level, i.e., even at strain level is possible.

It is generally argued that molecular methods, being genetic, are of a higher and better level than conventional methods, which are phenotypic. Isolated nucleic acids no longer function in the experiments as genetic material because they are immobilised in gels, cut into pieces, etc. and their analysis is also a purely phenotypic one. Some possible drawbacks of molecular biological methods are given below:

1. Specificity, sensitivity and reproducibility have been tested to only a limited extent.
2. Negative influence of conditions and biochemicals may lead to experimental errors.
3. Inability to discriminate between viable and non-viable cells and free nucleic acid in a sample.
4. It is difficult to verify false negatives and false positives.
5. Changes in probes/primers/enzymes/chemicals/methods may yield different or conflicting patterns or no patterns at all.
6. As only a small part of structural elements of an organism is used, this may lead to a sampling error.
7. Koch's postulates cannot be fulfilled.

The merits and demerits of modern serological and molecular techniques are given here. A positive view of these techniques is to acknowledge these as a welcome addition to the already existing ones. These should be carefully checked for their specificity, reproducibility and limitations because false positive reactions with unknown bacteria are generally possible. Genomic techniques are quick and have more specificity and sensitivity, but these have not been adopted in all the laboratories due to certain limitations. For some laboratories, the cost of chemicals and equipment, and the requirement of trained personnel for DNA extraction are the limitations for using PCR in the routine. Therefore, keeping in view the merits and demerits of all the available methods, Alvarez (2004) has rightly advocated the **adoption of integrated approaches** for the detection of plant pathogenic bacteria and diagnosis of bacterial plant diseases.

5.3.2.6 Bacteriophages

The host range of the bacteriophages (phages) varies substantially. It may be narrow, confined to a particular species or a strain or a group of strains within a species. The non-specific or polyvirulent phages have a wide range of hosts, which can infect bacteria belonging to different species, genera, or even families.

The phages have been used to detect plant pathogenic bacteria from diseased plant material. An increase in the population of phages, added to the substrate, indicates the presence of its host bacterium in the substrate. The propagated phage is separated from the microbial complex either by centrifugation or by chloroform treatment. In the case of a low population of the target bacterium in the substrate, the required enrichment of the bacterium is achieved by growing it in a suitable medium.

Phages have been successfully used to detect phytopathogenic bacteria from seeds of cowpea and green gram (Soni and Thind, 1991), rice, soybean, pea and kidney bean. Two phages, ϕ RSM1 and ϕ RSS1, of *Ral. solanacearum* are highly strain-specific and can be effectively used for the detection of specific strains of this bacterium (Yamada *et al.*, 2007). Kawasaki *et al.* (2007) reported that bacterial cells of *Ral. solanacearum*, transformed with a green fluorescent protein (GFP)-expressing plasmid pRSS12, constructed from a filamentous phage ϕ RSS1, emitted a strong green fluorescence, and the presence of these cells was easily monitored in infected potato tissues and soil samples. These results show that this plasmid can serve as an easy-to-use GFP-tagging tool for any given strain of *Ral. solanacearum* in cytological and field studies. Dobozi-King *et al.* (2005) developed a novel technique called "sensing of phage-triggered ion cascades" for the rapid detection and identification of live pathogenic bacteria on the scale of minutes, with an unparalleled specificity. The method has a potential ultimate sensitivity of 1 bacterium/microlitre (1 bacterium/mm^3) and is applicable to only physiological viable cells infected by mycophages and siphophages. The method combines the specificity and avidity of bacteriophages with fluctuation analysis of electrical noise. The method is based on the massive transitory ion leakage that occurs at the moment of phage DNA injection into the host cell. The ion fluxes require only that the cells should be physiologically viable (i.e., having energised membranes) and that they can occur within seconds after mixing the cells with a sufficient concentration of phage particles.

The phages can also be used for the identification of bacteria, including phytopathogenic ones. The phages used for this purpose should have a high specificity for the species or the strains to be identified. Phages attacking different groups of bacteria differ in their host range. Phages infecting *Pec. carotovorum* subsp. *carotovorum* generally have a narrow host range. Similarly, ϕ RSM1 and ϕ RSS1 of *Ral. solanacearum* have a narrow host range, while two other phages, namely, ϕ RSA1 and ϕ RSL1, of the same bacterium have a relatively wide host range, and the phage ϕ RSA1 was able to infect all the 15 strains of the bacterium belonging to different races and biovars tested in the study (Yamada *et al.*, 2007). The ability of two phages, namely, Cp1 and Cp2, to infect roughly 98% strains of *Xan. citri* subsp. *citri* (syn. *Xan. axonopodis* pv. *citri*) occurring in Japan, indicates the wide host range of the phages of this pathogen (Wakimoto, 1967). Wu *et al.* (1993) also obtained similar results in Taiwan.

In some phages, phage virulence is correlated to pathogenic variability or certain biochemical properties of the host bacterium. These phages can be used for the quick identification of pathogenic races. Certain phages of *Ral. solanacearum* attack strains of the bacterium characterised by lactose oxidation and pathogenicity to tobacco. Phages specific to a lactose-utilising strain of *Xan. citri* subsp. *malvacearum* (syn. *Xan. axonopodis* pv. *malvacearum*) and a mannitol-utilising strain of *Xan. citri* subsp. *citri* occur in nature and are used for the quick identification of these bacteria. Du Plessis *et al.* (1981) reported another bacteriophage successfully used for strain identification.

Three bacteriophages of *Xan. arboricola* pv. *pruni* isolated from soil beneath the diseased plum trees were used to type six South African and four other *Xan. arboricola* pv. *pruni* isolates, which were previously shown to be serologically closely related. On the basis of patterns of lysis and efficiency of plaque formation by the phages, the bacterial isolates formed three distinct groups. Isolates from South Africa, the United States and Argentina were lysed by the three phages with a high degree of efficiency, while one isolate from New Zealand showed no lysis and the other isolate showed lysis with a low degree of efficiency. Both the New Zealand isolates were lysogenic, liberating phages spontaneously or following UV irradiation. The phage typing also provided evidence that peach, plum and apricot trees in South Africa were infected by the same strain of *Xan. arboricola*. pv. *pruni*.

5.3.3 Pathogenicity Test

The pathogenicity test is the most important step in the fulfilment of Koch's postulates because there are saprophytic bacteria that almost completely resemble pathogenic bacteria in phenotypic or genetic characteristics. In spite of many molecular and serological tests available to detect the presence of a pathogen in host tissue, it is still a **most reliable test** to prove that a particular organism is the causal agent of the disease. It is the final proof and confirmation that a particular bacterium isolated from a plant with or without symptoms is really the cause of the disease and can be proved by a pathogenicity test using a susceptible host of the suspected pathogen. Janse (2005) has rightly concluded, "**The confirmatory pathogenicity test is still valid, still going strong and important to fulfil the Koch's postulates**." This test is still indispensable and obligatory in many official testing and diagnostic schemes and also in legal cases between government and growers or importers/exporters and between countries, when a pathogen is detected or described for the first time. An example is the diagnostic scheme for detection of latent infections, identification pathways and the confirmatory pathogenicity test for *Ral. solanacearum* specified in an official European Union (EU) and European Plant Protection Organisation (EPPO) testing scheme. The pathogenicity tests for quarantine organisms should be performed in a special quarantine greenhouse, where insects are excluded and a special quarantine protocol and regime for workers is in place.

However, the pathogenicity test requires time, the availability of host plants and optimum environmental conditions for disease development. Moreover, for some bacterial pathogens, the time required to produce the disease symptoms is too long. Some strains of *Cla. michiganensis* require 21–28 days to produce the symptoms. On the other hand, the confirmation by pathogenicity tests is often quick for leaf-spotting pathogens as they take only 3–4 days to produce the symptoms under optimum/favourable conditions.

5.3.3.1 Inoculation of Host

The isolated bacterium must be inoculated on the host to prove its pathogenicity. The bacterial cell suspension (made in buffer) used for inoculation should contain $10^{8\text{-}9}$ cells mL^{-1}. Different methods of inoculation are used depending on the pathogen and the host involved. For leaf spot diseases, the leaves are first rubbed with carborundum powder to make small wounds, and then the bacterial suspension is smeared or sprayed onto the leaf surface. In the case of xylem-inhabiting bacteria, like *Xyl. fastidiosa*, cell suspension is infiltrated into the xylem under vacuum through the injured root, stem, or leaf. Inoculation by cutting of leaf tips with scissors dipped in bacterial cell suspension works very well for bacterial blight of rice, another vascular disease. To avoid pitfalls, it is essential to keep a negative control (plants inoculated with sterile buffer solution only) in a pathogenicity test. If necessary, a positive control, i.e., using a known pathogenic strain of the pathogen, should also be included. To avoid any contamination among controls and test plants in the greenhouse, all the three sets of plants should be kept clearly separate from each other.

For some bacterial plant pathogens, a hypersensitivity test on tobacco, as an alternative to the pathogenicity test, is quite useful, while it does not work with others. The hypersensitivity test on tobacco is a specific plant test, which differentiates most plant pathogenic *Pseudomonas* spp. (fluorescent), and some *Xanthomonas* and *Erwinia* spp. from non-plant pathogenic bacteria. A dense bacterial cell suspension containing 10^8 cells mL^{-1} is infiltrated in the mesophyll of a tobacco leaf. A pathogenic bacterium will cause a hypersensitive reaction, in which the cells leak and collapse within 24 h after infiltration, rendering the tissue glassy and later turning necrotic. Infiltration with non-pathogenic bacteria only causes some yellowing after several days or no visible change at all.

5.3.3.2 Reisolation of Pathogen and its Identification

To be certain, that only the inoculated bacterium caused the symptoms observed in the test plants, reisolation from inoculated plants, at a distance from the point of original inoculation, is necessary. Reisolated bacterium should be purified and the pure culture obtained should be identical in all respects to the original inoculated culture. Some rapid confirmatory positive tests based on serology and PCR will make it a conclusive diagnosis. When a bacterium is found for the first time in a country or imported material, the tests for its identification should be repeated to confirm its identity and declare it as causal agent of the disease. In such cases, the original sample, sample extract, pure culture and other test material obtained should be kept for reference purposes. Lelliot and Stead (1987) have given a detailed account for the diagnosis of bacterial diseases of plants by providing their detailed symptoms, diagnostic procedures, host inoculation tests, composition of culture media and procedures for staining techniques and biochemical tests.

5.4 DIAGNOSIS REPORT

Finally, a diagnosis report containing the following information is sent to the original correspondent.

1. The origin and description of the sample.
2. Nature of the damage.
3. Name of the bacterium isolated.
4. Name of disease caused by the bacterium and description of disease symptoms.
5. Nature (biology, epidemiology) of the pathogen.
6. Preventive and management measures to be adopted.

Mahlein (2016) has highlighted the potential of optical techniques, such as RGB (Red Green Blue) imaging, multi- and hyperspectral sensors, thermography, or chlorophyll fluorescence in the detection, identification and quantification of plant diseases at an early stage of epiphytotics. Later on, 3D scanning has also been added as an optical analysis that gives additional information on vitality of plants. Different platforms from proximal to remote sensing are also available for multiscale monitoring of single crop organs or entire fields. Accurate and reliable detection of diseases is facilitated by highly sophisticated and innovative methods of data analysis, which lead to new insights derived from sensor data for complex plant pathogen systems. Non-destructive, sensor-based methods support and expand upon visual and/or molecular approaches to plant disease assessment. The most relevant areas for application of sensor-based analyses are precision agriculture and plant phenotyping.

Miller *et al.* (2009) stressed the importance of early and accurate diagnoses and pathogen surveillance on local, regional and global scales to predict outbreaks and allow time for the development and application of mitigation strategies. Plant disease diagnostic networks have developed worldwide to address the problems of efficient and effective disease diagnosis and pathogen detection, engendering cooperation of institutes and experts within countries and across national borders. The internet has opened global avenues for access to databases, communication and cooperation not thought possible decades ago. Without this innovation, the organisation of effective networks to meet the challenge of invasive pathogens was improbable. Case studies of regional, national and international diagnostic networks are presented. The authors have also emphasised the importance of morphological identification and systematics for the detection and identification of the pathogens in spite of frequent use of molecular methods in developed countries.

The National Plant Diagnostic Network (NPDN) has become a critical component of the plant biosecurity infrastructure of the United States. In this country, the target set in 2002 for the establishment of a distributed, but coordinated system of plant diagnostic laboratories at land grant universities and state departments of agriculture has been achieved. The NPDN has become a model for the cooperation necessary among the public and private entities to protect our natural and agricultural plant resources. Established at five regional networks, NPDN laboratories upload diagnostic data records into a National Data Repository at Purdue University. By facilitating early detection and providing triage and surge support during plant disease outbreaks, the NPDN has become an important partner among federal, state and local plant protection agencies and with the industries that support plant protection (Stack *et al.* 2014).

REFERENCES

Alvarez, A. M. (2001). Differentiation of bacterial populations in seed extracts by flow cytometry. In: S. H. De Boer (ed.). *Plant Pathogenic Bacteria*. Kluwer, Dordrecht, pp. 393–96.

Alvarez, A. M. (2004). Integrated approaches for detection of plant pathogenic bacteria and diagnosis of bacterial diseases. *Annu. Rev. Phytopathol.* 42: 339–66.

Alvarez, A. M. and Adams, P. D. (1999) Flow cytometry: A promising tool for seed health testing. *Proc. Third Int. Seed Health Symp*. Int. Seed Test. Assoc., Ames, Iowa, pp. 110–14.

Alvarez, A. M., Benedict, A. A. and Gananamanickam, S. S. (1993a). Identification of seed-borne bacterial pathogens of rice with monoclonal antibodies. *Proc. Symposium on Seed Health Testing 1st, Ottawa, Canada.*

Alvarez, A. M., Benedict, A. A. and Mizumoto, C. Y. (1985). Identification of xanthomonads and groupings of *Xanthomonas campestris* pv. *campestris* strains with monoclonal antibodies. *Phytopathology* 75: 722–28.

Alvarez, A. M., Berestecky, J., Stiles, J. I., Ferreira, S. J. and Benedict, A. A. (1992). Serological and molecular approaches to identification of *Pseudomonas solanacearum* strains from *Heliconia*. *Presented at Bacterial Wilt International Conference, Kaohsiung, Taiwan.*

Alvarez, A. M., Derie, M., Benedict, A. A. and Gabrielson, R. (1993b). Characteristics of a monoclonal antibody to *Clavibacter michiganensis* subsp. *michiganensis*. *Phytopathology* 83: 1405.

Alvarez, A. M., Rehman, F. U. and Leach, J. E. (1997). Comparison of serological and molecular methods for detection of *Xanthomonas oryzae* pv. *oryzae* in rice seed. Presented at *Seed Health Test: Progr. Towards the 21st Century, Cambridge, UK.*

Alvarez, A. M., Schenck, S. and Benedict, A. A. (1996). Differentiation of *Xanthomonas albilineans* strains with monoclonal antibody reaction patterns and DNA fingerprints. *Plant Pathol.* 45: 358–66.

Alvarez, E., Pardo, J. M. and Truke, M. J. (2015). A portable field pathogen-detection system for disease-free planting materials. *Phytopathology* 105: S1.5–S1.6.

Anonymous. (1998). Interim testing scheme for the diagnosis, detection and identification of *Ralstonia solanacearum* (Smith) Yabuuchi *et al.* in potatoes. Annex. II to the Council Directive 98/57/EC of 20 July 1998 on the control of *Ralstonia solanacearum* (Smith) Yabuuchi *et al.* Publication 97/647/EC, Official Journal European Communities No. L 235, 8–39.

Benedict, A. A., Alvarez, A. M., Berestecky, J., Imanaka, J., Mizumoto, C. Y. *et al.* (1989). Monoclonal antibodies specific to *Xanthomonas campestris* pv. *oryzae* and pv. *oryzicola*. *Phytopathology* 79: 322–28.

Benedict, A. A., Alvarez, A. M. and Pollard, L. W. (1990). Pathovar specific antigens of *Xanthomonas campestris* pv. *begoniae* and *X. campestris* pv. *pelargonii* detected with monoclonal antibodies. *Appl. Environ. Microbiol.* 56: 572–74.

Bertolini, E., Olmas, A., Lopez, M. M. and Cambra, M. (2003). Multiplex nested RT-PCR in a single closed tube for sensitive and simultaneous detection of four RNA viruses and *Pseudomonas savastanoi* pv. *savastanoi* in olive trees. *Phytopathology* 93: 286–92.

Bishop, A. L., Burr, T. J., Millak, V. L. and Katz, B. H. (1989). A monoclonal antibody specific to *Agrobacterium tumefaciens* biovar 3 and its utilization for indexing grapevine indexing material. *Phytopathology* 79: 995–98.

Bragard, C., Mehta, Y. R. and Maraile, H. (1993). Serodiagnostic assays versus the routine techniques to detect *Xanthomonas campestris* pv. *undulosa* in wheat seeds. *Fitopathol. Bras.* 18: 42–50.

Caruso, P., Bertolini, E., Cambra, M. and Lopez, M. M. (2003). A new and sensitive Co-operational polymerase chain reaction (Co-PCR) for rapid detection of *Ralstonia solanacearum* in water. *J. Microbiol. Methods* 55: 257–72.

Chitarra, L. G., Langerak, C. J., Bergervoet, J. H. and van den Bulk, R. W. (2002). Detection of the plant pathogenic bacterium, *Xanthomonas campestris* pv. *campestris* in seed extracts of *Brassica* sp. applying fluorescent antibodies and flow cytometry. *Cytometry* 47: 118–26.

Cubero, J. and Graham, J. H. (2002). Genetic relationship among worldwide strains of *Xanthomonas* causing canker in citrus species and design of new primers for their identification by PCR. *Appl. Environ. Microbiol.* 68: 157–64.

Danks, C. and Barker, I. (2000). On-site detection of plant pathogens using lateral flow devices. *Bull. OEPP/EPPO* 30: 421–26.

De Boer, S. H. (1987). Use of monoclonal antibodies to identify and detect plant pathogenic bacteria. *Can. J. Plant Pathol.* 9: 182–87.

De Boer, S. H. and Wieczorek, A. (1984). Production of monoclonal antibodies to *Corynebacterium sepedonicum*. *Phytopathology* 4: 1431–34.

Dobozi-King, M., Kim, J. U., Young, R., Cheng, M. and Kish, L. B. (2005). Rapid detection and identification of bacteria: Sensing of Phage-Triggered Ion Cascade (SEPTIC). *J. Biol. Phys. Chem.* 5: 3–7.

Du Plessis, H. J., Loos, M. A. and Matthee, F. N. (1981). Bacteriophage typing of *Xanthomonas campestris* var. *pruni* isolates from different stone fruit species. *Phytophylactica* 13: 57–62.

Franken, A. A. J. M. (1990). Production, characterization and application of monoclonal antibodies for detecting *Xanthomonas campestris* pv. *campestris* in cabbage seeds. *Proceedings of the Symposium "Perspectives for Monoclonal Antibodies in Agriculture." PUDOC, Wageningen, the Netherlands.*

Fuller, S. L., Savory, E. A., Weisberg, A. J., Buser, J. Z., Gordon, M. I. *et al.* (2017). Isothermal amplification and lateral-flow assay for detecting crown-gall-causing *Agrobacterium* spp. *Phytopathology* 107: 1062–68.

Getaz, M., Buhlmann, A., Schneeberger, P. H. H., Van Malderghem, C., Duffy, B. *et al.* (2017). A diagnostic tool for improved detection of *Xanthomonas fragariae* using a rapid and highly specific LAMP assay designed with comparative genomics. *Plant Pathol.* 66: 1094–1102.

Gorris, M. T., Cambra, M. and Lopez, M. M. (1989). Production of monoclonal antibodies specific to *Xylophilus ampelinus*. *Proceedings of the 7th International Conference on Plant Pathogenic Bacteria*, Budapest, Hungary, pp. 913–21.

Griep, R. A., Van Twisk, C., Van Beckhoven, J. R. C. M., Van der Wolf, J. M. and Schots, A. (1998). Development of specific recombinant monoclonal antibodies against the lipopolysaccharides of *Ralstonia solanacearum* race 3. *Phytopathology* 88: 795–803.

Horst, R. K. (2001). *Westcott's Plant Disease Handbook*. 6th Ed. Kluwer Academic Publishers, Boston, MA.

Hu F. P., Young, J. M., Triggs, C. M., Park, D.-C. and Saul, D. J. (2001). Relationships within the Proteobacteria of plant pathogenic *Acidovorax* species and subspecies, *Burkholderia* species and *Herbaspirillum rubrisubalbicans* by sequence analysis of 16S rDNA, numerical analysis and determinative tests. *Antonie Van Leeuwenhoek J. Microbiol. Serol.* 80: 201–14.

Janse, J. D. 2005. *Phytobacteriology: Principles and Practice*. CABI Publishing, Wallingford, 360 pp.

Kawasaki, T., Satsuma, H., Fujie, M., Usami, S. and Yamada, T. (2007). Monitoring of phytopathogenic *Ralstonia solanacearum* cells using green fluorescent protein-expressing plasmid derived from bacteriophage ϕRSS1. *J. Biosci. Bioeng.* 104: 451–56.

Koh, Y. J. and Nou, I. S. (2002). DNA markers for identification of *Pseudomonas syringae* pv. *actinidiae*. *Mol. Cells* 13: 309–14.

Lamka, G. L., Hill, J. H., McGee, D. C. and Braun, E. J. (1991). Development of an immunosorbent assay for seedborne *Erwinia stewartii* in corn seeds. *Phytopathology* 81: 839–46.

Lelliot, R. A. and Stead, D. E. (1987). *Methods for the Diagnosis of Bacterial Diseases of Plants*. Blackwell Scientific Publications, Oxford, 215 pp.

Lipp, R. L., Alvarez, A. M., Benedict, A. A. and Berestecky, J. (1992). Use of monoclonal antibodies and pathogenicity tests to characterize strains of *Xanthomonas campestris* pv. *dieffenbachiae* from aroids. *Phytopathology* 82: 677–82.

Lopez, M. M., Bertolini, E., Olmos, A., Caruso, P., Gorris, M. T. *et al.* (2003). Innovative tools for detection of plant pathogenic viruses and bacteria. *Int. Microbiol.* 6: 233–43.

Mahlein, A.-K. (2016). Plant disease detection by imaging sensors – Parallels and specific demands for precision agriculture and plant phenotyping. *Plant Dis.* 100: 241–51.

Manulis, S., Valinsky, L., Lichter, A. and Gabriel, D. W. (1994). Sensitive and specific detection of *Xanthomonas campestris* pv. *pelargonii* with DNA primers and probes identified by random amplified polymorphic DNA analysis. *Appl. Environ. Microbiol.* 60: 4094–99.

Miller, S. A., Beed, F. D. and Harmon, C. L. (2009). Plant disease diagnostic capabilities and networks. *Annu. Rev. Phytopathol.* 47: 15–38.

Moter, A. and Gobel, U. B. (2000). Fluorescence in situ hybridization (FISH) for direct visualization of microorganisms. *J. Microbiol. Methods* 41: 85–112.

Norman, D. J. and Alvarez, A. M. (1994). Rapid detection of *Xanthomonas campestris* pv. *dieffenbachiae* in anthurium plants with a miniplate enrichment/ELISA system. *Plant Dis.* 78: 954–58.

Ovod, V., Ashorn, P. Yakovleva, L. and Krohn, K. (1995). Classification of *Pseudomonas syringae* with monoclonal antibodies against the core and O-side chains of the lipopolysaccharide. *Phytopathology* 85: 226–32.

Permar, T. A. and Gottwald, T. R. (1989). Specific recognition of *Xanthomonas campestris* Florida nursery strain by a monoclonal antibody probe in a microfiltration enzyme immunoassay. *Phytopathology* 79: 780–83.

Rico, A., Lopez, R., Asensio, C., Aizpun, M. T., Asensio-S.-Manzanera, M. C. *et al.* (2003). Nontoxigenic strains of *Pseudomonas syringae* pv. *phaseolicola* are a main cause of halo blight of beans in Spain and escape current detection methods. *Phytopathology* 93: 1553–59.

Riley, M. B., Williamson, M. R. and Maloy, O. (2006). Plant disease diagnosis. APSnet Feature Story, [http://www.apsnet.org/online/feature/plant disease/].

Sakthivel, N., Mortensen, C. N. and Mathur, S. B. (2001). Detection of *Xanthomonas oryzae* pv. *oryzae* in artificially inoculated and naturally infected rice seeds and plants by molecular techniques. *Appl. Microbiol. Biotechnol.* 56: 435–41.

Sanders, G. M., Verschoor, J. A., van Wyngaard, S., Korsten, L. and Kotze, J. M. (1994). Production of monoclonal antibodies against *Xanthomonas campestris* pv. *mangiferaeindicae* and their use to investigate differences in virulence. *J. Appl. Bacteriol.* 77: 509–18.

Schaad, N. W., Frederick, R. D., Shaw, J., Schneider, W. L., Hickson, R. *et al.* (2003). Advances in molecular-based diagnostics in meeting crop biosecurity and phytosanitary issues. *Annu. Rev. Phytopathol.* 41: 305–24.

Singh, U., Trevors, C. M., De Boer, S. H. and Janse, J. D. (1999). Fimbrial-specific monoclonal antibody-based ELISA for European potato strains of *Erwinia chrysanthemi* and comparison to PCR. *Plant Dis.* 84: 443–48.

Soni, P. S. and Thind, B. S. (1991). Detection of *Xanthomonas campestris* pv. *vignaeradiatae* (Sabet *et al.*) Dye from green gram seeds and *X. campestris* pv. *vignicola* (Burkh.) Dye from cowpea seeds with the help of bacteriophages. *Plant Dis. Res.* 6: 6–11.

Stack, J. P., Bostock, R. M., Hammerschmidt, R., Jones, J. B. and Luke, E. (2014). The National Plant Diagnostic Network: Partnering to protect plant systems. *Plant Dis.* 98: 708–15.

Tsai, C. C., Lin, C. P. and Chen, C. T. (1990). Characterization and monoclonal antibody production of *Xanthomonas albilineans* (Ashby) Dowson, the causal agent of sugarcane leaf scald disease. *Plant Prot. Bull.* 32: 125–35.

Van der Wolf, J. M. and Schoen, C. D. (2004). *Bacterial Pathogens: Detection and Identification Methods*. Marcel Dekker, New York, pp. 1–5.

Van der Wolf, J. M., Sledz, V., Van Elsas, J. D., Van Overbeek, L. and Van Bergervoet, J. H. W. (2004). Flow cytometry to detect *Ralstonia solanacearum* and to assess viability. In: C. Allen, P. Prior and A. C. Hayward (eds.). *Bacterial Wilt: The Disease and the Ralstonia solanacearum Species Complex*. APS Press, St. Paul, MN, p 6.

Van Overbeek, L. S., Cassidy, M., Kozdroz, J., Trevors, J. T. and Van Elsas, J. D. (2002). A polyphasic approach for studying the interaction between *Ralstonia solanacearum* and potential control agents in the tomato phytosphere. *J. Microbiol. Methods*, 48: 69–86.

Veena, M. S. and van Vuurde, J. W. (2002). Indirect immunofluorescence colony staining method for detecting bacterial pathogens of tomato. *J. Microbiol. Methods* 49: 11–17.

Vernon-Shirley, M. and Burns, R. (1992). The development and use of monoclonal antibodies for detection of *Erwinia*. *J. Appl. Bacteriol.* 72: 97–102.

Wakimoto, S. (1967). Some characteristics of citrus canker bacteria, *Xanthomonas citri* (Hasse) Dowson, and the related phages isolated from Japan. *Ann. Phytopathol. Soc. Japan* 33: 301–10.

Walcott, R. R. and Gitaitis, R. D. (2000). Detection of *Acidovorax avenae* subsp. *citrulli* in watermelon seed using immunomagnetic separation and the polymerase chain reaction. *Plant Dis.* 84: 470–74.

Wong, W. C. (1990). Production of monoclonal antibodies to *Pseudomonas syringae* pv. *phaseolicola* and *Xanthomonas campestris* pv. *phaseoli*. *Lett. Appl. Microbiol.* 10: 241–44.

Wu, W. C., Lee, S. T., Kuo, H. F. and Wang, L. Y. (1993). Use of phages for identifying the citrus canker bacterium *Xanthomonas campestris* pv. *citri* in Taiwan. *Plant Pathol.* 42: 389–95.

Wullings, B. A., Van Beuningen, A. R., Janse, J. D. and Akkermans, A. D. L. (1998). Detection of *Ralstonia solanacearum*, which causes brown rot of potato, by fluorescent in situ hybridization with 23S rRNA targeted probes. *Appl. Environ. Microbiol.* 64: 4546–54.

Yamada, T., Kawasaki, T., Nagata, S., Fujiwara A., Usami S. *et al.* (2007). New bacteriophages that infect the phytopathogen *Ralstonia solanacearum*. *Microbiology* 153: 2630–39.

6 Classification of Bacteria

Classification is an arrangement of organisms into groups, called taxonomic ranks, on the basis of similarities or relationships. Two other areas closely related with the classification are nomenclature and identification. Nomenclature is the assignment of names to the taxonomic ranks according to the nomenclature code. Identification is the practical use of classification scheme to determine the identity of an organism as a member of an established taxon or as a member of a previously unidentified species.

6.1 NOMENCLATURE CODE

For several years, the Botanical Code of Nomenclature was the basis for bacterial nomenclature. In the International Congress of Microbiology held in London in 1936, it was decided that there should be a separate code for bacterial nomenclature. Therefore, the first separate code for bacteria, the **International Bacteriological Code of Nomenclature**, was published in 1948. The code has been emended on many occasions, but a major revision occurred in 1976 (Lapage *et al.*, 1975). The name of the Code has also been changed on many occasions; since 2000, however now, it is known as the **International Code of Nomenclature of Procaryotes**, governed by the **International Committee on Systematics of Procaryotes** (ICSP). Before the adoption of the above-mentioned last name of the code, it was called the **International Code of Nomenclature of Bacteria** and was governed by the **International Committee on Systematic Bacteriology**.

The 1976 revision of the International Code of Nomenclature of Bacteria (Lapage *et al.*, 1975) provides that after January 1, 1980, priority of publication shall date from January 1, 1980 and on that date, all names published prior to January 1, 1980 and included in the **Approved Lists of Bacterial Names** of ICSB (International Committee on Systemic Bacteriology) shall be treated validly published as having been published for the first time on that date. Those names validly published prior to January 1, 1980 but not included in the Approved Lists will have no further standing in nomenclature. After January 1, 1980, the priority of publication shall date from the date of their publication in the ***International Journal of Systematic Bacteriology*** (from 2000 in the ***International Journal of Systematic and Evolutionary Microbiology***). Since 1980, the names effectively published outside the above-mentioned two journals can only make a claim to priority, if they are subsequently published in validation lists in these two journals. Priority shall date from the date of their publication in the validation lists and not from the date of their original publication.

The Bacteriological Code (1990 Revision) did not cover taxa above the rank of a class. Therefore, in the second edition of *Bergey's Manual of Systematic Bacteriology*, other codes of nomenclature were used for naming the higher ranks. As there was no recognition of a rank above kingdom, Bergey's Trust, after considerable deliberations, concluded to incorporate the rank of **domain** into the hierarchy. Further, the rank of kingdom has not been used in this edition to avoid possible conflicts with other codes of nomenclature.

The rank of phylum is extensively used for groups of procaryotes. It has been estimated that ~1,300 bacterial phyla exist. As of May 2020, 41 bacterial phyla are formally accepted by the List of Procaryotic names with Standing in Nomenclature (LPSN), 89 bacterial phyla are recognised on the Silva database, dozens more have been proposed, and hundreds likely remain to be discovered. As of 2017, approximately 72% of widely recognised bacterial phyla were candidate phyla (i.e. have no cultured representatives) (from Wikipedia: text is available under the Creative Commons Attribution-Share Alike 4.0 License; page was last edited on May 28, 2025).

In addition, there are many deep lineages equivalent to phyla that are not yet represented by organisms available in pure culture. Oren *et al.*, (2015) proposed the inclusion of the rank of phylum in the International Code of Nomenclature of Procaryotes. The rank of phylum was added to the ranks covered by the rules of the Code. This important change was approved by the ICSP in a separate ballot held in 2021 after a public discussion was held on this topic (Oren *et al.*, 2021). Whitman *et al.*, (2018) proposed the suffix *-ota* to denote phyla, replacing the somewhat awkward *-aeota*, as an addendum to the earlier proposal of Oren *et al.*, (2015) to include the rank of phylum in the International Code of Nomenclature of Prokaryotes. The last revision (2022 Revision) of the code took place in 2022 (Oren *et al.*, 2023).

6.2 SPECIES

The bacterial species is the basic and most important taxonomic rank in the bacterial systematics. The concept of a bacterial species is poorly understood in comparison to higher organisms. This is mainly due to the reason that bacteria, being procaryotic organisms, differ markedly from higher organisms. Contrary to higher organisms, sexuality is not used in bacterial species classification because relatively few bacteria undergo conjugation. Moreover, most of the procaryotic microorganisms are relatively simple and similar in terms of their morphological character. Hence, the morphological features alone are usually of little classificatory significance in comparison to higher organisms.

DOI: 10.1201/9781003681878-7

Brenner *et al.*, (2001) defined the bacterial species as "a distinct group of strains that have certain distinguishing features and that generally bear a close resemblance to one another in the more essential features of organisation". Each species differs considerably and can be distinguished from all other species. The above-given definition of the species was generally loosely followed. It was mainly due to the reasons that one cannot accurately determine 'a close resemblance', 'essential features' or how many 'distinguishing features' are sufficient to create a species.

6.3 CLASSIFICATION

Von Naegeli (1857) placed bacteria in the class **Schizomycetes**. In the seventh edition of *Bergey's Manual of Determinative Bacteriology*, class Schizomycetes was placed in one of the five divisions, namely, Protophyta, recognised under the kingdom Plantae (Breed *et al.* 1957). Kingdom Plantae was proposed by Carolus Linnaeus in 1753. Chatton (1937), on the basis of cellular organisation, divided the living organisms into two groups, **procaryotes** and **eucaryotes**. Later on, R. G. E. Murray (1968) proposed two kingdoms, namely Procaryotae and Eucaryotae, to include these two groups. During the 1970s, the comparison of oligonucleotide catalogues of the 16S ribosomal (r) RNA of procaryotes and of 18S rRNA of eucaryotes of a broad range of living organisms revealed that there are two different kinds of procaryotes: the ***Archaea*** (also called archaebacteria or archaeobacteria) and the ***Bacteria***. On the basis of these findings, Woese *et al.*, (1990) opined that there are three fundamentally different kinds of living organisms: the *Archaea*, the *Bacteria* and the eucaryotes (now called the ***Eucarya***).

Different criteria have been followed to classify bacteria. First of all, there was the morphological approach in which the major emphasis was placed on morphological characters. Later on, the emphasis was also placed on physiological and biochemical characters, along with the morphological characters. As a result, the classification of bacteria became tedious and time-consuming because a large number of tests were to be carried out and some of these tests required long time for completion.

With the rapid advances in DNA sequencing technology, sequence analysis of highly conserved regions of the bacterial genome, most notably genes coding for the RNA of the small ribosomal subunit, has now led to a natural classification that reflects the evolutionary history of bacteria. One important development was the discovery by Doi and Igarashi (1965) and Dubnau *et al.*, (1965) that the rRNA citrons in bacterial species were conserved (slower to change) to a greater extent than the bulk of the genome, probably because of their critical role in the life of a cell. This function would allow any slow changes in nucleotide sequence to occur over a long period of time, relative to other genes that were not so critical to cell functions. This, in turn, led to the idea that rRNA–DNA hybridisation might be useful for deducing the broader relationships that DNA–DNA hybridisation could not reveal.

Genotyping (whole-genome sequence comparisons) is the most powerful procedure that will likely stabilise the taxonomy of bacteria. Current taxonomic classifications utilise DNA–DNA hybridisation values, guanine-plus-cytosine content in mole percent, 16S rDNA sequences, and, to a lesser extent, intergenic spacer sequences (16S–23S) and sequence comparisons of a few characterised genes, such as ATPase, the β subunit of F1F0 elongation factor Tu, *recA*, and *gyrA*, to generate new taxonomic proposals. The sequence analyses of genomes and plasmids have revealed complete profiles of gene families and identified pathogenicity islands and specific genes or gene sets that proved more useful and reliable than physiological and biochemical tests. The growing number of whole-genome sequences available provides the needed data to analyse evolutionary trends and define the age of each genus or bacterial family by using molecular clocks.

At present, the polyphasic approach used in the taxonomy and systematics of the bacteria and archaea includes the use of phenotypic, chemotaxonomic and genotypic data. The use of 16S rRNA gene sequence data has revolutionised our understanding of the microbial world and led to a rapid increase in the number of descriptions of novel taxa, especially at the species level. In many cases, it has allowed for the demarcation of taxa into distinct species, but its limitations in a number of groups have resulted in the continued use of DNA–DNA hybridisation. As technology has improved, next-generation sequencing has provided a rapid and cost-effective approach to obtain whole-genome sequences of microbial strains. The incorporation of genomics into the taxonomy and systematics of the bacteria and archaea coupled with computational advances will boost the credibility of taxonomy in the genomic era (Chun and Rainey, 2014).

Bergey's *Manuals of Determinative Bacteriology* and Bergey's *Manuals of Systematic Bacteriology* are the most comprehensive and authoritative publications in the field of bacterial taxonomy. The first edition of *Bergey's Manual of Determinative Bacteriology* was published in 1923 and the subsequent editions (up to the seventh) were published in 1925, 1930, 1934, 1939, 1948 and 1957, respectively. As the new information became available it was included in the subsequent editions. The eighth edition of *Bergey's Manual of Determinative Bacteriology* (Buchanan and Gibbons, 1974) is different in format, presentation of information and approach to classification from the seventh and earlier editions. This manual is meant mainly for the identification of bacteria and the bacteria are divided into 19 groups. The ninth edition of *Bergey's Manual of Determinative Bacteriology* (Holt *et al.*, 1994), published in 1994, is a departure from the past editions, in that it is intended solely for the identification of those bacteria that have been described and cultured. In this edition, the bacteria have been divided into 35 groups, including 5 groups of Archaeobacteria (archaea). These groups are not formal taxonomic ranks but easily recognised phenotypic groups useful for identification purposes.

The second edition of *Bergey's Manual of Systematic Bacteriology* is the latest edition of the manual. It represents a major departure from the first edition, as well as from the eighth and ninth editions of *Bergey's Manual of Determinative Bacteriology*, in that the organisation of the contents follows a phylogenetic framework based on analyses of the nucleotide sequence of the ribosomal small subunit RNA, rather than one based on phenotypic characters. The classification in the eighth and ninth editions of *Bergey's Manual of Determinative Bacteriology*, and also in the first edition of *Bergey's Manual of Systematic Bacteriology*, was organised in a non-hierarchical scheme because information about higher taxa was insufficient for the construction of a formal hierarchical classification such as those used in all the previous editions. Instead, the genera were organised into phenotypic groupings, and these groupings were called parts, sections or groups.

It is generally agreed that procaryotes (also called ***Procarya***) fall into two major lines of evolutionary descents, the *Archaea* and the *Bacteria*. These have been recognised as two different domains in the second edition of *Bergey's Manual of Systematic Bacteriology*. The domains are subdivided into phyla that represent the main organisational units in this edition. With the exception of two phyla, Cyanobacteria and Actinobacteria, phyla are further subdivided into classes, classes into orders, orders into families, families into genera and genera into species. In the case of Cyanobacteria, subsections replace the families. In the case of Actinobacteria, subclasses and suborders are included. Different taxonomic ranks and their numbers recognised till 2003 are given in Table 6.1 (Garrity *et al.*, 2005).

As of May 2020, 41 bacterial phyla are formally accepted by the LPSN, 89 bacterial phyla are recognised on the Silva database, dozens more have been proposed, and hundreds likely remain to be discovered. As of 2017, approximately 72% of widely recognised bacterial phyla were candidate phyla (i.e. have no cultured representatives).

The domain *Archaea* is divided into two phyla and nine classes. The domain *Bacteria* is divided into 24 phyla and 32 classes. The domain *Bacteria* is much larger and more complex than the other procaryotic domain, *Archaea*. The sizes of various phyla of domain *Bacteria* differ considerably. The phylum *Proteobacteria* was the largest; at the end of October 2003, it contained 2279 validly named species belonging to 521 genera. The phylum *Dictyoglomi* is the smallest and contains only a single recognised species. The phyla of the domain *Bacteria* along with their phyla numbers are listed in Table 6.2 (Garrity *et al.*, 2005).

Since the publication of Garrity *et al.*, (2005), Oren *et al.*, (2015) have listed seven more phylla names in use currently, i.e., *Armatimonadetes*, *Elusimicrobia*, *Ignavibacteriae*, *Lentisphaerae*, *Euryarchaeota*, *Thaumarchaeota*, and *Crenarchaeota*, but they omitted the phylum name *Cyanobacteria* already recognised.

6.3.1 Classification of Plant Pathogenic Bacteria

The classification of plant pathogenic bacteria has seen many revisions and rearrangements. Before 1967, i.e., prior to the discovery of role of mycoplasma-like organisms in causing plant diseases, there were only six genera of bacteria, which included plant pathogens. These genera were *Agrobacterium, Corynebacterium, Erwinia, Pseudomonas, Streptomyces* and *Xanthomonas*. The 1976 revision of the International Code of Nomenclature of Bacteria (Lapage *et al.*, 1975) had a major effect on the taxonomy of plant pathogenic bacteria. Due to the revision of the code, the names of many plant pathogenic bacteria were not accepted and, hence, were not included in the approved lists. Some of these were not accepted because of their poor descriptions or the absence of suitable type cultures. The others were not accepted because some of the plant pathogenic species reported were similar in morphological, cultural and biochemical characters and could not be differentiated from each other on the basis of these characters. However, these could be separated from each other on the basis of their

TABLE 6.1
Different taxonomic ranks used in classification of bacteria

Taxonomic Rank	Archaea	Bacteria	Total
Domain	1	1	2
Phylum	2	24	26
Class	9	32	41
Subclass	0	5	5
Order/subsection	13	75	88
Suborder	0	17	17
Family	23	217	240
Genus	79	1115	1194
Species	281	6185	6466

TABLE 6.2
Phyla of the domain *bacteria*

Phylum number	Phylum name	Phylum number	Phylum name
B1	*Aquificae*	B13	*"Firmicutes"*
B2	*Thermotogae*	B14	*"Actinobacteria"*
B3	*Thermodesulfobacteria*	B15	*"Planctomycetes"*
B4	*"Deinococcus-Thermus"*	B16	*"Chlamydiae"*
B5	*Chrysiogenetes*	B17	*"Spirochaetes"*
B6	*Chloroflexi*	B18	*"Fibrobacteres"*
B7	*Thermomicrobia*	B19	*"Acidobacteria"*
B8	*"Nitrospirae"*	B20	*"Bacteroidetes"*
B9	*Deferribacteres*	B21	*"Fusobacteria"*
B10	*Cyanobacteria*	B22	*"Verrucomicrobia"*
B11	*"Chlorobi"*	B23	*"Dictyoglomi"*
B12	*Proteobacteria*	B24	*Gemmatimonadetes*

Note: " " Used for provisional arrangement, the rest are established.

pathogenicity on different plant hosts. Most of the names in the latter category were nomenspecies of *Xan. campestris* and *Psm. syringae* groups. More than 40 species included in the genus *Pseudomonas* were identical to *Psm. syringae* on the basis of cultural, morphological and biochemical characters. According to the revised code, the specific epithets of these species (not included in the approved lists) would, thereafter, be available for reuse to name new taxa. Loss of species designations for recognised plant pathogens and subsequent reuse of these names for new taxa was bound to create an unnecessary confusion and loss of the knowledge previously published about these pathogens.

6.3.1.1 Pathovar Concept

The executive committee of the **International Society for Plant Pathology** (ISPP) set up a **Committee on Taxonomy of Plant Pathogenic Bacteria** (CTPPB) to supervise the consequences of the 1976 revision of the code for plant pathologists and to make necessary recommendations for retaining the names of plant pathogenic bacteria not included in the approved lists. The infrasubspecific taxon "pathovar" records most accurately the differences exhibited by plant pathogenic bacteria in terms of host range or in producing different symptoms (diseases) on the same host. The standards define a pathovar as "a strain or a group of strains with the same or similar characteristics, differentiated at the infrasubspecific level from other strains of the same species or subspecies on the basis of distinctive pathogenicity to one or more plant hosts". These differences in pathogenicity can be in either symptoms or host range. Pathovars may also differ in other characteristics as well as pathogenicity but such differences must be of minor importance; otherwise there would be justification for separation at subspecies or even species level (Bradbury, 1982). CTPPB of ISPP requested the Judicial Commission of International Committee on Systematic Bacteriology to alter the rules of the code to include the taxon pathovar. However, the Judicial Commission denied the request because it could set a precedent for the inclusion of other infrasubspecific subdivisions.

The consensus of opinion of ISPP members and its CTPPB was that the pathovar designation most accurately records the known status of nomenspecies of plant pathogenic bacteria not included in the approved lists and that formal recognition is essential to avoid confusion in the nomenclature. Young *et al.*, (1978) proposed that bacterial plant pathogens which do not have a claim to the species rank should be classified as pathovars and also listed all validly published names of plant pathogens having authentic cultures as species or pathovars. In order to formalise the use of pathovar names, the ISPP, through its executive committee and its CTPPB, decided to publish a list of pathovar names and standards, to be known as the International Standards for Naming Pathovars of Phytopathogenic Bacteria.

6.3.1.2 International Standards for Naming Pathovars of Phytopathogenic Bacteria

The rules for the Standards were adopted from the Code and were so written that if the Code ever does expand to assume authority over the infrasubspecific subdivisions, the pathovar standards would be transferred easily to the Code. The Standards were adapted to the Code in a manner that allows both the Standards and the Code to be applied without conflict. Principal requirements of the pathovars for legitimate names are essentially the same as for Code.

Unlike the valid publication of species names, no journal is specified for the valid publication of pathovar names. Although there is a requirement for the submission of pathotype culture to one international culture collection, the ISPP-CTPPB suggests that researchers should follow the Code and submit the cultures of new pathotype strains at least to two international culture collections located in two different countries in order to ensure stability and access. Priority for pathovars included in the list of pathovars names published with the standards (Dye *et al.*, 1980) and also those published before the publication of this list is assigned in a manner similar to those of species. These are treated as though they had been validly published for the first time on January 1, 1980. Priority for pathovar names published after January 1, 1980 is determined according to the date of effective publication of legitimate pathovar names. For more details on nomenclature of bacterial plant pathogens, the readers may consult the review articles by Young (2008) and Bull *et al.*, (2008).

Plant pathologists or plant bacteriologists have to follow two different but complementary sets of rules: the rules given in the International Code of Nomenclature of Procaryotes for naming down to subspecies level, and the standards in the International Standards for Naming Pathovars of Plant Pathogenic Bacteria to name pathovars. A list of pathovar names and pathotype strains is maintained by the CTPPB of the ISPP. The first such list was an appendix to the Standards and only contained the names of microorganisms for which pathovar names and pathotype or neopathotype strains needed to be assigned (Dye *et al.*, 1980). Names of plant pathogenic species in the approved lists were not included in the ISPP-CTPPB list. These two documents together contained the names of all bacterial plant pathogens validly published as of 1980. Since then, the ISPP-CTPPB has periodically published lists of recently published legitimate names and updates an annotated comprehensive list of names of plant pathogenic bacteria. The subsequent lists published by ISPP-CTPPB are given below:

1. Names of plant pathogenic bacteria 1864–1995 (Young *et al.*, 1996) is the first comprehensive list of names published by the ISPP-CTPPB in which all the names proposed for plant pathogenic bacteria were included in a single list. It was originally published in the *Review of Plant Pathology* (Vol. 75: pp. 721–863) and in 2000 was placed on the ISPP website, http://www.isppweb.org/names_bacterial_revised.asp.
2. Names of plant pathogenic bacteria published since 1995 (Young *et al.*, 2004; http://www.isppweb.org/names_bacterial.asp) is a list of names published between 1995 and March 2004. It is published only online and includes all the names that were published during this period.
3. Names of plant pathogenic bacteria, 1864–2004 (Young *et al.*, 2004; http://www.isppweb.org).
4. Names of plant pathogenic bacteria, 1864–2006 (Bull *et al.*, 2008; http://www.isppweb.org/about_tppb.asp).
5. Comprehensive list of names of plant pathogenic bacteria, 1980–2007 (Bull *et al.*, 2010).
6. List of new names of plant pathogenic bacteria, 2008–2010 (Bull *et al.*, 2012).
7. List of new names of plant pathogenic bacteria, 2011–2012 (Bull *et al.*, 2014).

In the second edition of *Bergey's Manual of Systemic Bacteriology*, phytopathogenic bacteria have been placed in three phyla, namely, *Proteobacteria*, "*Actinobacteria*" and "*Firmicutes*". Among these phyla, *Proteobacteria* is the largest and is divided into five classes, namely, "*Alphaproteobacteria*", "*Betaproteobacteria*", "*Gammaproteobacteria*", "*Deltaproteobacteria*" and "*Epsilonproteobacteria*". Out of these five classes, the first three classes contain the plant pathogenic bacteria. The phyla "*Firmicutes*" and "*Actinobacteria*" contain three classes and one class, respectively, and all these classes contain plant pathogenic bacteria. The important characteristics of the phyla containing plant pathogenic bacteria are given in Table 6.3.

Salam *et al.*, (2020) described the rearrangement of the higher taxonomic ranks of the members of the phylum '*Actinobacteria*', using the phylogeny of 16S rRNA gene sequences and supported by the phylogeny of the available genome sequences. Based on the refined phylogeny of the 16S rRNA gene sequences, the authors arranged all the members of the 425 genera of the phylum '*Actinobacteria*', with validly published names currently in use, into 6 classes, 46 orders and 79 families, including 16 new orders and 10 new families. The order *Micrococcales* Prévot 1940 (Approved Lists, 1980; emend. Nouioui *et al.* 2018) is now split into 11 monophyletic orders: the emended order *Micrococcales* and ten proposed new orders *Aquipuribacterales*, *Beutenbergiales*, *Bogoriellales*, *Brevibacteriales*, *Cellulomonadales*, *Demequinales*, *Dermabacterales*, *Dermatophilales*, *Microbacteriales* and *Ruaniales*. Further, the class '*Actinobacteria*' Stackebrandt *et al.* 1997; emend. Nouioui *et al.* 2018 has been described without any nomenclature type, and therefore the name '*Actinobacteria*' is deemed illegitimate. In accordance to Rule 8 of the International Code of Nomenclature of Procaryotes, Parker *et al.* 2019, the authors proposed the name *Actinomycetia* which is formed by using the stem of the name *Actinomycetales* (Buchanan 1917* (Approved Lists 1980); emend. Zhi *et al.*, (2009)), to replace the name '*Actinobacteria*'. The nomenclature type of the proposed new class *Actinomycetia* is the order *Actinomycetales* (Buchanan 1917* (Approved Lists 1980); emend. Zhi *et al.*, (2009)).

6.3.1.3 Genera of Phytopathogenic Bacteria

Till now, 72 genera of phytopathogenic bacteria have been reported in the literature. These include 70 established genera and two *Candidatus* categories, whose relatedness has been determined and verified by *in-situ* probing. A brief description of these genera, along with their systematic position, is given below. The authenticated species (and their subspecies and pathovars) of these genera are also listed alphabetically.

* References marked with an * are not cited in the bibliography.

TABLE 6.3
Phyla of the domain *Bacteria* encompassing phytopathogenic bacteria

Phylum name/properties	Class	Genera containing phytopathogenic bacteria
Proteobacteria Largest phylum of the domain *Bacteria* and 51 genera of this phylum contain plant pathogenic bacteria. It is divided into five classes.	*"Alphaproteobacteria"*	*Acetobacter, Agrobacterium, Allorhizobium, Liberibacter, Gluconobacter, Rhizobium, Rhizorhapis* and *Sphingomonas*
	"Betaproteobacteria"	*Achromobacter, Acidovorax, Burkholderia, Herbaspirillum, Janthinobacterium, Paraburkholderia, Paracidovorax, Ralstonia, Robbsia* and *Xylophilus*
	"Gammaproteobacteria"	*Acinetobacter, Affinibrenerria, Arsenophonus, Brenneria, 'Candidatus Phlomobacter', Citrobacter, Dickeya, Enterobacter, Erwinia, Ewingella, Gibbsiella, Klebsiella, Kosakonia, Lelliottia, Lonsdalea, Musicola, Pantoea, Pectobacterium, Phaseolibacter, Providencia, Pseudoalteromonas, Pseudomonas, Rahnella, Raoultella, Rhizobacter, Rouxiella, Salmonella, Samsonia, Serratia, Stenotrophomonas, Tatumella, Xanthomonas* and *Xylella*
	"Deltaproteobacteria"	None
	"Epsilonproteobacteria"	None
"Firmicutes" Contains phenotypically diverse groups of bacteria such as hyperthermophilic and thermophilic bacteria; anaerobic, straight, curved and helical Gram-stain-negative rods; phototrophic and non-phototrophic bacteria; Gram-stain-negative and Gram-stain-positive cocci; Gram-stain-positive and Gram-stain-negative rods; mycoplasmas and thermoactinomycetes.	*"Bacilli"*	*Bacillus, Paenibacillus, Lactobacillus, Lactococcus, Leuconostoc* and *Staphylococcus*
	"Clostridia"	*Clostridium*
	Mollicutes	'*Candidatus* Phytoplasma' and *Spiroplasma*
"Actinobacteria" A very high level of morphological, physiological and genomic diversity is found among its members. Members are divided into two major phenotypic groups: unicellular non-sporulating actinobacteria and filamentous sporulating sporoactinomycetes. Unicellular non-sporulating group contains Gram-stain-negative aerobic rods and cocci; budding and/or appendaged bacteria; Gram-stain-positive cocci; regular and irregular non-sporulating Gram-stain-positive rods and mycobacteria. Sporoactinomycetes include nocardioform actinomycetes, actinomycetes having multi-locular sporangia, *Streptomyces* and other related genera.	*Actinomycetia*	*Arthrobacter, Clavibacter, Corynebacterium, Curtobacterium, Janibacter, Leifsonia, Leucobacter, Nocardia, Rathayibacter, Rhodococcus, Sanguibacter* and *Streptomyces*
Phylum *Bacteroidota* The phylum *Bacteroidota* is composed of three large classes of Gram-negative, non-sporing, anaerobic or aerobic, and rod-shaped bacteria that are widely distributed in the environment, including in soil, sediments and sea water, as well as in the guts and on the skin of animals.	*Sphingobacteria*	*Sphingobacterium*

Domain: *Bacteria*

1. **Phylum: *Proteobacteria***
   Class: "*Alphaproteobacteria*"
   Order: *Rhodospirillales*
   Family: *Acetobacteraceae*
   Genera: *Acetobacter* and *Gluconobacter*

6.3.1.3.1 Genus Acetobacter *Beijerinck 1898*

The genus was established by Beijerinck (1898). The important characteristics of the genus are given below.

Cells are Gram-stain-negative, non-sporing, straight or slightly curved rods and measure 0.6–0.8 × 1.0–4.0 µm. Cells may occur singly, in pairs, or in chains and may be motile or non-motile; if motile, the motility is by means of peritrichous flagella. Colonies of most strains are non-pigmented, while some produce pale colonies and a small percentage of strains produces brown water-soluble pigments. Obligately aerobic; metabolism is strictly respiratory, never fermentative, with oxygen as the terminal electron acceptor. Tests for oxidase, gelatin liquefaction, indole and H_2S production are negative. Catalase test is usually positive and some strains reduce nitrates to nitrites. Chemoorganotrophs and some strains require growth factors such as thiamine, niacin, *p*-aminobenzoic acid, or pantothenic acid. Optimal temperature for growth is 30 °C and optimal pH is in the acidic range, i.e., 4.0–6.0. Acetate and lactate are oxidised to CO_2 and H_2O, and ethanol to acetic acid. Major quinone is Q-9. *Acetobacter* strains are found in fermented foods, fruits, flowers, vinegar and palm wine, and cause spoilage of cocoa and grape wine, beer, cider and fermented foods. The genus does not contain any human and animal pathogen. G + C content of the DNA varies from 50.5–60.3 mol%.

Type species: *Acetobacter aceti*

Beijerinck and Folpmers (1916) described two species, namely, *Ace. aceti* and *Ace. pasteurianus*, isolated mainly from flowers, fruits, palm wine, vinegar, kefir and fermented food. Later on, these species were also found to cause brown rot of apple and pear fruits and pink disease of pine apple (Kontaxis and Hayward, 1978; Dhanvantari *et al.*, 1978).

Plant pathogenic species: *Ace. aceti* and *Ace. pasteurianus*.

6.3.1.3.2 Genus Gluconobacter Asai 1935

The genus was established by Asai (1935). At present, it contains only one plant pathogenic species, the causal agent of bacterial rot of mature fruits (pear, apple and grape) and pink disease of pineapple.

The members of the genus have the following characteristic features. The cells are Gram-stain-negative, ellipsoidal to rod-shaped and measure 0.5–1.0 × 2.6–4.2 µm. Cells may be motile or non-motile and motile cells have 3–8 polar flagella. Cells may occur singly and/or in pairs but rarely in chains. Non-sporing. Obligately aerobic. Metabolism is strictly respiratory type in which the oxygen acts as a terminal acceptor. The best growth occurs at 25–30 °C. Optimum pH for growth is 5.0–6.0, though most strains can even grow at pH 3.5. Oxidase test is negative but catalase test is positive. Indole and hydrogen sulphide are not produced, gelatin is not liquefied and nitrates are not reduced. Ethanol is oxidised to acetic acid but acetic acid is not further oxidised to carbon dioxide and water, as the bacterium does not have a full set of Krebs cycle enzymes. Acid is generally produced from glucose, maltose, fructose and xylose and growth can occur even up to concentration of 10.0% glucose. Major quinone possessed is ubiquinone of Q-10 type. *Gluconobacter* strains occur in sugar-rich environments like flowers, fruits, soft drinks, grapes, wine, palm sap, coca wine, cider, beer etc. G + C content of the DNA varies from 52.0–64.0 mol%.

In addition to vinegar production, *Gluconobacter* species have many biotechnological applications. In recent decades, many bioconversion routes for rare and special sugars involving *Gluconobacter* spp. have been developed. Among the recent ones are the biotransformations involved in the production of L-ribose and miglitol, both very promising pharmaceutical lead molecules. Most of these processes make use of *Gluconobacter*'s membrane-bound polyol dehydrogenases. Other enzymes of recent interest in industrial biotechnology include dextran-dextrinase, capable of transglucosylating substrate molecules and intracellular NAD-dependent polyol dehydogenases involved in coenzyme regeneration.

Type species: *Gluconobacter oxydans*

Plant pathogenic species: *Glu. oxydans*

Order: "Rhizobiales"

Family: *Rhizobiaceae*

Genera: *Agrobacterium*, *Allorhizobium* and *Rhizobium*

6.3.1.3.3 Genus Agrobacterium Conn 1942

The genus was established by Conn (1942) to include *Agr. tumefaciens* and *Agr. rhizogenes*, which cause hypertrophies of the affected plant parts. Pathogenicity character is encoded by genes present on the transmissible plasmids. The ease with which the capacity to induce hypertrophies, is transferred between species of *Agrobacterium*, suggests that a stable taxonomy based on pathogenicity may not sound well. The use of names of *Agr. tumefaciens* and *Agr. rhizogenes* with natural classification recognising pathogenicity of strains according to their 'tumorigenic' or 'rhizogenic' and 'non-pathogenic state' has been supported earlier. According to Young *et al.*, (2001), this also posed a problem as the epithet *'tumefaciens'* was too closely associated with tumorigenicity to stand independently as a species name. In 1993, it was proposed to replace the name of *Agr. tumefaciens* with *Agr. radiobacter* to resolve this problem but it could not be adopted because *Agr. tumefaciens* was conserved as the type species of this proposed combination (Sawada *et al.*, 1993). Young *et al.*, (2001) found that three genera of family *Rhizobiaceae*, namely, *Rhizobium*, *Agrobacterium* and *Allorhizobium*, do not differ phenotypically and also do not have unique phenotypic and genetic circumscription. Moreover, their phylogenetic differentiation on the basis of algorithm and sequences of 16S rRNA is also not strong. Therefore, they amalgamated these three genera in a single genus *Rhizobium* as the senior subjective synonym. However, Garrity *et al.*, (2005) have recognised both the genera, *Rhizobium* and *Agrobacterium* in the family *Rhizobiaceae* for the time being. Therefore, the *Agrobacterium* is treated as a separate genus in this book also. Bouzar and Jones (2001) have included the causal agent of aerial tumours of *Ficus benjamina* in this genus.

The genus has the following important characteristics. Cells are Gram-stain-negative, non-sporing rods, measuring 0.6–1.0 × 1.5–3.0 µm. Cells are found singly or in pairs and are motile by means of 1–6 peritrichous flagella. Aerobic and chemoorganotrophs. Metabolism is strictly respiratory type in which oxygen is used as terminal electron acceptor. The colonies are generally circular, smooth, convex and colourless to light beige. Growth on culture media containing carbohydrates is usually copious and slimy. Optimal temperature for growth is 25–28 °C. Catalase and urease are always positive while oxidase is generally positive. Growth in mineral salts media containing mannitol and other carbohydrates is accompanied by an acidic reaction. Common inhabitants of soil. Tumour-forming strains occur in soils where diseased plants grew earlier or, are growing. G + C content of the DNA ranges from 57.0 to 63.0 mol%.

Type species: *Agrobacterium tumefaciens*

Agr. tumefaciens causes crown gall of plants. Virulent strains of the bacterium contain a tumour-inducing plasmid, called Ti-plasmid. For successful infection of the host, T-DNA, called transfer DNA, is cut from the Ti-plasmid and is transferred to the host cell genome. Penetration of *Agr. tumefaciens* into the host is through wounds. Although *Agrobacterium* species generally infect plants, they have also been shown to cause opportunistic infections in humans with weakened immune systems (Hulse *et al.*, 1993), but have not been proven to be a primary pathogen in otherwise healthy persons.

The DNA transmission capabilities of *Agrobacterium* species have been extensively exploited in biotechnology, as a means of inserting foreign genes into plants. The plasmid T-DNA that is transferred to the plants is an ideal vehicle for genetic engineering. Tumour-inducing genes are deleted from the plasmid and the desired gene sequence is cloned into the T-DNA.

Two plant-tumorigenic strains KFB 330^T and KFB 335 isolated from galls on raspberry (*Rubus idaeus*) in Serbia, and a non-pathogenic strain AL51.1 recovered from a cherry plum (*Prunus cerasifera*) tumour in Poland, were genotypically and phenotypically characterised by Kuzmanović *et al.*, (2015). Phylogenetic reconstruction based on 16S rDNA placed them within the genus *Agrobacterium*, with *Agr. nepotum* as their closest relative. Multilocus sequence analysis (MLSA) based on the partial sequences of *atpD*, *glnA*, *gyrB*, *recA* and *rpoB* housekeeping genes suggested that these three strains represent a new *Agrobacterium* species, that clustered with type strains of *Agr. nepotum*, *Agr. radiobacter*, *Agr. fabrum* and *Agr. pusense*. This was further supported by average nucleotide identity values (<92%) between the whole-genome sequences of strain KFB 330^T and related *Agrobacterium* species. The major cellular fatty acids of the novel strains were 18:1 w7c (72.8–77.87%) and 16:0 (6.82–8.58%). Phenotypic features allowed their differentiation from closely related species. Polyphasic characterisation showed that the three strains represent a novel species of the genus *Agrobacterium*, for which the name *Agrobacterium arsenijevicii* sp. nov. is proposed. The type strain of *Agr. arsenijevicii* is KFB 330^T (=CFBP 8308^T = LMG 28674^T).

Kuzmanović *et al.*, (2018) characterised a plant tumorigenic strain NCPPB 1650^T isolated from *Rosa* × *hybrida*, and four non-pathogenic strains isolated from tumours on grapevine (strain 384), raspberry (strain 839) and blueberry (strains B20.3 and B25.3) using polyphasic taxonomic methods. Based on 16S rRNA gene phylogeny, strains were clustered within the genus *Agrobacterium*. Furthermore, multilocus sequence analysis based on the partial sequences of *atpD*, *recA* and *rpoB* housekeeping genes indicated that the five strains studied form a novel *Agrobacterium* species and their closest relatives were *Agrobacterium* sp. R89-1, *Agr. rubi* and *Agr. skierniewicense*. Authenticity of the novel species was confirmed by average nucleotide identity and *in silico* DNA–DNA hybridisation comparisons between strains NCPPB 1650^T and B20.3, and their closest relatives, since obtained values were considerably below the proposed thresholds for the species delineation. Whole-genome-based phylogeny further supported distinctiveness of the novel species, which forms together with *Agr. rubi*, *Agr. skierniewicense* and *Agrobacterium* sp. R89-1, a well-delineated sub-clade of *Agrobacterium* spp. named "rubi". As for other species of the genus *Agrobacterium*, the major fatty acid of the strains studied was 18:1 w7c (73.42–78.12%). The five strains studied were phenotypically distinguishable from other species of the genus *Agrobacterium*. Overall, polyphasic characterisation showed that the five strains studied represent a novel species of the genus *Agrobacterium*, for which the name *Agrobacterium rosae* sp. nov. is proposed. The type strain of *Agr. rosae* is NCPPB 1650^T (=DSM 30203^T = LMG 230^T = CFBP 4470^T = IAM 13558^T = JCM 20915^T).

Mafakheri *et al.*, (2022) reported that whole-genome sequence-based core-genome phylogeny followed by average nucleotide identity (ANI) and digital DNA–DNA hybridisation (dDDH) calculations suggest that genomospecies G19 and G20 can be described as two novel and standalone species of genus *Agrobacterium*. In the phylogenetic tree, these two new genomospecies were clustered separately from other genomospecies/species of *Agr. tumefaciens* species complex. Moreover, both ANI and dDDH indices between the G19/G20 strains and other *Rhizobiaceae* members are clearly below the accepted thresholds for procaryotic species description. Hence, *Agrobacterium burrii* sp. nov. is proposed to encompass the G19 strains, with Rnr^T = CFBP 8705^T = DSM 112541^T as type strain. *Agr. shirazense* sp. nov. is also proposed to include G20 strains, with OT33^T = CFBP 8901^T = DSM 112540^T as type strain.

Basavand *et al*., (2022) reported for the first time the occurrence of crown and stem gall on pistachio caused by *Agr. radiobacter*, occurring world-wide. Xiao *et al*., (2023) isolated 45 pathogenic bacterial strains from *Prunus humilis* Bunge, from five regions of China. Using phenotypic and phylogenetic analysis of the 16S rRNA gene, they identified 12 strains as *Rhizobium rhizogenes* and 33 strains as members of the *Agr. tumefaciens* species complex. MLSA of three housekeeping genes (*rpoB*, *atpD* and *recA*) revealed that 11 strains belonged to *Agr. fabacearum* and 22 strains belonged to *Agr. radiobacter* within the *Agr. tumefaciens* species complex.

Warabieda *et al*., (2023) assessed three plant rhizogenic strains O132T, O115 and O34 isolated from *Cucumis* sp. L. for taxonomic affiliation by using polyphasic taxonomic methods. The physiological and biochemical properties of these strains allowed to distinguish them from their closest related species of the genus *Agrobacterium*. As a result of the performed overall characterisation, the authors proposed it as a new species, *Agrobacterium cucumeris* sp. nov., with O132T (=CFBP 8997T = LMG 32451T) as the type strain.

Plant pathogenic species: *Agr. arsenijevicii*, *Agr. burrii*, *Agr. cucumeris*, *Agr. fabacearum*, *Agr. larrymoorei*, *Agr. radiobacter Agr. rhizogenes*, *Agr. rosae*, *Agr. rubi*, *Agr. shirazense*, *Agr. tumefaciens* and *Agr. vitis*.

6.3.1.3.4 *Genus* Allorhizobium *gen. nov.*

Cells are Gram-stain-negative, non-sporing rods that are 2–4 µm long and 0·5–0·7 µm wide. Strains grow fast and form colonies of 0·5–3 mm diameter within 1–2 days on yeast mannitol mineral salts agar. Pronounced turbidity develops after 1–2 days in agitated broth media. Aerobic, chemo-organotrophic, utilise a wide range of carbohydrates, organic acids and amino acids as sole sources of carbon. 3-Ketolactose is not produced from lactose. Growth on carbohydrate media is usually accompanied by extracellular polysaccharide production. Members are typically able to invade the root hairs of some temperate zone leguminous plants (*Medicago sativa*) and some tropical zone plants (*Neptunia natans*, *Acacia senegal*, *A. seyal*, *A. tortilis* subsp. *raddiana*, *Lotus arabicus* and *Faidherbia albidd*) and induce the production of root nodules, wherein the bacteria occur as intracellular symbionts. All strains exhibit host specificity. No strain was found to nodulate *Sesbania rostrata*, *S. pubescens*, *S. grandiflora*, *Vigna unguiculata* or *Macroptilium atropurpureum*. G + C content of the DNA is 60·1 mol% (by T_m). At the molecular level, the genus can be recognised by SDS-PAGE whole-cell protein analysis, ITS PCR-RFLP and 16S rRNA gene sequencing.

Type species: *Allorhizobium undicola* sp. nov.

Kawaguchi (2022) reported that *All. vitis* causes crown gall of grapevine, an important bacterial disease of grapevine throughout the world. Gall incidence significantly increases inferior growth and death of grapevines.

Plant pathogenic species: *All. vitis*

6.3.1.3.5 *Genus* Rhizobium *Frank 1889*

The genus was created by Frank (1889) and the salient characteristics of the genus are given below. Cells are Gram-stain-negative, non-sporing, small to medium-sized rods measuring 0.5–1.0 × 1.2–3.0 µm. Motile by 1–6 flagella, which may be subpolar, bipolar, or peritrichous. Cells contain characteristic granules of polymerised β-hydroxybutyrate, which stain with Sudan Black and appear as highly refractive bodies under phase contrast illumination. Optimal temperature for growth is between 25 and 30 °C. Optimum pH lies between 6.0 and 7.0. Aerobic, having a respiratory type of metabolism with oxygen as the terminal electron accepter. Colonies are circular, convex, opaque or semi-translucent, raised and mucilaginous on yeast mannitol mineral salt agar. Colony colour is generally white or beige. Growth on carbohydrate media is usually accompanied by abundant amounts of extracellular polysaccharide slime. *Rhizobium* species can grow on relatively simple synthetic media. Chemoorganotrophs and utilise a wide range of carbohydrates and salts of organic acids as sole carbon sources, without production of gas. Produce an acidic reaction in mineral salts medium containing mannitol or other carbohydrates. Glutamate is more superior than nitrate or the ammonium ions as nitrogen source. Vitamin requirements differ for different species and strains. Thiamine, biotin and calcium pantothenate are required for growth while nicotinic acid, folic acid, *p*-aminobenzoic acid, inositol, vitamin B12 and riboflavin do not promote growth.

Casein, starch, chitin and agar are not hydrolysed. Starch and cellulose are not utilised. An acid reaction is produced in mineral salts medium having mannitol or other carbohydrates. Strains of all species cause hypertrophies in plants resulting in formation of root nodules with or without symbiotic nitrogen fixation. In root nodules, bacteria live endophytically and fix atmospheric nitrogen into a form utilisable by the host plant. Plant-host specificity is generally for a few legume genera but, in some strains, may extend to many legume genera and is largely determined by the chemical structure of lipochito-oligosaccharide. Nod factors are produced. The genes, which code for nodulation (*nod*) and nitrogen fixation (*nif*), are clustered on large plasmids or megaplasmids (pSyms) present in the bacteria. Plasmid transfer occurs between species and results in expression and stable inheritance of characters possessed by the plasmid-donor species. G + C content of the DNA varies from 57.0–66.0 mol%. Members are common inhabitants of arable land.

Three isolates of Gram-stain-negative, rod-shaped, nonsporing bacteria isolated from galls on chrysanthemum and cherry plum were able to cause crown galls on various plant species. On the basis of phylogenetic and phenotypic evidence, Pulawska *et al.*, (2012) considered that the isolates represent a single novel species of the genus *Rhizobium*, for which they proposed the name *Rhizobium skierniewicense* sp. nov.

Kuzmanović *et al.*, (2018) reported *Rhizobium tumorigenes* sp. nov., a novel plant tumorigenic bacterium, causing cane gall tumours on thornless blackberry.

Kuzmanović *et al.*, (2023) reported that the tumorigenic strains isolated from rhododendron represent a novel species of the genus *Rhizobium* for which the name *Rhizobium rhododendri* sp. nov. is proposed.

Type species: *Rhizobium leguminosarum*

Plant pathogenic species: *Rhiz. nepotum*, *Rhiz. rhizogenes* (Refer to reference of Xiao *et al.*, (2023) under the genus *Agrobacterium*), *Rhiz. rhododendri*, *Rhiz. skierniewicense* and *Rhiz. tumorigenes*

Both *Agrobacterium* and *Rhizobium* have been recognised as separate genera in this book, as done by Garrity *et al.*, (2005). The plant pathogenic species included under *Rhizobium* by Young *et al.*, (2001) are included under the genus *Agrobacterium* in this book.

Family: "*Phyllobacteriaceae*"

Genus: *Liberibacter*

6.3.1.3.6 Liberibacter *Fagen* et al. *2014*

Citrus huanglongbing (earlier called citrus greening) disease was initially considered to be caused by **mycoplasma-like organisms** (MLOs) and **rickettsia-like organisms** (RLOs). Later studies revealed that its causal agent is enclosed by 25 nm thick cell envelope and was called bacterium-like organism (BLO). Jagoueix-Eveillard *et al.*, (1994) characterised two uncultured phloem-restricted bacteria causing citrus huanglongbing in Asia and Africa and named them '*Candidatus* Liberobacter asiaticum' and '*Candidatus* Liberobacter africanum', respectively. Following the rules of the International Code of Nomenclature of Bacteria, these two pathogens have been renamed '*Candidatus* Liberibacter asiaticus' and '*Candidatus* Liberibacter africanus'. Teixeira *et al.*, (2005) have reported '*Candidatus*' Liberibacter americanus' associated with citrus huanglongbing in the Sao Paulo state of Brazil. Later on, '*Candidatus*' status was removed from the name of this genus.

The below-given description of the genus *Liberibacter* is based mainly on the publication of Jagoueix-Eveillard *et al.*, (1994). Bacteria are Gram-stain-negative, filamentous in shape and occur in phloem sieve tubes of plants. Round forms also occur but they are considered to be degenerating bacteria. These are transmitted by two insect vectors (Psyllidae), namely, *Trioza erytriae* and *Diaphornia citri*. These also multiply in the haemolymph and cells of salivary glands of both the vectors. The above-given description of the genus is based on the bacteria present in the phloem sieve tubes of sweet orange trees affected by huanglongbing (ex. greening disease) in Pune, India and in Nelspruit, South Africa. *Lib. asiaticus* requires higher temperature than *Lib. africanus* for optimum expression of the disease symptoms.

Sechler *et al.*, (2009) designed a medium called Liber A to culture all the three '*Candidatus* Liberibacter spp.' which cause citrus huanglongbing. The medium containing citrus vein extract and a growth factor supported the growth of '*Can.* Liberibacter spp.' for four or five single colony transfers before the decline of viability. The colonies of '*Can.* Lib. asiaticus' were irregularly shaped, convex and 0.1–0.3 mm in diam after 3–4 days of incubation. Two strains of '*Can.* Lib. asiaticus' and one of '*Can.* Lib. americanus' grown on this medium proved pathogenic to citrus and were isolated from uninoculated tissues of inoculated trees and seedlings 9 and 2 months later, respectively. The identity of the bacteria was confirmed by RT-PCR and 16S rDNA sequencing. This is the first report of the culturing and pathogenicity of '*Can.* Lib. asiaticus' and '*Can.* Lib. americanus' associated with citrus huanglongbing.

Roberts *et al.*, (2015) reported three novel lineages of '*Can.* Liberibacter africanus' associated with native rutaceous hosts of *Trioza erytreae* in South Africa.

Type species: *Liberibacter asiaticus*

Roberts and Pietersen (2017) reported a novel subspecies of '*Can.* Liberibacter africanus' infecting native *Teclea gerrardii* (Family: Rutaceae) from South Africa and named it '*Candidatus* Liberibacter africanus subsp. tecleae'.

Plant pathogenic species: *Lib. africanus*, ***Lib. africanus***: subsp. *capensis*, subsp. *clausenae*, subsp. *tecleae*, subsp. *vepridis* and subsp. *zanthoxyli*; *Lib. americanus*, *Lib. asiaticus*, '*Can.* Lib. psyllaurous' and *Lib. solanacearum*.

Order: "*Sphingomonadales*"

Family: *Sphingomonadaceae*

Genera: *Sphingomonas* and *Rhizorhapis*

6.3.1.3.7 *Genus* Sphingomonas *Yabuuchi* et al. *1990 emend. Yabuuchi* et al. *1999 emend. Takeuchi* et al. *2001 emend. Yabuuchi* et al. *2002 emend. Feng* et al. *2017*

Yabuuchi *et al.*, (1990) created the genus *Sphingomonas* to include strictly aerobic, chemoheterotrophic, yellow-pigmented and Gram-stain-negative rod-shaped bacteria that contain glycosphingolipids as cell envelope components. Later on, Yabbuchi *et al.*, (1999) emended the genus and transferred *Rhizomonas suberifaciens*, the causal agent of corky root disease of lettuce, to this genus along with two other species. On the basis of morphological, physiological and chemotaxonomic characteristics, along with DNA–DNA hybridisation and 16S rDNA sequences comparison data, Takeuchi *et al.*, (1995) included four new species in this genus. These species were *Sph. rosa* (isolated from rose plants and formerly named as *Agr. rhizogenes*), *Sph. pruni* (isolated from *Prunus persica*), *Sph. asaccharolytica* and *Sph. mali* (isolated from apple trees). They also identified two strains, isolated from *Psychotria nairobiensis* and formerly named as *Chromobacterium lividum*, as *Sph. yanoikuyae*. Buonaurio *et al*., (2002) described a novel pathogen causing brown spot on yellow Spanish melon (*Cucumis melo* var. *indodorus*) fruits and identified it as *Sph. melonis*.

The below-given description of the genus is based mainly on the emended description of the genus proposed by Yabuuchi *et al*., (2002). The cells are Gram-stain-negative, non-sporing, straight or slightly curved rods or ovoid-shaped, which measure 0.2–1.4 × 0.5–4.0 µm. Reproduction is generally by binary fission but budding or asymmetric division as visualised by electron microscopy has been found in two species, namely, *Sph. natatoria* and *Sph. ursincola*. Motility is usually by a single polar flagellum but *Sph. trueperi* is peritrichously flagellate. Many species are non-motile, as they do not show spreading growth on 0.3% semi-solid agar plates. Bipolar staining in Gram-stain reaction (*Sph. ursincola*) and a rosette-like aggregation caused by polar-fimbriae occurs in some species. Aerobic, having a strictly respiratory type of metabolism with oxygen as the terminal electron accepter. Colony colour is variable, for instance, deep-yellow, orange, lemon-yellow or non-pigmented. In certain cases, the pigmentation varies with the medium used. The colonies of some non-pigmented strains such as *Sph. yanoikuyae* become lemon-yellow after incubation of more than three days at room temperature. Esculin is hydrolysed. Two species, namely, *Sph. ursincola* and *Sph. natatoria* containing bacteriochlorophyl A, are facultative photoorganotrophs. Catalase test is positive. Oxidase positive or negative. Cell walls contain sphingolipids instead of lipopolysaccharides. Respiratory quinone is ubiquinone 10. Glucuronosyl-(1→1)-ceramide (SGL-1), galacturonosyl-β(1→1)-ceramide (in several species), and 2-hydroxymyristic acid, but not 3-hydroxy fatty acids are found. Major long-chain bases of SGL-1 are C-18:0, C-20:1 and C-21 cyclopropane. G + C content of the DNA ranges from 59.0–68.0 mol%. Inhabitants of natural and man-made environments. Some species are opportunistic pathogens of humans and a few are plant pathogens.

Type species: *Sphingomonas paucimobilis*

Liu *et al*., (2023) reported *Sph. spermidinifaciens* causing leaf spot of plum for the first time from China.

Plant pathogenic species: *Sph. asaccharolytica*, *Sph. mali*, *Sph. melonis*, *Sph. pruni*, *Sph. rosa*, *Sph. spermidinifaciens* and *Sph. yanoikuyae*.

6.3.1.3.8 *Genus* Rhizorhapis *Francis* et al. *2014*

The previous illegitimate name of the genus was *Rhizomonas* (van Bruggen *et al*., 1990).

The below-given description of the genus is based mainly on the publication of Francis *et al*., (2014). Cells are Gram-stain-negative, straight or slightly curved rods and motile by one lateral, subpolar or polar flagellum, or are non-motile (van Bruggen *et al*., 1990). Resting stages are unknown. Cells are obligate aerobes having an oxidative metabolism and multiply by binary fission. Cells accumulate polyhydroxybutyrate (PHB) granules and are arginine dihydrolase negative. Colonies are non-fluorescent, white or creamy white, and are smooth or wrinkled. Optimum growth temperature is 28–32 °C and maximum is 37 °C. Ethanol is not converted to acetic acid. Oxidase and catalase are produced. Denitrification to N_2 gas does not occur. Ubiquinone Q10 is present. Whole-cell fatty acids consist mainly of even-numbered unsaturated ($C_{18:1}$ and $C_{16:1}$) and saturated ($C_{16:0}$) straight-chain fatty acids, as well as $C_{14:0}$ 2-OH. The $C_{18:1}$ fatty acids represent at least 50% of the total fatty acids. G + C content of the DNA is in the range 58–60 mol%. The type species is *Rhizorhapis suberifaciens*, formerly known as *Rhizomonas suberifaciens* (van Bruggen *et al*., 1990).

Type species: *Rhizorhapis suberifaciens*

The bacterium causing corky root of lettuce was first classified as *Rhizomonas suberifaciens*. As the genus name *Rhizomonas* was found illegitimate, the bacterium was reclassified as *Sphingomonas suberifaciens* based on 16S rRNA gene sequences and the presence of sphingoglycolipid in the cell envelope. As the genus *Sphingomonas* is so diverse, the further reclassification of corky root of lettuce pathogen was deemed necessary. Twenty new isolates obtained from lettuce or sow thistle roots, or from soil using lettuce seedlings as bait, and the previously reported isolates were characterised in a polyphasic study, including 16S rRNA gene sequencing, DNA–DNA hybridisation, DNA G + C content, whole cell fatty acid composition, morphology, substrate oxidation, temperature, pH sensitivity and pathogenicity to lettuce. The isolates causing lettuce corky root belonged to the genera *Rhizorhapis* gen. nov., *Sphingobium*, *Sphingopyxis* and *Rhizorhabdus* gen. nov. Francis *et al*., (2014) specifically proposed to reclassify *Rhizomonas suberifaciens* as *Rhizorhapis suberifaciens* gen. nov., comb. nov., and also proposed the novel species *Rhizorhabdus argentea* gen. nov., sp. nov., *Sphingobium xanthum* sp. nov. and *Sphingobium mellinum* sp. nov. Several strains isolated from lettuce roots belonged to the genus *Sphingomonas*, but none of them was pathogenic.

Plant pathogenic species: *Rhir. suberifaciens*.

Class: “Betaproteobacteria”
Order: “*Burkholderiales*”
Family *Alcaligenaceae*
Genus *Achromobacter*

6.3.1.3.9 *Genus* Achromobacter

Cells are Gram-stain-negative, straight rods and are motile by 1–20 peritrichous flagella. Cells are strictly aerobic and are found in fresh and marine water and soils (Garrity *et al.*, 2005). The members have also been identified as contaminants in laboratory cell cultures (Gray *et al.*, 2010). The members have also been found as opportunistic human pathogens in people with certain immunosuppressive conditions such as cystic fibrosis, cancer and kidney failure (Swenson and Sadikot, 2015).

Type species: *Achromobacter xylosoxidans*

Wei *et al.*, (2023) reported *Achr. xylosoxidans* causing stem soft rot on *Amorphophallus konjac* from China.

Oksel *et al.*, (2024) reported for the first time *Achr. xylosoxidans* causing bacterial stem and leaf blight on *Cyrilla arida* from Tennessee and the United States.

Hua *et al.*, (2024) reported *Achr. xylosoxidans* causing wilt disease of ginger for the first time from China.

Plant pathogenic species: *Achr. xylosoxidans*

Class: "*Betaproteobacteria*"

Family: *Comamonadaceae*

Genera: *Acidovorax*, *Paracidovorax* and *Xylophilus*

6.3.1.3.10 Genus Acidovorax *Willems* et al. *1990 emend. Willems* et al. *1992 emend. Du* et al. *2023*

Willems *et al.*, (1990) created this genus to include bacteria isolated from soil, water, activated sludge and clinical samples. Two years later, Willems *et al.*, (1992) transferred four plant pathogenic species and subspecies of the genus *Pseudomonas* to this genus. Two more pathogens, i.e., causal agents of bacterial leaf spot of anthurium (*Anthurium andraeanum)* and bacterial leaf spot of lamb's lettuce (*Valerianella locusta*), were included in this genus (Gardan *et al.*, 2000, 2003a). Schaad *et al.*, (2009), on the basis of genetic and phenotypic data, proposed an emendation of the species *Aci. avenae* and elevated *Aci. avenae* subsp. *citrulli* and *Aci. avenae* subsp. *cattleyae* to species rank as *Acidovorax citrulli* and *Acidovorax cattleyae*, respectively. They also proposed a new taxon, *Acidovorax oryzae* sp. nov. with FC-143^T = International Collection of Phytopathogenic Bacteria (ICPB), 30003^T= International Collection of Microorganisms from Plants (ICMP) and 3960^T= American Type Culture Collection (ATCC) 19882^T as the type strain.

The below-given description of the genus is as proposed by Willems *et al.*, (1990) and emended by Willems *et al.*, (1992). Cells are Gram-stain-negative, straight to slightly curved rods and measure 0.2–1.2 × 0.8–5.0 µm. Cells occur singly, in pairs, or in short chains and are motile by means of one or rarely two or three polar flagella. Most strains produce non-pigmented growth on nutrient agar but some phytopathogenic strains produce a yellow to slightly brown diffusible pigment. Optimal temperature for growth is 30–35 °C. Oxidase test is positive while urease activity is variable among strains. Aerobic, having a strictly oxidative type of metabolism with oxygen as the terminal electron acceptor. However, some strains of *Aci. temperans* and *Aci. delafieldii* are capable of heterotrophic denitrification of nitrate. Most strains are chemoorganotrophs, although strains of *Aci. facilis* and *Aci. delafieldii* can grow lithoautotrophically, using the oxidation of H_2 as an energy source. Organic acids, amino acids and peptone support good growth of the bacterium but only a limited number of sugars support the growth. Two fatty acids, namely, 3-hydroxyoctanoic acid and 3-hydroxydecanoic acid, are always present while 2-hydroxy-substituted fatty acids are absent. G + C content of the DNA varies from 62.0–70.0 mol% (T_m). *Acidovorax* strains can be obtained from infected plants, soil, water and clinical samples and activated sludge.

Type species: *Acidovorax facilis*

Du *et al.*, (2023), on the basis of genome analyses, proposed the division of the current genus *Acidovorax* into an emended genus *Acidovorax* and the two novel genera *Paracidovorax* gen. nov. and *Paenacidovorax* gen. nov.

Sawada *et al.*, (2025) reported *Aci. sacchari* sp. nov., a pathogen causing red stripe of sugarcane in Japan.

Plant pathogenic species: *Aci. anthurii*, *Aci. avenae*, ***Aci. avenae***: subsp. *avenae*; *Aci. cattleyae*, *Aci. citrulli*, *Aci. konjaci*, *Aci. oryzae*, *Aci. sacchari* and *Aci. valerianellae*.

6.3.1.3.11 Genus Paracidovorax

Cells are straight to slightly curved rods and motile with a polar flagellum. The major fatty acids are summed feature 3 (C16:1 *ω7c* and/or C16:1 *ω6c*), C16:0, and summed feature 8 (C18:1 *ω7c* and/or C18:1 *ω6c*). The respiratory quinone is ubiquinone-8. The main polar lipids are phosphatidylethanolamine, diphosphatidylglycerol, and phosphatidylglycerol. The genomic sizes and DNA G + C contents of the sequenced species are 4.85–5.61 Mbp and 65.59–68.94%, respectively. The genus contains many plant pathogenic species.

Type species: *Paracidovorax avenae*

Duan *et al.*, (2024) reported the complete genome sequence of *Parac. avenae* causing red stripe of sugarcane.

Plant pathogenic species: *Parac. anthurii*, *Parac. avenae*, *Parac. cattleyae*, *Parac. citrulli*, *Parac. konjaci*, *Parac. oryzae*, *Parac. valerianellae*, and *Parac. wautersii*.

6.3.1.3.12 *Genus* Xylophilus *Willems* et al. *1987*

Willems *et al.*, (1987) found that the causal agent of bacterial necrosis and canker of grapevines, earlier classified as *Xanthomonas ampelina*, was not related to the genus *Xanthomonas*. Therefore, they created a new genus *Xylophilus* and transferred *Xan. ampelina* to the new genus as *Xylophilus ampelinus*. This, being the only species in the genus, was designated as type species.

The below-given description of the genus is taken from Willems *et al.*, (1987). Cells are Gram-stain-negative, straight to slightly curved rods, measure 0.4–0.8 × 0.6–3.3 µm and are motile by a single polar flagellum. Long filamentous cells (length ≥ 30 µm) may appear in older cultures. Cells occur singly, in pairs, or in short chains. Strictly aerobic, having a respiratory type of metabolism using oxygen as the terminal electron acceptor. The bacteria are chemoorganotrophs with oxidative carbohydrate metabolism; oxidase test negative and catalase test positive. The growth is generally very slow and poor even at the optimal temperature of 24 °C. G + C content of the DNA ranges from 68.0 to 69.0 mol%. The genus belongs to the acidovorans rRNA complex in rRNA superfamily III and is equidistantly related to the other taxa in this complex. *Xylophilus* strains can grow on L-glutamine but not on calcium lactate, while the reverse is true for *Xanthomonas* strains.

Type species: *Xylophilus ampelinus*

Plant pathogenic species: *Xylo. ampelinus*.

Family: "*Burkholderiaceae*"

Genera: *Burkholderia*, *Paraburkholderia*, *Ralstonia* and *Robbsia*

6.3.1.3.13 *Genus* Burkholderia *Yabuuchi* et al. *1999 emend. Gillis* et al. *1995*

The genus was created by Yabuuchi *et al.*, (1999) to include some species of the then genus *Pseudomonas*. Gillis *et al.*, (1995) emended this genus on the basis of polyphasic taxonomy and also proposed a new species for this genus. All the 19 species of this genus belong to the ribosomal RNA similarity group II, which can be differentiated from the other groups of aerobic pseudomonads by rRNA/DNA hybridisation or by rDNA sequencing.

Oliveira *et al.*, (2017) reported *Bur. cenocepacia* causing sour skin of onion (*Allium cepa*) in Brazil. As one plant pathogenic species of this genus has been transferred to a newly created genus *Ralstonia*, it contains only seven plant pathogenic species.

The genus was named after W. H. Burkholder, an American bacteriologist, who first discovered the etiological agent of sour skin of onions. The following description of the genus is based on its proposal by Yabuuchi *et al.*, (1992) and emended by Gillis *et al.*, (1995).

Cells are Gram-stain-negative, non-sporing, straight or curved rods but not helical and measure generally 0.5–1.0 × 1.5–4.0 µm. Cells may occur singly or in pairs and are motile by one or generally several polar flagella. One species, namely, *Bur. mallei*, lacks flagella and is non-motile. Most strains accumulate poly-β-hydroxybutyrate as carbon reserve material and are capable of *ortho* cleavage of protocatechuate. Catalase is produced but oxidase activity varies among species. Chemoorganotrophs. Metabolism is strictly respiratory type with oxygen as the terminal electron acceptor. Some species are capable of anaerobic respiration using nitrate. Strains of some species such as *Bur. vietnamiensis* and *Bur. cepacia* are capable of fixing atmospheric N_2. Most species grow at 40 °C. The species are characterised by the presence of hydroxy fatty acids having 14, 16 and 18 carbon atoms and most characteristic of these fatty acids is the $C_{16:0\ 3OH}$. The type strains of several species are characterised by the presence of two types of ornithine lipids. All species can utilise glucose, galactose, glycerol, sorbitol, inositol and mannitol as sole sources of carbon. More than half of the species are pathogenic on plants, animals and humans. Can be isolated from plant materials, soil and clinical samples. G + C content of the DNA ranges from 59.0–69.6 mol%.

The emended description of the genus by Sawana *et al.*, (2015) is as below. The members of the genus are Gram-stain-negative, straight or slightly curved rods, which exhibit motility mediated by one or more polar flagella. Only, *Bur. mallei* lacks flagella and is non-motile. The members do not produce sheaths or prosthecae and do not pass through any resting stages. Most species are able to accumulate and utilise poly-β-hydroxybutyrate for growth. The species are mostly aerobic, chemoorganotrophic, but some species are capable of anaerobic respiration using nitrate as the terminal electron acceptor. The members of the genus form a distinct monophyletic clade in phylogenetic trees, and these are distinguished from all other bacteria by the conserved sequence indels reported in this work in the following proteins: periplasmic amino acid-binding protein, 4-hydroxybenzoate 3-monooxygenase, 6-phosphogluconate dehydrogenase, sarcosine oxidase subunit alpha, a putative lipoprotein and a putative lyase. G + C content for the members ranges from 65.7 to 68.5%.

Han *et al.*, (2021) reported shot-hole on flowering cherry caused by *Bur. contaminans* and *Psm. syringae* pv. *syringae* for the first time from South Korea.

Paudel *et al.*, (2024) reported that in addition to the common onion pathogenic species, *Bur. cepacia*, *Bur. orbicola*, *Bur. gladioli* and other *Burkholderia* species have the potential to cause onion disease.

Type species: *Burkholderia cepacia*

Plant pathogenic species: *Bur. ambifaria*, *Bur. andropogonis*, *Bur. caryophylli*, *Bur. cenocepacia*, *Bur. cepacia*, *Bur. contaminans*, *Bur. gladioli*, ***Bur. gladioli***: pv. *agaricicola*, pv. *alliicola*, pv. *gladioli*; *Bur. glumae*, *Bur. lata*, *Bur. orbicola* and *Bur. plantarii*.

6.3.1.3.14 Genus Paraburkholderia *Sawana et al. 2015, gen. nov. emend. Dobritsa and Samadpour 2016*

Sawana *et al.*, (2014), on the basis of molecular signatures and phylogenomic analysis of the genus *Burkholderia*, proposed the amendment of genus *Burkholderia* into the emended genus *Burkholderia* containing pathogenic organisms and a new genus *Paraburkholderia* gen. nov. harbouring environmental species. The cells are Gram-stain-negative, straight or slightly curved rods that are motile with one or more polar flagella. Other morphological and metabolic characteristics are similar to those of the members of genus *Burkholderia*. G + C content of the members of the genus varies from 61.4% to 65.0%. Unlike *Burkholderia* species, *Paraburkholderia* members are not commonly associated with human infections. *Paraburkholderia* members form a monophyletic clade within the family *Burkholderiaceae*, and this prompted their separation as a genus independent from *Burkholderia* species, in combination with the finding of robust conserved signature indels (CSIs), which are unique to *Paraburkholderia* species, and are lacking in the members of genus *Burkholderia*. These CSIs distinguish the members of the genus from all other bacteria. Moreover, the CSIs that were found to be shared by *Burkholderia* species are absent in *Paraburkholderia species*, providing an evidence of separate lineages.

The CSIs found within the genus *Paraburkholderia* are in parallel with phylogenomic analyses that indicate two monophyletic clades within the genus; one clade harbours unnamed and '*Candidatus* Paraburkholderia', while the other clade is inclusive of environmental *Paraburkholderia*, commonly used for agricultural purposes. CSIs have been found exclusive to each of these clades, and have not been found specific for any other combination of *Paraburkholderia* species, providing an additional level of phylogenetic resolution within the genus level. At present, only one species has been found to be plant pathogenic. Sugiyama *et al.*, (2017) reported *Para. Andropogonis* causing bacterial leaf streak of *Strelitzia reginae* in Hawaii detected from imported plants.

Type species: *Paraburkholderia graminis*

Plant pathogenic species: *Para. Andropogonis*

6.3.1.3.15 Genus Ralstonia *Yabuuchi* et al. *1996*

Yabuuchi *et al.*, (1995) created this genus to accommodate bacteria, from ecological diverse niches, that were previously classified in genera *Burkholderia* and *Alcaligenes*. The genus was established on the basis of phenotypic characterisation, cellular lipid and fatty acids analysis, phylogenetic analysis of 16S rDNA sequences and rRNA–DNA hybridisation within the homology group II of pseudomonads. The genus contains plant and human pathogens and metal-resistant bacteria. Yabuuchi *et al.*, (1995) transferred *Bur. solanacearum* (syn. *Psm. solanacearum*) at the time of creation of the genus but *Psm. syzygii*, causal agent of Sumatra disease of clove, was transferred later on by Vaneechoutte *et al.*, (2004).

Ral. solanacearum species complex has long been recognised as a group of phenotypically diverse strains that can be subdivided into four phylotypes. Using a polyphasic taxonomic approach on an extensive set of strains, Safni *et al.*, (2014) have provided enough evidence for a taxonomic and nomenclatural revision of members of this complex. Data obtained from phylogenetic analysis of 16S–23S rRNA intergenic spacer (ITS) gene sequences, 16S–23S rRNA ITS region sequences and partial endoglucanase (egl) gene sequences and DNA–DNA hybridisations demonstrate that *Ral. solanacearum* species complex comprises three genospecies. One of these genospecies includes the type strain of *Ral. solanacearum* and consists only of strains of *Ral. solanacearum* phylotype II. Second genospecies includes the type strain of *Ral. syzygii* and contains only phylotype IV strains. This genospecies is further divided into three distinct groups, namely *Ral. syzygii*, the causal agent of Sumatra disease on clove trees in Indonesia, *Ral. solanacearum* phylotype IV strains isolated from different host plants mostly from Indonesia, and strains of the blood disease bacterium (BDB), the causal agent of blood disease of banana, a bacterial wilt in Indonesia affecting bananas and plantains. The last genospecies consists of *Ral. solanacearum* strains that belong to phylotypes I and III. As these genospecies are also supported by phenotypic data, which allow the differentiation of the three genospecies, the following taxonomic proposals are made:

1. Emendation of the descriptions of *Ral. solanacearum* and *Ral. syzygii*.
2. Descriptions of *Ral. syzygii* subsp. *syzygii* subsp. nov., *Ral. syzygii* subsp. *indonesiensis* subsp. nov., *Ral. syzygii* subsp. *celebesensis* subsp. nov. and *Ral. pseudosolanacearum* sp. nov.

The below-given description of the genus is based on Yabuuchi *et al.*, (1995). The genus was named after E. Ralston, an American bacteriologist, who first described *Psm. pickettii* and suggested a taxonomic relationship of this species to *Psm. solanacearum* based on DNA homology between these species. The cells are Gram-stain-negative, motile or non-motile rods. When motile, the motility is by means of a single polar flagellum or peritrichous flagella. Oxidase activity varies with species. Lysine and ornithine decarboxylases are not produced. Major quinone is ubiquinone Q-8. Cellular lipids of the members contain two kinds of phosphatidyl ethanolamines, PE-1 and PE-2. PE-2 contains 2-hydroxy fatty acid at *sn*-2 position of the glycerol moiety. Monosaccharides, disaccharides and polyalcohols are oxidised and assimilated as a sole source of carbon and energy. Inability of *Ral. pickettii, Ral. eutropha* and *Ral. solanacearum* to assimilate galactose, mannose, mannitol and sorbitol distinguishes these species from *Burkholderia* species. Two-dimensional thin layer chromatography failed to detect two kinds of ornithine lipids and C19 cyclopropanoic acid in cellular fatty acids in the type strains of the above-mentioned three *Ralstonia* species while the presence of these compounds was clearly demonstrated in six species of *Burkholderia*, including *Bur. cepacia*. Species have different flagellar morphology. Members of the genus are plant and human pathogens, and knallgas bacteria and metal-resistant bacteria. Strains of *Ral. pickettii* are found in various clinical specimens and hospital environments. *Ral. solanacearum* is taxonomically heterogeneous and infects a large number of host plants. G + C content of DNA ranges from 64.0–68.0 mol%.

Type species: *Ralstonia pickettii*

Liu *et al*., (2023) isolated a Gram-negative, aerobic, rod-shaped, motile bacterium (strain RS^T) with multiple flagella, causing bacterial wilt of tobacco in Yuxi city of Yunnan province, China. They proposed the name *Ralstonia nicotianae* sp. nov. for this bacterium. The type strain of *Ral. nicotianae* sp. nov. is RS^T (=GDMCC 1.3533^T=JCM 35814^T).

Plant pathogenic species: *Ral. nicotianae*, *Ral. pseudosolanacearum*, *Ral. solanacearum*, *Ral. solanacearum* complex: banana (*Musa* sp.) and ***Ral. syzygii***: subsp. *celebesensis*, subsp. *indonesiensis* and subsp. *syzygii*.

6.3.1.3.16 *Genus* Robbsia *Lopes-Santos* et al. *2017*, gen. nov.*

Robbsia (Robb's.i.a. N.L. fem. n. *Robbsia* named after Charles Frederick Robbs, a bacteriologist who pioneered phytobacteriology in Brazil). In addition to the description by Palleroni (1984a) and Gillis *et al*., (1995), of *Burkholderia andropogonis*, *Robbsia* strains are gelatin hydrolysis, oxidase, arginine dihydrolase, propionate, cellobiose, maltose, salicin, saccharose and erythritol negative. The members are positive for citrate, malonate, succinate, acetate, fructose, arabinose, galactose, mannitol, mannose, sorbitol and hypersensitive reaction. Its members can be differentiated from *Burkholderia*, *Paraburkholderia* and *Caballeronia* species by its single polar sheathed flagellum (Fuerst and Hayward, 1969a), rhizobitoxine production (Okazaki *et al*., 2004), 16S rRNA and housekeeping genes sequences, genome comparison by average nucleotide identity, tetranucleotide frequency signature and percentage similarity of conserved proteins. Pathogenic to a wide range of host plants, with an extensive geographical distribution (EPPO, 2014).

The phylogenetic classification of the species *Burkholderia andropogonis* within the *Burkholderia* genus was reassessed using 16S rRNA gene phylogenetic analysis and MLSA. Both phylogenetic trees revealed two main groups, named A and B, strongly supported by high bootstrap values (100%). Group A included all of the *Burkholderia* species complex, while Group B comprised only *Bur. andropogonis* species, with low percentage similarities with other species of the genus, from 92 to 95% for 16S rRNA gene sequences and 83% for conserved gene sequences. It is evident that *Bur. andropogonis* exhibits genotypic differences from other species of *Burkholderia* and represents a new genus described herein as *Robbsia andropogonis* gen. nov., comb. nov.

Type species: *Robbsia andropogonis* (Smith 1911) Lopes-Santos *et al*. 2017*

Cui *et al*., (2022) reported bacterial leaf spot of *Pueraria montana* var. *thomsonii* caused by *Robbsia andropogonis* from China for the first time

Plan pathogenic species: *Rob. andropogonis*
Family: "*Oxalobacteraceae*"
Genera: *Herbaspirillum* and *Janthinobacterium*

6.3.1.3.17 *Genus* Herbaspirillum *Baldani* et al. *1986 emend. Baldani* et al. *1996*

The genus was created in 1986 to include a root-associated nitrogen-fixing bacterium. Later, a plant pathogen, earlier described as *Psm. rubrisubalbicans* (causal agent of mottled stripe disease of sugarcane), was transferred to this genus.

The below-given description of the genus is based on the proposal of Baldani *et al*., (1986) and emended by Baldani *et al*., (1996). Cells are Gram-stain-negative, generally vibroid and sometimes helical, measuring 0.6–0.7 µm in width while the cell length varies from 1.5 to 5.0 µm depending on the culture media. Cells are motile, having one to three or more flagella on one or both the poles. The bacterium has a typical respiratory metabolism and sugars may be oxidised, but are not fermented. Among the species included in the genus, three species are capable of fixing atmospheric N_2 under microaerobic conditions and grow well with N_2 as the sole nitrogen source, even in the presence of 10% sucrose. The oxidase and urease tests are positive while the catalase test is weak or variable. The salts of certain organic acids such as malate, fumarate, succinate, pyruvate, citrate and transaconitate are favoured carbon sources for both NH_4^+- and N_2-dependent growth. Other carbon sources, such as galactose, glycerol, mannitol, sorbitol, D-glucose, L-arabinose, ribose, D-xylose, D-fructose, D-mannose, but not sucrose, are also utilised. Starch and gelatin are not hydrolysed. Optimal temperature for growth is between 30 and 34 °C and optimal pH between 5.3 and 8.0. The genus

* References marked with an * are not cited in the bibliography.

constitutes a separate rRNA cluster within the β-subclass of the *Proteobacteria*. G + C content of the DNA ranges from 60.0–65.0 mol% (T_m). The bacterium occurs mainly in association with graminaceous plants, endophytically colonising roots, stems and leaves, and also in soil.

Type species: *Herbaspirillum seropadicae*

Plant pathogenic species: *Her. rubrisubalbicans*

6.3.1.3.18 *Genus* Janthinobacterium *De Ley* et al. *1978 emend. Lincoln* et al. *1999*

The genus was created in 1978 to include some Gram-stain-negative, psychotrophic and violacein producing bacterial species associated with food spoilage. Lincoln *et al*., (1999) described a new species causing soft rot of mushrooms (*Agaricus bisporus*) and assigned it to this genus.

The description of the genus given below is as proposed by De Ley *et al.*, (1978) and emended by Lincoln *et al.*, (1999). Cells are Gram-stain-negative, straight and sometimes slightly curved rods, having rounded ends and measure 0.8–1.5 × 1.8–6.0 μm. Cells occur mostly singly, but sometimes in pairs or short chains. Non-sporing. Motile. Strictly aerobic, having a respiratory type of metabolism with oxygen as the terminal electron acceptor. Temperature range for growth is 4–30 °C but the best growth occurs at 25 °C. Optimal pH lies between 7.0 and 8.0. Many strains produce the violet pigment, violacein, hence the genus name *Janthinobacterium*, which is derived from janthinus, meaning violet-coloured. Chemoorganotrophs. Citrate is used as the sole source of carbon and ammonium as a sole source of nitrogen. The major phospholipids are phosphatidyl ethanolamine, phosphatidylglycerol and diphosphatidylglycerol as polar lipids. Major respiratory lipoquinone is Q-8. In Hugh and Leifson medium, acid is always produced from galactose, xylose, L-arabinose and *m*-inositol, and often from lactose, sucrose, mannitol, cellobiose, inulin and glycerol. Growth occurs on L-leucine but not on ornithine and phenylalanine. G + C content of the DNA ranges from 61.0 to 67.0 mol%. The bacterium is commonly found in cold and temperate climates in soil and water but can also be isolated from diseased mushrooms.

Type species: *Janthinobacterium lividum*

Plant pathogenic species: *Jant. agaricidamnosum*

Class: "*Gammaproteobacteria*"

Order: "*Enterobacteriales*"

Adeolu *et al.*, (2016) have revised the genome-based phylogeny and taxonomy of the *Enterobacteriales*. The authors proposed a new order "*Enterobacteriales*" and divided it into seven families. Of these seven families, five, namely, *Enterobacteriaceae*, *Erwiniaceae* fam. nov., *Pectobacteriaceae* fam. nov., *Yersiniaceae* fam. nov. and *Morganellaceae*, contain plant pathogenic bacteria.

Family: *Enterobacteriaceae*

Genera: *Citrobacter*, *Enterobacter*, *Klebsiella*, *Kosakonia*, *Lelliottia*, *Raoultella*, *Salmonella* and '*Candidatus* Phlomobacter'

6.3.1.3.19 *Genus* Citrobacter *Werkman and Gillen 1932*

The below-given description of the genus is as given by Sakazaki (1984). Cells are straight rods, measure 1.0 × 2.0–6.0 μm, and occur singly and in pairs. Gram-stain-negative, usually motile by peritrichous flagella, but generally not encapsulated. Facultatively anaerobic, having both a respiratory and a fermentative type of metabolism. Grow readily on ordinary media. The colonies on nutrient agar are generally 2–4 mm in diameter, smooth, low convex, moist, translucent or opaque and grey with shiny surface having entire edges. Mucoid or rough forms may occur occasionally. Catalase positive, but oxidase negative. Chemoorganotrophs. Nitrate is reduced to nitrite. Citrate can be used as a sole source of carbon. Lysine is not decarboxylated. Methyl red test is positive, but Voges–Proskauer test is negative. Acid and gas are produced from glucose. Phenylalanine deaminase, lipase, gelatinase and deoxyribonuclease are not produced. Alginate and pectate are not decomposed. The bacterium is found in the faeces of man and other animals; probably a normal intestinal inhabitant. Often isolated from clinical specimens as opportunistic pathogens. Also found in food, water, soil and sewage. Mol% G + C content of the DNA is 50–52 (*Tm*).

Type species: *Citrobacter freundii*

Allahverdi *et al*., (2016) reported *Cit. freundii* as the cause of bacterial canker of mulberry from Iran.

Plant pathogenic species: *Cit. freundii*.

6.3.1.3.20 *Genus* Enterobacter *Hormaeche and Edwards 1960*

The genus was established in 1960 by Hormaeche and Edwards. Earlier, the genus did not include any plant pathogen but now it contains four plant pathogenic species, transferred from *Erwinia*. In order to infer an encompassing phylogeny of this genus, 16S rDNA sequences of additional members of *Enterobacter* are required.

The genus has the following important characteristics. Cells are Gram-stain-negative, non-sporing, straight rods and measure 0.6–1.0 × 1.2–3.0 μm. Some strains are encapsulated. Motile by peritrichous flagella. Facultative anaerobes, and some *Enterobacter* species, as facultative anaerobes, ferment both glucose and lactose as a carbon source. Both acid and gas are produced from glucose. *Enterobacter* species can be differentiated from other Gram-stain-negative rods (*Escherichia coli* and

Klebsiella spp.) by virtue of being fast fermenter of lactose. Gas is produced in metabolic processes. Nitrate is reduced to nitrite. H_2S is not produced from thiosulphate. Methyl red and Voges–Proskauer tests are positive for many strains. Tween 80 and gelatin are either very slowly hydrolysed or not hydrolysed. Optimal temperature for growth is 30 °C.

Enterobacter are found on plants and human skin, in soil, water, sewage, and intestinal tracts of humans and animals. Some species of *Enterobacter*, such as *Entb. sakajakii*, are opportunistic human pathogens. They cause infections in immunocompromised (usually hospitalised) hosts and, rarely, in otherwise, healthy individuals. *Entb. sakajakii* is well known for infections in infants, who are fed milk-based powdered infant formulae. The bacterium survives when the contaminated powdered formula is heated and prepared. Since it has caused several outbreaks of infection in the past, baby formulae are effectively screened for *Entb. sakajakii* before they are sold. *Entb. cloacae* A-11 and similar bacteria are found on cucumber, radish, peas, soybean, sunflower and sweet corn seeds. *Entb. cloacae* is also used for biological control of plant diseases. G + C content of the DNA is 52.0–60.0 mol%.

Two isolates of Gram-stain-negative, facultatively anaerobic rods, 0.3–1.0 µm wide and 0.8–2.0 µm long, with peritrichous flagella, showing a DNA G + C content of 55.1 ± 0.5 mol%, isolated from diseased mulberry roots were investigated. Biochemical data revealed that the isolates could be differentiated from their nearest neighbours by the presence of lysine decarboxylase activity and their ability to utilise D-arabitol. DNA–DNA relatedness also distinguished the two isolates from phylogenetically closely related *Enterobacter* strains. Hence, these isolates were named *Enterobacter mori* sp. nov. (Zhu *et al*., 2011).

Furtado *et al*., (2012) isolated a facultative anaerobic, Gram-stain-negative bacterium causing brownish, necrotic, irregular spots on leaves of *Mabea fistulifera* Mart. in Brazil and on the basis of phenotypic, biochemical and molecular tests identified it as *Enterobacter cowanii*. This was the first report of a member of the *Enterobacteriaceae* family causing disease in *M. fistulifera*. Retana *et al*., (2017) reported *Entb. hormaechei* causing stem rot on *Hylocereus* spp. in Costa Rica.

Cao *et al*., (2021) reported the complete genome of *Entb. asburiae* strain SD4L, which represents the first whole-genome sequence of an isolate from rice seedlings in China. The assembled genome consisted of two contigs, with a circular chromosome of 4,574,166 bp and a plasmid of 85,271 bp, respectively.

Guo *et al*., (2023) reported for the first time *Entb. roggenkampii* and *Entb. cloacae* complex sp. causing bacterial pith necrosis of tomato.

Abedinzadeh *et al*., (2023) reported *Entb. cloacae* subsp. *cloacae, Entb. ludwigii* and *Entb. mori* causing decline and offshoot rot of the date palm. To the best of the author's knowledge, this is the first report of these three *Enterobacter* species as causal agents of decline and offshoot rot of the date palm in the world.

Tao *et al*., (2024) reported bacterial leaf spot on ginger caused by *Entb. quasiroggenkampii* from China. To the best of the author's knowledge, it is the first report of *Entb. quasiroggenkampii* causing bacterial leaf spot of ginger. In Guangxi, *Entb. quasiroggenkampii* was identified as one of the pathogens causing mulberry wilt (Jiao, 2022).

Wu *et al*., (2025) reported *Entb. kobei* and *Entb. mori* causing bacterial wilt of ginger for the first time from China.

Type species: *Enterobacter cloacae*

Plant pathogenic species: *Entb. asburiae*, *Entb. cancerogenus*, ***Entb. cloacae***: *Entb. cloacae* subsp. *cloacae*, *Entb. cloacae* subsp. *dissolvens*; *Entb. cowanii*, *Entb. hormaechei*, *Entb. kobei*, *Entb. ludwigii*, *Entb. mori*, *Entb. nimipressuralis*, *Entb. pyrinus*, *Entb. quasiroggenkampii* and *Entb. roggenkampii*

6.3.1.3.21 *Genus* Klebsiella *Trevisan 1885, emend. Carter* et al., *1999 emend. Drancourt* et al., *2001 emend. Ma* et al. *2016, emend.*

The cells are typically straight rods with rounded or slightly pointed ends and can be found singly, in pairs, or in short chains or linked end to end. Cells are Gram-stain-negative, usually non-motile, generally measure 0.3–1.5 µm wide and 0.5–5.0 µm long, and have a prominent polysaccharide-based capsule. Cells tend to be shorter and thicker when compared to other members of the family *Enterobacteriaceae*. Diplobacillary forms are commonly found *in vivo*. The members have no specific growth requirements and grow well on standard laboratory media, but grow best between 35 and 37 °C and at pH 7.2. The species are facultative anaerobes and most strains can survive with citrate and glucose as their sole carbon sources and ammonia as their sole nitrogen source. *Klebsiella* species are found everywhere in nature.

Members of the genus produce a prominent capsule, or slime layer, which can be used for serological identification, but molecular serotyping may replace this method. Members of the genus *Klebsiella* typically express two types of antigens on their cell surfaces. The first, O antigen, is a component of the lipopolysaccharide, of which nine varieties exist. The second is K antigen, a capsular polysaccharide with more than 80 varieties. Both contribute to pathogenicity and form the basis for serogrouping. Based on these two major antigenic determinants, several vaccines have been designed.

The genus was named after German-Swiss microbiologist Edwin Klebs (1834–1913). Members of the genus are a part of the human and animal's normal flora in the nose, mouth and intestines. This is thought to be due to distinct sublineages developing specific niche adaptations, with associated biochemical adaptations which make them better suited to a particular environment. The members can be found in water, soil, plants, insects and animals, including humans. *Kle. variicola* has been identified as one of the causes of bovine mastitis.

In plant systems, *Klebsiella* species can be found in a variety of plant hosts. *Kle. pneumoniae* and *Kle. oxytoca* are able to fix atmospheric nitrogen into a form that can be used by plants, and are therefore called **associative nitrogen fixers** or **diazotrophs**. The bacteria attach strongly to root hairs and less strongly to the surface of the zone of elongation and the root cap mucilage. These are of economic importance due to their ability to increase the yields of agricultural crops. Their high numbers in plants are thought to be at least partly attributable to their lack of a flagellum, as flagella are known to induce plant defences. Additionally, *Kle. variicola* is known to associate with a number of different plants including banana trees and sugarcane, and has been isolated from the fungal gardens of leaf-cutter ants.

Type species: *Klebsiella pneumoniae* (Schroeter, 1886) Trevisan, 1887

Loganathan *et al*., (2021) reported rhizome rot of banana caused by *Kle. variicola* for the first time from India. Earlier, *Kle. variicola* causing soft rot has been reported on banana in China (Fan *et al*., 2016), plantain soft rot in Haiti (Fulton *et al*., 2020) and carrot soft rot in India (Chandrashekar *et al*., 2018).

Basavand *et al*., (2021) observed a wet wood disease on *Morus* trees (Black mulberry, White mulberry and Red mulberry) in municipal lands and parks of Rafsanjan city (Kerman Province), Iran, in summer and spring of 2016. The causal bacterium was identified as *Kle. oxytoca* based on colony morphology, phenotypic and biochemical tests. Malik *et al*., (2022) reported the stem rot of pearl millet incited by *Kle. aerogenes* in southwestern regions of Haryana, India.

Plant pathogenic species: *Kle. aerogenes*, *Kle. oxytoca* and *Kle. variicola*.

6.3.1.3.22 *Genus* Kosakonia

The genus was named after Y. Kosako, for his contribution to the bacterial taxonomy. The cells are Gram-stain-negative, facultatively anaerobic and motile rods. Species are mesophilic. Optimum growth temperature ranges from 30–35 °C. Carbon sources such as glucose, cellobiose, fructose, galactose, gentiobiose, maltose, mannose and *N*-acetylglucosamine are readily assimilated. Species form a monophyletic clade when phylogenetic analysis is based on multilocus sequence analysis or whole-genome sequences. The major fatty acids are $C_{16:0}$ and summed features 2 (iso-$C_{16:1}$ and/or $C_{14:0}$ 3-OH) and 3 ($C_{16:1}$ ω7*c* and/or iso-$C_{15:0}$ 2-OH).

The species within the genus *Kosakonia* are primarily plant-growth-promoting bacteria, able to fix nitrogen and improve crop yield. However, several species have also been isolated from clinical samples and are thought to cause rare incidences of disease. Some species are also pathogenic on plants. DNA G+C content (mol%) ranges from 52.5–56.2 (genome analysis).

Type species: *Kosakonia cowanii* Brady *et al*. 2013

Berger *et al*., (2017) reported that plant growth-promoting bacterium *Kos. radicincitans* (DSM 16656) improved shoot and root growth of young tomato plants, but data on yield increase and fruit quality are lacking.

Chen *et al*., (2024) reported *Kos. cowanii* causing bacterial fruit blotch on *Trichosanthis fructus* from Jiangsu Province, China.

Plant pathogenic species: *Kos. cowanii*

6.3.1.3.23 *Genus* Lelliottia *Brady* et al. *2013*

Brady *et al*., (2013) taxonomically evaluated the genus *Enterobacter* based on MLSA and proposed to reclassify *Entb. nimipressuralis* and *Entb. amnigenus* into *Lelliottia* gen. nov. as *Lelliottia nimipressuralis* comb. nov. and *Lelliottia amnigena* comb. nov., respectively.

The genus *Lelliottia* includes species regularly isolated from water, soil, spoiled fruit and vegetables, elm trees suffering from "wetwood" disease and also from clinical samples. Cells are Gram-stain-negative, facultatively anaerobic and motile rods. Species are mesophilic, favouring growth temperatures at 30–35 °C. Carbon sources such as glucose, cellobiose, fructose, galactose, maltose, mannose, *N*-acetylglucosamine and *N*-acetyl galactosamine are readily assimilated. Species form a monophyletic clade when phylogenetic analysis is based on MLSA or whole-genome sequences. The major fatty acids are $C_{16:0}$, $C_{17:0}$ cyclo, $C_{18:1}$ ω7*c* and summed features 2 (iso-$C_{16:1}$ and/or $C_{14:0}$ 3-OH) and 3 ($C_{16:1}$ ω7*c* and/or iso-$C_{15:0}$ 2-OH).

The type species of this genus has alternatively been placed in the genus *Enterobacter* (Hormaeche and Edwards 1960; Approved lists 1980) (Brenner *et al*., 1986). The type species of this genus has also been alternatively placed in the genus *Erwinia* (Winslow *et al*. 1920; Approved Lists 1980) (Dye, 1969) and in the genus *Erwinia* Winslow *et al*. 1917 (Carter, 1945). DNA G+C content (mol%): 52.8–56.0 (genome analysis).

Thakur and Gauba (2022) reported the genome sequence of *Lell. amnigena* PTJIIT1005, isolated from a polluted site on the Yamuna River, Delhi. The genome is complete and consists of ~4.5 Mbp with a GC content of 52.62%, 4259 protein-coding genes, 76 tRNAs and 4 rRNAs. Strain PTJIIT1005 shows 98.89% average nucleotide identity with *Lell. amnigena*. El Zowalaty *et al*., (2023) reported the first draft genome sequences of a rare *Lell. nimipressuralis* strain MEZLN61.

Type species: *Lelliottia nimipressuralis* (Carter 1945) Brady *et al.* 2013

Osei *et al.*, (2022) reported *Lell. amnigena* causing soft rot of potato tuber from China.

Li *et al.*, (2025) reported for the first time *Lell. amnigena* causing soft rot on purple stem mustards from China. Zhao *et al.*, (2022) identified and characterised the pathogenicity of *Lell. nimipressuralis* causing soft rot of *Codonopsis pilosula* (dangshen) roots from China.

Plant pathogenic species: *Lell. amnigena* and *Lell. nimipressuralis*.

6.3.1.3.24 *Genus* Raoultella *Drancourt* et al. *2001*

Raoultella spp. are Gram-stain-negative, aerobic and non-motile rods. The members of the genus can be distinguished from those of genus *Klebsiella*, in that these use histamines as the only source of carbon in the medium. Also, *Raoultella* species grow at 4 °C and do not produce gas from lactose at 44.5 °C. *Raoultella* species are known to inhabit natural environments like water, soil and plants. The reservoir of *Raoultella* species is the gastrointestinal tract and upper respiratory tract. *Raoultella* spp. are opportunistic bacteria, which usually cause infections of the biliary tract, pneumonia and bacteraemia in oncologic and with lower-immunity patients. *Rao. planticola* and *Rao. ornithinolytica* are the most frequently encountered human pathogens among the members of this genus. In his review article, Sêkowska (2017) has summarised the information on infections caused by *Raoultella* species.

The members grow at 10 °C consistent with their recovery from plants, soil and water, whereas members of *Klebsiella* genus do not grow at 10 °C (Murray and Jo Baron, 2007) and are mainly recovered from mammals' mucosae. In human infections, *Raoultella* species are generally sensitive to treatment with carbapenems. In one series, 92% were sensitive to ciprofloxacin. More than 10% produced extended-spectrum beta-lactamases (Sêkowska *et al.*, 2020).

Type species: *Raoultella planticola* (Bagley *et al.*, 1981) Drancourt *et al.*, (2001)

Brady *et al.*, (2024) isolated 17 Gram-stain-negative, facultatively anaerobic bacterial strains from bleeding cankers of various broadleaf hosts and oak rhizosphere soil in Great Britain. Whole-genome sequencing comparisons confirmed these two clusters of strains belonging to two novel *Raoultella* species, namely *Raoultella scottia* sp. nov. (type strain = BAC 10a-01-01T = LMG 33072T = CCUG 77096T) and *Raoultella lignicola* sp. nov. (type strain = TW_WC1a.1T = LMG 33073T = CCUG 77094T).

Plant pathogenic species: *Rao. lignicola* and *Rao. scottia*

6.3.1.3.25 *Genus* Salmonella *Lignières 1900*

Joseph Leon Lignières named genus *Salmonella* in honour of Daniel Salmon, a veterinary pathologist of the US Department of Agriculture. *Salmonella* was first visualised in 1880 by Karl Eberth in the Peyer's patches and spleens of typhoid patients. In 1884, Georg Theodor Gaffky successfully grew the pathogen in pure culture. In 1985, Theobald Smith discovered what would be later known as *Salmonella enterica* (var. Choleraesuis). Initially, *Salmonella choleraesuis* was thought to be the causal agent of hog cholera, so Salmon and Smith named it "Hog-cholerabacillus". The name *Salmonella* was not used until 1900, when Joseph Leon Lignières proposed that the pathogen discovered by Daniel Salmon's group be named *Salmonella* in his honour (Lignières, 1900).

Cells are rod-shaped, measure 0.7–1.5 × 2.0–5.0 µm and are Gram-stain-negative. Cells are motile having peritrichous flagella and do not form endospores. These are chemotrophs and obtain their energy from oxidation and reduction reactions using organic sources. The members are also facultative anaerobes, capable of surviving with or without oxygen. Most subspecies of *Salmonella* produce H_2S, which can readily be detected by growing them on culture media containing ferrous sulphate. Most isolates exist in two phases, i.e., a motile phase I and a non-motile phase II. Cultures that are non-motile in primary culture may be switched to the motile phase using a Craigie tube or ditch plate. Bacterial cells are not killed by freezing, but UV light and heat accelerate their destruction. The cells are killed after being heated to 55 °C (131 °F) for 90 min. or to 60 °C (140 °F) for 12 min. The heating of food for at least 10 min. to an internal temperature of 75 °C (167 °F) is recommended to protect against *Salmonella* infection.

Genus *Salmonella* consists of only two species, *Sal. bongori* and *Sal. enterica*, and the latter is divided into six subspecies, namely, *enterica*, *salamae*, *arizonae*, *diarizonae*, *houtenae* and *indica*. *Sal. enterica* subsp. *enterica* includes more than 1,500 serotypes (while all the six subspecies include over 2,500 serotypes), which, despite their high genetic similarity, vary greatly in their host range and disease outcome, ranging from enteritis to typhoid fever (Ohl and Miller, 2001). *Sal. enterica* subsp. *enterica* is an important economic and public health problem throughout the world.

The members lead predominantly host-associated lifestyles, but can be found to persist in a bathroom setting for weeks following contamination, and are frequently isolated from water sources, which act as bacterial reservoirs and may help to facilitate transmission between hosts. *Salmonella* species can be found in the digestive tracts of humans and animals, especially reptiles. Food and water can also be contaminated with the bacteria if these come in contact with the faeces of infected people or animals. *Sal. enterica* subspecies are found worldwide in all warm-blooded animals and in the environment. *Sal. bongori* is restricted to cold-blooded animals, particularly reptiles. Strains of *Salmonella* cause illnesses such as typhoid fever, paratyphoid fever and food poisoning (salmonellosis).

Type species: *Salmonella enterica*

Strictly speaking, *Salmonella* species are not plant pathogens. Two reports, by Klerks *et al.*, (2007) and Schikora *et al.*, (2008), have suggested that plants show chlorosis or reduced vigour following *Sal. enterica* inoculation, but Koch's postulates were not proved in either case. Therefore, *Sal. enterica* remains characterised as plant-associated bacteria. However, Kado (2010) has listed *Sal. enterica* as a plant pathogen, responsible for causing systemic colonisation of thale cress (*Arabidopsis thaliana* (L.) Heynh.). There are numerous produce-implicated outbreaks, which are testament to *Sal. enterica* and pathogenic *Escherichia coli*'s ability to persist and use plants as vectors to humans (Taormina *et al.*, 1999).

Plant pathogenic species: *Sal. enterica.*

***Candidatus* category**: There are many bacteria that cannot be cultured by currently available methods. Hence, these cannot be investigated in ways that provide the description for proposal as legitimate taxa. Keeping this in view, Murray and Schleifer (1994) proposed the rank or category of *Candidatus* for classification of such procaryotes. The International Committee on Systematic Bacteriology accepted the proposal of Murray and Schleifer (1994) and recommended the category '*Candidatus*' as a possible taxonomic rank for the unculturable procaryotic organisms (Murray and Stackebrandt, 1995). In addition to genomic information, such as sequences apt to determine the phylogenetic position of the organism, all information, including structural, metabolic and reproductive features, should be included in the description of a provisional taxon, together with the natural environment in which the organism can be identified by *in situ*-hybridisation or other similar techniques for cell identification. The *Candidatus* species are differentiated primarily on the basis of the comparative analyses of 16S rRNA sequences, other sequence data, ecological information and metabolic data when available. To make clear the identification of organisms as *Candidatus*, names are preceded by the word *Candidatus* (italicised) but the name itself is not italicised and the entire designation is written between single quotation marks e.g., '*Candidatus* Liberibacter asiaticus'. *Candidatus* nomenclature is governed neither by the Code nor by the Standards for Naming Pathovars. However, to prevent any nomenclature confusion, the Phytoplasma/Spiroplasma Working Team of the International Research Project for Comparative Mycoplasmology (IRPCM) has suggested rules for the description of organisms as novel taxa within '*Candidatus* Phytoplasma' (IRPCM, 2004) and, by extension, to other plant pathogenic '*Candidatus* genera' (Young, 2008). As the *Candidatus* microorganisms cannot be cultured, their type strains cannot be isolated or maintained for reference. The reference material for *Candidatus* is either a DNA sample or a specific rDNA sequence. Two *Candidatus* genera are described below.

6.3.1.3.26 'Candidatus Phlomobacter' *Zreik* et al., *(1998)*

Zreik *et al.*, (1998) proposed '*Candidatus* Phlomobacter' to include bacterium-like organisms (BLOs) responsible for causing marginal chlorosis of strawberry. At present, it contains only one plant pathogenic species.

The below-given description of the '*Candidatus* Phlomobacter' is as published by Zreik *et al.*, (1998). Gram-stain-negative, walled bacterium, restricted to the sieve tubes of the phloem tissue of infected strawberry plants. Cells are 0.20–0.27 μm in diameter and 4.0 μm long. The bacterium is unable to grow in culture media, could not be transmitted to periwinkle plants through dodder, and insect vectors are not known. Sequence analysis of the cloned rDNA revealed that it is a new bacterium within group 3 of class "Gammaproteobacteria" while citrus huanglongbing agents belong to class "*Alphaproteobacteria*". '*Candidatus* Phlomobacter fragariae' name was proposed for strawberry marginal chlorosis agent.

Type species: '*Candidatus* Phlomobacter fragariae'

Garnier (2015) included this genus in the family *Enterobacteriaceae*, but Adeolu *et al.*, (2016) excluded it from this family.

Plant pathogenic species: '*Can.* Phl. fragariae'.

Family: Morganellaceae

Genera: *Arsenophonus* and *Providencia*

6.3.1.3.27 Genus Arsenophonus

Arsenophonus bacteria are among the most biologically diverse and widespread endosymbionts of arthropods. *Arsenophonus phytopathogenicus* has been characterised as an emerging insect-vectored plant pathogen. Bressan

(2014) has reviewed the current understanding on the lifestyle transition, evolution, host interaction and infection cycles of these emerging plant pathogens.

Presently, the plant pathogenic *Arsenophonus* bacteria have been reported to infect sugar beet and strawberry plants in different countries. They have been associated with two diseases: marginal chlorosis disease in strawberry plants and low sugar content disease in sugar beet plants. Symptoms of these diseases are convergent with those associated with phytoplasmas (Mollicutes) infections. There is evidence that plant pathogenic *Arsenophonus* bacteria have a complex relationship with their insect hosts.

Sequence analysis of the 16Sr RNA and spoT-spoU-recG genes of selected isolates, surprisingly, revealed the presence of two Arsenophonus bacteria: '*Can.* Phlomobacter fragrariae' and '*Can.* Arsenophonus phytopathogenicus', reported to be the causal agents of the "basses richesses" syndrome of sugar beet plants in eastern France. This is the first detection of two different protobacteria associated with strawberry marginal chlorosis and the first visualisation in Italy of the proteobacterium by electron microscopy. Further investigations will be necessary to better understand the etiology and the epidemiology of this disease in Italy.

Zübert and Kube (2021) introduced a TaqMan qPCR for diagnostics of '*Can.* Arsenophonus phytopathogenicus' targeting genes encoding a heat shock protein of the Hsp20 family and mannose-6-phosphate isomerase. Quantitation with synthetic oligonucleotides as standard showed that the developed TaqMan qPCR assays enable the detection up to 100 target copies. A comparison between the TaqMan qPCR and end-point PCR for '*Can.* Arsenophonus phytopathogenicus' detection was carried out on 78 sugar beet samples from different locations in southern Germany. The newly developed assays enable the fast, reliable and sensitive detection of '*Can.* Arsenophonus phytopathogenicus' in sugar beet.

Type species: *Arsenophonus nasoniae* Gherna *et al.* 1991

Plant pathogenic species: *Ars. phytopathogenicus*

6.3.1.3.28 Genus Providencia *Ewing 1962*

It was named after Providence, Rhode Island, where Stuart *et al.*, (1943) studied these bacteria at Brown University. The cells are Gram-stain-negative, motile, urease-producing bacilli that do not ferment lactose and are distinguished from other members of the family *Morganellaceae* by their ability to deaminate phenylalanine and lysine. *Providencia* species can deaminate aromatic amino acids, including tryptophan and phenylalanine, and can influence the formation of indole and indoxyl sulphate, which are metabolites of tryptophan. Produce urease.

Pro. stuartii cells are straight rods and measure 0.6–0.8 × 1.5–2.5 µm (Hanson *et al.*, 2019). This species is ubiquitous, being commonly found in water, soil and animal reservoirs (Ryan and Ray, 2004). Some species are opportunistic pathogens of humans. *Pro. stuartii* can cause urinary tract infections, particularly in patients with long-term indwelling urinary catheters or extensive severe burns (Ryan and Ray, 2004)

Many strains have been isolated from the haemolymph of fruit fly, *Drosophila melanogaster.* These strains display different levels of virulence. For example, *Pro. sneebia* is highly virulent, and infection always results in fly mortality (Juneja and Lazzaro, 2009). Alternatively, *Pro. rettgeri* displays an intermediate virulence level, wherein some individuals survive infection while others perish. The susceptibility to *Pro. rettgeri* is strongly tied to an allele of the antimicrobial peptide gene Diptericin (Unckless *et al.*, 2016). Surprisingly, the fly's defence against *Pro. rettgeri* seems to rely almost exclusively on Diptericin, as deletion of Diptericin leads to complete mortality. Meanwhile, deletion of multiple other antimicrobial peptides has no effect on *Pro. rettgeri* virulence. Yet defence against *Pro. burhodogranariea* is determined by multiple antimicrobial peptides beyond just Diptericin (Hanson *et al.*, 2019).

Type species: *Providencia alcalifaciens* (de Salles Gomes, 1944) Ewing (1962)

Zhou *et al.*, (2020) reported *Pro. rettgeri* causing brown slime flux on *Populus tomentosa* probably by inhibiting wound healing. To their knowledge, this was the first report of *Pro. rettgeri* as a pathogen of plants.

Wang *et al.*, (2023) reported *Pro. rettgeri* causing soft rot on calla lily from China. To their knowledge, this was the first report of the occurrence of *Pro. rettgeri* on calla lily in China.

Plant pathogenic species: *Pro. rettgeri*

Soutar and Stavrinides (2020) reported that genera *Gibbsiella* and *Samsonia* are currently assigned to incorrect families. Genus *Gibbsiella* should be placed in family *Yersiniaceae* and *Samsonia* in family *Pectobacteriaceae.*

Family: *Yersiniaceae*

Genera: *Gibbsiella*, *Serratia*, *Ewingella*, *Rahnella* and *Rouxiella*

6.3.1.3.29 Genus Gibbsiella *Brady* et al. *2011 emend. Kim* et al. *2013*

The genus was named in honour of British forest pathologist John N. Gibbs for his contributions to forest pathology. The below-given description of the genus is based on the publication of Brady *et al.*, (2010).

Cells are Gram-stain-negative, short rods (0.9 × 1–1.5 µm), facultatively anaerobic, oxidase negative and catalase positive. Cells occur singly, in pairs, or in groups of four and possess very fine fimbriae, but no flagella.

Colonies on nutrient agar are white to creamy, round, convex and smooth having entire margins. Strains can grow at temperatures between 10 and 40 °C. β-galactosidase activity is positive, but H_2S, urease, indole, acetoin and gelatinase production are negative. Citrate is utilised and nitrate is reduced to nitrite. Arginine dihydrolase, lysine decarboxylase, ornithine decarboxylase and tryptophan deaminase activity are negative. G + C content ranges from 56.0–56.4 mol%.

The emended description of the genus is same as given by Brady *et al.*, (2010), except that the DNA G + C content ranges from 56.0–58.7 mol% (Kim *et al.*, 2013).

Type species: *Gibbsiella quercinecans*

Brady *et al.*, (2010) reported *Gib. quercinecans* gen. nov., sp. nov., the cause of acute oak decline from Britain and Spain.

Plant pathogenic species: *Gib. quercinecans.*

6.3.1.3.30 Genus Serratia *Bizio 1823*

The genus was established by Bizio (1823). The genus is characterised by Gram-stain-negative, straight rods (0.5–0.8 × 0.9–2.0 µm) with rounded ends and generally motile by peritrichous flagella. Facultatively anaerobic. Growth factors are generally not required. Temperature range for growth lies between 10 and 36 °C. Colonies on nutrient agar are generally opaque, somewhat iridescent and appear white, pink or red in colour. Red colour of the colonies is due to the production of a characteristic red colour, prodigiosin. Members of the genus can be distinguished from other members of the family *Enterobacteriaceae* by their unique production of three enzymes, namely, DNase, lipase and gelatinase.

Glucose is fermented and gluconate is oxidised to 2-ketogluconate. The bacterium ferments and utilises maltose, fructose, D-galactose, D-mannose, D-mannitol, ribose and trehalose as sole carbon sources. Esculin is hydrolysed by most strains. Tween 80 is hydrolysed by all species except one. Acetoin is produced from pyruvate by all species except *Ser. fonticola*. Nitrate and chlorate are reduced anaerobically. Generally, the members of this genus occur on plant surfaces, in soil and water, and as opportunistic human pathogens. *Ser. marcescens* is mainly the human pathogen and causes nosocomial infections. The genus did not contain any plant pathogen till 1919, when Paine and Stansfield (1919) suggested the role of *Ser. proteamaculans* in causing leaf spot disease of king protea (*Protea cynaroides*), which was later on confirmed by Grimont *et al.*, (1978). Mol% G + C of the DNA is 52.0–60.0.

Manzo-Valencia *et al.*, (2023) isolated *Pseudomonas* sp., along with *Pantoea agglomerans* and *Serratia liquefaciens*, from broccoli heads showing head soft rot symptoms. Different phytopathogenic assays confirmed the ability of the isolated *Ser. liquefaciens* strain to produce soft rot disease in stems and florets of broccoli.

Mphande *et al.*, (2025) reported that bacteria that cause cucurbit yellow vine disease fall within the *Ser. ureilytica* species of the *Ser. marcescens* complex and can be vectored by cucumber beetles.

Type species: *Serratia marcescens*

Plant pathogenic species: *Ser. liquefaciens*, *Ser. marcescens*, *Ser. proteamaculans* and *Ser. ureilytica*.

6.3.1.3.31 Genus Ewingella *Grimont* et al. *1983*

Cells are Gram-stain-negative and rod-shaped. Genus *Ewingella* is phenotypically distinct from all other groups of *Enterobacteriaceae*. Members of the genus are anaerogenic, lipase and deoxyribonuclease negative; lysine, ornithine and arginine decarboxylase negative. Voges–Proskauer test positive. The members produce acid from glucose in the presence (and absence) of iodoacetate, but not from L-arabinose, melibiose, raffinose, D-sorbitol or sucrose. DNA-relatedness studies (S1-nuclease method) showed that the 10 *Ewingella* strains studied, form a single DNA-hybridisation group, has less than 21% relatedness to other members of the family *Enterobacteriaceae*. This single DNA-hybridisation group is named *Ewingella americana* sp. nov. The type strain of *Ewi. americana* is CDC 1468-78 (= ATCC 33852 = CIP 8194). Although the 10 strains of *Ewi. americana* were isolated from clinical sources in the United States, the clinical significance of these organisms is not known. *Ewi. americana* is the only species in the genus *Ewingella* and was first identified and characterised by Grimont *et al.*, (1983).

The organism is rarely reported as a human pathogen, though it has been isolated from a variety of clinical specimens, including wounds, sputum, urine, stool, blood, synovial fluid, conjunctiva and peritoneal dialysate. The bacterium was named in honour of William H. Ewing, an American biologist who contributed to modern taxonomy. *Ewi. americana* is an organism with simple nutritional needs that can survive in water and citrate solution. Domestic sources of water, including air conditioning units and ice baths, have been cited as sources of infection.

Type species: *Ewingella americana*

Although *Ewi. americana* has been reported from Korea causing internal stipe necrosis of *Agaricus bisporus* (Lee *et al.*, 2009), it has not been reported from shiitake mushrooms. It has also been reported from China, causing bacterial brown rot in *Flammulina velutipes* (Liu *et al.*, 2018). Na *et al.*, (2021) observed a bacterial brown rot on shiitake mushrooms (*Lentinula edodes*) at a mushroom farm in Daejeon, Korea during August 2019 and the incidence ranged from 5 to 15%. All the isolates were Gram-stain-negative and formed whitish, round, semi-transparent colonies on nutrient agar plates. On the basis of morphological and biochemical characteristics, polymerase chain reaction

(PCR) amplifications of the partial 16S rRNA and *gyrB* genes performed using primers 27F/1492R and gyrB-fseq/gyrB-rseq, the bacterium was identified as *Ewingella americana*. To their knowledge, this is the first report on the occurrence of bacterial brown rot on shiitake mushrooms caused by *Ewi. americana* in Korea.

Yang *et al.*, (2023) made the first worldwide report of *Ewi. americana* infecting and causing brown rot disease of the cultivated jelly fungus *Naematelia aurantialba*.

Plant pathogenic species: *Ewi. americana*

6.3.1.3.32 Genus Rahnella *Izard* et al. *1981**

Rahnella is a diverse genus of Gram-negative, rod-shaped, facultatively anaerobic bacteria, found widely in soil, water, plants, insects (like bark beetles), and food; capable of causing opportunistic human infections (like UTIs and bacteremia) and acting as plant growth promoters, with notable species including *Rah. aquatilis* and *Rah. contaminans*; known for antibiotic resistance genes and ecological adaptability. Ubiquitous, found in soil, water, plant roots/seeds, human clinical samples, and insects. Resistant to various stresses (acids, salts, metals) and capable of degrading complex substances. Some strains (e.g., *Rah. aquatilis*) act as biocontrol agents, reducing plant diseases and mitigating pesticide toxicity in honeybees. Found in the gut of bark beetles, aiding in nutrient processing and detoxification. Frequently found on vegetables, carrying antibiotic resistance genes. Can cause human infections like urinary tract infections, bacteremia, and wound infections, especially in immunocompromised individuals. Known to possess antibiotic resistance genes, though often susceptible to newer antibiotics like cephalosporins.

Type species: *Rahnella aquatilis* Izard *et al.* 1981*
Plant pathogenic species: *Rahnella victoriana* Brady *et al.*, (2017)*

6.3.1.3.33 Genus Rouxiella *Le Flèche-Matéos* et al. *2015 emend. Le Flèche- Matéos* et al. *2017 emend. Lee* et al. *2021*

The cells are Gram-stain-negative, rod-shaped, non-motile and do not form spores. Colonies are yellow or white coloured. The members are facultatively anaerobic and can grow in the temperature range of 4–30 °C. The genus was named to honour Émile Roux, a French physician, bacteriologist and immunologist, who was one of the closest collaborators of Louis Pasteur and a co-founder of the Pasteur Institute (Le Fleche-Mateos *et al.*, 2015, 2017). Le Flèche-Matéos *et al.*, (2017) emended the description of genus *Rouxiella* to include *Roux. badensis* sp. nov. and *Roux. silvae* sp. nov., isolated from peat bog soil or swampy meadow in Baden-Württemberg (Germany), in the genus *Rouxiella*.

The type species, *Rouxiella chamberiensis*, was originally isolated from parenteral nutrition bags used for premature newborns in neonatal intensive care units in Chambéry Hospital, and originally published in 2015 (Le Fleche-Mateos *et al.*, 2015). *Roux. badensis* and *Roux. silvae* were originally isolated from a peat bog (Le Fleche-Mateos *et al.*, 2017) and *Roux. aceris* was isolated from tree sap (Lee *et al.*, 2021)

Paul *et al.*, (2023) reported complete genome sequences of the type strains *Roux. badensis* DSM 100043 and *Roux. chamberiensis* DSM 28324, using nanopore long-read sequencing.

Lee *et al.*, (2021) performed polyphasic taxonomic studies for the seven strains, which were isolated from sap extracted from *Acer pictum* in Mt. Halla in Jeju, Republic of Korea. Cells of all the isolates were Gram-stain-negative, facultatively anaerobic, short rods and contained the major isoprenoid quinone of Q-8, the predominant fatty acids of $C_{16:0}$ and $C_{17:0}$ cyclo and the major polar lipids, including phosphatidylethanolamine, phosphatidylglycerol and an unidentified aminophospholipid. The G+C contents of the genomic DNAs were 50.6–51.3%. On the basis of data obtained here, the authors proposed *Rouxiella aceris* sp. nov. (type strain, SAP-1^T=KCTC 72599^T=CCM 9078^T) for these strains and emended the description of the genus *Rouxiella*.

Le Flèche-Matéos *et al.*, (2017) emended the description of genus *Rouxiella* to include *Roux. badensis* sp. nov. and *Roux. silvae* sp. nov., isolated from peat bog soil or swampy meadow in Baden-Württemberg (Germany), in the genus *Rouxiella*.

Zhao *et al.*, (2022) reported *Roux. badensis*, a new bacterial pathogen of onion causing bulb rot. This is the first report of *Roux. badensis* as a plant pathogen.

Plant pathogenic species: *Roux. badensis*

Family: Erwiniaceae

Genera: *Erwinia*, *Pantoea*, *Phaseolibacter* and *Tatumella*

6.3.1.3.34 Genus Erwinia *Winslow* et al. *1920 emend. Hauben* et al. *1999*

The genus was coined to contain all plant pathogenic bacteria that were Gram-stain-negative, peritrichously flagellate and facultatively anaerobic rods. The genus was named in honour of Erwin F. Smith, a pioneer plant bacteriologist. The heterogenicity of this genus was generally acknowledged right from its inception. Many proposals were put forth to emend or revise this genus. Waldee (1945) suggested that the genus *Erwinia* should be emended to contain pathogens that cause only necrosis or wilt diseases (*Erw. amylovora, Erw. salicis* and *Erw. tracheiphila*), and utilise a restricted range of carbon compounds for growth;

* References marked with an * are not cited in the bibliography.

and that the biochemically more active soft-rotting pathogens (*Erw. carotovora* and *Erw. chrysanthemi*) should be placed in a new genus, *Pectobacterium*. Several workers supported Waldee's suggestion. However, in the eighth edition of *Bergey's Manual of Determinative Bacteriology*, three clusters of organisms, namely, *amylovora, herbicola* and *carotovora* groups, were recognised under *Erwinia*. According to Kim *et al*., (1999), the phylogenetic group of seven species of the then genus *Erwinia* cluster *amylovora* (cluster-I), based on the complete 16S rDNA sequences, represents the true authentic erwinias. Some of the species and pathovars of *Erwinia* have been transferred to other genera, namely, *Brenneria, Dickeya, Enterobacter, Pantoea* and *Pectobacterium*. Gardan *et al*., (2004) added one species, the causal agent of pink canker of papaya, to this genus.

The below-given description of the genus is based on the emended description of the genus published by Hauben *et al*., (1999). Cells are Gram-stain-negative, non-sporing, straight rods and measure 0.5–1.0 × 1.0–3.0 µm. Cells occur mostly singly or in pairs and are peritrichously flagellate. These are facultative anaerobes but anaerobic growth of some species is poor. Optimal temperature for growth is between 27 and 30 °C. Strains do not produce indole and do not oxidise gluconate. Catalase is positive but oxidase is negative. Strains utilise L-alanine, L-serine, glycylglycine and L-glutaminic acid but not kynureninic acid and trigonelline as nitrogen sources. Members are sensitive to chloramphenicol, furazolidone, oxytetracycline, tetracycline and nalidixinic acid. Strains do not possess arginine dihydrolase, caseinase, urease and phenylalanine deaminase.

The species of the genus comprise a distinct phylogenetic group, as determined by 16S rRNA gene sequence comparisons, and are characterised by 14 generic signature nucleotides. The major fatty acids include 12:0, 14:0 and 16:0. G + C content of the DNA ranges from 49.8 to 54.1 mol%. The members of the genus occur as saprophytes, epiphytes and plant pathogens. Plant pathogens cause plant diseases that include mainly wilts and blights. The entry of the pathogens into host is mainly through wounds and natural openings.

The phylogenetic analyses of the 16S rRNA gene, *gyrB* and *rpoD* gene sequences of a bacterium, isolated from black lesions on shoots of European pear trees (*Pyrus communis* L.) in an orchard in Japan, indicated that it could not be assigned to any recognised species of the genus *Erwinia*. DNA–DNA hybridisation confirmed that the bacterial strains represented a novel species. Hence, the name *Erwinia uzenensis* sp. nov. was proposed (Matsuura *et al*., 2012).

Phenotypic analyses of eight *Erwinia* strains, isolated from necrotic pear blossoms in València, Spain, clustered them into one phenon, distinct from other species of the genus *Erwinia*. Phylogenetic analysis of 16S rRNA, *gpd* and *recA* gene sequences showed that the eight novel strains could not be assigned to any recognised species. As these novel isolates represent a novel species of the genus *Erwinia*, the name *Erwinia piriflorinigrans* sp. nov. is proposed for these isolates (Lopez *et al*., 2011). Uribe-Lorio and Wang (2017) identified *Erw. billingiae* as a causal agent of bacterial canker on mango (*Mangifera indica*) trees in Costa Rica. Luo *et al*., (2018) reported for the first time *Erw. aphidicola* causing pepper fruit brown spot from China.

Type species: *Erwinia amylovora*

Tao *et al*., (2023) isolated two Gram-stain-negative, facultatively anaerobic, motile and rod-shaped strains from the feces of ruddy shelducks in the Tibet Plateau of China. Distinct phylogenetic, genomic, phenotypic, biochemical and chemotaxonomic characters identified these as novel species belonging to the genus *Erwinia*, for which the name *Erwinia sorbitola* sp. nov. was proposed. The type strain was J780T (= CGMCC 1.17334T = GDMCC 1.1666T = JCM 33839T). Its virulence tests, causing blight and rot on the leaves and pear fruits, confirmed it as a phytopathogen.

Nie *et al*., (2024) reported complete genome sequence of *Erw. pyrifoliae* strain YKB12327, isolated from an apple tree in Korea.

Plant pathogenic species: *Erw. amylovora, Erw. aphidicola, Erw. billingiae, Erw. cacticida, Erw. mallotivora, Erw. papayae, Erw. persicina, Erw. piriflorinigrans, Erw. psidii, Erw. pyrifoliae, Erw. rhapontici, Erw. sorbitola, Erw. tracheiphila* and *Erw. uzenensis*.

6.3.1.3.35 *Genus* Pantoea *Gavini* et al. *1989 emend. Mergaert* et al. *1993*

Gavini *et al*., (1989) proposed this genus to include some species isolated from diverse geographical and ecological sources. Kageyama *et al*., (1992) described a new species from citrus fruits and transferred it to this genus. Later, some species of the then genus *Erwinia*, based mainly on the differences in DNA–DNA hybridisation, were also transferred to this genus. Brady *et al*., (2009), on the basis of MLSA and amplified fragment length polymorphism analysis, identified the bacteria isolated from eucalyptus leaves and shoots, showing symptoms of blight and dieback, collected in Uganda, Uruguay and Argentina, and from maize displaying brown stalk rot symptoms in South Africa as *Pantoea vagans* sp. nov., *Pantoea eucalypti* sp. nov. and *Pantoea deleyi* sp. nov. The below-given description of the genus is as given by Gavini *et al*., (1989) and emended by Mergaert *et al*., (1993).

Cells are Gram-stain-negative, non-encapsulated, non-sporing, straight rods and measure 0.5–1.3 × 1.0–3.0 µm. Cells are peritrichously flagellate or non-motile. Cultures grow well on nutrient agar at 30 °C. Colonies on nutrient agar are smooth, translucent and more or less convex having entire margins. Colonies are yellow, pale beige to pale reddish-yellow, or non-pigmented. Facultatively anaerobic and oxidase test is negative. No lysine decarboxylase, tryptophan deaminase, or urease activity. Pectate is not degraded and hydrogen sulphide is not produced from thiosulphate. Glucose dehydrogenase and gluconate dehydrogenase are produced. The bacterium has been isolated from plant surfaces, seeds, fruits, soil and water, as well as from humans (wounds, blood, urine and internal organs) and animals in several parts of the world. G + C content of the DNA is in the range 49.7–60.6 mol% (T_m).

Type species: *Pantoea agglomerans*

Ma *et al.*, (2016) has reported *Pan. beijingensis* as an emerging pathogen of *Pleurotus eryngii*, causing soft rot disease in China and Korea.

Koirala *et al.*, (2021) characterised phenotypically and genotypically 17 *Pan. stewartii* subsp. *indologenes* strains from onion and other sources (pearl millet, foxtail millet, guar pulse, verbena and corn). Based on their host range evaluation, they proposed two pathovars: *Pan. stewartii* subsp. *indologenes* pv. *cepacicola* pv. nov. and *Pan. stewartii* subsp. *indologenes* pv. *setariae* pv. nov. *Pan. stewartii* subsp. *indologenes* pv. *cepacicola* causes symptoms on *Allium* spp. (leek, onion, chive and Japanese bunching onion) and on foxtail millet, pearl millet and oats. However, *Pan. stewartii* subsp. *indologenes* pv. *setariae* can only infect the members of *Poaceae* family (foxtail millet, pearl millet and oats). They also proposed that the type strain of *Pan. stewartii* subsp. *indologenes* (LMG 2632^T) should be designated as a pathotype strain of *Pan. stewartii* subsp. *indologenes* pv. *setariae* and recommended that the strain PNA 14-12 be designated as the pathotype strain of *Pan. stewartii* subsp. *indologenes* pv. *cepacicola*. The digital DNA–DNA hybridisation (dDDH), average nucleotide identity (ANI) and MLSA study showed that the two pathovars are genotypically closely related. Their study also showed that *Pan. stewartii* subsp. *indologenes* pathovars and *Pan. stewartii* subsp. *stewartii* share high genotypic relatedness and cannot be differentiated by dDDH and ANI values. Although the newly proposed pathovars are not clearly distinguishable by their fatty acid and methyl esterase profiles and substrate use patterns, a fatty acid (unknown with retention time: 10.9525) and a few metabolites (3-methyl glucose, Na butyrate and fusidic acid) can be potentially used to distinguish them.

Ilyas *et al.*, (2021) reported *Pan. endophytica* causing bacterial rot of tobacco from Liuyang, China. To their knowledge, this was the first report of *Pan. endophytica* as a plant pathogen, which was first isolated from tobacco planted in southern China.

Pan. anthophila CL1 causes soft rot disease in *Clausena lansium* (wampee) in China and has inhibitory activity against the bacterial wilt pathogen *Ralstonia solanacearum*. He *et al.*, (2023) reported the genome sequencing and analysis of *Pan. anthophila* CL1, representing the first complete genome resource of the species. The CL1 genome consists of four circular replicons (one chromosome and three plasmids), with a total size of 4,594,065 bp, and contains 4,109 protein-coding genes and 106 RNA genes. The authors' bioinformatic analysis of CL1 predicted 228 virulence factors, two Type VI Secretion Systems and six secondary metabolite biosynthesis gene clusters producing saccharides, siderophores and terpene. The complete genome sequence of *Pan. anthophila* CL1 provides a solid foundation for further investigation of its pathogenesis and antimicrobial activity, and also represents a valuable resource for the comparative genomics of *Pantoea*.

Ferreira *et al.*, (2025) reported *Pan. graminicola* sp. nov. causing foliar chlorosis and streaking of sweet corn plants from Florida, USA.

Plant pathogenic species: *Pan. agglomerans*, ***Pan. agglomerans***: pv. *betae*, pv. *gypsophilae*, pv. *millettiae*; *Pan. allii*, *Pan. ananas*, *Pan. ananatis*, *Pan. ananatis* pv. *uredovora*; *Pan. anthophila*, *Pan. beijingensis*, *Pan. cypripedii*, *Pan. deleyi*, *Pan. dispersa*, *Pan. endophytica Pan. eucalypti*, *Pan. graminicola*, *Pan. stewartii*, ***Pan. stewartii***: subsp. *indologenes*, subsp. *indologenes* pv. *cepacicola*, subsp. *indologenes* pv. *setariae*, subsp. *stewartii* and *Pan. vagans*

6.3.1.3.36 Genus Phaseolibacter *Halpern* et al. *2013*

Cells are Gram-stain-negative, straight rods, motile by means of one or two polar flagella. Fermentative, facultative anaerobes. Oxidase test negative but catalase positive. The following results are obtained for the type species from API 20E strips after 48 h of incubation at 28 °C. D-glucose, sucrose and melibiose are fermented, acetoin is produced but H_2S and indole are not produced. Gelatin and urea are not hydrolysed but nitrate is reduced to nitrogen. Major fatty acids are $C_{16:0}$, summed feature 2 (one or more of $C_{14:0}$ 3-OH, iso-$C_{16:1}$ I and unknown ECL 10.928) and summed feature 3 ($C_{16:1}$ v7c and/or iso-$C_{15:0}$ 2-OH). The genus-specific signature regions in the 16S rRNA gene are 59-AGGCGTGAGTGTTAATATCACTTGCG and 59-ACCGAGCAGAGATGCTTGGG (the positions of the 59 nucleotides in the 16S rRNA gene of *Escherichia coli* are 451 and 1004, respectively). The type strain of *Phaseolibacter flectens* is ATCC 12775^T 5CFBP 3281^T 5ICMP 745^T 5LMG 2187^T 5NCPPB 539^T.

Type species: *Phaseolibacter flectens*, formerly *Pseudomonas flectens*, originally identified from French bean by Johnson (1956).

Plant Pathogenic species: *Phas. flectens*

6.3.1.3.37 Genus Tatumella *Hollis* et al. *1981* *emend. Brady* et al. *2010*

Hollis *et al.*, (1982) named a new genus *Tatumella* in honour of Harvey Tatum, an American bacteriologist who made significant contributions to our understanding of the classification and identification of fermentative and non-fermentative bacteria of medical importance. They proposed the name *Tatumella ptyseos* gen. nov., sp. nov., for a group of organisms (previously called group EF-9) isolated from clinical sources in the United States, Canada and Puerto Rico. A total of 68% of these isolates were from sputum specimens.

Members are Gram-stain-negative, oxidase negative, fermentative rods and grow on MacConkey agar. The distinctive biochemical characteristics of 44 *Tat. ptyseos* isolates were as follows: acid, but no gas from d-glucose, sucrose and usually (71%) from d-xylose (62% delayed); no acid from lactose, maltose, or d-mannitol; negative tests for indole, urea, methyl red, gelatin, l-lysine decarboxylase and l-ornithine decarboxylase; l-arginine dihydrolase variable; phenylalanine deaminase positive; Voges–Proskauer test positive by the Coblentz method, but negative by the O'Meara method; non-motile at 36 °C, but 66% weakly motile (30% delayed) at 25 °C; Simmons citrate test positive at 25 °C (89%) but negative at 36 °C. DNA–DNA relatedness studies with 26 *Tat. ptyseos* strains showed that they were 80–100% related at 60 °C, indicating that they comprise a single species. DNA relatedness to other species within the *Enterobacteriaceae* was 7–38%. Therefore, it is evident that this species belongs to this family, is distinct from all described species and needs to be placed in a new genus.

Tat. ptyseos isolates studied were susceptible to antimicrobial agents, namely, ampicillin, amikacin, cephalothin, chloramphenicol, gentamicin, kanamycin, tetracycline and tobramycin tested by broth dilution. Three striking differences between *Tat. ptyseos* and other members of the *Enterobacteriaceae* were its large zones of inhibition around penicillin (mean diameter 24 mm), its tendency to die on some laboratory media (such as blood agar) within seven days and its lesser number of flagella, usually one. Strain H36 (=ATCC 33301, =CDC D6168, =CDC 9591–78) is the type strain of this new species.

The below-given emended description of genus *Tatumella* is based on the data provided by Hollis *et al.*, (1982), Kageyama *et al.*, (1992) and Brady *et al.*, (2010). Gram-stain-negative, non-encapsulated, non-sporing, small rods that are 0.6–1.2 × 0.9–3.0 µm in size. Cells are motile by means of polar, subpolar, or lateral flagella or non-motile at 36 °C. Facultatively anaerobic, fermentative, catalase positive (weak and slow) and oxidase negative. Glucose dehydrogenase, gluconate dehydrogenase and 2-ketogluconate dehydrogenase are produced. Nitrates are reduced to nitrites.

Tests for indole, urease, gelatin, H_2S (TSI (triple sugar iron) agar media), ornithine decarboxylase, lysine decarboxylase, tryptophan deaminase, potassium cyanide, lipase and DNase are negative, but positive for Voges-Proskauer (Coblentz), methyl red and citrate (Simmons) and variable for phenylalanine, arginine dihydrolase and ortho-nitrophenyl-β-galactose. The growth is inhibited by many antibiotics. The bacterium was isolated from human clinical samples, fruit and soil. G + C content of the DNA ranges from 49.8 to 53 mol%. Brady *et al.*, (2010) transferred *Pantoea citrea*, *Pan. punctata* and *Pan. terrea* to the genus *Tatumella* emend. as *Tat. citrea* comb. nov., *Tat. punctata* comb. nov. and *Tat. terrea* comb. nov., and described *Tat. morbirosei* sp. nov.

Type species: *Tatumella ptyseos*

Plant pathogenic species: *Tat. citrea*, *Tat. morbirosei*, *Tat. punctate* and *Tat. terrea.*

Family: *Pectobacteriaceae*

Genera: *Affinibrenerria*, *Brenerria*, *Dickeya*, *Lonsdalea*, *Musicola*, *Pectobacterium* and *Samsonia*

6.3.1.3.38 Genus Affinibrenneria

Genus *Affinibrenneria* was proposed by Bian *et al.*, (2021) on the basis of a bacterial strain isolated from the symptomatic (cankered) bark of *Salix matsudana* in China. 16S rRNA gene analysis revealed that the novel strain shares the highest sequence similarity with *Brenneria goodwinii* FRB141^T (95.5%). In phylogenetic trees, based on four housekeeping genes (*gyr*B, *rpo*B, *atp*D and *inf*B) and the 16S rRNA gene sequence, the novel strain formed a separate branch from the five genera (*Lonsdalea*, *Brenneria*, *Dickeya*, *Pectobacterium* and *Sodalis*), of the family *Pectobacteriaceae*, suggesting that the novel strain should belong to a novel species of a novel genus within the family *Pectobacteriaceae.* The result was also supported by phylogenomics, amino acid identity and average nucleotide identity. Cells are Gram-stain-negative, facultatively anaerobic and motile. The major fatty acids are $C_{14:0}$, $C_{16:0}$, $C_{17:0}$ cyclo and $C_{19:0}$ cyclo *ω8c*. It has a large genome (5.89 Mb) with 5,052 coding genes, including 181 virulence genes, indicating that the novel strain is a potential pathogen of plants and animals. Based on phenotypic and genotypic characteristics, the L3-3HA^T strain represents a novel species of a novel genus in the *Pectobacteriaceae* family, for which the name *Affinibrenneria salicis* gen nov. sp. nov. is proposed. The strain type is L3-3HA^T (=CFCC 15588^T=LMG 31209^T).

Type species: *Affinibrenneria salicis*

Plant pathogenic species: *Aff. Salicis*

6.3.1.3.39 Genus Brenneria *Hauben* et al. *1999*

Hauben *et al.*, (1999) created this genus by transferring the species of cluster III of the then genus *Erwinia*. These species were transferred on the basis of complete 16S rDNA sequence analysis. One newly described species, *Erw. Alni*, was also transferred. However, Samson *et al.*, (2005) transferred one species, namely, *Bre. paradisiaca*, to a newly created genus *Dickeya*.

The genus was named after Don J. Brenner, an American bacteriologist. The below-given description of the genus is based on the data of Hauben *et al.*, (1999) and of Wilson *et al.*, (1957), Dye (1968), Lelliott and Dickey (1984), Verdonck (1992), Rijckaert (1994) and Surico *et al.*, (1996). The cells are Gram-stain-negative, non-sporing rods having rounded ends and measure 0.5–1.0 × 1.3–3.0 µm. Cells occur singly, rarely in pairs and are motile by means of peritrichous flagella. Facultatively anaerobic, catalase positive and oxidase negative. Optimal temperature for growth is 27–30 °C. The bacterium does not produce arginine dicarboxylases, arginine dihydrolase, lysine dicarboxylases and ornithine dicarboxylases. Members do not hydrolyse starch. Acid is produced from D-glucose, D-mannose, D-galactose, fructose, sucrose and salicin but not from adonitol, inositol and dulcitol. Metabolism is of fermentative type.

The major fatty acids include 12:0, 14:0, unknown 14.503, 16:0 and 17:0 cyclopropane. The species of the genus comprise a distinct phylogenetic group, as determined by 16S rRNA gene sequence comparisons, and are characterised of having 12 signature nucleotides. G + C contents of the members range from 50.1–56.1 mol%. The species of the genus cause tree diseases like wilts, necrosis, rots, blights and cankers. The entry of the bacteria into the host tissue generally occurs through wounds and natural openings.

The emended description of the genus based on the data from Hauben and Swings (2005) and Brady *et al.*, (2010) is given below.

Cells measure 0.5–1.0 × 1.3–3.0 µm, have rounded ends, occur singly or rarely in pairs, Gram-stain-negative and motile by means of peritrichous flagella. Facultatively anaerobic, but anaerobic growth of some species is weak. Optimum temperature for growth is 27–30 °C and maximum is 40 °C. Oxidase is negative and catalase positive. The tests for arginine dihydrolase, lysine decarboxylase, ornithine decarboxylase, tryptophan deaminase and starch hydrolysis are negative. Do not utilise citrate and do not produce H_2S, urease, indole or gelatinase. Nitrates are not reduced to nitrites. Major fatty acids are C12:0, C14:0, C16:0, and C17:0. Members of the genus *Brenneria* form a phylogenetic clade as determined by both 16S rRNA gene sequence analysis and multi-locus sequence analysis (MLSA) based on four housekeeping genes.

Brady *et al.*, (2014) isolated ten strains of Gram-negative, facultatively anaerobic bacteria from symptomatic oak tissue in the UK and USA. Their partial *gyrB* sequencing placed these strains in the genus *Brenneria*, with *Bre. goodwinii* as the closest phylogenetic relative. They further investigated these strains using a polyphasic approach, including MLSA (based on partial *gyrB*, *rpoB*, *infB* and *atpD* gene sequences), 16S rRNA gene sequencing, DNA–DNA relatedness studies and both phenotypic and chemotaxonomic assays. The MLSA and 16S rRNA gene analyses separated these strains into two groups based on origin, suggesting that they belong to *Brenneria* as two novel species. However, the DNA–DNA relatedness values revealed a closer relationship between the groups and indicated that they should belong to the same species. As the two groups of strains from the UK and USA can be differentiated from each other phenotypically and by ERIC PCR fingerprints, it is proposed to classify them as novel subspecies of a novel *Brenneria* species. The name *Brenneria roseae* sp. nov. (FRB 222^T = LMG 27714^T = NCPPB 4581^T) is proposed, with *Brenneria roseae* subsp. *roseae* ssp. nov. (FRB 222^T = LMG 27714^T = NCPPB 4581^T) for the strains from the UK and *Brenneria roseae* subsp. *americana* ssp. nov. (FRB 223^T = LMG 27715^T = NCPPB 4582^T) for the strains from the USA.

On the basis of phylogenetic and phenotypic data, Denman *et al.*, (2012) concluded that a group of nine Gram-stain-negative, facultatively anaerobic bacterial strains isolated from native oak trees displaying symptoms of acute oak decline in the UK represent a novel species in the genus *Brenneria*, for which the name *Brenneria goodwinii* sp. nov. is proposed.

Li *et al.*, (2015) isolated five Gram-stain-negative, facultatively anaerobic, motile, bacterial strains from symptomatic bark tissue of *Populus* × *euramericana* canker and on the basis of analysis of 16S rRNA gene sequences and MLSA classified these strains as a novel species of the genus *Brenneria* with the proposed name *Brenneria populi* sp. nov.

Kile *et al.*, (2022) isolated several strains of a previously undescribed bacterial species from mature *Tilia* hybrid trees suffering from bleeding cankers in the UK. Partial sequencing of the *gyrB* gene revealed that the strains belong to the genus *Brenneria*. Phylogenies based on the 16S rRNA gene and MLSA of partial housekeeping gene sequences of *gyrB*, *rpoB*, *infB* and *atpD* revealed that the strains formed an independent cluster within the genus *Brenneria*. As the data suggested that the strains belong to a novel taxon that can be differentiated from the closest relatives, the authors proposed that the strains represent a novel species, *Brenneria tiliae* sp. nov. (type strain WC1b.1^T=LMG 32575^T=NCPPB 4697^T) in the genus *Brenneria*.

Gašić *et al.*, (2022) reported that, on the basis of overall polyphasic characterisation, the strains isolated from walnut with symptoms of deep bark canker represent a novel species of the genus *Brenneria* for which the name *Brenneria izbisi* sp. nov. is proposed. The type strain of *Bre. izbisi* is KBI 423^T (= CFBP 9035^T = LMG 32479^T). To facilitate rapid identification of the newly described species, a conventional PCR protocol and primers targeting the putative gene *hrpP* were developed.

Type species: *Brenneria salicis*

Plant pathogenic species: *Bre. alni*, *Bre. goodwinii*, *Bre. izbisi*, *Bre. nigrifluens*, *Bre. populi*, ***Bre. quercina***: pv. *lupnicola*, pv. *quercina*; ***Bre. roseae***: subsp. *americana*, subsp. *roseae*; *Bre. rubrifaciens*, *Bre. salicis* and *Bre. tiliae*.

6.3.1.3.40 Genus *Dickeya* Samson et al. 2005

The genus was created by Samson *et al.*, (2005). The bacterium causing hollow stalk and blight of *Chrysanthemum morifolium* was described and identified as *Erwinia chrysanthemi*. Subsequently, similar bacteria were isolated from several plants causing typical soft rot and wilt symptoms. After extensive biochemical testes, all these pathogens were finally grouped into a single species, *Erw. chrysanthemi*, and designated as *Erw. chrysanthemi* complex. However, plant bacteriologists, for their convenience, divided the members of this complex into six pathovars, namely pv. *chrysanthemi*, pv. *dianthicola*, pv. *dieffenbachiae*, pv. *parthenii*, pv. *zeae* and pv. *paradisiaca*. Later on, Hauben *et al.*, (1999) renamed *Erw. chrysanthemi* pv. *paradisiaca* as *Brenneria paradisiaca* and transferred the other members to *Pectobacterium chrysanthemi*. Although these pathogens were grouped together as pathovars of *Pec. chrysanthemi*, a lot of diversity existed among these pathogens. Samson *et al.*, (2005) resolved this problem by using DNA–DNA hybridisation and analysis of 16S rRNA gene sequences. They found that on the basis of analysis of 16S rRNA gene sequences, *Pec. chrysanthemi* strains and *Bre. paradisiaca* formed a clade distinct from the genera *Pectobactrium* and *Brenneria*. Therefore, they transferred all these pathogens to a newly created genus *Dickeya*. On the basis of DNA–DNA hybridisation, all the strains of *Pec. chrysanthemi* formed six genomic species and hence were divided into five species, with one species having two biovars. Parkinson *et al.*, (2009) developed a phylogenetic method, using sequences from *recA* locus, to differentiate species of genus *Dickeya*. The method provides simple means for identification to species and intraspecies level.

The genus was named in honour of an American plant pathologist, Robert S. Dickey, for his contributions to the research on the *Erw. chrysanthemi* complex. The below-given description of the genus is taken from Samson *et al.*, (2005), which is based on their data and those of Waldee (1945) and Hauben *et al.*, (1999).

The cells are Gram-stain-negative rods, measuring 0.5–1.0 × 1.0–3.0 μm having rounded ends. Cells occur mostly singly or in pairs but sometimes in chains, and are usually motile by means of peritrichous flagella. Facultatively aerobic/anaerobic bacteria that catabolise glucose by a fermentative pathway and reduce nitrates to nitrites. The bacterium produces pectolytic enzymes and indole, and grows at 36 °C. Catabolises (+)-L-arabinose, myo-inositol, (+)-D-malate, malonate, D-mannose, mucate, saccharate and meso-tartrate but does not catabolise (+)-D-trehalose, methyl α-glucoside, (+)-D-arabitol or sorbitol. Members of the genus form a distinct clade based on the sequence analyses of 16S rRNA gene. G + C contents of its members range from 56.4–59.5 mol%. The members cause vascular wilts or soft rots on a range of host plants.

Eight isolates, seven from potato (plants exhibiting blackleg and slow wilt symptoms) and one from hyacinth, of Gram-stain-negative, motile, rod-shaped bacteria were isolated. Their characterisation by whole-cell matrix-assisted laser desorption ionisation time-of-flight (MALDI-TOF) mass spectrometry, pulsed field gel electrophoresis after digestion of whole-genome DNA with rare-cutting restriction enzymes, average nucleotide identity analysis and DNA–DNA hybridisation studies showed that although related to *Dic. dadantii*, these isolates represent a novel species within the genus *Dickeya*, for which the name *Dickeya solani* sp. nov. is proposed (van der Wolf *et al.*, 2014).

Tian *et al.*, (2016) reported that three strains of Gram-stain-negative, pectinolytic bacteria, isolated from pear trees displaying symptoms of bleeding canker in China, had identical 16S rRNA gene sequences that shared 99% similarity to the type strain of *Dic. dadantii*. Phylogenetic analysis of these strains with isolates representing all species of the genus *Dickeya* and related *Pectobacterium* species supported their affiliation to *Dickeya*. Multilocus sequence typing employing concatenated sequences encoding *recA*, *fusA*, *gapA*, *purA*, *rplB*, *dnaX* and the intergenic spacer illustrated a phylogeny which placed these strains as a distinct clade, separate from all other species of the genus *Dickeya*. The name *Dickeya fangzhongdai* sp. nov. was proposed to accommodate the pear strains.

Zaczek-Moczydłowska *et al.*, (2019) isolated *Dic. aquatica* causing soft rot of vegetables from Northern Ireland.

Hugouvieux-Cotte-Pattat and Van Gijsegem (2021) proposed a new species, *Dickeya parazeae* sp. nov., to characterise a clade distinct from those containing the *Dic. zeae* type strain NCPPB2538^T. Strain S31^T (CFBP 8716^T=LMG 32070^T) isolated from water in France is proposed as the type strain of the new species. Phenotypic analysis of eight publically available strains revealed traits common to the five tested *Dic. oryzae* members but apparently not shared by the *Dic. oryzae* type strain. Genomic analyses indicated that a simple distinction among the species *Dic. zeae*, *Dic. parazeae* and *Dic. oryzae* can be obtained on the basis of the *recA* sequence. *Dic. oryzae* can be distinguished from the two other species by growth on l-tartaric acid. Based on the *recA* marker, several strains previously identified as *Dic. zeae* were re-assigned to the species *Dic. parazeae* or *Dic. oryzae*. This study also highlighted the broad host range diversity of these three species.

Dobhal *et al.*, (2025) studied the genomes of strains isolated from bacterial heart rot of pineapple (*Ananas comosus* var. *comosus*). Based on polyphasic approaches, including ANI, dDDH, biochemical, physiological and phylogenomic analyses, they proposed the name *Dickeya ananatis* sp. nov. for this taxon, represented by the type strain A5410^T (=ICMP 25020^T=LMG 33197^T).

Type species: *Dickeya chrysanthemi*

Plant pathogenic species: *Dic. ananatis*, *Dic. aquatica*, *Dic. chrysanthemi*, ***Dic. chrysanthemi***: pv. *chrysanthemi*, pv. *parthenii*; *Dic. dadantii*, ***Dic. dadantii***: subsp. *dadantii*, subsp. *dieffenbachiae*; *Dic. dianthicola*, *Dic. fangzhongdai*, *Dic. oryzae*, *Dic. parazeae*, *Dic. solani* and *Dic. zeae*.

6.3.1.3.41 *Genus* Lonsdalea *Brady* et al. *2012 gen. nov.*

The genus was named in honour of David Lonsdale for his contributions to British forest pathology. The below-given description is based on the data of Hildebrand and Schroth (1967) and Brady *et al.*, (2012). Cells are Gram-stain-negative, short rods (0.5–1.0 × 1.0–2.0 µm) occurring singly, in pairs, or in groups, and motile by peritrichous flagella. Facultatively anaerobic, oxidase negative but catalase positive. Optimum temperature for growth is 28–30 °C. The colonies on tryptone soya agar are white to cream, round, convex and smooth with entire margins. The tests for beta-galactosidase, arginine dihydrolase, lysine decarboxylase, ornithine decarboxylase and tryptophan deaminase activities are negative. Citrate is utilised, but no production of H_2S, urease, indole, or gelatinase. Nitrates are not reduced to nitrites. Major fatty acids are $C_{14:0}$, $C_{16:0}$, $C_{18:1}$v7*c*, $C_{17:0}$ cyclo and summed features 2 (iso-$C_{16:1}$ and/or $C_{14:0}$ 3-OH) and 3 (C_{16}:$_1$v7*c* and/or iso-$C_{15:0}$ 2-OH). The members of the genus form a phylogenetic clade as determined by both 16S rRNA gene sequence analysis and MLSA based on four housekeeping genes. Members of the genus *Lonsdalea* may cause diseases on oak trees, including drippy nut disease and bark canker, and may be associated with acute oak decline. DNA G + C contents range from 54.5 to 55.1 mol%.

Type species: *Lonsdalea quercina*

Brady *et al.*, (2012) carried out taxonomic studies of bacterial isolates from oak trees in Spain and Britain, showing symptoms of bark canker and acute oak decline, respectively, by a polyphasic approach. Both 16S rRNA gene sequencing and MLSA, based on partial sequences of *gyrB*, *rpoB*, *infB* and *atpD* genes, separated the isolates into two genetic groups according to their origin. *Bre. quercina*, the causal agent of drippy nut disease of oak, was their closest phylogenetic relative, which clustered distant to the other species of the genus *Brenneria*. MLSA data for species of the genera *Brenneria*, *Pectobacterium*, *Dickeya*, *Erwinia*, *Pantoea* and *Samsonia* confirmed the polyphyletic nature of the genus *Brenneria* and indicated synonymy of *Dic. dadantii* and *Dic. dieffenbachiae*. DNA–DNA hybridisation experiments confirmed this synonymy and also revealed DNA–DNA relatedness values of 58–73% between the new oak isolates and *Bre. quercina*. Phenotypic and or chemotaxonomic methods allowed *Bre. quercina* and the two genetic groups of new oak isolates to be discriminated from other recognised species of the genus *Brenneria* and from members of the closely related genera *Dickeya*, *Pectobacterium* and *Samsonia*. Based on the data obtained, authors made the following taxonomic proposals:

1. Reclassification of *Bre. quercina* as the type species of a novel genus, *Lonsdalea* gen. nov., as *Lonsdalea quercina* comb. nov.
2. Classification of the oak isolates as *Lonsdalea quercina* subsp. *iberica* subsp. nov. and *Lonsdalea quercina* subsp. *britannica* subsp. nov.; leading to the automatic creation of *Lonsdalea quercina* subsp. *quercina* subsp. nov.
3. Emendation of the description of the genus *Brenneria*.
4. Reclassification of *Dic. dieffenbachiae* as *Dickeya dadantii* subsp. *dieffenbachiae* comb. nov. with the automatic creation of *Dickeya dadantii* subsp. *dadantii* subsp. nov.

Toth *et al.*, (2013) reported that seven Gram-stain-negative bacterial strains isolated from oozing bark canker of poplar (*Populus* × *euramericana*) trees in Hungary showed high (>98.3 %) 16S rRNA gene sequence similarity to *Lon. quercina*; but they differed from this species in several phenotypic characteristics. MLSA based on three housekeeping genes (*gyrB*, *atpD* and *infB*) revealed, and DNA–DNA hybridisation analysis confirmed, that this group of bacterial strains formed a distinct lineage within the species *Lon. quercina*. A detailed study of phenotypic and physiological characteristics confirmed the separation of poplars' isolates from other subspecies of *Lon. quercina*. Therefore, a novel subspecies, *Lonsdalea quercina* subsp. *populi*, was proposed.

Li *et al.*, (2017) studied four subspecies of *Lon. quercina*, namely, *Lon. quercina* subsp. *quercina*, *Lon. quercina* subsp. *britannica*, *Lon. quercina* subsp. *iberica* and *Lon. quercina* subsp. *populi*, by genome sequence-derived average nucleotide identity, phylogenetic analysis based on 16S rRNA gene sequences, and MLSA and phenotypic characteristics, and elevated three subspecies to the species level as *Lon. britannica* sp. nov., *Lon. iberica* sp. nov. and *Lon. populi* sp. nov.

Plant pathogenic species: *Lon. britannica*, *Lon. iberica*, *Lon. populi*, *Lon. quercina* and *Lon. quercina* subsp. *quercina*.

6.3.1.3.42 *Genus* Musicola *gen. nov. Hugouvieux-Cotte-Pattat* et al. *2021*

Hugouvieux-Cotte-Pattat *et al.*, (2021) sequenced the atypical strain A3967, registered as CFBP 722, isolated from *Solanum lycopersicum* growing in southern France in 1965. The genome of this strain shared digital DNA–DNA hybridisation and average nucleotide identity (ANI) values of 68 and 96%, respectively, with the *Dic. paradisiaca* type strain CFBP 4178^T. However, ANI analysis showed that *Dic. paradisiaca* strains are significantly dissimilar from the other *Dickeya* species, such that less than one-third of their genomes align to any other *Dickeya* genome. On phenotypic, phylogenetic and genomic grounds, they proposed a reassignment of *Dic. paradisiaca* to the genus level, and proposed the name *Musicola* gen. nov., with *Musicola paradisiaca* as the type species and CFBP 4178^T (NCPPB 2511^T) as the type strain. Phenotypic analysis showed differences between strain A3967^T and CFBP 4178^T, such as for the assimilation of melibiose, raffinose and *myo*-inositol. These results support the description of two novel species, namely *Musicola paradisiaca* comb. nov. and *Musicola keenii* sp. nov., with CFBP 4178^T (NCPPB 2511^T=LMG 2542^T) and A3967^T (CFBP 8732^T=LMG 31880^T) as the type strains, respectively.

Musicola members are Gram-stain-negative, non-sporing, facultatively anaerobic and pectinolytic bacteria. Cells have average dimensions of 0.6 by 1.7 μm. Cells are motile with peritrichous flagella. After 48 h of incubation at 30 °C on LB medium (tryptone 5 g L^{-1}, yeast extract 3 g L^{-1}, NaCl 5 g L^{-1} and agar 15 g L^{-1}), these form pale cream-coloured colonies of 0.5–2.5 mm in diameter with translucent appearance. The optimum temperature for bacterial growth is 25–36 °C and they can grow up to 40–41 °C. The members produce extracellular pectinases, a cellulase but no proteases. Cells do not produce the blue pigment indigoidine. *Musicola paradisiaca* was previously described as *Dickeya paradisiaca* (Samson *et al.*, 2005).

Type species: *Musicola paradisiaca*

Plant pathogenic species: *Mus. paradisiaca* and *Mus. keenii.*

6.3.1.3.43 Genus *Pectobacterium* Waldee 1945 emend. Hauben et al. 1999

Waldee (1945) proposed the emendment of the genus *Erwinia* and the creation of a new genus *Pectobacterium* to include biochemically more active and soft rotting bacteria like *Erw. carotovora* and *Erw. chrysanthemi*. Graham (1964), while reviewing the taxonomy of soft rot coliform bacteria, supported Waldee's suggestion, followed by many other workers. However, the genus *Pectobacterium* was not recognised in the eighth edition of *Bergey's Manual of Determinative Bacteriology* but it appeared in the *Approved Lists of Bacterial Names* (Skerman *et al.*, 1980). In 1998, the species of genus *Erwinia* were further divided into different phylogenetic groups based on complete 16S rDNA sequences. Hauben *et al.*, (1999) united the species constituting the cluster II group and transferred them to this genus. Gardan *et al.*, (2003b) elevated three subspecies of *Pec. carotovorum* to the species level.

The below-given description of the genus is taken from the emended description of the genus given by Hauben *et al.*, (1999). Cells are Gram-stain-negative, straight rods with rounded ends and measure 0.5–1.0 × 1.0–3.0 μm. Cells occur singly or in pairs and are motile by peritrichous flagella. Optimum temperature for growth is between 27 and 30 °C. Facultative anaerobes. Metabolism fermentative. Catalase positive and oxidase negative. Esculin is hydrolysed but not starch. No decarboxylases are formed for arginine, lysine, or ornithine. Strains do not possess tryptophan deaminase or urease. Strains utilise fructose, D-galactose, D-glucose, ribose, sucrose, mannose, acetate, fumarate, succinate, salicin, arbutin, glycerol, D-mannitol and β-methylglucoside but not betaine, benzoate, gallate, oxalate, propionate, butanol, methanol, or sorbose as carbon- and energy-yielding sources. Utilise L-alanine, arginine, asparagine, citrulline, glucosamine, glutamine, glycine, histidine, leucine, L-serine, tyrosine, L-tryptophan, ammonium chloride, asparaginic acid, urea, L-methionine and phenylalanine but not betaine, choline, anthranilic acid, quinolinic acid, spermine, spermidine, trigonelline, trimethyl-ammonium, hydroxyproline and sarcosine as nitrogen sources.

Major fatty acids of the type strains include 12:0, 14:0, 15:0, 16:0, 17:1ω8C and 17:0. The species of the genus comprise a distinct phylogenetic group, as determined by 16S rRNA gene sequence comparisons and are characterised by 17 signature nucleotides. G + C content of the DNA ranges from 50.5 to 56.1 mol%. *Pectobacterium* species produce pectolytic enzymes and cause plant diseases, mainly the soft rots and also blights, cankers, necroses and wilts. The entry of the bacteria into the host plant is mainly through wounds and natural openings.

Several pectolytic bacterial strains, mainly isolated from monocotyledonous plants and previously identified as *Pec. carotovorum*, were thought to belong to a novel species after several taxonomic analyses, including DNA–DNA hybridisation. In 16S rRNA gene sequence analyses, these strains had a similarity of >97.9 % to the 16S rRNA gene sequence of strains representing six other pectobacterial species and subspecies. These strains also showed some unique chemotaxonomic features and quantitative differences in polar lipids, lipoquinones and fatty acids. As these strains represented a novel species, Nabhan *et al.*, (2013) classified these strains as *Pectobacterium aroidearum* sp. nov.

Khayi *et al.*, (2016) reported that, on the basis of MLSA, *Pec. wasabiae* strains RNS 08-42-1A, SCC3193, CFIA1002 and WPP163, collected from the potato plant environment, constituted a separate clade from the original Japanese horseradish pathogen, *Pec. wasabiae*. Besides, these strains were phenotypically distinct from *Pec. wasabiae* strains by producing acids from (+)-raffinose, α-D(+)-α-lactose, D(+)-galactose and (+)-melibiose, but not from methyl α-D-glycopyranoside, (+)-maltose, or malonic acid. Hence, the name *Pectobacterium parmentieri* sp. nov. was proposed for this taxon. Wu *et al.*, (2017) reported *Pec. carotovorum* subsp. *actinidiae* on kiwifruit in China's Sichuan province. Sarfraz *et al.*, (2019) identified *Pectobacterium punjabense* sp. nov. as the causal agent of blackleg of potato from Punjab province, Pakistan. Sarfraz *et al.*, (2019) reported *Pec. parmentieri* and *Pec. polaris*, causing potato blackleg disease in Punjab province of Pakistan. Qin *et al.*, (2024) reported *Pec. polaris* causing soft rot on broccoli for the first time from China.

Portier *et al.*, (2019) proposed the elevation of *Pec. carotovorum* subsp. *odoriferum* to species level as *Pectobacterium odoriferum* sp. nov. (type strain CFBP 1878^T=LMG 5863^T=NCPPB 3839^T=ICMP 11533^T), proposed *Pectobacterium actinidiae* (type strain KKH3=LMG 26003 ^{T}=KCTC 23131^T) and *Pectobacterium brasiliense* as new species (type strain CFBP 6617^T= LMG 21371^T=NCPPB 4609^T) and emended the description of *Pec. carotovorum* (type strain CFBP 2046^T=LMG 2404^T=NCPPB 312^T=ICMP 5702^T). They also proposed a novel species, *Pectobacterium versatile* sp. nov. (type strain CFBP6051^T= NCPPB 3387^T=ICMP 9168^T), which includes the strains previously described as '*Candidatus* Pectobacterium maceratum'.

Based on average nucleotide identity, phylogenetic data and genome-to-genome distance, strains NIBIO1006^T (causal agent of soft rot of potato tubers in Norway), NIBIO1392 (from latently infected potato tubers) and NCPPB3395 are suggested to represent a novel species of the genus *Pectobacterium*, for which the name *Pectobacterium polaris* sp. nov. is proposed (Dees *et al.*, 2017).

Seven Gram-stain-negative, rod-shaped, pectinolytic bacterial strains isolated from potato tubers cultivated in Peru at high altitude (2400–3800 m) were subjected to polyphasic analyses that revealed their distinctiveness from the other *Pectobacterium* species (Waleron *et al.*, 2018). Phylogenetic analyses based on five housekeeping genes (*gyrA*, *recA*, *recN*, *rpoA* and *rpoS*) clearly showed their distinctiveness, simultaneously indicating *Pec. atrosepticum*, *Pec. wasabiae*, *Pec. parmentieri* and *Pec. betavasculorum* as the closest relatives. *In silico* DNA–DNA hybridisation of strain IFB5232^T with other *Pectobacterium*-type strains revealed a significant drop in DDH value below 70%, which is a prerequisite to distinguish *Pec. peruviense.* The ANI values supported the proposition of delineation of the *Pec. peruviense.* Genetic REP-PCR fingerprint and detailed MALDI-TOF MS proteomic profile sealed the individuality of the studied strains. The results of analyses performed for seven Peruvian strains formed the basis for novel species distinction and reclassification of the strains IFB5227-5232 and IFB5636, previously classified as *Pec. carotovorum* subsp. *carotovorum* into a new species named *Pectobacterium peruviense* sp. nov. and the IFB5232 isolate as a type strain (=PCM2893^T = LMG30269^T = SCRI179^T).

Lin *et al.*, (2022) reported *Pec. colocasium* causing bacterial soft rot of taro (*Colocasium esculenta*) in Taiwan for the first time.

Huang *et al.*, (2023) reported a high-quality, completely annotated genome sequence of *Pec. colocasium* strain F1-1. The 4,816,345 bp genome, which was assembled with Illumina and Nanopore reads with 217× and 311× coverage, respectively, consists of one chromosome and no plasmid. This completely circularised genome will aid future studies in comparative genomics, evolution and pathogenicity of *Pec. colocasium.* This genome resource will also be helpful for developing strategies to control *Pec. colocasium* causing soft rot of taro (*Colocasium esculenta*).

Currently, there are 21 *Pectobacterium* species and studies on some of them suggested that different species could exhibit distinct host preferences and phenotypes (Tang *et al.*, 2023).

Wu *et al.*, (2023) reported, for the first time, *Pectobacterium parvum* causing disease on potato in inner Mongolia, China.

Hong *et al.*, (2023) identified a *Pectobacterium*-like strain isolated from a diseased cucumber stem tissue, collected on Jeju Island, South Korea, as *Pectobacterium jejuense.* The type strain is 13-115^T (=KCTC 92800^T=JCM 35940^T).

Sawada *et al.*, (2024) isolated a phytopathogenic bacterium from lesions on Japanese angelica trees affected by bacterial soft rot in Yamanashi Prefecture, Japan. On the basis of phenotypic, chemotaxonomic and genotypic data, they classified the bacterium as *Pectobacterium araliae* sp. nov., designating MAFF 302110^T (=ICMP 25161^T) as the type strain.

Type species: *Pectobacterium carotovorum*

Plant pathogenic species: *Pec. actinidiae*, *Pec. araliae*, *Pec. aroidearum*, *Pec. atrosepticum*, *Pec. betavasculorum*, *Pec. brasiliense*, *Pec. cacticida*, *Pec. carotovorum*, *Pec. colocasium Pec. jejuense*, *Pec. odoriferum*; *Pec. parmentieri*, *Pec. parvum*, *Pec. peruviense*, *Pec. polaris*, *Pec. punjabense*, *Pec. versatile* and *Pec. wasabiae*.

6.3.1.3.44 Genus Samsonia *Sutra* et al. *2001*

The genus was established in 2001 and named in honour of R. Samson, a French phytobacteriologist, who did pioneering work on pectolytic species of genus *Erwinia*. At present, the genus contains only one plant pathogenic species, *Sam. erythrinae*, which causes bark necrotic lesions on *Erythrina* species. The strains of this species isolated from *Erythrina* species formed a homogenous phenon among plant pathogenic species of *Enterobacteriaceae* on the basis of phenotypic characters, 16S rDNA sequence analyses and DNA–DNA hybridisation. The below given description of the genus is based on the data of Sutra *et al.*, (1999) and (2001).

Cells are Gram-stain-negative, non-sporing, catalase positive and oxidase negative rods having rounded ends. Motile by means of peritrichous flagella. The strains are β-galactosidase positive, produce acetoin but not hydrogen sulphide, and do not possess arginine dihydrolase, lysine decarboxylase, ornithine decarboxylase, urease or tryptophan deaminase. Glucose is fermented without gas production. Nitrate is reduced. Strains do not utilise citrate and malonate as carbon sources and do not produce indole from tryptophan. These hydrolyse esculin but not gelatin, casein, lecithin, or Tween 80.

Type species: *Samsonia erythrinae*

Plant pathogenic species: *Sam. erythrinae*

Order: "Alteromonadales"

Family: *Alteromonadaceae*

Genus: *Pseudoalteromonas*

6.3.1.3.45 Genus Pseudoalteromonas *Gauthier* et al. *1995 emend. Beurmann* et al. *2017*

The genus was created by Gauthier *et al.*, (1995). They determined the sequences of small-subunit rDNA of 17 strains of genera *Alteromonas, Shewanella, Vibrio* and *Pseudomonas*, and these sequences were analysed by phylogenetic methods. On the basis of data obtained, they emended the existing genus *Alteromonas* and created a new genus *Pseudoalteromonas* to accommodate 11 species that were previously included in *Alteromonas* and one species from genus *Pseudomonas.* Later on, an aerobic, Gram-stain-negative, marine plant pathogenic bacterium, the causal agent of red spot disease of 'makonbu' (*Laminaria japonica*), identified as *Pseudoalteromonas bacteriolytica* sp. nov., was transferred to this genus (Sawabe *et al.*, 1998). Beurmann *et al.*, (2017) gave an emended description of the genus *Pseudoalteromonas.*

The following description of the genus is based on the publication of Gauthier *et al.*, (1995). Cells are Gram-stain-negative, non-sporing, straight rods, which measure 0.2–1.5 × 1.8–3.0 µm and occur singly or in pairs. Motile by a single polar flagellum, either unsheathed or sheathed. Strains do not accumulate poly-β-hydroxybutyrate. Non-bioluminescent. Several species produce pigments. Chemoorganotrophs. Aerobic, possessing a strictly respiratory type of metabolism but not fermentative, with oxygen as the terminal electron acceptor. Oxidase test is positive but catalase is generally weak and irregular. None of the strains has a constitutive arginine dihydrolase system. One species, namely, *Psa. denitrificans*, is capable of denitrification. Mostly psychrotolerant, with growth occurring between 0 and 40 °C but all species grow at 20 °C. Many strains require amino acids for growth. All species require sea water base for growth. The major quinone is ubiquinone-8 (Q-8). All the above-mentioned 12 species, initially included in the genus, are positive for gelatinase, lipase, lecithinase and DNase activities and utilisation of D-glucose as a sole source of carbon; and negative for utilisation of D-ribose, L-rhamnose, salicin, erythritol, sorbitol, *meso*-inositol and adonitol. Mol% G + C of the DNA is 38.0–48.0 (T_m).

Type species: *Pseudoalteromonas haloplanktis*

Plant pathogenic species: *Psa. bacteriolytica.*

Order: "*Pseudomonadales*"

Family: *Moraxellaceae*

Genus: *Acinetobacter*

6.3.1.3.46 Genus Acinetobacter *Brisou and Prevot 1954*

The cells are rods (0.9–1.6 × 1.5–2.5 µm) becoming spherical in the stationary phase of growth. Cells are Gram-stain-negative, but occasionally difficult to destain. Generally, occur in pairs and in chains of variable length, and non-sporing. Colonies are generally non-pigmented and are mucoid when the cells are encapsulated. Swimming motility is not observed, but the cells display twitching motility presumably due to the presence of fimbriae. Cells grow aerobically, having a strictly respiratory type of metabolism with oxygen as the terminal electron acceptor. Most strains do not reduce nitrate to nitrite. Oxidase test is negative but catalase is positive. Most strains grow between 20 and 37 °C with optimum temperature of 33–35 °C. Growth is good in most complex media. Most strains can grow in defined media having a single carbon or energy source, such as acetate or lactate, utilising ammonium or nitrate salts, or one of several common amino acids, as a source of nitrogen. For majority of strains, glutamic acid or aspartic acid can serve as a single source of carbon, energy and nitrogen in a defined mineral medium. Growth factor requirements are displayed rarely. Most members are saprophytes, occurring naturally in soil, water, sewage and foods such as raw vegetables. Can also occur on the human skin and in the human respiratory tract. Can be the cause of nosocomial infections such as bacteremia, secondary meningitis, pneumonia and urinary tract infections in humans. Mol% G + C of the DNA is 38–47.

Type species: *Acinetobacter calcoaceticus*

Patro *et al.*, (2006) reported top-rot phase of red stripe of sugarcane from Srikakulam district of Andhra Pradesh, India and identified the bacterium associated with the disease as *Acin. baumannii.*

Li *et al.*, (2015), on the basis of phenotypic, genotypic and phylogenetic characteristics, identified five Gram-stain-negative, non-motile and rod-shaped bacterial strains isolated from cankers of *Populus* × *euramericana* collected from different locations in Puyang city, Henan province, China as *Acinetobacter populi* sp. nov. These strains could be distinguished from members of most species of the genus *Acinetobacter* by their inability to assimilate l-arginine and benzoate.

Li *et al.*, (2014), on the basis of phenotypic, genotypic and phylogenetic characteristics, identified two Gram-stain-negative, non-motile, rod-shaped bacterial strains, isolated from canker bark of *Populus* × *euramericana* collected from different locations in Puyang City, Henan province, China, as *Acinetobacter qingfengensis* sp. nov.

Wang *et al.*, (2025) published first report of bacterial leaf spot on sweet potatoes caused by *Acin. seifertii* from China.

Plant pathogenic species: *Acin. baumannii, Acin. populi* and *Acin. qingfengensis.*

Family: *Pseudomondaceae*

Genera: *Pseudomonas* and *Rhizobacter*

6.3.1.3.47 Genus Pseudomonas *Migula 1894*

The genus was established by Migula (1894). It contains different types of bacteria, i.e., epiphytic and saprophytic species, animal, human and plant pathogens. The genus is characterised by straight or slightly curved rods but not helical, measuring 0.5–1.0 × 1.5–5.0 µm. Cells are motile by one to several polar flagella. Xanthomonadins are not produced. No resting stages are known, Gram-stain-negative, aerobic, having a strictly respiratory type of metabolism with oxygen as the terminal electron acceptor. However, in some cases, nitrate can be used as an alternate electron acceptor, allowing anaerobic growth. Most species do not accumulate polyhydroxybutyrate granules. Catalase positive, oxidase positive or negative. Chemoorganotrophs. Most species do not require organic growth factors. Mol% G + C of the DNA is 58.0–69.0.

Yang *et al.*, (2013) has emended the description of the genus, as given below. The emended description is as given previously (Migula, 1894; Palleroni, 2005) with the modification that most species bear one or several polar flagella, but some species have peritrichously flagellated cells.

Type species: *Pseudomonas aeruginosa*

In the first edition of *Bergey's Manual of Systematic Bacteriology*, Palleroni (1984b) enlisted 92 species in this genus, of which 23 were plant pathogens. Taxonomy of this genus has been revised many times. Several phytopathogenic species of this genus have been transferred to other genera, namely, *Acidovorax* (Willems *et al.*, 1992), *Herbaspirillum* (Baldani *et al.*, 1996), *Burkholderia* and *Ralstonia* (Urakami *et al.*, 1994; Yabuuchi *et al.*, 1993, 1995) and *Xanthomonas*. Many phytopathogenic species of the genus have been designated as pathovars of *Psm. savastanoi* and *Psm. syringae*. New species, namely, *Psm. salomonii* and *Psm. palleroniana* (Gardan *et al.*, 2002), *Psm. syringae* pv. *broussonetiae* (Takahashi *et al.*, 1996) and *Psm. costantinii* (Munsch *et al.*, 2002), have been added to this genus.

Cunty *et al.*, (2015) performed a multilocus sequence analysis based on four housekeeping genes (*gapA, gltA, gyrB* and *rpoD*) on 72 strains, representatives of the French outbreak that occurred since the beginning of 2010, and found that all the strains fell into two phylogenetic groups; one clonal corresponding to biovar 3 and the other corresponding to biovar 4. Strains of biovar 4 were found substantially different from those of other biovars as these were less aggressive and caused only leaf spots, whereas *Psm. syringae* pv. *actinidiae* (*psa*) biovars 1, 2 and 3 caused also canker and shoot die-back. Based on these pathogenic differences, supported by phenotypic, genetic and phylogenetic differences, they proposed that *Psa* biovar 4 be renamed as *Psm. syringae* pv. *actinidifoliorum* pv. nov.

Kałużna *et al.*, (2017) reported that, based on an analysis of 16S rDNA sequences, cherry isolates shared the highest (over 99.9%) similarity with *Psm. ficuserectae* and *Psm. congelans*. Phylogenetic analysis using housekeeping genes *gyrB*, *rpoD* and *rpoB* revealed that these isolates formed a separate cluster and confirmed their closest relation to *Psm. syringae* and *Psm. congelans*. On the basis of this polyphasic taxonomic study, eight strains from cherry were considered to represent a novel species of the genus *Pseudomonas* for which the name *Psm. cerasi* sp. nov. (Non Griffin, 1911*) was proposed. Plasencia-Marquez *et al.*, (2017) reported *Psm. aeruginosa* causing leaf infection in soybean (*Glycine max*) from Cuba.

Timilsina *et al.*, (2018) isolated an unusual fluorescent pseudomonad from tomato showing leaf spot symptoms similar to bacterial speck. The bacterium elicited a hypersensitive reaction on tobacco and caused soft rot of potato slices. Based on several tests coupled with Biolog GEN III tests, fatty acid methyl ester profiles and phylogenetic analysis using 16S rRNA and multiple housekeeping gene sequences, the authors identified the bacterium as *Psm. floridensis* sp. nov. Zhang *et al.*, (2018) reported *Psm. qessardii* causing ginseng bacterial root rot in China. Harmon *et al.*, (2018), based on molecular tests, identified the causal agent of stem galls of *Loropetalum chinense* as *Psm. amygdali* pv. *loropetali* pv. nov.

Vinatzer *et al.*, (2017) proposed to use genome similarity-based codes, called life identification numbers (LINs), to describe and name bacterial taxa. Using 93 genomes of *Psm. syringae sensu lato*, LINs were compared with a *Psm. syringae* genome tree whereby the assigned LINs were found to be informative of a majority of phylogenetic relationships. LINs also reflected host range and outbreak association for strains of *Psm. syringae* pv. *actinidiae*, a pathovar for which many genome sequences are available. The authors concluded that LINs could provide the basis for a new taxonomic framework to address the shortcomings of the current pathovar system and to complement the current taxonomic system of bacteria in general.

In 2018, a severe panicle blight (disease incidence ~15%, cultivar Tianyouhuazhan) was observed in rice fields in Fuyang, Hangzhou, China. This is the first report of *Psm. oryzihabitans* causing rice panicle blight and grain discolouration in China (Hou *et al.*, 2020).

Li *et al.*, (2021) isolated two Gram-stain-negative, aerobic, motile bacterial strains from leaf spot of *Quercus mongolica*. The authors proposed that these strains be named as *Pseudomonas quercus* sp. nov. The type strain is hsmgli-8^T (=CFCC 15739^T=LMG 31544^T).

Sawada *et al.*, (2021) isolated five phytopathogenic bacterial strains from bud blight lesions of cyclamen (*Cyclamen persicum* Mill.) in Miyagi, Japan, and subjected them to polyphasic taxonomic characterisation. Based on the phenotypic, chemotaxonomic and genotypic data, the authors concluded that the strains represent a novel *Pseudomonas* species, and proposed the name *Pseudomonas cyclaminis* sp. nov. for it. The type strain is MAFF 301449^T (=ICMP 23720^T).

Sawada *et al.*, (2021) isolated six phytopathogenic bacterial strains from soft rot lesions of onion (*Allium cepa* L.) in Japan. They concluded that on the basis of phenotypic, chemotaxonomic and genotypic data, the strains represent a

* References marked with an * are not cited in the bibliography.

novel *Pseudomonas* species, and proposed the name *Pseudomonas allii* sp. nov., with MAFF 301514^T (=ICMP 23680^T) being the type strain. Dutta *et al.*, (2018) classified strains from onions, responsible for yellow bud disease of onions, as *Psm. coronafaciens*. The authors identified it as a new species in genus *Pseudomonas*, which is diverse from *Psm. syringae*.

Sawada *et al.*, (2020) isolated five Gram-stain-negative, aerobic, motile with one to three polar flagella, rod-shaped bacterial strains (MAFF 212408^T, MAFF 212409, MAFF 212410, MAFF 301498 and MAFF 730085) from diseased Welsh onion (*Allium fistulosum* L.) in Japan. Based on the genetic and phenotypic evidence, the authors named the bacterium as *Pseudomonas kitaguniensis* sp. nov., representing a novel species. The type strain is MAFF 212408^T (=ICMP 23530^T).

Khanal *et al.*, (2022) isolated a Gram-stain-negative, aerobic and non-sporing bacterial strain, designated 20TX0172^T, from a rotting onion bulb in Texas, US. On the basis of phenotypic and genotypic characteristics, strain 20TX0172^T represented a novel species of the genus *Pseudomonas*, for which they proposed the name *Pseudomonas uvaldensis* sp. nov. (type strain 20TX0172^T (=NCIMB 15426^T=CIP 112022^T)). Khanal *et al.* (2023a) reported *Psm. alliivorans* pathogenic on onion bulbs from Texas.

Anurat *et al.*, (2019) isolated an aerobic, Gram-stain-negative, rod-shaped bacterium from bark of mango tree. On the basis of phenotypic characteristics, chemotaxonomic data and phylogenetic analysis, the bacterium was considered to represent a novel species in the genus *Pseudomonas*. The authors proposed the name *Pseudomonas mangiferae* sp. nov. for this novel species. The type strain of the novel species is DMKU BBB3-04^T (=TBRC 7080^T=JCM 32061^T).

Queslati *et al.*, (2019) isolated *Pseudomonas* strains from different regions of Tunisia during 2016–17 from the fruits and leaves of *Citrus sinensis* cv. 'Valencia Late' and *C. limon* cv. 'Eureka' plants showing symptoms of blast and black pit disease. A phylogenetic analysis of the housekeeping gene *rpoD* was used for strain identification at the species level. The results demonstrated the affiliation of these strains with the genus *Pseudomonas* and revealed that the 11 strains represent two putative new species in two monophyletic branches. The authors concluded that the strains KC12^T, KC17, KC20, KC22, KC24A, KC25 and KC26 represent a novel species of *Pseudomonas*, for which the name *Pseudomonas kairouanensis* is proposed. The type strain is KC12^T (=CECT9766 and CFBP 8662). The strains E10B^T, E10AB, E10CB1 and Iy3BA represent another novel species of *Pseudomonas* for which the name *Pseudomonas nabeulensis* is proposed; the type strain is E10B^T (=CECT9765 and CFBP 8661).

Bueno-Gonzalez *et al.*, (2019) isolated 22 cream-coloured bacterial strains from oak trees affected by acute oak decline in Southern England. The cells were Gram-stain-negative, motile, slightly curved rods, aerobic, non-sporing and catalase and oxidase positive. On the basis of the polyphasic approach, the authors proposed two novel *Pseudomonas* species, *Pseudomonas daroniae* sp. nov. (type strain FRB 228^T=LMG 31088^T=NCPPB 4672^T) and *Pseudomonas dryadis* sp. nov. (type strain FRB 230^T=LMG 31087^T=NCPPB 4673^T).

Tohya *et al.*, (2020), on the basis of genome analysis-based reclassification, reclassified *Psm. fuscovaginae* as a later heterotypic synonym of *Psm. asplenii*. Zhang *et al.*, (2021) reported that *Psm. amygdali* pv. *lachrymans* caused root decay and damaged the cucumber plants. Following root infection, the bacteria were transmitted through stems to the leaves. Cucumber seedlings in a moist planting substrate showed lodging, leaf chlorosis and wilting, and rotted roots.

Lu *et al.*, (2022) isolated a rod-shaped, Gram-stain-negative and motile bacterium, having amphi-lophotrichous flagella, from black spots on lower leaves of flue-cured tobacco in Yunnan, PR China and identified it as *Pseudomonas lijiangensis* sp. nov. as it represented a novel species of the genus *Pseudomonas*. Based on their distinctive pathogenicity profile, Caballo-Ponce *et al.*, (2021) proposed a *de novo* pathovar for *Psm. savastanoi* isolates obtained from dipladenia (*Mandevilla* spp.) as *Pseudomonas savastanoi* pv. *mandevillae* pv. nov., for which strain Ph3 (CFBP 8832^{PT}) has been designated as the pathotype strain. It is closely related to *Psm. savastanoi* pv. *nerii*.

Sawada *et al.*, (2022) subjected phytopathogenic bacterial strains (MAFF 311094^T, MAFF 311095, MAFF 311096 and MAFF 311097), isolated from rot lesions of parsley (*Petroselinum crispum*) sampled in Miyagi, Japan, to polyphasic characterisation to determine their taxonomic position. Cells were Gram-stain-negative, aerobic, non-sporing, motile with one or two polar flagella and rod-shaped. The 16S rRNA gene sequences analyses revealed that the strains belong to the genus *Pseudomonas*. Phenotypic characteristics, pathogenicity toward parsley and cellular fatty acid composition could differentiate the strains from their closest relatives. The phenotypic, chemotaxonomic and genotypic data presented in this study revealed that the strains constitute a novel *Pseudomonas* species, for which they proposed the name *Pseudomonas petroselini* sp. nov., with MAFF 311094^T (=ICMP 24279^T) being the type strain.

Sawada *et al.*, (2021) isolated two phytopathogenic bacteria, MAFF 301380^T and MAFF301381, from rot lesions of lettuce (*Lactuca sativa* L. var. *capitata* L.) in Japan and proposed the name *Pseudomonas lactucae* sp. nov. for these pathogens, with the type strain MAFF 301380^T (=ICMP 23838^T). Sawada *et al.*, (2022) isolated three phytopathogenic bacterial strains (MAFF 301350^T, MAFF 302030^T and MAFF 302046) from lettuce (*Lactuca sativa* var. *capitata*) showing bacterial rot symptoms in Japan, and subjected them to polyphasic characterisation to determine their taxonomic affiliations. The phenotypic, chemotaxonomic and genotypic data obtained showed that the strains represented two novel species of the genus *Pseudomonas*, *Pseudomonas aegrilactucae* sp. nov. for MAFF301350^T and *Pseudomonas morbosilactucae* sp. nov. for MAFF302030^T and

MAFF302046. The respective type strains are MAFF301350^T (= ICMP 23989^T) and MAFF302030^T (= ICMP 24377^T).

Wang *et al.*, (2021) reported the occurrence of stem blight of *Pinellia ternata* caused by *Psm. extremorientalis* from China. To their knowledge, this was the first report of bacterial stem blight of *P. ternata* caused by *Psm. extremorientalis* in China. Stem blight caused by *Psm. extremorientalis* poses a significant threat to yield and marketability of *P. ternata*.

Chu *et al.*, (2023) reported that *Psm. bijieensis* causes soft rot of roots of yam bean. The authors also reported the high-quality and complete genome sequence for *Psm. bijieensis* strain SP1, created by hybrid assembly using both short reads and long reads. The genomic data will be a valuable resource for exploring the virulence mechanism and antifungal activity of this strain.

Li *et al.*, (2023) reported bacterial leaf spot of tobacco caused by *Psm. psychrotolerans* from China for the first time. Huang *et al.*, (2023) reported *Psm. mosselii* causing white blotch disease in *Pleurotus pulmonarius* from China for the first time. Head soft rot is the major disease of broccoli caused by *Psm. fluorescens* and *Pec. carotovorum* (Manzo-Valencia *et al.*, 2023). Rodrigues *et al.*, (2023) reported *Psm. coronafaciens* pv. *garcae* causing bacterial halo blight and *Psm. amygdali* pv. *tabaci* causing bacterial leaf spots of Arabica coffee.

Manna *et al.*, (2024) reported *Psm. amygdali* pv. *morsprunorum* as the causal agent of bacterial canker in sweet cherry orchards in Washington state. Marin *et al.*, (2024) reported *Psm. fragariae* sp. nov., a novel bacterial species causing leaf spots on strawberry (*Fragaria*×*ananassa*).

Fullem *et al.*, (2024) isolated two bacterial species from watermelon in Florida and identified these as *Pseudomonas boreofloridensis* sp. nov. and *Pseudomonas citrulli* sp. nov. based on the combination of genetic and phenotypic data.

Qiu *et al.*, (2024) identified a Gram-stain-negative, aerobic, rod-shaped and non-endospore-forming bacterium isolated from branches of wilted pepper plants (*Capsicum annuum*) as *Pseudomonas machongensis* sp. nov.

Chu *et al.*, (2025) reported bacterial gall disease of *Myrica rubra* caused by *Psm. amygdali* pv. *myricae* from Taiwan.

Plant pathogenic species: *Psm. aegrilactucae*, *Psm. aeruginosa*, *Psm. agarici*, *Psm. allii*, *Psm. alliivorans*, *Psm. amygdali*, ***Psm. amygdali***, pv. *lachrymans*, pv. *loropetali*, pv. myricae, pv. *morsprunorum*, pv. *tabaci*; *Psm. asplenii*, *Psm. avellanae*, *Psm. beteli*, *Psm. bijieensis*, *Psm. boreofloridensis*, *Psm. cannabina*, ***Psm. cannabina***: pv. *alisalensis*, pv. *cannabina*; *Psm. caricapapayae*, *Psm. cerasi*, *Psm. cichorii*, *Psm. cissicola*, *Psm. citrulli*, *Psm. coronafaciens*, *Psm. coronafaciens* pv. *garcae*, *Psm. corrugata*, *Psm. costantinii*, *Psm. cyclaminis*, *Psm. daroniae*, *Psm. dryadis*, *Psm. extremorientalis Psm. ficuserectae*, *Psm. flectens*, *Psm. floridensis*, *Psm. fluorescens*, *Psm. fragariae*, *Psm. hibiscicola*, *Psm. kairouanensis*, *Psm. kitaguniensis*, *Psm. lactucae*, *Psm. lijiangensis*, *Psm. machongensis*, *Psm. mangiferae*, *Psm. marginalis*, ***Psm. marginalis***: pv. *alfalfae*, pv. *marginalis*, pv. *pastinacae*; *Psm. mediterranea*, *Psm. meliae*, *Psm. morbosilactucae, Psm. mosselii*, *Psm. nabeulensis*, *Psm. oryzihabitans*, *Psm. palleroniana*, *Psm. petroselini*, *Psm. polaris*, *Psm. psychrotolerans*, *Psm. qessardii*, *Psm. quercus*, *Psm. rubrilineans*, *Psm. rubrisubalbicans*, *Psm. salomonii*, *Psm. savastanoi*, ***Psm. savastanoi***: pv. *fraxini*, pv. *glycinea*, pv. *mandevillae*, pv. *nerii*, pv. *phaseolicola*, pv. *retacarpa*, pv. *savastanoi*; *Psm. syringae*, ***Psm. syringae***: pv. *aceris*, pv. *actinidiae*, pv. *actinidifoliorum*, pv. *aesculi*, pv. *allii*, pv. *antirrhini*, pv. *apii*, pv. *aptata*, pv. *atrofaciens*, pv. *atropurpurea*, pv. *averrhoi*, pv. *avii*, pv. *berberidis*, pv. *broussonetiae*, pv. *castaneae*, pv. *cerasicola*, pv. *ciccaronei*, pv. *coriandricola*, pv. *coronafaciens*, pv. *coryli*, pv. *cunninghamiae*, pv. *daphniphylli*, pv. *delphinii*, pv. *dendropanacis*, pv. *eriobotryae*, pv. *garcae*, pv. *helianthi*, pv. *hibisci*, pv. *japonica*, pv. *lachrymans*, pv. *lapsa*, pv. *maculicola*, pv. *mellea*, pv. *mori*, pv. *morsprunorum*, pv. *myricae*, pv. *oryzae*, pv. *panici*, pv. *papulans*, pv. *passiflorae*, pv. *persicae*, pv. *philadelphi*, pv. *photiniae*, pv. *pisi*, pv. *porri*, pv. *primulae*, pv. *rhaphiolepidis*, pv. *ribicola*, pv. *sesami*, pv. *solidagae*, pv. *spinaceae*, pv. *striafaciens*, pv. *syringae*, pv. *tabaci*, pv. *tagetis*, pv. *theae*, pv. *tomato*, pv. *ulmi*, pv. *viburni*, pv. *zizaniae*; *Psm. tolaasii*, *Psm. tremae*, *Psm. uvaldensis* and *Psm. viridiflava*.

6.3.1.3.48 Genus *Rhizobacter* Goto and Kuwata 1988

The genus was created by Goto and Kuwata (1988) to classify a bacterium causing galls on carrots. The following description of the genus is based on the publication of Goto and Kuwata (1988).

The cells are Gram-stain-negative, non-sporing, encapsulated, straight to slightly curved rods, measuring 0.9–1.3 × 2.1–2.5 µm. Cells are motile by polar or lateral flagella or by both, but motile cells are rare in the populations. Poly-β-hydroxybutyrate granules are formed. Chemoorganotrophic. Aerobic, having respiratory type of metabolism of glucose. D-glucose is utilised as a sole source of carbon and energy. On agar plates, the colonies are white or yellowish-white, plicated, tough or viscid; if the colour is yellow, the pigment differs from xanthomonadins. In liquid media, floccular growth consisting of globular units occurs without finger-like projections. Oxidase and catalase tests are positive. Nitrates are reduced to nitrites. H_2S is produced from cysteine. Growth is inhibited by KCN. *o*-nitrophenyl-β-D-galactopyranoside test is positive. Arginine dihydrolase, nitrogenase, fluorescent pigment, indole and acetoin are not produced and the tests for methyl red and denitrification are also negative. Starch, dextrin and glycogen are hydrolysed. The bacterium is susceptible to 10 µg vibriostatic agent O/129 phosphate. Does not require NaCl or growth factors. Contains ubiquinone Q-8. Benzene derivatives are not utilised as carbon sources. G + C content of the DNA is 67–71 mol%. The type species is a plant pathogen that causes galls on carrot roots and is found in soil.

Type species: *Rhizobacter dauci*

Plant pathogenic species: *Rhib. dauci*

Order: "Xanthomonadales"

Family: "*Xanthomonadaceae*"

Genera: *Stenotrophomonas*, *Xanthomonas* and *Xylella*

6.3.1.3.49 *Genus:* Stenotrophomonas *Palleroni and Bradbury 1993 emend. Ouattara* et al. *2017*

Palleroni and Bradbury (1993) proposed a new genus *Stenotrophomonas* and gave its below-given description. Gram-stain-negative, nonsporulating cells, about 0.5 by 1.5 µm in size, motile with several polar flagella (Hugh and Ryschenkow, 1961). They may produce fimbriae (Fuerst and Hayward, 1969b). Poly-β-hydroxybutyrate granules are not accumulated, and exogenous polymer is not hydrolysed. The colonies are smooth, glistening, with entire margin, white, greyish, or pale yellow. Strains require methionine for growth, but this is not a universal property (Ikemoto *et al.*, 1980). Nitrate is reduced, but it is not used as a nitrogen source. Denitrification and egg yolk reaction are negative. Liquefaction of gelatin and lipolytic reaction (Tween 80 hydrolysis) are positive. Optimal growth temperature is 35 °C. The nutritional spectrum is restricted. Of 145 organic compounds tested as the sole sources of carbon and energy, only 23 were used for growth (Stanier *et al.*, 1966). These include the disaccharides maltose, lactose, trehalose, and cellobiose and the glucoside salicin. Growth on lactose is poor, probably because of the inability to use galactose. Few fatty acids and amino acids were used for growth. None of the polyalcohols, aromatic compounds, or amines were used (Stanier *et al.*, 1966). Freshly isolated strains were found to use hydrocarbons, but the property was lost after subculturing on common laboratory media (Iizuka and Komagata, 1964a, 1964b). Isolated from assorted natural materials, human infections and clinical materials. The members are the most frequently isolated pseudomonads, after *Psm. aeruginosa* in the clinical laboratory (Hugh and Gilardi, 1980). Characteristics useful for differentiating them from species of *Xanthomonas* are given by Palleroni and Bradbury (1993). The type strain is ATCC 13637 (Hugh 810-2, NCIB 9203, NCTC 10257, NRC 729).

The main reservoirs of *Stenotrophomonas* members are soil and plants. *Stenotrophomonas* species range from common soil organisms (*Stn. nitritireducens*) to opportunistic human pathogens (*Stn. maltophilia*). The most common species, *Stn. maltophilia*, is very versatile and can be beneficial for plant growth and health; and hence can be used in agriculture, biocontrol, bioremediation and phytoremediation strategies as well as the production of biomolecules of economic value. Some *Stn. maltophilia* strains are pathogenic to humans with multidrug resistant profile (Ryan *et al.*, 2009). *Stn. indologenes* can also cause or be part of polymicrobial infections in humans, especially small children (*Aykac et al.*, 2016). Some species are phytopathogenic and cause plant diseases. Members of the genus have an important ecological role in the nitrogen and sulphur cycles. *Stenotrophomonas* species, especially *Stn. maltophilia* and *Stn. rhizophila* are often found in association with plants, such as cucumber, oilseed rape, potato, strawberry, alfalfa, sunflower, maize, rice, wheat, various weeds, willow and poplar.

Stenotrophomonas spp. metabolise a large range of organic compounds present in the rhizosphere, including phenolic compounds found in plant root exudates. *Stenotrophomonas* spp. produce plant growth hormone indole-3-acetic acid. The members can also promote plant growth due to nitrogen fixation and the oxidation of elemental sulphur, which in turn provides sulphate for the plant growth. Most *Stn. maltophilia* isolates produce antifungal compounds, such as maltophilin and xanthobaccin or volatile organic compounds with antifungal activity. *Stn. maltophilia* strains have an extraordinarily high hydrolytic potential; they produce diverse proteases, chitinases, glucanases, DNases, RNases, lipases and laccases. *Stn. maltophilia* cells are equipped for iron uptake, as they produce the siderophore enterobactin and use many TonB-dependent receptors for the active transport of iron-siderophore complexes (Ryan *et al.*, 2009).

Full genome sequence of an environmental isolate, *Stn. maltophilia* R551-3, and a clinical isolate, *Stn. maltophilia* K279a, are available (Ryan *et al.*, 2009). Both strains contain genes that encode type I pili, which have been implicated in adhesion and the early stages of biofilm formation, and type IV pili, which have been implicated in adherence, auto-aggregation, twitching motility and biofilm formation. Conserved distribution of pili-coding gene clusters in sequenced genomes may indicate similarities in the plant and animal colonisation strategies.

Type species: *Stenotrophomonas maltophilia* (Hugh 1981* ex Hugh and Ryschenkow 1961) Palleroni and Bradbury 1993

Singh *et al.*, (2023) proposed a taxonomic amendment of *Psm hibiscicola* and *Psm. beteli*, reported as pathogens of *Hibiscus rosa-sinensis* and Betel vine (*Piper betle*) plants, respectively, to be reclassified as *Stn. hibiscicola* and *Stn. beteli*, respectively. A novel species of the genus, *Stn. cyclobalanopsidis*, has also been reported as a leaf spot pathogen

* References marked with an * are not cited in the bibliography.

of the oak tree genus *Cyclobalanopsis*. They also provided deep phylo-taxonogenomic evidence that *Stn. maltophilia* strain JZL8, reported as a plant pathogen, is a misclassified strain of *Stn. geniculata*, making it the fourth pathogenic species of this genus.

Ouattara *et al.*, (2017) transferred *Pseudomonas pictorum* to the genus *Stenotrophomonas* as *Stenotrophomonas pictorum* comb. nov. and emended the description of the genus to match better with the characteristics of the eleven novel species assigned to this genus.

Chuang *et al.*, (2023) proposed three new species: *Stenotrophomonas aracearum* sp. nov. and *Stn. oahuensis* sp. nov. for two anthurium strains; and *Xan. hawaiiensis* sp. nov. for the spathiphyllum and colocasia strains.

Qiu *et al.*, (2024) identified a Gram-stain-negative, aerobic, rod-shaped and non-endospore-forming bacterium isolated from branches of wilted pepper plants (*Capsicum annuum*) as *Stn. capsici* sp. nov.

Plant pathogenic species: *Stn. beteli*, *Stn. capsici*, *Stn. cyclobalanopsidis*, *Stn. hibiscicola* and *Stn. geniculata.*

6.3.1.3.50 Genus Xanthomonas *Dowson 1939 emend. van den Mooter and Swings 1990 emend. Vauterin* et al. *1995*

The genus was established by Dowson in 1939 to include plant pathogens which cause leaf spots, cankers, blights etc. on various crop plants. Cells are Gram-stain-negative straight rods, which measure 0.4–0.6 × 0.8–2.0 µm, occur mostly singly or in pairs, occasionally in short chains and rarely as filaments. Cells are motile by means of a single polar flagellum. Usually produce yellow colonies, which are smooth and butyrous, mucoid or viscid. The exopolysaccharide xanthan, which is responsible for the mucoid or viscous growth, is a characteristic of the genus. No growth occurs at pH 4.5 or 4.0, and at 37 °C. Bacterial growth is inhibited by 6% NaCl, 30% glucose, 0.01% methyl green, 0.01% lead acetate or 0.1% triphenyl tetrazolium chloride. Chemoorganotrophs and may require one or more growth factors. Glycine, L-asparagine and L-glutamine are not utilised as sole sources of both carbon and nitrogen. Obligately aerobic and metabolism is strictly respiratory type with oxygen as the terminal electron acceptor. Catalase is produced. Oxidase production is negative or weak but urease and indole or acetoin are not produced. Nitrate is not reduced and litmus milk is not acidified.

Small amounts of acids are produced from many carbohydrates but not from L-rhamnose, adonitol, D-sorbitol, *meso*-erythritol, *meso*-inositol, or sorbose. In Biolog GN microplate tests, D-fructose, α-D-glucose, D-mannose, methylpyruvate and α-ketoglutaric acid are oxidised, and these reactions are the characteristics of the genus. α-cyclodextrin, adonitol, D-arabitol, *meso*-erythritol, *meso*-inositol, xylitol, D-glucosaminic acid, γ-hydroxybutyric acid, itaconic acid, sebacic acid, L-ornithine, L-pyroglutamic acid, DL-carnitine, D-serine, γ-aminobutyric acid, phenylethylamine, putrescine, 2-aminoethanol, and 2, 3 butanediol are not oxidised. Among the nine fatty acids that predominate in whole-cell preparations, three acids, namely, 9-methyl decanoic acid, 3-hydroxy-9-methyl decanoic acid and 3-hydroxy-11-methyl dodecanoic acid, are highly characteristic of the genus and serve as useful criteria for differentiating *Xanthomonas* strains from other bacteria. The ubiquinone that is present has eight isoprene units. Spermidine is the major polyamine present in the *Xanthomonas* strains. Spermine is usually present in detectable amounts, while 2-hydroxyputrescine and 1, 3-diaminopropane are never present. According to Saddler and Bradbury (2015), species so far described are either plant pathogens or associated with plants.

Xanthomonas is a large genus of plant-associated and plant-pathogenic bacteria. Collectively, members cause diseases on over 392 plant species. Individually, they exhibit marked host- and tissue-specificity. The determinants of this specificity are unknown (Lu *et al.* 2008). The mol% G + C of the DNA is 63.3–69.7 (T_m).

Type species: *Xanthomonas campestris*

In the eighth edition of *Bergey's Manual of Determinative Bacteriology*, only five species, namely, *Xan. albilineans, Xan. ampelina, Xan. axonopodis, Xan. campestris* and *Xan. fragariae*, were included in this genus. The number of species increased with the passage of time but with the introduction of the pathovar system (Young *et al.*, 1978), many species were downgraded and designated as pathovars of *Xan. campestris* and *Xan. axonopodis.*

Vauterin *et al.*, (1995), on the basis of a DNA–DNA hybridisation study of 183 strains of the genus, divided it into 20 DNA homology groups and proposed a comprehensive reclassification of the genus. They found that a previously described species, *Xan. campestris* was heterogeneous; it comprised of 16 DNA homology groups and one of these groups exhibited a high level of DNA homology with *Xan. axonopodis.* They emended *Xan. campestris* to include only the pathovars infecting crucifers and transferred its 34 pathovars to *Xan. axonopodis.* They recognised a total of 16 species in the genus, leaving 66 former pathovars of *Xan. campestris* unassigned to any species of the genus. Other changes in the taxonomy of the genus include reclassification of *Psm. cissicola* as *Xan. cissicola*, transfer of the bacterium isolated from passion fruit plants (*Passiflora* spp.) and originally classified as *Xan. campestris* pv. *passiflorae* to *Xan. axonopodis* pv. *passiflorae* on the basis of DNA–DNA hybridisation data (Goncalves and Rosato, 2000) and inclusion of *Xan. cynarae*, the causal agent of bacterial bract spot of artichoke (*Cynara scolymus*), as a new species (Trebaol *et al.*, 2000). Roumagnac *et al.*, (2004), on the basis of DNA–DNA hybridisation, thermal stability of DNA reassociation and fluorescent amplified fragment length polymorphism analysis, identified the causal agent of bacterial blight of onion, Welsh onion (*Allium spp.*) and garlic as *Xan. axonopodis* pv. *allii*. The pathogens causing bacterial spot of tomato, pepper or both have been divided into four species, namely, *Xan. vesicatoria, Xan. euvesicatoria, Xan. perforans* and *Xan. gardneri* (Jones *et al.*, 2004). Parkinson *et al.*, (2007) showed that phylogenetic analysis of *Xanthomonas* species based on comparison of partial gyrase B gene sequences was a useful method to determine inter-species relatedness and for rapid and accurate species-level identification. Ah-You *et al.*, (2009) used amplified fragment length polymorphism, MLSA and DNA–DNA hybridisation for genotypic classification of *Xanthomonas* pathovars associated with the plant family *Anacardiaceae* and concluded that the causal agents of mango bacterial canker and cashew bacterial spot should be classified as pathovars of *Xan. citri*, namely, *Xan. citri* pv. *mangiferaeindicae* and *Xan. citri* pv. *anacardii*, respectively. J. P. Euzeby has listed 30 species under this genus (List of Prokaryotes with Standing in Nomenclatur: http://www.bacterio.cict.fr/). Another addition to the genus is the inclusion of *Xan. dyei* with its three pathovars, namely, *eucalypti*, *laureliae* and *dysoxyli*, classified on the basis of MLSA (Young *et al.*, 2010). The strains of this species have a unique fatty acid methyl ester profile.

On the basis of the MLSA and virulence phenotypes, Huang *et al.*, (2013) classified a xanthomonad causing a bacterial spot of rose in Florida and Texas as *Xan. alfalfae* subsp. *citrumelonis* pv. *rosa*. Vicente *et al.*, (2017) reported *Xan. nasturtii* sp. nov. causing a disease on leaves of watercress (*Nasturtium officinale*) from Florida, US.

Lopez *et al.*, (2018), on the basis of a polyphasic approach that included both phenotypic and genomic methods, identified three isolates obtained from symptomatic nectarine trees (*Prunus persica* var. *nectarina*) cultivated in Murcia, Spain as *Xan. prunicola*.

Martins *et al.*, (2020) described a novel species isolated from walnut (*Juglans regia*) which comprised non-pathogenic and pathogenic strains on walnut. The isolates, obtained from a single ornamental walnut tree showing disease symptoms, grew on yeast extract dextrose carbonate agar as mucoid yellow colonies, characteristic of *Xanthomonas* species. Pathogenicity assays showed that while strain CPBF 424^T caused disease in walnut, strain CPBF 367 was non-pathogenic on walnut leaves. Two strains showed over 98% similarity to each other. Genomic analysis showed that strain CPBF 424^T harbours a complete type III secretion system and several type III effector proteins, in contrast to strain CPBF 367, shown to be non-pathogenic in plant bioassays. Based on these data, the authors found that strains CPBF 367 and CPBF 424^T belong to a new species and hence named these strains as *Xanthomonas euroxanthea* sp. nov., with CPBF 424^T (=LMG 31037^T=CCOS 1891^T=NCPPB 4675^T) as the type strain.

Mafakheri *et al.*, (2022) isolated five *Xanthomonas* strains from *Agrobacterium*-induced crown gall tissues of amaranth (*Amaranthus* sp.) and weeping fig (*Ficus benjamina*) plants in Iran. Phenotypic characteristics (i.e. biochemical tests and pathogenicity features) and whole-genome sequence-based core-genome phylogeny followed by average nucleotide identity and digital DNA–DNA hybridisation calculations suggested that these gall-associated strains belong to two new species within the genus *Xanthomonas*. *Xanthomonas bonasiae* sp. nov. is proposed for the strains isolated from weeping fig with $FX4^T$ (=CFBP 8703^T=DSM 112530^T) as type strain. *Xanthomonas youngii* sp. nov. is proposed for the strains isolated from amaranth with $AmX2^T$ (=CFBP 8902^T=DSM 112529^T) as type strain.

Dia *et al.*, (2021) isolated a bacterium in 2011 and 2012 from nursery-grown *Hydrangea arborescens* cultivars in Flanders, Belgium. Analysis using MALDI-TOF MS identified the *Hydrangea* strains as belonging to the genus *Xanthomonas* but excluded them from the species *Xan. hortorum*. A phylogenetic tree based on gyrB confirmed its close relation to *Xan. hortorum*. Phylogenetic characterisation based on MLSA and phylogenomic characterisation revealed that the strains are close to, yet distinct from, *Xan. hortorum*. The genome sequences of the strains had average nucleotide identity values ranging from 94.35–95.19% and *in silico* DNA–DNA hybridisation values ranging from 55.70 to 59.40% in the genomes of the *Xan. hortorum* pathovars. A genomics-based loop-mediated isothermal amplification assay was developed which was specific to the *Hydrangea* strains for its early detection. A novel species, *Xanthomonas hydrangeae* sp. nov., is proposed with strain LMG 31884T (=CCOS 1956T) as the type strain.

Studholme *et al.*, (2020) transferred *Xan. campestris* pv. *arecae* and *Xan. campestris* pv. *musacearum* to *Xan. vasicola* as *Xan. vasicola* pv. *arecae* comb. nov. and *Xan. vasicola* pv. *musacearum* comb. nov. The authors also amended description of the bacterial species *Xan. vasicola* to include the causative agent of banana *Xanthomonas* wilt, as well as strains that cause disease on *Areca* palm, *Tripsacum* grass, sugarcane and maize. Genome-sequence data reveal that all these strains share more than 98% average nucleotide with each other and with the type strain.

Campigli and Rizzo (2023) reported for the first time *Xan. hydrangeae* causing leaf spot on oakleaf hydrangea (*Hydrangea quercifolia*) in Tuscany, Italy. Chuang *et al.*, (2023) proposed three new species: *Stenotrophomonas aracearum* sp. nov. and *Stn. oahuensis* sp. nov. for two anthurium strains; and *Xan. hawaiiensis* sp. nov. for the spathiphyllum and colocasia strains. Ferraz *et al.*, (2024) proposed the transfer of *Xan. axonopodis* pv. *eucalyptorum* to *Xan. citri* as *Xanthomonas citri* pv. *eucalyptorum*.

Choudhary *et al.*, (2024) sequenced the whole genomes of eight strains isolated from two *Eucalyptus* species between 2007 and 2015. Based on average nucleotide identities, phylogenetic relationships and pathogenicity, they designated these *Eucalyptus* strains as *Xanthomonas euvesicatoria* pv. *eucalypti*.

Tambong *et al.*, (2024) reported that based on phenotypic, biochemical and whole-genome sequence data, taxonogenomic analysis of the *Xanthomonas translucens* complex leads to the descriptions of *Xanthomonas cerealis* sp. nov. and *Xanthomonas graminis* sp. nov.

Zhao *et al.*, (2025) reported for the first time *Xan. hortorum* pv. *vitians* as the causal agent of bacterial leaf spot on lettuce from China.

Wang *et al.*, (2025) reviewed the occurrence, host–pathogen interactions and management of bacterial pustule disease of soybean caused by *Xan. citri* pv. *glycines*. Duin *et al.*, (2025) isolated strains of *Xanthomonas* from angular leaf spot lesions on crape myrtle in North Carolina from 2014 to 2023. They studied the full reference genomes and fulfilled the Koch's postulates. Genomes were obtained with hybrid whole-genome sequencing using Illumina and Nanopore and assembled to develop robust genomic resources for these disease-causing strains. The completed genomes support the inclusion of the strains in the *Xan. citri* species group; however, both phylogenetic analysis and the identification of a novel plant host suggest the creation of the new pathovar *Xanthomonas citri* pv. *lagerstroemium*.

Plant pathogenic species: *Xan. albilineans*, *Xan. alfalfae*, ***Xan. alfalfae***: subsp. *alfalfae*, subsp. *citrumelonis*, subsp. *citrumelonis* pv. *rosa*; *Xan. arboricola*, ***Xan. arboricola***: pv. *celebensis*, pv. *corylina*, pv. *fragariae*, pv. *juglandis*, pv. *populi*, pv. *pruni*, *pv. zantedeschiae*; *Xan. axonopodis*, ***Xan. axonopodis***: pv. *allii*, pv. *anacardii*, pv. *axonopodis*, pv. *bauhiniae*, pv. *begoniae*, pv. *betlicola*, pv. *biophyti*, pv. *cajani*, pv. *cassiae*, pv. *clitoriae*, pv. *coracanae*, pv. *cyamopsidis*, pv. *desmodii*, pv. *desmodiigangetici*, pv. *desmodiilaxiflori*, pv. *desmodiirotundifolii*, pv. *dieffenbachiae*, pv. *erythrinae*, pv. *fascicularis*, pv. *glycines*, pv. *khayae*, pv. *lespedezae*, pv. *maculifoliigardeniae*, pv. *mangiferaeindicae*, pv. *manihotis*, pv. *martyniicola*, pv. *melhusii*, pv. *nakataecorchori*, pv. *patelii*, pv. *pedalii*, pv. *phaseoli*, pv. *phyllanthi*, pv. *physalidicola*, pv. *poinsettiicola*, pv. *punicae*, pv. *rhynchosiae*, pv. *ricini*, pv. *rosa*, pv. *sesbaniae*, pv. *spondiae*, pv. *tamarindi*, pv. *vasculorum*, pv. *vignaeradiatae*, pv. *vignicola*, pv. *vitians*; *Xan. bonasiae*, *Xan. bromi*, *Xan. campestris*, ***Xan. campestris***: pv. *aberrans*, pv. *alangii**, pv. *amaranthicola**, pv. *amorphophalli**, pv. *aracearum**, pv. *argemones**, pv. *armoraciae*, pv. *arracaciae**, pv. *asclepiadis**, pv. *azadirachtae**, pv. *badrii**, pv. *barbareae*, pv. *betae**, pv. *bilvae**, pv. *blepharidis**, pv. *boerhaaviae**, pv. *brunneivaginae**, pv. *campestris*, pv. *cannabis**, pv. *cannae**, pv. *carissae**, pv. *centellae**, pv. *citrumelo*, pv. *clerodendri**, pv. *convolvuli**, pv. *coriandri**, pv. *daturae**, pv. *durantae**, pv. *esculenti**, pv. *euphorbiae**, pv. *fici**, pv. *guizotiae**, pv. *gummisudans**, pv. *heliotropii**, pv. *incanae*, pv. *ionidii**, pv. *lantanae**, pv. *lawsoniae**, pv. *leeana**, pv. *leersiae**, pv. *malloti**, pv. *merremiae**, pv. *mirabilis**, pv. *mori**, pv. *nigromaculans**, pv. *obscurae**, pv. *olitorii**, pv. *papavericola**, pv. *parthenii**, pv. *passiflorae*, pv. *paulliniae**, pv. *pennamericanum**, pv. *phormiicola*, pv. *physalidis**, pv. *plantaginis*, pv. *raphani*, pv. *sesami**, pv. *spermacoces**, pv. *syngonii**, pv. *tardicrescens**, pv. *thespesiae**, pv. *thirumalacharii**, pv. *tribuli**, pv. *trichodesmae**, pv. *uppalii**, pv. *vernoniae**, pv. *viegasii**, pv. *vitians*, pv. *viticola**, pv. *vitiscarnosae**, pv. *vitistrifoliae**, pv. *vitiswoodrowii**, pv. *zantedeschiae**, pv. *zingibericola**, pv. *zinniae**; *Xan. cassavae*, *Xan. cerealis*, ***Xan. citri***: pv. *anacardii*, pv. *eucalyptorum*, pv. *glycines*, pv. *lagerstroemium*, ***Xan. citri***: subsp. *citri*, subsp. *malvacearum*; *Xan. codiaei*, *Xan. cucurbitae*, *Xan. cynarae*, *Xan. dyei*, ***Xan. dyei***: pv. *dysoxyli*, pv. *eucalypti*, pv. *laureliae*; *Xan. euroxanthea*, *Xan. euvesicatoria*, *Xan. euvesicatoria* pv. *eucalypti*, *Xan. fragariae*, *Xan. fuscans*, ***Xan. fuscans***: subsp. *aurantifolii*, subsp. *fuscans*; *Xan. gardneri*, *Xan. graminis*, *Xan. hawaiiensis*, *Xan. hortorum*, ***Xan. hortorum***: pv. *carotae*, pv. *hederae*, pv. *pelargonii*, pv. *taraxaci*, pv. *vitians*; *Xan. hyacinthi*, *Xan. hydrangeae*, *Xan. maltophilia*, *Xan. melonis*, *Xan. nasturtii*, *Xan. oryzae*, ***Xan. oryzae***: pv. *oryzae*, pv. *oryzicola*; *Xan. perforans*, *Xan. pisi*, *Xan. populi*, *Xan. prunicola*, *Xan. sacchari*, *Xan. theicola*, *Xan. translucens*, ***Xan. translucens***: pv. *arrhenatheri*, pv. *cerealis*, pv. *graminis*, pv. *phlei*, pv. *phleipratensis*, pv. *pistaciae*, pv. *poae*, pv. *secalis*, pv. *translucens*, pv. *undulosa*; *Xan. vasicola*, ***Xan. vasicola***: pv. *arecae*, pv. *holcicola*, pv. *musacearum*, pv. *vasculorum*; *Xan. vesicatoria* and *Xan. youngii*.

*These pathovars were not examined by Vauterin *et al.*, (1995) nor reclassified by others since 1995 and therefore are *Xan. campestris sensu lato* and may not belong to the species as emended by Vauterin *et al.*, (1995) and Bull *et al.*, (2010).

6.3.1.3.51 Genus *Xylella* Wells et al. 1987

The genus was established by Wells *et al.*, (1987) to include xylem-limited procaryotes responsible for causing Pierce's disease of grapevines, phony peach disease, leaf scorches of oak, mulberry, sycamore, elm, almond and plum, and periwinkle wilt. The below-given description of the genus is as given by Wells *et al.*, (1987).

Cells are Gram-stain-negative, predominantly single straight rods measuring 0.25–0.35 × 0.9–3.5 μm. Under certain culturing conditions, long filamentous strands are formed. Non-motile and non-sporing. Colonies are creamy to white and are generally of two types: 1) convex to pulvinate, smooth, opalescent with entire margins; 2) umbonate, rough with finely undulated margins. The bacteria are strictly aerobic and the growth is inhibited by 2.5% carbon dioxide. Metabolism is respiratory in which the oxygen is used as terminal electron acceptor. Non-fermentative, non-halophilic and non-pigmented; oxidase test negative and catalase test positive. Optimal temperature for growth is 26–28 °C and optimum pH is 6.5–6.9. The bacteria are nutritionally fastidious, requiring a specialised medium such as buffered cysteine-yeast extract agar containing charcoal or glutamine-peptone medium containing serum albumin. G + C content of the DNA is 51.0–52.4 mol% (T_m) or 52.0–53.1 mol% (Bd). Parasitic members are found mainly in the xylem of plant tissue.

Type species: *Xylella fastidiosa*

Although Wells *et al.*, (1987) recognised only one species to include all the strains, there are definitely different pathotypes, most of which have not been characterised and compared with each other (Hopkins and Purcell, 2002). Schaad *et al.*, (2004) proposed three new subspecies, namely, *Xyl. fastidiosa* subsp. *piercei, Xyl. fastidiosa* subsp. *multiplex* and *Xyl. fastidiosa* subsp. *pauca*. *Xyl. fastidiosa* subsp. *pauca* was not validated due to the lack of deposition of a culture of the type strain. As the type strain designated for subsp. *piercei* is the type strain of the species, the name *Xyl. fastidiosa* subsp. *fastidiosa* takes priority over *Xyl. fastidiosa* subsp. *piercei*.

Su *et al.*, (2016) proposed a novel species, *Xylella taiwanensis* sp. nov. for a Gram-stain-negative fastidious bacterium causing pear leaf scorch in Taiwan.

Surano *et al.*, (2023) reported *Xyl. fastidiosa* subsp. *pauca* ST53 as a pathogenic bacterium causing the olive quick decline syndrome, one of the most severe plant diseases currently threatening the olive growing areas of the Mediterranean.

Plant pathogenic species: *Xyl. fastidiosa*, ***Xyl. fastidiosa***: subsp. *fastidiosa*, subsp. *multiplex*, subsp. *pauca* and *Xyl. taiwanensis*.

2. **Phylum: "*Actinobacteria*"**

   Class: *Actinobacteria*

   Order: *Corynebacteriales* ord. nov.

   Family: *Corynebacteriaceae*

   Genus: *Corynebacterium*

6.3.1.3.52 Genus Corynebacterium *Lehmann and Neumann 1896*

The genus was created by Lehmann and Neumann in 1896 with *Cor. diphtheriae* as the type species. It contained human, animal and plant pathogens. All the plant pathogenic species of this genus except *Cor. ilicis* have been transferred to different genera, namely, *Clavibacter, Curtobacterium, Rhodococcus* and *Rathayibacter.* The Judicial Commission of the International Committee on Systematics of Prokaryotes ruled (opinion 87) that the name *Cor. ilicis* Mandel *et al.* 1961 is represented by the type strain ICMP 2608 =ICPB CI144 and is reported to be a plant pathogenic species (*Int. J. Syst. Evol. Microbiol.* Vol. 58: pp. 1976–78).

Type species: *Corynebacterium diphtheriae*

Plant pathogenic species: *Cor. ilicis*.

Family: *Nocardiaceae*

Genera: *Nocardia* and *Rhodococcus*

6.3.1.3.53 Genus Nocardia *Trevisan 1889*

The genus was created by Trevisan (1889). In 1952, one phytopathogenic bacterium causing galls, abundant proliferation of buds and witches' broom effect on blue berry was included in this genus.

The genus was named in an honour to Edmond Nocard, a French veterinarian, who discovered the first member of the genus in 1889 as the cause of bovine infection. *Nocardia* are strict aerobes and catalase positive bacteria, which stain from Gram-stain-positive to Gram-stain-variable. Non-sporing, non-motile, chemoorgantrophs, having an oxidative type of metabolism. Rudimentary to extensively branched vegetative hyphae measuring 0.5–1.2 μm in diameter are produced. These hyphae often fragment into rod-shaped and coccoid forms. The conidia are occasionally produced in chains on the aerial hyphae. Colonies have a variable shape and may be pigmented or white. Colonies are also of variable appearance and most species appear to have aerial hyphae when viewed with a dissecting microscope. Some strains are partially acid-fast, i.e., these require less concentration of sulphuric or hydrochloric acid during the staining procedure due to the presence of mycolic acids of intermediate length in their cell walls. *Nocardia* species grow under a wide range of temperature, but grow slowly on non-selective culture media. Colonies become evident in 3–5 days on most commonly used media. However, sometimes a prolonged incubation period of 2–3 weeks may be required.

The members are distributed around the world in soils that are rich in organic matter and are also found in aquatic habitats. These are involved in the degradation of hydrocarbons and waxes and can contribute to the biodeterioration of rubber joints in water and sewage pipes. Although most members are free-living saprophytes, some species, such as *Noc. asteroides*, are opportunistic pathogens, which cause nocardiosis in humans and animals. Persons with a weak immune system, like children, the elderly and the immunocompromised, are the most at risk. The most commonly infected organs are lungs but the central nervous system and other organs may also be involved. Some *Nocardia* species also cause a variety of cutaneous infections like actinomycetoma. Virulence factors are the catalase and superoxide dismutase, which inactivate reactive oxygen species that would otherwise prove toxic to the bacteria, and a cord factor (trehalose 6-6′ dimycolate). The cord factor interferes with fagocytosis by macrophages by preventing the fusion of the phagosome with the lysosome. G + C content of the DNA varies from 64.0–72.0 mol%.

Type species: *Nocardia asteroides*

Plant pathogenic species: *Noc. vaccinii.*

6.3.1.3.54 Genus Rhodococcus *Zopf 1891*

The genus was described by Zopf (1891) to include a saprophytic Gram-stain-positive bacterium producing red pigment and forming irregular masses of cells. Later on, Goodfellow (1984) transferred *Cor. fascians*, the causal agent of fasciation of sweet pea, and leafy galls and shoot proliferations on many plants, to this genus.

The genus is characterised by Gram-stain-positive, nonsporing, non-motile and aerobic bacteria. Cells are nocardioform, meaning they may have mycelial growth with fragmentation into rod-shaped or coccoid forms. The cocci may germinate into short rods, form filaments with side projections, show elementary branching, or, in some cases, produce extensively branched substrate hyphae. The fragmentation of rods, filaments and hyphae produces next generation of short rods or cocci. In some strains, feeble aerial hyphae are produced. No conidia are produced. Colony appearance may be rough, smooth or mucoid and pigmented but rarely colourless. Catalase positive. Chemoorganotrophs, having an oxidative type of metabolism. A wide range of organic compounds is utilised as sole sources of carbon for growth and energy. The growth is generally good at 30 °C on laboratory media. Cell walls contain mycolic acids.

Genus *Rhodococcus* is genetically diverse and well adapted to the soil environment. As a group, the rhodococci have tolerance for starvation, which may aid in their persistence in soil. The other habitats are aquatic habitat and herbivore dung, and a few reside in eucaryotic cells. The members of the genus are important due to their ability to metabolise/degrade harmful environmental pollutants such as toluene, naphthalene, polychlorinated biphenyls, petroleum hydrocarbons, detergents, benzene and various herbicides/pesticides. Strains of *Rhodococcus* are also important due to their ability to catabolise a wide range of compounds and produce bioactive steroids, acrylamide and acrylic acid, and their involvement in fossil fuel biodesulfurisation. *Rho. qingshengii* is capable of degrading carbendazim, a fungicide widely used for plant disease control. The genome of *Rhodococcus* sp. strain RHA 1 was completely sequenced in October 2006 (McLeod *et al.*, 2006). The genome is 9.7 mega base pairs long and is one of the largest bacterial genomes completely sequenced to date. The genetic and catabolic diversity of the genus is not due only to the large bacterial chromosome, but also due to the presence of three large linear plasmids. *Rho. equi* contains a circular plasmid while *Rho. fascians* harbours a linear plasmid (about 200 kb). Bacterial isolates identified as *Rho. fascians* have been found in diverse habitats, including polar seas and ice sediments in Greenland. Two isolates have been reported to fix nitrogen. G + C content of the DNA ranges from 63.0–72.0 mol% (T_m).

Type species: *Rhodococcus rhodochrous*

Kolozsváriné Nagy (2024) published the first report of *Rho. fascians* causing leafy gall on *Iberis sempervirens* from Hungary.

Plant pathogenic species: *Rho. fascians.*
Order: *Micrococcales*
Family: *Micrococcaceae*
Genus: *Arthrobacter*

6.3.1.3.55 Genus Arthrobacter *Conn and Dimmick 1947*

The genus was created by Conn and Dimmick (1947). The bacterium causing bacterial blight of leaves and twigs of American hollyhock, and earlier classified as *Corynebacterium ilicis*, was transferred to this genus. However, according to Young *et al.*, (2004), *Art. ilicis* is a validly named species, but probably is not a plant pathogen. The Judicial Commission ruled (opinion 87) that the name *Art. ilicis* is represented by the type strain DSM 20138 =ATCC 14264 =NCPPB 1228 and is not a homotypic synonym of *Cor. ilicis* Mandel *et al.* 1961, and is reported not to be a plant pathogen (*Int. J. Syst. Evol. Microbiol.* Vol. 58: pp. 1976–78).

Arthrobacter (from the Greek, "**jointed small stick**") are coryneform bacteria having pleomorphism and showing Gram-stain variable reaction. In young cultures, slender rods are Gram-stain-negative and in about 30 h-old cultures, very short rods and coccoids are Gram-stain-positive. Coccoid cells measure ~ from 0.6–1.0 µm in diameter. Colonies on blood agar are small, measure 2 mm in diam and range in colour from yellow to white. The bacterium can be readily isolated on sheep blood agar. Moderate to good growth occurs on yeast extract peptone medium. Optimal temperature for growth is between 25 and 30 °C. These have a complex life cycle composed of two distinct stages when grown in complex media. It is also called a **rod-coccus growth cycle** and is a most distinctive feature of the genus. In the exponential phase, the bacteria are irregular, branched rods and may reproduce by a process called **snapping division**. As the bacteria enter the stationary phase, the cells change to coccoid form. After transfer to fresh complex medium, the coccoid cells swell and produce one or more overgrowths and again form actively growing irregular rods. When viewed under the microscope, some cells are seen forming V-shaped arrangements and also more complex angular arrangements.

Most species of arthrobacteria are obligate aerobes, but all exhibit a pure respiratory but never a fermentative metabolism. However, two species, namely, *Art. globiformis* and *Art. nicotianae*, exhibit anaerobic metabolism. Both the species utilise nitrate as an electron acceptor at the end of their respiratory chain, reducing it to ammonia via nitrite. Small amount or no acid is produced from glucose and other sugars in peptone media. Non-sporulating. Many species exhibit a weak motility that often goes unnoticed. Members are nutritionally versatile and use a variety of substrates, including nicotine, nucleic acids and various herbicides and pesticides, in their oxidative metabolism. Pyridone is used as a sole source of carbon. Chemoorganotrophs and catalase positive.

Although arthrobacters are often isolated from sewage, fish and plant surfaces, soil is their most important habitat, where they form a significant component of the microbial flora. These are well adapted to this habitat because the cocci are resistant to desiccation and starvation. Due to their diverse metabolic activity, they degrade a variety of chemicals, including agricultural pesticides. Hexavalent chromium is 100 times more toxic than trivalent chromium due to its oxidative state and is also much more soluble in water, thus allowing it to seep into ground water very easily. Arthrobacteria cannot only grow in its presence; it can also reduce it to trivalent chromium, a much less toxic form. *Arthrobacter* species in association with several strains of *Streptomyces* can completely degrade an organophosphate insecticide, diazinon. In combination, both can use this as the only source of carbon and energy but individually none can grow on this compound. Genome of *Arthrobacter* species FB 24 has been sequenced and it contains 5011598 base pairs. G + C content of the DNA ranges from ~59.0–70.0 mol% (T_m).

Busse (2016) reviewed the taxonomy of the genus *Arthrobacter* and emended the genus *Arthrobacter sensu lato*. The emphasis was also placed on intrageneric phylogeny and chemotaxonomic characteristics, concentrating on quinone systems, peptidoglycan compositions and polar lipid profiles.

Kämpfer *et al.*, (2020) emended the description of genus *Arthrobacter*, which is given below.

Emended description of the genus *Arthrobacter* Conn and Dimmick 1947, emend. Koch *et al.* 1995, emend. Busse 2016 *sensu lato.*

Cells exhibit a rod-coccus cycle but exclusively coccoid cells may also occur. The peptidoglycan type is A3α and comprises the variations Lys-Ala_{1-4}, Lys-Thr-Ala_{1-3}, Lys-Ala-Ser-Ala_3, Lys-Gly-Ala_3 and Lys-Ser-Thr-Ala. The peptidoglycan type A4α with the variations Lys-D-Asp and Lys-Ala-D-Glu may be present in some species. The quinone system usually contains MK-9(H_2) as a predominant component but almost equal amounts of MK-9(H_2) and MK-8(H_2), as major components, may also occur. The polar lipid profile contains predominantly diphosphatidylglycerol, phosphatidylglycerol, phosphatidylinositol and dimannosylglyceride. The fatty acid profile is dominated by anteiso- and iso-methyl-branched acids. The major fatty acid is usually anteiso-$C_{15:0}$. Fatty acids, iso-$C_{15:0}$, iso-$C_{16:0}$ and anteiso-$C_{17:0}$ often contribute significantly to the profile.

Type species: *Arthrobacter globiformis*.

Huang *et al.*, (2023) published the first report of *Arthrobacter woluwensis* causing brown blotch disease on white button mushroom (*Agaricus bisporus*).

Plant pathogenic species: *Art. ilicis* and *Art. woluwensis*

Family: *Intrasporangiaceae*

Genus: *Janibacter*

6.3.1.3.56 *Genus* Janibacter *Martin* et al. *1997*

Originally, the genus was created to accommodate a Gram-stain-positive actinomycete, isolated from environmentally polluted samples and whose cell walls contained meso-diaminopimelic acid. Yoon *et al.*, (2004) described a bacterium isolated from abnormally spoiled oriental melons and transferred it to this genus on the basis of 16S rRNA gene sequence analyses and genomic relatedness.

The below-given description of the genus is as proposed by Martin *et al.*, (1997). The name of the genus is derived from a Latin word, *Janus*, a God in Roman mythology, who is said to have had two faces. Cells are Gram-stain-positive, coccoid to rod-shaped and non-motile. Cells occur singly, in pairs or occasionally in irregular clumps. Non-sporing and not acid-fast. Aerobic, chemoorganotrophs. Catalase is produced while oxidase test is negative. Cell wall type is A1γ (directly cross-linked *meso*-A_2 pm). Mycolic acids are absent; the menaquinone is M2K-8(H_4). The complex fatty acid profile contains straight-chain saturated and monounsaturated fatty acids as the major components and the polar lipids are diphosphatidylglycerol, phosphatidylglycerol and phosphatidylinositol. The strains of the genus are inhabitants of sludge and sewage waste. G + C content of the DNA is 70.0 mol%.

Type species: *Janibacter limosus*

Plant pathogenic species: *Jani. melonis.*

Family: *Microbacteriaceae*

Genera: *Clavibacter*, *Curtobacterium*, *Leifsonia*, *Leucobacter* and *Rathayibacter*

6.3.1.3.57 *Genus* Clavibacter *Davis* et al. *1984*

The genus was created by Davis *et al.*, (1984) to accommodate some phytopathogenic coryneform bacteria whose cell walls contain diaminobutyric acid. Two newly described plant pathogens, *Cla. xyli* subsp. *xyli* and *Cla. xyli* subsp. *cynodontis*, causal agents of ratoon stunting disease of sugarcane and Bermuda grass stunting disease, respectively, were classified in this genus. The creation of this genus was considered necessary, because no other genus as circumscribed for coryneform bacteria was suitable for the accommodation of these two phytopathogens. Later on, both *Cla. xyli* subsp. *xyli* and *Cla. xyli* subsp. *cynodontis* were transferred to *Leifsonia*. Three other species of this genus, namely, *Cla. toxicus, Cla. rathayi* and *Cla. tritici*, have been transferred to genus *Rathayibacter.*

The below-given description of the genus is based on the publication of Davis *et al.*, (1984). Cells are Gram-stain-positive, not acid-fast, pleomorphic rods, which are often arranged at an angle to give V-shaped arrangements; but a marked rod-coccus cycle is not observed. Cells are non-motile and endospores are not formed. Obligately aerobic. Optimal temperature for growth is ca. 20–29 °C and maximum is ca. 29–35 °C. The bacterium has exact nutritional requirements and the growth occurs on suitable solid media. Slow and weak oxidative production of acids occurs from some carbohydrates. Some organic acids are utilised. Neither nitrate is reduced nor utilised. Oxidase, urease, lipase and tyrosinase are not generally produced.

Cell wall peptidoglycan contains 2, 4-diamino butyric acid. Rhamnose is the major component of the cell wall while arabinose is absent. Mycolic acids are also absent. Non-hydroxylated fatty acids consist predominantly of *anteiso*- and *iso*-methyl branched chains; small amounts of straight-chain saturated acids are found. Respiratory quinones are menaquinones. Polar lipids comprise diphosphatidylglycerol, phosphatidylglycerol and lipids. G + C content of the DNA is about 70 ± 5 mol%.

Gonzalez and Trapiello (2014) identified a yellow Gram-stain-positive bacterium, causal agent of a newly identified bacterial disease named as bacterial bean leaf yellowing, isolated from bean seeds (*Phaseolus vulgaris* L.), and classified it as *Cla. michiganensis* based on 16S rRNA gene sequencing. The results of molecular methods employed such as amplification of specific sequences by PCR, 16S amplified rDNA restriction analysis, restriction fragment length polymorphism and MLSA as well as the analysis of biochemical and phenotypic traits, including API 50CH and API ZYM results, showed that the bacterium did not represent any known subspecies of *Cla. michiganensis.* Therefore, the name *Clavibacter michiganensis* subsp. *phaseoli* subsp. nov. was proposed.

Based on multilocus sequence typing and other biochemical and physiological characteristics, including colony colour, utilisation of carbon sources, and enzyme activities, Oh *et al.*, (2016) observed that strain PF008T isolated from pepper was categorically different from eight known subspecies of *Cla. michiganensis* and proposed a novel subspecies, *Clavibacter michiganensis* subsp. *capsici*, for the bacterium causing bacterial canker of pepper.

Li *et al.*, (2018) reported that whole-genome analysis based on average nucleotide identity and digital DNA–DNA hybridisation, as well as MLSA of seven housekeeping genes, support raising each of the *Cla. michiganensis* subspecies to species status. On the basis of whole-genome and MLSA data, they proposed the establishment of two new species and three new combinations, namely, *Clavibacter capsici* sp. nov., comb. nov., *Clavibacter tessellarius* sp. nov., comb. nov., *Clavibacter insidiosus* comb. nov., *Clavibacter nebraskensis* comb. nov. and *Clavibacter sepedonicus* comb. nov.

Tian *et al.*, (2021) examined the biochemical, physiological, fatty acids and genetic characteristics of strains DM1^T and DM3 isolated from barley. On the basis of results from multilocus sequence typing and other biochemical and physiological features, including colony colour, carbon source utilisation and enzyme activities, DM1^T and DM3 are categorically differentiated from eight species and subspecies of the genus *Clavibacter.* The results of genomic analysis reveal that the DNA G + C contents of DM1^T and DM3 are 73.7 and 73.5%, respectively, and the average nucleotide identity (ANI) values between DM1^T and DM3 and other species and subspecies range from 90.4 to 92.0%. The ANI value between DM1^T and DM3 is 98.0%. These results indicate that DM1^T and DM3 are distinct from other known species and subspecies of the genus *Clavibacter.* Therefore, they proposed a novel species, *Clavibacter zhangzhiyongii*, with DM1^T (=CFCC 16553 ^{T}=LMG 31970^T) as the type strain.

To complete the valid publication of the new species names resulting from the reclassification of the genus *Clavibacter*, Li and De Boer (2019) have provided descriptions of *Cla. insidiosus* sp. nov. and *Cla. tessellarius* sp. nov.

Arizala *et al.*, (2022), based on complete genome *in silico* analyses, elevated *Clavibacter michiganensis* subsp. *californiensis* to species level as *Clavibacter californiensis* sp. nov., and merged and re-classified *Cla. michiganensis* subsp. *chilensis* and *Cla. michiganensis* subsp. *phaseoli* as *Clavibacter phaseoli* sp. nov.

Nasiri *et al.*, (2024) reported *Cla. tessellarius* causing bacterial mosaic of wheat from southern Iran.

Type species: *Clavibacter michiganensis*

Plant pathogenic species: *Cla. californiensis*, *Cla. capsici*, *Cla. insidiosus*, *Cla. michiganensis*, *Cla. michiganensis* subsp. *phaseoli*, *Cla. nebraskensis*, *Cal. phaesoli*, *Cla. sepedonicus*, *Cla. tessellarius* and *Cla. zhangzhiyongii.*

6.3.1.3.58 Genus *Curtobacterium* Yamada and Komagata 1972

The genus was proposed by Yamada and Komagata (1972). Collins and Jones (1983) proposed the transfer of *Cor. flaccumfaciens* to *Curtobacterium* as *Cur. flaccumfaciens* and transfer of *Cor. betae*, *Cor. oortii* and *Cor. poinsettiae* as pathovars of *Cur. flaccumfaciens.*

The characteristics of the genus as described by Yamada and Komagata (1972) are given below. The cells are weakly Gram-stain-positive and small short rods. Coccoid cells are found in old cultures and old cells frequently lose the character of Gram-stain-positive reaction. Generally motile by means of lateral flagella. Cells multiply by bending type of cell division. Pleomorphism is slightly recognised. Metachromatic granules are not recognised. Ornithine is a principal amino acid found in the cell walls. Acid production is slow and weak from several sugars. Chemoheterotroph. Various kinds of organic acids are assimilated in addition to acetate, lactate and pyruvate. Gelatin is hydrolysed slowly. G + C content of the DNA is 66.0–71.0 mol%. The bacterium is widely distributed in soil and plant material.

Type species: *Curtobacterium citreum*

Khanal *et al.*, (2023b) isolated a Gram-stain-positive, aerobic and non-spore-forming bacterial strain 20TX0166^T from a diseased onion bulb in Texas, USA. Upon testing its pathogenicity on onion bulbs, it produced pathogenic response. Using a polyphasic approach based on phenotypic and genotypic analyses, strain 20TX0166^T was classified as a novel species of the genus *Curtobacterium* and the name *Curtobacterium allii* sp. nov. was proposed. The type strain is 20TX0166^T (=LMG 32517^T=CIP112023^T=NCIMB 15427^T).

Osdaghi *et al.*, (2024) elevated the poinsettia pathogen *Cur. flaccumfaciens* pv. *poinsettiae* to the species level as *Cur. poinsettiae* sp. nov.

Plant pathogenic species: *Cur. allii*, *Cur. flaccumfaciens*, ***Cur. flaccumfaciens***: pv. *betae*, pv. *beticola*, pv. *flaccumfaciens*, pv. *ilicis* and pv. *oortii* and *Cur. poinsettiae*.

6.3.1.3.59 Genus Leifsonia *Evtushenko* et al. *2000 emend. Dastager* et al. *2009*

The genus was created by Evtushenko *et al.*, (2000) to accommodate two new species, namely, *Lef. poae* and *Lef. aquatica* found in *Poa annua* root galls induced by the nematode *Subanguina radicicola* and *Corynebacterium aquaticum*. *Cor. aquaticum* was reclassified as *Lef. aquatica*. *Cla. xyli* was reclassified with two subspecies, namely, *Lef. xyli* subsp. x*yli* and *Lef. xyli* subsp. c*ynodontis* in this genus. The members of the genus are characterised by coryneform morphology. The species clearly differ from each other phylogenetically and phenotypically. Species can be distinguished on the bases of their morphologies, the composition of cell wall sugars, the requirement of complex media for growth and many physiological characteristics, including the oxidase reaction. Currently, the genus contains only one phytopathogenic species having two subspecies.

The below-given description of the genus is based on the publication of Evtushenko *et al.*, (2000). The colonies are yellow or white, circular, somewhat convex, glistening, opaque and butyrous. The cells are Gram-stain-positive, non-sporing, irregular rods or filaments, which usually fragment into shorter rods or coccoid forms. Primary branching occurs in young cultures of some species. The cells are usually motile, not acid-fast, obligately aerobic and mesophilic. Catalase test is negative while oxidase test is variable among species. Cell wall peptidoglycan of B-type contains glycine, glutamic acid, diaminobutyric acid (DAB) and alanine in a molar ratio close to 1:1:2:1; both isomers, L-DAB and D-DAB, are usually present in almost equal proportions. The major menaquinone is MK-11; MK-10 is also present. Rhamnose is a major component of cell wall sugars with small amounts of galactose, glucose and mannose, while some species also contain fucose. Cell wall teichoic acids and mycolic acids are absent. The principal phospholipids are phosphatidylglycerol and diphosphatidylglycerol. Among fatty acids, *anteiso*-15:0, *anteiso*-17:0 and *iso*-16:0 predominate. The concentrations of polyamines are low while putrescine is the predominant compound. G + C content of the DNA ranges from 66.0–73.0 mol%. Some species can be distinguished by their tolerance to antibiotics. The genus forms a coherent phylogenetic cluster attached to the branch of *Agromyces* spp.

Type species: *Leifsonia aquatica*

Plant pathogenic species: *Lef. xyli*, ***Lef. xyli***: subsp. *cynodontis* and subsp. *xyli*.

6.3.1.3.60 Genus Leucobacter *Takeuchi* et al. *1996*

The below-given description of the genus is based primarily on the publication of Takeuchi *et al.*, (1996). The cells are Gram-stain-positive, non-motile rods and non-sporulating. The mycelium is not produced. Catalase is produced, but oxidase, arginine dihydrolase, hydrogen sulphide and urease are not produced. The cell wall peptidoglycan contains 2, 4-diaminobutyric acid as the characteristic diamino acid. The peptidoglycan acyl type is acetyl. The quinone system contains either menaquinone MK-10 and MK-11 or MK-11 only as major isoprenoid quinine. Mycolic acids are not present. The major cellular fatty acids are C15:0 anteiso, C16:0 iso and C17:0 anteiso. Diphosphatidylglycerol, phosphatidylglycerol and one unknown glycolipid are present. G + C content of the DNA is 62.8–69.5 mol%.

Type species: *Leucobacter komagatae*

Fang *et al.*, (2016) isolated a Gram-stain-positive, aerobic, nonmotile and rod-shaped bacterium from the symptomatic bark of a *Populus* × *euramericana* canker. On the basis of molecular data and physiological and biochemical characteristics, the isolated bacterium was considered to represent a novel species of the genus *Leucobacter* and they named it as *Leucobacter populi* sp. nov.

Plant pathogenic species: *Leb. populi*.

6.3.1.3.61 Genus Rathayibacter *Zgurskaya* et al. *1993*

The genus was created by Zgurskaya *et al.*, (1993) to accommodate three species of Gram-stain-positive, aerobic, coryneform bacteria previously classified in the genus *Clavibacter*, namely, *Cla. rathayi, Cla. tritici* and *Cla. iranicus* as well as six strains that were isolated from annual cereal grasses, responsible for ryegrass toxicity and are very similar to *Cla. toxicus*. Later on, Sasaki *et al.*, (1998) transferred *Cla. toxicus* to this genus.

The below-given description of the genus is based on the publication of Zgurskaya *et al.*, (1993). The genus was named after E. Rathay, an Australian plant pathologist who first isolated members of this genus. The cells are irregular, pleomorphic, short rods and measure 0.4–0.8 × 0.5–1.8 µm. V-forms and sometimes short cell chains are formed but a clear rod-coccus cycle is never observed. Gram-stain-positive, non-sporing and not acid-fast. Obligately aerobic. Chemoorganotrophs, having an oxidative type of catabolism. Nutritionally exacting. Optimal growth occurs between 24 and 28 °C; no growth occurs at 10 or 37 °C. Cell wall peptidoglycan contains glutamic acid, alanine, glycine and diaminobutyric acid (1:1:1:2). Cell wall sugars include glucose, mannose and rhamnose. Cell walls of some strains also contain xylose and galactose. Menaquinones are of the MK-10 type. Mycolic acids are absent. Branching saturated fatty acids of the *anteiso* type (*anteiso*-15:0 and *anteiso*-17:0) are predominant. Diagnostic phospholipids are represented by phosphatidylglycerol, diphosphatidylglycerol and unknown glycosyldiacylglycerols; several strains contain lyso-phosphatidylglycerol.

Phenylalanine and methionine are assimilated but most amino acids are not used as sole sources of nitrogen. Casein, guanine, starch, Tween 60, Tween 80 and urea are not hydrolysed or decomposed. Nitrate is not reduced to nitrite. Methyl red and Voges–Proskauer tests are variable. G + C content of the DNA ranges from 63.0–72.0 mol%. The bacterium is isolated mainly from cereal crops and annual cereal grasses.

In July 2013, inflorescences of bluebunch wheatgrass, *Pseudoroegneria spicata* subsp. *spicata* (Pursh) Á. Löve, growing along a roadside were observed exhibiting signs of bacterial head blight caused by *Rathayibacter agropyri* (non O'Gara, 1916) comb. nov., nom. rev. (Schroeder *et al.*, 2018). The primary symptoms associated with this disease are dry, resinous droplets (ooze) on the inflorescence. To authors' knowledge, this is the first report of *Rat. agropyri* on bluebunch wheatgrass, which is a widely occurring native perennial wheatgrass in the western United States.

Type species: *Rathayibacter rathayi*

Plant pathogenic species: *Rat. agropyri, Rat. iranicus, Rat. rathayi, Rat. toxicus* and *Rat. tritici.*

Order: *Streptomycetales* ord. nov.
Family: *Streptomycetaceae*
Genus: *Streptomyces*

6.3.1.3.62 Genus Streptomyces *Waksman and Henrici 1943*

The genus was described by Waksman and Henrici (1943). The members of the genus are characterised by the production of **coenocytic mycelium** and Gram-stain-positive reaction. The aerial mycelium produces chains of three to many spores, which may be round, oval or cylindrical in shape. More than 500 species of *Streptomyces* have been described but most of them are saprophytes. However, some species attack tuber and root crops such as potato, sweet potato, radish, turnip, carrot and beet, and cause scabbing. Till 2006, 24 phytopathogenic species of this genus have been validly published (Bouchek-Mechiche *et al.*, 2000, 2006; Goyer *et al.*, 1996; Miyajima *et al.*, 1998; Park *et al.*, 2003). A few species are also pathogens of animals, including humans. Guo *et al.*, (2008) reported the results of an MLSA scheme developed to address the phylogeny of *Str. griseus* 16S rRNA gene clade. Sequence fragments of six housekeeping genes, *atpD, gyrB, recA, rpoB, trpB* and 16S rRNA, were obtained for 53 reference strains representing 45 valid species and subspecies. The analysis of each individual locus confirmed the suitability of loci and the congruence of single-gene trees for concatenation. Concatenated trees of three, four, five and all six genes were constructed, and the stability of the topology and discriminatory power of each tree were analysed. It was concluded from the results that phylogenetic analysis based on multilocus sequences is more accurate and robust for species delineation within *Streptomyces.* A multilocus phylogeny of six genes proved to be optimal for elucidating the interspecies relationships within the *Str. griseus* 16S rRNA gene clade. Hence, this multilocus sequence analysis scheme provides a valuable tool that can be applied to other *Streptomyces* clades for refining the systematic framework of this genus.

The characteristic feature of the genus is the production of an extensively branched mycelium resembling fungi. The diameter of vegetative hyphae ranges from 0.5–2.0 µm. The chains of three to many spores, called **conidia**, are formed on mature aerial hyphae. In a few species, substrate mycelium produces conidia in short chains. The conidia are often pigmented having smooth, hairy, or spiny outer surface. Conidia are non-motile and responsible for the reproduction and dispersal of *Streptomyces* species. The colonies are discrete and may be leathery or butyrous. Initially, the colonies have a smooth surface but later on these appear powdery, granular, or velvet-like due to the formation of a weft of aerial mycelium. Produce a wide variety of pigments responsible for the colour of colonies. Gram-stain-positive, strict aerobes and not acid-fast. Chemoorganotrophs having an oxidative type of metabolism. Use many organic compounds as sole source of carbon for energy and growth. Catalase positive. Nitrates are generally reduced to nitrites and, gelatin, casein and starch are hydrolysed. Members of the genus are widely distributed and found predominantly in soil and decaying vegetation. The members play an important role in the degradation of organic matter, most commonly observed in compost piles. In fact, the odour of the moist earth is largely the result of production of volatile substances such as **geosmin** by *Streptomyces* spp. Members are a great source for the production of many antibacterial antibiotics. These include erythromycin, neomycin, streptomycin, vancomycin, rifamycin, lincomycin, tetracycline and chloramphenicol. The complete genome sequence of *Str. coelicolor* strain A 3 was published in 2002.

The genome sequences of *Str. scabiei*, the cause of potato common scab, and of *Str. avermetilis* have also been completed. In all these species, the chromosome is a linear structure, in contrast to most bacterial chromosomes, which are circular in shape. G + C content of the DNA rages from 69.0 to 78.0 mol% (T_m).

Labeda (2011) reported that phylogenetic analysis of single gene alignments and a concatenated four-gene alignment demonstrated that the phytopathogenic species are taxonomically distinct from each other in spite of high 16S rRNA gene sequence similarities and also provided a tool for the identification of unknown putative phytopathogenic *Streptomyces* strains at the species level.

Huguet-Tapia *et al.*, (2016) reported that genome sequencing, comparative genomics and phylogenetic analysis of genomes of *Streptomyces* species, including four plant pathogenic species, enabled to discriminate pathogenic strains from saprophytic *Streptomyces* strains. The pathogen-specific genome contained 4,662 orthologs. Phylogenetic reconstruction suggested that *Str. scabies* and *Str. ipomoeae* share an ancestor, but their biosynthetic clusters encoding the required virulence factor thaxtomin have diverged. In contrast, *Str. turgidiscabies* and *Str. acidiscabies*, two relatively unrelated pathogens, possess highly similar thaxtomin biosynthesis clusters, which suggests that the acquisition of these genes was through lateral gene transfer.

Nguyen *et al.*, (2022) described 14 strains of *Streptomyces*, isolated from scab lesions on potato, as members of a novel species based on genetic distance, morphological observation and biochemical analyses. They proposed the name *Streptomyces caniscabiei* sp. nov. for strain NE06-02D^T (=DSM111602^T=ATCC TSD-236^T) and other members of this novel species group.

Streptomyces strain NRRL B-2795^T (DSM 112329^T=NRRL B-2795^T) is described as the type strain of *Str. griseiscabiei* sp. nov. using whole-genome average nucleotide identity and MLSA in addition to plant pathogenicity and antibiotic resistance. A second *Streptomyces* strain, NRRL B-16521, previously named *Streptomyces scabiei*, and also previously suggested as a potential novel species, is assigned to *Str. acidiscabies* based on whole-genome average nucleotide identity. Morphological and biochemical characterisations also support this designation for NRRL B-16521. Both *Streptomyces* sp. strain NRRL B-2795^T and NRRL B-16521 cause common scab on multiple cultivars of potato (Nguyen *et al.*, 2022).

Type species: *Streptomyces albus*

Wu *et al.*, (2024) reported *Str. brasiliscabiei* causing common scab of potato from China.

Str. hayashii and *Str. hilarionis* has been associated with potato common scab in Brazil (Vitor *et al.*, 2023).

Plant pathogenic species: *Str. acidiscabies*, *Str. albidoflavus*, *Str. aureofaciens*, *Str. bottropensis*, *Str. brasiliscabiei*, *Str. candidus*, *Str. caniscabiei*, *Str. caviscabies*, *Str. collinus*, *Str. diastatochromogenes*, *Str. dmorensis*, *Str. europaeiscabiei*, *Str. galilaeus*, *Str. griseiscabiei*, *Str. hayashii*, *Str. hilarionis* *Str. intermedius*, *Str. ipomoeae*, *Str. luridiscabiei*, *Str. niveiscabiei*, *Str. puniciscabiei*, *Str. reticuliscabiei*, *Str. rhizophilus*, *Str. scabiei*, *Str. setonii*, *Str. steliiscabiei* and *Str. turgidiscabies*.

Phylum: "*Firmicutes*"

Class: "*Bacilli*"
Order: *Bacillales*
Family: *Bacillaceae*
Genus: *Bacillus*

6.3.1.3.63 Genus Bacillus *Cohn 1872*

It is a very old genus and was created by Cohn (1872). The important characteristics of the genus are given below.

Cells are Gram-stain-positive, endospore-forming rods that are usually motile and peritrichously flagellate. Presence of spores makes the cells (rods) to bulge at one end. The spores may be thin- or thick-walled. The bacterium is heterotrophic, aerobic or sometimes facultative and catalase positive. Ubiquitous in nature, the genus includes both free-living and pathogenic species. *Bac. subtilis* is one of the best-understood procaryotes, in terms of molecular biology and cell biology. The cell wall is composed of teichoic and teichuronic acids. *Bac. subtilis* and *Bac. megaterium* are strict aerobes and do not require growth factors. The former is a powerful producer of extracellular amylases and proteases and serves as a source of commercial production of these enzymes.

Many *Bacillus* species are of considerable importance. Some species produce antibiotics namely, bacitracin, gramicidin, polymyxin etc. *Bac. cereus* causes various types of food poisonings and can infect humans. *Bac. anthracis* is the causal agent of anthrax disease that afflicts both farm animals and humans. *Bac. polymyxa* fixes atmospheric nitrogen. *Bac. thuringiensis* and *Bac. sphaericus* have insecticidal values. Both the species form a solid protein crystal, the parasporal body, next to their spores during endospore formation. The parasporal body of *Bac. thuringiensis* contains proteinaceous toxins that can kill larvae of more than 100 lepidopterous insect species. Most of the *Bac. thuringiensis* toxin genes are carried on large plasmids. The genes of the bacterium, which encode toxin production (BT genes), have been incorporated into cotton, corn and brinjal plants. These genetically modified crops are called BT cotton, BT corn and BT brinjal, respectively and possess the insecticidal properties. The parasporal body of *Bac. sphaericus* contains proteins toxic to mosquito larvae and may prove useful in controlling mosquitoes that carry the malaria parasite, *Plasmodium*. The 4.2 million base pairs genome of *Bac. subtilis* has been sequenced. The genome contains genes for the catabolism of many diverse carbon sources, and for protein secretion and antibiotic synthesis. G + C content of the DNA ranges from 32.0–69.0 mol% (T_m).

Gupta *et al.*, (2020) have given an emended description of the genus *Bacillus* to restrict it to only the members of the Subtilis and Cereus clades. The authors also reported 31 conserved signature indels that are unique characteristics of either the members of the Subtilis clade (containing the type species *Bac. subtilis*) or the Cereus clade (containing *Bac. anthracis* and *Bac. cereus*). As most *Bacillus* species, which are not part of these two clades, can now be assigned to other genera.

Type species: *Bacillus subtilis*

Generally, the members of this genus are not active plant pathogens. However, some of its members are found associated with plant material and cause its breakdown, particularly at or after the crop harvest. This breakdown also occurs at relatively higher temperatures in comparison to the temperature required by other plant pathogenic bacteria for their pathogenic action. In 1982, a pathogen of cereals, namely, *Bac. megaterium* pv. *cerealis*, was transferred to this genus. Peng *et al.*, (2013) identified a Gram-stain-positive, rod-shaped, spore-forming bacterium, causing ginger rhizome rot in Shandong province, China, as *Bac. pumilus*. Elbanna *et al.*, (2014) reported for the first time, *Bac. altitudinis* causing soft rot of apple and pear fruits.

Jin *et al.*, (2022) collected five isolates of bacteria causing bacterial leaf spot of peach. Phylogenetic analysis of the partial 16S rRNA genes of these isolates showed that these belong to *Bacillus cereus* species complex. Three of these isolates were identified as *Bac. cereus*, with the remaining isolates being *Bac. thuringiensis* and *Bac. pacificus*, based on multilocus sequence analysis (16S rDNA, *adk*, *ccpA*, *ftsA*, *glpT*, *pyrE*, *recF* and *sucC*) and biochemical characteristics. All the five isolates were pathogenic on *Prunus persica* as well as on *Cerasus pseudocerasus*, but none of these was pathogenic on *C. humilis*, plum (*Prunus salicina*) or *Armeniaca vulgaris*. To their knowledge, this is the first report on the occurrence of *Bac. cereus* species complex as causal agents of bacterial leaf spot of peach, which will benefit the development of future control measures.

Kiyan *et al.*, (2024) reported that *Bacillus subtilis*, *Bac. megaterium*, *Bac. flexus* and *Paenibacillus peoriae* could cause leaf mottle disease in common sunflower (*Helianthus annuus* L.).

Plant pathogenic species: *Bac. altitudinis*, *Bac. cereus*, *Bac. flexus*, ***Bac. megaterium***, *Bac. megaterium* pv. *cerealis*, *Bac. pacificus*, *Bac. pumilus*, *Bac. subtilis* and *Bac. thuringiensis*.

Family: *Paenibacillaceae*

Genus: *Paenibacillus*

6.3.1.3.64 *Genus* Paenibacillus *Ash* et al. *1994 emend. Shida* et al. *1997 emend. Behrendt* et al. *2010*

Ash *et al.*, (1994) proposed that members of "group 3" within the genus *Bacillus* should be transferred to the genus *Paenibacillus*, for which they proposed *Paenibacillus polymyxa* as the type species. Latin *paene* means almost, so the paenibacilli are literally "**almost bacilli**".

Cells are rod-shaped and motile by means of peritrichous flagella. Gram-stain reaction is positive, but usually stain variable or negative in the laboratory. Oval endospores are formed that distend the sporangium. Strictly aerobic or facultatively anaerobic. Most species are catalase positive. Colonies are generally smooth and translucent, light brown, white, or sometimes light pink in colour. Optimal temperature for growth is 28–40 °C. Optimal pH is 7.0, although strains of some species are alkaliphilic. Growth is inhibited by 10% sodium chloride. Major fatty acid is C15:0 anteiso. Additional important fatty acid fractions often contain C16:0, C15:0 iso and C17:0 anteiso. DNA G + C content (mol%) ranges from 40–59.

The bacteria belonging to this genus have been detected in various environments, such as water, soil, rhizosphere, vegetable matter, forage and insect larvae, as well as clinical samples. The genus includes *Paen. larvae*, which causes American foulbrood in honeybees, and *Paen. polymyxa*, which fixes nitrogen in soil, and is thus used in agriculture and horticulture. Various *Paenibacillus* species also produce antimicrobial substances that affect a wide spectrum of microorganisms such as fungi, soil bacteria, plant pathogenic bacteria and even important anaerobic pathogens such as *Clostridium botulinum*. *Paen. vortex* and *Paen. dendritiformis* form complex colonies with intricate architectures.

Type species: *Paenibacillus polymyxa*.

Zhang *et al.*, (2017) reported *Paen. polymyxa* as the causal agent of bacterial stem rot of *Hylocereus undulates* from China.

Plant pathogenic species: *Paen. peoriae* and *Paen. polymyxa*.

Family: Staphylococcaceae

Genus: *Staphylococcus*

6.3.1.3.65 *Genus* Staphylococcus *Rosenbach 1884*

The genus name *Staphylococcus* was coined in 1880 by Scottish surgeon and bacteriologist Alexander Ogston (1844–1929), following the pattern established five years earlier with the naming of *Streptococcus*. It combines the prefix "staphylo" (from Ancient Greek: σταφυλή, romanised: staphylē, lit. 'bunch of grapes'), and suffixed by the New Latin: coccus, lit. 'spherical bacterium' (from Ancient Greek: κόκκος, romanised: kókkos, lit. 'grain, seed, berry').

Staphylococci are a group of spherical bacteria, the best-known species of which are universally present in great numbers on the mucous membranes and skin of humans and other warm-blooded animals. The term staphylococcus, generally used for all the bacterial species, refers to the cells' habit of aggregating in grapelike clusters. Staphylococci are microbiologically characterised as Gram-stain-positive (in young cultures), non-sporing, non-motile and facultative anaerobes (not requiring oxygen).

Of significance to humans are various strains of the species *Staphylococcus aureus* and *Stp. epidermidis*. While *Stp. epidermidis* is a mild opportunistic pathogen in people with lowered resistance, strains of *Stp. aureus* are major agents of wound infections, boils and other human skin infections and are one of the most common causes of food poisoning. *Stp. aureus* also causes meningitis, pneumonia, urinary tract infections and mastitis, an infection of the breast in women or of the udder in domestic animals. In addition, local staphylococcal infections can lead to toxic shock syndrome, a disease associated with the liberation of a toxin into the bloodstream from the site of infection.

Staphylococcus was one of the leading infections in hospitals and many strains of this bacterium have become antibiotic-resistant. Despite strong attempts to get rid of them, staphylococcus bacteria stay present in hospitals, where they can infect people who are most at risk of infection.

Staphylococcus includes at least 45 species. Many species cannot cause disease and reside normally on the skin and mucous membranes of humans and other animals. *Staphylococcus* species have been found to be nectar-inhabiting microbes. (Jacquemyn *et al*., 2013). They are also a small component of the soil microbiome (Madigan and Martinko, 2005).

Not only salt-resistant, *Staphylococcus* spp. are always facultatively anaerobic. When stained, the cells are seen in small clusters (staphylo = cluster). *Staphylococcus* is usually either beta-hemolytic or not hemolytic at all (called gamma hemolysis). Pathogenic *Staphylococci* spp. can produce a variety of virulence factors, including toxins, coagulase, leucocidins and hydrolytic enzymes that can damage host tissues.

One strain of great concern to humans is methicillin-resistant *Stp. aureus* (MRSA), which is characterised by the act of a mutation that renders it resistant to methicillin, a semisynthetic penicillin used to treat staphylococcus infections that are resistant to mold-derived penicillin. This strain of *Stp. aureus* was first isolated in the early 1960s, shortly after methicillin came into wide use as an antibiotic. Today methicillin is no longer used, but the strain of MRSA to which it gave rise is commonly found on the skin, in the nose, and in the blood and urine of humans.

An estimated two in every 100 people carry MRSA, which is readily passed by skin contact but rarely causes infection in healthy individuals. However, very young children and elderly or sick patients in hospitals and nursing homes are particularly susceptible to MRSA infection, which is difficult to treat because of its resistance to most antibiotics. The treatment of MRSA infections with vancomycin, an antibiotic often considered as a last line of defence against MRSA, has led to the emergence of vancomycin-resistant *Stp. aureus* (VRSA), against which few agents are effective. In 2005 in the United States, deaths from MRSA (approximately 18,000) surpassed deaths from HIV/AIDS (approximately 17,000). While the rate of MRSA bloodstream infections in the United States slowed in the following years, the organism still caused 20,000 deaths in 2017. Globally, MRSA was responsible for more than 100,000 deaths in 2019, underscoring the need for improved surveillance to prevent and control the spread of this potentially lethal organism.

Asadi *et al*., (2024) reported *Staphylococcus warneri* causing bacterial canker on almond and other prunus species. It appears to be a newly emerging disease of apricot, peach and almond trees.

Plant pathogenic species: *Stp. warneri*

Order: *Lactobacillales*

Family: *Lactobacillaceae*

Genus: *Lactobacillus*

6.3.1.3.66 *Genus* Lactobacillus *Beijerinck 1901*

Cells are Gram-stain-positive and non-sporing. Cells vary from long and slender, sometimes bent rods, to short, often coryneform coccobacilli; chain formation is common. Motility is uncommon, but when it does occur, it is by peritrichous flagella.

Metabolism is fermentative; obligately saccharoclastic, at least half of end-product carbon is lactate. Lactate is usually not fermented. Microaerophilic, surface growth on solid media is generally enhanced by anaerobiosis or reduced oxygen pressure and 5–10% CO_2. Some members are anaerobes on isolation. Pigment production is rare; if present, it is yellow or orange to rust or brick red. Nitrate reduction is highly unusual, occurs only when terminal pH is above 6.0. Gelatin is not liquefied. Casein is not digested. Indole and H_2S are not produced. Catalase and cytochrome tests and benzidine reaction are negative.

Growth temperature ranges from 2–53 °C, while the optimal lies between 30 and 40 °C. Optimal pH ranges from 5.5–6.2, growth generally occurs at 5.0 or lower pH. Cell multiplication is often reduced at neutral or initially alkaline pH. The members are found in dairy, meat, fish and grain products, sewage, water, beer, wine, fruits and fruit products. They are also found in vegetable pickles, silage, sour dough and mash. Members are a part of the normal flora of the mouth, intestinal tract and vagina of many homothermic animals, including humans. Pathogenic action is rare. Mole% G + C of the DNA ranges from 32–53 (Bd, *Tm*).

Based on the polyphasic approach, Zheng *et al.*, (2020) proposed reclassification of the genus *Lactobacillus* into 25 genera, including the emended genus *Lactobacillus* which includes host-adapted organisms that have been referred to as the *Lactobacillus delbrueckii* group, *Paralactobacillus* and 23 other novel genera. They also proposed to emend the description of the family *Lactobacillaceae* to include all genera that were previously included in families *Lactobacillaceae* and *Leuconostocaceae*.

Type species: *Lactobacillus delbrueckii*

Based on growth characteristics and sequence analysis of 16S rRNA and *rpoA* genes, Bonasera *et al.*, (2017) identified lactic acid bacteria isolated from diseased onion plants and bulbs as *Lactobacillus plantarum*. *Lactob. plantarum* caused symptoms in both leaves and bulbs. This is the first report of a Gram-stain-positive bacterium to cause a disease on onions.

Plant pathogenic species: *Lactob. plantarum*

Family: *Streptococcaceae*

Genus: *Lactococcus*

6.3.1.3.67 Genus Lactococcus *Schleifer* et al. *1986 gen. nov.*

Lactococcus is a genus of lactic acid bacteria that were earlier included in the genus *Streptococcus* Group N1. The members are known as homofermentors because these produce a single product, i.e., lactic acid in this case, as the major or only product of glucose fermentation. However, this homofermentative character can be altered by adjusting environmental conditions such as pH, glucose concentration and nutrient limitation. The cells are Gram-stain-positive, cocci, found singly, in pairs or in chains, non-motile and catalase negative. The genus contains strains known to grow at or below 7 °C.

These microorganisms are commonly used in the dairy industry for manufacturing fermented dairy products such as cheeses. These can be used as single-strain starter cultures, or as mixed-strain starter cultures with other lactic acid bacteria such as *Lactobacillus* and *Streptococcus*. Special emphasis is placed in the study of *Lactoc. lactis* subsp. *lactis* and *Lactoc. lactis* subsp. *cremoris*, as these are the strains generally used as starter cultures in industrial dairy fermentations. Their main role in dairy fermentation is the rapid acidification of milk, which lowers the pH of the fermented product and prevents the growth of spoilage bacteria. These bacteria also play a role in the flavour of the final product. Lactococci are also currently used in the biotechnology industry. The members are easily grown at industrial scale on whey-based media. As food-grade bacteria, they are used in the production of foreign proteins that are used in the food industry.

Type species: *Lactococcus lactis*

On the basis of growth characteristics and sequence analysis of 16S rRNA and *rpoA* genes, Bonasera *et al.*, (2017) identified lactic acid bacteria isolated from diseased onion bulbs as *Lactococcus lactis*. *Lactoc. lactis* caused scale discolouration in bulbs, but failed to cause lesions on leaves. This is the first report of *Lactoc. lactis* to cause a disease in onion bulbs.

Plant pathogenic species: *Lactoc. lactis*

Family: *Leuconostocaceae*

Genus: *Leuconostoc*

6.3.1.3.68 Genus Leuconostoc *van Tieghem 1878*

Cells are Gram-stain-positive, generally ovoid cocci and often form chains. *Leuconostoc* spp. are intrinsically resistant to vancomycin and are catalase negative (which distinguishes them from staphylococci). All species within this genus are heterofermentative and are able to produce dextrin from sucrose.

The members generally form the slime. Blamed for causing the "stink" when used for creating a sourdough starter. Some species are also capable of causing human infection. As the members are not a common cause of disease in humans, standard commercial identification kits are often unable to identify the organism. *Leuconostoc*, along with other lactic acid bacteria such as *Pediococcus* and *Lactobacillus*, is responsible for the fermentation of cabbage, making it **sauerkraut**. In this process, fresh cabbage is fermented in light brine, where the sugars in the cabbage are transformed by lacto-fermentation to lactic acid which gives the cabbage a sour flavour and good keeping qualities. Similarly, *Leuconostoc* is a part of the symbiotic colonies of bacteria and yeast or symbiotic culture of bacteria and yeast (SCOBY) involved in the fermentation of kefir, a fermented milk beverage.

Type species: *Leuconostoc mesenteroides*

Bonasera *et al.*, (2017), on the basis of growth characteristics and sequence analysis of 16S rRNA and *rpoA* genes, identified the strains of lactic acid bacteria isolated from diseased onion plants and bulbs as *Leuconostoc citreum*,

Len. mesenteroides and *Len. pseudomesenteroides*. All the three *Leuconostoc* species caused symptoms in both leaves and bulbs. *Len. citreum* caused bulb decay in 7 days at 18 °C as well as 37 °C. This is the first report of *Leuconostoc* species to cause a disease in onion leaves and bulbs.

Plant pathogenic species: *Len. citreum*, *Len. mesenteroides* and *Len. pseudomesenteroides*.

Class: *Mollicutes*

Order: *Entomoplasmatales*

Family: *Spiroplasmataceae*

Genus: *Spiroplasma*

6.3.1.3.69 Genus Spiroplasma *Saglio* et al. *1973*

The genus was first described by Saglio *et al.*, (1973) to classify a mycoplasma-like organism associated with citrus stubborn disease. Later on, two more species, namely, *Spr. kunkelii* and *Spr. phoeniceum*, causal agents of corn stunt and stubborn of periwinkles, respectively, were added to this genus.

The salient features of the genus are given below. The cells are pleomorphic and vary in shape and size. Different shapes found are helical and branched non-helical filaments, and spherical and ovoid forms. Helical forms are generally 100–200 nm in width and 3–5 µm long. Spherical cells are approximately 300 nm in diameter. The colonies have a typical **fried-egg appearance**. Colony margin is frequently diffuse due to the movement of growing cells. Colony size varies from 0.1–4.0 mm in diam. Cells divide by binary fission. Facultatively anaerobes. Flagella, periplasmic fibrils or other organs of locomotion are not found. However, helical filaments show motility due to flexional and twitching movements. Chemoorganotrophs. Acid is produced from glucose, but fermentation of other carbohydrates is variable. Arginine is generally hydrolysed, but hydrolysis of urea, esculin fermentation, and serum digestion are negative. Cholesterol or possibly other sterols are required for growth. Sensitive to erythromycin, tetracycline and amphotericin B, but resistant to penicillin. Genome size is very small and G + C content of the DNA varies from 25.0–31.0 mol%. The members can be isolated from phloem fluid of vascular plants and insects that feed on this fluid, and from surfaces of flowers and other plant parts. *Spr. citri* causes stubborn disease of citrus and corn stunt.

Type species: *Spiroplasma citri*

Plant pathogenic species: *Spr. citri*, *Spr. kunkelii* and *Spr. phoeniceum*.

Genus: '*Candidatus* Phytoplasma': The order and family are not yet established.

6.3.1.3.70 Genus 'Candidatus *Phytoplasma*'

Several wall-less procaryotes are associated with plants and cause yellow-type diseases. These were earlier called **mycoplasma-like organisms** (MLOs) due to their morphological similarities to mycoplasmas. Some of these microorganisms, which are spiral-shaped and can grow on artificial culture media, have been assigned to the genus *Spiroplasma*. The remaining wall-less, non-helical and not yet cultured procaryotes that colonise plant phloem and sap feeding insects are called phytoplasmas. The trivial name "Phytoplasma" was adopted by the Phytoplasma Working Team at the 10th Congress of International Organisation of Mycoplasmology, held in 1994, replacing the term "mycoplasma-like organisms" or MLOs. Zreik *et al.*, (1995) proposed the genus '*Candidatus* Phytoplasma' to include '*Candidatus* Phytoplasma aurantifolia', the causal agent of witches' broom disease of lime. Since then, many other '*Candidatus* phytoplasmas' causing different plant diseases have been added to this genus (Jung *et al.*, 2002, 2003, 2003; Verdin *et al.*, 2003; Hiruki and Wang, 2004; Lee *et al.*, 2004, 2004, 2006, 2011; Marcone *et al.*, 2004, 2004; Schneider *et al.*, 2005; Seemuller and Schneider, 2004; Arocha *et al.*, 2005, 2007; Valiunas *et al.*, 2006; Al-Saady *et al.*, 2008; Zhao *et al.*, 2009, Malembic-Maher *et al.*, 2011; Davis *et al.* 2012, 2013, 2017; Martini *et al.*, 2012; Nejat *et al.*, 2013; Quaglino *et al.*, 2013; Win *et al.*, 2013; Harrison *et al.*, 2014; Šafarova *et al.*, 2016; Naderali *et al.*, 2017; Miyazaki *et al.*, 2018). Most plant pathogenic *Candidatus* species are members of this genus.

'*Can.* Phytoplasma' cells lack rigid cell walls, are surrounded by a single unit membrane and are pleomorphic in shape. When observed by transmission electron microscopy, they appear as round to filamentous pleomorphic bodies with a mean diameter of 200–800 nm. Some of the filamentous forms are branched. Phytoplasmas are sensitive to tetracyclines, but not sensitive to penicillin and other cell wall inhibiting antibiotics. They are resistant to digitonin and sensitive to hypotonic salt solutions, and are, therefore, similar to those mollicutes that do not require sterols. The members use UGA as a stop codon and not as a tryptophan codon, as several other mycoplasmas do. Phytoplasmas are closest relatives of the members of the genus *Acholeplasma*. The genome size of different phytoplasmas varies considerably, ranging from 530–1,350 kilo base pairs. The Bermuda grass white leaf phytoplasma has the smallest genome size of 530 kbp found in phytoplasma to date and may represent the smallest chromosome known for any living cell. Like other members of the mollicutes, phytoplasmas contain one circular double-stranded chromosomal DNA molecule. Short, circular extra chromosomal DNAs or plasmids were found in all the members of the aster yellows group and stolbur group and in some members of the X-disease and clover proliferation groups. Large plasmids may also be present in some phytoplasmas. G + C contents of phytoplasma chromosomal DNA are estimated to be between 23 and 29 mol% (Bd) (Lee *et al.* 2000). The members inhabit the phloem sieve elements of vascular plants, and the gut, haemolymph, salivary glands and other organs of sap-sucking insects (leafhoppers and planthoppers). They may cause premature mortality in insect hosts. In plant hosts, they cause a wide range of symptoms, such as virescence, phyllody, sterility of flowers, proliferation of auxiliary shoots, abnormal elongation of internodes, and many other, often less specific symptoms.

Classification and taxonomic assignment of phytoplasmas have been based primarily on molecular analysis of 16S rRNA gene sequences because of the inaccessibility of measurable phenotypic characters suitable for conventional microbial characterisation. In the recent past, many molecular techniques have been developed for the identification and classification of phytoplasmas. Wei *et al.*, (2007), on the basis of distinctive virtual RFLP patterns and calculated similarity coefficients (obtained by using computer-simulated RFLP analysis of 16S rRNA genes), classified phytoplasma strains into 28 groups. They were able to classify hundreds of previously unclassified phytoplasmas and delineate ten new phytoplasma groups representing three formerly described and seven novel putative '*Candidatus* Phytoplasma' taxa. Wei *et al.*, (2008), extending their above-mentioned efforts in the exploitation of computer-simulated 16S rRNA gene RFLP analysis and virtual gel plotting for rapid classification of phytoplasmas, developed a **Perl program** for automated RFLP pattern comparison and similarity coefficient calculation. By streamlining virtual RFLP pattern analysis, the program has led to the establishment of a criterion for phytoplasma 16Sr subgroup classification and to the delineation of new and distinct sub group lineages in the clover proliferation phytoplasma group (16SrVI).

Martini *et al.*, (2007) showed that ribosomal protein (rp) genes, rplV (rpl22)- and rpsC (rps3)-based phylogeny allowed for finer resolution of distinct lineages within the phytoplasma 16Sr groups. RFLP analysis of rp gene sequences permitted finer differentiation of phytoplasma strains in a given 16Sr group. Hodgetts *et al.*, (2008) highlighted the utility of virtual RFLP analysis of secA gene sequences to differentiate 16Sr groups and subgroups, suggesting that this gene may provide an informative alternative molecular marker for pathogen identification and diagnosis of phytoplasma diseases. Zhao *et al.*, (2009) designed an interactive online tool, **iPhyClassifier**, to expand the efficacy and capacity of the current 16S rRNA gene sequence-based phytoplasma classification system. iPhyClassifier performs sequence similarity analysis, simulates laboratory restriction enzyme digestions and subsequent gel electrophoresis, and generates virtual restriction fragment length polymorphism (RFLP) profiles. iPhyClassifier makes instant suggestions on tentative phytoplasma 16Sr group/subgroup classification status and '*Candidatus* Phytoplasma' species assignment based on calculated RFLP pattern similarity coefficients and overall sequence similarity scores. Using iPhyClassifier, the authors have revised and updated the classification of strains affiliated with the peach X-disease phytoplasma group. The online tool is available at http://www.ba.ars.usda.gov/data/mppl/iPhyClassifier.html.

As the DNA-dependent RNA polymerase (DpRp) β-subunit gene (*rpoB*) exists as a single copy in the phytoplasma genome, Valiunas *et al.*, (2013) explored the use of *rpoB* for phytoplasma classification and phylogenetic analysis. They sequenced a clover phyllody phytoplasma genetic locus containing ribosomal protein genes, a complete *rpoB* gene, and a partial *rpoC* gene encoding the β′-subunit of DpRp. *rpoB* gene sequences from phytoplasmas classified in groups 16SrI, 16SrII, 16SrIII, 16SrX and 16SrXII were subjected to sequence similarity and phylogenetic analyses. The *rpoB* gene sequences were more variable than 16S rRNA gene sequences, more clearly distinguishing among phytoplasma lineages. Phylogenetic trees based on 16S rRNA and *rpoB* gene sequences had similar topologies, and branch lengths in the *rpoB* tree facilitated distinctions among closely related phytoplasmas. Virtual RFLP analysis of *rpoB* gene sequences also improved distinctions among closely related lineages. Results indicated that as the *rpoB* gene provides a useful additional marker for phytoplasma classification, it should facilitate studies of disease etiology and epidemiology.

Lee *et al.*, (2010) emphasised the importance of *secY* gene sequence for delineating phytoplasma strains within each 16Sr group. Phylogenetic analysis based on the *secY* gene permitted finer differentiation of phytoplasma strains. *secY* gene-based phylogeny not only readily resolved 16Sr sub groups within a given 16Sr group, but also delineated distinct lineages irresolvable by 16S rRNA gene-based phylogeny. Such high resolving power makes the *secY* gene a more useful genetic marker than the 16S rRNA gene for finer differentiation of closely related phytoplasma strains based on RFLP analysis with selected restriction enzymes.

Wei *et al.*, (2015) have emphasised the importance of occurrence, distribution and possible functional roles of simple sequence repeats (SSRs) in phytoplasma genomes. SSRs of microbes are mutation-prone, phase-variable short DNA tracts that function as "evolutionary rheostats" and enhance rapid adaptations to vastly different environments. Overall density of SSRs in phytoplasma genomes was higher than in representative strains of other procaryotes. While mono- and tri-nucleotide SSRs were significantly over-represented in the phytoplasma genomes, dinucleotide SSRs and other higher-order SSRs were under-represented. Occurrence and distribution of long SSRs in the prophage islands and phytoplasma-unique genetic loci indicated that SSRs played a role in compounding the complexity of sequence mosaics in individual genomes and in increasing allelic diversity among genomes.

Rao (2021) has published a review article on the phytoplasma research scenario in India, highlighting the various aspects dealt and pointing out the areas on which the future research is to be focused.

Type species: '*Candidatus* Phytoplasma aurantifolia'

Zhao *et al.*, (2021), on the basis of ecological, molecular and genomic evidences, identified the phytoplasma causing wheat blue dwarf disease as '*Can.* Phytoplasma tritici'.

Based on phylogenetic analyses and OGRI values obtained from the genome sequences, Kirdat *et al.*, (2021) proposed a novel taxon '*Can.* Phytoplasma sacchari' for the causal agent of sugarcane grassy shoot disease.

In Australia, Jardim *et al.*, (2021) detected *Stylosanthes* little leaf phytoplasma in *Stylosanthes scabra*, *Arachis pintoi*, *Saccharum officinarum*, *Carica papaya*, *Medicago sativa* and *Solanum tuberosum*. The authors proposed its name as '*Can.* Phytoplasma stylosanthis'. Jones *et al.*, (2021) proposed a novel taxon '*Can.* Phytoplasma dypsidis' associated with a lethal wilt disease of palms in Australia. Mou *et al.*, (2022) reported lethal bronzing of palm caused by the phytoplasma '*Can.* Phytoplasma aculeata' and investigated the acquisition and transmission of this phytoplasma by *Haplaxius crudus*.

Pilet *et al.*, (2023) reported association of '*Can.* Phytoplasma palmae' (16SrIV-A Subgroup) with palm lethal yellowing disease on *Cocos nucifera* and *Pritchardia* sp. in Guadeloupe, French West Indies.

Rios *et al.*, (2024) reported the occurrence of a '*Candidatus* Phytoplasma sacchari'-related strain with yellowing and decline of silver bluestem for the first time from Texas, U.S.A.

Plant pathogenic species: '*Can.* Phy. aculeata' *Phy. allocasuarinae*, *Phy. americanum*, *Phy. asteris*, '*Can.* Phy. aurantifolia', '*Can.* Phy. australasia', *Phy. australiense*, '*Can.* Phy. balanitae', *Phy. brasiliense*, *Phy. caricae*, *Phy. castaneae*, *Phy. cirsii*, *Phy. convolvuli*, '*Phy. costaricanum*, *Phy. cynodontis*, '*Can.* Phy. dypsidis', '*Can.* Phy. fragariae', *Phy. fraxini*, *Phy. graminis*, *Phy. hispanicum*, *Phy. japonicum*, *Phy. luffae*, *Phy. lycopersici*, *Phy. malaysianum*, *Phy. mali*, *Phy. meliae*, *Phy. noviguineense*, *Phy. omanense*, *Phy. oryzae*, '*Can.* Phy. palmae', *Phy. palmicola*, *Phy. phoenicium*, *Phy. pini*, *Phy. pruni*, *Phy. prunorum*, *Phy. pyri*, *Phy. rhamni*, *Phy. rubi*, '*Can.* Phy. sacchari', *Phy. solani*, *Phy. spartii*, '*Can.* Phy. stylosanthis' '*Can.* Phy. sudamericanum', *Phy. tamaricis*, *Phy. trifolii*, '*Can.* Phy. tritici' *Phy. ulmi*, *Phy. wodyetiae* and *Phy. ziziphi*.
Class: '*Clostridia*'
Order: *Clostridiales*
Family: *Clostridiaceae*
Genus: *Clostridium*

6.3.1.3.71 Genus Clostridium *Prazmowski 1880*

It is a very old genus established way back in 1880. At present, it contains only one plant pathogenic species associated with spoilage of potatoes and carrots.

The genus has the following important characteristics. The cells are Gram-stain-positive and spore forming rods. Obligate anaerobes. Most species are obligate fermenters and they ferment by pathways that generate organic solvents such as butantol. No doubt, clostridial cells are rod-shaped, but, when producing spores, they appear more like drum sticks with a bulge at one end. Do not carry out dissimilatory sulphate reduction. An exposure of vegetative cells to oxygen for 5–10 minutes may cause their massive death. However, the spores are not killed by the oxygen. In many cases, the death results from the production of hydrogen peroxide, a highly toxic gas by the bacteria. Most species completely lack heme proteins, including catalase and are incapable of decomposing hydrogen peroxide.

Clostridium spp. are commonly found in the environment. The members inhabit sewage, soil, marine sediments and the intestines of both animals and humans. Some species produce toxins and cause diseases. *Clo. tetani* causes tetanus. The bacterium releases an exotoxin called tetanus toxin, which blocks the release of neurotransmitters from the presynaptic membrane of inhibitory interneurons of spinal cord and brainstem of mammals that regulate muscle contraction. *Clo. botulinum* is the causal agent of botulism and it is contracted usually through the ingestion of botulinum toxin. *Clo. perfringens* is one of the several species of the genus that causes gas gangrene and is the causative agent in 95% gas gangrene cases. *Clo. sordellii* has been linked to the death of a few women after childbirth.

Apart from causing diseases and harmful effects, *Clostriduim* species have commercial uses and industrial value. *Clo. thermocellum* can utilise lignocellulosic waste and generates ethanol. *Clo. ljungdahlii* produces ethanol from single-carbon sources, including synthesis gas, a mixture of carbon monoxide and hydrogen that can be generated from the partial combustion of either fossil fuels or biomass. *Clo. butyricum* and *Clo. pasteurianum* fix nitrogen. The genomes of *Clo. tetani, Clo. acetobutylicum* and *Clo. perfringens* have been sequenced. *Clo. tetani* also contains a plasmid that measures 74082 bp long, having 61 open reading frames, and encodes the tetanus toxin. The genome of *Clo. perfringens* encodes the typical anaerobic fermentation enzymes leading to gas production but no enzymes for the tricarboxylic acid cycle of respiratory chain (*Clostridium* – Microbe Wiki, accessed in December, 2025). G + C content of the DNA ranges from 22.0–55.0 mol% (T_m).

Type species: *Clostridium butyricum*

Plant pathogenic species: *Clo. puniceum*.

3. **Phylum: Bacteroidota**
Class: Sphingobacteriia
Order: Sphingobacteriales
Family: Sphingobacteriaceae

6.3.1.3.72 Genus Sphingobacterium *Yabuuchi* et al. *1983 emend. Wauters* et al. *2012*

The genus *Sphingobacterium* was proposed originally by Yabuuchi *et al.*, (1983) to include Gram-stain-negative rods that are positive for catalase and oxidase, negative for heparinase, gelatinase and indole production and have iso-C15:

0, iso-C15: 0 2-OH, C16: 1v7c and C17: 0 3-OH as the main fatty acids.

The genus *Sphingobacterium* includes organisms previously classified as *Flavobacterium* species. The organisms that were transferred to this genus contain large amounts of sphingophospholipid compounds in their cell membranes and have other taxonomic features that distinguish them from flavobacteria.

Sphingobacterium spp. are characterised by colonies that develop a yellow pigment after a few days at room temperature. *Sph. multivorum*, *Sph. spiritivorum* and *Sph. thalpophilum* have been isolated from a variety of infections, including BSIs, peritonitis, wound infections, UTI and abscesses. *Sphingobacterium* spp. *in vitro* are usually resistant to aminoglycosides and polymyxin B and susceptible to fluoroquinolones and TMP-SMX.

Sphingobacterium species are ubiquitous and are isolated from natural environments, such as soil and water. However, these rarely cause infections in humans. Only a limited number of cases have been reported in elderly and immunocompromised patients with underlying diseases and predisposing factors. *Sph. spiritivorum* is a glucose non-fermenting Gram-negative rod, formerly classified as one of the *Flavobacterium* species. It is characterised by a large number of cellular membrane sphingophospholipids.

Chen *et al.*, (2024) reported *Sphingobacterium multivorum* causing bacterial rot of cauliflower mushroom (*Sparassis latifolia*) from China. This is the first report of *Sph. multivorum* causing bacterial rot of mushroom.

Type species: *Sphingobacterium spiritivorum*

Plant pathogenic species: *Sph. multivorum*

4. **Phylum: Actinomycetota**
   Class: Actinomycetes
   Order: Micrococcales
   Family: Sanguibacteraceae

6.3.1.3.73 Genus Sanguibacter *Fernández-Garayzábal* et al. *1995*

Fernández-Garayzábal *et al.*, (1995) performed 16S rRNA gene sequencing studies on some Gram-positive coryneform bacteria of unknown taxonomic position isolated from blood and milk of healthy cows. Their comparative sequence analysis demonstrated that the milk isolates corresponded to *Oerskovia xanthineolytica* whereas those from blood consisted of two distinct, albeit highly related species, within the family Cellulomonadaceae. Based on the phylogenetic and phenotypic distinctiveness of the blood isolates, the authors proposed that these be classified in a new genus *Sanguibacter.*

Pikuta *et al.*, (2017) emended the descriptions of the family Sanguibacteraceae, the genus *Sanguibacter* and species *S. antarcticus*, *S. inulinus*, *S. kedieii*, *S. marinus*, *S. soli* and *S. suarezii.*

Ivanova *et al.*, (2009) reported complete genome sequence of *Sanguibacter keddieii* type strain (ST-74^T). This is the first complete genome sequence of a member of the family *Sanguibacteraceae*, and the 4,253,413 bp long single replicon genome with its 3735 protein-coding and 70 RNA genes.

Kiel *et al.*, (2024) isolated a bacterial strain from pathogenic lesions of *Acer campestre* tree leaves from the Teutoburg Forest in North Rhine-Westphalia, Germany, by culturing on a non-selective agar medium. Bacterial cells were coryneform motile rods, with a length of 1.1–3.3 µm and a constant diameter of 0.5 µm. Cells did not form spores under the tested conditions and were Gram-stain-positive. Growth occurred between 0.5 and 4% NaCl concentration (optimal: 1%), at pH 5.5–9.5 (optimal: 8.0–9.0). The bacterium was classified as *Sanguibacter biliveldensis* sp. nov.

Plant pathogenic species: *San. biliveldensis*

BIBLIOGRAPHY

Abedinzadeh, A., Falahi Charkhabi, N., Aeini, M. and Amani, M. (2023). *Enterobacter* species, emerging plant-pathogenic bacteria, associated with decline and offshoot rot of date palm in Iran. *Eur. J. Plant Pathol.* 166: 341–51. https://doi.org/10.1007/s10658-023-02666-3

Adeolu, M., Alnajar, S., Naushad, S. and Gupta, R. S. (2016). Genome based phylogeny and taxonomy of the "Enterobacteriales": proposal for Enterobacterales ord. nov. divided into the families Enterobacteriaceae, Erwiniaceae fam. nov., Pectobacteriaceae fam. nov., Yersiniaceae fam. nov., Hafniaceae fam. nov., Morganellaceae fam. nov., and Budviciaceae fam. nov. *Int. J. Syst. Evol. Microbiol.* 66: 5575–99.

Ah-You, N., Gagnevin, L., Grimont, P. A. D. Brisse, S., Nesme, X. *et al.* (2009). Polyphasic characterization of xanthomonads pathogenic to members of the Anacardiaceae and their relatedness to species of *Xanthomonas. Int. J. Syst. Evol. Microbiol.* 59: 306–18.

Allahverdi, T., Rahimian, H. and Ravanlou, A. (2016). First report of bacterial canker in mulberry caused by *Citrobacter freundii* in Iran. *Plant Dis.* 100: 1774.

Al-Saady, N. A., Khan, A. M., Calari, A. Al-Subhi, A. M. and Bertaccini, A. (2008). '*Candidatus* Phytoplasma omanense', associated with witches'-broom of *Cassia italica* (Mill.) Spreng. in Oman. *Int. J. Syst. Evol. Microbiol.* 58: 461–66.

Anurat, P., Duangmal, K. and Srisuk, N. (2019). *Pseudomonas mangiferae* sp. nov., isolated from bark of mango tree in Thailand. *Int. J. Syst. Evol. Micro.* 69: 3537–43.

Arizala, D., Dobhal, S., Alvarez, A. M. and Arif, M. (2022). Elevation of *Clavibacter michiganensis* subsp. *californiensis* to species level as *Clavibacter californiensis* sp. nov., merging and re-classification of *Clavibacter michiganensis* subsp. *chilensis* and *Clavibacter michiganensis* subsp. *phaseoli* as *Clavibacter phaseoli* sp. nov. based on complete genome in silico analyses. *Int. J. Syst. Evol. Microbiol.* 72(9). https://doi.org/10.1099/ijsem.0.005427. PMID: 36107753.

Arocha, Y., Antesana, O., Montellano, E., Franco, P., Plata, G. *et al.* (2007). '*Candidatus* Phytoplasma lycopersici', a phytoplasma associated with 'hoja de perejil' disease in Bolivia. *Int. J. Syst. Evol. Microbiol.* 57: 1704–10.

Arocha, Y., Lopez, M., Pinol, B. Fernández, M. Picornell, B. *et al.* (2005). '*Candidatus* Phytoplasma graminis' and '*Candidatus* Phytoplasma caricae', two novel phytoplasmas associated with diseases of sugarcane, weeds and papaya in Cuba. *Int. J. Syst. Evol. Microbiol.* 55: 2451–63.

Asadi, E., Rahimian, H., Babaeizad, V. and Basavand, E. (2024). *Staphylococcus warneri*, an unconventional plant pathogen involved in canker disease of almond and other *Prunus* species. *Plant Pathol.* 73: 1719–29. https://doi.org/10.1111/ppa.13920

Asai, T. (1935). Taxonomic studies on acetic acid bacteria and allied oxidative bacteria isolated from fruits. A new classification of the oxidative bacteria. *J. Agri. Chem. Soc. Japan.* 11: 674–708.

Ash, C., Priest, F. G. and Collins, M. D. (1994). List of new names and new combinations previously effectively, but not validly, published. *Int. J. Syst. Bacteriol.* Vol. 44: p. 852. Effective publication: Ash, C., Priest, F. G., and Collins, M. D. (1993). Molecular identification of rRNA group 3 bacilli (Ash, Farrow, Wallbanks and Collins) using a PCR probe test. *Antonie van Leeuwenhoek*, 64: 253–60.

Aykac, K., Ozsurekci, Y. Tuncer, O., Sancak, B., Cengiz, A. B. *et al.* (2016). Six cases during 2012–2015 and literature review of Chryseobacterium indologenes infections in paediatric patients. *Canadian J. Microbiol.* 62: 812–19. https://doi.org/10.1139/cjm-2015-0800.

Bagley, S., Seidler, R. J. and Brenner, D. J. (1981). *Klebsiella planticola* sp. nov.: A new species of Enterobacteriaceae found primarily in nonclinical environments. *Curr. Microbiol.* 6: 105–09.

Baldani, J. I., Baldani, V. L. D., Seldin, L. and Döbereiner, J. (1986). Characterization of *Herbaspirillum seropedicae* gen. nov., sp. nov., a root-associated nitrogen-fixing bacterium. *Int. J. Syst. Bacteriol.* 36: 86–93.

Baldani J. I., Pot, B., Kirchhof, G. Falsen, E., Baldani, V. L. D. *et al.* (1996). Emended description of *Herbaspirillum*; inclusion of [*Pseudomonas*] *rubrisubalbicans*, a mild plant pathogen, as *Herbaspirillum rubrisubalbicans* comb. nov.; and classification of a group of clinical isolates (EF Group 1) as *Herbaspirillum* species 3. *Int. J. Syst. Bacteriol.* 46: 802–10.

Basavand, E., Khodaygan, P., Moradi, M., Mousavi, S. A., Lindström, K. *et al.* (2022). Characterization of *Agrobacterium radiobacter*, a new pathogen of pistachio. *Austr. Plant Pathol.* 51: 167–74. https://doi.org/10.1007/s13313-021-00831-y

Basavand, E., Khodaygan, P., Rahimian, H., Ganjeh, A. Molaei, S. *et al.* (2021). Identification and characterization of *Klebsiella oxytoca* strains associated with wetwood disease of *Morus* trees. *Indian Phytopath.* 74: 1123–27. https://doi.org/10.1007/s42360-021-00360-0

Behrendt, U., Schumann, P., Stieglmeier, M., Pukall, R., Augustin, J., *et al.* (2010). Characterization of heterotrophic nitrifying bacteria with respiratory ammonification and denitrification activity – Description of *Paenibacillus uliginis* sp. nov., an inhabitant of fen peat soil and *Paenibacillus purispatii* sp. nov., isolated from a spacecraft assembly clean room. *Syst. Appl. Mirobiol.* 33: 328–36.

Beijerinck, M. W. (1898). Uber die Arten der Essigbakterien. *Zentralbl Bakteriol Parasitenkd Infektionskr Hyg.* Abt II 4: 209–16.

Beijerinck, M. W. (1901). Sur les ferments lactiques de l'industrie. *Archives Néerlandaises des Sciences Exactes et Naturelles (Section 2)*, 6: 212–43.

Beijerinck, M. W. and Folpmers, T. (1916). Verslagen van de gewone vergadering der wis-en naturkundige afdeeling. *Konenklijke Akademie van Westen-chappen te Amsterdam*, 18: 1198–1200.

Berger, B., Baldermann, S. and Ruppel, S. (2017). The plant growth-promoting bacterium *Kosakonia radicincitans* improves fruit yield and quality of *Solanum lycopersicum*. *J. Sci. Food & Agri.* 97: 4865–71. https://doi.org/10.1002/jsfa.8357

Beurmann, S., Ushijima, B., Svoboda, C. M., Videau, P., Smith, A. M. *et al.* (2017). *Pseudoalteromonas piratica* sp. nov., a budding, prosthecate bacterium from diseased *Montipora capitata*, and emended description of the genus *Pseudoalteromonas*. *Int. J. Syst. Evol. Microbiol.* 67: 2683–88.

Bian, D. R., Xue, H., Wang, G. M., Piao, C. G. and Li, Y. (2021). *Affinibrenneria salicis* gen. nov. sp. nov. isolated from *Salix matsudana* bark canker. *Arch. Microbiol.* 203: 3473–81.

Bizio, B. (1823). Lettera di Bartolomeo Bizio al chiarissimo canonico Angelo Bellani sopra il fenomeno della polenta porporina Biblioteca Italiana o sia Giornale di letteratura, *Scienze e Arti (Anno VIII)*, 30: 275–95.

Bonasera, J. M., Asselin, J. A. E. and Beer, S. V. (2017). Lactic acid bacteria cause a leaf blight and bulb decay of onion (*Allium cepa*). *Plant Dis.* 101: 29–33.

Bouchek-Mechiche, K., Gardan, L., Andrivon, D. and Normand, P. (2006). *Streptomyces turgidiscabies* and *Streptomyces reticulsiscabiei*: One genomic species, two pathogenic groups. *Int. J. Syst. Evol. Microbiol.* 56: 2771–76.

Bouchek-Mechiche, K., Gardan, L., Normand, P. and Jouan, B. (2000). DNA relatedness among strains of *Streptomyces* pathogenic to potato in France: description of three new species, *S. Europaeiscabiei* sp. Nov. And *S. Steliiscabiei* sp. Nov. Associated with common scab, and *S. Reticulsiscabiei* sp. Nov. Associated with netted scab. *Int. J. Syst. Evol. Microbiol.* 50: 91–99.

Bouzar, H. and Jones, J. B. (2001). *Agrobacterium larrymoorei* sp. nov., a pathogen isolated from aerial tumours of *Ficus benjamina*. *Int. J. Syst. Evol. Microbiol.* 51: 1023–26.

Bradbury, J. F. (1982). The new bacterial nomenclature – What to do? *Trop. Pest Manag.* 28: 42–44.

Bradbury, J. F. (1984). Genus 11. *Xanthomonas*. In: N. R. Krieg and J. G. Holt (eds.), *Bergey's Manual of Systematic Bacteriology*, 1: 199–210. The Williams & Wilkins Co., Baltimore.

Brady, C., Cleenwerck, I., Venter, S., Coutinho, T. and De Vos, P. (2013). Taxonomic evaluation of the genus *Enterobacter* based on multilocus sequence analysis (MLSA): Proposal to reclassify *E. nimipressuralis* and *E. amnigenus* into *Lelliottia* gen. nov. as *Lelliottia nimipressuralis* comb. nov. and *Lelliottia amnigena* comb. nov., respectively, *E. gergoviae* and *E. pyrinus* into *Pluralibacter* gen. nov. as *Pluralibacter gergoviae* comb. nov. and *Pluralibacter pyrinus* comb. nov., respectively, *E. cowanii, E. radicincitans, E. oryzae* and *E. arachidis* into *Kosakonia* gen. nov. as *Kosakonia cowanii* comb. nov., *Kosakonia radicincitans* comb. nov., *Kosakonia oryzae* comb. nov. and *Kosakonia arachidis* comb. nov., respectively, and *E. turicensis, E. helveticus* and *E. pulveris* into *Cronobacter* as *Cronobacter zurichensis* nom. nov., *Cronobacter helveticus* comb. nov. and *Cronobacter pulveris* comb. nov., respectively, and emended description of the genera *Enterobacter* and *Cronobacter*. *Syst. Appl. Microbiol.* 36: 309–19.

Brady, C., Crampton, B., Kaur, S., Maddock, D., Kile, H. *et al.* (2024). Two novel *Raoultella* species associated with bleeding cankers of broadleaf hosts, *Raoultella scottia* sp. nov. and *Raoultella lignicola* sp. nov. *Front. Microbiol.* 15: 1386923. https://doi.org/10.3389/fmicb.2024.1386923

Brady, C., Denman, S., Kirk, S., Venter, S., Rodriguez-Palenzuela, P. *et al.* (2011). Validation list no. 139. *Int. J. Syst. Evol. Microbiol.* 61: 1011–13.

Brady, C., Hunter, G., Kirk, S., Arnold, D. and Denman, S. (2014). Description of *Brenneria roseae* sp. nov. and two subspecies, *Brenneria roseae* subspecies *roseae* ssp. nov. and *Brenneria roseae* subspecies *americana* ssp. nov. isolated from symptomatic oak. *Syst. Appl. Microbiol.* 37: 396–401.

Brady, C. L., Cleenwerck, I., Denman, S., Venter, S. N., Rodriguez-Palenzuela, P. *et al.* (2012). Proposal to reclassify *Brenneria quercina* (Hildebrand and Schroth 1967) Hauben *et al.* 1999 into a new genus, *Lonsdalea* gen. nov., as *Lonsdalea quercina* comb. nov., descriptions of *Lonsdalea quercina* subsp. *quercina* comb. nov., *Lonsdalea quercina* subsp. *iberica* subsp. nov. and *Lonsdalea quercina* subsp. *britannica* subsp. nov., emendation of the description of the genus *Brenneria*, reclassification of *Dickeya dieffenbachiae* as *Dickeya dadantii* subsp. *dieffenbachiae* comb. nov., and emendation of the description of *Dickeya dadantii*. *Int. J. Syst. Evol. Microbiol.* 62: 1592–1602.

Brady, C. L., Venter, S. N., Cleenwerck, I. Engelbeen, K., Vancanneyt, M. *et al.* (2009). *Pantoea vagans* sp. nov., *Pantoea eucalypti* sp. nov., *Pantoea deleyi* sp. nov. and *Pantoea anthophila* sp. nov. *Int. J. Syst. Evol. Microbiol.* 59: 2339–45.

Brady, C. L., Venter, S. N., Cleenwerck, I., Vandemeulebroecke, K., De Vos, P. *et al.* (2010). Transfer of *Pantoea citrea*, *Pantoea punctata* and *Pantoea terrea* to the genus *Tatumella* emend. as *Tatumella citrea* comb. nov., *Tatumella punctata* comb. nov. and *Tatumella terrea* comb. nov. and description of *Tatumella morbirosei* sp. nov. *Int. J. Syst. Evol. Microbiol.* 60: 484–94.

Breed, R. S., Murray E. G. D. and Smith, N. R. (eds.) (1957). *Bergey's Manual of Determinative Bacteriology*. 7th Ed. Williams and Wilkins Co. Baltimore, MD, 1094 pp.

Brenner, D. J., McWhorter, A. C., Kai, A., Steigerwalt, A. G. and Farmer, J. J. 3rd. (1986). *Enterobacter asburiae* sp. nov., a new species found in clinical specimens, and reassignment of *Erwinia dissolvens* and *Erwinia nimipressuralis* to the genus *Enterobacter* as *Enterobacter dissolvens* comb. nov. and *Enterobacter nimipressuralis* comb. nov. *J. Clin. Microbiol.* 23: 1114–20.

Brenner, D. J., Staley, J. T. and Krieg, N. R. (2001). Classification of prokaryotic organisms and the concept of bacterial speciation. In: D. R. Boone and R. W. Castenholz (volume editors), G. M. Garrity (editor-in-chief). *Bergey's Manual of Systematic Bacteriology*. 2nd Ed. Vol. 1, *The Archaea and the Deeply Branching and Phototrophic Bacteria*. Springer-Verlag, New York, pp. 27–31.

Bressan, A. (2014). Emergence and evolution of *Arsenophonus* bacteria as insect-vectored plant pathogens. *Sci. Direct*, 22: 81–90.

Brisou, J. and Prevot, A. R. (1954). Etudes de systematique bacterienne. X. Revision des especes reunies dans le genre *Achromobacter*. *Annales de l'Institut Pasteur (Paris)*, 86: 722–28.

Buchanan, R. E. and Gibbons, N. E. (eds.) (1974). *Bergey's Manual of Determinative Bacteriology*. 8th Ed. Williams and Wilkins Co., Baltimore, MD, 1268 pp.

Bueno-Gonzalez, V., Brady, C., Denman, S., Plummer, S., Allainguillaume, J. *et al.* (2019). *Pseudomonas daroniae* sp. nov. and *Pseudomonas dryadis* sp. nov., isolated from pedunculate oak affected by acute oak decline in the UK. *Int. J. Syst. Evol. Micro.* 69: 3368–76.

Bull, C. T., Coutinho, T. A., Denny, T. P., Firrao, G. and Fischer-Le Saux, M. (2014). List of new names of plant pathogenic bacteria (2011–2012). *J. Plant Pathol.* 96: 223–26.

Bull, C. T., De Boer, S. H., Denny, T. P. Firrao, G., Fischer-Le Saux, M. *et al.* (2008). Demystifying the nomenclature of bacterial plant pathogens. *J. Plant Pathol.* 90: 403–17.

Bull, C. T., De Boer, S. H., Denny, T. P., Firrao, G., Fischer-Le Saux, M. *et al.* (2012). List of new names of plant pathogenic bacteria (2008–2010). *J. Plant Pathol.* 94: 21–27.

Bull, C. T., De Boer, S. H., Denny, T. P., Firrao, G., Fischer-Le Saux, M. *et al.* (2010). Comprehensive list of names of plant pathogenic bacteria, 1980–2007. *J. Plant Pathol.* 92: 551–92.

Buonaurio, R., Stravato, V. M., Kosako, Y. Fujiwara, N., Naka. T. *et al.* (2002). *Sphingomonas melonis* sp. nov., a novel pathogen that causes brown spots on yellow Spanish melon fruits. *Int. J. Syst. Evol. Microbiol.* 52: 2081–87.

Busse H-J. (2016). Review of the taxonomy of the genus *Arthrobacter*, emendation of the genus *Arthrobacter* sensu lato, proposal to reclassify selected species of the genus *Arthrobacter* in the novel genera *Glutamicibacter* gen. nov., *Paeniglutamicibacter* gen. nov., *Pseudoglutamicibacter* gen. nov., *Paenarthrobacter* gen. nov. and *Pseudarthrobacter* gen. nov., and emended description of *Arthrobacter roseus*. *Int. J. Syst. Evol. Microbiol.* 66: 9–37.

Caballo-Ponce, E., Pintado, A., Moreno-Pérez, A., Murillo, J., Smalla, K., *et al.* (2021). *Pseudomonas savastanoi* pv. *mandevillae* pv. nov., a clonal pathogen causing an emerging, devastating disease of the ornamental plant *Mandevilla* spp. *Phytopathology*, 111: 1277–88.

Campigli, S. and Rizzo, D. (2023). First report of *Xanthomonas hydrangeae* causing leaf spot on oakleaf hydrangea (*Hydrangea quercifolia*) in Tuscany, Italy. *Plant Dis.* 107: 2514.

Cao, P., Li, C., Tan, K., Zhao, J., Wang, X. *et al.* (2021). Complete genomic data of *Enterobacter asburiae* strain SD4L associated with bacterial palea browning of rice in China. *Plant Dis.* 105: 4138–40.

Carter, J. C. (1945). Wetwood of elms. *Illinois Nat. Hist. Surv. Bull.* 23: 407–48.

Carter, J. S., Bowden, F. J., Bastian, I., Myers, G. M., Sriprakash, K. S. *et al.* (1999). Phylogenetic evidence for reclassification of *Calymmatobacterium granulomatis* as *Klebsiella granulomatis* comb. nov. *Int. J. Syst. Bacteriol.* 49: 1695–1700.

Chandrashekar, B. S., Prasannakumar, M. K., Puneeth, M. E., Teli, K., Priyanka, K. *et al.* (2018). First report of bacterial soft rot of carrot caused by *Klebsiella variicola* in India. *New Disease Reports.* 37: 21.

Chatton, E. (1937). *Titres et Travaux Scientifiques*. Sottano, Sete.

Chen, J., Lin, J., Gan, Y., Tong, Z., Han, X. *et al.* (2024). First report of *Sphingobacterium multivorum* causing bacterial rot of cauliflower mushroom (*Sparassis latifolia*) in China. *Plant Dis.* 108: 1881. https://doi.org/10.1094/PDIS-01-24-0022-PDN

Chen, X., Tang, B., Laborda, P., Zhao, Y. and Liu, F. (2024). First report of bacterial fruit blotch caused by *Kosakonia cowanii* on *Trichosanthis fructus* in Jiangsu Province, China. *Plant Dis.* 108: 782.

Choudhary, M., Minsavage, G. V., Goss, E. M., Timilsina, S., Coutinho, T. A. *et al.* (2024). Whole-genome-sequence-based classification of *Xanthomonas euvesicatoria* pv. *eucalypti* and computational analysis of the type III secretion system. *Phytopathology*, 114: 47–60. https://doi.org/10.1094/PHYTO-05-23-0150-R

Chu, H., Peng, S., Li, Z., Gao, Y. and Zhang, Y. (2023). Complete genome sequence data of *Pseudomonas bijieensis* SP1 associated with root soft rot of yam bean. *Plant Dis.* 107: 1617–19.

Chu, H.-H., Tang, W.-C. K. Iamnok, K., Goh, R.-P. and Chu, C.-C. (2025). A bacterial gall disease of Myrica rubra caused by Pseudomonas amygdali pv. myricae in Taiwan. *Plant Dis.* Vol.109, p. 1781.

Chuang, S. C., Dobhal, S., Alvarez, A. M. and Arif, M. (2023). Three new species, *Xanthomonas hawaiiensis* sp. nov., *Stenotrophomonas aracearum* sp. nov., and *Stenotrophomonas oahuensis* sp. nov., isolated from Araceae. *Phytopathology*, 113: S3.143.

Chun, J. and Rainey, F. A. (2014). Integrating genomics into the taxonomy and systematics of the *Bacteria* and *Archaea*. *Int. J. Syst. Evol. Microbiol.* 64: 316–24.

Cohn, F. (1872). Untersuchungen uber Bakterien. *Beitr. Biol. Pflanz.* 1(Heft 2): 127–224.

Collins, M. D. and Jones, D. (1983). Reclassification of *Corynebacterium flaccumfaciens*, *Corynebacterium betae*, *Corynebacterium oortii* and *Corynebacterium poinsettiae* in the genus *Curtobacterium*, as *Curtobacterium flaccumfaciens* comb. nov. *J. Gen. Microbiol.* 129: 3545–48.

Conn, H. J. (1942). Validity of the genus *Acaligenes*. *J. Bacteriol.* 44: 353–60.

Conn, H. J. and Dimmick, I. (1947). Soil bacteria similar in morphology to mycobacterium and *Corynebacterium*. *J. Bacteriol.* 54: 291–303.

Cui, X. D., Cai, Y. P., Chen, R. Z., Li, M. H. and Liu, Q. G. (2022). First report of bacterial leaf spot disease on *Pueraria montana* var. *thomsonii* caused by *Robbsia andropogonis* in China. *Plant Dis.* 106: 2985.

Cunty, A., Poliakoff, F., Rivoal, C., Cesbron, S., Fischer-Le Saux, M. *et al.* (2015). Characterization of *Pseudomonas syringae* pv. *actinidiae* (Psa) isolated from France and assignment of Psa biovar 4 to a *de novo* pathovar: *Pseudomonas syringae* pv. *actinidifoliorum* pv. nov. *Plant Pathol.* 64: 582–96.

Dastager, S. G., Lee, J. C., Ju, Y. J. Park, D.-J. and Kim, C.-J. (2009). *Leifsonia kribbensis* sp. nov., isolated from soil. *Int. J. Syst. Evol. Microbiol.* 59: 18–21.

Davis, M. J., Gillaspie, A. G., Jr., Vidaver, A. K. and Harris, R. W. (1984). *Clavibacter*: a new genus containing some phytopathogenic coryneform bacteria, including *Clavibacter xyli* subsp. *xyli* sp. nov., subsp. nov. and *Clavibacter xyli* subsp. *cynodontis* subsp. nov., pathogens that cause ratoon stunting disease of sugarcane and Bermudagrass stunting disease. *Int. J. Syst. Bacteriol.* 34: 107–17.

Davis, R. E., Zhao, Y., Dally, E. L., Jomantiene, R., Lee, I.-M. *et al.* (2012). "*Candidatus* Phytoplasma sudamericanum," a novel taxon, and strain PassWB-Br4, a new subgroup 16SrIII-V phytoplasma, from diseased passion fruit (*Passiflora edulis* f. *flavicarpa* Deg.). *Int. J. Syst. Evol. Microbiol.* 62: 984–89.

Davis, R. E., Zhao, Y., Dally, E. L., Lee, I.-M., Jomantiene, R. *et al.* (2013). "*Candidatus* Phytoplasma pruni," a novel taxon associated with X-disease of stone fruits, *Prunus* spp.: Multilocus characterization based on 16S rRNA, *secY*, and ribosomal protein genes. *Int. J. Syst. Evol. Microbiol.* 63: 2766–76.

Davis, R. E., Zhao, Y., Wei, W., Dally, E. L. and Lee, I.-M. (2017). "*Candidatus* Phytoplasma luffae," a novel taxon associated with witches' broom disease of loofah, *Luffa aegyptica* Mill. *Int. J. Syst. Evol. Microbiol.* 67: 3127–33.

De Ley, J., Segers, P. and Gillis, M. (1978). Intra- and intergeneric similarities of *Chromobacterium* and *Janthinobacterium* ribosomal ribonucleic acid cistrons. *Int. J. Syst. Bacteriol.* 28: 154–68.

de Salles Gomes, L. (1944). Sobre una nova especie do genero Eberthella Buchanan, isolada de fezes patologicas de crianca. *Rev. Inst. Lutz*, 4: 182–95.

Dees, M. W., Lysøe, E., Rossmann, S., Perminow, J. and Brurberg, M. B. (2017). *Pectobacterium polaris* sp. nov., isolated from potato (*Solanum tuberosum*). *Int. J. Syst. Evol. Mirobiol.* 67: 5222–29. https://doi.org/10.1099/ijsem.0.002448.

Denman, S., Brady, C., Kirk, S., Cleenwerck, I., Venter, S. *et al.* (2012). *Brenneria goodwinii* sp. nov., associated with acute oak decline in the UK. *Int. J. Syst. Evol. Microbiol.* 62: 2451–56.

Dhanvantari, B. N., Dye, D. W. and Young, J. M. (1978). *Pseudomonas pomi* Cole 1959 is a latter subject synonym of *Acetobacter pasteurianus* (Hansen 1879) Beijerinck 1898 and *Pseudomonas melophthora* Allen and Riker 1932 is a nomen dubium. *Int. J. Syst. Bacteriol.* 28: 532–37.

Dia, N. C., Vaerenbergh, J. V., Malderghem, C. V., Blom, J., Smits, T. H. M. *et al.* (2021). *Xanthomonas hydrangeae* sp. nov., a novel plant pathogen isolated from *Hydrangea arborescens*. *Int. J. Syst. Evol. Microbiol.* 71. https://doi.org/10.1099/ijsem.0.005163. PMID: 34913859.

Dobhal, S., Hugouvieux-Cotte-Pattat, N., Arizala, D., Sari, G. B. Chuang, S.-C. *et al.* (2025). *Dickeya ananatis* sp. nov., pectinolytic bacterium isolated from pineapple (*Ananas comosus*). *Int. Jour. Syst. Evol. Microbiol.* 75. https://doi.org/10.1099/ijsem.0.006822

Dobritsa, A. P. and Samadpour, M. (2016). Transfer of eleven species of the genus *Burkholderia* to the genus *Paraburkholderia* and proposal of *Caballeronia* gen. nov. to accommodate twelve species of the genera *Burkholderia* and *Paraburkholderia*. *Int. J. Syst. Evol. Microbiol.* 66: 2836–46.

Doi, R. H. and Igarashi, R. T. (1965). Conservation of ribosomal and messenger ribonucleic acid citrons and *Bacillus* species. *J. Bacteriol.* 90: 384–90.

Dowson, W. J. (1939). On the systematic position and generic names of the gram negative bacterial plant pathogens. *Zentralbl. Bakteriol. Parasitenk. Infektionskr. Hyg. Abt. II* 100: 177–93.

Drancourt, M., Bollet, C., Carta, A. and Rousselier, P. (2001). Phylogenetic analyses of Klebsiella species delineate Klebsiella and Raoultella gen. nov., with description of *Raoultella ornithinolytica* comb. nov., *Raoultella terrigena* comb. nov. and *Raoultella planticola* comb. nov. *Int. J. Syst. Evol. Microbiol.* 51: 925–32.

Du, J., Liu, Y. and Zhu, H. (2023). Genome-based analyses of the genus *Acidovorax*: proposal of the two novel genera *Paracidovorax* gen. nov., *Paenacidovorax* gen. nov. and the reclassification of *Acidovorax antarcticus* as *Comamonas antarctica* comb. nov. and emended description of the genus *Acidovorax*. *Arch. Microbiol.* 205: Article No. 42. https://doi.org/10.1007/s00203-022-03379-7

Duan, H. H., Zhao, J. Y., Liu, W. T., Xie, Q., Zheng, C. Y. *et al.* (2024). Complete genome sequence of *Paracidovorax avenae* causing red stripe in sugarcane. *BMC Genom. Data* 25: Article No. 89. https://doi.org/10.1186/s12863-024-01271-5

Dubnau, D., Smith, I., Morel, P. and Marmur, J. (1965). Gene conservation in *Bacillus* species. 1. Conserved genetic and nucleic acid base sequence homologies. *Proc. Natl. Acad. Sci. USA*, 54: 491–98.

Duin, I. M., Ritchie, D. F., Braswell, E. and Fagen, J. R. (2025). *Xanthomonas citri* pv. *lagerstroemium*, description of a new pathovar causing leaf spot on crape myrtle. *Plant Dis.* Vol. 109: pp. 1843–47. https://doi.org/10.1094/PDIS-09-24-2009-SC

Dutta, B., Gitaitis, R., Agarwal, G., Coutinho, T. and Langston, D. (2018). *Pseudomonas coronafaciens* sp. nov., a new phytobacterial species diverse from *Pseudomonas syringae*. *PLoS ONE*, 13(12): e0208271.

Dye, D. W. (1968). A study of the genus *Erwinia*. I. The '*amylovora*' group. *N. Z. J. Sci.* 11: 590–607.

Dye, D. W. (1969). A taxonomic study of the genus *Erwinia*. IV. "Atypical" erwinias. *New Zealand J. Sci.* 12: 833–39.

Dye, D. W., Bradbury, J. F., Goto, M. Hayward, A. C., Lelliott, R. A. *et al.* (1980). International standards for naming pathovars of phytopathogenic bacteria and a list of pathovar names and pathotype strains. *Rev. Plant. Pathol.* 59: 153–68.

El Zowalaty, M. E., Falgenhauer, L., Forsythe, S. and Helmy, Y. A. (2023). Draft genome sequences of rare *Lelliottia nimipressuralis* strain MEZLN61 and two *Enterobacter kobei* strains MEZEK193 and MEZEK194 carrying mobile colistin resistance gene mcr-9 isolated from wastewater in South Africa. *Jour. Glob. Antimicrob. Resis.* 33: 231–37. https://doi.org/10.1016/j.jgar.2023.03.007

Elbanna, K., Elnaggar, S. and Bakeer, A. (2014). Characterization of *Bacillus altitudinis* as a new causative agent of bacterial soft rot. *J. Phytopath.* 162: 712–22.

European and Mediterranean Plant Protection Organization (2014). PQR-EPPO database on quarantine pests. European and Mediterranean Plant Protection Organization, Paris, France. https://www.eppo.int/DATABASES/pqr/pqr.htm

Evtushenko, L. I., Dorofeeva, L. V., Subbotin, S. A. Cole, J. R. and Tiedje, J. M. (2000). *Leifsonia poae* gen. nov., sp. nov., isolated from nematode galls on *Poa annua*, and reclassification of '*Corynebacterium aquaticum'* Leifson 1962 as *Leifsonia aquatica* (ex Leifson 1962) gen. nov., nom. rev., comb. nov. and *Clavibacter xyli* Davis *et al.* 1984 with two subspecies as *Leifsonia xyli* (Davis *et al.* 1984) gen. nov., comb. nov. *Int. J. Syst. Evol. Microbiol.* 50: 371–80.

Ewing, W. H. (1962). The Tribe Proteae: Its nomenclature and taxonomy. *Int. Bull. Bacteriol. Nomen. Taxon.* 12: 93–102.

Fagen, J. R., Leonard, M. T., Coyle, J. F., McCullough, C. M., Davis-Richardson, A. G. *et al.* (2014). *Liberibacter crescens* gen. nov., sp. nov., the first cultured member of the genus *Liberibacter*. *Int. J. Syst. Evol. Microbiol.* 64: 2461–66. https://doi.org/10.1099/ijs.0.063255-0. PMID 24786353.

Fan, H. C., Zeng, L., Yang, P. W., Guo, Z. X. and Bai, T. T. (2016). First report of banana soft rot caused by *Klebsiella varicola* in China. *Plant Dis.* 100: 517.

Fang, W., Li, X., Tan, X.-M., Wang, L.-F., Piao, C.-F. *et al.* (2016). *Leucobacter populi* sp. nov. isolated from a symptomatic bark of *Populus* × *euramericana* canker. *Int. J. Syst. Evol. Microbiol.* 66: 2254–58.

Feng, G.-D., Yang, S.-Z., Xiong, X., Li, H.-P. and Zhu, H.-H. (2017). *Sphingomonas spermidinifaciens* sp. nov., a novel bacterium containing spermidine as the major polyamine, isolated from an abandoned lead-zinc mine and the emended description of the genus *Sphingomonas* and the species *Sphingomonas yantingensis* and *Sphingomonas japonica*. *Int. J. Syst. Evol. Microbiol.* 67: 2160–65.

Fernández-Garayzábal, J. F., Dominguez, L., Pascual, C. Jones, D. and Collins, M. D. (1995). Phenotypic and phylogenetic characterization of some unknown coryneform bacteria isolated from bovine blood and milk: description of *Sanguibacter* gen. nov. *Lett. Appl. Microbiol.* 20: 69–75. https://doi.org/10.1111/j.1472-765X.1995.tb01289.x

Ferraz, H. G. M., Badel, J. L., Neves, Y. F., Eloi, A. C. L., Vidigal, P. M. P. *et al.* (2024). *Xanthomonas* species causing leaf blight on eucalypt plants in Brazil and transfer of *Xanthomonas axonopodis* pv. *eucalyptorum* to *Xanthomonas citri* pv. *eucalyptorum* comb. nov. *Plant Pathol.* 73: 677–91.

Ferreira, L. C., de Moura, V. A. S., Poudel, M., Fullem, K. R., Jones, J. B. *et al.* (2025). *Pantoea graminicola* sp. nov., a Gram-negative bacterium isolated from sweet corn (*Zea mays* L.) in the USA. *Int. Jour. Syst. and Evol. Microbiol.* Vol. 75: No. 11. https://doi.org/10.1099/ijsem.0.006952

Francis, I. M., Jochimsen, K. N., De Vos, P. and van Bruggen, A. H. C. (2014). Reclassification of rhizosphere bacteria including strains causing corky root of lettuce and proposal of *Rhizorhapis suberifaciens* gen. nov., comb. nov., *Sphingobium mellinum* sp. nov., *Sphingobium xanthum* sp. nov. and *Rhizorhabdus argentea* gen. nov., sp. nov. *Int. J. Syst. Evol. Microbiol.* 64: 1340–50.

Frank, B. (1889). Über die Pilzsymbiose der Leguminosen. *Ber. Deutsch. Bot. Ges.* 7: 332–46.

Fuerst, J. A. and Hayward, A. C. (1969a). The sheathed flagellum of *Pseudomonas stizolobii*. *J. Clin. Microbiol.* 59: 239–45.

Fuerst, J. A. and Hayward, A. C. (1969b). Surface appendages similar to fimbriae (pili) on *Pseudomonas* species. *J. Gen. Microbiol.* 58: 227–37.

Fullem, K. R., MacLellan, M. P., Poudel, M., Goss, E. M., Potnis, N. *et al.* (2024). *Pseudomonas boreofloridensis* sp. nov. and *Pseudomonas citrulli* sp. nov., isolated from watermelon in Florida. *Int. J. Syst. and Evol. Microbiol.* 74: https://doi.org/10.1099/ijsem.0.006596

Fulton, J. C., Bec, S., Fayette, J., Ploetz, R. C., Garrett, K. A. *et al.* (2020). First report of plantain soft rot caused by *Klebsiella varicola* in Haiti. *Plant Dis.* 104: 1851.

Furtado, G. Q., Guimaraes, L. M. S., Lisboa, D. O., Cavalcante, G. P., Arriel, D. A. A. *et al.* (2012). First report of *Enterobacter cowanii* causing bacterial spot on *Mabea fistulifera*, a native forest species in Brazil. *Plant Dis.* 96: 1576.

Gardan, L., Bella, P., Meyer, J. M. Christen, R., Rott, P. *et al.* (2002). *Pseudomonas salomonii* sp. nov., pathogenic on garlic, and *Pseudomonas palleroniana* sp. nov., isolated from rice. *Int. J. Syst. Evol. Microbiol.* 52: 2065–74.

Gardan, L., Christen, R., Achouak, W. and Prior, P. (2004). *Erwinia papayae* sp. nov., a pathogen of papaya (*Carica papaya*). *Int. J. Syst. Evol. Microbiol.* 54: 107–13. https://doi.org/10.1099/ijs.0.02718-0.

Gardan, L., Dauga, C., Prior, P. Gillis, M. and Saddler, G. S. (2000). *Acidovorax anthurii* sp. nov., a new phytopathogenic bacterium which causes bacterial leaf-spot of anthurium. *Int. J. Syst. Evol. Microbiol.* 50: 235–46. https://doi.org/10.1099/00207713-50-1-235

Gardan, L., Gouy, C., Christen, R. and Samson, R. (2003a). Elevation of three subspecies of *Pectobacterium carotovorum* to species level: *Pectobacterium atrosepticum* sp. nov., *Pectobacterium betavasculorum* sp. nov. and *Pectobacterium wasabiae* sp. nov. *Int. J. Syst. Evol. Microbiol.* 53: 381–91. https://doi.org/10.1099/ijs.0.02423-0.

Gardan. L., Stead, D. E., Dauga, C. and Gillis, M. (2003b). *Acidovorax valerianellae* sp. nov., a novel pathogen of Lamb's lettuce [*Valerianella locusta* (L.) Laterr.]. *Int. J. Syst. Evol. Microbiol.* 53: 795–800.

Garnier, M. (2015). *Candidatus* Phlomobacter. In: *Bergey's Manual of Systematics of Archaea and Bacteria*. John Wiley and Sons, Hoboken, NJ, pp. 1–3.

Garrity, G. M., Bell, J. A. and Lilburn, T. (2005). The revised road map to the manual. In: D. J. Brenner, N. R. Krieg and J. T. Staley (volume editors), G. M. Garrity (editor-in-chief),

Bergey's Manual of Systematic Bacteriology. 2nd Ed. Vol. 2: *The Proteobacteria*, Part A, *Introductory Essays*. Springer-Verlag, New York, pp. 159–200.

Garrity, G. M., Brenner, D. J., Krieg, N. R. and Staley, J. T. (eds.) (2005). *Bergey's Manual of Systematic Bacteriology*, Vol. 2: *The Proteobacteria*, Part C: The Alpha-, Beta-, Delta- and Epsilonproteobacteria. New York: Springer. ISBN 978-0-387-24145-6.

Gašić, K., Zlatković, N. and Kuzmanović, N. (2022). Polyphasic study of phytopathogenic bacterial strains associated with deep bark canker of walnut in Serbia revealed a new species, *Brenneria izbisi* sp. nov. *Front. Plant Sci.* 13: https://doi.org/10.3389/fpls.2022.1055186

Gauthier, G., Gauthier, M. and Christen, R. (1995). Phylogenetic analysis of the genera *Alteromonas*, *Shewanella*, & *Moritella* using genes coding for small-subunit rRNA sequences and division of the genus *Alteromonas* into two genera, *Alteromonas* (emended) and *Pseudoalteromonas* gen. nov., and proposal of twelve new species combinations. *Int. J. Syst. Bacteriol.* 45: 755–61.

Gavini, F., Mergaert, J., Beji, A. Mielcarek, C., Izard, D. *et al.* (1989). Transfer of *Enterobacter agglomerans* (Beijerinck 1888) Ewing and Fife 1972 to *Pantoea* gen. nov. as *Pantoea agglomerans* comb. nov. and description of *Pantoea dispersa* sp. nov. *Int. J. Syst. Bacteriol.* 39: 337–45.

Gherna, R. L., Werren, J. H., Weisburg, W., Cote, R., Woese, C. R. *et al.* (1991). *Arsenophonus nasoniae* gen. nov., sp. nov., the causative agent of the son-killer trait in the parasitic wasp *Nasonia vitripennis*. *Int. J. Syst. Bacteriol.* 41: 563–65.

Gillis, M., van Van T., Bardin, R., Goor, M., Hebbar, P. *et al.* (1995). Polyphasic taxonomy in the genus *Burkholderia* leading to an emended description of the genus and proposition of *Burkholderia vietnamiensis* sp. nov. for N_2-fixing isolates from rice in Vietnam. *Int. J. Syst. Evol. Microbiol.* 45: 274–89.

Goncalves, E. R. and Rosato, Y. B. (2000). Genotypic characterization of xanthomonad strains isolated from passion fruit plants (*Passiflora* spp.) and their relatedness to different *Xanthomonas* species. *Int. J. Syst. Evol. Microbiol.* 50: 811–21.

Gonzalez, A. J. and Trapiello, E. (2014). *Clavibacter michiganensis* subsp. *phaseoli* subsp. nov., pathogenic in bean. *Int. J. Syst. Evol. Microbiol.* 64: 1752–55.

Goodfellow, M. (1984). Reclassification of *Corynebacteirum fascians* (Tilford) Dowson in the genus *Rhodococcus*, as *Rhodococcus fascians* comb. nov. *Syst. Appl. Microbiol.* 5: 225–29.

Goto, M. and Kuwata, H. (1988). *Rhizobacter daucus* gen. nov., sp. nov., the causal agent of carrot bacterial gall. *Int. J. Syst. Bacteriol.* 38: 233–39.

Goyer, C., Faucher, E. and Beaulieu, C. (1996). *Streptomyces caviscabies* sp. nov., from deep-pitted lesions in potatoes in Quebec, Canada. *Int. J. Syst. Bacteriol.* 46: 635–39.

Graham, D. C. (1964). Taxonomy of the soft rot coliform bacteria. *Annu. Rev. Phytopathol.* 2: 13–42.

Gray, J. S., Birmingham, J. M. and Fenton, J. I. (2010). Got black swimming dots in your cell culture? Identification of *Achromobacter* as a novel cell culture contaminant. *Biologicals.* 38: 273–77. https://doi.org/10.1016/j.biologicals.2009.09.006. PMC 2849847. PMID 19926304.

Grimont, P. A., Farmer, J. J. 3rd, Grimont, F., Asbury, M. A., Brenner, D. J. *et al.* (1983). *Ewingella americana* gen. nov., sp. nov., a new Enterobacteriaceae isolated from clinical specimens. *Ann. Microbiol. (Paris)*, 134A(1): 39–52. https://doi.org/10.1016/0769-2609(83)90102-3.

Grimont, P. A. D., Grimont, F. and Starr, M. P. (1978). *Serratia proteamaculans* (Paine and Stansfield) comb. nov., a senior subjective synonym of *Serratia liquefaciens* (Grimes and Hennerty) Bascomb *et al. Int. J. Syst. Bacteriol.* 28: 503–10.

Guo, Y., Zheng, W., Rong, X. and Huang, Y. (2008). A multilocus phylogeny of the *Streptomyces griseus* 16S rRNA gene clade: use of multilocus sequence analysis for streptomycete systematics. *Int. J. Syst. Evol. Microbiol.* 58: 149–59.

Guo, Z., Mai, Y., Mo, Q., Liang, J., Qi, X. *et al.* (2023). First report of *Enterobacter roggenkampii* and *E. cloacae* complex sp. causing tomato pith necrosis in Guangdong Province, China. *Plant Dis.* 107: 2212.

Gupta, R. S., Patel, S., Saini, N. and Chen, S. (2020). Robust demarcation of 17 distinct *Bacillus* species clades, proposed as novel *Bacillaceae* genera, by phylogenomics and comparative genomic analyses: description of *Robertmurraya kyonggiensis* sp. nov. and proposal for an emended genus *Bacillus* limiting it only to the members of the Subtilis and Cereus clades of species. *Int. J. Syst. Evol. Micro.* 70: 5753–98. https://doi.org/10.1099/ijsem.0.004475.

Halpern, M., Fridman, S., Aizenberg-Gershtein, Y. and Izhaki, I. (2013). Transfer of *Pseudomonas flectens* Johnson 1956 to *Phaseolibacter* gen. nov., in the family Enterobacteriaceae, as *Phaseolibacter flectens* gen. nov., comb. nov. *Int. J. Syst. Evol. Micro.* 63: 268–73.

Han, V.-C., Yu, N. H., Park, A. R., Yoon, H., Son, Y. K. *et al.* (2021). First report of shot-hole on flowering cherry caused by *Burkholderia contaminans* and *Pseudomonas syringae* pv. *syringae*. *Plant Dis.* 105: 3795–3802.

Hanson, M. A., Dostálová, A., Ceroni, C., Poidevin, M., Kondo, S. *et al.* (2019). Synergy and remarkable specificity of antimicrobial peptides *in vivo* using a systematic knockout approach. *eLife.* 8. https://doi.org/10.7554/eLife.44341. PMC 6398976. PMID 30803481

Harmon, C. L., Timilsina, S., Bonkowski, J., Jones, D. D., Sun, X. *et al.* (2018). Bacterial gall of *Loropetalum chinense* caused by *Pseudomonas amygdali* pv. *loropetali* pv. nov. *Plant Dis.* 102: 799–806.

Harrison, N. A., Davis, R. E., Oropeza, C., Helmick, E. E., Narvaez, M. *et al.* (2014). "*Candidatus* Phytoplasma palmicola," associated with a lethal yellowing-type disease of coconut (*Cocos nucifera* L.) in Mozambique. *Int. J. Syst. Evol. Microbiol.* 64: 1890–99.

Hauben, L., Moore, E. R. B., Vauterin, L. Steenackers, M., Mergaert, J. *et al.* (1999). Validation of publication of new names and new combinations previously effectively published outside the IJSB. *Int. J. Syst. Bacteriol.* Vol. 49: pp. 1–3. Effective publication: Hauben, L., Moore, E. R. B., Vauterin, L. Steenackers, M., Mergaert, J. *et al.* (1998). Phylogenetic position of phytopathogens within the *Enterobacteriaceae*. *Syst. Appl. Microbiol.* 21: 384–97.

Hauben, L. and Swings, J. (2005). Genus *Brenneria*. In: D. J. Brenner, N. R. Krieg and J. T. Staley (eds.). *Bergey's Manual of Systematic Bacteriology*. 2nd ed. Vol. 2: The Proteobacteria, Part B, The Gammaproteobacteria, Springer, New York, pp. 628–33.

He, Q., Li, J., Liu, Q., Li, Y., Liao, X. *et al.* (2023). Complete genome sequence resource of *Pantoea anthophila* CL1 causing soft rot disease in *Clausena lansium* (Wampee) in China. *Plant Dis.* 107: 1613–16.

Hildebrand, D. C. and Schroth, M. N. (1967). A new species of *Erwinia* causing the drippy nut disease of live oaks. *Phytopathology* 57: 250–53.

Hiruki, C. and Wang, K. (2004). Clover proliferation phytoplasma: '*Candidatus* Phytoplasma trifolii'. *Int. J. Syst. Evol. Microbiol.* 54: 1349–53.

Hodgetts, J., Boonham, N., Mumford, R. Harrison, N. and Dickinson, M. *et al.* (2008). Phytoplasma phylogenetics based on analysis of *secA* and 23S rRNA gene sequences for improved resolution of candidate species of '*Candidatus* Phytoplasma'. *Int. J. Syst. Evol. Microbiol.* 58: 1826–37.

Hollis, D. G., Hickman, F. W., Fanning, G. R., Farmer, III, J. J., Weaver, R. E. *et al.* (1982). Validation List No. 8. *Int. J. Syst. Bacteriol.* 32: 266–68. Effective publication: Hollis, D. G., Hickman, F. W., Fanning, G. R., Farmer, III, J. J., Weaver, R. E. *et al.* (1981). *Tatumella ptyseos* gen. nov., sp. nov., a member of the family Enterobacteriaceae found in clinical specimens. *J. Clin. Microbiol.* 14: 79–88.

Holt, J. G., Krieg, N. R., Sneath P. H. A., Staley, J. T. and Williams, S. T. (eds.) (1994). *Bergey's Manual of Determinative Bacteriology*. 9th Ed. Williams and Wilkins, Baltimore, MD, 787 pp.

Hong, S. M., Ten, L. N., Park, K. T. Back, C. G., Waleron, M. *et al.* (2023). *Pectobacterium jejuense* sp. nov. isolated from cucumber stem tissue. *Curr. Microbiol.* 80: 308. https://doi.org/10.1007/s00284-023-03419-5

Hopkins, D. L. and Purcell, A. H. (2002). *Xylella fastidiosa*: Cause of Pierce's disease of grapevine and other emergent diseases. *Plant Dis.* 86: 1056–66.

Hormaeche, E. and Edwards, P. R. (1960). A proposed genus *Enterobacter*. *Int. Bull. Bacteriol. Nomen. Taxon.* 10: 71–74.

Hou, Y., Zhang, Y., Yu, L., Ding, X., Liu, L. *et al.* (2020). First report of *Pseudomonas oryzihabitans* causing rice panicle blight and grain discoloration in China. *Plant Dis.* 104: 3055.

Hua, Y., Zheng, Z., Wang, J., Zhu, L. and Zhou, C. (2024). First report of wilt disease of ginger caused by *Achromobacter xylosoxidans* in China. *Plant Dis.* 108: 778. https://doi.org/10.1094/PDIS-05-23-0891-PDN

Huang, C.-H., Vallad, G. E., Adkison, H., Summers, C., Margenthaler, E. *et al.* (2013). A novel *Xanthomonas* sp. causes bacterial spot of rose (*Rosa* spp.). *Plant Dis.* 97: 1301–07.

Huang, C.-J., Wu, T.-L., Zheng, P.-X., Ou, J.-Y., Ting, C.-L. *et al.* (2023). Complete genome sequence resource of *Pectobacterium colocasium* strain F1-1 that causes soft rot disease of taro. *Plant Dis.* 107: 1911–15.

Huang, Z., Huang, H., Ding, C., Liu, Y., Liu, Y. *et al.* (2023). First report of *Arthrobacter woluwensis* causing brown blotch disease on white button mushroom (*Agaricus bisporus*). *Plant Dis.* 107: 2840.

Huang, Z., Wei, Z., Chen, H., Liu, Y., Liu, Y. *et al.* (2023). First report of *Pseudomonas mosselii* causing white blotch disease in *Pleurotus pulmonarius* in China. *Plant Dis.* 107: 549.

Hugh, R. and Gilardi, G. L. (1980). *Pseudomonas*. In: E. H. Lennette, A. Balows, W. J. Hausler, Jr. and J. P. Truant (eds.). *Manual of Clinical Microbiology*, 3rd Ed. American Society for Microbiology, Washington, DC, pp. 288–317.

Hugh, R. and Ryschenkow, E. (1961). *Pseudomonas maltophilia*, an Alcaligenes-like species. *J. Gen. Microbiol.* 26: 123–32.

Hugouvieux-Cotte-Pattat, N., des-Combes, C. J., Briolay, J. and Pritchard, L. (2021). Proposal for the creation of a new genus *Musicola* gen. nov., reclassification of *Dickeya paradisiaca* (Samson *et al.* 2005) as *Musicola paradisiaca* comb. nov. and description of a new species *Musicola keenii* sp. nov. *Int. J. Syst. Evol. Micro.* 71: 5037.

Hugouvieux-Cotte-Pattat, N. and Van Gijsegem, F. (2021). Diversity within the *Dickeya zeae* complex, identification of *Dickeya zeae* and *Dickeya oryzae* members, proposal of the novel species *Dickeya parazeae* sp. nov. *Int. J. Syst. Evol. Microbiol.* 71: https://doi.org/10.1099/ijsem.0.005059

Huguet-Tapia, J. C., Lefebure, T., Badger, J. H., Guan, D., Pettis, G. S. *et al.* (2016). Genome content and phylogenomics reveal both ancestral and lateral evolutionary pathways in plant pathogenic *Streptomyces* species. *Appl. Environ. Microbiol.* 82: 2146–55.

Hulse, M., Johnson, S. and Ferrieri, P. (1993). *Agrobacterium* infections in humans: experience at one hospital and review. *Clin. Infect. Dis.* 16: 112–17.

Iizuka, H. and Komagata, K. (1964a). Microbiological studies on petroleum and natural gas. I. Determination of hydrocarbon utilizing bacteria. *J. Gen. Appl. Microbiol.* 10: 207–21.

Iizuka, H. and Komagata, K. (1964b). Microbiological studies on petroleum and natural gas. II. Determination of pseudomonads isolated from oil-brines and related materials. *J. Gen. Appl. Microbiol.* 10: 223–31.

Ikemoto, S., Suzuki, K., Kaneko, T. and Komagata, K. (1980). Characterization of strains of *Pseudomonas maltophilia* which do not require methionine. *Int. J. Syst. Bacteriol.* 30: 437–47.

Ilyas, N., Yang, Y., Liu, W., Li, X., Pu, W. *et al.* (2021). First report of bacterial rot caused by *Pantoea endophytica* on tobacco in Liuyang, China. *Plant Dis.* 105: 4147.

IRPCM Phytoplasma/Spiroplasma Working Team – Phytoplasma taxonomy group (2004). '*Candidatus* Phytoplasma', a taxon for the wall-less, non-helical prokaryotes that colonize plant phloem and insects. *Int. J. Syst. Evol. Microbiol.* 54: 1243–55.

Ivanova, N., Sikorski, J., Sims, D., Brettin, T., Detter, J. C. *et al.* (2009). Complete genome sequence of *Sanguibacter keddieii* type strain (ST-74^{T}). *Stand Genomic Sci.* 1: 110–18. https://doi.org/10.4056/sigs.16197

Jacquemyn, H., Lenaerts, M., Brys, R., Willems, K., Honnay, O. *et al.* (2013). Among-population variation in microbial community structure in the floral nectar of the bee-pollinated forest herb *Pulmonaria officinalis* L. *Plos ONE.* **8** (3): e56917.

Jagoueix-Eveillard, S., Bove, J. M. and Garnier, M. (1994). The phloem-limited bacterium of greening disease of citrus is a member of the ∝ subdivision of the *Proteobacteria. Int. J. Syst. Bacteriol.* 44: 379–86.

Jardim, B. R., Kinoti, W. M., Tran-Nguyen, L. T. T., Gambley, C., Rodoni, B. *et al.* (2021). '*Candidatus* Phytoplasma stylosanthis', a novel taxon with a diverse host range in Australia, characterised using multilocus sequence analysis of 16S rRNA, *secA*, *tuf*, and *rp* genes. *Int. J. Syst. Evol. Microbiol.* 71: 004589.

Jiao, S. (2022). *Isolation, identification and genomic analysis of bacteriophage of mulberry bacterial wilt.* MD dissertation. Huazhong Agricultural University, Wuhan.

Jin, S., Wang, N., Zhou, Z., Ji, Z. and Zhang, J. (2022). Identification of *Bacillus* species causing bacterial leaf spot of peach in China. *J. Phytopath.* 170: 811–19.

Johnson J. C. (1956). Pod twist: a previously unrecorded bacterial disease of French bean (*Phaseolus vulgaris* L.). *Queensland J. Agricultural Sci.* 13: 127–58.

Jones, J. B., Lacy, G. H., Bouzar, H. Stall, R. E., Schaad, N. W. *et al.* (2004). Reclassification of the *xanthomonads* associated with bacterial spot disease of tomato and pepper. *Syst. Appl. Microbiol.* 27: 755–62.

Jones, L. M., Pease, B., Perkins, S. L., Constable, F. E., Kinoti, W. M. *et al.* (2021). '*Candidatus* Phytoplasma dypsidis', a novel taxon associated with a lethal wilt disease of palms in Australia. *Int. J. Syst. Evol. Micro.* 71: 10.1099/ijsem.0.004818

Judicial Commission of the International Committee on Systematics of Prokaryotes. (2008). *Corynebacterium ilicis* is typified by ICMP 2608 =ICPB CI144, *Arthrobacter ilicis* is typified by DSM 20138 =ATCC 14264 =NCPPB 1228 and the two are not homotypic synonyms, and clarification of the authorship of these two species. Opinion 87. *Int. J. Syst. Evol. Microbiol.* 58: 1976–78.

Juneja, P. and Lazzaro, B. P. (2009). *Providencia sneebia* sp. nov. and *Providencia burhodogranariea* sp. nov., isolated from wild *Drosophila melanogaster*. *Int. J. Syst. Evol. Microbiol.* 59: 1108–11. https://doi.org/10.1099/ijs.0.000117-0. PMID 19406801

Jung, H. Y., Sawayanagi, T., Kakizawa, S. Nishigawa, H., Miyata, S.-I. *et al.* (2002). '*Candidatus* Phytoplalsma castaneae', a novel phytoplasma taxon associated with chestnut witches' broom disease. *Int. J. Syst. Evol. Microbiol.* 52: 1543–49.

Jung, H. Y., Sawayanagi, T., Kakizawa, S. Nishigawa, H., Wei, W. *et al.* (2003). '*Candidatus* Phytoplasma ziziphi', a novel phytoplasma taxon associated with jujube witches'-broom disease. *Int. J. Syst. Evol. Microbiol.* 53: 1037–41.

Jung, H. Y., Sawayanagi, T., Wongkaew, P., Kakizawa, S., Nishigawa, H. *et al.* (2003). '*Candidatus* Phytoplasma oryzae', a novel phytoplasma taxon associated with rice yellow dwarf disease. *Int. J. Syst. Evol. Microbiol.* 53: 1925–29.

Kado, C. I. (2010). *Plant Bacteriology*. The American Phytopathological Society, St. Paul, MN, 336 pp.

Kageyama, B., Nakae, M., Yagi, S. and Sonoyama, T. (1992). *Pantoea punctata* sp. nov., *Pantoea citrea* sp. nov., and *Pantoea terrea* sp. nov. isolated from fruit and soil samples. *Int. J. Syst. Bacteriol.* 42: 203–10.

Kałużna, M., Willems, A., Pothier, J. F., Ruinelli, M., Sobiczewski, P. *et al.* (2017). Validation list No. 173. *Int. J. Syst. Evol. Microbiol.* 67: 1–3. Effective publication: Kałużna, M., Willems, A., Pothier, J. F., Ruinelli, M., Sobiczewski, P. *et al.* (2016). *Pseudomonas cerasi* sp. nov. (non Griffin, 1911) isolated from diseased tissue of cherry. *Syst. Appl. Microbiol.* 39: 370–77.

Kämpfer, P., Busse, H.-J., Schumann, P., Criscuolo, A., Clermont, D. *et al.* (2020). *Arthrobacter ulcerisalmonis* sp. nov., isolated from an ulcer of a farmed Atlantic salmon (*Salmo salar*), and emended description of the genus *Arthrobacter sensu lato*. *Int. J. Syst. Evol. Micro.* 70: 1963–68.

Kawaguchi, A. (2022). Risk assessment of inferior growth and death of grapevines due to crown gall. *Eur. J. Plant Pathol.* 164: 613–18. https://doi.org/10.1007/s10658-022-02577-9

Khanal, M., Bhatta, B. P. and Malla, S. (2023a). Isolation and characterization of bacteria associated with onion and first report of onion diseases caused by five bacterial pathogens in Texas, USA. *Plant Dis.* 107: 1721–29.

Khanal, M., Bhatta, B. P., Timilsina, S., Ghimire, S., Cochran, K. *et al.* (2023b). *Curtobacterium allii* sp. nov., the actinobacterial pathogen causing onion bulb rot. *Antonie van Leeuwenhoek* 116: 83–96. https://doi.org/10.1007/s10482-022-01775-z

Khanal, M., Timilsina, S., Bhatta, B. P., Bophela, K., Coutinho, T. *et al.* (2022). *Pseudomonas uvaldensis* sp. nov., a bacterial pathogen causing onion bulb rot. *Int. J. Syst. Evol. Microbiol.* 72: https://doi.org/10.1099/ijsem.0.005311. PMID: 35442877.

Khayi, S., Cigna, J., Chong, T. M., Quetu-Laurent, A., Chan, K.-G. *et al.* (2016). Transfer of the potato plant isolates of *Pectobacterium wasabiae* to *Pectobacterium parmentieri* sp. nov. *Int. J. Syst. Evol. Microbiol.* 66: 5379–83.

Kiel, A., Helweg, L. P., Kaltschmidt, B. P., Wohllebe, M. P., Hitz, F. *et al.* (2024). *Sanguibacter biliveldensis* sp. nov., a Gram-positive mesophilic bacterium isolated from plant lesions. *Int. J. Syst. and Evol. Microbiol.* 74. https://doi.org/10.1099/ijsem.0.006560

Kile, H., Arnold, D., Allainguilaume, J., Denman, S. and Brady, C. (2022). *Brenneria tiliae* sp. nov., isolated from symptomatic *Tilia × moltkei* and *Tilia × europaea* trees in the UK. *Int. J. Syst. Evol. Microbiol.* 72: https://doi.org/10.1099/ijsem.0.005515

Kim, P. S., Shin, N.-R., Kim, J. Y., Yun, J.-H., Hyun, D.-W. *et al.* (2013). *Gibbsiella papilionis* sp. nov., isolated from the intestinal tract of the butterfly *Mycalesis gotama*, and emended description of the genus *Gibbsiella*. *Int. J. Syst. Evol. Microbiol.* 63: 2607–11.

Kim, W. S., Gardan, L., Rhim, S. L. and Geider, K. (1999). *Erwinia pyrifoliae* sp. nov., a novel pathogen that affects Asian pear trees (*Pyrus pyrifolia* Nakai). *Int. J. Syst. Bacteriol.* 49: 899–906.

Kirdat, K., Tiwarekar, B., Thorat, V., Sathe, S., Shouche, Y. *et al.* (2021). '*Candidatus* Phytoplasma sacchari', a novel taxon-associated with Sugarcane Grassy Shoot (SCGS) disease. *Int. J. Syst. Evol. Micro.* 71. https://doi.org/10.1099/ijsem.0.004591.

Kiyan, V., Smagulova, A., Kukhar, Y., Savin, T., Bekenova, A. *et al.* (2024). Morphological and molecular characterization of bacterial pathogens associated with leaf mottle of sunflower in Northern Kazakhstan. *Plant Dis.* 108: 264–69. https://doi.org/10.1094/PDIS-07-23-1352-SR

Klerks, M. M., van Gent-Pelzer, M., Franz, E., Zijlstra, C. and van Bruggen, A. H. C. (2007). Physiological and molecular responses of *Lactuca sativa* to colonization by *Salmonella enterica* serovar Dublin. *Appl. Environ. Microbiol.* 73: 4905–14.

Koch, C., Schumann, P. and Stackebrandt, E. (1995). Reclassification of *Micrococcus agilis* (Ali-Cohen 1889) to the genus *Arthrobacter* as *Arthrobacter agilis* comb. nov. and emendation of the genus *Arthrobacter*. *Int. J. Syst. Bacteriol.* 45: 837–39. Anonymous. Notification list. Notification that new names and new combinations have appeared in volume 45, part 4 of the IJSB. *Int. J. Syst. Bacteriol.* 46: 364–65.

Koirala, S., Zhao, M., Agarwal, G., Gitaitis, R., Stice, S. *et al.* (2021). Identification of two novel pathovars of *Pantoea stewartii* subsp. *indologenes* effecting *Allium* sp. and millets. *Phytopathology*, 111: 1509–19.

Kolozsváriné Nagy, J., Fodor, J., Bozsó, Z., Ágoston, J., Dlauchy, D. *et al.* (2024). First report of *Rhodococcus fascians* causing leafy gall on *Iberis sempervirens* in Hungary. *Phytopath. Mediterr.* 63: 465–73.

Kontaxis, D. G. and Hayward, A. C. (1978). The pathogen and symptomatology of pink disease of pineapple fruit in the Philippines. *Plant Dis. Reptr*, 62: 446–50.

Kuzmanović, N., diCenzo, G. C., Bunk, B., Spröer, C., Frühling, A. *et al.* (2023). Genomics of the "tumorigenes" clade of the family *Rhizobiaceae* and description of *Rhizobium rhododendri* sp. nov. *Microbiology* 12: e1352. https://doi.org/10.1002/mbo3.1352

Kuzmanović, N., Puławska, J., Prokić, A., Ivanović, M., Zlatković, N. *et al.* (2015). *Agrobacterium arsenijevicii* sp. nov., isolated from crown gall tumors on raspberry and cherry plum. *Syst. Appl. Microbiol.* 38: 373–78.

Kuzmanović, N., Puławska, J., Smalla, K. and Nesme, X. (2018). Agrobacterium rosae sp. nov., isolated from galls on different agricultural crops. Syst. *Appl. Microbiol.* 41: 191–97.

Kuzmanović, N., Smalla, K., Gronow, S. and Puławska, J. (2018). *Rhizobium tumorigenes* sp. nov., a novel plant tumorigenic bacterium isolated from cane gall tumors on thornless blackberry. *Sci. Rep.* 8: Article No. 9051. https://doi.org/10.1038/s41598-018-27485-z

Labeda, D. P. (2011). Multilocus sequence analysis of phytopathogenic species of the genus *Streptomyces*. *Int. J. Syst. Evol. Microbiol.* 61: 2525–31.

Lapage, S. P., Sneath, P. H. A., Lessel, E. F., Skerman, V. B. D., Seeliger, H. P. R. *et al.* (1975). *International Code of Nomenclature of Bacteria: 1976 Revision*. American Society for Microbiology, Washington, DC.

Le Flèche-Matéos, A., Kügler, J. H., Hansen, S. H., Syldatk, C., Hausmann, R. *et al.* (2017). *Rouxiella badensis* sp. nov. and *Rouxiella silvae* sp. nov. isolated from peat bog soil and emendation description of the genus *Rouxiella*. *Int. J. Syst. Evol. Microbiol.* 67: 1255–59. https://doi.org/10.1099/ijsem.0.001794.

Le Flèche-Matéos, A., Levast, M., Lomprez, F., Arnoux, Y., Andonian, C. *et al.* (2015). *Rouxiella chamberiensis* gen. nov., sp. nov., a member of the family Enterobacteriaceae isolated from parenteral nutrition bags. *Int. J. Syst. Evol. Microbiol.* 65: 1812–18. https://doi.org/10.1099/ijs.0.000179. PMID 25747423.

Lee, C. J., Jhune, C. S., Cheong, J. C., Yun, H. S. and Cho, W. D. (2009). Occurrence of internal stipe necrosis of cultivated mushrooms (*Agaricus bisporus*) caused by *Ewingella americana* in Korea. *Mycobiology*, 37: 62–66. https://doi.org/10.4489/MYCO.2009.37.1.062

Lee, I. M., Bottner, K. D., Secor, G. and Rivera-Vara, V. (2006). '*Candidatus* Phytoplasma americanum', a phytoplasma associated with a potato purple top wilt disease complex. *Int. J. Syst. Evol. Microbiol.* 56: 1593–97.

Lee, I.-M., Bottner-Parker, K. D., Zhao, Y., Villalobos, W. and Moreira, L. (2011). "*Candidatus* Phytoplasma costaricanum": a novel phytoplasma associated with an emerging disease in soybean (*Glycine max*). *Int. J. Syst. Evol. Microbiol.* 61: 2822–26.

Lee, I-M., Bottner-Parker, K. D., Zhao, Y., Davis, R. E. and Harrison, N. A. (2010). Phylogenetic analysis and delineation of phytoplasmas based on secY gene sequences. *Int. J. Syst. Evol. Microbiol.* 60: 2887–97. DOI: 10.1099/ijs.0.019695-0

Lee, I. M., Davis, R. E. and Gundersen-Rindal, D. E. (2000). Phytoplasma: Phytopathogenic mollicutes. *Annu. Rev. Microbiol.* 54: 221–55.

Lee, I. M., Gundersen-Rindal, D. E., Davis, R. E., Bottner, K. D., Marcone, C. *et al.* (2004). '*Candidatus* Phytoplasma asteris', a novel phytoplasma taxon associated with aster yellows and related diseases. *Int. J. Syst. Evol. Microbiol.* 54: 1037–48.

Lee, I. M., Martini, M., Marcone, C. and Zhu, S. F. (2004). Classification of phytoplasma strains in the elm yellows group (16SrV) and proposal of '*Candidatus* Phytoplalsma ulmi' for the phytoplasma associated with elm yellows. *Int. J. Syst. Evol. Microbiol.* 54: 337–47.

Lee, S. D., Yang, H. L. and Kim, I. S. (2021). *Rouxiella aceris* sp. nov., isolated from tree sap and the emended description of the genus *Rouxiella*. *Anton. Van Leeuwen.* 114: 1013–24.

Lehmann, K. B. and Neumann, R. (1896). *Atlas und Grundriss der Bakteriologie und Lehrbuch der speciellen bakteriologischen D bniagnostik*, 1st Ed., J. F. Lehmann, Munchen.

Lelliott, R. A. and Dickey, R. S. (1984). Genus VII. *Erwinia* Winslow, Broadhurst, Buchanan, Krumwiede, Rogers and Smith 1920. In: N. R. Krieg, and J. G. Holt (eds.), *Bergey's Manual of Systematic Bacteriology*. I. Williams and Wilkins, Baltimore, MD, MD, 469–76.

Li, J.-H., Qin, S.-Y., Shi, Y.-X., Xie, X.-W., Chai, A.-L. *et al.* (2025). First report of *Lelliottia amnigena* causing soft rot on purple stem mustards in China. *Plant Dis.* 109: 228. https://doi.org/10.1094/PDIS-01-24-0180-PDN

Li, X. and De Boer, S. H. (2019). Descriptions of *Clavibacter insidiosus* sp. nov. and *Clavibacter tessellarius* sp. nov. *Int. J. Syst. Evol. Micro.* 69: 2069.

Li, X., Tambong, J., Yuan, K. X., Chen, W., Xu, H. *et al.* (2018). Reclassification of *Clavibacter michiganensis* subspecies on the basis of whole-genome and multilocus sequence analyses. *Int. J. Syst. Evol. Microbiol.* 68: 234–40.

Li, Y., Chang, J., Guo, L.-M., Wang, H.-M., Xie, S.-J. *et al.* (2015). Description of *Acinetobacter populi* sp. nov. isolated from symptomatic bark of *Populus* × *euramericana* canker. *Int. J. Syst. Evol. Microbiol.* 65: 4461–68.

Li, Y., Fang, W., Xue, H., Liang, W-x., Wang, L.-F. *et al.* (2015). *Brenneria populi* sp. nov., isolated from symptomatic bark of *Populus* × *euramericana* canker. *Int. J. Syst. Evol. Microbiol.* 65: 432–37.

Li, Y., He, W., Wang, T., Piao, C.-G., Guo, L.-M. *et al.* (2014). *Acinetobacter qingfengensis* sp. nov., isolated from canker bark of *Populus* × *euramericana*. *Int. J. Syst. Evol. Microbiol.* 64: 1043–50.

Li, Y., Wang, D., Cao, S., Wang, X. and Ren, G. (2023). First report of bacterial leaf spot on tobacco caused by *Pseudomonas psychrotolerans* in China. *Plant Dis.* 107: 935.

Li, Y., Wang, S., Bian, D., Sun, S. and Ma, T. (2021). *Pseudomonas quercus* sp. nov, associated with leaf spot disease of *Quercus mongolica*. *Int. J. Syst. Evol. Micro.* 71

Li, Y., Xue, H., Guo, L.-M., Koltay, A., Palacio-Bielsa, A. *et al.* (2017). Elevation of three subspecies of *Lonsdalea quercina* to species level: *Lonsdalea britannica* sp. nov., *Lonsdalea iberica* sp. nov. and *Lonsdalea populi* sp. nov. *Intr. J. Syst. Evol. Microbiol.* 67: 4680–84.

Lignieres, J. (1900). Maladies du porc. *Bull. Soc. Cent. Med. Vet.*, 18: 389–431.

Lin, P.-R., Ting, C.-L., Sun, Y.-C. and Huang, C.-J. (2022). First report of *Pectobacterium colocasium* causing bacterial soft rot of taro in Taiwan. *J. Plant Pathol.* 104: 1577–78. https://doi.org/10.1007/s42161-022-01204-0

Lincoln, S. P., Fermor, T. R. and Tindall, B. J. (1999). *Janthinobacterium agaricidamnosum* sp. nov., a soft rot pathogen of *Agaricus bisporus*. *Int. J. Syst. Bacteriol.* 49: 1577–89.

Liu, J-Y, Zhang, J-F, Wu, H-L, Chen, Z, Li, S-Y *et al.* (2023). Proposal to classify *Ralstonia solanacearum* phylotype I strains as *Ralstonia nicotianae* sp. nov., and a genomic comparison between members of the genus *Ralstonia*. *Front. Microbiol.* 14: 1135872. https://doi.org/10.3389/fmicb.2023.1135872

Liu, Y., Sun, W., Chen, X., Huang, S., Tang, L. *et al.* (2023). Bacterial leaf spot of plum caused by *Sphingomonas spermidinifaciens* in Guangxi, China. *Plant Dis.* 107: 2839.

Liu, Z. H., Sossah, F. L., Li, Y. and Fu, Y. P. (2018). First report of *Ewingella americana* causing bacterial Brown Rot Disease on cultivated needle mushroom (*Flammulina velutipes*) in China. *Plant Dis.* 102: 2633.

Loganathan, M., Thangavelu, R., Pushpakanth, P., Muthubharathi, K., Ramesh, R. *et al.* (2021). First report of rhizome rot of banana caused by *Klebsiella variicola* in India. *Plant Dis.* 105: 2011.

Lopez, M. M., Lopez-Soriano, P., Garita-Cambronero, J., Beltran, C., Taghouti, G. *et al.* (2018). *Xanthomonas prunicola* sp. nov., a novel pathogen that affects nectarine (*Prunus persica* var. *nectarina*) trees. *Int. J. Syst. Evol. Microbiol.* 68: 1857–66.

Lopez, M. M., Rosello, M., Llop, P., Ferrer, S., Christen, R. *et al.* (2011). *Erwinia piriflorinigrans* sp. nov., a novel pathogen that causes necrosis of pear blossoms. *Int. J. Syst. Evol. Microbiol.* 61: 561–67.

Lu, C.-H., Han, T.-H., Jiang, N., Gai, X.-T., Cao, Z.-H. *et al.* (2022). *Pseudomonas lijiangensis* sp. nov., a novel phytopathogenic bacterium isolated from black spots of tobacco. *Int. J. Syst. Evol. Microbiol.* 72: https://doi.org/10.1099/ijsem.0.005591

Lu, H., Patil, P., Van Sluys, M.-A., White, F. F., Ryan, R. P. *et al.* (2008). Acquisition and evolution of plant pathogenesis-associated gene clusters and candidate determinants of tissue-specificity in *Xanthomonas*. *PLoS ONE* 3: e3828. https://doi.org/10.1371/journal.pone.0003828

Luo, M., Sheng, Q., Wang, C. L. and Zhang, X. L. (2018). First report of fruit spot on pepper caused by *Erwinia aphidicola* in China. *Plant Dis.* 102: 1445.

Ma, Y., Rong, C., Wang, S., Liu, Y., Wang, J. *et al.* (2016). Development of a species-specific real-time PCR assay for diagnosing *Pantoea beijingensis*. *J. Plant Pathol.* 98: 365–68.

Madigan, M. and Martinko, J. eds. (2005). *Brock Biomlogy of Microorganisms* (11th Ed.). Prentice Hall, Hoboken, NJ. ISBN 978-0-13-144329-7.

Mafakheri, H., Taghavi, S. M., Khezerpour, K., Kuzmanović, N. and Osdaghi, E. (2022). Genomic analyses of rose crown gall-associated bacteria revealed two new Agrobacterium species: Agrobacterium burrii sp. nov. and Agrobacterium shirazense sp. nov. *Phytopathology*, 112: 1208–13.

Mafakheri, H., Taghavi, S. M., Zarei, S., Portier, P., Dimkić, I. *et al.* (2022). Xanthomonas bonasiae sp. nov. and Xanthomonas youngii sp. nov., isolated from crown gall tissues. *Int. J. Syst. Evol. Microbiol.* 72: https://doi.org/10.1099/ijsem.0.005418

Malembic-Maher, S., Salar, P., Filippin, L., Carle, P., Angelini, E. *et al.* (2011). Genetic diversity of European phytoplasmas of the 16SrV taxonomic group and proposal of "*Candidatus* Phytoplasma rubi". *Int. J. Syst. Evol. Microbiol.* 61: 2129–34.

Malik, V. K., Sangwan, P., Singh, M., Punia, R., Yadav, D. V. *et al.* (2022). First report of *Klebsiella aerogenes* inciting stem rot of pearl millet in Haryana, India. *Plant Dis.* 106: 754.

Manna, S., Santander, R. D. and Zhao, Y. (2024). First report of *Pseudomonas amygdali* pv. *morsprunorum* causing bacterial canker in sweet cherry orchards in Washington state. *Plant Dis.* 108: 2560. https://doi.org/10.1094/PDIS-04-24-0718-PDN

Mandel, M., Guba, E. F. and Litsky, W. (1961). The causal agent of bacterial blight of American holly. *Bacteriol. Proc. (1961)*, 61: [Google Scholar].

Manzo-Valencia, M. K., Cervantes-Juan, M. M., Mozqueda-Cervantes, M. d. S., Olalde-Portugal, V., Arias-Padró, M. D. *et al.* (2023). Amaranth cystatin inhibits the growth of broccoli (*Brassica oleracea* var. *italica* L.) pathogenic bacteria. *J. Phytopathol.* 171: 152–64.

Marcone, C., Gibb, K. S., Streten, C. and Schneider, B. (2004). '*Candidatus* Phytoplasma spartii', '*Candidatus* Phytoplasma rhamni' and '*Candidatus* Phytoplasma allocasuarinae', respectively associated with spartium witches'-broom, buckthorn witches'-broom and allocasuarina yellows diseases. *Int. J. Syst. Evol. Microbiol.* 54: 1025–29.

Marcone, C., Schneider, B. and Seemuller, E. (2004). '*Candidatus* Phytoplasma cynodontis', the phytoplasma associated with Bermuda grass white leaf disease. *Int. J. Syst. Evol. Microbiol.* 54: 1077–82.

Marin, M. V., Carvalho, R., Paret, M. L., Jones, J. B. and Peres, N. A. (2024). *Pseudomonas fragariae* sp. nov., a novel bacterial species causing leaf spots on strawberry (*Fragaria×ananassa*). *Int. J. Syst. Evol. Microbiol.* Vol. 74: https://doi.org/10.1099/ijsem.0.006476

Martin, K., Schumann, P., Rainey, F. A. Schuetze, B. and Groth, I. (1997). *Janibacter limosus* gen. nov., sp. nov., a new actinomycete with *meso*-diaminopimelic acid in the cell wall. *Int. J. Syst. Bacteriol.* 47: 529–34.

Martini, M., Lee, I. M., Bottner, K. D., Zhao, Y., Botti, S. *et al.* (2007). Ribosomal protein gene-based phylogeny for finer differentiation and classification of phytoplasmas. *Int. J. Syst. Evol. Microbiol.* 57: 2037–51.

Martini, M., Marcone, C., Mitrović. J., Maixner, M., Delić, D. *et al.* (2012). "*Candidatus* Phytoplasma convolvuli," a new phytoplasma taxon associated with bindweed yellows in four European countries. *Int. J. Syst. Evol. Microbiol.* 62: 2910–15.

Martins, L., Fernandes, C., Blom, J., Dia, N. C., Pothier, J. F. *et al.* (2020). *Xanthomonas euroxanthea* sp. nov., a new xanthomonad species including pathogenic and non-pathogenic strains of walnut. *Int. J. Syst. Evol. Micro.* 70: 6024–31.

Matsuura, T., Mizuno, A., Tsukamoto, T., Shimizu, Y., Saito, N. *et al.* (2012). *Erwinia uzenensis* sp. nov., a novel pathogen that affects European pear trees (*Pyrus communis* L.). *Int. J. Syst. Evol. Microbiol.* 62: 1799–1803.

McLeod, M. P., Warren, R. L., Hsiao, W. W. L., Zhao, Y., Botti, S. *et al.* (2006). The complete genome of *Rhodococcus* sp. RHA1 provides insights into a catabolic powerhouse. *Proc. Natl. Acad. Sci.* USA, 103: 15582–87.

McNeill, J., Barrie, F. R., Buck, W. R., Demoulin, V., Greuter, W. *et al.* (eds.) (2012). *International Code of Nomenclature for algae, fungi, and plants (Melbourne Code) adopted by the Eighteenth International Botanical Congress Melbourne, Australia, July 2011. Regnum Vegetabile 154.* Koeltz Scientific Books, Koenigstein.

Mergaert, J., Verdonck, L. and Kersters, K. (1993). Transfer of *Erwinia ananas* (synonym, *Erwinia uredovora*) and *Erwinia stewartii* to the genus *Pantoea* emend. as *Pantoea ananas* (Serrano 1928) comb. nov. and *Pantoea stewartii* (Smith 1898) comb. nov., respectively, and description of *Pantoea stewartii* subsp. *indologenes* subsp. nov. *Int. J. Syst. Bacteriol.* 43: 162–73.

Migula, W. (1894). Uber ein neues System der Bakterien. *Arb Bakteriol Inst Karlsruhe* 1: 235–38 (in German).

Miyajima, K., Tanaka, F., Takeuchi, T. and Kuninaga, S. (1998). *Streptomyces turgidiscabies* sp. nov. *Int. J. Syst. Bacteriol.* 48: 495–502.

Miyazaki, A., Shigaki, T., Koinuma, H., Iwabuchi, N., Rauka, G. B. *et al.* (2018). "*Candidatus* Phytoplasma noviguineense," a novel taxon associated with Bogia coconut syndrome and banana wilt disease on the island of New Guinea. *Int. J. Syst. Evol. Microbiol.* 68: 170–75.

Mou, D.-F., Di Lella, B., Halbert, S. E., Bextine, B., Helmick, E. E. *et al.* (2022). Acquisition and transmission of the lethal bronzing Phytoplasma by *Haplaxius crudus* using infected palm spear leaves and artificial feeding media. *Phytopathology* 112: 2052–61.

Mphande, K., Beattie, G. A., & Gleason, M. L. (2025). Bacteria that cause cucurbit yellow vine disease fall within the *Serratia ureilytica* species of the *S. marcescens* complex and can be vectored by cucumber beetles. *Bacteriology* 115: 606–17.

Munsch, P., Alatossava, T., Marttinen, N., Meyer, J.-M., Christen, R. *et al.* (2002). *Pseudomonas costantinii* sp. nov. another causal agent of brown blotch disease, isolated from cultivated mushroom sporophores in Finland. *Int. J. Syst. Evol. Microbiol.* 52: 1973–83.

Murray, P. R. and Jo Baron, E. (2007). *Manual of Clinical Microbiology*. 9th ed. Washington, DC, 705 pp. ISBN 978-1-55581-587-5. OCLC 647917473.

Murray, R. G. E. (1968). Microbial structure as an aid to microbial classification and taxonomy. *Spisy (Faculté des Sciences de l' Université J. E. Purkyne Brno)* 43: 249–52.

Murray, R. G. E. and Schleifer, K. H. (1994). Taxonomic notes: A proposal for recording the properties of putative taxa of procaryotes. *Int. J. Syst. Bacteriol.* 44: 174–76.

Murray, R. G. E. and Stackebrandt, E. (1995). Taxonomic Note: Implementation of the provisional status *Candidatus* for incompletely described procaryotes. *Int. J. Syst. Bacteriol.* 45: 186–87.

Murray, T. D., Barrantes-Infante, B. and Schroeder, B. K. (2020). First report of bacterial head blight of *Pseudoroegneria spicata* subsp. *spicata* caused by *Rathayibacter agropyri* in Idaho. *Plant Dis.* 104: 1534.

Na, W., Luo, H. and Yu, J. M. (2021). First report of bacterial brown rot disease on shiitake mushroom (*Lentinula edodes*) caused by *Ewingella americana* in Korea. *J. Plant Pathol.* 103: 1325–26. https://doi.org/10.1007/s42161-021-00897-z

Nabhan, S., De Boer, S. H., Maiss, E. and Wydra, K. (2013). *Pectobacterium aroidearum* sp. nov., a soft rot pathogen with preference for monocotyledonous plants. *Int. J. Syst. Evol. Microbiol.* 63: 2520–25.

Naderali, N., Nejat, N., Vadamalai, G., Davis, R. E., Wei, W. *et al.* (2017). "*Candidatus* Phytoplasma wodyetiae," a new taxon associated with yellow decline disease of foxtail palm (*Wodyetia bifurcata*) in Malays. *Int. J. Syst. Evol. Microbiol.* 67: 3765–72.

Nasiri, M., Faghihi, M. M., Rahimian, H. and Osdaghi, E. (2024). *Clavibacter tessellarius* causing bacterial mosaic of wheat establishes in the Old World. *Plant Pathol.* 73: 1363–71.

Nejat, N., Vadamalai, G., Davis, R. E., Harrison, N. A., Sijam, K. *et al.* (2013). "*Candidatus* Phytoplasma malaysianum," a novel taxon associated with virescence and phyllody of Madagascar periwinkle (*Catharanthus roseus*). *Int. J. Syst. Evol. Microbiol.* 63: 540–48.

Nguyen, H. P., Shelley, B. A., Mowery, J. and Clarke, C. R. (2022). Description of *Streptomyces griseiscabiei* sp. nov. and reassignment of *Streptomyces* sp. strain NRRL B-16521 to *Streptomyces acidiscabies*. *Int. J. Syst. Evol. Microbiol.* 72: https://doi.org/10.1099/ijsem.0.005574

Nguyen, H. P., Weisberg, A. J., Chang, J. H. and Clarke, C. R. (2022). *Streptomyces caniscabiei* sp. nov., which causes potato common scab and is distributed across the world. *Int. J. Syst. Evol. Microbiol.* 72. https://doi.org/10.1099/ijsem.0.005225

Nie, H., Lee, S., Ko, S.-R., Lee, Y. K., Lee, M-H. *et al.* (2024). Complete genome sequence of *Erwinia pyrifoliae* strain YKB12327, isolated from an apple tree in Korea. *Jour. Plant Pathol.* 106: 1843–46.

Nouioui, I., Carro, L., García-López, M., Meier-Kolthoff, J. P., Woyke, T. *et al.* (2018). Genome-based taxonomic classification of the phylum *Actinobacteria*. *Front Microbiol.* 9: 02007

O'Gara, P. J. (1916). A bacterial disease of western wheat grass, *Agropyron smithii*. *Phytopathology*, 6: 341–50. [Google Scholar].

Oh, E.-J., Bae, C., Lee, H.-B., Hwang, I. S., Lee, H.-I. *et al.* (2016). *Clavibacter michiganensis* subsp. *capsici* subsp. nov., causing bacterial canker disease in pepper. *Int. J. Syst. Evol. Microbiol.* 66: 4065–70.

Ohl, M. E. and Miller, S. I. (2001). *Salmonella*: A model for bacterial pathogenesis. *Annu. Rev. Med.* 52: 259–74.

Okazaki, S., Nukui, N., Sugawara, M. and Minamisawa, K. (2004). Rhizobial strategies to enhance symbiotic interaction: rhizobitoxine and 1-aminocyclopropane-1-carboxylate deaminase. *Microbes Environ.* 19: 99–111.

Oksel, C., Avin, F. A., Liyanapathiranage, P., Shreckhise, J. and Baysal-Gurel, F. (2024). First report of *Achromobacter xylosoxidans* causing bacterial stem and leaf blight on *Cyrilla arida* in Tennessee and the United States. *Plant Dis.* 108: 1094. https://doi.org/10.1094/PDIS-01-24-0005-PDN

Oliveira, W. J., Silva, W. A., Silva, A. M. F., Candeia, J. A., Souza, E. B. *et al.* (2017). First report of *Burkholderia cenocepacia* causing sour skin of onion (*Allium cepa*) in Brazil. *Plant Dis.* 101: 1950.

Oren, A., Arahal, D. R., Göker, M., Moore, E. R. B., Rossello-Mora, R. *et al.* (2023). ICSP Matters. International Code of Nomenclature of Prokaryotes. Prokaryotic Code (2022 Revision). *Int. J. Syst. Evol. Microbiol.* 73: 005585.

Oren, A., Arahal, D. R., Rosselló-Móra, R., Sutcliffe, I. C. and Moore, E. J. B. (2021). Emendation of Rules 5b, 8, 15, and 22 of the International Code of Nomenclature of Prokaryotes to include the rank of phylum. *Int. J. Syst. Evol. Microbiol.* 71: 004851.

Oren, A., da Costa, M. S., Garrity, G. M., Rainey, F. A. and Rossello-Mora, R. (2015). Proposal to include the rank of phylum in the International Code of Nomenclature of Prokaryotes. *Int. J. Syst. Evol. Microbiol.* 65: 4284–87.

Oren, A. and Garrity, G. M. (2017). Validation list no. 177. List of new names and new combinations previously effectively, but not validly, published. *Int. J. Syst. Evol. Microbiol.* 67: 3140–43. Effective publication: Lopes-Santos, L., Castro, D. B. A., Ferreira-Tonin, M., Corrêa, D. B. A., Weir, B. S. *et al.* (2017). Reassessment of the taxonomic position of *Burkholderia andropogonis* and description of *Robbsia andropogonis* gen. nov., comb. nov. *Antonie van Leeuwenhoek*, 110: 727–36.

Oren, A. and Garrity, G. M. (2019). Validation list No. 187. *Int. J. Syst. Evol. Microbiol.* 69: 1247–50.

Osdaghi, E., Taghavi, S. M., Hamidizade, M., Kariminejhad, M., Fazliarab, A. *et al.* (2024). Multiphasic investigations imply transfer of orange-/red-pigmented strains of the bean pathogen *Curtobacterium flaccumfaciens* pv. *flaccumfaciens* to a new species as *C. aurantiacum* sp. nov., elevation of the poinsettia pathogen *C. flaccumfaciens* pv. *poinsettiae* to the species level as *C. poinsettiae* sp. nov., and synonymy of *C. albidum* with *C. citreum*. *Syst. Appl. Microbiol.* 47: 126489.

Osei, R., Yang, C., Cui, L., Ma, T., Li, Z. *et al.* (2022). Isolation, identification, and pathogenicity of *Lelliottia amnigena* causing soft rot of potato tuber in China. *Microb. Pathogen.* 164. https://doi.org/10.1016/j.micpath.2022.105441

Ouattara, A. S., Le Mer, J., Joseph, M. and Macarie, H. (2017). Transfer of *Pseudomonas pictorum* Gray and Thornton 1928 to genus *Stenotrophomonas* as *Stenotrophomonas pictorum* comb. nov., and emended description of the genus *Stenotrophomonas*. *Int. J. Syst. Evol. Microbiol.* 67: 1894–1900. https://doi.org/10.1099/ijsem.0.001880

Paine, S. G. and Stansfield, H. (1919). Studies in bacteriosis III. A bacterial leaf-spot disease of *Protea cynaroides* exhibiting a host reaction of possibly bacteriolytic nature. *Ann. Appl. Boil.* 6: 27–39.

Palleroni, N. J. (1984a). Genus I. *Pseudomonas* Migula. In: N. R. Krieg and J. G. Holt (eds.). *Bergey's Manual of Systematic Bacteriology*, Vol. 1. Williams and Wilkins, Baltimore, MD, pp. 141–99.

Palleroni, N. J. (1984b). Family I. Pseudomonadaceae. In: N. R. Krieg and J. G. Holt (eds). *Bergey's Manual of*

Systematic Bacteriology, 1st Ed. I. Williams and Wilkins, Baltimore, MD, pp. 144–99.

Palleroni, N. J. (2005). Genus I. *Pseudomonas* Migula 1894, 237AL (nom. cons., Opin. 5 of the Jud. Comm. 1952, 121). In: D. J. Brenner, N. R. Krieg, J. T. Staley and G. M. Garrity (eds.) *Bergey's Manual of Systematic Bacteriology*. 2nd Ed. 2B. Springer, New York.

Palleroni, N. J. and Bradbury, J. F. (1993). *Stenotrophomonas*, a new bacterial genus for *Xanthomonas maltophilia* (Hugh 1980) Swings *et al*. 1983. *Int. J. Syst. Bacteriol.* 43: 606–09. https://doi.org/10.1099/00207713-43-3-606 [PubMed].

Park, D. H., Kim J. S., Kwon S. W., Wilson, C., Yu, Y. M. *et al.* (2003). *Streptomyces luridiscabiei* sp. nov., *Streptomyces puniciscabiei* sp. nov. and *Streptomyces niveiscabiei* sp. nov., which cause potato common scab disease in Korea. *Int. J. Syst. Evol. Microbiol.* 53: 2049–54.

Parker, C. T., Tindal, B. J. and Garrity, G. M. (eds.) (2015). *International code of Nomenclature of Prokaryotes. Prokaryotic Code. 2008 Revision*. Published for the International Committee on Systematics of Prokaryotes. *Int. J. Syst. Evol. Microbiol.* https://doi.org/10.1099/ijsem.0.000778.

Parker, C. T., Tindall, B. J. and Garrity, G. M. (2019). International code of nomenclature of prokaryotes. *Int. J. Syst. Evol. Microbiol.* 69: S1–S111.

Parkinson, N., Aritua, V., Heeney, J. Cowie, C., Bew, J. *et al.* (2007). Phylogenetic analysis of *Xanthomonas* species by comparison of partial gyrase B gene sequences. *Int. J. Syst. Evol. Microbiol.* 57: 2881–87.

Parkinson, N., Stead, D., Bew, J., Heeney, J., Tsror Lahkim, L. *et al.* (2009). *Dickeya* species relatedness and clade structure determined by comparison of *recA* sequences. *Int. J. Syst. Evol. Microbiol.* 59: 2388–93.

Parvin, S. M. R., Taghavi, S. M. and Osdaghi, E. (2023). Field surveys indicate taxonomically diverse *Pectobacterium* species inducing soft rot of vegetables and annual crops in Iran. *Plant Pathol.* 72: 1260–71.

Patro, T. S. S. K., Nageswara Rao, G. V. and Gopalakrishnan, J. (2006). Association of *Acinetobacter baumannii* with a top rot phase of sugarcane red stripe disease in India. *Indian Phytopathol.* 59: 501–02.

Paudel, S., Dutta, B. and Kvitko, B. (2024). Onion-pathogenic *Burkholderia* species: Role and regulation of characterized virulence determinants. *Plant Pathol.* 73: 2281–97.

Paul, S., Anderson, P. J., Maynard, G. J., Dyall-Smith, M. and Kudinha, T. (2023). Complete genome sequences of the type strains *Rouxiella badensis* DSM 100043 and *R. chamberiensis* DSM 28324, resolved using nanopore long-read sequencing. *Microbiol. Resour. Announc.* 12: e00156-23. https://doi.org/10.1128/mra.00156-23

Peng, Q., Yuan, Y. and Gao, M. (2013). *Bacillus pumilus*, a novel ginger rhizome rot pathogen in China. *Plant Dis.* 97: 1308–15.

Pikuta, E. V., Lyu, Z., Williams, M. D., Patel, N. B., Liu, Y. *et al.* (2017). *Sanguibacter gelidistatuariae* sp. nov., a novel psychrotolerant anaerobe from an ice sculpture in Antarctica, and emendation of descriptions of the family Sanguibacteraceae, the genus *Sanguibacter* and species *S. antarcticus*, *S. inulinus*, *S. kedieii*, *S. marinus*, *S. soli* and *S. suarezii*. *Int. J. Syst. Evol. Microbiol.* 67: 1442–50. https://doi.org/10.1099/ijsem.0.001838. PMID: 28150571.

Pilet, F., Loiseau, M., Boyer, C., Cavalier, A. and Diman, C. (2023). First report of '*Candidatus* Phytoplasma palmae' (16SrIV-A Subgroup) associated with palm lethal yellowing disease on *Cocos nucifera* and *Pritchardia* sp. in Guadeloupe, French West Indies. *Plant Dis.* 107: 1621.

Plasencia-Marquez, O., Corzo, M., Martinez-Zubiaur, Y., Rivero, D., Devescovi, G. *et al.* (2017). First report of soybean (*Glycine max*) disease caused by *Pseudomonas aeruginosa* in Cuba. *Plant Dis.* 101: 1950.

Portier, P., Pédron, J., Taghouti, G., Fischer-Le Saux, M., Caullireau, E. *et al.* (2019). Elevation of *Pectobacterium carotovorum* subsp. *odoriferum* to species level as *Pectobacterium odoriferum* sp. nov., proposal of *Pectobacterium brasiliense* sp. nov. and *Pectobacterium actinidiae* sp. nov., emended description of *Pectobacterium carotovorum* and description of *Pectobacterium versatile* sp. nov., isolated from streams and symptoms on diverse plants. *Int. J. Syst. Evol. Micro.* 69: 3207–16.

Prazmowski, A. (1880). *Untersuchungen über die Entwickelungsgeschichte und Firment-wirking einiger Bacterien-Arten*. Inaugural Dissertation, Hugo Voigt, Leipzig, Germany.

Pulawska, J., Willems, A. and Sobiczewski, P. (2012). *Rhizobium skierniewicense* sp. nov., isolated from tumours on chrysanthemum and cherry plum. *Int. J. Syst. Evol. Microbiol.* 62: 895–99.

Qin, S.-Y., Xie, X.-W., Chen, R.-X., Shi, Y.-X., Chai, A.-L. *et al.* (2024). First report of *Pectobacterium polaris* causing soft rot on broccoli in China. *Plant Dis.* 108: 2215. https://doi.org/10.1094/PDIS-08-23-1475-PDN

Qiu, M., Li, Y., Liu, Q., Zhang, X., Huang, Y. *et al.* (2024). *Pseudomonas machongensis* sp. nov. and *Stenotrophomonas capsici* sp. nov., isolated from wilted pepper plants. *Int. J. Syst. and Evol. Microbiol.* 74: https://doi.org/10.1099/ijsem.0.006550

Quaglino, F., Zhao, Y., Casati, P., Bulgari, D., Bianco, P. A. *et al.* (2013). "*Candidatus* Phytoplasma solani," a novel taxon associated with stolbur- and bois noir-related diseases of plants. *Int. J. Syst. Evol. Microbiol.* 63: 2879–94.

Queslati, M., Mulet, M., Gomila, M., Berge, O., Hajlaoui, M. R. *et al.* (2019). New species of pathogenic *Pseudomonas* isolated from citrus in Tunisia: Proposal of *Pseudomonas kairouanensis* sp. nov. and *Pseudomonas nabeulensis* sp. nov. *Syst. Appl. Micro.* 42: 348–59.

Rao, G. P. (2021). Our understanding about phytoplasma research scenario in India. *Indian Phytopathol.* 74: 371–401. https://doi.org/10.1007/s42360-020-00303-1

Retana, K., Castro, O., Quesada, A. and Blanco, M. (2017). Etiology of stem rot on *Hylocereus* spp. cause by *Enterobacter hormaechei* in Costa Rica. (Abstr.) *Phytopathology*, 107: S4.18–19.

Rijckaert, C. (1994). *Taxonomische studie van de Erwinia amylovora groep (Dye, 1968)*. PhD thesis, University Gent, Belgium.

Rios, D., Ueckert, J., Ong, K., Barillas, J. R. and Costanzo, S. (2024). First report of a '*Candidatus* Phytoplasma sacchari'-related strain associated with yellowing and decline of silver bluestem in Texas, USA. *Plant Dis.* 108: 2216. https://doi.org/10.1094/PDIS-03-24-0524-PDN

Roberts, R. and Pietersen, G. (2017). A novel subspecies of '*Candidatus* Liberibacter africanus' found on native *Teclea gerrardii* (Family: Rutaceae) from South Africa. *Antonie Van Leeuwenhoek.* 110: 437–44.

Roberts, R., Steenkamp, E. T. and Pietersen, G. (2015). Three novel lineages of '*Candidatus* Liberibacter africanus' associated with native rutaceous hosts of *Trioza erytreae* in South Africa. *Int. J. Syst. Evol. Microbiol.* 65: 723–31.

Rodrigues, L. M. R., Andrade, V. T., Destéfano, S. A. L., Braghini, M. T., Caixeta, L. B. *et al.* (2023). Inheritance of resistance to bacterial halo blight and bacterial leaf spot in Arabica coffee. *Plant Dis.* 107: 3810–16.

Rosenbach, F. J. (1884). *Microorganismen bei den Wund-Infections-Krankheiten des Menschen.* J.F. Bergmann, Wiesbaden.

Roumagnac, P., Gagnevin, L., Gardan, L. Sutra, L., Manceau, C. *et al.* (2004). Polyphasic characterization of xanthomonads isolated from onion, garlic and Welsh onion (*Allium spp.*) and their relatedness to different *Xanthomonas* species. *Int. J. Syst. Evol. Microbiol.* 54: 15–24.

Ryan K. J. and Ray C. G. eds. (2004). *Sherris Medical Microbiology.* 4th ed. McGraw Hill. ISBN 0-8385-8529-9

Ryan, R. P., Monchy, S., Cardinale, M., Taghavi, S., Crossman, L. *et al.* (2009). The versatility and adaptation of bacteria from the genus *Stenotrophomonas. Nature Rev. Microbiol.* 7: 514–525. https://doi.org/10.1038/nrmicro2163

Saddler, G. S. and Bradbury, J. F. (2015). *Xanthomonas. Bergey's Manual of Systematics of Archaea and Bacteria.* pp.1–53. Šafarova, D., Zemanek, T., Valova, P. and Navratil, M. (2016). "*Candidatus* Phytoplasma cirsii," a novel taxon from creeping thistle [*Cirsium arvense* (L.) Scop.]. *Int. J. Syst. Evol. Microbiol.* 66: 1745–53.

Šafarova, D., Zemanek, T., Valova, P. and Navratil, M. (2016). '*Candidatus* Phytoplasma cirsii,' a novel taxon from creeping thistle [*Cirsium arvense* (L.) Scop.]. *Int. J. Syst. Evol. Microbiol.* 66: 1745–53.

Safni, I., Cleenwerck, I., De Vos, P., Fegan, M., Sly, L. *et al.* (2014). Polyphasic taxonomic revision of the *Ralstonia solanacearum* species complex: Proposal to emend the descriptions of *Ralstonia solanacearum* and *Ralstonia syzygii* and reclassify current *R. syzygii* strains as *Ralstonia syzygii* subsp. *syzygii* subsp. nov., *R. solanacearum* phylotype IV strains as *Ralstonia syzygii* subsp. *indonesiensis* subsp. nov., banana blood disease bacterium strains as *Ralstonia syzygii* subsp. *celebesensis* subsp. nov. and *R. solanacearum* phylotype I and III strains as *Ralstonia pseudosolanacearum* sp. nov. *Int. J. Syst. Evol. Microbiol.* 64: 3087–3103.

Saglio, P., Lhospital, M., Lafleche, D. Dupont, G., Bove, J. M. *et al.* (1973). *Spiroplasma citri* gen. and sp. n.: A mycoplasma-like organism associated with "Stubborn" disease of citrus. *Int. J. Syst. Bacteriol.* 23: 191–204.

Sakazaki, R. (1984). Genus IV. *Citrobacter* Werkman and Gillen 1932. In: N. R. Krieg and J. G. Holt (eds.). *Bergey's Manual of Systematic Bacteriology*, I ed. Vol. 1: The Williams and Wilkins Co, Baltimore, MD, pp. 458–61.

Salam, N., Jiao, J.-Y., Zhang, X.-T. and Li, W.-J. (2020). Update on the classification of higher ranks in the phylum Actinobacteria. *Int. J. Syst. Evol. Microbiol.* 70: 131–35.

Samson, R., Legendre, J. B., Christen, R., Saux M.F-L., Achouak, W. *et al.* (2005). Transfer of *Pectobacterium chrysanthemi* (Burkholder *et al.* 1953) Brenner *et al.* 1973 and *Brenneria paradisiaca* to the genus *Dickeya* gen. nov. as *Dickeya chrysanthemi* comb. nov. and *Dickeya paradisiaca* comb. nov. and delineation of four novel species, *Dickeya dadantii* sp. nov., *Dickeya dianthicola* sp. nov., *Dickeya dieffenbachiae* sp. nov. and *Dickeya zeae* sp. nov. *Int. J. Syst. Evol. Microbiol.* 55: 1415–27.

Sarfraz, S., Riaz, K., Oulghazi, S., Cigna, J., Sahi, S. T. *et al.* (2019). First report of *Pectobacterium parmentieri* and *Pectobacterium polaris* causing potato blackleg disease in Punjab, Pakistan. *Plant Dis.* 103: 1405.

Sasaki, J., Chijimatsu, M. and Suzuki, K.-I. (1998). Taxonomic significance of 2, 4-diaminobutyric acid isomers in the cell wall peptidoglycan of actinomycetes and reclassification of *Clavibacter toxicus* as *Rathayibacter toxicus* comb. nov. *Int. J. Syst. Bacteriol.* 48: 403–10.

Sawabe, T., Makino, H., Tatsumi, M., Nakano, K., Tajima, K. *et al.* (1998). *Pseudoalteromonas bacteriolytica* sp. nov., a marine bacterium that is the causative agent of red spot disease of *Laminaria japonica. Int. J. Syst. Bacteriol.* 48: 769–74.

Sawada, H., Fujikawa, T., Nishiwaki, Y. and Horita, H. (2020). *Pseudomonas kitaguniensis* sp. nov., a pathogen causing bacterial rot of Welsh onion in Japan. *Int. J. Syst. Evol. Micro.* 70: 3018–26.

Sawada, H., Fujikawa, T., Osada, S. and Satou, M. (2021). *Pseudomonas cyclaminis* sp. nov., a pathogen causing bacterial bud blight of cyclamen in Japan. *Int. J. Syst. Evol. Micro.* 71: https://doi.org/10.1099/ijsem.0.004723

Sawada, H., Fujikawa, T., Osada, S. and Satou, M. (2022). *Pseudomonas petroselini* sp. nov., a pathogen causing bacterial rot of parsley in Japan. *Int. J. Syst. Evol. Microbiol.* 72: https://doi.org/10.1099/ijsem.0.005424

Sawada, H., Fujikawa, T. and Satou, M. (2021). *Pseudomonas lactucae* sp. nov., a pathogen causing bacterial rot of lettuce in Japan. *Int. J. Syst. Evol. Micro.* 71: https://doi.org/10.1099/ijsem.0.004917

Sawada, H., Fujikawa, T. and Satou, M. (2022). *Pseudomonas aegrilactucae* sp. nov. and *Pseudomonas morbosilactucae* sp. nov., pathogens causing bacterial rot of lettuce in Japan. *Int. J. Syst. Evol. Microbiol.* 72. https://doi.org/10.1099/ijsem.0.005599

Sawada, H., Fujikawa, T., Tsuji, M. and Satou, M. (2021). *Pseudomonas allii* sp. nov., a pathogen causing soft rot of onion in Japan. *Int. J. Syst. Evol. Micro.* 71. https://doi.org/10.1099/ijsem.0.004582

Sawada, H., Ieki, H., Oyaizu, H. and Matsumoto, S. (1993). Proposal for rejection of *Agrobacterium tumefaciens* and revised descriptions for the genus *Agrobacterium* and for *Agrobacterium radiobacter* and *Agrobacterium rhizogenes. Int. J. Syst. Bacteriol.* 43: 694–702.

Sawada, H., Shinohara, H., Takashima, Y., Naito, K. and Satou, M. (2025). *Acidovorax sacchari* sp. nov., a pathogen causing red stripe of sugarcane in Japan. *Int. Jour. Syst. and Evol. Microbiol.* 75 (2). https://doi.org/10.1099/ijsem.0.006575

Sawada, H., Someya, N., Morohoshi, T., Ono, M. and Satou, M. (2024). *Pectobacterium araliae* sp. nov., a pathogen causing bacterial soft rot of Japanese angelica tree in Japan. *Int. J. Syst. Evol. Microbiol.* 74: https://doi.org/10.1099/ijsem.0.006326

Sawana, A., Adeolu, M. and Gupta, R. S. (2015). List of new names and new combinations previously effectively, but not validly, published. *Int. J. Syst. Evol. Microbiol.* 65: 2017–25. Effective publication: Sawana, A., Adeolu, M. and Gupta, R. S. (2014). Molecular signatures and phylogenomic analysis of the genus *Burkholderia*: Proposal for division of this genus into the emended genus *Burkholderia* containing pathogenic organisms and a new genus *Paraburkholderia* gen. nov. harboring environmental species. *Front. Genet.* 5: 429.

Schaad, N. W., Postnikova, E., Lacy, G., M'Barck, F., Chang, C.-J. *et al.* (2004). *Xylella fastidiosa* subspecies: *X. fastidiosa* subsp. *piercei*, subsp. nov., *X. fastidiosa* subsp. *multiplex*, subsp. nov., and *X. fastidiosa* subsp. *pauca* subsp. nov. *Syst. Appl. Microbiol.* 27: 290–300.

Schaad, N. W., Postnikova, E., Sechler, A., Claflin, L. E., Vidaver, A. K. *et al.* (2009). List of new names and new combinations previously effectively, but not validly, published. *Int. J. Syst. Evol. Microbiol.* Vol. 59: pp. 923 -25. Effective publication: Schaad, N. W., Postnikova, E., Sechler, A., Claflin, L. E., Vidaver, A. K. *et al.* (2008). Reclassification of subspecies of *Acidovorax avenae* as *A. avenae* (Manns 1905) emend., *A. cattleyae* (Pavarino 1911) comb. nov., *A. citrulli* (Schaad *et al.* 1978) comb. nov., and proposal of *A. oryzae* sp. nov. *Syst. Appl. Microbiol.* 31: 434–46.

Schikora, A., Carreri, A., Charpentier, E. and Hirt, H. (2008). The dark side of the salad: *Salmonella typhimurium* overcomes the innate immune response of *Arabidopsis thaliana* and shows an endopathogenic lifestyle. *PLoS One* 3: e2279.

Schleifer, K. H., Kraus, J., Dvorak, C., Kilpper-Balz, R., Collins M. D. *et al.* (1986). Validation list No. 20. *Int. J. Syst. Bacteriol.* 36: 354–56. Effective publication: Schleifer, K. H., Kraus, J., Dvorak, C., Kilpper-Balz, R., Collins M. D. *et al.* (1985). Transfer of *Streptococcus lactis* and related streptococci to the genus *Lactococcus* gen. nov. *Syst. Appl. Microbiol.* 6: 183–95.

Schneider, B., Torres, E., Martin, M. P., Schroder, M., Behnke, H. D. *et al.* (2005). '*Candidatus* phytoplasma pini', a novel taxon from *Pinus silvestris* and *Pinus halepensis*. *Int. J. Syst. Evol. Microbiol.* 55: 303–07.

Schroeter J. *Schizomycetes* (1886), pp. 131–256. In: Cohn F. (ed.), *Kryptogamen-Flora von Schlesien, Band 3, Heft 3, Pilze*, Max Müller, Breslau, 1889, pp. 1–814.

Schroeder, B. K., Schneider, W. L., Luster, D. G., Sechler, A. and Murray, T. D. (2018). *Rathayibacter agropyri* (non-O'Gara 1916) comb. nov., nom. rev., isolated from western wheatgrass (*Pascopyrum smithii*). *Int. Jour. Syst. Evol. Microbiol.* 68: 1519–25. doi:10.1099/ijsem.0.002708. Epub 2018 Mar 20. PMID: 29557775.

Sechler, A., Schuenzel, E. L., Cooke, P., Donnua, S., Thaveechai, N. *et al.* (2009). Cultivation of '*Candidatus* Liberibacter asiaticus,' '*Ca.* L. africanus' and '*Ca.* L. americanus' associated with huanglongbing. *Phytopathology* 99: 480–86.

Seemuller, E. and Schneider, B. (2004). '*Candidatus* Phytoplasma mali', '*Candidatus* Phytoplasma pyri' and '*Candidatus* Phytoplasma prunorum', the causal agents of apple proliferation, pear decline and European stone fruit yellows, respectively. *Int. J. Syst. Evol. Microbiol.* 54: 1217–26.

Sêkowska, A. (2017). *Raoultella* spp.-clinical significance, infections and susceptibility to antibiotics. *Folia Microbiol. (Praha).* 62(3): 221–27.

Sêkowska, A., Bogiel, T., Woźniak, M. and Gospodarek-Komkowska, E. (2020). *Raoultella* spp.–reliable identification, susceptibility to antimicrobials and antibiotic resistance mechanisms. *J. Med. Microbiol.* 69 (2): 233–38. https://doi.org/10.1099/jmm.0.001150. ISSN 0022-2615. PMID 31971501.

Shida, O., Takagi, H., Kadowaki, K., Nakamura, L. K. and Komagata, K. (1997). Transfer of *Bacillus alginolyticus*, *Bacillus chondroitinus, Bacillus curdlanolyticus, Bacillus glucanolyticus, Bacillus kobensis* and *Bacillus thiaminolyticus* to the genus *Paenibacillus* and emended description of the genus *Paenibacillus*. *Int. J. Syst. Bacteriol.* 47: 289–98.

Singh, A., Kumar, S., Bansal, K. and Patil, P. B. (2023). Taxonomic and phylogenomic assessment identifies phytopathogenicity potential of *Stenotrophomonas maltophilia* complex. *Phytopathology*, 113: 1833–38.

Skerman, V. B. D., McGowan, V. and Sneath, P. H. A. (eds.) (1980). Approved lists of bacterial names. *Int. J. Syst. Bacteriol.* 30: 225–420.

Smith, E. F. (1911). Bacteria in relation to plant diseases. *Carnegie Inst. Wash. Publ.* 2: 1–368.

Soutar, C. D. and Stavrinides J. (2020). Phylogenetic analysis supporting the taxonomic revision of eight genera within the bacterial order *Enterobacterales*. *Int. J. Syst. Evol. Micro.* 70: 6524–30.

Stackebrandt, E., Rainey, F. A. and Ward-Rainey, N. L. (1997). Proposal for a new hierarchic classification system, *Actinobacteria* classis nov. *Int. J. Syst. Bacteriol.* 47: 479–91.

Stanier, R. Y., Palleroni, N. J. and Doudoroff. M. (1966). The aerobic pseudomonads: a taxonomic study. *J. Gen. Microbiol.* 43: 159–271.

Stuart, C. A., Wheeler, K. M., Rustigian, R. and Zimmerman, A (1943). Biochemical and antigenic relationships of the paracolon bacteria. *J. Bacteriol.* 45: 101–19. https://doi.org/10.1128/jb.45.2.101-119.1943. PMC 373720. PMID 16560614

Studholme, D. J., Wicker, E., Abrare, S. M., Aspin, A., Bogdanove, A. *et al.* (2020). Transfer of *Xanthomonas campestris* pv. *arecae* and *X. campestris* pv. *musacearum* to *X. vasicola* (Vauterin) as *X. vasicola* pv. *arecae* comb. nov. and *X. vasicola* pv. *musacearum* comb. nov. and Description of *X. vasicola* pv. *vasculorum* pv. nov. *Phytopathology* 110: 1153–60.

Su, C.-C., Deng, W.-L., Jan, F.-J., Chang, C.-J., Huang, H. *et al.* (2016). *Xylella taiwanensis* sp. nov., causing pear leaf scorch disease. *Int. J. Syst. Evol. Microbiol.* 66: 4766–71.

Sugiyama, L. S., Bushe, B. C., Heller, W. P. and Keith, L. M. (2017). First report of *Paraburkholderia andropogonis* causing bacterial leaf streak of *Strelitzia reginae* in Hawaii detected from imported plants. *Plant Dis.* 101: 1030.

Surano, A., del Grosso, C., Musio, B., Todisco, S., Giampetruzzi, A. *et al.* (2023). Exploring the xylem-sap to unravel biological features of *Xylella fastidiosa* subspecies *pauca* ST53 in immune, resistant and susceptible crop species through metabolomics and *in vitro* studies. *Front. Plant Sci.* 14: https://doi.org/10.3389/fpls.2023.1343876

Surico, G., Mugnai, L., Pastorelli, R., Giovannetti, L. and Stead, D. E. (1996). *Erwinia alni*, a new species causing bark canker of alder (*Alnus* Miller) species. *Int. J. Syst. Bacteriol.* 46: 720–26.

Sutra, L., Christen, R., Bollet, C. Simoneau, P. and Gardan, L. (2001). *Samsonia erythrinae* gen. nov., sp. nov., isolated from bark necrotic lesions of *Erythrina* sp., and discrimination of plant-pathogenic *Enterobacteriaceae* by phenotypic features. *Int. J. Syst. Evol. Microbiol.* 51: 1291–1304.

Sutra, L., Prior, P., Perlemoine, K., Rise'de, J. M., Cao-Van, P. *et al.* (1999). Description of a new disease on *Erythrina* sp. in Martinique (French West Indies) and preliminary characterization of the causal agent as a novel *Erwinia* species. *Plant Pathol.* 48: 253–59.

Swenson, C. E. and Sadikot, R. T. (2015). *Achromobacter* respiratory infections. *Ann. Am. Thorac. Soc.* 12: 252–58. https://doi.org/10.1513/AnnalsATS.201406-288FR. ISSN 2329-6933. PMID 25706494.

Takahashi, K., Nishiyama, K. and Sato, M. (1996). *Pseudomonas syringae* pv. *broussonetiae* pv. nov., the causal agent of bacterial blight of paper mulberry (*Broussonetia kazinoki* x *B. papyrifera*). *Ann. Phytopathol. Soc. Japan*, 62: 17–22.

Takeuchi, M., Hamana, K. and Hiraishi, A. (2001). Proposal of the genus *Sphingomonas* sensu stricto and three new genera, *Sphingobium*, *Novosphingobium* and *Sphingopyxis*, on the basis of phylogenetic and chemotaxonomic analyses. *Int. J. Sys. Evol. Microbiol.* 51: 1405–17.

Takeuchi, M., Sakane, T., Yanagi, M., Yamasato, K., Hamana, K. *et al.* (1995). Taxonomic study of bacteria isolated from plants: Proposal of *Sphingomonas rosa* sp. nov., *Sphingomonas pruni* sp. nov., *Sphingomonas asaccharolytica* sp. nov. and *Sphingomonas mali* sp. nov. *Int. J. Syst. Bacteriol.* 45: 334–41.

Takeuchi, M., Weiss, N., Schumann, P. and Yokota, A. (1996). *Leucobacter komagatae* gen. nov., sp. nov., a new aerobic gram-positive, nonsporulating rod with 2,4-diaminobutyric acid in the cell wall. *Int. J. Syst. Bacteriol.* 46: 967–71.

Tambong, J. T., Xu, R., Fleitas, M. C. and Kutcher, R. (2024). Taxonogenomic analysis of the *Xanthomonas translucens* complex leads to the descriptions of *Xanthomonas cerealis* sp. nov. and *Xanthomonas graminis* sp. nov. *Int. J. Syst. and Evo. Micro.* 74: https://doi.org/10.1099/ijsem.0.006523

Tang, W. C., Wang, L. H. and Chu, C. C. (2023). Phylogenetic placements and characteristics of *Pectobacterium* strains obtained from different hosts in Taiwan. *Phytopathology* 113 (S3): 55.

Tao, B., Li, X., Zhou, S. and Yuan, G. (2024). First report of bacterial leaf spot disease on ginger caused by *Enterobacter quasiroggenkampii* in China. *Plant Dis.* 108: 3646. https://doi.org/10.1094/PDIS-06-24-1254-PDN

Tao, Y., Ge, Y., Yang, J., Song, W., Jin, D. *et al.* (2023). A novel phytopathogen *Erwinia sorbitola* sp. nov., isolated from the feces of ruddy shelducks. *Front. Cell. Infect. Microbiol.* 13: https://doi.org/10.3389/fcimb.2023.1109634.

Taormina, P. J., Beuchat, L. R. and Slutsker, L. (1999). Infections associated with eating seed sprouts: An international concern. *Emerg. Infect. Dis.* 5: 626–34.

Teixeira, D. C., Saillard, C., Eveillard, S., Danet, J. L., da Costa, P. I. *et al.* (2005). '*Candidatus* Liberibacter americanus', associated with citrus huanglongbing (greening disease) in Sao Paulo State, Brazil. *Int. J. Syst. Evol. Microbiol.* 55: 1857–62.

Thakur, P. and Gauba, P. (2022). Genomic characterization of *Lelliottia amnigena* PTJIIT1005, a nitrate tolerant strain isolated from water sample of Yamuna River, Delhi. *Microbiol. Resour. Announc.* https://doi.org/10.1128/mra.00229-22

Tian, Q., Chuan, J., Sun, X., Zhou, A., Wang, L. *et al.* (2021). Description of *Clavibacter zhangzhiyongii* sp. nov., a phytopathogenic actinobacterium isolated from barley seeds, causing leaf brown spot and decline. *Int. J. Syst. Evol. Micro.* 71: 004786.

Tian, Y., Zhao, Y., Yuan, X., Yi, J., Fan, J. *et al.* (2016). *Dickeya fangzhongdai* sp. nov., a plant-pathogenic bacterium isolated from pear trees (*Pyrus pyrifolia*). *Int. J. Syst. Evol. Microbiol.* 66: 2831–35.

Timilsina, S., Minsavage, G. V., Preston, J., Newberry, E. A., Paret, M. L. *et al.* (2018). *Pseudomonas floridensis* sp. nov., a bacterial pathogen isolated from tomato. *Int. J. Syst. Evol. Microbiol.* 68: 64–70.

Tohya, M., Watanabe, S., Tada, T., Tin, H. H. and Kirikae, T. (2020). Genome analysis-based reclassification of *Pseudomonas fuscovaginae* and *Pseudomonas shirazica* as later heterotypic synonyms of *Pseudomonas asplenii* and *Pseudomonas asiatica*, respectively. *Int. J. Syst. Evol. Micro.* 70: 3547–52.

Toth, T., Lakatos, T. and Koltay, A. (2013). *Lonsdalea quercina* subsp. *populi* subsp. nov., isolated from bark canker of poplar trees. *Int. J. Syst. Evol. Microbiol.* 63: 2309–13.

Trevisan V. (1885). Carratteri di alcuni nuovi generi di *Batteriacee. Atti della Accademia Fisica-Medica-Stastistica in Milano (ser 4).* 3: 92–107.

Trebaol, G., Gardan, L., Manceau, C. Tanguy, J. L., Tirilly, Y. *et al.* (2000). Genomic and phenotypic characterization of *Xanthomonas cynarae* sp. nov., a new species that causes bacterial bract spot of artichoke (*Cynara scolymus* L.). *IntJ. Syst. Evo. Microbiol.* 50: 1471–78.

Trevisan, V. (1887). Sul Micrococco della rabbia e sulla possiblità di riconoscere durante il periode d'incubazione, dall'esame del sangue della persona moricata, se ha contratta l'infezione rabbica. *Rendiconti dell'Instituto Lombardo di Scienze e Lettere (Ser 2).* 20: 88–105.

Trevisan, V. (1889). *I generie le specie delle bacteriacee.* Zanaboni and Gabuzzi, Milan, pp. 1–33.

Unckless R. L., Howick, V. M. and Lazzaro, B. P. (2016). Convergent balancing selection on an antimicrobial peptide in *Drosophila. Curr. Biol.* 26: 257–62. https://doi.org/10.1016/j.cub.2015.11.063.

Urakami, T., Ito-Yoshida, C., Araki, H., Kijima, T., Suzuki, K.-I. *et al.* (1994). Transfer of *Pseudomonas plantarii* and *Pseudomonas glumae* to *Burkholderia* as *Burkholderia* spp. and description of *Burkholderia vandii* sp. nov. *Int. J. Syst. Bacteriol.* 44: 235–45.

Uribe-Lorio, L. and Wang, W. A. (2017). Identification of *Erwinia billingiae* as a causal agent of bacterial canker on mango (*Mangifera indica*) trees in Costa Rica (Abstr.) *Phytopathology*, 107 (S4).12–13.

Valiunas, D., Staniulis, J. and Davis, R. E. (2006). '*Candidatus* Phytoplasma fragariae', a novel phytoplasma taxon discovered in yellows diseased strawberry, *Fragaria* x *ananassa. Int. J. Syst. Evol. Microbiol.* 56: 277–81.

Valiunas, D., Jomantiene, R. and Davis, R. E. (2013). Evaluation of the DNA-dependent RNA polymerase b-subunit gene (rpoB) for phytoplasma classification and phylogeny. *Int. J. Syst. Evol. Microbiol.* 63: 3904–14.

Van Bruggen, A. H. C., Jochimsen, K. N. and Brown, P. R. (1990). *Rhizomonas suberifaciens* gen. nov., sp. nov., the causal agent of corky root of lettuce. *Int. J. Syst. Bacteriol.* 40: 175–88.

van den Mooter, M. and Swings J. (1990). Numerical analysis of 295 phenotypic features of 266 *Xanthomonas* strains and related strains and an improved taxonomy of the genus. *Int. J. Syst. Bacteriol.* 40: 348–69.

van der Wolf, J. M., Nijhuis, E. H., Kowalewska, M. J., Saddler, G. S., Parkinson, N. *et al.* (2014). *Dickeya solani* sp. nov., a pectinolytic plant-pathogenic bacterium isolated from potato (*Solanum tuberosum*). *Int. J. Syst. Evol. Microbiol.* 64: 768–74.

Van Tieghem, P. (1878). Sur la gomme de sucrerie (*Leuconostoc mesenteroides*). *Ann. Sci. Nat. Bot.* 7: 180–203.

Vaneechoutte, M., Kampfer, P., Baere, T. D., Falsen, E. and Verschraegen, G. (2004). *Wautersia* gen. nov., a novel genus accommodating the phylogenetic lineage including *Ralstonia eutropha* and related species, and proposal of *Ralstonia* [*Pseudomonas*] *syzygii* (Roberts *et al.* 1990) comb. nov. *Int. J. Syst. Evol. Microbiol.* 54: 317–27.

Vauterin, L., Hoste, B., Kersters, K. and Swings, J. (1995). Reclassification of *Xanthomonas. Int. J. Syst. Bacteriol.* 45: 472–89.

Verdin, E., Salar, P., Danet, J. L., Choueiri, E., Jreijiri, F. *et al.* (2003). '*Candidatus* Phytoplasma phoenicium' sp. nov., a novel phytoplasma associated with an emerging lethal disease of almond trees in Lebanon and Iran. *Int. J. Syst. Evol. Microbiol.* 53: 833–38.

Verdonck, L. (1992). *Taxonomische studie van de "soft rot" verwekkende soorten van het bacteriengeslacht Erwinia.* PhD thesis, University Gent, Belgium.

Vicente, J. G., Rothwell, S., Holub, E. B. and Studholme, D. J. (2017). Pathogenic, phenotypic and molecular characterisation of *Xanthomonas nasturtii* sp. nov. and *Xanthomonas floridensis* sp. nov., new species of *Xanthomonas* associated with watercress production in Florida. *Int. J. Syst. Evol. Microbiol.* 67: 3645–54.

Vinatzer, B. A., Weisberg, A. J., Monteil, C. L., Elmarakeby, H. A., Sheppard, S. K. *et al.* (2017). A proposal for a genome similarity-based taxonomy for plant-pathogenic bacteria that is sufficiently precise to reflect phylogeny, host range and outbreak affiliation applied to *Pseudomonas syringae sensu lato* as a proof of concept. *Phytopathology* 107: 18–28.

Vitor, L., Amaral, D. T., Corrêa, D. B. A., Ferreira-Tonin, M., Lucon, E. T. *et al.* (2023). *Streptomyces hilarionis* sp. nov. and *Streptomyces hayashii* sp. nov., two new strains associated with potato scab in Brazil. *Int. J. Syst. Evol. Microbiol.* 73 (6): https://doi.org/10.1099/ijsem.0.005916. ISSN 1466-5034. PMID 37319004.

Von Naegeli, C. (1857). In Caspary, Bericht uber die Ver handlungen der 33, Ver sammlung Deutscher Naturforsches und Aertzte gehalten in Bonn von 18 bis 24 September 1857. *Artz Bot Ztg*, 15: 749–92.

Waksman, S. A. and Henrici, A. T. (1943). The nomenclature and classification of actinomycetes. *J. Bacteriol.* 46: 337–41

Waldee, E. L. (1945). Comparative studies of some peritrichous phytopathogenic bacteria. *Iowa State Coll. J. Sci.* 19: 435–84.

Waleron, M., Misztak, A., Waleron, M., Franczuk, M., Wielgomas, B. *et al.* (2018). Transfer of *Pectobacterium carotovorum* subsp. *carotovorum* strains isolated from potatoes grown at high altitudes to *Pectobacterium peruviense* sp. nov. *Syst. Appl. Microbiol.* 41: 85–93.

Wang, F., Tang, T., Guo, J., Guo, X., Duan, Y. *et al.* (2021). Occurrence of stem blight caused by *Pseudomonas extremorientalis* on *Pinellia ternata* in China. *Plant Dis.* 105: 1851.

Wang, S.-Y. Jiang, Y.-H. and Huo, Z-Y. (2025). Bacterial pustule disease in soybean: occurrence, host–pathogen interactions and management. *Plant Pathol.* Vol. 74: pp. 1941–57. https://doi.org/10.1111/ppa.70000

Wang, X., Wang, D., Pun, M. K., Gou, R., Liang, H. *et al.* (2023). First report of *Providencia rettgeri* causing soft rot on calla lily in China. *Plant Dis.* 107: 2837.

Warabieda, M., Kuzmanović, N., Trzciński, P. and Puławska, J. (2023). *Agrobacterium cucumeris* sp. nov. isolated from crazy roots on cucumber (*Cucumis sativus*). *Sys. Appl. Microbiol.* 46: https://doi.org/10.1016/j.syapm.2023.126402

Wauters, G., Janssens, M., De Baere, T., Vaneechoutte, M. and Deschaght, P. (2012). Isolates belonging to CDC group II-i belong predominantly to *Sphingobacterium mizutaii* Yabuuchi *et al.* 1983: emended descriptions of *S. mizutaii* and of the genus *Sphingobacterium*. *Int. J. Syst. Evol. Microbiol.* 62: 2598–2601. Euzeby, J. P. (2013). Notification list. Notification that new names and new combinations have appeared in volume 62, part 11 of the IJSEM. *Int. J. Syst. Evol. Microbiol.* 63: 399–400.

Wei, W., Davis, R. E., Lee, I.-M. and Zhao, Y. (2007). Computer-simulated RFLP analysis of 16S rRNA genes: identification of ten new phytoplasma groups. *Int. J. Syst. Evol. Microbiol.* 57: 1855–67.

Wei, W., Davis, R. E., Suo, X. and Zhao, Y. (2015). Occurrence, distribution and possible functional roles of simple sequence repeats in phytoplasma genomes. *Int. J. Syst. Evol. Microbiol.* 65: 2748–60.

Wei, W., Lee, I.-M., Davis, R. E. Suo, X. and Zhao, Y. (2008). Automated RFLP pattern comparison and similarity coefficient calculation for rapid delineation of new and distinct phytoplasma 16Sr subgroup lineages. *Int. J. Syst. Evol. Microbiol.* 58: 2368–77.

Wei, W., Zhong, L. W., Chen, Z. B., Fan, H. C., Xu, S. G. *et al.* (2023). First report of *Achromobacter xylosoxidans* causing stem soft rot on *Amorphophallus konjac* in China. *Plant Dis.* 107: 1935.

Wells, J. M., Raju, B. C., Hung, H. Y., Weisburg, W. G., Mandelco-Paul, L. *et al.* (1987). *Xylella fastidiosa* gen. nov., sp. nov.: Gram-negative, xylem-limited, fastidious plant bacteria related to *Xanthomonas* spp. *Int. J. Syst. Bacteriol.* 37: 136–43.

Werkman, C. H. and Gillen, G. F. (1932). Bacteria producing trimethylene glycol. *J. Bacteriol.* 23: 167–82.

Whitman, W. B., Oren, A., Chuvochina, M., da Costa, M. S., Garrity, G. M. *et al.* (2018). Proposal of the suffix *-ota* to denote phyla. Addendum to "Proposal to include the rank of phylum in the International Code of Nomenclature of Prokaryotes". *Int. J. Syst. Evol. Microbiol.* 68: 967–69.

Willems, A., Falsen, E., Pot, B., Jantzen, E., Hoste, B. *et al.* (1990). *Acidovorax*, a new genus for *Pseudomonas facilis, Pseudomonas delafieldii*, E. Falsen (EF) group 13, EF group 16, and several clinical isolates, with the species *Acidovorax facilis* comb. nov., *Acidovorax delafieldii* comb. nov., and *Acidovorax temperans* sp. nov. *Int. J. Syst. Bacteriol.* 40: 384–98.

Willems, A., Gillis, M., Kersters, K., Van Den Broecke, L. and De Ley, J. (1987). Transfer of *Xanthomonas ampelina* Panagopoulos 1969 to a new genus, *Xylophilus* gen. nov., as *Xylophilus ampelinus* (Panagopoulos 1969) comb. nov. *Int. J. Syst. Bacteriol.* 37: 422–30.

Willems, A., Goor, M., Thielemans, S. Gillis, M., Kerster, K. *et al.* (1992). Transfer of several phytopathogenic *Pseudomonas* species to *Acidovorax* as *Acidovorax avenae* subsp. *avenae* subsp. nov., comb. nov., *Acidovorax avenae* subsp. *citrulli, Acidovorax avenae* subsp. *cattleyae*, and *Acidovorax konjaci*. *Int. J. Syst. Bacteriol.* 42: 107–19.

Wilson, E. E., Starr, M. P. and Berger, J. A. (1957). Bark canker, a bacterial disease of the Persian walnut tree. *Phytopathology*, 47: 669–73.

Win, N. K. K., Lee, S.-Y., Bertaccini, A., Namba, S. and Jung, H.-Y. (2013). '*Candidatus* Phytoplasma balanitae' associated with witches' broom disease of *Balanites triflora*. *Int. J. Syst. Evol. Microbiol.* 63: 636–46.

Winslow, C. E. A., Broadhurst, J., Buchanan, R. E., Krumwiede, C. Jr., Rogers, L. A. et al. (1917). The Families and Genera of the Bacteria: Preliminary Report of the Committee of the Society of American Bacteriologists on Characterization and Classification of Bacterial Types. *J. Bacteriol.* 2: 505.

Winslow, C. E. A., Broadhurst, J., Buchanan, R. E., Krumwiede, C. J., Rogers, L. A. *et al.* (1920). The families and genera of the bacteria. Final report of the Committee of the Society of American Bacteriologists on characterization and classification of bacterial types. *J. Bacteriol.* 5: 191–229.

Woese, C. R., Kandler, O. and Wheelis, M. L. (1990). Towards a natural system of organisms: Proposal for the domains *Archaea*, *Bacteria* and *Eubacteria*. *Proc. Natl. Acad. Sci. USA* 87: 4576–79.

Wu, C., Zhang, Z., Qiao, C., Zhu, L., Wang, J. *et al.* (2025). First report of *Enterobacter kobei* and *Enterobacter mori* causing ginger bacterial wilt in China. *Plant Dis.* 109: 1374.

Wu, J., Duan, J., Zhang, C., Zhang, R. F. and Handique, U. (2024). First report of *Streptomyces brasiliscabiei* causing common scab on potato in China. *Pl. Dis.* 108: 2559. https://doi.org/10.1094/PDIS-03-24-0607-PDN

Wu, J., Lv, J., Zeng, M., Zhang, S., Feng, Z. *et al.* (2023). First report of *Pectobacterium parvum* causing disease on potato in inner Mongolia, China. *Plant Dis.* 107: 4016.

Wu, W. X., Liu, Y., Huang, X. Q. and Zhang, L. (2017). First report of summer canker caused by *Pectobacterium carotovorum* subsp. *actinidiae* on kiwifruit in China's Sichuan province. *Plant Dis.* 101: 1540.

Xiao, R., Mu, X.-P., Zhang, J.-C., Zhang, S., Zhang, C.-F. *et al.* (2023). Isolation and characterization of tumorigenic bacteria associated with crown gall disease of *Prunus humilis* Bunge in China. *Eur. J. Plant Pathol.* 166: 463–83.

Yabuuchi, E., Kaneko, T., Yano, I., Moss, C. W. and Miyoshi, N. (1983). *Sphingobacterium* gen. nov., *Sphingobacterium spiritivorum* comb. nov., *Sphingobacterium multivorum* comb. nov., *Sphingobacterium mizutae* sp. nov., and *Flavobacterium indologenes* sp. nov.: glucose-nonfermenting gram-negative rods in CDC groups IIK-2 and IIb. *Int. J. Syst. Bacteriol.* 33: 580–98.

Yabuuchi, E., Kosako, Y., Fujiwara, N., Naka, T., Matsunaga, I. *et al.* (2002). Emendation of the genus *Sphingomonas* Yabuuchi *et al.* 1990 and junior objective synonymy of the species of three genera, *Sphingobium, Novosphingobium* and *Sphingopyxis*, in conjuction with *Blastomonas ursincola. Int. J. Syst. Evol. Microbiol.* 52: 1485–96.

Yabuuchi, E., Kosako, Y., Naka, T. Suzuki, S. and Yano, I. (1999). Validation of publication of new names and new combinations previously effectively published outside the IJSB. *Int. J. Syst. Bacteriol.* 49: 935–36. Effective publication: Yabuuchi, E., Kosako, Y., Naka, T. *et al.* (1999). Proposal of *Sphingomonas suberifaciens* (van Bruggen, Jochimsen and Brown 1990) comb. nov., *Sphingomonas natatoria* (Sly 1985) comb. nov., *Sphingomonas ursincola* (Yurkov *et al.* 1997) comb. nov., and emendation of the genus *Sphingomonas*. *Microbiol. Immunol.* 43: 339–49.

Yabuuchi, E., Kosako, Y., Oyaizu, H., Yano, I., Hotta, H. *et al.* (1993). Validation of the publication of new names and new combinations previously effectively published outside the IJSB. *Int. J. Syst. Bacteriol.* 43: 398–99. Effective publication: Yabuuchi, E., Kosako, Y., Oyaizu, H. *et al.* (1992). Proposal of *Burkholderia* gen. nov. and transfer of seven species of the genus *Pseudomonas* homology group II to the new genus, with the type species *Burkholderia cepacia* (Palleroni and Holmes 1981) comb. nov. *Microbiol. Immunol.* 36: 1251–75.

Yabuuchi, E., Kosako, Y., Yano, I., Hotta, H. and Nishiuchi, Y. (1996). Validation of the publication of new names and new combinations previously effectively published outside the IJSB. *Int. J. Syst. Bacteriol.* 46: 625–26. Effective publication: Yabuuchi, E., Kosako, Y., Yano, I. *et al.* (1995). Transfer of two *Burkholderia* and an *Alcaligenes* species to *Ralstonia* gen. nov.: Proposal of *Ralstonia pickettii* (Ralston, Palleroni and Doudoroff 1973) comb. nov., *Ralstonia solanacearum* (Smith 1896) comb. nov. and *Ralstonia eutropha* (Davis 1969) comb. nov. *Microbiol. Immunol.* 39: 897–904. https://doi.org/10.1111/j.1348-0421.1995.tb03275.x.

Yabuuchi, E., Yano, I., Oyaizu, H., Hashimoto, Y., Ezaki, T. *et al.* (1990). Validation of the publication of new names and new combinations previously effectively published outside the IJSB. *Int. J. Syst. Bacteriol.* 40: 320–21. Effective publication: Yabuuchi, E., Yano, I., Oyaizu, H. *et al.* (1990). Proposals of *Sphingomonas paucimobilis* gen. nov. and comb. nov., *Sphingomonas parapaucimobilis* sp. nov., *Sphingomonas yanoikuyae* sp. nov., *Sphingomonas adhaesiva* sp. nov., *Sphingomonas capsulata* comb. nov. and two genospecies of the genus *Sphingomonas*. *Microbiol. Immunol.* 34: 99–119.

Yamada, K. and Komagata, K. (1972). Taxonomic studies on coryneform bacteria V. Classification of coryneform bacteria. *J. Gen. Appl. Microbiol.* 18: 417–31.

Yang, G., Han, L., Wen, J. and Zhou, S. (2013). *Pseudomonas guangdongensis* sp. nov., isolated from an electroactive biofilm, and emended description of the genus *Pseudomonas* Migula. 1894. *Int. J. Syst. Evol. Microbiol.* 63: 4599–4605.

Yang, Y., Liu, Q., Zhang, Z. and Dong, C. (2023). First report of bacterial brown rot disease caused by *Ewingella americana* on cultivated *Naematelia aurantialba* in China. *Plant Dis.* 107: 2513.

Yoon, J. H., Lee, H. B., Yeo, S. H. and Choi, J. E. (2004). *Janibacter melonis* sp. nov., isolated from abnormally spoiled oriental melon in Korea. *Int. J. Syst. Evol. Microbiol.* 54: 1975–80.

Young, J. M. (2008). An overview of bacterial nomenclature with special reference to plant pathogenic bacteria. *Syst. Appl. Microbiol.* 31: 405–24.

Young, J. M., Bull, C. T., De Boer, S. H., Firrao, G., Saddler, G. E. *et al.* (2004). Names of plant pathogenic bacteria, 1864–2004. Online [http://www.isppweb.org/names_bacterial_revised.asp].

Young, J. M., Dye, D. W., Bradbury, J. F. Panagopoulos, C. G. and Robbs, C. F. (1978). A proposed nomenclature and classification for plant pathogenic bacteria. *N. Z. J. Agric. Res.* 21: 153–77.

Young, J. M., Kuykendall, L. D., Martinez-Romero, E., Kerr, A. and Swada, H. (2001). A revision of *Rhizobium* Frank 1889, with an emended description of the genus, and the inclusion of all species of *Agrobacteriun* Conn 1942 and *Allorhizobium undicola* de Laudie *et al.* 1998 as new combinations: *Rhizobium radiobacter*, *R. rhizogenes*, *R. rubi*, *R. undicola*, and *R. vitis*. *Int. J. Syst. Evol. Microbiol.* 51: 89–103.

Young, J. M., Saddler, G. S., Takikawa, Y., De Boer, S. H., Vauterin, L. *et al.* (1996). Names of plant pathogenic bacteria 1864–1995. *Rev. Plant Pathol.* 75: 721–63.

Young, J. M., Wilkie, J. P., Park, D. C. and Watson, D. R. W. (2010). List of new names and new combinations previously effectively, but not validly, published. *Int. J. Syst. Evol. Microbiol.* Vol. 60: pp. 1009–1010. Effective Publication: Young, J. M., Wilkie, J. P., Park, D. C. *et al.* (2010). New Zealand strains of plant pathogenic bacteria classified by multilocus sequence analysis; proposal of *Xanthomonas dyei* sp. nov. *Plant Pathol.* 59: 270–81.

Zaczek-Moczydłowska, M. A., Fleming, C. C., Young, G. K., Campbell, K. and O'Hanlon, R. (2019). *Pectobacterium* and *Dickeya* species detected in vegetables in Northern Ireland. *Eur. J. Plant Pathol.* 154: 635–47. https://doi.org/10.1007/s10658-019-01687-1

Zgurskaya, H. I., Evtushenko, L. I., Akimov, V. N. and Kalakoutskii, L. V. (1993). *Rathayibacter* gen. nov., including the species *Rathayibacter rathayi* comb. nov., *Rathayibacter tritici* comb. nov., *Rathayibacter iranicus* comb. nov., and six strains from annual grasses. *Int. J. Syst. Bacteriol.* 43: 143–49.

Zhang, A. H., Zhang, X. X., Lei, F. J. and Zhang, L. X. (2018). First report of bacterial soft rot of ginseng caused by *Pseudomonas qessardii* in Jilin Province of China. *Plant Dis.* 102: 437.

Zhang, R. Y., Zhao, S. X., Tan, Z. Q. and Zhu, C. H. (2017). First report of bacterial stem rot disease caused by *Paenibacillus polymyxa* on *Hylocereus undulatus* in China. *Plant Dis.* 101: 1031.

Zhang, S., Meng, X., Cheng, Y., Wang, Y., Hu, T. *et al.* (2021). Infection of cucumber seedling roots by *Pseudomonas amygdali* pv. *lachrymans* following artificial substrate inoculation. *Eur. J. Plant Pathol.* 160: 385–95. https://doi.org/10.1007/s10658-021-02251-6

Zhao, M., Tyson, C., Gitaitis, R., Kvitko, B. and Dutta, B. (2022). *Rouxiella badensis*, a new bacterial pathogen of onion causing bulb rot. *Front. Microbiol.* 13: https://doi.org/10.3389/fmicb.2022.1054813

Zhao, X., Tian, Y., Yue, L., Liu, Y., Yan, Y. *et al.* (2022). Identification and characterization of pathogenicity of Lelliottia nimipressuralis causing soft rot of *Codonopsis pilosula* (dangshen) roots in China. *Plant Pathol.* 71: 1801–11. https://doi.org/10.1111/ppa.13606

Zhao, Y., Sun, Q., Wei, W., Davis, R. E., Wu, W. *et al.* (2009). '*Candidatus* Phytoplasma tamaricis', a novel taxon discovered in witches'-broom-diseased salt cedar (*Tamarix chinensis* Lour.). *Int. J. Syst. Evol. Microbiol.* 59: 2496–2504.

Zhao, Y., Wei, W., Davis, R. E., Lee, I.-M. and Bottner-Parker. K. D. (2021). The agent associated with blue dwarf disease in wheat represents a new phytoplasma taxon, '*Candidatus* Phytoplasma tritici'. *Int. J. Syst. Evol. Micro.* 71: https://doi.org/10.1099/ijsem.0.004604

Zhao, Y., Wei, W., Lee, I. M., Shao, J., Suo, X. *et al.* (2009). Construction of an interactive online phytoplasma classification tool, *i*PhyClassifier, and its application in analysis of the peach X-disease phytoplasma group (16SrIII). *Int. J. Syst. Evol. Microbiol.* 59: 2582–93.

Zhao, Z.-X., Wang, Q., Shi, Y.-X., Xie, X.-W., Chai, A.-L. *et al.* (2025). First report of *Xanthomonas hortorum* pv. *vitians* as the causal agent of bacterial leaf spot on lettuce in China. *Plant Dis.* 109: 228. https://doi.org/10.1094/PDIS-02-24-0321-PDN

Zheng, J., Wittouck, S., Salvetti, E., Franz, C. M. A. P., Harris, H. M. B. *et al.* (2020). A taxonomic note on the genus *Lactobacillus*: Description of 23 novel genera, emended description of the genus *Lactobacillus* Beijerinck 1901, and union of *Lactobacillaceae* and *Leuconostocaceae*. *Int. J. Syst. Evol. Micro.* 70: 2782–2858.

Zhi, X.-Y., Li, W.-J. and Stackebrandt, E. (2009). An update of the structure and 16S rRNA gene sequence-based definition of higher ranks of the class *Actinobacteria*, with the proposal of two new suborders and four new families and emended descriptions of the existing higher taxa. *Int. J. Syst. Evol. Microbiol.* 59: 589–608.

Zhou, J., Xia, F., Che, S., Wang, J. and Zhang, G. (2020). *Providencia rettgeri* as the causal agent of the brown slime flux of *Populus tomentosa*. *Forest Pathol.* 50: e12576.

Zhu, B., Lou, M.-M., Xie, G.-L., Wang, G.-F., Zhou, Q. *et al.* (2011). *Enterobacter mori* sp. nov., associated with bacterial wilt on *Morus alba* L. *Int. J. Syst. Evol. Microbiol.* 61: 2769–74.

Zopf, W. (1891). Uber Ausscheidung von Fellfarbstoffen (Lipochromen) seitens gewisser Spattpilze. *Berichte der Deut Schen Botanischen Gesellschaft.* 9: 22–28.

Zreik, L., Bove, J. M. and Garnier, M. (1998). Phylogenetic characterization of the bacterium-like organism associated with marginal chlorosis of strawberry and proposition of a *Candidatus* taxon for the organism, '*Candidatus* Phlomobacter fragariae'. *Int. J. Syst. Bacteriol.* 48: 257–61.

Zreik, L., Carle, P., Bove, J. M. and Garnier, M. (1995). Characterization of the mycoplasma like organism associated with witches'-broom disease of lime and proposition of a "*Candidatus*" taxon for the organism, '*Candidatus* Phytoplama aurantifolia'. *Int. J. Syst. Bacteriol.* 45: 449–53.

Zübert, C. and Kube, M. (2021). Application of TaqMan Real-Time PCR for detecting ' *Candidatus* arsenophonus phytopathogenicus' infection in sugar beet. *Pathogens*, 10: 1466. https://doi.org/10.3390/pathogens10111466

7 Molecular Mechanisms of Virulence and Pathogenesis

The rapid emergence of new bacterial diseases negatively affects both human health and agricultural productivity. Although the molecular mechanisms underlying these disease emergences are shared between both human and plant pathogenic bacteria, little effort has been made to date to understand the emergence of diseases caused by plant pathogenic bacteria. In particular, there is a paucity of information in the literature about the role of environmental habitats in which plant pathogenic bacteria evolve and on the stress factors to which these microbes are unceasingly exposed. Bartoli *et al.* (2016), in their micro review, focus on three molecular mechanisms underlying pathogenicity in bacteria: mutations, genomic rearrangements, and the acquisition of new DNA sequences through horizontal gene transfer. They briefly discuss the role of these mechanisms in bacterial disease emergence and elucidate how the environment can influence the occurrence and regulation of these molecular mechanisms by directly impacting such processes. The understanding of such molecular evolutionary mechanisms and their environmental drivers will represent an important step towards predicting the emergence of bacterial disease and developing sustainable management strategies for crops.

Emerging diseases can be defined either as infections that have either newly appeared in a given population or have historically existed and are once more spreading in incidence or geographical range (Morse, 1995). Over the past four decades, 335 new infectious diseases have emerged in humans, the majority of which are caused by multidrug-resistant (MDR) bacteria (Jones *et al.*, 2008). Similarly, the number of plant disease outbreaks, some caused by plant pathogenic bacteria, has also increased in recent decades. For example, since the beginning of this century, *Pseudomonas syringae* has caused over 55 disease outbreaks in perennial plants (Lamichhane *et al.*, 2014) and more than 70 disease outbreaks in annual plants (Lamichhane *et al.*, 2015). Such an increasing number of disease outbreaks raises concerns about the origin of the genetic lines causing such epidemics, as well as the evolutionary processes involved.

A large number of infectious diseases are caused by host-adapted microbial agents, which evolve a pathogenic phenotype through the acquisition of new pathogenic determinants (Morens and Fauci, 2013). In addition, most of the disease outbreaks are caused by different bacterial genetic lineages. The understanding of the molecular mechanisms underlying the emergence of such bacterial genetic lines, therefore, appears to be a fundamental step towards predicting disease emergence.

Overall, three mechanisms are seen to be behind the origin of DNA modifications that lead to the evolution of genomic traits underlying pathogenicity in bacteria: (i) point mutations (small local sequence change); (ii) the rearrangements of DNA segments (such as gene duplications or insertion/deletion of entire or portions of genes); and (iii) the acquisition of new DNA components from other organisms via horizontal gene transfer (Arber, 2008). The result of these processes is an altered bacterial phenotype; this could be either beneficial or harmful for bacteria depending on the environment they encounter. In addition, whether or not the new alleles coming from genomic modifications will be fixed into a given bacterial population is strictly related to the effect of genetic drift and natural selection, both of which are strongly influenced by environmental fluctuations (Ohta, 2000).

However, the emergence of new diseases is also affected by a combination of ecological, environmental and socio-economic factors (Morens *et al.*, 2004). Most of the emerging diseases seem to be caused by existing pathogens which, through adaptation to new habitats, evolve the capability to cause a disease in new hosts. As in the case for human pathogens, both abiotic and biotic environments can act as potential reservoirs of plant pathogenic bacteria. These bacterial lineages can evolve under pressures that are different from those exerted by the final host plants. Some studies have emphasised the role of water habitats in the evolution of *Psm. syringae* lineages pathogenic to kiwifruit and tomato plants (Bartoli *et al.*, 2015; Monteil *et al.*, 2013). In addition, plants are likely to play an important role in disease emergence by acting as reservoirs of pathogenic lineages, as observed, for example, in *Xan. campestris* strains (Toussaint *et al.*, 2012). More generally, under the selective pressure of resistant and/or tolerant plants, bacteria can evolve new allelic forms of virulence which, in turn, may trigger disease on new hosts. In particular, tolerant plants seem to be much more suitable for bacterial evolution, as tolerance is attained by limiting pathogen damage without reducing bacterial growth (i.e. without reducing population size). The authors have summarised the role of point mutations, genomic rearrangements and horizontal gene transfer in the emergence of bacterial lineages pathogenic to either plants or humans, with particular emphasis on the role of the environment in triggering these mechanisms. An understanding of the molecular evolutionary mechanisms underlying bacterial disease emergence and the role of the environment in this evolutionary scenario can shed light on the possible development of sustainable disease management strategies.

DOI: 10.1201/9781003681878-8

Gene activation is essential for plant pathogens to sense, swim towards, colonise and infect their hosts. Many plant pathogenic bacteria recognise the presence of their plant hosts through the perception of plant-generated chemical signals. The signals perceived by the bacterial cells are transduced through the bacterial membranes to initiate specific motility events. The pathogen reaches its host plant following the chemical gradient of the signal. The movement of the bacterial cells through the gradient occurs through either flagella or pili. Various chemicals serve as specific signals and are recognised by the bacteria. *Agrobacterium tumefaciens* recognises methoxyphenols, namely, sinapinic and cinnamic acids, acetosyringone and coniferyl alcohol, derived from the lignin biosynthetic pathway produced by the plant. A transmembrane-docked histidine kinase called VirA assists in perceiving the methoxyphenols potentiated by simple sugars and low pH and in activating the response regulator VirG in this bacterium. Yuan *et al.* (2008) reported that an acidic environment (around 5.5 pH) has been shown to induce as many as 78 genes and to repress 74 genes in *Ag. tumefaciens*. Venturi and Fuqua (2013) reviewed the knowledge in chemical signalling that takes place between plants and plant pathogenic bacteria.

7.1 QUORUM SENSING

Plant pathogens not only recognise plant-generated chemical signals, but also recognise self-generated signals. Quorum sensing (QS) (dependent on cell population size), an ability to communicate and coordinate behaviour via signalling molecules is widespread among both Gram-stain-positive and Gram-stain-negative bacteria. A given species may possess more than one QS system. In this process, as the bacteria grow, they regulate gene expression by sensing the concentration of their own diffusible signal molecule. Each bacterial cell produces a small amount of a QS signal molecule that becomes sufficiently concentrated as the cell population increases. There are certain traits that are only expressed when bacteria are crowded. Traits regulated in phytopathogenic bacteria by QS include the production of extracellular polysaccharides, degradative enzymes, pigments, antibiotics and siderophores, as well as hypersensitive reaction and pathogenicity protein secretion, Ti-plasmid transfer, motility, biofilm formation and epiphytic fitness. As QS regulatory systems are often required for pathogenesis, interference with QS signalling may offer a means of controlling bacterial plant diseases. A brief account of the QS mechanisms operating in some important bacterial plant pathogenic species is given below.

In Gram-stain-negative bacteria, the best-characterised signal molecules are the *N*-acyl-homoserine lactones (AHLs). QS systems are tuned to preferentially synthesise, and respond to, specific AHLs having an acyl side group that may range in length from four to six carbons, may contain a degree of unsaturation and may be modified with a carbonyl or hydroxyl functional group at carbon position 3. AHL is used and produced by many Gram-stain-negative plant pathogenic bacteria and endosymbionts, including *Pseudomonas syringae* pathovars, *Xanthomonas campestris* pathovars, *Agr. tumefaciens*, *Pantoea stewartii* subsp. *stewartii*, *Pectobacterium carotovorum*, *Erwinia amylovora* and *Tatumella citrea*.

A lot of information is available on the molecular and structural basis of QS control of Ti-plasmid conjugal transfer in *Agr. tumefaciens*. In the nopaline strain C 58, ApoTraR appears to be plasma membrane-associated, but disassociates from the membrane when complexed with AHL. AHLTraR dimerises and positions the C-terminal helix-turn-helix domain to attach to the major groove of the 18-bp *tra*-box recognition site; *tra*-boxes are located immediately upstream of the *traAFB*, *traCDG*, *traI-trb*, and *rep* operons. Activation of *traI-trb* operon by TraR results in the creation of a positive feedback loop for *traI*. TraR activity also stimulates *traM* gene expression, which encodes a small anti-activator with affinity for the C-terminal domain of TraR. It is assumed that TraM titrates TraR to prevent Ti-plasmid transfer until TraR levels exceed than those of TraM.

In young tumours, sufficient nutrients are present, which favour rapid growth of colonising agrobacteria over Ti-plasmid maintenance. At this stage, the galls contain a significant population of avirulent agrobacteria due to plasmid loss, and these bacteria appear to have a definite growth advantage. At later stages of gall formation, nutrients may be limiting and the bacterial cell density may be sufficiently high to trigger QS to result in Ti-plasmid transfer to more cells in the galls. These cells, then, would be capable of causing new infections. A higher plasmid copy number would also increase the expression of the opine catabolic genes resulting in more efficient use of the available opines.

QS in *Pec. carotovorum* subsp. *carotovorum* controls the population density-dependent expression of pathogenicity factors such as extracellular enzymes and the hypersensitive response, and pathogenicity (Hrp) secretion system, as well as carbapenem antibiotic production. Primary AHL, 3-oxoC6HL, is produced by the LuxI-like signal synthase CarI. Mutants defective in *carI* do not produce carbapenem, pectolytic enzymes, endoglucanases and proteases, and hence fail to secrete harpin. Therefore, these are completely non-pathogenic. The car QS system directly regulates the genes for the production of carbapenem, which constitute the *carA-H* biosynthetic operon. The QS mechanism of *Pectobacterium* species causing soft rot has been called the "**mob attack**" mechanism of pathogenesis. The bacteria avoid production of pectic enzymes in early stages of infection, possibly to avoid triggering a defence response. The early production of cell wall degrading enzymes can release pectic wall fragments that can elicit host defence responses prematurely. The pathogen, therefore, requires building a sufficient population before risking detection. In an experiment, *carI* was introduced into transgenic tobacco plants. These plants produced 3-oxoC6HL and presumably activated the *Pec. carotovorum* subsp. *carotovorum* to produce pectic enzymes early in infection, leading to the triggering of a defence response. This increased resistance of transgenic plants to the pathogen suggests that such an early enzyme production might be harmful to the pathogen. The production of carbapenem late in infection avoids the tissue colonisation by other bacteria and thus protects the food supply of the pathogen.

The QS signalling molecule *N*-(3-oxo-hexanoyl)-L-homoserine lactone, produced by *Pec. carotovorum* subsp. *Carotovorum*, is required for the production of a β-lactam antibiotic called carbapenem and for the production of pectin lyases, cellulases and proteases that are required for efficacious survival and for optimal virulence (McGowan *et al*., 2005). Cui *et al*. (2005) reported that *Pec. carotovorum* subsp. *carotovorum* encodes ExpR, a homolog of LuxR that activates the expression of a global negative regulator that binds RNA.

Ralstonia solanacearum causes vascular wilt in many solanaceous and a few non-solanaceous plants. After entering the vascular system, the bacterium colonises the entire plant, and its population exceeds 10^{10} cells cm^{-1} length of stem tissue. Most of its traits related to infection and virulence are regulated by the Phc (*ph*enotype *c*onversion) regulatory system in a population density-dependent manner. PhcA, a LysR-type transcriptional regulator, is the most important component of a complex regulatory system, and its activity is regulated by a unique, volatile signal molecule, 3-OH palmitic acid methyl ester (3-OH PAME). *phcA* mutants do not produce extracellular polysaccharide (EPS), pectin methyl esterase and endoglucanase, but they have an increased motility and higher production of polygalacturonase and siderophore. 3-OH PAME acts as a QS signal and is used to sense population density and to regulate the production of EPS and degradative enzymes such as pectin methyl esterase, polygalacturonase, endoglucanase, etc. Low levels of 3-OH PAME lead to reduced EPS and exoenzyme synthesis, but increased levels of 3-OH PAME promote PhcA activity, resulting in enhanced production of EPS and exoenzymes, but decreased motility and siderophore synthesis. Thus, the Phc regulatory system serves as a master control switch to turn off behaviours suited to free-living survival of the bacterium and to turn on these when needed for initial host contact, microcolony/biofilm formation and pathogenesis.

Pan. stewartii subsp. *stewartii* causes Stewart's wilt of sweet corn and maize. After entering the plant, the bacterium grows in the apoplast (intercellular spaces) of young leaves causing "water-soaked" lesions and colonises the xylem vessels, leading to subsequent wilting. Accumulation of large amounts of EPS on pit membranes causes vascular occlusion.

EPS synthesis in *Pan. stewartii* subsp. *stewartii* strain DC283 is regulated in part by the EsaI/EsaR QS system. The AHL synthase, EsaI, catalyses the production of 3-oxo-C6HL and small amounts of 3-oxoC8HL. The *esaI* gene is constitutively expressed and is not subjected to EsaR-mediated autoregulation. Mutations in the *esaI* gene cause pleiotropic effects, eliminating ALH production, EPS synthesis and virulence, while the mutations in the *esaR* gene result in constitutive growth-independent hypermucoidy. In contrast, the wild-type strain produces EPS in a population density-dependent manner, with measurable levels detected primarily at population densities $>10^8$ cells mL^{-1}. The inoculation of the *esaI* mutant strain on sweet corn seedlings fails to induce wilting and causes only slight water-soaking as this strain has a functional copy of EsaR that represses EPS synthesis and Hrp-mediated water-soaking. In contrast, the inoculation of wild-type strains results in the death of seedlings. The *esaR* mutant, and *esaI* and *esaR* double-deletion mutants are significantly less virulent than the wild-type strains in spite of the fact that they produce large amounts of EPS and express *hrp* genes. These findings, thus, clearly show that QS control of pathogenicity genes is important. The reduced virulence of the mutants indicates that the early production of some virulence factors, such as EPS, may be counterproductive unless these are produced at the proper time and location of infection. *In vitro* studies with the wild-type and QS mutants showed that EPS synthesis interferes with the ability of the pathogen to attach to plastic surfaces.

The same attachment deficiency could be artificially induced in the *esaI* mutant strain through exogenous addition of AHL, the degree of attachment being inversely proportional to the amount of AHL supplied. In plant essays, the similar behaviour for bacterial attachment to plants was observed; the *esaI* mutants remained localised at the site of infection, whereas the wild-type strains moved at an impressive rate through the vascular system. Another species of *Pantoea*, *Pan. agglomerans* pv. *gypsophilae*, the causal agent of gypsophyla galls, produces C4HL.

Erw. amylovora harbours *luxI* and *lusR* homologs of *expI* and *expR*, *hslI* and *hslR* genes of *Pec. carotovorum* and *expI* and *expR* genes of *Dickeya chrysanthemi*. These genes are used for AHL synthesis and activation in these plant pathogens (Molina *et al*. 2005). AHL in *Erw. amylovora* regulates the expression of virulence effectors and subsequent symptom development.

Cheng *et al*. (2018) studied the interactions between AHL-producing transgenic tobacco plants and *Psm. syringae* pv. *tabaci* 11528. Both a reduction in disease incidence and a decrease in the growth of bacterium were observed in AHL-producing plants compared to wild-type plants, indicating that plant-produced AHLs enhance disease resistance against this pathogen. Subsequent RNA sequencing analysis showed that the exogenous addition of AHLs up-regulated the expression of *Psm. syringae* pv. *tabaci* 11528 genes for flagella production. It is evident that plant-produced AHLs activated a wide spectrum of defence responses in plants following inoculation, including the oxidative burst, hypersensitive response, cell wall strengthening and the production of certain metabolites. Exogenous AHLs alter the gene expression patterns of pathogens, while the plant-produced AHLs, either directly or indirectly, enhance plant local immunity during the early stage of plant infection.

Xan. campestris requires the QS gene cluster *rpf*, which encodes sensors of the diffusible signal factor and is required for virulence. HD-GYP domain regulator RpfG in this organism regulates the synthesis of extracellular virulence factors and negatively regulates biofilm formation (Dow *et al.*, 2003). Recognition of the diffusible signal factor is somehow coordinated with the degradation of cyclic di-GMP via a two-component RpfC (histidine kinase) and the RpfG response regulator (Ryan *et al.*, 2006).

Xan. oryzae pv. *oryzae* does not produce AHL, but it does produce a modular protein called OryR which has an AHL domain and a helix-turn-helix domain, both of which are typical of the LuxR-family subgroup of QS regulators (Fuqua *et al.*, 2001; Ferluga *et al.*, 2007). The rice signal molecule that is present in high concentrations in the xylem sap of rice plants solubilises OryR protein (Ferluga and Venturi, 2009). The chemical structure of the rice signal molecule is unknown, but the molecular size is less than 1 kDa.

Counteracting of QS signals: Plant pathogenic bacteria depend greatly on QS regulation to coordinate their colonisation and infection of plant hosts. A significant percentage of soil bacteria do produce enzymes for degrading or inactivating AHLs. The best studied of these enzymes is AiiA, a lactonase from *Bacillus cereus* that opens the lactone ring of all AHL tested, reducing their potency as signals by about 1000-fold. Interestingly, when AiiA was expressed in transgenic tobacco and potato plants, the plants showed a high degree of resistance to *Pec. carotovorum* subsp. *carotovorum* infection, which depends heavily on 3-oxo-C6-HSL-mediated QS for pathogenicity. Hence, the destruction or inactivation of AHL signals *in situ* could be a useful strategy for engineering disease-resistant plants.

Information available to date suggests that AHLs are not degraded substantially by plant enzymes, at least in the rhizosphere of soil-grown plants. However, plants and animals have innate defence mechanisms for disrupting bacterial QS by producing compounds that mimic AHL signals (Givskov *et al.*, 1996; Gao *et al.*, 2003). Available information shows that some higher plants, namely, pea, rice, tomato, soybean and *Medicago truncatula*, secrete various compounds that act like (mimic) bacterial QS signals. Such "signal-mimic" compounds may provide plants with important tools to disrupt or manipulate QS regulation in associated bacteria. First of all, QS signal-mimic compounds were found in marine red alga, *Delisera pulchra*. The active compounds were a set of halogenated furanones that are structurally similar to AHLs and interact specifically with AHL receptors. These compounds also inhibit AHL-mediated QS in many bacteria, interfere with biofilm formation and considerably alter the structure of natural bacterial communities on the algal surface in marine environments. Higher plants also produce signal-mimic compounds that can inhibit bacterial responses to an added AHL signal. However, in contrast to the strictly inhibitory activities of the furanone mimics of the *D. pulchra*, many of the signal mimics from higher plants stimulate AHL-induced QS behaviours (Von Bodman *et al.*, 2003).

Williams *et al.* (2004) reported that the AHL signals of *Psm. aeruginosa* enter the eucaryotic cells and activate transcription factors. The chemical cross-talk between plants and their associated bacteria appears to reflect a highly evolved and complicated biological system that seems to benefit both parties. Therefore, it is evident that QS is used to synchronise the production of virulence factors in pathogens utilising this signalling system. More than 350 genes have been observed to be regulated in *Psm. aeruginosa* by QS (Dong *et al.*, 2005).

7.2 ROLE OF PATHOGENICITY FACTORS IN DISEASE DEVELOPMENT

7.2.1 Bacterial Secretion Systems

Most determinants of pathogenicity need to be secreted to interact with host cells. The successful infection by plant pathogenic bacteria utilises various influx and efflux systems as well as enzyme- and effector-protein translocation pathways. The influx systems are used for the uptake of essential components needed for survival, which are usually produced by the enzymatic breakdown of complex substrates into smaller compounds accessible by the influx system. The efflux systems protect the bacterial cells from toxic compounds and antimetabolites, including pH differences and induced innate host defence compounds, such as hydrogen peroxide, nitric oxide, phytoalexins and phenolics.

Six secretion systems, i.e., I to VI, are found in Gram-stain-negative bacteria, which deliver various chemicals, including proteins and nucleic acids, to perform a multitude of biochemical and physiological functions. The secretion systems of plant pathogenic bacteria play a major role in their pathogenicity as they transport bacterial proteins and other molecules, required for pathogenicity, into the host cells. As the secretion machineries of these secretion systems utilise the inner and outer membranes of the bacterial envelop, therefore, only the Gram-stain-negative bacteria can be equipped with these secretion pathways. *Pec. atrosepticum* strain 1043 contains genes encoding all the six secretion systems, suggesting a complex secretome and an impressive ability to interact with its environment(s). All the six secretion systems are found in plant pathogenic bacteria.

7.2.1.1 Type I Secretion System

The type I secretion system involves protein translocation from the cytoplasm across the inner and outer membranes without any interaction in the periplasm. The type I export mechanism requires only three proteins and does not generate periplasmic intermediates. The type I secretion apparatus is composed of membrane ATPase (the ATP-binding cassette [ABC] protein), a membrane fusion protein that belongs to the membrane fusion protein family and an outer membrane peptide.

ABC transporter containing these membrane proteins can translocate a wide variety of substrates across extra- and intra-cellular membranes. Polypeptides that are secreted via the type I secretion apparatus are transferred from the cytosol, without a periplasmic intermediate, through the inner and outer membranes out into the extracellular environment. The secreted polypeptides can be toxin or enzymes, such as proteases of *Dic. chrysanthemi* and *Psm. aeruginosa*. Based on genome sequence analysis, *Agr. tumefaciens*, *Leifsonia xyli*, *Pec. carotovorum*, *Ral. solanacearum*, *Xylella fastidiosa*, *Xan. oryzae* pv. *oryzae*, *Xan. campestris* and pathovars of *Xan. axonopodis* and *Psm. syringae* are equipped with type I secretion systems (TISSs).

Type I is present nearly in all plant pathogenic bacteria. It secretes proteins of various sizes, often with calcium-binding glycine rich repeats, from the cytoplasm to the extracellular matrix in a single step. Various toxins, such as hemolysins, rhizobiocin and cyclolysin, are also secreted. These consist of ATP-binding cassette proteins and perform in and out movement of a variety of compounds for which the energy is provided by the hydrolysis of ATP. In the tribe erwiniae, the TISS is required for the secretion of proteases, which apparently play only a minor role in pathogenesis. The genome sequence of *Pec. atrosepticum* strain 1043 has revealed additional TISSs. In summary, TISS plays only a minor role in pathogenicity.

7.2.1.2 Type II Secretion System

The type II secretion system (TIISS), conserved among Gram-stain-negative bacteria, is characterised by a two-step protein translocation process. The first step is protein processing through the inner (cytoplasmic) membrane, followed by protein modifications to exit the outer membrane. The major route of protein translocation across the inner membrane is a secretion pathway, to which substrates are targeted by aminoterminal signal peptides. For more details of the system, the readers may refer to Sandkvist (2001) and Kado (2010).

This is common among the Gram-stain-negative bacteria and shares many features with the type IV pilus system. TIISSs are found in several phytopathogenic bacteria, including xanthomonads, pseudomonads, *Xyl. fastidiosa*, *Ral. solanacearum* and the tribe erwiniae. In *Xan. campestris* pv. *campestris* and *Xan. oryzae* pv. *oryzae*, it is involved in the secretion of plant cell wall-degrading enzymes (PCWDEs), polysaccharide and proteases, and hence plays an important role in virulence. TIISS has been studied in detail in soft rot erwiniae (SRE). It is responsible for the secretion of PCWDEs and is, therefore, essential for soft rot disease development. TIISS is involved in the translocation of various virulence factors, including toxins, enzymes and proteins. The translocation of proteins involves a two-step process. In the first step, unfolded proteins move to the periplasm via the secretion (Sec) pathway across the inner membrane. In the second step, processed and folded proteins move through the periplasm and across the outer membrane via an apparatus consisting of 12–14 proteins encoded by a cluster of genes. *Xanthomonas* and *Ralstonia* species possess two TIISS per cell and use them for the secretion of virulence factors such as pectinolytic and cellulolytic enzymes. *Xylella* and *Agrobacterium* species possess only one TIISS per cell. *Agrobacterium* species have the genes only for the first step of protein transport across the inner membrane, and these use the type IV secretion system (TIVSS) for the remaining process. TIISS is regulated in part by a QS mechanism (Agrios, 2004).

7.2.1.3 Type III Secretion System

In plant pathogens, type III secretion apparatus is encoded by *hrp* genes. As the type III secretion system (TIIISS) in phytopathogens was first recognised through its association with the HR, the genes encoding it have been named as *hrp* genes. The *hrp* clusters from several phytopathogenic bacteria have been thoroughly studied. The first *hrp* cluster to be characterised was that of *Psm. syringae* pv. *phaseolicola* (Lindgren *et al.*, 1986), while for *Xanthomonas* species, the first one was of *Xan. campestris* pv. *campestris* (Arlat *et al.*, 1991). Later on, the *hrp* clusters from *Xan. campestris* pv. *vesicatoria, Xan. oryzae* pv. *oryzae* and *Xan. axonopodis* pv. *glycines* were also sequenced and characterised (Zhu *et al.*, 2000; Buttner and Bonas, 2002; Kim *et al.*, 2003). The widely studied *hrp* cluster from *Xan. campestris* pv. *vesicatoria* has a region of 23 kb and contains six operons, from hrpA to hrpF. Nine out of the twenty-two proteins encoded by the cluster, conserved among plant and animal pathogens, are named *hrc* (*hrp* conserved), and constitute the core of the type III translocon. Structurally, type III protein secretion system in phytopathogens is composed of a flagellum-the Hrp pilus and a basal body that crosses the two bacterial membranes and ensures the stabilisation of the whole structure. This complex structure facilitates the bacterial attachment to the host cell membrane to ensure the translocation of effector proteins to the interior of the plant cell (Hueck, 1998; Tampakaki *et al.*, 2004).

There are a number of genes in the *hrp* cluster, which show a high level of similarity to those in the animal pathogens, and, hence, these have been termed as *hrp*-conserved (*hrc*) genes. The *hrp* cluster in the soft rot erwiniae has been demonstrated to play an important role in pathogenesis. *Erw. amylovora* (*Eam*), a close relative of soft rot erwiniae, has been best-studied of the erwiniae in terms of TIIISS and its effectors. Three TIIISS secreted proteins, namely, DspE/AEam, HrpNEam (harpin) and HrpWEam, have been characterised in *Erw. amylovora*. DspE/AEam appears to be widely conserved in many plant pathogens and is currently the only demonstrated effector protein in the erwiniae. HrpWEam and HrpNEam, on the other hand, have been termed helper proteins that may assist the translocation of effectors (such as DspE/A), but are believed to remain outside the host cell (Alfano and Collmer, 2004).

A number of plant pathogenic bacteria use one or more TIIISSs to alter host cell machinery and cause disease. TIIISSs are found in a variety of bacterial pathogens (Desvaux *et al.*, 2006), including major human pathogens, such as *Salmonella*, *Shigella*, *Yersinia* and *Pseudomonas* species and enteropathogenic *Escherichia coli*; and are composed of at least 40 genes, making their temporal and spatial regulation very complex. Optimum temperature and pH are critical for the functioning of TIIISS. TIIISS plays a very important role in the pathogenicity of bacteria belonging to the genera *Xanthomonas*, *Pseudomonas* and *Ralstonia*. Its primary function is the transport of effector proteins across the bacterial membrane and further into the plant cell. In addition to the above-mentioned three genera, TIIISS is also found in most Gram-stain-negative bacteria (*Erwinia* and *Pantoea* species), including those causing diseases in humans and animals. The translocation of the bacterial effector proteins into the host cells interferes with host cell signal transduction and other cellular processes, leading to the enhancement of bacterial virulence in the susceptible hosts. Without the TIIISS, phytopathogens cannot defeat basal defences of the host, grow in plants, produce disease symptoms on host, or elicit hypersensitive response in non-hosts. TIIISS mutants fail to suppress the basal plant defences. Some effectors also act as virulence proteins that are recognised by corresponding host resistance (R) proteins, triggering the HR that restricts pathogen spread. Alfano and Collmer (2004) have presented a conceptual overview of TIIISS effectors functioning as both elicitors and suppressors of plant defences.

Type III secretion apparatus is usually encoded by a set of genes often on a plasmid or within a chromosomal pathogenicity island. The secretion apparatus is composed of a macromolecular complex spanning both the inner and outer membranes. All TIIISSs mediate injection of virulence proteins, called type III effectors as well as nodulation outer proteins.

Psm. syringae utilises TIIISS to inject effector proteins into plant host cells. The bacterium has a wide host range, and this is caused, in part, due to the diversity of effectors employed by this phytopathogen. More than 60 different effector families exist in *Psm. syringae*, and one such family is HopF, which contains over 100 distinct alleles. However, despite this diversity, researchers have focused on only two members of this family: HopF1 from *Psm. syringae* pv. *phaseolicola* 1449B and HopF2 from *Psm. syringae* pv. *tomato* DC3000. Lo *et al.* (2017) have reviewed the work done on HopF family members, including their host targets and molecular mechanisms of immunity suppression and their enzymatic function; and they have also provided a phylogenetic analysis of this expanding effector family which provides a basis for a proposed nomenclature to guide the future research. Extensive genetic diversity within the HopF family presents a great opportunity to study how the functional diversification of an effector family contributes to host specialisation.

Chakravarthy *et al.* (2018) constructed *Psm. syringae* pv. *tomato* DC3000 derivatives, differentially producing coronatine (COR), the TIIISS machinery and subsets of key effectors, and assayed in leaves of *Nicotiana benthamiana*. The inoculations were made by dipping the whole plants in bacterial cell suspension and the plants assayed for bacterial population and the production of chlorotic spots on leaves. The strains fell into three classes. Class I strains were TIIISS+, but functionally effector-less; they grew poorly *in planta* and produced faint chlorotic spots only if COR+. Class II strains were TIIISS–, or, if TIIISS+, produced effectors avirulence (Avr) PtoB and HopM1. However, class II strains grew better than class I strains *in planta* and, if COR+, also produced robust chlorotic spots. Class III strains were TIIISS+ and minimally produced AvrPtoB, HopM1 and three other effectors encoded in the *Psm. syringae* conserved effector locus. These strains grew better than class II strains *in planta* and produced chlorotic spots without COR, if the precursor coronafacic acid was produced. COR made no significant contribution to the bacterial colonisation of the apoplast, but enabled a gratuitous, semi-quantitative surface indication of bacterial growth, which is determined by the strain's effector composition.

Fan *et al.* (2017) screened 56 plant phenolic compounds and derivatives for their effects on the *Xan. oryzae* pv. *oryzae* TIIISS. Ten of these compounds significantly inhibited the promoter activity of a harpin gene, *hpa1*. Further testing of these inhibitors for their impact on the HR caused by *Xan. oryzae* pv. *oryzae* on non-host tobacco plants revealed that pre- treatment of *Xan. oryzae* pv. *oryzae* with TS006 (*o*-coumaric acid, OCA), TS010, TS015 and TS018 resulted in significantly attenuated HR without affecting bacterial growth or survival. Additionally, Cya translocation assays demonstrated that the translocation of two TIII effectors was suppressed by the four inhibitors. Quantitative reverse transcription-polymerase chain reaction analysis showed that mRNA levels of representative genes in the *hrp* cluster, as well as the regulatory genes *hrpG* and *hrpX*, were reduced by treatment with the four inhibitors, suggesting that expression of the *Xan. oryzae* pv. *oryzae* TIIISS was suppressed. The expression of other virulence factors was not suppressed, which indicated possible TIIISS-specific inhibition. These inhibitors reduced the disease symptoms of *Xan. oryzae* pv. *oryzae* and *Xan. oryzae* pv. *oryzicola* on the rice cultivar IR24 to varying extents.

Bismerthiazol provides an excellent control of bacterial blight of rice caused by *Xan. oryzae* pv. *oryzae*, but does not inhibit considerable *in vitro* growth of the bacterium. According to RNA-sequencing analysis, the transcription of the histidine utilisation (Hut) pathway genes of *Xan. oryzae* pv. *oryzae* ZJ173 was inhibited after 4.5 and 9.0 h of bismerthiazol treatment. Functional studies of *hutG* and *hutU* indicated that the Hut pathway had little effect on the growth and bismerthiazol sensitivity of the bacterium *in vitro*, but significantly reduced the aggregation of the cells. Deletion mutants of *hutG* or *hutU* were more motile, produced less biofilm and were less virulent than the wild-type, indicating that the Hut pathway is involved in QS and contributes to virulence. Over-expression of the *hutG-U* operons in ZJ173 reduced inhibition of the bacterium by bismerthiazol. Bismerthiazol did not inhibit the transcription of Hut pathway genes, QS, or virulence of the bismerthiazol-resistant strain 2-1-1. It is evident that bismerthiazol reduces virulence of the bacterium by inhibiting the Hut pathway and QS (Liang *et al.*, 2018).

Gottig *et al.* (2010) have given a detailed account of the type III protein secretion system involved in citrus canker caused by *Xan. axonopodis* pv. *citri* (recent name *Xan. citri* subsp. *citri*). Hajri *et al.* (2009) predicted several effector proteins to be secreted by this system in *Xan. axonopodis* pv. *citri*; among them is the well-characterised PthA, a member of the *avrBs3* gene family which was first identified as the elicitor of the symptoms of citrus canker (Swarup *et al.*, 1992). *Hrp* cluster is critical for the development of the disease. Dunger *et al.* (2005) reported that *Xan. axonopodis* pv. *citri* mutants in the *hrp* cluster, the one in the gene that codifies for HrcN, the ATPase that may be providing the energy for the transport system or for the assembly of the secretion apparatus; another in an operon bearing genes, which codify for proteins of the core of the translocon (HrcQ, HrcR and HrcS); and the last one in a gene that codifies for a pore-forming protein that mediates translocation of effector proteins into the plant cell (HrpF), showed no virulence when infiltrated in citrus leaves, showing consistency with the crucial role of this system in the pathogenesis. In contrast to the compatible interactions produced in citrus canker, *Xan. axonopodis* pv. *citri* induces non-host resistance with HR in cotton, tobacco, tomato, bean and pepper, and a functional *hrp* cluster is required for the induction of HR on these non-host plants, since the above-mentioned mutants produced no HR lesions in these plants.

Xan. citri subsp. *citri* uses transcription activator-like effectors (TALEs) as major pathogenicity factors. TALEs, which are delivered into plant cells through TIIISS, interact with effector binding elements (*EBE*s) in host genomes to activate the expression of downstream susceptibility genes to promote disease. Predictably, TALEs bind *EBE*s in host promoters via known combinations of TALE amino acids to DNA bases, known as the TALE code.

Shantharaj *et al.* (2017) fused *ProBs31EBE* and *ProBs314EBE* to the *Xanthomonas* gene, *avrGf1*, which encodes a bacterial effector that elicits an HR in grapefruit and sweet orange, and demonstrated, in transient assays, that activation of *ProBs314EBE* by *Xan. citri* subsp. *citri* TALEs is TIIISS-dependent, and that the expression of AvrGf1 triggers HR and correlates with decreased bacterial growth. All tested virulent *Xan. citri* subsp. *citri* strains from diverse geographical locations activated *ProBs314EBE*. TALEs are essential for the virulence of *Xan. citri* subsp. *citri* strains and, because the engineered promoter traps are activated by multiple TALEs, this concept has the potential to confer broad-spectrum, durable resistance to citrus canker in stably transformed plants.

Xan. fuscans subsp. *aurantifolii* group C strains possess a type III effector, AvrGf2, belonging to the *Xanthomonas* outer protein (Xop)AG effector gene family, which restricts the host range among citrus species. Gochez *et al.* (2017) dissected the modular nature and mode of action of AvrGf2 in grapefruit resistance. Mutation of GPLL to AASL in AvrGf2 abolished the elicitation of the HR, while the mutation in only the first amino acid to SPLL delayed the HR in grapefruit. Yeast two hybrid experiments showed strong interaction of AvrGf2 with grapefruit cyclophilin (GfCyp), while the AvrGf2-SPLL and AvrGf2-AASL mutants exhibited weak and no interaction, respectively. Molecular modelling and *in silico* docking investigations for the cyclophilin–AvrGf2 interaction predicted the binding of citrus cyclophilins (CsCyp, GfCyp) to hexameric peptides spanning the cyclophilin-binding domain of AvrGf2 and AvrGf2 mutants (VAGPLL, VASPLL and VAAASL) with affinities equivalent to or better than a positive control peptide (YSPSA) previously demonstrated to bind CsCyp. Additionally, the C-terminal domain of XopAG family effectors contains a highly conserved motif, CLNA × YD, which was identified to be crucial for the induction of HR based on site-directed mutagenesis (CLNA × YD to CASA × YD). These findings suggest a model in which grapefruit cyclophilin promotes a conformational change in AvrGf2, thereby triggering the resistance response.

Xan. axonopodis pv. *manihotis* strain CIO151 possesses 17 predicted type III effectors (TIIIEs) belonging to the Xop class. Medina *et al.* (2018) characterised nine Xop effectors present in strain CIO151 for their role in virulence and modulation of plant immunity and demonstrated the importance of XopZ, XopX, XopAO1 and AvrBs2 for full virulence, as well as a redundant function in virulence between XopN and XopQ in susceptible cassava plants. Their role in pathogen-associated molecular pattern (PAMP)-triggered immunity (PTI) and effector-triggered immunity (ETI) using heterologous systems revealed that AvrBs2, XopR and XopAO1 were capable of suppressing PTI. ETI suppression activity was only detected for XopE4 and XopAO1. These results show the overall importance and diversity in functions of major virulence effectors AvrBs2 and XopAO1 in *Xan. axonopodis* pv. *manihotis* during cassava infection.

Nissan *et al.* (2018) generated draft genome sequences of *Pan. agglomerans* pv. *betae* and pv. *gypsophilae* and employed these sequences in combination with a machine learning approach and a translocation assay into beet roots to identify the pools of TIIIEs in these pathovars. Mutational analysis showed that in pv. *betae* 4188, eight TIIIEs (HsvB, HsvG, PseB, DspA/E, HopAY1, HopX2, HopAF1 and HrpK) contribute to pathogenicity on beet and gypsophila. In pv. *gypsophilae* 824-1, nine TIIIEs (HsvB, HsvG, PthG, DspA/E, HopAY1, HopD1, HopX2, HopAF1 and HrpK) contribute to pathogenicity on gypsophila, while the PthG effector triggers HR on beet. HsvB, HsvG, PthG and PseB appear to endow pathovar specificities to pv. *betae* and pv. *gypsophilae*, and no homologous TIIIEs were identified for these proteins in other plant pathogenic bacteria. Conversely, the remaining TIIIEs contribute to the virulence of both the pathovars, and homologous TIIIEs were found in other phytopathogenic bacteria. Remarkably, HsvG and HsvB, which act as host-specific transcription factors, contributed largely to the disease development.

Sang *et al.* (2018) found that the *Ral. solanacearum* type III effector RipAY suppresses plant immune responses triggered by bacterial elicitors and by the phytohormone salicylic acid. Biochemical analysis indicated that RipAY associates *in planta* with thioredoxins from *N. benthamiana* and Arabidopsis. RipAY displays γ-glutamyl cyclotransferase activity to degrade glutathione in plant cells, which is required for the reported suppression of immune responses. Seeing the significance of thioredoxins and glutathione as major redox regulators in eukaryotic cells, RipAY activity may constitute a novel and powerful virulence strategy employed by *Ral. solanacearum* to suppress immune responses and potentially alter general redox signalling in host cells.

Duan *et al.* (2018) obtained a gene expression profile of *Ral. solanacearum* PO41 under the root secretions environment of *Solanum tuberosum* at 8, 16 and 24 h after infection, with RNA microarray technology. It was found that the virulence factors of *Ral. solanacearum* mainly focused on the output pathways of toxic protein [Sec pathway, twin-arginine translocation (Tat) pathway and TIIISS], the aggregation and transfer of exopolysaccharides and the chemotactic movement, and the adhesion of flagellum in the potato root secretion ecological niche, whereas the virulence factors in the atypical output pathway were mainly distributed in Sec (*secB*, *secDF*, *yidc*) and Tat (*tatA*, *tatC*) pathways to promote the output of folded and unfolded toxic proteins. The *fliI* ATPase was obviously upregulated 8 h after inoculation, showing that type III secretion system was only active at the early stage of *Ral. solanacearum* infection. Upregulated expression of phosphoglucomutase and epimerase showed that the virulence factor of exopolysaccharides was synthesised at the early stage of bacterial infection. Chemotactic receptor and motor protein were obviously upregulated within 24 h of inoculation. It was evident that *Ral. solanacearum* PO41 had already colonised to the roots within 24 h due to the stimulation of root secretion. Some pathogenic genes were upregulated during this period.

7.2.1.4 Type IV Secretion System

This secretion system transports macromolecules between bacteria and from bacteria into their eucaryotic hosts. These transferred proteins are very similar to those which mobilise plasmids among bacteria. The *virB* operon of *Agr. tumefaciens* encodes 11 proteins that form an organised structure. These proteins are involved in the transfer of T-DNA strand from the bacterium to the cytoplasm of the plant cell. The transporting structure originating from bacterial inner membrane stretches through the outer membrane and terminates in a pilus-like structure that protrudes from the bacterial cell (Agrios, 2004). Three other phytopathogenic bacteria, namely, *Pec. atrosepticum* strain 1043, *Xan. citri* subsp. *citri* and *Xyl. Fastidiosa*, possess gene clusters homologous to *VirB*. The presence of *virB* cluster of *Pec. atrosepticum* strain 1043 within horizontally acquired island 7 suggests that it was acquired through horizontal gene transfer.

The type IV secretion system (TIVSS) transports proteins, nucleoproteins and nucleic acid substrates through specific channelling complexes incorporated into the outer membrane and cell envelop. Highly conserved TIVSSs have been identified in both plant and mammalian pathogens. This secretion system can also include pathogen-based effectors that are injected into eucaryotic cells. Like the TIIISSs, TIVSSs are used by various bacterial pathogens to transfer bacterial proteins and DNA into eucaryotic cells. Among the bacterial plant pathogens, *Ag. tumefaciens* harbours the well-characterised prototypical TIVSS used to transfer the T-DNA into the plant cell to cause crown gall. Genome and plasmid sequences of *Agr. tumefaciens*, *Agr. rhizogenes*, *Agr. vitis*, *Xyl. fastidiosa* and *Xanthomonas* species have revealed the presence of type IV secretion components. Their functional presence, and the role of the Ti-plasmid-based TIVSS in *Agr. Tumefaciens*, are relatively well characterised. *Xan. citri* pv. *citri* contains two TIVSS loci; one on the chromosome and the other on a 64-kb pXAC64 plasmid (da Silva *et al.*, 2002). Alegria *et al.* (2005) reported that these loci are VirB orthologs, including aVirB7 protein discovered in *Xan. citri* pv. *citri*.

More than one type of IV secretion pathway can occur in a given pathogen, and a pathogen can also be equipped with type II and type III as well as type IV secretion systems, e.g., *Burkholderia cepacia*. Genes encoding the TIVSS component can be found on plasmids and on chromosomes such as that of *Bur. cepacia*. The available information shows that TIVSS loci are highly conserved and that more than one TIVSS has been widely dispersed among Gram-stain-negative bacteria via lateral gene transfer. *Spiroplasma kunkelii* harbours a 14.6-kb plasmid, pSKU146, which apparently encodes a TIVSS. Plasmid pSKU146 contains open reading frames that encode proteins having similarity to conjugative elements and components of the bacterial type IV secretion apparatus (Davis *et al.*, 2005). Basically, TIVSS are used to either transfer protein effectors into host cells or to take up protein and DNA (retrotransfer) for environmental adaptation (Kado, 2002). Interestingly, TIVSSs operate to move substrates in a bidirectional manner, i.e., certain TIVSSs function to deliver or take up protein or nucleic acid substrates.

7.2.1.5 Type V Secretion System

The type V secretion system (TVSS), known as the autotransported protein system, is composed of proteins that are sub-classified as either an autotransporter or a two-partner secretion pathway family member. The characteristic of this family is the capacity of each member to mediate its own translocation across the bacterial outer membrane. The Sec system is used to mediate translocation across the inner membrane. Autotransporter proteins use a *C*-terminal translocater domain to mediate the secretion of an *N*-terminal effector domain across the outer membrane to the bacterial surface in a process requiring no accessory proteins. Proteins of two-partner secretion pathway differ from autotransporters in that the effector and translocater domains are expressed as separate proteins (Henderson *et al.*, 2004). In contrast to type I to IV secretion systems, only a single gene encodes the secreted protein, and the complete machinery is required for secretion. TVSS is found in a wide range of Gram-stain-negative bacteria. *Xylella* and *Xanthomonas* species contain genes for TVSS (autotransporters), which encode surface associated adhesins.

Two-partner system also occurs in *Pec. atrosepticum* strain 1043, *Erw. amylovora* and *Dic. chrysanthemi* (Toth *et al.*, 2006).

7.2.1.6 Type VI Secretion System

In *Vibrio cholerae*, the virulence-associated secretion (VAS) system transports proteins to the exterior of the bacterial cell by a mechanism that does not require hydrophobic *N*-terminal signal sequences. Several *vas* genes products constitute the structural components of the VAS apparatus that delivers effector proteins via the type VI secretion system (TVISS) (Pukatzki *et al.*, 2006). Several Gram-stain-negative bacterial plant pathogens and endosymbionts, such as *Agr. tumefaciens* and *Rhizobium leguminosarum*, respectively, harbour genes homologous to *vas* genes. Wu *et al.* (2008) reported that the TVISS is used by *Agr. tumefaciens* to secrete a hemolysin-regulated protein that is required for efficient tumorigenesis. TVISS has been found associated with secretion of hemolysins and cytolysins in *Serratia marcescens.* A homologous system is present in *Xyl. fastidiosa* and *Ral. solancearum.*

It is hoped that TVISS will also be discovered in other Gram-stain-negative bacterial plant pathogens.

7.2.1.7 Twin-Arginine Translocation Protein Secretion System

Besides the six secretion pathways (types I–VI) described above, there is another pathway used by the bacteria, the Tat secretion system, , It is found in the cytoplasmic membranes of most bacteria and in the energy-transducing membranes of plant organelles (Berks *et al.*, 2005; Muller and Klosgen, 2005). This pathway transports proteins in a Sec-independent manner. Such proteins are synthesised with amino-terminal signal peptides that contain a distinctive invariant peptide sequence of SRR × FLK, which contains the twin arginine amino acid motif. Berks *et al.* (2000) reported that proteins bearing twin-arginine signal peptides are directed post-translationally to the membrane-embedded Tat secretion system that transports fully folded proteins across the inner membrane using energy provided by the transmembrane pH gradient. The core of the Tat transporter is a large oligomeric complex of TatA, TatB and TatC proteins. TatA protein provides the protein-conducting channel within the complex through which folded proteins are transported.

The studies on a Tat secretion system in *Agr. tumefaciens* showed that the inactivation of this system abolishes anaerobic respiration and chemotaxis, and attenuates virulence (Ding and Christie, 2003). Ferreira *et al.* (2006) found substrates of Tat secretion system in *Psm. syringae* pv. *tomato.* Phospholipases were found to be substrates that are translocated across the outer membrane by a *Psm. syringae* TIISS encoded by the *gsp* operon (Bronstein *et al.*, 2005). It indicates that the Tat secretion system and the TIISS coordinate the secretion of these substrates. With intensification of research work on this aspect in the future, plant pathogenic bacteria like *Psm. syringae* (Caldelari *et al.*, 2006) and *Ral. solanacearum* (Gonzalez *et al.*, 2007) will be found to be using this pathway to deliver effectors used in virulence and pathogenesis.

7.3 CIRCUMVENTING HOST RESISTANCE BY PLANT PATHOGENIC BACTERIA

Successful plant pathogenic bacteria are equipped with an arsenal of products that help evade innate and induced host defence mechanisms. Innate immunity depends on germline encoded receptors that have evolved to recognise highly conserved PAMPs. These receptors are called pattern recognition receptors (PRRs). Two distinct types of elicitors, namely, general elicitors and specific elicitors, activate host defences. General elicitors are comprised of bacterial lipopolysaccharides, phospholipids, flagellins, peptidoglycans, proteins such as elongation factors, double-stranded nucleic acids, cell wall pieces and methylated DNA that can bind to host PRRs and trigger host defences (Kunze *et al.*, 2004; Ingle *et al.*, 2006). Very specific receptors that are often host cultivar-specific, perceive specific elicitors.

During the infection process, the pathogenic bacteria inject several proteins into the host tissue, via the secretion systems described above. The effector proteins, which include proteins that suppress defence reactions and promote bacterial survival, include race-specific elicitor proteins encoded by *avr* genes. If the host plant carries the corresponding resistant (*R*) gene, such Avr proteins usually induce defence responses such as HR. This is known as an incompatible host–pathogen interaction, in which the pathogen produces an Avr effector protein that is recognised by a cognate *R* gene product in the resistant cultivar (Keen, 1990; Martin *et al.*, 2003). R proteins are thought to recognise the modification of the virulence target by the Avr protein. R proteins elicit HR in response to cell death suppression activity, which is only observed in the absence of the cognate R protein, is a simple model. It has been replaced by a more complex model, the **trump model**, proposed by Abramovitch and Martin (2005). The trump model includes a trump (T) factor that would act on the cognate R protein to generate an HR-based immunity in response to a cell death suppressor. Certain effector proteins, such as HopPtoN, suppress cell death in both compatible and incompatible interactions (Lopez-Solanilla *et al.*, 2004).

Morel and Dangl (1997) reported that HR is observed in resistant plants challenged with a pathogen as well as being a consequence of non-host resistance in non-host plants. This response is associated with cell death at the initial stage of infection that results in rapid localised necrosis that initially involves the activation of pre-existing compounds, followed by the induction of plant defence mechanisms. This is accompanied by the production of reactive oxygen species leading to plant cell death, and, consequently, the elimination of the pathogen (Mysore and Ryu, 2004).

Chisholm *et al.* (2006) reported that the activation of R protein-mediated resistance also suppresses microbial growth, but after the limited proliferation of the pathogen. These results challenge the gene-for-gene hypothesis of Flor (1971), in which disease resistance occurs only with the simultaneous presence of *R* gene resistance in the plant and corresponding avirulence (*Avr*) gene in the pathogen, suggesting that a direct Avr–R interaction activates the resistance. However, despite many investigations, the physical interaction between Avr and R proteins has been observed in only a few cases. As many Avr are known to increase the virulence of the pathogen in susceptible plants, an alternative model has been proposed in which the Avr protein is able to interact and modify a target specific to the plant to promote disease in the absence of the corresponding R protein. The function of R protein would be to guard or protect these targets and activate the resistance signalling pathways when these plant targets are modified by a specific Avr (Chisholm *et al.*, 2006).

PAMPs trigger innate immunity defences in the host (Abramovitch *et al.*, 2006; Jones and Dangl, 2006). These defences are regulated by interconnected chemical signalling networks that utilise ethylene, jasmonic acid and salicylic acid. PAMPs activate mitogen-activated protein kinases (MAPKs) and the production of reactive oxygen species (ROSs) by the host. Superoxide dismutases produced by the pathogen help reduce the effects of ROSs. Espinosa and Alfano (2004) reported that such virulence factors can inhibit PAMP-elicited basal defences. Recruitment of host factors by the pathogen is a highly evolved means of circumventing innate defences. Djamei *et al.* (2007) reported that the *Agrobacterium* VirE2-binding protein VIPI, produced by the plant host, appears to be a substrate of MAPK3 that phosphorilates VIPI at serine position 79 of the VIPI peptide in order to translocate the T-DNA strand-VirE2 complex into the plant cell nucleus. Thus, *Agr. tumefaciens* has recruited MAPK3 to facilitate nuclear localisation and the eventual integration of the T-DNA strand in the host chromosome.

Cysteine proteases play an important role in plant pathogenesis. Hotson and Mudgett (2004) reported that the effectors AvrXv4, AvrPphb, AvrRpt2 and XopD have cysteine protease functions. As cysteine proteases have been identified as effectors injected into plant cells by the TIIISS, their role is likely to be targeting defence-related proteins.

7.4 NEUTRALISATION OF REACTIVE OXYGEN SPECIES

When perturbed, signalled, or damaged lysosomes in the host produce ROSs as a host defence response. The pathogens equipped with a ROS detoxification pathway can effectively dodge this defence. Among plant pathogenic bacteria, *Agr. tumefaciens* harbours the *katG* gene that encodes catalase to hydrolyse toxic peroxides, and, as reported by Yuan *et al.* (2005), it is induced by low phosphate concentration. A catalase peroxidase encoded by *katG* in *Xan. campestris* pv. *campestris* appears to be required for virulence by conferring resistance to the detrimental effects of the hydrogen peroxide defence system (Jittawuttipoka *et al.* 2009). Sukchawalit *et al.* (2001) reported that ROS is also detoxified by an alkyl-hydroperoxide reductase encoded by the *ahpC* gene in *Xan. campestris* pv. *phaseoli*. The hydrogen peroxide-induced transcription factor OxyR regulates the expression of both *katG* and *ahpC* genes (Zheng *et al.*, 2001). Hydrogen peroxide activates OxyR by the formation of an intramolecular disulphide bond. Oxidised OxyR, in turn, activates transcription of *katG* and *ahpC* among other reductases. Saenkham *et al.* (2007) reported that *Agr. tumefaciens* possesses three iron-containing superoxide dismutases called SodBI, SodBII and SodBIII, all of which play a role in virulence.

7.5 ROLE OF PHYTOALEXINS AND THEIR DETOXIFICATION

Phytoalexins are antimicrobial compounds produced by the plants in response to pathogen invasion. Some of these compounds restrict or inhibit the multiplication of bacteria by hydrogen bonding interactions with vital proteins, such as enzymes. Virulent pathogens trigger plant defence responses at later stages of infection. These are able to detoxify phytoalexins using various mechanisms that include modification of these compounds followed by their metabolism as well as efficient efflux systems, which prevent intracellular accumulation of toxic compounds. Due to their broad substrate specificity, efflux pumps expel a wide range of harmful compounds from the cell directly into the extracellular environment, bypassing the periplasm. Palumbo *et al.* (1998) reported that an efflux pump induced by isoflavonoids was found in *Agr. tumefaciens*. The pump greatly reduces the cellular accumulation of isoflavonoids produced by alfalfa roots as exudates. The efflux system is encoded by *ifeABR* locus and is induced by various isoflavonoids and a chalcone. The mutations in *ifeABR* reduced the ability of *Agr. tumefaciens* to colonise the root system. Burse *et al.* (2004) reported the presence of an efflux pump, encoded by a putative *acrB* operon, in *Erw. amylovora*. Disruption of the efflux pump by mutation in *acrB* dramatically decreases the virulence of this bacterium.

Plants have the ability to counteract the bacterial efflux system by producing inhibitors of the multidrug resistance pumps. Glazebrook and Ausubel (1994) found the accumulation of phytoalexin camalexin in *Arabidopsis thaliana* on inoculation with *Psm. syringae* pv. *maculicola*. Arabidopsis mutants deficient in camalexin biosynthesis enhanced the sensitivity of host to *Psm. syringae* pv. *maculicola*. They further reported that camalexin is not required for resistance to avirulent *Psm. syringae* strains in *A. thaliana*.

Bacterial pathogens that incorporate parts of their genome into plant cells are equipped with mechanisms to suppress the small interfering RNAs (siRNAs) produced by the plant's defence reaction. Dunoyer *et al.* (2006) reported that *Agr. tumefaciens* suppressed siRNAs, whose production was induced by the incorporation of multiple copies of the T-DNA into the plant genome. The tumour-inducing pathogen triggers the formation of siRNAs 3 days after inoculation, but these siRNAs are not found in mature crown gall tumours. The formation of microRNAs that are required for tumour formation, presumably by a degradation pathway of target mRNAs, induces formation of siRNAs in the tumour that are subsequently degraded by uncharacterised *Agrobacterium*-mediated factor(s).

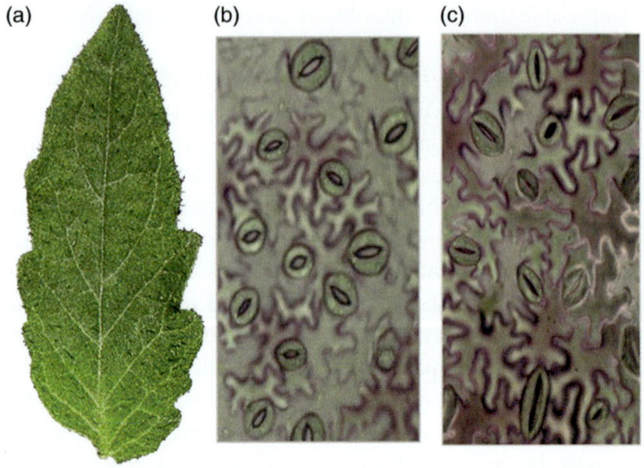

FIGURE 7.1 Light-conditioned tomato leaf (a) and leaf surface under the microscope prior to exposure to *Psm. syringae* pv. *tomato* (*Pst*) DC3000 showing mostly open stomata (b). The same leaf was exposed to *Pst* DC3000, and after 1 h of exposure most stomata were closed (c). (Reproduced with thanks from Baker, C. M. *et al.* (2010), *Braz. J. Med. Biol. Res.*, 43, 698–704, CCAL.)

7.6 OTHER VIRULENCE FACTORS

There is considerable information on virulence and avirulence genes and the contribution of their protein/peptide products in pathogenicity. However, there are also other chemicals produced by bacterial plant pathogens that are injurious and lethal to the host plants. Many bacterial plant pathogens produce peptide- or protein-containing toxins, such as coronatine (abbreviated as COR), tabtoxin, phaseolotoxin, syringomycin, tagetitoxin and rhizobitoxine. Tagetitoxin specifically inhibits the action of RNA polymerase III in eucaryotic 5S and tRNA gene transcription and of chloroplast RNA polymerase.

The foliar infection of plants by bacteria such as *Psm. syringae* pv. *tomato* occurs through stomata (Underwood *et al.*, 2007), which serve as critical entry sites and allow bacteria to transition from epiphytic to endophytic lifestyle. Earlier, it was assumed that the entry of bacteria into leaf tissues through natural openings was a passive process, where the plant lacked mechanisms for preventing bacterial entry, and the bacterium lacked active virulence mechanisms to promote entry. Later on, it was found that the entry of bacteria into leaf tissue through stomata is more complex and dynamic than the simple act of swimming into the leaf through passive openings (Gudesblat *et al.*, 2009). Several lines of evidence suggest that stomata actively close in response to plant pathogenic and human pathogenic bacteria or when exposed to conserved molecules found on the surface of bacterial cells known as pathogen/microbe-associated molecular patterns (Melotto *et al.*, 2006).

Bacterium-induced stomatal closure (also known as stomatal defence) is one part of plant immune defences (Figure 7.1). Thus, stomatal closure is an integral basal plant defence mechanism to restrict the invasion of pathogenic bacteria into plant tissues.

The production of phytotoxins and the type III secretion system has emerged as important factors to overcome stomatal defence by bacterial pathogens. COR is produced by several pathovars of *Psm. syringae* including *tomato*, *maculicola*, *glycinea* and *atropurpurea*, where it is known to function as a virulence factor promoting chlorosis in several host plants (Mittal and Davis, 1995; Brooks *et al.*, 2004). Bender *et al.* (1999) reported that COR induces modifications in the plant's physiology, such as anthocyanin production, alkaloid accumulation, ethylene emission, tendril coiling and root inhibition. This toxin acts as a virulence factor and contributes to disease development. Ishiga *et al.* (2009) reported that COR induced effects on photosynthetic machinery and on ROS in modulating necrotic cell death in bacterial speck of tomato.

COR is one of the most extensively studied bacterial phytotoxins and is a non-host-specific phytotoxin. The available evidence suggests that COR plays multiple roles in bacterial pathogenesis, including promoting the entry of bacteria through stomata at the initial stages of infection (Melotto *et al.*, 2006) and the suppression of defences mediated by the plant hormone salicylic acid later in the infection process (Uppalapati *et al.*, 2007).

The possibility that COR could suppress early defence responses during *Psm. syringae* pv. *tomato* DC3000 infection of Arabidopsis and tomato was suggested by Mittal and Davis (1995) and later on confirmed by Melotto *et al.* (2006). The discovery that COR is required for re-opening stomata by *Psm. syringae* pv. *tomato* DC3000 represents the first identification of a bacterial virulence factor that suppresses stomatal closure. COR-defective mutants have reduced multiplication and symptom production *in planta* compared to the wild-type (Mittal and Davis, 1995; Brooks *et al.*, 2004). Investigations of host–pathogen interactions, involving both *Psm. syringae* pv. *tomato* and *Psm. syringae* pv. *glycinea*, have suggested that COR may be important for bacterial invasion of plant tissue.

The discovery of this virulence mechanism of *Psm. syringae* pv. *tomato* DC3000 has created a lot of interest in elucidating the mode of action of COR at the molecular level. Mino *et al.* (1987) showed that 10 μM COR promoted the opening of dark closed stomata of broad bean and Italian ryegrass. The effect of COR on the stomatal aperture was more pronounced on Italian ryegrass. The authors also reported that COR activates membrane-bound ATPase activity, inducing stomatal opening. Liu *et al.* (2009) reported that the plant protein RIN4 (RPM1 interacting protein 4) and the activation of H^+-ATPase have been found to be necessary for COR to re-open stomata. Coronatine insensitive 1 (COI1) is another plant protein necessary for COR functioning in the guard cell (Melotto *et al.*, 2006). In fact, COI1 has been shown to be a receptor for COR (Yan *et al.*, 2009).

Figure 7.2 illustrates signalling components and interactions between molecules to be involved in stomatal defence and bacterial counter defence. In the stomatal guard cell, microbe-associated molecular patterns (MAMPs), e.g., flagellin and lipopolysaccharide (LPS) are perceived by cognate immune receptors (e.g., flagellin receptor FLS2). The perception of MAMPs triggers stomatal closure, which requires the phytohormones salicylic acid (SA) and abscisic acid (ABA), as well as ABA signalling components listed in the grey rectangle (components are listed in order where the top one is the most upstream, their localisation in the cell is not shown). Flagellin also prevents stomatal opening by inhibiting inwardly rectifying K^+ channels (K^+ in) through ABA signalling components (GPA1 and possibly others). COR-mediated inhibition of MAMP-triggered stomatal closure requires the plant proteins COI1 (a COR receptor) and RIN4. While COI1 physically binds to COR, RIN4 binds and activates the proton pump (H^+ ATPase) causing the membrane hyperpolarisation and activation of K^+ influx (K^+ in), a condition that promotes stomatal opening.

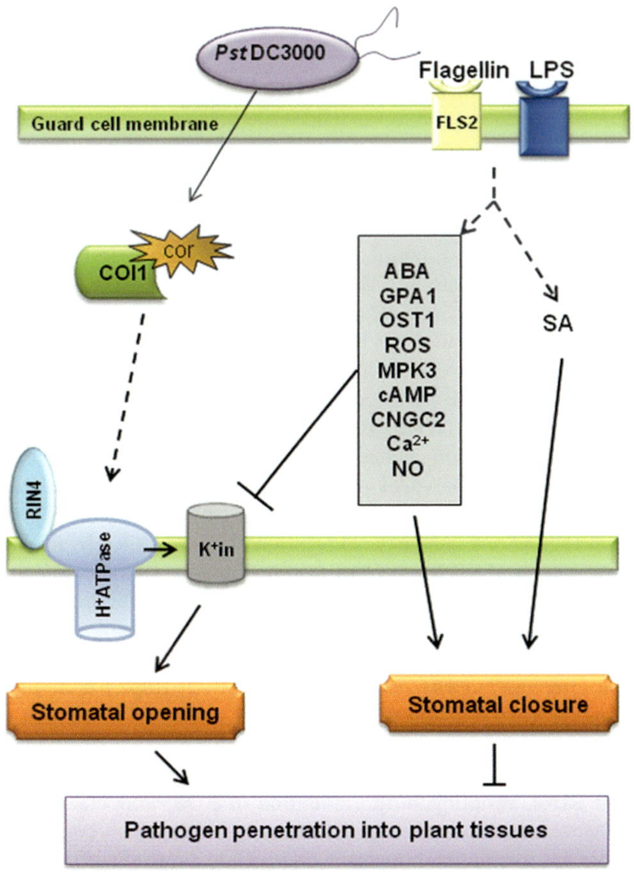

FIGURE 7.2 Model illustrating the signalling components and interactions between molecules that have been experimentally demonstrated to be involved in stomatal defence and bacterial counter defence. Dashed arrows indicate possibly indirect pathways. (Reproduced with thanks from Baker, C. M. *et al.* (2010), *Braz. J. Med. Biol. Res.*, 43, 698–704, CCAL.)

Molecular action of coronatine in plant cells: The molecular mechanisms through which COR may operate in the plant cell to cause disease are given below.

Staswick and Tiryaki (2004) reported that coronatine is a structural and functional mimic of the plant hormone jasmonate (JA) conjugated to the amino acid isoleucine (JA-Ile). Biological concentrations of COR activate the JA signalling pathway in the plant. cDNA microarray analysis indicated that the induction of JA-responsive genes in the tomato-*Psm. syringae* pv. *tomato* DC3000 interaction depends on the bacterial production of COR (Zhao *et al.* 2003). For more details, the readers may refer to Baker *et al.* (2010).

Yan *et al.* (2009) demonstrated COI1 to be a receptor for JA-Ile and COR. Based on this, Baker *et al.* (2010) developed a plausible model for the entire set of interactions where COR produced by the bacterium binds to COI1 and leads to the degradation of jasmonate ZIM-domain (JAZ) proteins through the SCF^{COI1} complex. In this model, JAZ proteins and other adaptor proteins act as repressors of JA signalling. The degradation of JAZ proteins allow for the expression of JA-responsive genes in the plant cell (Chini *et al.*, 2007; Thines *et al.*, 2007; Chung and Howe, 2009; Pauwels *et al.*, 2010) blocking plant innate immune responses including stomatal defence (Figure 7.3).

In their update on stomatal defence, Melotto *et al.* (2017) have given a perspective of the basic mechanisms underlying the process, and their implications in the understanding of plant–microbe interactions. They have also focused on significant advances towards a mechanistic understanding of stomatal defence.

Bacterium-triggered stomatal closure is a fast response (<1 h), and the basic mechanism underlying this process includes the following. ABA has long been recognised to induce stomatal closure under drought stress, thereby minimising water loss through the leaves. However, the role of this hormone in Arabidopsis defence against *Psm. syringae*

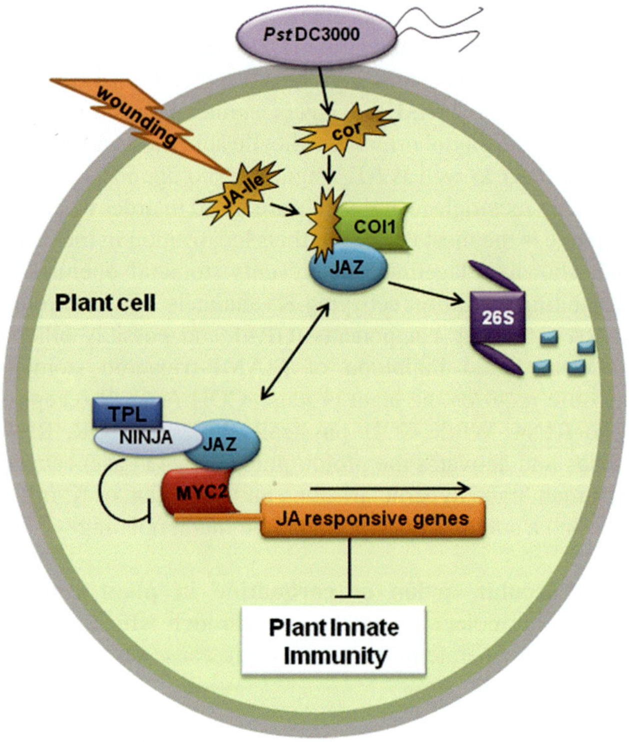

FIGURE 7.3 A model depicting the molecular action of coronatine in plant cells (possibly for all cell types). COR is secreted by *Psm. syringae* pv. *tomato* DC3000 into the plant cell and increases the affinity of the COI1 protein (as part of the SCFCOI1 ubiquitin-ligase complex, not shown here) towards the JAZ repressor. The SCFCOI1 complex catalyses ubiquitination of JAZ, which is then degraded through the 26S proteasome (denoted as "26S"). JAZ protein is part of a repressor complex that also contains NINJA and TPL, and physically binds to transcriptional activators (such as MYC2) of jasmonate response genes. Upon degradation of JAZ, JA response genes are activated blocking plant innate immune responses, including stomatal closure. COR = coronatine; JA = jasmonate; JAZ = jasmonate ZIM-domain; NINJA = novel interactor of JAZ; TPL = TOPLESS. (Reproduced with thanks from Baker, C. M. *et al.* (2010), *Braz. J. Med. Biol. Res.*, 43, 698–704, CCAL.)

differs depending on the stage of infection. The current experimental evidence suggests a prominent role of ABA signalling in stomatal defence (Eisenach and de Angeli, 2017; Inoue and Kinoshita, 2017; Vialet-Chabrand *et al.*, 2017).

Melotto *et al.* (2017) have elucidated a putative signalling cascade by which COR prevents bacterium triggered stomatal closure (Figure 7.4). COR, a molecular mimic of JA-isoleucine, promotes physical interaction between the F-box protein COI1 and the transcriptional repressor JAZ, leading to ubiquitination and proteasome-mediated degradation of JAZ proteins (Chini *et al.* 2007; Thines *et al.* 2007; Yan *et al.* 2007). Degradation of JAZ proteins derepresses bHLH transcription factors, such as MYC2, MYC3 and MYC4, leading to the activation of downstream transcriptional responses (Zhang *et al.* 2017). Zheng *et al.* (2012) reported that COR activation of JA signalling induced the expression of three homologous *NAC* transcription factor genes, *ANAC019*, *ANAC055* and *ANAC072*, which are the direct targets of MYC2. Genetic characterisation of *nac* null mutants reveals that NACs mediate COR-induced stomatal reopening and bacterial multiplication in plant tissues by inhibiting the accumulation of SA. Specifically, these NACs exert an inhibitory

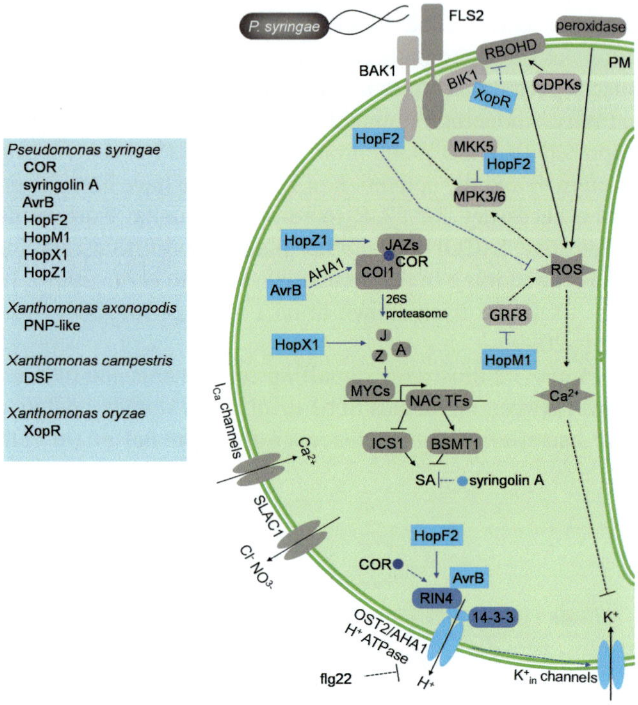

FIGURE 7.4 A simplified diagram of microbial virulence factors that manipulate MAMP-induced stomatal closure. NADPH oxidase RBOHD mediates flg22-induced ROS production and stomatal defence through BIK1-/CDPKs-regulated phosphorylation. Type III peroxidase also contributes to flg22-induced ROS production. MAMP-induced stomatal closure includes accumulation of ROS and NO, cytosolic calcium oscillations, the activation of S-type anion channels and the inhibition of K^{+}_{in} channels. COR induces COI1–JAZ interaction, mediates JAZ degradation and activates the expression of NAC TFs, which inhibit the expression of *ICS1* and induce the expression of *BSMT1*, thereby leading to a decreased SA level. COR also may induce stomatal opening through RIN4, which interacts with H^+-ATPase AHA1 and AHA2, resulting in the induction of K^{+}_{in} channels. Syringolin A inhibits SA signalling via its proteasome inhibitor activity. Multiple bacterial effectors HopM1, HopF2, HopX1, AvrB, HopZ1 and XopR could disrupt stomatal defence. HopM1 disrupts the function of GRF8 (a 14-3-3 protein), resulting in the reduction of MAMP-induced ROS production. HopF2 directly targets BAK1, MKK5 and RIN4, resulting in the inhibition of ROS production. AvrB enhances the interaction of RIN4 and AHA1, which could promote the interaction between COI1 and JAZs, leading to stomatal opening. HopX1 and HopZ1 induce stomatal opening after PAMP-triggered immunity-mediated stomatal closure, through a COI1-independent and COI1-dependent manner, respectively. XopR interacts with BIK1 and suppresses stomatal closure possibly by inhibition of RBOHD. (Reproduced with thanks from Melotto, M. *et al.* (2017), *Plant Physiol.*, 174, 561–71. With permission from American Society of Plant Biologists.)

effect by repressing the expression of genes involved in SA biosynthesis and activating the expression of genes involved in SA metabolism, resulting in the overall reduction of SA in infected plants (Zheng *et al.*, 2012; Gimenez-Ibanez *et al.*, 2017).

Du *et al.* (2014) showed that JA and COR activate the expression of the NAC tomato homolog, *JASMONIC ACID2-LIKE*. This transcription factor binds to and activates the expression of *SAMT1* and *SAMT2* that encode enzymes that deactivate SA by methylation, thereby suppressing the accumulation of SA and promoting stomatal opening.

ANAC032, another member of the NAC family of transcription factors, acts as both a positive regulator of SA signalling and a negative regulator of JA signalling. ANAC032 directly binds to the promoter of *MYC2* (positive regulator of JA signalling) and *NIMIN1* (negative regulator of SA signalling) and concomitantly suppresses their transcription within 6 h of *Pst* DC3000 infection. Accordingly, over expression of ANAC032 in Arabidopsis inhibits COR-dependent reopening of stomata (Allu *et al.*, 2016).

COR has also been shown to trigger cellular responses that do not depend on the canonical COI1-JAZ signalling pathway to manipulate guard cell movement. For example, RIN4, a negative regulator of plant innate immunity, is required by COR to open the stomatal pore as evidenced by the facts that neither *Pst* DC3000 nor COR was able to reopen stomata of *rpm1/rps2/rin4* mutants (Liu *et al.*, 2009; Lee *et al.*, 2015; Zhou *et al.*, 2015). The perception of flg22 via the FLS2 receptor leads to phosphorylation of the plasma membrane H^+-ATPases and the subsequent alkalisation of the apoplast that, along with the induction of ROS via NADPH oxidase RbohD, triggers stomatal closure (Liu *et al.*, 2009; Li *et al.*, 2014). RIN4 interacts with the plasma membrane H^+-ATPases AHA1 and AHA2, resulting in their inhibition and acidification of the apoplast through the hyperpolarisation of the plasma membrane and the subsequent induction of inward K^+ channels, promoting stomatal opening (Zhang *et al.*, 2008; Liu *et al.*, 2009). Zhang *et al.* (2008) showed that COR reversed the inhibitory effects of flg22 on K^+ currents to promote stomatal opening.

Syringolin A, a peptide toxin produced by *Psm. syringae* pv. *syringae*, is another virulence factor that has a proteasome inhibitory function and impacts stomatal defence (Groll *et al.*, 2008). Syringolin A promotes stomatal opening (Schellenberg *et al.*, 2010). In contrast to *Pst* DC3000, *Psm. syringae* pv. *syringae* does not induce an initial stomatal closure on either its bean host or Arabidopsis (Schellenberg *et al.*, 2010; Panchal *et al.*, 2016). These finding show that this bacterium constitutively produces syringolin A, which is a stronger signal than the MAMPs produced by *Psm. syringae* pv. *syringae* and/or that these plants do not recognise *Psm. syringae* pv. *syringae* MAMPs efficiently. Misas-Villamil *et al.* (2013) demonstrated that syringolin A diffuses from the site of infection and suppresses SA signalling, thereby decreasing immune responses in adjacent tissues. Schellenberg *et al.* (2010) reported that this might be the mechanism involved in syringolin A inhibition of MAMP-induced stomatal closure because syringolin A inhibits the proteasome-mediated turnover of NPR1, an important component of SA signalling to induce stomatal defence.

Xan. campestris pv. *campestris* is also capable of interfering with stomatal closure induced by MAMP or ABA signalling. The bacterium does so by inducing the production of a diffusible signal factor that is involved in bacterium to bacterium signalling (Gudesblat *et al.*, 2009). The mechanism of stomatal reopening by diffusible signal factor upon bacterial invasion is still unknown. *Xan. axonopodis* pv. *citri* produces a compound, known as plant natriuretic peptide-like, that can open stomata during plant infection that correlates with enhanced disease symptoms. Plant natriuretic peptide-like controls stomatal aperture in a cGMP-dependent manner (Gottig *et al.*, 2008).

In addition to phytotoxins, plant pathogenic bacteria also produce type-III-secretion-system effectors (T3SSEs) that can counter stomatal defence by either inhibiting MAMP-triggered stomatal closure or actively inducing stomatal opening. Lozano-Duran *et al.* (2014) reported that the T3SSE HopM1 produced by *Psm. syringae* disrupts the function of a 14-3-3 protein, GRF8, leading to reduction in MAMP-triggered ROS burst and stomatal defence. However, it is not clear in this case whether stomatal opening is a consequence of HopM1-mediated ROS suppression. Similarly, the *Psm. syringae* effector HopF2 inhibits flg22-induced ROS and stomatal defence (Hurley *et al.*, 2014). The XopR effector from *Xan. oryzae* pv. *oryzae* strain PXO99A also inhibits flg22-induced stomatal closure; however, the mechanism remains elusive (Wang *et al.*, 2016).

Benali *et al.* (2014) have reviewed the development in virulence strategies of phytopathogenic bacteria and their role in plant disease pathogenesis. They have dealt with secretion systems (type I–VI) employed by phytopathogenic bacteria, and the role of pectin-degrading enzymes, adhesins and extracellular polysaccharides, siderophores, COR, phaseolotoxins, syringomycins, syringopeptins, QS and biofilm production in causing different plant diseases.

Leigh and Coplin (1992) have reviewed the role of exopolysaccharides in plant–bacterial interactions. The genetic regulation of exopolysaccharide synthesis in many plant pathogens is complex, perhaps reflecting the various niches, free-living and in planta, in which exopolysaccharides function. In some cases, exopolysaccharide synthesis is regulated co-ordinately with other virulence factors, and in other cases separately. Regulatory genes that have homology to the two-component sensor and transcriptional effector systems are a common motif. In *Rhizobium* species, exopolysaccharide synthesis is regulated by transcriptional as well as post-translational mechanisms. Exopolysaccharides function differently in the root-nodule symbiosis versus plant pathogenesis. Specific *Rhizobium* exopolysaccharide structures promote nodule development and invasion in legumes that form indeterminate nodules. In plant pathogenesis, less specific mechanisms of pathogenesis occur: exopolysaccharides cause wilting by blocking xylem vessels, are partly responsible for water-soaked lesions and may also aid in invasion, growth, and survival in plant tissues.

The virulence of phytopathogenic bacteria is often correlated with their ability to produce extracellular polysaccharides (EPSs). The composition and amount of lipopolysaccharide O-antigens and low molecular weight, cell-associated 13-linked glucans may also affect the virulence of Gram-negative pathogens. For a few species of *Agrobacterium*, *Clavibacter*, *Erwinia*, *Pseudomonas*, and *Xanthomonas*, sufficient biochemical and genetical data has accumulated to permit critical evaluation of possible functions of polysaccharides during pathogenesis. It is clear that EPSs are necessary for several pathogens to cause normal disease symptoms such as water-soaking and wilting. Evidence is accumulating that EPSs and cell-associated polysaccharides also promote colonisation and enhance survival of some bacteria within host tissues. Further progress will require the use of thoroughly characterised polysaccharide-minus mutants, "natural" inoculation procedures, and careful monitoring of the fate of bacteria within plants (Denny, 1995).

Rudolph *et al.* (1994) investigated the role of bacterial exopolysaccharides (EPS) as virulence factors for leaf spot diseases caused by pseudomonads and xanthomonads. The capacity of these bacteria to induce persistent water-soaking in leaves plays a crucial role during pathogenesis that seems to be accomplished by a synergistic interaction between bacterial EPS and plant polymers. Under conditions of low EPS production (e.g. in continuously darkened plants), the bacteria were not able to cause typical water-soaking symptoms. The main EPS components were alginate and levan (*Pseudomonas*), xanthan (*Xanthomonas*), as well as lipopolysaccharides (LPS) and a small amount of proteins. It is suggested that alginate, which is very similar to plant pectate, is required for establishing bacterial infections in later stages of disease development. This concept was confirmed by evaluating transposon mutants with EPS deficiencies. LPS may be involved in specific interactions with plant polymers, leading to agglutination and precipitation (incompatibility) or gel-formation (compatibility). Bacteria which are embedded in a gel-like matrix *in planta* are not easily recognised by the plant and are protected against bacteriostatic compounds and desiccation.

Pathogenicity islands (PAIs) are now defined as regions of DNA that contain virulence genes and are present in the genome of pathogenic strains, but absent from – or only rarely present – in non-pathogenic variants of the same or related strains. Other features include a variable G+C content, distinct boundaries from the rest of the genome and the presence of genes related to mobile elements such as insertion sequences, integrases and transposases. Although PAIs have now been described in a wide range of both plant and animal pathogens, it has become evident that the general features of PAIs are displayed by a number of regions of DNA with functions other than pathogenicity, such as symbiosis and antibiotic resistance, and the general term genomic islands has been adopted. The review by Arnold *et al.* (2003) describes a range of genomic islands in plant pathogenic bacteria, including those that carry effector genes, phytotoxins and the type III protein secretion cluster. The review also considers some medically important bacteria in order to discuss the range, acquisition and stabilisation of genomic islands.

REFERENCES

Abramovitch, R. B., Anderson, J. C. and Martin, G. B. (2006). Bacterial elicitation and evasion of plant innate immunity. *Nat. Rev. Mol. Cell. Biol.* 7: 601–11.

Abramovitch, R. B. and Martin, G. B. (2005). AvrPtoB: A bacterial type III effector that both elicits and suppresses programmed cell death associated with plant immunity. *FEMS Microbiol. Lett.* 245: 1–8.

Agrios, G. N. (2004). *Plant Pathology*. 5th Ed. Elsevier Academic Press, San Diego, CA, 922 pp.

Alegria, M. C., Souza, D. P., Andrade, M. O., Docena, C., Khater, L. *et al.* (2005). Identification of new protein–protein interactions involving the products of the chromosome- and plasmid-encoded type IV secretion loci of the phytopathogen *Xanthomonas axonopodis* pv. *citri*. *J. Bacteriol.* 187: 2315–25.

Alfano, J. R. and Collmer, A. (2004). Type III secretion system effector proteins: Double agents in bacterial disease and plant defense. *Annu. Rev. Phytopathol.* 42: 385–414.

Allu, A. D., Brotman, Y., Xue, G. P. and Balazadeh, S. (2016). Transcription factor ANAC032 modulates JA/SA signalling in response to *Pseudomonas syringae* infection. *EMBO Rep.* 17: 1578–89.

Arber, W. (2008). Molecular mechanisms driving Darwinian evolution. *Math. Comput. Model.* 47: 666–74.

Arlat, M., Gough, C. L., Barber, C. E., Boucher, C. and Daniels, M. J. (1991). *Xanthomonas campestris* contains a cluster of hrp genes related to the larger hrp cluster of *Pseudomonas solanacearum*. *Mol. Plant-Microbe Interact.* 4: 593–601.

Arnold, D. L., Pitman, A. and Jackson, R. W. (2003). Pathogenicity and other genomic islands in plant pathogenic bacteria. *Mol. Plant Pathol.* 4: 407–20. https://doi.org/10.1046/J.1364-3703.2003.00187.X

Baker, C. M., Chitrakar, R., Obulareddy, N., Panchal, S. Williams, P. *et al.* (2010). Molecular battles between plant and pathogenic bacteria in the phyllosphere. *Braz. J. Med. Biol. Res.* 43: 698–704.

Bartoli, C., Lamichhane, J. R., Berge, O., Guilbaud, C., Varvaro, L. *et al.* (2015). A framework to gauge the epidemic potential of plant pathogens in environmental reservoirs: The example of kiwifruit canker. *Mol. Plant Pathol.* 16: 137–49.

Bartoli, C., Roux, F. and Lamichhane, J. R. (2016). Molecular mechanisms underlying the emergence of bacterial pathogens: An ecological perspective. *Mol. Plant Pathol.* 17: 303–10.

Benali, S., Mohamed, B. and Eddine, H. J. (2014). Virulence strategies of phytopathogenic bacteria and their role in plant disease pathogenesis. *Afr. J. Microbiol. Res.* 8: 2809–15.

Bender, C. L., Alarcon-Chaidez, F. and Gross, D. C. (1999). *Pseudomonas syringae* phytotoxins: Mode of action, regulation, and biosynthesis by peptide and polyketide synthetases. *Microbiol. Mol. Biol. Rev.* 63: 266–92.

Berks, B. C., Palmer, T. and Sargent, F. (2005). Protein targeting by the bacterial twin-arginine translocation (Tat) pathway. *Curr. Opin. Microbiol.* 8: 174–81.

Berks, B. C., Sargent, F. and Palmer, T. (2000). The Tat protein export pathway. *Mol. Microbiol.* 35: 260–274.

Bronstein, P. A., Marrichi, M., Cartinhour, S., Schneider, D. J. and DeLisa, M. P. (2005). Identification of a twin-arginine translocation system in *Pseudomonas syringae* pv. *tomato* DC3000 and its contribution to pathogenicity and fitness. *J. Bacteriol.* 87: 8450–61.

Brooks, D. M., Hernandez-Guzman, G., Kloek, A. P., Alarcon-Chaidez, F., Sreedharan, A. *et al.* (2004). Identification and characterization of a well-defined series of coronatine biosynthetic mutants of *Pseudomonas syringae* pv. *tomato* DC3000. *Mol. Plant-Microbe Interact.* 17: 162–74.

Burse, A., Weingart, H. and Ullrich, M. S. (2004). The phytoalexin-inducible multidrug efflux pump AcrAB contributes to virulence in the fire blight pathogen, *Erwinia amylovora. Mol. Plant-Microbe Interact.* 17: 43–54.

Buttner, D. and Bonas, U. (2002). Getting across-bacterial type III effector proteins on their way to the plant cell. *EMBO J.* 21: 5313–22.

Caldelari, I., Mann, S., Crooks, C. and Palmer, T. (2006). The Tat pathway of the plant pathogen *Pseudomonas syringae* is required for optimal virulence. *Mol. Plant–Microbe Interact.* 19: 200–12.

Chakravarthy, S., Worley, J. N., Montes-Rodriguez, A. and Collmer, A. (2018). *Pseudomonas syringae* pv. *tomato* DC3000 polymutants deploying coronatine and two type III effectors produce quantifiable chlorotic spots from individual bacterial colonies in *Nicotiana benthamiana* leaves. *Mol. Plant Pathol.* 19: 935–47.

Cheng, F., Ma, A., Zhuang, G. and Fray, R. G. (2018). Exogenous *N*-acyl-homoserine lactones enhance the expression of flagella of *Pseudomonas syringae* and activate defence responses in plants. *Mol. Plant Pathol.* 19: 104–15.

Chini, A., Fonseca, S., Fernandez, G., Adie, B., Chico, J. M. *et al.* 2007. The JAZ family of repressors is the missing link in jasmonate signalling. *Nature* 448: 666–71.

Chisholm, S. T., Coaker, G., Day, B. and Staskawicz, B. J. (2006). Host–microbe interactions: Shaping the evolution of the plant immune response. *Cell* 124: 803–14.

Chung, H. S. and Howe, G. A. (2009). A critical role for the TIFY motif in repression of jasmonate signaling by a stabilized splice variant of the jasmonate ZIM-domain protein JAZ10 in *Arabidopsis*. *Plant Cell* 21: 131–45.

Cui, Y., Chatterjee, A., Hasegawa, H., Dixit, V., Leigh, N. *et al.* (2005). ExpR, a LuxR homolog of *Erwinia carotovora* subsp. *carotovora*, activates transcription of *rsmA*, which specifies a global regulatory RNA-binding protein. *J. Bacteriol.* 187: 4792–4803.

da Silva, A. C., Ferro, J. A., Reinach, F. C., Farah, C. S., Furlan, L. R. *et al.* (2002). Comparison of the genomes of two *Xanthomonas* pathogens with differing host specificities. *Nature* 417: 459–63.

Davis, R. E., Dally, E. L., Jomantiene, R., Zhao, Y., Roe, B. *et al.* (2005). Cryptic plasmid pSKU146 from the wall-less plant pathogen *Spiroplasma kunkelii* encodes an adhesion and components of a type IV translocation-related conjugation system. *Plasmid* 3: 179–90.

Denny, T. P. (1995). Involvement of bacterial polysaccharides in plant pathogenesis. *Annu. Rev. Phytopathol.* 33: 173–97.

Desvaux, M., Hebraud, M., Henderson, I. R. and Pallen, M. J. (2006). Type III secretion: What's in a name? *Trends Microbiol.* 14: 157–60.

Ding, Z. and Christie, P. J. (2003). *Agrobacterium tumefaciens* twin-arginine-dependent translocation is important for virulence, flagellation, and chemotaxis but not type IV secretion. *J. Bacteriol.* 185: 760–71.

Djamei, A., Pitzschke, A., Nakagami, H., Rajh, I. and Heribert, H. (2007). Trojan horse strategy in *Agrobacterium* transformation: Abusing MAPK defense signaling. *Science* 318: 453–56.

Dong, Y. H., Zhang, X.-F., Xu, J.-L., Tan, A.-T. and Zhang, L.-H. (2005). VqsM, a novel AraC-type global regulator of quorum sensing signaling and virulence in *Pseudomonas aeruginosa*. *Mol. Microbiol.* 58: 552–64.

Dow, J. M., Crossman, L. C., Findlay, K., He, Y. Q., Feng, J. X. *et al.* (2003). Biofilm dispersal in *Xanthomonas campestris* is controlled by cell–cell signaling and is required for full virulence to plants. *Proc. Natl. Acad. Sci. USA* 100: 10995–11000.

Du, M., Zhai, Q., Deng, L., Li, S., Li. H. *et al.* (2014). Closely related NAC transcription factors of tomato differentially regulate stomatal closure and reopening during pathogen attack. *Plant Cell* 26: 3167–84.

Duan, T. T., Luo, Y. P., Kong, C. Y., Gao, X. D., Zhao, X. *et al.* (2018). Gene expression profile of *Ralstonia solanacearum* for the rhizosphere ecological niche of *Solanum tuberosum*. *J. Phytopathol.* 166: 151–60.

Dunger, G., Arabolaza, L. N., Gottig, N., Orellano, E. G. and Ottado, J. (2005). Participation of *Xanthomonas axonopodis* pv. *citri hrp* cluster in citrus canker and in non-host plants responses. *Plant Pathol.* 54: 781–88.

Dunoyer, P., Himber, C. and Voinnet, O. (2006). Induction, suppression and requirement of RNA silencing pathways in virulent *Agrobacterium* infections. *Nat. Genet.* 38: 258–63.

Eisenach, C. and de Angeli, A. (2017). Ion transport at the vacuole during stomatal movements. *Plant Physiol.* 174: 520–30.

Espinosa, A. and Alfano, J. R. (2004). Disabling surveillance: Bacterial type III secretion system effectors that suppress innate immunity. *Cell Microbiol.* 6: 1027–40.

Fan, S., Tian, F., Li, J., Hutchins, W., Chen, H. *et al.* (2017). Identification of phenolic compounds that suppress the virulence of *Xanthomonas oryzae* on rice via the type III secretion system. *Mol. Plant Pathol.* 18: 555–68.

Ferluga, S., Bigirimana, J., Hőfte, M. and Venturi, V. (2007). A LuxR homologue of *Xanthomonas oryzae* pv. *oryzae* is required for optimal rice virulence. *Mol. Plant Pathol.* 8: 529–38.

Ferluga, S. and Venturi, V. (2009). OryR is a LuxR-family protein involved in interkingdom signaling between pathogenic *Xanthomonas oryzae* and rice. *J. Bacteriol.* 191: 890–97.

Ferreira, A. O., Myers, C. R., Gordon, J. S., Martin, G. B., Vencato, M. *et al.* (2006). Whole-genome expression profiling defines the HrpL regulon of *Pseudomonas syringae* pv. *tomato* DC3000, allows de novo reconstruction of the Hrp *cis* element and identifies novel coregulated genes. *Mol. Plant-Microbe Interact.* 19: 1167–79.

Flor, H. H. (1971). Current status of the gene-for-gene concept. *Annu Rev Phytopathol.* 9: 275–98.

Fuqua, C., Parsek, M. R. and Greenberg, E. P. (2001). Regulation of gene expression by cell-to-cell communication: Acylhomoserine lactone quorum sensing. *Annu. Rev. Genet.* 35: 439–68.

Gao, M., Teplitski, M., Robinson, J. B. and Bauer, W. D. (2003). Production of substances by *Medicago truncatula* that affect bacterial quorum sensing. *Mol. Plant-Microbe Interact.* 16: 827–34.

Gimenez-Ibanez, S., Boter, M., Ortigosa, A., Garcia-Casado, G., Chini, A. *et al.* (2017). JAZ2 controls stomata dynamics during bacterial invasion. *New Phytol.* 213: 137892.

Givskov, M., De Nys, R., Manefield, M., Gram, L., Maximilien, R. *et al.* (1996). Eukaryotic interference with homoserine lactonemediated prokaryotic signaling. *J. Bacteriol.* 178: 6618–22.

Glazebrook, J. and Ausubel, F. M. (1994). Isolation of phytoalexin-deficient mutants of *Arabidopsis thaliana* and characterization of their interactions with bacterial pathogens. *Proc. Natl. Acad. Sci. USA* 91: 8955–59.

Gochez, A. M., Shantharaj, D., Potnis, N., Zhou, X., Minsavage, G. V. *et al.* (2017). Molecular characterization of XopAG effector AvrGf2 from *Xanthomonas fuscans* ssp. *aurantifolii* in grapefruit. *Mol. Plant Pathol.* 18: 405–19.

Gonzalez, E. T., Brown, D. G., Swanson, J. K. and Allen, C. (2007). Using the *Ralstonia solanacearum* Tat secretome to identify bacterial wilt virulence factors. *Appl. Environ. Microbiol.* 73: 3779–86.

Gottig, N., Garavaglia, B. S., Daurelio, L. D., Valentine, A., Gehring, C. *et al.* (2008). *Xanthomonas axonopodis* pv. *citri* uses a plant natriuretic peptide-like protein to modify host homeostasis. *Proc. Natl. Acad. Sci. USA* 105: 18631–36.

Gottig, N., Garavaglia, B. S., Garofalo, C. G., Zimaro, T., Sgro, G. G. *et al.* (2010). Mechanisms of infection used by *Xanthomonas axonopodis* pv. *citri* in citrus canker disease. In: A. Mendez-Vilas (ed.). *Current Research, Technology and Education Topics in Applied Microbiology and Microbial Biotechnology*. Formatex Research Center, Badajoz, pp. 196–204.

Groll, M., Schellenberg, B., Bachmann, A. S., Archer, C. R., Huber, R. *et al.* (2008). A plant pathogen virulence factor inhibits the eukaryotic proteasome by a novel mechanism. *Nature* 452: 755–58.

Gudesblat, G. E., Torres, P. S. and Vojnov, A. A. (2009). *Xanthomonas campestris* overcomes Arabidopsis stomatal innate immunity through a DSF cell-to-cell signal-regulated virulence factor. *Plant Physiol.* 149: 1017–27.

Hajri, A., Brin, C., Hunault, G., Lardeux, F., Lemaire, C. *et al.* (2009). A "repertoire for repertoire" hypothesis: Repertoires of type three effectors are candidate determinants of host specificity in *Xanthomonas*. *PLoS ONE.* 4: e6632.

Henderson, I. R., Navarro-Garcia, F., Desvaux, M., Fernandez, R. C. and Ala'Aldeen, D. (2004). Type V protein secretion pathway: The autotransporter story. *Microbiol. Mol. Biol. Rev.* 68: 692–744.

Hotson, A. and Mudgett, M. B. (2004). Cysteine proteases in phytopathogenic bacteria: Identification of plant targets and activation of innate immunity. *Curr. Opin. Plant Biol.* 7: 384–90.

Hueck, C. J. (1998). Type III protein secretion systems in bacterial pathogens of animals and plants. *Microbiol. Mol. Biol. Rev.* 62: 379–433.

Hurley, B., Lee, D., Mott, A., Wilton, M., Liu, J. *et al.* (2014). The *Pseudomonas syringae* type III effector HopF2 suppresses Arabidopsis stomatal immunity. *PLoS One* 9: e114921.

Ingle, R. A., Carstens, M. and Denby, K. J. (2006). PAMP recognition and the plant–pathogen arms race. *Bioessays* 28: 880–89.

Inoue, S. and Kinoshita, T. (2017). Blue light regulation of stomatal opening and the plasma membrane H+-ATPase. *Plant Physiol.* 174: 531–38.

Ishiga, Y., Uppalapati, S. R., Ishiga, T., Elavarthi, S., Martin, B. *et al.* (2009). The phytotoxin coronatine induces light-dependent oxygen species in tomato seedlings. *New Phytol.* 181: 147–60.

Jittawuttipoka, T., Buranajitpakorn, S., Vattanaviboon, P. and Mongkolsuk, S. 2009. The catalase-peroxidase KatG is required for virulence of *Xanthomonas campestris* pv. *campestris* in a host plant by providing protection against low levels of H2O2. *J. Bacteriol.* 191: 7372–77.

Jones, J. D. G. and Dangl, J. L. (2006). The plant immune system. *Nature* 444: 323–29.

Jones, K. E., Patel, N. G., Levy, M. A., Storeygard, A., Balk, D. *et al.* (2008). Global trends in emerging infectious diseases. *Nature* 451: 990–93.

Kado, C. I. (2002). Horizontal transmission of genes by *Agrobacterium* species. In: M. Syvanen and C. I. Kado (eds.). *Horizontal Gene Transfer*. 2nd Ed. Academic Press, San Diego, CA, pp. 45–50.

Kado, C. I. (2010). *Plant Bacteriology*. American Phytopathological Society, St. Paul, MN, 336 pp.

Keen, N. T. (1990). Gene-for-gene complementarity in plant–pathogen interactions. *Annu. Rev. Genet.* 24: 447–63.

Kim, J. G., Park, B. K., Yoo, C. H., Jeon, E., Oh, J. *et al.* 2003. Characterization of the *Xanthomonas axonopodis* pv. *glycines* Hrp pathogenicity island. *J. Bacteriol.* 185: 3155–66.

Kunze, G., Zipfel, C., Robatzek, S., Niehaus, K., Boller, T. *et al.* (2004). The N terminus of bacterial elongation factor TU elicits innate immunity in *Arabidopsis* plants. *Plant Cell* 16: 3496–3507.

Lamichhane, J. R., Messéan, A. and Morris, C. E. (2015). Insights into epidemiology and control of diseases of annual plants caused by Pseudomonas syringae. *J. Gen. Plant. Pathol.* https://doi.org/10.1007/s10327-015-0605-z

Lamichhane, J. R., Varvaro, L., Parisi, L., Audergon, J.-M. and Morris, C. E. (2014). Disease and frost damage of woody plants caused by *Pseudomonas syringae*: Seeing the forest for the trees. *Adv. Agron.* 126: 235–95.

Lee, D., Bourdais, G., Yu, G., Robatzek, S. and Coaker, G. (2015). Phosphorylation of the plant immune regulator RPM1-INTERACTING PROTEIN4 enhances plant plasma membrane H+-ATPase activity and inhibits flagellin triggered immune responses in Arabidopsis. *Plant Cell* 27: 2042–56.

Leigh, J. A. and Coplin, D. L. (1992). Exopolysaccharides in plant–bacterial interactions. *Annu. Rev. Microbiol.* 46: 307–46.

Li, L., Li, M., Yu, L., Zhou, Z., Liang, X. *et al.* (2014). The FLS2-associated kinase BIK1 directly phosphorylates the NADPH oxidase RbohD to control plant immunity. *Cell Host Microbe*, 15: 329–38.

Liang, X., Yu, X., Pan, X., Wu, J., Duan, Y. *et al.* (2018). A thiadiazole reduces the virulence of *Xanthomonas oryzae* pv. *oryzae* by inhibiting the histidine utilization pathway and quorum sensing. *Mol. Plant Pathol.* 19: 116–28.

Lindgren, P. B., Peet, R. C. and Panopoulos, N. J. (1986). Gene cluster of *Pseudomonas syringae* pv. "*phaseolicola*" controls pathogenicity of bean plants and hypersensitivity of nonhost plants. *J. Bacteriol.* 168: 512–22.

Liu, J., Elmore, J. M., Fuglsang, A. T., Palmgren, M. G., Staskawicz, B. J. *et al.* (2009). RIN4 functions with plasma membrane H+-ATPases to regulate stomatal apertures during pathogen attack. *PLoS Biol.* 7: e1000139.

Lo, T., Koulena, N., Seto, D., Guttman, D. S. and Desveaux, D. (2017). The HopF family of *Pseudomonas syringae* type III secreted effectors. *Mol. Plant Pathol.* 18: 457–68.

Lopez-Solanilla, E., Bronstein, P. A., Schneider, A. R. and Collmer, A. (2004). HopPtoN is a *Pseudomonas syringae* Hrp (type III secretion system) cysteine protease effector that suppresses pathogen-induced necrosis associated with both compatible and incompatible plant interactions. *Mol. Microbiol.* 54: 353–65.

Lozano-Duran, R., Bourdais, G., He, S. Y. and Robatzek, S. (2014). The bacterial effector HopM1 suppresses PAMP-triggered oxidative burst and stomatal immunity. *New Phytol.* 202: 259–69.

Martin, G. B., Bogdanova, A. J. and Sessa, G. (2003). Understanding the functions of plant disease resistance proteins. *Annu. Rev. Plant Biol.* 54: 23–61.

McGowan, S. J., Barnard, A. M. L., Bosgelmez, G., Sebalhia, M., Simpson, N. J. L. *et al.* (2005). Carbapenem antibiotic biosynthesis in *Erwinia carotovora* is regulated by physiological and genetic factors modulating the quorum sensing-dependent control pathway. *Mol. Microbiol.* 55: 526–45.

Medina, C. A., Reyes, P. A., Trujillo, C. A., Gonzalez, J. L., Bejarano, D. A. *et al.* (2018). The role of type III effectors from *Xanthomonas axonopodis* pv. *manihotis* in virulence and suppression of plant immunity. *Mol. Plant Pathol.* 19: 593–606.

Melotto, M., Underwood, W., Koczan, J., Nomura, K. and He, S. Y. (2006). Plant stomata function in innate immunity against bacterial invasion. *Cell*, 126: 969–80.

Melotto, M., Zhang, L., Oblessuc, P. R. and He, S. Y. (2017). Stomatal defense a decade later. *Plant Physiol.* 174: 561–71.

Mino, Y., Matsushita, Y. and Sakai, R. (1987). Effect of coronatine on stomatal opening in leaves of broadbean and Italian ryegrass. *Ann. Phytopathol. Soc. Jpn.* 53: 53–55.

Misas-Villamil, J. C., Kolodziejek, I., Crabill, E., Kaschani, F., Niessen, S. *et al.* (2013). *Pseudomonas syringae* pv. *syringae* uses proteasome inhibitor syringolin A to colonize from wound infection sites. *PLoS Pathog* 9: e1003281.

Mittal, S. and Davis, K. R. (1995). Role of the phytotoxin coronatine in the infection of *Arabidopsis thaliana* by *Pseudomonas syringae* pv. *tomato*. *Mol. Plant-Microbe Interact.* 8: 165–71.

Molina, L., Rezzonico, F., Defago, G. and Duffy, B. (2005). Autoinduction in *Erwinia amylovora*: Evidence of an acylhomoserine lactone signal in the fire blight pathogen. *J. Bacteriol.* 187: 3206–13.

Monteil, C. L., Cai, R., Liu, H., Llontop, M. E. M., Leman, S. *et al.* (2013). Nonagricultural reservoirs contribute to emergence and evolution of *Pseudomonas syringae* crop pathogens. *New Phytol.* 199: 800–11.

Morel, J. B. and Dangl, J. L. (1997). The hypersensitive response and the induction of cell death in plants. *Cell Death Differ.* 4: 671–83.

Morens, D. M. and Fauci, A. S. (2013). Emerging infectious diseases: Threats to human health and global stability. *PLoS Pathog.* 9: e1003467.

Morens, D. M., Folkers, G. K. and Fauci, A. S. (2004). The challenge of emerging and re-emerging infectious diseases. *Nature*, 430: 242–49.

Morse, S. S. (1995). Factors in the emergence of infectious diseases. *Emerg. Infect. Dis.* 1: 7–15.

Muller, M. and Klosgen, R. B. (2005). The Tat pathway in bacteria and chloroplasts. *Mol. Membr. Biol.* 22: 113–21.

Mysore, K. S. and Ryu, C. M. (2004). Nonhost resistance: How much do we know? *Trends Plant Sci.* 9: 97–104.

Nissan, G., Gershovits, M., Morozov, M., Chalupowicz, L., Sessa, G. *et al.* (2018). Revealing the inventory of type III effectors in *Pantoea agglomerans* gall-forming pathovars using draft genome sequences and a machine-learning approach. *Mol. Plant Pathol.* 19: 381–92.

Ohta, T. (2000). Mechanisms of molecular evolution. *Philos. Trans. R. Soc. London, B: Biol. Sci.* 355: 1623–26.

Palumbo, J. D., Kado, C. I. and Phillips, D. A. (1998). An isoflavonoid-inducible efflux pump in *Agrobacterium tumefaciens* is involved in competitive colonization of roots. *J. Bacteriol.* 180: 3107–13.

Panchal, S., Chitrakar, R., Thompson, B. K., Obulareddy, N., Roy, D. *et al.* (2016). Regulation of stomatal defense by air relative humidity. *Plant Physiol.* 172: 2021–32.

Pauwels, L., Barbero, G. F., Geerinck, J., Tilleman, S., Grunewald, W. *et al.* (2010). NINJA connects the co-repressor TOPLESS to jasmonate signalling. *Nature* 464: 788–91.

Pukatzki, S., Ma, A. T., Sturtevant, D., Krastins, B., Sarracino, D. *et al.* (2006). Identification of a conserved bacterial protein secretion system in *Vibrio cholerae* using the *Dictyostelium* host model system. *Proc. Nat. Acad. Sci. USA* 103: 1528–33.

Rudolph, K. W. E., Gross, M., Ebrahim-Nesbat, F., Nöllenburg, M., Zomorodian, A. *et al.* (1994). The role of extracellular polysaccharides as virulence factors for phytopathogenic pseudomonads and xanthomonads. In: Kado, C. I. and Crosa, J. H. (eds.) *Molecular Mechanisms of Bacterial Virulence. Developments in Plant Pathology*, Vol. 3. Springer, Dordrecht. https://doi.org/10.1007/978-94-011-0746-4_25

Ryan, R. P., Fouhy, Y., Lucey, J. F. and Dow, J. M. (2006). Cyclic di-GMP signaling in bacteria: Recent advances and new puzzles. *J. Bacteriol.* 188: 8327–34.

Saenkham, P., Eiamphungporn, W., Farrand, S. K., Vattanaviboon, P. and Mongkolsuk, S. (2007). Multiple superoxide dismutases in *Agrobacterium tumefaciens*: Functional analysis, gene regulation, and influence on tumorigenesis. *J. Bacteriol.* 189: 8807–17.

Sandkvist, M. (2001). Biology of type II secretion. *Mol. Microbiol.* 40: 271–83.

Sang, Y., Wang, Y., Ni, H., Cazale, A.-C., She, Y.-M. *et al.* (2018). The *Ralstonia solanacearum* type III effector RipAY targets plant redox regulators to suppress immune responses. *Mol. Plant Pathol.* 19: 129–42.

Schellenberg, B., Ramel, C. and Dudler, R. (2010). *Pseudomonas syringae* virulence factor syringolin A counteracts stomatal immunity by proteasome inhibition. *Mol. Plant-Microbe Interact.* 23: 1287–93.

Shantharaj, D., Romer, P., Figueiredo, J. F. L., Minsavage, G. V., Kronauer, C. *et al.* (2017). An engineered promoter driving expression of a microbial avirulence gene confers recognition of TAL effectors and reduces growth of diverse *Xanthomonas* strains in citrus. *Mol. Plant Pathol.* 18: 976–89.

Staswick, P. E. and Tiryaki, I. (2004). The oxylipin signal jasmonic acid is activated by an enzyme that conjugates it to isoleucine in Arabidopsis. *Plant Cell*, 16: 2117–27.

Sukchawalit, R., Loprasert, S., Atichartpongkul, S. and Mongkolsuk, S. (2001). Complex regulation of the organic hydroperoxide resistance gene (*ohr*) from *Xanthomonas* involves Ohr, a novel organic peroxide-inducible negative regulator, and posttranscriptional modifications. *J. Bacteriol.* 183: 4405–12.

Swarup, S., Yang, Y., Kingsley, M. T. and Gabriel, D. W. (1992). An *Xanthomonas citri* pathogenicity gene, *pthA*, pleiotropically encodes gratuitous avirulence on nonhosts. *Mol. Plant-Microbe Interact.* 5: 204–13.

Tampakaki, A. P., Fadouloglou, V. E., Gazi, A. D., Panopoulos, N. J. and Kokkinidis, M. (2004). Conserved features of type III secretion. *Cell Microbiol.* 6: 805–16.

Thines, B., Katsir, L., Melotto, M., Niu, Y., Mandaokar, A. *et al.* (2007). JAZ repressor proteins are targets of the SCF(COI1) complex during jasmonate signalling. *Nature*, 448: 661–65.

Toth, I. K., Pritchard, L. and Birch, P. R. J. (2006). Comparative genomics reveals what makes an enterobacterial plant pathogen. *Annu. Rev. Phytopathol.* 44: 305–36.

Toussaint, V., Benoit, D. L. and Carisse, O. (2012). Potential of weed species to serve as a reservoir for Xanthomonas campestris pv. vitians, the causal agent of bacterial leaf spot of lettuce. *Crop Prot.* 41: 64–70.

Underwood, W., Melotto, M. and He, S. Y. (2007). Role of plant stomata in bacterial invasion. *Cell Microbiol.* 9: 1621–29.

Uppalapati, S. R., Ishiga, Y., Wangdi, T., Kunkel, B. N., Anand, A. *et al.* (2007). The phytotoxin coronatine contributes to pathogen fitness and is required for suppression of salicylic acid accumulation in tomato inoculated with *Pseudomonas syringae* pv. *tomato* DC3000. *Mol. Plant-Microbe Interact.* 20: 955–65.

Venturi, V. and Fuqua, C. (2013). Chemical signaling between plants and plant-pathogenic bacteria. *Annu. Rev. Phytopathol.* 51: 17–37.

Vialet-Chabrand, S., Hills, A., Wang, Y., Griffiths, H., Lew, V. L. *et al.* (2017). Global sensitivity analysis of OnGuard models identifies key hubs for transport interaction in stomatal dynamics. *Plant Physiol.* 174: 680–88.

Von Bodman, S. B., Bauer, W. D. and Coplin, D. L. (2003). Quorum sensing in plant-pathogenic bacteria. *Annu. Rev. Phytopathol.* 41: 455–82.

Wang, S., Sun, J., Fan, F., Tan, Z., Zou, Y. *et al.* (2016). A *Xanthomonas oryzae* pv. *oryzae* effector, XopR, associates with receptor like cytoplasmic kinases and suppresses PAMP-triggered stomatal closure. *Sci. China Life Sci.* 59: 897–905.

Williams, S. C., Patterson, E. K., Carty, N. L., Grisold, J. A., Hamood, A. N. *et al.* (2004). *Pseudomonas aeruginosa* autoinducer enters and functions in mammalian cells. *J. Bacteriol.* 186: 2281–87.

Wu, H.-Y., Chung, P.-C., Shih, H.-W., Wen, S.-R. and Lai, E.-M. (2008). Secretome analysis uncovers an Hcp-family protein secreted via a type VI secretion system in *Agrobacterium tumefaciens*. *J. Bacteriol.* 190: 2841–50.

Yan, J., Zhang, C., Gu, M., Bai, Z., Zhang, W. *et al.* (2009). The Arabidopsis coronatine insensitive 1 protein is a jasmonate receptor. *Plant Cell*, 21: 2220–36.

Yan, Y., Stolz, S., Chetelat, A., Reymond, P., Pagni, M. *et al.* (2007). A downstream mediator in the growth repression limb of the jasmonate pathway. *Plant Cell*, 19: 2470–83.

Yuan, Z.-C., Liu, P., Saenkham, P., Kerr, K. and Nester, E. W. (2008). Transcriptome profiling and functional analysis of *Agrobacterium tumefaciens* reveals a general conserved response to acidic conditions (pH 5.5) and a complex acid mediated signaling involved in *Agrobacterium*-plant interactions. *J. Bacteriol.* 190: 494–507.

Yuan, Z.-C., Zaheer, R. and Finan, T. M. (2005). Phosphate limitation induces catalase expression in *Sinorhizobium meliloti*, *Pseudomonas aeruginosa* and *Agrobacterium tumefaciens*. *Mol. Microbiol.* 58: 877–94.

Zhang, L., Zhang, F., Melotto, M., Yao, J and He, S. Y. (2017). Jasmonate signaling and manipulation by pathogens and insects. *J. Exp. Bot.* 68: 1371–85.

Zhang, W., He, S. Y. and Assmann, S. M. (2008). The plant innate immunity response in stomatal guard cells invokes G-protein dependent ion channel regulation. *Plant J.* 56: 984–96.

Zhao, Y., Thilmony, R., Bender, C. L., Schaller, A., He, S. Y. *et al.* (2003). Virulence systems of *Pseudomonas syringae* pv. *tomato* promote bacterial speck disease in tomato by targeting the jasmonate signaling pathway. *Plant J.* 36: 485–99.

Zheng, M., Wang, X., Doan, B., Lewis, K. A., Schneider, T. D. *et al.* (2001). Computation-directed identification of OxyR DNA binding sites in *Escherichia coli*. *J. Bacteriol.* 183: 4571–79.

Zheng, X.-Y., Spivey, N. W., Zeng, W., Liu, P.-P., Fu, Z. Q. *et al.* (2012). Coronatine promotes *Pseudomonas syringae* virulence in plants by activating a signaling cascade that inhibits salicylic acid accumulation. *Cell Host Microbe* 11: 587–96.

Zhou, Z., Wu, Y., Yang, Y., Du, M., Zhang, X. *et al.* (2015). An Arabidopsis plasma membrane proton ATPase modulates JA signaling and is exploited by the *Pseudomonas syringae* effector protein AvrB for stomatal invasion. *Plant Cell* 27: 2032–41.

Zhu, W., MaGbanua, M. M. and White, F. F. (2000). Identification of two novel hrp-associated genes in the hrp gene cluster of *Xanthomonas oryzae* pv. *oryzae*. *J. Bacteriol.* 182: 1844–53.

8 Plants as Carriers of Human Enteric Bacterial Pathogens

The cases of human illness associated with foodborne diseases caused by fresh produce have been noticed since the early 1970s and reported outbreaks associated with this commodity have grown steadily. Earlier it was thought that most cases of illness linked to fresh produce were the result of post-harvest microbial contamination rather than contamination in the field. However, with the recognition of pre-harvest crop contamination being the cause of foodborne illness, new ecological concepts regarding the fitness of enteric pathogens on crop plants need to be answered. Various aspects, such as bacterial epiphytic fitness, endophytic growth and survival, resource utilisation in the phyllosphere and the rhizosphere, plant–microbe interactions and microbe–microbe interactions will provide information for a better understanding of the ecology of enteric pathogens on plants. The review article by Brandl (2006) presents an overview of issues related to the microbial contamination of fresh produce with a focus on the behaviour of enteric pathogens on plant surfaces in the context of concepts in food safety and plant microbiology. Tyler and Triplett (2008), in their review article, have stated that enteric bacteria are common inhabitants of the interior of plants and *Salmonella* and *Escherichia coli* O157:H7 have been shown to enter into plants. The extent of endophytic colonisation by strains is regulated by plant defences and several genetic determinants necessary for this interior colonisation in endophytic bacteria have been identified. Common virulence factors in plant and animal pathogens have also been described in bacteria that can infect both plant and animal models.

8.1 ASSOCIATION OF HUMAN ENTERIC BACTERIAL PATHOGENS WITH THE PLANT ENVIRONMENT

Several bacterial pathogens, including *Sal. enterica*, pathogenic *Esc. coli*, *Shigella* spp., *Campylobacter* spp., *Listeria monocytogenes*, *Staphylococcus aureus*, *Yersinia* spp. and *Bacillus cereus*, have caused fresh produce-associated epidemics of enteric illness. The association of human enteric viruses with foodborne illness has also been reported by Hirneisen and Kniel (2013).

Sal. enterica and pathogenic *Esc. coli* are not plant pathogens. Schikora *et al.* (2008) and Klerks *et al.* (2007b) reported that plants showed chlorosis or reduced vigour following inoculation with *Sal. enterica*, but Koch's postulates were not fulfilled in either case. Therefore, *Sal. enterica* and pathogenic *Esc. coli* remain characterised as plant-associated bacteria. Certain bacterial species that commonly inhabit plant surfaces, including *Pseudomonas aeruginosa*, *Burkholderia cepacia* complex members, *Enterobacter cloacae*, *Enterococcus faecalis* and *Erwinia* spp., have been shown to infect both plant and human tissue. The traditional view that human enteric bacterial pathogens have a defined ecological niche has been challenged by several investigators reporting outbreaks of foodborne illness associated with these pathogens. The available evidence that enteric bacterial pathogens have a high incidence on fresh fruits and vegetables, and that their presence on produce can lead to epidemics of foodborne illness, has caused a further change in the concept regarding the niche specificity of these pathogens.

Unexpected increase in produce-associated bacterial infections is further indicative of a much more important role of plants as a secondary habitat for enteric pathogens than previously thought. The role of plants to their human hosts, as a vector of these cross-domain pathogens, has remained overlooked in the past. Barak and Schroeder (2012), in their review article, have highlighted the significant steps throughout the life cycle of human enteric bacterial pathogens to focus the role of plants as their vectors.

Epidemics of foodborne illness linked to fresh produce in the United States increased from 1973 to 1997, both in absolute number and in the proportion of total food-linked outbreaks with a known etiologic agent (Sivapalasingam *et al.*, 2004). During this period, seed sprouts, lettuce, melon, and fruit juice were the four most important single-produce items implicated in epidemics of foodborne illness in the United States. A survey by Center for Science in the Public Interest (2005) revealed that fresh produce caused 28,315 cases of foodborne illness and was the second most common single food vehicle linked to outbreaks in the United States during the period 1990–2003. In England and Wales, salad, vegetables and fruit caused 6.4% and 10.1% of all outbreaks with a known food vehicle during the periods of 1993–1998 (WHO, 2000) and 1999–2000 (WHO, 2003), respectively. Other fresh produce commonly associated with outbreaks of enteric bacterial diseases include tomato, cilantro, parsley, spinach, green onions, carrot and cabbage.

Earlier outbreaks of foodborne disease epidemics linked to fresh produce were initially suspected to result from the cross-contamination of vegetables or fruits during food preparation with other food items, particularly meat products. Surveys conducted by the United States (US) Food and Drug Administration (2001a, 2001b) of samples obtained from major distributors indicated that 1.6% of the

DOI: 10.1201/9781003681878-9

domestic and 4.4% of the imported produce were contaminated with human enteric pathogens. A study by the Microbial Data Programme of the US Department of Agriculture found that, in 2002, 0.62% of samples of domestic or imported celery, cantaloupe, lettuce and tomatoes were contaminated with *Esc. coli* strains harbouring virulence factors (US Department of Agriculture, 2004). In other surveys, Harris *et al.* (2003) reported the presence of pathogenic enteric bacteria on produce and in unpasteurised fruit or vegetable juices sampled during production or at retail markets. The above-mentioned surveillance studies support the hypotheses of possible contamination of fresh produce with human pathogens in the pre-harvest environment, or in post-harvest during washing and minimal processing, and, consequently, of outbreaks linked to fresh produce not necessarily being the outcome of cross-contamination events with other foods. The occurrence of foodborne illness from contaminated fresh produce has changed the belief that such disease was linked only to the consumption of foods of animal origin, including poultry, eggs, meat and milk.

Several outbreaks of foodborne illness of humans mainly due to the pre-harvest contamination of food have been mentioned above, but the major outbreak caused by *Esc. coli* O157:H7 in Japan in 1996 is of great significance. It caused sickness among more than 10,000 people, the majority being schoolchildren. Furthermore, the food was not of animal origin; it was in radish sprouts contaminated with Shiga toxin-producing *Esc. coli* (STEC) (Watanabe *et al.*, 1999). This Japanese outbreak highlighted the fact that plant and animal pathogens are similar in many ways, something which was not appreciated by many people before 1990. The numerous produce-implicated outbreaks are testament to *Sal. enterica* and pathogenic *Esc. coli*'s ability to persist and use plants as vectors to humans (Taormina *et al.*, 1999). A number of the important outbreaks associated with foodborne illness are listed in Table 8.1.

In US, there are more chances of *Sal. enterica* contamination from fresh plant produce than animal products. Sivapalasingam *et al.* (2004) reported that *Sal. enterica*, the most frequent etiological agent of outbreaks of fresh produce, caused 48% of such outbreaks in the US recorded between 1973 and 1997. However, of special concern was the occurrence of outbreaks caused by multidrug-resistant (MDR) strains of *Sal. enterica* serovar Typhimurium DT104 and linked to the consumption of lettuce (Horby *et al.*, 2003; Takkinen *et al.*, 2005). It also commonly occurs on tomato and cantaloupe. Since 2002, yearly salmonellosis outbreaks have occurred in the United States resulting from contaminated tomato fruits produced on the eastern shore of Virginia (Bidol *et al.*, 2007). Pathogenic *Esc. coli* is the second most important causal agent of outbreaks from fresh produce, and the pathogenic strains include the enterotoxigenic serotype O157:H7 which causes hemolytic uremic syndrome leading to death, particularly in children, the elderly and the immunocompromised. *Esc. coli* O157:H7 caused 21% of all produce-linked outbreaks during 1982–2002 in the US (Rangel *et al.*, 2005). In the US during 1990–2004, 76% of the epidemics from contaminated fruits were caused by *Sal. enterica* and 48% of the epidemics from contaminated fresh leafy vegetables were caused by *Esc. coli*. It is apparent, therefore, that America's salad bowl contains *Esc. coli* O157:H7 and *Sal. enterica*, which can lead to leafy green contamination and sometimes to outbreaks of human illness.Data are also available for various plant parts linked to human illness caused by *Sal. enterica*, pathogenic *Esc. coli*, *Shigella* spp. and *Campylobacter* spp. for United States for the period from 1990 - 2004 (Table 8.2). During this period of 14 years, fruits contaminated with *Sal. enterica* caused 76% of the epidemics, whereas pathogenic *Esc. coli* strains were the

TABLE 8.1

Foodborne outbreaks caused by human enteric pathogens linked to contaminated seed

Seed	Outbreak year	Pathogen	Clinical cases	References
Mung	1988	*Salmonella enterica* Saint Paul	143	O' Mahony *et al.* (1990)
Alfalfa	1994	*Sal. enterica* Bovismorbificans	210	Puohiniemi *et al.* (1997)
Alfalfa	1995	*Sal. enterica* Stanley	242	Mahon *et al.* (1997)
Alfalfa	1996	*Sal. enterica* Meleagridis	75	Mohle-Boetani *et al.* (2001)
Alfalfa	1997	*Escherichia coli* O157:H7	85	Como-Sabetti *et al.* (1997)
Alfalfa	1998	*Sal. enterica* Cubana	22	Mohle-Boetani *et al.* (2001)
Alfalfa	1999	*Sal. enterica* Muenchen	157	Proctor *et al.* (2001)
Alfalfa	2001	*Sal. enterica* Kottbus	31	Winthrop *et al.* (2003)
Alfalfa	2009	*Sal. enterica* Bovismorbificans	42	Rimhanen-Finne *et al.* (2011)
Radish	1996	*Esc. coli* O157:H7	10,000	Nathan (1997)
Clover	1999	*Sal. enterica* Typhimurium	112	Brenner (1984)
Fenugreek	2011	*Esc. coli* O104:H4	4,321	Buchholz *et al.* (2011)

TABLE 8.2
Number of outbreaks[a] linked to single items of fresh produce in the United States during 1990–2004

Produce type	*Salmonella enterica*	Pathogenic *Escherichia coli*	*Shigella* spp.	*Campylobacter* spp.
Fruits and fruit juices	32 (76)	8 (19)	1 (2)	1 (2)
Leafy vegetables	8 (30)	13 (48)	3 (11)	3 (11)
Seed sprouts	9 (60)	6 (40)	0	0
Total	49	27	4	4

Source: Brandl, M. T. (2006), *Annu. Rev. Phytopathol.*, 44, 367–92.

Note: Figures in parentheses indicate the percentage of outbreaks caused by a given pathogen within a given produce category.

[a] Data are compiled from the publications: Center for Science in the Public Interest (2002), Harris *et al.* (2003), The US Centers for Disease Control and Prevention (2003, 2004, 2005). The authors of these publications did not report outbreaks from single items of fresh produce caused by *Staphylococcus aureus*, *Listeria monocytogenes*, or *Yersinia enterocolitica* in the United States during the given time period.

causal agent of the highest proportion of 48% foodborne illness outbreaks associated with fresh leafy vegetables. These leafy vegetables consisted mostly of lettuce, while the pathogenic *Esc. coli* was predominantly serotype O157:H7.

Apart from enteric bacteria, agricultural produce can also become contaminated with human viral pathogens in the field through soil, faeces, or water used for irrigation; through application of manure, biosolids, pesticides and fertilisers; and through animals, insects and dust. The Norovirus (more commonly known as the winter vomiting bug in the United Kingdom) was originally named the 'Norwalk agent' after Norwalk, Ohio in the United States, where an outbreak of acute gastroenteritis occurred among children at Bronson Elementary School in November 1968. The genus name *Norovirus* is derived from Norwalk virus, the only species of the genus. The species causes approximately 90% of epidemic outbreaks of non-bacterial gastroenteritis around the world (Lindersmith *et al.*, 2003) and may be responsible for 50% of all foodborne outbreaks of gastroenteritis in the United States (Widdowson *et al.*, 2005; Anonymous, 2007). The virus affects around 267 million people and causes more than 200,000 deaths each year. These deaths are usually in less developed countries and in the very young, the elderly and immunosuppressed persons (Debbink *et al.*, 2012).

Hirneisen and Kniel (2013) studied the survival and stability of human noroviruses and norovirus surrogates, namely, *Murine norovirus* (MNV) and *Tulane virus* (TV), on foliar surfaces of spinach plants in pre-harvest growth stages. The site of virus inoculation greatly impacted the virus survival, as viruses inoculated on adaxial leaf surfaces had lower decimal reduction time (D values) than the viruses inoculated on abaxial leaf surfaces. Under certain conditions, spinach type also influenced virus survival, with greater D values observed for survival on semi-savoy spinach leaves compared to those on smooth spinach leaves. Additional UVA and UVB exposure to mimic sunlight influenced virus survival on adaxial surfaces of both semi-savoy and smooth spinach plants for both the viruses. The average D value of human genogroup II norovirus inoculated onto semi-savoy spinach was not statistically significant from those of MNV and TV, suggesting that these surrogates may have similar survival on spinach leaves compared with human noroviruses. The information on the behaviour of enteric viruses on spinach leaves can help the growers for a better risk assessment under certain growing conditions. The following factors are responsible for the frequent occurrence of produce-linked outbreaks:

1. Introduction of minimally processed produce.
2. Increased importation of fresh produce.
3. Increased consumption of meals outside the home.
4. Increased popularity of salad bars.
5. Increased consumption of fresh fruits and vegetables, and fresh fruit juices.
6. Improved methods to identify and track pathogens.

8.2 SOURCES OF CONTAMINATION

Common sources of contamination of fresh plant produce with these pathogens are contaminated manure (enteric pathogens can survive for prolonged periods in animal faeces), contaminated water used for irrigation, pesticide and fertiliser application and soil. In general, enteric bacterial pathogens enter agricultural environments via animal faeces. The verified routes of crop produce contamination from faeces are soil, compost, seed and water.

8.2.1 Animals

After the occurrence of outbreaks caused by contaminated fresh produce, animals found in agricultural production areas have been identified as carriers of Shiga toxin-producing *Esc. coli* and *Sal. enterica.* These may be freely wandering animals in the fields, but also include those contained in pastures or high-density feeding operations. The faeces of these contained animals are a common source of human bacterial pathogens, as evidenced by many reports (Brashears *et al.*, 2003; Hutchison *et al.*, 2005a, 2005b; Thurston-Enriquez *et al.*, 2005; Soupir *et al.*, 2006). The pathogens associated with these faeces are moved by flood waters and rain or aerosolised by high-velocity winds. After mixing with water, these pathogens can move along with surface water, commonly used for irrigation and also as pesticide and fertiliser diluents. In addition to direct flow into field crops, surface water can also percolate through the soil column into ground water.

8.2.2 Soil

The survival of human enteric bacterial pathogens in agricultural fields depends on many characteristics of the soil, including soil structure, clay content, organic matter and resident microflora. In temperate climates, the populations of both *Sal. enterica* and pathogenic *Esc. coli* in soil decline with time following entry of pathogen either via water or manure (Natvig *et al.*, 2002; Semenov *et al.*, 2008).

The survival of *Salmonella* and/or pathogenic *Esc. coli* in manure-emended soil ranges from several weeks to more than 8 months (Islam *et al.*, 2004a, 2004b, 2004c; Nicholson *et al.*, 2005; Johannessen *et al.*, 2005; Holley *et al.*, 2006; You *et al.*, 2006; Semenov *et al.*, 2008). The survival of these pathogens in manure, compost and, ultimately in soil, leading to contamination of fresh produce, is influenced by the prevailing environment. The form of manure applied to fields as fertiliser amendments can influence the survival of *Esc. coli* O157:H7 and *Sal. enterica* in soil and on plants. Semenov *et al.* (2009) reported that, in general, slurry application led to more pathogen cells percolating to greater depths than manure. After mixing farmyard manure or injecting slurry into the top 10 cm of soil, *Sal. enterica* survived approximately 25 days in soil treated with manure or slurry, and *Esc. coli* O157:H7 survived longer (~30 days) in soil when injected as slurry compared to mixing of manure (~18 days). More pathogens are found in the rhizosphere in comparison to bulk soil after a slurry application, but no difference is observed after addition of manure.

Sal. enterica survived for more than 8 months in soil of a radish or carrot crop following an application in October of contaminated manure or irrigation water and was isolated from radishes and carrots up to 84 and 203 days, respectively, after the seeds were sown in the contaminated manure or irrigated soil (Islam *et al.*, 2004b). *Esc. coli* O157:H7 persisted for 217 days in soils emended with contaminated compost and was detected on lettuce and parsley up to 77 and 177 days, respectively, after planting (Islam *et al.*, 2004a). *Esc. coli* O157:H7 persisted longer in soil in association with plants, similar to leaf spot pathogens such as *Xanthomonas campestris* pv. *vitians* (Barak *et al.*, 2001) and *Psm. syringae* (Hirano and Upper, 1983). Plant pathogenic bacteria that lack the ability to survive in bulk soil can utilise plant debris to persist between crops or for overwintering/oversummering. Although *Esc. coli* O157:H7 can survive in the bulk soil for more than 8 months under certain conditions, its populations do decline. However, its survival in soil is enhanced by the presence of plant roots (Gagliardi and Karns, 2002) and crop debris.

Samaddar *et al.* (2021) have provided an overview of pathogens in soil, focusing on human and crop pathogens, and discussed general strategies and examples, of how soils' extraordinarily diverse microbial communities regulate soilborne pathogens. The ecological principles underpinning the regulation of soil pathogens, as well as relationships between pathogen suppression and soil health, are reviewed. Mechanisms and specific examples are presented of how soil and soil biota are involved in regulating pathogens of humans and plants. How specific agricultural management practices can either promote or interfere with soil's ability to regulate pathogens are also evaluated. The authors conclude that how integrating soil, plant, animal and human health through a 'One Health' framework could lead to more integrated, efficient and multifunctional strategies for regulating detrimental organisms and processes.

8.2.3 Water

The most common route of human enteric bacterial pathogens' entry into agricultural environments is via contaminated faecal matter deposited directly into surface water as reported by Jay *et al.* (2007) or animal faeces transported into surface water following rain or flood events. On the basis of a survey conducted in Salinas Valley, California, Cooley *et al.* (2007) found that the frequency of isolation of *Esc. coli* O157:H7 from surface water, e.g., creeks and streams, was more likely in areas where cattle were present. The isolation frequency was also higher following rain or flooding events.

Besides *Esc. coli* O157:H7, *Sal. enterica* is often isolated from surface waters commonly used for irrigation water in some agricultural production areas (Winfield and Groisman, 2003; Haley *et al.*, 2009). In a survey for *Sal. enterica* in and around Monterey County, California, Gorski *et al.* (2011) isolated several strains from surface water over a period of several months, indicating their persistence in the environment. However, the occurrence of these human pathogens in the fresh produce production environment is not solely limited to California, but could be anywhere where animals, including humans, deposit faeces.

8.2.4 Seed

The contaminated seed is a confirmed route of contamination in foodborne illness outbreaks. Cooley *et al.* (2003) reported that both *Sal. enterica* and *Esc. coli* O157:H7 moved from sites of inoculation on roots to leaves and flowers. This extensive movement was along the outside of the plant, and the flagella-mediated motility leads to seed contamination following *Sal. enterica* introduction into the rhizosphere. The US Food and Drug Administration has recommended soaking of seed for 20 minutes in a 20,000 mg L^{-1} of calcium hypochlorite solution. This treatment does reduce, but does not eliminate the contaminating bacteria from the seeds.

Sprouts are a common fresh produce vehicle in causing foodborne illness (CDC, 2009). Alfalfa seeds are identified as the most common contaminated seed, but it is unknown whether or not this is due to the popularity of this type of sprout with consumers, to seed production practices, or to sprout production practices. As only a small quantity of alfalfa seed is used for human consumption, therefore, no specific production practices are followed to check its contamination with human enteric pathogens. Mohle-Boetani *et al.* (2001) reported that alfalfa seeds which tested positive for an outbreak strain were fertilised with un-composted chicken manure and irrigated with non-potable canal water. As the seed treatment recommended by US Food and Drug Administration is not fully effective against human enteric pathogens, their use leads to occurrence of severe outbreaks. Brooks *et al.* (2001) have given an interesting account of clover sprouts responsible for a 1999 salmonellosis outbreak, which were grown by two independent producers using a common seed lot. Producer X did not treat the clover seeds prior to sprouting, while producer Y soaked seeds for 20 min in a 20,000 mg L^{-1} calcium hypochlorite solution (based on the recommendations of US Food and Drug Administration). The seed produced by producer X showed four times more illnesses than those of producer Y. It is evident that seed treatment reduces, but does not eliminate the bacterium responsible for illness from the sprouts. Subsequently, Proctor *et al.* (2001) confirmed these findings in multistate outbreaks resulting from alfalfa sprouts.

Perez-Lewis *et al.* (2022) studied the survival of *Esc. coli* O157:H7 and *Sal. enterica* serovar Typhimurium on tomato skins as a result of sanitisation treatment, under three differing contamination and sanitisation scenarios. Sanitising treatments consisted of the plant-derived antimicrobial geraniol (0.5 wt.%) emulsified in the polymeric surfactant Pluronic F-127 (GNP), 0.5 wt.% unencapsulated geraniol (UG), 200 mg/L hypochlorous acid at pH 7.0 (HOCl), and a sterile distilled water wash. Experimental contamination and sanitisation scenarios tested were: (1) pathogen inoculation preceded by treatment' (2) the pathogen was inoculated onto samples twice with a sanitising treatment applied in between inoculations, and (3) pathogen inoculation followed by sanitising treatment. Reductions in the counts of surviving pathogens depended on the sanitising treatment, the storage period, or the interaction of these independent/main effects. GNP treatment yielded the greatest reductions in pathogen counts on tomato skins; pathogen survivor counts following GNP treatment were consistently statistically lower than those achieved by HOCl or UG treatments ($p < 0.05$). GNP treatment provided the greatest pathogen reduction under differing conditions of pre- and/or post-harvest cross-contamination and reduced hygiene, indicating microbes the most of all treatments on non-inoculated samples. Encapsulated geraniol can reduce the risk of pathogen transmission on tomato fruit, thereby reducing food safety hazard risks for tomato consumers.

8.3 COLONISATION OF PLANTS BY HUMAN ENTERIC BACTERIAL PATHOGENS

8.3.1 Attraction by Plants

The rhizosphere is an ideal reservoir of plant-associated bacteria, including human enteric pathogens. Human enteric pathogenic bacteria are capable of sensing available plant nutrients and use these signals to actively move towards plants for subsequent attachment and colonisation. The rhizosphere of plants attracts *Salmonella* spp. This active movement of *Sal. enterica* to the plant roots is different from the previously reported passive contamination due to soil splashing, irrigation, insect transmission, or even passive uptake through roots (Solomon *et al.*, 2002). *Sal. enterica* is attracted to the lettuce rhizosphere by a sugar-like carbon source in root exudates (Klerks *et al.*, 2007b). Many bacteria colonise the roots of plants or persist in the rhizosphere by using these exudates as a nutritional source. *Salmonella*'s behaviour is similar to that of *Ralstonia solanacearum*, which demonstrates a preferential chemotaxis towards exudates of host plants (Yao and Allen, 2006). Whether the numerous human pathogens associated with plants or different serovars of specific pathogens differentially chemotax to various plants remains to be determined.

Due to chemotaxis, *Sal. enterica* in the soil can find the rhizosphere and, subsequently, colonises the phyllosphere. *Sal. enterica* differentially colonises agricultural plants growing in the contaminated soil, which suggests a case of differential chemotaxis towards specific plants (Barak *et al.*, 2008). The bacterium preferred radish, turnip and broccoli for colonisation over carrot, lettuce and tomato. Although the bacterium showed no preference among lettuce cultivars, it preferred the specific leafy greens radicchio and endive in comparison to lettuce. This preference for specific crop plants may reflect differences in the root exudates or a more favourable rhizosphere microflora, which is also influenced by root exudates. The steep gradient of root exudates in soil is continually modulated by the indigenous soil microflora.

After attraction by the root exudates, *Sal. enterica* actively moves towards the roots and colonises them. After the colonisation of roots, *Sal. enterica* and *Esc. coli* O157:H7 can move into the phyllosphere by flagella-mediated motility. It is not known whether human pathogens, such as *Sal. enterica*, require directed motility to locate roots, although niche exploration on the root has been shown to be flagella-mediated (Cooley *et al.*, 2003). Root exudates, therefore, have a dual role, i.e., the induction of chemotaxis of *Sal. enterica* to the roots and the simultaneous conditioning of its cells for plant attachment and subsequent colonisation (Klerks *et al.*, 2007a; Barak *et al.*, 2008).

Like rhizosphere, the bacteria are also attracted to the phyllosphere. Depending on the plant species, *Sal. enterica* preferentially colonises specific areas of leaves, i.e., cell junctions (Cooley *et al.*, 2003), along veins (Brandl and Mandrell, 2002), stomata (Kroupitski *et al.*, 2009), and/or trichomes (Barak *et al.*, 2011). It is unknown whether any attractant, other than a preferred microclimate, shelter from UV exposure, or a place of free water accumulation, exists for bacteria at cell junctions or along veins. The attractant produced by trichomes is also unknown, although it is suspected that the nutrient-rich secretions common of glandular trichomes (Wagner, 1991) could be alluring. Nutrients produced by actively photosynthesising guard cells of lettuce attract *Sal. enterica* to stomata and the addition of exogenous sugars disrupts the attraction (Kroupitski *et al.*, 2009). Tropism towards stomata is shared by *Sal. enterica* (Kroupitski *et al.*, 2009). Bacteria aggregate near open stomata for subsequent entry to the leaf apoplast. For plant-associated bacteria, entry to the leaf apoplast serves two purposes; (1) bacteria escape the harsh climate of the leaf surface and (2) apoplast entry is the first step in leaf infection.

8.3.2 Attachment and Persistence

The colonisation of plants by bacteria requires their adhesion to the plant surface and subsequent multiplication. Flagella, pili and extracellular polysaccharides are commonly employed by bacteria for plant colonisation. Flagella are best known for their role in bacterial motility; however, other uses of the flagella that could be useful during plant colonisation do exist. Wang *et al.* (2005) reported the sensing of external wetness by flagella of *Sal. enterica*. The sensing of the external environment by flagella leads to the regulation of its own biogenesis as well as a type III secretion system (TIIISS) in the *Salmonella* pathogenicity island-1. In addition to chemotaxis to roots or movement along plants, *Sal. enterica* and *Esc. coli* O157:H7 use flagella for adhesion to leaves, as reported by Kroupitski *et al.* (2009) and Saldana *et al.* (2011), respectively.

The enteric plant-associated bacteria, both animal and plant pathogens, require extracellular polysaccharide, cellulose and protein filaments for biofilm formation. Biofilm formation by plant-associated bacteria is a multicellular and aggregative behaviour, and it is used by them to successfully colonise plants (Danhorn and Fuqua, 2007). AgfD is the master regulator of aggregative behaviour in *Sal. enterica* (Gibson *et al.*, 2006), while the CsgD is the *Esc. coli* homolog of AgfD (Romling *et al.*, 1998). AgfD-regulated aggregative components, namely, thin aggregative fimbriae (but curli in *Esc. coli*), cellulose and O-antigen capsules, are involved in adhesion and colonisation to alfalfa sprouts (Barak *et al.*, 2005, 2007; Jeter and Matthysse, 2005). *Sal. enterica* and *Esc. coli* O157:H7 differ in terms of their colonisation of plants. *Esc. coli* O157:H7 utilises only curli fimbriae, and not cellulose, during leaf colonisation (Saldana *et al.*, 2011), whereas *Sal. enterica* requires both type of fimbriae, although curli to a lesser extent. These AgfD-regulated aggregative components have no role in pathogenicity studies carried out with animals (van der Velden *et al.*, 1998). Hence, AgfD is the first enteric bacterial regulator identified outside an animal host. The fact that AgfD and those yet uncharacterised regulators that govern plant fitness factors are necessary for these enteric pathogens to complete their life cycles on plants, gives support to the better understanding that plants have a crucial role in the life cycle of these pathogens. Brandl *et al.* (2005) reported that 54% of the cell population of *Sal. enterica* was present in aggregates composed of at least 128 cells, while some were comprised of as many as 4096 cells, 3 days after inoculation onto moist cilantro leaves. These bacterial cells, while in the exopolymer matrix of aggregates, may remain protected from the detrimental effect of rapid and frequent fluxes in water activity on plants.

Although thin aggregative fimbriae (curli), cellulose, O-antigen and flagella are required for aggregative behaviour in plant-associated bacteria, these have distinct roles in their persistence on plant tissue, such as bacterial cell–cell or bacterial cell–plant cell interactions. Thin aggregative fimbriae and cellulose protect *Sal. enterica* from desiccation and the capsule delays the oxidative degradation (Gibson *et al.*, 2006; Jonas *et al.*, 2007). For persistence on plants, it is paramount to ward off desiccation due to the rapid and daily fluctuations in free water on leaf surfaces (Lindow and Leveau, 2002). *Sal. enterica* strains with stronger biofilm-forming ability *in vitro* have stronger adhesion and persistence on lettuce leaves (Kroupitski *et al.*, 2009). Biofilm formation is equally important for the colonisation of plant roots (Barak *et al.*, 2007). *Sal. enterica*'s tolerance to desiccation stress, coupled with its ability to multiply under subsequent wet conditions, helps in its survival on plant surface (Kroupitski *et al.*, 2009).

Kroupitski *et al.* (2013), using recombinase-based *in vivo* expression technology, identified *Sal. enterica* genes involved in the persistence of this organism on post-harvest lettuce during cold storage. Knockout mutations in eight upregulated genes revealed that four of them have a role in the persistence of the pathogen in this system. These genes were *stfC*, *bcsA*, *misL* and *yidR*, encoding a fimbrial outer membrane usher, a cellulose synthase catalytic subunit, an adhesin of the autotransporter family expressed from the *Salmonella* pathogenicity island-3 and a putative ATP/GTP-binding protein, respectively. *misL*, *bcsA* and *yidR*, but not *stfC* mutants, were impaired also in attachment and biofilm formation, suggesting that these functions are required for survival of *Sal. enterica* on post-harvest lettuce. This being the first report that gene *misL*, which has a role in *Sal. enterica* binding to fibronectin in animal hosts, is involved also in adhesin to post-harvested lettuce.

In addition to biofilm formation, *Sal. enterica* has a quorum-sensing system and the cells can swarm. Swarming is a coordinated, flagella-mediated motility carried out by both Gram-stain-positive and Gram-stain-negative bacteria across semi-solid surfaces (Kearns, 2010). The swarming cells of *Sal. enterica* are morphologically and physiologically distinct from swimming or stationary cells. During swarm state induction, *Sal. enterica* cells elongate, produce more flagella (Harshey and Matsuyama, 1994), produce either a surfactant (Mireles *et al.* 2001) or osmolite (Chen *et al.*, 2007), exhibit adaptive resistance to multiple antibiotics (Butler *et al.*, 2009) and initiate *de novo* biosynthesis (Kim *et al.*, 2004). Root exudates induce swarming in plant-associated bacteria such as *Serratia liquefaciens* (Teplitski *et al.*, 2000) as well as *Sal. enterica* (Barak *et al.*, 2009). Bacteria preferentially colonise the zone of root elongation and sites of lateral root expansion (Walker *et al.*, 2003), a nutrient-rich environment due to root exudates and, hence, colonised by large, diverse bacterial communities. For survival in this niche, bacteria capable of inducing swarming may avoid antimicrobial compounds produced by the plant to defend against pathogens or compete with rival microbes for nutrients, and propel themselves through the semi-solid sloughing of root cellular debris and mucilage to ideal niches. The inability of *Sal. enterica* to swarm results in its failure to colonise the roots (Barak *et al.*, 2009). Non-flagellated and non-motile strains neither move beyond the point of inoculation on roots nor invade lateral roots with a small initial inoculum (10 cfu mL^{-1}) (Cooley *et al.*, 2003). Swarming is the only one type of flagella-mediated motility used by *Sal. enterica* to locate preferential niches.

Plant-associated bacteria also use swimming to colonise the plants. Cooley *et al.* (2003) reported that *Sal. enterica* travelled from the roots into the phyllosphere by swimming. However, the ability to swim requires coordinated behavior. A hypermotile strain of *Ral. solanacearum* poorly colonised roots (Meng *et al.*, 2011). Therefore, colonisation is a balanced act, and it requires the coordination of cells for flagella-mediated motility. Continuous association of a bacterium with a single host or an environment may reduce its fitness for other hosts. Perpetual passage from animal to soil and/or water/to plants/to animal may push pathogens to retain and gain fitness for each environment. Analysis of *Esc. coli* O157:H7 spinach outbreak that occurred in 2006 revealed that disease severity increased compared with the 1996 sprout outbreak (Manning *et al.*, 2008). Genomic sequencing of the strain causing the spinach outbreak showed substantial genomic differences that contributed to more severe disease. The fitness factors important for survival on plants, or at least outside the animal host, may have selected for such a deadly strain.

8.3.3 Growth and Multiplication

Human enteric bacterial pathogens grow and multiply on plants before their entry into the intestinal tract of humans along with seed and other plant products. Their multiplication on plants in the phyllosphere, rhizosphere and spermosphere is influenced by many biological and environmental factors. The phyllosphere is a comparatively harsh environment for the multiplication and survival of microorganisms compared to the rhizosphere. A brief account of the multiplication and survival of human enteric bacterial pathogens on plants is given below.

8.3.3.1 In the Phyllosphere

For an effective colonisation of plants, the bacteria should grow and multiply on the plants. The plant phyllosphere and rhizosphere are highly heterogeneous environments and vary in nutrient availability. Depending on the plant species, physiological state and plant growth conditions; sugars, organic acids and amino acids are available to bacteria in the phyllosphere. *Esc. coli* O157:H7 and *Sal. enterica* persist on tomato, lettuce and cilantro plants with minimal growth (Islam *et al.*, 2004a, 2004b).

There are an increasing number of outbreaks of human pathogens related to fresh produce. Therefore, the growth of human pathogens on plants should be explored. Human pathogens can survive under the harsh environments in plants, and can adhere and actively invade plants. Plant-associated microbiota or insects contribute to the survival and transmission of enteric pathogens in plants. Human enteric pathogens also trigger plant innate immunity, but some pathogens, such as *Salmonella*, can overcome this defence mechanism (Lim *et al.*, 2014).

Koukkidis *et al.* (2017) reported that traces of juices released from damaged salad leaves can significantly enhance colonisation of salad leaves by *Sal. enterica*. Salad juices in water increased *Salmonella* growth by 110% and in host-like serum-based media by more than 2,400-fold compared to control. Other aspects relevant to salad leaf colonisation and retention, such as motility and biofilm formation, increased over control levels by >220% and 250%, respectively where as the direct attachment to salad leaves increased by >350% when the salad leaf juice was present. Salad leaf juices also enhanced pathogen attachment to the salad bag plastic. Over 5 days of refrigeration (a standard storage time for bagged salad leaves), even traces of juice within the salad bag fluids, when mixed in water, increased *Salmonella* growth 280 times more than the control, as well as enhanced salad bag colonisation, which could lead to an infection of the consumers.

Mishra *et al.* (2017) made an attempt to develop a mathematical system model to understand the pathway of *Esc. coli* O157:H7 in the production of leafy greens. The results indicate that the seasonality of outbreaks of *Esc. coli* O157:H7-associated contamination of leafy greens was in good agreement with the prevalence of this pathogen in cattle and wild pig faeces in a major leafy green-producing region in California. On the basis of results obtained in different scenarios, it is evident that the concentration of *Esc. coli* O157:H7 in leafy greens can be reduced considerably, if contamination of soil with wild pig and cattle faeces is mitigated.

Leaf surface is a harsh site for bacterial growth due to rapid fluctuations in temperature and availability of free water, the effect of UV radiation and the diversity of available nutrients (Hirano and Upper, 1983; Lindow and Brandl, 2003). Many authors have reported the persistence of enteric bacteria, namely, *Sal. enterica*, *Entb. cloacae* and *Esc. coli* O157:H7, in the phyllosphere for extended periods, although their populations may decline with time (Pederson and Leser, 1992; Takikawa *et al.*, 2002; Cooley *et al.*, 2003; Islam *et al.*, 2004a, 2004b, 2004c). The colonisation of leaf tissue by *Sal. enterica* led to its further colonisation of fruit and seed (Cooley *et al.*, 2003; Barak *et al.*, 2011). A free film of water enhances bacterial colonisation and survival rates. Relative humidity of 85%–95% sustained *Psm. aeruginosa* (Green *et al.*, 1974) and *Sal. enterica* (Cooley *et al.*, 2003; Barak *et al.*, 2011) populations on leaves. However, survival on leaves at lower humidity levels requires fitness factors like UV-mediated DNA repair systems (Fernandez *et al.*, 2006), pigmentation (Poplawsky *et al.*, 2000), and extracellular polysaccharide production (Kemp *et al.*, 2004) commonly expressed by phytopathogenic bacteria. Escape from such an environment by bacteria lacking these fitness factors increases their survival in the phyllosphere.

Esc. coli O157:H7 occurs to a greater extent on the external tissue on plants, but it can also occur in the internal tissue of plants. Wright *et al.* (2013) using high-resolution microscopy determined the extent of internalisation of *Esc. coli* O157:H7 in live spinach and lettuce plants. The bacterium was found in the internal tissue of both the plant species. Colonisation occurred within the apoplast between the plant cells. Furthermore, colonies were detected inside the cell wall of epidermal and cortical cells of spinach and *Nicotiana benthamiana* roots. Internal colonisation of epidermal cells resembled that of *Pec. atrosepticum* on potato. In contrast, only sporadic cells of *Esc. coli* strain K-12 were found on spinach, but there was no occurrence of bacteria inside the cells. It appears that the internal colonisation of plants by bacteria is limited to a specific group of plant-interacting bacteria, including *Esc. coli* O157:H7.

O'Brien and Lindow (1989) demonstrated that the population sizes of *Sal. enterica* Typhimurium and *Esc. coli* reached the same level of those of *Psm. syringae* in the bean and corn phyllospheres under constant wet conditions, while their population sizes were significantly less than those of the epiphytic bacteria after incubation under dry conditions. The colonisation studies of a strain of *Sal. enterica* serovar Thompson, isolated during an outbreak linked to cilantro, on the cilantro phyllosphere revealed that the pathogen was less of a fit than the common epiphytic bacteria *Psm. chlororaphis* and *Pantoea agglomerans* under wet conditions at 22 °C (Brandl and Mandrell, 2002). However, the competitiveness of *Sal. enterica* serovar Thompson on wet cilantro leaves significantly increased when the plants were incubated at 30 °C and 37 °C, enabling it to achieve rapidly higher population sizes than at 24 °C. In growth chamber experiments, Natvig *et al.* (2002) detected *Sal. enterica* at harvest from the leaves of arugula and the roots of radish plants (raised from seeds planted in soil cores amended with contaminated manure) grown under conditions simulating a wet and warm summer environment, but not under those simulating a comparatively drier and cooler spring environment. The higher multiplication rates of the enteric pathogen at warm temperatures likely allow it to use a larger share of the nutrient resources that it can assimilate on the plant surface. Hence, it is probable that free water from rainfall, dew and irrigation, and accompanying warm temperatures enable *Sal. enterica* to undergo bursts of growth on plants in the field, thereby reaching the required population sizes to cause foodborne illness. However, *Sal. enterica* and other enteric pathogens are likely not as well adapted as plant-colonising bacteria to utilise the full spectrum of nutrients present on plant surfaces and, therefore, are limited by the amount and range of available nutrients that they can assimilate to thrive on plants. Lin (1996) reported that, in contrast to many plant-associated bacteria and *Esc. coli*, *Sal. enterica* was typically unable to assimilate sucrose, one of the main sugars present in leaf and root exudates. This fact may explain the relatively lower number of outbreaks caused by *Sal. enterica* on leaves in comparison to fruit, as shown in Table 8.2. Gu and associates (2013) investigated the effects of soil management on movement of *Sal. enterica* Typhimurium in tomato plants and found that more *Salmonella* spp. survived in inoculated leaves on plants grown in conventional than in organic soil. Only tomato plants grown in conventional soil produced contaminated fruit (1%) and approximately 5% of the seed from infested fruit were internally contaminated. No *Salmonella* sp. was detected in plants grown from contaminated seed.

Low water availability is considered to be one of the main limitations of bacterial survival on plant surfaces. However, *Sal. enterica* has a high tolerance to long-term desiccation stress in non-host environments. This is the reason why *Sal. enterica* is the most common etiologic agent of foodborne illness linked to spices (Vibha *et al.*, 2006) and also responsible for the occurrence of two salmonellosis outbreaks associated with dry raw almonds (The Centers for Disease Control and Prevention, 2004; Isaacs *et al.*, 2005). It was hypothesised that at least in one of these almond-linked outbreaks, *Sal. enterica* multiplied on the moist almond fruits on the orchard floor at harvest and survived on the dry kernels to which it had migrated. It shows the pathogen's ability to cope with low water availability as the moisture content of dried nuts in storage is less than 6%. Brandl and Mandrell (2002) reported that the growth of *Sal. enterica* declined at rates similar to those of epiphytic bacteria on the dry leaf surface of cilantro plants under conditions of low relative humidity imposed for 24 h, but it resumed at significant rates upon reoccurrence of wet conditions on the leaves. Therefore, the tolerance of *Sal. enterica* to desiccation stress, combined with its potential to multiply under subsequent wet conditions may confer on this pathogen the ability to persist despite the high fluctuations in water availability on plant surfaces in the field. It shows that even small populations of the pathogen, capable of survival on crop plants before harvest, may increase to infectious dose levels during growth-conducive conditions while the produce is stored or processed for consumption.

Generally, *Sal. enterica* has comparatively low epiphytic fitness on plants. Poza-Carrion *et al.* (2013) reported that viable population sizes of *Sal. enterica* applied to plants pre-inoculated with *Psm. syringae* or either of two *Erw. herbicola* strains, were ≥10-fold higher than that on control plants that were not pre-colonised by such indigenous bacteria when assessed 24–72 h after the imposition of desiccation stress. Hence, the survival of immigrant cells of *Sal. enterica* on plants appears to be strongly influenced by the epiphytic bacteria, and the presence of common epiphytic bacteria on plants can protect such immigrants from a key stress like desiccation encountered on leaf surfaces.

The occurrence of large numbers of cases of human illness linked to fresh produce, spices and nuts contaminated with non-typhoidal *Salmonella* supports the hypothesis that the colonisation of plants is an important part of the life cycle of this human pathogen. Although plant responses to human enteric pathogens are distinct from the more specific responses to phytopathogenic bacteria, plants appear to recognise *Salmonella*, likely by detecting conserved microbial patterns, which subsequently activate basal defences. Numerous *Salmonella* genes have been identified for playing a role in its colonisation of plant surfaces and tissues, and in its various interactions with other members of the plant-microbial community. Importantly, *Salmonella* species employ diverse and overlapping strategies to interact with plants and their microflora and to successfully colonise its vertebrate hosts (Brandl *et al.*, 2013).

The phyllosphere contamination on agricultural plants is influenced by the traits unique to the phyllosphere of different plants. Barak *et al.* (2008) reported that *Sal. enterica* failed to colonise some cultivars of tomato following their sowing in contaminated soil. This differential persistence in the phyllosphere was also observed following dip inoculation of the plants in a *Sal. enterica* suspension, as reported by Barak *et al.* (2011). Leaf characteristics that varied among the tomato cultivars might have contributed to the differential survival of *Sal. enterica*. Physical structures, such as leaf waxes, are known to influence the bacterial colonisation of leaves. Marcell and Beattie (2002) showed that the reduction in the density of crystalline waxes of corn increased the epiphytic populations of *Clavibacter michiganensis* and *Pan. agglomerans*. The changes in cuticular wax biosynthesis that produce glossy corn exhibit differential effects on populations of *Cla. michiganensis* and *Pan. agglomerans*. Those leaf characteristics that encourage the colonisation of non-pathogens may also benefit plant-associated human pathogens, such as *Sal. enterica* and pathogenic *Esc. coli*. The trichome, a known preferential site of bacterial colonisation on the leaf surface identified by Beattie and Lindow (1995), also influences the persistence of *Sal. enterica*. Barak *et al.* (2011) identified glandular trichomes as the preferential site of *Sal. enterica* on tomato leaves.

The bacteria, including the phytopathogenic bacteria, lack the ability to directly penetrate the intact plant surface. Instead, they rely entirely on natural openings or accidental wounds to enter the internal plant tissues. High humidity or wet weather results in the opening of stomata to avoid leaf water-soaking. The bacteria that fail to open stomata, namely, *Psm. aeruginosa* (Starkey and Rahme, 2009), *Sal. enterica* (Kroupitski *et al.*, 2009) and *Bur. cepacia* (Kawamoto and Lorbeer, 1974), can enter the leaf apoplast. Entry in the substomatal chamber places the bacteria in a protected niche that facilitates their persistence and is the first step towards plant infection by phytopathogenic bacteria.

The route of human enteric bacterial pathogens internalisation into plant tissue has been a subject for extensive discussion (Warriner and Namvar, 2010). Human bacterial pathogens' internalisation through plant stomata has been studied for *Esc. coli* O157:H7 (Melotto *et al.*, 2006) and *Sal. enterica* serovar Typhimurium (Kroupitski *et al.*, 2009). Thilmony *et al.* (2006) reported that *Esc. coli* O157:H7 triggers stomatal closure, but it is not able to overcome this plant immune response when inoculated as pure cultures under laboratory conditions. However, Kroupitski *et al.* (2009) documented a remarkable ability of *Sal. enterica* serovar Typhimurium to migrate towards stomata and enter plant tissues without triggering stomatal immune response. This finding suggests the possibility that not only plant pathogens, but also some human enteric bacterial pathogens have evolved mechanisms to subvert plant stomata-based defence to enter plant tissues. The underlying mechanism of this observation is not understood and is a topic of active research (Baker *et al.*, 2010).

Several researchers have examined the role of plant diseases in the contamination of produce with foodborne pathogens. Although *Salmonella* spp. and *Esc. coli* O157:H7 fail to infect plants or grow in the phyllosphere unaided, these grow in plant cell lysates, root exudates and nutrients liberated by plant pathogens (Aruscavage *et al.*, 2006; Kyle *et al.*, 2010) and persist as infective doses, documented by the numerous foodborne illness outbreaks. Wells and Butterfield (1997) observed twice the incidence of *Salmonella* species on fruit and vegetables affected by soft rot in comparison to healthy produce. They found that the population of *Salmonella* species on potato, carrot and bell pepper discs co-inoculated with the soft rot pathogen, *Erw. carotovora* or *Psm. viridiflava* was 10- and 3-fold higher, respectively, than when inoculated with the enteric pathogen alone. The hypothesis that *Salmonella* spp. exchange AI-2 signals with *Pec. carotovorum* to increase its competitive fitness was tested by Cox *et al.* (2013), using mutants involved in AI-2 production (*luxS*) or perception (*lsrACDBF* or *lsrG*). Combined infections of a wild-type *Salmonella* sp. and its AI-2 mutants (at ≈3–10^4 cfu mL^{-1}) were established in green or red tomato ("FL 47" or "Campari") for 3 or 5 days, as well as tomato co-infected with *Pectobacterium* (at 10^9 cfu mL^{-1}) or its *luxS* mutant. The results of this study showed that AI-2 signalling offers no significant benefit to *Salmonella* spp. in this model of colonisation of tomato or soft rots.

Carlin *et al.* (1995) reported that soft rot also increased the population of *Lis. monocytogenes* on endive leaves. It is presumed that the damaged cells in rotten plant tissue leak abundant nutrients, which otherwise are unavailable to enteric pathogens on healthy plant surfaces. This assumption is supported by the fact that the growth of enteric pathogens is also enhanced in plant tissue that is simply mechanically injured. The degradation of plant tissue by plant pathogens may offer enteric pathogens a broad spectrum of nutrients for growth. However, not all plant pathogens affect the colonisation of plant tissue by enteric pathogens to the same extent. Brandl (2004) reported that co-inoculation of *Sal. enterica* with *Erw. chrysanthemi* or *Psm. viridiflava* revealed the better correlation in population increase of *Sal. enterica* on the diseased leaves of cilantro plants with those of *Erw. chrysanthemi* than with those of *Psm. viridiflava*. Both the growth rate and the final population sizes of *Sal. enterica* were greater in *Erw. chrysanthemi*-infected leaves than in leaves infected with *Psm. viridiflava*. In another study, Liao and Sapers (1999) found that the presence of *Psm. fluorescens* and *Psm. viridiflava* strongly inhibited the growth of *Lis. monocytogenes*, while the presence of *Erw. carotovora* or *Xan. campestris* did not affect its growth, either negatively or positively, after co-inoculation with one of these four pathogens onto potato tuber slices. The researcher suggested the competition for iron via production of siderophores as a possible mechanism for the antagonism by the pseudomonads in this study. Conversely, in other situations, the growth of enteric pathogens may be enhanced by their heterologous utilisation of siderophores produced by plant-associated microorganisms or by the plants themselves. Some plant pathogenic bacteria promote the growth of enteric pathogens even without causing disease. *Xan. vesicatoria* promotes *Sal. enterica* growth in the phyllosphere in the absence of plant disease (Barak and Liang, 2008). The populations of *Sal. enterica* grew several hundred times higher in the phyllosphere in the presence of *Xan. vesicatoria* than *Sal. enterica* populations on leaves without xanthomonads. It shows that apart from benefiting the enteric pathogens via nutrient liberation, plant pathogens benefit the enteric pathogens also in other ways.

8.3.3.2 Inhabitation of Spermosphere and Rhizosphere

Both spermosphere and rhizosphere are rich in plant exudates that can be utilised by the surrounding microbial community. Bacterial multiplication varies in spermospheres of different plants and is higher when seed exudates include essential nutrients and lack microbial inhibitors. Roberts *et al.* (2009) reported that *Entb. cloacae* exhibited a higher metabolic activity in the spermosphere of pea than that of cucumber, and the difference might be due to sugar concentration or composition, as replication could be restored in the cucumber spermosphere by addition of xylose and fructose. Colonisation of the corn, radish and wheat spermospheres was not related to the total peptide catabolic activity of *Entb. cloacae* as reported by Roberts *et al.* (1996).

The rhizosphere is a reservoir for plant-associated bacteria, including human enteric pathogens. Nutrient availability in the rhizosphere is also active and dynamic. Germinating seeds and subsequent roots release soluble and insoluble products into the rhizosphere via root exudates. After seed inoculation, *Sal. enterica* preferred to colonise the roots over the phyllosphere, most likely due to root exudates as reported by Charkowski *et al.* (2002). The composition and distribution of root exudates vary with many factors, including plant age, different root zones, nutritional status and soil types. These factors influence the microbial community associated with the root system, as bacterial species differ in their ability to utilise and compete for various nutrient sources (Petra *et al.*, 2004; Haichar *et al.*, 2008). Populations of *Bur. cepacia* complex members and *Esc. coli* O157:H7 in the rhizosphere were generally up to 100 times more than those in bulk soil (Pallud *et al.*, 2001; Gagliardi and Karns, 2002; Jacobs *et al.*, 2008).

8.4 MULTIPLE SEROVAR INFECTIONS

In a trace back investigation of the 1988 salmonellosis outbreak from seeds or sprouts in the United Kingdom, *Sal. enterica* ser. SaintPaul, the outbreak strain; ser. Virchow, a second outbreak strain; ser. Muenchen, ser. Lancing, ser. Litchfield and ser. Arizona were isolated by O'Mahony *et al.* (1990). In another salmonellosis outbreak from alfalfa seed lot occurring in 1995, *Sal. enterica* ser. Newport, the outbreak strain; ser. Albany and ser. Schwarzengrund were isolated. In these outbreaks, different serovars were associated with the plants involved. When the produce is contaminated with multiple serovars, it is not known why one or two particular serovars cause human illness and not the others. The following possibilities may be there: (1) the outbreak strain had a higher population on the seed and thus had more chances to infect, (2) some *Sal. enterica* serovars are more virulent or infectious than others and (3) the outbreak strain was more easily isolated in the clinical examination than those isolated in the research laboratory.

8.5 PLANT RESPONSE/REACTION

Two of the human enteric pathogens, namely, *Sal. enterica* and *Esc. coli*, have generally been reported to cause outbreaks on different fresh produce commodities. *Sal. enterica* was found on tomato and cantaloupe, pathogenic *Esc. coli* on lettuce, and *Sal. enterica* and pathogenic *Esc. coli* on sprouted seeds. Different *Sal. enterica* serovars were found to colonise the plants in these outbreaks. It is not known whether this differential serovar colonisation was due to the pathogen involved or the host. Some *Sal. enterica* serovars have caused repeated outbreaks in different years on the same fresh produce commodity, e.g., Poona serovar on cantaloupe and Newport and Javiana serovars on tomato. The reports of outbreaks of foodborne illness occurred to date have not identified a single dominant serovar on all the plants tested.

8.6 DEFENCE MECHANISMS OF PLANTS

The plants rely on the innate immunity of each cell, which acts as a first line of defence. FLS2 is one of the best-described molecular pattern receptors, and it recognises flagellin (Zipfel *et al.*, 2004). FLS2 acts as a first line of defence by recognising the most conserved amino acid sequence in flagellin, which is common to all plant-associated bacteria, including *Sal. enterica* (Felix *et al.*, 1999). On their arrival on leaves, the bacteria are detected by plants and are normally prevented from entering the apoplast by the immediate closure of stomata and left on the leaf surface to survive the harsh and unfavourable conditions.

Roy *et al.* (2013) reported that *Sal. enterica* serovar Typhimurium SL1344, like *Psm. syringae* pv. *tabaci* DC3000, causes a transient stomatal closure in lettuce. The mechanisms underlying stomatal closure and reopening mediated by SL1344 are not understood, but it appears that not only plant pathogens, but also some human pathogens may have evolved mechanisms to modulate plant stomatal movements as part of their colonisation strategy of the phyllosphere (Roy *et al.*, 2013; Melotto *et al.*, 2014). As such, the study of stomatal defence may have implications beyond plant diseases.

The role played by plant defence mechanisms in restricting colonisation and/or the persistence of human pathogens has been elucidated by bacterial mutational analysis and plant response assays. Iniguez *et al.* (2005) reported that the elimination of the microbe-associated molecular pattern and flagella increased endophytic colonisation of *Sal. enterica*. This finding suggests that *Medicago truncatula* recognises *Sal. enterica* on arrival via flagellin and mounts a defence response, leading to a reduction in the bacterial population. *Sal. enterica* induced both salicylic acid (SA)-dependent and -independent plant defences as reported by Iniguez *et al.* (2005). Ethylene, a signal molecule for induced systemic resistance in plants, decreases endophytic colonisation of *Sal. enterica*. It shows that the plants use at least two different defence pathways to recognise the human pathogens and mount a defence to keep their population in check.

Phytopathogenic bacteria utilise a TIIISS and inject effectors to overcome plant innate immunity and facilitate pathogenic action. Plant and human pathogens share TIIISSs and their effectors, as well as transcriptional regulators. Two TIIISSs required for animal infection by *Salmonella* occur within *Salmonella* pathogenicity island (SPI) 1 or 2. *SPI*-1 promotes invasion of mammalian cells. Iniguez *et al.* (2005) reported that mutations that eliminate the structural component of the *SPI*-1 TIIISS needle apparatus are required for translocation of other effectors and increase endophytic populations of *Sal. enterica* in the roots. TIIISS mutants of *Sal. enterica* had even higher populations on plants that fail to accumulate SA or are disrupted in both SA-mediated and SA-independent defence responses (Iniguez *et al.*, 2005). These results show that both SA-dependent and -independent pathways restrict populations of *Sal. enterica* in wild-type plants, and that the TIIISSs, including flagella, trigger either innate immunity and/or effector-triggered immunity. Hirano *et al.* (1999) reported that TIIISS mutants of *Psm. syringae* had no fitness defect in the spermosphere, although their populations in the phyllosphere were small. The role of TIIISS in plant pathogenic bacteria appears to facilitate bacterial replication because without growth the disease does not occur.

8.7 FUTURE STRATEGIES

Roy *et al.* (2013) investigated plant defence responses induced by the fully pathogenic bacteria *Esc. coli* O157:H7 and *Sal. enterica* serovar Typhimurium SL1344 in both *Arabidopsis thaliana* and lettuce (*Lactuca sativa*). Unlike SL1344, O157:H7 induced strong plant immunity at both pre-invasion and post-invasion steps of infection. O157:H7 triggered stomatal closure even under high relative humidity. It indicates that plants may recognise and respond to some human pathogens more effectively than others. It is very important to understand how plant responses can decrease the contamination of human bacterial pathogens for preventing disease outbreaks and improving the safety of food supplies.

Several investigators have demonstrated that the outcome of interactions of *Salmonella* with plants depends on the plant species and genotype. Internalisation of *Salmonella* into plant tissue varies greatly among plant species, and plant colonisation also varies among cultivars of a species. Barak *et al.* (2011) reported that there was an approximately 100-fold variation in the phyllosphere populations of *Salmonella* on four tomato cultivars, and *Solanum pimpinellifolium* variety WVa700 supported the lowest number of bacteria. Interestingly, WVa700 was also significantly less susceptible to *Psm. syringae* pv. *tomato*, the causal agent of bacterial speck in tomato. Quilliam *et al.* (2012) also showed the important role of the plant genotype in the proliferation of *Esc. coli* O157:H7 in the lettuce phyllosphere. These findings suggest that specific genetic factors governing the plant response to microbial colonisation reflect the outcome of the interaction of enteric pathogens with plants, or that the differences in physicochemical properties, such as the availability of certain nutrients or surface morphology, associated with different crop genotypes impact the proliferation of *Salmonella* and *Esc. coli* on and inside plants. These findings give a clue to the potential of breeding for resistance to colonisation by enteric pathogens, keeping in view the economic feasibility of such breeding programmes. However, it is unknown whether or not there is any correlation between plant basal immune responses to phytopathogens and to human pathogens. A positive correlation would provide an opportunity to integrate breeding for increased basal resistance of crops to both plant and human enteric pathogens.

In addition to competition for nutrients on plant surfaces, enteric pathogens may be exposed to phages and antibacterial compounds producing bacteria. Janisiewicz *et al.* (1999) reported that a strain of *Psm. syringae*, previously shown having fungicidal properties, reduced growth of *Esc. coli* O157:H7 on wounded apples by 10- to 1,000-fold. This and other such findings prompted further experiments on the biological control of human enteric pathogens associated with fresh produce. Fett (2006) showed that a well-characterised biocontrol strain of *Psm. fluorescens* 2–79 effectively reduced *Salmonella* populations on alfalfa sprouts. Some phytobacteria, including those involved in the biocontrol of plant diseases, significantly inhibit attachment and plant colonisation by non-typhoidal *Salmonella* and enterovirulent *Esc. coli* by producing antibiotics or competing for nutrients in the phyllosphere (Teplitski *et al.*, 2011). Similarly, bacteriophages can also be used to reduce the contamination of enteric pathogens on the produce. With the development of host range mutant (h-mutants) phages by Jackson (1989), the efficacy of phages for controlling bacteria has increased manifold, and these have been used successfully for controlling some bacterial diseases of plants in the field. It is evident that phages and certain bacteria have a proven antibacterial activity and can be used as potential tools for controlling enteric pathogens throughout the produce production cycle. However, it is important to consider that zero tolerance for most human pathogens on fresh fruit and vegetables implies that even the most effective biocontrol agent would need to be integrated into a general control strategy.

It is evident from the above-given account that the plants play a critical role in the life cycle of human enteric bacterial pathogens. These enteric pathogens use plants as vectors between animal hosts, all while following the life cycle script of plant-associated bacteria (Barak and Schroeder, 2012). Human/animal bacterial pathogens, and also bacterial plant pathogens, have some common mechanisms, such as type III secretion systems and their effectors, and transcriptional regulators, which function in both animal and plant hosts. It is just possible that during the process of evolution, *Sal. enterica* and pathogenic *Esc. coli*, presently characterised as plant-associated bacteria, may acquire pathogenic capabilities and become plant pathogens. It will be a matter of great concern for plant bacteriologists and plant pathologists.

Fletcher *et al.* (2013) reported that the frequent occurrences of outbreaks of foodborne illness associated with fresh fruits and vegetables have revived interest among public health agencies and sparked a new wave of research on food safety issues related to the microbial contamination of fresh produce. Recent efforts to address concerns about microbial contamination of food plants and resulting foodborne human illness have prompted new collaborations and interactions between scientific communities of plant pathology and food safety. They have emphasised the need for active collaboration among members of the plant pathology discipline, public health researchers and food microbiologists, to address the problems of human pathogens on plants.

The repeated outbreaks of human illness resulting from the pre-harvest contamination of plant products by human enteric pathogens such as Shiga toxin-producing *Esc. coli* and *Salmonella* spp. have led some plant pathologists to broaden the application of their science in the past two decades to address problems of human enteric pathogens on plants. New collaborations between members of plant pathology and food safety communities have led to enhanced research activity and an understanding of the issues about which research is needed. Therefore, continued interactions and communication among plant pathologists and food safety scientists is essential and can be achieved by the creation of an interdisciplinary research coordination network. Brandl and Sundin (2013) have emphasised that the critical aspect of produce safety is relevant to research in plant microbial ecology and intersects with numerous concepts that are explored in plant pathology.

REFERENCES

Anonymous (2007). *Norovirus: Technical fact sheet.* National Center for Infectious Diseases, CDC. http://www.cdc.gov/Ncidod/dvrd/revb/gastro/norovirus-factsheet.htm

Aruscavage, D., Lee, K., Miller, S. A. and LeJeune, J. T. (2006). Interactions affecting the proliferation and control of human pathogens on edible plants. *J. Food Sci.* 71: 89–98.

Baker, C. M., Chitrakar, R., Obulareddy, N., Panchal, S., Williams, P. *et al.* (2010). Molecular battles between plant and pathogenic bacteria in the phyllosphere. *Braz. J. Med. Biol. Res.* 43: 698–704.

Barak, J. D., Gorski, L., Liang, A. S. and Narm, K.-E. (2009). Previously uncharacterized *Salmonella enterica* genes required for swarming play a role in seedling colonization. *Microbiology* 155: 3701–09.

Barak, J. D., Gorski, L., Naraghi-Arani, P. and Charkowski, A. O. (2005). *Salmonella enterica* virulence genes are required for bacterial attachment to plant tissue. *Appl. Environ. Microbiol.* 71: 5685–91.

Barak, J. D., Jahn, C. E., Gibson, D. L. and Charkowski, A. O. (2007). The role of cellulose and O-antigen capsule in the colonization of plants by *Salmonella enterica. Mol. Plant-Microbe Interact.* 20: 1083–91.

Barak, J. D., Koike, S. T. and Gilbertson, R. L. (2001). Role of crop debris and weeds in the epidemiology of bacterial leaf spot of lettuce in California. *Plant Dis.* 85: 169–78.

Barak, J. D., Kramer, L. C. and Hao, L.-Y. (2011). Colonization of tomato plants by *Salmonella enterica* is cultivar dependent and type I trichomes are preferred colonization sites. *Appl. Environ. Microbiol.* 77: 498–504.

Barak, J. D., Liang, A. and Narm, K.-E. (2008). Differential attachment to and subsequent contamination of agricultural crops by *Salmonella enterica. Appl. Environ. Microbiol.* 74: 5568–70.

Barak, J. D. and Liang, A. S. (2008). Role of soil, crop debris, and a plant pathogen in *Salmonella enterica* contamination of tomato plants. *PLoS One*, 3: e1657.

Barak, J. D. and Schroeder, B. K. (2012). Interrelationships of food safety and plant pathology: The life cycle of human pathogens on plants. *Annu. Rev. Phytopathol.* 50: 241–66.

Beattie, G. A. and Lindow, S. E. (1995). The secret life of foliar bacterial pathogens on leaves. *Annu. Rev. Phytopathol.* 33: 145–72.

Bidol, S. A., Daly, E. R., Ricker, R. E., Hill, T. A., Taylor, T. H. *et al.* (2007). Multistate outbreaks of *Salmonella* infections associated with raw tomatoes eaten in restaurants: United States, 2005–2006. *CDC Morb. Mortal. Wkly. Rep.* 56: 909–11.

Brandl, M. T. (2004). Fitness of *Salmonella enterica* in the phyllosphere. *Phytopathology*, 94(Suppl.): 128.

Brandl, M. T. (2006). Fitness of human enteric pathogens on plants and implications for food safety. *Annu. Rev. Phytopathol.* 44: 367–92.

Brandl, M. T., Cox, C. E. and Teplitski, M. (2013). *Salmonella* interactions with plants and their associated microbiota. *Phytopathology*, 103: 316–25.

Brandl, M. T. and Mandrell, R. E. (2002). Fitness of *Salmonella enterica* serovar Thompson in the cilantro phyllosphere. *Appl. Environ. Microbiol.* 68: 3614–21.

Brandl, M. T., Miller, W. G., Bates, A. H. and Mandrell, R. E. (2005). Production of autoinducer 2 in *Salmonella enterica* serovar Thompson contributes to its fitness in chickens but not on cilantro leaf surfaces. *Appl. Environ. Microbiol.* 71: 2653–62.

Brandl, M. T. and Sundin, G. W. (2013). Focus on food safety: Human pathogens on plants. *Phytopathology*, 103: 304–05.

Brashears, M. M., Galyean, M. L., Loneragan, G. H., Mann, J. E. and Killinger-Mann, K. M. (2003). Prevalence of *Escherichia coli* O157 and performance of beef feedlot cattle given *Lactobacillus* direct-fed antimicrobials. *J. Food. Prot.* 66: 748–54.

Brenner, D. J. (ed.) (1984). Family 1. Enterobacteriaceae Rahn 1937, Nom. fam. cons. In: *Bergey's Manual of Systematic Bacteriology*. 1st Ed., Vol. 1. Williams & Wilkins Co., Baltimore, MD, pp. 408–20.

Brooks, J. T., Rowe, S. Y., Shillam, P., Heltzel, D. M., Hunter, S. B. *et al.* (2001). *Salmonella typhimurium* infections transmitted by chlorine-pretreated clover sprout seeds. *Am. J. Epidemiol.* 154: 1020–28.

Buchholz, U., Bernard, H., Werber, D., Bohmer, M. M., Remschmidt, C. *et al.* (2011). German outbreak of *Escherichia coli* O104:H4 associated with sprouts. *N. Engl. J. Med.* 365: 1763–70.

Butler, M. T., Wang, Q. and Harshey, R. M. (2009). Cell density and mobility protect swarming bacteria against antibiotics. *Proc. Natl. Acad. Sci. USA*, 107: 3776–81.

Carlin, F., Nguyen-The, C. and Abreu da Silva, A. (1995). Factors affecting the growth of *Listeria monocytogenes* on minimally processed fresh endive. *J. Appl. Bacteriol.* 78: 636–46.

Center for Science in the Public Interest. (2002). Outbreak Alert! Closing the Gaps in Our Federal Food Safety Net. Washington, DC. http://www.cspinet.org/Report.

Center for Science in the Public Interest. (2005). Outbreak Alert! Closing the Gaps in Our Federal Food Safety Net. Washington, DC. http://www.cspinet.org/Report.

Charkowski, A. O., Barak, J. D., Sarreal, C. Z. and Mandrell, R. E. (2002). Differences in growth of *Salmonella enterica* and *Escherichia coli* O157:H7 on alfalfa sprouts. *Appl. Environ. Microbiol.* 68: 3114–20.

Chen, B. G., Turner, L. and Berg, H. C. (2007). The wetting agent required for swarming in *Salmonella enterica* serovar Typhimurium is not a surfactant. *J. Bacteriol.* 189: 8750–53.

Como-Sabetti, K., Reagan, S., Allaire, S., Parrott, K., Simonds, C. M. *et al.* (1997). Outbreaks of *Escherichia coli* O157:H7 infection associated with eating alfalfa sprouts: Michigan and Virginia, June–July 1997. *Morb. Mortal. Wkly. Rep.* 46: 741–44.

Cooley, M., Carychao, D., Crawford-Miksza, L., Jay, M. T., Myers, C. *et al.* (2007). Incidence and tracking of *Escherichia coli* O157:H7 in a major produce production region in California. *PLoS ONE* 2: e1159.

Cooley, M. B., Miller, W. G. and Mandrell, R. E. (2003). Colonization of A*rabidopsis thaliana* with *Salmonella enterica* and enterohemorrhagic *Escherichia coli* O157:H7 and competition by *Enterobacter asburiae*. *Appl. Environ. Microbiol.* 69: 4915–26.

Cox, C. E., McClelland, M. and Teplitski, M. (2013). Consequences of disrupting *Salmonella* AI-2 signaling on interactions within soft rots. *Phytopathology* 103: 352–61.

Danhorn, T. and Fuqua, C. (2007). Biofilm formation by plant-associated bacteria. *Annu. Rev. Microbiol.* 61: 401–22.

Debbink, K., Lindesmith, L. C., Donaldson, E. F. and Baric, R. S. (2012). Norovirus immunity and the great escape. *PLoS Pathog* 8 (10): e1002921.

Felix, G., Duran, J. D., Volko, S. and Boller, T. (1999). Plants have a sensitive perception system for the most conserved domain of bacterial flagellin. *Plant J.* 18: 265–76.

Fernandez Zenoff, V., Sineriz, F. and Farias, M. E. (2006). Diverse responses to UV-B radiation and repair mechanisms of bacteria isolated from high-altitude aquatic environments. *Appl. Environ. Microbiol.* 72: 7857–63.

Fett, W. F. (2006). Inhibition of *Salmonella enterica* by plant-associated pseudomonads *in vitro* and on sprouting alfalfa seed. *J. Food Prot.* 69: 719–28.

Fletcher, J., Leach, J. E., Eversole, K. and Tauxe, R. (2013). Human pathogens on plants: Designing a multidisciplinary strategy for research. *Phytopathology*, 103: 306–15.

Gagliardi, J. V. and Karns, J. S. (2002). Persistence of *Escherichia coli* O157:H7 in soil and on plant roots. *Environ. Microbiol.* 4: 89–96.

Gibson, D. L., White, A. P., Snyder, S. D., Martin, S., Heiss, C. *et al.* (2006). *Salmonella* produces an O-antigen capsule regulated by AgfD and important for environmental persistence. *J. Bacteriol.* 188: 7722–30.

Gorski, L., Parker, C. T., Liang, A., Cooley, M. B., Jay-Russell, M. T. *et al.* (2011). Prevalence, distribution, and diversity of *Salmonella enterica* in a major produce region of California. *Appl. Environ. Microbiol.* 77: 2734–48.

Green, S. K., Schroth, M. N., Cho, J. J., Kominos, S. D. and Vitanza-Jack, V. B. (1974). Agricultural plants and soil as a reservoir for *Pseudomonas aeruginosa*. *Appl. Microbiol.* 28: 987–91.

Gu, G., Cevallos-Cevallos, J. M., Vallad, G. E. and van Bruggen, A. H. C. (2013). Organically managed soils reduce internal colonization of tomato plants by *Salmonella enterica* serovar Typhimurium. *Phytopathology*, 103: 381–88.

Haichar, F. Z., Marol, C., Berge, O., Rangel-Castro, J. I., Prosser, J. I. *et al.* (2008). Plant host habitat and root exudates shape soil bacterial community structure. *Int. Soc. Microbiol. Ecol. J.* 2: 1221–30.

Haley, B. J., Cole, D. J. and Lipp, E. K. (2009). Distribution, diversity and seasonality of waterborne *Salmonella* in a rural watershed. *Appl. Environ. Microbiol.* 75: 1248–55.

Harris, L. J., Farber, J. N., Beuchat, L. R., Parish, M. E., Suslow, T. V. *et al.* (2003). Outbreaks associated with fresh produce: Incidence, growth, and survival of pathogens in fresh and fresh-cut produce. *Comp. Rev. Food Sci. Food Saf.* 2 (Suppl.): 78–141.

Harshey, R. M. and Matsuyama, T. (1994). Dimorphic transition in *Escherichia coli* and *Salmonella typhimurium*: Surface induced differentiation into hyperflagellate swarmer cells. *Proc. Natl. Acad. Sci. USA*, 91: 8631–35.

Hirano, S. S., Charkowski, A. O., Collmer, A., Willis, D. K. and Upper, C. D. (1999). Role of the Hrp type III protein secretion system in growth of *Pseudomonas syringae* pv. *syringae* B728a on host plants in the field. *Proc. Natl. Acad. Sci. USA*, 96: 9851–56.

Hirano, S. S. and Upper, C. D. (1983). Ecology and epidemiology of foliar bacterial plant pathogens. *Annu. Rev. Phytopathol.* 21: 243–70.

Hirneisen, K. A. and Kniel, K. E. (2013). Norovirus surrogate survival on spinach during preharvest growth. *Phytopathology*, 103: 389–94.

Holley, R. A., Arrus, K. M., Ominski, K. H., Tenuta, M. and Blank, G. (2006). *Salmonella* survival in manure-treated soils during simulated seasonal temperature exposure. *J. Environ. Qual.*, 35: 1170–80.

Horby, P. W., O'Brien, S. J., Adak, G. K., Graham, C., Hawker, J. I. *et al.* (2003). A national outbreak of multi-resistant *Salmonella enterica* serovar Typhimurium definitive phage type (DT) 104 associated with consumption of lettuce. *Epidemiol. Infect.* 130: 169–78.

Hutchison, M. L., Walters, L. D., Avery, S. M., Munro, F. and Moore, A. (2005a). Analyses of livestock production, waste storage, and pathogen levels and prevalences in farm manures. *Appl. Environ. Microbiol.* 71: 1231–36.

Hutchison, M. L., Walters, L. D., Moore, T., Thomas, D. J. I. and Avery, S. M. (2005b). Fate of pathogens present in livestock wastes spread onto fescue plots. *Appl. Environ. Microbiol.* 71: 691–96.

Iniguez, A. L., Dong, Y., Carter, H. D., Ahmer, B. M., Stone, J. M. *et al.* 2005. Regulation of enteric endophytic bacterial colonization by plant defenses. *Mol. Plant-Microbe Interact.* 18: 169–78.

Isaacs, S., Aramini, J., Ciebin, B., Farrar, J. A., Ahmed, R. *et al.* (2005). An international outbreak of salmonellosis associated with raw almonds contaminated with a rare phage type of *Salmonella enteritidis*. *J. Food Prot.* 68: 191–98.

Islam, M., Doyle, M. P., Phatak, S. C., Millner, P. and Jiang, X. (2004a). Persistence of enterohemmorrhagic *Escherichia coli* O157:H7 in soil and on leaf lettuce and parsley grown in fields treated with contaminated manure composts or irrigation water. *J. Food Prot.* 67: 1365–70.

Islam, M., Morgan, J., Doyle, M. P., Phatak, S. C., Millner, P. *et al.* (2004b). Fate of *Salmonella enterica* serovar Typhimurium on carrots and radishes grown in fields treated with contaminated manure composts or irrigation water. *Appl. Environ. Microbiol.* 70: 2497–2502.

Islam, M., Morgan, J., Doyle, M. P., Phatak, S. C., Millner, P. *et al.* (2004c). Persistence of *Salmonella enterica* serovar Typhimurium on lettuce and parsley and in soils on which they were grown in fields treated with contaminated manure composts or irrigation water. *Foodborne Pathog. Dis.* 1: 27–35.

Jackson, L. E. (1989). Bacteriophage prevention and control of harmful plant bacteria. U. S. Patent No. 4828999.

Jacobs, J. L., Fasi, A. C., Ramette, A., Smith, J. J., Hammerschmidt, R. *et al.* (2008). Identification and onion pathogenicity of *Burkholderia cepacia* complex isolates from the onion rhizosphere and onion field soil. *Appl. Environ. Microbiol.* 74: 3121–29.

Janisiewicz, W. J., Conway, W. S. and Leverentz, B. (1999). Biological control of postharvest decays of apple can prevent growth of *Escherichia coli* O157: H7 in apple wounds. *J. Food Prot.* 62: 1372–75.

Jay, M. T., Cooley, M., Carychao, D., Wiscomb, G. W., Sweitzer, R. A. *et al.* (2007). *Escherichia coli* O157:H7 in feral swine near spinach fields and cattle, central California coast. *Emerg. Infect. Dis.* 13: 1908–11.

Jeter, C. and Matthysse, A. G. (2005). Characterization of the binding of diarrheagenic strains of *E. coli* to plant surfaces and the role of curli in the interaction of the bacteria with alfalfa sprouts. *Mol. Plant-Microbe Interact.* 18: 1235–42.

Johannessen, G. S., Bengtsson, G. B., Heier, B. T., Bredholt, S., Wasteson, Y. *et al.* (2005). Potential uptake of *Escherichia coli* O157:H7 from organic manure into crisphead lettuce. *Appl. Environ. Microbiol.* 71: 2221–25.

Jonas, K., Tomenius, H., Kader, A., Normark, S., Romling, U. *et al.* (2007). Roles of curli, cellulose and BapA in *Salmonella* biofilm morphology studied by atomic force microscopy. *BMC Microbiol.* 7: 70.

Kawamoto, S. O. and Lorbeer, J. W. (1974). Infection of onion leaves by *Pseudomonas cepacia*. *Phytopathology* 64: 1440–45.

Kearns, D. B. (2010). A field guide to bacterial swarming motility. *Nat. Rev. Microbiol.* 8: 634–44.

Kemp, B. P., Horne, J., Bryant, A. and Cooper, R. M. (2004). *Xanthomonas axonopodis* pv. *manihotis gumD* gene is essential for EPS production and pathogenicity and enhances epiphytic survival on cassava (*Manihot esculenta*). *Physiol. Mol. Plant Pathol.* 64: 209–18.

Kim, J. W., Kim, Y. S. and Kyung, K. H. (2004). Inhibitory activity of essential oils of garlic and onion against bacteria and yeasts. *J. Food Prot.* 67: 499–504.

Klerks, M. M., Franz, E., van Gent-Pelzer, M., Zijlstra, C. and van Bruggen, A. H. C. (2007a). Differential interaction of *Salmonella enterica* serovars with lettuce cultivars and plantmicrobe factors influencing the colonization efficiency. *Int. Soc. Microbiol. Ecol. J.* 1: 620–31.

Klerks, M. M., van Gent-Pelzer, M., Franz, E., Zijlstra, C. and van Bruggen, A. H. C. (2007b). Physiological and molecular responses of *Lactuca sativa* to colonization by *Salmonella enterica* serovar Dublin. *Appl. Environ. Microbiol.* 73: 4905–14.

Koukkidis, G., Haigh, R., Allcock, N., Jordan, S. and Freestone, P. (2017). Salad leaf juices enhance *Salmonella* growth, colonization of fresh produce, and virulence. *Appl. Environ. Microbiol.* 83: e02416–e02516.

Kroupitski, Y., Brandl, M. T., Pinto, R., Belausov, E., Tamir-Ariel, D. *et al.* (2013). Identification of *Salmonella enterica* genes with a role in persistence on lettuce leaves during cold storage by recombinase-based *in vivo* expression technology. *Phytopathology*, 103: 362–72.

Kroupitski, Y., Golberg, D., Belausov, E., Pinto, R., Swartzberg, D. *et al.* (2009). Internalization of *Salmonella enterica* in leaves is induced by light and involves chemotaxis and penetration through open stomata. *Appl. Environ. Microbiol.* 75: 6076–86.

Kyle, J. L., Parker, C. T., Goudeau, D. and Brandl, M. T. (2010). Transcriptome analysis of *Escherichia coli* O157:H7 exposed to lysates of lettuce leaves. *Appl. Environ. Microbiol.* 76: 1375–87.

Liao, C. H. and Sapers, G. M. (1999). Influence of soft rot bacteria on growth of *Listeria monocytogenes* on potato tuber slices. *J. Food Prot.* 62: 343–48.

Lim, J.-A., Lee, D. H. and Heu, S. (2014). The interaction of human enteric pathogens with plants. *Plant Pathol. J.* 30(2): 109–16.

Lin, E. (1996). Dissimilatory pathways for sugars, polyols, and carboxylates. In: F. Neidhardt (ed.). *Escherichia coli and Salmonella: Cellular and Molecular Biology*. ASM Press, Washington, DC, pp. 307–42.

Lindesmith, L., Moe, C., Marionneau, S., Ruvoen, N., Jiang, X. *et al.* (2003). Human susceptibility and resistance to Norwalk virus infection. *Nat. Med.* 9: 548–53.

Lindow, S. E. and Brandl, M. T. (2003). Microbiology of the phyllosphere. *Appl. Environ. Microbiol.* 69: 1875–83.

Lindow, S. E. and Leveau, J. H. J. (2002). Phyllosphere microbiology. *Curr. Opin. Biotechnol.* 13: 238–43.

Mahon, B. E., Ponka, A., Hall, W. N., Komatsu, K., Dietrich, S. E. *et al.* (1997). An international outbreak of *Salmonella* infections caused by alfalfa sprouts grown from contaminated seeds. *J. Infect. Dis.* 175: 876–82.

Manning, S. D., Motiwala, A. S., Springman, A. C., Qi, W., Lacher, D. W. *et al.* (2008). Variation in virulence among clades of *Escherichia coli* O157:H7 associated with disease outbreaks. *Proc. Nat. Acad. Sci. USA* 105: 4868–73.

Marcell, L. M. and Beattie, G. A. (2002). Effect of leaf surface waxes on leaf colonization by *Pantoea agglomerans* and *Clavibacter michiganensis*. *Mol. Plant-Microbe Interact.* 15: 1236–44.

Melotto, M., Panchal, S. and Roy, D. (2014). Plant innate immunity against human bacterial pathogens. *Front. Microbiol.* 5: 411.

Melotto, M., Underwood, W., Koczan, J., Nomura, K. and He, S. Y. (2006). Plant stomata function in innate immunity against bacterial invasion. *Cell* 126: 969–80.

Meng, F., Yao, J. and Allen, C. (2011). A MotN mutant of *Ralstonia solanacearum* is hypermotile and has reduced virulence. *J. Bacteriol.* 193: 2477–86.

Mireles, J. R., Toguchi, A. and Harshey, R. M. (2001). *Salmonella enterica* serovar Typhimurium swarming mutants with altered biofilm-forming abilities: Surfactin inhibits biofilm formation. *J. Bacteriol.* 183: 5848–54.

Mishra, A., Pang, H., Buchanan, R. L., Schaffner, D. W. and Pradhan, A. K. (2017). A system model for understanding the role of animal feces as a route of contamination of leafy greens before harvest. *Appl. Environ. Microbiol.* 83: e02775–e02816.

Mohle-Boetani, J. C., Farrar, J. A., Werner, S. B., Minassian, D., Bryant, R. *et al.* (2001). *Escherichia coli* O157 and *Salmonella* infections associated with sprouts in California, 1996–1998. *Ann. Intern. Med.* 135: 239–47.

Nathan, R. (1997). American seeds suspected in Japanese food poisoning epidemic. *Nat. Med.* 3: 705–06.

Natvig, E. E., Ingham, S. C., Ingham, B. H., Cooperband, L. R. and Roper, T. R. (2002). *Salmonella enterica* serovar Typhimurium and *Escherichia coli* contamination of root and leaf vegetables grown in soils with incorporated bovine manure. *Appl. Environ. Microbiol.* 68: 2737–44.

Nicholson, F. A., Groves, S. J. and Chambers, B. J. (2005). Pathogen survival during livestock manure storage and following land application. *Bioresour. Technol.* 96: 135–43.

O'Brien, R. D. and Lindow, S. E. (1989). Effect of plant species and environmental conditions on epiphytic population sizes of *Pseudomonas syringae* and other bacteria. *Phytopathology* 79: 619–27.

O'Mahony, M., Cowden, J., Smyth, B., Lynch, D., Hall, M. *et al.* (1990). An outbreak of *Salmonella* saint-paul infection associated with bean sprouts. *Epidemiol. Infect.* 104: 229–35.

Pallud, C., Viallard, V., Balandreau, J., Normand, P. and Grundmann, G. (2001). Combined use of a specific probe and PCAT medium to study *Burkholderia* in soil. *J. Microbiol. Methods* 47: 25–34.

Pederson, J. C. and Leser, T. D. (1992). Survival of *Enterobacter cloacae* on leaves and in soil detected by immunofluorescence microscopy in comparison with selective plating. *Microbiol. Releases* 1: 95–102.

Perez-Lewis, K. L., Yegin, Y., Oh, J. K., Castillo, A., Cisneros-Zevallos, L. *et al.* (2022). Reduction of bacterial enteric pathogens and hygiene indicator bacteria on tomato skin surfaces by a polymeric nanoparticle-loaded plant-derived antimicrobial. *Microorganisms* 10(2): 448. https://doi.org/10.3390/microorganisms10020448

Petra, M., David, C. and Ching Hong, Y. (2004). Development of specific rhizosphere bacterial communities in relation to plant species, nutrition and soil type. *Plant Soil* 261: 199–208.

Poplawsky, A. R., Urban, S. C. and Chun, W. (2000). Biological role of xanthomonadin pigments in *Xanthomonas campestris* pv. *campestris*. *Appl. Environ. Microbiol.* 66: 5123–27.

Poza-Carrion, C., Suslow, T. and Lindow, S. (2013). Resident bacteria on leaves enhance survival of immigrant cells of *Salmonella enterica*. *Phytopathology* 103: 341–51.

Proctor, M. E., Hamacher, M., Tortorello, M. L., Archer, J. R. and Davis, J. P. (2001). Multistate outbreak of *Salmonella* serovar Muenchen infections associated with alfalfa sprouts grown from seeds pretreated with calcium hypochlorite. *J. Clin. Microbiol.* 39: 3461–65.

Puohiniemi, R., Heiskanen, T. and Siitonen, A. (1997). Molecular epidemiology of two international sproutborne *Salmonella* outbreaks. *J. Clin. Microbiol.* 35: 2487–91.

Quilliam, R. S., Williams, A. P. and Jones, D. L. (2012). Lettuce cultivar mediates both phyllosphere and rhizosphere activity of *Escherichia coli* O157:H7. *PLoS ONE* 7(3): e33842.

Rangel, J. M., Sparling, P. H., Crowe, C., Griffin, P. M. and Swerdlow, D. L. (2005). Epidemiology of *Escherichia coli* O157:H7 outbreaks, United States, 1982–2002. *Emerg. Infect. Dis.* 11: 603–09.

Rimhanen-Finne, R., Niskanen, T., Lienemann, T., Johansson, T., Sjoman, M. *et al.* (2011). A nationwide outbreak of *Salmonella* Bovismorbificans associated with sprouted alfalfa seeds in Finland, 2009. *Zoonoses Public Health* 58: 589–96.

Roberts, D. P., Baker, C. J., McKenna, L., Liu, S., Buyer, J. S. *et al.* (2009). Influence of host seed on metabolic activity of *Enterobacter cloacae* in the spermosphere. *Soil Biol. Biochem.* 41: 754–61.

Roberts, D. P., Dery, P. D. and Hartung, J. S. (1996). Peptide utilization and colonization of corn, radish and wheat spermospheres by *Enterobacter cloacae*. *Soil Biol. Biochem.* 28: 1109–11.

Romling, U., Sierralta, W. D., Eriksson, K. and Normark, S. (1998). Multicellular and aggregative behaviour of *Salmonella typhimurium* strains is controlled by mutations in the *agf*D promoter. *Mol. Microbiol.* 28: 249–64.

Roy, D., Panchal, S., Rosa, B. A. and Melotto, M. (2013). *Escherichia coli* O157:H7 induces stronger plant immunity than *Salmonella enterica* Typhimurium SL1344. *Phytopathology* 103: 326–32.

Saldana, Z., Sanchez, E., Xicohtencatl-Cortes, J., Puente, J. L. and Giron, J. A. (2011). Surface structures involved in plant stomata and leaf colonization by Shiga-toxigenic *Escherichia coli* O157:H7. *Front. Microbiol.* 2: 119.

Samaddar, S., Karp, D. S., Schmidt, R., Devarajan, N., McGarvey, J. A. *et al.* (2021). Role of soil in the regulation of human and plant pathogens: Soils' contributions to people. *Phil. Trans. R. Soc.* B376: 20200179. https://doi.org/10.1098/rstb.2020.0179

Schikora, A., Carreri, A., Charpentier, E. and Hirt, H. (2008). The dark side of the salad: *Salmonella typhimurium* overcomes the innate immune response of *Arabidopsis thaliana* and shows an endopathogenic lifestyle. *PLoS ONE* 3: e2279.

Semenov, A. V., Franz, E., van Overbeek, L., Termorshuizen, A. J. and van Bruggen, A. H. C. (2008). Estimating the stability of *Escherichia coli* O157:H7 survival in manure-amended soils with different management histories. *Environ. Microbiol.* 10: 1450–59.

Semenov, A. V., van Overbeek, L. and van Bruggen, A. H. C. (2009). Percolation and survival of *Escherichia coli* O157:H7 and *Salmonella enterica* serovar Typhimurium in soil amended with contaminated dairy manure or slurry. *Appl. Environ. Microbiol.* 75: 3206–15.

Sivapalasingam, S., Friedman, C. R., Cohen, L. and Tauxe, R. V. (2004). Fresh produce: A growing cause of outbreaks of foodborne illness in the United States, 1973 through 1997. *J. Food Prot.* 67: 2342–53.

Solomon, E. B., Yaron, S. and Matthews, K. R. (2002). Transmission of *Escherichia coli* O157:H7 from contaminated manure and irrigation water to lettuce plant tissue and its subsequent internalization. *Appl. Environ. Microbiol.* 68: 397–400.

Soupir, M., Mostaghimi, S., Yagow, E., Hagedorn, C. and Vaughan, D. (2006). Transport of fecal bacteria from poultry litter and cattle manures applied to pastureland. *Water Air Soil Pollut.* 169: 125–36.

Starkey, M. and Rahme, L. G. (2009). Modeling *Pseudomonas aeruginosa* pathogenesis in plant hosts. *Nat. Protoc.* 4: 117–24.

Takikawa, Y., Mori, H., Otsu, Y., Matsuda, Y., Nonomura, T. *et al.* (2002). Rapid detection of phylloplane bacterium *Enterobacter cloacae* based on chitinase gene transformation and lytic infection by specific bacteriophages. *J. Appl. Microbiol.* 93: 1042–50.

Takkinen, J., Nakari, U. M., Johansson, T., Niskanen, T., Siitonen, A. *et al.* (2005). A nationwide outbreak of multiresistant *Salmonella* Typhimurium va. Copenhagen DT104B infection in Finland due to contaminated lettuce from Spain. *Eurosurveill. Wkly.* http://www.eurosurveillance.org. Accessed on June 23, 2025.

Taormina, P. J., Beuchat, L. R. and Slutsker, L. (1999). Infections associated with eating seed sprouts: An international concern. *Emerg. Infect. Dis.* 5: 626–34.

Teplitski, M., Robinson, J. B. and Bauer, W. D. (2000). Plants secrete substances that mimic bacterial n-acyl homoserine lactone signal activities and affect population density-dependent behaviours in associated bacteria. *Mol. Plant-Microbe Interact.* 13: 637–48.

Teplitski, M., Warriner, K., Bartz, J. and Schneider, K. R. (2011). Untangling metabolic and communication networks: Interactions of enterics with phytobacteria and their implications in produce safety. *Trends Microbiol.* 19: 121–27.

The Centers for Disease Control and Prevention (CDC). (2003). 2002 – Foodborne disease outbreaks due to bacterial etiologies. http://www.cdc.gov/foodborneoutbreaks/us_outb/fbo2002./bacterial02.htm.

The Centers for Disease Control and Prevention (CDC). (2004). Outbreaks of *Salmonella* serotype Enteritidis infections associated with raw almonds-United States and Canada, 2003–2004. *MMWR* 53: 1–3.

The Centers for Disease Control and Prevention (CDC). (2009). Surveillance for food borne disease outbreaks: United States, 2006. *Morb. Mortal. Wkly. Rep.* 58: 609–15.

The Centers for Disease Control and Prevention (CDC). (2005). Foodborne outbreaks due to bacterial etiologies, 2004. http://www.cdc.gov/foodborneoutbreaks/us_outb/fbo2004./summary04.htm

Thilmony, R., Underwood, W. and He, S. Y. (2006). Genome-wide transcriptional analysis of the *Arabidopsis thaliana* interaction with the plant pathogen *Pseudomonas syringae* pv. *tomato* DC3000 and the human pathogen *Escherichia coli* O157:H7. *Plant J.* 46: 34–53.

Thurston-Enriquez, J. A., Gilley, J. E. and Eghball, B. (2005). Microbial quality of runoff following land application of cattle manure and swine slurry. *J. Water Health* 3: 157–71.

Tyler, H. L. and Triplett, E. W. (2008). Plants as a habitat for beneficial and/or human pathogenic bacteria. *Annu. Rev. Phytopathol.* 46: 53–73.

US Department of Agriculture. (2004). Progress update and 2002 data summary. Microbial data program. http://www.ams.usda.gov/(Cited by Brandl, M. T. (2006). Fitness of human enteric pathogens on plants and implications for food safety. *Ann. Rev. Phytopath.* 44: 367–92.

US Food and Drug Administration (FDA). (2001a). FDA survey of imported fresh produce. FY 1999 Field Assignment. US Food and Drug Administration. Center for Food Safety and Applied Nutrition. Office of Plant and Dairy Foods and Beverages. January 30. http://www.cfsan.gov/~dms/prodsur6.html. Accessed on June 24, 2025.

US Food and Drug Administration (FDA). (2001b). FDA Survey of domestic fresh produce: Interim results. US Food and Drug Administration. Center for Food Safety and Applied Nutrition. June 30. http://vm.cfsan.gov/~dms/prodsur8.html. Accessed on June 24, 2025.

van der Velden, A. W., Baumler, A. J., Tsolis, R. M. and Heffron, F. (1998). Multiple fimbrial adhesins are required for full virulence of *Salmonella typhimurium* in mice. *Infect. Immun.* 66: 2803–08.

Vibha, V., Ailes, E., Wolyniak, C., Angulo, F. and Klontz, K. (2006). Recalls of spices due to bacterial contamination monitored by the US Food and Drug Administration: The predominance of Salmonellae. *J. Food Prot.* 69: 233–37.

Wagner, G. J. (1991). Secreting glandular trichomes: More than just hairs. *Plant Physiol.* 96: 675–79.

Walker, T. S., Bais, H. P., Grotewold, E. and Vivanco, J. M. (2003). Root exudation and rhizosphere biology. *Plant Physiol.* 132: 44–51.

Wang, Q., Suzuki, A., Mariconda, S., Porwollik, S. and Harshey, R. M. (2005). Sensing wetness: A new role for the bacterial flagellum. *EMBO J.* 24: 2034–42.

Warriner, K. and Namvar, A. (2010). The tricks learnt by human enteric pathogens from phytopathogens to persist within the plant environment. *Curr. Opin. Biotechnol.* 21: 131–36.

Watanabe, Y., Ozasa, K., Mermin, J. H., Griffin, P. M., Masuda, K. *et al.* (1999). Factory outbreak of *Escherichia coli* O157:H7 infection in Japan. *Emerg. Infect. Dis.* 5: 424–28.

Wells, J. M. and Butterfield, J. E. (1997). *Salmonella* contamination associated with bacterial soft rot of fresh fruits and vegetables in the market place. *Plant Dis.* 81: 867–72.

WHO. (2000). Seventh report of WHO surveillance program for control of foodborne infections and intoxications in Europe. *FAO/WHO Collab. Cent. Res. Train. Food Hygiene Zoonoses.*

WHO. (2003). Eighth report of WHO surveillance program for control of foodborne infections and intoxications in Europe. *Collab. Cent. Res. Train. Food Hygiene Zoonoses.*

Widdowson, M-A., Sulka, A., Bulens, S. N., Beard, R. S., Chaves, S. S. *et al.* (2005). Norovirus and foodborne disease, United States, 1991–2000. *Emerg. Infect. Dis.* 11: 95–102.

Winfield, M. D. and Groisman, E. A. (2003). Role of nonhost environments in the lifestyles of *Salmonella* and *Escherichia coli*. *Appl. Environ. Microbiol.* 69: 3687–94.

Winthrop, K. L., Palumbo, M. S., Farrar, J. A., Mohle-Boetanim, J. C., Abbott, S. *et al.* (2003). Alfalfa sprouts and *Salmonella* Kottbus infection: A multistate outbreak following inadequate seed disinfection with heat and chlorine. *J. Food Prot.* 66: 13–17.

Wright, K. M., Chapman, S., McGeachy, K., Humphris, S., Campbell, E. *et al.* (2013). The endophytic lifestyle of *Escherichia coli* O157:H7: Quantification and internal localization in roots. *Phytopathology* 103: 333–40.

Yao, J. and Allen, C. (2006). Chemotaxis is required for virulence and competitive fitness of the bacterial wilt pathogen *Ralstonia solanacearum*. *J. Bacteriol.* 188: 3697–3708.

You, Y., Rankin, S. C., Aceto, H. W., Benson, C. E., Toth, J. D. *et al.* (2006). Survival of *Salmonella enterica* serovar Newport in manure and manure-amended soils. *Appl. Environ. Microbiol.* 72: 5777–83.

Zipfel, C., Robatzek, S., Navarro, L., Oakeley, E. J., Jones, J. D. G. *et al.* (2004). Bacterial disease resistance in *Arabidopsis* through flagellin perception. *Nature* 428: 764–67.

Section II

Specific Plant Diseases

9 Description of Plant Diseases

9.1 BACTERIAL BLIGHT OF RICE

Bacterial blight (hereafter referred to as BB) is one of the most devastating diseases of rice (*Oryza sativa* L.) due to its high epiphytotic potential and its destructive nature to nitrogen-responsive high-yielding cultivars in both temperate and tropical climates. Its occurrence in 1970 in Africa has added a new dimension to concerns about the transmission and dissemination of its causal bacterium. It has also been a serious problem on hybrid rice. In the 1980s, when hybrid rice was first released in China, epiphytotics occurred widely because of the lack of resistance in the hybrid rice, resulting in severe yield losses.

It is probably one of the most extensively studied diseases and a large number of research publications, dealing with different aspects of the pathogen and the disease, have been published. Mew *et al.* (1993), in their paper "Focus on bacterial blight of rice", have provided information on various aspects of the pathogen and the disease. Several review papers containing valuable information on the disease have also been published by workers from different countries. Two such review papers have also appeared from India (Srivastava, 1972; Thind, 2002). A review paper by Niño-Liu *et al.* (2006) provides very useful information on different aspects of the causal organism of the disease. For additional information, the readers may also see the annual reports of the International Rice Research Institute, Manila, Philippines and Central Rice Research Institute, Cuttack, India.

The disease was first noticed by farmers in the Fukuoka Prefecture of Japan in 1884 and it was called **white withering disease**. Initially, it was believed to be caused due to a reaction of an acidic soil. In 1908, Takaishi found bacterial masses in dewdrops of infected rice leaves, successfully isolated a bacterium and proved its pathogenicity but did not name the causal agent. Bokura, in 1911, also isolated a bacterium, proved its pathogenicity and named it *Bacillus oryzae* Hori and Bokura. Reitsma and Schure (1950) studied a rice disease called '**Kresek**' in Indonesia and considering it different from bacterial blight, named the pathogen as *Xanthomonas kresek*. Subsequently, Goto (1964) reported the **pale-yellow leaf phase** of the disease, which occurs at a later stage of plant growth. Kresek phase and pale-yellow leaf phase are not different diseases; rather, they are different phases of bacterial blight.

The disease is widely spread and occurs in the Philippines, Thailand, China, Indonesia, Taiwan, Malaysia, Cambodia, Vietnam, Myanmar, Laos, Bangladesh, Mexico, South East Russia, Australia, West Africa, Latin America, the Caribbean, Pakistan and India. However, the occurrence of the disease has not been confirmed in the United States and has not yet been found in Europe. Quarantines for the BB pathogen are in place in the United States and other rice-growing countries where the disease is not endemic but also in places where it is present, to prevent the introduction of new virulent strains. It is also acknowledged as a European and Mediterranean Plant Protection Organisation (EPPO) A1 quarantine pest.

In India, the disease was first reported by Sreenivasan *et al.* (1959) from Maharashtra, but the symptoms described were not exactly those of bacterial blight. Bhapkar *et al.* (1960) reported correct symptoms of the disease, but they did not isolate the bacterium. Later, Srivastava and Rao (1963) reported its epiphytotic on cv. BR 34 from Shahabad district of Bihar. During this period, a Formosan rice variety, Taichung Native 1, a high nitrogen-responsive but resistant to lodging was introduced into India to increase rice production. This variety was highly susceptible to BB. With the introduction of this variety, BB, which had been unknown in all but two states of India to that date, became pandemic within a couple of years. Rao and Shekhawat (1972) have given a brief account of this disease dealing with its different aspects.

Although the disease occurs in many countries, it is most destructive in Southeast Asia. The yield losses vary with host cultivar, age and nutrition of the host and environmental conditions. These are usually high in early infected crops and the occurrence of kresek phase may lead to a total failure of the crop. Losses due to BB increased significantly following the widespread cultivation of nitrogen-responsive, high-yielding, semi-dwarf rice cultivars derived from Taichung Native 1 and its derivatives in the 1960s. Prior to the introduction of resistant varieties and the implementation of strict quarantine measures in Japan, losses generally ranged from 20–30% but in some cases rose as high as 50% (Ou, 1972). Reports from the Philippines, Indonesia and India estimate that losses due to kresek phase have reached 60–75%, depending on weather, location and rice variety (Ou, 1985; Reddy *et al.*, 1979). Ray and Sengupta (1970) also reported losses up to 50% from semi-dwarf high-yielding varieties. Singh *et al.* (1977) reported that four rice cultivars with a similar leaf score of 9 showed variable losses in grain yield. Reddy *et al.* (1979) established a relationship between the severity of BB at different growth stages of host plant and rice yield and found that early epiphytotics reduced grain yield, panicle fertility and kernel weight while late epiphytotics had no measurable effect on grain yield. They also devised a scale for calculating BB-associated losses. In addition to reduction in yield, the disease may also affect grain quality by interfering with grain maturation.

Pantoea ananatis is a major pathogen that causes the new bacterial blight in rice, and its symptoms are very

DOI: 10.1201/9781003681878-11

similar to those of rice bacterial blight. Aiying *et al.* (2023) devised an early and rapid visual detection method for *Pan. ananatis*. Using *GyrB* gene as the target sequence, an innovative recombinase-aided amplification detection system integrated with a lateral flow dipstick (RAA-LFD) was constructed. The specificity and sensitivity tests showed that the RAA-LFD method could detect *Pan. ananatis*, whereas other common plant pathogens were not detected, and its detection sensitivity for *Pan. ananatis* DNA reached 100 copies/μL. The detection of diseased tissues indicated that this method could accurately detect *Pan. ananatis* in artificially inoculated rice tissues in the early stages of infection before the appearance of symptoms. The RAA-LFD detection system established in this study is simple and fast, with visible results, excellent specificity and high sensitivity. It is semi-quantitative and should be used for the early detection and rapid field diagnosis of new leaf blight, which provides technical support for the early warning and real-time detection of field samples.

9.1.1 Symptoms

It is a typical vascular disease. The below-given three different types of symptoms appear.

9.1.1.1 Leaf Blight Phase

These are the most common symptoms of the disease and generally appear 4–6 weeks after transplanting. Initially, the symptoms start as small, green, water-soaked spots at the tips and margins of fully developed leaves. The spots expand forming linear, yellow to straw-coloured stripes having wavy inner margins. The stripes appear generally on both the margins and only rarely on just one margin of the leaves. The stripes progress rapidly, extending both lengthwise and crosswise. The drying and twisting of the leaf tips occur (Figure 9.1a). Occasionally, linear stripes may develop anywhere on the leaf lamina or midrib with or without marginal stripes. The blighting extends to leaf sheaths, culms, killing the tillers. In wet weather, yellow- or amber-coloured drops of bacterial ooze appear on young lesions at the margins of the leaves. On drying, these form yellowish hard beads. The symptoms on panicles appear as grey to light brown lesions on glumes, resulting in poor fertility and low-quality grains. The symptoms of BB often occur simultaneously with those of bacterial leaf streak and individual leaves may show symptoms of both the diseases.

9.1.1.2 Kresek (Wilt) Phase

This is the most destructive phase of the disease and results from the systemic infection of plants in the nursery or an early infection in the main field. In an infected plant, the bacterium spreads through the xylem of the infected leaf to the culm and infects the crown (base of other leaves). Subsequently, all the tillers of the plants are infected, resulting in the rapid wilting of the entire plant. The leaves of the affected plants roll up completely, droop and turn yellow or grey. Ultimately, the diseased tillers wither away and, in severely infected plants, stools are killed. The plants that survive, exhibit arrested growth of tillers and stunted appearance (Figure 9.1b). Kresek phase is common in tropical countries. Temperature range of 28–34° C favours kresek development. The kresek phase is generally confused with attacks by stem borers. However, in the latter case, affected tillers/panicles can be easily pulled out and the detached part of the tiller shows the signs of insect feeding at the broken end.

9.1.1.3 Pale Yellow Leaf Phase

The youngest leaves in a clump turn pale yellow or whitish. The diseased leaves later turn yellowish brown and the tillers do not grow fully.

The disease can be easily diagnosed on the basis of the above-given symptoms. However, the confirmation of the diagnosis can be done by observing the bacterial ooze in the laboratory. In the field, diagnosis of the disease can be done by placing infected leaf bits in a glass containing clean water. If the water turns turbid after some time, presence of the disease is confirmed. However, further confirmation of the diagnosis can be done by observing the bacterial ooze in the laboratory.

The symptoms of bacterial blight on rice leaves, similar characteristics to those caused by *Xan. oryzae* pv. *oryzae*, have been reported to be caused by *Pantoea ananatis* and *Pan. stewartii* in Togo (Kini *et al.*, 2017).

(a) (b)

FIGURE 9.1 Bacterial blight of rice: (a) Leaf blight phase and (b) Kresek phase.

9.1.2 Causal Organism

Xanthomonas oryzae pv. *oryzae* (Ishiyama 1922) Swings *et al.* 1990
Syn. *Xan. oryzae*, *Xan. campestris* pv. *oryzae*

The bacterium was first named as *Bacillus oryzae* Hori & Bokura. Ishiyama (1922) later renamed it as *Pseudomonas oryzae* Uyeda & Ishiyama according to Migula's system of classification. In 1927, it was renamed *Bacterium oryzae* (Uyeda & Ishiyama) Nakata according to E. F. Smith's concept and subsequently *Xanthomonas oryzae* (Uyeda & Ishiyama) Dowson with the creation of *Xanthomonas* genus. With the introduction of the pathovar concept, the name *Xan. campestris* pv. *oryzae* (Ishiyama) Dye was adopted (Dye, 1978). Swings *et al.* (1990), on the basis of phenotypic, genotypic and chemotaxonomic data, proposed *Xanthomonas oryzae* (ex Ishiyama, 1922) sp. nov., nom. rev. and renamed the blight pathogen as *Xanthomonas oryzae* pv. *oryzae* (Ishiyama) Swings *et al.* (hereafter referred to as *Xoo*). The following description of *Xan. oryzae* (Ishiyama 1922) sp. nov., nom. rev. is taken primarily from Swings *et al.* (1990).

The cells are straight rods (0.4–0.8 × 1.5–2.9 μm) with rounded ends, Gram-stain-negative, non-sporulating and motile with a single polar flagellum. Cells occur singly, in pairs or sometimes in chains; filaments may also occur. The bacterium is obligately aerobic and catalase positive. The colonies on solid media containing glucose are round, convex, mucoid and yellow in colour due to the production of xanthomonadin. Xanthomonadin, though not required for virulence, protects the bacterium from UV light and probably contributes to its survival in the field. Cells produce copious, capsular, extracellular polysaccharide (EPS). EPS plays a role in virulence and *Xoo* mutants lacking this ability are severely attenuated in virulence. The tests for indole and 2-ketogluconate formation, urease, egg yolk hydrolysis, nitrate reduction and oxidase are negative. Starch is hydrolysed and carbon sources are used only oxidatively. Growth occurs on many carbon compounds. Acid is produced from D-glucose, D-fructose, D-galactose, trehalose and cellobiose but not from D-ribose, L-rhamnose, maltose, lactose, salicin, inulin, sorbitol, *meso*-inositol, dulcitol and adonitol. Litmus milk is not acidified but H_2S is produced. Esculin, Tween 40 and 80 are hydrolysed. No growth occurs at 4 or 35° C or in the presence of 3% NaCl, but poor growth occurs at 10 and 32° C. Optimum temperature for growth is between 25 and 30° C. The G + C contents of the DNA range from 64.6 to 65.0 mol%.

Xoo differs from *Xan. oryzae* pv. *oryzicola* in causing different types of symptoms on rice. *Xoo* also differs from pathovar *oryzicola* in not producing acetoin, in its inability to use L-alanine as a sole source of carbon and in being unable to grow on 0.2% vitamin-free casamino acids. *Xan. oryzae* pv. *oryzicola* is positive for all the above- given three characteristics. *Xoo* is able to grow in the presence of 0.001% cupric nitrate while *Xan. oryzae* pv. *oryzicola* is unable.

Xu *et al.* (2010) obtained 534 single-colony isolates of *Xoo* from rice leaves showing bacterial blight symptoms, collected from southern China in 2007 and 2008, and tested these isolates on plates for sensitivity to streptomycin. Four strains (0.75%) isolated from the same county of Yunnan province were resistant to streptomycin, and the resistance factor (the ratio of the mean median effective concentration inhibiting growth of resistant isolates to that of sensitive isolates) was approximately 226. The resistant isolate also showed resistance to streptomycin *in vivo*. Besides resistant isolates, isolates of less sensitivity were also present in the population of *Xoo* from Yunnan province. Mutations in the *rpsL* (encoding S12 protein) and *rrs* genes (encoding 16S rRNA) and the presence of the *strA* gene accounting for streptomycin resistance in other phytopathogenic or animal and human pathogenic bacteria, were examined on sensitive and resistant strains of *Xoo* by polymerase chain reaction amplification and sequencing. Neither the presence of *strA* gene nor mutations in *rpsL* or *rrs* genes were found, suggesting that different resistance mechanisms are involved in the resistant isolates of *Xoo*.

Elshakh *et al.* (2016) reported that the culture filtrates of three *Bacillus* strains, namely, rice seed-associated strain *Bac. subtilis* A15, rhizobacterial strains *Bac. amyloliquefaciens* D29 and *Bac. methylotrophicus* H8 significantly suppressed the growth and biofilm formation of *Xoo*, along with changes in bacterial cell morphology such as the swelling of cells and severe cell wall alterations. Polymerase chain reaction (PCR) analysis revealed that all three strains harbour the antimicrobial-associated genes that are responsible for biosynthesis of bacillomycin, fengycin, iturin and surfactin. Subsequent real-time quantitative (q)PCR analysis revealed the upregulated expression of *fenD* and *srfAA* genes in D29 and H8, and *fenD* and *ituC* genes in A15 during their *in vitro* interaction with *Xoo*. It indicates that the antibacterial mechanisms of these strains may be at least partially associated with their ability to secrete corresponding lipopeptides. Further, the application of these strains under greenhouse conditions was found to be effective in controlling bacterial blight, which was achieved through the activation of inducing systemic resistance, resulted from the enhanced activities of defence-related enzymes. This is the first report of demonstration of the mode of antibacterial effect of *Bacillus* strains against *Xoo*.

Lee *et al.* (2005) and Ochiai *et al.* (2005) sequenced the genomes of two *Xoo* strains, namely, Japanese strain MAFF 311018 (also called T 7174) and Korean strain KACC 10331 (also called KXO 85), respectively. Salzberg *et al.* (2008) have reported the sequencing of genome of a third strain of the bacterium, PXO 99^A. The genome of PXO 99^A is a single circular chromosome of 5,240,075 bp and considerably longer than the genomes of other two strains (4,941,439 bp for KACC 10331 and 4,940,217 bp for MAFF 311018). The genome of PXO 99^A contains 5083 protein coding genes, including 87 not found in the genomes of other two strains. The findings of Salzberg *et al.* (2008) provide striking evidence of **genome plasticity and rapid evolution** within *Xoo*. The comparisons point to sources of genomic variation and candidates for the strain-specific adaptation of this pathogen that help to explain the extraordinary diversity of *Xoo* genotypes and races that have been isolated throughout the world.

The interactions between *Xan. oryzae* pv. *oryzae* and the rice plant follow the gene-for-gene hypothesis to some extent, making the *Xan. oryzae* pv. *oryzae*-rice pathosystem an excellent model platform in plant sciences (Zhang and Wang, 2013). Among the identified *Xan. oryzae* pv. *oryzae* strains, PXO99^A, also known as Philippine race 6, is widely acknowledged as the most representative type strain of *Xan. oryzae* pv. *oryzae* and important for differentiating *Xan. oryzae* pv. *oryzae* strains in laboratories and in rice disease resistance breeding. In 2008, Salzburg and his associates published the first version of the complete genome of PXO99^A, which contains a unique, near-perfect tandem repeat of 212 kb close to the replication terminus (Salzberg *et al.*, 2008). In 2015, Booher and his colleagues confirmed that there are two genomic structures in the progenies from the original PXO99^A, with or without the 212-kb duplication, by using long-range sequencing and specific PCR verification (Booher *et al.*, 2015). The genome of the PXO99^A duplication version, designated the PXO99^A-s accession, has widely been used as a benchmarking reference genome in relevant research around the world (Booher *et al.* 2015; Salzberg *et al.* 2008). Li *et al.* (2022) obtained a substrain of PXO99^A, named *Xan. oryzae* pv. *oryzae* PXO99^A-GX (shortened to GX99), from the College of Plant Protection at Nanjing Agriculture University, the first institute which had imported the strain PXO99^A used in China. In their mutational gene manipulations in PXO99^A-GX, they found that the phenotypic changes in mutants of genes in the putative 212-kb duplication region had no significant differences from that of the genes in the other genomic regions, which suggested that PXO99^A-GX is not a PXO99^A-s accession (Booher *et al.*, 2015). In order to clarify the relationship between this substrain and the original strain PXO99^A, Li *et al.* (2022) sequenced the complete genome of PXO99^A-GX using PacBio and Illumina sequencing.

Xan. oryzae pv. *oryzae* PXO99^A-GX was grown in PSG medium (peptone 10 g, sucrose 10 g and glutamic acid 1 g L^{-1}) for 36 h, approximately up to the early stationary phase. Genomic DNA was isolated using HiPure Bacterial DNA Kits (Magen, Guangzhou, China). The quality of the isolated genomic DNA was assessed using Qubit and Nanodrop accordingly and was fragmented with G-tubes. The fragment size (>10 kb) was selected by the blue pippin system to prepare SMRTbell DNA template libraries. DNA sequencing was performed on the Pacific Biosciences Sequel (PacBio, Menlo Park, CA, USA) by Genedenovo Biotechnology Co., Ltd. (Guangzhou, China). In all, 145,826 reads and 1,162,120,421 bp were obtained and the N_{50} length was 9,258 bp. The sequencing depth of PXO99^A-GX was around 230×. Continuous long reads obtained from SMRT sequencing were assembled *de novo* using Hifiasm software (version 0.16.0) (Cheng *et al.*, 2021). The reads from the Illumina platform were used to correct the genome sequences to improve the quality of the assembly by Pilon software (version 1.23) (Walker *et al.*, 2014). A circular chromosome of 5,028,051 bp was attained with 63.65% G + C content. The genome completeness and contamination were evaluated by CheckM v1.1.3 (Parks *et al.*, 2015). The genome annotation of PXO99^A-GX was carried out by the NCBI prokaryotic genome annotation pipeline (Tatusova *et al.*, 2016) and GeneMarkS-2 software (Lomsadze *et al.*, 2018). In total, 4,493 protein coding genes, 6 ribosomal RNAs, 53 transfer RNAs and 147 soluble RNAs were predicted by the pipeline.

Zhao *et al.* (2007) have developed a novel and highly sensitive real-time method for the identification of *Xoo* based on a TaqMan probe, which is designed to recognise the sequence of a putative siderophore receptor gene cds specific to *Xoo* and can be identified from either a bacterial culture or naturally infected rice seeds and leaves in about 2 h. The sensitivity of the method is 100 times higher than that of the currently used polymerase chain reaction gel electrophoresis method for diagnosis.

Song *et al.* (2012) developed a triplex PCR method for the race-specific detection of *Xoo*, designing three primer sets: for the specific internal regions of two genes (*hpaA* and *XorII* very-short-patch-repair endonuclease) and for a genomic locus derived from an amplified fragment length polymorphism fragment specific for the K3 and K5 races. The pathogen was quickly detected in the asymptomatic rice leaf three days after inoculation and at a distance of 6 cm from the lesion site. This simple and rapid assay will be useful for the detection and identification of *Xoo* as well as for disease forecasting in paddy fields.

Yu *et al.* (2020) reported that the RpoN2-PilRX regulatory system controls bacterial motility and virulence by regulating type IV pilus gene transcription in *Xan. oryzae* pv. *oryzae*.

Mishra *et al.* (2025) reported that PthXo2B orthologue Tal7 of *Xan. oryzae* pv. *oryzae* strain IX-221 acts as a major virulence factor in *Indica* rice without activating a Clade III *SWEET* gene.

9.1.2.1 Pathogenic Variation

A high degree of race-cultivar specificity occurs in the host–pathogen interactions. The first indication of host specialisation in *Xoo* came when Asakaze, a resistant rice cultivar, was unexpectedly found to have been severely attacked in the Kyushu Island areas of Japan. This prompted active research on this aspect. Sato *et al.* (1976) analysed the distribution of the bacterial population in Kyushu area and divided it into two races: one attacked all differential cultivars, including Wase Aikoku 3; the other attacked Kinmaze and Wase Aikoku 3 but not the other cultivars. In 1977, Yamamoto *et al.* tested 71 isolates from Indonesia on Kozaka's differential cultivars (included Rantaj-emas along with others) and found no isolates belonging to Kozaka's groups (Kozaka, 1969) I or II but 46 isolates with virulence similar to group III in Japan, i.e., virulent to Rantaj-emas. They also identified two more groups, IV and V; the last two groups showed pathogenicity similar to the two races identified by Sato *et al.* (1976). These results not only demonstrated pathogenic specialisation/variation but also showed a strong interaction between rice cultivars and strains of the pathogen. At present, there are more than 30 races of the pathogen reported from different countries (Niño-Liu *et al.*, 2006). A set of races identified in the Philippines using five differential rice cultivars (Mew, 1987), has been used widely for identifying and classifying resistance to the pathogen in other cultivars.

Lore *et al.* (2011) found that the current *Xoo* population from the Punjab state in India is highly variable and can be classified into seven distinct pathotypes (PbXo-1 to PbXo-7) on the basis of differential reactions on a set of near-isogenic lines in the background of IR24 and some international, national and regional cultivars. PbXo-7, the most dominant pathotype, was found to be virulent and induced susceptible/moderately susceptible reaction on 22 of the 40 test genotypes, followed by PbXo-1, PbXo-5 and PbXo-6; PbXo-2 was the least virulent pathotype. Chen *et al.* (2016) reported a new race that was highly virulent on a near-isogenic line CBB23 (*Xa23*) derived from a cross between a wild rice *O. rufipogon* accession (RBB16) and a susceptible *indica* rice variety Jingang 30. Wang *et al.* (2014) reported this line as being resistant to earlier races of the pathogen in the province.

Kogeethavani *et al.* (2021) used nine near-isogenic rice Lines (NILs) carrying single bacterial blight (BB) resistance genes (*Xa*) to characterise the virulence of 40 *Xan. oryzae* pv. *oryzae* isolates collected from Malaysian rice granaries. The results revealed a highly diverse pathogenic profile, consisting of 21 distinct *Xan. oryzae* pv. *oryzae* races named R1 to R21. Race R20, followed by R16 were identified as dominant, showing the highest frequency. Although no *Xa* genes have been used in Malaysian rice granaries to date, unexpectedly race R20 was found virulent to most of the NILs carrying the *Xa* genes. However, the wide distribution of avirulence gene *avrxa5* and *avrXa7* among the isolates indicated that the population was most effective to NILs carrying the *xa5* and *Xa7* genes.

9.1.3 Disease Cycle

9.1.3.1 Perpetuation

It has been carefully studied in temperate regions, particularly in Japan (Tagami and Mizukami, 1962; Mizukami, 1966). The following sources help in the survival of the bacterium in different regions of the world.

9.1.3.1.1 Weed Hosts

In Japan, natural infection is found on weeds such as *Leersia* spp. and *Zizania latifolia*; of these, *L. sayanuka* is the most important source of perpetuation. The bacterium survives in the rhizosphere of this weed and multiplies in early spring when new shoots of the weed begin to emerge. From this weed, the bacterium is dispersed through irrigation water to the nursery or transplanted fields of rice. Gonzalez et al. (1991) reported that L. hexandra acts as a symptomless carrier of the bacterium. In India, Brar and Thind (1994) reported *Echinochloa colonum* to be a natural host of the bacterium and also proved that seeds from infected plants transmit the bacterium to the seedlings. In the tropics, *Leptochloa* spp., *Cyperus difformis* and *C. rotundus* also act as natural hosts of the bacterium. Other hosts of the bacterium include several species of wild rice (*O. rufipogon*, *O. australiensis*) and a number of weeds, such as *Cynodon dactylon*, *Cenchrus ciliaris*, *Brachiaria mutica*, *E. crusgalli*, *Panicum maxicum* and *Paspalum scrobiculatum*.

9.1.3.1.2 Rice Straw

The bacterium can perpetuate in piles of rice straw kept outdoors in temperate regions (Isaka, 1962). In tropical areas like Punjab, India, the bacterium has been found to survive in rice straw kept in big heaps protected from weathering agencies, but this source does not play any role in the perpetuation of the bacterium (personal observation).

9.1.3.1.3 Rice Stubbles

The bacterium survives in the stubbles (base of stems and roots) of the plants left behind in the field after harvest until the following spring in Japan. It is the most important source of infection in Thailand (Tabei and Eamchit, 1974).

9.1.3.1.4 Volunteer Plants

Infected seeds from the previous crop shed in the field give rise to diseased seedlings; these serve as a source of infection in double-cropped areas. In North India, off-type tall plants, usually infected with bacterial blight, are generally found in rice fields. As these plants mature before harvesting of the crop, they shed their seeds and these infected seeds initiate the disease in the subsequent year (Durgapal, 1985).

9.1.3.1.5 Rhizosphere of Succeeding Crops

Thind and Brar (1989), on the basis of two-year field trials, demonstrated that the bacterium survived in the rhizosphere of berseem, wheat and winter maize (sown in fields after harvesting bacterial blight infected rice) till the sowing of next rice crop. The survival in the rhizospheres of wheat and rapeseed has also been reported by Wakimoto (1956) and Mohiuddin *et al.* (1977). The survival in the rhizosphere of wheat and berseem in Punjab and other North Indian states, where rice is generally sown after these crops, has an important bearing because most of the rice acreage in these areas is sown after harvesting wheat and berseem.

9.1.3.1.6 Seed

The seed-borne nature of the bacterium was first reported by Fang *et al.* (1956) and, according to them, the bacterium was present mainly in glumes and occasionally in the endosperm. Since then, the seed-borne nature of the bacterium has remained controversial, with some workers reporting positive results while others reported negative findings. Different methods, like the isolation of the bacterium from seed, the inoculation of seed macerate on young seedlings, bacterial streaming from seed parts and the seedling symptom test, have been used to prove the seed-borne nature. Of these, the seedling symptom test is considered best, since it is both natural and reliable.

Srivastava and Rao (1963, 1964) and Singh (1972) reported positive seed-plant transmission of the bacterium. Reddy (1983), using nine-month-old seeds, observed typical blight symptoms 60 days after sowing. Brar and Thind (1995) developed a technique called, **"Nitrogen-Supplemented Seedling Symptom Test"** (NSSST) using irrigation water containing 300 μg of nitrogen mL^{-1}, to prove seed-plant transmission of the bacterium. The typical symptoms of the disease appear around 10–12 days after sowing in NSSST. Thind and Brar (1998) reported that the bacterium survived in all the seed samples, till the sowing of the next rice crop. These seed samples were collected from naturally infected crop from different places in Punjab, India and tested at monthly intervals. However, the number of seeds showing positive seed-plant transmission decreased with storage period. The seed-borne inoculum is a major source of perpetuation of the bacterium in Punjab and possibly in other northwestern states of India where only one crop of rice is taken in a year.

Tian *et al.* (2014), based on 16S–23S internal transcribed spacer ribosomal DNA sequence data, designed padlock probe (PLP), P-Xoo and tested to detect *Xoo* in rice seed by combining with dot-blot hybridisation. With this technique, a detection sensitivity of 1 pg of *Xoo* genomic DNA was observed. The technique facilitated the detection of *Xoo* in rice seed lots with 2% artificially infested seed. PLP assay detected *Xoo* in 39.3% (13 of 33) of naturally infested commercial rice seed lots. In contrast, conventional PCR using OSF1/OSR1 primers sets detected *Xoo* in 9.1% (3 of 33) of the same rice seed lots. The combining of both the padlock probes (P-Xoo for pv. *oryzae* and P-Xoc for pv. *oryzicola*) with reverse dot-blot hybridisation enabled to simultaneously detect *Xoo* and *Xan. oryzae* pv. *oryzicola* in naturally infested commercial rice seed lots. This technique can serve as an important tool to detect multiple pathogens in seed and thereby prevent the spread of important pathogens.

9.1.3.2 Disease Cycle

The bacterium enters the host through hydathodes, wounds, and growth cracks formed due to the development of new roots. In terms of infection, new wounds are more conducive than old wounds. The bacteria multiply in the intercellular spaces of the underlying epitheme, then enter and spread into the plant through the xylem. The pathogen moves vertically through the leaves through primary veins, but also progresses laterally through commissural veins. Within a few days, bacterial cells and extracellular polysaccharide (EPS) produced by the bacterium fill the xylem vessels and the bacteria ooze out from hydathodes, forming beads or strands of exudate on the leaf surface, which acts as a source of secondary inoculum. The EPS protects the bacterial cells from desiccation.

Temperature range of 25–31 °C is highly favourable for disease development. The kresek phase appears in 20 days when rice plants are kept at 31 °C, while it takes 40 or more days at 21 °C. Younger plants, i.e., those less than 21 days old, are the most susceptible and the susceptibility decreases with increase in age. The common practice of cutting the leaf tips of seedlings before transplanting, particularly when the seedlings are quite tall, plays an important role in the development of kresek symptoms because the cut ends serve as infection courts for the penetration of the pathogen. In addition, broken roots resulting from pulling the seedlings off the seed-bed also serve as entry points for bacteria present in flood-irrigated fields.

The disease is favoured by high levels of relative humidity. The epiphytotics of the disease are common in the rainy or monsoon season in India. The same varieties, which show high disease severity in rainy season, either escape infection or show traces to moderate infection in the *Rabi* season. Secondary spread of the pathogen occurs through wind-splashed rain, irrigation water and the contact of healthy leaves with diseased leaves. Wind-driven rains and persistent warm weather are highly favourable for the spread of pathogen and disease development. Severe epiphytotics often occur following typhoons, since the fierce winds, wind-blown rain and hail, all cause wounds on rice plants and disperse bacteria.

Heavy doses of nitrogenous fertilisers influence the disease incidence; even the source of nitrogen also influences disease severity. However, increase in disease intensity with increased levels of fertilisers differs with different cultivars. In general, nitrogen-induced susceptibility is expressed by susceptible cultivars while this is not the case with resistant cultivars. Reddy *et al.* (1979) showed that high levels of nitrogen (> 100 kg/ha) caused a reduction in yield due to increased bacterial blight incidence in susceptible variety IET 2895, whereas the resistant variety IET 4141 was the least affected by all the nitrogen levels. The increase in disease severity due to application of excess doses of nitrogen results from enhanced vegetative growth leading to high humidity.

The disease severity also increases under shade. In Punjab, India, it is a common sight to see high disease severity on plants growing under shade in transplanted rice (personal observation). The high disease severity results from high relative humidity and favourable temperatures, which are maintained for longer periods under shade.

Attempts have been made to forecast the disease by different methods based on climatic factors and counting bacterial and bacteriophage populations. The unusual changes in climate affect forecasting by all these methods. In temperate regions, the monitoring of bacteriophage strains specific to *Xoo* has been used for forecasting since the 1960s. The time when the phage population in irrigation water exceeds 200 particles mL^{-1} corresponds to the period 10 days before the initial development of the disease (Yoshimura *et al.*, 1960). In order to achieve an accurate forecast, the water sample should be taken in the morning at the out flow. However, the bacteriophage forecasting system is not practised extensively in tropical Asia, because rice cultivation is mostly rain fed, limiting the use of phage detection in paddy fields. In general also, disease forecasting has been of limited utility because no practical control measures are available.

Mondal *et al.* (2016) investigated the role of XopF, one of the conserved effectors in *Xoo*, using a null mutant developed through a PCR-based homologous recombination strategy. XopF regulated the growth of *Xoo* in plants and suppressed pathogen-associated molecular patterns-triggered immune (PTI) response in rice. *Xoo* wild, but not mutant, *Xoo* Δ*xopF*, produced intense blight lesions after inoculation by leaf clipping method, while *Xoo* Δ*xopF* showed significant reduction in colonisation of host in comparison to the wild strain. The relative expression analysis of PTI marker genes *PR10*, *OsWRKY13*, *OsRLCK16*, and *OsFLS2* showed that these genes were up-regulated 1.5- to 5-fold upon challenge inoculation with *Xoo*Δ*xopF*, indicating the role of XopF in suppressing PTI in rice. The findings show that XopF suppresses basal PTI response in plants and favours growth and pathogenic action of *Xoo* in rice.

Wu *et al.* (2017) identified 15 potential Tat (twinarginine translocation)-dependent translocation proteins (TDTPs) by using comparative proteomics and bioinformatics analyses. Combining systematic mutagenesis, phenotypic characterisation and gene expression, they further found that multiple TDTPs play key roles in *Xoo* adaption or virulence. In particular, four TDTPs were involved in virulence, three contributed to colonisation *in planta*, one had a key role in attachment to leaf surface, four were involved in tolerance to multiple stresses, and two were required for full swarming motility. These findings suggest that multiple TDTPs may have differential contributions to the involvement of the Tat pathway in *Xoo* adaption, physiology and pathogenicity.

Talreja and Nerurkar (2018) found that cinnamic acid, and its derivatives like ferulic acid, caffeic acid and p-coumaric acid, showed a significant reduction in promoter activity of xopQ, an effector secreted through type third secretion system by *Xoo* BXO43. They also found significant reduction in disease symptoms, confirming cinnamic acid, ferulic acid, caffeic acid and p-coumaric acid as effective virulence-attenuating agents of *Xoo*.

9.1.3.3 Host Resistance

The development of resistant cultivars carrying major resistance (*R*) genes has been found to be the most effective approach for managing BB. According to Wang *et al.* (2009), 31 genes imparting resistance to the BB pathogen have been identified. Most of these genes have been found in cultivars of *O. sativa* subspecies *indica* but some are also from *japonica* cultivars and from wild species of rice, such as *O. longistaminata*, *O. rufipogon*, *O. nivara*, *O. minuta* and *O. officinalis*. Some of these *R* genes, like *Xa21*, are effective only in adult plants, although most of these do not seem to be related to any specific plant growth stage. However, *Xa3* is typically effective only in adult plants, bearing at least one race, against which it is effective at all stages of growth. Some genes, like *Xa21* and *Xa31*, impart resistance against a wide range of *Xoo* races while others are effective against only one or a few races of the pathogen. Most *R* genes are dominant while some are recessive (*xa5, xa8, xa13, xa19, xa20, xa24* and *xa28*) while still others are semi-dominant.

The effectiveness of disease resistance genes is sometimes influenced by the plant developmental stage. Webb *et al.* (2010) reported that gene *Xa7* was not fully functional in very young seedlings, but it was completely effective 21 days after sowing. However, if a partial or fully functional *avrXa7* allele was present, *Xa7* resistance was effective at all growth stages tested after the transplanting stage, i.e., more than 21 days after sowing.

Near-isogenic lines (NILs) have been developed by introducing single *R* genes into the background of susceptible *indica* cultivar IR 24 to serve as differential lines. These differential lines are used for the identification of races of the bacterium. Rice differentials recommended by the scientists of the All India Coordinated Rice Improvement Programme of Directorate of Rice Research; Hyderabad are given in Table 9.1. Rice workers also include commercially grown rice varieties in their regions, as additional differentials, to have a better assessment of the pathogenic variation in the region.

The resistance of rice to specific *Xoo* races is governed by major *R* genes with a qualitative effect that impart complete resistance and polygenes with a quantitative effect, which impart partial resistance. It was believed that gene *Xa21*, derived from *O. longistaminata*, conferred resistance to all the pathotypes/races of *Xoo* from India and Philippines. However, Saini and Goel (1998) reported that even this growth stage-specific gene was ineffective in 70-day-old plants against some isolates of the bacterium from Punjab, India. Priyadarisini and Gnanamanickam (1999) also found that of 140 isolates of *Xoo* isolated from an epiphytotic-affected plants that occurred in Kerala, South India in 1998, 20 isolates were virulent on rice line IRBB 21 (carrying *Xa21* gene) but the rice line NH 56, which carried four resistance genes (*Xa4*, *xa5*, *xa13* and *Xa21*), was resistant to all the isolates. Sirisha *et al.* (2004) again reported the breakdown of resistance of *Xa21* against *Xoo* due to the possible gain of virulence by the pathogen in the process of co-evolution. Swamy *et al.* (2006) also reported that Pusa Basmati 1 transgenic lines expressing the *Xa21* gene were susceptible to all the six isolates collected from different regions of India while another line, MH 2R, into which three resistant genes, *xa5*, *xa13* and *Xa21*, were introduced through marker-assisted selection (MAS), showed resistant reaction to all the isolates. Sanchez

TABLE 9.1
Rice Differentials Recommended by the Scientists of All India Coordinated Rice Improvement Programme of Directorate of Rice Research, Hyderabad for Field Monitoring Virulences in *Xanthomonas oryzae* pv. *oryzae*

Sr. No.	Line/Differential	Gene Combination
1	IRBB 1	*Xa1*
2	IRBB 3	*Xa3*
3	IRBB 4	*Xa4*
4	IRBB 5	*xa5*
5	IRBB 7	*Xa7*
6	IRBB 8	*xa8*
7	IRBB 10	*Xa10*
8	IRBB 11	*Xa11*
9	IRBB 13	*xa13*
10	IRBB 14	*Xa14*
11	IRBB 21	*Xa21*
12	IRBB 50	*Xa4+xa5*
13	IRBB 51	*Xa4+xa13*
14	IRBB 52	*Xa4+Xa21*
15	IRBB 53	*xa5+xa13*
16	IRBB 54	*xa5+Xa21*
17	IRBB 55	*xa13+Xa21*
18	IRBB 56	*Xa4+xa5+xa13*
19	IRBB 57	*Xa4+xa5+Xa21*
20	IRBB 58	*Xa4+xa13+Xa21*
21	IRBB 59	*xa5+xa13+Xa21*
22	IRBB 60	*Xa4+xa5+xa13+Xa21*
23	DV 85	*xa5+Xa7*
24	Ajaya (IET 4141/CR 98-7216)	
25	RP B10-226	
26	TN 1	*Xa14*

Note: 1–9 scale of Standard Evaluation System for Rice (IRRI, 1996) is used to record the disease reaction.

et al. (2000) has already emphasised the usefulness of MAS for the pyramiding of bacterial blight resistance genes, particularly for recessive genes like *xa5* and *xa13* that are difficult to select through conventional breeding in the presence of dominant genes, such as *Xa21*. They also reported that MAS showed accuracies of 95% and 96% for identifying homozygous-resistant plants containing *xa5* and *xa13* genes, respectively. Host resistance conferred by single genes against *Xoo* is generally not stable due to considerable variation in the pathogen population. Therefore, pyramiding of BB resistance genes in rice is based on pathogen population response to these genes. Hence, the selected resistance genes are combined in a cultivar through MAS. This helps to develop varieties with broader-spectrum and longer-lasting resistance. Shanti *et al.* (2001) also found that three gene combinations, i.e., *Xa4 +xa5*, *xa5 + Xa21* and *Xa4 + xa5 + Xa21*, conferred a broad-spectrum resistance to all the strains evaluated (collected from Orissa, Madhya Pradesh and Uttar Pradesh of India), thereby supporting the strategy of pyramiding appropriate resistance genes.

Kanipriya *et al.* (2024) reported that the resistant genotypes not only harbour multiple bacterial blight resistance genes but also manifest up regulated defence responses against *Xoo*, possibly substantiating their elevated level of disease resistance.

Pradhan *et al.* (2016) developed pyramided lines carrying two BB resistance gene combinations (*Xa21 + xa13* and *Xa21 + xa5*) in a lowland cultivar Jalmagna background through backcross breeding by integrating molecular markers. The genotypes with a combination of two BB resistance genes conferred high levels of resistance to the predominant *Xoo* isolates prevalent in Odisha, India. The pyramided lines showed similarity with the recipient parent with respect to major agromorphological traits.

Yugander *et al.* (2017) collected 400 isolates of *Xoo* from diverse rice-growing regions of India and carried out their virulence profiling on a set of differentials consisting of 22 near-isogenic lines of IR24, possessing different BB resistance genes and their combinations along with the checks. They found that none of the single BB resistance genes could provide broad spectrum resistance in India. However, two-gene combinations like *xa5 + xa13* and different three- or four-gene combinations like *Xa4 + xa5 + xa13*, *Xa4 + xa13 + Xa21*, *xa5 + xa13 + Xa21* and *Xa4 + xa5 + xa13 + Xa21* are broadly effective throughout India.

Xa21 is the first resistance gene cloned in the monocots. It is a member of a multigene family in rice and of a larger family of genes in plants and animals involved in the recognition of pathogens and developing innate immunity. In addition to *Xa21*, other genes, such as *Xa1*, *xa5*, *xa13* and *Xa27*, have been cloned and represent gene families. The *avrXa7* and *avrXa10* corresponding to *R* genes *Xa7* and *Xa10*, respectively are the first avirulence genes of *Xoo* cloned from race 2 strain PXO 86. These avirulence genes are members of the large gene family of *avrBs3* and share an extensive sequence homology with *avrBs3* gene cloned from the pepper and tomato pathogen, *Xan. vesicatoria*. This gene family is widely distributed in *Xanthomonas* species, and *Xoo* possesses 12–14 copies. Type III secretion system-mediated secretion and the delivery of *avrXa10* into rice cells by *Xoo* have been demonstrated using a reporter protein fusion.

Dossa *et al.* (2016) reported that African rice, *O. glaberrima*, accessions contain bacterial blight resistance different from the *Xa* gene type and high temperature enhances the resistance of this species. Dossa *et al.* (2017) found that drought stress affected rice response to *Xoo*, and the response varied according to the rice genotype. They concluded that the evaluation of rice varieties under combined abiotic and biotic stresses will be the best strategy to determine biotic stress resistance durability under climate change.

Zhang *et al.* (2015) reported that a rice introgression line, FF329, derived from the cross between donor PSBRC66 and recipient Huang-Hua-Zhan, exhibited a typical hypersensitive response when inoculated with all 21 representative *Xoo* strains and found that BB resistance in FF329 was controlled by a single dominant gene. As FF329 showed broad-spectrum BB resistance, which differed from that of both the parents, the gene identified was novel and was designated *Xa39*. This gene can provide rice breeders with a new option to incorporate BB resistance into newly developed cultivars.

Chen *et al.* (2019) reported that *Xa21*-mediated resistance can be fully activated when the juvenile plants are grown at 27 °C under fluorescent light. Once transferred back to 31 °C under the same light intensity, *Xa21* seedlings lose the gained resistance. Temperature shift experiments indicate that high temperature treatment can suppress *Xa21*-activated resistance. It is also evident that the abundance of the *Xa21* protein is not significantly influenced by the temperature changes. These results highlight an interplay between development and temperature in this immune system. Full activation of resistance at juvenile stage will greatly facilitate the studies of *Xa21* immunity.

Li *et al.* (2020) identified PilN, a T4P (type IV pilus) alignment subcomplex protein, which is involved in the regulation of swimming motility. They analysed its contribution to bacterial surface-associated behaviours and virulence. They found that the *pilN* deletion mutant exhibited dramatically reduced twitching motility and scarcely detectable levels of T4P major pili PilA, as well as enhanced biofilm formation and exopolysaccharide (EPS) production. In addition, deletion of the *pilN* gene in *Xoo* resulted in impaired virulence in host rice and attenuated type III secretion system (T3SS) genes expression, which is independent of PilA assembly. Expression of the relevant *pilN* gene in trans was capable of restoring twitching motility and biofilm formation to the wild-type levels in the *pilN* mutant, but it resulted in only partially recovered EPS production and virulence. Moreover, the expression of *trh* and *xrvA* genes, which encode the HrpG-positive regulators, was decreased in the *pilN* mutant. The results suggest that PilN executes versatile functions in bacterial virulence and cell surface-associated behaviours.

Chen *et al.* (2020) reported that the treatment of rice plants with exogenous melatonin (20 μg mL^{-1}) increased nitrate reductase, nitric oxide synthase and peroxidase activity, enabling high intracellular concentrations of melatonin, nitric oxide and H_2O_2. The expression of *NPR1*, a key regulator in the salicylic acid signalling pathway, was upregulated more than 10-fold when the plants were challenged with melatonin. Similarly, the messenger RNA level of *PDF1.2*, a jasmonic acid-induced defence marker, was 15 times higher in the treated plants than in control plants. Moreover, three pathogenesis-related proteins, *PR1b*, *PR8a* and *PR9*, were upregulated 20-fold in the presence of melatonin. The application of melatonin (100 μg mL^{-1}) to soil-grown rice reduced the incidence of bacterial blight by 86.21%. These results not only provide a better understanding of melatonin-mediated innate immunity to *Xan. oryzae* pv. *oryzae* in rice but also offer a promising cultivation strategy to protect rice from *Xoo* infection. Zhang *et al.* (2019) reported that the loss of the *rocF* gene caused a significant decrease in the virulence of *Xoo* in the susceptible rice cultivar IR24. Bioinformatics analysis demonstrated that *rocF* encodes arginase. Quantitative real-time PCR and Western blot assays revealed that *rocF* expression was significantly induced by rice and arginine. The *rocF* deletion mutant strain showed elevated sensitivity to hydrogen peroxide, reduced extracellular polysaccharide production and reduced biofilm formation, all of which are important determinants for the full virulence of *Xoo*, compared with the wild-type strain. This study reveals a mechanism by which a bacterial arginase is required for the full virulence of *Xoo* on rice because of its contribution to tolerance to reactive oxygen species, extracellular polysaccharide production and biofilm formation.

Liu *et al.* (2021) developed two functional markers [Tetra primer amplification refractory mutation system polymerase chain reaction (ARMS-PCR) and competitive allele specific PCR (KASP)] based on the variation of two bases in the coding region of the *xa5* allele. The specificity and applicability of the two functional markers were verified in a breeding segregating population and a rice micro-core germplasm resource population. The authors concluded that these two newly-developed functional markers can effectively determine the homozygosity and heterozygosity of the target gene under different laboratory conditions. Hence, these can be used for genotyping the *xa5* gene in molecular breeding programmes, thereby improving breeding efficiency.

Rashid *et al.* (2021) collected 118 isolates of *Xoo* from 8 rice-growing regions of Bangladesh and inoculated these isolates on 13 near-isogenic lines of rice. Each line contained a single bacterial blight resistant gene, viz. *Xa1, Xa2, Xa3, Xa4, xa5, Xa7, xa8, Xa10, Xa11, xa13, Xa14, Xa21* and *Xa23*. Based on reaction patterns of these 118 *Xoo* isolates on near-isogenic lines, 12 pathotypes/races were identified for the first time in Bangladesh. Race 1, Race 2 and Race 3 were predominantly distributed in most areas of Bangladesh and considered as major races containing the maximum number of isolates (48%, 14% and 11%, respectively). Among these, Race 1 was the most prevalent and widely distributed while Race 5 was the most virulent, circumventing all of the resistance genes tested. Bacterial blight resistant gene *Xa21* showed a resistant reaction against 11 of 12 races (i.e. 94.91% of the isolates tested). Effective genes and information of races generated from this study can be deployed for the development of race-specific bacterial blight resistant varieties in Bangladesh.

Liang *et al.* (2015) reported that the thiadiazoles of zinc thiazole and bismerthiazol induced H_2O_2 accumulation, up-regulated defence-related genes, callose deposition and hypersensitive response-like cell death in rice leaves infected with *Xoo* strain ZJ173, but not in non-infected leaves. These defence responses in *Xoo*-infected leaves were suppressed by the exogenous application of catalase, which reduces H_2O_2 accumulation. The application of extracellular polysaccharides (EPSs) extracted from strain ZJ173 significantly compromised rice defence against ZJ173 with or without thiadiazole treatment. The EPS-deficient *Xoo* mutant Δ*gumH* triggered a stronger defence than its parent strain ZJ173. Thiadiazole treatments reduced EPS production by strain ZJ173, but not by the thiadiazole-resistant strain 2-1-1, which is thiadiazole-resistant *in vivo*, but not *in vitro*. However, enhanced defence was not detected in thiadiazole-treated rice inoculated with strain 2-1-1. Based on these data, it can be inferred that zinc thiazole and bismerthiazol promote rice defence against *Xoo* by inhibiting the production of bacterial EPS.

Nguyen *et al.* (2016) researched a knockout strain (*detR*⁻) of *Xoo*, lacking the *PXO_04659* gene which shows dramatically reduced virulence relatively to the wild-type. The elucidation of *detR* function in *Xoo* pathogenesis revealed a reduction in extracellular polysaccharide (EPS), intolerance to reactive oxygen species (ROS) and the deregulation of iron homeostasis in the *detR*⁻ strain. Moreover, gene expression of regulatory factors, including other response regulators and transcription factors, was altered in the absence of DetR protein, as determined by reverse transcription-PCR and/or real-time quantitative PCR analyses. It was concluded that DetR is essential for *Xoo* virulence through the regulation of the *Xoo* defence system, including EPS synthesis, ROS detoxification and iron homeostasis, solely or cooperatively with other regulatory factors.

Verma *et al.* (2018) showed that *Xoo* displays a positive chemotaxis response to components of rice xylem sap: glutamine, xylose and methionine. The authors performed a detailed deletion mutant analysis to know the role of chemotaxis components involved in the promotion of chemotaxis, entry and virulence. Analysis of mutants defective in chemotaxis components, flagellar angbiogenesis, expression analysis and assays of virulence-associated functions indicated that chemotaxis-mediated signalling in *Xoo* is involved in the regulation of several virulence-associated functions, such as motility, attachment and iron homeostasis. The Δ*cheY1* mutant of *Xoo* exhibited a reduced expression of genes involved in motility, adhesins and iron uptake and its metabolism. The expression of *Xoo* chemotaxis and motility components was induced under *in planta* conditions and it is required for entry, colonisation and virulence. Moreover, deletion analysis of a putative chemoreceptor *mcp2* gene revealed that chemoreceptor Mcp2 is involved in the sensing of xylem sap and constituents of xylem exudate, including methionine, serine and histidine, and that it plays an important role in epiphytic entry and virulence. This is the first report of the role of chemotaxis in the virulence of *Xoo*.

Cellobiosidase (CbsA) is one of several cell wall-degrading enzymes secreted by *Xoo* via the type II secretion system (T2SS) and it is considered a fundamental virulence factor for vascular pathogenesis. CbsA has an N-terminal glycosyl hydrolase domain and a C-terminal fibronectin type III (FnIII) domain. Interestingly, the secreted form of CbsA lacks the FnIII domain during *in planta* growth. Nathawat *et al.* (2022) showed that the presence of the FnIII domain inhibits the enzyme activity of CbsA on polysaccharide substrates like carboxymethylcellulose. The FnIII domain is required for the interaction of CbsA with SecB chaperone, and this interaction is crucial for the stability and efficient transport of CbsA across the inner membrane. Deletion of the FnIII domain reduced virulence similar to Δ*cbsA* Xoo, which corroborates the importance of the FnIII domain in CbsA. This finding elucidates a hitherto unknown function of the FnIII domain in enabling the virulence-promoting activity of CbsA.

Xue *et al.* (2021) isolated two enterobacteria from blighted rice leaves obtained from Sichuan and Guangdong provinces of China and identified these as, *Enterobacter asburiae* and *Pantoea ananatis*, as the causal agents. Of the two, *Entb. asburiae* was isolated from samples of both the provinces and *Pan. ananatis* was isolated only from the Sichuan samples. Different from rice foot rot pathogen *Dickeya zeae* EC1 and rice bacterial blight pathogen *Xoo* PXO99A, strains SC1, RG1 and SC7 produced rare cell wall-degrading enzymes but more extra polysaccharides. *Entb. asburiae* strains SC1 and RG1 produced bacteriostatic substances while *Pan. ananatis* strain SC7 produced none. Pathogenicity tests indicated that all of them infected monocotyledonous rice and banana seedlings, but not dicotyledonous potato, radish, or cabbage. Moreover, strain RG1 was the most virulent, while strains SC1 and SC7 were similarly virulent on rice leaves, even though strain SC1 propagated significantly faster in rice leaf tissues than strain SC7. This is the first record of *Entb. asburiae* as a new pathogen of rice bacterial blight, and, in some cases, *Pan. ananatis* could be a companion pathogen. Analysis on production of virulence factors suggested that both pathogens probably employ a different mechanism to infect hosts other than using cell wall-degrading enzymes to break through host cell walls.

Protein glycosylation is an essential process that plays an important role in proteome stability, protein structure and protein function modulation in eucaryotes. Wang *et al.* (2021) investigated the relationship between protein glycosylation and pathogenicity by using *Xoo*. In their previous work, the authors identified a virulence-related hypothetical protein, PXO_03177, but how this protein regulates *Xan. oryzae* pv. *oryzae* virulence remained unclear. They found that both PXO_03177 and PXO_02523 (a related OmpA family-like protein) play important roles in regulating pathogen virulence and viability under stressful environments. This work provides the first evidence that protein glycosylation is necessary for the virulence of plant pathogenic bacteria.

Li *et al.* (2023) investigated the role of HpaP homologues in *Xoo* and showed that HpaP is required for the full

virulence of *Xoo*. Deletion of *hpaP* in *Xoo* led to a reduction in virulence and alteration in the production of virulence factors, including extracellular polysaccharide and cell motility. Comparative transcriptomics and reverse transcription-quantitative PCR assays revealed that in the XVM2 medium, a mimic medium of the plant environment, the expression levels of *hrp* genes (for HR and pathogenicity) were enhanced in the *Xoo hpaP* deletion mutant compared to the wild type. The expression of *hrp* genes of *Xoo hpaP* mutant was reduced *in planta*. These findings indicate that HpaP plays a role in *Xoo* infection.

Hou *et al.* (2023) investigated the vitamin B6 (VB6) biosynthesis pathway in *Xoo* and its role in host–pathogen interactions. PdxA and PdxJ are key components for deoxyxylose 5-phosphate (DXP)-dependent VB6 biosynthesis. It was found that *Xoo* mutants containing mutations in these two putative genes (*pdxA* and *pdxJ*) were fully auxotrophic for VB6, demonstrating that *Xoo* VB6 *de novo* biosynthesis is DXP-dependent. The authors also found that mutation in these key genes (*pdxA* and *pdxJ*) significantly reduced *Xoo* virulence in rice. Hence, it is evident that VB6 de novo biosynthesis is essential for *Xoo* virulence.

Sun *et al.* (2023) identified and characterised *OsGLP3-7*, a member of the germin-like protein (GLP) family in rice. Expression of *OsGLP3-7* was significantly induced by pathogen infection, jasmonic acid (JA) treatment, and hydrogen peroxide (H_2O_2) treatment. *OsGLP3-7* was highly expressed in leaves and sub-localised in the cytoplasm. Overexpression of *OsGLP3-7* increased plant resistance to leaf blast, panicle blast and bacterial blight, whereas disease resistance in *OsGLP3-7* RNAi-silenced plants was remarkably compromised, suggesting this gene is a positive regulator of disease resistance in rice. Further work showed that *OsGLP3-7* has superoxide dismutase activity and can influence the accumulation of H_2O_2 in transgenic plants. Many genes involved in JA and phytoalexin biosynthesis were strongly induced, accompanied with elevated levels of JA and phytoalexins in *OsGLP3-7*-overexpressing plants, while expression of these genes was significantly suppressed and the levels of JA and phytoalexins were reduced in *OsGLP3-7* RNAi plants compared to control plants, both before and after pathogen inoculation. It also showed that *OsGLP3-7*-dependent phytoalexin accumulation may, at least partially, be attributed to the elevated JA levels observed after pathogen infection. These findings show that *OsGLP3-7* positively regulates rice disease resistance by activating JA and phytoalexin metabolic pathways, thus providing novel insights into the disease resistance mechanisms conferred by GLPs in rice.

Xoo modulates the rice SWEET expression by directly injecting transcription activator-like effector (TALEs) via a type III secretion system (TIIISS). However, *Xoo* TALEs have specificity for the particular rice SWEETs. The specificity is determined by the RVD region of the TALE. Thus, understanding of the rice *Xoo* interactions, which is essential for formulating disease management strategy, largely depends on the regulation of rice SWEETs proteins by the *Xoo* TALEs. The review article by Mondal *et al.* (2021) brings an overview of the rice SWEETs proteins and their regulation by the *Xoo* TALEs during bacterial blight infection.

Rashmi *et al.* (2024) examined genetic variations in the promoter region of *OsSWEET11*, *OsSWEET13* and *OsSWEET14* in 20 genotypes grown in India. Further sequence analysis of three genotypes having extreme response (resistance and susceptibility) to bacterial blight revealed that the promoter of the resistant genotype Pusa basmati 1692 is more closely clustered with the *xa13* and *xa25* resistance genes when compared with susceptible genotype Pusa basmati 1. Relative gene expression studies showed that, upon *Xoo* inoculation, the fold expression of clade III *OsSWEET* genes, namely *OsSWEET11*, *OsSWEET13*, *OsSWEET14* and *OsSWEET15*, increases in cultivars IR24, Pusa Basmati 1 and Improved Pusa Basmati 1. Intriguingly, after *Xoo* challenge inoculation, *OsSWEET15* expression was found to be increased along with other major *OsSWEET* genes, implying that it could be a novel target for a bacterial blight pathogen.

Sun *et al.* (2023) observed that the *Oscds5* mutant shows enhanced resistance to rice blast, bacterial blight and bacterial leaf streak. Mutation of *OsCDS5* gene promotes production of ROS and increases the expression level of multiple defence-related genes. Transcriptomic analyses indicate that genes involved in responses to stress, biotic/abiotic stimuli and metabolic processes are highly upregulated and enriched in mutant *Oscds5*. Metabolomic analyses showed that differential metabolites were enriched in the lipid metabolic and tryptophan metabolic pathways. The decreased level of phosphatidylinositol and increased level of serotonin probably contribute to enhanced disease resistance of the *Oscds5* mutant. Hence, *OsCDS5* may be a promising target for genetic engineering to enhance the resilience of rice to abiotic and biotic stresses simultaneously.

9.1.4 Management

The measures adopted for the management of BB vary depending on the location and disease incidence. The following measures can help in reducing the disease severity.

1. Destruction of diseased plant debris, including rice straw and stubbles.
2. Removal of weeds from canals and ridges.
3. Seed treatment before sowing is practised in tropical Asia. Seed can be treated with Streptocycline (100 µg mL^{-1}) + Agallol (0.1%) or Aretan/Tafasan (0.05%) followed by hot water treatment at 52–54° C for 30 minutes. Alternately, use the seed from a disease-free crop.
4. Nursery should be raised away from diseased fields and avoid its flooding with contaminated water. It should also not be sown under shade. Raising the nursery under upland dry conditions has been found effective in Sri Lanka.

5. Keep proper plant spacing and avoid high doses of nitrogenous fertilisers.
6. Chemical control of BB began in the 1950s with the prophylactic sprays of Bordeaux mixture. Since then, many fungicides, including copper and mercurial compounds, and several antibiotics have been tried singly and in combination but with limited success. In temperate regions, probenazole is applied in the irrigation water in nursery and in the transplanted rice to prevent the bacterial multiplication. The chemicals used for spraying include tecloftalam, phenazine oxide and nickel dimethyl dithiocarbamate. In the tropical monsoon climate of Asia, chemical sprays are not generally given due to the non-availability of effective commercial chemicals and difficulty in spraying due to uncertain weather. However, the following three publications suggest the possibility of managing the disease with chemical sprays. Foliar sprays of Blitox- 50WP (0.25%) + Streptocycline (100 μg mL^{-1}) have been found to give significant disease control along with significant increase in grain yield (Thind and Mehra, 1992; Sinha and Sinha, 2000). Tilak *et al.* (2001) obtained significant disease control along with significant increase in grain yield with 2 sprays of Bacterimycin–2000 (0.04%) in a multilocation trial.
7. Bacterial antagonists of *Xoo* have received special attention as biocontrol candidates. As a result, some success has been achieved in this respect. In India, more than 40 bacterial isolates antagonistic to the bacterium have been identified through *in vitro* and *in vivo* testing. Among them, rice-associated rhizobacteria, *Psm. fluorescens* and *Psm. putida* strain V14i, significantly suppressed the disease severity when sprayed on leaves (Sivamani *et al.*, 1987). The application of *Bacillus* species, as seed treatment before sowing, root-dip prior to transplanting and foliar sprays in the field, has also given encouraging results. In at least one study, the antagonists suppressed the disease by almost 60% and increased the plant height and grain yield by two-fold (Vasudevan *et al.*, 2002).

   Thind and Ahmad (1994) reported that, out of 32 microorganisms, isolated from phylloplane and spermosphere of 4 rice cultivars and from phylloplane of 4 rice weeds, 9 proved antagonistic to the bacterium in *in vitro* evaluation and these antagonists also gave significant control of the disease under field conditions for two consecutive years, along with an increase in grain yield. Of these nine antagonists, three antagonists, namely, *Bac. subtilis*, *Erwinia herbicola* and *Penicillium oxalicum*, gave significantly higher grain yield in addition to significantly better disease control in comparison to Streptocycline (100 μg mL^{-1}). *Bac. subtilis*, the most promising antagonist, gave 35.4% disease control and 10.55% increase in grain yield.

   Vidhyasekaran *et al.* (1996) reported that the treatment of rice seeds with talc-based formulation of *Psm. fluorescens* at 10 g kg^{-1} of seed-induced systemic resistance and gave protection up to 60 days against bacterial blight pathogen while the foliar sprays imparted resistance up to 15 days. Kamala and Vidhyasekaran (1998) tested talc-based formulation of *Psm. fluorescens* str. Pf 1 as seed treatment, root-dip application and foliar sprays individually and in combination, and found that the combination treatment induced resistance and gave 60–70% disease control in the field trials, along with an increase in grain yield. Despite these encouraging results, biocontrol agents have not attained a widespread use for the control of BB.
8. Ogunyemi *et al.* (2019) reported that, out of five *Xoo* bacteriophages, namely X1, X2, X3, X4 and X5 studied, phage X3 had the broadest host range, lysing 22 out of the 23 *Xoo* strains tested; it also had the longest latent period of 40 min with a burst size of 50-plaque forming units per infected cell. It caused 53% degradation of exopolysaccharide production, 43% degradation of biofilm, while the highest rate of bacteriophage-resistant bacterial colonies emerged at log 4 plaque-forming units mL^{-1} of strain Gz 0011. Application of this phage also decreased disease severity *in vivo*. It was more effective (83.1% decrease) if applied by spraying before pathogen inoculation rather than after pathogen inoculation (28.9–73.9% decrease). Seed treatment was also very effective and it caused 95.4% decrease in disease severity.
9. Use of resistant cultivars is the most effective and economical method to control bacterial blight of rice. However, the development of resistant varieties and durability of the resistance is complicated by the high diversity in the pathogen population. Resistant varieties, based on the pathogen population structure of different regions, have been developed and released for cultivation. Rice cultivars, namely, PR113, PR114, PR121, PR122, PR126, PR127, PR128, PR129, PR130, PR131 and PR132, developed by Punjab Agricultural University, Ludhiana, Punjab have been released for cultivation in Punjab, India (Anonymous, 2024). Pusa 1460, a variety developed by pyramiding resistance genes *Xa5* + *xa13* + *xa21* in the background of Pusa Basmati 1 through marker-assisted backcross breeding, has been released for cultivation in Punjab, Haryana, J & K, Delhi and Uttarakhand of India. Transgenic rice developed from cultivar IR 72 showed a high level of resistance to existing races of the pathogen in the Philippines (Mew and Vera Cruz, 2001). The growing of cultivar mixtures containing appropriate single resistance genes has also been tried for the management of the disease.

REFERENCES

Aiying, W., Ju, L., Cilin, W., Yuxuan, H., Baojun, Y. *et al.* (2023). Establishment and application of the Recombinase-Aided Amplification-Lateral Flow Dipstick detection method for *Pantoea ananatis* on rice. *Australas. Plant Pathol.* 52: 283–91.

Anonymous (2024). *Package of Practices for Kharif Crops.* Punjab Agricultural University, Ludhiana, pp. 2–3.

Bhapkar, D. G., Kulkarni, N. B. and Chavan, V. M. (1960). Bacterial blight of paddy. *Poona Agri. Coll. Mag.* 51: 36–46.

Booher, N. J., Carpenter, S. C. D., Sebra, R. P., Wang, L., Salzberg, S. L. *et al.* (2015). Single molecule real-time sequencing of *Xanthomonas oryzae* genomes reveals a dynamic structure and complex TAL (transcription activator-like) effector gene relationships. *Microb. Genom.* 1: e000032.

Brar, J. S. and Thind, B. S. (1994). A new weed host of *Xanthomonas oryzae* pv. *oryzae*, the causal agent of bacterial leaf blight of rice. *Ann. Plant Prot. Sci.* 2: 79–80.

Brar, J. S. and Thind, B. S. (1995). Nitrogen supplemented seedling symptom test- a new method for proving seed transmission of *Xanthomonas oryzae* pv. *oryzae*. *Oryza* 32: 42–45.

Chen, Q., Huang, X., Chen, X., Shamsunnaher and Song, W.-Y. (2019). Reversible activation of XA21-mediated resistance by temperature. *Eur. J. Plant Pathol.* 153: 1177–84. https://doi.org/10.1007/s10658-018-01634-6

Chen, X., Laborda, P. and Liu, F. (2020). Exogenous melatonin enhances rice plant resistance against *Xanthomonas oryzae* pv. *oryzae*. *Plant Dis.* 104: 1701–08.

Chen, X. L., Yan, Q., Li, R. F., Li, K. H. and Gao, L. J. (2016). First report of a new and highly virulent race of *Xanthomonas oryzae* pv. *oryzae*, the causal agent of bacterial leaf blight of rice in Guangxi province, China. *Plant Dis.* 100: 1492.

Cheng, H., Concepcion, G. T., Feng, X., Zhang, H. and Li, H. (2021). Haplotype-resolved de novo assembly using phased assembly graphs with hifiasm. *Nat. Methods* 18: 170–75.

Dossa, G. S., Oliva, R., Maiss, E., Cruz, C. V. and Wydra, K. (2016). High temperature enhances the resistance of cultivated African rice, *Oryza glaberrima*, to bacterial blight. *Plant Dis.* 100: 380–87.

Dossa, G. S., Torres, R., Henry, A., Oliva, R., Maiss, E. *et al.* (2017). Rice response to simultaneous bacterial blight and drought stress during compatible and incompatible interactions. *Eur. J. Plant Pathol.* 147: 115–27.

Durgapal, J. C. (1985). Self-sown plants from bacterial blight infected rice seeds- a possible source of primary infection in North West India. *Curr. Sci.* 54: 1283–84.

Dye, D. W. (1978). Genus IX *Xanthomonas* Dowson 1939. In: Young, J. M., Dye, D. W., Bradbury, J. F. *et al.* A proposed nomenclature and classification for plant pathogenic bacteria. *N. Z. J. Agric. Res.* 21: 162–66.

Elshakh, A. S. A., Anjum, S. I., Qiu, W., Almoneafy, A. A., Li, W. *et al.* (2016). Controlling and defence-related mechanisms of *Bacillus* strains against bacterial leaf blight of rice. *J. Phytopathol.* 164: 534–46.

Fang, C. T., Lin, C. F. and Chu, C. L. (1956). A preliminary study on the disease cycle of the bacterial blight of rice. *Acta Phytopathol. Sin.* 2: 173–85 (Ch. en).

Gonzalez, C. F., Xu, D. W. and Li, H. L. (1991). *Leersia hexandra*, an alternate host for *Xanthomonas campestris* pv. *oryzae* in Texas. *Plant Dis.* 75: 159–62.

Goto, M. (1964). 'Kresek' and pale yellow leaf systemic symptoms of bacterial leaf blight of rice caused by *Xanthomonas oryzae* (Uyeda et Ishiyama) Dowson. *Plant Dis. Reptr.* 48: 858–61.

Hou, Y., Cai, J., Lin, Y., Mi, D., Tao, J. *et al.* (2023). Vitamin B6 biosynthesis is essential for the virulence of *Xanthomonas oryzae* pv. *oryzae*. *Plant Pathol.* 72: 861–67.

IRRI (1996). *Standard evaluation system for rice.* 4th Edition, IRRI, The Philippines.

Isaka, M. (1962). Overwintering of bacterial leaf blight organism in damaged leaf in paddy field. *Proc. Assoc. Plant Prot.* Hokuriku 10: 90.

Ishiyama, S. (1922). Studies on the white leaf disease of rice plants. *Rept. Agri. Expt. Sta. Tokyo* 45, 233–51 [Ja.].

Kamala, N. and Vidhyasekaran, P. (1998). Evaluation of new biopesticide formulation for the management of some foliar diseases of rice. In: G. S. Dhaliwal, N. S. Randhawa, R. Arora and A. K. Dhawan (eds.). *Ecological Agricultural and Sustainable Development*, Vol. 2, pp. 382–88. Indian Society of Ecological Agriculture and Sustainable Development.

Kanipriya, R., Natarajan, S., Gopalakrishnan, C., Ramalingam, J., Saraswathi, R. *et al.* (2024). Screening for disease resistance and profiling the expression of defense-related genes contributing to resistance against bacterial blight (*Xanthomonas oryzae* pv. *oryzae*) in rice genotypes. *Physiol. Mol. Plant Pathol.* 131: 102286. https://doi.org/10.1016/j.pmpp.2024.102286

Kini, K., Agnimonhan, R., Afolabi, O., Milan, B., Soglonou, B. et al. (2017). First report of a new bacterial leaf blight of rice caused by *Pantoea ananatis* and *Pantoea stewartii* in Benin. *Plant Dis.* 101: 242.

Kogeethavani, R., Fatin, N. A., Suzianti, I. V. and Erwan, S. S. (2021). Characterization of pathogenic variability of *Xanthomonas oryzae* pv. *oryzae* isolates causing bacterial leaf blight disease in Malaysian rice granaries. *Australasian Plant Pathol.* 50: 293–98. https://doi.org/10.1007/s13313-020-00771-z

Kozaka, T. (1969). Control of rice diseases with resistant varieties. *Agric. Hortic.* 44: 208–12.

Lee, B. M., Park, Y. J., Park, D. S., Kang, H.-W., Kim, J.-G. *et al.* (2005). The genome sequence of *Xanthomonas oryzae* pathovar *oryzae* KACC 10331, the bacterial blight pathogen of rice. *Nucleic Acids Res.* 33: 577–86.

Li, R.-F., Ren, P.-D., Zhang, D.-P., Cui, P., Zhu, G.-N. *et al.* (2023). HpaP divergently regulates the expression of hrp genes in *Xanthomonas oryzae* pathovars *oryzae* and *oryzicola*. *Mol. Plant Pathol.* 24: 44–58.

Li, Y., Dan, X., Li, T., Wu, Q., Wang, Z. *et al.* (2022). Complete genome resource of a commonly used laboratory substrain of *Xanthomonas oryzae* pv. *oryzae* PXO99A. *Plant Dis.* 106: 1045–48.

Li, Y., Yan, Y., Deng, S., Zhang, C., Haq, F. *et al.* (2020). The *Xanthomonas oryzae* pv. *oryzae* type IV pilus alignment subcomplex protein PilN contributes to regulation of bacterial surface-associated behaviours and T3SS system. *Plant Pathol.* 69: 744–55.

Liang, X., Yu, X., Dong, W., Guo, S., Xu, S. *et al.* (2015). Two thiadiazole compounds promote rice defence against *Xanthomonas oryzae* pv. *oryzae* by suppressing the bacterium's production of extracellular polysaccharides. *Mol. Plant Pathol.* 16: 882–92.

Liu, Y., Wang, F., Zhang, A. Kong, D., Liu, G. *et al.* (2021). Development and validation of functional markers (Tetra-primer ARMS and KASP) for the bacterial blight resistance gene *xa5* in rice. *Australasian Plant Pathol.* 50: 323–27. https://doi.org/10.1007/s13313-021-00776-2

Lomsadze, A., Gemayel, K., Tang, S. and Borodovsky, M. (2018). Modeling leaderless transcription and atypical genes results in more accurate gene prediction in prokaryotes. *Genome Res.* 28: 1079–89.

Lore, J. S., Vikal, Y., Hunjan, M. S., Goel, R. K., Bharaj, T. S., *et al.* (2011). Genotypic and pathotypic diversity of *Xanthomonas oryzae* pv. *oryzae*, the cause of bacterial blight of rice in Punjab state of India. *J. Phytopathol.* 159: 479–87.

Mew, T. W. (1987). Current status and future prospects of research on bacterial blight of rice. *Annu. Rev. Phytopathol.* 25: 359–82.

Mew, T. W., Alvarez, A. M., Leach, J. E. and Swings, J. (1993). Focus on bacterial blight of rice. *Plant Dis.* 77: 5–12.

Mew, T. W. and Vera Cruz, C. M. (2001). Bacterial blight of rice. In: O. C. Maloy and T. D. Murray (eds.). *Encyclopedia of Plant Pathology*. Vol. 1, John Wiley and Sons, Inc., New York, pp. 71–74.

Mishra, P., Shakespear, S., Carpenter, S. C. D., Hamsa, S, Vigi, S. et al. (2025). PthXo2B orthologue Tal7 of *Xanthomonas oryzae* pv. *oryzae* strain IX-221 acts as a major virulence factor in indica rice without activating a clade III SWEET Gene. *Mol. Plant. Pathol.* 2025 Mar;26(3): e70067. https://doi.org/10.1111/mpp.70067

Mizukami, T. (1966). Epidemiology of bacterial leaf blight of rice and use of phages for forecasting. *Eleventh Pacific Science Congress, Symposium on Plant Diseases in the Pacific*, pp. 15–32.

Mohiuddin, M. S., Rao, Y. P., Mohan, S. K. and Verma, J. P. (1977). Survival of *Xanthomonas oryzae*, the casual incitant of bacterial blight of rice in the rhizosphere of wheat. *Sci. & Cult.* 43: 124–25.

Mondal, K. K., Rashmi E. R. and Ghoshal, T. (2021). Rice SWEET proteins: The key targets by the bacterial blight pathogen, *Xanthomonas oryzae* pv. *oryzae* to ensure nutrition in the rice apoplast during infection. *Indian Phytopathol.* 74: 323–31.

Mondal, K. K., Verma, G., Junaid, M. A. and Mani, C. 2016. Rice pathogen *Xanthomonas oryzae* pv. *oryzae* employs inducible *hrp*-dependent XopF type III effector protein for its growth, pathogenicity and for suppression of PTI response to induce blight disease. *Eur. J. Plant Pathol.* 144: 311–23.

Nathawat, R., Maku, R. V., Patel, H. K., Sankaranarayanan, R. and Sonti, R. V. (2022). Role of the FnIII domain associated with a cell wall-degrading enzyme cellobiosidase of *Xanthomonas oryzae* pv. *oryzae*. *Mol. Plant Pathol.* 23: 1011–21.

Nguyen, M.-P., Park, J., Cho, M.-H. and Lee, S.-W. (2016). Role of DetR in defence is critical for virulence of *Xanthomonas oryzae* pv. *oryzae*. *Mol. Plant Pathol.* 17: 601–13.

Niño-Liu, D. O., Ronald, P. C. and Bogdanove, A. J. (2006). *Xanthomonas oryzae* pathovars: Model pathogens of a model crop. *Mol. Plant Pathol.* 7: 303–24.

Ochiai, H., Inoue, Y., Takeya, M., Sasaki, A. and Kaku, H. (2005). Genome sequence of *Xanthomonas oryzae* pv. *oryzae* suggests contribution of large numbers of effector genes and insertion sequences to its race diversity. *Jpn. Agr. Res. Q.* 39: 275–87.

Ogunyemi, S. O., Chen, J., Zhang, M., Wang, L., Masum, M. M. I. *et al.* (2019). Identification and characterization of five new OP2-related *Myoviridae* bacteriophages infecting different strains of *Xanthomonas oryzae* pv. *oryzae*. *J. Plant Pathol.* 101: 263–73. https://doi.org/10.1007/s42161-018-0188-6

Ou, S. H. (1972). *Rice Diseases*. 1st Ed. Commonwealth Mycol. Int. Kew, Surrey, 368 pp.

Ou, S. H. (1985). *Rice Diseases*. 2nd Ed. C.A.B. International, Slough, 380 pp.

Parks, D. H., Imelfort, M., Skennerton, C. T., Hugenholtz, P. and Tyson, G. W. (2015). Assessing the quality of microbial genomes recovered from isolates, single cells, and metagenomes. *Genome Res.* 25: 1043–55.

Pradhan, S. K., Nayak, D. K., Pandit, E., Behera, L., Anandan, A. *et al.* (2016). Incorporation of bacterial blight resistance genes into lowland rice cultivar through marker-assisted backcross breeding. *Phytopathology* 106: 710–18.

Priyadarisini, V. B. and Gnanamanickam, S. S. (1999). Occurrence of a subpopulation of *Xanthomonas oryzae* pv. *oryzae* with virulence to rice cv. IRBB 21 (*Xa21*) in Southern India (Abstr.). *Plant Dis.* 83: 781.

Rao, Y. P. and Shekhawat, G. S. (1972). Bacterial blight of rice. In: P. N. Patel (ed.). *Plant Bacteriology*. Vol. I, *Bacterial Diseases of Plants in India*. Indian Agriculture Research Institute, New Delhi, pp. 1–12.

Rashid, M. M., Nihad, S. A. I., Khan, M. A. I., Haque, A., Ara, A. *et al.* (2021). Pathotype profiling, distribution and virulence analysis of *Xanthomonas oryzae* pv. *oryzae* causing bacterial blight disease of rice in Bangladesh. *J. Phytopathol.* 169: 438–46.

Rashmi, E. R., Mondal, K. K., Ghoshal, T. and Lakshmi, M. A. (2024). Comparative gene expression of OsSWEETs from various rice genotypes during the development of bacterial blight as induced by the most virulent Race 4 of *Xanthomonas oryzae* pv. *oryzae*. *Australasian Plant Pathol.* 53: 401–12. https://doi.org/10.1007/s13313-024-00990-8

Ray, P. R. and Sengupta, T. K. (1970). A study on the extent of loss of yield in rice due to bacterial blight of rice. *Indian Phytopathol.* 23: 713–14.

Reddy, A. P. K., Katyal, J. C., Rouse, D. I. and MacKenzie, D. R. *et al.* (1979). Relationship between nitrogen fertilization, bacterial leaf blight severity and yield of rice. *Phytopathology* 69: 970–73.

Reddy, A. P. K., MacKenzie, D. R., Rouse, D. I. and Rao, A. V. (1979). Relationship of bacterial leaf blight severity to grain yield of rice. *Phytopathology* 69: 967–69.

Reddy, P. R. (1983). Evidence for seed transmission of *Xanthomonas campestris* pv. *oryzae*. *Curr. Sci.* 52: 265–66.

Reitsma, J. and Schure, P. S. J. (1950). 'Kresek', a bacterial disease of rice. Contributions of the Central Agricultural Research Station, Bogor No. 117, 17 pp.

Saini, R. G. and Goel, R. K. (1998). Disease resistance in rice (*Oryza sativa* L.)- an appraisal of the Indian work. *Crop Improv.* 25: 141–54.

Salzberg, S. L., Sommer, D. D., Schatz, M. C., Phillippy, A. M., Rabinowicz, P. D. *et al.* (2008). Genome sequence and rapid evolution of the rice pathogen *Xanthomonas oryzae* pv. *oryzae* PXO99^{A}. *BMC Genom.* 9: 204–19.

Sanchez, A. C., Brar, D. S., Huang, N., Li, Z. and Kush, G. S. (2000). Sequence tagged site marker-assisted selection for three bacterial blight resistance genes in rice. *Crop Sci.* 40: 792–97.

Sato, T., Choi, Y. S., Iwasaki, M. and Watanabe, B. (1976). Distribution of races of *Xanthomonas oryzae* in Kyushu (Abstr.). *Ann. Phytopathol. Soc. Jpn.* 17: 42.

Shanti, M. L., George, M. L. C., Vera Cruz, C. M., Bernardo, M. A., Nelson, R. G. *et al.* (2001). Identification of resistance genes effective against rice bacterial blight pathogen in Eastern India. *Plant Dis.* 85: 506–12.

Singh, G. P., Srivastava, M. K., Singh, R. V. and Singh, R. M. (1977). Variation in quantitative and qualitative losses caused by bacterial blight in different rice varieties. *Indian Phytopathol.* 30: 180–85.

Singh, R. N. (1972). Perpetuation of the bacterial blight disease of rice in North India. *Indian Phytopathol.* 25: 148–50.

Sinha, B. B. P. and Sinha, R. K. P. (2000). Integrated disease management of bacterial blight of rice with special reference to neem product. *Ann. Plant Prot. Sci.* 8: 260–61.

Sirisha, C., Reddy, J. N., Mishra, D., Das, K. M., Bernardo, M. A. *et al.* (2004). Susceptibility of IRBB 21 carrying the resistance gene *Xa21* to bacterial blight. *Rice Genet. Newslet.* 21: 75–77.

Sivamani, E., Anuratha, C. S. and Gnanamanickam, S. S. (1987). Toxicity of *Pseudomonas fluorescens* towards bacterial plant pathogens of banana (*Pseudomonas solanacearum*) and rice (*Xanthomonas campestris* pv. *oryzae*). *Curr. Sci.* 56: 547–48.

Song, E.-S., Lee, B.-M., Lee, C.-S. and Park, Y.-J. 2012. PCR-based rapid assay for discriminative detection of latent infections of rice bacterial blight. *J. Phytopathol.* 160: 195–200.

Sreenivasan, M. C., Thirumalachar, M. J. and Patel, M. K. (1959). Bacterial blight of rice. *Curr. Sci.* 28: 469–70.

Srivastava, D. N. (1972). Bacterial blight of rice. *Indian Phytopathol.* 25: 1–16.

Srivastava, D. N. and Rao, Y. P. (1963). Epidemic of bacterial blight disease of rice in North India. *Indian Phytopathol.* 16: 393–94.

Srivastava, D. N. and Rao, Y. P. (1964). Seed transmission and epidemiology of bacterial blight of rice in North India. *Indian Phytopathol.* 17: 77–78.

Sun, B., Li, W., Ma, Y., Yu, T., Huang, W. *et al.* (2023). *OsGLP3-7* positively regulates rice immune response by activating hydrogen peroxide, jasmonic acid, and phytoalexin metabolic pathways. *Mol. Plant Pathol.* 24: 248–61.

Swamy, P., Panchbhai, A. N., Dodiya, P., Naik, V., Panchbhai, S. D. *et al.* (2006). Evaluation of bacterial blight resistance in rice lines carrying multiple resistance genes and *Xa21* transgenic lines. *Curr. Sci.* 90: 818–24.

Swings, J., Van den Moorter, M., Vauterin, L., Hoste, B., Gillis, M. *et. al.* (1990). Reclassification of the casual agents of bacterial blight (*Xanthomonas campestris* pv. *oryzae*) and bacterial leaf streak (*Xanthomonas campestris* pv. *oryzicola*) of rice as pathovars of *Xanthomonas oryzae* (ex Ishiyama 1922) sp. nov., nom. rev. *Int. J. Syst. Bacteriol.* 40: 309–11.

Tabei, H. and Eamchit, S. (1974). Infection source of bacterial leaf blight of rice in Thailand. *Jpn. Agric. Res. Q.* 8: 123–25.

Tagami, Y. and Mizukami, T. (1962). Historical review of the researches on bacterial leaf blight of rice caused by *Xanthomonas oryzae* (Uyeda et Ishiyama) Dowson. *Special Report of the Plant Disease and Insect Pests Forecasting Service* No. 10, 112 pp.

Talreja, S. S. and Nerurkar, A. S. 2018. Small molecules cause virulence attenuation of *Xanthomonas oryzae* pv. *oryzae*, the pathogen causing bacterial blight of rice. *Eur. J. Plant Pathol.* 151: 229–41.

Tatusova, T., DiCuccio, M., Badretdin, A., Chetvernin, V., Nawrocki, E. P. *et al.* (2016). NCBI prokaryotic genome annotation pipeline. *Nucleic Acids Res.* 44: 6614–24.

Thind, B. S. (2002). Bacterial blight of rice- an overview. *Plant Dis. Res.* 17: 227–51.

Thind, B. S. and Ahmad, M. (1994). Biological control of *Xanthomonas oryzae* pv. *oryzae*. In: Lemattre, M., Freigoun, S., Rudolph, K. *et al.* (eds). *Proc. 8th Int. Conf. Plant Path. Bact.* Versailles, pp. 867–73.

Thind, B. S. and Brar, J. S. (1989). Perpetuation of *Xanthomonas campestris* pv. *oryzae* (Ishiyama) Dye in the rhizosphere of succeeding crops. In: Z. Klement (ed.). *Proc. 7th Int. Conf. Plant Path. Bact.* Part A, Budapest, pp. 1021–27.

Thind, B. S. and Brar, J. S. (1998). Perpetuation of *Xanthomonas oryzae* pv. *oryzae* under North Indian conditions. In: R. Mahadevan (ed.). *Proc. 9th Int. Conf. Plant Path. Bact.* Chennai, pp. 409–19.

Thind, B. S. and Mehra, R. K. (1992). Chemical control of bacterial blight of rice. *Plant Dis. Res.* 7: 226–34.

Tian, Y., Zhao, Y., Xu, R., Liu, F., Hu, B., *et al.* (2014). Simultaneous detection of *Xanthomonas oryzae* pv. *oryzae* and *X. oryzae* pv. *oryzicola* in rice seed using a padlock probe-based assay. *Phytopathology* 104: 1130–37.

Tilak, K. V. B. R., Udaiyan, K. and Ramarethinam, S. (2001). Bioefficacy of bacterimycin–2000, a plant immunomodulator in the control of bacterial leaf blight (BLB) in rice. *Pestology* 25: 1–10.

Vasudevan, P., Kavitha, S., Priyadarisini, V. B., Babujee, L. and Gnanamanickam, S. (2002). Biological control of rice diseases. In: S. Gnanamanickam (ed.). *Biological Control of Crop Diseases*. Marcel Dekker, New York, pp. 11–32.

Verma, R. K., Samal, B. and Chatterjee, S. (2018). *Xanthomonas oryzae* pv. *oryzae* chemotaxis components and chemoreceptor Mcp2 are involved in the sensing of constituents of xylem sap and contribute to the regulation of virulence-associated functions and entry into rice. *Mol. Plant Pathol.* 19: 2397–2415.

Vidhyasekaran, P., Kamala, N. and Ramanathan, A. (1996). Induced systemic resistance against rice bacterial blight. Paper presented in *9th Int. Conf. Plant Path. Bact.* held during August 26–29, 1996, Chennai, India, p. 98.

Wakimoto, S. (1956). Considerations of the overwintering and the infection mechanisms of *Xanthomonas oryzae*. *Plant Prot. Jpn.* 10: 421–24.

Walker, B. J., Abeel, T., Shea, T., Priest, M., Abouelliel, A. *et al.* (2014). Pilon: An integrated tool for comprehensive microbial variant detection and genome assembly improvement. *PLoS ONE* 9: e112963.

Wang, B., Wu, G., Li, K., Ling, J., Zhao, Y. *et al.* (2021). A glycoside hydrolase family 99-like domain-containing protein modifies outer membrane proteins to maintain *Xanthomonas* pathogenicity and viability in stressful environments. *Phytopathology* 111: 929–39.

Wang, C., Wen, G., Lin, X., Liu, X. and Zhang, D. (2009). Identification and fine mapping of the new bacterial blight gene, *Xa31* (t) in rice. *Eur. J. Plant Pathol.* 123: 235–40.

Wang, C.-L., Qin, T.-F., Yu, H.-M., Zhang, X.-P., Che, J.-Y., *et al.* (2014). The broad bacterial blight resistance of rice line CBB23 is triggered by a novel transcription activator-like (TAL) effector of *Xanthomonas oryzae* pv. *oryzae*. *Mol. Plant Pathol.* 15: 333–41.

Webb, K. M., Garcia, E., Vera Cruz, C. M. and Leach, J. E. (2010). Influence of rice development on the function of bacterial blight resistance genes. *Eur. J. Plant Pathol.* 128: 399–407.

Wu, G., Su, P., Wang, B., Zhang, Y., Qian, G., *et al.* (2017). Novel insights into Tat pathway in *Xanthomonas oryzae* pv. *oryzae* stress adaption and virulence: Identification and characterization of Tat-dependent translocation proteins. *Phytopathology* 107: 1011–21.

Xu, Y., Zhu, X.-F., Zhou, M.-G., Kuang, J., Zhang, Y., *et al.* (2010). Status of streptomycin resistance development in *Xanthomonas oryzae* pv. *oryzae* and *Xanthomonas oryzae* pv. *oryzicola* in China and their resistance characters. *J. Phytopathol.* 158: 601–08.

Xue, Y., Hu, M., Chen, S., Hu, A., Li, S. *et al.* (2021). *Enterobacter asburiae* and *Pantoea ananatis* causing rice bacterial blight in China. *Plant Dis.* 105: 2078–88.

Yoshimura, S., Aoyagi, K., Morihasni, T. Yoshino, M., Nishimura, H. *et al.* (1960). Not explicitly stated in your message, but likely related to plant protection or rice pathology based on the journal. *Proc. Assoc. Plant Prot.* Hokuriku 8: 31–41.

Yu, C., Nguyen, D.-P., Ren, Z., Liu, J., Yang, F. *et al.* (2020). The RpoN2-PilRX regulatory system governs type IV pilus gene transcription and is required for bacterial motility and virulence in *Xanthomonas oryzae* pv. *oryzae*. *Mol. Plant Pathol.* 21: 652–66.

Yugander, A., Sundaram, R. M., Ladhalakshmi, D., Hajira, S. K., Prakasam, V., *et al.* (2017). Virulence profiling of *Xanthomonas oryzae* pv. *oryzae* isolates, causing bacterial blight of rice in India. *Eur. J. Plant Pathol.* 149: 171–91.

Zhang, F., Zhuo, D.-L., Zhang, F., Huang, L.-Y., Wang, W.-S., *et al.* (2015). *Xa39*, a novel dominant gene conferring broad-spectrum resistance to *Xanthomonas oryzae* pv. *oryzae* in rice. *Plant Pathol.* 64: 568–75.

Zhang, H. and Wang, S. (2013). Rice versus *Xanthomonas oryzae* pv. *oryzae*: A unique pathosystem. *Curr. Opin. Plant Biol.* 16: 188–95.

Zhang, Y., Wu, G., Palmer, I., Wang, B., Qian, G. *et al.* (2019). The role of a host-induced arginase of *Xanthomonas oryzae* pv. *oryzae* in promoting virulence on rice. *Phytopathology* 109: 1869–77.

Zhao, W.-J., Zhu, S.-F., Liao X.-L., Chen, H.-Y. and Tan, T.-W. (2007). Detection of *Xanthomonas oryzae* pv. *oryzae* in seeds using a specific TaqMan probe. *Mol. Biotechnol.* 35: 119–27.

9.2 BACTERIAL LEAF STREAK OF RICE

Bacterial leaf streak of rice is another important disease of rice, but it is comparatively less destructive and less widespread than bacterial blight (hereafter referred to as BB). Reinking (1918) was the first to report bacterial leaf streak of rice from the Philippines, but he gave it the name bacterial stripe. However, for several years the disease was erroneously referred to as BB. Fang *et al.* (1957) studied the disease in southern China, distinguished it from BB and renamed it as bacterial leaf streak (hereafter referred to as BLS). He also found its causal agent different from that of BB and named it as *Xanthomonas oryzicola*.

The occurrence of the disease is restricted largely to tropical and subtropical Asia, including Thailand, Indonesia, Bangladesh, Malaysia, Vietnam and India in addition to southern China and the Philippines. It is one of the major rice diseases in southern China. It is also present in regions of northern Australia and has become a significant problem in parts of West Africa. It has also been reported from Kenya, East Africa (Onaga *et al.* 2018).

The quarantines for its causal pathogen are in place in the United States and other rice-growing countries where the disease is not present, but also in places where the disease occurs, to guard against the introduction of new virulent strains. Its causal agent is also listed in the 2002 Agricultural Bioterrorism Protection Act of the USA as a potential bioterrorism agent, necessitating strict biosecurity measures in addition to standard biosafety measures. The available data suggest that yield losses range from 0–20% depending upon the rice variety, climatic conditions and severity of the disease (Ou, 1985). Under conditions favourable for disease development, a reduction in grain weight up to 32% has been reported. The incidence of BLS is increasing in parts of Asia, where hybrid rice is grown, such as China, because hybrid rice cultivars are susceptible to the pathogen.

In 2022, Han *et al.* (2024) observed BLS on the leaves of the rice variety Meixiangzhan 2 in Guangdong Province, China. The disease incidence was up to 30% in each field. Infected rice leaves exhibited distinctive symptoms at the boundary between diseased and healthy tissue, featuring dark green to yellow-brown streaks. Disease progression from the leaf tip inwards revealed grey-white or dehydrated lesions with a bluish-grey colour. Some leaves exhibited wrinkling at the edges, and severe symptoms at the leaf tips resembled those of bacterial blight of rice. MLSA analysis using concatenated sequences of *gyrB*, *leuS* and *rpoB* genes indicated that the isolated strain LZ1 belongs to *Pantoea ananatis*. Subsequent 16S rDNA, *gyrB*, *leuS* and *rpoB* gene sequencing results further confirmed the identity of the pathogen as *Pan. ananatis*, thereby fulfilling Koch's postulates.

9.2.1 Symptoms

Although it is mainly a foliage disease infecting the parenchymatous tissue exclusively, infection of the seed coat is common. The symptoms start as minute, water-soaked, translucent, interveinal pinheads that can be better seen when the leaf is held against light. These pinheads enlarge gradually into 1–10 cm long, water-soaked, dark green, translucent, discrete vein-limited streaks, which are typically yellow in colour. The width of the streaks varies from 0.5–1.00 mm depending on the rice cultivar, but is generally more in broad-leaf varieties. At a later stage, infected leaves turn greyish-white and die. Under humid conditions, yellow- to amber-coloured bacterial exudate is deposited in the form of beads in linear rows all along the streaks on both the leaf surfaces. This bacterial ooze is a typical sign of the disease. Irregular, diffused, cholorotic areas appear around the lesions in highly susceptible cultivars (Figure 9.2). In severe cases, streaks may coalesce resulting in blotchy lesions and the blightening of the leaves. Sheath infection is rare, but the symptoms are similar to that of leaf lamina.

In resistant cultivars, water-soaked pinheads turn brown without elongation or enlarge into 5–10 mm long reddish-brown streaks. Glume infection occurs in the form of water-soaked lesions. Under conditions favourable for the spread of the pathogen, the disease may affect the entire field.

When infection results from the entry of the bacterium through breaks in the leaf, which might occur due to high wind velocity, symptoms may expand length-wise, killing most or whole of the leaf. The symptoms of BB and BLS are quite distinct in the initial stage but they become less distinguishable in the advanced stage. Moreover, the severe infection of BLS may also result in partial or total leaf blightening resembling BB symptoms. However, even in such cases, both the diseases can be correctly and conveniently diagnosed on the basis of diagnostic hints given below.

BB infected leaves, when viewed against light, appear uniformly opaque. In the case of BLS, numerous interveinal, translucent streaks are seen. In case of BB, bacterial

FIGURE 9.2 Bacterial leaf streak of rice. (Courtesy of Donald Groth, Louisiana State University AgCenter, Bugwood.org.)

exudate appears in the form of isolated and infrequently formed beads near the margins on the upper surface of the leaves. In the case of BLS, several beads of yellow to amber colour, arranged closely in a linear fashion, appear all along the streak length on both the leaf surfaces (Shekhawat and Rao, 1972).

The symptoms of BB and BLS often occur simultaneously in rice fields and the symptoms of both the diseases may be present on the same leaves.

9.2.2 Causal Organism

Xanthomonas oryzae pv. *oryzicola* (Fang *et al.* 1957) Swings *et al.* 1990

Syn. *Xan. oryzicola*, *Xan. campestris* pv. *oryzicola*

Fang *et al.* (1957) first named the bacterium as *Xan. oryzicola*. Later on, it was changed to *Xan. translucens* f. sp. *oryzae* (Fang *et al.*) Pordesimo (Pordesimo, 1958) and then to *Xan. translucens* f. sp. *oryzicola* (Fang *et al.*) Bradbury (Bradbury, 1971). With the introduction of the pathovar concept, it was renamed as *Xan. campestris* pv. *oryzicola* (Fang *et al.*) Dye (Dye, 1978). Swings *et al.* (1990) reclassified it as *Xan. oryzae* pv. *oryzicola* (hereafter referred to as *Xoc*).

Xoc, being a pathovar of *Xan. oryzae*, shares the characteristics of *Xan. oryzae*. The description of *Xanthomonas oryzae* (Ishiyama 1922) sp. nov., nom. rev. is given under bacterial blight of rice. *Xoc* differs from *Xan. oryzae* pv. *oryzae* (hereafter referred to as *Xoo*) in terms of its ability to produce acetoin, to utilise L-alanine as a sole source of carbon, to grow on 0.2% vitamin-free casamino acids and also for its sensitivity to 0.001% cupric nitrate. The growth of *Xoc* on solid media is fast in comparison to *Xoo*. Variation in *Xoc* has not been studied as extensively as in *Xoo*. However, the analysis of strains isolated from different locations in the Philippines indicates the diversity in the pathogen (Raymundo *et al.*, 1999). The clusters of isolates were not robust, and there was no correlation between clusters and the reactions of a set of diverse rice varieties. Like *Xoo, Xoc* also serves as a model for understanding fundamental aspects of bacterial interaction with plants (Nino-Liu *et al.*, 2006).

The genome of *Xoc* strain BLS 256 has been sequenced (available in the Comprehensive Microbial Resource, http://www.tigr.org). The genome has overall similarity to *Xoo* genomes sequenced with respect to size, G + C content, abundance of IS elements and TAL effector genes and the presence of inversions and rearrangements. However, the genome also has some notable unique features like a gene cluster with predicted functions in the synthesis and transport of lipopolysaccharides, is bordered by two highly conserved genes and is distinct both in gene order and in gene content from the corresponding locus in the *Xoo* genomes. Half of the cluster is orthologous with *Xan. citri* subsp. *citri* cluster present at this locus while half is apparently unique.

The genome of *Xoc* annotates one uncharacterised gene, *XOC_3841*, only one open reading frame (ORF) in this strain is annotated to encode phosphohexosemutase (XanA), which reversibly converts glucose 1-phosphate to glucose 6-phosphate that implicates in the carbon metabolism pathways. Guo *et al.* (2014) reported that *xanA* is required for extracellular polysaccharides production, cell motility and the full virulence of *Xoc*.

Xu *et al.* (2010) obtained 827 single-colony isolates of *Xoc*, from rice leaves showing bacterial leaf streak symptoms, collected from southern China in 2007 and 2008, and tested these isolates on plates for sensitivity to streptomycin. All the isolates examined were sensitive to streptomycin and none showed decreased sensitivity to the antibiotic.

Su *et al.* (2018) reported that *his* operon was involved in histidine biosynthesis and two genes within this operon, *trpR* and *hisB*, were required for both virulence and bacterial growth *in planta*. Hence, these two genes play an important role in the pathogenicity of *Xoc* Rs105.

9.2.3 Disease Cycle

Infected seed is the main source of the perpetuation of the bacterium. The bacterium can also survive on wild rice species. In tropical areas, where only one rice crop is taken in a year, plant debris, stubbles and soil play no role in its survival. In India, where two or three crops of rice are taken in a year, the bacterium can survive on the crop itself. The problem of survival is only in areas like North or Northwestern India, where only one crop of rice is taken in a year.

Tian *et al.* (2014), based on 16S-23S internal transcribed spacer ribosomal DNA sequence data, designed padlock probe, P-Xoc to detect *Xoc* in rice seed by combining it with dot-blot hybridisation. Through the application of this technique, a detection sensitivity of 1 pg of *Xoc* genomic DNA was found. The technique facilitated the detection of *Xoc* in rice seed lots with 0.2% artificially infested seed. The padlock probe assay detected *Xoc* in 21.3% (10 of 47) of naturally infested commercial rice seed lots. In contrast, conventional PCR using XoocF/XoocR primers sets detected *Xoc* in 8.5% (4 of 47) of the same rice seed lots. The combining of both the padlock probes (P-Xoo for pv. *oryzae* and P-Xoc for pv. *oryzicola*) with reverse dot-blot hybridisation enabled the simultaneous detection of *Xoo* and *Xoc* in naturally infested commercial rice seed lots. This technique can serve as an important tool to detect multiple pathogens in seed and thereby prevent the spread of important pathogens.

During the germination of the seed, the bacterium present in the inner surface of the glumes infects the developing plumule. The first leaf carries the bacterium to the subsequent leaves, which are also infected successively as they emerge. The bacterium penetrates the leaf mainly through stomata, multiplies in the substomatal cavity and then colonises the intercellular spaces of parenchyma. Leaf veins act as a barrier to the lateral spread of the pathogen. The penetration can also occur through wounds, but it remains restricted to the apoplast of the mesophyll tissue and does not invade the xylem. In humid weather, masses of bacterial cells ooze out in the form of bacterial exudate from stomata and act as an inoculum for the secondary dispersal of the pathogen through wind, rain, irrigation water, insects and contact. Wind-driven rain is the most important agent for the dissemination of the bacterium. Bacterial exudate on the flag leaf provides abundant inoculum for the infection of seeds in the panicle. Without strong winds and rain, the secondary spread of the pathogen remains limited and the effect of disease dwindles rapidly as the growth of new leaves compensates for damage to infected leaves.

The disease is more severe in the rainy season but disappears during short periods of dry spell or bright sunshine. Relative humidity of 90% or above or dew during morning hours for 2–3 days is necessary for rapid infection. Temperature is more important than relative humidity for the enlargement of lesions. Optimal temperature for disease development is 26–30 °C; below 22 °C, the enlargement of lesions is checked. Younger plants are more susceptible; after 60 days of age, only the top young leaves remain susceptible.

Different methods, like the spraying of bacterial cell suspension without injury, spraying the cell suspension containing carborundum powder and smearing the cell suspension on the underside of the leaves with a piece of muslin cloth, are used for the artificial inoculation of plants to evaluate the germplasm. Of these methods, the smear method gives better and consistent results even under dry conditions, provided water is sprayed twice a day on inoculated plants. The best results are obtained up to one month of age of plants.

Qian *et al.* (2012) detected six putative hypothetical secreted proteins from *Xoc* strain BLS 256, but these were absent in *Xoo*. A disruption-based mutagenesis study revealed that one of them, *Xoc_15235*, named the extracellular polysaccharide and virulence-related gene (*epv*), were required for the optimal virulence in rice host, but not for the induction of a hypersensitive reaction in non-host tobacco. Sequence analysis showed that *epv* was highly conserved in *Xanthomonas* spp., with the exception of *Xoo*. The in-frame deletion of *epv* in *Xoc* dramatically impaired pathogen virulence and extracellular polysaccharide production, one of the important known virulence-associated functions in *Xanthomonas* spp. The quantitative real-time reverse-transcription PCR revealed that the expressions of both encoding exopolysaccharide xanthan biosynthesis export protein (*gumB*) gene and a known virulence-related encoding phosphoglycerate kinase (*pgk*) gene were obviously reduced in the *epv*-deletion mutant compared with the wild-type strain Rs 105. Additionally, *epv* was positively regulated by both diffusible signal factor and global regulator Clp in *Xoc*. This is the first investigation concerning the role and genetics of *epv* in the extracellular polysaccharide production and virulence of *Xoc*.

Zhao *et al.* (2012) reported that a novel pathogenicity-related gene, *Xoryp_010100018570* (named *hshB*) of *Xoc* encodes a hydrolase with a putative signal peptide, which is a homolog of imidazolonepropionase. Reverse-transcription PCR showed that *hshB* and its upstream gene, *Xoryp_010100018565* (named *hshA*), are cotranscribed in *Xoc*. Subsequent experimental results indicated that the mutation of *hshB* remarkably impaired the virulence, extracellular protease activity, extracellular polysaccharide production, growth in minimal medium, and resistance to oxidative stress and bismerthiazol of *Xoc*. The mutation of *clp*, encoding a global regulator, resulted in similar phenotypes. Real-time PCR assays showed that *hshB* transcription is positively regulated by *clp* and diffusible signal factor (DSF) and induced by poor nutrition. The finding not only discovered a novel gene *hshB* regulated by DSF-dependent quorum-sensing system and *clp*, but also showed that *hshB* was required for virulence of *Xoc*.

Ren *et al.* (2025) reported that EmvR (an orphan single-domain RR) plays a vital role in modulating *Xan. oryzae* pv. *oryzicola* (*Xoc*) T4P synthesis in the early stage of *Xoc* infection through rice stomata. Moreover, their data also revealed that EmvR can also interact with the histidine sensor kinase of the two-component signalling systems (TCS) $ColS_{XOCgx\_4036}/ColR_{XOCgx\_4037}$, which positively and negatively affects *Xoc* spreading and twitching, respectively. They proposed a ‘one-to-two’ TCS working model for the role of $ColS_{XOCgx\_4036}$, $ColR_{XOCgx\_4037}$ and EmvR in modulating *Xoc* motility.

9.2.3.1 Host Resistance

No major genes resistant to *Xoc* have been identified in rice. In contrast to resistance to BB, available genetic resistance to BLS in rice is quantitative. Several quantitative trait loci conditioning resistance to BLS have been identified in *indica* rice. Zhao *et al.* (2005) cloned *Rxol* gene from maize and showed that it provided resistance against several strains of *Xoc*, as a transgene in rice. As a transgene, *Rox*1 holds a great promise for rice breeding as a simply inherited source of resistance to BLS. An avirulent gene *avrRxo1* has also been cloned; it has been found to be present in all the strains of *Xoc* tested. Hence, the indication is that it may play a critical role in fitness, including the virulence of the bacterium. The activity of *avrRxo1* is dependent on the type III secretion system but its product is not TAL effector and lacks similarity to any known protein.

BLS resistance in rice is quantitatively inherited and is controlled by multiple quantitative trait loci (QTLs). Several QTLs conditioning resistance to BLS have been identified in *indica* rice. A QTL with relatively large effect, *qBlsr5a*, has been mapped in a ~380 kb region on chromosome 5. To further refine the mapping of *qBlsr5a*, Xie *et al.* (2014) developed overlapping sub-chromosome segment substitution lines (sub-CSSLs) from a secondary F2 population of 7000 plants, in which only the chromosomal region harbouring *qBlsr5a* segregated. *qBlsr5a* was delimited to a 30-kb interval, in which only three genes were predicted. These three putative genes did not show significant response to the infection of BLS pathogen in both resistant and susceptible parent lines. Two nucleotide substitutions in the coding sequence of gene LOC_Os05g01710, which encodes the gamma chain of transcription initiation factor IIA (TFIIAγ), resulted in a change of the 39th amino acid from valine (in the susceptible parent) to glutamic acid (in the resistant parent). The resistant parent allele of LOC_Os05g01710 is identical to *xa5*, a major gene resistant to pathogen of BB of rice. LOC_Os05g01710 is very possibly the candidate gene of *qBlsr5a*.

Song *et al.* (2015) identified two genes, *hshA* and *hshC*, within *nodB-rghB* locus which were unique to *Xoc* Rs 105 compared with *Xoo* PXO99A, and found that the transcription of these genes was regulated by DSF signalling in *Xoc*. When the low-inoculation density of *Xoc* was used, the mutation of these genes impaired the virulence of the wild-type Rs 105. However, the mutation of *hshA* or *hshC* significantly reduced the *in planta* growth ability and the level of epiphytic survival of *Xoc* cells, which were the probable mechanisms of involvement of these two genes in virulence. The identification of these two novels DSF-controlled virulence associated genes, 5*hshA* and *hshC*, is an addition to our knowledge of the regulatory *in planta* mechanisms of conserved DSF virulence signalling in *Xanthomonas* species.

Li *et al.* (2023) investigated the role of HpaP homologues in *Xoc* and showed that HpaP is required for the full virulence of *Xoc*. Deletion of *hpaP* in *Xoc* led to a reduction in virulence and alteration in the production of virulence factors, including extracellular polysaccharide and cell motility. Comparative transcriptomics and reverse transcription-quantitative PCR assays revealed that in XVM2 medium, a mimic medium of the plant environment, the expression levels of *hrp* genes (for HR and pathogenicity) were decreased in the *Xoc hpaP* deletion mutant compared to the wild type. The expression of *hrp* genes of *Xoc hpaP* mutant increased *in planta*. These findings indicate that HpaP plays a role in *Xoc* infection.

9.2.4 Management

Not many investigations have been conducted for the management of the disease but many of the measures used for managing BB can also be expected to work for this disease. As in the case of BB, host genetic resistance is also important for the control of BLS. A brief information on this aspect is given above under host resistance. The other management measures include:

1. Use of pathogen-free seed or treatment of seed by soaking in Streptocycline (250 µg mL^{-1}) solution for 12 h or in hot water at 50 °C for 30 minutes.
2. As the disease is exclusively parenchymatous, it can be effectively managed by spraying Streptocycline (100 µg mL^{-1}). The number of sprays will vary with disease severity; in general, however, 2–3 sprays at 10–15-day intervals should give satisfactory results.
3. Yuan *et al.* (2017) identified two compounds, namely, berberine and jatrorrhizine extracted from stems of *Mahonia fortunei*, as having a strong inhibitory effect on *Xoc*. The antibacterial activity of berberine was stronger, with a half-maximal inhibitory concentration (IC50) of 2.9008 mg L^{-1}. At the concentration of 0.5 g L^{-1}, its control efficacy for BLS of rice was more than 84%. Furthermore, berberine was absorbed by rice leaves, it translocated up and down in the rice plant, and the effective period was also long, but its lateral translocation inside the blade was poor.
4. Chen *et al.* (2019) made the initial report that the treatment of rice leaves with exogenous melatonin reduced the incidence of bacterial leaf streak by 17%. It is possible, therefore, that melatonin may serve as a bactericide for controlling BLS of rice.

REFERENCES

Bradbury, J. F. (1971). Nomenclature of the bacterial leaf streak pathogen of rice. *Int. J. Syst. Bacteriol.* 21: 72.

Chen, X., Sun, C., Laborda, P., He, Y., Zhao, Y. *et al.* (2019). Melatonin treatments reduce the pathogenicity and inhibit the growth of *Xanthomonas oryzae* pv. *oryzicola*. *Plant Pathol.* 68: 288–96.

Dye, D. W. (1978). Genus IX *Xanthomonas* Dowson 1939. In: Young, J. M., Dye, D. W., Bradbury, J. F. *et al.* A proposed nomenclature and classification for plant pathogenic bacteria. *N. Z. J. Agric. Res.* 21: 162–66.

Fang, C. T., Ren, H. C., Chen. T. Y., Chu, Y. K., Faan, H. C., *et al.* (1957). A comparison of the rice bacterial leaf blight organism with the bacterial leaf streak organism of rice and *Leersia hexandra* Swartz. *Acta Phytopathol. Sin.* 3: 99–124 (Ch. en).

Guo, W., Chu, C., Yang, X.-X., Fang, Y., Liu, X., *et al.* (2014). Phosphohexosemutase of *Xanthomonas oryzae* pv. *oryzicola* is negatively regulated by HrpG and HrpX, and required for the full virulence in rice. *Eur. J. Plant Pathol.* 140: 353–64.

Han, Z., Huang, D., Wang, M., Xie, H. and Wang, J. (2024). First report of bacterial leaf streak of rice caused by *Pantoea ananatis* in Guangdong Province, China. *Plant Dis.* 108: 1881. https://doi.org/10.1094/PDIS-02-24-0421-PDN

Ishiyama, S. (1922). Studies of bacterial leaf blight of rice. *Rep. Imp. Agric. Stn. Konosu* 45: 233–61.

Li, R.-F., Ren, P.-D., Zhang, D.-P., Cui, P., Zhu, G.-N. *et al.* (2023). HpaP divergently regulates the expression of hrp genes in *Xanthomonas oryzae* pathovars *oryzae* and *oryzicola*. *Mol. Plant Pathol.* 24: 44–58.

Nino-Liu, D. O., Ronald, P. C. and Bogdanove, A. J. (2006). *Xanthomonas oryzae* pathovars: Model pathogens of a model crop. *Mol. Plant Pathol.* 7: 303–24.

Onaga, G., Murori, R., Habarugira, G., Nyongesa, O., Bigirimana, J. *et al.* (2018). First report of *Xanthomonas oryzae* pv. *oryzicola* causing bacterial leaf streak of rice in Kenya. *Plant Dis.* 102: 1025.

Ou, S. H. (1985). *Rice Diseases*. 2nd Ed. C.A.B. International, Slough, 380 pp.

Pordesimo, A. N. (1958). Bacterial blight of rice. *Philippine Agri.* 42: 115–28.

Qian, G., Zhang, Y., Zhou, Y., Liu, C., Zhao, Y., *et al.* (2012). *epv*, encoding a hypothetical protein, is regulated by DSF-mediating quorum sensing as well as global regulator Clp and is required for optimal virulence in *Xanthomonas oryzae* pv. *oryzicola*. *Phytopathology* 102: 841–47.

Raymundo, A. K., Briones, A. M. Jr., Ardales, E. Y., Perez, M. T., Fernandez, L. C. *et al.* (1999). Analysis of DNA polymorphism and virulence in Philippine strains of *Xanthomonas oryzae* pv. *oryzicola*. *Plant Dis.* 83: 434–40.

Reinking, O. A. (1918). Philippine economic plant diseases. *Oryza sativa* L. Rice bacterial leaf stripe. *Philippine J. Sci.* Sec. A 13: 225–26.

Ren, P.-D., Ma, Z.-F., Liu, Q.-Q., Xia, X.-Q., Zhu, G.-N. *et al.* (2025). *Xanthomonas oryzae* orphan response regulator EmvR is involved in virulence, extracellular polysaccharide production and cell motility. *Mol. Plant Pathol.* 26: e70083. https://doi.org/10.1111/mpp.70083

Shekhawat, G. S. and Rao, Y. P. (1972). Bacterial leaf streak of rice. In: P. N. Patel (ed.). *Plant Bacteriology*. Vol. I, *Bacterial Diseases of Plants in India*. Indian Agricultural Research Institute, New Delhi, pp. 13–21.

Song, Z., Zhao, Y., Zhou, X., Wu, G., Zhang, Y., *et al.* (2015). Identification and characterization of two novel DSF-controlled virulence-associated genes within the *nodB-rhgB* locus of *Xanthomonas oryzae* pv. *oryzicola* Rs105. *Phytopathology* 105: 588–96.

Su, P., Song, Z., Wu, G., Zhao, Y., Zhang, Y., *et al.* (2018). Insights into the roles of two genes of the histidine biosynthesis operon in pathogenicity of *Xanthomonas oryzae* pv. *oryzicola*. *Phytopathology* 108: 542–51.

Swings, J., Van den Moorter, M., Vauterin, L., Hoste, B., Gillis, M. *et al.* (1990). Reclassification of the causal agents of bacterial leaf blight (*Xanthomonas campestris* pv. *oryzae)* and bacterial leaf streak (*Xanthomonas campestris* pv. *oryzicola)* of rice as pathovars of *Xanthomonas oryzae* (ex Ishiyama 1922) sp. nov., nom. rev. *Int. J. Syst. Bacteriol.* 40: 309–11.

Tian, Y., Zhao, Y., Xu, R., Liu, F., Hu, B., *et al.* (2014). Simultaneous detection of *Xanthomonas oryzae* pv. *oryzae* and *X. oryzae* pv. *oryzicola* in rice seed using a padlock probe-based assay. *Phytopathology* 104: 1130–37.

Xie, X., Chen, Z., Cao, J., Guan, H., Lin, D., *et al.* (2014). Toward the positional cloning of *qBlsr5a*, a QTL underlying resistance to bacterial leaf streak, using overlapping sub-CSSLs in rice. *PLoS ONE* 9: e95751.

Xu, Y., Zhu, X.-F., Zhou, M.-G., Kuang, J., Zhang, Y., *et al.* (2010). Status of streptomycin resistance development in *Xanthomonas oryzae* pv. *oryzae* and *Xanthomonas oryzae* pv. *oryzicola* in China and their resistance characters. *J. Phytopathol.* 158: 601–08.

Yuan, G., Chen, Y., Li, F., Zhou, R., Li, Q., *et al.* (2017). Isolation of an antibacterial substance from *Mahonia fortunei* and its biological activity against *Xanthomonas oryzae* pv. *oryzicola*. *J. Phytopathol.* 165: 289–96.

Zhao, B., Lin. X., Poland, J., Trick, H., Leach, J. *et al.* (2005). A maize resistance gene functions against bacterial streak disease in rice. *Proc. Natl. Acad. Sci.* USA 102: 15383–88.

Zhao, Y., Qian, G., Fan, J., Yin, F., Zhou, Y. *et al.* (2012). Identification and characterization of a novel gene, *hshB*, in *Xanthomonas oryzae* pv. *oryzicola* co-regulated by quorum sensing and *clp*. *Phytopathology* 102: 252–59.

9.3 BACTERIAL FOOT ROT OF RICE

A bacterial disease of rice, "bacterial foot rot", was found in Japan in July 1977 (Goto, 1979). It was characterised by a dark brown decay of the tillers. In the early stages of the disease, browning of sheath rot occurred that seemed to spread from the ligule regions. The lesions quickly extended down to nodes and culms and, finally, to crowns. Neighbouring tillers of the same crown were invaded systemically, causing foot rot symptoms. A soft rot, with an unpleasant odour, developed in young tissues of infected tillers. In the advanced stage, many tillers decayed and the entire diseased plants could be removed easily with a slight pull. The syndrome was similar to the damage caused by the rice stem borer or the "**kresek**" symptoms of bacterial blight of rice. The phenotypic characters of the rice strains were very similar to those of the corn pathovar of *Erwinia chrysanthemi*. The rice strain produced top soft rot or stalk rot on corn seedlings within one week. The corn strain was less virulent on rice, however, suggesting that the rice strain may be a distinct pathovar of the bacterium. Pu *et al.* (2012) reported *Dickeya zeae* causing foot rot disease on rice for the first time from China.

9.3.1 Symptoms

The first symptom of infection is the development of small, water-soaked lesions on the leaf sheath. The bacteria spread into nodes, stems and roots. Finally, the stems turn black, rot and may collapse (Figures 9.3a and 9.3b). The leaf blade also turns yellow and produces a rotten smell, which is typical of the disease and can help to distinguish foot rot from other pests such as stem borers and brown planthoppers. The yield loss caused by foot rot may reach 30–40%.

9.3.2 Causal Organism

Dickey zeae Samson *et al.*

Synonym: *Erwinia chrysanthemi* Burkholder *et al.*

(a)

FIGURE 9.3A Bacterial foot rot of rice: Dead tillers.

(b)

FIGURE 9.3B Bacterial foot rot of rice: Rotting of leaf sheath and stem.

Zhang *et al.* (2020) identified a new strain, *Erw. chrysanthemi* 2, from rice in Guangdong province, China, which differed from the previously identified strain from rice in its biochemical characteristics, pathogenicity and genomic constituents. To explore genomic discrepancies between *Erw. chrysanthemi* 2 and previously identified strains from rice, the authors obtained a complete genome sequence of *Erw. chrysanthemi* 2 and used for comparative genomic analyses. The complete genome sequence of *Erw. chrysanthemi* 2 is 4,575,125 bp in length. *Erw. chrysanthemi* 2 was phylogenetically closest to previously identified *Dickeya* strains from rice, but not within their subgroup. In terms of secretion systems, genomic comparisons revealed that *Erw. chrysanthemi* 2 harboured only type I (TISS), type II (TIISS) and type VI (TVISS) secretion systems. The flagella cluster of this strain possessed specific genomic characteristics like other *Dic. zeae* strains from Guangdong and from rice; within this locus, the genetic diversity among strains from rice was much lower than that of within strains from non-rice hosts. Unlike other strains from rice, *Erw. chrysanthemi* 2 lost the zeamine cluster, but retained the clustered regularly interspaced short palindromic repeats-1 (CRISPR-1) array. Compared to the other *Dic. zeae* strains containing both exopolysaccharide (EPS) and capsular polysaccharide (CPS) clusters, *Erw. chrysanthemi* 2 harboured only the CPS cluster, while the other strains from rice carried only the EPS cluster. Furthermore, the authors found strain MS1 from banana, carrying both EPS and CPS clusters, produced significantly more EPS than the strains from rice, and exhibited different biofilm-associated phenotypes. Comparative genomics analyses suggest *Erw. chrysanthemi* 2 likely evolved through a pathway different from the other *Dic. zeae* strains from rice, producing a new type of rice foot rot pathogen. These findings emphasise the emergence of a new type of *Dic. zeae* strain causing rice foot rot, an essential step in the early prevention of this rice bacterial disease.

Bertani *et al.* (2013) presented the draft genome sequence of *Dic. zeae* DZ2Q, an isolate obtained from rice grown in Italy.

9.3.3 Disease Cycle

Foot rot is more severe at high temperatures, high humidity and high levels of nitrogen. The disease is also more common in fields with a poor drainage system and soil oxidation. Hybrid varieties can be more susceptible to foot rot than local varieties. The studies on bacterial plant diseases have thus far been focused on the single bacterial species causing the disease, with very little attention given to the many other microorganisms present in the microbiome. The investigation by Bez *et al.* (2021) intends to use pathobiome analysis of the rice foot rot disease, caused by *Dic. zeae*, as a case study to investigate the effects of this bacterial pathogen to the total resident microbiome and to highlight possible interactions between the pathogen and the members of the community involved in the disease process. The microbiome of asymptomatic and the pathobiome of foot-rot symptomatic field-grown rice plants over two growing periods and belonging to two rice cultivars were determined via 16S rRNA gene amplicon sequencing. Results showed that the presence of *Dic. zeae* is associated with an alteration of the resident bacterial community in terms of species composition, abundance and richness, leading to the formation of microbial consortia linked to the disease state. Several bacterial species were significantly co-presented with the pathogen in the two growing periods, suggesting that they could be involved in the disease process. Besides, culture-dependent isolation and *in planta* inoculation studies of a bacterial member of the pathobiome, identified as positively correlated with the pathogen in our *in silico* analysis, indicated that it benefits from the presence of *Dic. zeae*. A similar microbiome/pathobiome experiment was also performed in a symptomatically different rice disease, evidencing that not all plant diseases have the same consequence/relationship with the plant microbiome. This study moves away from a pathogen-focused stance and goes towards a more ecological perception considering the effect of the entire microbial community which could be involved in the pathogenesis, persistence, transmission and evolution of plant pathogens. In summary, this study highlights that the plant microbiome shows marked changes in its composition and structure during the foot rot disease and that this disease can be viewed in part as the result of the interactions between a primary plant pathogen with members of the pathobiome.

9.3.4 Management

The following measures will help to reduce or control the disease.

1. To reduce the spread of disease from field to field, irrigation systems must be controlled so that water from the infested fields does not cross over to the non-infested fields.
2. To increase soil ventilation, inter-cultivation is needed. This should be done approximately 25 days after transplanting
3. Potash (56 lb/acre) and Gypsum (6 viss/acre) should be applied to improve soil conditions.
4. Goto (1978) screened rice cultivars from different Asian countries for their responses to the bacterial foot rot pathogen. Inoculation by spraying with bacterial suspension was not useful for screening; the injection method proved more effective. With high inoculum doses, a majority of the cultivars tested showed severe infection and the symptoms produced were very similar. As the dose of inoculum was decreased, varying degrees of resistance became apparent depending on the cultivars. With inoculum density of 1×10^4 cfu per tiller, a few varieties showed a high resistance. With inoculum density of 2×10^3 cfu per tiller, varietal differences in resistance became more distinct. Taichung 65, which was severely attacked in the field, showed a high level of susceptibility resulting in decay of stems at the base. However, some other varieties were slightly infected with this level of inoculum or were completely free from the disease. When the inoculum dose was further decreased to 2×10^1 cfu, many varieties became tolerant to the disease development though some showed a moderate infection.
5. Li *et al.* (2020) worked out a safer and more effective biocontrol method for soft rot disease caused by *Dic. zeae*. The spot-on-lawn assay was used to screen antagonistic bacteria, and three strains, including SC3, SC11 and 3-10, revealed strong antagonistic activity and were identified as *Pseudomonas fluorescens*, *Psm. parafulva* and *Bacillus velezensis*, respectively, using multilocus sequence analysis (MLSA) based on the sequences of 16S rRNA and other housekeeping genes. *In vitro* antimicrobial activity showed that two *Pseudomonas* strains, SC3 and SC11, were only antagonistic to some pathogenic bacteria, while strain 3-10 had broad-spectrum antimicrobial activity on both pathogenic bacteria and fungi. Evaluation of control efficacy in greenhouse trials showed that they all restrained the occurrence and development of soft rot disease caused by *Dic. zeae* MS2 or EC1. Among them, strain SC3 had the most impressive biocontrol efficacy on alleviating the soft rot symptoms on both monocotyledonous and dicotyledonous hosts, and strain 3-10 additionally reduced the occurrence of banana wilt disease caused by *Fusarium oxysporum* f. sp. *cubensis*. This is the first report of *Psm. fluorescens*, *Psm. parafulva* and *Bac. velezensis* as potential bioagents for controlling soft rot disease caused by *Dic. zeae*.

REFERENCES

Bertani, I., Passos da Silva, D., Abbruscato, P., Piffanelli, P. and Venturi, V. (2013). Draft genome sequence of the plant pathogen *Dickeya zeae* DZ2Q, isolated from rice in Italy. *Genome Announc.* 1 (6): e00905-13. https://doi.org/10.1128/genomeA.00905-13

Bez, C., Esposito, A., Thuy, H. D., Hong, M. N., Valè, G. *et al.* (2021). The rice foot rot pathogen *Dickeya zeae* alters the in-field plant microbiome. *Environ Microbiol.* 23: 7671–87. https://doi.org/10.1111/1462-2920.15726 PMCID: PMC9292192, PMID: 34398481

Goto, M. (1978). Resistance of rice cultivars to bacterial foot rot caused by a strain of *Erwinia chrysanthemi*. *Ann. Phytopath. Soc. Japan* 44: 646–51.

Goto, M. (1979). Bacterial foot rot of rice caused by a strain of *Erwinia chrysanthemi*. *Phytopathology* 69: 213–16.

Li. J., Hu, M., Xue, Y., Chen, X., Lu, G. *et al.* (2020). Screening, identification and efficacy evaluation of antagonistic bacteria for biocontrol of soft rot disease caused by *Dickeya zeae*. *Microorganisms* 8: 697. https://doi.org/10.3390/microorganisms8050697

Pu, X. M., Zhou, J. N., Lin, B. R. and Shen, H. F. (2012). First report of bacterial foot rot of rice caused by a *Dickeya zeae* in China. *Plant Dis.* 96: 1818. https://doi.org/10.1094/PDIS-03-12-0315-PDN. PMID: 30727269.

Zhang, J., Arif, M., Shen, H., Hu, J., Sun, D. *et al.* (2020) Genomic divergence between *Dickeya zeae* strain EC2 isolated from rice and previously identified strains, suggests a different rice foot rot strain. *PLoS ONE* 15 (10): e0240908. https://doi.org/10.1371/journal.pone.0240908

9.4 YELLOW EAR ROT OF WHEAT

The disease is also known by other names, such as **yellow slime rot** and **spike blight** and in India as ***Tundu* disease**. Hutchinson (1917) reported this disease for the first from Punjab, India but did not establish Koch's postulates. According to him, the disease had been present since 1908. The disease is present in the entire wheat belt of North India (Swarup, 1972). In general, the losses in India range from 1–2%; in extreme cases, however, losses up to

50% have also been reported. Paruthi and Gupta (1987) reported an infection level of 60.8% from India. The disease is more severe in Pakistan and the yield losses ranging from 69.6–93.5% (depending upon the cultivars) have been reported by Anwar and Inam-ul-Haq (1992). They also reported drastic reductions in other plant parameters, i.e., 15.1–21.7% in plant height, 20.8–27.7% in tillering, 10.9–14.5% in ear length and 28.6–84.3% in the number of grains. Besides India and Pakistan, the disease has been reported from China, Egypt, Australia, Iran, Iraq, Cyprus and Ethiopia.

Yellow ear rot is actually a **disease complex**, i.e., it is caused by a bacterium, *Rathayibacter tritici* and a nematode, *Anguina tritici*. *A. tritici* alone causes another disease, namely, **ear cockle of wheat**. Milne (1919) gave the first indication that the disease is caused by a bacterium and a nematode when he observed that some ear cockle galls could be obtained from certain ear heads showing yellow ear rot. However, the experimental proof for this was provided by Fahmy and Mikhail (1925) by showing that bacterium alone could not produce the disease. Later on, Chaudhuri (1935) claimed that the bacterium alone can cause the disease, but his claim has not been confirmed by the subsequent workers. Vasudeva and Hingorani (1952) confirmed the work of Fahmy and Mikhail (1925) and the latest confirmation is by Sahebani *et al.* (2006).

9.4.1 Symptoms

A. tritici, being a part of the causal complex, initial symptoms of ear cockle and yellow ear rot are the same. The first symptom is the enlargement of the stem near the soil line in 20- to 25-day-old seedlings. The leaves emerging from such seedlings are twisted and wrinkled. Affected plants are dwarfed and spreading type, and many die in the case of severe infection.

The symptoms of yellow ear rot become evident at the time of ear emergence and are characterised by the presence of a bright yellow slime or gum on abortive ears and also on the leaves in contact with such ears in the boot leaf. The affected ears fail to produce any grain. In humid weather, yellow slimy mass in the infected ears trickles down and may fall onto the lower leaves. On drying, the slimy mass becomes hard, brittle and brown. Affected culms either die in young stage or grow until heading. In the latter case, the emerging spike is narrow and short, partially filled with grains or completely filled with bacterial mass. Sometimes, the spike may fail to come out of the boot leaf. The stalk below the ear is always distorted when the bacterial mass is produced in the ears. If the conditions favourable for the multiplication of the bacteria continue, the slimy mass completely fills the ear, resulting into a **complete *Tundu* ear**. However, if the conditions are not that favourable, then partial ear cockle and partial yellow ear rot symptoms develop. In the latter case, ear cockle galls carry the bacterial inoculum for initiating the disease in the next year. Bright yellow streaks or a slimy mass also appear on the leaf surface of young plants.

9.4.2 Causal Organisms

1. *Rathayibacter tritici* (ex Hutchinson 1917) Zgurskaya *et al.* 1993
   Syn. *Clavibacter tritici*, *Corynebacterium michiganense* pv. *tritici*, *Cor. tritici*
2. *Anguina tritici*

Rat. tritici: The bacterium was first named *Pseudomonas tritici* by Hutchinson (1917); Burkholder (1948) then changed it to *Corynebacterium tritici*. With the introduction of the pathovar concept for the classification of plant pathogenic bacteria, it was changed to *Cor. michiganense* pv. *tritici*. Later on, Davis *et al.* (1984) reclassified it as *Clavibacter tritici*. Zgurskaya *et al.* (1993) created a new genus *Rathayibacter* and renamed it as *Rat. tritici*. The below given description of the bacterium is taken from Zgurskaya *et al.* (1993).

The cells are irregular, pleomorphic, short rods and measure 0.4–0.8 × 0.5–1.8 μm. V-forms; sometimes short cell chains are formed but a clear rod-coccus cycle is not observed. Nonmotile, non-sporulating, Gram-stain-positive, not acid-fast and obligately aerobic. Chemoorganotrophs, having an oxidative type of metabolism and catalase test positive. The optimum growth occurs between 24 and 28° C. The cell walls contain galactose and xylose. The bacterium is capable of utilising xylose, lactose, L-arabinose, sorbitol, mannitol, inulin, citrate, tartrate and sebacinate. Acid is produced from glucose, galactose, fructose, mannose, maltose, sucrose and xylose. Glutamic acid and DL-ornithine are utilised but not DL-valine. The bacterium is able to hydrolyse Tween 21, 40 and 85 and grow in the presence of 5% NaCl and 0.03% potassium tellurite. Voges-Proskauer test is negative but methyl red is positive. The bacterium contains a 50 kb plasmid.

Park *et al.* (2017) reported the first complete genome sequence of *Rat. tritici* NCPPB 1953 with diverse features of this strain. The whole genome consists of one circular chromosome of 3,354,681 bp having a G + C content of 69.48%. In all 2,979 genes, comprising 2,866 protein coding genes and 49 RNA genes, were predicted. The comparison of genomic analyses of *Rat. tritici* NCPPB 1953 and *Rat. toxicus* strains identified 1,052 specific genes in *Rat. tritici* NCPPB 1953. Using the BlastKOALA database, the authors found that the flexible genome of *Rat. tritici* NCPPB 1953 is highly enriched in the "Environmental Information Processing" system and metabolic processes for diverse substrates. Moreover, many specific genes of *Rat. tritici* NCPPB 1953 are distributed in substrate-binding proteins for extracellular signals, including saccharides, lipids, phosphates, amino acids and metallic cations. These findings provide clues for rapid and stable colonisation of *Rat. tritici* for disease mechanism and nematode association.

It is generally believed that both the bacterium and the nematode are equally important in causing the disease. However, the opinion is changing in favour of a major role played by the bacterium. Now it is believed that the disease is caused mainly by the bacterium; the nematode helps in carrying the bacterium to the ears and in its entry into the host tissue by causing wounds in addition to the survival of the bacterium in or on the nematode galls.

9.4.3 Disease Cycle

The primary source of inoculum for both the bacterium and the nematode are cockles or nematode galls. These galls may be present in the soil or in the seed. The bacterium is present inside the galls and also on the surface of galls. Under dry conditions, the nematode can survive inside the galls for several years, while the bacterium survives for at least five years.

The nematode galls present in the soil, or brought to the soil along with the contaminated seed, absorb moisture, swell and burst open to release 2nd stage larvae. These bacteria-laden larvae find their way to the growing point of the seedlings and feed ectoparasitically. With the natural growth of the plants, the larvae are carried to the inflorescence, where they enter the floral primordia and feed endoparasitically. The bacteria also gain entry into the developing ear heads, along with the nematode. Under favourable weather conditions, the bacteria multiply profusely inside the ears, producing an abundant slimy mass which envelops the developing spikelets. Poorly drained soils, early sowing and a greater number of irrigations increase the disease incidence.

9.4.4 Management

As the nematode galls of *A. tritici* help in the survival of the bacterium and its larvae initiate the disease, the control of this nematode will automatically eliminate the disease. The following measures are adopted for the control of *A. tritici* and, ultimately, the yellow ear rot.

1. Use the gall-free seed. If the seed is not gall-free, it should be made gall-free by adopting the following measures.
    a. Soak the seed in 20% brine solution (NaCl or KCl). Vigorously stir the seed and remove the galls, which float on the surface of water. Destroy these galls by burning. After soaking in brine solution, wash the seed 2–3 times with water to avoid reduction in seed germination.
    b. Soak the seed in water for 2 h and heat the water to raise its temperature to 51° C and maintain it for 30 minutes. The heat treatment activates the dormant larvae and then kills these.
    c. Galls can also be removed by sieving, fanning or winnowing. In China, nematode eliminators are used to remove the galls.
2. Practise the crop rotation for 2–3 years with non-graminaceous crops.
3. Grow the disease-resistant or -tolerant cultivars.
4. Use of chemicals for seed/soil treatment.

REFERENCES

Anwar, M. J. and Inam-ul-Haq, M. (1992). Effect of *Tundu* disease on growth and yield of wheat. *Pakistan J. Phytopathol.* 4: 66–69.

Burkholder, W. H. (1948). Genus I. *Corynebacterium* Lehmann & Neumann. In: R. S. Breed, E. G. D. Murray and A. P. Hitchens (eds.). *Bergey's Manual of Determinative Bacteriology*, 6th ed. Bailliere, Tindall & Cox, London and Williams & Wilkins Co., Baltimore, MD, pp. 381–408.

Chaudhuri, H. (1935). A bacterial disease of wheat in Punjab. *Proc. Indian Acad. Sci.* 1(B): 579–85.

Davis, M. J., Gillaspie, A. J. Jr., Vidaver, A. K. and Harris, R. W. (1984). *Clavibacter*: A new genus containing some phytopathogenic coryneform bacteria, including *Clavibacter xyli* subsp. *xyli* sp. nov., subsp. nov. and *Clavibacter xyli* subsp. *cynodontis* subsp. nov., pathogens that cause ratoon stunting disease of sugarcane and bermudagrass stunting disease. *Int. J. Syst. Bacteriol.* 34: 107–17.

Fahmy T. and Mikhail, T. (1925). The bacterial disease of wheat caused by *Pseudomonas tritici* Hutchinson. *Agri. J. Egypt* 1: 64–72.

Hutchinson, C. M. (1917). A bacterial disease of wheat in Punjab. *Mem. Dept. Agric. India Bact. Ser.* 1: 169–75.

Milne, D. (1919). Ear-cockle in wheat. *Bull. Dept. Agri. Punjab Bot.* 1: 1–5.

Park, J., Lee, P. A., Lee, H. H., Choi, K., Lee, S. W. *et al.* (2017). Comparative genome analysis of *Rathayibacter tritici* NCPPB 1953 with *Rathayibacter toxicus* strains can facilitate studies on mechanisms of nematode association and host infection. *Plant Pathol. J.* 33 (4): 370–81.

Paruthi, I. J. and Gupta, D. C. (1987). Incidence of "tundu" in barley and kanki in wheat field infested with *Anguina tritici*. *Haryana Agric. Univ. J. Res.* 17: 78–79.

Sahebani, N., Kheiri, A., Rahimian, H. *et al.* (2006). A study on interaction between wheat seed gall nematode *Anguina tritici* and wheat yellow ear rot *Rathayibacter tritici*. *Iranian J. Agric. Sci.* 36: 1355–60.

Swarup, G. (1972). Tundu disease of wheat. In: P. N. Patel (ed.). *Plant Bacteriology*. Vol. I, Bacterial Diseases of Plants in India. Indian Agricultural Research Institute, New Delhi, pp. 33–37.

Vasudeva, R. S. and Hingorani, M. K. (1952). Bacterial disease of wheat caused by *Corynebacterium tritici* (Hutchinson) Bergey *et al. Phytopathology* 42: 291–93.

Zgurskaya, H. I., Evtushenko, L. I., Akimov, V. N. *et al.* (1993). *Rathayibacter* gen. nov., including the species *Rathayibacter rathayi* comb. nov., *Rathayibacter tritici* comb. nov., *Rathayibacter iranicus* comb. nov., and six strains from annual grasses. *Int. J. Syst. Bacteriol.* 43: 143–49.

9.5 BACTERIAL STRIPE AND BLACK CHAFF OF BARLEY

The disease was first reported from the USA by Jones *et al.* (1917) as bacterial blight of barley (*Hordeum vulgare* L.). Shekhawat and Patel (1972) have also given a brief account of bacterial blight of barley. The currently used name of the disease is bacterial stripe and black chaff of barley. It is prevalent in Mexico, the USA, South America, Belgium, France, Spain and India. In India, the disease was first reported by Patel and Shekhawat (1971), as bacterial blight of barley from Ludhiana, Punjab, India. Later on, Maharshi (1996) reported it from Rajasthan in 1989, where it is has become more or less endemic (Maharshi and Sharma, 2006). The disease incidence may reach nearly 100% in overhead-irrigated fields in contrast to fields irrigated with the conventional flood irrigation system. Sharma and Maharshi (2012) have given a brief account of the occurrence of the disease in Rajasthan, along with information on seed-plant transmission of the bacterium and the environmental conditions favouring the disease development. However, there is little information available on losses caused by the disease.

Beutler *et al.* (2023) reported bacterial leaf streak of barley caused by *Xanthomonas translucens* pv. *translucens* from British Columbia, Canada.

9.5.1 Symptoms

All the above-ground parts of the plants are affected. There are two principal types of symptoms, i.e., stripes and black chaff. The stripes appear on leaf sheath, leaf lamina and stems, while the black chaff appears on ears and neck. Initially, water-soaked, dark green flecks/lesions appear on leaf sheaths and leaf blades, which elongate to linear stripes parallel to the veins. The stripes also appear on the stem internodes. Soon the stripes lose their greenish-yellow and water-soaked appearance and become necrotic, brown and translucent. The length of the stripes varies from 15–25 cm, or they may even cover the entire length of the leaf blade (Figure 9.4a and b). Severe infection on susceptible varieties may cause chlorosis of the leaves. Severe infection of flag leaf and its sheath results in the sealing of sheath lumen with bacterial mass, resulting in the distortion of emerging spike and its rachis. Often, the emergence of spike is prevented completely. Frequently, the stripes coalesce to from yellowish-brown blotches, leading to the shrivelling and drying of the infected leaves/sheaths. In the field, the symptoms are more pronounced at the tillering stage. In the early morning hours, copious amounts of a viscous slimy mass of bacterial ooze appears on the surface of stripes. This then dries up into numerous beads or crusty translucent scales under dry conditions.

The seed-borne nature of the bacterium leads to the infection of young seedlings. Severely affected plants remain weak and stunted, and their ears often fail to emerge. If the ears do emerge, they are dwarfed, twisted and even sterile, having compressed and closed spikelets. Young infected plants at the two- to three-leaf stage are sometimes so severely stunted and may be killed; if they grow further; additionally, the grain formation is poor and is accompanied by deformed spikes. Sometimes, only the awns emerge from the ears followed by the death of plants. The symptoms of the disease can be confused with those of barley stripe caused by *Helminthosporium graminieum*, but it can be clearly diagnosed on the basis of the bacterial ooze/turbidity test in the laboratory and field, respectively.

The initiation of leaf lesions from the apex and their downward extension and the starting of the lesions primarily from the central part of the blade where the leaf bends have been reported by different workers from different regions. It appears that the symptom development and further progress may vary in different geographical regions.

9.5.2 Causal Organism

Xanthomonas translucens pv. *translucens* (Jones *et al.* 1917) Vauterin *et al.* 1995

Synonym: *Xan. campestris* pv. *translucens*

(a) (b)

FIGURE 9.4 Bacterial stripe and black chaff of barley: (a) Infected plant (b) Stripes present on leaves.

Jones *et al.* (1917) originally described the bacterium as *Bacterium translucens*; with the establishment of genus *Xanthomonas*, it was reclassified as *Xanthomonas translucens*. Later on, it was designated as *Xan. translucens* f. sp. *hordei*. With the introduction of pathovar system, it was named as *Xan. campestris* pv. *translucens*. Vauterin *et al.* (1995), while reclassifying the species of *Xanthomonas*, changed it to *Xan. translucens* pv. *translucens*.

The colonies on nutrient agar are yellow, mucoid, convex and smooth, having entire margins. Cells measure 0.4 × 1.0 µm and are motile by a single polar flagellum. The bacterium produces hydrogen sulphide but does not reduce nitrate, does not produce indole, does not hydrolyse starch and does not utilise lactose; catalase test is positive while oxidase test is negative. The optimum temperature for its growth ranges from 28–30 °C.

The following four pathogens, often grouped under the name **"translucens group"**, cause bacterial stripe and streak disease on cereals.

Xan. translucens pv. *translucens*
Xan. translucens pv. *cerealis*
Xan. translucens pv. *secalis*
Xan. translucens pv. *undulosa*

Jones *et al.* (1917), who first isolated the bacterium from barley, found that the bacterium was pathogenic only to barley and did not infect wheat, oat, rye, timothy, spelt emmer or einkorn (all *Triticum* spp.). Maharshi (1996) also reported that the Rajasthan isolate of the bacterium (CABIMIB 12929) was also non-pathogenic to wheat. Following the rules of bacterial nomenclature (International Society for Plant Pathology, Young *et al.*, 1996), the name *Xan. translucens* pv. *translucens* should be reserved for strains pathogenic only to barley, as originally reported by Jones *et al.* (1917). As *Xan. translucens* pv. *undulosa* strains infect wheat and triticale but can also be isolated from barley and rye, their host range is not only broader than that of *Xan. translucens* pv. *translucens* but also covers the host range of *Xan. translucens* pv. *cerealis*. Based on amplified fragment length polymorphism and fatty acid methyl ester analysis, Bragard *et al.* (1997) showed that strains pathogenic on barley and not on wheat clustered together in a genetically different group. They further reported that the pathovars *cerealis, translucens* and *undulosa* corresponded to true biological entities.

Roman-Reyna *et al.* (2020) sequenced the complete genome of *Xan. translucens* pv. *translucens* strain UPB886 to better understand the biology of *Xan. translucens* subgroup.

Fu *et al.* (2024) developed two probe-based quantitative PCR (qPCR) systems, namely P-Xtt and P-Xtu, to diagnose cereal bacterial leaf streak pathogens, *Xan. translucens* pv. *translucens* and *Xan. translucens* pv. *undulosa*, respectively. P-Xtt is specific to pv. *translucens* and P-Xtu is specific to pv. *undulosa*, pv. *cerealis*, pv. *secalis* and pv. *pistaciae*. P-Xtt and P-Xtu worked on all accessible strains of pv. *translucens* and pv. *undulosa*, respectively. Both systems could detect 100 copies of the target gBlock DNA.

9.5.3 Disease Cycle

The bacterium survives in the infected seed and it is both externally and internally seed-borne. Using the paper towel and seedling system test, Maharshi (1996) reported 30% seed-plant transmission at harvesting time, which decreased to 4% after 7 months, i.e., up to the sowing time of the next crop. *Xan. translucens* specific medium is better than nutrient glucose agar for the detection of the pathogen from the seeds. No inoculum threshold is known for infected barley seeds but wheat seeds, the washings of which contained approximately 1000 cfu of the bacterium per mL^{-1} per gram of seed, produced the diseased seedlings (Schaad, 1988).

Jones *et al.* (1917) reported that the bacterium from the seeds enters into lemma and leaf through stomata and further moves into intercellular spaces of thin-walled parenchymatous tissue. Sharma (2003) traced the movement of the bacterium from husk (glumes) and pericarp, but not from endosperm or embryo, to the coleoptile, first true leaf and then into the aerial parts of the seedlings. The transmission from seed to seedlings is facilitated by the presence of free water, i.e., it is more in high rainfall or under overhead-irrigation system.

The survival of the bacterium in infected crop debris and soil is only for a limited period and hence of no significance. The bacterium enters the host through stomata and wounds. The dissemination of the bacterium occurs through wind-splashed rain, overhead irrigation, plant contact and insects. In Montana, USA, under the sprinkler irrigation system, the bacterium has been found to spread from a single infection locus over an area of 28 m^2 in 39 days. Sprinkler irrigation during the daytime also promotes the penetration of the bacterium, as the stomata are open during the daytime.

The bacterium multiplies under a wide range of temperature and moisture conditions, but the disease is more severe in warm and humid conditions. The bacterium multiplies rapidly in the host tissue in tissue creases and intercellular spaces. The incubation period is short, lasting approximately 10 days. The bacterium multiplies in all the aerial plant parts including spikes, which contaminate the developing seeds.

Butchacas *et al.* (2020) observed that cbsA, a mutant of *Xan. transluscens* pv. *translucens* (*Xtt*) developed by them, displayed wild-type vascular blight development. However, this mutant significantly increased apoplastic water-soaking symptoms downstream of vascular blight symptoms, similar to the mesophyll pathogen *Xan. translucens* pv. *undulosa*. This indicates that the loss of CbsA enhances non-vascular leaf colonisation. Further co-inoculation of the mutant and the wild-type suppressed water-soaking, suggesting that presence of this secreted enzyme in the plant dictates pathogen tissue specificity. These findings demonstrate that CbsA plays a major role in *Xtt* regulation of mesophyll colonisation behaviour.

Shoaf *et al.* (2020) developed primers based on whole-genome differences for two multiplex-PCR reactions: one to distinguish between pathovars of *Xan. translucens* and the other to differentiate *Xan. translucens* pv. *translucens* (*Xtt*) subgroups. The primers efficacy was examined using seeds and leaves from infected barley plants. In the case of the infected leaf tissue PCR, water was used from leaf suspensions to ensure simplicity, avoiding both genomic extraction and bacterial culture. *Xtt* subgroups A and B were identified from US field samples. As far as the authors are aware, these are the first primers to distinguish *Xan. translucens* pv. *undulosa* and *Xtt*, and *Xtt* subgroups. These molecular tools will support disease management strategies by enabling the detection and pathovar incidence analysis of *Xan. translucens*.

9.5.4 Management

1. Use of pathogen-free seed is the best approach for the management of the disease. The seed certification programmes and seed assays are necessary for the production of pathogen-free seed.
2. Dry heat treatment of barley seeds at 72 °C for 5 to 7 days has been found to eradicate the pathogen. Fourest *et al.* (1990) found that dry heat treatment of heavily infested barley seed at 71, 75 or 84 °C for 11 days eliminated the pathogen while treatment at 72 °C for 4 days apparently eliminated the pathogen from moderately infested seed. Reduction in seed germination was practically negligible for seed treated at 71 or 72 °C for 7 days or less while the treatment at 71 °C for 11 days caused an average reduction of 8%.
3. Avoid the overhead-irrigation or limit its use.
4. Use of disease-resistant cultivars is the most effective approach for disease management, but no disease-resistant cultivars are available. However, the available cultivators differ in their level of susceptibility to the pathogen.

REFERENCES

Beutler, J., Li, T., Román-Reyna, V., Fleitas, M. C., Bamrah, R. *et al.* (2023). First report of bacterial leaf streak of barley (*Hordeum vulgare*) caused by *Xanthomonas translucens* pv. *translucens* in British Columbia, Canada. *Plant Dis.* 107: 2215.

Bragard, C., Singer, E., Alizadeh, A., Vauterin, L., Maraite, H. *et al.* (1997). *Xanthomonas translucens* from small grains: Diversity and phytopathological relevance. *Phytopathology* 87: 1111–17.

Butchacas, J., Roman-Reyna, V. and Jacobs, J. M. (2020). Loss of a single gene enhances non-vascular *Xanthomonas* streak (Abstr.). *Phytopathology* 110: S2.30–2.31.

Fourest, E., Rehms, L. D., Sands, D. C., Bjarko, M. and Lund, R. E. (1990). Eradication of *Xanthomonas campestris* pv. *translucens* from barley seed with dry heat treatments. *Plant Dis.* 74: 816–18.

Fu, H., Fleitas, M. C., Sarkes, A., Wang, L., Yang, Y. *et al.* (2024). Detection and differentiation of *Xanthomonas translucens* pathovars *translucens* and *undulosa* from wheat and barley by duplex quantitative PCR. *Plant Dis.* 108: 270–77.

Jones, L. R., Johnson, A. G. and Reddy, C. S. (1917). Bacterial blight of barley. *J. Agric. Res.* 11: 625–43.

Maharshi, R. P. (1996). Bacterial stripe and black chaff disease of barley caused by *Xanthomonas campestris* pv. *translucens*: A report on the natural occurrence in Rajasthan, India. *Proc. 9th Int. Conf. Plant Path. Bact., Chennai, India*, pp. 367–70.

Maharshi. R. P. and Sharma, H. (2006). Bacterial stripe and black chaff of barley (*Hordeum vulgare* L.) caused by *Xanthomonas campestris* pv. *translucens* in Rajasthan, India. *Ann. Exptl. Agri. Al. Sci.* 1: 103–10.

Patel, P. N. and Shekhawat, G. S. (1971). Occurrence of *Xanthomonas translucens* f. sp. *hordei* in India and its pathogenicity to rice. *Plant Dis. Reptr* 55: 365–68.

Roman-Reyna, V., Luna, E. K., Pesce, C., Vancheva, T., Chang, C. *et al.* (2020). Genome resource of barley bacterial blight and leaf streak pathogen *Xanthomonas translucens* pv. *translucens* strain UPB886. *Plant Dis.* 104: 13–15.

Schaad, N. W. (1988). Bacteria. In: Symposium on Inoculum Thresholds of Seed-borne Pathogens. *Phytopathology* 78: 872–75.

Sharma, H. (2003). Role of Seed Transmission and Environmental Factors on Development of Bacterial Stripe Disease of Barley (*Hordeum vulgare* L.) Caused by *Xanthomonas campestris* pv. *translucens* (Jones, Johnson & Reddy) Dye. Ph.D. Thesis. Rajasthan Agricultural University Bikaner, 68 pp.

Sharma, H. and Maharshi, R. P. (2012). *Bacterial Stripe Disease of Barley: A Research Investigation*. Lambert Academic Publishing, Riga, Latvia, 96 pp.

Shekhawat, G. S. and Patel, P. N. (1972). Bacterial blight of barley. In: P. N. Patel (ed.). *Plant Bacteriology*. Vol. I, *Bacterial Diseases of Plants in India*. Indian Agriculture Research Institute, New Delhi, pp. 38–42.

Shoaf, W., Mezache, N., Luna, E. K., Lang, J. M., Carpenter, S. *et al.* (2020). Development and application of multiplex-PCR to specifically detect *Xanthomonas translucens* pvs. *undulosa* and *translucens* in infected field samples (Abstr.) *Phytopathology* 110: S2.173–74.

Vauterin, L. Hoste, B., Kerster, K. and Swings, J. (1995). Reclassification of *Xanthomonas*. *Int. J. Syst. Bacteriol.* 45: 472–89.

Young, J. M., Saddler, G. S., Takikawa, Y., De Boer, S. H., Vauterin, L. *et al.* (1996). Names of plant pathogenic bacteria 1864–1995. *Rev. Plant Pathol.* 75: 721–63.

9.6 HALO BLIGHT OF OATS

There are two types of bacterial diseases which infect the foliage of oats (*Avena sativa* L.): halo blight and bacterial stripe blight. Formerly, it was thought that *Pseudomonas syringae* pv. *coronafaciens* causes halo blight of oats and *Psm. syringae* pv. *striafaciens* causes bacterial stripe blight of oats and barley. However, Schaad and Cunfer (1979) proved that the same organism causes both diseases.

Personal observations in Wisconsin by Elliott (1920) and specimens of diseased plants from Ohio, Illinois, Indiana, Minnesota, Tennessee, California and Virginia led to the conclusion that halo blight was present in oat fields every season. Scattered lesions occur on the lower leaves more or less throughout the season and, occasionally, the panicles are also attacked. These lesions on the lower leaves are generally hidden by the upper new leaves and so escape observation. Only under favourable weather conditions does the blight develop sufficiently to attract attention and cause serious losses.

Halo blight of oats caused by *Psm. coronafaciens* occurs in most oat-growing regions of the world (Paul and Smith, 1989). Reid (1938) reported its first occurrence in New Zealand and attributed it to *Psm. coronafaciens* (Elliott, 1920; Stevens, 1925). He described the disease symptoms on oats, but did not publish a full description of the pathogen. Wilkie (1972) again reported the disease from New Zealand, along with a full description of the pathogen. Harder and Harris (1973) reported it from Kenya.

In 1996, it was detected in Norway and Sweden for the first time. The damage was most serious in areas where the weather was cool and humid early in the summer, before changing to warmer temperatures with heavy rain and showers (Persson and Sletten, 2001). Halo blight occurs quite commonly on oats in North America; infections just before panicle emergence can reduce yield. Early season infections, however, have little impact on yield (Dickson, 1947).

Wang *et al.* (2024) reported halo blight of oats caused by *Psm. syringae* pv. *coronafaciens* from Huan county, Gansu Province of China. Average leaf disease incidence was between 1 and 14%. The symptoms were mainly observed on oat leaves and appeared as white-to-yellow streaks. After 20–30 days, the streaks changed to yellowish necrotic lesions and spread over the whole leaves, eventually causing them to wilt and senesce. The inoculation of the bacterium on plants in the glasshouse caused 19.44, 11.49, 11.34, 9.23, 11.61, 7.57, 6.11, 9.73, 9.31 and 4.15% reduction in number of tillers, plant height, flag-leaf length, flag-leaf width, fresh weight, dry weight, panicle length, grains per spike, number of reproductive branches, and thousand grains weight, respectively.

Halo blight of rye was widespread in the state of Georgia during 1974 and 1975, marking the first-known report of this pathogen on large acreages of rye in the USA. The bacterium was identified by pathogenicity and biochemical tests. Disease symptoms were most abundant in March and April, but the pathogen could still be isolated from lesions on a few rye cultivars in late May. *Psm. syringae* pv. *coronafaciens* caused extensive head blight and all of the leaves were killed on many plants. The disease has probably been present in Georgia, but was not detected for many years (Cunfer and Schaad, 1976).

Chocolate spot, caused by *Psm. syringae* pv. *coronafaciens*, has been reported on maize from Wisconsin and Minnesota (when reported, it was not found in the other states of the USA). It occurs sporadically, and is of no economic importance (Ribeiro *et al.*, 1977; Bowden and Stromberg, 1982).

9.6.1 Host Range

The main hosts are *Avena sativa* (oats) and *Hordeum vulgare* (barley). *Secale cereale* (rye), *Triticum aestivum* (wheat), *Avena byzantina* (red oat, Algerian oat), *Agropyron repens* (quack grass), *Bromus* spp. (brome grass) and *Phleum pratense* (timothy) have also been reported as natural hosts (Janse, 2005).

9.6.2 Symptoms

9.6.2.1 Oats

Halo blight is most conspicuous on the leaves, although it may also occur on leaf sheaths and glumes. Typical well-developed lesions of the disease are oval, chlorotic spots, 1/2 to 2 cm or more in diameter, around points of infection; the latter consist of grey-brown collapsed tissue, measuring from 1 to several millimetres in length. In the beginning, the halo-like border is only slightly lighter green than the surrounding tissue; as it becomes older, however, it loses more of its green colour and forms oval yellow halos around the centrally infected areas.

The lesions are first visible as light green oval spots (4 to 5 mm in diameter) with central sunken points of infection, initially evident on only one side of the leaf. This centre of infection increases slowly in size, penetrates the leaf tissue, and, within a day or two, forms a grey or brown dry tissue from 1 to several millimetres in diameter, evident on both sides of the leaf blade. The halo-like margin spreads rapidly, becoming uniformly lighter green to yellow or showing concentric markings of different shades of green and yellow. Occasionally, these halo-like margins prolong at one end, ranging from 1 to several centimetres in length. They may extend as yellow streaks through the centre or along the margin to the tip of the leaf; however, they appear as oval spots, measuring 1 cm or more in diameter. Marginal infections are common, forming crescent-shaped lesions. These halo-like lesions are conspicuous and characteristic. Except in the central infected area, the tissues remain turgid and have a normal appearance, except for the paler yellowish colour (Figure 9.5). Exudate does not occur in connection with the lesions.

When several lesions occur on the same leaf, these lesions often coalesce and produce a general yellowing, followed by a breaking across of leaf blades or a shrivelling and drying of tips and margins. During periods of warm and dry weather, yellow-haloed leaf tissue loses its turgidity and colour, forming oval, greyish-brown dead spots which, on some leaves, have narrow, brown margins, and on others narrow, yellow halo-like margins. Very rarely, the dead tissue may be pinkish or reddish-brown in colour. In separated lesions, the oval outline of the dead halo persists; even when the whole leaf becomes dry and brown, the original halo outlines can still be distinguished. On chaff, small spots with a yellow halo may also occur and on glumes yellow translucent spots between the veins can be observed.

FIGURE 9.5 Halo blight of oats.

9.6.2.2 Rye

Several types of symptom were caused by *Psm. syringae* pv. *coronafaciens* on rye in Georgia, USA. Elliptical to linear, brown lesions, surrounded by mild chlorosis, often in association with frost injury, were observed on leaves during the winter when temperatures rarely exceeded 15 °C. Halo symptoms were prominent in early spring when the weather conditions were mild (maximum temperature 15–22 °C). Chlorotic lesions were less common after heading. Halo lesions became buff within 1–2 days, giving the lesions a 'scalded' appearance. Scald symptoms were most common from flowering until maturity, but easily recognisable halo symptoms persisted on leaf sheaths almost until maturity. Sometimes, minute halos were found on the glumes, but more commonly the entire floret blighted, dried, and became white (Cunfer *et al.*, 1978).

9.6.3 Causal Organism

Pseudomonas coronafaciens (Elliott 1920) Schaad and Cunfer 1979

Synonym: *Psm. syringae* pv. *coronafaciens*, *Psm. syringae* pv. *striafaciens*

Cells are Gram-stain-negative, non-capsulated, motile and non-sporing short rods. The colonies are cream-coloured, smooth, round and convex with entire margins on nutrient agar supplemented with 0.5% yeast extract. Growth occurs at 24–30 °C, but not at 4 or 40 °C. The bacterium is strictly aerobic and negative for indole activity. It produces levan, negative for oxidase, negative for arginine dihydrolase, LOPAT reaction: +—+. The bacterium is ice-nucleation positive and is variable for the production of a water-soluble, fluorescent pigment when grown on King's medium B. The bacterium can hydrolyse aesculin and gelatin and utilises many carbon and nitrogen sources at 24 °C. The DNA G + C % for *Psm. coronafaciens* type strain (NCPPB 600^{PT}) was 58.2 and 57.8 mol%.

The bacterium does not cause rot in potato tubers, but it does produce a hypersensitive reaction in tobacco.

Kim (2020) reported the complete genome sequence of *Psm. coronafaciens* X-1, containing a circular chromosome and one circular megaplasmid, isolated from halo blight lesions on oat leaves in Korea. This is the first genome sequence of the oat halo blight pathogen and will contribute to our understanding of the pathogenicity and ecological aspects of the halo blight pathogen and the interactions between *Psm. coronafaciens* and oats.

9.6.4 Disease Cycle

The bacterium survives in crop residues and also in/on seed. There is evidence from the following three facts that the bacterium causing halo blight over-winters in/on the seed: 1) Presence of typical halo lesions on the glumes and lemmas of maturing spikelets, 2) Early appearance of the disease on seedlings grown in soil where oats were not sown previously and 3) Higher disease severity in plots sown with untreated seed compared to those sown with treated seed.

Short distance or secondary spread of the bacterium occurs by wind-splashed rain. It can also occur through direct contact of healthy plant parts with infected plant parts, through aphids or other insects.

Psm. coronafaciens is a plant-quarantined bacterium in many countries, including the Republic of Korea and the use of certified seed is important for the control of this pathogen. As effective detection method for *Psm. coronafaciens* are not available, An *et al.* (2015) developed PCR and TaqMan PCR assays for the detection of this bacterium from seed. PCR primers were designed from the draft genome sequence of *Psm. coronafaciens* LMG 5060, which was obtained by the next-generation sequencing in this study. The PCR primer set Pc-12-F/Pc-12-R specifically amplified 498 bp from the 13 strains of the bacterium isolated from Canada, Japan, the United Kingdom, Zimbabwe, Kenya, Germany and New Zealand. The nested primer set Pc-12-ne-F/Pc-12-ne-R specifically amplified 298 bp from these strains. The target-size PCR product was not amplified from the non-target bacteria with the PCR and nested primer sets. TaqMan PCR with Pc-12-ne-F/Pc-12-ne-R and a TaqMan probe, Pc-TaqMan, which were designed inside of the nested PCR amplicon, generated Ct values which, in a dose-dependent manner to the amount of the target DNA and the Ct values of all the *Psm. coronafaciens* strains, were above the threshold Ct value for positive detection. TaqMan PCR generated positive Ct values from the seed extracts of the artificially inoculated oat seeds were above the 10 cfu/ml inoculation level. Assays developed in this study will serve as useful tools to detect and identify this pathogen.

Kim and Kim (2023), on the basis of comparative genomic analysis of *Psm. coronafaciens* B2017 with various *Psm. syringae* pathogens, showed that B2017 possesses the mgo operon encoding mangotoxin production. They used RT-PCR analysis and marker exchange mutagenesis to study the regulation of transcription from mgoB and mgoC genes in the cluster and the importance of mangotoxin production in the development of halo blight disease. The mgo operon in *Psm. coronafaciens* B2017 is similar to those in other *Psm. syringae* pathovars with four genes (mgoBCAD) in a transcription unit, as shown by RT-PCR analysis. Mutants defective in mgoB and mgoC were generated by marker exchange mutagenesis. Both the disease incidence and the severity of halo blight on oat seedlings were reduced in these mutants. Full virulence was observed when the genes were complemented. The results indicate that production of mangotoxin is a key factor contributing to the virulence of a Korean isolate *Psm. coronafaciens* B2017 in the development of oat halo blight.

Lee *et al.* (2023) reported that the increase in temperature from 15 °C to above 20 °C significantly enhanced the development of halo blight. The foliar symptoms also developed sooner and were more severe when RH was increased from 80 to 100%; low RH (< 60%) limited symptom development. Disease incidence and RH had a strong linear regression association. Elevated spring temperatures with higher humidity due to climate warming could increase the disease.

9.6.4.1 Host Resistance

Oat varieties having resistance to halo blight exist. Porter, for example, is a selection from the cross Purdue 623A1-1-9-1/Stout and was developed by using modified pedigree breeding with single plant selection in the F2, F5 and F7 generations. It has moderate resistance to *Psm. syringae* pv. *coronafaciens* (Ohm *et al.*, 1982). Symptoms in the halo blight-resistant oat line Cc 4146 and in barley were similar to those in leaves of the susceptible oat cultivar Milford after infiltration with suspensions (10^8 cells mL^{-1}) of *Psm. syringae* pv. *coronafaciens.* However, resistance of barley and Cc 4146 was expressed if less concentrated inocula were used: symptoms either developed more slowly or were absent. Susceptibility was characterised by normosensitive collapse of infiltrated areas on the third day from inoculation after the rapid multiplication of bacteria. Wheat leaves were resistant, this being expressed not by a hypersensitive reaction but by a slight yellowing at the inoculation site after 4 days (Smith and Mansfield, 1981).

9.6.5 Management

1. Use healthy or treated seed for sowing. Seed treatment with 1 to 320 formalin for three hours, as for smut control, reduces, but does not entirely control, the disease. A hot-air treatment for 30 hours at 100 °C does control the blight.
2. Burn, bury or graze oat stubbles. Deep ploughing helps in burying and destruction of the stubbles leading to reduction in pathogen population.
3. Use disease-resistant/-tolerant varieties.
4. Follow crop rotation.

REFERENCES

An, J. H., Noh, Y. H., Kim. Y. E., Lee, H. I. and Cha, J. S. (2015). Development of PCR and TaqMan PCR assays to detect *Pseudomonas coronafaciens*, a causal agent of halo blight of oats. *Plant Pathol. J.* 31: 25–32.

Bowden, R. L. and Stromberg, E. L. (1982). Chocolate spot of corn in Minnesota. *Plant Dis*. 66: 744.

Cunfer, B. M. and Schaad, N. W. (1976). Halo blight of rye. *Plant Dis Rept*. 60: 61–64.

Cunfer, B. M., Schaad, N. W. and Morey, D. D. (1978). Halo blight of rye: Multiplicity of symptoms under field conditions. *Phytopathology* 68: 1545–48.

Dickson, J. G. (1947). *Diseases of Field Crops*. McGraw Hill Book Company, Inc., New York.

Elliott, C. (1920). Halo blight of oats. *J. Agric. Res.* 19: 139–72.

Harder, D. E. and Harris, D. C. (1973). Halo blight of oats in Kenya. *East Afr. Agric. Forest. J.* 38: 241–45.

Janse, J. D. (2005). *Phytobacteriology: Principles and Practice*. Cabi Publishing, Wallingford, 360 pp.

Kim, Y. C. (2020). Complete genome resource of *Pseudomonas coronafaciens* x-1 causing halo blight disease in oats. *Mol. Plant-Micro. Inter*. 33: 1100–02.

Kim, Y. N. and Kim, Y. C. (2023). Production of mangotoxin is a key virulence factor in *Pseudomonas coronafaciens* causes halo blight in oats. (Abstr.) *Phytopathology* 113: S3.1.

Lee, J., Kim, Y. N. and Kim, Y. C. (2023). Effect of climate on development of halo blight in oats caused by *Pseudomonas coronafaciens*. *Phytopathology* 113: S3.41.

Ohm, H. W., Patterson, F. L., Shaner, G. E., Foster, J. E., Finney, R. E. *et al*. (1982). Registration of Porter spring oat (Reg. No. 299). *Crop Sci*. 22: 447–48.

Paul, V. H. and Smith, I. M. (1989). Bacterial pathogens of Gramineae: Systematic review and assessment of quarantine for the EPPO region. *EPPO Bull*. 19: 33–42.

Persson, P. and Sletten, A. (2001). Halo blight of oats in Scandinavia. In: S. H. De Boer (ed.). *Plant Pathogenic Bacteria*. Springer, Dordrecht, pp. 265–68. https://doi.org/10.1007/978-94-010-0003-1_61

Reid, W. D. (1938). Halo blight of oats. *N. Z. J. Sci. Techn*. 20A: 266–68.

Ribeiro, R. de L. D., Durbin, R. D., Arny, D. C. and Uchytil, T. F. (1977). Characterization of the bacterium inciting chocolate spot of corn. *Phytopathology* 67: 1427–31.

Schaad, N. W. and Cunfer, B. M. (1979). Synonymy of *Pseudomonas coronafaciens*, *Pseudomonas coronafaciens* pathovar *zeae*, *Pseudomonas coronafaciens* subsp. *atropurpurea* and *Pseudomonas striafaciens*. *Int. J. Syst. Bacteriol*. 29: 213–21.

Smith J. J. and Mansfield, J. W. (1981). Interactions between pseudomonads and leaves of oats, wheat and barley. *Physiol. Plant Pathol*. 18: 345–56.

Stevens, F. L. (1925). *Plant Disease Fungi*. MacMillan Company, New York, 469 pp.

Wang, J., Zhang, R., Deng, H., Han, X., Wei, X. *et al*. (2024). *Pseudomonas syringae* pv. *coronafaciens*, causing a new bacterial halo blight disease on oat (*Avena sativa* L.) in China. *Eur. J. Plant Pathol*. 170: 631–47.

Wilkie, J. P. (1972). Halo blight of oats in New Zealand. *N. Z. J. Agric. Res*. 15: 461–68.

9.7 BACTERIAL LEAF STREAK OF CEREAL CROPS

Bacterial leaf streak (hereafter, referred to as BLS) of wheat and barley had been noticed or recorded around the world even before the 20th century, but official reports and the characterisation of the causal bacterial pathogens were not made until the 1910s (Jones *et al.*, 1916; Smith *et al.*, 1919). Later, similar diseases were also identified and characterised on other small grains and related grasses (reviewed by Sapkota *et al.*, 2020). BLS of wheat and barley has a global distribution with the exception of some particular regions, for example Western Europe, where environmental conditions are not favourable for the development of the disease (Duveiller *et al.*, 1997). Yield losses due to BLS on wheat and barley can be variable, rising even as high as 40% (Forster and Schaad, 1988; Shane *et al.*, 1987). BLS on maize is a new and emerging disease constraining maize production in the United States. Liu *et al.* (2023) have given a brief account of the disease, covering its various aspects and also highlighting the gaps in the information. Duveiller *et al.* (2002) have given a detailed account of bacterial leaf streak and black chaff of wheat. Similarly, Kaur *et al.* (2020) have also published a detailed account of bacterial leaf streak of wheat, highlighting various aspects of the disease.

Most studies concerning yield losses have been based on the single-tiller approach (Forster and Schaad, 1988; Shane *et al.*, 1987; Tillman *et al.*, 1999). The yield losses in wheat due to BLS are dependent on disease pressure and genotype, but losses as high as 30–40% have been reported (Duveiller *et al.*, 1997). For an individual genotype, a linear relationship is usually observed between flag leaf severity and yield losses (Duveiller *et al.*, 1997). Friskop *et al.* (2023), on the basis of field experiments conducted over multiple years and locations, estimated yield and economic losses in North Dakota spring wheat due to BLS. Their study revealed that highly susceptible spring wheat cultivars suffered yield losses up to 60%. In combination with disease survey and variety acreage data, the authors estimated direct economic losses of $4.2 million and $8 million for the years 2019 and 2020, respectively, in the state.

In their review, Osdaghi *et al.* (2023) have provided an updated overview of the detection and diagnosis of the bacterial leaf streak pathogens. First, a short history of the leaf streak pathogens is given; this is then followed by the symptomology and host range of the causal agents. Then, the utility of conventional methods and high-throughput molecular approaches in the precise detection and identification of the pathogens are given. Finally, the authors highlight the role of quarantine inspections and early detection of the pathogen in combating the risk of bacterial leaf streak in the 21st century's small-grains cereals' industry.

Friskop *et al.* (2023) reported that BLS-affected hectares of hard red spring wheat (HRSW) in North Dakota grew massively, from 747 to 141,680 between 2015 and 2021. Yield loss observed at multiple performance trial locations was as high as 60% on susceptible varieties. The number of BLS-affected hectares was the highest in 2019 and 2020, and direct economic losses for North Dakota HRSW producers were estimated to be as high as $4.7 and $8.0 million, respectively. These data highlight the importance of loss due to BLS in HRSW and the need to procure resources for breeding efforts and growers' education on management of BLS.

9.7.1 Host Range

Xan. translucens and its pathovars has a very wide host range, including cereals like wheat, barley, rice, oats, rye, sorghum and maize; and many graminaceous (grass family) weedy grasses.

9.7.2 Symptoms

Typical symptoms on the leaf consist of elongated, light brown lesions, several centimetres long, which are initially distinct but later coalesce to cover larger areas. Early symptoms are characterised by translucent stripes that are easily seen under incident light. Initially, the lesions are water-soaked and produce honey-like exudates, producing a milky slime under humid conditions (Smith, 1917). If undisturbed, the exudates harden into yellowish, resinous granules or scales studding the surface of the lesions and are easily detachable.

On the glumes, BLS is characterised by black, longitudinal, more or less parallel stripes that are more numerous and conspicuous on the upper parts (Smith, 1917). Bacterial leaf streak can be recognised by a greasy appearance or alternating bands of diseased and healthy areas on the awns. Purple-black symptoms may extend to the peduncle between the inflorescence and the flag leaf, and may sometimes show a yellow centre (Forster *et al.*, 1986).

Sharp, discoloured interveinal streaks on the glumes suggest the presence of *Xanthomonas*, particularly if also

FIGURE 9.6 Bacterial leaf streak of cereal crops.

irregularly distributed on the spike and if there is abundant BLS on the leaves. In contrast, melanosis on the peduncle, which occurs on the same side of most culms in a field as a result of exposure to sunshine, is indicative of brown melanosis.

9.7.3 Causal Organisms

The bacterial pathogens causing BLS on small grains and bacterial wilt on forage crops are related and have been classified into a single species, i.e., *Xan. translucens*, with several pathovars described according to the results from DNA–DNA hybridisation and host pathogenicity tests (Bragard *et al.*, 1997; Vauterin *et al.*, 1992, 1995). Recent molecular and genomic data have confirmed and further refined this classification system (Goettelmann *et al.*, 2022; Langlois *et al.*, 2017; Peng *et al.*, 2016). The availability of genome sequences for this group of bacterial pathogens has allowed us to catalogue bacterial effector genes that might be involved in virulence and host–pathogen interaction (Sapkota *et al.*, 2020); however, their roles in virulence, as well as the underlying molecular mechanisms, remain largely unknown. The taxonomy of *Xan. translucens* has undergone many changes in history and has gradually become clearer with the results of molecular phylogenetic and genomic studies. A total of 11 pathovars had been described after the establishment of the species, i.e., *translucens*, *hordei*, *undulosa*, *secalis*, *cerealis*, *pistaciae*, *graminis*, *arrhenatheri*, *poae*, *phlei* and *phleipratensis*. The first five pathovars cause BLS on cereal crops and are closely related, forming the "translucens" group, while the last five pathovars cause bacterial wilt diseases on forage crops constituting the "graminis" group (Sapkota *et al.*, 2020). Strains in pv. *pistaciae* cause dieback on pistachio (*Pistacia vera* L.) and are genetically closer to the translucens group with two phylogenetic subclades: Pistaciae A and Pistaciae B (Giblot-Ducray *et al.*, 2009).

The names of four BLS-inducing pathovars of *Xan. translucens* included in the International Society for Plant Pathology (ISPP) list of plant pathogenic bacteria (Young *et al.*, 1996) are:

1. *Xanthomonas translucens* pv. *translucens* (Jones *et al.* 1917) Vauterin, Hoste, Kersters and Swings 1995;
2. *Xanthomonas translucens* pv. *cerealis* (Hagborg 1942) Vauterin, Hoste, Kersters and Swings 1995;
3. *Xanthomonas translucens* pv. *secalis* (Reddy *et al.* 1917*) Vauterin *et al.* 1995;
4. *Xanthomonas translucens* pv. *undulosa* (Smith *et al.* 1919) Vauterin *et al.* 1995.

Xan. translucens pathovars, namely, *arrhenatheri*, *graminis*, *hordei*, *phlei*, *poae* and *phleipratensis*, also cause bacterial leaf streak of cereal crops and graminaceous grassy weeds.

The bacteria are non-sporing, Gram-stain-negative and motile by a single polar flagellum. Cells are rods, measure 0.4–0.8 × 1.0–2.5 μm, occur singly or in pairs, except in peptonised nutrient broth with 2 percent sodium chloride (NaCl) in which long non-motile chains are formed (Jones *et al.*, 1917; Dowson, 1939). *Xan. translucens* is oxidative and there is no reduction of nitrates to nitrites. Hypersensitivity reaction on tobacco is positive (Mohan and Mehta, 1985), but this reaction is not always clear and a cool incubation temperature is required.

Heiden *et al.* (2023a) found that average nucleotide identity (ANI)-based approaches organise *Xan. translucens* pv. *translucens* (hereafter, referred to as *Xtt*) strains differently from the previous standard multilocus sequencing analysis approach. According to ANI, *Xtt* forms a separate clade from *Xan. translucens* pv. *undulosa* (hereafter, referred to as *Xtu*) and consists of three main groups which are represented on multiple continents. Some virulence factors, such as 17 Type III-secreted effectors, are highly conserved and offer potential targets for the elicitation of broad resistance. However, there is a high degree of variation in virulence factors, meaning that germplasm should be screened for resistance with a diverse panel of *Xtt*.

Hong *et al.* (2023) developed DNA markers to differentiate *Xtu* and *Xtt* and used these to characterise a collection of *Xan. translucens* strains with diverse origins, followed by confirmation and characterisation with pathogenicity tests and multilocus sequence analysis/typing (hereafter, referred to as MLSA and MLST, respectively). The authors first developed cleaved amplified polymorphic sequence markers based on the single-nucleotide polymorphisms within a cereal pathovar-specific DNA sequence. In addition, two *Xtt*-specific markers, designated *Xtt*-XopM and *Xtt*-SP1, were developed from comparative genomics among the sequenced *Xtt*/*Xtu* genomes. Using the developed markers, a collection of *Xan. translucens* strains were successfully identified as *Xtu* or *Xtt*. Pathogenicity tests on wheat and barley plants and MLSA of four housekeeping genes validated the pathovar assignation of those strains. Furthermore, MLSA revealed distinct subclades within both *Xtu* and *Xtt* groups. Seven and three sequence types were identified from MLST for *Xtu* and *Xtt* strains, respectively. The establishment of efficient *Xtt*/*Xtu* differentiation methods and the characterisation of those strains will be useful in studying disease epidemiology and host–pathogen interactions and breeding programmes when screening for sources of resistance for these two important bacterial pathogens.

Román-Reyna *et al.* (2023) developed a multiplex PCR based on whole-genome differences to distinguish *Xtt* and *Xtu*. They validated the primers across different *Xanthomonas* and non-*Xanthomonas* strains. To their knowledge, this is the first multiplex PCR to distinguish *Xtt* and *Xtu*. These molecular tools will support disease management strategies enabling detection and pathovar incidence analysis of *Xan. translucens*.

Alvandi *et al.* (2023) used the comparative genomics approach to design a pathovar-specific primer pair for the

* References marked with an * are not cited in the bibliography.

detection of *Xtu* in naturally infected seeds and its differentiation from other pathovars of the species. For this purpose, complete genome sequences of strains of different *Xan. translucens* pathovars were compared and the specific PCR primer pair XtuF/XtuR was designed. These primers were strictly specific to *Xtu* because the expected 229-bp DNA fragment was not amplified in the closely related pathovars or in other xanthomonads, wheat-pathogenic bacteria and other plant-pathogenic bacteria. High sensitivity of the primer pair XtuF/XtuR allowed the detection of pure DNA of the pathogen in a concentration as low as 4.5 pg μL^{-1}. The pathogen was also detected in water suspension at a concentration of 8.6×10^2 CFU mL^{-1}. The PCR test was capable of detecting the pathogen in extracts of naturally infected wheat seeds at a concentration of 3.5×10^4 CFU/g while a culture-plate method was able to detect the pathogen at a concentration of 50×10^5 CFU/g from the same seeds. The PCR test developed in this study is a step forward for precise detection and identification of *Xtu* to prevent outbreaks of the bacterial leaf streak disease.

Bankole *et al.* (2023) developed several DNA markers to differentiate *Xtu* and *Xtt* and used them to characterise a diverse collection of *Xan. translucens* strains including strains from historical collections and contemporary strains collected in North Dakota. In this study, MLSA of four house-keeping genes (rpoD, dnaK, fyuA and gyrB) was used to infer identification and to examine the phylogenetic relationship of these strains. MLSA assigned all strains to the previously described *Xan. translucens* pathovars and correlated with the results from molecular marker analyses and pathogenicity tests. The inferred phylogenetic tree clearly separated *Xtu* and *Xtt* strains and grouped the *Xtu* strains into two clades, while separated *Xtt* strains into three clades. This work further aids the accurate identification and classification of *Xan. translucens* strains, which will be important for the breeding of durable disease resistance to BLS in wheat and barley.

Tambong *et al.* (2024) developed a TaqMan real-time PCR for specific detection/identification of *Xan. translucens* pv. *undulosa* targeting the recombination mediator gene (*rec*F). Analysis of the complete *rec*F (1,117 bp) sequences identified the gene as a reliable phylogenetic marker for the identification of *Xan. translucens* pv. *undulosa*, differentiating it from the other pathovars; *rec*F-based sequence homology values among the 11 pathovars correlated well with genome-based DNA–DNA hybridisation values. The discriminatory power of *rec*F to differentiate *Xan. translucens* pv. *undulosa* from the other pathovars is due to nucleotide polymorphic positions. They used these nucleotide polymorphisms to develop a TaqMan PCR for the specific detection of *Xan. translucens* pv. *undulosa*. The specificity of the assay was validated using 67 bacterial and fungal (oomycetes) strains. The selected primers and the double-quenched FAM-labelled TaqMan probe were specific for the detection of 11 *Xan. translucens* pv. *undulosa/secalis* strains. Fifty-six strains of other *Xan. translucens* pathovars ($n = 39$) and non-*Xanthomonas* spp. ($n = 17$) did not exhibit any detectable fluorescence. Also, greenhouse-inoculated and naturally infected wheat leaf samples showed positive reactions for the presence of *Xan. translucens* pv. *undulosa* DNA but not healthy control plants. The TaqMan assay reliably detected as low as 1-pg DNA amount and 10 colony-forming units of the target pathogen per reaction. This TaqMan assay could be useful to regulatory agencies with economic benefits to wheat growers.

Complete genome sequences of *Xan. translucens* and bioinformatic tools have helped reveal bacterial secretion systems and profile their effector contents for virulence (Charkhabi *et al.*, 2017; Goettelmann *et al.*, 2022; Peng *et al.*, 2016; Shah *et al.*, 2021). Among them, the most important are the type III secretion system (TIIISS) and TIII effectors. In *Xanthomonas*, TIII effectors can be categorised into *Xanthomonas* outer proteins (Xops) and transcription activator-like effectors (TALEs). A highly conserved TIIISS has been shown to be present in all sequenced *Xtu* and *Xtt* strains and has also been shown to be important for disease induction (Peng *et al.*, 2016; Pesce *et al.*, 2017; Shah *et al.*, 2019). The number of Xops in *Xtu* and *Xtt* strains ranges from 21 to 36 and these appear to be highly conserved between strains within each pathovar and across pathovars, but a few unique Xops have been identified for specific pathovars (Goettelmann *et al.*, 2022; Khojasteh *et al.*, 2020). In contrast, TALEs are highly diverse between strains even within each pathovar (Goettelmann *et al.*, 2022; Khojasteh *et al.*, 2019; Peng *et al.*, 2016, 2019). Three TALEs in *Xtu* and one in *Xan. translucens* pv. *cerealis* (hereafter, referred to as *Xtc*) have been shown to play a significant role in disease development (Charkhabi *et al.*, 2017; Peng *et al.*, 2019; Shah *et al.*, 2019). No study has been done to functionally test the role of Xops in virulence. In addition to TIIISS, types II (TIISS), IV and VI have also been identified in all or some pathovars of *Xan. translucens* (Goettelmann *et al.*, 2022). Shah *et al.* (2023) generated *Xtc* mutants deficient for TIISS, TIIISS and TALEs and compared wheat gene expressions in response to the wild-type strain and those mutants. The study revealed for the first time that the TIISS contributes to virulence for *Xtc* on wheat and different sets of host genes were regulated by each secretion system and effector. From a broader perspective, Heiden et al. (2023b) provided a short commentary on BLS diseases on small grains, rice and maize aiming to gain an understanding of the convergent evolution of the three diseases and common virulence mechanisms by *Xan. translucens*, *Xan. oryzae* and *Xan. vasicola*.

9.7.4 Disease Cycle

A general disease cycle for wheat and barley BLS has been proposed by Duveiller *et al.* (1997), but many aspects of the disease cycle and epidemiology remain untested or undetermined (Sapkota *et al.*, 2020).

Earlier studies have indicated that perennial grasses such as smooth brome (*Bromus inermis* Leyss.), quack grass (*Elymus repens* (L.) Gould.) and timothy (*Phleum pratense* L.) could be potential hosts for *Xan. translucens* (Boosalis, 1952; Wallin, 1946). Recently, *Xtu* strains have been successfully isolated from cultivated wild rice (*Zizania palustris*) and annual and perennial weedy grasses around wheat fields, capable of causing water-soaking symptoms on some grass hosts by artificial inoculations (Curland *et al*., 2021; Ledman *et al*., 2021). Ledman *et al.* (2023b) provided complete genome sequences of four *Xtu* strains isolated from cultivated wild rice and annual and perennial grasses. Phylogenetic analysis based on average nucleotide identity did not reveal a clear separation of these four strains from *Xtu* strains isolated from wheat. This strongly suggests that *Xtu* can survive in grasses as a potential pathogen or as an epiphytic population, and that these grasses play a significant role in the life cycle of this bacterial pathogen.

Seed is the most important source of primary inoculum, and large-scale transmission of the BLS pathogen is due to its seed-borne nature (Smith *et al*., 1919). Depending on storage conditions, it is estimated that the bacterium will die in between 63 and 81 months (Forster and Schaad, 1990). The bacterium survives poorly in soil, but does better when crop debris is present (Boosalis, 1952). Also, plant stubbles usually decay very fast in warm and humid climates and wheat pathogenic bacteria cannot survive in decomposing debris. *Xtu* can survive on weeds and grasses due to its broad host range; however, this is probably not significant on annual hosts.

BLS outbreaks are characterised by sporadic epiphytotics and are usually observed by farmers relatively late in the growing season (Forster *et al*., 1986). Moisture facilitates the pathogen's release from the seed and contributes to leaf colonisation and invasion of leaf tissue. Free water allows the pathogen to spread in the field and to disperse on the leaf, thereby increasing the number of lesions. Bacteria enter through the stomata and multiply in large masses in the parenchyma. This causes elongated streaks limited by the veins, which act as barriers. Later milky or yellow exudates form on the surface of lesions. Rain and wind greatly influence the spread of these exudates and of the disease from leaf to leaf throughout the field. Micro-injuries to awns and leaves caused by hail or wind may contribute to bacterial penetration.

BLS-causing pathogens tolerate a wide range of temperatures i.e., 15–30 °C (Duveiller *et al*., 1991). Duveiller and Maraite (1995) have shown that temperature has a major impact on epidemics. Pathogen multiplication in leaf tissue is directly dependent on temperature and dry air conditions (relative humidity less than 30%) do not limit disease progress. Symptoms only occur when temperature allows the bacterial population to reach an estimated threshold of 10^8 cfu/leaf. Low temperatures retard the multiplication of the pathogen and disease progress.

Epiphytic populations may be important for understanding the aetiology of disease and discovering why the disease is sporadic. In Mexico, it was possible to monitor a *Xtu* population in plots of symptomless genotypes contrasting in their field resistance to the pathogen (Duveiller, 1994a). The population of pathogenic bacteria decreased after a heavy rainfall, which suggests that epiphytic *Xtu* cells are present on wheat leaves before they actually penetrate the parenchyma.

The local dissemination of the pathogen occurs by rain and dew and plant-to-plant contact (Boosalis, 1952). In addition, visitors to demonstration plots, particularly in the morning when dew is at its maximum, increase the spread of bacteria. In Brazil, Mehta (1990) reported that the spread of BLS from one field to another is limited, and that pathogen spread through splashing rain is restricted to distances as short as 4 to 5 metres. The role of aphids in long-distance transmission of the pathogen is probably limited.

9.7.4.1 Host Resistance

The sources of resistance to BLS in wheat and barley are very limited, and the genetics of resistance is complicated and mostly controlled by multiple genes (Adhikari *et al*., 2012; Duveiller *et al*., 1993; Sapkota *et al*., 2020; Tillman and Harrison 1996).

A **Pathogen Spotlight Issue** was dedicated to corn BLS in 2020, which greatly advanced our understanding of the disease system in several areas (Ortiz-Castro *et al*., 2020). The Pathogen Spotlight Issue (*Phytopathology*, year 2023, Vol. 113, No. 11, pages 2020–23) mainly focuses on BLS diseases of wheat and barley and comprehensively integrates all three bacterial species. In this issue, two reviews, one perspective, one short communication and eight research articles are included, covering the economic impact of the disease, pathogen genomics and classification, pathogen detection, pathogen virulence and the genetics of host resistance.

Duveiller *et al.* (1997), at the International Maize and Wheat Improvement Centre (CIMMYT), published a manual on wheat bacterial diseases, which included a literature review on wheat BLS. The main purpose of the manual was to provide principles and guidelines for breeding and the management of the disease. The first comprehensive review of BLS on small grains was conducted by Sapkota *et al.* (2020), who summarised the general features of the disease systems and up-to-date information regarding the taxonomy and genomics of the bacterial pathogens, pathogen virulence and the genetic mapping of host resistance. Over the last few years, BLS of wheat and barley has received greater attention due to their increasing importance in many countries, and research publications on these two diseases have proliferated. This calls for new reviews for these two diseases. The first review in this Pathogen Spotlight Issue by Osdaghi *et al.* (2023) primarily focuses on pathogen detection and disease diagnosis in both field and laboratory. In this mini-review, high-resolution and high-quality photos from fresh field samples at different growth stages are provided in order to facilitate accurate and timely field disease diagnosis. Additionally, conventional and modern methods for bacterial isolation and detection are described in detail, which is useful for researchers in a variety of laboratory settings. The second review by Ledman *et al.* (2023a) highlights new research and the current understanding of BLS epidemiology, diagnostic tools, pathogen genomics and virulence, and host resistance and screening methods. In particular, new developments in genome sequences have expanded our understanding of *Xan. translucens* diversity, taxonomy and effector content and improved the tools for pathogen diagnostics (Ledman *et al.* 2023a).

9.7.5 Management

Knowledge and tools are lacking for the management of BLS on wheat and barley despite it being a major disease in many wheat- and barley-growing areas around the world for over a century. The following measures are suggested to minimise the disease incidence.

1. Cultural practices: As straw from a previously infected crop or weed hosts serves as a source of overwintering/oversummering inoculum, field infestation can be reduced if the straw is incorporated into the soil. The pathogen is short-lived in soil, so excluding weed hosts and rotating to a non-host crop will contribute to the reductions of inoculum in the field.
2. Pathogen-free seed: As seed-borne inoculum is the most frequent source of primary infections, sowing pathogen-free seed is critical for the management of BLS. Use of clean and pathogen-free seed is important, and seed multiplication should take place in disease-free areas under dry conditions without overhead irrigation. Tests can be conducted to determine the presence of the pathogen in the seeds. For pathogen-free certification, a seed wash test on a semi-selective medium after dilution plating, is widely used.

   Discarding infected seed, prior to sowing, should be the primary control measure, as sowing pathogen-free seed is the first logical step in avoiding an outbreak of the disease. Foundation seed should be multiplied in disease-free areas where climatic conditions are unfavourable for the development of epiphytotics. Seed should be disinfected before sowing, even if currently available seed treatments are not fully satisfactory. Nevertheless, even if not completely satisfactory, seed treatment with dry heat or a product such as Panoctine Plus is recommended.
3. Chemical control: Seed treatment can reduce seed-borne inoculum, but it cannot prevent outbreaks of BLS in the field, if conditions are conducive for disease development. Seed treatment with acidified cupric acetate (0.5%) at 45 °C for 20 minutes can reduce the incidence of BLS in the field due to seed-borne infections. Studies have shown some reductions in disease incidence with seed treatments including guazatine 300 g/L and imazalil 20 g/L, formaline and dry heat, but these are not sufficient to prevent disease outbreaks when contaminated seed is sown.

   Foliar-applied products with active ingredients based on mono- and dipotassium salts of phosphorous acid are labelled and commercially available for the control of BLS. These products, which are considered to have both fungicidal and bactericidal activity, are labelled for the control of *Xanthomonas* diseases in wheat. However, the efficacy of these products with regard to the control of BLS has not been demonstrated, and they are not typically recommended for management of this disease.

   As controlling black chaff through seed treatment is not easy, the breeding of resistant genotypes appears to be the best way to reduce the risk of yield losses. Therefore, evaluating resistance based on adult plant response in the field is recommended. Screening for resistance should be encouraged in areas where pathogen populations present the maximum variation.

   Shohanipor *et al.* (2025) reported that copper oxychloride and Cobre at the rate of 3000 μg mL^{-1} as preventive treatment decreased the disease severity up to 68.90 and 68.87% in the first *in planta* bioassay, and 75.67 and 74.98% in the second *in planta* bioassay, respectively.
4. Host resistance: Though commercial cultivars of wheat with high levels of BLS resistance are unavailable, some cultivars are less susceptible to the disease than others. Resistance to BLS is quantitative in nature and single-gene resistance to BLS has not been identified. Major quantitative trait loci (QTL) associated with BLS resistance have been reported across the wheat genome from at least 9 of the 42 chromosomes (chromosomes 1A, 3A, 4A, 4B, 5A, 5D, 6B and 7D). Wheat germplasm can be screened for resistance in either greenhouse assays or under field conditions. In greenhouse screening, seedlings and young plants can be evaluated by inoculating leaves with a concentration of 10^3 cfu/ml to determine the level of resistance that exists in different germplasm accessions. To promote infection and disease development following inoculation, plants are incubated in a humidity chamber (relative humidity >85%) for 5–7 days at 24–26 °C.

   For field screening, highly concentrated (10^9 cfu/mL^{-1}) bacterial suspensions are sprayed onto plants at the tillering stage. Carborundum is added as an abrasive to facilitate entry of the pathogen into the plant. Disease incidence and disease severity can be measured at the flowering stage. Disease severity is assessed based on the percent leaf area with lesions and position of the lesions (i.e. height) within the plant canopy. By screening germplasm, additional QTLs associated with BLS resistance can be identified and used in breeding programmes to select wheat varieties with improved disease resistance.

Genetic resistance is the preferable way to control the disease and research has been conducted to screen sources of resistance and also to map genetic resistance in wheat. The overall conclusion is that the sources of resistance are relatively limited, even though some lines have partial resistance (Sapkota *et al.*, 2020). Very limited research has been done for barley BLS worldwide. Ritzinger *et al.* (2022) screened 2,094 accessions of barley germplasm and identified 32 accessions with high level of resistance and a desirable maturity date. Ritzinger *et al.* (2023) reported a genome-wide association study to reveal genomic locations that are associated with BLS resistance in barley. A major and robust marker-trait association was identified on barley 6H across the panels and could be the target for marker-assisted selection in breeding programmes. Screening and genetic mapping of host resistance have been conducted for some maize commercial hybrids and diverse inbred lines (Qiu *et al.*, 2020; Robaina *et al.*, 2020). Mullens *et al.* (2023) reported the identification of quantitative trait locus (QTL) associated with BLS resistance in maize by using a recombinant inbred line population derived from the cross between B73 and CML228. In this study, three QTLs were identified, with two derived from CML228 and one from B73. Interestingly, the reaction to BLS had a significant correlation with the reaction to other fungal diseases suggesting common pathways underlying plant reactions to bacterial and fungal diseases.

REFERENCES

Adhikari, T. B., Gurung, S., Hansen, J. M., Jackson, E. W. and Bonman, J. M. (2012). Association mapping of quantitative trait loci in spring wheat landraces conferring resistance to bacterial leaf streak and spot blotch. *Plant Genome* 5: 1–16. https://doi.org/10.3835/plantgenome2011.12.0032

Alvandi, H., Taghavi, S. M., Khojasteh, M., Rahimi, T., Dutrieux, C. *et al.* (2023). Pathovar-specific PCR method for detection and identification of *Xanthomonas translucens* pv. *undulosa*. *Plant Dis.* 107: 2279–87.

Bankole, I. A., Hong, E., Curland, R., Chapara, V. and Liu, Z. (2023). Multilocus sequence typing and phylogenetic analysis of *Xanthomonas translucens* strains causing bacterial leaf streak in wheat and barley. (Abstr.) *Phytopathology* 113: S2.3.

Boosalis, M. G. (1952). The epidemiology of *Xanthomonas translucens* (J. J. and R.) Dowson on cereals and grasses. *Phytopathology* 42: 387–95. ISI.

Bragard, C., Singer, E., Alizadeh, A., Vauterin, L., Maraite, H. *et al.* (1997). *Xanthomonas translucens* from small grains: Diversity and phytopathological relevance. *Phytopathology* 87: 1111–17. https://doi.org/10.1094/PHYTO.1997.87.11.1111

Charkhabi, N. F., Booher, N. J., Peng, Z., Wang, L., Rahimian, H. *et al.* (2017). Complete genome sequencing and targeted mutagenesis reveal virulence contributions of Tal2 and Tal4b of *Xanthomonas translucens* pv. *undulosa* ICMP11055 in bacterial leaf streak of wheat. *Front. Microbiol.* 8: 1488. https://doi.org/10.3389/fmicb.2017.01488

Curland, R. D., Hallada, K. R., Ledman, K. E. and Dill-Macky, R. (2021). First report of bacterial leaf streak caused by *Xanthomonas translucens* pv. *undulosa* on cultivated wild rice (*Zizania palustris*) in Minnesota. *Plant Dis.* 105: 2711. https://doi.org/10.1094/PDIS-02-21-0407-PDN

Dowson, W. J. (1939). On the systematic position and generic names of the Gram negative bacterial plant pathogens. *Zentralblatt für Bakt.* etc. II. Abt. Bd. 100 9/13: 177–93.

Duveiller, E. (1994a). A study of *Xanthomonas campestris* pv. *undulosa* populations associated with symptomless wheat leaves. *Parasitica* 50: 109–17.

Duveiller, E., Bragard, C. and Maraite, H. (1991). Bacterial diseases of wheat in the warmer areas- reality or myth? In: D. Saunders (ed.). *Wheat for the Non-traditional Warm Areas. Proc. Int. Conf., Iguazu Falls, Brazil.* Mexico, DF, CIMMYT, pp. 189–202.

Duveiller, E., Bragard, C. and Maraite, H. (2002). Bacterial leaf streak and black chaff. Food and Agriculture Organization of the United Nations, Rome. https://www.fao.org

Duveiller, E., Fucikovsky, L. and Rudolph, K. (eds.). (1997). *The Bacterial Diseases of Wheat: Concepts and Methods of Disease Management.* CIMMYT, Mexico, D.F.

Duveiller, E. and Maraite, H. (1995). Effect of temperature and air humidity on multiplication of *Xanthomonas campestris* pv. *undulosa* and symptom expression in susceptible and field-tolerant wheat genotypes. *J. Phytopathol.* 143: 227–32.

Duveiller, E., van Ginkel, M. and Thijssen, M. (1993). Genetic analysis of resistance to bacterial leaf streak caused by *Xanthomonas campestris* pv. *undulosa* in bread wheat. *Euphytica* 66: 35–43. https://doi.org/10.1007/BF00023506

Forster, R. L., Mihuta-Grimm, L. and Schaad, N. W. (1986). Black chaff of wheat and barley. *Current Information Series*, No. 784, p. 2. University of Idaho, College of Agriculture.

Forster, R. L. and Schaad, N. W. (1988). Control of black chaff of wheat with seed treatment and a foundation seed health program. *Plant Dis.* 72: 935–38. https://doi.org/10.1094/PD-72-0935

Forster, R. L. and Schaad, N. W. (1990). Longevity of *Xanthomonas campestris* pv. *translucens* in wheat seed under two storage conditions. In: Z. Klement (ed.). *Proc. 7th Int. Conf. Plant Pathogenic Bacteria*, Budapest, Part A. Akadémiai Kiadó, Budapest, pp. 329–31.

Friskop, A., Green, A., Ransom, J., Liu, Z., Knodel, J. *et al.* (2023). Increase of bacterial leaf streak in hard red spring wheat in North Dakota and yield loss considerations. *Phytopathology* 113: 2103–09. https://doi.org/10.1094/PHYTO-08-22-0282-SA

Giblot-Ducray, D., Marefat, A., Gillings, M. R., Parkinson, N. M., Bowman, J. P. *et al.* (2009). Proposal of *Xanthomonas translucens* pv. *pistaciae* pv. nov., pathogenic to pistachio (*Pistacia vera*). *Syst. Appl. Microbiol.* 32: 549–57. https://doi.org/10.1016/j.syapm.2009.08.001

Goettelmann, F., Roman-Reyna, V., Cunnac, S., Jacobs, J. M., Bragard, C. *et al.* (2022). Complete genome assemblies of all *Xanthomonas translucens* pathotype strains reveal three genetically distinct clades. *Front. Microbiol.* 12: 817815. https://doi.org/10.3389/fmicb.2021.817815

Hagborg, W. A. F. (1942). Classification revision in *Xanthomonas translucens*. *Can. J. Res.* 20: 312–26.

Heiden, N., Broders, K. A., Hutin, M., Ortiz Castro, M., Roman-Reyna, V. *et al.* (2023a). Bacterial leaf streak diseases of plants: Symptom convergence in monocot plants by distant pathogenic *Xanthomonas* species. *Phytopathology* 113: 2048–55. https://doi.org/10.1094/PHYTO-05-23-0155-IA

Heiden, N., Roman-Reyna, V., Curland, R. D., Dill-Macky, R. and Jacobs, J. M. (2023b). Comparative genomics of barley-infecting *Xanthomonas translucens* shows overall genetic similarity but globally distributed virulence factor diversity. *Phytopathology* 113: 2056–61. https://doi.org/10.1094/PHYTO-04-22-0113-SC

Hong, E., Bankole, I. A., Zhao, B., Shi, G., Buck, J. W. *et al.* (2023). DNA markers, pathogenicity test, and multilocus sequence analysis to differentiate and characterize cereal-specific *Xanthomonas translucens* strains. *Phytopathology* 113: 2062–72. https://doi.org/10.1094/PHYTO-10-22-0381-SA

Jones, L. R., Johnson, A. G. and Reddy, C. S. (1916). Bacterial blights of barley and certain other cereals. *Science* 44: 432–33. https://doi.org/10.1126/science.44.1134.432

Jones, L. R., Johnson, A. G. and Reddy, C. S. (1917). Bacterial blight of barley. *J. Agric. Res.* 11: 625–43.

Kaur, N., Ishimaru, C., Vinatzer, B. A. and Mehl, H. L. (2020). Bacterial leaf streak of wheat. *The Plant Health Instructor*.

Khojasteh, M., Shah, S. M. A., Haq, F., Xu, X., Taghavi, S. M. *et al.* (2020). Transcription activator-like effectors diversity in Iranian strains of *Xanthomonas translucens*. *Phytopathology* 110: 758–67. https://doi.org/10.1094/PHYTO-11-19-0428-R

Khojasteh, M., Taghavi, S. M., Khodaygan, P., Hamzehzarghani, H., Chen, G. *et al.* (2019). Molecular typing reveals high genetic diversity of *Xanthomonas translucens* strains infecting small-grain cereals in Iran. *Appl. Environ. Microbiol.* 85: e01518–19. https://doi.org/10.1128/AEM.01518-19

Langlois, P. A., Snelling, J., Hamilton, J. P., Bragard, C., Koebnik, R. *et al.* (2017). Characterization of the *Xanthomonas translucens* complex using draft genomes, comparative genomics, phylogenetic analysis, and diagnostic LAMP assays. *Phytopathology* 107: 519–27. https://doi.org/10.1094/PHYTO-08-16-0286-R

Ledman, K. E., Curland, R. D., Ishimaru, C. A. and Dill-Macky, R. (2021). *Xanthomonas translucens* pv. *undulosa* identified on common weedy grasses in naturally infected wheat fields in Minnesota. *Phytopathology* 111: 1114–21. https://doi.org/10.1094/PHYTO-08-20-0337-R

Ledman, K. E., Osdaghi, E., Curland, R. D., Liu, Z. and Dill-Macky, R. (2023a). Epidemiology, host resistance, and genomics of the small grain cereals pathogen *Xanthomonas translucens*: New advances and future prospects. *Phytopathology* 113: 2037–47. https://doi.org/10.1094/PHYTO-11-22-0403-SA

Ledman, K. E., Roman-Reyna, V., Curland, R. D., Heiden, N., Jacobs, J. M. *et al.* (2023b). Comparative genomics of *Xanthomonas translucens* pv. *undulosa* strains isolated from weedy grasses and cultivated wild rice. *Phytopathology* 113: 2083–90.

Liu, Z., Friskop, A., Jacobs, J. M. and Dill-Macky, R. (2023). Bacterial leaf streak: A persistent and increasingly important disease problem for cereal crops. *Phytopathology* 113: 2020–23.

Mehta, Y. R. (1990). Management of *Xanthomonas campestris* pv. *undulosa* and *hordei* through cereal seed testing. *Seed Sci. Tech.* 18: 467–76.

Mohan, S. K. and Mehta, Y. R. (1985). Studies on *Xanthomonas campestris* pv. *undulosa* in wheat and triticale in Paraná State. *Fitopatologia Brasileira* 10: 447–53 (in Portuguese).

Mullens, A., Lipka, A. E., Balint-Kurti, P. and Jamann, T. (2023). Exploring the relationship between pattern-triggered immunity and quantitative resistance to *Xanthomonas vasicola* pv. *vasculorum* in maize. *Phytopathology* 113: 2127–33. https://doi.org/10.1094/PHYTO-09-22-0357-SA

Ortiz-Castro, M., Hartman, T., Coutinho, T., Lang, J. M., Korus *et al.* (2020). Current understanding of the history, global spread, ecology, evolution, and management of the corn bacterial leaf streak pathogen, *Xanthomonas vasicola* pv. *vasculorum*. *Phytopathology* 110: 1124–31. https://doi.org/10.1094/PHYTO-01-20-0018-PER

Osdaghi, E., Taghavi, S. M., Aliabadi, A. A., Khojasteh, M., Abachi, H. *et al.* (2023). Detection and diagnosis of bacterial leaf streak on small grain cereals: From laboratory to field. *Phytopathology* 113: 2024–36.

Peng, Z., Hu, Y., Xie, J., Potnis, N., Akhunova, A. *et al.* (2016). Long read and single molecule DNA sequencing simplifies genome assembly and TAL effector gene analysis of *Xanthomonas translucens*. *BMC Genom* 17: 1–19. https://doi.org/10.1186/s12864-015-2348-9

Peng, Z., Hu, Y., Zhang, J., Huguet-Tapia, J. C., Block, A. K. *et al.* (2019). *Xanthomonas translucens* commandeers the host rate-limiting step in ABA biosynthesis for disease susceptibility. *Proc. Natl. Acad. Sci.* 116: 20938–46. https://doi.org/10.1073/pnas.1911660116

Pesce, C., Jacobs, J. M., Berthelot, E., Perret, M., Vancheva, T. *et al.* (2017). Comparative genomics identifies a novel conserved protein, HpaT, in proteobacterial type III secretion systems that do not possess the putative translocon protein HrpF. *Front. Microbiol.* 8: 1177. https://doi.org/10.3389/fmicb.2017.01177

Qiu, Y., Kaiser, C., Schmidt, C., Broders, K., Robertson, A. E. *et al.* (2020). Identification of quantitative trait loci associated with maize resistance to bacterial leaf streak. *Crop Sci.* 60: 226–37. https://doi.org/10.1002/csc2.20099

Ritzinger, M., Sallam, A. H., Smith, K. P., Case, A. J., Wodarek, J. *et al.* (2023). Genome-wide association mapping of bacterial leaf streak resistance in two elite barley breeding panels. *Phytopathology* 113: 2119–26. https://doi.org/10.1094/PHYTO-10-22-0402-SA

Ritzinger, M. G., Smith, K. P., Case, A. J., Wodarek, J. R., Dill-Macky, R. *et al.* (2022). Sources of bacterial leaf streak resistance identified in a diverse collection of barley germplasm. *Plant Dis.* 107: 802–8. https://doi.org/10.1094/PDIS-04-22-0751-RE

Robaina, R. R., Longhi, T. V., Zeffa, D. M., Gonçalves, L. S. A. and Leite, R. P., Jr. (2020). Development of a protocol and a diagrammatic scale for quantification of bacterial leaf streak disease on young plants of maize. *Plant Dis.* 104: 2921–27. https://doi.org/10.1094/PDIS-01-20-0041-RE

Román-Reyna, V., Curland, R. D., Velez-Negron, Y., Ledman, K. E., Gutierrez-Castillo, D. E. *et al.* (2023). Development of genome-driven, lifestyle-informed markers for identification of the cereal-infecting pathogens *Xanthomonas translucens* pathovars *undulosa* and *translucens*. *Phytopathology* 113: 2110–18.

Sapkota, S., Mergoum, M. and Liu, Z. (2020). The translucens group of *Xanthomonas translucens*: Complicated and important pathogens causing bacterial leaf streak on cereals. *Mol. Plant Pathol.* 21: 291–302. https://doi.org/10.1111/mpp.12909

Shah, S. M. A., Haq, F., Ma, W., Xu, X., Wang, S. *et al.* (2019). Tal1NXtc01 in *Xanthomonas translucens* pv. *cerealis* contributes to virulence in bacterial leaf streak of wheat. *Front. Microbiol.* 10: 2040. https://doi.org/10.3389/fmicb.2019.02040

Shah, S. M. A., Khojasteh, M., Wang, Q., Haq, F., Xu, X. *et al.* (2023). Comparative transcriptomic analysis of wheat cultivars in response to *Xanthomonas translucens* pv. *cerealis* and its T2SS, T3SS, and TALEs deficient strains. *Phytopathology* 113: 2073–82. https://doi.org/10.1094/PHYTO-02-23-0049-SA

Shah, S. M. A., Khojasteh, M., Wang, Q., Taghavi, S. M., Xu, Z. *et al.* (2021). Genomics-enabled novel insight into the pathovar-specific population structure of the bacterial leaf streak pathogen *Xanthomonas translucens* in small grain cereals. *Front. Microbiol.* 12: 674952. https://doi.org/10.3389/fmicb.2021.674952

Shane, W. W., Baumer, J. S. and Teng, P. S. (1987). Crop losses caused by *Xanthomonas* streak on spring wheat and barley. *Plant Dis.* 71: 927–30. https://doi.org/10.1094/PD-71-0927

Shohanipor, T., Sharifi, R. and Hosseini, S. (2025). Efficacy assessment of some copper compounds on the bacterial leaf streak pathogen, *Xanthomonas translucens*. *Eur. Jour. Plant Pathol.* Vol. 173: pp. 877–92. https://doi.org/10.1007/s10658-025-03099-w

Smith, E. F. (1917). A new disease of wheat. *J. Agric. Res.* 10: 51–54.

Smith, E. F., Jones, L. R. and Reddy, C. S. (1919). The black chaff of wheat. *Science* 50: 48. https://doi.org/10.1126/science.50.1280.48.a

Tambong, J. T., Xu, R., Fleitas, M. C., Wang, L., Akuma, M. *et al.* (2024). TaqMan Real-Time PCR assay for specific detection and differentiation of *Xanthomonas translucens* pv. *undulosa* from other pathovars targeting a recombination mediator gene, *recF*. *Plant Dis.* 108: 1869–78. https://doi.org/10.1094/PDIS-09-23-1827-RE

Tillman, B. L. and Harrison, S. A. (1996). Heritability of resistance to bacterial streak in winter wheat. *Crop Sci.* 36: 412–18. https://doi.org/10.2135/cropsci1996.0011183X003600020033x

Tillman, B. L., Kursell, W. S., Harrison, S. A. and Russin, J. S. (1999). Yield loss caused by bacterial streak in winter wheat. *Plant Dis.* 83: 609–14. https://doi.org/10.1094/PDIS.1999.83.7.609

Vauterin, L., Hoste, B., Kersters, K. and Swings, J. (1995). Reclassification of *Xanthomonas*. *Int. J. Syst. Bacteriol.* 45: 472–89. https://doi.org/10.1099/00207713-45-3-472

Vauterin, L., Yang, P., Hoste, B., Pot, B., Swings, J. *et al.* (1992). Taxonomy of xanthomonads from cereals and grasses based on SDS-PAGE of proteins, fatty acid analysis and DNA hybridization. *Journal of General Microbiology* 138: 1467–77. https://doi.org/10.1099/00221287-138-7-1467

Wallin, J. R. (1946). Seed and seedling infection of barley, bromegrass, and wheat by *Xanthomonas translucens* var. *cerealis*. *Phytopathology* 36: 445–57.

Young, J. M., Saddler, G. S., Takikawa, Y., De Boer, S. H., Vauterin, L. *et al.* (1996). Names of plant pathogenic bacteria 1864–1995. *Rev. Plant Pathol.* 75: 721–63.

9.8 BACTERIAL STALK ROT OF MAIZE

Four bacterial plant pathogens, namely, *Dickeya zeae* (syn. *Erwinia chrysanthemi* pv. *zeae*, *Pectobacterium chrysanthemi* pv. *zeae*), *Enterobacter dissolvens* (syn. *Erw. dissolvens*), *Pseudomonas syringae* pv. *lapsa* (syn. *Psm. lapsa*) and *Acidovorax avenae* subsp. *avenae* (syn. *Psm. alboprecipitans*), have been reported to cause stalk rot in maize (*Zea mays* L.). Among the diseases caused by the above-mentioned four bacterial pathogens, stalk rot caused by *Dic. zeae* is the most important and widespread disease of maize and, hence, is described below. It was first reported from the United Arab Republic by Sabet (1954), who identified the bacterium as *Erw. carotovora* f. sp. *zeae*. Since then, it has been reported from many countries, including the USA, where it is called **top rot**. Martinez-Cisneros *et al.* (2014) reported it from Mexico and also hinted that the control of causal bacterium's vector, *Chilo partellus*, can aid in the management of the disease.

Hingorani *et al.* (1959) reported the disease for the first time from India. In India, it is one of the major diseases of maize and is prevalent in Himachal Pradesh, Uttar Pradesh, Delhi, Punjab, Rajasthan, Bihar, Andhra Pradesh and West Bengal. During *Kharif* 1969, an outbreak of the disease occurred in the Sundar Nagar area of Mandi district of Himachal Pradesh on hybrid Ganga 5; the disease incidence reached as high as 80–85% in several fields (Thind, 1970). The work done on the disease in India has been reviewed by Thind and Payak (1985). Kumar *et al.* (2017b), in their review article on bacterial stalk rot of maize, have presented information on nomenclature; biochemical, physiological and molecular characterisation; host range and survival of the pathogen; etiology and epidemiology of the disease; germplasm evaluation; and disease management. A brief account of Erwinia stalk rot of maize is also given by Payak (1972).

Prokić *et al.* (2020) reported that phenotypic tests and the amplification of the specific 420-bp fragment in PCR assay of the strains (isolated from maize plants showing bacterial stalk soft rot symptoms) revealed that the strains belong to genus *Dickeya*. Using ERIC-PCR analysis, seven different genetic profiles were obtained, suggesting the presence of genetic diversity in the population of this pathogen in Serbia. Additionally, phylogenetic analysis based on the *recA* gene sequence analysis indicated that the strains isolated from soft rotted maize plants belong to *Dickeya zeae*, leading to the conclusion that this bacterium was the causal agent of stalk soft rot of maize in Serbia.

9.8.1 Host Range

The bacterium has a very wide host range. It infects many fleshy fruits and vegetables and solanaceous plants such as brinjal, tomato and tobacco. The bacterium also infects sorghum, pearl millet and sugarcane.

9.8.2 Symptoms

In nature, the symptoms usually appear at the flowering stage. Partially, the seriousness of the disease lies in its often-sudden and dramatic appearance and the swiftness with which it can kill the plants (within 2–4 days) under favourable weather conditions. The first symptoms are the water-soaking and rotting of basal leaves, especially the leaf sheaths. The close examination of infected plants reveals water-soaking, and the rotting and discolouration of basal internodes. The rind loses its healthy green colour and resembles pale straw. The diseased stalks appear as they have been boiled in water. Infected plants wilt.

Infection advances rapidly covering the considerable length or the entire length of the stalk. Two most characteristic symptoms of the disease are: 1) Infected plants emit a sweet fermenting smell and 2) Infected plants usually topple down from 2nd or 3rd basal internode (Figure 9.7a). It is mainly a parenchymatous disease. In advanced infection, the pith is completely destroyed, leaving the vascular bundles in a disorganised state, but intact. Longitudinally splitted stems show discolouration of the pith (Figure 9.7b).

Occasionally, developing ears, shanks and stalks get infected at a level higher than the ground level. In the majority of these cases, the borer attack is found to be associated. In such cases, ears droop down and hang limply. This occurs due to the softening of shanks. The husk leaves, the grains, if any, as well as the cob become soft and appear water-soaked. In the USA, the disease is known as **top rot** because the symptoms usually appear on apical portion of the plants including the leaves. It is primarily due to the use of sprinkler irrigation system.

(a)

FIGURE 9.7A Bacterial stalk rot of maize: Stalks appear as boiled in water and on the right side, a toppled plant.

(b)

FIGURE 9.7B Bacterial stalk rot of maize: Longitudinally split stem shows discolouration of the pith. (Source: www.cropdisease.cropsci.illinois.edu).

The symptoms of stalk rots caused by other bacteria are given below to distinguish these from those described above. *Psm. syringae* pv. *lapsa* causes the rapid rotting of stalks and leaves. Affected plants collapse, but do not emit a fermenting smell. *Aci. avenae* subsp. *avenae* causes bacterial leaf blight and stalk rot of maize. The symptoms include leaf blight accompanied by the shredding of leaves and the rotting of stalks. Infection occurs at the top, leading to rotting and bleaching of the top. Multiple earing is also a common feature. Infection caused by *Entb. dissolvens* is generally localised and may not cover the whole circumference or length of the stalk.

9.8.3 Causal Organism

Dickeya zeae (Burkholder *et al.* 1953) Samson *et al.* 2005

Syn. *Erw. chrysanthemi* pv. *zeae*, *Pec. chrysanthemi* pv. *zeae*

Sabet (1954), on the basis of similarity of stalk rot pathogen to *Erw. carotovora* in cultural, morphological and biochemical characters and its ability to attack maize and other graminaceous hosts in addition to other known hosts of *Erw. carotovora*, named it as *Erw. carotovora* f. sp. *zeae*. However, the assignment of the epithet of *forma specialis* to this bacterium by Sabet (1954) was not correct, as there was no specialisation or restriction in attacking the hosts by the maize pathogen. On the contrary, it attacks more hosts than *Erw. carotovora*. Graham (1964), while reviewing the taxonomy of soft rot coliform bacteria, disfavoured the use of *forma specialis* for maize pathogen, as there was no specialisation in pathogenicity, similar to that found in rust fungi. In the eighth edition of *Bergey's Manual of Determinative Bacteriology*, it was named *Erw. chrysanthemi* Burk. e*t al.* corn pathotype (Buchanan and Gibbons, 1974). Victoria *et al.* (1975) named it *Erw. chrysanthemi* pv. *zeae*. Hauben *et al.* (1998) renamed it *Pectobacterium chrysanthemi*. Samson *et al.* (2005) renamed it *Dickeya zeae* sp. nov.

Cells are Gram-stain-negative rods (0.6–0.8 × 1.5–3.0 µm) with rounded ends, motile by peritrichous flagella, facultative anaerobes and metabolism is fermentative. The bacterium produces indole and H_2S, reduces nitrates, liquefies gelatin and degrades pectin. Oxidase, urea hydrolase, phenylalanine deaminase and urease tests are negative. Citrate, formate, lactate and malonate salts are used as sources of carbon and energy. G + C content of the DNA is 56.4 mol %.

Kumar *et al.* (2015) reported that 59 isolates of *D. zeae* obtained from maize-growing areas of Punjab during *Kharif* 2012 and 2013 showed variation in virulence when tested on susceptible cv. Punjab Sweet Corn-1 and were grouped as highly virulent, moderately virulent and less virulent. All the isolates produced cavities on crystal violet pectate medium. Characterisations of these isolates on the basis of 27 biochemical tests showed differential reactions for utilisation of starch and other carbohydrates, gelatin liquefication and growth at high salt concentration. Multiple antibiotic resistance was also observed in all the isolates tested. Numerical analysis of phenotypic features revealed two major groups of isolates associated at 60% similarity coefficient.

The bacterium is highly sensitive to 1 µg mL^{-1} of chlorine and 1, 10, 25 and 50 µg mL^{-1} of Streptocycline. Highly motile cultures are also highly virulent while non-motile cultures are non-infective. Phages specific to the bacterium occur in nature (Thind, 1970).

9.8.4 Disease Cycle

Infected plant debris seems to be the only known source of survival of the bacterium. The bacterium can survive for 9 months in soil containing naturally and artificially infected stalks. Maize–potato–maize rotation increases the disease incidence as potato also acts as a host for the bacterium. The bacterium enters the host through wounds. Maize borer (*Chilo partellus*) have been shown to transmit the bacterium (Thind and Singh, 1976). Dalmacio *et al.* (2007) have also reported the association of Asian corn borer (*Ostrinia furnacalis*) with the occurrence of the disease. Optimum temperature for disease development is 28–30° C and relative humidity above 90% also favours the disease development. High doses of nitrogenous fertilisers increase the incidence of the disease. Nitrogen doses of 120 and 180 kg per hectare, without any level of potash and phosphorus, significantly increased the disease incidence than 60 kg of nitrogen (Saxena and Lall, 1981). They recommended 60, 60 and 40 kg of nitrogen, phosphorus and potash, respectively per hectare to keep the disease under check.

A positive correlation exists between early infection/disease incidence and loss in grain yield. Early infection leads to heavy losses or death of plants. Screening of 32 maize genotypes, including inbred lines, hybrids, composites and open-pollinated varieties, revealed the disease incidence ranging from 42.9–100%, yield losses ranging from 21.3–98.8% and high susceptibility of most of the hybrids and composites (Thind and Payak, 1978). High levels of phenols in hybrid Ganga 2 plants after 60 and 75 days of planting, i.e., flowering stage, in comparison to 40 and 95 days after planting were implicated with relatively more resistance at flowering.

Kumar *et al.* (2017a) analysed activities of phenylalanine ammonia lyase 1 (PAL), peroxidase (POX), and polyphenoloxidase (PPO) from three sets of maize lines showing moderately resistant, moderately susceptible and highly susceptible reaction to *D. zeae* and found an elevated activity of these enzymes in all the three sets. Moderately resistant lines showed significantly more activity of these enzymes than the highly susceptible lines. The activity of PAL and PPO peaked after 48 h and of POX after 72 h of challenge inoculation with *D. zeae* in all the maize lines. The activity of these enzymes was also negatively correlated with disease development. It is evident that PAL, POX and PPO play an important role in imparting resistance in maize against *D. zeae*.

9.8.5 Management

1. Removal and destruction of diseased plant debris are of the utmost importance for the management of the disease.
2. Avoid water logging in the field.
3. Number of plants should not exceed the recommended population density, i.e. 55,000/ha.
4. Grow resistant cultivars. Dalmacio *et al.* (2007) reported that transgenic hybrids showed significantly less bacterial stalk rot mortality and ear rot incidence and gave higher grain yield in comparison to their conventional counterparts.
5. Chlorinated water has been tried and found effective in the USA (Thompson, 1965) and India (Thind and Payak, 1972; Sharma *et al.*, 1982). Thind and Payak (1972), on the basis of three-year field trials, reported that soil drenching with chlorinated

water (100 μg mL^{-1} chlorine) at two-week intervals, starting from knee-high stage to flowering stage, reduced the disease incidence by 75.92%. One pre-planting application was also given. Lall and Saxena (1978) recommended two applications of bleaching powder @ 25 kg/ha per application. The first application should be given at flowering and the second, 10 days after the first application.

REFERENCES

Buchanan, R. E. and Gibbons, N. E. (eds) (1974). *Bergey's Manual of Determinative Bacteriology*. 8th Ed. The Williams and Wilkins Co. Baltimore, MD. 1268 pp.

Burkholder, W. H., McFadden, L. A. and Dimock, A. W. (1953). A bacterial blight of chrysanthemums. *Phytopathology* 43: 522–26.

Dalmacio, S. C., Lugod, T. R., Serrano, E. M. and Munkvold, G. P. (2007). Reduced incidence of bacterial stalk rot on transgenic insect-resistant maize in the Philippines. *Plant Dis*. 91: 346–51.

Graham, D. C. (1964). Taxonomy of the soft rot coliform bacteria. *Annu. Rev. Phytopathol*. 2: 13–42.

Hauben, L., Moore, E. R. B., Vauterin, L. Steenackers, M., Mergaert, J.*et al*. (1998). Phylogenetic position of phytopathogens within the Enterobacteriaceae. *Syst. Appl. Microbiol*. 21: 384–97.

Hingorani, M. K., Grant, U. J. and Singh, N. J. (1959). *Erwinia carotovora* f. sp. *zeae, a* destructive pathogen of maize in India. *Indian Phytopathol*. 12: 151–57.

Kumar, A., Hunjan, M. S., Kaur, H., Dhillon, H. K. and Singh, P. P. (2017a). Biochemical responses associated with resistance to bacterial stalk rot caused by *Dickeya zeae* in maize. *J. Phytopathol*. 165: 822–32.

Kumar, A., Hunjan, M. S., Kaur, H., Rawal, R., Kumar, A., *et al*. (2017b). A review on bacterial stalk rot disease of maize caused by *Dickeya zeae*. *J. Appl. Natural Sci*. 9: 1214–25.

Kumar, A., Hunjan, M. S., Kaur, H. and Singh, P. P. (2015). Characterization of *Dickeya zeae* isolates causing stalk rot of maize based on biochemical assays and antibiotic sensitivity. *Indian Phytopathol*. 68: 375–79.

Lall, S. and Saxena, S. C. (1978). Bacterial stalk rot of maize: Loss assessment and possibility of disease control by application of calcium hypochlorite (Abstr.). *Indian Phytopathol*. 31: 120.

Martinez-Cisneros, B. A., Juarez-Lopez, G., Valencia-Torres, N., Duran-Peralta, E. and Mezzalama, M. (2014). First report of bacterial stalk rot of maize caused by *Dickeya zeae* in Mexico. *Plant Dis*. 98: 1267.

Payak, M. M. (1972). Erwinia stalk rot of maize. In: P. N. Patel (ed.). *Plant Bacteriology*. Vol. I, *Bacterial Diseases of Plants in India*. Indian Agriculture Research Institute, New Delhi, pp. 22–28.

Prokić, A., Zlatković, N., Kuzmanović, N., Ivanović, M., Gašić, K. *et al*. (2020). Identification and characterization of *Dickeya zeae* strains associated with maize stalk soft-rot in northern Serbia. *Eur. J. Plant Pathol*. 157: 685–91. https://doi.org/10.1007/s10658-020-02019-4

Sabet, K. A. (1954). A new bacterial disease of maize in Egypt. *Empire J. Exptl. Agric*. 22: 65–67.

Samson, R., Legendre, J. B., Christen, R., Fischer-Le Saux, M., Achouak, W. *et al*. (2005). Transfer of *Pectobacterium chrysanthemi* (Burkholder *et al*. 1953) Brenner *et al*. 1973 and *Brenneria paradisiaca* to the genus *Dickeya* gen. nov. as *Dickeya chrysanthemi* comb. nov. and *Dickeya paradisiaca* comb. nov. and delineation of four novel species, *Dickeya dadantii* sp. nov., *Dickeya dianthicola* sp. nov., *Dickeya dieffenbachiae* sp. nov. and *Dickeya zeae* sp. nov. *Int. J. Syst. Evol. Microbiol*. 55: 1415–27.

Saxena, S. C. and Lall, S. (1981). Effect of fertilizer application on the incidence of bacterial stalk rot of maize. *Indian J. Mycol. Plant Pathol*. 11: 164–68.

Sharma, S. C., Randhawa, P. S., Thind, B. S. and Khera, A. S. (1982). Use of Klorocin for the control of bacterial stalk rot and its absorption, translocation and persistence in maize tissue. *Indian J. Mycol. Plant Pathol*. 12: 185–90.

Thind, B. S. (1970). Investigations on Bacterial Stalk Rot of Maize (*Erwinia carotovora* var. *zeae* Sabet). Ph.D. Thesis, Postgraduate School, Indian Agricultural Research Institute, New Delhi, 113 pp.

Thind, B. S. and Payak, M. M. (1972). Antibiotics and bleaching powder in the control of bacterial stalk rot of maize. *Hindustan Antibiot. Bull*. 15: 9–13.

Thind, B. S. and Payak, M. M. (1978). Evaluation of maize germplasm and estimation of losses due to maize stalk rot. *Plant Dis. Reptr* 62: 319–23.

Thind, B. S. and Payak, M. M. (1985). A review of bacterial stalk rot of maize in India. *Trop. Pest Manag*. 31: 311–16.

Thind, B. S. and Singh, N. J. (1976). Maize borer (*Chilo partellus* Swinhoe) as carrier of *Erwinia carotovora* var. *zeae*, the causal agent of bacterial stalk rot of maize. *Curr. Sci*. 45: 117–18.

Thompson, D. L. (1965). Control of bacterial stalk rot of corn by chlorination of water in sprinkler irrigation. *Crop Sci*. 5: 369–70.

Victoria, J. I., Arboleda, F. and Munoz, S. (1975). La pudricion suave del tallo de maiz (*Zea mays* L.) en Colombia. *Noticas Fitopatologicas* 4: 136–47.

9.9 BACTERIAL LEAF STRIPE OF MAIZE

Six bacterial plant pathogens have been reported to attack maize foliage in different parts of the world. Important among them are *Acidovorax avenae* subsp. *avenae* (syn. *Pseudomonas rubrilineans*), *Burkholderia andropogonis* (syn. *Psm. andropogonis*) and *Psm. alboprecipitans*.

Ullasa *et al.* (1967) observed, for the first time, a severe leaf stripe disease in some inbred lines of maize during the 1965 and 1966 *Kharif* seasons at Delhi, India and identified the bacterium as *Xanthomonas rubrilineans*. Later, Dange (1972), while working on this disease, found that it is caused by *Psm. rubrilineans*, the pathogen that causes red stripe of sugarcane and has already been reported on that crop by Chona and Rao (1963). The first natural occurrence of this pathogen on maize was observed by Orian (1957) in Mauritius. The disease has also been reported from Jiangsu Province (Gao *et al.*, 2007) and Guangdong province, China (Ji *et al.*, 2014). As the disease being described here causes symptoms mainly on the leaves, it is appropriate to call it **'Bacterial Leaf Stripe of Maize'** rather than bacterial stripe disease of maize.

Bacterial leaf spot disease of maize due to *Xanthomonas maydis*, reported by Rangaswami *et al.* (1961), has not been found subsequently and the identity of the pathogen is also in doubt (Dye, 1966).

Since 1965, bacterial leaf stripe has been regularly observed at Delhi and Pant Nagar, but has never caused significant losses. The disease has also been seen at Ludhiana (Punjab) and in Rajasthan and Himachal Pradesh and probably occurs in other maize growing areas of India also.

9.9.1 Host Range

In addition to maize, sugarcane and teosinte, the bacterium attacks sorghum, barley, *Pennisetum purpureum, Setaria verticillata* and *Sorghum halepanse* (Ullasa *et al.*, 1967; Dange, 1972).

9.9.2 Symptoms

Initially, water-soaked spots appear on both the leaf surfaces, which develop rapidly into long stripes. The stripes are distributed throughout the leaf lamina, but are generally more concentrated towards the leaf base. The length of stripes varies from 1–25 cm, or perhaps more, and these may cover the whole length of the leaf. The stripes often coalesce, forming elongated broad bands of necrotic tissue. Later, the stripes turn dark brown and papery. On old papery stripes, many saprophytic fungi grow.

Under humid weather, bacterial exudate appears on the stripes, which on drying forms a dark brown crust. In later stages, stripes split and disintegrate, giving the leaves a shredded or tattered appearance. Occasionally, narrow, blackish, longitudinal stripes appear on the midribs. Generally, leaf sheaths, stalks and ears are not infected in nature (Figure 9.8).

FIGURE 9.8 Bacterial leaf stripe of maize

9.9.3 Causal Organism

Acidovorax avenae subsp. *avenae* (Manns 1909) Willems *et al.* 1992

Syn. *Pseudomonas rubrilineans*

It is primarily a pathogen of sugarcane. Hence, its taxonomy and cultural, morphological and biochemical characters are given under red stripe of sugarcane.

Dhkal *et al.* (2019) studied the genetic diversity and pathogenic variation in Punjab populations of *Aci. avenae* subsp. *avenae* using 25 isolates collected from infected maize crop grown in different geographical regions of Punjab, India. All the primers showed amplification, with a total of 1769 amplified fragments. Among RAPD primers, S112 was found to be highly polymorphic with PIC value of 0.95 while OPT-4 was the least polymorphic. The size of amplified DNA fragments ranged from 59 bp to 3.9 kbp. From a total of 25 isolates, 20 were used for their pathological characterisation and, based on the multivar cluster analysis of pathogenicity data, these were divided into three groups, each producing a distinct range of disease score on particular maize inbred/hybrid used in the study. Isolate Aaa2 was found to be the most virulent with an average disease score of 8.6, whereas Aaa17 was found to be the least virulent, with an average disease score of 5.1. The most virulent isolates of this pathogen were present in central Punjab. This was the first attempt to study the genetic diversity and pathogenic variation in Punjab populations of *Aci. avenae* subsp. *avenae*.

Maize, sugarcane and teosinte isolates have the same cultural, morphological and biochemical characters. These cross-infect the hosts of each other on artificial inoculation. A phage isolated from diseased maize tissues typed all the 12 isolates of maize but did not type sugarcane isolates in spot testing. It also caused lysis of teosinte isolates but later on resistant colonies appeared in the lysed areas (Dange, 1972).

9.9.4 Disease Cycle

The bacterium is internally and/or externally seed-borne. It enters the host through stomata or wounds. Secondary spread of the pathogen occurs mainly through wind-splashed rain. Young plants are more susceptible than older plants.

Dhkal *et al.* (2016b) investigated the potential role of four antioxidant enzymes in imparting resistance against *Aci. avenae* subsp. *avenae* in six inbreds/hybrids of maize, namely, G-5414, PMH-1, CM-139, LM-13, CM-600 and CML-25. Antioxidant enzyme activity was estimated spectrophotometrically from the leaf samples collected at 0, 24, 48, 72, 96, 120 and 144 h after inoculation. A spike in the expression of various antioxidant enzymes was recorded following the pathogen challenge, with maximum spike after 48 or 72 h post-inoculation in different maize lines. Inbred line CML-25 showed maximum enzyme activity, while G-5414 showed the least. Significantly high values of Pearson's correlation coefficient obtained between antioxidant enzyme activity and lesion length development suggest a possible involvement of these enzymes in imparting disease resistance against bacterial leaf stripe of maize.

Although the maize, sugarcane and teosinte isolates artificially cross-infect the hosts of each other, it appears that in nature they do not cross infect other hosts. In fields, the pathogen may be present on maize but absent on sugarcane crop growing adjacent to it, suggesting that maize strain does not infect sugarcane in nature.

Inbred lines, namely, CM 104, CM 105, CM 106, CM 109, CM 112, CM 201, CM 300 and Eto group, were found to be resistant on artificial screening and most of the commercial grown hybrids and composites are not severely infected in nature (Dange, 1972).

To develop artificial epiphytotics of the disease, Dhkal *et al.* (2016a) field evaluated four inoculation methods during the 2014 *Kharif* season. The whorl prick inoculation method was the most effective in producing the disease followed by pin-prick inoculation and cotton swab inoculation methods.

9.9.5 Management

1. Use of pathogen-free seed is the most effective method of disease management. In the absence of pathogen-free seed, there is an urgent need to devise the methods for the eradication of seed-borne inoculum.
2. As the pathogen attacks only the parenchyma of the host tissue, it can be effectively checked with Streptocycline. Thind (1978) obtained 67% control of the disease with two sprays of Streptocycline (100 µg mL^{-1}) given at an interval of 10 days.

REFERENCES

Chona, B. L. and Rao, Y. P. (1963). Association of *Pseudomonas rubrilineans* (Lee *et al.*) Stapp with the red stripe disease of sugarcane in India. *Indian Phytopathol.* 16: 392–93.

Dange, S. R. S. (1972). Studies on bacterial stripe disease of maize caused by *Pseudomonas rubrilineans*. Ph.D. Thesis, Postgraduate School, Indian Agricultural Research Institute, New Delhi.

Dhkal, M., Hunjan, M. S. and Kaur, H. (2016a). Molecular identification of the pathogen causing bacterial leaf streak disease of maize and standardization of inoculation technique. *Plant Disease Research* 31: 79–85.

Dhkal, M., Hunjan, M. S., Kaur, H. and Kaur, R. (2016b). Biochemical basis of bacterial leaf streak resistance in maize. *Indian Phytopath.* 69: 373–80.

Dhkal, M., Hunjan, M. S., Kaur, H. and Pannu, P. S. P. (2019). Characterization of *Acidovorax avenae* subsp. *avenae* causing bacterial leaf streak of maize in Punjab state of India. *J. Plant Pathol.* 101: 71–79. https://doi.org/10.1007/s42161-018-0138-3

Dye, D. W. (1966). A comparative study of some atypical 'Xanthomonads'. *N. Z. J. Sci.* 9: 843–54.

Gao, Y., Hu, B.-S., Wang, F.-X., Liu, F.-Q. and Xu, Z-G. (2007). Identification of causal agent of maize bacterial leaf stripe. *Jiangsu J. Agric. Sci.* 23: 22–25.

Ji, C., Ou, J., Xu, D. and Pan, R. (2014). First report of corn bacterial leaf stripe caused by *Acidovorax avenae* in Guangdong Province, China. *Plant Dis.* 98: 1424.

Manns, T. F. (1909). The blade blight of oats. A bacterial disease. *Bull. Ohio Agric. Exp. Stn.* 210: 91–167.

Orian, G. (1957). Plant pathology division. *Report of the Department of Agriculture, Mauritius*, 1995, pp. 90–93.

Rangaswami, G., Prasad, N. N. and Easwaran, K. S. S. (1961). A new bacterial leaf spot disease of maize. *Madras Agric. J.* 48: 392–93.

Thind, B. S. (1978). Chemical control of bacterial leaf stripe of maize caused by *Pseudomonas rubrilineans*. *Indian J. Plant Prot.* VI: 70–71.

Ullasa, B. A., Mehta, Y. R., Payak, M. M. and Renfro, B. L. (1967). *Xanthomonas rubrilineans* on *Zea mays* in India. *Indian Phytopathol.* 20: 77–78.

Willems, A., Goor, M., Thielemans, S., Gillis, M., Kersters, K. *et al.* (1992). Transfer of several phytopathogenic *Pseudomonas* species to *Acidovorax* as *Acidovorax avenae* subsp. *avenae* subsp. nov., comb. nov., *Acidovorax avenae* subsp. *citrulli*, *Acidovorax avenae* subsp. *cattleyae*, and *Acidovorax konjaci*. *Int. J. Syst. Bacteriol.* 42: 107–19.

9.10 STEWART'S WILT OF CORN

In 1895, F. C. Stewart first observed a bacterial wilt of sweet corn from Long Island, New York (Stewart, 1897). He described the symptoms of the disease, isolated the bacterium and proved Koch's postulates. In 1898, E. F. Smith described and named the causal bacterium (of the disease), the culture of which was sent to him by Stewart, as *Pseudomonas stewartii*, in honour of F. C. Stewart. Prior to Stewart's report, T. J. Burrill reported a new bacterial disease of field corn from southern Illinois. However, T. J. Burrill and an entomologist, S. A. Forbes, attributed this disease to dry conditions and the large population of clinch bugs in spite of the presence of bacteria in the diseased tissue. Moreover, Burrill could not prove Koch's postulates despite isolating the bacterium from diseased plants and insects, and also described many symptoms of the disease, presently not attributed to Stewart's wilt. Hence, the credit of reporting Stewart's wilt correctly for the first time goes to F. C. Stewart. The disease was also called **bacterial wilt** or **bacterial leaf blight**.

The disease is generally more severe on sweet corn than on dent corn or popcorn, but some hybrids and inbred lines of popcorn and field corn are also highly susceptible. Economic losses are caused due to the reduction in yield and restrictions on the import of seed, as the causal bacterium is a quarantined pathogen in several countries (EPPO A2 quarantine pest).

The disease has been a major threat to corn producers in the eastern corn belt of the United States for more than a century. Severe epiphytotics of the disease occurred in Nebraska in 1999 and 2000. However, the economic losses due to the disease have been reduced as a result of the development of resistant hybrids, adequate warning systems and effective control measures. Golden Cross Bantam, a single-cross sweet corn hybrid developed by Glenn Smith and released in 1932, reduced the enormous levels of losses from the disease. It was so widely grown that within a few years of its release, it accounted for 70–80% of all the sweet corn canned in the USA. The disease also occurs in other countries of South and Central America, Europe and Asia. Some countries, including South Africa, are believed to be free from the pathogen. Orio *et al.* (2012) reported the disease from Argentina for the first time. Cui *et al.* (2021) reported *Pantoea stewartii* subsp. *stewartii* as having caused bacterial leaf wilt of sugarcane from Zhanjiang, Guangdong province of China.

9.10.1 Host Range

Gamma grass (*Tripsacum dactyloides*) and teosinte (*Zea maxicana*) are also natural hosts of the bacterium. Choi and Kim (2013) reported the occurrence of the disease on *Dracaena sanderiana* (common name lucky bamboo) plants in greenhouses from Seongam, Gyeonggi Province, South Korea, with an incidence of 35–50%. The artificially infected hosts include sorghum, millet, Sudan grass and some weed species like yellow foxtail.

9.10.2 Symptoms

There are two phases of the disease, namely, **wilt phase** and **leaf blight phase**. In both phases, the initial symptoms appear as leaf lesions originating from corn flea beetle feeding wounds.

9.10.2.1 Wilt Phase

In general, this phase is more severe as the plants may be killed outrightly. The symptoms usually appear on young seedlings, but can also occur on older plants of very susceptible sweet corn cultivars. Initially, the leaf tissue surrounding flea beetle feeding wounds becomes water-soaked. Later on, pale-green to white-yellow linear streaks, with irregular or wavy margins, develop parallel to leaf veins. In susceptible cultivars, these lesions may extend along the whole length of the leaves. With age, water-soaked and chlorotic tissues become necrotic and may result in leaf blight, if the weather conditions remain favourable.

The wilt phase is associated with systemic infection. After entering the host tissue, the bacteria multiply in the xylem vessels and spread throughout the plant but occasionally infect kernels. The multiplication of bacteria and production of large quantities of extracellular polysaccharides inside the xylem vessels results in the plugging of xylem vessels and a reduction in the flow of water and nutrients. Bacterial slime may exude in the form of droplets from the cob husks. Bacteria penetrate the seed but do not enter the embryo. The systemically infected plants become stunted, wilt and die (Figure 9.9a). Infected plants produce dwarfed and bleached tassels, which ultimately shrivel and die. The bacterial multiplication in the stalk tissue causes browning and the necrosis of vascular bundles. Yellow bacterial ooze comes out from the cut ends when systemically infected stalks are pressed. In severely infected plants, open cavities may develop in the stalks at the base.

9.10.2.2 Leaf Blight Phase

These symptoms usually appear after tasselling, although they can appear any time during the crop season. Foliar symptoms of this phase are similar to those which occur in the wilt phase. The blighted leaves often wither and die, but these are replaced by new leaves in young seedlings. On adult plants, this phase becomes more obvious because no new leaves are produced. In severe attacks, the entire leaves may be blighted due to coalescence of lesions (Figure 9.7b). Secondary fungi often grow on the dead tissue.

The leaf symptoms, similar to those of Stewart's wilt, are also produced by other leaf blights caused by *Acidovorax avenae* subsp. *avenae*, *Burkholderia andropogonis* and *Aspergillus* species. Therefore, it is advisable to confirm the identity of the disease by isolating the causal agent and proving its pathogenicity. Stewart's wilt lesions on plants after tasselling can also be confused with those of Goss's wilt. However, these two diseases can be distinguished by using a selective growth medium for Goss's wilt bacterium and an ELISA antibody technique for Stewart's wilt bacterium.

9.10.3 Causal Organism

Pantoea stewartii subsp. *stewartii* (Smith 1898) Mergaert *et al.* 1993

Synonym: *Erwinia stewartii*

Cells are Gram-stain-negative, non-flagellate, non-sporing, facultatively anaerobic rods and measure approximately 0.4–0.8 × 0.9–2.2 µm. Capsules are formed. The colonies on nutrient glucose agar are cream-yellow to orange-yellow. The tests for methyl red, Voges–Proskauer, H_2S and indole production, gelatin and esculin hydrolysis and nitrate reduction are negative. The bacterium also does not produce oxidase, urea hydrolase, arginine dihydrolase or ornithine dicarboxylase but catalase is produced. Acid is produced from glucose but not gas. Acid production occurs from L-arabinose, D-mannitol, D-mannose, melibiose, cellobiose, raffinose, D-sorbitol, trehalose and D-xylose but not from D-adonitol, dulcitol, glycerol, myo-inositol, maltose, lactose, L-rhamnose, salicin and α-methyl D-glucoside. The metabolism is fermentative. Pathogenicity is related to the production of extracellular polysaccharides, which require quorum-sensing regulatory proteins. Maltose, D-arabitol, quinate, citrate, betaine, D-saccharate and 3-*0*-methyl-D-glucose are not used as sole source of carbon and energy. The mol% G + C of the DNA is 54.6–55.1 (*Tm*, melting temperature).

FIGURE 9.9 Stewart's wilt of corn: (a) Wilted plant (b) Leaf blight phase. (Courtesy: J. K. Pataky, University of Illinois at Urbana-Champaign, Bugwood.org.)

9.10.4 Disease Cycle

The bacterium primarily survives (overwinters) in the digestive tracts of hibernating adult corn flea beetles (*Chaetocnema pulicaria* Melsheimer, order Coleoptera, family *Chrysomelidae*). The beetles hibernate and pass the winter in soil of grassy areas near the corn fields. The beetles emerge from soil in the spring when the soil temperature is around 18–21 °C; they feed on young corn plants and deposit the bacteria in wounds as faecal contamination. The bacteria enter into the leaf veins and cause infection. These beetles are also the primary mode of secondary spread of the pathogen. The beetles feed on the infected plants, acquire the bacterium and spread it to the healthy plants. An infested beetle can carry and transmit the bacterium throughout its life. Other insects, like root worms and flea beetle species, can be vectors of the bacterium; however, their role is negligible in comparison to the corn flea beetles. Corn fields near to grassy areas and winter wheat are generally more severely affected by the disease. Menelas and associates (2006) reported that after 48 h of a feeding period on diseased plants, corn flea beetles required a minimum period of 3 h to successfully transmit the bacterium; the study found that 50% transmission had occurred by 7.6 ± 0.87 h. They further concluded that the beetles are highly efficient vectors, both in acquiring and transmitting the bacterium; hence only those insecticides should be used for seed treatment and foliar sprays, which act quickly to prevent beetles from acquiring and transmitting the bacterium.

The bacterium can also survive in the infected seed, but the chances of seed-plant transmission are extremely low or almost non-existent under commercial conditions. Only severely and systemically infected plants produce infected seed. As most of this seed is of poor quality, it will be removed during routine seed processing. Moreover, among infected seeds, the frequency of seed-plant transmission is extremely low, ranging from 0.022 to 0.038% (Block *et al*., 1998; Michener *et al*., 2002). In addition, the seed-borne infection is also influenced by certain disease parameters. It is more common in systemically infected plants. The ratio of infected seeds in systemically infected susceptible plants is 1 in 50,000 seeds, whereas it is 1 in 20,000,000 seeds in resistant plants showing leaf blight symptoms. The seed-plant transmission of the bacterium is least significant in North America, where the disease is endemic. But more than 60 countries outside North America enforce quarantine restrictions on import of corn seed to prevent the entry of the pathogen (Pataky, 2003).

The severity of disease is influenced by many factors, such as host nutrition, host reaction to the pathogen, temperature and the number of over-wintering corn flea beetles. High levels of ammonium nitrogen and phosphorus tend to increase the susceptibility while high levels of calcium and potassium impart resistance. Meyer *et al*. (2010), on the basis of 741 observations made from 79 field trials conducted at 15 locations throughout Illinois and one each in Kentucky and Delaware, found that the host resistance affected the incidence of systemic seedling wilt within ranges of winter temperatures used for forecasts of Stewart's wilt. In the case of moderately resistant and resistant hybrids, frequency distributions of Stewart's wilt incidence and mean incidence ranging from 0.7 to 1.8% did not differ among three winter temperature ranges above –2.8 °C. On the other hand, distributions of Stewart's wilt incidence on susceptible hybrids differed among each of the four ranges of winter temperature from Stevens-Boewe forecast (i.e., >0.6, –1.1 to 0.6, –2.8 to –1.1 and < -2.8), with mean incidence ranging from 0.5 % to 8.5 %. They also observed that levels of host resistance and winter temperature influenced the probability of exceeding economic thresholds of 1 or 5% incidence of systemic Stewart's wilt. Stewart's wilt is unlikely to exceed economic thresholds when the mean winter temperature is below –4.4° C. When the mean winter temperature was above –2.8 °C, the probability of exceeding 1% incidence of systemic Stewart's wilt was 0.59 for susceptible sweet corn hybrids and 0.28 for moderate and resistant hybrids. With mean winter temperature below –2.8 °C, the probability of exceeding 1% incidence of systemic Stewart's wilt was 0.22% for susceptible hybrids and 0.04 for moderate and resistant sweet corn hybrids. The probability of exceeding 5% incidence was less than 0.1, except when the mean winter temperature was above –2.8 °C and susceptible hybrids were grown. High temperatures also favour the disease development. The temperature during the winter indirectly affects the disease by influencing the survival of corn flea beetles. Stevens (1934) developed a forecasting system based on winter temperature to predict the occurrence of the disease. Later on, Boewe (1949) modified Stevens's method to predict more accurately the leaf blight phase of the disease. The simplified version of the Stevens–Boewe method uses the average of mean monthly temperatures for December, January and February. If the mean daily temperature for December, January and February is above 1 °C, the flea beetles will survive and the disease is likely to be severe. If this average daily temperature is less than –3 °C, flea beetles are unlikely to survive and it is less probable that Stewart's wilt will be severe. Stevens-Boewe's forecasting system seems to be a reliable predictor of the disease, because severe epiphytotics of the disease, which occurred in the early 1930s, the early 1950s and the 1990s in the Midwest and the eastern United States, followed winters that were among the warmest of that century.

Pan. stewartii subsp. *stewartii* is a remarkable bacterium for laboratory studies due to its relative ease of propagation and genetic manipulation, and the fact that it appears to employ a minimal number of pathogenicity mechanisms. Moreover, *Pan. stewartii* subsp. *stewartii* produces copious amounts of its quorum sensing (QS) signal, AHL, making it an excellent organism for studying QS-controlled gene regulation in a plant pathogenic bacterium. Hence, *Pan. stewartii* subsp. *stewartii* has become the microbial paradigm for QS control of gene expression by both repression and activation via a QS regulator that binds DNA in the absence and dissociates in the presence of the signal ligand. As *Pan. stewartii* subsp. *stewartii* is a member of the family *Enterobacteriaceae*, the lessons learned from its interaction with plants may be extrapolated to other plant-associated enterobacteria, such as *Erwinia*, *Dickeya* and *Pectobacterium* spp., or enteric human pathogens associated with plants, such as *Escherichia* coli and *Salmonella* spp. Studies of this bacterium have provided useful insight into how xylem-dwelling bacteria establish themselves and incite disease in their hosts (Roper, 2011).

Pan. stewartii subsp. *stewartii* is an important bacterial pathogen of sweet corn that colonises both the apoplast and xylem tissues in which reactive oxygen species (ROS) are produced. The genome of *Pan. stewartii* predicts the presence of two redox-sensing transcriptional regulators, OxyR and SoxR, which activate gene expression in response to oxidative stress. ROS exposure, in the form of hydrogen peroxide and the superoxide-generating compound paraquat, initiates an induced stress response through OxyR and SoxR that includes the activation of the ROS-detoxifying enzymes alkyl hydroperoxide reductase and superoxide dismutase. *Pan. stewartii* ΔsoxR was more sensitive to paraquat and was compromised in the ability to form water-soaked lesions, whereas ΔoxyR was more sensitive to hydrogen peroxide treatment and was deficient in exopolysaccharide production and the elicitation of wilting symptoms. It is evident that both SoxR and OxyR play an important role in the virulence of *Pan. stewartii* during the infection process while colonising different niches (Burbank and Roper, 2014).

Yield losses vary with host cultivars, the growth stage of the plant at the time of infection and whether or not the infection is systemic or non-systemic. The losses may range from 40–100% when the infection of susceptible sweet corn hybrids occurs before 5-leaf stage. In sweet corn, the yield loss is about 0.8% for every 1% incidence of systemically infected seedlings.

9.10.5 Management

1. Cultivation of resistant hybrids of sweet corn is the best control measure. Many sweet corn hybrids contain sufficient resistance to the wilt pathogen (Michener *et al*., 2003; Meyer *et al*., 2010).
2. Another equally effective control measure is the control of corn flea beetles with insecticides. The insecticides can be applied as seed treatment, furrow application before sowing or foliar application but seed treatment is most effective. Use of insecticides is more viable in sweet corn and seed corn production. Treatment of sweet corn seed of susceptible variety 'Sprint' with imidacloprid @ of 4 ml/kg caused 37–83% reduction in disease incidence (Kuhar *et al*., 2002).
3. Use of pathogen-free seed. The seed-borne infection can be detected using ELISA.

REFERENCES

Block, C. C., Hill, J. H. and McGee, D. C. (1998). Seed transmission of *Pantoea stewartii* in field and sweet corn. *Plant Dis.* 82: 775–80.

Boewe, G. H. (1949). Late season incidence of Stewart's disease on sweet corn and winter temperatures in Illinois, 1944-1948. *Plant Dis. Reptr* 33: 192–94.

Burbank, L. and Roper, M. C. (2014). OxyR and SoxR modulate the inducible oxidative stress response and are implicated during different stages of infection for the bacterial phytopathogen *Pantoea stewartii* subsp. *stewartii*. *Mol. Plant-Microbe Interact.* 27: 479–90.

Choi, O. and Kim, J. (2013). *Pantoea stewartii* causing Stewart's wilt on *Dracaena sanderiana* in Korea. *J. Phytopathol.* 161: 578–81.

Cui, D., Huang, M.-T., Hu, C.-Y., Su, J.-B., Lin, L.-H. *et al.* (2021). First report of *Pantoea stewartii* subsp. *stewartii* causing bacterial leaf wilt of sugarcane in China. *Plant Dis.* 105: 1190.

Kuhar, T. P., Stivers-Young, L. J., Hoffmann, M. P. and Taylor, A. G. (2002). Control of corn flea beetle and Stewart's wilt in sweet corn with imidacloprid and thiamethoxam seed treatments. *Crop Prot.* 21: 25–31.

Menelas, B., Block, C. C., Esker, P. D. and Nutter, F. W. Jr. (2006). Quantifying the feeding periods required by corn flea beetles to acquire and transmit *Pantoea stewartii*. *Plant Dis.* 90: 319–24.
Mergaert, J. Verdonck, L. and Kersters, K. (1993). Transfer of *Erwinia ananas* (synonym, *Erwinia uredovora*) and *Erwinia stewartii* to the genus *Pantoea* emend. as *Pantoea ananas* (Serrano 1928) comb. nov. and *Pantoea stewartii* (Smith 1898) comb. nov., respectively, and description of *Pantoea stewartii* subsp. *indologenes* subsp. nov. *Int. J. Syst. Bacteriol.* 43: 162–73.
Meyer, M. D., Pataky, J. K., Joos, D. K., Esgar, R. W. and Henry, B. R. (2010). Influence of host resistance on Stewart's wilt forecasts and probability of exceeding thresholds for use of seed-treatment insecticides on sweet corn. *Plant Dis.* 94: 1111–17.
Michener, P. M., Freeman, N. D. and Pataky, J. K. (2003). Relationship between reactions of sweet corn hybrids to Stewart's wilt and incidence of systemic infection by *Erwinia stewartii*. *Plant Dis.* 87: 223–28.
Michener, P. M., Pataky, J. K. and White, D. G. (2002). Rates of transmitting *Erwinia stewartii* from seed to seedlings of a sweet corn hybrid susceptible to Stewart's wilt. *Plant Dis.* 86: 1031–35.
Orio, A. G. A., Brucher, E., Plazas, M. C., Sayago, P., Guerra, F. *et al.* (2012). First report of Stewart's wilt of maize in Argentina caused by *Pantoea stewartii*. *Plant Dis.* 96: 1819.
Pataky, J. K. (2003). Stewart's wilt of corn. APSnet Feature Story, July-August 2003. http://www.apsnet.org/feature/stewarts
Roper, M. C. (2011). *Pantoea stewartii* subsp. *stewartii*: Lessons learned from a xylem-dwelling pathogen of sweet corn. *Mol. Plant Pathol.* 12: 628–37.
Smith, E. F. (1898). Notes on Stewart's sweet-corn germ. *Pseudomonas stewarti*, n. sp. *Proc. American Assoc. Advancement of Sci.* 47: 422–26.
Stevens, N. E. (1934). Stewart's disease in relation to winter temperatures. *Plant Dis. Reptr* 18: 141–49.
Stewart, F. C. (1897). A bacterial disease of sweet corn. *N. Y. Agric. Exp. Sta. Bull.* 130: 422–39.

9.11 BACTERIAL LEAF STREAK AND BACTERIAL LEAF STRIPE OF SORGHUM

More than ten bacterial diseases have been reported on sorghum (*Sorghum bicolor* (L.) Moench), of which bacterial leaf streak, abbreviated as BLS, caused by *Xanthomonas vasicola* pv. *holcicola* (Elliott) Vauterin *et al.*, bacterial leaf stripe caused by *Paraburkholderia andropogonis* (Gillis *et al.*) Sawana *et al.* and bacterial spot caused by *Pseudomonas syringae* pv. *syringae* van Hall are considered to be economically significant (Claflin *et al.*, 1992).

In southern Africa, BLS of sorghum has been reported from Angola, Lesotho, Malawi and Tanzania but the pathogen's identity was not confirmed; in contrast, in the Republic of South Africa, BLS has been reported with confirmed identity (de Milliano, 1992). Occurrence of BLS has also been reported from Argentina, Australia, Iran, Japan, Mexico, New Zealand, Nigeria, North America, the Philippines, South America, the USA and West Africa. Noble and Richardson (1968) reported the seed-borne nature of BLS pathogen; however, the experimental data were not provided. Rao *et al.* (1990) detected *Xan. campestris* pv. *holcicola* in sorghum leaves by dot immunobinding assay in a sorghum sample imported to India from Yemen in 1987.

Xan. campestris pv. *holcicola* is a quarantined pathogen. In several countries, this bacterium is used as a trade barrier. BLS has not previously been reported from India. Navi *et al.* (2002) reported the occurrence of this disease on sorghum under field conditions at the ICRISAT research farm, Patancheru, Andhra Pradesh, India, in several farmers' fields surveyed in Karnataka, India, from August 1999 to March 2001, and in elite germplasm accessions evaluated for resistance to foliar and panicle diseases during the 2001 rainy season at ICRISAT, Patancheru.

Hale and Wilkie (1972) reported BLS of sorghum caused by *Psm. rubrisubalbicans* from New Zealand. This was the first record of this pathogen from New Zealand and of the natural occurrence of *Psm. rubrisubalbicans* on sorghum. The disease is also called bacterial stripe of sorghum.

9.11.1 Symptoms

9.11.1.1 Bacterial Leaf Streak

Small, water-soaked, reddish-brown, necrotic streaks appear in early stages that later elongate and darken. Later, the lesions broaden and develop tan centres with a narrow red margin. These symptoms are delimited by the veins. Occasionally, tiny, yellow, bead-like exudations are found on lesions. The narrow, long or short translucent lesions coalesce, forming large patches. In severe cases, affected leaf parts acquire a burnt appearance. At this stage, the leaves wither and turn brown. The most prominent symptoms are elongate lesions or areas of discoloration, usually of limited length, and parallel to the leaf veins.

9.11.1.2 Bacterial Leaf Stripe

Initial symptoms are small (1 cm), linear, interveinal lesions. The lesions on leaves and sheaths are purple, red, yellow or tan, depending on the host reaction. Under favourable conditions, lesions may exceed 20 cm in length (as much as 42 cm has been reported) and they usually coalesce along the width of the leaf. Moderately severe infections may produce shorter lesions which closely mimic symptoms of bacterial leaf streak (*Xan. campestris* pv. *holcicola*). Water-soaking of tissue adjacent to a lesion is usually not observed under field conditions. Bacterial exudate is usually observed on infected portions of the leaf (Figure 9.10). Lesions may also occur on the kernel, peduncle, and rachis and in the interior of the stalk.

9.11.2 Causal Organisms

Bacterial leaf streak of sorghum: *Xanthomonas vasicola* pv. *holcicola* (Elliott) Vauterin *et al.*

Syn. *Xan. campestris* pv. *holcicola* (Elliott) Dye, *Xan. holcicola*

FIGURE 9.10 Bacterial leaf streak and stripe of sorghum.

(Elliott) Starr and Burkholder

Bacterial leaf stripe of sorghum: *Paraburkholderia andropogonis*

(Smith) Sawana *et al.*

Synonym: *Burkholderia andropogonis* (Smith) Gills *et al.*

Pseudomonas andropogonis (Smith) Stapp

9.11.3 Disease Cycle

Xan. vasicola pv. *holcicola* and *Para. andropogonis* are seed-borne and can also survive in infected plant debris, soil and left-over diseased plants. Local dissemination of both the bacteria occurs through wind-splashed rain and insects. Long-range dissemination of bacteria occurs through infected seed. The penetration of the bacteria occurs through injuries caused by insects and wind. Disease development is favoured by warm and wet weather such as the cloudy humid days following rain.

Out of 1014 elite *germplasm* accessions (originating from 45 countries) evaluated for various foliar and panicle diseases during the 2001 rainy season at ICRISAT, Patancheru, Navi *et al.* (2002) observed BLS in seven accessions (IS 21977 (India), IS 14008, IS 14305, IS 24497 (South Africa), IS 29389 (Lesotho), IS 13177 (Argentina) and IS 25428 (Kenya)). BLS incidence was 4–10% and the severity was 2–20%. BLS was also observed in several farmers' fields in the state of Karnataka, during rainy and post-rainy seasons (August 1999 to March 2001) while conducting disease surveys. The locations where BLS was observed were Ainapur [16° 49' 40" (N), 75° 46' 19" (E)], Kannolli [16° 50' 84" (N), 76° 08' 24" (E)] and Bijapur [16° 49' 40" (N), 75° 46' 19" (E)] in Bijapur district; Hunagund [16° 18' 33" (N), 75° 54' 57" (E)] in Bagalakot district, Attigullapura [11° 50' 00" (N), 76° 59' 94" (E)], and Seenappanadoddi [11° 48' 88" (N), 77° 00' 17" (E)] in Chamaraj Nagar district; Dandinakuru-barahatti [14° 13' 82" (N), 76° 27' 70" (E)] in Chitradurga district; and Sege [13° 06' 69" (N), 76° 04' 73" (E)] in Hassan district. BLS incidence varied from 0.5 to 20% and its severity from 5 to 60%. The occurrence of BLS in various farmers' fields and on the research farm appears to be the first record from India. However, attempts to prove pathogenicity, using several methods, were not successful.

9.11.4 Management

The following measures can be helpful in minimising or checking the disease incidence.

1. Use pathogen-free seed or treat the seed with Streptocycline 100 μg ml^{-1}.
2. Deep plough and bury the diseased plant debris.
3. Follow crop rotation for three years with non-host crops.
4. Sow disease-resistant varieties.

REFERENCES

Claflin, L. E., Ramundo, B. A., Leach, J. E. and Qhobela, M. (1992). Bacterial diseases of sorghum. In: W. A. J. de Milliano, R. A. Frederiksen, and G. D. Bengston (eds.). *Sorghum and Millets Diseases: A Second World Review*, International Crops Research Institute for the Semi-Arid Tropics (ICRISAT, Andhra Pradesh, pp. 135–151.

de Milliano, W. A. J. (1992). Sorghum diseases in southern Africa. In: W. A. J. de Milliano, R. A. Frederiksen, and G. D. Bengston (eds.). *Sorghum and Millets Diseases: A Second World Review*, International Crops Research Institute for the Semi-Arid Tropics (ICRISAT, Andhra Pradesh, pp. 9–19.

Hale, C. N. and Wilkie, J. P. (1972). Bacterial leaf stripe of sorghum in New Zealand caused by *Pseudomonas rubrisubalbicans*. *N. Z. J. Agric. Res.* 15: 457–60.

Navi, S. S., Bandyopadhyay, R., Devi, K. T. and Reddy, D. V. R. (2002). Bacterial Leaf Streak of Sorghum- a new report from India. *ISMN* 43: 61–63.

Noble, M. and Richardson, M. J. (1968). An annotated list of seed-borne diseases. 2nd ed. *Commonwealth Mycological Institute Phytopathological Papers* 8: 138.

Rao, R. D. V. J. R., Chakrabarty, S. K., Ratna, R. S. and Reddy, J. R. (1990). Detection of *Xanthomonas campestris* pv. *holcicola* in sorghum by dot immunobinding assay. *Indian Phytopath.* 43: 213–14.

9.12 RED STRIPE OF SUGARCANE

Lyon (1922) was the first to observe red stripe on tip canes in Hawaii. Shortly afterwards, Lee and Martin (1925) established the bacterial nature of the disease. Another kind of symptom/condition, called **top-rot of sugarcane**, was reported from Queensland by Tryon (1923). Cottrell-Dormer (1932) showed that red stripe and top-rot are two different phases of the same disease and that they are caused by the same bacterium. Initially, **red stripe** and **top-rot** were considered two different diseases. Since these early studies, the disease has been reported from most of the sugarcane-growing countries of the world, for which the readers can refer to chapter XXII of the book entitled, *Diseases of Sugarcane*: *Major Diseases*, edited by Ricaud *et al.* (1989).

In India, the disease was first observed simultaneously by Mc Rae (1933) and Desai (1933). Since then, several reports of severe infections have appeared infecting varieties like Co 419 in Maharashtra (Albuquerque and Arakeri, 1956; Bhide *et al.*, 1956; Summanwar and Bhide, 1962), Co 449, Co 527 and Fiji B in Tamil Nadu (Rangaswami, 1960a, 1960b) and Co 312 in Delhi and eastern districts of Punjab (Chona and Rao, 1963). In Hawaii, the disease was of great economic importance in commercial plantings of Tip canes and occasionally on other varieties.

The losses can occur if susceptible varieties are grown under environmental conditions favourable for the development and dissemination of the pathogen. Losses up to 15% or more have been reported due to the top-rot phase (Martin and Wismer, 1961). The red stripe phase alone generally does not affect yield and the symptoms usually disappear with the onset of the new growth.

Girard *et al.* (2014) reported *Acidovorax avenae* subsp. *avenae* causing sugarcane red stripe for the first time from Gabon. This was also the first recorded description of the occurrence of the top-rot form of the disease in R570, a cultivar that is grown in several locations of Africa, the Mascarene Islands and the French West Indies. Li *et al.* (2018) reported the widespread occurrence of the disease in China. Sawada *et al.* (2025) reported *Aci. sacchari* sp. nov., a pathogen causing red stripe of sugarcane in Japan. Patro *et al.* (2006) reported the top-rot phase of the disease from Srikakulam district of Andhra Pradesh, India. In this case, however, the bacterium associated with this disease was identified as *Acinetobacter baumannii*.

9.12.1 Host Range

In addition to sugarcane, the bacterium naturally infects maize (Orian, 1957; Ullasa *et al.*, 1967), maize and teosinte (Dange and Payak, 1973) and two grasses, namely, *Paspalum nutans* and *P. paniculatum* (Orian, 1956). A few more hosts have been shown to be artificially infected. Bhide *et al.* (1956) reported the successful inoculation of sorghum and Summanwar and Bhide (1962) of *Pennisetum typhoides* (Burm. f.) Stapf and Hubb. with red stripe bacterium. Ullasa *et al.* (1967) could only successfully inoculate sorghum and barley with *Xan. rubrilineans* isolated from maize, but not from several other graminaceous hosts. Dange (1972) reported that maize isolates infected *bajra*, barley, sorghum, elephant grass (*Pennisetum purpureum*) and *Setaria verticillata*. Martin and Wismer (1961), who could infect only several varieties of *Sorghum bicolor* out of many wild and cultivated grasses, stated that the common field grasses were of little or no importance as a source of inoculum for sugarcane crop.

9.12.2 Symptoms

There are two types/phases of symptoms of red stripe: leaf stripe and top-rot. These may occur either singly or simultaneously; under field conditions, both are favoured by high relative atmospheric humidity.

9.12.2.1 Leaf Stripe

In the beginning, the symptoms appear as watery green stripes on the leaf lamina. These usually appear in the middle of the leaf and near the midrib, but sometimes they may concentrate near the leaf base. The stripes spread rapidly upward and downward and develop a reddish colour, which later on changes to dark red; hence the name **red stripe**. These are straight and uniform in width and, in the early stages of infection, they remain confined to the parenchymatous tissue. Two or more stripes often coalesce to form broad bands of diseased leaf tissue. The width of the stripes varies from 0.5 to 4.0 mm and the length from a few centimetres to the entire length of the leaf lamina (Figure 9.12a). However, in some varieties, the length of the stripes may also extend to the leaf sheath. The stripes may also appear on the lower surface of the midrib. During the warm humid weather, the bacterial ooze comes out through the stomata on the lower leaf surface of the affected tissue, which often dries to form whitish flakes.

The disease appears mostly on the young and middle-aged leaves and may also attack the partially unrolled youngest leaves. In young cane, the entire stool may be killed while in older cane, the individual stalks die. Usually the young ratoons are more susceptible to infection than plant cane of the same age.

9.12.2.2 Top-rot

As the name indicates, this is the rotting of the top portion of the plant. The losses from this phase are much higher than those from leaf stripes. The yellowing and wilting of older leaves occurs and the affected plants may show the typical reddish leaf stripes. Top-rot may result from bud or stem infection without showing leaf symptoms as well as from infection of the leaves. The reddish discolouration on the outside and reddish splashes on the inner surfaces often appear on the leaf sheaths attached to the affected internodes. The sunken areas, initially water-soaked in appearance and later turning reddish-brown, often appear on the affected internodes (Figure 9.11b). As the rotting advances, large cavities are formed inside the internodes. In advanced stage, the leaf spindle can be easily pulled out of enclosing sheaths.

A characteristic strong unpleasant odour, an important **diagnostic feature** of the disease, is emitted by the rotted

(a) (b)

FIGURE 9.11 Red stripe of sugarcane: (a) Red stripes on leaf (b) Top-rot phase.

spindles. Sometimes, the odour is so strong that it may be felt from a distance. Stalks exhibiting top-rot are stunted in growth and usually die; their tops often break and fall down. Sometimes, the uppermost healthy buds of top-rot affected stalks give rise to side shoots, and the leaves of these shoots may exhibit stripe symptoms.

A new disease called **false red stripe**, different from the above-described disease of sugarcane, has been reported from central and southern areas of Brazil (Mantovani *et al.*, 2006). It does not occur in North and northeastern areas of Brazil or any other country. The causal bacterium was identified as a species of *Xanthomonas*. Artificial inoculation, by injecting and spraying of bacterial cell suspension, of 31 host species comprising crops, grasses and weeds revealed that only sorghum, maize and oats expressed disease symptoms 15 days after inoculation.

Occasionally, the leaf symptoms of red stripe and those of gumming (*Xan. axonopodis* pv. *vasculorum*) and mottled stripe (*Pseudomonas rubrisubalbicans*) can be confused with each other. Gumming can be distinguished due to its some specific symptoms like formation of gum pockets. There is no top rot in mottled stripe and in general its symptoms are paler in colour.

9.12.3 Causal Organism

Paracidovorax avenae

Synonym: *Pseudomonas avenae* subsp. *avenae, Psm. rubrilineans*, *Acidovorax avenae* subsp. *avenae*

The taxonomy of bacterium causing red stripe has undergone many changes. It was first classified as *Phytomonas rubrilineans* by Lee *et al.* (1925). It was renamed as *Pseudomonas rubrilineans* by Stapp (1928) and as *Bacterium rubrilineans* by Elliott (1930). Following the establishment of genus *Xanthomonas* by Dowson in 1939, Starr and Burkholder (1942) transferred the bacterium to this genus as *Xan. rubrilineans*. However, they pointed out that as far as lipolytic activity is concerned, red stripe pathogen does not agree with a typical *Xanthomonas*. Wernham (1948) also included it in doubtful species of *Xanthomonas*. In spite of these doubts, Breed *et al.* (1957) listed it as a species of *Xanthomonas* in the seventh edition of *Bergey's Manual of Determinative Bacteriology*.

After a detailed systematic study of bacteria causing sugarcane diseases, Hayward (1962) found that red stripe pathogen has many characteristics of genus *Pseudomonas* and hence proposed that the name *Psm. rubrilineans* (Lee *et al.*) Stapp should replace *Xan. rubrilineans* (Lee *et al.*) Starr and Burkholder. Later on, Dye (1963) and Bradbury (1967) supported Hayward's proposal. Dange (1972) reported that the red stripe of sugarcane and bacterial stripe disease of maize were caused by the same pathogen. He further reported that earlier reports from India identifying *Xan. rubrilineans* as the causal agent of red stripe of sugarcane (Desai, 1933; Albuquerque and Arakeri, 1956; Bhide *et al.*, 1956; Rangaswami, 1960a, 1960b; Summanwar and Bhide, 1962; Chona and Rao, 1963) and of bacterial stripe disease of maize (Ullasa *et al.*, 1967) were probably erroneous. Willems *et al.* (1992), on the basis of DNA–DNA hybridisation and other studies, grouped *Psm. avenae, Psm. rubrilineans* and *Psm. setariae* into *Acidovorax avenae* subsp. *avenae* subsp. nov., comb. nov. The below given description of *Psm. rubrilineans* (recent name, *Aci. avenae* subsp. *avenae*) is based on the publication of Hayward (1962).

Cells are straight rods and motile by one polar flagellum. Poly-β-hydroxybutyrate is accumulated. Colonies are non-mucoid on 2% glucose peptone agar. No pigments are produced. Oxidase positive; H_2S produced. Nitrates are reduced to nitrites, but denitrification is negative. Tween 80 is hydrolysed; starch hydrolysis is negative and gelatin liquefaction is weak. Capable of growth at 40 °C. Acid is produced from glucose, fructose, galactose, arabinose, mannitol, sorbitol and glycerol. No acid from sucrose, raffinose, salicin, lactose, maltose, cellobiose or m-inositol.

Fontana *et al.* (2013), using randomly amplified polymorphic DNA (RAPD)-PCR, detected the presence of at least four different biotypes among the analysed isolates of *Aci. avenae* subsp. *avenae* collected from sugarcane producing areas of Tucuman and Salta (northwest of Argentina).

Bertani *et al.* (2021) reported a significant increase in the occurrence of red stripe in Argentina in the past decade and employed molecular markers to analyse 112 isolates from Tucumán. By using repetitive element polymorphism-based polymerase chain reaction (rep-PCR) almost all the isolates were differentiated and grouped into ten clusters, revealing a high genetic diversity. Cluster analysis showed no clear association between isolate clustering, sugarcane host genotype, crop age, type of tissue sampled, fertilisation, or year of sampling. Results showed a differential level of aggressiveness among the isolates on a resistant sugarcane variety. This study confirmed that rep-PCR is an adequate tool for genetic analysis and population structure characterisation in bacteria, and revealed both high genetic diversity and clonal population structure of *Aci. avenae* subsp. *avenae* in Tucumán, Argentina.

Duan *et al.* (2024) reported the complete genome sequence of *Paracidovorax avenae* causing red stripe of sugarcane.

9.12.4 Disease Cycle

As the sugarcane crop is of 10 months duration or is taken as a ratoon crop, undecomposed plant debris or the infected crop may serve as primary source of inoculum. In Java, Bolle (1929) showed that the bacterium isolated from old withered leaves, showing red stripes, was virulent after four months. Cottrell-Dormer (1932) was able to isolate the pathogen from leaf stripe-infected plant debris kept in a dry cardboard box for seven months.

The bacterium is disseminated mainly through wind-splashed rain. After multiplying in large numbers in parenchyma of leaves, the bacteria ooze out on the surface of leaf lesions in moist warm weather. Then, with the help of wind and rain, these bacteria spread from plant to plant, from one field to another or fall down on the lower plant parts or may run down the leaves to cause the stem infection. The bacterium is rarely disseminated through cane knives and mechanical equipment.

The bacterium enters the host through stomata and wounds. The leaf wounds are caused due to the scratches made by the marginal spines of the leaves through the action of the wind. The bacterium enters through the stomata present on both the leaf surfaces, but the symptoms from natural infection usually appear first on the lower leaf surfaces. Barnum and Martin (1925) found that the bacterium artificially infected all the plant parts, but the youngest internodes and the youngest leaves showed more susceptibility. Vigorously growing plants were also much more susceptible than the less vigorous plants. Pal *et al.* (1996) showed that out of four methods tried for screening of sugarcane cultivars, whorl inoculation method was the most effective as it gave highest disease incidence and also induced both the phases of the disease.

Johnson *et al.* (2016) found a negative correlation between yield components and red stripe incidence, with the strongest relationship between sucrose per metric ton and disease incidence. The disease incidence was positively correlated with soil nutrition, i.e., phosphorus, potassium, zinc and calcium. The incidence also increased with increasing nitrogen rate showing maximum effect in heavy soils. The data also indicated that the use of infected canes as a seed source can significantly decrease shoot emergence, stalk population and subsequent cane and sugar yields.

9.12.5 Management

1. The use of resistant varieties is the most effective and economical approach for managing the disease. The replacement of susceptible varieties with resistant varieties has relegated red stripe to be of little economic importance. Before release, the varieties must be screened using artificial inoculation techniques. In North Queensland, the top-rot phase of the disease was markedly reduced by altering the planting date. The autumn (March–April) sown crop showed much less disease incidence compared to spring (August–September) sown crop.
2. Quarantine measures are very helpful in restricting the movement of the pathogen to disease-free areas.

REFERENCES

Albuquerque, M. J. and Arakeri, H. R. (1956). Sugarcane red stripe disease on Co 419. *Indian Sugar* 6: 323–24.

Barnum, C. C. and Martin, J. P. (1925). *The Susceptibility of Roots, Stalks, Leaf Sheath and Leaf Blades to Red-stripe Disease, and the Relationship of Maturity of Tissues to Increasing Resistance to Red Stripe*. Red-Stripe Disease Studies. Hawaiian Sugar Planters' Association Experiment Station, Pathology Department, Honolulu, pp. 35–48.

Bertani, R. P., Perera, M. F., Joya, C. M., Henriquez, D. D., Funes, C., *et al.* (2021). Genetic diversity and population structure of *Acidovorax avenae* subsp. *avenae* isolated from sugarcane in Argentina. *Plant Pathol.* 70: 1719–32.

Bhide, V. P., Hedge, P. K. and Desai, M. K. (1956). Bacterial red-stripe disease of sugarcane caused by *Xanthomonas rubrilineans* in Bombay state. *Curr. Sci.* 25: 330.

Bolle, P. C. (1929). De roodestrepenziekte. *Arch. Suikerind.* 37: 1147–1218.

Bradbury, J. F. (1967). *Pseudomonas rubrilineans. C. M. I. Description of Patho-genic Fungi and Bacteria.* Set 13, No. 127.

Breed, R. S., Murray, E. G. D. and Smith, N. R. (eds.) (1957). *Bergey's Manual of Determinative Bacteriology*. 7th Ed. Williams and Wilkins Co., Baltimore, MD, 1094pp.

Chona, B. L. and Rao, Y. P. (1963). Association of *Pseudomonas rubrilineans* (Lee *et al.*) Stapp with the red stripe disease of sugarcane in India. *Indian Phytopathol.* 16: 392–93.

Cottrell-Dormer, W. (1932). Red-Stripe disease of sugar-cane in Queensland. *Bull. Bur. Sugar Exp. Stn. Div. Pathol.* 3: 25–59.

Dange, S. R. S. (1972). Studies on Bacterial Stripe Disease of Maize Caused by *Pseudomonas rubrilineans*. Ph.D. Thesis, Postgraduate School, Indian Agricultural Research Institute, New Delhi.

Dange, S. R. S. and Payak, M. M. (1973). The taxonomic position of the pathogen of red stripe of sugar-cane. *Sugarcane Pathol. Newsl.* 10: 25–28.

Desai, S. V. (1933). Occurrence of the red stripe disease of sugarcane in India. (Abstr.). *Proc. 20th Indian Sci. Congress, Patna*, pp. 61–62.

Duan, H. H., Zhao, J. Y., Liu, W. T., Xie, Q., Zheng, C. Y. *et al.* (2024). Complete genome sequence of *Paracidovorax avenae* causing red stripe in sugarcane. *BMC Genom. Data* 25: Article No. 89. https://doi.org/10.1186/s12863-024-01271-5

Dye, D. W. (1963). Comparative study of the biochemical reactions of additional *Xanthomonas* spp. *N. Z. J. Sci.* 6: 483–86.

Elliott, C. (1930). *Manual of Bacterial Plant Pathogens*. Bailliere, Tindall and Cox Publ., London, 349 pp.

Fontana, P. D., Rago, A. M., Fontana, C. A., Vignolo, G. M., Cocconcelli, P. S. *et al.* (2013). Isolation and genetic characterization of *Acidovorax avenae* from red stripe infected sugarcane in Northwestern Argentina. *Eur. J. Plant Pathol.* 137: 525–34.

Girard, J.-C., Noell, J., Larbre, F., Roumagnac, P. and Rott, P. (2014). First report of *Acidovorax avenae* subsp. *avenae* causing sugarcane red stripe in Gabon. *Plant Dis.* 98: 684.

Hayward, A. C. (1962). Studies on bacterial pathogens of sugar cane. II. Differentiation, taxonomy and nomenclature of the bacteria causing red stripe and mottled stripe diseases. *Mauritius Sugar Ind. Res. Inst., Occas. Pap.* 13: 13–27.

Johnson, R. M., Grisham, M. P., Warnke, K. Z. and Maggio, J. R. (2016). Relationship of soil properties and sugarcane yields to red stripe in Louisiana. *Phytopathology* 106: 737–44.

Lee, H. A. and Martin, J. P. (1925). *The Cause of Red-stripe Disease of Sugar Cane. Red-stripe Disease Studies*. Hawaiian Sugar Planters' Association Experiment Station, Pathology Department, Honolulu, pp. 83–93.

Lee, H. A., Purdy, H. A., Barnum, C. C. and Martin, J. P. (1925). *A Comparison of Red-stripe Disease with Bacterial Diseases of Sugar Cane and Other Grasses*. Red-Stripe Disease Studies. Hawaiian Sugar Planters' Association Experiment Station, Pathology Department, Honolulu, pp. 64–74.

Li, X.-Y., Sun, H.-D., Rott, P. C., Wang, J.-D., Huang, M.-T. *et al.* (2018). Molecular identification and prevalence of *Acidovorax avenae* subsp. *avenae* causing red stripe of sugarcane in China. *Plant Pathol.* 67: 929–37.

Lyon, H. L. (1922). A leaf disease of the Tip canes. *Proc. Hawaii. Sugar Plant. Assoc.* 246.

Mantovani, E. S., Marini, D. C. and Giglioi, E. A. (2006). Host range of *Xantho-monas* sp., causal agent of the false red stripe of sugarcane, among grasses. *Summa Phytopathol.* 32: 124–30.

Martin, J. P. and Wismer, C. A. (1961). In: J. P. Martin, E. V. Abbott and C. G. Hughes (eds.). *Sugar-Cane Diseases of the World*. Vol. I, Elsevier, Amsterdam, pp. 109–26.

Mc Rae, W. (1933). New diseases reported during the year 1932. *Int. Bull. Pl. Prot.* 7: 79–80.

Orian, G. (1956). Occurrence of a disease similar to red stripe of sugar cane in Mauritius. *Proc. Int. Soc. Sugar Cane Technol.* 9: 1042–48.

Orian, G. (1957). Plant pathology division. *Rep. Dep. Agric. Mauritius* 1955: 90–93.

Pal, V., Kumar, A. and Vir, S. (1996). Evaluation of inoculation methods for red stripe induction in sugarcane. *Proc. 9th Int. Conf. Plant Path. Bact.* Chennai, pp. 430–33.

Patro, T. S. S. K., Nageswara Rao, G. V. and Gopalakrishnan, J. (2006). Association of *Acinetobacter baumannii* with a top rot phase of sugarcane red stripe disease in India. *Indian Phytopathol.* 59: 501–02.

Rangaswami, G. (1960a). Studies on two bacterial diseases of sugarcane. *Curr. Sci.* 29: 318–19.

Rangaswami, G. (1960b). Further studies on bacterial gummosis and red stripe disease of sugarcane. *J. Annamalai Univ.* 22: 135–50.

Ricaud, C., Egan, B. T., Gillaspie, A. G. Jr. and Hughes, C. G. (eds.) (1989). *Diseases of sugarcane: Major diseases*. Elsevier, Amsterdam, 399 pp.

Sawada, H., Shinohara, H., Takashima, Y., Naito, K. and Satou, M. (2025). *Acidovorax sacchari* sp. nov., a pathogen causing red stripe of sugarcane in Japan. *Int. Jour. Syst. and Evol. Microbiol.* 75 (2). https://doi.org/10.1099/ijsem.0.006575

Stapp, C. (1928). In: *Sorauer's Handb. Pflanzenkr.* 5. Aufl. 2: 35.

Starr, M. P. and Burkholder, W. H. (1942). Lipolytic activity of phytopathogenic bacteria determined by means of spirit blue agar and its taxonomic significance. *Phytopathology* 32: 598–604.

Summanwar, A. S. and Bhide, V. P. (1962). Bacterial red stripe disease of sugar-cane (*Saccharum officinarum*) caused by *Xanthomonas rubrilineans* var. *indicus* in Maharashtra. *Indian J. Sugarcane Res. Dev.* 6: 65–68.

Tryon, H. (1923). Top rot of the sugar cane. *Bull. Bur. Sugar Exp. Stn., Div. Pathol.* 1.

Ullasa, B. A., Mehta, Y. R., Payak, M. M. and Renfro, B. L. (1967). *Xanthomonas rubrilineans* on *Zea mays* in India. *Indian Phytopathol.* 20: 77–78.

Wernham, C. C. (1948). The species value of pathogenicity in the genus *Xanthomonas*. *Phytopathology* 38: 283–291.

Willems, A., Goor, M., Thielemans, S., Gillis, M., Kersters, K. *et al.* (1992). Transfer of several phytopathogenic *Pseudomonas* species to *Acidovorax* as *Acidovorax avenae* subsp. *avenae* subsp. nov., comb. nov., *Acidovorax avenae* subsp. *citrulli, Acidovorax avenae* subsp. *cattleyae*, and *Acidovorax konjaci*. *Int. J. Syst. Bacteriol.* 42: 107–19.

9.13 RATOON STUNTING DISEASE OF SUGARCANE

Ratoon stunting disease was first observed by Steindl in Queensland, Australia in the summer of 1944–45 on ratoon crops of sugarcane cultivar Q 28. He also coined the term '**ratoon stunting**' as the ratoon crops were more severely infected. Due to the seed-borne nature of the pathogen, the disease continued to spread from one country to another along with the movement of the germplasm; it is now present in most of the cane-growing countries of the world. The names of 52 countries from where the disease has been reported, are given in the list compiled in 1986, by the International Society of Sugar Cane Technologists (ISSCT's) Standing Committee on Sugarcane Diseases (Ricaud *et al.*, 1989).

In a review article, Young (2016) traced the origin of ratoon stunting disease to the release of modern commercial hybrids and provided evidence that *Saccharum officinarum*, the major progenitor of modern sugarcane cultivars, is not the natural host of *Leifsonia xyli* subsp. *xyli* (hereafter referred to as *Lxx*). Rather, it is proposed that the wild relative, *S. spontaneum*, is more likely to be the original host, and that *Lef. xyli* subsp. *xyli* was acquired during interspecific hybridisation work undertaken in Java during the 1920s.

In India, the disease was first observed by Prof. S. J. P. Chilton in 1956 in CoS 510 from Golagokarnath, Uttar Pradesh (Chona, 1956). However, the presence of the disease was reportedly questioned as most of the experiments conducted subsequently failed to confirm the presence of the disease. Later on, Singh (1966) confirmed the presence of the disease in Co 290 obtained from Jaora, Madhya Pradesh.

The name ratoon stunting disease (hereafter called RSD) is a misnomer as the disease also retards the growth of the plant crop and causes considerable losses. Cumulative losses due to RSD are generally more than the losses caused by any other disease of sugarcane. The losses caused by the disease are influenced by many factors, such as the susceptibility of the variety cultivated, weather conditions, the amount of irrigation water available and whether the crop is freshly planted or a ratoon crop. Overall losses ranging from 5–10% have been reported by Hughes (1974). Losses up to 30% or more have also been reported in certain years and in highly susceptible cultivars, these may go as high as 50%. Urashima and Marchetti (2013) reported disease incidence (based on dot blot enzyme immunoassay of sap) of 23.6%, 27.1% and 25.8% in 2009, 2010 and 2011, respectively, from Brazil. RB867515, the major cultivar in Sao Paulo, had field incidence up to 70% in 2009, 48% in 2010 and 88% in 2011.

In spite of the reduction in cane yield and sugar per unit area, the sugar content of diseased canes may be increased. This results from the reduced growth and lower water content of the infected canes. This is the only disease of sugarcane in which the sugar content of the infected canes increases.

In their review article, Chakraborty *et al.* (2024) have summarised the historical background and the current knowledge of sugarcane ratoon stunting disease, including some aspects related to transmission, pathosystem, and management practices; and the drawbacks of current diagnostic methods and the potential for application of advanced diagnostics to improve disease management.

9.13.1 Host Range

No natural hosts other than sugarcane have been reported. Gillaspie and Teakle (1989) have listed 14 hosts, which were artificially infected by the bacterium; from these hosts, the bacterium was successfully reinoculated to the sugarcane. These hosts included maize, sorghum, sweet Sudan grass and many other grasses commonly found in sugarcane fields. Symptoms

There are two types of symptoms: external and internal symptoms.

9.13.1.1 External Symptoms

The stunting and unthrifty growth of the plants are the only external symptoms of the disease, which can occur not only due to RSD but due to several other factors like nutritional deficiency, lack of moisture, poor agronomic practices, etc. Stunting and unthriftiness due to RSD are generally more pronounced when the crop grows in reduced soil moisture. The slow growth of the infected crop, accompanied with the production of small and thinner canes, causes a reduction in yield in spite of the total numbers of millable canes remaining unchanged. However, in a drought year, the total number of stalks may also be reduced, along with the variation in the length of individual canes in a stool. Stunting is not uniform in all the plants, even if all the stools in a field are infected. There is a characteristic **'up and down'** appearance, which means that not all the plants are of uniform height.

Varieties differ in their response to infection and show different degrees of stunting. Ratoon crops usually show more disease severity than plant crops. The losses in clone Q 28 are in the range of 12–37% in plant crops while these usually exceed 60% in ratoon crops. The roots and underground parts of the diseased canes do not show any abnormality but there is general reduction in root mass.

9.13.1.2 Internal Symptoms

In contrast to generalised external symptoms, there are well-defined and characteristic internal symptoms of the disease. The discolouration of individual vascular bundles in fully differentiated nodes of mature stalks is generally observed in diseased canes. The discolouration occurs in the lower part of the node, just below the region of leaf sheath attachment, and is confined to the nodes.

In the longitudinally split, diseased stalks, the discolouration appears as dots, commas or short straight or bent lines of 2–3 mm in length. Discolouration varies in type and intensity with the severity of disease and the variety involved. Some varieties do not show any discolouration and those which show discolouration, it may vary in the same variety from time to time. The discolouration may be of yellow, orange, pink-red and reddish-brown shade (Figure 9.12).

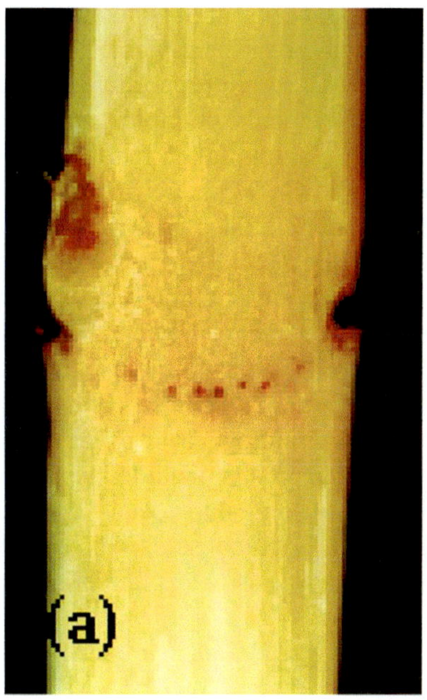

FIGURE 9.12 Ratoon stunting disease of sugarcane: Internal stalk symptoms.

The discolouration is due to the plugging of many large xylem vessels with a coloured gummy substance. For a reliable diagnosis, all the nodes in the fully developed stalks should show discolouration. Only the undamaged stalks should be examined for the diagnosis of RSD as the damage to the internal tissues of the sugarcane due to mechanical/pest injuries also leads to the discolouration of tissues. Diseased stalks should be examined immediately after splitting because the symptoms become inconspicuous as the tissues darken or dry. The nodes of young canes infected with RSD also show a pinkish colour below the apical meristem, which may extend downward about a centimetre. However, these symptoms are not as reliable as the mature node symptoms for diagnosing the disease, but in some varieties they are more helpful in detecting the disease at an early stage.

The vascular discolouration also occurs in leaf scald and gumming disease of sugarcane caused by *Xanthomonas albilineans* and *Xan. axonopodis* pv. *vasculorum*, respectively. In these diseases, the intensity of discolouration is greater and the discolouration also extends from nodes to internodes in comparison to RSD. Moreover, in gumming disease, the formation of gum pockets and the exudation of gum form cut ends of the canes are other distinguishing features.

Viswanathan (1996) found that internal/nodal symptoms were often unreliable for diagnosing the disease as these were not produced in all the susceptible varieties and these also varied in varieties in which these were produced. He also found that indirect ELISA and dot-blot techniques were more reliable for the detection of the disease, the former being more sensitive than the latter. Duttamajumder (1996) reported that many workers have questioned the reliability of nodal symptoms for diagnosing the disease and, at present, it is taken as an indication, but not the confirmation of the disease. The author further reported that Co 975, a genotype universally susceptible to RSD with 100% nodal symptoms, did not yield any RSD bacterium in extracts taken from most of the samples. This shows that Co 975 is not susceptible to the disease and that it may be a false positive case for RSD. He also screened 25 commercially grown genotypes for the presence of RSD bacterium using dark field, phase contrast and fluorescence microscopy and found that no genotype was completely free from the bacterium. Dark field microscopy was superior to phase contrast microscopy. Other techniques commonly used to detect RSD bacterium in the infected plants include tissue-blot enzyme immunological assay and fluorescent antibody staining (FAS); among these techniques, also including the above-mentioned ones, FAS is considered more effective than others.

In nature, RSD is often found in association with mosaic. Simultaneous occurrence of both the diseases has been more common in certain cultivars. The combined infection of both the diseases causes marked reduction in germination and yield.

The diagnosis of RSD based on visual symptomatology is highly difficult as this disease does not exhibit distinct external and internal symptoms. Sun *et al.* (2019) designed a novel *Lxx*-specific primer pair Lxx-F1/Lxx-R1 to detect *Lxx* using a conventional PCR assay. These primers were then compared with four published *Lxx*-specific primers and one universal *Leifsonia* generic primer pair LayF/LayR. Sugarcane leaf samples were collected from *Saccharum* spp. hybrids in commercial fields (315 samples) and from germplasm clones of five *Saccharum* species and *Erianthus arundinaceus* (216 samples). These samples were used for comparative field diagnosis with six conventional PCR assays. Sensitivity tests suggested that the PCR assay with primers Lxx-F1/Lxx-R1 had the same detection limit (1 pg of *Lxx* genomic DNA) as the primer pairs Cxx1/Cxx2 and CxxITSf#5/CxxITSr#5 and had 10-fold higher sensitivity than the primer pairs Pat1-F2/Pat1-R2, LayF/LayR and C2F/C2R. Comparison of PCR assays revealed that natural *Lxx*-infection incidence (6.1%) in field sample evaluation identified by Lxx-F1/Lxx-R1 primers was higher than the incidences (0.7 to 3.0%) determined by other primer pairs. Moreover, no nonspecific DNA amplification occurred within these field samples with Lxx-F1/Lxx-R1 primers, in contrast to with the primer pairs Cxx1/Cxx2 and LayF/LayR. Diverse *Leifsonia* strains were identified by PCR detection with LayF/LayR primers in the field samples, but whether these *Leifsonia* strains were pathogenic to sugarcane required further testing. Our findings reveal that the PCR assay with the newly designed primers, Lxx-F1/Lxx-R1 can be widely used for RSD diagnosis and *Lxx*-pathogen detection with satisfactory sensitivity and specificity.

9.13.2 Causal Organism

Leifsonia xyli subsp. *xyli* (Davis *et al.* 1984) Evtushenko *et al.* 2000

Synonym: *Clavibacter xyli* subsp. *xyli*

For around three decades, the causal agent of RSD was believed to be a virus because of its mode of transmission and the failure of the workers to either observe or isolate the pathogen. Gillaspie *et al.* (1973), Teakle *et al.* (1973) and Maramorosh *et al.* (1973) independently reported the association of a microorganism with the disease. Davis *et al.* (1980) isolated the bacterium and proved its pathogenicity.

The colonies on SC medium after two weeks of incubation at 28° C are circular with entire margin, non-pigmented, convex and measure 0.1–0.3 mm in diameter. Cells measure 0.25–0.35 × 1.0–4.0 μm and are straight or slightly curved rods, some cells swollen in the middle or at the tip. Variations observed in cell size have been attributed to different methods of sample preparation. Cells are nonmotile. Gram-stain-positive, non-sporing, aerobic, not acid-fast, oxidase negative and catalase positive. Cell walls contain fucose and rhamnose as the major sugars and 2, 4-diamino-butyric acid, glutamic acid, glycine and alanine as the major amino acids. Guanine + cytosine content of the bacterium is 66 mol% and the whole cell fatty acid extracts contain 17–24% 15:0-anteiso, 5–14% 16:0-iso and 62–72% 17:0-anteiso acids. All the coryneform plant pathogenic bacteria that contain 2, 4-diaminobutyric acid were placed in the genus *Clavibacter.* Evtushenko *et al.* (2000) transferred *Cla. xyli* subsp. *xyli* to a newly created genus *Leifsonia* as *Leifsonia xyli* subsp. *xyli.*

9.13.3 Disease Cycle

The bacterium perpetuates in the diseased canes and infected stubbles and the latter give rise to an infected ratoon crop. There is no evidence of its survival in the true seed or soil.

As the bacterium can be easily inoculated mechanically to the healthy plants through the sap of diseased canes, cutting knives/cutter-planter machines and harvesting equipment play an important role in the dissemination of the bacterium. Hughes and Steindl (1955) found that, in one case, the cutter machine spread the bacterium from one diseased cane of clone Q 28 to 60 consecutive plants and, thereafter, to some scattered plants in Queensland. In Louisiana, the dissemination by harvester equipment increased the infected plants in the plant crop from 16 to 47%, and then in the subsequent ratoon crops from 50 to 80–90% (Steib *et al.*, 1957). The exchange of germplasm from one country to another country also plays a major role in the spread of disease. The field rats, dogs and foxes play a limited role in disseminating the pathogen.

There seems to be no possibility of the dissemination of the pathogen due to wind-splashed rain, as the susceptible varieties growing adjacent to the infected cane fields remain free from infection. It was also the case in the varietal evaluation trials that the rows of healthy canes, planted adjacent to the rows of inoculated canes, remained free from infection for three years when proper sanitation measures were observed (Hughes and Steindl, 1955).

The bacterium penetrates the host through wounds and then enters the xylem vessels. After penetration, it colonises the xylem vessels of roots, stalks and leaves. It seems that the bacterium generally multiplies and spreads in dysfunctional vessels. The enzymatic activity of the bacterium dissolves pit membranes to facilitate its movement through xylem vessels. The extent and nature of anatomical barriers, which restrict the movement of the bacterium in the host, are considered mainly responsible for the susceptible/resistant reaction of the cultivars. The susceptibility of the host is influenced more by the accessibility of vascular tissue to the pathogen than the pathogen ability to multiply within the host.

Quecine *et al.* (2016) studied the colonisation of sugarcane tissues using a mutant strain of *Lef. xyli* subsp. *xyli* that stably expressed the *gfp* gene in sugarcane tissues. In the leaves, *Lef. xyli* subsp. *xyli*-tagged cells were found in a new niche, i.e., in the mesophyll and in the bundle sheath cells surrounding the vascular system, in addition to within the xylem vessels, as reported earlier. This finding indicates that the bacterium is able to move from the xylem to the parenchyma of the leaf cells and the colonisation of sugarcane plants by this pathogen is not limited to the xylem vessels only as commonly reported.

Zhu *et al.* (2018) inoculated cuttings of two sugarcane cultivars, Badila and GT11, with *Lef. xyli* subsp. *xyli* to investigate the effects of infection on their growth, physiological characteristics and associated gene differential expression. The bacterial population gradually increased with plant growth after inoculation. Plant height, stalk diameter, single stalk weight and water potential decreased as a result of infection, while membrane permeability and amino acid content increased in infected plants compared to the control. The expression of *PAL*, *ZFP* and *NBS-LRR* genes also increased in plants subjected to ratoon stunt stress, indicating that these genes are involved in sugarcane responses to *Lef. xyli* subsp. *xyli* infection.

Farahani *et al.* (2015) compared conventional PCR, nested PCR and real-time PCR for the detection of *Lef. xyli* subsp. *xyli* from infected sugarcane plants and reported that conventional PCR lacked considerable sensitivity, nested PCR had acceptable sensitivity, but real-time PCR was the most efficient technique to be used as a diagnostic tool for the disease.

Young *et al.* (2016) compared four diagnostic techniques, namely, quantitative PCR on pooled leaf sheath biopsies (LSB-qPCR), conventional PCR on the same templates (LSB-PCR), evaporative-binding enzyme immunoassay (EB-EIA) coupled with phase contrast microscopy (PCM) on expressed xylem sap from the same fields, and conventional PCR on the same xylem sap samples for the detection of causal agent from 100 fields of sugarcane (*Saccharum* interspecific hybrids) of unknown infection status. LSB-qPCR and LSB-PCR detected the causal agent in 27 and 18 fields, respectively, whereas, from samples of expressed xylem sap from the same fields, conventional PCR identified 12 infections and EB-EIA/PCM detected the bacterium from 3 fields. The sensitivities of qPCR and PCR, determined from plate counts of a dilution series, were approximately 10^3 and 10^4 cfu mL^{-1}, respectively. In further tests conducted on 139 LSB samples from across the Australian industry, qPCR and PCR diagnosed the disease in 31 and 25 fields, respectively. Using qPCR and PCR on LSB samples, the disease was diagnosed in a range of cultivars throughout the year, and these methods could detect the pathogen in sugarcane plants aged three months to more than one year.

9.13.4 Management

1. Use of pathogen-free seed and prevention of the spread of the pathogen from the diseased plants to the healthy plants are the two most effective control measures. The selection of apparently healthy and thick canes for seed greatly reduces the disease

incidence. However, the selection of healthy canes is hampered due to the absence of clearly defined external symptoms of the disease.

2. In the absence of pathogen-free seed, the diseased canes/setts should be treated with heat therapy to eliminate the seed-borne inoculum. Four methods that use heat therapy for treating sugarcane setts are given below.

   a. **Hot water treatment**: This method is used globally. In India, the hot water treatment is normally carried out at 50 °C for ½ h. The same is done at 50 °C for 3 h in Queensland (Australia), Florida and Hawaii (USA). The seed setts of mature stalks are usually less affected and give better germination after hot water treatment than those from immature or senescing stalks.
   b. Andreato *et al.* (2019) reported that hot water treatment of sugarcane setts at 50 °C for 2 h was more effective for controlling ratoon stunting disease compared to 52 °C for 30 min. and also found that this treatment was also effective for controlling leaf scald.
   c. **Hot air treatment**: In India, it is used to a lesser extent than hot water treatment. It is carried out at 54 °C for 8 h. In this treatment, the bud damage is less than that of hot water treatment. It is more suitable for soft, succulent and immature canes, which cannot withstand hot water treatment. After the treatment, dip the seed cane in water at ambient temperatures to reduce the loss in germination.
   d. **Moist hot air treatment**: The method was developed in India to check the loss of moisture in hot air treatment. Within 1 h, the required temperature of 54 °C is obtained and it is maintained for 7 h in a sealed chamber (Shukla *et al.*, 1974).
   e. **Aerated-steam method**: It is the most recent method in which an aerated steam, a mixture of air and steam, is used. In India, the recommended temperature is 54 °C for 4 h while in the USA the treatment is done at 51 °C for 4 h. The smooth and uniform distribution of heat in the treatment chamber makes this method more effective. It is also economical than the other methods. In addition to RSD, the treatment is also effective for grassy shoot, leaf scald and red rot of sugarcane.

   The major drawbacks with heat therapy are the reduction in germination, the high cost of equipment and the failure to eradicate the pathogen completely. The percentage of disease-free setts never exceeds more than 90–95, especially in hot water treatment. In the remaining 5–10% setts, the number of bacteria is very low, making it difficult to detect with the commonly available methods. Hence, there is an urgent need to develop more sensitive and accurate methods to detect this low population of the bacterium.

3. Sanitation, including disinfection of cutting knives, cultivators and harvester blades, is very useful in preventing the spread of the pathogen. The commonly used disinfectants include 5–15% Lysol (neutralised cresylic acid), 1% Dettol, 50% ethanol or 10% formalin. A minimum contact of 5 min. of the disinfectant with the cutting surface is essential for proper disinfection.
4. Strict quarantine measures should be followed to check the spread of the pathogen to the disease-free areas.
5. Give proper irrigation to the crop to minimise the losses.
6. Use of resistant varieties is the most viable method of disease control. Though most commercial sugarcane clones are susceptible to the pathogen, substantial resistance has been found in clones such as CP 29–116, Q 50 and H 60–6909.

The Australian sugar industry has never pursued genetic resistance to RSD, despite it being widely considered to be one of the most important diseases of sugarcane. Young (2018), in his review, traces the factors that have led to an industry stance that is apparently without any scientific justification, and which has tended to downplay the significance of RSD on Australian sugarcane productivity, and thus has led to a significant loss in production. Based on the available information, the author calls on the Australian sugar industry to prioritise selection for RSD resistance in the plant improvement programme.

REFERENCES

Andreato, C., Gazaffi, R. and Urashima, A. S. (2019). Effect of thermotherapy for ratoon stunt on leaf scald control (Abstr.). *Phytopathology* 109: S2.71–2.72.

Chakraborty, M., Soda, N., Strachan, S., Ngo, C. N., Bhuiyan, S. A. *et al.* (2024). Ratoon stunting disease of sugarcane: A review emphasizing detection strategies and challenges. *Phytopathology* 114: 7–20. https://doi.org/10.1094/PHYTO-05-23-0181-RVW

Chona, B. L. (1956). Presidential address for the pathology section. *Proc. Int. Soc. Sugar Cane Technol., 9th Congress*, pp. 975–80.

Davis, M. J., Gillaspie, A. G., Vidaver, A. K. and Harris, R. W. (1984). *Clavibacter*: A new genus containing some phytopathogenic coryneform bacteria, including *Clavibacter xyli* subsp. *xyli* sp. nov., subsp. nov. and *Clavibacter xyli* subsp. *cynodontis* subsp. nov., pathogens that cause ratoon stunting disease of sugarcane and Bermudagrass stunting disease. *Int. J. Syst. Bacteriol.* 34: 107–17.

Davis, M. J., Gillaspie, A. G. Jr., Harris, R. W. and Lawson, R. H. (1980). Ratoon stunting disease of sugarcane: Isolation of the causal bacterium. *Science* 240: 1365–67.

Duttamajumder, S. K. (1996). Ratoon stunting disease in subtropical India: Prevalence and control. *Proc. 9th Int. Conf. Plant Path. Bac.* Chennai, pp. 333–41.

Evtushenko, L. I., Dorofeeva, L. V., Subbotin, S. A., Cole, J. R. and Tiedje, J. M. (2000). *Leifsonia poae* gen. nov., sp. nov., isolated from nematode galls on *Poa annua*, and reclassification of '*Corynebacterium aquaticum*' Leifson 1962 as *Leifsonia aquatica* (ex Leifson 1962) gen. nov., nom. rev., comb. nov. and *Clavibacter xyli* Davis *et al.* 1984 with two subspecies as *Leifsonia xyli* (Davis *et al.* 1984) gen. nov., comb. nov. *Int. J. Syst. Evol. Microbiol.* 50: 371–80.

Farahani, A. S., Taghavi, S. M. and Taher-Khani, K. (2015). Comparison of conventional, nested and real-time PCR for detection of the causal agent of ratoon stunt in Iran. *J. Plant Pathol.* 97: 259–63.

Gillaspie, A. G. Jr., Davis, R. E. and Worley, J. F. (1973). Diagnosis of ratoon stunting disease based on the presence of a specific microorganism. *Plant Dis. Reptr* 57: 987–90.

Gillaspie, A. G. Jr. and Teakle, D. S. (1989). Ratoon stunting disease. In: C. Ricaud, B. T. Egan, A. G. Gillaspie, Jr. and C. G. Hughes (eds.). *Diseases of Sugarcane: Major Diseases*. Elsevier, Amsterdam, pp. 59–80.

Hughes, C. G. (1974). The economic importance of ratoon stunting disease. *Proc. Int. Soc. Sugar Cane Technol.* 15: 213–17.

Hughes, C. G. and Steindl, D. R. L. (1955). Ratoon stunting disease of sugarcane. *Queensland Bureau of Sugar Experiment Stations Technical Communications* No. 2, 54 pp.

Maramorosh, K., Plavsic-Banjac, B., Bird, J. and Liu, L. J. (1973). Electron microscopy of ratoon stunted sugarcane: Microorganisms in xylem. *Phytopathol. Z.* 77: 270–73.

Quecine, M. C., Silva, T. M., Carvalho, G., Saito, S., Mondin, M., *et al.* (2016). A stable *Leifsonia xyli* subsp. *xyli* GFP-tagged strain reveals a new colonization niche in sugarcane tissues. *Plant Pathol.* 65: 154–62.

Ricaud, C., Egan B. T., Gillaspie, A. G. Jr. and Hughes, C. G. (eds.) (1989). *Diseases of Sugarcane: Major Diseases*. Elsevier, Amsterdam, 399 pp.

Shukla, U. S., Ram R. S. and Tripathi, R. C. (1974). Effect of moist hot air treatments on the control of GSD and RSD. *Annu. Rep. Indian Inst. Sugarcane Res.* Lucknow, pp. 72–74.

Singh, K. (1966). Ratoon stunting disease of sugarcane in India. *Indian Sugar* 16: 335–37.

Steib, R. J., Forbes, I. L. and Chilton, S. J. P. (1957). A report on further studies on the ratoon stunting disease of sugarcane in Louisiana. *Sugar J.* 19: 35–37.

Sun, S.-R., Chen, J.-L., Duan, Y.-Y., Chu, N., Huang, M.-T., *et al.* (2019). Improved primers for the specific detection of *Leifsonia xyli* subsp. *xyli* in sugarcane using a conventional PCR assay. *Plant Dis.* 103: 3251–58.

Teakle, D. S., Smith, P. M. and Steindl, D. R. L. (1973). Association of small coryneform bacterium with the ratoon stunting disease of sugarcane. *Aust. J. Agric. Res.* 24: 869–74.

Urashima, A. S. and Marchetti, L. B. L. (2013). Incidence and severity of *Leifsonia xyli* subsp. *xyli* infection of sugarcane in Sao Paulo State, Brazil. *J. Phytopathol.* 161: 478–84.

Viswanathan, R. (1996). Serological techniques for the detection of a bacterium causing ratoon stunting disease (RSD) in sugarcane. *Proc. 9th Int. Conf. Plant Path. Bac.* Chennai, India, pp. 162–68.

Young, A. J. (2016). Possible origin of ratoon stunting disease following interspecific hybridization of *Saccharum* species. *Plant Pathol.* 65: 1403–10.

Young, A. J. (2018). Turning a blind eye to ratoon stunting disease of sugarcane in Australia. *Plant Dis.* 102: 473–82.

Young, A. J., Kawamata, A., Ensbey, M. A., Lambley, E. and Nock, C. J. (2016). Efficient diagnosis of ratoon stunting disease of sugarcane by quantitative PCR on pooled leaf sheath biopsies. *Plant Dis.* 100: 2492–98.

Zhu, K., Yuan, D., Zhang, X.-Q., Yang, L.-T. and Li, Y.-R. (2018). The physiological characteristics and associated gene expression of sugar cane inoculated with *Leifsonia xyli* subsp. *xyli*. *J. Phytopathol.* 166: 44–52.

9.14 GUMMOSIS OF SUGARCANE

Gummosis, also known as **gumming or Cobb's disease**, was the first disease of sugarcane reported in the literature. Dranert (1869) reported this disease for the first time from Brazil, but its bacterial nature was first established by Cobb (1893) while he was working in Australia. The occurrence of the disease was reported from Mauritius in 1894, from Fiji in 1895 and, later on, from Réunion, the Caribbean, Argentina, Belize, Columbia, the Dominican Republic, French Guiana, Jamaica, Ghana, Malawi, Madagascar, Mozambique, Mexico, Panama, Puerto Rico, South Africa, Swaziland and Zimbabwe (Ricaud and Autrey, 1989; Janse, 2006). In India, Rangaswami (1960) reported it from Tamil Nadu on Co 449 and Co 527. To date, there is no report of its occurrence from North India.

The disease caused heavy losses till 1930s where highly susceptible noble canes where grown. However, the development of resistant interspecific hybrids brought the disease under control and it was eradicated from Australia, Brazil, Fiji and the West Indies (Ricaud and Autrey, 1989). At present, it is a minor disease in most countries, except in those where climatic conditions are highly favourable for disease development and dissemination of the pathogen.

The losses from the disease occur due to reductions in the yield and sugar content of the canes. Besides the loss in yield, the presence of gum (polysaccharides) in the cane juice interferes with the milling process leading to low sugar recovery. The race of the pathogen involved, and the degree of tolerance of the varieties affected, influence the extent of loss. In the same variety, the loss from systemic infection of the stalks will be more than from stalks showing foliar infection only. The losses as high as 30–40% in cane yield and 9–70% in sugar content have been reported under epiphytotic conditions in Australia.

9.14.1 Host Range

Ricaud and Autrey (1989) have listed six natural hosts of the bacterium, in addition to sugarcane, reported by different workers. These are maize, three palms, namely, princess palm (*Dictyosperma album*), royal palm (*Roystonia regia*) and betel nut palm (*Areca catechu*), broom bamboo (*Thysanolaena maxima*) and Guatemala grass (*Tripsacum laxum*). The infection was rarely observed on maize, *A. catechu* and Guatemala grass.

T. maxima often shows a systemic infection, but the causal bacterium is quite different from the bacterium infecting the sugarcane. Isolates identical to the races infecting sugarcane obtained from *T. maxima* cause only foliar symptoms on this host. It is evident from the above-given account that sugarcane is the only main host of gumming pathogen. The other hosts are only temporary and accidental hosts and do not harbour the pathogen for long periods in the absence of sugarcane.

A more or less similar disease occurs on maize and Guatemala grass. Strains of *Xanthomonas* causing this disease differ in several aspects from the gumming pathogen and probably these belong to a group of strains reclassified as *Xan. vasicola* pv. *vasculorum*.

9.14.2 Symptoms

Two types of symptoms, namely, **leaf-streaking phase** and **systemic phase**, are produced.

In the leaf-streaking phase, yellow to orange, 3–5 mm wide streaks appear on the leaves and progress along the affected veins. With age, their colour changes to ashy grey and the affected tissue becomes necrotic. Usually, these originate from the leaf margin and extend towards the base, but they can arise from any area of leaf surface when the infection occurs through the wounds and develop in either direction. The streaks may cover the entire length of leaf lamina and may even extend to the leaf sheath, in the latter case leading to the systemic infection of the stalks (Figure 9.13a). The streaks of gummosis differ in colour and patterns of appearance form those caused by leaf scald and chlorotic streak.

FIGURE 9.13 Gummosis of sugarcane. (a) Yellow to orange streaks on leaf lamina. (b) Reddish discolouration of vascular bundles.

The severity of leaf infection is influenced by the age of leaf, environmental conditions and varietal reaction of the cane. The intensity of the streaks is at its highest on the mature leaves. Under highly favourable conditions for disease development, many streaks may appear on a leaf and several leaves of a stalk may show symptoms.

The characteristic symptoms of the disease are observed, when transversely cut and longitudinally split stalks are examined. Longitudinally split canes may show reddish discolouration of vascular bundles, which is more prominent at the nodes (Figure 9.13b). The breakdown of the internal stalk tissue results in the formation of pockets/cavities, particularly below the apical meristem. These pockets contain gum consisting of bacterial cells and extracellular polysaccharide.

The bacteria ooze out in the form of yellow shiny droplets from infected bundles of the transversally cut surface of stumps or stalks having systemic infection. The bacterial ooze can be seen more clearly by performing a **sweating test**. In this test, cut pieces of affected stalks kept in sealed polythene bags or moist chambers are incubated at 30–35° C. Within a few hours, yellowish bacterial ooze can be seen on the cut ends. This test is more helpful in detecting the bacterial ooze from doubtful cases.

Systematic infection may also result in malformation of the stalk, making it flat on one side and bulging on the opposite side. Transverse splits appearing as 'knife cuts' can be seen near the apex of affected stalks. Systemically infected plants remain stunted. Ratoon crop suffers less than the plant crop. Systemic infection is usually, but not always associated with the chlorosis of leaves. The leaves of mature stalks may show chlorosis but it is usually more common in young ratoon shoots. In highly susceptible varieties, the death of plants may occasionally occur after the chlorotic symptoms while in tolerant varieties, the affected stalks may completely recover from these symptoms, especially under highly favourable growth conditions.

A part or whole of the lamina may show chlorosis and from one to all leaves of a stalk may be affected. In case of partial chlorosis, white patches appear on lamina; starting from the base of the oldest leaf and moving to the tip of the youngest leaf. The chlorotic symptoms of gummosis and leaf scald are indistinguishable from each other. As the pathogen is absent in the affected tissue in both the diseases, the involvement of a toxin is suspected.

Narrow red stripes, indistinguishable from those of red stripe disease, may also occur on the leaves of systemically infected stalks. A film of dried bacterial exudate is often seen on these stripes. The bacterium infects both xylem vessels and mesophyll cells.

A new bacterial disease of sugarcane caused by *Pantoea stewartii* subsp. *indologenes* has been reported by Silva-Rojas *et al.* (2017) from Mexico. Affected plants show chlorotic streaks in the leaves, and these symptoms frequently appear in the middle of the leaves as fine lines with irregular borders. The disease may be confused with gummosis as it has some overlapping symptoms.

9.14.3 Causal Organism

Xanthomonas axonopodis pv. *vasculorum* (ex Cobb 1893) Vauterin *et al.* 1995

Synonym: *Xan. vasculorum, Xan. campestris* pv. *vasculorum*

Cells are straight rods (0.4–0.5 × 1.0–1.5 μm) Gram-stain-negative; occur predominantly singly, motile by a single polar flagellum, non-capsular, non-sporing and not acid-fast. Xanthomonadins are produced. Hydrolysis of starch and esculin is positive. Growth on nutrient agar is poor. Amino acids are required for growth. Acid is produced within 21 days from glucose, sucrose and trehalose in Dye's medium C but not from arabinose, galactose, mannose, fructose and cellobiose. Catalase test and H_2S production from peptone are positive but nitrate reduction and lysine dicarboxylase tests are negative. The PCR-RFLP of the 16S–23S rDNA spacer region has been used to differentiate *Xanthomonas* species attacking sugarcane. The method also allows differentiation of Type A from Type B of *Xan. axonopodis* pv. *vasculorum* (Destefano *et al.*, 2003).

9.14.4 Disease Cycle

The bacterium perpetuates in diseased canes and infected stubbles. The diseased setts disseminate the bacterium to short as well as long distances. The diseased setts and contaminated setts give rise to primary infection of the crop. Secondary spread of the bacteria occurs through the cutting knives, harvesting implements, wind-blown rain and cyclones; the spread is very fast through the last two means. Cutting knives and implements are important agents for disseminating the bacterium at harvest time.

During humid weather and periods of heavy dew, bacterial cells ooze out from wounds of infected stalks and leaves and may be also through stomata of leaves and leaf sheaths. Then these are splashed away/carried with the help of wind/rain and enter the host through wounds and stomatal openings. The bacteria may also enter the stalks via leaf scars and wounds caused due to farm operations. The wounds on the leaves are caused due to the shearing action of the wind and the serrated margins of leaves. For the successful infection of the host, high humidity or free moisture on host surface is essential. From vascular infection, leaf streaks develop within 1–3 weeks under favourable weather conditions.

In Mauritius, the severity of epiphytotics was positively correlated with the number and intensity of cyclones in the growing season. Epiphytotics were more severe when cyclones immediately preceded the early beginning of a dry and cool maturing season (Ricaud, 1969).

9.14.5 Management

1. Use of resistant varieties is the most effective method of disease management. In Australia and Mauritius, legislative measures were followed for the replacement of susceptible varieties with resistant varieties. Varieties having high degree of resistance to foliar infection should be recommended in areas where environmental conditions favour frequent epiphytotics of the disease.
2. The other management measures, like sanitation, have a limited success due to the rapid spread of the pathogen. However, these can be helpful in areas where conditions for rapid aerial spread occur very infrequently. The solution of an iodophor (250 μg mL^{-1} available iodine) can be used to disinfect cutting knives and blades of harvesters.
3. Sow pathogen-free seed. The seed should not be taken form fields showing systemic infection.
4. Rouging of diseased plants particularly, those showing systemic infection (chlorosis), will reduce the disease incidence.
5. Do not grow the collateral host like maize, pearl millet and sorghum near the sugarcane fields.

REFERENCES

Cobb, N. A. (1893). Plant diseases and their remedies-Diseases of the sugarcane. *Agric. Gaz. N. S. W. for 1893* 4: 777–98.

Destefano, S. A. L., Almeida, I. M. G., Rodrigues, N. J. Ferreira, M. and Balani, D. M. (2003). Differentiation of *Xanthomonas* species pathogenic to sugarcane by PCR-RFLP analysis. *Eur. J. Plant Pathol.* 109: 283–88.

Dranert, F. M. (1869) Bericht uber die Krankheit des Zuckerrohres. *Z. Parasi-tenkd. Bd.* 1: 13–17.

Janse, J. D. (2006). *Phytobacteriology: Principles and Practice*. CABI Publishing, Wallingford, 360 pp.

Rangaswami, G. (1960). Studies on two bacterial diseases of sugarcane. *Curr. Sci.* 29: 318–19.

Ricaud, C. (1969). Investigation on the systemic infection of gumming disease. *Proc. Int. Soc. Sugar Cane Technol.* 13: 1159–69.

Ricaud, C. and Autrey, L. J. C. (1989). Gumming disease. In: C. Ricaud, B. T. Egan, A. G. Gillaspie Jr. and C. G. Hughes (eds.). *Diseases of Sugarcane: Major Diseases*. Elsevier, Amsterdam, pp. 21–31.

Silva-Rojas, H., Aguilar-Granados, A., Rebollar-Alviter, A., Valdes-Balero, A., Sanchez-Pale, J., *et al.* (2017). *Pantoea stewartii* subsp. *indologenes* is responsible for a new bacterial disease in sugarcane. *Phytopathology* 107: S2.4.

Vauterin, L., Hoste, B., Kerster, K. and Swings, J. (1995). Reclassification of *Xanthomonas*. *Int. J. Syst. Bacteriol.* 45: 472–79.

9.15 LEAF SCALD OF SUGARCANE

Leaf scald of sugarcane was first reported by Wilbrink (1920) from Java and later on by North (1926) from Australia and Fiji. Since then, it has been reported from many countries, and Ricaud and Ryan (1989) have listed 44 countries from where it has been reported. Later on, Rott and Davis (2000) reported that the disease occurs in at least 66 countries. In India, the disease was first observed by Egan in 1960 at the Indian Agricultural Research Institute, New Delhi during his visit to India (Egan, 1979) but the first detailed report of the disease was published by Satyanarayana (1974) from Andhra Pradesh. Since then, the incidence of the disease has increased manifold and now it has become a major disease of sugarcane in the country, occurring in Punjab, Uttar Pradesh and Bihar. Among the susceptible cultivars, prominent ones include CoJ 64, CoS 767 and CoLK 8001.

Mensi *et al.* (2013) reported the occurrence of leaf scald for the first time from Gabon, describing an unusual highly virulent and aggressive strain of *Xanthomonas albilineans*.

Leaf scald was a serious disease of noble canes, which were widely grown in the early years of the 20th century. North (1926) reported substantial losses due to the disease from Australia and in severe outbreaks; the losses were more than 25%. Egan (1971) reported that outbreaks of the disease in the early 1900s caused heavy losses in highly susceptible varieties in Queensland. The replacement of susceptible noble canes with varieties containing *Saccharum spontaneum* genes, resistant to the pathogen, has considerably reduced the incidence of the disease. However, it is still a major constraint in the successful cultivation of sugarcane in many countries like Mauritius, Fiji, Cuba, Brazil, Taiwan, Indonesia and the USA.

Apart from reducing cane yield, the disease also lowers the quality of the juice by reducing Brix, Pol and Purity. The slow growth of affected stalks, the reduction in number of millable canes per stool and poor ratooning reduce the yield. The acute stage causes the death of stools and a reduction in sugar yield. These factors, which influence the growth of sugarcane plants, also influence the disease severity. The onset of adverse conditions soon after the maturity of the canes give rise to large number of side shoots. The number and size of side shoots are directly related to deterioration of juice quality.

9.15.1 Host Range

In addition to sugarcane, the bacterium causes natural infection on *Paspalum dilatatum*, *P. congugatum*, *Imperata cylindrica* var. *major* (blady grass) and *Brachiaria piligera*.

The possibility of survival of the bacterium in blady grass for some time can make this host as a source of inoculum for the sugarcane crop. *Zea mays*, *Rottboellia cochinchinensis*, *I. cylindrica*, *Panicum maximum*, *Pennisetum purpureum*, *Paspalum* sp. and *Sorghum* sp. also act as natural hosts of the bacterium. It seems that *Xan. albilineans* has a narrow host range, restricted to sugarcane and some other members of the family Graminae.

9.15.2 Symptoms

There are two types of symptoms, namely, the **chronic phase** and the **acute phase**. The latent infection is also an important characteristic feature of the disease.

9.15.2.1 Chronic Phase

These are the most commonly observed symptoms. The most characteristic symptom of the disease is the appearance of **'white pencil-line'** streaks, which are narrow, about 1–2 mm wide and sharply defined, on the leaf lamina. The streaks develop parallel to the midrib and often extend to the base of leaf lamina or the leaf sheath (Figure 9.14a). The specific epithet of the causal bacterium has been derived from the description of this streak.

Initially, the streaks are generally well-defined but these become diffused and broader with the passage of time. However, the close examination of these broader and diffused stripes will show the white pencil-line at the centre. As the disease progresses, the stripes start drying causing withering away of the leaves. The drying always starts from the tip and extends downward, giving a scalded appearance to the leaves. The streaks also appear on leaves in **gummosis** and **chlorotic streak of sugarcane** but these streaks are devoid of central white pencil-line. The partial or complete chlorosis of the leaf blade also occurs. White pencil-line or its reddish necrotic patches can be seen on close examination of these chlorotic blades.

Generally, the side shoots, varying in number, develop on the mature infected stalks (Figure 9.12b). Some of the side shoots die when they are still young while the surviving shoots generally exhibit typical white pencil-line or chlorosis of leaves.

The necrosis of the vascular bundles in the form of bright to dark red streaks can be seen on the longitudinally

(a) (b)

FIGURE 9.14 Leaf scald of sugarcane: (a) Pencil-line streaks on leaf lamina (b) Infected plant bearing adventitious shoots.

split canes. These streaks are generally more prominent at the nodes than in the internodes and also appear in the side shoots. Inside the severely infected canes, longitudinal lysigenous cavities may be formed in the nodal and internodal areas, particularly near the shoot apex. These cavities are relatively dry compared to those of gumming disease as these contain less amount of bacterial slime or gum.

9.15.2.2 Acute Phase

The sudden wilting and death of mature stalks, often without any previous symptoms, mark the onset of the acute phase. Occasionally, the small shoots having typical streaks appear at the base of stalks. The other symptoms, like proliferation of buds, bunchy top appearance and reductions of the root systems, were observed by Agnihotri (1986). The acute phase generally appears after a stress resulting from a prolonged period of dry weather, or dry weather after a rainy season. To confirm the occurrence of the acute phase, one should look for canes showing chronic phase symptoms among the affected plants or wait till the new tillers come out showing the white pencil-line symptoms.

In India, the symptoms of the acute phase have not been observed. However, the gradual wilting of plants showing chronic-phase symptoms, especially during the summer months and after the first monsoon shower, is commonly observed in some susceptible varieties like CoLK 8001, CoJ 81 and Co 62399 (Agnihotri and Duttamajumder, 1992).

9.15.2.3 Latent Infection

The disease is characterised by a latent phase/infection in which the infected plants do not show any symptoms. The plants suffering from the chronic phase recover from the disease under certain environmental conditions. The recovery is more common in tolerant varieties or when the conditions are suitable for plant growth. The latency may involve a few shoots or stools or an entire field. The symptoms may appear on such plants after ratooning and on the plants growing from setts taken from the infected stools. The latent infection can also lead to the development of the acute phase. Various techniques like the fluorescent antibody technique, ELISA and the use of indicator plants are used to detect the latent infection. Among these, fluorescent antibody technique is more effective

9.15.3 Causal Organism

Xanthomonas albilineans (Ashby 1929) Dowson 1943
Synonym: *Phytomonas albilineans*

In 1929, Ashby named the bacterium as *Bacterium albilineans*. Later on, it was transferred to genus *Phytomonas* as *Pht. albilineans*. After the establishment of genus *Xanthomonas*, it was transferred to it as *Xan. albilineans*. Wilbrink's agar, now known as sucrose peptone agar (SPA), was earlier used for isolation and culturing of this bacterium. Glutamic acid and methionine are required for growth. Optimum temperature for growth on SPA is 25 °C and the maximum is 37 °C. The isolation of the bacterium can be made by using the macerated infected tissue or juice of infected plants after diluting it 1000-fold. On SPA, the colonies appear after an incubation of 5–7 days, and these are honey-yellow to Naples-yellow, minute, glistening and viscid, having an entire margin.

A selective medium (*Xanthomonas albilineans* selective [XAS] medium) has been devised by supplementing Wilbrink's medium with 5 g of KBr, 100 mg of cycloheximide, 2 mg of benomyl, 25 mg of cephalexin, 30 mg of novobiocin and 50 mg of Kasugamycin per litre. The isolation on this medium gives more than 98% recovery from symptomatic plants. The growth rate, morphology and pigmentation of colonies on the medium serve as useful differential characteristics to identify the bacterium.

Cells are straight rods (0.25–0.3 × 0.6–1.0 µm), predominantly single, Gram-stain-negative and motile by a single polar flagellum. Xanthomonadins are produced. The bacterium neither liquefies gelatin nor reduces nitrate and does not produce H_2S, ammonia and indole. Esculin is hydrolysed but not starch. Acid is produced within 21 days on Dye's medium C from glucose, sucrose and mannose. Ammonium salts, nitrates or asparagine are not used as a sole source of nitrogen.

Variation in genetic, physiological and serological characteristics, and the pathogenicity of the bacterium, indicating the existence of different pathotypes in the species, has been found but to date there is no evidence of existence of races. According to Silva *et al.* (2007), there is evidence for the occurrence of pathogenic variation in the bacterium, as a particular cultivar gives a susceptible reaction to the pathogen in one region but a different reaction in another region. Rep-PCR analysis of 50 isolates, collected by them from different regions of Sao Paulo state in Brazil, revealed genetic diversity among these isolates and they divided them into three groups. The pathogenicity test of these isolates revealed the variation in aggressiveness among the isolates, suggesting a possible relationship between the occurrence of bacterial variants and the reaction of sugarcane cultivars in Sao Paulo State. It can be inferred from these findings that pathogenic races of the bacterium occur in nature and that there is an urgent need to develop suitable differentials for the identification of these races.

9.15.4 Disease Cycle

The main sources of the perpetuation of the bacterium are infected stubbles and infected setts used for sowing. Maximum dissemination of the bacterium occurs at the time of harvesting and planting, and it occurs mainly through the infected cuttings and implements used for the harvesting and cutting of canes. Infected cuttings are an important mode of spread over long distances. Duttamajumder (1996) showed that setts obtained from the upper half of the infected canes were much more efficient (48%) in producing the disease than those from lower half (21%).

The cane knives and base cutters of mechanical harvesters also play a role in the secondary dissemination of the bacterium within a crop. The harvesting implements carry the inoculum after cutting an infected cane and transmit the same up to the third stool, after which the transmission becomes irregular. Fluff (seed)-plant transmission of the bacterium has also been proved. Aerial dissemination of the pathogen has been reported to occur in several geographical areas of the world, and was associated with outbreaks of the disease (Comstock, 2001; Daugrois *et al.*, 2003). The role of phyllosphere colonisation by the bacterium in the epidemiological cycle of the disease in Guadeloupe was also shown by Daugrois *et al.* (2003). The role of insects such as grasshoppers, leafhoppers, leaf beetles and cane borers, and rodents in transmission is suspected but not yet proved.

Xan. albilineans is a systemic and xylem-invading pathogen and colonises only the xylem of the vascular strands of the affected parts. However, occasionally the bacterium may break the xylem vessels and destroy the surrounding cells, producing cavities. Generally, this situation is manifested by the appearance of red areas on the white pencil-line. The bacterium produces gum, a xanthan-like polysaccharide composed of glucose, mannose and glucuronic acid. Gum occludes phloem and xylem vessels and partially enters mesophyll cells (Solas *et al.*, 2003). The chlorosis-inducing isolates of the bacterium produce toxic compounds, of which the **albicidin** being the most important, plays a key role in pathogenicity. Albicidin inhibits DNA replication of protoplastids, thereby blocking plastid development. There is no correlation between toxin production and pathogenicity of *Xan. albilineans*. Therefore, it is apparent that some other virulence factors play a key role concurrently with albicidin in pathogenicity (Champoiseau *et al.*, 2006).

Garces *et al.* (2014) reported that assays with SYBR Green primers and TaqMan probe and primers derived from the albicidin toxin biosynthesis gene cluster efficiently and reproducibly amplified *Xan. albilineans*. Detection was more sensitive with qPCR than with conventional PCR. Assays were specific for *Xan. albilineans*, and sap extracts did not inhibit the qPCR reaction. Populations of *Xan. albilineans* varied in different tissues and differences were the greatest within tissues in resistant cultivars. The bacterial populations in systemically infected young, but not yet fully emerged leaves exhibited the greatest differences between resistant and susceptible cultivars. As the qPCR is a highly sensitive method for the detection of *Xan. albilineans*, it will provide a reliable method for leaf scald resistance screening. Gutierrez *et al.* (2016) compared qPCR with the visual symptom rating in 31 clones at different times after inoculation and found the qPCR assay more consistent than the visual symptom rating at 8 weeks after inoculation. They also suggested that qPCR can provide an improved method to evaluate resistance to leaf scald in sugarcane.

Daugrois *et al.* (2014) reported that the resistance of sugarcane to leaf scald pathogen appears to involve several traits, including limiting the size of epiphytic *Xan. albilineans* populations and limiting the ability of the pathogen to produce leaf necrotic symptoms by invading the leaf vascular system or to move from the leaf into the stalk.

The disease is favoured by wet weather, particularly in combination with cyclones. Usually, the appearance of disease after the first rain shower indicates the involvement of wind-blown rain in disease development. Stress caused by drought and water logging usually increases the disease severity.

Shi *et al.* (2021) developed a sensitive and robust quantitative PCR (qPCR) assay for the detection and quantification of *Xan. albilineans* using TaqMan probe and primers targeting a putative adenosine triphosphate-binding cassette (ABC) transporter gene (*abc*). This novel qPCR assay was highly specific to the 43 tested *Xan. albilineans* strains belonging to different pulsed-field gel electrophoresis groups. The detection thresholds were 100 copies^{ul-} of plasmid DNA, 100 fg$^{\mu l-}$ of bacterial genomic DNA and 100 CFU^{ml-} of bacterial suspension prepared from pure culture. The assay was 100 times more sensitive than a conventional PCR assay. The pathogen was detected by qPCR in 75.1% (410/546) of symptomless stalk samples, whereas only 28.4% (155/546) of samples tested positive by conventional PCR. Based on qPCR data, population densities of *Xan. albilineans* in symptomless stalks of the same varieties differed between two sugarcane production areas in China, Beihai (Guangxi Province) and Zhanjiang (Guangdong Province), and no significant correlation between these populations was identified. Furthermore, no relationship was found between these populations of the pathogen in asymptomatic stalks and the resistance level of the sugarcane varieties to leaf scald. The newly developed qPCR assay proved to be highly sensitive and reliable for the detection and quantification of *Xan. albilineans* in sugarcane stalks.

9.15.5 Management

An integrated approach consisting of the following measures can prove very effective for the management of the disease.

1. **Resistant varieties**: The use of resistant varieties is the most effective, practical and economical method of disease management. Before release, the resistant varieties must be tested by artificial inoculation, and both plant and ratoon crops should be screened. For assessing the reaction of varieties, take into account only the systemic infection on the non-inoculated leaves and ignore the initial symptoms on the inoculated plants.
2. **Pathogen-free seed**: As the bacterium is transmitted mainly through the diseased setts, the use of pathogen-free setts is of utmost importance for the management of the disease. In areas free from severe

epiphytotics, pathogen-free setts can be obtained by regular inspection of seed plots, rouging of diseased plants and then protecting the crop by adopting phytosanitary measures. Duttamajumder (1996) showed that a directed-selection for an apparently healthy cane progeny yielded symptomless plants in the fourth progeny with no trace of leaf scald bacterium on isolation. From 25% of the apparently healthy canes in the first year, it increased to about 90% at the end of the 4th year.

3. **Heat therapy**: Heat therapy have been tried to eradicate the bacterium from the diseased planting material. Hot air treatment at 54 °C for 8 h is not effective. Long hot water (50 °C for 3 h) and short hot water (52 °C for 20 min) treatments are effective but these considerably reduce the germination of buds. Nowadays, moist-hot-air treatment (54 °C for 4 h) is mostly used. Agnihotri (1984) obtained an effective control of the disease with this method. It is superior to all other heat therapy methods used in India and it also controls grassy shoot and ratoon stunting diseases of sugarcane.
4. **Chemotherapy**: Dip the single bud sets after desiccating for 12 h, in 100 μg mL^{-1} solution of streptomycin for 1 day. However, its higher concentrations are injurious to plants.
5. **Phytosanitary measures**: The disinfection of cane knives, mechanical harvesters and planters reduces the dissemination of the bacterium. Ten percent commercial formalin, 1% Dettol or 5% Lysol can be used for the disinfection of implements. While cutting setts, the knives should be disinfected at frequent intervals. The destruction of infected debris, stubbles and of weeds like *I. cylindrica* helps in reducing the inoculum load. *Sorghum* sp. and *Z. mays*, which act as collateral hosts of the bacterium, should not be grown near the sugarcane crop.
6. **Quarantine measures**: Strict quarantines should be enforced to check the dissemination of the pathogen to the disease-free areas. However, the latent infection of the bacterium is a big hurdle in the effective implementation of the quarantines. All the quarantine stations should be equipped with adequate facilities for the detection of the bacterium in the seed material.
7. **Cultural practices**: High doses of nitrogenous fertilisers, water stress and water logging increase the incidence of the disease. The correction or removal of these factors will reduce the incidence of the disease. The early ploughing of heavily infected fields considerably reduces the source of inoculum.

REFERENCES

Agnihotri, V. P. (1984). Control of leaf scald of sugarcane, a bacterial disorder, by heat therapy and streptomycin. *Indian Sugar Crops J*. 10: 13–14.

Agnihotri, V. P. (1986). Some new or interesting symptoms of leaf scald of sugarcane. *Indian Sugar Crops J*. 11: 21.

Agnihotri, V. P. and Duttamajumder, S. K. (1992). Leaf scald disease of sugarcane. In: A. N. Mukhopadhyay, J. Kumar, H. S. Chaube and U. S. Singh (eds.). *Plant Diseases of International Importance*. Vol. IV, *Diseases of Sugar, Forests and Plantation Crops*. Prentice Hall, Englewood Cliffs, NJ, pp. 27–40.

Ashby, S. F. (1929). The bacterium which causes gumming disease of sugar canes with notes on two other bacterial diseases of the same host. *Trop. Agric*. 6: 135–38.

Champoiseau, P., Daugrois, J. H., Pieretti, I., Cociancich, S., Royer, M. *et al*. (2006). High variation in pathogenicity of genetically closely related strains of *Xanthomonas albilineans*, the sugarcane leaf scald pathogen, in Guadeloupe. *Phytopathology* 96: 1081–91.

Comstock, J. C. (2001). Foliar symptoms of sugarcane leaf scald. *Sugar J*. 64: 23–32.

Daugrois, J. H., Boisne-Noc, R. and Rott, P. (2014). Leaf surface colonization of sugarcane by *Xanthomonas albilineans* and subsequent disease progress vary according to the host cultivar. *Plant Dis*. 98: 191–96.

Daugrois, J. H., Dumont, V., Champoiseau, P., Costet, L., Boisne, R. *et al*. (2003). Aerial contamination of sugarcane in Guadeloupe by two strains of *Xanthomonas albilineans*. *Eur. J. Plant Pathol*. 109: 445–58.

Dowson, W. J. (1943). On the generic names *Pseudomonas*, *Xanthomonas*, and *Bacterium* for certain bacterial plant pathogens. *Trans. Brit. Mycol. Soc*. 26: 1–14.

Duttamajumder, S. K. (1996). Dynamics of field spread of *Xanthomonas albilineans* (Ashby) Dowson causing leaf scald disease in sugarcane. *Proc. 9th Int. Conf. Plant Path. Bact*., Chennai, pp. 324–30.

Egan, B. T. (1971). The decline of leaf scald as a major disease in northern Queensland. *Proc. Queensland Soc. Sugarcane Technol*. 38: 157–61.

Egan, B. T. (1979). Leaf scald disease in India: Letters to the editor. *Sugarcane Pathol. News*. 22: 53.

Garces, F. F., Gutierrez, A. and Hoy, J. W. (2014). Detection and quantification of *Xanthomonas albilineans* by qPCR and potential characterization of sugarcane resistance to leaf scald. *Plant Dis*. 98: 121–26.

Gutierrez, A., Garces, F. F. and Hoy, J. W. (2016). Evaluation of resistance to leaf scald by quantitative PCR of *Xanthomonas albilineans* in sugarcane. *Plant Dis*. 100: 1331–38.

Mensi, I., Girard, J.-C., Pieretti, I., Larbre, F., Roumagnac, P., *et al*. (2013). First report of sugarcane leaf scald in Gabon caused by a highly virulent and aggressive strain of *Xanthomonas albilineans*. *Plant Dis*. 97: 988.

North, D. S. (1926). Leaf scald, a bacterial disease of sugarcane. Colonial Sugar Refining Co. Ltd. Sydney Agric. Rep. No. 8, 80 pp.

Ricaud, C. and Ryan, C. C. (1989). Leaf scald. In: C. Ricaud, B. T. Egan, A. G. Gillaspie, Jr. and C. G. Hughes (eds.). *Diseases of Sugarcane: Major Diseases*. Elsevier, Amsterdam, pp. 39–58.

Rott, P. and Davis, M. J. (2000). Leaf scald. In: P. Rott, R. A. Bailey, J. C. Comstock, B. J. Croft and A. S. Saumtally (eds.). *A Guide to Sugarcane Diseases*. La Librairie du Cirad, Montpellier, pp. 38–44.

Satyanarayana, Y. (1974). Leaf scald of sugarcane: A new disease in Andhra Pradesh. *Indian Sugar* 24: 23–26.

Shi, Y., Zhao, J.-Y., Zhou, J. R., Ntambo, M. S., Xu, P.-Y. *et al*. (2021). Molecular detection and quantification of *Xanthomonas albilineans* in juice from symptomless

sugarcane stalks using a real-time quantitative PCR assay. *Plant Dis*. 105: 3451–58.
Silva, M. de S. e, Bedendo, I. P. and Casagrande, M. V. (2007). Molecular and pathogenic characterization of isolates of *Xanthomonas albilineans* (Ashby) Dow-son, causal agent of sugarcane leaf scald. *Summa Phytopathol*. 33: 341–47.
Solas, M. T., Pinon, D., Acevedo, R., Fontaniella, B., Legaz, M. E. *et al*. (2003). Ultrastructural changes and production of xanthan like polysaccharide associated with scald of sugarcane leaves caused by *Xanthomonas albilineans*. *Eur. J. Plant Pathol*. 109: 351–59.
Wilbrink, G. (1920). De Gomziekte van het Suikerriet, hare Oorzaak en hare Bestrijding. *Arch. Suikerind, Ned. Indie* 28: 1399–1525.

9.16 BACTERIAL WILT OF SOLANACEOUS PLANTS

Bacterial wilt caused by *Ralstonia solanacearum* is a highly destructive disease of many solanaceous and some non-solanaceous crops. It is considered the world's single most destructive bacterial plant disease. It is also considered the second most important disease of potato worldwide after late blight of potato. It is also known by several other names, such as **brown rot of potato, southern bacterial wilt of tomato** (due to its prevalence in the southern states of the USA), **Moko disease of banana** and **Granville wilt of tobacco**. The bacterium infects many solanaceous plants, namely, potato, tomato, tobacco, eggplant and pepper, and some non-solanaceous hosts such as banana, peanut and Heliconia; it also impacts geraniums and other ornamental plants and weeds. More than 200 plant species belonging to 50 families are attacked by the bacterium.

Tjou-Tam-Sin *et al*. (2017) reported, for the first time, bacterial wilt of cut roses (ornamental *Rosa* sp. cv. Armando) caused by *Ral. solanacearum*.

The disease was first described by E. F. Smith (1896), who also described the bacterium and named it *Bacillus solanacearum*. The disease occurs in tropical, sub-tropical and temperate regions of the world. In the past, the disease has caused serious economic losses to different crops in different countries. The damage to the potato crop occurs on account of premature wilting of the standing crop and of the rotting of potato tubers both in field and in stores. In Granville county, North Carolina, USA, the tobacco crop suffered losses in the range 25–100%; in 1908 alone, the loss was above $100,000 for the entire state of North Carolina. During 1935–45, the disease caused an annual loss of US$2 million to the tobacco crop (Kelman, 1953). Around 1890, the disease almost eliminated the moko plantain, a relative of banana, in Trinidad; later on, it destroyed one million plants on the pacific coast of Costa Rica. The cultivation of tomato and brinjal was abandoned due to this disease in some parts of the Madhya Pradesh, India. In India, losses ranging from 5–6% for tobacco, 50% for potato and 62–65% for brinjal have been reported (Rao, 1972). According to Champoiseau and Allen (2009), the bacterium is responsible for an estimated loss of US$1 billion worldwide every year, mostly to poor small-scale farmers in the tropical high lands.

In India, the disease was first reported by Cappel in 1892 from Pune district of the Maharashtra as bangle blight, or bungdi disease of potato; but its bacterial nature was established in 1909 by Coleman. In 1913, Hutchinson reported it as Rangpur wilt of tobacco, making it the first bacterial disease recorded from India. The disease has not been noticed in the Northwestern high hills (excluding Kumaon hills), where the mean monthly minimum temperatures during December–February remain below 0 °C and in the Northwestern and North central plains having mean monthly maximum summer temperatures above 40 °C coupled with a water deficit in soil for a minimum continuous period of 45 days. In India, race 1 is more predominant and it has a wide genetic diversity. The strains of bacterium prevalent in India are also more virulent than those found in other countries particularly the USA and South America.

Sakthivel *et al*. (2016) characterised 14 *Ral. solanacearum* strains obtained from tomato, brinjal and chili pepper on the Andaman Islands, India, using a polyphasic approach. Strains were identified as *Ral. Solanacearum*, based on carbon substrate utilisation profiling with Biolog similarity coefficients >0.82. The species identity was further confirmed by 16S ribosomal RNA and *recN* gene sequence analysis. The intraspecific identification of strains revealed the presence of race 1 biovar 3 and biovar 4. Both biovars caused the wilting of plants with similar aggression. All strains were identified as phylotype I, and multilocus sequence typing revealing that the strains belong to a small number of clonal complexes that also comprise strains from mainland India, especially West Bengal and Kerala states.

Sagar *et al*. (2014) classified 75 bacterial strains, isolated from wilt-infected potato plants from the various potato-growing regions of India, using traditional and molecular methods. The identity of all the strains was confirmed as *Ral. Solanacearum*, as expected single 280-bp fragment resulted in all the strains following PCR amplification using *Ral. Solanacearum*-specific universal primer pair 759/760. It was found that the Indian potato strains of *Ral. solanacearum* belong to three of four phylotypes, namely the Asian phylotype I, the American phylotype II and the Indonesian phylotype IV. It is the first study regarding the diversity of *Ral. solanacearum* from potato in India using the phylotype and sequevar schemes. It is also the first report of the occurrence of the phylotype IV sequevar 8 (bv2T) strain of *Ral. solanacearum* causing potato bacterial wilt in the middle hills of Meghalaya in India.

9.16.1 Symptoms

As the bacterium infects many plant hosts, the symptoms differ from host to host, although some of the symptoms are common to all hosts.

9.16.1.1 Potato

Foliage symptoms include the rapid wilting of leaves and stems, usually first visible at the warmest time of the day. Initially, plants recover from wilt at night. The leaves may have bronze tint and the petioles may develop epinasty. Eventually, plants fail to recover, become yellow and brown, and then becoming necrotic and dying. As the disease progresses, a streaky brown discolouration appears on the stems above the soil line and also on the branches of the plants. The brown discolouration of the vascular system of the stems occurs, which can be observed by cutting the infected stems longitudinally or transversely. A white, slimy bacterial mass oozes from vascular bundles of the cut or broken stems. Bacterial ooze in the form of threads can be seen when the cut potato stems are suspended in water. Under a cool, growing climate, wilting and other foliage symptoms may not occur.

On tubers, external symptoms may or may not appear. However, in damp soils, the bacterial ooze often emerges from the eyes and the stolen-end attachment of infected tubers. Soil may adhere to bacterial mass exuded from the tubers. Infected cut tubers reveal a browning and eventual necrosis of the vascular ring and immediately surrounding tissues. A few minutes after cutting, a creamy fluid exudate usually appears spontaneously on the vascular ring of the cut surface (Figures 9.15a). In the case of ring rot of potato, an infected tuber has to be squeezed in order to press out a mass of yellowish macerated vascular tissue and bacterial slime. Plants with foliar symptoms of wilt may produce healthy and diseased tubers, while those showing no symptoms of the disease may sometimes produce diseased tubers. Water-soaked lesions, 0.5–1.0 cm in diameter and 5 mm deep, develop at the site of infected lenticels. Tuber skin remains intact but becomes shrivelled. A tuber cut through the lesion exposes a dirty white bacterial mass below the skin.

9.16.1.2 Tomato

The youngest leaves show the first sign of the disease as they become flaccid in appearance, usually at the warmest time of the day. The disease develops slowly under less favourable conditions. Stunting may occur (Figure 9.15b) and large numbers of adventitious roots develop on the stem. The vascular tissues of the stem show a brown discolouration (Figure 9.15c). Drops of white or yellowish bacterial ooze come out from the transverse sections of the stems.

9.16.1.3 Tobacco

On young seedlings, the wilt develops rapidly. The leaves of infected plants become light green and gradually turn yellow. The midrib and veins become flaccid and large leaves droop in an umbrella-like fashion. Necrotic areas may frequently develop between the veins. One of the main symptoms is the **unilateral wilting**. This wilting of the entire plants and of half-leaves is a common feature of the disease. In severe cases, the leaves wilt without changing colour, remaining attached to the stem. When the stems are cut open, the vascular tissue shows a brown discolouration.

9.16.1.4 Banana

The most characteristic symptom is the vascular discolouration in the inner leaves and fruit stalk (peduncle). On young and fast-growing plants, the youngest leaves turn pale-green or yellow and collapse. Lower leaves become yellow, droop and dry out. Sometimes the first symptom is in the fruit, which appears distorted, with pulp exhibiting a dark brown discolouration. Premature yellowing and cracking occurs in infected banana fruits. Brown discolouration of the vascular tissue is seen in cross-sections of the stem. The extensive breakdown of pith occurs in the lower parts of the stem and taproot, and the stems become hollow. Young suckers may be blackened, stunted or twisted. The pseudostems show brown vascular discolouration.

(a) (b) (c)

FIGURE 9.15 (a) Brown rot of potato: Cross section of an infected potato tuber shows greyish discolouration of vascular tissue and bacterial ooze. (Courtesy: Patrice Champoiseau, UF Guatemala.) (b) Bacterial wilt of plants: Wilted tomato plants. (Courtesy: AVRDC- The World Vegetable Center.) (c) Bacterial wilt of plants: Longitudinally splitted tomato stem shows reddish brown discolouration. (Courtesy: AVRDC- The World Vegetable Center.)

9.16.1.5 Geranium

The first symptoms are wilting followed by yellowing of leaves, often sectorial. Plants develop a systemic wilt but initially recover during cooler temperatures or at night. Lower leaves tend to wilt first, becoming chlorotic. Brown discolouration of the vascular tissue is often visible internally. In the later stages, leaves become brown and necrotic; plants never recover from wilt, desiccate and die.

9.16.1.6 Eggplant

The most characteristic symptom is leaf drop, which occurs before there is much wilting.

9.16.1.7 Peanut

Foliage loses its green colour, wilts suddenly and turns a dull black. Leaf yellowing is quite conspicuous.

9.16.1.8 Pepper

Plants rarely show true wilting. Leaves turn yellow and roll or curl before dropping. Vascular browning is the characteristic symptom of the disease.

Shekhawat and associates (2000) found that vascular browning in the host depends on the ability of the infecting bacterial strains to produce brown pigment. The strains, which produce brown pigment in cultures, produce vascular browning. The strains, which do not produce vascular browning, have been reported from Portugal, Kenya and Australia, in addition to India.

Apart from the symptoms, the streaming test is a valuable diagnostic tool for the quick identification of the disease in the field. It is performed by placing cut stem pieces of infected plants in clear water. After a few minutes, threads of a viscous white slime can be seen streaming from the cut end of the stem. These threads are of bacterial ooze exuding from the infected xylem vascular bundles.

9.16.2 Causal Organism

Ralstonia solanacearum species complex (Smith 1896) Yabuuchi *et al.* 1996 emend Safni *et al.* 2014

Synonym: *Burkholderia solanacearum*, *Pseudomonas solanacearum*

Ral. Pseudosolanacearum, previously known as *Ral. solanacearum* phylotypes I and III, is one of the causal agents of bacterial wilt, a devastating disease that affects more than 250 plant species (Bocsanczy *et al.*, 2022).

The bacterium can be easily isolated from the symptomatic material on Yeast Peptone Glucose Agar (YPGA), a non-selective medium or on Kelman's tetrazolium chloride medium. For isolating from symptomless material, Semiselective Medium South Africa (SMSA) medium modified by Elphinstone *et al.* (1996) is useful and has been used successfully in Europe.

On solid agar media, individual bacterial colonies are usually visible after 36–48 h of incubation at 28 °C. The bacterium is Gram-stain-negative, rod-shaped, non-fluorescent, non-sporing, non-encapsulated, strictly aerobic with a polar tuft of flagella. The bacterium is very sensitive to desiccation and is inhibited in culture by 2.0% sodium chloride. On solid agar media, individual bacterial colonies are usually visible after 36–48 h of incubation at 28 °C. The colonies of the normal or virulent type are white or cream-coloured, irregularly shaped, highly fluidal and opaque. The colonies of the mutant or non-virulent type, which appear occasionally, are uniformly round, smaller, butyrous or dry.

The colonies of virulent strains on tetrazolium chloride medium are irregularly shaped, fluidal and white having pink centres while those of non-virulent strains are round, smaller and dark red. This medium is also used for the differentiation of virulent and weakly virulent/avirulent strains (Kelman, 1954). The bacterium can be stored for many years at room temperature in sterilised tap, distilled, or deionised water. It can also survive for a long time at minus 80 °C in liquid culture broth amended to 40% glycerol. Repeated transfers of cultures on agar media result in rapid loss of virulence and cultures stored at 4 °C also lose viability.

Polyhydroxy butyrate is accumulated. Gelatin hydrolysis is negative or weak, nitrate reduced by nearly all strains and many produce gas. Levan is not produced from sucrose; starch and esculin are not hydrolysed, ammonia is produced and only glucose is metabolised oxidatively. The tests for oxidase and catalase are positive while those for arginine dihydrolase and lipase are negative. For most strains, the optimum growth temperature is between 28 and 32 °C. The bacterium can utilise glucose, sucrose and n-propanol but not D-arabinose, L-rhamnose, D-fucose and D-raffinose.

The bacterium undergoes spontaneous conversion from fluidal wild-type to afluidal, avirulent or weakly virulent state when stored in water or cultured in media. Such afluidal variants can be easily recognised by their characteristic colony morphology on tetrazolium agar medium. This conversion was considered unidirectional until the report of Shekhawat *et al.* (1992), which showed that the reversion back to fluidal virulent forms also occurs. The average of reversion rate was 3.45%. This reversion to virulent fluidal form may be playing a role in maintaining the virulence of the bacterium. The exposure of the bacterium to Terramycin and tetracycline increased the number of fluidal revertants 5- to 16-fold. The incubation at higher temperatures of 35 and 37 °C, in comparison to optimum temperature of 30 °C, also increased the reversion rate by 4.0 to 6.4-fold (Gadevar *et al.*, 1999).

Li *et al.* (2024)), in their microreview, have presented and discussed the current data of the SolI/R and RasI/R quorum-sensing (QS) systems in the *Ral. solanacearum* species complex (RSSC). They have also presented the distribution and frequency of these *N*-acyl homoserine lactone (AHL) QS systems in the RSSC, discussed possible ecological roles and evolutive implications. The complex QS hierarchical networks emphasises the crucial role of cell–cell signalling in the virulence of the RSSC.

Dewberry *et al.* (2024) reported that genotypic and phenotypic analyses show *Ralstonia solanacearum* cool virulence is a quantitative trait not restricted to "race 3 biovar 2".

The species *Ral. solanacearum* is a complex taxonomic unit with a great degree of both phenotypic and genetic diversity. Thousands of genetically distinct strains of *Ral. solanacearum* have been characterised worldwide, revealing significant variation within the group. Therefore, *Ral. solanacearum* is considered a "**species complex**". Unlike other phytopathogenic bacteria, the race system of *Ral. solanacearum* is not based on gene-for-gene interactions, i.e., different cultivars carrying different *R* gene(s). Instead, the races are based on their pathogenicity on different host species. Although the biovar and race systems are widely accepted for the classification of *Ral. solanacearum*, there is no correlation between biovar and race; each race transects the biovars and each biovar contains various races. Only positive correlation between the biovar and race systems exists for biovar 2 and race 3. The differentiation of biovars is given in Tables 9.16a and 9.16b.

Safni *et al.* (2014) emended the description of *Ral. solanacearum* given by Yabuuchi *et al.* (1995) to include only strains belonging to phylotype II. The below-given emended description of the bacterium is based on the data of Yabuuchi *et al.* (1995) and Safni *et al.* (2014). Cells are Gram-stain-negative and rod-shaped, and may be motile or non-motile, but the type strain is motile. Catalase and oxidase tests are positive. Bacterium is able to grow at 28 °C on casamino acid peptone glucose (CPG) medium. Nitrate is reduced to nitrite and some strains also reduce nitrite to gas. Most strains can utilise citrate, hydrolyse Tween 80, and produce urease. Tests for arginine dihydrolase and lysine, and ornithine decarboxylases are negative. Acid is produced oxidatively from glucose, sucrose, mannose, fructose, glycerol and galactose. The major cellular fatty acids are summed feature 3 (C16: 1v7c/iso-C15: 0 2-OH), C16: 0, C18: 1v7c, summed feature 2 (iso-C16: 1 I/C14: 0 3-OH), C17: 0 cyclo, C18: 1 2-OH, C16: 1 2-OH, and C14: 0. The strains contain a signature nucleotide sequence, 59-AAGTTATGGACGG TGGAAGTC (Fegan and Prior, 2005; Prior and Fegan, 2005), in the 16S–23S rRNA ITS gene sequence. The DNA G + C content of the type strain is 66.6 mol% (HPLC method).

TABLE 9.16A

Tests for Differentiation of *Ral. solanacearum* Biovars

	Biovar				
	1	2	3	4	5
Maltose	–	+	+	–	+
Lactose	–	+	+	–	+
Cellobiose	–	+	+	–	+
Mannitol	–	–	+	+	+
Sorbitol	–	–	+	+	–
Dulcitol	–	–	+	+	–

Note: + = Utilised, – = Not utilised. After EPPO Bulletin 34 (2004).

TABLE 9.16B

Tests for Differentiation of Subphenotypes of Biovar 2 of *Ral. solanacearum*

Test	Biovar 2A RFLP Group26	Biovar 2A RFLP Group 27	Biovar 2T or 2N
Utilisation of trehalose	–	+	+
Utilisation of inositol	+	–	+
Utilisation of D-ribose	–	–	+
Pectolytic activity	Low	Low	High

Note: + = Utilised, – = Not utilised

Using a polyphasic taxonomic approach on an extensive set of strains, Safni *et al.* (2014) provided evidence for a taxonomic and nomenclatural revision of members of *Ral. solanacearum* species complex. The data obtained from phylogenetic analysis of 16S–23S rRNA ITS gene sequences, 16S–23S rRNA ITS region sequences and partial endoglucanase (egl) gene sequences, and DNA–DNA hybridisations demonstrate that the *Ral. solanacearum* species complex comprises three genospecies. One of these includes the type strain of *Ral. solanacearum* and consists of strains of *Ral. solanacearum* phylotype II only. Second genospecies includes the type strain of *Ral. syzygii* and contains only phylotype IV strains. This genospecies is further divided into three distinct groups: *Ral. syzygii*, the causal agent of Sumatra disease on clove trees in Indonesia; *Ral. solanacearum* phylotype IV strains isolated from different host plants, mostly from Indonesia; and strains of the blood disease bacterium (BDB) that cause the banana blood disease, a bacterial wilt disease in Indonesia that affects bananas and plantains. The third genospecies is composed of *Ral. solanacearum* strains that belong to phylotypes I and III. As the differentiation of these genospecies is also supported by phenotypic data, the authors made the following taxonomic proposals: emendation of the descriptions of *Ral. solanacearum* and *Ral. syzygi*i, and descriptions of *Ral. syzygii* subsp. nov. for the current *Ral. syzygii* strains, *Ral. syzygii* subsp. *indonesiensis* subsp. nov. for the current *Ral. solanacearum* phylotype IV strains, *Ral. syzygii* subsp. *celebesensis* subsp. nov. for the BDB strains and *Ral. pseudosolanacearum* sp. nov. for the strains of *Ral. solanacearum* phylotypes I and III.

Genome of 3 strains of *Ral. solanacearum* has been sequenced. The first genome sequenced was that of strain GMI 1000, which has a wide host range (Salanoubat *et al.*, 2002). It is organised into two replicons, a chromosome of 3.7 Mb and a mega plasmid of 2.1 Mb, with a very similar G + C % (about 67%). The strains Mo IK2 (a strain of race 2, isolated from a banana tree) and 1609 (a race 3 strain, isolated from potato) have genomes, which are comparable to that of GMI 1000 (about 6 Mb, G + C % estimated at 67%).

The bacterium belongs to RNA homology group II (non-fluorescent) within the pseudomonads and within this group it is readily distinguished from other members by its failure to grow at 40 °C. On the basis of RFLP and other genetic fingerprinting studies (Hayward, 2000), the species is divided into Division I called the Asiaticum division (biovars 3, 4 and 5 originating in Asia) and Division II called the Americanum division (biovars 1, 2A and 2T, originating in South America).

To date, five races of the bacterium have been identified on the basis of their ability to attack different host species. Race 1 occurs in tropical areas all over the world and is endemic in the southern USA. It has a wide host range and attacks tobacco, many other solanaceous crops and many other hosts and floriculture crops. It infects hundreds of different plant species in 50 families.

Race 2 occurs mainly in tropical areas of South America, in the Caribbean and Brazil, and infects bananas and *Heliconia* species, causing so-called Moko disease. In the Philippines, it causes bugtok disease on plantains.

Race 3 (biovar 2) is the most important race; it occurs in subtropical and temperate regions and at higher altitudes in the tropics. It is of special concern to the potato industry in the USA and Canada because, being a cold-temperature tolerant race, it has caused serious disease problems in potato crop in other temperate countries of the world. It is also an extremely destructive pathogen of potato in the highland tropics of Africa, Asia and Latin America. This race was introduced into the USA from Guatemala in geranium cuttings on several occasions in 1999. In the United States, it was listed in 2002 as a Select Agent plant pathogen, and is now subjected to the strictest biosecurity regulations. Apart from its status as Select Agent, race 3 is considered a serious quarantine pest by the Animal and Plant Health Inspection Service of the US Department of Agriculture and by the European Plant Protection Organisation (Champoiseau and Allen, 2009). Government regulations in the USA include a zero tolerance for this race. It attacks potato, tomato, occasionally *Pelargonium zonale*, aubergine (eggplant) and capsicum, some solanaceous weeds like black nightshade (*Solanum nigrum*) and bitter sweet or climbing nightshade (*S. dulcamara*). Race 3 also infects a number of non-solanaceous weeds, which often do not show symptoms. This is the most predominant race and is mostly comprised of biovar 2A or RFLP group 26 with a worldwide distribution; RFLP group 27 found in Chile and Colombia or biovar 2T, sometimes also called 2N, found in tropical areas in South America. Race 3 is not found in Canada. It is on the North American Plant Protection Organization (NAPPO) list of A1 quarantine pests as well as Canada's list of quarantine pests. Milling *et al.* (2009) compared different traits of tropical, temperate (race 3 biovar 2) and warm-temperate strains of the bacterium in *in vitro* and *in vivo* and found that interaction with plants is required for expression of the temperate epidemiological trait of race 3 biovar 2. Race 4 affects ginger in most parts of Asia. Race 5 (biovar 5) infects mulberries in China.

Nowadays, the identification of *Ral. solanacearum* into races and biovars is considered unsatisfactory because it is not predictive and some groups (e.g. race 1) contain very large variation (Champoiseau and Allen, 2009). A phylogenetically meaningful classification scheme was developed based on DNA sequence analysis (Fegan and Prior, 2005; Prior and Fegan, 2005). In this classification, twenty-three sequevars and four phylotypes have been recognised within the species complex; these broadly reflect the ancestral relationships and geographical origins of the strains.

Molecular studies have shown *Ral. solanacearum* as a species complex consisting of highly diverse strains (Poussier *et al.* 2000; Fegan and Prior, 2005). Molecular analyses using the nucleotide sequences of genes *egl* (encoding endoglucanase, a conserved virulence factor), *mut S* (a DNA mismatch repair enzyme), *hrp B* (a regulator of type III secretion) and the ITS region between the 16S and 23S ribosomal RNA genes have generated a phylogenetically meaningful subdivision of the species complex into four phylotypes, which group strains by ancestral geographic origin. Phylotype I strains come from Asia, phylotype II strains from the Americas, phylotype III strains from Africa and phylotype IV strains from Indonesia, which is the probable origin of the group (Poussier *et al.*, 2000; Fegan and Prior, 2005). These four groups are distinguished by a multiplex PCR reaction. Within phylotypes, strains are further clustered into sequevars based on nucleotide sequences of the *egl* and *mut S* genes (Poussier *et al.*, 2000; Fegan and Prior, 2005).

Schaad *et al.* (2001) developed Real-time PCR for the detection and identification of *Ral. solanacearum* race 3 biovar 2. Ozakman and Schaad (2003) further refined this technique by developing a Real-time Bio-PCR. Real-time Bio-PCR was more sensitive (detected 20 cells/ml) than Real-time PCR and it detected *Ral. solanacearum* race 3 biovar 2 from latently infected potato tubers. Serological field kits, which incorporate highly specific monoclonal antibodies, are also commercially available (immunostrips-AgDia, CSL Pocket Diagnostics) for rapid on-site diagnosis. However, a confirmed diagnosis should always be obtained through laboratory tests on a pure culture.

Priou *et al.* (2010) developed two sensitive, specific and user-friendly serological methods, namely, double-antibody sandwich (DAS)-ELISA and indirect ELISA on the nitrocellulose membrane after the enrichment of the plant extract in a semi-specific broth for detection of *Ral. solanacearum* (biovar 2A) in stems of symptomless plants 20 days before harvest of potato crop. Detection probabilities were higher with DAS-ELISA. Being a reliable and user-friendly technique, DAS-ELISA can easily be used by national plant protection services and seed certification programmes in developing countries.

Chandrasekhar and Umesha (2015) developed a species-specific PCR (sPCR) method to identify and detect isolates of *Ral. solanacearum*, causing bacterial wilt in chilli. PCR primers for *Ral. solanacearum* were identified by alignment of *hrpB* gene sequences and selection of sequences specific for *Ral. solanacearum* at their 3′ ends. These primers were found specific for *Ral. solanacearum*, as no PCR product was obtained when genomic DNA from other bacterial species, including closely related *Ralstonia* species, were used as test species. Lone pair of primers (RshrpBF and RshrpBR) was designed using *hrpB* gene sequence, unique to *Ral. solanacearum* which amplified a predicted PCR product of 810 bp from 20 different isolates. To understand the evolutionary divergence of Indian *Ral. solanacearum* isolates, phylogenetic analysis was also attempted. Based on phylogenetic analysis, Indian isolates showed homology with the standard reference isolates from other countries; however, one new isolate showed complete evolutionary divergence by forming an out-group.

Stulberg *et al.* (2015) developed a multiplex PCR assay that identified *Ral. solanacearum* species complex strains, signalled whether the strain detected was a select agent (race 3 biovar 2 strains are considered select agents by the US government), and controlled false negatives associated with PCR inhibition or unsuccessful DNA extractions in one reaction. They identified unique sequences of non-phage-related DNA for the *Ral. solanacearum* species complex strains, and for Select Agent strains, using *in silico* genome subtraction. An internal plant DNA control assay was also designed and included. The multiplex PCR assay correctly identified 90 *Ral. solanacearum* species complex strains and 34 Select Agent strains, but did not recognise five out-group bacterial species. Moreover, the multiplex PCR assay facilitated the detection of plant DNA and *Ral. solanacearum* from infected tomato, potato, geranium and tobacco plants. This rapid, accurate and reliable detection assay can help US government officials to make timely and appropriate recommendations to exclude this bacterium from the United States.

Ral. solanacearum race 3 biovar 2 (R3bv2) sub group is a high-concern quarantine pathogen, while the related sequevar 7 group is endemic to the southeastern United States. Tran *et al.* (2016) compared the sensitivity, cost, and technical complexity of several assays to detect and distinguish R3bv2 and sequevar 7 strains of *Ral. solanacearum* in geranium, tomato and potato tissue in the laboratory and in naturally infected tomato plants from the field. The sensitivity of PCR-based methods in infected geranium tissues was significantly improved by the use of Kapa3G Plant, a polymerase with enhanced performance in the presence of plant inhibitors. The cells of R3bv2 were killed within 60 minutes of application to Whatman FTA(R) nucleic acid-binding cards, suggesting that samples on FTA cards can be safely transported for diagnosis. On the whole, culture enrichment followed by dilution plating was the most sensitive detection method (10^1 cfu mL^{-1}), but it was also most laborious. Conducting PCR from FTA cards was faster, easier, and sensitive enough to detect approximately 10^4 cfu mL^{-1}, levels similar to those found in latently infected geranium plants.

Massart *et al.* (2014) developed a new multiplex real-time PCR assay to simultaneously detect *Clavibacter michiganensis* subsp. *sepedonicus* and *Ral. solanacearum*, listed as A2 pests in the EPPO region and as zero tolerance quarantine organisms in the European Union, in a single assay. Additionally, the reliability of this molecular diagnostic test has been improved by the simultaneous amplification of an internal control, corresponding to a potato gene co-extracted from the sample. The polyvalence and the specificity of each set of bacterial primers and probes were evaluated on more than 90 bacterial strains. Detection limit of this triplex real-time protocol was similar to those observed with other molecular protocols previously developed for the individual detection of these bacteria. A concordance of 100% was obtained in a blind test mimicking the routine application of the technology. The new protocol represents a straight forward and convenient method potentially adapted to primary screening of potato tubers.

Albuquerque *et al.* (2015) selected 17 novel markers based on *Ral. solanacearum* specific protein domains and thoroughly validated for specificity and stability, both *in silico* and using "wet lab" assays. Validation assays based on PCR and hybridisation revealed that the DNA regions selected as markers were unevenly distributed amongst the tested strains, with nine markers present throughout the species complex. The distribution of the remaining eight markers was highly variable between the different analysed strains and allowed the attainment of strain-specific dot blot hybridisation patterns, particularly informative for typing. Average probability value of each strain being positive for each of the 17 markers was calculated by an algorithm and used to obtain a dendrogram representing hierarchical clustering analysis of *Ral. solanacearum*, according to the similarity of their hybridisation patterns. This method should prove to be a robust and straight forward procedure for genotyping members of the *Ral. solanacearum* species complex. Moreover, this quantitative hybridisation approach will allow the construction of informative databases to determine new *Ral. solanacearum* genotypes and infer epidemiological patterns.

Saini *et al.* (2025) reported that while the strains of *Ral. solanacearum* species complex remain major pathogens, emerging evidence reveals a broader spectrum of bacterial pathogens from the Enterobacteriaceae family, including *Entb. cloacae*, *Entb. mori*, *Kle. oxytoca*, *Kosa. cowanii* and *Ser. marcescens*, as well as *Pro. vermicola* from the Morganellaceae family, all implicated in causing comparable wilt symptoms in plant hosts traditionally associated with *Ral. solanacearum* species complex. Although much remains to be understood about their pathogenicity and infection mechanisms, recent advancements through whole-genome sequencing have provided limited but valuable insights into their potential virulence factors.

9.16.3 Disease Cycle

Ral. solanacearum is primarily a soil-borne and also water-borne pathogen. *Ralstonia* spp. survive remarkably well in water, with some isolates surviving in water stocks for decades (Cope-Arguello *et al.*, 2023).

The primary inoculum comes from soil, but there is no conclusive evidence that it is a persistent soil inhabitant. The top 30 cm of soil contains the high population of the bacterium. The bacterium also survives in infected tubers, including symptomless tubers, infected stock, reservoir plants and weeds. Though air-borne spread has never been reported, there is some evidence of epiphytic survival of the bacteria on the plant surfaces under conditions of high relative humidity. Depending upon biological and environmental conditions, the bacterium can survive from days to years in contaminated water, wet soils or deep soil layers forming a reservoir of inoculum from which it can disperse. In European countries, bittersweet or climbing nightshade (*S. dulcamara*), a solanaceous weed growing along waterways, is considered to be a source of primary inoculum. *Ral. solanacearum* in contaminated water in waterways infects the roots of this weed. The source of bacteria in waterways may come from waste generated by the processing of infected potatoes. Growers may pull *Ralstonia*-contaminated water from the waterways for irrigation, thereby inoculating their potato and tomato crops. Outbreaks of brown rot of potato in the United Kingdom were traced to *Ral. solanacearum*-infected *S. dulcamara* plants, which release the pathogen from roots into river waters that were subsequently used for potato irrigation. Hong *et al.* (2008) also showed the survival of the bacterium in irrigation ponds in Gadsden county, North Florida (USA) and in common aquatic weeds growing in these ponds, including *Hydrocotyle ranunculoides* (dollar weed) and *Polygonum pennsylvanicum* (Pennsylvania smart weed). Both weeds were latently infected and showed no symptoms of wilt. There was a positive correlation between ambient temperature and *Ral. solanacearum* population in irrigation water, as shown by previous research workers. In Australia, England, Kenya and Sweden, race 3 biovar 2 was not detected in previously diseased fields after two years, suggesting that long-term survival in temperate regions is reduced.

Bacterium spreads through infected tubers, banana rhizomes and transplants, contaminated equipment, irrigation water, insects and also, in banana, through pruning knives. Its movement from plant to plant occurs through soil. Race 2 is transmitted by insects and has a high potential for natural spread. Race 3 spreads more easily with surface water when infected *S. dulcamara* grows with its roots floating in water. The long-distance spread of the bacterium occurs through infected (latently) seed potatoes and other vegetative propagating planting materials. Among the hosts of the bacterium producing true seed, only groundnut seed has been shown to be naturally infected.

Bacterium enters the host through wounds caused by agricultural operations and pruning knives, and through the rifts caused by the emergence of lateral roots. The feeding injuries caused by root and spiral nematodes also help in the penetration of the pathogen. In addition, certain gall-forming insects, cut worms, white grubs and root invading parasitic fungi also provide avenues for the entry of bacterium. The secondary spread of the bacterium is of little significance, as the bacterium does not become air-borne. It occurs, to a limited extent, through irrigation water, insects and contaminated equipment. The bacterial cells released in the soil following the death of diseased roots invade the nearby roots.

Soil and air temperature, and soil moisture, play important roles in disease development. Temperature and soil moisture have a synergistic effect on disease development. High temperature and high soil moisture alone cannot induce the disease. Light intensity, day length and mineral nutrition also influence disease development. Race 3 has comparatively low optimum temperature of 27 °C while the range for maximum disease development is between 24 and 35 °C. The virulence of the bacterium decreases when temperature exceeds 35 °C or falls below 12 °C. The optimum temperature for disease development for races 1, 2, 4 and 5 is 35 °C. Favourable temperature facilitates rapid multiplication of the bacterium in the host, its release and infection of neighbouring plants. High soil moisture (50–100% water-holding capacity) favours rapid disease development. It increases survival of the bacterium in soil, enhances multiplication and disease development and spread through soil. The disease is found in diverse soil types and in both acidic (pH 5.0–5.5) and alkaline (pH 7.0–8.5) soils.

The bacterium first enters the large xylem vessels, and, in succulent stems, it multiplies in the intercellular spaces of the cortex and the pith, causing lysigenous cavities. In potato tubers, the pathogen may produce an internal decay emanating from the vascular bundles and causing a breakdown in the bud tissue of the eyes.

Physiology of the host–parasite interaction has been excellently reviewed by Buddenhagen and Kelman (1964). Various biochemical and physiological changes occur in the host during pathogenesis, which are manifested in the form of internal and external symptoms on the host.

After entering the xylem vessels, the bacteria multiply and travel rapidly to the upper parts of the plants. This results in the production of large amounts of exopolysaccharide (EPS) slime. The presence of this slime, along with large number of bacterial cells in the xylem vessels, leads to a reduction in sap flow, resulting in the wilting of the plants. The wilting can also occur due to rupturing of vessel elements on account of excessive hydrostatic pressure. The main virulence factor of *Ral. solanacearum* is EPS I, a long (>10^6 Da) polymer. Many diverse isolates of the bacterium produce EPS I (or a very similar polymer) and 85% of the EPS I exists as released, cell-free slime while 15% remains in a cell surface-bound capsular form. EPS I accounts for >90% of its total exopolysaccharide produced in the plants. Mutants of *Ral. solanacearum* specifically blocked in the synthesis of only EPS I rarely cause wilting or killing of plants, even if a large number of bacterial cells are directly injected into the stem. EPS I also promotes stem colonisation, because EPS I-deficient mutants colonise stems much more slowly and to a much lesser extent in comparison to wild-types.

The bacterium produces ~10 major exoproteins that are readily detectable, in addition to many others that are less abundant and/or conditionally produced. The bacterium uses two different systems to export polypeptides from its cytoplasm to the outside environment. These systems consist of a Type II secretion system encoded by *eep* (enhanced expression of pheromone) genes, which exports most major exoproteins, including all the cell wall-degrading exoenzymes, and a Type III secretion system encoded by *hrp* (hypersensitive reaction and pathogenicity) genes, which appears to deliver toxic proteins directly into plant cells. Mutants lacking either system are severely handicapped in colonisation ability and/or multiplication in plants. These exoproteins are absolutely essential for the disease development.

Ral. solanacearum produces pectin methyl esterase (PME) that removes methoxy groups from pectin, converting it into methanol and polygalacturonic acid units. An endopolygalacturonase, PehA (also called PglA), and two exopolygalacturonases, PehB (PglB) and PehC (PglC), attack the polygalacturonic acid units. In pectin-rich regions of plants and also in the soil, these four enzymes break down pectin into oligomers and also provide monogalacturonic acid for the growth of the bacterium. PehB-deficient mutants show only a slight reduction in virulence, while PehA-deficient mutants take at least twice as long as the wild-type strains to wilt and kill the plants. This reduction in virulence is even more pronounced when soil infestation is used for inoculation in comparison to stem or petiole inoculation. The role of PME in wilt development is not clear.

Ral. solanacearum produces two extracellular glucanases, Egl and CbhA, *in vitro* and these enzymes attack β-1-4 glycosidic linkages of cellulose. Egl releases cellobiose, cellotriose and cellotetraose from soluble cellulose and CbhA releases cellobiose from the reducing ends of crystalline cellulose. Mutants deficient in Egl are slower to cause the disease than wild-type strains. It seems that Egl facilitates invasion of roots and/or the penetration of xylem vessels by degrading cellulosic glucans in the cell walls.

Schell (2000) has reviewed the information on the control of virulence and pathogenicity genes of the bacterium by an elaborate sensory network. Many of the key extracytoplasmic virulence and pathogenicity factors are transcriptionally controlled by an extensive network of distinct, interacting signal transduction pathways. A five-gene Phc (*ph*enotype *c*onversion) system, which is the core of this sensory network, regulates exopolysaccharide, cell wall-degrading exoenzymes, and other factors in response to self-produced signal molecule that monitors the pathogen's growth status and environment. Four additional environmentally responsive two-component systems work independently, and with the Phc system to fine-tune virulence gene expression. Another critical system is Prh (*pro*line-rich *h*omeodomain), which transduces plant cell-derived signals through a six-gene cascade to activate deployment of the Type III secretion pathway encoded by the *hrp* genes.

Aliye *et al.* (2015) evaluated 28 cultivars of potato under greenhouse and field conditions and found that, under greenhouse conditions, all the cultivars were resistant to *Ral. solanacearum* with scores ranging from 0.77 to 1.17 out of 5. However, under field conditions, only 2 of 28 cultivars were found to be resistant under greenhouse conditions, showed adequate resistance to the pathogen, indicating the significant impact of the environment on the activity of the pathogen and reaction of the crop. Percentage wilt incidence and latent infection showed significant ($P <$ 0.05) positive correlation, with $r = 0.9438$. Hence, the evaluation of a crop's performance based on the combination of the parameters like field wilt incidence and proportion of latent infection give us a better picture of the overall crop feat, than simply using wilt incidence as a sole parameter of evaluation as has been done in most studies. Moreover, the established correlation of latent infection with field wilt incidence will also be helpful in understanding the disease epidemiology and design effective management measures, accordingly.

Kumar *et al.* (2017) and Singh *et al.* (2018) performed virulence and pathogenicity assays on 6- to 7-day-old tomato seedlings using wild and mutant strains of *Ral. solanacearum* and found the technique as an easy and useful approach for investigating virulence functions of the pathogen at the seedling stage of host. The technique used was economical in terms of space, labour and cost as well as the amount of bacterial inoculum required.

Du *et al.* (2017) evaluated *Ral. solanacearum* infection dynamics in pepper lines using bioluminescence imaging. The utility of the bioluminescence assay was validated by comparing *Ral. solanacearum* infection dynamics in real-time *in vivo* between resistant line BVRC 1 and susceptible line BVRC 25. Luminescence intensity was strongly correlated to bacterial population in planta ($R^2 = 0.88$). The distribution and multiplication of BL-Rs7 strain in resistant line BVRC 1 was conspicuously limited in plants inoculated in either roots or stem in comparison to susceptible line BVRC 25. These results suggest that pepper line BVRC 1 may resist colonisation by interfering with *Ral. solanacearum* multiplication in the roots and stem.

Baichoo and Jaufeerally-Fakim (2017) reported that salicylic acid and ethylene pathway marker genes were expressed at higher levels with a statistically significant difference ($P < 0.05$) in fold change expression compared to very little activation of the jasmonic acid pathway marker genes among the *Solanum* lines. The wild *Lycopersicon cerasiforme*, resistant lines CRA 66 and Hawaii 7996 and *S. commersonii* showed stronger activation of the salicylic acid and ethylene marker genes than the moderately resistant cultivar (MST 32/1) and the susceptible lines (Quatre, carrees and Spunta). The marker genes reached their highest expression levels 4 h post-inoculation in the resistant and moderately resistant lines; in contrast, in the susceptible lines, it occurred 48 h after inoculation. The results indicate that salicylic acid and ethylene signalling pathways have a significant role in defence against *Ral. solanacearum*. The timing and magnitude of the upregulation of gene expression may determine the plant ability to put up a defence response against the pathogen.

Nakaho *et al.* (2017) found that on infiltration of leaves of resistant tomato cultivar LS-89 or susceptible cultivar Ponderosa with 10^6 cells mL^{-1} of *Ral. solanacearum*, cell death occurred in LS-89 at 18 and 24 h after infiltration. The proliferation of bacteria in the infiltrated leaf tissues of LS-89 was suppressed approximately 10–30% of that of Ponderosa leaf tissues, and expression of the defence-related gene *PR-2* and HR marker gene *hsr203J* was induced in the infiltrated tissues. These results show that the response of LS-89 is a true HR, and the induction of vascular HR in xylem parenchyma and pith cells surrounding xylem vessels seems to be associated with the quantitative resistance of LS-89 to *Ral. solanacearum*.

Caldwell *et al.* (2017) used a combination of scanning electron microscopy and light microscopy to investigate *Ral. solanacearum* colonisation in roots of soil-grown resistant and susceptible tomato cultivars at multiple time points after inoculation. It was found that colonisation of the root vascular cylinder was delayed in resistant cultivar "Hawaii 7996" and that, once the bacteria entered the root vascular tissues, colonisation in the vasculature was spatially restricted. The data suggest that resistance is due, in part, to the ability of the resistant cultivar to restrict bacterial root colonisation in space and time.

Jiang *et al.* (2013) investigated the effect of calcium nutrition on tomato bacterial wilt and the regulation of resistance mechanisms. Plants grown in nutrient solution with calcium concentrations of 0.5, 5.0 and 25.0 mM were inoculated with *Ral. solanacearum* by the root dip method. Disease severities of 100, 77.1 and 56.8 per cent were found in Ca concentrations of 0.5, 5.0 and 25.0 mM, respectively. Plant growth, i.e., height, stem diameter and biomass, was significantly better in high Ca concentration than in low Ca concentration. Tomato plants absorbed significantly more Ca in roots and shoots as the level of Ca in the nutrient solution increased. A negative correlation was found between Ca concentration, levels of hydrogen peroxide (H_2O_2), PO and PPO in tomato plants, and disease severity, indicating that these chemicals played an important role in tomato resistance to bacterial wilt. The results suggested that Ca was involved in the regulation of H_2O_2 concentration, and activity of PO and PPO in tomato.

Youssef and Tartoura (2013) reported that compost provides effective protection against *Ral. solanacearum* via the upregulation of the capacity of the ascorbate-glutathione redox cycle and the modulation of the cellular redox status, thereby eliminating reactive oxygen species damage and sustaining membrane stability.

Okiro *et al.* (2019) developed a loop-mediated isothermal amplification (LAMP) assay using the UDP-(3-*O*-acyl)-*N*-acetylglucosamine deacetylase gene (*IpxC*) to screen seed potato for *Ral. solanacearum* strains. The assay was assessed using DNA extracted from *Ral. solanacearum* and other soil and potato pathogens to demonstrate specificity and sensitivity. The LAMP assay was validated using field samples from different potato growing regions of Kenya collected over two growing seasons and compared with established nucleic acid and protein-based assays. The *IpxC* LAMP assay was found to be specific and sensitive to *Ral. solanacearum*, detecting as low as 2.5 pg/µl of *Ral. solanacearum* DNA. Of the 47 potentially infected field samples collected, both *IpxC* LAMP and quantitative polymerase chain reaction (PCR) detected *Ral. solanacearum* DNA in 90% of the samples, followed by conventional PCR (86%) and ELISA (75%). This *IpxC* LAMP assay is a promising diagnostic tool to rapidly screen for *Ral. solanacearum* in seed potato with high sensitivity in Kenya.

Kumbar *et al.* (2021) found that the root dip method of artificial inoculation was the most reliable method compared to media drenching and stem inoculation methods. Maximum plant spread, stem girth, number of primary branches, yield per plant (6.69 kg), number of fruits per plant (94.80), fruit length (22.22 cm), root length (63.65 cm) and root spread (87.05 cm) were exhibited by Green Long Hybrid scion when grafted onto *Solanum melongena* cv. Haritha rootstock. Taller plants with high fruit girth (10.97 cm) and average fruit weight (78.00 g) of scion were recorded on SM 116 rootstock. The highest levels of dry matter content (11.12%) and total phenolic content (113.30 mg/100 g) of fruits were observed in scion when SM 398 and SM 3 were used as rootstocks, respectively. Overall, 'Haritha' cultivar was found to be the best rootstock for grafting in the current study.

Mori *et al.* (2016) reported that *Ral. solanacearum* strain OE1-1 produces mature biofilms on the surfaces of tomato cells after invasion into intercellular spaces. Mutation of *lecM* encoding a lectin, RS-IIL, may contribute to biofilm formation by OE1-1, which is required for OE1-1 virulence.

Cheng *et al.* (2017) identified and characterised a novel LRR protein from pepper, CaLRR51, structurally distinct from the previously characterised CaLRR1. It was localised to the plasma membrane and transcriptionally up-regulated by *Ral. solanacearum* infection, as well as by exogenously applied salicylic acid, jasmonic acid and ethephon. The data of loss- and gain-of-function assays indicated that CaLRR51 acts as a positive regulator in the response of pepper to *Ral. solanacearum* infection. In addition, the signal peptide and transmembrane region were found to be required for plasma membrane targeting of CaLRR51, which is required for the role of CaLRR51 in pepper immunity.

Sang *et al.* (2018) found that the *Ral. solanacearum* type III effector RipAY suppresses plant immune responses triggered by bacterial elicitors and by the phytohormone salicylic acid. The biochemical analysis indicated that RipAY associates *in planta* with thioredoxins from *Nicotiana benthamiana* and Arabidopsis. RipAY displays γ-glutamyl cyclotransferase activity to degrade glutathione in plant cells, which is required for the reported suppression of immune responses. Seeing the importance of thioredoxins and glutathione as major redox regulators in eukaryotic cells, RipAY activity may constitute a novel and powerful virulence strategy employed by *Ral. solanacearum* to suppress immune responses and potentially alter general redox signalling in host cells.

Mori *et al.* (2018) reported that integrated signalling via ralfuranones influences the quorum sensing and virulence of *Ral. solanacearum*.

Sharma *et al.* (2022) identified two *Ral. solanacearum* species complex (RSSC) phylotypes, phylotype II (98.4%, $n = 506$ [sequevar 1 ($n = 505$) and sequevar 2 ($n = 1$)]) and phylotype I (1.6%, $n = 30$ [sequevar 13 ($n = 9$) and a new sequevar ($n = 21$)]). The phylotype II sequevar 1 strains were haplotyped using multilocus tandem repeat sequence typing (TRST) schemes. The TRST scheme identified 51 TRST profiles within the phylotype II sequevar 1 strains with a modest diversity index (HGDI = 0.87), confirming the epidemic clonality of RSSC phylotype II sequevar 1 strains in Kenya. A minimum spanning tree and mapping of the TRST profiles revealed that TRST27 '8-5-12-7-5' is the primary founder of the clonal complex of RSSC phylotype II sequevar 1 and is widely distributed via latently infected seed tubers.

Shen *et al.* (2020) demonstrated that L-glutamic acid from host plants is the key active component associated with increased extracellular polysaccharide production, cellulase activity, swimming motility and biofilm formation in *R. solanacearum* GMI1000. In addition, L-glutamic acid also promoted colonisation of *Ral. solanacearum* cells in the roots and stems of tomato plants and accelerated disease incidence. Furthermore, genetic screening and biochemical analysis suggested that *RS01577*, a hybrid sensor histidine kinase/response regulator, is involved in L-glutamic acid signalling in *Ral. solanacearum*. Mutations in *RS01577* and exogenous addition of L-glutamic acid to the GMI1000 wild-type strain had overlapping effects on both the transcriptome and biological functions of *Ral. solanacearum*, including on motility, biofilm formation and virulence. These findings have shown a new interaction mechanism between *Ral. solanacearum* and its host plants that highlights the complexity of the virulence regulation mechanism and may provide new insight into disease control.

Landry *et al.* (2020), in their micro review, have summarised and discussed the current knowledge on the characterised *Ral. solanacearum* species complex type III effectors.

Moon *et al.* (2021) reported that the nucleocytoplasmic type III-secreted effector (T3E) RipAZ1 (*Ralstonia* injected protein AZ1) confers *Ral. solanacearum* avirulence in *Solanum americanum*. Few avirulence genes are known in vascular bacterial phytopathogens and *ripAZ1* is the first one in *Ral. solanacearum* that is recognised in black nightshades. This work thus opens the way for the identification of disease resistance genes responsible for the specific recognition of RipAZ1, which can be a source of resistance against the devastating bacterial wilt disease.

Morel *et al.* (2018) demonstrated that the *Ral. solanacearum* species complex (RSSC) type III effector (T3E) *ripAX2* triggers specific resistance in eggplant AG91-25, which carries the major resistance locus *EBWR9*. The eggplant accession AG91-25 is resistant to the wild-type *Ral. pseudosolanacearum* strain GMI1000, whereas a *ripAX2* defective mutant of this strain can cause wilt. Notably, the addition of *ripAX2* from GMI1000 to PSS4 suppresses wilt development, demonstrating that RipAX2 is an elicitor of AG91-25 resistance. RipAX2 has been shown previously to induce effector-triggered immunity (ETI) in the wild relative eggplant *Solanum torvum*, and its putative zinc (Zn)-binding motif (HELIH) is critical for ETI. The authors showed that, in their model, the HELIH motif is not necessary for ETI on AG91-25 eggplant. The *ripAX2* gene was present in 68.1% of 91 screened RSSC strains, but in only 31.1% of a 74-genome collection comprising *Ral. solanacearum* and *Ral. syzygii* strains. Overall, it is preferentially associated with *Ral. pseudosolanacearum* phylotype I. $RipAX2_{GMI1000}$ appears to be the dominant allele, prevalent in both *Ral. pseudosolanacearum* and *Ral. solanacearum*, suggesting that the deployment of AG91-25 resistance could efficiently control bacterial wilt in the Asian, African and American tropics. This finding advances the understanding of the interaction between RipAX2 and the resistance genes at the *EBWR9* locus, and paves the way for both functional genetics and evolutionary analyses.

Qiu *et al.* (2023) showed that RipBT is a novel type III secretion system (T3SS)-secreted effector by using a cyaA reporter system. Transient expression of RipBT in *Nicotiania benthamiana* induced strong cell death in a plasma membrane-localisation dependent manner. Notably, mutation of RipBT in *Ral. solanacearum* showed attenuated virulence on potato, while RipBT transgenic potato plants exhibited enhanced susceptibility to *Ral. solanacearum*. Interestingly, transcriptomic analyses suggest that RipBT may interfere with plant reactive oxygen species (ROS) metabolism during the *Ral. solanacearum* infection of potato roots. In addition, the expression of RipBT remarkably suppressed the flg22-induced pathogen-associated molecular pattern-triggered immunity responses, such as the ROS burst. These findings show that RipBT acts as a T3SS effector, promoting *Ral. solanacearum* infection on potato and presumably disturbing ROS homeostasis.

Zhang *et al.* (2023) used exogenous salicylic acid to improve the immunity of tomato plants, followed by transposon insertion sequencing analysis to identify the *Ral. solanacearum* genes associated with tolerance to plant immunity. Three lipopolysaccharide genes, and two genes with unknown function, were deleted to validate the transposon insertion sequencing results, revealing that lipopolysaccharide genes are essential for *Ral. solanacearum* tolerance to plant immunity and that these genes with unknown function may contribute to the trade-off between tolerance to plant immunity and the fitness cost.

Castillo (2023) used comparative genomic tools to investigate the speciation process of *Ral. solanacearum* species complex (RSSC) by studying the essential and flexible gene content and the openness of the pangenome of 314 genomic sequences. The results indicate that the vast majority of genes in RSSC pangenome are dispensable with only 228 essential genes. The pangenome of RSSC is open, which is characteristic of sympatric species. *Ral.*

pseudosolanacearum, *Ral. solanacearum*, and *Ral. syzygii* pangenomes are also open, however, the pangenome of *Ral. syzygii* is near to turn closed. Phylogenetic and statistical evidence reported in the literature and the results obtained in this study suggest that both allopatric and sympatric lifestyles have shaped the speciation process of RSSC.

Using growth in *ex vivo* xylem sap as a proxy for host xylem, Hamilton *et al.* (2024) found that *Ralstonia* strain GMI1000 grows in sap from both healthy plants and *Ralstonia*-infected susceptible plants. However, sap from *Ralstonia*-infected Hawaii 7996 plants inhibited *Ralstonia* growth, suggesting that, in response to *Ralstonia* infection, resistant plants increase inhibitors in their xylem sap. Consistent with this, reciprocal grafting and defence gene expression experiments indicated that Hawaii 7996 wilt resistance acts in both above- and below-ground plant parts.

Hawaii 7996 resistance is broken by *Ralstonia* strain UW551 of the pandemic lineage that threatens highland tropical agriculture. Unlike other *Ralstonia*, UW551 grew well in sap from *Ralstonia*-infected Hawaii 7996 plants. Moreover, other *Ralstonia* strains could grow in sap from H7996 plants previously infected by UW551. Thus, UW551 overcomes H7996 resistance in part by detoxifying inhibitors in xylem sap. Testing a panel of xylem sap compounds identified by metabolomics revealed that no single chemical differentially inhibits *Ralstonia* strains that cannot infect H7996. However, sap from *Ralstonia*-infected H7996 contained more phenolic compounds, which are known to be involved in plant antimicrobial defence. Culturing UW551 in this sap reduced total phenolic levels, indicating that the resistance-breaking *Ralstonia* strain degrades these chemical defences. Overall, these results suggest that Hawaii 7996 tomato wilt resistance depends, in part, on inducible phenolic compounds in xylem sap.

Huang *et al.* (2024) demonstrated that RSc2734, hereafter named PrhX, positively regulated *hrp* gene expression via a PrhA-PrhIR-PrhJ-HrpG signalling cascade. Moreover, PrhX was crucial for *Ral. solanacearum* to invade host roots and grow *in planta* naturally. *prhX* expression was independent of the PhcB-PhcA quorum sensing system. PrhX did not affect the expression of *phcB* and *phcA* and the QS-dependent phenotypes, such as extra polysaccharide production and biofilm formation. These results provide novel insights into the complex regulatory network of the TIIISS and pathogenesis in *Ral. solanacearum.*

9.16.4 Management

The management of the disease is very difficult due to the soil-borne nature and wide host range of the pathogen, its high pathogenic variability and the limited host resistance. However, the following measures can prove helpful in reducing the disease incidence and severity.

1. Crop rotation of 5–7 years with non-host crops is recommended. Elhalag *et al.* (2015) reported that cowpea and maize are suitable for crop rotation with potato as the root exudates, containing asparagines, glutamine and sulphur-containing methionine, from these crops enhance the sustainability of *Stenotrophomonas maltophilia*, a biocontrol agent of *Ral. solanacearum.*

   Deberdt *et al.* (2015) reported that the incidence of tomato wilt decreased by 86% and 60%, when tomato crop was sown after *Raphanus sativus* cv. Melody and *Crotalaria spectabilis*, respectively. Hence, *C. spectabilis* and *R. sativus* cv. Melody can be used as previous crops to control bacterial wilt in ecological management strategies without any drastic suppression of *Ral. solanacearum* population in stem tissues and in the rhizosphere.
2. Altering soil pH based on the cropping period can be effective. For potatoes, lowering the pH to 4–5 in the summer and raising it to 6 in autumn helped to eradicate the pathogen (Smith *et al.*, 1997). He *et al.* (2014) reported that the addition of organic fertiliser and $CaCO_3$ increased soil pH and lowered the *Ral. solanacearum* population nearly 100-fold and increased the Ca^{2+} content in tobacco plants significantly compared to treatments without $CaCO_3$. It suggests that $CaCO_3$ can serve as a potential soil amendment for the control of bacterial wilt caused by *Ral. solanacearum*
3. Use of resistant varieties. Development of resistant varieties saved the tobacco industry in the United States from near extinction around the middle of the 20th century. However, host plant resistance has not been very effective for the management of bacterial wilt on potato and tomato, as host resistance often varies among locations due to the strain variation of the pathogen. In potato, resistance often fails at warmer temperatures. Grafting susceptible tomato cultivars onto resistant tomato or other solanaceous rootstock is effective on the commercial scale in Japan, Bangladesh and the Philippines, but grafting has not been tested for use against race 3 biovar 2. Lin *et al.* (2008) obtained a significant control of tomato wilt by integrating the use of a resistant tomato rootstock (Hawaii 1996) with soil amendment consisting of urea and slaked lime.
4. Destruction and removal of weed hosts like *S. dulcamara* with herbicides.
5. Many attempts to manage the disease through soil fumigation and antibiotics have not given satisfactory results. However, dipping of tomato seedlings in Streptocycline solution (1 g in 30 litres of water) proved effective and it gave better results than soil drench and soil drench + foliar sprays. Ji *et al.* (2007) obtained very effective control of tomato wilt and a significant increase in yield with the combined application of thymol (applied as preplant fumigation through drip irrigation @ 73 kg/ha) and acibenzolar-S-methyl (foliar spray @ 25

mg/l). Deberdt *et al.* (2018) observed no incidence of bacterial wilt in greenhouse tomato plants grown in soil treated with chemotype 3 (clove-scented) of *Pimenta racemosa* var. *racemosa* essential oil at a concentration of 0.14%. In the untreated control soil, 62% of plants showed symptoms of bacterial wilt. Treatment with chemotype 3 significantly increased the growth of tomato plants in comparison to untreated controls. These findings suggest that chemotype 3 of *P. racemosa* var. *racemosa* essential oil is a good candidate for further development as a soil biofumigant for the control of tomato bacterial wilt.

6. Imada *et al.* (2016) reported that magnesium oxide nanoparticles (MgO NP) induced systemic resistance in tomato plants against *Ral. solanacearum.* Salicylic acid-inducible *PR1*, jasmonic acid inducible lipoxygenase (*LoxA*), ethylene-inducible *Osm*, and systemic resistance-related *GluA* were up-regulated in both the roots and the hypocotyls of tomato plants after treatment of the plant roots with MgO NP. Histochemical analyses revealed that *β*-1, 3-glucanase and tyloses accumulated in the xylem and apoplast of pith tissues of the hypocotyls after MgO NP treatment
7. *Psm. fluorescens* isolates and vascular arbuscular mycorrhizae, *Glomus mosseae* have given encouraging results for the control of the disease in the preliminary trials. Messiha *et al.* (2007a) obtained a significant control of brown rot of potato with *Stenotrophomonas maltophilia* in the Nile Delta area of Egypt. Sood and associates (personal communication) have suggested the integrated management of tomato wilt utilising various cultural (soil amendments, soil solarisation, adjusting time of transplanting and biofumigation) and biological methods (host resistance, antagonistic rhizobacteria and use of VAM) in Himachal Pradesh, India, where race 1 biovar 3 of the bacterium is prevalent. Ambadar and Sood (2010) obtained 43% and 63% reduction in terminal wilt incidence of tomato through soil solarisation for 10 and 8 weeks, respectively, under Palampur (H. P., India) conditions.

Soliman *et al.* (2024) reported that the mixture of *Alternaria burnsii*, *Aspergillus niveus* and *Aspergillus niger* was the most effective treatment for reducing the percent disease index, with an 18% decrease compared to the infected control.

Abo-Elyousr *et al.* (2012) investigated the induced defence responses caused by acibenzolar-*S*-methyl (ASM) and *Psm. fluorescens* alone or in combination against tomato bacterial wilt. Treatments of tomato seedlings with *Psm. fluorescens* or ASM caused 58% and 56% disease reduction, respectively. The highest disease reduction (72%) resulted from a combined application of both *Psm. fluorescens* and ASM. The application of ASM alone increased 64.3% seedlings biomass compared to infected control. Significant changes ($P \leq 0.05$) in the activities of polyphenol oxidase, β-glucosidase and peroxidase were found. These findings show that the management of tomato bacterial wilt can be enhanced by integrating foliar sprays and soil drench of ASM and *Psm. fluorescens* into the existing management schedule. This is the first report of the use of both ASM and *Psm. fluorescens* to control the tomato bacterial wilt disease under field conditions.

Kunwar *et al.* (2017) reported that grafting (susceptible tomato variety BHN 602 grafted onto a resistant rootstock BHN 998) alone or in combination with drip applications of ASM (178.6 μM) significantly reduced disease incidence and increased total marketable yield compared to non-grafted treatments. There were no significant differences between grafted plants with or without drip ASM applications in terms of bacterial wilt incidence or total marketable yield. However, it was demonstrated for the first time that foliar applications of ASM on grafted plants negatively affected the total marketable yield compared with drip applications of ASM on grafted plants or non-treated grafted control.

Erjavec *et al.* (2016) tested 150 protein extracts from 94 different basidiomycete and ascomycete wild mushroom species for antibacterial activity against *Ral. solanacearum.* In *in vitro* microtiter plate assays, 11 extracts completely inhibited bacterial growth and 4 extracts showed partial inhibition. On further testing of 5 of these 15 extracts, 3 extracts slowed disease progression and reduced disease severity in artificially inoculated tomato and potato plants. This is the first *in vitro* and *in vivo* study to demonstrate that mushroom protein extracts can be promising for treatment of bacterial wilt caused by *Ral. solanacearum.*

Khan *et al.* (2020) reported that 30 g powder of leaves and succulent shoots of *Rhazya stricta*, mixed with 1 kg of infested soil, 20 days before transplanting was most effective in controlling bacterial wilt. It increased root length (cm), shoot length (cm), and plant fresh biomass (g) by 55, 42 and 40%, respectively, compared to control plants. Mixing of plant powder with the artificially infested pot soil [35 ml (10^8 CFU/ml) per kilogram of soil] was better than surface mulching. Mixing of 30 grams dose per kilogram of infested soil increased root length (cm), shoot length (cm) and plant fresh biomass (g) of treated plants by 67, 36 and 46%, respectively, over control plants. One drench application of 30 g of powder per kilogram of soil, applied 20 days before transplanting, decreased the disease severity by 37% over the control. The results show that the dried powder (30 g/kg of soil) of leaves or succulent shoots of *R. stricta*, thoroughly mixed with soil, 20 days before transplanting, could act as an effective control measure against bacterial wilt.

Ganiyu *et al.* (2020) reported that sole or combined application of thymol and acibenzolar-S-methyl (ASM) significantly reduced incidence of bacterial wilt in both seasons, from the second to eight weeks after transplanting. At the eighth week after transplanting, the effect of combined application of thymol and ASM translated to 100% and 90.3% disease reduction in early and late planting seasons, respectively. Grafted tomato, thymol and ASM combination produced an average yield of 74.3 and 86.9 ton/ha in early and 65.3 and 58.1 ton/ha in late planting seasons, respectively. The findings reveal that integrated use of grafting, thymol (0.7% w/v) and ASM (30 μg/ml) can be of great help for controlling bacterial wilt of tomato.

8. Biological soil disinfestation (BSD, that is, induction of anaerobic soil conditions by increasing microbial respiration through incorporation of fresh organic amendments and by reducing the resupply of oxygen by covering with airtight plastic sheets) has also been used to attempt to manage the disease. Messiha *et al.* (2007b) reported that BSD reduced the soil population of *Ral. solanacearum* race 3 biovar 2 by 92.5 to 99.9% in the field and that, therefore, it has the potential to become an important element in a sustainable and effective management strategy to control brown rot of potato especially in areas where the disease is endemic.
9. Hot air treatment of ginger rhizomes for 30 min at 50 °C is also successful. Spafford *et al.* (2021) recommended use of rootstocks, namely High Power, Ganbarune Vect and Ganbarune No. 37 for fields having high incidence of bacterial wilt. Barik *et al.* (2023) recommended brinjal cultivar CARI-1 as root stock for the management of wilt. An increase of yield by 69.23 % and 33.46% under laterite acidic soils along with the reduced severity of bacterial wilt was observed in grafted eggplant on CARI-1 rootstock over both the non-grafted scions viz., VNR-212 and Arka Neelachal Shyama, respectively.

Shekhawat *et al.* (2000), after reviewing the exhaustive work done on bacterial wilt of potato in India, have suggested the following measures for its management:

1. Use of pathogen-free whole tubers for sowing.
2. Application of stable bleaching powder @ 12 kg/ha mixed with fertilisers in furrows at the time of sowing.
3. Application of urea @ 1000 kg/ha.
4. Bacterisation of seed tubers with *Bacillus subtilis*.
5. Use of resistant varieties.
6. Two- to three-year crop rotation with wheat, lupin, maize, paddy and bean.
7. Pre-treatment of seed tubers with 250 μg/ml of Streptocycline solution, followed by 2 foliar sprays of 100 μg/ml concentration.
8. Adjustment of date of planting and minimum tillage in the standing crop. Post-emergence earthing up in potatoes causes the maximum injury ranging from 10–50%.
9. One or two deep ploughings of infested soil during May–June on plains or plateau and in October–November in the hills.

REFERENCES

Abo-Elyousr, K. A. M., Ibrahim, Y. E. and Balabel, N. M. (2012). Induction of disease defensive enzymes in response to treatment with acibenzolar-S-methyl (ASM) and *Pseudomonas fluorescens* Pf2 and inoculation with *Ralstonia solanacearum* race 3, biovar 2 (phylotype II). *J. Phytopath.* 160: 382–89.

Albuquerque, P., Marcal, A. R. S., Caridade, C., Costa, R., Mendes, M. V. *et al.* (2015). A quantitative hybridization approach using 17 DNA markers for identification and clustering analysis of *Ralstonia solanacearum*. *Pl. Pathol. 64*: 1270–83.

Aliye, N., Dilbo, C. and Pillay, M. (2015). Understanding reaction of potato (*Solanum tuberosum*) to *Ralstonia solanacearum* and relationship of wilt incidence to latent infection. *J. Phytopath.* 163: 444–55.

Ambadar, V. K. and Sood, A. K. (2010). Effect of solarization on tomato wilt incidence and population dynamics of *Pseudomonas fluorescens* and *Ralstonia solanacearum*. *J. Mycol. Plant Pathol.* 40: 120–23.

Baichoo, Z. and Jaufeerally-Fakim, Y. (2017). *Ralstonia solanacearum* upregulates marker genes of the salicylic acid and ethylene signaling pathways but not those of the jasmonic acid pathway in leaflets of *Solanum* lines during early stage of infection. *Eur. J. Plant Pathol.* 147: 615–25.

Barik, S., Ponnam, N., Acharya, G. C., Kumari, M., Adamala, A. K. *et al.* (2023). Assessment of Bacterial wilt-resistant *Solanum* genetic resources as rootstocks for yield and fruit quality traits in eggplant. *Australasian Plant Pathol.* 52: 253–69.

Bocsanczy, A. M., Bonants, P., van der Wolf, J., Bergsma-Vlami, M. and Norman, D. J. (2022). Identification of candidate type 3 effectors that determine host specificity associated with emerging *Ralstonia pseudosolanacearum* strains. *Eur. J. Plant Pathol.* 163: 35–50. https://doi.org/10.1007/s10658-021-02455-w

Buddenhagen, I. W. and Kelman, A. (1964). Biological and Physiological aspects of bacterial wilt caused by *Pseudomonas solanacearum*. *Annu. Rev. Phytopathol.* 2: 203–30.

Caldwell, D., Kim, B-S. and Iyer-Pascuzzi, A. S. (2017). *Ralstonia solanacearum* differentially colonizes roots of resistant and susceptible tomato plants. *Phytopathology* 107: 528–36.

Castillo, J. A. (2023). Analysis of the speciation process suggests a dual lifestyle in the plant pathogen *Ralstonia solanacearum* species complex. *Eur. J. Plant Pathol.* 166: 251–57. https://doi.org/10.1007/s10658-023-02670-7

Champoiseau, P. G. and Allen, C. (2009). *Ralstonia solanacearum* race 3 biovar 2: Tropical losses, temperate anxieties. APSnet Feature Story. http://www.apsnet.org/online/feature/ralstonia/.

Chandrasekhar, B. and Umesha, S. (2015). Detection, identification and phylogenetic analysis of Indian *Ralstonia solanacearum* isolates by comparison of partial *hrpB* gene sequences. *J. Phytopath.* 163: 105–15.

Cheng, W., Xiao, Z., Cai, H., Wang, C., Hu, Y. (2017). A novel leucine-rich repeat protein, CaLRR51, acts as a positive regulator in the response of pepper to *Ralstonia solanacearum* infection. *Mol. Plant Pathol.* 18: 1089–1100.

Cope-Arguello, M. L., Ingel, B., Deutschbauer, A. and Lowe-Power, T. (2023). What genes are necessary for the *Ralstonia solanacearum* species complex to survive in water? *Phytopathology* 113: S3.137.

Deberdt, P., Davezies, I., Coranson-Beaudu, R. and Jestin, A. (2018). Efficacy of leaf oil from *Pimenta racemosa* var. *racemosa* in controlling bacterial wilt of tomato. *Plant Dis.* 102: 124–31.

Deberdt, P., Gozé, E., Coranson-Beaudu, R., Perrin, B., Fernandes, P. *et al.* (2015). *Crotalaria spectabilis* and *Raphanus sativus* as previous crops show promise for the control of bacterial wilt of tomato without reducing bacterial populations. *J. Phytopath.* 163: 377–85.

Dewberry, R. J., Sharma, P., Prom, J. L., Kinscherf, N. A., Lowe-Power, T. *et al.* (2024). Genotypic and phenotypic analyses show *Ralstonia solanacearum* cool virulence is a quantitative trait not restricted to "race 3 biovar 2". *Phytopathology* 114: 2468–80. https://doi.org/10.1094/PHYTO-06-24-0187-R

Du, H., Chen, B., Zhang, X., Zhang, F., Miller, S. A. *et al.* (2017). Evaluation of *Ralstonia solanacearum* infection dynamics in resistant and susceptible pepper lines using bioluminescence imaging. *Plant Dis.* 101: 272–78.

Elhalag, K. M., Emara, H. M., Messiha, N. A. S., Safwat, A., Elhadad, S. A. *et al.* (2015). The relation of different crop roots exudates to the survival and suppressive effect of *Stenotrophomonas maltophilia* (PD4560), biocontrol agent of bacterial wilt of potato. *J. Phytopath.* 163: 829–40.

Elphinstone, J. G., Hennessey, J., Wilson, J. K. and Stead, D. E. (1996). Sensitivity of different methods for the detection of *Pseudomonas solanacearum* in potato tuber extracts. *Bulletin OEPP/EPPO Bulletin* 26: 663–78.

Erjavec, J., Ravnikar, M., Brzin, J., Grebenc, T., Blejec, A. *et al.* (2016). Antibacterial activity of wild mushroom extracts on bacterial wilt pathogen *Ralstonia solanacearum. Plant Dis.* 100: 453–64.

Fegan, M. and Prior, P. (2005). How complex is the "*Ralstonia solanacearum* species complex"? In: C. Allen, P. Prior and A. C. Hayward (eds.). *Bacterial Wilt: The Disease and the Ralstonia solanacearum Species Complex.* APS Press, St. Paul, MN, pp. 449–61.

Gadewar, A. V., Shekhawat, G. S., Chakrabarti, S. K. and Birhman, R. K. (1999). Temperature induced non-directed heritable changes in virulence of an unstable avirulent strain of *Ralstonia solanacearum. J. Indian Potato Assoc.* 26: 11–18.

Ganiyu, S. A., Popoola, A. R., Enikuomehin, O. A. and Bodunde, J. G. (2020). Evaluation of integrated management of bacterial wilt of tomato using grafting, biofumigant and plant resistance activator under field conditions. *Australasian Plant Pathol.* 49: 249–55. https://doi.org/10.1007/s13313-020-00702-y

Hamilton, C. D., Zaricor, B., Dye, C. J., Dresserl, E., Michaels, R. *et al.* (2024). *Ralstonia solanacearum* pandemic lineage strain UW551 overcomes inhibitory xylem chemistry to break tomato bacterial wilt resistance. *Mol. Plant Pathol.* 25: e13395. https://doi.org/10.1111/mpp.13395

Hayward, A. C. (2000). *Ralstonia solancearum. Encyclopedia of Microbiology*. 2nd Ed. Vol. 4. Academic Press, London, pp. 32–42.

He, K., Yang, S.-Y., Li, H., Wang, H. and Li, Z-L. (2014). Effects of calcium carbonate on the survival of *Ralstonia solanacearum* in soil and control of tobacco bacterial wilt. *Eur. J. Plant Pathol.* 140: 665–75. https://doi.org/10.1007/s10658-014-0496-4

Hong, J. C., Momol, M. T., Jones, J. B. Ji, P., Olson, S. M.*et al.* (2008). Detection of *Ralstonia solanacearum* in irrigation ponds and aquatic weeds associated with ponds in North Florida. *Plant Dis.* 92: 1674–82.

Huang, J., Wang, R., Zhang, Q., Wang, C., Liang, T. *et al.* (2024). Positive regulation of the *PhcB* neighbouring regulator *PrhX* on expression of the type III secretion system and pathogenesis in *Ralstonia solanacearum. Mol. Plant Pathol.* 25: e13398. https://doi.org/10.1111/mpp.13398

Imada, K., Sakai, S., Kajihara, H., Tanaka, S. and Ito, S. (2016). Magnesium oxide nanoparticles induce systemic resistance in tomato against bacterial wilt disease. *Plant Pathol.* 65: 551–60.

Ji, P., Momol, M. T., Rich, J. R. and Olson, S. M. (2007). Development of an integrated approach for managing bacterial wilt and root-knot on tomato under field conditions. *Plant Dis.* 91: 1321–26.

Jiang, J.-F., Li, J.-G. and Dong, Y.-H. (2013). Effect of calcium nutrition on resistance of tomato against bacterial wilt induced by *Ralstonia solanacearum. Eur. J. Plant Pathol.* 136: 547–55.

Kelman, A. (1953). The Bacterial wilt caused by *Pseudomonas solanacearum*. A literature review and bibliography. *Tech. Bull. N. C. Agr. Exp. Stn.* No. 99, 194 pp.

Kelman, A. (1954). The relationship of pathogenicity of *Pseudomonas solanacearum* to colony appearance in a tetrazolium medium. *Phytopathology* 44: 693–95.

Khan, R. A. A., Alam, S. S., Hayat, S., Ahmad, M., Ali, A. *et al.* (2020). Eco-friendly management of bacterial wilt in tomato using dried powder of the wild arid-land medicinal shrub *Rhazya stricta. Plant Dis.* 104: 2613–21.

Kumar, R., Barman, A., Phukan, T., Kabyashree, K., Singh, N. *et al.* (2017). *Ralstonia solanacearum* virulence in tomato seedlings inoculated by leaf clipping. *Plant Pathol.* 66: 835–41.

Kumbar, S., Narayanankutty, C., Sainamole Kurian, P., Sreelatha, U. and Barik, S. (2021). Evaluation of eggplant rootstocks for grafting eggplant to improve fruit yield and control bacterial wilt disease. *Eur. J. Plant Pathol.* 161: 73–90. https://doi.org/10.1007/s10658-021-02305-9

Kunwar, S., Paret, M. L., Freeman, J. H., Ritchie, L. Olson, S. M. *et al.* (2017). Foliar applications of acibenzolar-s-methyl negatively affect the yield of grafted tomatoes in fields infested with *Ralstonia solanacearum. Plant Dis.* 101: 890–94.

Landry, D., González-Fuente, M., Deslandes, L. and Peeters, N. (2020). The large, diverse, and robust arsenal of *Ralstonia solanacearum* type III effectors and their *in planta* functions. *Mol. Plant Pathol.* 21: 1377–88.

Li, P., Bez, C., Zhang, Y., Deng, Y. and Venturi, V. (2024). *N*-acyl homoserine lactone cell-cell diffusible signalling in the

Ralstonia solanacearum species complex. *Mol. Plant Pathol.* 25: e13467. https://doi.org/10.1111/mpp.13467

Lin, C.-H., Hsu, S. T., Tzeng, K. C. and Wang, J.-F. (2008). Application of a preliminary screen to select locally adapted resistant rootstock and soil amendment for integrated management of tomato bacterial wilt in Taiwan. *Plant Dis.* 92: 909–16.

Massart, S., Nagy, C. and Jijakli, M. H. (2014). Development of the simultaneous detection of *Ralstonia solanacearum* race 3 and *Clavibacter michiganensis* subsp. *sepedonicus* in potato tubers by a multiplex real-time PCR assay. *Eur. J. Plant Pathol.* 138: 29–37.

Messiha, N., Diepeningen, A., Farag, N., Abdallah, S., Janse, J. *et al.* (2007a). *Stenotrophomonas maltophilia*: A new potential biocontrol agent of *Ralstonia solanacearum*, causal agent of potato brown rot. *Eur. J. Plant Pathol.* 118: 211–25.

Messiha, N., Diepeningen, A., Wenneker, M., Beuningen, A., Janse, J. *et al.* (2007b) Biological soil disinfestation (BSD), a new control method for potato brown rot, caused by *Ralstonia solanacearum* race 3 biovar 2. *Eur. J. Plant Pathol.* 117: 403–15.

Milling, A., Meng, F., Denny, T. P. and Allen, C. (2009). Interactions with hosts at cool temperatures, not cold tolerance, explain the unique epidemiology of *Ralstonia solanacearum* race 3 biovar 2. *Phytopathology* 99: 1127–34.

Moon, H., Pandey, A., Yoon, H., Choi, S., Jeon, H. *et al.* (2021). Identification of RipAZ1 as an avirulence determinant of *Ralstonia solanacearum* in *Solanum americanum*. *Mol. Plant Pathol.* 22: 317–33.

Morel, A., Guinard, J., Lonjon, F., Sujeeun, L., Barberis, P. *et al.* (2018). The eggplant AG91-25 recognizes the Type III-secreted effector RipAX2 to trigger resistance to bacterial wilt (*Ralstonia solanacearum* species complex). *Mol. Plant Pathol.* 19: 2459–72.

Mori, Y., Inoue, K., Ikeda, K., Nakayashiki, H., Higashimoto, C. *et al.* (2016). The vascular plant-pathogenic bacterium *Ralstonia solanacearum* produces biofilms required for its virulence on the surfaces of tomato cells adjacent to intercellular spaces. *Mol. Plant Pathol.* 17: 890–902.

Mori, Y., Ishikawa, S., Ohnishi, H., Shimatani, M., Morikawa, Y., *et al.* (2018). Involvement of ralfuranones in the quorum sensing signalling pathway and virulence of *Ralstonia solanacearum* strain OE1-1. *Mol. Plant Pathol.* 19: 454–63.

Nakaho, K., Seo, S., Ookawa, K., Inoue, Y., Ando, S. *et al.* (2017). Involvement of a vascular hypersensitive response in quantitative resistance to *Ralstonia solanacearum* on tomato rootstock cultivar LS-89. *Plant Pathol.* 66: 150–58.

OEPP/EPPO (2004). EPPO diagnostic protocols for regulated pests: *Ralstonia solanacearum. Bulletin OEPP/EPPO Bulletin* 34: 173–78.

Okiro, L. A., Tancos, M. A., Nyanjom, S. G., Smart, C. D. and Parker, M. L. (2019). Comparative evaluation of LAMP, qPCR, conventional PCR, and ELISA to detect *Ralstonia solanacearum* in Kenyan potato fields. *Plant Dis.* 103: 659–65.

Ozakman, M. and Schaad, N. W. (2003). A real-time Bio-PCR assay for detection of *Ralstonia solanacearum* race 3, biovar 2 in asymptomatic potato tubers. *Can. J. Plant Pathol.* 25: 232–39.

Poussier, S., Prior, P., Luisetti, J. Hayward, C. and Fegan, M. (2000). Partial sequencing of the *hrpB* and endoglucanase genes confirms and expands the known diversity within the *Ralstonia solanacearum* species complex. *Syst. Appl. Microbiol.* 23: 479–86.

Prior, P. and Fegan, M. (2005). Recent developments in the phylogeny and classification of *Ralstonia solanacearum. Acta Hort.* 695: 127–36.

Priou, S., Gutarra, L., Aley, P., De Mendiburu, F. and Lique, R. (2010). Detection of *Ralstonia solanacearum* (biovar 2A) in stems of symptomless plants before harvest of the potato crop using post-enrichment DAS-ELISA. *Plant Pathol.* 59: 59–67.

Qiu, H., Wang, B., Huang, M., Sun, X., Yu, L. *et al.* (2023). A novel effector RipBT contributes to *Ralstonia solanacearum* virulence on potato. *Mol. Plant Pathol.* 24: 947–60.

Rao, M. V. B. (1972). Bacterial wilt due to *Pseudomonas solanacearum*. In: P. N. Patel (ed.). *Plant Bacteriology*. Vol. I, *Bacterial Diseases of Plants in India*. Indian Agricultural Research Institute, New Delhi, pp. 94–121.

Safni, I., Cleenwerck, I., De Vos, P., Fegan, M., Sly, L. *et al.* (2014). Polyphasic taxonomic revision of the *Ralstonia solanacearum* species complex: Proposal to emend the descriptions of *Ralstonia solanacearum* and *Ralstonia syzygii* and reclassify current *R. syzygii* strains as *Ralstonia syzygii* subsp. *syzygii* subsp. nov., *R. solanacearum* phylotype IV strains as *Ralstonia syzygii* subsp. *indonesiensis* subsp. nov., banana blood disease bacterium strains as *Ralstonia syzygii* subsp. *celebesensis* subsp. nov. and *R. solanacearum* phylotype I and III strains as *Ralstonia pseudosolanacearum* sp. nov. *Int. J. Syst. Evol. Microb.* 64: 3087–3103.

Sagar, V., Jeevalatha, A., Mian, S., Chakrabarti, S. K., Gurjar, M. S. *et al.* (2014). Potato bacterial wilt in India caused by strains of phylotype I, II and IV of *Ralstonia solanacearum. Eur. J. Plant Pathol.* 138: 51–65.

Saini, M., Gupta, M., Kumar, P. and Saini, R. (2025). Expanding pathogenic diversity in bacterial wilt- insights beyond *Ralstonia solanacearum* species complex. *Austr. Plant Pathol.* 54: 443–52. https://doi.org/10.1007/s13313-025-01066-x

Sakthivel, K., Gautam, R. K., Kumar, K., Roy, S. D., Kumar, A. *et al.* (2016). Diversity of *Ralstonia solanacearum* strains on the Andaman Islands in India. *Plant Dis.* 100: 732–38.

Salanoubat, M., Genin, S., Artiguenave, F., Gouzy, J., Mangenot, S. *et al.* (2002). Genome sequence of the plant pathogen *Ralstonia solanacearum. Nature* 415: 497–502.

Sang, Y., Wang, Y., Ni, H., Cazalé, A.-C., She, Y.-M. *et al.* (2018). The *Ralstonia solanacearum* type III effector RipAY targets plant redox regulators to suppress immune responses. *Mol. Plant Pathol.* 19: 129–42.

Schaad, N. W., Gaush, P., Postnikova, E. and Frederick, R. (2001). On-site one hr PCR diagnosis of bacterial diseases. APS Congress.

Schell, M. A. (2000). Control of virulence and pathogenicity genes of *Ralstonia solanacearum* by an elaborate sensory network. *Annu. Rev. Phytopathol.* 38: 263–92.

Sharma, K., Iruegas-Bocardo, F., Abdurahman, A., Alcalá-Briseño, R. I., Garrett, K. A. *et al.* (2022). *Ralstonia* strains from potato-growing regions of Kenya reveal two phylotypes and epidemic clonality of phylotype II sequevar 1 strains. *Phytopathology* 112: 2072–83.

Shekhawat, G. S., Gadewar, A. V. and Chakrabarti S. K. (1992). Spontaneous phenotypic reversion from afluidal to fluidal state in strains of *Pseudomonas solanacearum. ACIAR Bacterial Wilt Newsletter*, No. 8, pp. 5–6.

Shekhawat, G. S., Gadewar, A. V. and Chakravarti, S. K. (2000). Potato bacterial wilt in India. Technical Bulletin No. 38, Central Potato Research Institute, Shimla.

Shen, F., Yin, W., Song, S., Zhang, Z., Ye, P. *et al.* (2020). *Ralstonia solanacearum* promotes pathogenicity by utilizing

L-glutamic acid from host plants. *Mol. Plant Pathol.* 21: 1099–1110.

Singh, N., Phukan, T., Sharma, P. L., Kabyashree, K. Barman, A. *et al.* (2018). An innovative root inoculation method to study *Ralstonia solanacearum* pathogenicity in tomato seedlings. *Phytopathology* 108: 436–42.

Smith, E. F. (1896). A bacterial disease of the tomato, eggplant and Irish potato (*Bacillus solanacearum* nov. sp.). *U. S. Dept. Agr. Div. Veg. Phys. and Path. Bull.* 12: 1–28.

Smith, I. M., McNamara, D. G., Scott, P. R. and Holderness, M. (1997). *Ralstonia solanacearum*. In: *Quarantine Pests for Europe*. 2nd Ed. EPPO/CABI.

Soliman, E. A., Abdelaziz, A. M., Dorry, M-A. E. and Attia, M. S. (2024). Protective role of biomass endophytic fungi to enhance potato plant responses against wilt disease caused by *Ralstonia solanacearum*. *Physiol. Mol. Plant Pathol.* https://doi.org/10.1016/j.pmpp.2024.102287

Spafford, J., Reeves, E. and Meadows, I. (2021). Variability in survival of grafted tomato plants exposed to the bacterial wilt pathogen in the field (Abstr.). *Phytopathology* 111: S1.6.

Stulberg, M. J., Shao, J. and Huang, Q. (2015). A multiplex PCR assay to detect and differentiate select agent strains of *Ralstonia solanacearum*. *Plant Dis.* 99: 333–41.

Tjou-Tam-Sin, N. N. A., van de Bilt, J. L. J., Westenberg, M., Bergsma-Vlami, M., Korpershoek, H. J. *et al.* (2017). First report of bacterial wilt caused by *Ralstonia solanacearum* in ornamental *Rosa* sp. *Plant Dis.* 101: 378.

Tran, T. M., Jacobs, J. M., Huerta, A., Milling, A., Weibel, J. *et al.* (2016). Sensitive, secure detection of race 3 biovar 2 and native U. S. strains of *Ralstonia solanacearum*. *Plant Dis.* 100: 630–39.

Yabuuchi, E., Kosako, Y., Yano, I., Hotta, H. and Nishiuchi, Y. (1996). Validation of the publication of new names and new combinations previously effectively published outside the IJSB. *Int. J. Syst. Bacteriol.* 46: 625–26. Effective publication: Yabuuchi, E., Kosako, Y., Yano, I. Hotta, H. and Nishiuchi, Y. (1995). Transfer of two *Burkholderia* and an *Alcaligenes* species to *Ralstonia* gen. nov.: Proposal of *Ralstonia pickettii* (Ralston, Palleroni and Doudoroff 1973) comb. nov., *Ralstonia solanacearum* (Smith 1896) comb. nov. and *Ralstonia eutropha* (Davis 1969) comb. nov. *Microbiol. Immunol.* 39: 897–904.

Youssef, S. A. and Tartoura, K. A. H. (2013). Compost enhances plant resistance against the bacterial wilt pathogen *Ralstonia solanacearum* via up-regulation of ascorbate-glutathione redox cycle. *Eur. J. Plant Pathol.* 137: 821–34.

Zhang, H., Xu, Y., Huang, Y., Xiong, X., Wu, X. *et al.* (2023). Tn-seq identifies *Ralstonia solanacearum* genes required for tolerance of plant immunity induced by exogenous salicylic acid. *Mol. Plant Pathol.* 24: 536–48.

9.17 SOFT ROT OF FLESHY VEGETABLES AND FRUITS

Fleshy tissues of fruits and vegetables often undergo rotting in field, transit and in storage. The bacteria are often associated with these rotting tissues. In the majority of these cases, the bacteria involved are not plant pathogens. These are either saprophytes, i.e., they do not attack the living cells, or secondary parasites, which grow in tissues already damaged or weakened by pathogens or adverse environmental conditions. However, some real plant pathogenic bacteria attack living plant tissues and cause soft rots. *Pectobacterium carotovorum* subsp. *carotovorum* (syn. *Erwinia carotovora* subsp. *carotovora*) and *Pec. atrosepticum* (syn. *Erw. carotovora* pv. *atroseptica, Pec. carotovorum* subsp. *atrosepticum*) are the main soft rotting bacteria responsible for enormous losses due to soft rots. This chapter describes soft rot of fleshy vegetables and fruits caused by *Pec. carotovorum* subsp. *carotovorum* and blackleg of potato caused by *Pec. atrosepticum*. The other soft rotting bacteria, of less economic importance, include:

1. *Pseudomonas* spp.: *Psm. fluorescens, Psm. cichorii, Psm. marginalis* and *Psm. viridiflava* are found to be associated with soft rot of fleshy vegetables and fruits, pink eye disease of potato, slippery skin disease of onion and sour skin of onion. These bacteria cause soft rots at lower temperatures than *Pectobacterium* and *Bacillus* spp.
2. *Bac. subtilis* and *Bac. megaterium* cause rotting of potato tubers and tobacco leaves. These cause the rotting of tissue at higher temperatures compared to other soft rotting bacteria.
3. *Clostridium* spp. cause rotting of potato tubers and tobacco leaves, and wet wood symptoms in poplar. These bacteria cause rotting under anaerobic conditions.

The disease is worldwide in occurrence, but is most prevalent in warm and humid climates. It occurs most commonly on fleshy vegetables such as potato tubers, onion bulbs, the roots of radish, carrot and turnip, cabbage and cauliflower heads, fleshy fruits such as tomato, eggplant, cucumber, watermelon, strawberry, peach and papaya, celery petioles and succulent leaves of lettuce and spinach. The losses may occur in the field, transit and storage/market. The cumulative losses due to this disease are more than any other bacterial disease. In 1901, L. R. Jones became the first person to report the soft rot of carrots from Vermont.

Based on multilocus sequence analysis (MLSA) of *gyrA*, *recA*, *recN*, *rpoA*, and *rpoS* genes, Waleron *et al.* (2019a) proposed that four strains from potatoes (IFB5169, IFB5266, IFB5206 and IFB5208) and two strains from bittersweet (IFB5213 and IFB5214) should be transferred to '*Candidatus* Pectobacterium maceratum'. According to authors, this is the first report of '*Can.* Pec. maceratum' causing soft rot on potato in Poland, and that this species was already present in the country in 1996.

Waleron *et al.* (2019b) found that the five strains isolated from potato (IFB5220, IFB5222, IFB5225, IFB5226 and IFB5252), and one from bittersweet (IFB5223), previously identified as *Pectobacterium carotovorum* subsp. *carotovorum*, should be renamed as *Pec. polaris*. To their knowledge, this was the first report of *Pec. polaris* causing soft rot on potato in Poland. Based on the original isolation date, this species was present in Poland as early as 1996.

Sawada *et al.* (2020) isolated five Gram-reaction-negative, aerobic, motile with one to three polar flagella, rod-shaped bacterial strains (MAFF 212408^T, MAFF 212409, MAFF 212410, MAFF 301498 and MAFF 730085) from diseased Welsh onion (*Allium fistulosum* L.) in Japan. Based on the genetic and phenotypic evidence, the authors named the bacterium *Pseudomonas kitaguniensis* sp. nov., representing a novel species. The type strain is MAFF 212408^T (=ICMP 23530^T).

Ma *et al.* (2022) reported *Pec. versatile* causing potato soft rot in Oregon and Washington for the first time. Wang *et al.* (2022) reported aerial stem rot of potato caused by *Pectobacterium polaris* for the first time from China.

9.17.1 Symptoms

As the name indicates, it is a characteristic soft decay of fleshy organs of fruits and vegetables.

Soft rot symptoms on potato tubers start as small, water-soaked, tannish lesions on the surface, which enlarge rapidly in diameter and in depth. The affected inner tissue becomes soft, mushy and watery, while its surface appears slightly depressed and wrinkled. The affected tissues become cream-coloured and slimy, leading to the disintegration and disorganisation of plant cells. In certain cases, such as potato tubers and tomato fruits, the outer surface remains intact while the inner entire contents change to a thin paste. The slimy mass comprising bacterial cells and cell debris oozes out onto the outer surface through cracks. The slimy mass turns tan, grey or dark brown due to its exposure to air. Within three or four days, a whole tuber or fruit may rot completely under favourable weather conditions (Figure 9.16a). Onion bulbs, potato tubers and cruciferous plants, when infected by soft rot bacteria, often emit an offensive sulphurous odour; in most other plants, however, the infected tissues remain almost odourless until the secondary invaders grow on the rotten tissues and produce offensive smell (Agrios, 2004).

Soft rot bacteria also invade swollen potato lenticels, which commonly swell when tubers are soaked in water or exposed to wet soil. Infected areas around the lenticels become slightly raised and darker in colour than the potato skin. The lesions are initially ¼ inch in diameter and ¼–½ inch deep, but under favourable conditions these lesions develop into soft rot (Figure 9.16b). When low temperature prevails, the lenticel infection dries out, leaving a shallow spot with a chalky-white deposit under the normal skin colour.

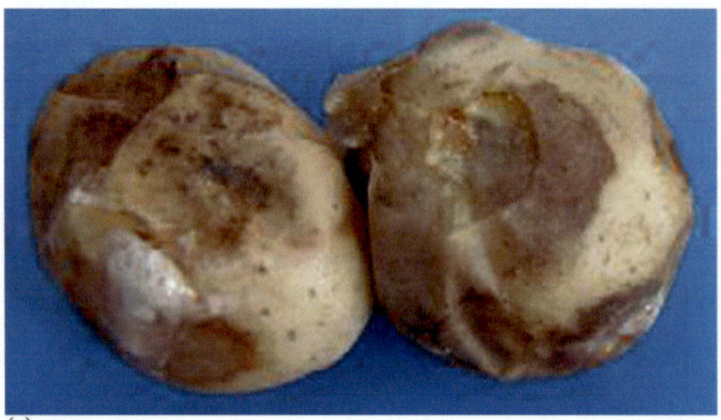
(a)

FIGURE 9.16A Bacterial soft rot: Advanced soft rot symptoms on potato tubers. (Courtesy: Queen's Printer for Ontario; www.omafra.gov.on.ca)

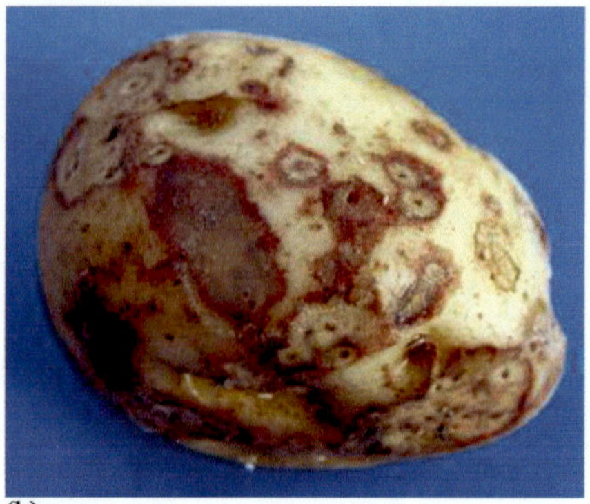
(b)

FIGURE 9.16B Bacterial soft rot: Lenticels show soft rot symptoms. (Courtesy: Queen's Printer for Ontario; www.omafra.gov.on.ca)

The aerial symptoms manifest as weak and chlorotic plants, with margins of the leaflets curled upward. Irregular, brownish to black (but not as black as in blackleg of potato) lesions appear on the stems. These lesions enlarge, causing a soft mushy rot of the stems. These symptoms are referred to as **aerial stem rot** or **bacterial stem rot**.

Succulent leaves are rarely infected in the field, but their infection in storage or in plastic containers leads to their rapid rotting and disintegration.

9.17.2 Causal Organisms

1. *Pectobacterium carotovorum* subsp. *carotovorum* (Jones 1901*) Waldee 1945 emend Hauben *et al.* 1999
   Synonym: *Erwinia carotovora* subsp. *carotovora*
2. *Pectobacterium jejuense* Hong *et al.* 2023

Waldee (1945) created a new genus *Pectobacterium* to include soft rotting bacteria and suggested that all the soft rot-causing species of the then genus *Erwinia* should be transferred to this genus. Graham (1964) also supported Waldee's proposal. However, Waldee's proposal did not get widespread acceptance, but the name of *Pectobacterium* appeared in the *Approved Lists of Bacterial Names* in 1980. Hauben and associates (1998) emended the description of genus *Pectobacterium* and its type species, *Pec. Carotovorum*, and also proposed five new subspecies combinations, including *Pec. carotovorum* subsp. *carotovorum*.

* References marked with an * are not cited in the bibliography.

The below-given description of this subspecies is taken from Hauben *et al.* (1999) and Gardan *et al.* (2003).

Cells are Gram-stain-negative rods with rounded ends and measure 0.5–1.0 × 1.0–3.0 µm. Cells occur mostly singly or in pairs, but chains are also formed and are usually motile by means of peritrichous flagella. The bacterium produces H_2S from cysteine, liquefies gelatin, reduces nitrate, produces catalase and degrades pectin. The tests for indole production, oxidase and starch hydrolysis are negative. Acid is produced from α-methylglucoside, glucanate, sorbitol and inulin. The bacterium assimilates 1-0-methyl α-galactopyranoside, 1-0-methyl β-galactopyranoside and L-glutamate, but does not assimilate palatinose, *meso*-tartrate or L-alanine. It can grow in the presence of 7% NaCl and can utilise citrate. The bacterium does not utilise hypoxanthin as a nitrogen source. The bacterium is able to grow at 36° C. It does not produce reducing substances from sucrose. G + C content of the DNA of the strains of the subspecies is in the range 50.5–53.1 mol%. The selective medium of Beraha (1968) is helpful in the isolation of soft rot bacteria, while the differential medium of Friedman (1964) is useful in separating virulent and avirulent colonies.

In addition to *Pec. carotovorum* subsp. *carotovorum*, other phytopathogenic bacteria have also been shown to cause soft rot symptoms. Nabhan *et al.* (2013) identified a pectolytic soft rot pathogen with preference for monocotyledonous plants as *Pec. aroidearum* sp. nov. after several taxonomic analyses, including DNA–DNA hybridisation. Moretti *et al.* (2016) made the first official report of *Pec. aroidearum* as a causal agent of soft rot on potato plants in Lebanon.

Pec. carotovorum subsp. *odoriferum* is generally considered to have a narrow host range and most often causes soft rot of chicory. Waleron *et al.* (2014) reported for the first time that *Pec. carotovorum* subsp. *odoriferum* occurs on a wide range of vegetables and has the ability to cause soft rot of potato during storage. All isolates of *Pec. carotovorum* subsp. *odoriferum* caused soft rot symptoms on chicory and potato. Moreover, *Pec. carotovorum* subsp. *odoriferum* isolates caused more severe symptoms on potatoes at temperatures, ranging from 15 °C to 37 °C compared to *Pec. atrosepticum* or *Pec. carotovorum* subsp. *carotovorum* isolates. Tissue maceration by *Pec. carotovorum* subsp. *odoriferum* isolates was highest at 28 °C; it was two times higher compared to that caused by *Pec. carotovorum* subsp. *carotovorum* isolates.

Pec. atrosepticum, *Pec. wasabiae* and *Pec. carotovorum* subsp. *brasiliense* have also been reported to cause soft rot symptoms (Pitman *et al.*, 2010; Barzic and Com, 2012; Onkendi and Moleleki, 2014). Meng *et al.* (2017) reported cucumber stem soft rot caused by *Pec. carotovorum* subsp. *brasiliense* for the first time from China as well as in the world, which has a significant economic impact on cucumber production.

Hong *et al.* (2023) identified a *Pectobacterium*-like strain isolated from a diseased cucumber stem tissue, collected on Jeju Island, South Korea, as *Pectobacterium jejuense*. The bacterium induced soft rot symptoms in cucumber plants.

Zoledowska *et al.* (2018) observed significant phenotypic differences among *Pec. parmentieri* isolates regarding the production of virulence factors and the ability to macerate plants. To establish the possible genomic basis of these differences, they sequenced 12 genomes of *Pec. parmentieri* strains (10 isolated in Poland, 2 in Belgium) with the combined use of Illumina and PacBio approaches. *De novo* genome assembly was performed with the use of SPAdes software, while annotation was conducted by the NCBI Prokaryotic Genome Annotation Pipeline.

Li *et al.* (2020) developed a conventional PCR and a quantitative PCR (qPCR) assay for the identification of *Pec. carotovorum* subsp. *odoriferum* (*Pco*) based on its specific *srlE* gene encoding sorbitol-specific phosphotransferase. These two methods could specifically amplify the expected products of 674 and 108 bp, respectively, from all of the *Pco* strains. The assays demonstrated high sensitivity and could detect as little as 1 and 100 pg/µl of bacterial genomic DNA, respectively. Both assays could also detect the pathogen directly from plant tissues infected with as little as 2.5×10^{-2} CFU/mg of *Pco*, even before external symptoms appeared. These assays constitute effective tools for disease diagnosis and the rapid identification of soft rot pathogens.

Boluk *et al.* (2020) developed a rapid and field-deployable recombinase polymerase amplification (RPA) assay coupled with a lateral flow device (LFD) that can accurately detect *Dickeya* spp. in infected plant tissues without the need for DNA isolation. A unique genomic region (*mgl*A/*mgl*C genes) conserved among *Dickeya* spp. was used to design highly specific robust primers and probes for an RPA assay. Assay specificity was validated with 34 representative strains from all *Dickeya* spp. and 24 strains from other genera and species; no false positives or negatives were detected. An RPA assay targeting the internal transcribed spacer region of the host genome was included to enhance the reliability and accuracy of the *Dickeya* assay. The detection limit of 1 fg was determined by both sensitivity and spiked sensitivity assays; no inhibitory effects were observed when 1 µl of host sap, macerated in Tris-EDTA buffer, was added to each reaction in the sensitivity tests. The developed RPA assay is rapid, highly accurate, sensitive and fully field-deployable. It has numerous applications in routine diagnostics, surveillance, biosecurity, and disease management. Dobhal *et al.* (2020) developed a multigene-based TaqMan qPCR multiplex assay for sensitive and reliable detection of *Dickeya solani*. Sensitivity of the developed assay was 1 pg for the detection of *Dic. solani*. The developed assay was also validated with plant samples infected naturally and artificially with *Dickeya* species and/or *Dic. solani*, and the assay detected the target in plant samples with 100% accuracy. The developed assay has applications in routine diagnostics, disease management and agriculture biosecurity.

Yasuhara-Bell *et al.* (2016) developed a loop-mediated isothermal amplification (LAMP) assay specifically designed to detect *Pec. carotovorum*. The assay allows the specific detection of *Pec. carotovorum* in the field and in the laboratory, as well as the differentiation from *Erwinia*, *Dickeya* and *Brenneria* spp. As a field assay for crops, such as potato, exhibiting blackleg or soft rot disease symptoms, it is extremely useful to have a rapid test that can identify/differentiate between pathogenic genera that could be causal agents. This assay can be used as a field test with a hand-held real-time monitoring device, such as the SMART-DART, or in the laboratory using standard real-time fluorescence monitoring equipment. The assay was tested against a broad range of non-target bacteria, but it showed specificity only for *Pec. carotovorum*. In addition, the gene target chosen for this LAMP assay can be applied to all molecular-based amplification technologies for the specific detection of *Pec. carotovorum*.

9.17.3 Disease Cycle

The bacterium perpetuates in infected plant debris, fleshy organs in storage, in ponds and streams used for irrigation water, in infested soil, in contaminated stores/warehouses and in pupae of several species of maggot fly. Contamination of potato tubers mostly occurs during harvest, handling and washing. Plant tissues weakened, invaded or killed by pathogens or by mechanical means are more susceptible to soft rot bacteria.

Penetration of bacteria mostly occurs through wounds caused by cultural operations and insects. The bacteria can also penetrate through lenticels of freshly harvested tubers. The dissemination of bacteria occurs through wind-splashed rain, irrigation water and insects. Several species of maggot fly are known to disseminate the bacteria.

Aerial stem rot infection usually results from bacteria present in the infested soil and brought by irrigation water, wind-blown rain and insects. Stem infections occur through wounds and leaf scar avenues. In the field, the disease first appears on plants grown from previously infected seed pieces or healthy seed pieces contaminated from infested soil. The further spread and inoculation of bacteria into fleshy organs are facilitated greatly by insects (maggot flies). These bacteria can live in all growth stages of the insect. Adult insects lay eggs on decaying tissues. When the eggs hatch, larvae feed on rotting tissue and their bodies become contaminated externally and internally with bacteria. These contaminated larvae inoculate the bacteria into healthy plant parts while feeding on them. The pupae and adult flies developing from internally contaminated larvae also become internally infested. These adult flies lay eggs smeared with soft rot bacteria. The larvae coming out from these eggs become contaminated; when they feed on healthy tissue, they not only cause wounds but also introduce the pathogen. By causing wounds, the maggots also break the natural barrier of host resistance, thereby making even the resistant plants or plant parts susceptible.

Soft rot bacteria are typical wound parasites. After entering wounds, these bacteria feed and multiply on cell sap released by wounded cells. During multiplication, the bacteria produce a considerable amount of pectolytic enzymes and a lesser amount of cellulolytic enzymes. Pectolytic enzymes break down the pectic substances of middle lamella and cell walls. The pectic substances are present mainly in the middle lamella as calcium pectate. Pectolytic enzymes produced move ahead of bacteria and cause maceration of the host tissue by breaking the pectic substances of middle lamella and cell walls. Due to high osmotic pressure of the macerated tissue, water from the cells comes into the intercellular spaces, leading to plasmolysis, collapse and the death of cells. Bacteria move into intercellular spaces of macerated tissue, multiply, produce pectolytic enzymes, which move ahead of bacteria and prepare that tissue for invasion. The invasion of host cells occurs at later stages when cell walls are dissolved partially. Infected plant parts become soft and are converted into a slimy mass containing numerous bacteria present in the cell sap. Slimy mass containing bacteria exudes through cracks or wounds formed on the outer surface of infected plant parts and spreads the bacteria to other fleshy organs present in the field or in stores.

A fairly high relative humidity is essential for the progress of the disease. When a decaying fleshy organ is placed in a dry atmosphere, the rotted organ dehydrates rapidly and the further progress of the disease may be checked completely. The progress of the disease is best at temperatures above 24° C. Oxygen depletion in potato tubers also favours soft rot. A film of water on potato tubers in wet soil or in stores depletes the oxygen from tubers.

DuRaan *et al.* (2016) determined the cardinal growth temperatures for the species responsible for soft rot of potato *in vitro*. The optimal temperature of 31 °C and the temperature range of 20 °C to 38 °C for *Pec. carotovorum* subsp. *brasiliense* were similar to those recorded for *Pec. carotovorum* subsp. *carotovorum*. *Pec. wasabiae* grew at an optimal temperature of 29 °C and a range of 20 °C–34 °C. Authors suggested that the minimal variation in optimal temperatures between different species might be an indication that temperature ranges, rather than optimal temperatures, play an important role in disease development.

Kubheka *et al.* (2013) generated an mCherry-*Pec. carotovorum* subsp. *brasiliense*-tagged strain to study interactions between *Pec. carotovorum* subsp. *brasiliense* and potato plants. Confocal laser-scanning microscopy and *in vitro* viable cell counts showed that *Pec. carotovorum* subsp. *brasiliense* was able to penetrate roots of a susceptible potato cultivar as early as 12 h post-inoculation and migrated upward into aerial stem parts. A comparison of *Pec. carotovorum* subsp. *brasiliense* colonisation patterns in tolerant and susceptible cultivars revealed that in the susceptible cultivar, *Pec. carotovorum* subsp. *brasiliense* cells colonised the xylem tissue, forming "biofilm-like" aggregates that led to the occlusion of some of the vessels. In contrast, in the tolerant cultivar, *Pec. carotovorum* subsp.

brasiliense appeared as free-swimming planktonic cells with no specific tissue localisation. It suggests that resistance mechanisms in the tolerant cultivar limit aggregation of *Pec. carotovorum* subsp. *brasiliense in planta*, leading to a lack of symptom development.

Koh *et al.* (2013) investigated the effect of bromelain gene (*BL1*) of pineapple (*Ananas comosus* L. Merrill) for enhancing resistance to soft rot in transgenic *Brassica rapa* "Seoulbaechu." On the inoculation of three homozygous T2 lines with *Pec. carotovorum* subsp. *carotovorum*; BL8-2 line showed the lowest number of infected leaves in both wound- and non-wound-inoculated categories, while the non-infected line showed 100% infected leaves in both cases. The highest expression of *BL1* gene was also observed in the BL8-2 homozygous line. Thus, the over-expressed *BL1* gene conferred enhanced resistance to soft rot in *B. rapa*.

Yang *et al.* (2011) characterised 2-aminobenzoic acid (2-AB) as a main induced systemic resistance (ISR) determinant from *Bacillus* sp. BS107 (BS107). The application of 2-AB at 2.3 mM significantly suppressed disease development, but exhibited no direct contact inhibition of pathogen. Root drench treatment of tobacco seedlings with BS107 suppressed disease development caused by *Pec. carotovorum* subsp. *carotovorum* SCC1. Reverse transcriptase-PCR analyses of tobacco leaves revealed the up-regulation of the induced resistance marker genes such as *PR1a*, *PR1c*, *PR2* and *PR4* by the application of 2-AB on the root. It is evident that BS107 can play a role in promoting plant defences by the secretion of bacterial determinants, including 2-AB, for the elicitation of ISR.

9.17.4 Management

The combination of sanitary measures and cultural practices can provide a very effective management of soft rot.

1. Remove all the debris from the warehouses and disinfect the floor and walls of the warehouses with copper sulphate (1 lb in 5 gal of water) or formaldehyde (1 pint in 10 gal of water).
2. All efforts should be made to avoid the wounding of plants and their storage organs.
3. Do not wash potatoes and other products before shipping or storage. If absolutely necessary, use chlorinated water for washing and the product must be dried before shipping or storage.
4. Maintain temperature of 4 °C and low humidity in stores.
5. Use pathogen-free potato seed and, if possible, use whole tubers instead of seed pieces.
6. Harvest potato crop only after the vines are completely dead to ensure skin maturity of tubers.
7. Insecticides should be sprayed in the fields and in warehouses to kill the insects responsible for the spread of bacteria.
8. Chemical sprays are not generally recommended except for the control of soft rot of tomatoes, for which Bordeaux mixture (0.8%) should be sprayed repeatedly. Currently, chlorine compounds are used to control post-harvest soft rot of fleshy fruits and vegetables. Mahovic *et al.* (2007) reported that wounded harvested tomato fruits exposed to 88 mg ClO_2 produced over 24 h or 99 mg ClO_2 produced over 2 h were free from soft rot decay.
9. Rotate the susceptible crops with cereals and other non-host crops.
10. da Silva *et al.* (2017) found that $Ca(NO_3)_2$, applied as leaf spray and soil drench, was effective in controlling soft rot of Chinese cabbage caused by *Pec. carotovorum* subsp. *carotovorum*, and it caused disease reduction of up to 48.5% when sprayed onto the leaves at the rate of 0.15 g L^{-1}. A significant increase in the leaf calcium content was observed only in the plants that were sprayed with higher doses of $Ca(NO_3)_2$ and $CaCl_2$. Light microscopy examination revealed an increased number of chloroplasts and the improved structuring of the palisade parenchyma in all the treatments, while transmission electron microscopy examination revealed an increased cell wall thickness that was especially evident for the 0.15 g L^{-1} $Ca(NO_3)_2$ treatment applied by leaf spraying and soil drenching. Abdallah *et al.* (2024) reported the use of silver nanoparticles (AgNPs) for the control of postharvest diseases of potato caused by *Pec. carotovorum* subsp. *carotovorum*.
11. *Pantoea agglomerans* (*Erw. herbicola* 252) has shown promising results in controlling both soft rot and blackleg diseases. Potato seed pieces/tubers are treated with the antagonist before sowing. Lapidot *et al.* (2015) reported that *Paenibacillus dendritiformis* significantly reduced the maceration area of tuber slices infected by *Pec. carotovorum* subsp. *carotovorum*, significantly reduced disease indices in greenhouse experiments and significantly increased tuber yield of infected plants in the field. This finding demonstrates the potential of preliminary screening based on genome analysis to identify effective biocontrol agents.

    Gerayeli *et al.* (2018) reported that of 235 *Bacillus* strains isolated from the rhizosphere of potato plants, 15 strains inhibited *in vitro* growth of *Pec. carotovorum* subsp. *carotovorum*. Antagonistic strains reduced the maceration capacity of soft rotting strains by 1.6–4.0 times. *Bac. pumilus* IrB8 and *Bac. amyloliquefaciens* IrB12 suppressed tuber maceration by 63.7 % and 47.8 %, respectively. The high levels of phenylalanine ammonia-lyase, polyphenol oxidase, peroxidase and total phenols were obtained in a single application of IrB8 or IrB12 after 8 h. It is evident that IrB8 and IrB12 strains have the potential to be used for the control of potato soft rot.

Aghdam *et al.* (2023) reported that strain EnPb of *Psm. brassicacearum* reduced soft rot development by 62.2% *in vivo* and significantly suppressed symptoms of *Pec. carotovorum* subsp. *carotovorum* (*Pcc*) in both preventative (58%) and curative (54.6%) applications under storage conditions. In addition, volatile organic compounds emitted by EnPb inhibited the growth of *Pcc in vitro* and were identified as sulphur-containing compounds consisting of dimethyl disulphide and dimethyl trisulfide. This is the first application of an endophytic *Psm. brassicacearum* stains as a biocontrol agent in the potato-*Pcc* patho-system, decreasing a post-harvest disease.

REFERENCES

Abdallah, Y., Ali, E. F., Abo-Elyousr, K. A. M. and Bagy, H. M. K. (2024). Biosynthesized silver nanoparticles using Jojoba oil as a novel antibacterial agent for controlling *Pectobacterium carotovorum* in potato. *Eur. J. Plant Pathol.* 170: 337–47. https://doi.org/10.1007/s10658-024-02902-4

Aghdam, N. M.-N., Baghaee-Ravari, S. and Shiri, A. (2023). Antimicrobial capacity of *Pseudomonas brassicacearum* strain EnPb against potato soft rot agent. *Eur. J. Plant Pathol.* 165: 215–31.

Agrios, G. N. (2004). *Plant Pathology*. 5th Ed. Elsevier Academic Press, San Diego, CA, 922 pp.

Barzic, M. R. and Com, E. (2012). Proteins involved in the interaction of potato tubers with *Pectobacterium atrosepticum*: A proteomic approach to understanding partial resistance. *J. Phytopath.* 160: 561–75.

Beraha, L. (1968). A rapid method for the preparation of a semi-solid agar medium for detection of pectolytic enzyme activity in *Erwinia carotovora*. *Plant. Dis. Reptr.* 52: 167.

Boluk, G., Dobhal, S., Crockford, A. B., Melzer, M., Alvarez, A. M. *et al.* (2020). Genome-informed recombinase polymerase amplification assay coupled with a lateral flow device for in-field detection of *Dickeya* species. *Plant Dis.* 104: 2217–24.

da Silva Felix, K. C., da Silva, C. L., de Oliveira, W. J., Mariano, R. de L. R. and de Souza, E. B. (2017). Calcium-mediated reduction of soft rot disease in Chinese cabbage. *Eur. J. Plant Pathol.* 147: 73–84.

Dobhal, S., Santillana, G., Stulberg, M. J., Boluk G., Rascoe, J. *et al.* (2020). Multigene based TaqMan qPCR multiplex assay for sensitive and reliable detection of *Dickeya solani* (Abstr.). *Phytopathology*, 110: S2.8–2.9.

duRaan, S., Coutinho, T. A. and van der Waals, J. E. (2016). Cardinal temperature differences, determined *in vitro*, between closely related species and subspecies of pectinolytic bacteria responsible for blackleg and soft rot on potatoes. *Eur. J. Plant Pathol.* 144: 361–69.

Friedman, B. A. (1964). Carbon source and tetrazolium agar to distinguish virulence in colonies of *Erwinia carotovora*. *Phytopathology* 54: 494–95.

Gardan, L., Gouy, C., Christen, R. and Samson, R. (2003). Elevation of three subspecies of *Pectobacterium carotovorum* to species level: *Pectobacterium atrosepticum* sp. nov., *Pectobacterium betavasculorum* sp. nov. and *Pectobacterium wasabiae* sp. nov. *Int. J. Syst. Evol. Microbiol.* 53: 381–91.

Gerayeli, N., Baghaee-Ravari, S. and Tarighi, S. (2018). Evaluation of the antagonistic potential of *Bacillus* strains against *Pectobacterium carotovorum* subsp. *carotovorum* and their role in the induction of resistance to potato soft rot infection. *Eur. J. Plant Pathol.* 150: 1049–63.

Graham, D. C. (1964). Taxonomy of the soft rot coliform bacteria. *Annu. Rev. Phytopathol.* 2: 13–42.

Hauben, L., Moore, E. R. B., Vauterin, L., Steenackers, M., Mergaert, J. *et al.* (1999). Validation of publication of new names and new combinations previously effectively published outside the IJSB. *Int. J. Syst. Bacteriol.* 49: 1. Effective Publication: Hauben, L., Moore, E.R.B., Vauterin, L. *et al.* (1998). Phylogenetic position of phytopathogens within the enterobacteriaceae. *Syst. Appl. Microbiol.* 21: 384–97.

Hong, S. M., Ten, L. N., Park, K. T., Back, C. G., Waleron, M. *et al.* (2023). *Pectobacterium jejuense* sp. nov. isolated from cucumber stem tissue. *Curr. Microbiol.* 80: 308. https://doi.org/10.1007/s00284-023-03419-5

Koh, Y.-J., Park, J.-I., Ahmed, N. U., Jung, H.-J. Kang, K.-K. *et al.* (2013). Enhancement of resistance to soft rot (*Pectobacterium carotovorum* subsp. *carotovorum*) in transgenic *Brassica rapa*. *Eur. J. Plant Pathol.* 136: 317–22.

Kubheka, G. C., Coutinho, T. A., Moleleki, N. and Moleleki, L. N. (2013). Colonization patterns of an mcherry-Tagged *Pectobacterium carotovorum* subsp. *brasiliense* strain in potato plants. *Phytopathology* 103: 1268–79.

Lapidot, D., Dror, R., Vered, E., Mishli, O., Levy, D., *et al.* (2015). Disease protection and growth promotion of potatoes (*Solanum tuberosum* L.) by *Paenibacillus dendritiformis*. *Plant Pathol.* 64: 545–51.

Li, X., Fu, L., Chen, C., Sun, W., Tian, Y. *et al.* (2020). Characteristics and rapid diagnosis of *Pectobacterium carotovorum* ssp. associated with bacterial soft rot of vegetables in China. *Plant Dis.* 104: 1158–66.

Ma, X., Brazil, J., Rivedal, H., Perry, K. L., Frost, K. *et al.* (2022). First report of *Pectobacterium versatile* causing potato soft rot in Oregon and Washington. *Plant Dis.* 106: 1292.

Mahovic, M. J., Tenney, J. D. and Bartz. J. A. (2007). Applications of chlorine dioxide gas for control of bacterial soft rot of tomatoes. *Plant Dis.* 91: 1316–20.

Meng, X., Chai, A., Shi, Y., Xie, X., Ma, Z. *et al.* (2017). Emergence of bacterial soft rot in cucumber caused by *Pectobacterium carotovorum* subsp. *brasiliense* in China. *Plant Dis.* 101: 279–87.

Moretti, C., Fakhr, R., Cortese, C., De Vos, P., Cerri, M. *et al.* (2016). *Pectobacterium aroidearum* and *Pectobacterium carotovorum* subsp. *carotovorum* as causal agents of potato soft rot in Lebanon. *Eur. J. Plant Pathol.* 144: 205–11.

Nabhan, S., De Boer, S. H., Maiss, E. and Wydra, K. (2013). *Pectobacterium aroidearum* sp. nov., a soft rot pathogen with preference for monocotyledonous plants. *Int. J. Syst. Evol. Micro.* 63: 2520–25.

Onkendi, E. M. and Moleleki, L. N. (2014). Characterization of *Pectobacterium carotovorum* subsp. *carotovorum* and *brasiliense* from diseased potatoes in Kenya. *Eur. J. Plant Pathol.* 139: 557–66.

Pitman, A. R., Harrow, S. A. and Visnovsky, S. B. (2010). Genetic characterisation of *Pectobacterium wasabiae* causing soft rot disease of potato in New Zealand. *Eur. J. Plant Pathol.* 126: 423–35.

Sawada, H., Fujikawa, T., Nishiwaki, Y. and Horita, H. (2020). *Pseudomonas kitaguniensis* sp. nov., a pathogen causing bacterial rot of Welsh onion in Japan. *Int. J. Syst. Evol. Micro*. 70: 3018–26.

Waldee, E. L. (1945). Comparative studies of some peritrichous phytopathogenic bacteria. *Iowa St. Coll. J. Sci*. 19: 435–84.

Waleron, M., Misztak, A., Jońca, J., Furmaniak, M., Waleron, M. M. *et al*. (2019a). First report of '*Candidatus* Pectobacterium maceratum' causing soft rot of potato in Poland. *Plant Dis*. 103: 1409.

Waleron, M., Misztak, A., Jońca, J. and Waleron, K. (2019b). First report of *Pectobacterium polaris* causing soft rot of potato in Poland. *Plant Dis*. 103: 144.

Waleron, M., Waleron, K. and Lojkowska, E. (2014). Characterization *of Pectobacterium carotovorum* subsp. *odoriferum* causing soft rot of stored vegetables. *Eur. J. Plant Pathol*. 139: 457–69.

Wang, J., Han, W., Li, Z., Cheng, J., Pan, Y. *et al*. (2022). First report of *Pectobacterium polaris* causing aerial stem rot of potato in China. *Plant Dis*. 106: 755.

Yang, S. Y., Park, M. R., Kim. I. S., Kim, Y. C., Yang, J. W. *et al*. (2011). 2-Aminobenzoic acid of *Bacillus* sp. BS107 as an ISR determinant against *Pectobacterium carotovorum* subsp. *carotovotrum* SCC1 in tobacco. *Eur. J. Plant Pathol*. 129: 371–78.

Yasuhara-Bell, J., Marrero, G., De Silva, A. And Alvarez, A. M. (2016). Specific detection of Pectobacterium carotovorum by loop-mediated isothermal amplification. *Mol. Plant Pathol*. 17: 1499–1505.

Zoledowska, S., Motyka-Pomagruk, A., Sledz, W., Mengoni, A. and Lojkowska, E. (2018). High genomic variability in the plant pathogenic bacterium *Pectobacterium parmentieri* deciphered from de novo assembled complete genomes. *BMC Genom*. 19: Article No. 751.

9.18 BLACKLEG OF POTATO

Blackleg of potato has many similarities with soft rot disease. However, it affects mainly potatoes, in contrast to the wide host range of soft rot bacterium. Its occurrence is worldwide. In the first decade of the 20th century, the disease was independently studied/described by Appel in Germany, van Hall in the Netherlands and Harrison in Canada. It used to cause considerable economic losses in all potato-growing regions; in recent years, however, its incidence has greatly decreased in countries where micropropagated plantlets are used instead of seed tubers, thereby reducing the number of field generations.

Severe stand losses and blackleg symptoms were observed in Maine, USA during the 2014 potato crop season. Again in 2015, severe stand losses and blackleg symptoms were present in a number of states, including Maine. PCR confirmed *Dickeya* spp. as being present in number of these instances. It is probable that *Dickeya* spp. have been in the Maine seed system, albeit at low levels, for more than five years (Johnson, 2016). Hao *et al*. (2016), on the basis of molecular analysis, identified *Dic. dianthicola* as the cause of the outbreak of blackleg of potato that occurred in Maine in 2015.

Dic. solani have been isolated most frequently from potato plants with the symptoms of blackleg and soft rot. *Dic. solani* strains were shown to cause more severe disease symptoms on potato plants than *Dic. Dianthicola*, especially at higher temperatures. These are also able to cause blackleg disease from lower inoculum levels. Polish *Dic. solani* strains showed higher activities of cell wall-degrading enzymes than the Finnish and Israeli strains at all the three tested temperatures, i.e., 18 °C, 27 °C and 37 °C. This observation is correlated with the higher ability of Polish *Dic. solani* strains to cause soft rot. In addition, *Dic. solani* strains exhibited higher activity of the above-mentioned enzymes and caused more severe potato tuber maceration in laboratory tests than the tested *Dic. dianthicola* strains (Golanowska *et al*., 2017). In Western Europe, *Pec. carotovorum* subsp. *brasiliense* is emerging as a causal agent of blackleg disease (van der Wolf *et al*., 2017).

Czajkowski *et al*. (2011) have briefly reviewed research on the causal agents of blackleg and soft rot of potato, namely, *Dickeya* and *Pectobacterium* species, and the disease syndrome, including epidemiological and aetiological aspects. The review includes critical evaluation of control methods used in practice, i.e., the avoidance of the contamination of plants, in particular the use of seed-testing programs and the application of hygienic procedures during crop production. The perspective of breeding and genetic modification to introduce resistance is dealt with, and the review also includes the application of physical and chemical tuber treatments to reduce inoculum load and an examination of the possibility of biocontrol using antagonistic bacteria and bacteriophages.

Sarfraz *et al*. (2019) reported, for the first time, *Pectobacterium parmentieri* and *Pec. polaris* as the causal agents of black leg of potato from Punjab state of Pakistan. Marković *et al*. (2022) reported the occurrence of *Pec. versatile* causing blackleg of potato in Serbia for the first time.

Voronina *et al*. (2021) reported soft rot and black leg of potato caused by *Pec. polaris* for the first time from Russia.

Ge *et al*. (2021) reported an outbreak of blackleg and soft rot of potato, caused primarily by the bacterial pathogen *Dic. dianthicola*, which has resulted in significant economic losses in the northeastern United States since 2015. Fingerprinting with BOX-A1R-based repetitive extragenic palindromic PCR and phylogenetic analysis based on sequences of the 16S rRNA and gapA genes indicated that *Dic. dianthicola* isolates were divided into three genotypes, denoted types I, II and III.

In Finland, *Dic. solani* was first found in 2004; it has now been the cause of major outbreaks of blackleg in Finland for more than a decade. Currently, its incidence has declined significantly because of a stringent, concerted engagement strategy adopted to combat the pathogen that is described in detail in this article. This strategy implemented in Finland could serve as a model system to combat other emerging and re-emerging species of *Dickeya* and *Pectobacterium* and serve as useful support information for the formulation of a policy framework for the management of emerging plant health risks.

9.18.1 Symptoms

The symptoms of blackleg usually originate from a contaminated seed piece or tuber. Under favourable weather conditions, entire seed pieces or tubers, and sprouts emerging from it, may rot in the soil, resulting in a poor stand. The non-emergence of plants, also called **blanking**, is a serious economic consequence of the disease. Under less favourable conditions, plants emerging from infected tubers show blackleg symptoms.

Symptoms usually start on the stems at the point of attachment to the tubers and extend upward. In the early growing season, a light brown to inky decay originating from the seed piece extends to the stem portions which are entirely below ground and the plant shows yellowing and stunting, and present a stiffened and upright habit. However, the typical phase of the disease, i.e., blackleg, usually develops after the plants have emerged from the soil. When there is sufficient moisture in the soil, the shoots arising from the infected tubers show watery decay of the cortical tissue, and the shoots wilt. If the soil moisture is reduced, the base of the shoot develops a blackened, shrivelled cortex, the **blackleg phase** (Figure 9.17), and the growth of the shoot is stunted. The leaves turn yellow, the margins of the leaflets roll upward, and, later, the entire plant may wilt and often dies. Stem decay and associated blackening is not always uniform throughout the entire length of the stem. Sometimes, the blackening of the tissue occurs near the soil level and then at a node or stem, distant from the previous one while the intervening tissue shows no external symptoms.

The tubers of the infected plants also become infected by the movement of pathogen through stolons. The dark-coloured decay of the tubers starts at the stolon end, extends inward, causing decay of pith tissue and turning the tuber into a hollow shell. These decaying tubers are often invaded by the secondary soil microorganisms, which modify the original disease symptoms. In improperly stored tubers, the pathogen causes decay at the sites of tuber lenticels and wounds. Under dry conditions, lenticel infections result in the formation of restricted, small, sunken lesions around the lenticels. In literature, these restricted lesions are referred to as **"hard rot"**. Nevertheless, under moist conditions and in association with weakly pathogenic or saprophytic microorganisms, lenticel infections can lead to the decay of entire tubers. The flesh of the affected tubers first turns creamy, then greyish and then, finally, black. Irregular cavities having black walls may extend through the centre of the potato tubers. Near the mid-season, inky-black, slightly sunken lesions are formed at the stem end of the tubers.

FIGURE 9.17 Blackleg of potato caused by *Dickeya* sp. (Courtesy: Margaret T. McGrath.)

9.18.2 Causal Organism

Pectobacterium atrosepticum (van Hall 1902) Gardan *et al.* 2003

Synonym: *Erwinia carotovora* pv. *atroseptica, Pec. carotovorum* subsp. *atrosepticum*

Van Hall in 1902 first named the bacterium as *Erw. atroseptica*. Later, Dye, in 1969, renamed it *Erw. carotovora* subsp. *atroseptica* and then, in 1978, *Erw. carotovora* pv. *atroseptica*. Hauben *et al.* (1999), on the basis of 16S rDNA sequences reclassified it as *Pec. carotovorum* subsp. *atrosepticum*. Gardan *et al.* (2003) raised it to the species level, naming it *Pec. atrosepticum*. The below-given description of the bacterium is based on the publications of Hauben *et al.* (1999) and Gardan *et al.* (2003).

The cells are Gram-stain-negative rods having round ends and measuring 0.5–1.0 × 1.0–3.0 μm. These occur mostly singly or in pairs, but chains also occur, and are usually motile by means of peritrichous flagella. Acid is produced from gluconate, α-methyl-glucoside, lactose, cellobiose, raffinose and melibiose but not from inulin, sorbitol and D(+)-arabitol. The bacterium assimilates lactulose, platinose, L-alanine, 1-*0*-methyl β-galactopyranoside and D-glucuronate but does not assimilate *meso*-tartrate or D(-)-tartrate. The bacterium is unable to grow at 36 °C and produces reducing compounds from sucrose. G + C content of the DNA ranges from 51.3 to 53.1 mol%. Bell *et al.* (2004) sequenced the genome of the bacterium and demonstrated the presence of novel pathogenicity determinants in it.

Cigna *et al.* (2017) developed a PCR-sequencing tool to easily characterise the different *Pectobacterium* and *Dickeya* taxa. The *gapA* gene sequences from 53 published genomes were aligned and a phylogeny tree was constructed. A set of 35 signature nucleotides was discovered to distinguish genera, species and subspecies of *Pectobacterium* and *Dickeya*. Then, a PCR-primer couple was designed for amplifying the *gapA* gene in pectinolytic enterobacteria. The primers were tested on 22 isolates recovered from blackleg symptoms in several potato fields. Amplicons were sequenced and signature-nucleotides were analysed. A phylogeny that includes *gapA* sequence specimens confirmed the taxonomical identification of these environmental isolates.

Zijlstra *et al.* (2020) developed two TaqMan assays for the generic detection of *Dickeya* species in plant material. The assays enabled detection of all strains of *Dic. chrysanthemi*, *Dic. dadantii*, *Dic. dianthicola*, *Dic. solani* and *Dic. zeae* tested, but only weak reactions were found with strains of *Dic. paradisiaca*. The assays enabled detection of as low as 100 fg of target DNA in a background of potato DNA. The assays were evaluated with DNA extracted from potato tuber extracts, in which different *Dickeya* species can be present. Results of both the assays were highly similar, indicating that these can be used reliably in potato seed testing programmes.

De Boer *et al.* (1987) recognised four different serological groups on the basis of antigenicity to the lipopolysaccharide. Serogroup I accounts for about 95 strains of this bacterium associated with blackleg in Canada, the United States and several countries in Western Europe. Monoclonal antibodies specific to this group are available commercially and have been used to index seed tubers for the presence of the bacterium. Fragments of genomic DNA to be used as primers in the PCR have been identified for the detection of the bacterium.

The production of pectolytic enzymes, namely, pectin methyl esterase, pectate lyase and polygalacturonase, is the primary but not the sole pathogenicity factor. Additionally, cellulase, xylanase and protease activity is present, but their role in tissue maceration has not been studied in detail.

Plant pathogenic bacteria, other than *Pec. atrosepticum*, have also been shown to be the cause of blackleg of potato. De Boer *et al.* (2012) detected *Pec. carotovorum* subsp. *brasiliensis*, *Pec. carotovorum* subsp. *carotovorum*, *Pec. wasabiae* and *Pec. atrosepticum* in potato stems showing blackleg symptoms, collected from Canada, using species and subspecies specific PCR. The tests included a new assay for *Pec. wasabiae* based on the phytase gene sequence. As mentioned above, Hao *et al.* (2016) identified *Dic. dianthicola* as the cause of blackleg of potato in Maine, USA. *Dic. solani* also causes blackleg and soft rot of potatoes.

Lin *et al.* (2023) reported a first high-quality genome assembly and the announcement of the *Pec. polaris* strain QK413-1, which causes blackleg disease in potatoes in China. The QK413-1 genome was sequenced and assembled using the PacBio Sequel II and Illumina sequencing platform. The assembled genome has a total size of 5,005,507 bp with a G + C content of 51.81%, encoding 4,782 open reading frames, including 639 virulence genes, 273 drug resistance genes and 416 secreted proteins. The QK413-1 genome sequence provides a valuable resource for the control of potato blackleg and research into its mechanism.

Sledz *et al.* (2023) undertook broad characterisation on both genotypic and phenotypic levels of the *Pec. atrosepticum* (*Pa*) isolates acquired during monitoring of potato fields in Poland in 2013, 2014 and 2016. Of REP, ERIC and BOX-based genomic fingerprinting, the BOX-based approach turned out to be the most informative and allowed for the classification of 118 *Pa* strains into six groups. The study indicated that IV and I BOX profiles dominated among the investigated population. Finally, 23 *Pa* strains representing all BOX groups and originating from various growing seasons were selected for genotypic, phylogenetic and phenotypic studies. *rpoS*-based phylogeny revealed intraspecies variation (16 SNP sites) among the studied *Pa* strains, in contrast to the analyses relying on *gyrA* (1 SNP) and *recA* (0 SNP) sequences. *Pa* strains showed higher potency to macerate potato tissue at 20 °C than 28 °C. These isolates exhibited rather uniform production of plant cell wall-degrading enzymes (pectinases, cellulases and proteases), lipases, siderophores and biofilm; however, in the majority of cases inferior in comparison to *Dic. solani* and *Pec. carotovorum* strains. The study revealed genotypic and phenotypic uniformity in addition to high virulence potential of *Pa* strains isolated from potato fields during several growing seasons in Poland.

9.18.3 Disease Cycle

The bacterium mainly survives in or on the contaminated potato tubers in the lenticels and in the damaged periderm tissue (Perombelon and Kelman, 1980). It does not survive well in soil for more than one year; however, its survival in infected debris is possible. Dissemination of the bacterium is mainly through contaminated potato tubers. Other means of dispersal are wind-blown rain, seed planting equipment, cutting knives and dipterous insects. The bacterium enters mainly through wounds and bruises caused on the potato tubers.

Temperature and moisture are the critical factors in the initiation and development of the disease. The extent of seed lot contamination and seed handling techniques also influence the severity of the disease. Most of the blackleg outbreaks have resulted from contaminated seed lots. Toth *et al.* (2003) showed that the disease development was directly related to seed tuber contamination and the threshold level of seed tuber contamination remains an important factor in predicting both disease development and progeny tuber contamination. High soil moisture and high temperature favour seed-piece decay and pre-emergence blackleg. Cool and wet soils at planting and high temperature after the emergence of plants are favourable for the blackleg phase. Even in the absence of disease, population of the bacterium may increase considerably in the root zone of the plants and may contaminate the progeny tubers. Hence, visual evaluation of blackleg incidence is not a proper assessment of the level of contamination on progeny tubers.

Skelsey *et al.* (2016) applied spatial point pattern analysis to blackleg occurrence in seed potato crops in Scotland during the 2010–2013 crop seasons to assess whether its distribution was random, regular, or aggregated, and the spatial scales at which these patterns occurred. These analyses provided the first quantitative evidence of localised and large-scale spatial clustering of potato blackleg.

Skelsey *et al.* (2023) conducted the first national-scale analysis of spatial and spatiotemporal patterns of blackleg incidence rates and associated risk factors for blackleg at the landscape scale. This was achieved through a combination of ArcGIS and interpretable machine learning applied to a longitudinal dataset of naturally infected seed potato crops from across Scotland. They found striking differences in long-term disease outcomes across the country and identified that features (variables) related to the health status and management of mother crops (seed stocks), matching features in daughter crops, and the characteristics of surrounding potato crop distributions were the most important predictors of disease, followed by field, bioclimatic and soil features.

Blackleg and soft rot of potato can be caused by individual species or a complex of *Dickeya* and *Pectobacterium* spp. To determine the role of these two bacterial species on disease expression, Ge *et al.* (2020) conducted a field trial in Maine. Seed pieces of the potato varieties 'Shepody', 'Lamoka' and 'Atlantic' were inoculated with *Dic. dianthicola*, *Pec. parmentieri*, or a mixture of both the species using a vacuum infiltration method. The tubers were planted in a randomised block designed field trial. Blackleg symptoms appeared six weeks after planting. Potato inoculated with *Dic. dianthicola* had higher disease incidence (16.22%) than that of *Pec. parmentieri* (10.4%), but the highest disease incidence was observed when the two species were co-inoculated (25.43%). These results were consistent across all three potato varieties in two-year trials. It is evident that co-infection of two bacterial species synergistically promoted disease development at a higher level.

Taylor *et al.* (2021) found that relative susceptibility of five potato cultivars namely, Lamoka, Classic Russet, Alegria, Yukon Gem and Chipeta to *Pec. atrosepticum* depended on the assay used and the laboratory and greenhouse results differed from field observations.

9.18.4 Management

1. As the bacterium does not survive well in soil, the rotation with nonhost crops is not very helpful in managing the disease.
2. Use of pathogen-free seed, and seed handling techniques which cause minimum injuries to the seed, are very effective.
3. The disposal and burial of infected plant refuse and cull piles and disinfection of tools used in the cutting and planting of potato tubers reduce the disease's severity (De Boer, 2001).
4. Proper wound healing and drying of potato tubers before storage, and storage under proper conditions, reduce the spread of the pathogen during storage.

Hadizadeh *et al.* (2019) reported that the potato endophyte *Serratia plymuthica* A30 protected potato plants by reducing blackleg development on average by 58.5% and transmission to tuber progeny as latent infection by 47–75%. These results suggest that treatment of potato tubers with biocontrol agents after harvest can reduce the severity of soft rot disease during storage and affect the transmission of soft rot bacteria from mother tubers to progeny tubers during field cultivation.

REFERENCES

Bell, K. S., Sebaihia, M., Pritchard, L., Holden, M. T. G., Hyman, L. J. *et al.* (2004). Genome sequence of the enterobacterial phytopathogen *Erwinia carotovora* subsp. *atroseptica* and characterization of virulence factors. *Proc. Natl. Acad. Sci.* USA, 101: 11105–10.

Cigna, J., Dewaegeneire, P., Beury, A., Gobert, V. and Faure, D. (2017). A *gapA* PCR-sequencing assay for identifying the *Dickeya* and *Pectobacterium* potato pathogens. *Plant Dis.* 101: 1278–82.

Czajkowski, R., Pérombelon, M. C. M., van Veen, J. A. and van der Wolf, J. M. (2011). Control of blackleg and tuber soft rot of potato caused by *Pectobacterium* and *Dickeya* species: A review. *Plant Pathol.* 60: 999–1013.

De Boer, S. H. (2001). Black leg/soft rot of potato. In: O. C. Maloy and T. D. Murray (eds.). *Encyclopedia of Plant Pathology*. Vol. 1. John Wiley and Sons, Inc., New York, pp. 146–48.

De Boer, S. H., Li, X. and Ward, L. J. (2012). *Pectobacterium* spp. associated with bacterial stem rot syndrome of potato in Canada. *Phytopathology* 102: 937–47.

De Boer, S. H., Verdonck, L., Vruggink, H., Harju, P., Bang, H. O. *et al.* (1987). Serological and biochemical variation among potato strains of *Erwinia carotovora* subsp. *atroseptica* and their taxonomic relationship to other *E. carotovora* strains. *J. Appl. Bacteriol.* 63: 487–95.

Gardan, L., Gouy, C., Christen, R. and Samson, R. (2003). Elevation of three subspecies of *Pectobacterium carotovorum* to species level: *Pectobacterium atrosepticum* sp. nov., *Pectobacterium betavasculorum* sp. nov. and *Pectobacterium wasabiae* sp. nov. *Int. J. Syst. Evol. Microbiol.* 53: 381–91.

Ge, T., Jiang, H., Johnson, S. B., Larkin, R. P., Charkowski, A. O. *et al.* (2021). Genotyping *Dickeya dianthicola* causing potato blackleg and soft rot outbreak associated with inoculum geography in the United States. *Plant Dis.* 105: 1976–83.

Ge, T., Johnson, S., Larkin, R. P., Charkowski, A. O. and Hao, J. (2020). Pathogen synergism of blackleg disease on potato (Abstr.). *Phytopathology* 110: S2.211.

Golanowska, M., Kielar, J. and Lojkowska, E. (2017). The effect of temperature on the phenotypic features and the maceration ability of *Dickeya solani* strains isolated in Finland, Israel and Poland. *Eur. J. Plant Pathol*. 47: 803–17.

Hadizadeh, I., Peivastegan. B., Hannukkala, A., van der Wolf, J. M., Nissinen, R., *et al*. (2019). Biological control of potato soft rot caused by *Dickeya solani* and the survival of bacterial antagonists under cold storage conditions. *Plant Pathol*. 68: 297–311.

Hao, J., Jiang, H. and Johnson, S. (2016). Detection and characterization of *Dickeya* species in the outbreak of blackleg disease of potato in Maine (Abstr.). *Phytopathology*, 106: S2.2.

Hauben, L., Moore, E. R. B., Vauterin, L., Steenackers, M., Mergaert, J. *et al*. (1999). Validation of publication of new names and new combinations previously effectively published outside the IJSB. *Int. J. Syst. Bacteriol*. 49: 1. Effective Publication: Hauben, L., Moore, E.R.B., Vauterin, L. *et al*. (1998). Phylogenetic position of phytopathogens within the enterobacteriaceae. *Syst. Appl. Microbiol*. 21: 384–97.

Johnson, S. (2016). Dickeya, a new potato pathogen in Maine and elsewhere (Abstr.). *Phytopathology*, 106: S2.3.

Lin, Z., Li, H., Luo, W., Xu, Y., Xu, G. *et al*. (2023). Genome sequence resource of *Pectobacterium polaris* QK413-1 that causes blackleg on potato in Fujian Province, China. *Plant Dis*. 107: 2201–04.

Marković, S., Komić, S. M., Jelušic, A., Iličić, R., Bagi, F. *et al*. (2022). First report of *Pectobacterium versatile* causing blackleg of potato in Serbia. *Plant Dis*. 106: 312.

Perombelon, M. C. M. and Kelman, A. (1980). Ecology of the soft rot erwinias. *Annu. Rev. Phytopathol*. 18: 361–87.

Sarfraz, S., Riaz, K., Oulghazi, S., Cigna, J. Sahi, S. T. *et al*. (2019). First report of *Pectobacterium parmentieri* and *Pectobacterium polaris* causing potato blackleg disease in Punjab, Pakistan. *Plant Dis*. 103: 1405.

Skelsey, P., Civita, F. and Humphris, S. (2023). Landscape epidemiology of potato blackleg. *Phytopathology* 113: 1474–82.

Skelsey, P., Elphinstone, J. G., Saddler, G. S., Wale, S. J. and Toth, I. K. (2016). Spatial analysis of blackleg-affected seed potato crops in Scotland. *Plant Pathol*. 65: 570–76.

Sledz, W., Motyka-Pomagruk, A., Zukowska, D., Babinska-Wensierska, W., Zoledowska, S. *et al*. (2023). Genotypic and phenotypic uniformity among the population of *Pectobacterium atrosepticum* strains isolated during three growing seasons from potato fields in Poland. *Eur. J. Plant Pathol*. 167: 99–121. https://doi.org/10.1007/s10658-023-02687-y

Taylor, D., Charkowski, A. O. and Zeng, Y. (2021). Laboratory assays used to rank potato cultivar tolerance to blackleg showed that tuber vacuum infiltration results correlate with field observations. *Plant Dis*. 105: 585–91.

Toth I. K., Sullivan, L., Brierley, J. L., Avrova, A. O., Hyman, L. J.*et al*. (2003). Relationship between potato seed tuber contamination by *Erwinia carotovora* ssp. *atroseptica*, black leg disease development and progeny tuber contamination. *Plant Pathol*. 52: 119–26.

van der Wolf, J. M., de Haan, E. G., Kastelein, P., Krijger, M., de Haas, B. H., *et al*. (2017). Virulence of *Pectobacterium carotovorum* subsp. *brasiliense* on potato compared with that of other *Pectobacterium* and *Dickeya* species under climatic conditions prevailing in the Netherlands. *Plant Pathol*. 66: 571–83.

van Hall, C. J. J. (1902). Bijdragen tot de kennis der Bakterieele Plantenzeikten. Coöperatieve Drukkerij-vereeniging "Plantijn", Inaugural dissertation. Amsterdam.

Voronina, M. V., Lukianova, A. A., Shneider, M. M., Korzhenk A. A., Toschakov, S. V. *et al*. (2021). First report of *Pectobacterium polaris* causing soft rot and blackleg of potato in Russia. *Plant Dis*. 105: 1851.

Zijlstra, C., De Haas, L. G., Krijger, M., Verstappen, E., Warris, S. *et al*. (2020). Development and evaluation of two Taqman assays for generic detection of *Dickeya* species. *Eur. J. Plant Pathol*. 156: 311–16. https://doi.org/10.1007/s10658-019-01876-y

9.19 COMMON SCAB OF POTATO

Common scab of potato was first discovered in 1825 in the United States, but its causal agent was identified as *Oospora scabiei* by Thaxter (1891). In 1914, Gussow transferred the causal agent to *Actinomycetes*. Later on, Waksman and Henrici (1943) transferred it to the genus *Streptomyces* as *Streptomyces scabies*. In 1953, potato scab was found in Maine, in soils having pH as low as 4.3; it was named **"acid scab"**. Common scab of potato occurs in potato-growing regions throughout the world. It is more prevalent in neutral or slightly alkaline soils and the severity increases during the dry years. Generally, the disease does not cause a reduction in yield but significant losses do result from reduction in the market price of potato tubers. Severe infection may also cause a reduction in yield, and deep-pitted lesions increase losses resulting from wastage due to peeling of tubers. Additionally, deep-pitted lesions, which are common in diseased tubers, affect the processing quality of potatoes, where the normal steam-peeling processes do not remove lesioned material. The disease is rated among the top five destructive diseases of potato by potato seed producers in the USA. It is also considered one of the most economically important diseases to affect potato production around the world. According to Wilson (2004), the disease annually costs the Tasmanian (Australia) potato industry in excess of A$ 3.66 million, or approximately 4% of the industry value.

Wei *et al*. (2022) isolated *Streptomyces* strain 6-2-1(1) from scabby potatoes in Keshan County, Heilongjiang Province, China. PCR analysis confirmed that the strain harboured the characteristic *Streptomyces* pathogenicity island genes (*txtA*, *txtAB*, *necl*, and *tomA*). Phylogenetic analysis based on 16S rRNA gene sequences revealed that strain 6-2-1(1) shared 99.86% sequence similarity with *Streptomyces rhizophilus* JR-41^T, isolated initially from bamboo in rhizospheric soil in Korea. PCR amplification followed by Sanger sequencing of the 16S rRNA gene of 164 scabby potato samples collected in Heilongjiang Province from 2019 to 2020 demonstrated that approximately 2% of the tested samples were infected with *Str. rhizophilus*. *Str. rhizophilus* may pose a potential challenge to potato production in Heilongjiang Province of China. *Str. galilaeus* has been recently identified as a dominant species causing potato common scab in Inner Mongolia, China.

9.19.1 Symptoms

The symptoms appear on the underground plant parts, mainly on the tubers. The symptoms are usually noticed late in the growing season or at harvest. Corky or scabby lesions of three types, namely, surface (shallow), raised (erumpent) and pitted (deep), appear on the surface of infected potato tubers. Sometimes surface lesions are also referred to as russeting **(russet scab)**, particularly on white-skinned potato varieties, as the general appearance resembles the skin of a russet-type tuber. The lesions start as small, superficial, roughened, circular spots, sometimes raised above and often slightly below the plane of the healthy skin. The spots enlarge, merging with other spots, forming large scabby lesions. Lesions become corky due to the abnormal proliferation of cells of the periderm of tubers. Lesions vary in shape and size and are slightly darker in colour than the healthy skin (Figure 9.18).

In **deep** or **pitted scab**, the lesions extend deep into the tuber and, on average, may be 3–4 mm deep. These lesions are darker in colour than the surface (shallow) lesions. Association of insects is suggested in the formation of pitted lesions (Walker, 1969). The type of lesion formed is considered to be the outcome of the interaction of host response, aggressiveness of the pathogen strain, time of infection and environmental conditions. Another scab, called **acid scab**, is caused by *Str. acidiscabies*. The symptoms of acid scab are similar, if not identical, to those of common scab. However, the acid scab occurs in soils having pH below 5.2, as well at higher pH levels while the common scab is found in soils having a pH range of 5.2 to 8.0.

Netted scab, caused by *Str. reticulsiscabiei*, is another disease of potato, which differs from common scab in terms of the type of symptoms produced, the causal agent (Bouchek-Mechiche *et al.* 2000a), the host range, and the varietal reaction and optimum soil temperature required for disease development (Bouchek-Mechiche *et al.*, 2000b). Netted scab is reported only from European countries. The symptoms consist of superficial brown lesions on the tuber's surface and on the roots, resulting in yield reductions. Several potato cultivars are immune to this pathogen. However, the most popular cultivars for commercial cropping in continental Europe, such as Bintje, are susceptible. Sugarbeet is another host, but carrot and radish are not infected. The differences in climatic requirements, i.e., soil temperature and moisture, for the development of common and netted scabs suggest that netted scab may become a problem in areas, or in seasons unfavourable for common scab, if susceptible cultivars are grown.

FIGURE 9.18 Common scab of potato: scab lesions on potato tubers.

Gouws and McLeod (2012) observed a distinct atypical symptom on potato in isolated production regions (Western Free State and Mpumalanga) in South Africa since early 2010. Deep longitudinal fissures (3–12 mm) containing scab-like lesions were observed on the surface area of tubers of several potato cultivars (Mondial, BP1 and Buffelspoort). The lesions on cv. Mondial were interesting since this cultivar is tolerant to typical common scab in South Africa. *Str. scabiei* reference isolate caused typical, circular, raised, brown, corky common scab lesions. A previously unreported *Streptomyces* sp. is the causal agent of a new lesion type, **fissure scab**, on potato in South Africa.

9.19.2 Causal Organisms

To date, the following species have been reported as causal agents of potato scabs:

Streptomyces acidiscabies Lambert and Loria 1989a
Str. bottropensis Waksman 1961*
Str. diastatochromogenes
Str. europaeiscabiei Bouchek-Mechiche *et al.* 2000a
Str. galilaeus Ettlinger *et al.* 1958*
Str. griseus (Krainsky 1914*) Waksman and Henrici 1948*
Str. hayashii sp. nov. Vitor *et al.* 2023
Str. hilarionis sp. nov. Vitor *et al.* 2023
Str. luridiscabiei Park *et al.* 2003
Str. niveiscabiei Park *et al.* 2003
Str. puniciscabiei Park *et al.* 2003
Str. reticulsiscabiei Bouchek-Mechiche *et al.* 2000a
Str. rhizophilus Lee and Whang 2014*
Str. sampsonii (Millard and Burr 1926*) Waksman 1953*
Str. scabiei (syn. *Str. scabies*) (ex Thaxter 1892) Lambert and Loria 1989b
Str. steliiscabiei Bouchek-Mechiche *et al.* 2000a
Str. turgidiscabies Miyajima *et al.* 1998
Str. venezuelae (Millard and Burr 1926*) Waksman 1953*

Leiminger *et al.* (2013) reported *Str. bottropensis* causing potato common scab for the first time from Germany and Guo *et al.* (2014) reported *Str. galilaeus* from China. Yang *et al.* (2017) reported *Str. diastatochromogenes* causing potato common scab in China. Hiltunen *et al.* (2014) reported *Str. europaeiscabiei* causing common scab on potato from Finland for the first time.

Chalupowicz *et al.* (2022) isolated seven *Streptomyces* species from potato tubers and peanut pods showing scab symptoms in Israel. These pathogenic species were identified as *Str. bottropensis*, *Str. europaeiscabiei*, *Str. griseus*,

* References marked with an * are not cited in the bibliography.

Str. scabiei, *Str. sampsonii*, *Str. turgidiscabies* and *Str. venezuelae*. *Str. europaeiscabiei* was the most dominant among isolates from imported seed tubers, while *Str. scabiei* and *Str. turgidiscabies* were isolated from tubers grown in Israel.

Vitor *et al.* (2023) isolated two strains, IBSBF 2807^T and IBSBF 2953^T, from scab lesions on potato tubers grown in the southern Brazilian states of Rio Grande do Sul and Santa Catarina, respectively, and characterised and identified these strains through a polyphasic approach. As these strains represented two new *Streptomyces* species related to potato scab, they proposed new names for these strains as *Streptomyces hilarionis* sp. nov. (IBSBF 2807^T=CBMAI 2674^T=ICMP 24297^T=MUM 22.66^T) and *Streptomyces hayashii* sp. nov. (IBSBF 2953^T=CBMAI 2675^T=ICMP 24301^T=MUM 22.68^T).

Str. acidiscabies and *Str. reticulsiscabiei* cause acid scab and netted scab, respectively while the remaining species cause common scab. Among these species, *Str. scabiei* is the most frequently isolated species from common scab lesions, and is worldwide in distribution. Unlike most other plant pathogenic procaryotes, *Streptomyces* species consist of slender, branched, prostrate mycelium containing few or no septa. The diameter of the hyphae is about 1 micrometre.

Str. scabiei: The formation of grey-coloured spores in spiral chains is a characteristic feature of the species. The spores are smooth, cylindrical and measure 0.5 × 0.9–1.0 μm. About 20 or more spores are formed in a chain. Brown melanoid pigment is formed in spores on tyrosine nutrient agar and peptone iron agar. Septa formation on sporogenous hyphae starts from the tip and proceeds towards the base. As the spores mature, the septa constrict, eventually breaking the cells at the tip. On germination, the spores produce one or two germ tubes, which grow into mycelium.

The organism is tyrosinase-positive and gives Gram-stain-positive reaction. It utilises L-arabinose, D-fructose, D-glucose, D-mannitol, rhamnose, sucrose, D-xylose and raffinose but not xanthine as a sole source of carbon. It is sensitive to 20 μg mL^{-1} streptomycin and 0.5 μg mL^{-1} crystal violet. The lowest pH for its growth is 5.0. G + C content of the DNA is 70% mol (Goto, 1992).

Str. acidiscabies: Spores are formed in flexuous chains, consisting about 20 or more spores in a chain. Spores are smooth, white but reddish on certain media, having high pH and measure 0.4 × 0.5–1.1 μm. The organism utilises L-arabinose, D-fructose, D-glucose, D-mannose, D-rhamnose, D-xylose and sucrose but not raffinose as a sole source of carbon. The lowest pH for its growth is 4.0. It is tolerant to 20 μg mL^{-1} streptomycin and 0.5 μg mL^{-1} crystal violet. G + C content of the DNA is 71% mol. The species is distinguished from other pathogenic *Streptomyces* species by the production of red pigment, white to reddish spore mass and resistance to high acidic reactions.

Despite the morphological differences and genetic diversity in scab-causing species, it seems that these species share pathogenicity genes and mechanism(s) of pathogenicity. These pathogenicity genes are clustered in a large block called pathogenicity island. Pathogenicity island described in *Str. turgidiscabies* is a single 425 kb restriction endonuclease fragment which contains genes for synthesis of the pathogenicity determinant thaxtomin *(txtA, txtB, txtC* and *nos*) near one end and genes for pathogenicity factors tomatinase (*tomA*) and a necrosis-inducing protein (*nec1*) near the other end. These three loci are also found in a similar large contiguous chromosomal region of *Str. acidiscabies* and *Str. scabiei* (Wanner, 2006).

Pathogenicity island genes have been found to be transferred horizontally from pathogenic species to saprophytic species, resulting in the emergence of new pathogenic forms. The independent emergence of *Str. acidiscabies* and *Str. turgidiscabies* in recent years in the USA and Japan is the result of this phenomenon (Bouchek-Mechiche *et al.*, 2006).

Nitrated dipeptide phytotoxin, thaxtomin, inhibits cellulose biosynthesis in expanding plant tissues, stimulates Ca^{2+} spiking and causes cell death. Thaxtomin production by *Streptomyces* species, which cause common, acid and pitted scabs, the quantitative relationship between toxin production and virulence and the lack of its production by saprophytic species support the hypothesis that this toxin is a pathogenicity determinant on potato tubers (Loria *et al.*, 2006). Thaxtomin is not associated with the development of netted scab lesions, suggesting that *Str. reticulsiscabiei* might possess some other pathogenicity determinants. Thaxtomin production is also not found to be associated with the species which causes russet scab.

Lerat *et al.* (2010) reported that the addition of both cellobiose (0.5%) and suberin (0.1%) in a starch-containing minimal medium doubled bacterial growth and triggered thaxtomin A production, which correlated with the upregulation (up to 342-fold) of genes involved in thaxtomin synthesis. The addition of either suberin or cellobiose alone did not affect these parameters. Suberin appeared to stimulate the onset of secondary metabolism, which is a prerequisite to the production of molecules such as thaxtomin A, while cellobiose induced the biosynthesis of this secondary metabolite.

Flores-Gonzalez *et al.* (2008) developed a PCR-based method to directly detect pathogenic *Streptomyces*, causing common scab of potato using primers designed to amplify a fragment of the *txtA* and *txtB* genes, which are pathogenicity determinants in the main pathogenic *Streptomyces* species. They detected pathogenic *Streptomyces* from tuber lesions in 70 samples out of 84 naturally infected samples comprising 19 potato cultivars collected between 2000 and 2004 from the Netherlands, the UK, France, Germany and Spain by PCR. Results of pathogen detection by PCR were confirmed by isolation. All the pathogenic isolates showed the basic general phenotypic traits of the *Str. scabiei* phenetic cluster. The RFLP analysis of amplified rRNA sequences, together with carbon sources utilisation and repetitive BOX profiles, allowed most isolates to be assigned to *Str. europaeiscabiei*, which emerged as the main cause of potato common scab in Western Europe. Another PCR assay (SYBR green quantitative real-time) developed by Qu *et al.* (2008) using primers designed from the *txtAB* (genes encoding thaxtomin synthetase) operon allows rapid, accurate and cost-effective quantification of pathogenic

Streptomyces strains in potato tubers and in soil. The assay was specific for pathogenic *Streptomyces* strains and the detection limit was 10 fg of the target DNA, or one genome equivalent. The amount of pathogenic *Streptomyces* DNA in total DNA extracts from 1 g asymptomatic and symptomatic tubers quantified ranged from 10^1 to 10^6 pg.

Correa *et al.* (2015) showed that the PCR-RFLP of the *atpD* gene technique can be used for the diagnosis and identification of phytopathogenic *Streptomyces* strains, representing a rapid and inexpensive tool. Moreover, this gene can be useful for the phylogenetic analysis of *Streptomyces* spp. associated with potato scab.

Streptomyces species cause many plant diseases, resulting in great economic losses, as given above. However, many *Streptomyces* species have antimicrobial activity against many plant pathogens. The review by Al-Quwaie (2024) briefly summarises the role, mechanisms, advantages and disadvantages of using *Streptomyces* species in plant disease management as an alternative method. They produce bioactive-rich antimicrobial metabolites and enzymes that can kill or inhibit the growth of plant pathogens. *Streptomyces* species are widely distributed in nature, but are especially abundant in the rhizosphere, the soil region surrounding plant roots. *Streptomyces* can be used as bioinoculants to protect plants from diseases. In addition to their disease-fighting abilities, they can promote plant growth in many ways. They produce plant growth-promoting substances, such as indole-3-acetic acid, cytokinin, and siderophores. They also suppress diseases through antibiosis, mycoparasitism and nutrient competition. *Streptomyces* can also supply plants with essential minerals, i.e., iron, copper, phosphorus and sulphur. Therefore, the authors conclude that *Streptomyces* species can be used as an alternative to chemicals to control plant diseases.

9.19.3 Disease Cycle

Str. scabiei can survive indefinitely, as a saprophyte, in most soils except the most acidic ones. Its population in soil in the absence of hosts can be reduced but cannot be eliminated completely, as it multiplies to some extent on organic matter present in the soil. In contrast, *Str. acidiscabies* is a poor soil inhabitant and does not survive as well as *Str. scabiei*. *Str. scabiei* can also survive on infected potato tubers and in infected plant debris.

The pathogenic species are spread through infected potato seed tubers, soil, water and soil carried by wind and implements. The penetration into the host occurs through lenticels, wounds, stomata and directly through the cuticle in freshly harvested and young tubers. Young tubers are more susceptible to infection than older ones. None of the potato cultivars is immune, although a very few show highly resistant reaction. Other hosts include radish, carrot, turnip, beet and sweet potato.

Khatri *et al.* (2010) reported that the direct inoculation of tubers with a spore suspension of *Str. scabiei* resulted in disease development, demonstrating that infection could be initiated in a soilless media. Tubers were most susceptible to infection between 3 and 20 days after tuber initiation, confirming that this early period of tuber formation is critical to disease development.

Infection by *Str. scabiei* usually starts with the onset of tuberisation. Optimum temperature for tuber infection is 20–22 °C, while the infection range varies from 10 to 31 °C. After penetration, the pathogen grows between or through a few layers of peridermal cells, these cells die due to the pathogenic action and the bacterium then feeds saprophytically on them, producing a metabolite. The metabolite promotes the rapid division of living cells surrounding the lesion. This results in the formation of several layers of corky (suberised) cells that isolate the pathogen and surrounding tuber cells. After the first layer of suberised cells is penetrated by the bacterium, the second layer is formed; it goes on till several such layers are formed in the growing season. As these layers are pushed outward and sloughed off, the bacterium grows and multiplies in additional dead cells, thereby leading to the formation of large scab lesions. The depth of lesion is influenced by the host variety, time of infection, soil pH, soil moisture and invasion of the scab lesion by secondary invaders, including insects. Insect attack usually results in the breakdown of the corky layers and allows the pathogen to invade the tuber deeply. The following factors influence the severity of the disease.

9.19.3.1 Soil pH

Soil pH from 5.2 to 8.0 favours the development of common scab, but the acid scab can also occur up to pH 4.0.

9.19.3.2 Soil Moisture

Maintaining soil moisture near field capacity (above–0.4 bars) in the two-to-six-week period following tuber initiation will reduce the incidence of the disease. Bacteria that grow at high soil moisture outgrow *Str. scabiei* on the tuber surface.

9.19.3.3 Soil Type and nutrition

Light-textured soils, and those containing high organic matter, are conducive to disease development.

The disease cycle of *Str. acidiscabies* is similar to that of *Str. scabiei*, except that the former can grow in the more acidic soils, i.e., up to pH 4.0.

Nahar *et al.* (2018) reported that the genetic background of potato cultivars influenced the abundance of pathogenic *Streptomyces* spp., having five to six times more abundant *Streptomyces* spp. in rhizosphere soil of susceptible cultivars compared with tolerant cultivars. This would result in substantially more inoculum being left in the field after harvest.

Wilson and Tegg (2010) observed no association between resistance to common scab disease and tolerance to thaxtomin A toxicity. Disease-resistant cultivars "Russet Burbank" and "Atlantic" were sensitive and tolerant to thaxtomin A toxicity, respectively. Similarly, disease-susceptible cultivars "Bismark" and "Tasman" showed susceptibility and tolerance to thaxtomin A. It is obvious that while thaxtomin A is critical to disease expression, reaction to this toxin is only one component influencing resistance to common scab disease; and many other anatomical, physiological, or biochemical factors are critical to defence against this disease. Fyans *et al.* (2016) also provided evidence that phytotoxins other than thaxtomin A may also contribute to the development of common scab by *Streptomyces* spp.

Lapaz *et al.* (2017) analysed the presence of *Streptomyces* pathogenicity island (PAI) genes, including genes encoding for *txtA*, *txtB* synthetase, *tomA* and a necrosis protein (*nec1*). Among the pathogenic isolates, 50% isolates contained four pathogenicity genes, 33% had an atypical composition of PAI marker genes, and 17% did not contain any gene. The absence of the genes reported to be involved in thaxtomin biosynthesis (*txtA*, *txtB*) was confirmed by whole-genome sequencing of two representative strains of this group. The finding suggests the participation of other virulence factors in plant pathogenicity.

Hiltunen *et al.* (2011) reported that 18 genotypes selected based on high sensitivity or tolerance, using shoot growth as the criterion, were multiplied *in vitro*, and the evaluation of their progeny of ca. 6500 tubers for resistance to common scab caused by *Str. turgidiscabies* and *Str. scabies* in a glasshouse and in three different fields showed that these genotypes differed in scab indices and disease severity ($P < 0.0001$). The relative shoot height *in vitro* (thaxtomin A used at 0.5 µg mL^{-1}) and the scab index in the field showed significant correlation ($rs = -0.463$, $P = 0.0528$, $n = 18$), which is also consistent with the results obtained under controlled conditions in the glasshouse. Therefore, the *in vitro* bioassay may be used to discard scab susceptible genotypes and elevate the overall levels of common scab resistance in the potato-breeding populations.

Andrade *et al.* (2019) developed standard area diagrams (SADs) to assess the common scab severity in potato tubers. A set of 115 tuber images, with and without symptoms, was used to develop the scale, which contained 12 images of tubers with different symptoms. The SADs were then tested by 12 evaluators with no experience in appraising plant diseases, who assessed a set of 50 images on four occasions, first without the aid of the SADs, then twice using the proposed scale, and then again using the scale of James (*Canadian Plant Disease*, 51(2), 39–65, 1971). The use of SADs increased the accuracy (coefficient of bias (Cb) = 0.99, 0.95 and 0.95 with SADs, without SADs and with the James SADs, respectively) and agreement (Lin's coefficient (ρc) = 0.97, 0.90 and 0.88 with SADs, without SADs and with James SADs, respectively) of the severity estimates. Severity estimates were also more reproducible with the use of the scale (determination coefficient (R^2) = 0.93 using SADs and R^2 = 0.84 without the use of SADs). The results obtained with the scale were also repeatable (intra-rater determination coefficient (R^2) = 0.94). The proposed SADs have been shown to be useful in evaluating common scab in potatoes, increasing the accuracy, precision and reliability of disease severity estimates.

Li *et al.* (2019) investigated the importance of thaxtomin (Txt) H during thaxtomin biosynthesis in *Str. scabies.* Biochemical studies showed that TxtH is required for promoting the soluble expression of both the TxtA and TxtB A-domains in *Escherichia coli*, and amino acid residues essential for this activity were identified. The deletion of *txtH* in *Str. scabies* significantly reduced thaxtomin A production, and the deletion of one of the two additional MbtH-like protein (MLP) homologues in *Str. scabies* completely abolished production. Engineered expression of all three *Str. scabies* MLPs could restore thaxtomin A production in a triple MLP-deficient strain, while the engineered expression of MLPs from other *Streptomyces* spp. could not. Furthermore, the constructed MLP mutants were reduced in virulence compared to wild-type *Str. scabies*. The results of this study confirm that TxtH plays a key role in thaxtomin A biosynthesis and plant pathogenicity in *Str. scabies*.

Da Silva Pereira *et al.* (2021) evaluated a mapping population (~150 full sibs) derived from a cross between two tetraploid potatoes ('Atlantic' × B1829-5) in three environments/locations (MN11, PA11, ME12) under natural common scab pressure. Three measures to common scab reaction, namely percentage of scabby tubers, diseased area and lesion indices, were found to be highly correlated (>0.76). Because of the large environmental effect, heritability values were zero for all the three traits in MN11, but moderate to high in PA11 and ME12 (~0.44 to 0.79). The authors identified a single quantitative trait locus (QTL) for lesion index in PA11, ME12 and joint analyses on linkage group 3, explaining ~22 to 30% of the total variation. The identification of QTL haplotypes and candidate genes contributing to disease resistance can support genomics-assisted breeding approaches in the crop.

Liu *et al.* (2023) characterised a multifunctional regulatory protein, SCAB_Lrp, which controls secondary metabolism, pathogenicity and sporulation in *Str. scabies*. These findings provide new insights into the complex regulatory network that modulates thaxtomin phytotoxins in pathogenic *Streptomyces*.

Isayenka *et al.* (2023) reported that somaclone RB9 regenerated from thaxtomin A-habituated potato Russet Burbank calli produced tubers more resistant to common scab than the original variety. Comparison of the RB9 tuber proteome with that of Russet Burbank using label-free quantitative proteomic analysis revealed changes in the accumulation of defence-related proteins from the patatin and lipoxygenase (LOX) families, which are involved in the metabolism of lipids, and of two miraculins of the Kunitz-type protease inhibitors family. The implication of LOX during common scab infection was studied using synchronised mini tubers developed from leaf-bud cuttings. *Str. scabiei* infection stimulated the accumulation of LOX in both Russet Burbank and RB9 mini tubers, but this accumulation was intensified in RB9 mini tubers. Infection also increased LOX activity in Russet Burbank and RB9 mini tubers. However, LOX activity measured in noninfected RB9 mini tubers was similar to that of infected Russet Burbank mini tubers, indicating endogenous activation of LOX activity in RB9 mini tubers. Ther is also a discussion of how increased LOX abundance and activity in the somaclone RB9 may contribute to improving tuber defence against common scab.

Two somaclones, YG8 and YG32, were regenerated from thaxtomin A (TA)-habituated potato var. Yukon Gold calli and infected with *Str. scabiei* in controlled conditions. The severity of common scab and lesion depth were significantly decreased in both the somaclones compared to the original variety, indicating increased resistance to the disease. Proteomic analysis comparing YG8 and YG32 tuber proteomes with that of Yukon Gold tubers showed changes in the abundance of proteins involved in various processes, including response to oxidative stress and iron homeostasis. Abundance of the iron storage protein ferritin was considerably enhanced in YG8 and YG32 tubers, correlating with increased iron accumulation in tubers. Ferritin gene expression was induced in response to *Str. scabiei* in YG32 minitubers but not in YGP tubers, suggesting the reprogramming of ferritin gene expression in TA-habituated somaclone minitubers. The treatment of *Str. scabiei*-infected YG32 minitubers with the iron chelator deferoxamine restored their susceptibility to the disease. These results suggest that changes in the regulation of iron homeostasis are involved in increasing resistance to common scab in the Yukon Gold TA-habituated somaclones (Labidi et al., 2023).

9.19.4 Management

1. Use of resistant varieties is the most effective approach for the management of scab. There is a good correlation between resistance to common and acid scab among potato cultivars. Wilson *et al.* (2009) have suggested a somatic cell selection approach using the phytotoxin, thaxtomin A, as a positive selection agent. The best disease-resistant variant, among the 13 disease-resistant variants obtained from cv. Iwa, showed on average a 85–86% lower disease score (or 91–92% less estimated tuber surface cover) than the parent cultivar. Wilson *et al.* (2010) further advocated the use of somatic cell selection with thaxtomin A to obtain stable and extremely resistant variants of potato cv. Russet Burbank to common scab.
2. Maintaining soil moisture levels near field capacity by light and frequent irrigation during 2 to 6 weeks period following tuberisation will inhibit/reduce infection.
3. Severity of common scab is significantly reduced in soils by maintaining pH at 5.2 or below (by using fertilisers like ammonium sulphate which produce acidic reaction), but this measure is associated with other drawbacks. Firstly, the number of crops, which can be grown in soils having such a low pH, is greatly reduced. Secondly, plant nutrients are best available in soils having pH levels near 6.5.
4. Rotation with small grain crops, such as corn and alfalfa, reduces the soil-borne inoculum. In Punjab, India, potato rice rotation almost eliminated the disease, which at one time was very severe. Hunjan and Sabhikhi (2020) found that *Str. scabies* population in soil can be effectively reduced either by keeping the field fallow after potato or by including mungbean or sunhemp as green manure, in a single year cropping sequence over a period of four years. Three crop rotations, namely, fallow-rice-potato, mungbean-rice-potato and sunhemp-rice-potato, led to a maximum decrease in the population of *Str. scabies* in soil, resulting in reduced scab severity. The mungbean- and sunhemp-based crop rotations also enhanced rhizospheric soil microflora, especially fluorescent pseudomonads and *Trichoderma* spp.
5. The use of pathogen-free seed and the treatment of infected seed will prevent the introduction of the pathogen into disease-free areas. It also reduces disease severity in infested fields.
6. Biological control with *Streptomyces* phages has shown promising results.
7. Green manuring reduces soil-borne inoculum by encouraging antagonistic microorganisms.
8. Tegg *et al.* (2008) have suggested a novel approach for managing the disease by foliar application of 2, 4–D @ 0.90 mM (0.20 g/litre). 2, 4–D is not inhibitory to the growth of *Str. scabiei* but it reduces the toxicity of phytotoxin, thaxtomin A, essential for induction of common scab. The application of 2, 4–D can be incorporated into the integrated management strategy of the disease. Clarke (2019), on the basis of a two-year field trial, reported that low-dose treatment of 2, 4–D mitigated tuber damage from common scab.
9. *Pseudomonas fluorescens* LBUM223 controls common scab of potato caused by *St. scabies*, through phenazine-1-carboxylic acid (PCA) production. Arseneault *et al.* (2013) reported that, under soil conditions, PCA production by *Pseudomonas* sp. LBUM223 does not control common scab by inhibiting growth of *Str. scabies* in the geocaulosphere; instead, it reduces *Str. scabies* thaxtomin A production, required for pathogenesis, leading to a reduced virulence of *Str. scabies*. (Arseneault *et al.* 2016) reported that *Psm. fluorescens* LBUM223 must colonise the potato geocaulosphere at high levels (10^7 bacteria g^{-1} of soil) in order to achieve biocontrol of common scab through increased PCA production.
10. Tomihama *et al.* (2016) found that rice bran (RB) amendment reduced potato scab by repressing the pathogenic *Streptomyces* population in young tubers. Amplicon sequencing analyses of

16S ribosomal RNA genes from the rhizosphere microbiome revealed that RB amendment dramatically changed bacterial composition and led to an increase in the relative abundance of Gram-stain-positive bacteria such as *Streptomyces* spp., and this was negatively correlated with the disease severity. Most actinomycete isolates derived from the RB-amended soil showed antagonistic activity against pathogenic *Str. scabiei* and *Str. turgidiscabies* on R2A medium. Some of the *Streptomyces* isolates suppressed the disease on inoculation onto potato plants in a field experiment. These findings suggest that RB amendment increases the levels of antagonistic bacteria against the scab pathogens in the potato rhizosphere.

11. Hiltunen *et al.* (2017) investigated the effect of repeated applications of an antagonistic *Streptomyces* strain 272 on common scab suppressiveness in a long-term field trial over five years. After a single application of strain 272, efficient disease suppression did not persist in the soil to the following growing season. However, when strain 272 was applied for three or more consecutive years, the soil remained suppressive to scab for at least two years beyond the last application, suggesting that, with repeated applications, it may be possible to enhance development of scab suppression in soil.
12. Zhang *et al.* (2020) isolated a bacterial strain PBSH9 from a potato tuber and identified it as a *Streptomyces* species on the basis of morphological, physiological and biochemical characteristics, as well as DNA sequence analysis. PBSH9 inhibited the growth of *Str. galilaeus* on agar plates and produced inhibition zones of 19.8 mm in diameter. The extracellular filtrate of PBSH9 also inhibited *Str. galilaeus* growth and produced inhibition zones of 10.0 mm in diameter. The level of disease control achieved was up to 81.88% in greenhouse trials, and from 47.64 to 73.97% in three-year field trials. Among the inoculation methods tested, seed treatment was more effective than soil drenching for PBSH9 application. PBSH9 not only effectively controlled potato common scab but also promoted potato sprouting and emergence, leading to increased potato yield.

REFERENCES

Al-Quwaie, D. A. (2024). The role of *Streptomyces* species in controlling plant diseases: A comprehensive review. *Australas. Plant Pathol.* 53: 1–14. https://doi.org/10.1007/s13313-023-00959-z

Andrade, M. H. M. L., Niederheitmann, M., de Paula Ribeiro, S. R. R., Oliveira, L. C., Pozza, E. A. *et al.* (2019). Development and validation of a standard area diagram to assess common scab in potato tubers. *Eur. J. Plant Pathol.* 154: 739–50. https://doi.org/10.1007/s10658-019-01697-z

Arseneault, T., Goyer, C. and Filion, M. (2013). Phenazine production by *Pseudomonas* sp. LBUM223 contributes to the biological control of potato common scab. *Phytopathology* 103: 995–1000.

Arseneault, T., Goyer, C. and Filion, M. (2016). Biocontrol of potato common scab is associated with high *Pseudomonas fluorescens* LBUM223 populations and phenazine-1-carboxylic acid biosynthetic transcript accumulation in the potato geocaulosphere. *Phytopathology* 106: 963–70.

Bouchek-Mechiche, K., Gardan, L., Andrivon, D. and Normand, P. (2006). *Streptomyces turgidiscabies* and *Streptomyces reticulsiscabiei*: One genomic species, two pathogenic groups. *Int. J. Syst. Evol. Microbiol.* 56: 2771–76.

Bouchek-Mechiche, K., Gardan, L., Normand, P. and Jouan, B. (2000a). DNA relatedness among strains of *Streptomyces* pathogenic to potato in France: Description of three new species, *S. europaeiscabiei* sp. nov. and *S. steliiscabiei* sp. nov. associated with common scab, and *S. reticulsiscabiei* sp. nov. associated with netted scab. *Int. J. Syst. Evol. Microbiol.* 50: 91–99.

Bouchek-Mechiche, K., Pasco, C., Andrivon, D. and Jouan, B. (2000b). Differences in host range, pathogenicity to potato cultivars and response to soil temperature among *Streptomyces* species causing common and netted scab in France. *Plant Pathol.* 49: 3–10.

Chalupowicz, L., Tsror, L., Hazanovsky, M., Erlich, O., Reuven, M. *et al.* (2022). Isolation and characterization of *Streptomyces* spp. from potato and peanut in Israel. *Plant Pathol.* 71: 1870–79.

Clarke, C. (2019). Streptomyces in potato production: Common scab pathogens and potential biocontrol strains (Abstr.). *Phytopathology* 109: S1.17.

Corrêa, D. B. A., Salomão, D., Rodrigues-Neto, J., Harakava, R. and Deståfano, S. A. L. (2015). Application of PCR-RFLP technique to species identification and phylogenetic analysis of *Streptomyces* associated with potato scab in Brazil based on partial *atpD* gene sequences. *Eur. J. Plant Pathol.* 142: 1–12.

da Silva Pereira, G., Mollinari, M., Qu, X., Thill, C., Zeng, Z.-B., *et al.* (2021). Quantitative trait locus mapping for common scab resistance in a tetraploid potato full-sib population. *Plant Dis.* 105: 3048–54.

Flores-Gonzalez, R., Velasco, I. and Montes, F. (2008). Detection and characterization of *Streptomyces* causing potato common scab in Western Europe. *Plant Pathol.* 57: 162–69.

Fyans, J. K., Bown, L. and Bignell, D. R. D. (2016). Isolation and characterization of plant-pathogenic *Streptomyces* species associated with common scab-infected potato tubers in Newfoundland. *Phytopathology* 106: 123–31.

Goto, M. (1992). *Fundamentals of Bacterial Plant Pathology*. Academic Press, Inc. San Diego, CA, 342 pp.

Gouws, R. and McLeod, A. (2012). Fissure scab, a new symptom associated with potato common scab caused by a *Streptomyces* sp. in South Africa. *Plant Dis.* 96: 1223.

Guo, F. L., Zhang, H. Y., Yu, X. M., Zhao, W. Q., Liu, D. Q. *et al.* (2014). First report of *Streptomyces galilaeus* associated with common scab in China. *Plant Dis.* 98: 683.

Hiltunen, L. H., Alanen, M., Laakso, I., Kangas, A., Virtanen, E., *et al.* (2011). Elimination of common scab sensitive progeny from a potato breeding population using thaxtomin A as a selective agent. *Plant Pathol.* 60: 426–35

Hiltunen, L. H., Kelloniemi, J. and Valkonen, J. P. T. (2014). First report of *Streptomyces europaeiscabiei* causing common scab on potato in Finland. *Plant Dis.* 98: 1267.

Hiltunen, L. H., Kelloniemi, J. and Valkonen, J. P. T. (2017). Repeated applications of a nonpathogenic *Streptomyces* strain enhance development of suppressiveness to potato common scab. *Plant Dis*. 101: 224–32.

Hunjan, M. S. and Sabhikhi, H. S. (2020). Designing a crop rotation strategy to manage *Streptomyces scabies* causing potato scab in North India. *J. Phytopathol*. 168: 469–77.

Isayenka, L., Duque-Yate, J., Goulet, M.-C., Michaud, D., Beaulieu, C. *et al*. (2023). Increased abundance of patatins, lipoxygenase and miraculins in a thaxtomin A-habituated potato Russet Burbank somaclone with enhanced resistance to common scab. *Plant Pathol*. 72: 100–11.

Khatri, B. B., Tegg, R. S., Brown, P. H. and Wilson, C. R. (2010). Infection of potato tubers with the common scab pathogen *Streptomyces scabiei* in a soilless system. *J. Phytopathol*. 158: 453–55.

Labidi, S., Sandhu, R. K., Beaulieu, C. and Beaudoin, N. (2023). Increased ferritin and iron accumulation in tubers of thaxtomin A-habituated potato var. Yukon Gold somaclones with enhanced resistance to common scab. *J. Plant Pathol*. 105: 107–19. https://doi.org/10.1007/s42161-022-01289-7

Lambert, D. H. and Loria, R. (1989a). *Streptomyces scabies sp. nov., nom. rev. Int. J. Syst. Bacteriol*. 39: 387–92.

Lambert, D. H. and Loria, R. (1989b). *Streptomyces acidiscabies sp. nov. Int. J. Syst. Bacteriol*. 39: 393–96.

Lapaz, M. I., Huguet-Tapia, J. C., Siri, M. I., Verdier, E., Loria, R. *et al*. (2017). Genotypic and phenotypic characterization of *Streptomyces* species causing potato common scab in Uruguay. *Plant Dis*. 101: 1362–72.

Leiminger, J., Frank, M., Wenk, C., Poschenrieder, G., Kellermann, A. *et al*. (2013). Distribution and characterization of *Streptomyces* species causing potato common scab in Germany. *Plant Pathol*. 62: 611–23.

Lerat, S., Simao-Beaunoir, A.-M., Wu, R., Beaudoin, N. and Beaulieu, C. (2010). Involvement of the plant polymer suberin and the disaccharide cellobiose in triggering thaxtomin A biosynthesis, a phytotoxin produced by the pathogenic agent *Streptomyces scabies*. *Phytopathology* 100: 91–96.

Li, Y., Liu, J., Adekunle, D., Bown, L., Tahlan, K. *et al*. (2019). TxtH is a key component of the thaxtomin biosynthetic machinery in the potato common scab pathogen *Streptomyces scabies*. *Mol. Plant Pathol*. 20: 1379–93.

Liu, J., Wang, Y., He, H., Dong, S., Tang, L. *et al*. (2023). The leucine-responsive regulatory protein SCAB_Lrp modulates thaxtomin biosynthesis, pathogenicity, and morphological development in *Streptomyces scabies*. *Mol. Plant Pathol*. 24: 167–78.

Loria, R., Kers, J. and Joshi, M. (2006). Evolution of plant pathogenicity in *Streptomyces*. *Annu. Rev. Phytopathol*. 44: 469–87.

Miyajima, K., Tanaka, F., Takeuchi, T. and Kuninaga, S. (1998). *Streptomyces turgidiscabies* sp. nov. *Int. J. Syst. Bacteriol*. 48: 495–502.

Nahar, K., Goyer, C., Zebarth, B. J., Burton, D. L. and Whitney, S. (2018). Pathogenic *Streptomyces* spp. Abundance affected by potato cultivars. *Phytopathology* 108: 1046–55.

Park, D. H., Kim, J. S., Kwon, S. W., Wilson, C., Yu, Y. M. *et al*. (2003). *Streptomyces luridiscabiei*, sp. nov., *Streptomyces puniciscabiei* sp. nov. and *Streptomyces niveiscabiei* sp. nov., which cause potato common scab disease in Korea. *Int. J. Syst. Bacteriol*. 53: 2049–54.

Qu, X., Wanner, L. A. and Christ, B. J. (2008). Using the *txtAB* operon to quantify pathogenic *Streptomyces* in potato tubers and soil. *Phytopathology* 98: 405–12.

Tegg, R. S., Gill, W. M., Thompson, H. K., Davies, N. W., Ross, J. J. *et al*. (2008). Auxin-induced resistance to common scab disease of potato linked to inhibition of thaxtomin A toxicity. *Plant Dis*. 92: 1321–28.

Tegg, R. S. and Wilson, C. R. (2008). Relationship of resistance to common scab disease and tolerance to thaxtomin A toxicity within potato cultivars. *Eur. J. Plant Pathol*. 128: 143–48.

Thaxter, R. (1891). Potato scab. Annual Report of the Connecticut Agricultural Experiment Station for 1891, pp. 153–60.

Tomihama, T., Nishi, Y., Mori, K., Shirao, T., Iida, T. *et al*. (2016). Rice bran amendment suppresses potato common scab by increasing antagonistic bacterial community levels in the rhizosphere. *Phytopathology* 106: 719–28.

Vitor, L., Amaral, D. T., Corrêa, D. B. A., Ferreira-Tonin, M., Lucon, E. T. *et al*. (2023). *Streptomyces hilarionis* sp. nov. and *Streptomyces hayashii* sp. nov., two new strains associated with potato scab in Brazil. *Int. J. Syst. Evol. Microbiol*. 73

Walker, J. C. (1969). *Plant Pathology*. 3rd Ed. McGraw-Hill Book Co., New York, 819 pp.

Wanner, L. A. (2006). A survey of genetic variation in *Streptomyces* isolates causing potato common scab in the United States. *Phytopathology* 96: 1363–71.

Wei, Q., Li, J., Yang, S., Wang, W., Min, F. *et al*. (2022). *Streptomyces rhizophilus* causes potato common scab disease. *Plant Dis*. 106: 266–74.

Wilson, C. R. (2004). A summary of common scab disease of potato research from Australia. In: S. Naito, N. Kondo, S. Akino *et al*. (eds). *Proc. Int. Potato Scab Symp*. Hokkaido Univ., Sapporo, pp. 198–214.

Wilson, C. R., Luckman, G. A., Tegg, R. S., Yuan, Z. Q., Wilson. A. J. *et al*. (2009). Enhanced resistance to common scab of potato through somatic cell selection in cv. Iwa with the phytotoxin thaxtomin A. *Plant Pathol*. 58: 137–44.

Wilson, C. R., Tegg, R. S., Wilson, A. J., Luckman, G. A., Eyles, A. *et al*. (2010). Stable and extreme resistance to common scab of potato obtained through somatic cell selection. *Phytopathology* 100: 460–67.

Yang, F. Y., Yang, D. J., Zhao, W. Q., Liu, D. Q. and Yu, X. M. (2017). First report of *Streptomyces diastatochromogenes* causing potato common scab in China. *Plant Dis*. 101: 243.

Zhang, X.-Y., Li, C., Hao, J.-J., Li, Y.-C., Li, D.-Z. *et al*. (2020). A novel *Streptomyces* sp. strain PBSH9 for controlling potato common scab caused by *Streptomyces galilaeus*. *Plant Dis*. 104: 1986–93.

9.20 RING ROT OF POTATO

The disease was first observed in Germany in 1906; seven years later, and working in the same country Spieckermann described it. In the early 1930s, the disease spread to Canada and the United States and by the 1940s, it was found in most major potato-growing regions of the North America. The rapid spread of the disease occurred mainly through the infected tubers. The disease caused heavy losses in Canada and the United States; in some instances, the losses were up to 100%. After the 1940s, yield losses due to the disease have been considerably reduced by the adoption of zero tolerance in seed certification standards in most seed-producing areas, including the United States, Canada and the European Union. Many countries, including the United States, have attempted, unsuccessfully, to achieve the eradication of the disease. The main reason for the persistence of the disease is the unnoticed presence of the pathogen in apparently healthy tubers due to a lack of disease symptoms. The disease still poses a continuous threat to the potato seed industry. The occurrence of the disease has also been reported from Turkey (Altundag *et al.*, 2009).

9.20.1 Symptoms

In continuous cool weather, the symptoms are not observed before harvest. In warm weather, the symptoms usually appear after the mid-season. Severely infected tubers may not sprout, and shoots may be killed before emerging from the soil. The shoots which emerge appear stunted, bear a lower number of leaves and, finally, die.

When the disease develops late in the season, often only one or two stems in a hill show symptoms. Affected stems show stunting and wilt during the day and recover at night. The wilting progresses slowly in the humid weather. Wilting and yellowing usually start from the lower leaves and progress upward. Interveinal areas of the leaflets turn yellowish; leaf margins become necrotic and roll upward (Figure 9.19a). Brown stripes appear on the stems, which may extend to leaf petioles. In advanced stages of disease development, longitudinally split stems show a brown discolouration of vascular tissue. When an infected stem is cut across at the base, a milky white to creamy bacterial exudate can be seen at the cut end by squeezing the stem. The deterioration of roots, particularly young roots, also occurs. The extent and type of symptoms can vary with the environment and host cultivar. However, in some cases, there are no above-ground symptoms.

Infection of tubers occurs through stolons. In tubers, sometimes the symptoms may not appear until harvest and appear later in storage. In most cases, there are no symptoms on the surface of tubers, except in severely infected tubers. When an infected tuber is cut across the stem end, a creamy-yellow to light brown discolouration of the vascular ring, about 1/4 inch below the skin, is seen along with bacterial ooze that may be increased by squeezing the tuber. As the disease progresses, a creamy-yellow or light brown crumbly or cheesy rot develops in the vascular ring; on squeezing the tuber, a soft pulpy exudate emerges from the diseased areas (Agrios, 2004). This is the most **characteristic symptom** of the disease and the name, **"ring rot"** (Figure 9.19b), has been derived from this diagnostic symptom.

As the disease advances further, bacterial exudate becomes more copious and a breakdown of the tissue of the vascular ring occurs, forming macroscopic cavities. Due to dehydration, the affected areas may become crumbly and powdery, or they may be soft, mushy and foul-smelling, if invaded by soft-rotting bacteria, which usually invade. In an advanced stage, the rot involves the pith of the tuber, leaving a hollow but firm shell of the tuber. The rot also extends to the cortex of tubers, leading to the formation of canker lesions on the surface of tubers and cracking of the skin (Figure 9.19c). In stores, the disease continuously progresses in tubers, occasionally forming cracks on the surface and finally causing an extensive rot. Symptoms of ring rot in the vascular tissue of infected tubers are not always as evident/prominent, as described above, and can appear as only a broken, sporadically appearing dark line, or as a continuous yellowish discolouration. Ring rot has some overlapping symptoms with brown rot.

9.20.1.1 Diagnosis

Infected tubers may grow for several generations in the field without producing any visible symptoms, making the diagnosis of the disease difficult. Tomato and eggplant have been used as indicator plants for the diagnosis of the disease, because foliar symptoms on these hosts appear within two to three weeks, in comparison with four to six weeks on potato plants. Since the late 1980s, serological techniques like latex agglutination test, enzyme-linked immunosorbent assay (ELISA) and immunofluorescent antibody staining (IFAS) are routinely used for the diagnosis of ring rot. Toda, polymerase chain reaction (PCR) methods developed for detecting specific DNA sequences found in the chromosome or plasmid of *Clavibacter sepedonicus* are being used for this purpose (Lee *et al.*, 2001; van Beckhoven *et al.*, 2002).

Smith *et al.* (2008) integrated an internal reaction control into a TaqMan PCR assay for the detection of this bacterium. The reaction control, cloned into plasmid pCmsC4, consisted of a sequence unrelated to *Cla. sepedonicus* flanked by primer sequences used in the TaqMan PCR, thus eliminating the need for multiplexing. The use of the reaction control plasmid permitted the validation of negative results, and thus facilitated the use of TaqMan real-time PCR in the routine testing of diagnostic samples for *Cla. sepedonicus*. This method takes care of drawbacks such as the occurrence of false negative results due to reagent or thermocyler failure and amplicon carry-over into samples or reagents leading to false positive results, associated with some of the earlier described methods. Another real-time PCR assay developed by Gudmestad *et al.* (2009) uses the cellulase A (CelA) gene sequence as the basis for primer design. CelA primers were specific to *Cla. Sepedonicus*, and did not react with any other coryneform bacteria or bacteria pathogenic to potato. The assay was more sensitive than Cms50/72A PCR primers, immunofluorescence antibody and enzyme-linked immunosorbent assays for the detection of the bacterium from infected and symptomless potato tubers. The assay proved to be a very robust detection tool with the added advantage of detecting only virulent strains of the bacterium.

9.20.2 Causal Organism

Clavibacter sepedonicus (Spieckermann and Kotthoff 1914) Li *et al.* 2018

Synonym: *Cla. michiganensis* subsp. *sepedonicus*, *Corynebacterium michiganense* subsp. *sepedonicum*, *Cor. michiganense* pv. *sepedonicum*, *Cor. sepedonicum*

(a) (b) (c)

FIGURE 9.19 (a) Ring rot of potato: Symptoms on aerial plant parts. (Courtesy: Solke H. De Boer, Agriculture & Agri-Food, Canada.) (b) Ring rot of potato: Cross section of potato tuber showing vascular discolouration. (Courtesy: Solke H. De Boer, Agriculture & Agri-Food, Canada.) (c) Ring rot of potato: Discolouration and cracking of potato tuber skin. (Courtesy: Solke H. De Boer, Agriculture & Agri-Food, Canada.)

For several years, the bacterium was classified in the genus, *Corynebacterium*. However, due to considerable differences in cell wall and DNA base composition, it was placed in a newly created genus, *Clavibacter.* The below-given description of the bacterium has been taken mainly from Davis *et al.* (1984).

As the bacterium grows slowly in culture media, its isolation is difficult from the infected plant material. It is rarely isolated when the samples contain less than 10^5 cells/cm^2 or a gram of host tissue. Colonies on nutrient agar appear within three to six days and are white to slightly yellow. Cells are Gram-stain-positive, club-shaped rods that vary from small coccoids through rod shapes and may appear as snapping, V-shaped cells. Cells are non-motile, non-sporing, obligate aerobes and measure 0.4–0.6 × 0.8–1.2 µm. Optimal temperature for growth is 18–21 °C while the maximum is 30 °C. Starch hydrolysis is positive, but neither galatin liquefaction nor production of pectinolytic enzymes occurs. Acid is produced from mannitol but not from lactose, trehalose, mannitol and inositol. Pathogen strains vary in the virulence and the production of extracellular slime. Both mucoid and non-mucoid colony types have been isolated from potato.

In nature, the bacterium mostly infects potato, although it can colonise the roots of sugar beet (Rowe *et al.*, 1995).

Ma *et al.* (2023) *de novo* assembled the genome sequence of *Cla. sepedonicus* K496 with 1,924,544,313 bp of Nanopore reads (N50 = 13,785 bp) using Flye v2.9 and polished it with 2 × 150 bp paired-end Illumina reads (855,788,703 bp in total). The resulting genome of K496 consists of a single circular chromosome 3,266,016 bp long and a linear plasmid of 135,489 bp. Using the NCBI PGAP v5.3, this genome was predicted to have 3,301 genes, encompassing 3,247 protein-coding genes, 90 pseudogenes, two 5S rRNA-coding, two 16S rRNA-coding, two 23S rRNA-coding sequences, 45 tRNAs and three noncoding RNAs.

9.20.3 Disease Cycle

Infected tubers are the main source of survival of the bacterium. The bacterium also survives in the infected plant refuse and volunteer plants growing from infected tubers left in the soil from the previous season. The survival also occurs for 10–24 months in dried bacterial slime spread from infected tubers and stems on crates, baskets, storage bins, sacks and warehouses and also on the surface of cultivating, harvesting and grading equipment. Survival is longest under cool and dry conditions.

Dissemination of the bacterium occurs through cutting knives, planters, particularly picker-type planters, chewing insects like Colorado potato beetles and flea beetles, and irrigation water. A knife that cuts an infected potato tuber can spread the bacterium to the subsequent 20–100 healthy seed pieces. Secondary spread occurs through the bacteria released from the wounds caused by chewing insects and cultural practices. The secondary spread is much less significant than primary spread from infected tubers.

Penetration of the bacterium into the host occurs only through wounds. The bacterium can remain viable in the host for a long time without producing visible symptoms. A symptomless tuber carrying the bacterium internally may produce symptomless plants and symptomless tubers; actually, however, all the three stages are infected. This is particularly true when the plants grow under continuous cool conditions. The favourable soil temperature for disease development is 19–28° C. In plants growing from infected tubers, the bacteria move slowly through the vascular tissue, reaching all the lower portions of the growing plant. Bacteria also move out of the vessels into the surrounding parenchyma tissues, where these multiply and cause their breakdown by forming cavities. From the parenchymatous tissue, the bacteria again move to the newly formed xylem vessels. Late in the season, the bacteria migrate from the stem to the newly formed tubers via stolons, and these infected tubers serve as a source of inoculum for the next crop.

Steinmoller *et al.* (2013) examined the risk of dissemination of *Cla. michiganensis* subsp. *sepedonicus* through potato residues from processing industries. The residues were categorised into different risk categories, ranging from category 0 (no risk of dissemination) to category 4 (high risk of dissemination). The residues not heated during processing and used in agriculture, e.g., as fertiliser, were pooled in the highest risk category 4. The residues that were sanitised before use in agriculture by composting or pasteurisation were still classified as being probably high risk (risk category 3) because no information was available concerning the inactivation of the pathogen by these treatments. The samples contaminated with *Cla. michiganensis* subsp. *sepedonicus* were composted for six days at 70° C and for 13 days at 55° C, and pasteurised for 90 min at 70° C. The viable bacteria detected by bioassay on eggplants, and culturing on semi-selective media from plant sap-forming characteristic colonies, were found in all the treatments. It is evident that these sanitation treatments are not effective in inactivating the bacterium, and the above-given classification of treated residues in category 3 (probably high risk) is thus confirmed.

Karim *et al.* (2019) developed a droplet digital PCR (ddPCR) assay for the accurate and sensitive detection of *Cla. michiganensis* subsp. *sepedonicus* (*Cms*). The assay was compared empirically to the existing methods, including conventional PCR, real-time PCR (RT-PCR), and ELISA for detection. The ddPCR assay improved detection capacity at least tenfold using genomic DNA dilution in sterile water compared to RT-PCR. The ddPCR assay is sensitive enough to consistently detect one infected potato tuber core among 800 uninfected tuber cores, whereas RT-PCR is limited to one infected core in 400. Therefore, large potato seed lots can be more comprehensively sampled using ddPCR. Because of the improved sensitivity of ddPCR, *Cms* was found on potatoes sold for consumption in grocery stores. This discovery highlights the improved detection capabilities of ddPCR and the need for expanded *Cms* monitoring.

9.20.4 Management

The management of the disease is mainly through the exclusion and eradication of the pathogen. The following measures are effective in controlling or minimising the disease.

1. Use of certified pathogen-free seed. In the USA and Canada, certified seed has a zero tolerance for ring rot.
2. Destruction of infected plant debris and volunteer plants.
3. Follow crop rotation with non-host plants for three years.
4. When ring rot is present on a farm, a through clean-up must be carried out. Dispose of all infected potato tubers away from potato production areas. Clean all surfaces of warehouses and farming equipment by removing all debris, dirt, trash and plant refuse and then washing with a strong detergent in hot water applied by a high-pressure washer. After cleaning, disinfect all warehouses and equipment with a disinfectant (1000–2000 μg mL^{-1} chlorine solution, 5% Lysol solution or 2 pounds of copper sulphate dissolved in 10 gallons of water). For proper disinfection, the disinfectant must remain in contact for at least 10 minutes, preferably 20–30 minutes, on any surface being treated. Disinfect also crates, bags, bins etc.
5. While cutting seed tubers, frequently disinfect the cutting knives with a disinfectant or use rotary knives. These are continuously run through a disinfectant solution, which should be frequently renewed.
6. Use assist- or cup-feed planters instead of picker-type planters.
7. Control insects such as Colorado potato beetles, aphids, plant bugs and leafhoppers, which may transmit the bacterium.
8. As the highly resistant cultivars also support sizable populations of the bacterium and act as potentially silent carriers of the pathogen, the use of resistant cultivars should be practised with caution (Ishimaru, 2001).

REFERENCES

Agrios, G. N. (2004). *Plant Pathology*. 5th Ed. Elsevier Academic Press, San Diego, CA, 922 pp.

Altundag, S., Karahan, A., Kilinc, A. O. and Ozakman, M. (2009). First report of *Clavibacter michiganensis* subsp. *sepedonicus* causing bacterial ring rot of potato in Turkey. *Plant Pathol*. 58: 794.

Davis, M. J., Gillaspie, A. G. Jr., Vidaver, A. K. and Harris, R. W. (1984). *Clavibacter*: A new genus containing some phytopathogenic coryneform bacteria, including *Clavibacter xyli* subsp. *xyli* sp. nov., subsp. nov. and *Clavibacter xyli* subsp. *cynodontis* subsp. nov., pathogens that cause ratoon stunting disease of sugarcane and Bermudagrass stunting disease. *Int. J. Syst. Bacteriol*. 34: 107–17.

Gudmestad, N. C., Mallik, I., Pasche, J. S. Anderson, N. R. and Kinzer, K. (2009). A real-time PCR assay for the detection of *Clavibacter michiganensis* subsp. *sepedonicus* based on the cellulase A gene sequence. *Plant Dis*. 93: 649–59.

Ishimaru, C. (2001). Ring rot of potato. In: O. C. Maloy and T. D. Murray (eds.). *Encyclopedia of Plant Pathology*. Vol. 2. John Wiley and Sons, Inc. New York., pp. 863–64.

Karim, S., Zink, F., Charkowski, A. O. and Tembrock, L. (2019). Improved detection of the potato pathogen *Clavibacter michiganensis* subsp. *sepedonicus* using droplet digital PCR (Abstr.). *Phytopathology* 109: S3.10.

Lee, I. M., Lukaesko, L. A. and Maroon, C. J. M. (2001). Comparison of dig-labeled PCR, nested PCR and ELISA for the detection of *Clavibacter michiganensis* subsp. *sepedonicus* in field grown potatoes. *Plant Dis*. 85: 261–66.

Li, X., Tambong, J., Yuan, K. X., Chen, W., Xu, H., *et al.* (2018). Reclassification of *Clavibacter michiganensis* subspecies on the basis of whole-genome and multi-locus sequence analyses. *Int. J. Syst. Evol. Microbiol*. 68: 234–40.

Ma, X., Perry, K. L. and Swingle, B. (2023). Complete genome sequence resource for a recently isolated potato ring rot pathogen, *Clavibacter sepedonicus* K496. *Plant Dis*. 107: 1202–06.

Rowe, R. C., Miller, S. A. and Riedel, R. M. (1995). Bacterial ring rot of potatoes. Extension Fact Sheet, Plant Pathology. Ohio State University.

Smith, D. S., De Boer, S. H. and Gourley, J. (2008). An internal reaction control for routine detection of *Clavibacter michiganensis* subsp. *sepedonicus* using a real-time TaqMan PCR-based assay. *Plant Dis*. 92: 684–93.

Spieckermann, A. and Kotthoff, P. (1914). Untersuchungen uber die Kartoffelpflanze und ihre Krankheiten. I. Die Bakterienringfa˙ule der Kartoffelpflanze. *Landwirtschaftliche Jahrbücher, Berlin* 46, 659–732.

Steinmoller, S., Möller, P., Bandte, M. and Büttner, C. (2013). Risk of dissemination of *Clavibacter michiganensis* ssp. *sepedonicus* with potato waste. *Eur. J. Plant Pathol*. 137: 573–84.

Van Beckhoven, J. R. C. M., Stead, D. E. and van der Wolf, J. M. (2002). Detection of *Clavibacter michiganensis* subsp. *sepedonicus* by AmpliDet RNA, a new technology based on real-time monitoring of NASBA amplicons with a molecular beacon. *J. Appl. Microbiol*. 93: 840–49.

9.21 BLACK ROT OF CRUCIFERS

Black rot (bacterial wilt) is one of the most destructive diseases of cruciferous crops. The disease was first reported by Garman in 1891 from Lexington Kentucky, USA on cabbage; later, Pammel first described the pathogen from rutabaga in Iowa in 1895. The pathogen was one of the first bacterium shown to be seed-borne by Harding and associates (1904). According to Williams (1980), it is the most important worldwide disease of crucifers, attacking all cultivated brassicas, radishes and numerous cruciferous weeds. It occurs on many cultivated and wild crucifers. Cabbage, cauliflower and kale are the most susceptible hosts while broccoli, mustards, rape, rutabaga and turnip are susceptible hosts. Radish, *gobhi sarson* (canola) and Chinese *sarson* are also infected. In a warm and wet climate, the losses due to black rot may exceed 50% on some crops. It is particularly more severe in developing countries, in the tropical and subtropical regions, where small-scale and uneducated farmers often use their self-produced seed to raise the nursery (Schaad, 2001).

The disease is worldwide in occurrence where cabbage is grown. In India, it was first reported by Patwardhan in 1928. According to Patel and Kulkarni (1953), the pathogen was not only introduced in India but was repeatedly brought with the imported seed. Several epiphytotics of the disease on cabbage and cauliflower have been reported from Rajasthan, West Bengal and Himachal Pradesh. In 1969, the National Seeds Corporation of India suffered an estimated loss of Rs. 5,00,000.00 from a 10-hectare seed production plot of cauliflower laid in Himachal Pradesh, due to the combined infection of black rot and stump rot (*Erwinia* sp.) (Trivedi and Patel, 1972).

Laala *et al*. (2021) recorded the occurrence of *Xanthomonas* (hereafter referred as *Xan*.) *campestris* pv. *campestris* for the first time from Algeria on the basis of diseased plants, showing typical symptoms of black rot, collected from cabbage and cauliflower fields during 2011 to 2014.

9.21.1 Symptoms

The most characteristic symptom of the disease is the formation of V-shaped chlorotic lesions, the base of V towards the midrib, on the margins of leaves, developing from hydathodal infections. Lesions enlarge and spread towards the midrib of the leaf and the infected areas later on turn brown and dry. In the chlorotic areas, veins and veinlets turn black; this network of black venation can be seen more clearly when the infected leaf is held against light (Figure 9.20a). The bacteria move through veins to the midrib of the leaf, from where they spread to stem through petiole. From the stem, the bacteria move upward and downward to other leaves and roots. The chlorotic lesions may appear anywhere on the leaf lamina from the systemic infection, moving from stem to midrib of the leaves. Infected leaves show stunted growth and may fall off prematurely. The cross-sections of infected stems and petioles show black rings due to the infection of water-conducting vessels. Chlorotic lesions also appear from invasions occurring through wounds caused by insects or other means. Under ideal conditions, the symptoms appear 10–14 days after the entry of the bacterium.

Cabbage and cauliflower heads are also invaded and become discoloured. The fleshy tissue of radish, turnip, knolkhol and rutabaga infected by black rot show vascular discolouration and, later, internal breakdown. Such fleshy tissues are often invaded by soft rot bacteria and undergo rotting, emitting an offensive foul smell.

On seedlings, developing from infected seeds or those developing from healthy seeds but infected by bacteria present in plant refuse or soil, the symptoms appear as black lesions on the margins of the cotyledons. The lesions enlarge and coalesce, and the cotyledons turn yellow to black. The infected cotyledons shrivel, hang down and drop prematurely; before this, however, the bacteria move into stem and young leaves. Affected seedlings wilt, collapse and die (Figure 9.20b).

In addition to *Xan. campestris* pv. *campestris*, four other pathovars of *Xan. campestris*, namely, *armoraciae*, *aberrans*, *barbareae* and *incanae*, have been reported to cause non-vascular leaf spots (some also cause sunken lesions on

(a)

FIGURE 9.20A Black rot of cabbage: V-shaped lesions on field plants. (Courtesy: Margaret T. McGrath, www.vegetablemdonline.com)

(b)

FIGURE 9.20B Black rot of cauliflower: V-shaped lesions on transplants. (Courtesy: Sally Miller, www.ohioonline.osu.edu)

stem and petioles) on many cruciferous and some non-cruciferous plants. McCulloch (1929) was the first to report *Xan. campestris* pv. *armoraciae* (syn. *Bacterium campestris* var. *armoraciae*) as the causal agent of non-vascular leaf spots on horseradish (*Armoracia rusticana*), but weekly pathogenic on cabbage and cauliflower (causing only leaf spots but no lesions on stems and petioles) and non-pathogenic on kale, mustard and radish. White (1930) reported, for the first time, *Xan. campestris* pv. *raphani* (syn. *Bac. vesicatorium* var. *raphani*) as the cause of bacterial leaf spot of radish, turnip, cabbage, cauliflower, kale, Brussels sprout and mustard, but it is non-pathogenic to horseradish. Although these pathovars generally cause non-vascular infections, vascular infection by *Xan. campestris* pv. *aberrans* (Vicente *et al.*, 2006) and by *Xan. campestris* pv. *armoraciae* and *Xan. campestris* pv. *incanae* (Vicente *et al.*, 2001) have also been reported.

9.21.2 Causal Organism

Xanthomonas campestris pv. *campestris* (Pammel 1895) Dowson 1939

Synonym: *Xan. campestris*

Cells are small, Gram-stain-negative, non-sporing rods having a single polar flagellum. It is catalase and hydrogen sulphide positive but negative for oxidase, nitrate reduction and indole production. Colonies on nutrient agar and yeast dextrose carbonate agar are yellow, convex, mucoid and shining. It produces a yellow extracellular polysaccharide called xanthan on media containing glucose. The bacterium hydrolyses starch and this character is used for its easy recognition on Schaad's selective medium.

The bacterium prefers a neutral or alkaline medium and its growth is inhibited or suppressed by acids. The optimum growth occurs at 30 °C while the lethal temperature is around 50 °C (Onsando, 1992). Acid is produced within 21 days in Dye's medium C from arabinose, mannose, galactose, trehalose, cellobiose and fructose. Complete genome sequencing has been done for two strains, namely, ATCC 33913 (CFBP 5241) and 8004 (CFBP 6650) of the bacterium.

Lugo *et al.* (2013) evaluated 56 native isolates of *Xan. campestris* pv. *campestris* collected from 12 farming districts of Trinidad, along with seven reference strains for resistance to copper in buffered (pH 7.0) and unbuffered (pH 5.6) nutrient agar media. Thirty-four and 33 native isolates were highly resistant to copper (growth on ≥200 µg mL^{-1} copper) in buffered and unbuffered media, respectively; by contrast, all the reference strains were highly susceptible to copper. Mean minimum inhibition concentration for the 56 native isolates was 224.6 µg mL^{-1}, copper indicating that high levels of copper resistance are present in *Xan. campestris* pv. *campestris* in Trinidad. A strong association between length of time of continuous applications of copper formulations and proportion of the native *Xan. campestris* pv. *campestris* with resistance to copper (Pearson's $r = 0.96$; Spearman's $r = 0.93$) was found. However, there was no association between resistance to copper and aggressiveness of the pathogen at ten days after inoculation.

The bacterium is composed of genetic and serological heterogeneous groups of strains. Variability in *Xan. campestris* pv. *campestris* continuously endangers cultivars with a narrow genetic base for resistance.

Chen *et al.* (2023) reported that the growth and virulence of *Xan. campestris* pv. *campestris* (*Xcc*) were not affected by deletion of RNase D but were severely attenuated by RNase D overexpression. However, the deletion of RNase D in *Xcc* resulted in a significant reduction in exopolysaccharide (EPS) production. In addition, either deletion or overexpression of RNase D in *Xcc* did not influence the tRNAs tested, inconsistent with the finding in *Escherichia coli* that the primary function of RNase D is to participate in tRNA maturation and its overexpression degrades tRNAs. More importantly, deletion, overexpression, and *in vitro* enzymatic analyses revealed that the *Xcc* RNase D degrades 5S rRNA but not 16S and 23S rRNAs that share an operon with 5S rRNA. Our results suggest that *Xcc* employs RNase D to realise specific modulation of the cellular 5S rRNA content after transcription and maturation whenever necessary. The finding expands our knowledge about the function of RNase D in bacteria.

Vicente *et al.* (2001), on the basis of reaction of the pathogen to cultivars of *Brassica oleracea*, *B. rapa*, *B. napus*, *B. carinata* and *B. juncea*, identified six races of the pathogen. Races 1 and 4 are predominant on *B. oleracea* crops. Based on four avirulence genes of *Xan. campestris* pv. *campestris* and four matching *Brassica* resistance genes, they also proposed a gene for gene hypothesis to explain the interaction of races and differential cultivars. Fargier and Manceau (2007) described three new races, hence making the total number of races to nine. However, they detected no new avirulence gene in bacterial genotypes and no resistance gene in plant genotypes to those proposed in the gene-for-gene model of Vicente *et al.* (2001).

Mulema *et al.* (2012) race-typed 60 of the 250 isolates of *Xan. campestris* pv. *campestris* obtained from black rot samples collected in 2009 mainly from *B. oleracea* crops (cabbage, cauliflower, broccoli and kales), and using a differential set of *Brassica* spp., found only two races: race 1 occurring in Kenya and Tanzania and race 4 occurring in Kenya, Tanzania and Uganda. Another race, race 5, was observed from one isolate obtained from a *B. rapa* sample collected from Tanzania in 2003. Genomic fingerprinting with repetitive-PCR revealed clusters that did not show significant correlations between isolates and geographical location, isolates and host adaptation, or isolates and race. However, it did demonstrate the existence of genetic differences within the Kenya, Tanzania and Uganda *Xan. campestris* pv. *campestris* populations, indicating that it is not a similar clonal population of the same genetic background.

Chidamba and Bezuidenhout (2012) determined genetic diversity in *Xan. campestris* pv. *campestris* using 28 isolates obtained from cabbage samples collected from the Gauteng, Mpumalanga and North West Provinces of South Africa. Four differential *Brassica* cultivars with known avirulence genes were used for race typing using the spray inoculation method. Four races, namely, 1, 3, 4 and 6, were identified. Of the 28 isolates, four were identified as race 1, two as race 3, 19 as race 4 and three as race 6. Repetitive DNA PCR-based fingerprinting using ERIC- and BOX-primers was used to assess the genetic diversity. Generated fingerprints of the bacterium were relatively similar. As the cluster analysis did not strictly group isolates based on their geographical origin, this suggests a limited diversity in *Xan. campestris* pv. *campestris* strains occurring in cabbage-producing regions of South Africa.

Lema *et al.* (2012) identified seven races of *Xan. campestris* pv. *campestris* based on race-typing tests of 75 isolates of the bacterium obtained from infected samples collected from fields of northwestern Spain. The identified races were 1, 4, 5, 6, 7, 8 and 9; race 4 was the most frequent in *B. oleracea* and race 6 in *B. rapa* crops. Cluster analysis derived from the combined fingerprints showed four groups, but no clear relationship to race, crop, or geographical origin was found. Rep-PCR analysis was not found to be a reliable method for the differentiation of these races. Hence, the race typing of this pathogen should be done by using the differentials of *Brassica* spp. or another alternative approach.

Singh *et al.* (2016) obtained 217 isolates of *Xan. campestris* pv. *campestris* from 12 different black rot-infected crucifer crops from 19 states of India and using a set of seven differential crucifer hosts, identified races 1, 4 and 6. Among these races, race 1 followed by race 4 dominated most of the states of India.

Cruz *et al.* (2017) reported the occurrence of races 4, 6 and 7 and also described two new races, i.e., race 10 and 11 from Portugal.

Bella *et al.* (2019) investigated the presence of physiological races in *Xan. campestris* pv. *campestris* (*Xcc*) for the first time in Italy. The race structure of *Xcc* was determined by inoculating eight differential *Brassica* lines belonging to five species. Race 4 was the most widespread, followed by races 1 and 6 in Italy. No correlation was found between allelic profiles, host of isolation, geographical origin and races, although a prevalent race was identified within the same clonal complex.

Singh *et al.* (2022) reported the whole-genome sequence of isolate *Xcc*-C7, race 4 of *Xan. campestris* pv. *campestris* isolated from cabbage (*Brassica oleracea* var. *capitata*) from Bengaluru, South India. Whole-genome sequence data were generated by the next-generation sequencing-based single-molecule real-time sequencing techniques. This information will improve our knowledge of genomic diversity in *Xan. campestris* pv. *campestris* and pave the way for research on host–pathogen interactions to develop resistance in cultivated *Brassicaceae* crops.

9.21.3 Disease Cycle

The bacterium survives internally and externally in infected seed, plant refuse and numerous weeds. Infected cruciferous weeds growing near to the crop are an important source of carryover and infection of the subsequent crops. Most of these weeds do not show typical disease symptoms and many show no symptoms at all under most favourable conditions for disease development. If the weed population around the seed production fields becomes infected, it becomes extremely difficult to produce the pathogen-free seed.

Silva *et al.* (2017) investigated the survival of *Xan. campestris* pv. *campestris* in the phyllosphere and rhizosphere of 26 weed species belonging to 14 families between August 2014 and October 2015 in Brazil. *Lepidium virginicum* and *Raphanus raphanistrum* (*Brassicaceae*) showed great potential for the survival of *Xa. campestris* pv. *campestris* in the phyllosphere, and the bacterium was isolated after 56 and 70 days, respectively. Low variation between maximum and minimum temperatures, high rainfall and high relative humidity at specific times of the year contributed to longer survival periods in the phyllosphere of some weed species. The bacterium survived in the rhizosphere of *R. raphanistrum* only, from where it was isolated up to 28 days. No relation was found between climatic factors and survival in the rhizosphere.

Krauthausen *et al.* (2018) reported that *Xanthomonas* isolates from shepherd's purse (*Capsella bursa-pastoris*) may not constitute a serious risk to serve as primary inoculum for black rot epidemics in Brassicas. However, *Xanthomonas* isolates from wallflower (*Erysimum cheiri*) have this potential. Most of the isolates from cruciferous weeds and the wallflower isolates do not fit into the actual classification of *Xan. campestris* pathovars.

Local spread of the pathogen is mainly through wind-splashed rain, while the distant spread is through infected seed and diseased nursery plants. As low as 0.03% of seeds being infected can result in the development of an epiphytotic. The other means of dissemination include wind-blown detached leaves, dust particles, insects, agricultural equipment and workers. The bacterium penetrates mainly through hydathodes and wounds caused by insects and cultural operations. The stomatal penetration occurs in cotyledons or very young leaves. In developed leaves, the penetration through stomata occurs in highly susceptible crops like cauliflower, in water-soaked tissue or when inoculum is applied with a suitable detergent.

Optimum temperature for disease development is 25–28 °C. The bacteria can multiply at 16–18 °C, but the progress of disease is very slow and that too with no visible symptoms. Free moisture in the form of rain or dew on plant surface is required for the infection and disease development. Under favourable conditions, the symptoms may appear 8–10 days after infection. Generally, the epiphytotics occur due to favourable weather conditions, but occasionally these can be traced to an infected seed lot or infected nursery bed.

After penetration through stomata and hydathodes, the bacteria multiply and progress intercellularly until they reach and invade the spiral vessels. Bacteria multiply in the vascular tissue and produce a mucilaginous extracellular polysaccharide, xanthan. The multiplication of bacterial cells and the accumulation of xanthan cause plugging of xylem vessels impeding the normal flow of water and nutrients. Bacteria then spread in the vessels throughout the plant reaching the seed via xylem of funiculus. In the infected seeds, the bacteria remain protected below the seed coat. External seed infestation occurs either through local lesions on siliques or during threshing. Infected xylem vessels disintegrate at places and then the bacteria move to the surrounding parenchyma cells, killing and disintegrating these cells and forming cavities. During wet weather, the bacteria often ooze to the surface of leaves through hydathodes and wounds from where they are disseminated by rain splashes, wind and insects. Unfavourable weather conditions and the absence of disseminating agents may result in the disappearance of the visible symptoms. In these plants, the bacteria progress slowly within the vascular tissue and the visible symptoms appear again on the return of favourable weather conditions.

Nunez *et al.* (2017) developed a diagrammatic scale, with eight levels of severity, ranging from 0.19% to 48.8%, to evaluate black rot severity on kale (*B. oleraceae* var. *acephala*) leaves. More than 95% of the leaves assessed showed severity levels ranging from 0.1% to 21%, and 5% of the leaves showed severities higher than 22%. The validation of the scale was performed by ten inexperienced evaluators, and the data were analysed using two methods: linear regression and Lin's statistics. In the absence of scale, most evaluators overestimated disease severity, but the use of the scale resulted in increased precision, accuracy, repeatability and reproducibility of the estimates according to both validation methods. The proposed diagrammatic scale proved useful for assessments of black rot severity in kale leaves, and it may be of interest to researchers performing studies on epidemiology or breeding for resistance.

Schaad and Kendrick (1975) developed a laboratory technique capable of detecting low number of *Xan. campestris* pv. *campestris* cells in crucifer seeds. The technique was capable of detecting a single diseased seed per 1000 inoculated seeds and was superior to several other methods tested. Schaad (1978) also demonstrated the use of immunofluorescence for rapid identification of the bacterium in crucifer seeds. Indirect immunofluorescence test (using antiserum to ribosomes) was better than the direct test for quick identification of the bacterium.

Zaccardelli *et al.* (2007) also developed a method for the detection and identification of the bacterium from crucifer seeds and from naturally infected and inoculated plants. The method is based on PCR amplification of fragments from the pathogenicity-associated gene *hrcC* (a PCR product of 519 bp). The method was very specific for *Xan. campestris* pv. *campestris* because no amplicons were obtained from 39 pure cultures of phytopathogenic bacteria, including four pathovars of *Xan. campestris*, two pathovars of *Xan. axonopodis* and species of *Pseudomonas* and *Pectobacterium*. Moreover, PCR reactions were also negative for 50 unidentified environmental isolates purified from the surface of crucifers.

Van Der Wolf and Van Der Zouwen (2010) found that the inoculation of cauliflower blossom with *Xan. campestris* pv. *campestris*, by pollination with blue bottle flies (*Calliphora vomitoria*) as a vector, can result in seed infestation. In two years of poly-tunnel experiments, fly inoculation of cauliflower resulted in high densities of *Xan. campestris* pv. *campestris* (>10,000 cfu g^{-1}) in approximately 30% of seed batches of a breeding line, both before and after seed treatment with warm water. Presence of the bacterium in seed derived from fly-inoculated plants was confirmed in a grow-out test. In a one-year trial, no seed infection was found in warm water-treated seed derived from cv. Opaal, an open-pollinated cultivar. No indication of internal seed infection was found after blossom inoculation with *Xan. campestris* pv. *armoraciae*.

van der Wolf *et al.* (2013) spray-inoculated cauliflower plants with *Xan. campestris* pv. *campestris* at the eight-leaf stage or during flowering. Examination of seed samples by dilution plating for deep-seated infection following hot water treatment detected *Xan. campestris* pv. *campestris* in 61% of the 23 seed samples harvested from plants inoculated during flowering. However, symptom development in seedlings raised from the seeds could not be confirmed in a grow-out test under favourable conditions for infection. The bacterium was not detected in 59 seed samples harvested from leaf-inoculated plants, with the exception of one sample from plants inoculated at peduncle formation. In another experiment, the spray inoculation of flowers resulted in the infection of 52% of flowers leading to 0.18% infected seeds, and 1–10% of the infected siliques contained infected seeds.

Laala *et al.* (2015) devised a technique named seed-qPCR for the detection of living bacterial cells of *Xan. campestris* pv. *campestris* in seed lots. The technique is based on an enrichment of bacterial population associated with infected seeds by seed germination coupled with real-time Taq-man PCR after extraction of the target DNA. The technique is inexpensive, and can detect even one contaminated seed among 10,000 healthy seeds. The seed-qPCR method combines an efficient extraction based on bacterial multiplication on seedlings with a sensitive technique qPCR for the detection of bacteria in seed lots.

Van der Wolf *et al.* (2019) conducted glasshouse experiments to study the colonisation of seedpods (siliques) and seeds of rapid cycling *Brassica oleracea* plants after spraying inoculum on clusters of recently opened flowers, with *Xan. campestris* pv. *campestris* (*Xcc*) at densities of $10^{7\text{-}8}$ cfu mL^{-1}. Authors found that *Xcc* was found in association with the seed coat and in a single seed; it was also found in the endosperm and embryo, indicating deep-seated seed infection. The estimated incidence of contaminated seeds in both years was ca. 7%. Estimated incidence of deep-seated infections, still detectable after warm water treatment of seeds, was also high (2.0–3.8%). It is concluded that spray-inoculation of flower clusters with *Xcc* can result in the infection of sepals and reproductive organs, and in deep-seated seed infections. Laala *et al.* (2015) devised a technique, named seed-qPCR, for the detection of living bacterial cells of *Xan. campestris* pv. *campestris* in seed lots. The technique is based on an enrichment of bacterial population associated with infected seeds by seed germination coupled with real-time Taq-man PCR after the extraction of the target DNA. The technique is inexpensive and can detect even one contaminated seed among 10,000 healthy seeds. The seed-qPCR method combines an efficient extraction based on the bacterial multiplication on seedlings with a sensitive technique qPCR for the detection of bacteria in seed lots.

Eichmeier *et al.* (2019) developed a real-time PCR assay based on a dual-labelled hydrolysis TaqMan® probe for the rapid and sensitive detection of *Xan. campestris* pv. *campestris* (*Xcc*) and related pathovars that affect mainly *Brassicaceae* crops and ornamentals. Primers were designed to specifically amplify a 152 bp fragment of the *Zur* gene from *Xan. campestris.* To confirm the specificity of the detection, primers targeting the *Zur* and *hrpF* genes were used for standard and real-time PCR with DNA samples from 13 *Xcc* strains, seven *Xanthomonas* species and pathovars, and five different bacterial endophytes, including *Bacillus*, *Erwinia*, *Klebsiella*, *Pantoea* and *Pseudomonas*, previously isolated from tissues of crucifers. Real-time PCR protocol was optimised to allow the detection at the level of ten copies of *Zur* PCR fragment per one microliter of DNA. Although the real-time assay based on the detection of *Zur* also detected *Xan. campestris* pvs. *raphani*, *armoraciae*, *incanae* and a strain of *Xan. hortorum* pv. *carotae*, it improved the specificity in relation to the previously published *hrpF*-based real-time method. A multiplex assay for *Zur* and *hrpF* genes further improved the specificity by excluding *Xan. hortorum* pv. *carotae.* Tests of brassica tissues and seeds artificially inoculated with *Xcc* showed that the real-time PCR based on the detection of *Zur* is an efficient and robust assay.

Sawada *et al.* (2020) isolated a bacterium from head rot lesions of broccoli (*Brassica oleracea* L. var. *italica* Plenck) in Hokkaido, Japan, and subjected it to polyphasic taxonomic characterisation. It was differentiated from their closest relatives by phenotypic characteristics, pathogenicity on broccoli, and whole-cell MALDI-TOF mass spectrometry profiles. The phenotypic, chemotaxonomic and genotypic data showed that the bacterium represents a novel *Pseudomonas* species, for which the name *Pseudomonas brassicae* sp. nov. is proposed. The type strain is MAFF 212427^{T} (=ICMP 23635^{T}).

Lin *et al.* (2021) investigated the role of extracellular amylase in *Xan. campestris* pv. *campestris* virulence and the control of its expression. Deletion of *XC3487* (here renamed *amyAXcc*), a putative amylase-encoding gene from the genome of *Xan. campestris* pv. *campestris* strain 8004, resulted in a complete loss of extracellular amylase activity and significant reduction in virulence. The extracellular amylase activity and virulence of the *amyAXcc* mutant could be restored to the wild-type level by expressing *amyAXcc* in trans. These results demonstrate that *amyAXcc* is responsible for the extracellular amylase activity of *Xan. campestris* pv. *campestris* and also indicate that extracellular amylase plays an important role in *Xan. campestris* pv. *campestris* virulence. They also found that the expression of *amyAXcc* is strongly induced by starch and requires activation by the global posttranscriptional regulator RsmA. RsmA binds specifically to the 5′-untranslated region of *amyAXcc* transcripts, suggesting that RsmA regulates *amyAXcc* directly at the posttranscriptional level. Unexpectedly, in addition to posttranscriptional regulation, the use of a transcriptional reporter demonstrated that RsmA also regulates *amyAXcc* expression at the transcriptional level, possibly by an indirect mechanism.

Li *et al.* (2022) examined the role of the SD-RR proteins encoded by *Xan. campestris* pv. *campestris* (*Xcc*). They described and characterised a SD-RR protein named McvR (motility, chemotaxis and virulence-related response regulator) that is required for virulence and motility regulation in *Xcc.* Deletion of the *mcvR* open reading frame caused reduction in motility, chemotactic movement and virulence of *Xcc.* Global transcriptome analyses revealed that the McvR had a broad regulatory role and that most motility and pathogenicity genes were down-regulated in the *mcvR* mutant. Bacterial two-hybrid and protein pull-down assays revealed that McvR did not physically interact with components of the bacterial flagellum, but interacted with other SD-RR proteins (like CheY) and the subset of DNA-binding proteins involved in gene regulation. Site-directed mutagenesis and phosphor-transfer experiments revealed that the aspartyl residue at position 55 of the receiver domain is important for phosphorylation and the regulatory activity of McvR protein. These findings describe a previously unrecognised class of SD-RR protein that contributes to the regulation of motility and virulence in *Xcc.*

Wang *et al.* (2023) identified a key sulforaphane (SFN) tolerance-related gene, *saxF*, in *Xan. campestris* pv. *campestris* (*Xcc*). They found that the transcription of *saxF* was regulated by the novel transcription factor SFN-sensing transcription factor (SstF). As a LysR family transcription factor, SstF can sense SFN and regulate the expression of *saxF* cluster genes to increase SFN resistance by directly binding to the promoter of *saxF*. The authors also found that SstF and *saxF* also play an important role in positively regulating the virulence of *Xcc*. Collectively, the results illustrate a previously unknown mechanism by which *Xcc* senses the host defence signal SFN and activates the expression of SFN tolerance-related genes to increase virulence. Hence, this study provides a remarkable result; that is, during pathogen-plant co-evolution, new functions of existing scaffolds are activated, thereby improving the proficiency of the pathogenic mechanism.

Xan. campestris pv. *campestris* (*Xcc*) contains an orphan LuxR regulator (*Xcc*R). Zhang *et al.* (2023) demonstrated that plant-produced glucose serves as a signal for *XccR/pip* regulation. Molecular docking shows that glucose is embedded into a novel sugar-binding pocket neighbouring the putative AHL pocket in the *Xcc*R autoinducer domain. The structure of the highly conserved sugar-binding pocket of several LuxR proteins generated by 3D homology models and their high affinity for glucose implied that glucose may be important for host–plant interaction. *Xcc*-induced expression of the specific sugar transporter and cell-wall invertase activity in host plants, which resulted in glucose accumulation at the bacterial infection site. These findings extend our knowledge about glucose; it can serve not only as a major carbon source for bacterial growth, but also as a signal to stimulate bacterial virulence gene expression, which enhances the ability of phytopathogens to deal with host plant immunity as well as bacterial survival. Understanding the chemical communication between plants and bacteria could help to design new strategies to control plant pathogen infection.

VemR is a response regulator of the two-component signalling systems (TCSs). It consists solely of a receiver domain. Previous studies have shown that VemR plays an important role in influencing the production of exopolysaccharides and exoenzymes, cell motility, and virulence of *Xan. campestris* pv. *campestris* (*Xcc*). However, whether VemR is involved in the essential pathogenicity determinant type III secretion system (TIIISS) is not clear. Li *et al.* (2023) found by transcriptome analysis that VemR modulates about 10% of *Xcc* genes, which are involved in various cellular processes, including the TIIISS. Further experiments revealed that VemR physically interacts with numerous proteins, including the TCS sensor kinases HpaS and RavA, and the TCS response regulator HrpG, which directly activates the transcription of HrpX, a key regulator controlling TIIISS expression. It has been demonstrated previously that HpaS composes a TCS with HrpG or VemR to control the expression of TIIISS or swimming motility, while RavA and VemR form a TCS to control the expression of flagellar genes. Mutation analysis and *in vitro* transcription assay revealed that phosphorylation might be essential for the function of VemR and phosphorylated VemR could significantly enhance the activation of *hrpX* transcription by HrpG. The authors infer that the binding of VemR to HrpG can modulate the activity of HrpG to the *hrpX* promoter, thereby enhancing *hrpX* transcription. Although further studies are required to validate this inference and explore the detailed functional mechanism of VemR, these findings provide some insights into the complex regulatory cascade of the HpaS/RavA-VemR/HrpG-HrpX signal transduction system in the control of TIIISS.

9.21.4 Management

The management of the disease is based mainly on the eradicative and preventive measures given below:

1. Use certified pathogen-free seed.
2. Hot water treatment (50 °C for 30 min) of all crucifer seeds is essential. However, Carisse and associates (1999) have recommended the treatment of cabbage, broccoli and Brussel's sprouts seeds at 50° C for 25 min and those of cauliflower, turnip, rutabaga and kale for 15 min. Immediately after the treatment, wash the seed in cold water and allow it to dry. Proper hot water treatment also helps to eliminate seed-borne pathogens of black leg, anthracnose, downy mildew, Alternaria leaf spot and Fusarium yellows. Another recommendation is soaking of seed in 0.5% cupric acetate dissolved in 0.005 N acetic acid at 40 °C for 20 min (Schaad *et al.*, 1980) or in 10% household bleach at 50 °C for 20 min.
3. Nursery beds should be at least ¼ mile away from the transplanted fields.
4. Three-year rotation should be practised for both nursery beds and transplanted fields.
5. Avoid sprinkler irrigation or use it in moderation.
6. Agricultural equipment and tools should be disinfected with bleach after use.
7. Destroy the plant refuse by burning or by deeply burying in the soil immediately after harvest. High soil moisture hastens the decomposition of plant refuse.
8. Control the insects, which spread the pathogen and destroy the cruciferous weeds, which serve as source of inoculum for the bacterium.
9. Maintain balanced soil fertility in both seedbed and main field.
10. Spray the crop with copper fungicides at ten-day intervals. Sripo-Ngam *et al.* (2024) reported that seed treatment and foliar spray application of high surface area ZnO NPs at 250 μg mL^{-1} significantly reduced the disease severity of black rot.
11. Use of biocontrol agents. Wulff *et al.* (2002) have shown the possibility of managing the disease with

an endophytic strain of *Bacillus subtilis* (strain BB). The control was better in dry and short rainy seasons when compared with the main rainy season and also in clay loam soil, in contrast to sandy loam soil. Liu *et al.* (2015) tested individual plant growth-promoting rhizobacterial (PGPR) strains in greenhouse assays, and mixtures of PGPR strains in greenhouse and field tests for the control of black rot in Chinese cabbage. In the field test, all the treatments significantly reduced the disease incidence at three weeks after transplanting, the head disease severity at the harvest time and increased the marketable yield compared to the non-bacterised control.

12. Grow resistant varieties. Two cabbage cultivars, namely, Early Fuji and Badger, are resistant while the cultivar, White rock, of cauliflower is tolerant to the pathogen.

Stenotrophomonas rhizophila CFBP13503 is a seed-borne commensal bacterial strain, which is efficiently transmitted to seedlings and can outcompete the phytopathogenic bacterium, *Xan. campestris* pv. *campestris* strain 8004. Garin *et al.* (2024) found that Type VI Secretion System of *Stn. rhizophila* CFBP13503 is involved in the antibiosis against *Xan. campestris* pv. *campestris* 8004 and reduces seedling transmission of this bacterium in radish. The distribution of this Type VI Secretion System cluster in the *Stn. rhizophila* complex could make it possible to exploit these strains as biocontrol agents against *Xan. campestris* pv. *campestris*.

REFERENCES

Bella, P., Moretti, C., Licciardello, G., Strano, C. P., Pulvirenti, A. *et al.* (2019). Multilocus sequence typing analysis of Italian *Xanthomonas campestris* pv. *campestris* strains suggests the evolution of local endemic populations of the pathogen and does not correlate with race distribution. *Plant Pathol.* 68: 278–87.

Carisse, O., Wellman-Desbiens, E., Toussaint, V. and Otis, T. (1999). *Preventing Black Rot.* Horticultural Research and Development Centre, Quebec.

Chen, X.-Y., Hu, S.-S., Xu, X.-C., Tang, J.-L. and Tang, D.-J. (2023). RNase D is involved in 5S rRNA degradation and exopolysaccharide production in *Xanthomonas campestris* pv. *campestris*. *Phytopathology*, 113: 1822–32.

Chidamba, L. and Bezuidenhout, C. C. (2012). Characterisation of *Xanthomonas campestris* pv. *campestris* isolates from South Africa using genomic DNA fingerprinting and pathogenicity tests. *Eur. J. Plant Pathol.* 133: 811–18.

Cruz, J., Tenreiro, R. and Cruz, L. (2017). Assessment of diversity of *Xanthomonas campestris* pathovars affecting cruciferous plants in Portugal and disclosure of two novel *X. campestris* pv. *campestris* races. *J. Plant Pathol.* 99: 403–14.

Dowson, W. J. (1939). On the systematic position and generic names of the Gram-negative bacterial plant pathogens. *Zentralblatt fur Bakteriologie, Parasitenkunde, Infektionskrankheiten und Hygiene*, No. 9–13, pp. 177–93.

Eichmeier, A., Peňázová, E., Pokluda, R. and Vicente, J. G. (2019). Detection of *Xanthomonas campestris* pv. *campestris* through a real-time PCR assay targeting the *Zur* gene and comparison with detection targeting the *hrpF* gene. *Eur. J. Plant Pathol.* 155: 891–902. https://doi.org/10.1007/s10658-019-01820-0

Fargier, E. and Manceau, C. (2007). Pathogenicity assays restrict the species *Xanthomonas campestris* into three pathovars and reveal nine races within *X. campestris* pv. *campestris*. *Plant Pathol.* 56: 805–18.

Garin, T., Brin, C., Préveaux, A., Brault, A., Briand, M. *et al.* (2024). The type VI secretion system of *Stenotrophomonas rhizophila* CFBP13503 limits the transmission of *Xanthomonas campestris* pv. *campestris* 8004 from radish seeds to seedlings. *Mol. Plant Pathol.* 25: e13412.

Harding, H. A., Stewart, F. C. and Prucha, M. J. (1904). Vitality of the cabbage black rot germ on cabbage seed. *N. Y. Agr. Exp. Sta. Bull.* 251: 177–194.

Krauthausen, H.-J., Horner, G., Zimmermann, S., Voegele, R. T. and Brandle, F. (2018). Competence of *Xanthomonas campestris* from cruciferous weeds and wallflower (*Erysimum cheiri*) to induce black rot in cabbage. *Eur. J. Plant Pathol.* 151: 275–89.

Laala, S., Bouznad, Z. and Manceau, C. (2015). Development of a new technique to detect living cells of *Xanthomonas campestris* pv. *campestris* in crucifers seeds: The seed-qPCR. *Eur. J. Plant Pathol.* 141: 637–46.

Laala, S., Cesbron, S., Kerkoud, M., Valentini, F., Bouznad, Z. *et al.* (2021). Characterization of *Xanthomonas campestris* pv. *campestris* in Algeria. *Phytopath. Mediterr.* 60: 51–62.

Lema, M., Cartea, M. E., Sotelo, T., Velasco, P. and Soengas, P. (2012). Discrimination of *Xanthomonas campestris* pv. *campestris* races among strains from northwestern Spain by *Brassica* spp. genotypes and rep-PCR. *Eur. J. Pl. Pathol.* 133: 159–69.

Li, R.-F., Peng, J.-L., Liu, Q.-Q., Chang, Z., Huang, Y.-X. *et al.* (2023). *Xanthomonas campestris* VemR enhances the transcription of the T3SS key regulator HrpX via physical interaction with HrpG. *Mol. Plant Pathol.* 24: 232–47.

Li, R.-F., Ren, P.-D., Liu, Q.-Q., Yao, J.-L., Wu, L. *et al.* (2022). McvR, a single domain response regulator regulates motility and virulence in the plant pathogen *Xanthomonas campestris*. *Mol. Plant Pathol.* 23: 649–63.

Lin, Y., Liao, Y.-Y., Huang, R.-X., Li, A.-Z., An, S.-Q. *et al.* (2021). Extracellular amylase is required for full virulence and regulated by the global posttranscriptional regulator RsmA in *Xanthomonas campestris* pathovar *campestris*. *Phytopathology*, 111: 1104–13.

Liu, K., Garrett, C., Kloepper, J. W. and McInroy, J. A. (2015). Induction of systemic resistance in Chinese cabbage against black rot by plant growth-promoting rhizobacteria. *Phytopathology*, 105: S2.7.

Lugo, A. J., Elibox, W., Jones, J. B. and Ramsubhag, A. (2013). Copper resistance in *Xanthomonas campestris* pv. *campestris* affecting crucifers in Trinidad. *Eur. J. Plant Pathol.* 136: 61–70.

McCulloch, L. (1929). A bacterial leaf spot of horseradish caused by *Bacterium campestris* var. *armoraciae*, n. var. *J. Agric. Res.* 38: 269–87.

Mulema, J. M. K., Vicente, J. G., Pink, D. A. C., Jackson. A., Chacha, D. O. *et al.* (2012). Characterization of isolates that cause black rot of crucifers in East Africa. *Eur. J. Plant Pathol.* 133: 427–38.

Nunez, A. M. P., Monteiro, F. P., Pacheco, L. P., Rodriguez, G. A. A., Carvalho, C. *et al.* (2017). Development and

validation of a diagrammatic scale to assess the severity of black rot of crucifers in kale. *J. Phytopathol.* 165: 195–203.
Onsando, J. M. (1992). Black rot of crucifers. In: H. S. Chaube, U. S. Singh, A. N. Mukhopadhyay and J. Kumar (eds.). *Plant Diseases of International Importance*. Vol. II, *Diseases of Vegetables and Oilseed Crops*. Prentice Hall, Englewood Cliffs, NJ, pp. 243–52.
Pammel, L. H. (1895). Bacteriosis of rutabaga. (*Bacillus campestris* n. sp.). *Bull. Iowa State College Agri. Expt. Sta.* 27: 130–34.
Patel, M. K. and Kulkarni, Y. S. (1953). A review of bacterial plant disease investigation of India. *Indian Phytopathol.* 6: 131–40.
Sawada, H., Fujikawa, T. and Horita, H. (2020). *Pseudomonas brassicae* sp. nov., a pathogen causing head rot of broccoli in Japan. *Int. J. Syst. Evol. Micro.* 70: 5319–29.
Schaad, N. W. (1978). Use of direct and indirect immunofluorescence tests for identification of *Xanthomonas campestris*. *Phytopathology* 68: 249–52.
Schaad, N. W. (2001). Black rot of crucifers. In: O. C. Maloy and T. D. Murray (eds.). *Encyclopedia of Plant Pathology*. Vol. 1. John Wiley and Sons, Inc. New York, pp. 145–46.
Schaad, N. W., Gabrielson, R. L. and Mulanax, M. W. (1980). Hot acidified cupric acetate soaks for eradication of *Xanthomonas campestris* from crucifer seeds. *Appl. Environ. Microbiol.* 39: 803–07.
Schaad, N. W. and Kendrick, R. (1975). A qualitative method for detecting *Xanthomonas campestris* in crucifer seed. *Phytopathology* 65: 1034.
Silva, J. C., Silva T. A. F. Jr., Soman, J. M., Tomasini, T. D., Sartori, M. M. P. *et al.* (2017). Survival of *Xanthomonas campestris* pv. *campestris* in the phyllosphere and rhizosphere of weeds. *Plant Pathol.* 66: 1517–26.
Singh, D., Kesharwani, A. K., Singh, K., Jaiswal, S., Iquebal, M. A. *et al.* (2022). Whole-genome sequence resource of Indian race 4 of *Xanthomonas campestris* pv. *campestris*, the causal agent of black rot disease of *Brassica oleracea* var. *capitate*. *Plant Dis.* 106: 1502–05.
Singh, D., Rathaur, P. S. and Vicente, J. G. (2016). Characterization, genetic diversity and distribution of *Xanthomonas campestris* pv. *campestris* races causing black rot disease in cruciferous crops of India. *Plant Pathol.* 65: 1411–18.
Sripo-Ngam, S., Khaengraeng, C., Kasem, S. and Chatnaparat, T. (2024). Antibacterial activity of high surface area zinc oxide nanoparticles for controlling black rot disease of Chinese kale. *Plant Pathol.* 73: 1957–68. https://doi.org/10.1111/ppa.13923
Trivedi, B. M. and Patel, P. N. (1972). Black rot of crucifers. In: P. N. Patel (ed.). *Plant Bacteriology*. Vol. I: *Bacterial Diseases of Plants in India*. Indian Agricultural Research Institute, New Delhi, pp. 122–33.
van der Wolf, J., Kastelein, P., da Silva Júnior, T. A. F., Lelis, F. V. and van der Zouwen, P. (2019). Colonization of siliques and seeds of rapid cycling *Brassica oleracea* plants by *Xanthomonas campestris* pv. *campestris* after spray-inoculation of flower clusters. *Eur. J. Plant Pathol.* 154: 445–61. https://doi.org/10.1007/s10658-019-01668-4
Van Der Wolf, J. M. and Van Der Zouwen, P. S. (2010). Colonization of cauliflower blossom (*Brassica oleracea*) by *Xanthomonas campestris* pv. *campestris*, via flies (*Calliphora vomitoria*) can result in seed infestation. *J. Phytopathol.* 158: 726–32.
van der Wolf, J. M., van der Zouwen, P. S. and van der Heijden, L. (2013). Flower infection of *Brassica oleracea* with *Xanthomonas campestris* pv. *campestris* results in high levels of seed infection. *Eur. J. Plant Pathol.* 136: 103–11.
Vicente, J. G., Conway, J., Roberts, S. J. and Taylor, J. D. (2001). Identification and origin of *Xanthomonas campestris* pv. *campestris* races and related pathovars. *Phytopathology* 91: 492–99.
Vicente, J. G., Everett, B. and Roberts, S. J. (2006). Identification of isolates that cause a leaf spot disease of brassicas as *Xanthomonas campestris* pv. *raphani* and pathogenic and genetic comparison with related pathovars. *Phytopathology* 96: 735–45.
Wang, B., Xu, Z., Zhao, Y., Wu, G., Li, K. *et al.* (2023). SstF, a novel sulforaphane-sensing transcription factor of *Xanthomonas campestris*, is required for sulforaphane tolerance and virulence. *Mol. Plant Pathol.* 24: 452–65.
White, H. E. (1930). Bacterial spot of radish and turnip. *Phytopathology* 20: 653–62.
Williams, P. H. (1980). Black rot: A continuing threat to world crucifers. *Plant Dis.* 64: 736–42.
Wulff, E. G., Mguni, C. M., Mortensen, C. N., Keswani, C. L. and Hockenhull, J. (2002). Biological control of black rot (*Xanthomonas campestris* pv. *campestris*) of brassicas with an antagonistic strain of *Bacillus subtilis* in Zimbabwe. *Eur. J. Plant Pathol.* 108: 317–25.
Zaccardelli, M., Campanile, F., Spasiano, A. and Merighi, M. (2007). Detection and identification of the crucifer pathogen, *Xanthomonas campestris* pv. *campestris*, by PCR amplification of the conserved Hrp/type III secretion system gene *hrcC*. *Eur. J. Plant Pathol.* 118: 299–306.
Zhang, S., Kan, J., Liu, X., Wu, Y., Zhang, M. *et al.* (2023). Phytopathogenic bacteria utilize host glucose as a signal to stimulate virulence through LuxR homologues. *Mol. Plant Pathol.* 24: 359–73.

9.22 BACTERIAL ZONATE SPOT OF CABBAGE

Bacterial zonate spot is primarily a disease of cabbage, although other members of the Cruciferae and Cucurbitaceae are also susceptible. It was first observed on Florida-grown cabbage in the Chicago market, USA, in 1953. Superficially, the lesions resemble those caused by *Alternaria brassicae* on cabbage. Infected areas are firm and pliable but only slightly soft. This is in contrast to the slimy type of decay produced by *Erwinia carotovora*. Lesions may occur on any part of the leaf, but there is no evidence that the disease involves the vascular system.

Spotting of cabbage heads is a disease that occurs every year in the Everglades, USA. The severity of the disease varies from field to field and from year to year. Sometimes only a small percentage of the cabbage heads is affected and the spots are so small in size and in number that the damage is negligible. Other times up to 50% of the cabbage heads are so heavily spotted that they are not fit to be harvested. In the fall and winter of 1962, it was not uncommon to see fields where almost every cabbage head was spotted to some degree. Under favourable conditions for its development, this disease can be as destructive to cabbage as black rot (Wehlburg, 1963).

9.22.1 Host Range

Pseudomonas cichorii is a multihost pathogen and infects a large number of plant species. Some of these hosts, along with their respective diseases, are listed below: bacterial spot of celery (Alippi, 1996), bacterial stem rot in greenhouse grown lettuce (Dhanvantari, 1990), bacterial rot of witloof chicory (Duveiller, 1986), bacterial leaf spot of cucumber (Fu *et al.*, 2019), bacterial blight of endive (Kotte, 1930), leaf spot of basil (Holcomb and Cox, 1998), bacterial blight of okra (Inou *et al.*, 2013), bacterial blight of gerbera (Marques *et al.*, 2016), bacterial leaf spot of witloof (Myung *et al.*, 2013), leaf spot of pumpkin (Newberry *et al.*, 2016), bacterial leaf spot of sweet basil (Patel *et al.*, 2019), stem melanosis of spring wheat (Piening and MacPherson, 1985), pith necrosis of tomato (Ruan *et al.*, 2019), bacterial leaf blight of *Plumeria pudica* (Sugiyama *et al.*, 2018), leaf spot of Rhododendron (Uddin and McCarter, 1996) and leaf spot of soybean (Yu and Lee, 2012). Bolick (1960) and McFadden (1961) described it as a disease of ornamental chrysanthemums.

9.22.2 Symptoms

The symptoms of bacterial zonate spot found in the field are as given below. On the cabbage heads, greyish-brown to dark brown, slightly sunken, round or oval spots appear, measuring 1 to 5 mm in diameter. If the spots are located on one of the big veins, they become elongated and sometimes irregular in shape. The spots may coalesce to form large areas of affected tissue. Sometimes the spots display concentric rings of light-coloured zones alternating with brown lines. These target-like symptoms have been described by Ramsey and Smith (1961) under the name of bacterial zonate spot of cabbage.

These target spots are larger than the plain spots and measure up to 15 mm in diameter. They may also coalesce and form still larger spots of irregular form. Furthermore, the affected wrapper leaf sometimes cracks and this offers the pathogen a convenient infection court for the invasion of new areas, so that often more than half of the top leaf is affected. Unless soft-rot bacteria become involved, the affected tissue remains firm and does not smell. Sometimes along the margins of partly expanded leaves of the cabbage head, dark brown to almost black, triangular spots, up to 5 mm in length, are formed.

When a wrapper leaf that is only lightly affected is peeled off, the underlying leaf generally does not display any spots at all. The lesions are often not even visible on the reverse side of the affected wrapper leaf. Unless extremely favourable conditions prevail for the development of the pathogen or when wounds or other mechanical damage permits the invasion of deeper layers, the lesions remain limited to the outside (or wrapper) leaf of the cabbage head. However, there are cases where the lesions can be found on the inner leaves of the cabbage head, sometimes as deep as the ninth leaf, counting from the wrapper leaf down. The spots on the inner leaves are irregular in form and almost black. The younger the leaf the fewer spots, until the last affected leaf shows only one or two lesions. On the younger affected inner leaves, the spots are often localised near or on the leaf margin. The spots on the inner leaves do not correspond with those on the leaf immediately above it.

The infection usually remains shallow, and often does not penetrate as deep as the thickness of the leaf. This is an indication that the pathogen grew systemically in those cases where several leaves inside the cabbage head are affected. Cross-sections through the stem of cabbage plants with systemic infections in most cases showed darkened vascular bundles. Microscopic examinations revealed bacteria in the xylem vessels and in the surrounding parenchyma cells.

The symptoms of bacterial spot of cabbage have been confused with those of downy mildew for a long time. This explains why fungicides that are able to control downy mildew, as it appears on young plants, do not prevent the cabbage heads from becoming spotted. Downy mildew lesions can be distinguished from bacterial spot by their irregular shape and greyish colour, whereas the spots caused by *Psm. cichorii* are usually round or oval and dark brown. However, the differences between the symptoms of the two diseases are not always clear enough, to identify with certainty the cause of the disease. In the Everglades, downy mildew is a disease that is generally serious only on young plants; on older plants, it may be present on the lower leaves, but this is rather the exception. In the majority of the cases, dark brown spots on the cabbage heads are bacterial spot and not downy mildew (Figure 9.21).

9.22.3 Causal Organism

Pseudomonas cichorii (Swingle 1925) Stapp 1928

Synonym: *Chlorobacter cichorii* (Swingle 1925) Patel and Kulkarni 1951, *Psm. papaveris* Luketina and Young 1979

Cells are rods, single or in short chains, motile by one or two polar flagella, Gram-stain-negative, not acid fast. On nutrient agar, the colonies are raised, glistening, circular with slightly irregular margin, greyish white, opalescent and a delicate light brown with almost colourless transparent

FIGURE 9.21 Bacterial zonate spot of cabbage: Greyish-brown to dark brown spots on cabbage heads.

margin, when viewed under transmitted light. On beef extract agar, growth is slow, and the colonies measure less than 1 mm in 6 days. Green fluorescent pigment is abundantly produced on proteose peptone medium. Nutrient broth turns very turbid, with a viscous white pellicle and abundant white sediment. Milk is turned alkaline, although with some isolates it remains neutral; no curd is produced. Some isolates reduce nitrate strongly, some feebly and some not at all. Ammonia is produced, but not hydrogen sulphide or indole. Starch is not hydrolysed, gelatin is not liquefied and corn seed oil is not decomposed. Acid is produced, but not gas from dextrose, levulose and mannitol; sucrose, maltose and lactose are not fermented.

These data agree closely with the description that Clara (1934) gave for *Phytomonas endiviae* (Kotte) Clara, and is identical with the organism that was originally described by Swingle (1925) under the name of *Phytomonas cichorii* (sic). Stapp (1928) transferred it to the genus *Pseudomonas* as *Psm. cichorii* (Swingle) Stapp.

9.22.4 Disease Cycle

Psm. cichorii is seed-borne. It also survives in diseased plant debris. Short distance spread occurs through the wind-splashed rain. Long-distance spread and spread to the new areas generally occurs through the transporting of infected seeds. Disease development is favoured by high humidity and warm temperature.

Psm. cichorii causes rot on lettuce leaves, distinct from the necrotic spots of infected eggplant leaves. Hikichi *et al.* (2013) worked out the mechanism of disease development of *Psm. cichorii* on lettuce leaves. On lettuce leaves, *Psm. cichorii* invades intercellular spaces through stomata and grows vigorously, causing rot on the leaves. Surprisingly, *Psm. cichorii* does not produce pectate lyase, the most important enzyme for degrading plant cell walls. Alternatively, infection with *Psm. cichorii* causes heterochromatin condensation and DNA laddering in lettuce cells, followed by induced cell death, which results in disease symptoms. Thus, apoptotic programmed cell death (PCD) is associated with the symptoms on lettuce leaves. Although PCD in *Psm. cichorii*-infected eggplant leaves is also associated with necrotic spots caused by the bacteria, there are differences in the responses of the two hosts. *Psm. cichorii* harbours the *N*-acetyltransferase family gene (*pat*) and the *hrp* genes (*hrp*) encoding a type III secretion system, adjacent to an aldehyde dehydrogenase gene (*aldH*). The virulence of *Psm. cichorii* is *hrp*-dependent in eggplant, but not in lettuce. Moreover, *hrp*, *pat* and *aldH* are implicated in the diversity of *Psm. cichorii* virulence on susceptible Asteraceae species. The involvement of *hrp*, *aldH* and *pat* in bacterial virulence on the respective species has no relationship with the phylogeny of the plant species. These results show that *Psm. cichorii* has multiple virulence determinants. The involvement of not only *hrp* but also *aldH* and *pat* in *Psm. cichorii* virulence arose after the species diversification of host plants. Host responses implicated in symptom development have been responsible for the development of virulence determinants of *Psm. cichorii*.

9.22.5 Management

Control measures listed by Kabeere (2010) are given below:

1. Use the pathogen-free seed for planting and make sure your seed is from a clean source.
2. Tolerant varieties should be grown where possible.
3. Do not to grow cabbages on the same land year after year, even if the crops were healthy.
4. Rouge out the infected plants carefully.
5. Infected plant material provides a large amount of inoculum to infect the next crop, so try to reduce the amount of infected material being returned to the soil.
6. Mulching the crop will reduce rain water splashes to the plants, thereby reducing the spread of the pathogen.
7. Avoiding weeding during damp conditions and accidental damage to the plants will avoid the bacteria gaining entry into the plants.
8. Use bleach or fire on the weeding tools regularly in order to kill any bacterial cells present on them.
9. Use copper as a dust or as a spray as soon as possible after the plants are thinned to a stand. If the copper is used, other fungicides for the control of downy mildew are no longer required.

REFERENCES

Alippi, A. M. (1996). First report of bacterial spot of celery caused by *Pseudomonas cichorii* in Argentina. *Plant Dis.* 80: 599.

Bolick, J. H. (1960). Bacterial bud blight of chrysanthemum. *Proc. Florida State Hort. Soc.* 73: 346–51.

Clara, F. M. (1934). A comparative study of the green-fluorescent bacterial pathogens. Cornell University Agricultural Experiment Station Memoirs, 159.

Dhanvantari, B. N. (1990). Occurrence of bacterial stem rot caused by *Pseudomonas cichorii* in greenhouse-grown lettuce in Ontario. *Plant Dis.* 74: 394.

Duveiller, E. (1986). Bacterial rot of witloof chicory caused by *Pseudomonas cichorii* in Burundi. *Parasitica* 42: 55–57.

Fu, X. Y., Zhang, R. Y., Tan, Z. Q., Liu, T. and Peng, Z. Q. (2019). First report of bacterial leaf spot of cucumber caused by *Pseudomonas cichorii* in China. *Plant Dis.* 103: 147–48.

Hikichi, Y., Wali, U. M., Ohnishi, K. and Kiba, A. (2013). Mechanism of disease development caused by a multihost plant bacterium, *Pseudomonas cichorii*, and its virulence diversity. *J. Gen. Plant Pathol.* 79: 379–89

Holcomb, G. and Cox, P. J. (1998). First report of basil leaf spot caused by *Pseudomonas cichorii* in Louisiana and cultivar screening results. *Plant Dis.* 82: 1283.

Inou, T., Kajihara, H., Muramoto, K., Yoshioka, R. and Sawada, H. (2013). Fruit rot, a new symptom of okra bacterial blight caused by a *Pseudomonas chicorii*. *Jap. J. Phytopath.* 79: 99–104.

Kabeere, F. (2010). Cabbage bacterial leaf spot. Factsheets for Farmers. www.plantwise.org.

Kotte, W. (1930). Eine bakterielle Blattfaule der Winter-Endivie (*Cichorium endiviae* L). *Phytopath. Zeitschr.* 1: 605–13.

Luketina, R. C. and Young, J. M. (1979). Nomenclatural status of *Pseudomonas oryzicola* Klement 1955, *Pseudomonas papaveris* Lelliott and Wallace 1955, and *Pseudomonas syringae* var. *capsici* (Orsini 1942) Klement 1956. *N. Z. J. Agric. Res.* 22: 349–53.

Marques, E., Borges, R. C. and Uesugi, C. H. (2016). Identification and pathogenicity of *Pseudomonas cichorii* associated with a bacterial blight of gerbera in the Federal District. *Horticultura Brasileira*, 34: 244–48. https://doi.org/10.1590/S0102-053620160000200015.

McFadden, L. A. (1961). A bacterial leaf spot of florists' chrysanthemums (*Chrysanthemum morifolium*). *Plant Dis. Reptr*, 45: 16–19.

Myung, I. S., Choi, J. K., Lee, J. Y., Yoon, M. J., Hwang, E. Y. *et al.* (2013). First report of bacterial leaf spot of witloof, caused by *Pseudomonas cichorii* in Korea. *Plant Dis.* 97: 1376.

Newberry, E. A., Paret, M. L., Jones, J. B. and Bost, S. C. (2016). First report of leaf spot of pumpkin caused by *Pseudomonas cichorii* in Tennessee. *Plant Dis.* 100: 2159–60.

Patel, M. K. and Kulkarni, Y. S. (1951). Nomenclature of bacterial plant pathogens. *Indian Phytopath.* 4: 74–84.

Patel, N., Kobayashi, D. Y., Noto, A. J., Baldwin, A. C., Simon, J. E. *et al.* (2019). First Report of *Pseudomonas cichorii* causing bacterial leaf spot on sweet basil (*Ocimum basilicum*) in New Jersey. *Plant Dis.* 103: 2666. https://doi.org/10.1094/PDIS-04-19-0895-PDN.

Piening, L. J. and MacPherson, D. J. (1985). Stem melanosis, a disease of spring wheat caused by *Pseudomonas cichorii*. *Can. J. Plant Pathol.* 7: 168–72.

Ramsey, G. B. and Smith, M. A. (1961). Bacterial zonate spot. Market diseases of cabbage, cauliflower, turnips, cucumbers, melons, and related crops. *Agriculture Handbook* No. 184. 78 pp.

Ruan, H., Shi, N., Du, Y., Chen, F., Yang, X. *et al.* (2019). First report of *Pseudomonas cichorii* causing tomato pith necrosis in Fujian Province, China. *Plant Dis.* 103: 145.

Stapp, C. (1928). Schizomycetes (Spaltpilze oder Bakterien). In: Sorauer (ed.). *Handbuch der Pflanzenkrankheiten*. 5th Ed., Vol. 2. Paul Parey, Berlin, pp. 1–295.

Sugiyama, L. S., Bushe, B. C., Heller, W. P. and Keith, L. M. (2018). First report of *Pseudomonas cichorii* causing bacterial leaf blight of *Plumeria pudica* in Hawaii. *Plant Dis.* 102: 1025. https://doi.org/10.1094/PDIS-11-17-1771-PDN.

Swingle, D. B. (1925). Center rot of French endive or wilt of Chicory (*Cichorium intybus* L). *Phytopathology* 15: 730.

Uddin, W. and McCarter, S. M. (1996). First report of Rhododendron leaf spot caused by *Pseudomonas cichorii*. *Plant Dis.* 80: 960.

Wehlburg, C. (1963). A bacterial spot of cabbage caused by *Pseudomonas cichorii*. *Proc. Florida State Hortic. Soc.* pp. 119–22.

Yu, S.-M. and Lee, Y. H. (2012). First report of *Pseudomonas cichorii* associated with leaf spot on soybean in South Korea. *Plant Dis.* 96: 142.

BACTERIAL DISEASES OF TOMATO

Many bacterial diseases occur on tomato (*Solanum lycopersicum* L.). Among them, the important ones are bacterial spot, which also occurs on pepper (*Capsicum annuum* L.), bacterial speck, bacterial canker and wilt, pith necrosis and bacterial wilt. The first three diseases are described below while the bacterial wilt is covered under the description of bacterial wilt of solanaceous plants.

9.23 BACTERIAL SPOT OF TOMATO AND PEPPER

Bacterial spot of tomato and pepper is a serious disease of both crops. Localised epiphytotics occur in areas where warm and moist conditions prevail during the crop season. It is a major problem in tropical and subtropical regions. Especially under conditions of overhead irrigation, the disease can cause enormous losses. The disease is much less damaging in greenhouses. Crop losses occur due to defoliation, which results in yield reductions and the severe spotting of fruits. This leads to a reduction in their market value, sometimes even making the fruits unsalable (Jones, 2001).

Bacterial leaf spot of chillies was first observed in the USA in 1912; it was first observed on tomatoes in 1917. The first detailed account of bacterial spot of tomato was published by Doidge in 1920; in this she named the causal bacterium as *Bacterium vesicatorium*. The specific name of the causal agent was derived in part from the development of vesicles in host cells during pathogenesis. In 1923, Gardner and Kendric showed that the same bacterium also caused leaf spot of pepper. At approximately the same time, Higgins described bacterial spot of pepper in detail. Since then, the disease has been reported in many countries and regions, including Canada, Central and South America, China, Australia, Austria, Bulgaria, France, Germany, Niger, Nigeria, the Philippines, Poland, Romania, Russia, Senegal, Spain, Sudan, Taiwan, Tanzania, Thailand, Turkey, Zimbabwe, Zambia, Israel, Hungary, New Zealand, Italy, the Caribbean, Yugoslavia and Southeast Asia. *Xanthomonas gardneri* (group D strains) is predominant in Ontario, Canada. Due to its efficient overwintering under the prevalent conditions in southern Ontario, it is considered to be an aggressive pathogen. In 2000, the bacterium was responsible for the epiphytotic of the disease in Ontario processing tomato fields. The occurrence of this species has also been reported from Brazil. In India, bacterial spot was first observed on pepper from Pune in 1948 by Patel *et al.* (1950). It is now present in most of the country's tomato- and pepper-growing areas.

9.23.1 Host Range

In addition to tomato and pepper, the bacterial spot pathogens infect chilli pepper (*C. frutescens*), currant tomato (*S. pimpinellifolium*), Aztec tobacco (*Nicotiana rustica*), *Datura tatula*, buffalo bur (*S. rostratum*), nightshade (*S. douglassii*), black nightshade (*S. nigrum*) and bittersweet or climbing nightshade (*S. dulcamara*). The survival of pathogens on these hosts may be an important source of the primary inoculum in specific localities.

Araújo *et al.* (2015) reported that plants of *Nicandra physaloides*, *S. americanum* and *Euphorbia heterophylla* showing leaf lesions were found to be growing naturally among tomato plants infected with bacterial spot in commercial tomato fields in Brazil. Isolates from *N. physaloides* and *S. americanum* were identified as *Xan. perforans* and those from *E. heterophylla* as *Xan. gardneri*. Each of these was able to artificially infect and cause symptoms on the three weed species and tomato plants.

Bophela *et al.* (2019) reported a leaf disease of *Eucalyptus pellita* caused by *Xan. perforans* from Indonesia; this bacterium also induced bacterial spot symptoms on tomato and pepper seedlings.

9.23.2 Symptoms

The symptoms appear on all the above-ground plant parts of tomato and pepper. On tomato leaves, small (up to 3 mm in diameter), irregular and dark brown to black, greasy lesions appear. After a few days, the lesions are often surrounded by yellow haloes. Centres of the lesions dry out and frequently tear. Lesions coalesce to form big irregular dead areas, which give a scorched appearance. On mature plants, more lesions are found on fully-expanded and older leaves. Leaves having numerous lesions turn yellow and may defoliate (Figure 9.22a). Defoliation begins from the lower leaves and it is generally more common on pepper plants. The shot-hole symptoms caused on tomato leaves are usually associated with *Xan. perforans*. Large black lesions are formed on stem and petioles. The infection of flower parts, particularly on pedicel, generally results in severe blossom drop/blossom abortion, resulting in yield loss and splitting of fruits.

On green tomato fruits, small, water-soaked, slightly raised spots appear. The spots increase in size up to 4–6 mm in diameter and become dark brown, scabby and slightly sunken. Initially, the spots may be surrounded by greenish-white haloes, which eventually disappear (Figure 9.22b). The symptoms do not appear on ripe fruits.

On pepper, the spots on leaves, petioles, stem and fruits are distinctly raised. Under conditions of high rainfall and high temperature, the lesions on pepper may take the form of blight.

The symptoms of bacterial spot and bacterial speck on tomato are similar, except on fruits, and hence they are often confused. Symptoms of bacterial speck on green fruits appear as black, slightly raised stippling, which causes lesions less than 1.5 mm in diameter.

9.23.3 Causal Organisms

Xanthomonas euvesicatoria Jones *et al.* 2004

Xanthomonas gardneri (ex Sutic 1957) Jones *et al.* 2004

Xanthomonas perforans Jones *et al.* 2004

Xanthomonas vesicatoria (ex Doidge 1920) Vauterin *et al.* 1995

Synonym: *Xa. campestris* pv. *vesicatoria*

The taxonomy of the bacteria causing bacterial spot of tomato and pepper has undergone many changes. In 1921, Ethel Doidge identified it as *Bacterium vesicatorium*. It was reclassified as *Pseudomonas vesicatoria* in 1925, as *Phytomonas vesicatoria* in 1930 and as *Xanthomonas vesicatoria* in 1939. With the introduction of the pathovar system in 1978, it was changed to *Xan. campestris* pv. *vesicatoria*. Vauterin *et al.* (1995) subdivided the bacterium into two species, placing the strongly amylolytic strains in *Xan. vesicatoria* and the non-amylolytic strains in *Xan. axonopodis* pv. *vesicatoria*. Jones *et al.* (2000) found that xanthomonads pathogenic to tomato, pepper or both consist of at least four phenotypic groups, namely, A, B, C and D. Later on, Jones *et al.* (2004) renamed group A strains as *Xan. euvesicatoria*, group B strains as *Xan. vesicatoria*, group C strains as *Xan. perforans* sp. nov. and group D strains as *Xan. gardneri* on the basis of DNA–DNA hybridisation

(a) (b)

FIGURE 9.22 (a) Bacterial spot of tomato: Symptoms on foliage. (Source: *Illinois Fruit and Vegetable News*, Vol. 13, No. 8.) (b) Bacterial spot of tomato: Symptoms on fruits. (Courtesy: M. Babadoost, *Illinois Fruit and Vegetable News*, Vol. 13, No. 8.)

data. Group A strains reclassified here as *Xan. euvesicatoria* represent the strains originally identified by Doidge in 1921 as *Bac. vesicatorium* referring to them as feebly amylolytic. The specific epithet "*gardneri*" was proposed by Sutic in 1957, who identified a bacterial disease of tomato from Yugoslavia. However, this bacterium has not been a major plant pathogen, except in Ontario, Canada, since its discovery. The strains of *Xan. vesicatoria* and *Xan. euvesicatoria* are widely distributed compared to those of other two species (Jones *et al.*, 1998).

Xan. vesicatoria and *Xan. euvesicatoria* differ from each other in the utilisation of cis-aconitic acid, reaction to monoclonal antibodies, pectolytic and amylolytic activity, whole-cell protein profiles, rep-PCR genomic fingerprinting and reaction on differential hosts (races).

The bacteria are Gram-stain-negative rods, measure 0.7–1.0 × 2.0–2.4 μm, motile by a single polar flagellum and strictly aerobic. These grow slowly on nutrient agar-forming colonies, which are circular, shinning, wet, entire and Naples yellow. Xanthomonadin, a water-insoluble yellow pigment, is produced.

A 420-bp fragment of the *hrpB7* gene was amplified by PCR from 75 strains representing *Xan. euvesicatoria*, *Xan. vesicatoria*, *Xan. perforans* and *Xan. gardneri*; the PCR products were sequenced and phylogenetic analysis revealed that *hrpB7* is highly conserved within each species, with a single nucleotide polymorphism among the *Xan. vesicatoria* strains. Four probes and two primer sets were designed to target the four bacterial spot pathogens based on their *hrpB7* gene sequences. For the simultaneous detection of four bacterial spot pathogens, four probes and two primer sets were optimised for a multiplex real-time TaqMan PCR assay and the optimised multiplex assay proved to be highly specific to the four bacterial spot pathogens. As the optimised multiplex assay facilitated the identification of each bacterial spot pathogen from pure cultures and infected plant tissue, it holds great potential as a diagnostic tool (Strayer *et al.* 2016a).

Thieme *et al.* (2005) presented the whole-genome sequence of the pepper-pathogenic *Xan. campestris* pv. *vesicatoria* strain 85-10, which comprises a 5.17-Mb circular chromosome and four plasmids. The genome has a high G+C content (64.75%) and signatures of extensive genome plasticity. Whole-genome comparisons revealed a gene order similar to both *Xan. axonopodis* pv. *citri* and *Xan campestris* pv. *campestris* and a structure completely different from *Xan. oryzae* pv. *oryzae*. A total of 548 coding sequences (12.2%) are unique to *Xan. campestris* pv. *vesicatoria*. In addition to a type III secretion system, which is essential for pathogenicity, the genome of strain 85-10 encodes all other types of protein secretion systems described to date in Gram-negative bacteria. Remarkably, one of the putative type IV secretion systems encoded on the largest plasmid is similar to the Icm/Dot systems of the human pathogens, *Legionella pneumophila* and *Coxiella burnetii*. Comparisons with other completely sequenced plant pathogens predicted six novel type III effector proteins and several other virulence factors, including adhesins, cell wall-degrading enzymes and extracellular polysaccharides.

Osdaghi *et al.* (2021) have listed the following three species as causal agents of the disease.

1. *Xanthomonas euvesicatoria*
2. *Xanthomonas hortorum*
3. *Xanthomonas vesicatoria*

Fan *et al.* (2023) reported *Xan. euvesicatoria* pv. *perforans* as the causative agent of bacterial leaf spot on *Ipomoea aquatica* from Fujian Province in China.

Due to the continuous use of copper-containing bactericides without effective alternative bactericides, copper resistance has become more prevalent in plant pathogens, including *Xanthomonas euvesicatoria* pv. *perforans* (formerly *Xan. perforans*). Previously, reports of copper resistance have been associated with a large conjugative plasmid. However, Bibi *et al.* (2023) characterised a copper resistance genomic island located within the chromosome of multiple *Xan. euvesicatoria* pv. *perforans* strains. The island is distinct from a previously described chromosomally encoded copper resistance island in *Xan. vesicatoria* strain XVP26. Computational analysis revealed the genomic island to contain multiple genes associated with genetic mobility, including both phage-related genes and transposase. Among copper-tolerant strains of *Xan. euvesicatoria* pv. *perforans* isolated from Florida, the majority of strains were found to have the copper resistance chromosomally encoded rather than plasmid-borne. These results suggest that this copper resistance island may have two modes of horizontal gene transfer and that chromosomally encoded copper resistance genes may provide a fitness advantage over plasmid-borne resistance.

9.23.3.1 Races

Stall *et al.* (2009) have given an excellent account of races of bacterial spot pathogens that infect pepper and tomato, resistance genes and their corresponding avirulence genes involved in imparting resistance to both the hosts and the occurrence of hypersensitive and quantitative forms of resistance. The authors also emphasised the need to intensify the research on quantitative or multigenic resistance and to transfer this resistance to commercial cultivars of tomato and pepper.

Pathogenic variation in bacterial spot pathogens is clearly established. Some strains are pathogenic only to tomato or pepper, while others infect both hosts. Both hypersensitive and quantitative forms of resistance to bacterial spot pathogens occur in tomato and pepper. To date, five resistance genes involved in hypersensitive reaction (HR) in pepper and four in tomato have been identified. The evolution of resistance in tomato and pepper to bacterial spot pathogens has been slow to develop. The genes in tomato and pepper that impart resistance against these pathogens are listed in Table 9.23a. The origin of these genes and their corresponding avirulence genes present in the bacteria are also given.

The occurrence of pathogenic races in bacterial spot pathogens is known. Some of these races attack both the hosts. There are 11 races (0 to 10), which attack pepper. The differential reaction of these races to four resistance genes is given in Table 9.23b. The resistance of plants of accession PI 163192 of pepper was the result of an HR to the bacterial spot pathogen. This was the first report of an HR occurring in resistant host plants to a bacterial plant pathogen. Initially, two races attacking pepper were identified. The bacterium pathogenic on plants (accession PI 163192) having resistance gene *Bs1* was designated as race 1, while the prevalent bacteria in the field that caused bacterial leaf spot were designated as race 2. Later on, an avirulence gene cloned from race 2 caused race 1 to be hypersensitive in pepper leaves containing the *Bs1* gene. This finding strongly supported the **gene-for-gene** hypothesis of disease resistance established by Flor in 1955 as a resistance gene *Bs1* in the host (cv. Early Calwonder 10R) and an avirulence gene (*avrBs1*) in the bacterium was necessary for resistance to bacterial leaf spot. Subsequently, *Bs2*, *Bs3* and *Bs4* resistance genes were identified. Each of these single dominant genes, i.e., *Bs1* to *Bs4*, can be overcome individually by specific races of the pathogens in the field-grown plants. To improve the level of resistance in the host, three genes, namely, *Bs1*, *Bs2* and *Bs3*, from *C. annuum* were transferred to a single plant of Early Calwonder (ECW) and designated as ECW 123. Each gene was inherited independently of the others. The combination of all the three resistance genes in a plant provided a level of disease less than on the universally susceptible control to strains of race 6, which is pathogenic on plants with all three genes. The combination of *Bs1* and *Bs2* provided some control against race 6, while the combination of *Bs1* and *Bs3* was ineffective. None of the combinations provided a satisfactory disease control.

A Hawaiian tomato accession (H 7998) showed a highly resistant reaction to the bacterial spot pathogen. Resistant plants exhibited an HR, but the resistance was not inherited as a single dominant gene as in most other hypersensitive resistances. Three genes were involved in this case; these genes were designated as *rx1*, *rx2* and *rx3*. The inoculation of pathogen strains from Brazil on H 7998 plants produced two types of reactions. Some strains produced the susceptible reaction (these strains were designated as race T2) and the strains giving HR on H 7998 were designated as race T1 (Table 9.23c). T2 strains were abundant in Brazil, Argentina, Asia and the Midwestern states of United States, but were never found in Florida. As T2 strains had many characteristics different from those of T1 strains, these were identified as *Xan. vesicatoria*. Race T1 remained the sole pathogen of bacterial spot of tomato in Florida for at least 30 years. In 1991, H 7998 showed susceptible reaction to the bacterial spot pathogen. This bacterium showed pectolytic and amylolytic activity and was similar to previously discovered T2 strains. However, unlike the T2 strains, it caused HR in many PI accessions of *S. pimpinellifolium*. Hence, this strain was designated as race T3. A near-isogenic line containing *Xv3* was developed by backcrossing plants with the resistance gene from PI 128216 into FL 7060 seven times. This line was designated FL 216 and is useful in determining races of the tomato bacterial spot pathogens, along with H 7998. In 1991, T3 strains comprised only a small percentage of the bacterial spot pathogen population in Florida. It gradually eliminated the T1 race in Florida; in 2006, no culture of T1 was isolated. The elimination of T1 strains by T3 strains was thought to be the result of an antagonism of T 3 strains to T1 strains, caused by bacteriocin production among the T3 strains (Tudor-Nelson *et al.*, 2003). At least three bacteriocins are produced by T3 strains, and at least one of them is antagonistic to all the T1 strains tested. The *avr* gene associated with T3 hypersensitivity was cloned and designated as *avrXv3*. Tomato strains with this *avr* gene also initiate hypersensitivity in pepper plants. Due to the genetic differences of race T3 strains from the other strains of xanthomonads that cause bacterial spot of tomato, T3 strains were identified as *Xan. perforans*.

TABLE 9.23A

Resistances in Tomato and Pepper and Bacterial Effectors that Interact with Them

Resistance gene	Plant		Bacterium	
	Source	Species	Effector	Location
Pepper				
Bs1	PI 163192	*Capsicum annuum*	AvrBs1	Plasmid
Bs2	PI 260435	*C. chacoense*	AvrBs2	Chromosome
Bs3	PI 271322	*C. annuum*	AvrBs3	Plasmid
Bs4	PI 235047	*C. pubescens*	AvrBs4	Plasmid
	PI 163192 or	*C. annuum*	ND	ND
bs5	PI 271322	*C. annuum*	ND	ND
	PI 163192 or	*C. annuum*	ND	ND
bs6	PI 271322	*C. annuum*	ND	ND
BsT	Commercial pepper	*C. annuum*	AvrBsT	Plasmid
Tomato				
rx1, *rx2*, *rx3*	Hawaii 7988	*Solanum lycopersicum*	AvrRxv	Chromosome
Xv3	PI 128216 & Hawaii 7981	*S. pimpinellifolium* & *S. lycopersicum*	AvrXv3	Chromosome
Xv4	716	*S. pennellii*	AvrXv4	Chromosome
Bs4	Commercial tomato	*S. lycopersicum*	AvrBs4	Plasmid

Source: Stall *et al.* (2009).
Note: ND = Not determined.

TABLE 9.23B
Differential Reaction of Races of Xanthomonads to Pepper Resistance Genes

	Resistance Gene[a]			
Pathogen Race	***Bs1***	***Bs2***	***Bs3***	***Bs4***
0	HR[b]	HR	HR	HR
1	Sus	HR	HR	HR
2	HR	HR	Sus	Sus
3	Sus	HR	Sus	HR
4	Sus	Sus	HR	HR
5	HR	Sus	Sus	Sus
6	Sus	Sus	Sus	HR
7	Sus	HR	HR	Sus
8	Sus	HR	Sus	Sus
9	Sus	Sus	HR	Sus
10	Sus	Sus	Sus	Sus

Source: Stall *et al.* (2009).

[a] Sources of resistance genes are listed in Table 9.23a.

[b] HR refers to the hypersensitive reaction which is involved in resistance, Sus refers to susceptible reaction.

TABLE 9.23C
Differential Reactions of Races of Xanthomonads on Tomato Resistance Genes

	Resistance Gene[b]		
Pathogen Race[a]	***rx1, rx2, rx3***	***Xv3***	***Xv4***
T1	HR[c]	Sus	Sus
T2	Sus	Sus	Sus
T3	Sus	HR	HR
T4	Sus	Sus	HR

Source: Stall *et al*., (2009).

[a] Letter T before each race indicates the reaction of races from tomato plants.

[b] Sources of resistance genes are listed in Table 9.23a.

[c] HR refers to the hypersensitive reaction which is involved in resistance, Sus refers to susceptible reaction.

Strains of *Xan. perforans* were found in Florida which proved pathogenic on plants having gene *Xv3*. These strains were designated as race T4. The cultivation of grape tomatoes in Florida apparently provided the selection pressure to cause the development of T4 strains or their increase over T3 strains. In a survey of races on tomato in 2006 in Florida, approximately 77% of the strains collected were of the T4 race; the rest were of T3 race. The *avr* gene in race T4 that incites HR in plants having *Xv4* gene was cloned and designated as *avrXv4*. Both the T3 and the T4 races contain *avrXv4*. Probably, race T4 originated by the mutation of *avrXv3* in race T3 because race T3 contains both *avrXv3* and *avrXv4* genes, and T3 and T4 are genetically similar. The durability of a resistance gene based on HR in a plant depends on the stability of the *avr* gene in the pathogen. Many mechanisms operate in bacterial spot pathogens that result in the instability of *avr* genes. Many *avr* genes of bacterial spot pathogens are contained in self-transmissible plasmids. The loss of such a plasmid can result in the loss of an *avr* gene. Another mechanism of inactivation of *avr* genes is the addition of an insertion element into the *avr* gene coding sequence. The total genomic sequencing of one strain of the bacterial spot pathogen revealed 58 insertion elements in the chromosome and another eight on four plasmids of the bacterium. One of these elements was *IS476*, which is associated with the inactivation of *avrBs1* in race 2 of pepper strains. The frequency of these changes affects race composition among pathogen populations and determines the durability of the corresponding plant resistance.

Adhikari *et al.* (2019) have given a detailed account of phenotypic and genetic diversity of *Xan. perforans* populations from tomato in North Carolina, suggesting that bacterial spot management practices in tomato should focus on deploying resistance genes to combat emerging pathogenic races of *Xan. perforans* and overcome the challenges currently posed by intense use of copper-based bactericides.

9.23.4 Disease Cycle

The causal bacteria survive in diseased plant debris, volunteer plants and as epiphytes on weed species. Epiphytic survival on weeds seems to be of minimal importance as a primary source of inoculum. These bacteria have also been found to persist as latent infections on host leaves and, furthermore, to survive there under unfavourable conditions. This is important for the spread of these pathogens on transplants that may not show visible symptoms. Thr role of plant debris in the survival of bacteria is more important in temperate regions where the crop is raised in the summer months and the debris does not decompose readily due to low temperature. The bacteria also survive on seed as contaminants and the seed contamination occurs during the seed extraction process.

Local (or short-distance) spread of the bacteria occurs through wind-splashed rain, rain splashes and overhead irrigation. The dissemination to distant places and new areas occurs through infested seed and diseased transplants.

The bacteria penetrate into the host through stomata and wounds caused by wind-blown sand, insects, mechanical means and the breaking of hairs. The entry of the bacteria into the fruits occurs through cracks in the epidermis and through fruit hairs, which are present only on immature fruits.

Disease development is favoured by high relative humidity and overhead irrigation. The optimum temperature for disease development is 24–30° C. The bacteria may occur in low populations as epiphytes on the surface of host plants. At the onset of favourable conditions, these low populations increase rapidly and cause infection. Marcuzzo *et al.* (2009) made the following observations on the epidemiological aspects of bacterial spot of tomato in Cacador, Brazil. Epiphytotic onset of the disease coincided with physiological maturation of the first fruit clusters. Bacterial leaf spot continued to develop even in the dry period; it is possible that leaf wetness from dewdrops favoured development of the disease. The bacterial population on leaf surface varied during the first ten weeks after planting depending upon the weather, but it remained stable after the appearance of the first symptoms of the disease.

Dutta *et al.* (2014) investigated the colonisation of pepper blossoms by *Xan. euvesicatoria*. The bacterial populations reached up to 10^5 cfu/blossom on stigmas and ranged from 10^5 to 10^6 cfu/blossom on stylar and ovary tissues 96 h after inoculation. The inoculation of blossoms resulted in asymptomatic pepper fruit, but a positive correlation was found between inoculum concentration and the percentage of infested seed lots. Real-time PCR detected *Xan. euvesicatoria* in 39% of the seed lots assayed and viable colonies were recovered from 35% of these seed lots. Successful transmission occurred in 16% of the seed lots tested.

Giovanardi *et al.* (2018) investigated the impact of disease outbreaks on the phytosanitary quality of seeds for two pathosystems, i.e., tomato-*Xan. vesicatoria* and pepper-*Xan. euvesicatoria*. Seed contamination levels in pepper ranged from 34–100 cfu g^{-1}, and the contamination rate ranged from 1.50% to 3.17% for *Xan. euvesicatoria*, while the tomato seeds produced were not infected by *Xan. vesicatoria*. During seedling grow-out assays and the second cropping year, no symptoms were observed in either tomato or pepper plants. It is evident that the calculated pepper seed contamination rate for *Xan. euvesicatoria* appeared to be less than the threshold necessary to initiate a disease outbreak. Finally, all seeds obtained during the second cropping year were uninfected.

Sun *et al.* (2014) investigated the feasibility of using image analysis to estimate the severity of foliar disease of bacterial spot in tomato. Five diseased leaves from each plant, spray-inoculated with bacterial spot race T3 (*Xan. perforans*), were randomly collected and scanned to obtain digital images 21 days after inoculation. Disease severity (% leaf area) was measured using image analysis. Marker-trait association analysis identified four quantitative trait loci conferring resistance to race T3 in resistant line, PI 114490 using data obtained from image analysis, the Horsfall–Barratt (HB) category scale data, and HB midpoint converted values. Nevertheless, the disease severity was slightly underestimated using the HB category scale, and the phenotypic variation explained by each marker was overestimated using the HB category data compared to the image analysis-measured disease severity data. It is evident that image analysis can provide a consistent, accurate, and reliable method compared to the HB scale to estimate disease severity for genetic studies of foliar bacterial spot in tomato.

Duan *et al.* (2015) prepared and evaluated a standard area diagram (SAD) set to assess bacterial spot severity on tomato leaves caused by *Xanthomonas* spp. The proposed SAD set contains illustrations of leaves with 12 distinct disease severity values ranging from 0.5% to 90% and was validated by 12 raters without any experience of plant disease evaluation. Regression analysis and Lin's concordance correlation (ρc) analysis of estimated (without the SADs) versus true disease severity (based on image analysis) showed that the precision and accuracy improved for all raters using the SADs. The SAD set improved accuracy (correction factor, $Cb = 0.91$ and 1.00, without and with the SADs, respectively) and agreement ($\rho c = 0.83$ and 0.96, without and with the SADs, respectively) of the severity estimates. Severity estimates using the SADs were also more reliable (coefficient of determination, $R2 = 0.74–0.94$ without the SADs, and $R2 = 0.82–0.95$ with the SADs, and intra-class correlation $\rho = 0.85$ without the SADs and $\rho = 0.91$ using the SADs). SAD set proposed in this study will improve the accuracy and reliability of estimates of bacterial spot severity on tomato leaves for inexperienced raters and can be used to assess the severity of other bacterial diseases on tomato leaves.

Araújo *et al.* (2017) reported that of 204 strains obtained from 33 counties (22 with processing tomatoes and 11 with fresh-market tomatoes) of Brazil, *Xan. perforans* predominated among the strains (92%) and was present in most counties. Moreover, this species was also prevalent in most areas of both fresh-market tomatoes (63.6% of counties surveyed) and processing tomatoes (95.4% of counties surveyed). High genetic diversity was found within *Xan. perforans*, with 137 BOX-PCR haplotypes. Fifteen strains (7.5%) were identified as *Xan. gardneri*, which was found mostly in fresh-market fields located at regions with altitude higher than 900 meters, and only one strain of *Xan. euvesicatoria* (0.5%) was found in a tomato processing field.

Safaie Farahani and Taghavi (2018) found that rutin, a polyphenolic substance, at a concentration of 2 mM induced resistance in tomato against *Xan. perforans* and reduced the disease severity of bacterial spot, but showed no antibacterial activity *in vitro* against the bacterium. Pre-treatment of tomato plants with rutin led to the induction of *PR-5*, *PAL*, and *LOX* genes compared to control and rutin-induced resistance against *Xa. perforans* in tomato might be mediated through the stimulation of some defence genes (such as *PR-5*, *PAL* and *LOX*).

Wang *et al.* (2020) developed a viability-qPCR assay for the specific quantification of viable *Xanthomonas* pathogens on tomato seed and compared the performance of this assay with the industry-standard culturing method for quantifying viable pathogen cells in seed samples. The Ct values from PMA-qPCR assay and the colony count results from the culturing method were compared using latent class analysis to determine their detection sensitivity and specificity. Results showed that the PMA-qPCR assay had a higher sensitivity for detecting the viable *Xanthomonas* pathogens on tomato seed.

Moura *et al.* (2020) investigated the role of crop rotation in the control of volunteer plants in commercial fields of processing tomato in the Central-West Region of Brazil and the occurrence of tomato diseases in these plants. The number of volunteer plants per hectare was estimated and compared between commercial fields with one and two years of crop rotation or without crop rotation. The rotation of processing tomato crops with soybean, maize, sweet corn and bean crops significantly reduced the numbers of volunteer plants while the consecutive tomato cultivation promoted a higher number of volunteer plants for the next crop. Thirty-nine isolates of *Xanthomonas perforans* were isolated from volunteer plants symptomatic for bacterial spot. All the isolates were identified as *Xan. perforans* by specifics primers, and all were pathogenic when inoculated in tomato plants. The authors concluded that crop rotation is an efficient practice to reduce volunteer plants in processing tomato crops. This method can reduce the inoculum sources of a pathogen transmitted by seeds, like *Xan. perforans*.

To improve the seed certification process, Wang *et al.* (2019) developed a viability-qPCR assay in conjunction with a crude DNA extraction procedure to selectively quantify viable xanthomonads in tomato seed. A DNA-binding dye, propidium monoazide (PMA), was used to treat seed extract; it selectively binds to the DNA of dead cells, thus preventing the DNA of dead cells from being amplified in downstream qPCR. The procedure was tested with tomato seed samples spiked with mixtures of viable and dead cells in different concentrations and proportions. The assay accurately quantified viable cells within mixtures of 1:10,000 (live:dead), and the use of the crude DNA extract provided results similar to those obtained using purified DNA prepared from commercial kits. This technique will provide a sensitive, specific and cost-effective way to certify tomato seed free from bacterial spot pathogens.

Wang and Zhao (2020) reported that a *Xan. euvesicatoria* type III-secreted effector, XopN, is required to trigger the water-soaking symptom on *Nicotiana benthamiana* and pepper plants infected with *Xan. euvesicatoria.* With the function of XopN, a higher apoplastic water content is maintained in *N. benthamiana* leaf infected by *Xan. euvesicatoria.* In addition, XopN positively regulates stomatal conductance, by directly or indirectly increasing the stomatal aperture, which can enhance the colonisation of a non-pathogenic *Xan. euvesicatoria* strain on tobacco and pepper leaves. These findings showed that a glycerol uptake facilitator and the type II secretion system of *Xan. euvesicatoria* also contribute to the development of the water-soaking phenotype. *Xan. euvesicatoria* utilises different mechanisms to promote an aqueous apoplast, which is vital for enhancing bacterial proliferation in both tobacco and pepper plants.

Strayer-Scherer *et al.* (2019a) designed novel recombinase olymerase amplification (RPA) probes and primers to amplify regions of the *hrcN* gene of *Xan. gardneri*, *Xan. euvesicatoria* and *Xan. perforans*. *Xan. gardneri* RPA is specific to *Xan. gardneri*, with a detection limit of 10^6 CFU/ml and detected *Xan. gardneri* in lesions from naturally (n = 6) or artificially (n = 18) infected plants. *Xan. euvesicatoria* RPA detects both *Xan. euvesicatoria* and *Xan. perforans* with a detection limit of 10^6 CFU/ml and detected both pathogens in plants artificially infected (n = 36) or naturally infected (n = 85) with either *Xan. euvesicatoria* or *Xan. perforans*. *Xan. perforans* RPA is specific to *Xan. perforans* with a detection limit of 10^7 CFU/ml. Although the *Xan. perforans* RPA assay was unable to detect *Xan. perforans* from lesions, the *Xan. euvesicatoria* RPA was successfully used in the field to detect *Xan. perforans* from symptomatic field samples (n = 31). *Xan. perforans* RPA was then used to confirm the pathogen in the laboratory. The *Xan. euvesicatoria* and *Xa. gardneri* RPA is promising for rapid, real-time in-field detection of bacterial spot pathogens and is one of the first developed among plant pathogenic bacteria.

In order to elucidate how an effector gene, avrBsT, confers fitness to *Xan. perforans* (*Xp*), Sharma *et al.* (2019) conducted a study with the objectives of quantifying pathogen dispersal and predicting the prevalence and movement of Xp strains in relation to the avrBsT gene. The authors developed a model to quantify the spread of *Xp* with or without avrBsT over spatial and temporal dimensions. According to the model, wild-type strains spread faster than mutant strains (~4 vs. ~1.25 meters/week). Preliminary results show that plants infected with wild-type strains exhibit higher disease severity than those with mutant strains. These findings show that the presence of a single gene can affect pathogen movement and significantly alter bacterial population dynamics.

Bernal *et al.* (2020) identified three independent sources of resistance with quantitative trait loci (QTL) mapping to the centromeric region on chromosome 11. These sources were derived from Hawaii 7998 (QTL-11A), PI 114490 (QTL-11B) and LA2533 (QTL-11C). To determine which QTL introgression from chromosome 11 provides the greatest resistance to multiple species, the authors developed near-isogenic lines (NILs) using marker-assisted backcrossing. In parallel, the authors also developed a NIL that contains *Rx-4/Xv3*, which provides major gene resistance to *Xan. perforans*. *Rx-4/Xv3* resistance was also combined with QTL-11A. These sources of resistance were independently introduced into the susceptible parent, OH88119. Over the course of a three-year period from 2016 to 2018, backcross-derived families and NILs from each source in independent field trials inoculated with *Xan. perforans*, *Xan. euvesicatoria*, or *Xan. gardneri* were evaluated. These results suggest that both QTL-11C and QTL-11A can combine with *Rx-4/Xv3* and provide effective genetic resistance against multiple *Xanthomonas* species. The evidence for additive to dominant genetic action for the QTL introgressions is also provided.

Although *Xan. perforans* and *Xan. euvesicatoria* are closely related and both attack tomato and pepper, these species differ in host specificity. *Xan. perforans* is the dominant pathogen of tomato and *Xan. euvesicatoria* that of pepper. This is, in part, due to the activity of avirulence proteins that are secreted by *Xan. perforans* strains and elicit effector-triggered immunity in pepper leaves, thereby restricting pathogen growth (Newberry *et al.*, 2023). Newberry *et al.* (2023) investigated the basal evolutionary processes underlying the host range of *Xan. perforans* using multiple genome-wide association analyses. They identified two novel gene candidates that were significantly associated with pepper-pathogenic *Xan. perforans* and *Xan. euvesicatoria*. Both candidates were predicted to be involved in the acquisition/transport of nutrients common to the plant cell wall or apoplast and included a TonB-dependent receptor, which was disrupted through independent mutations within the *Xan. perforans* lineage. The other included a symporter of protons/glutamate, *gltP*, enriched with pepper-associated mutations near the promoter and start codon of the gene. Functional analysis of these candidates revealed that only the TonB-dependent receptor had a minor effect on the symptom development and growth of *Xan. perforans* in pepper leaves, indicating that pathogenicity to this host might have evolved independently within the bacterial species and is likely a complex, multigenic trait.

9.23.5 Management

For the effective management of the disease, it is desirable to integrate the following management measures.

1. Use of pathogen-free seed. If the pathogen-free seed is not available, it must be treated by any of the following treatments.
   a. Hot water treatment at 50 °C for 25 minutes after pre-warming the seed at 38 °C in water for 10 minutes. Immediately after the treatment, put the seed in cold water for 5 minutes (Leboeuf *et al.*, 2005).
   b. Acetic acid (0.8%) steep for 24 h or 5% HCl for 5–10 h at 21 °C.
   c. Shahrtash and Lewis Ivey (2015) treated tomato and pepper seed artificially infested with a mixture of *Xan. vesicatoria*, *Xan. euvesicatoria* and *Xan. perforans* (~10^4 cfu mL^{-1}) with hot water (pepper seeds for 30 min at 51 ° C; tomato seeds for 25 min at 50 °C) or chlorine (20% for one minute), which represent industry standards for disinfesting seed, or sodium selenate (765 mg L^{-1} for 30 min). All the three treatments significantly reduced the incidence of infested pepper and tomato seed compared to the non-treated infested control. Incidence of infested pepper seeds following treatment with sodium selenate was significantly higher than that of seed treated with chlorine and hot water. There was no difference in the incidence of infested seed between sodium selenate-treated and chlorine-treated tomato seed. Hot water treatment of pepper and tomato seed was the most effective in reducing *Xanthomonas* spp. on seed. The presence of *Xanthomonas* spp. was confirmed using colony-pick PCR with *Xanthomonas*-specific PCR primers. None of the treatments significantly reduced or delayed germination.
2. Use of disease-free transplants; this is achieved by growing in the nursery where crop production does not occur.
3. Rotation with non-host plants for two or three years.
4. Destroy all the volunteer plants.
5. Avoid cull piles near greenhouse and field locations.
6. Use of chemicals: Use of copper compounds and streptomycin has created a serious problem of development of pathogen strains resistant to these chemicals. In order to control copper-resistant strains, a combination of copper and mancozeb should be used and these chemicals should be applied as preventive sprays after transplanting.

Strayer *et al.* (2016b) reported that in greenhouse studies, the application of Ag-dsDNA-GO, a silver-based nanocomposite, at either 75 or 100 µg mL^{-1} prior to the artificial inoculation of tomato plants with *Xan. perforans* significantly reduced disease severity when compared with copper-mancozeb and negative controls ($P = 0.05$). The study highlights the potential of Ag-dsDNA-GO as an alternative to copper in tomato transplant production. Ag-dsDNA-GO also exhibited antibacterial activity against copper-tolerant *Xan. vesicatoria*, *Xan. euvesicatoria* and *Xan. gardneri* strains.

Paret *et al.* (2013) reported that the application of TiO_2/Zn, AgriTitan (nanoscale titanium dioxide doped with zinc) at ≈500 to 800 µg mL^{-1} significantly reduced the severity of bacterial spot (*Xan. perforans*) compared with untreated and copper control in greenhouse studies on naturally and artificially infected transplants. Protection was similar to the grower standard, copper + mancozeb. Use of TiO2/Zn at ≈500 to 800 µg mL^{-1} also significantly reduced disease incidence in three of the four field trials compared with untreated and copper control, and was comparable to, or better than, the grower standard. These treatments did not cause any adverse effects on tomato yield in any of the field trials.

Integration of foliar sprays of ammonium lignosulfonate (2% or 4%) and potassium phosphate (25 mM) in the standard bactericide management programmes has also been suggested for improving the efficiency of disease management (Abbasi *et al.* 2002).

Qiao *et al.* (2021) evaluated the effect of N-acetylcysteine (NAC) alone, and also in combination with copper, on a copper-resistant *Xan. perforans* strain *in vitro* and its ability to control bacterial spot of tomato under greenhouse and field conditions. *In vitro*, the minimum inhibitory concentration of NAC against the *Xan. perforans* strain was 2,048 mg L^{-1}. NAC increased sensitivity of the copper-resistant *Xan. perforans* to copper *in vitro* when the application of NAC was followed by copper application after 6 h. In greenhouse assays, NAC applied alone or in combination with copper significantly ($P < 0.05$) reduced the disease severity of bacterial spot on tomato compared with the untreated control. NAC at 100 mg L^{-1} + copper at 300 mg L^{-1} consistently exhibited synergistic effects against bacterial spot. In the field trials, NAC at 1,000 mg L^{-1} + copper at 150 mg L^{-1} significantly reduced disease severity compared with the untreated control. Results from this study demonstrated that NAC significantly reduced the severity of bacterial spot of tomato and enhanced the efficacy of copper against copper-resistant *Xan. perforans*, suggesting that NAC could be applied for the effective management of bacterial spot of tomato.

Fan *et al.* (2019) investigated the antibacterial effect of three novel copper-composites, Core-shell copper (CS-Cu), Mixed-valence copper (MV-Cu) and Fixed-quat copper (FQ-Cu) in *in vitro* and in greenhouse experiments against *Xan. euvesicatoria* copper-tolerant strains. All concentrations (100, 200, 500 and 1000 μg mL^{-1}) of CS-Cu and FQ-Cu effectively killed bacteria within 1 h and MV-Cu killed all the cells within 24 h, *in vitro*. In the first greenhouse experiment, plants treated with all copper-composites, at all concentrations (100, 200 and 500 μg mL^{-1}) had disease severity (AUDPC) statistically equivalent to, or better than, copper-mancozeb, the growers' standard. All three concentrations of FQ-Cu, 200 and 500 μg mL^{-1} of CS-Cu and 100 and 200 μg mL^{-1} of MV-Cu showed significantly less AUDPC compared to Kocide 3000. In the second greenhouse experiment, all concentrations of FQ-Cu, and 100 μg mL^{-1} and 200 μg mL^{-1} of MV-Cu significantly reduced AUDPC compared to Kocide-3000. All concentrations of MV-Cu and FQ-Cu were statistically equivalent to copper-mancozeb. FQ-Cu and MV-Cu with low-to-no phytotoxicity and consistent efficacy highlight the potential of copper-nanocomposites for an effective management of bacterial spot of pepper.

Due to the presence of copper (Cu)-tolerant *Xan. perforans* strains, Cu bactericides are not considered effective in disease management. Liao *et al.* (2019) evaluated the antibacterial activity of magnesium oxide (Nano-MgO) and other metal oxide nanoparticles against a Cu-tolerant and -sensitive *Xan. perforans* strain. *In vitro* experiments demonstrated high antibacterial activity of Nano-MgO against both strains compared with the commercial Cu. The minimum inhibitory concentration of Nano-MgO was 25 μg mL^{-1} and the minimum bactericidal concentration was 100 μg mL^{-1} against a Cu-tolerant *Xan. perforans* strain after 4 h of exposure. In two greenhouse experiments with a Cu-tolerant strain, Nano-MgO at 200 μg mL^{-1} significantly reduced bacterial spot severity compared with Cu-ethylene bis-dithiocarbamate (grower standard), and the untreated control ($P = 0.05$). In three field experiments, Nano-MgO at 200 μg mL^{-1} also significantly reduced disease severity with no negative impact on yield compared with the untreated control. Inductively coupled plasma mass spectrometric analysis of the fruit confirmed that Nano-MgO application did not lead to the accumulation of Mg, Cu, Ca, K, Mn, P and S. This is the first study to demonstrate the potential of Nano-MgO against bacterial spot of tomato.

Liao *et al.* (2020) reported that in a field evaluation in the fall of 2019, applications of formulated Mg and Cu nanomaterials (i.e., Mg-Cu and Mg-DB) at 500 μg mL^{-1} significantly reduced the disease severity compared to untreated controls, whereas conventional Cu bactericide was not significantly different from untreated controls ($p = 0.05$). Additional field studies are required to determine the consistent efficacy of these formulated Mg nanomaterials across multiple production seasons.

Choudhary *et al.* (2024) reported that magnesium double-coated and magnesium-copper nanomaterials significantly reduced disease severity in two out of four field trials. Magnesium double-coated at 500 μg/mL^{-1} reduced severity by 34% compared with the nontreated control without affecting yield in fall, 2020. The use of hybrid nanomaterials at the highest concentrations (500 μg/mL^{-1}) evaluated in the field experiments can reduce copper use by 90% compared with the growers' standard. There was no phytotoxicity observed with the use of hybrid nanomaterials in the field.

Chien and Huang (2022) reported that weekly foliar sprays of potassium silicate at pH 10.0 rather than at pH 7.0 can effectively reduce bacterial spot of tomato and can provide an alternative disease management strategy.

Acibenzolar-S-methyl (ASM, trade names Actigard 50 WG and CGA 245704) is a plant activator that induces systemic acquired resistance in many crops against a number of pathogens. Louws *et al.* (2001) showed that ASM (35 g a.i. ha^{-1}) can be used as a viable alternative to copper-based bactericides for field management of bacterial spot of tomato, and also for speck of tomato, where copper-resistant populations of these pathogens predominate.

Huang and Vallad (2018) investigated the effects of soil application rates of ASM on bacterial spot of tomato and the expression levels of the two pathogenesis-related genes, *PR1a* and *PR1b*, in leaf tissues. ASM at 18.8 mg L^{-1}, corresponding to the labelled rate, was sprayed on tomato seedlings and soil application was made at 0.84 and 10 mg L^{-1}. The soil application of ASM at 10 mg L^{-1} consistently reduced the final disease severity and disease progress compared to the untreated control. The expression levels of *PR1a* and *PR1b* in the leaf tissues were significantly induced by both soil and foliar applications of ASM. The soil applications of ASM at 10 mg L^{-1} markedly reduced disease progress compared to the control and copper standard. Although the control efficiency of soil applications of ASM depends on the rates used, the study suggests that ASM can be used as soil applications to induce tomato resistance against bacterial spot.

7. Use of phages: Obradovic *et al.* (2004) obtained better control of the disease with the combination of phage and ASM than phage, ASM or copper-mancozeb alone.

Phage treatment, integrated with other practices, is currently widely used in greenhouse and production fields in Florida as an aspect of standard integrated management programme for the disease. On the basis of the results of greenhouse and field trials, Omniytics, Inc., in Salt Lake City, UT, is the first to obtain registration from the Environmental Protection Agency (EPA) of the USA for using host-specific phages on tomatoes in greenhouses and production fields in Florida as a part of standard integrated management programme for the control of the disease. Obradovic *et al.* (2005) also suggested the integrated use of acibezolar-S-methyl and phages as an alternative management strategy for the disease.

8. Biological control:
   Ji *et al.* (2006) obtained significant control of bacterial spot and speck of tomato with the combined application of *Pseudomonas syringae* str. Cit 7 and *Psm. fluorescens* str. 89B-61.

Felipe *et al.* (2021) reported that two novel strains, VMA11p and VM05, closely related to *Bacillus velezensis*, and VMAP1, closely related to *Pseudomonas soli*, antagonised *Xan. vesicatoria in vitro.* Bacterial suspensions of VMA11p and VMAP1, applied to the soil by irrigation, significantly ($P<0.05$) reduced bacterial spot severity by 53.9% and 44.2%, respectively, under greenhouse conditions. However, the most effective strategy to control bacterial spot was to use the cell-free supernatant produced by VMA11p, VM05 or VMAP1 applied as foliar spray, which significantly ($P<0.05$) reduced the severity of the disease by 98.5%, 94.2% and 75.2%, respectively. None of the treatments reduced the growth of tomato plants.

Chien and Huang (2020) evaluated the effects of the combined and individual use of *Bacillus amyloliquefaciens* and *Trichoderma asperellum* on the nutrient contents of tomato plants and bacterial spot on tomato caused by *Xan. perforans*. The application methods used for *Bac. amyloliquefaciens* and *T. asperellum* significantly influenced the P, K and Mg contents in tomato plants. All biocontrol treatments significantly reduced the final disease severity of bacterial spot on tomato by 25.6–56.0% and the area under the disease progress curve (AUDPC) by 33.9–57.2% compared with the untreated control. The application of either *Ba. amyloliquefaciens* or *T. asperellum* via growth medium produced statistically equivalent results in reducing the final disease severity compared with the standard, but the other biocontrol treatments were significantly better than the standard. Similarly, the application of either *Bac. amyloliquefaciens* or *T. asperellum* via growth medium produced statistically equivalent results in lowering the AUDPC compared with the standard. However, the other biocontrol treatments significantly reduced the AUDPC by 39.7–44.8% compared with the standard. It is evident that the combined use and individual application of *Bac. amyloliquefaciens* and *T. asperellum* via foliar spray or growth medium application or both have the potential to control bacterial spot on tomato.

Strayer-Scherer *et al.* (2019b) evaluated the efficacy of Leap™ ES (*Bacillus thuringiensis* subsp. *kurstaki* strain ABTS-351), LifeGard™ WG (*Bac. mycoides* isolate J), Actigard® 50WG (acibenzolar-S-methyl), Harbour® (streptomycin) and three experimental nanomaterials loaded with copper (Cu) and/or chitosan as potential alternatives to the grower standard, copper-mancozeb, under greenhouse and/or field conditions. In greenhouse studies, tomato plants treated with either Actigard® 50WG S1.37 at 0.4 g^{L-} alone, or in combination with Harbour® at 1.2 g^{L-} or Kocide® 3000 at 2.1 g^{L-} (copper hydroxide), and Leap™ ES at 2.6 ml^{L-} in combination with Harbour® at 1.2 g^{L-} significantly reduced bacterial spot disease severity when compared to the grower standard and water controls ($P < 0.05$). In field trials, LifeGard™ WG at 4.5 oz/100 gal, and nanomaterials + chitosan at 100 mg^{L-}, significantly reduced bacterial spot disease severity when compared to the non-treated control without a significant impact on yield ($P < 0.05$). This study highlights the potential of several different products for the management of bacterial spot of tomato.

Suazo Tejada and Vallad (2023) reported that irrigation with plasma-activated water statistically reduced the AUD progress curve by 19.2% and the final disease incidence by 17.6% ($P = <0.0001$), compared to the irrigation control. These findings support the use of plasma-activated water as a viable alternative to bactericides for the management of bacterial spot on tomato seedlings.

REFERENCES

Abbasi, P. A., Soltani, N., Cuppels, D. A. and Lazarovits, G. (2002). Reduction of bacterial spot disease severity on tomato and pepper plants with foliar applications of ammonium lignosulfonate and potassium phosphate. *Plant Dis.* 86: 1232–36.

Adhikari, P., Adhikari, T. B., Timilsina, S., Meadows, I., Jones, J. B. *et al.* (2019). Phenotypic and genetic diversity of *Xanthomonas perforans* populations from tomato in North Carolina. *Phytopathology* 109: 1533–43.

Araújo, E. R., Costa, J. R., Ferreira, M. A. S. V., Quezado-Duval, A. M. (2017). Widespread distribution of *Xanthomonas perforans* and limited presence of *X. gardneri* in Brazil. *Plant Pathol.* 66: 159–68.

Araújo, E. R., Costa, J. R., Pontes, N. C. and Quezado-Duval, A. M. (2015). *Xanthomonas perforans* and *X. gardneri* associated with bacterial leaf spot on weeds in Brazilian tomato fields. *Eur. J. Plant Pathol.* 143: 543–48.

Bernal, E., Liabeuf, D. and Francis, D. M. (2020). Evaluating quantitative trait locus resistance in tomato to multiple *Xanthomonas* spp. *Plant Dis.* 104: 423–29.

Bibi, S., Weis, K., Kaur, A., Bhandari, R., Goss, E. *et al.* (2023). A brief evaluation of a copper resistance mobile genetic island in the bacterial leaf spot pathogen *Xanthomonas euvesicatoria* pv. *perforans*. *Phytopathology* 113: 1394–98.

Bophela, K. N., Venter, S. N., Wingfield, M. J., Duran, A., Tarigan, M. *et al.* (2019). *Xanthomonas perforans*: A tomato and pepper pathogen associated with bacterial blight and dieback of *Eucalyptus pellita* seedlings in Indonesia. *Australasian Plant Pathol.* 48: 543–51. https://doi.org/10.1007/s13313-019-00657-9

Chien, Y. C. and Huang, C. H. (2020). Biocontrol of bacterial spot on tomato by foliar spray and growth medium application of *Bacillus amyloliquefaciens* and *Trichoderma asperellum*. *Eur. J. Plant Pathol.* 156: 995–1003. https://doi.org/10.1007/s10658-020-01947-5

Chien, Y. C. and Huang, C. H. (2022). Effects of pH values and application methods of potassium silicate on nutrient uptake and bacterial spot of tomato. *Eur. J. Plant Pathol.* 162: 119–30. https://doi.org/10.1007/s10658-021-02390-w

Choudhary, M., Liao, Y-Y., Huang, Z., Pereira, J., Santra, S. *et al.* (2024). Novel magnesium-copper hybrid nanomaterials for management of bacterial spot of tomato. *Plant Dis.* 108: 3234–42.

Doidge, E. M. (1920). A tomato canker. *J. Dept. Agric. Union S. Afr.* 1: 718–21.

Duan, J., Zhao, B., Wang, Y. and Yang, W. (2015). Development and validation of a standard area diagram set to aid estimation of bacterial spot severity on tomato leaves. *Eur. J. Plant Pathol.* 142: 665–75.

Dutta, B., Gitaitis, R., Sanders, H., Booth, C., Smith, S. *et al.* (2014). Role of blossom colonization in pepper seed infestation by *Xanthomonas euvesicatoria*. *Phytopathology* 104: 232–39.

Fan, Q., Liao, Y.-Y., Kunwar, S., Young, M., Santra, S. *et al.* (2019). Managing bacterial spot of pepper caused by *Xanthomonas euvesicatoria* using novel copper-composites. (Abstr.) *Phytopathology* 109: S1.5.

Fan, X., Zheng, H., Luo, H. Zhuo, T. and Chen, Y. (2023). *Xanthomonas euvesicatoria* pv. *perforans* is the causative agent of bacterial leaf spot on *Ipomoea aquatica* from Fujian Province in China. *Australas. Plant Pathol.* 52: 327–37.

Felipe, V., Bianco, M. I., Terrestre, M., Mielnichuk, N., Romero A. M. *et al.* (2021). Biocontrol of tomato bacterial spot by novel *Bacillus* and *Pseudomonas* strains. *Eur. J. Plant Pathol.* 160: 935–48.

Giovanardi, D., Biondi, E., Ignjatov, M., Jevtić, R. and Stefani. E. (2018). Impact of bacterial spot outbreaks on the phytosanitary quality of tomato and pepper seeds. *Plant Pathol.* 67: 1168–76.

Huang, C-H. and Vallad, G. E. (2018). Soil applications of acibenzolar-*S*-methyl induce defense gene expression in tomato plants against bacterial spot. *Eur. J. Plant Pathol.* 150: 971–81.

Ji, P., Campbell, H. L., Kloepper, J. W. Jones, J. B., Suslow, T. V. *et al.* (2006). Integrated biological control of bacterial speck and spot of tomato under field conditions using foliar biological control agents and plant growth-promoting rhizobacteria. *Biol. Control* 36: 358–67.

Jones, J. B. (2001). Bacterial spot of tomato. In: O. C. Maloy and T. D. Murray (eds.). *Encyclopedia of Plant Pathology*. Vol. 1: John Wiley and Sons, Inc. New York, pp. 83–84.

Jones, J. B., Bouzar, H., Stall, R. E. Almira, E. C., Roberts, P. D. *et al.* (2000). Systematic analysis of xanthomonads (*Xanthomonas* spp.) associated with pepper and tomato lesions. *Int. J. Syst. Evol. Microbiol.* 50: 1211–19.

Jones, J. B., Lacy, G. H., Bouzar, H. Stall, R. E. and Schaad, N. W. (2004). Reclassification of the xanthomonads associated with bacterial spot disease of tomato and pepper. *Syst. Appl. Microbiol.* 27: 755–62.

Jones, J. B., Stall, R. E. and Bouzar, H. (1998). Diversity among xanthomonads pathogenic on pepper and tomato. *Annu. Rev. Phytopathol.* 36: 41–58.

Leboeuf, J., Cuppels, D., Dick, J., Pitblado, R., Loewen, S. *et al.* (2005). Bacterial diseases of tomato: Bacterial spot, bacterial speck, bacterial canker. Ministry of Agri., Food and Rural Affairs, Ontario, Canada. http://www.omafra.gov.on.ca/english/crops/facts/05-069.htm

Liao, Y. Y., Huang, Z., Strayer-Scherer, A. L., Santra, S., White, J. C. *et al.* (2020). Magnesium-Copper nanomaterials as alternative to conventional copper for tomato and pepper bacterial spot disease management. (Abstr.) *Phytopathology* 110: S2.100.

Liao, Y.-Y., Strayer-Scherer, A. L., White, J., Mukherjee, A., De La Torre-Roche, R. *et al.* (2019). Nano-magnesium oxide: A novel bactericide against copper-tolerant *Xanthomonas perforans* causing tomato bacterial spot. *Phytopathology* 109: 52–62.

Louws, F. J., Wilson, M., Campbell, H. L. Cuppels, D. A., Jones, J. B. *et al.* (2001). Field control of bacterial spot and bacterial speck of tomato using a plant activator. *Plant Dis.* 85: 481–88.

Marcuzzo, L. L., Becker, W. F. and Fernandes, J. M. C. (2009). Some epidemiological aspects of bacterial spot (*Xanthomonas* spp.) of tomato in Cacador/ SC, Brazil. *Summa Phytopathol.* 35: 132–35.

Moura, D. R., Yamada, J. K., de Albuquerque, L. C. and de Carvalho Pontes, N. (2020). Crop rotation reduces the density of volunteer plants in processing tomato fields and the inoculum of bacterial spot. *Eur. J. Plant Pathol.* 156: 299–304. https://doi.org/10.1007/s10658-019-01868-y

Newberry, E. A., Minsavage, G. V., Holland, A., Jones, J. B. and Potnis, N. (2023). Genome-wide association to study the host-specificity determinants of *Xanthomonas perforans*. *Phytopathology* 113: 400–12.

Obradovic, A., Jones, J. B., Momol, M. T., Balogh, B. and Olson, S. M. (2004). Management of tomato bacterial spot in the field by foliar application of bacteriophages and SAR inducers. *Plant Dis.* 88: 736–40.

Obradovic, A., Jones, J. B., Momol, M. T., Olson, S. M., Jackson, L. E. *et al.* (2005). Integration of biological control agents and systemic acquired resistance inducers against bacterial spot on tomato. *Plant Dis.* 89: 712–16.

Osdaghi, E., Jones, J. B., Sharma. A., Goss, E. M., Abrahamian, P. *et al.* (2021). A centenary for bacterial spot of tomato and pepper. *Mol. Plant Pathol.* 22: 1500–19.

Paret, M. L., Vallad, G. E., Averett, D. R., Jones, J. B. and Olson, S. M. (2013). Photocatalysis: Effect of light-activated nanoscale formulations of TiO_2 on *Xanthomonas perforans* and control of bacterial spot of tomato. *Phytopathology* 103: 228–36.

Patel, M. K., Kulkarni, Y. S. and Dhande, G. W. (1950). Bacterial spot of chillies. *Indian Phytopathol.* 3: 95–97.

Qiao, K., Liu, Q., Xia, Y. and Zhang, S. (2021). Evaluation of a small-molecule compound, N-acetylcysteine, for the management of bacterial spot of tomato caused by copper-resistant *Xanthomonas perforans*. *Plant Dis.* 105: 108–13.

Safaie Farahani, A. and Taghavi, S. M. (2018). Rutin promoted resistance of tomato against *Xanthomonas perforans*. *Eur. J. Plant Pathol.* 151: 527–31.

Shahrtash, M. and Lewis Ivey, M. L. (2015). Evaluation of sodium selenate as a seed disinfectant to eradicate *Xanthomonas vesicatoria*, *X. euvesicatoria* and, *X. perforans* on pepper and tomato seeds. (Abstr.) *Phytopathology* 105 (Suppl. 2): S2.9–2.10.

Sharma, A., Timilsina, S., Abrahamian, P., Ojiambo, P. S., Colee, J. *et al*. (2019). Role of avrBsT in dispersal of *Xanthomonas perforans* and severity of bacterial spot of tomato. (Abstr.) *Phytopathology* 109: S2.24.

Stall, R. E., Jones, J. B. and Minsavage, G. V. (2009). Durability of resistance in tomato and pepper to xanthomonads causing bacterial spot. *Annu. Rev. Phytopathol*. 47: 265–84.

Strayer, A., Ocsoy, I., Tan, W., Jones, J. B. and Paret, M. L. (2016b). Low concentrations of a silver-based nanocomposite to manage bacterial spot of tomato in the greenhouse. *Plant Dis*. 100: 1460–65.

Strayer, A. L., Jeyaprakash, A., Minsavage, G. V., Timilsina, S., Vallad, G. E. *et al*. (2016a). A multiplex real-time PCR assay differentiates four *Xanthomonas* species associated with bacterial spot of tomato. *Plant Dis*. 100: 1660–68.

Strayer-Scherer, A., Jones, J. B. and Paret, M. L. (2019a). Recombinase polymerase amplification assay for field detection of tomato bacterial spot pathogens. *Phytopathology* 109: 690–700.

Strayer-Scherer, A. L., Sharpe, S., Henson, M. and Meadows, I. (2019b). Evaluation of alternative fungicide spray programs for the management of bacterial spot on fresh-market tomatoes. (Abstr.) *Phytopathology* 109: S1.36–1.37.

Suazo Tejada, A. K. and Vallad, G. E. (2023). Application of plasma activated water (PAW) for the management of *Xanthomonas euvesicatoria* pv. *perforans* on tomato seedlings. *Phytopathology* 113: S3.65.

Sun, H., Wei, J., Zhang, J. and Yang, W. (2014). A comparison of disease severity measurements using image analysis and visual estimates using a category scale for genetic analysis of resistance to bacterial spot in tomato. *Eur. J. Plant Pathol*. 139: 125–36.

Sutic, D. (1957). *Bakterioze crvenog patlidzana* (*Tomato bacteriosis*). Posebna Izd. Inst. Zasht. Bilja Beograd (Spec. Edit. Inst. Plant Prot. Beograd) 6: 1–65. English summary: *Rev. Appl. Mycol*. 36: 734–35.

Thieme, F., Koebnik, R., Bekel, T., Berger, C., Boch, J. *et al*. (2005). Insights into genome plasticity and pathogenicity of the plant pathogenic bacterium *Xanthomonas campestris* pv. *vesicatoria* revealed by the complete genome sequence. *J. Bacteriol*. 187: 7254–66. PMID: 16237009; PMCID: PMC1272972.

Tudor-Nelson, S., Minsavage, G. V., Stall, R. E. and Jones, J. B. (2003). Bacteriocin-like substances from tomato race 3 strains of *Xanthomonas campestris* pv. *vesicatoria*. *Phytopathology* 93: 1415–21.

Vauterin, L., Hoste, B. Kersters, K. and Swings, J. (1995). Reclassification of *Xanthomonas*. *Int. J. Syst. Bacteriol*. 45: 472–89.

Wang, H., Vallad, G. E., Jones, J. B. and Turechek, W. W. (2019). A viability-qPCR test for quantifying living pathogens of bacterial spot in tomato seed. (Abstr.) *Phytopathology* 109: S2.1. https://doi.org/10.1094/PHYTO-109-10-S2.1

Wang, H., Vallad, G. E., Jones, J. B. and Turechek, W. W. (2020). Comparison of methods for detecting bacterial spot pathogens on tomato seed (Abstr.). *Phytopathology* 110: S2.10.

Wang, Z. and Zhao, B. (2020). *Xanthomonas euvesicatoria* type III effector XopN promotes an aqueous apoplast and enhances bacterial proliferation in both tobacco and pepper. (Abstr.) *Phytopathology* 110: S1.36.

9.24 BACTERIAL CANKER AND WILT OF TOMATO

Bacterial canker of tomato was first observed by Erwin F. Smith in 1909 in Grand Rapids, Michigan (USA); hence the name he gave it: Grand Rapids disease. However, Smith soon renamed the disease as bacterial canker of tomato. By the 1920s, the occurrence of the disease was reported from Central and South America, Asia, Africa, Australia and Europe. Now the disease occurs around the globe, wherever tomatoes are grown. Although usually sporadic in its occurrence, it is so destructive in nature that both field and greenhouse crops, although especially the latter, suffer heavily from the disease. It is a quarantined pathogen and Canada, the USA, the European Union (EPPO A2 quarantine pest) and some other countries enforce zero tolerance for the import and export of plant material infected with the pathogen. Janse (2005) has listed 60 countries where the disease occurs. Yield losses, ranging from 20–70%, have been reported from France and the USA.

The review article by Nandi *et al.* (2018) presents a detailed investigation regarding the pathogen, bacterial canker infection, molecular interactions between pathogen and tomato, and current perspectives on improved disease management.

9.24.1 Host Range

In addition to tomato, the bacterium attacks *Solanum douglassii* (nightshade), *S. nigrum* (black nightshade) and *S. triflorum*.

9.24.2 Symptoms

Tomato plants are susceptible at all the growth stages. A wide variety of symptoms, resulting from superficial or localised and systemic infections, are produced. **Wilting** is the most common symptom in spite of different names of the disease.

Generally, there are no symptoms on the seedlings, but under favourable conditions, infected seedlings show weak and stunted growth or may be killed. The wilting is first seen on the older leaves and the leaves may curl downward before wilting. Wilting of leaflets often occurs on one side of the petiole, but soon the leaflets on the other side also wilt and wilting progresses, involving the whole plant. The progress of the wilt is more rapid in younger plants. A light brown discolouration of vascular tissues is visible when wilted stems, branches and petioles are split open vertically. The stem discolouration is more prominent at nodes and just above the soil line. In the advanced stages of disease development, cankers develop on stems, petioles and midribs in the form of brownish streaks. These streaks split open, exposing necrotic tissue in the pith and cortex. Pith turns granular to mealy and cavities are formed. In severely infected plants, yellow ooze may exude from the cut end of a stem when it is squeezed. The seed-borne inoculum, or invasion of the vascular system by the pathogen entering through wounds, results in systemic infection, leading to wilting and canker formation.

Superficial/localised infections resulting from the entry of epiphytic bacteria through stomata, hydathodes and broken trichomes (hairs) produce the following symptoms. Bacteria entering through hydathodes cause marginal necrosis of leaves; these brown necrotic areas (**leaf "firing"**) may extend until the entire leaflet is shrivelled. Round to irregular spots also appear on the leaves. Similar spots may also be noted on the stem and petioles (Figure 9.23a).

On green fruits, initially small, white, superficial spots appear. These spots enlarge up to 1–3 mm in diameter and develop a raised dark brown centre surrounded by a distinct white halo (Figure 9.23b). As these spots resemble a bird's eye, they are known as **bird's-eye spots**. However, their occurrence is erratic and these do not always develop in disease outbreaks. Their occurrence reduces the market value of the fruits, sometimes making them unsalable (Braun, 2001). Systemic infection results in a darkening of the vascular tissues within the fruit. The bacteria can multiply in the vascular bundles within the fruits up to the seed. This leads to the formation of yellowish strands from the stem to the seeds, and also the internal infection of the seed. The fruit may show a black peppering at the vascular bundles under the calyx scar.

9.24.3 Causal Organism

Clavibacter michiganensis (Smith 1910) Li *et al.* 2018

Synonyms: *Cla. michiganensis* subsp. *michiganensis*, *Corynebacterium michiganense* subsp. *michiganense*, *Cor. michiganense* pv. michiganense

FIGURE 9.23 (a) Bacterial canker and wilt of tomato: Wilted plant showing necrosis and death of leaves. (Courtesy: Diane Cuppels, Agriculture & Agri-Food, Canada.) (b) Bacterial canker of tomato: Bird's eye-like spots on fruit. (Courtesy: Diane Cuppels, Agriculture & Agri-Food, Canada.)

E. F. Smith was the first to isolate and characterise the bacterium; he named it *Aplanobacter michiganense*. Its name was changed several times before it was named *Corynebacterium michiganense*, a name that remained in use for several decades. Davis *et al.* (1984) suggested the removal of all the plant pathogenic bacteria from genus *Corynebacterium* and placed this bacterium in a new genus, *Clavibacter*, as *Cla. michiganensis*.

Davis *et al.* (1984) proposed the reclassification of *Cor. michiganense* and its five subspecies namely, *michiganense*, *insidiosum*, *sepedonicum*, *nebraskense* and *tessellarium*), *Cor. iranicum*, *Cor. tritici* and *Cor. rathayi* in the new genus *Clavibacter*. Li *et al.* (2018) reported that the whole-genome analysis, based on average nucleotide identity and digital DNA–DNA hybridisation as well as multilocus sequence analysis (MLSA) of seven housekeeping genes, support raising each of the *Cla. michiganensis* subspecies to species level. Therefore, the recent name of the causal organism of the disease is *Cla. michiganensis* and not *Cla. michiganensis* subsp. *michiganensis*, as proposed by Davis *et al.* (1984).

The bacterium grows well on nutrient agar containing 5% glucose and potato dextrose agar producing yellow mucoid colonies. Cells are Gram-stain-positive, non-motile, non-sporing and irregularly shaped rods. Gelatin is weakly liquefied and H_2S is produced. Acid is produced from mannose fructose and sucrose, but not from dulcitol, sorbitol, inulin and α-methyl-D-glucoside. Acetate, citrate, fumarate, malate and succinate are used as a sole source of carbon. The genes involved in pathogenicity have been identified and characterised. The following characteristics of this bacterium differentiate it from other species of *Clavibacter*: tolerance to sodium chloride up to 6%, H_2S production, weak liquefaction of gelatin and utilisation of mannose, cellobiose and succinate but not of mannitol and propionate as the sole source of carbon.

Ftayeh *et al.* (2011) developed a new selective and highly sensitive medium for isolation of *Cla. michiganensis* subsp. *michiganensis* (hereafter referred to as *Cmm*) from seed and latently infected plants. The new medium (BCT) proved superior to all published semi-selective media (D2, KBT, D2ANX, SCM, mSCM, CMM1, mCNS and EPPO) for *Cmm* detection. BCT is designated as a selective medium because: (i) its mean plating efficiency amounts to $\leq$89% within 7 days for all 30 *Cmm* strains from different sources tested; (ii) the high selectivity, because of inhibition to an extent of 98–100% of accompanying bacterial species occurring on tomato plants and seed, or bacteria obtained from culture collections; and (iii) the remarkable detection sensitivity. Because of its high selectivity, 8 cfu of *Cmm* in field plant homogenates containing 12,750 cfu of accompanying saprophytes were detected on BCT. Moreover, BCT also supported growth of the closely related *Cla. michiganensis* subsp. *insidiosus*, *nebraskensis* and *tessellarius*. The authors have recommended the new medium for *Cmm* detection in tomato seed and in symptomless tomato plantlets to improve the control of bacterial canker of tomato.

Yasuhara-Bell *et al.* (2014) used LAMP to specifically identify *Cmm*, using LAMP primers developed to detect *micA*, a chromosomally stable gene that encodes a type II lantibiotic, michiganin A. A total of 409 bacterial strains (351 *Cmm* and 58 non-*Cmm*) from a worldwide collection were tested with LAMP to determine its specificity. LAMP results were compared with genetic profiles established using PCR amplification of seven genes (*dnaA*, *ppaJ*, *pat-1*, *chpC*, *tomA*, *ppaA* and *ppaC*). *Cmm* strains produced eight distinct profiles. LAMP reaction identified all *Cmm* strains and distinguished them from other *Cla. michiganensis* subspecies and non-*Clavibacter* bacteria. LAMP has advantages over immunodiagnostic and other molecular detection methods due to its specificity and isothermal nature, which allows for easy field application. LAMP reaction is also not affected by as many inhibitors as PCR. Therefore, LAMP has a potential to provide an easy, one-step test for rapid identification of *Cmm*.

Yasuhara-Bell *et al.* (2014) reported that three genes, clavicidin (*clvA*) and two associated genes (*clvF* and *clvG*, encoding a putative ABC-type multi-drug transport system ATPase and permease component, respectively) are unique and highly conserved in *Cmm*. Loop-mediated amplification of *clvA* and PCR of *clvA*, *clvF* and *clvG* showed that these genes were present only in *Cmm* and not in other *Clavibacter* species/sub species, as well as other genera of plant-associated bacteria. Sequences of *clvA*, *clvF* and *clvG* from 48 geographically diverse strains of *Cmm* were analysed phylogenetically to determine sequence variation. Maximum parsimony, neighbour-joining and maximum likelihood analyses placed strains into sub groups irrespective of their geographical origins. Multi-locus sequence analysis of these three genes produced the same results. The data support the use of the *clvA* gene for identification of *Cmm*, using either a previously designed LAMP assay or PCR with additional primers developed in this study. The genes investigated here are novel targets for specific detection of *Cmm* and have numerous molecular diagnostic applications.

Sen *et al.* (2015) reported that *Cmm* strains show great variability in virulence and are usually described as being hypervirulent, hypovirulent, or non-virulent. The bacterium lacks a type III secretion system, and only a few virulence factors have been experimentally determined from the many putative virulence factors. As its molecular mode of infection is unknown, researchers have avoided intensive work on this pathogen. Genetic plant mechanisms conferring resistance to the bacterium are apparently complex, and breeders have yet to develop disease resistant cultivars.

Larrea-Sarmiento *et al.* (2019) sequenced three strains of *Cmm*: virulent and extracellular polysaccharide (EPS) producing, virulent and non-EPS-producing and non-virulent and EPS-producing using PacBio. Comparative genomic analyses revealed no truncated clusters associated with EPS production; however, large-scale rearrangement events were observed. Plasmid organisation was different for each strain and the plasmid sequences showed significant differences in their genomic constituents. Plasmids pCM1 and pCM2 were present in virulent EPS-producing and virulent non-EPS-producing strains, respectively, but showed high discrepancy when aligned with reference genome NCPPB 382. Non-pathogenic strains showed disrupted and/or truncated pathogenicity islands (tomA and chp clusters). Scanning electron microscopy and qPCR results confirmed that the EPS-producing virulent strain colonised the host more rapidly, whereas, non-virulent strain showed no colonisation of the stem vascular tissue (lacked the colonisation facilitator gene, chpG), although it was detected in the root samples.

Malliarakis *et al.* (2023) collected isolates of *Clavibacter michiganensis* subsp. *michiganensis* (*Cmm*) from infected tomato plants of 11 Greek Prefectures during the years 2003–18 and investigated the genetic variability. Multi-Locus Sequencing Analysis highlighted 4.1% variability in nucleotide sequences and classified the 93 *Cmm* strains into 38 haplotypes. The study also provides important epidemiological information on the disease outbreaks and the spread of *Cmm* in Greece.

9.24.4 Disease Cycle

An infected seed is an important source of perpetuation of the bacterium, surviving in the innermost layers of the seed coat and acting as an external contaminant. The infection of the seed results from the systemic infection of the plant while external contamination occurs during the extraction of the seed. One infected seed in 10,000 may be enough for the outbreak of the disease under favourable conditions. The pathogen can survive in the infected plant debris in soil up to two or three years. However, the survival in plant debris is related to the extent of decomposition of the debris in the soil. Woody stems buried deep in soil usually remain intact and help in the survival of the bacterium. The bacterium also survives on volunteer tomato plants and weed hosts.

Seed-borne inoculum plays an important role in the outbreak of the disease. The contamination of seeds with only a few bacterial cells, below the detection level limits, gives rise to a high number of infected seedlings. The seed lots, having very few infested seeds, pose a serious problem in greenhouse crops. Undoubtedly the certified seed reduces the chances of infection, but is no guarantee of a pathogen-free seed. When contaminated seeds germinate, the bacteria enter the cotyledons through wounds, probably through broken trichomes. The bacteria move systemically through the xylem, from which it enters the phloem, pith and cortex. The optimum conditions for disease development include high moisture/relative humidity and a temperature of 24–32° C. There is usually a long latent period of several weeks in mature plants.

Detection of pathogen in seed lots is required for seed certification. In this respect, PCR is considered the most sensitive method for the detection of plant pathogenic bacteria. Classical PR primers derived from the *pat-1* gene in plasmid pCM2 of *Cmm* were developed for its detection, but the primers reacted with 75% of the strains tested.

Brochu *et al.* (2024) developed a specific and sensitive multiplex TaqMan qPCR assay to detect *Cla. michiganensis* (*Cm*) and distinguish it from related bacterial species that affect tomato plants. Two *Cm* chromosomal virulence-related genes, *rhu*M and *tom*A, were used as specific targets. The plant internal control *tubulin alpha-3* was included in each of the multiplexes to improve the reliability of the assay. Specificity was evaluated with 37 bacterial strains, including other *Clavibacter* spp. and related and unrelated bacterial pathogens from different geographic locations affecting a wide variety of hosts. Results showed that the assay is able to discriminate *Cm* strains from other related bacteria. The assay was validated on tissue and seed samples following artificial infection, and all tested samples accurately detected the presence of *Cm*. The tool described here is highly specific, sensitive and reliable for the detection of *Cm*; it allows the quantification of *Cm* in seeds, roots, stems and leaves. The diagnostic assay can also be adapted for multiple purposes, such as seed certification programmes, surveillance, biosafety, the effectiveness of control methods, border protection and epidemiological studies.

Infected seedlings are one of the most important primary sources of inoculum of the bacterium in commercial production plots. It was found that the pathogen dispersed spatially from root-inoculated source seedlings and colonised the leaf surfaces of surrounding seedlings to distances of 65–75 cm. A subirrigation (also known as seepage *irrigation)* system reduced, but did not prevent, dispersal of the bacterium. Infections by epiphytic *Cmm* populations can occur under a wide range of temperature conditions, seedling ages and foliar wetness periods. There is a need to minimise the primary sources of infection as well as developing new strategies to reduce the dispersal during the nursery stage (Frenkel *et al.*, 2016).

The pathogen can survive in the infected plant debris in soil for two to three years. However, the survival in plant debris is related to the extent of decomposition of the debris in the soil. Woody stems buried deep in soil usually remain intact and help in the survival of the bacterium. Vega and Romero (2016) reported that *Cmm* survived in host debris left on the soil surface in greenhouses (La Plata, Buenos Aires Province, Argentina) for 120–260 days for crop production cycles that ended in winter and for 45–75 days for those that ended in summer. In stems or roots buried in winter, this period was 45–75 days. They concluded that host debris, including roots, might be an important primary inoculum source of the pathogen in greenhouses.

Bach *et al.* (2003) developed a real-time TaqMan PCR assay based on the primers specific to the internal transcribed spacer sequence between 16S and 23S rDNA, a common region in all the five subspecies of *Cla. michiganensis*, and a subspecies-specific TaqMan probe for the detection of this bacterium. However, both the above-mentioned PCR assays do not distinguish between the living and dead cells of the bacterium and false positive results obtained with dead cells limit the use of PCR in risk analysis of seed lots. However, a quantitative real-time PCR assay devised by binding the bacterial DNA with ethidium monoazide has taken care of this problem as the method allows the selective quantification of only viable cells (Luo *et al.*, 2008). The primers and TaqMan probe, based on 16S–23S rDNA spacer sequences, were highly specific for *Cmm* at the subspecies level. The detection limit of the direct real-time PCR was 10^3 cfu mL^{-1} in samples and with an apparent sensitivity of 2 cfu of target cells in PCR reaction solution.

Penetration of the bacterium into the host occurs through wounds caused by cultural operations, broken trichomes and natural openings such as stomata and hydathodes. On tomato fruits, the wounds caused by wind-blown sand, breaking of hairs and insect punctures also act as avenues for the entry of the pathogen. The secondary spread of the bacterium occurs mainly through wind-splashed rain and in greenhouses through sprinkler irrigation. Secondary spread also occurs through contaminated equipment and cultural operations like pruning, and the staking and training of vines. The bacteria growing epiphytically on healthy and diseased plants serve as a major source of inoculum for the secondary spread.

Lelis *et al.* (2014) investigated the colonisation and disease development of axenically grown tomato plants by *Cmm*, using a spontaneous rifampicin resistant strain of *Cmm* tagged with a marker that expressed a green fluorescent protein (GFP) in a stable way and which possessed a similar virulence to the parental strain. Symptoms such as canker, wilting and growth reduction, typical for *Cmm*, developed in *in vitro* plants. The presence of *Cmm* in the vascular and parenchymatic tissue of *in vitro* tomato plants was confirmed by epifluorescence stereo- and confocal laser-scanning microscopy. The study shows that *in vitro* tomato plants can be effectively used for detailed studies on interactions between *Cmm* and its host, in particular if a GFP-tagged strain of the pathogen is used.

Xu *et al.* (2012) reported that a virulent, stable, constitutively bioluminescent *Cmm* strain, BL-*Cmm* 17, coupled with an *in vivo* imaging system, allowed visualisation of the *Cmm* colonisation process in tomato seedlings in real time. The dynamics of bacterial infection in seedlings through wounds were compared under two humidity regimes, i.e., 45% and 83%. Bacteria multiplied rapidly in cotyledon petioles after clip inoculation, and moved in the stem toward both roots and shoots. Luminescent signals were also observed in the roots of tomato seedling over time, and root development was retarded in inoculated plants maintained under both humidity regimes. Wilting was more severe in seedlings at 83% relative humidity. A strong positive correlation between light intensity and bacterial population *in planta* suggests that bioluminescent *Cmm* strains will be useful in evaluating the efficacy of bactericides and host resistance.

The pathogenicity of *Cmm* is dependent on plasmid-borne virulence factors and serine proteases located on the chromosomal *chp/tomA* PAI. Chalupowicz *et al.* (2012) examined the colonisation patterns and movement of *Cmm* during tomato infection using a GFP-labeled strain. A plasmid expressing GFP in *Cmm* was constructed and found to be stable *in planta* for at least for one month. Confocal laser-scanning microscopy (CLSM) of inoculated stems showed that the pathogen extensively colonised the lumen of xylem vessels and preferentially attached to spiral secondary wall thickening of the protoxylem. Acropetal movement of the wild-type strain *Cmm* NCPPB382 (*Cmm*382) in tomato resulted in an extensive systemic colonisation of the whole plant reaching the apical region after 15 days, whereas *Cmm*100 (lacking the plasmids pCM1 and pCM2) or *Cmm*27 (lacking the *chp/tomA* PAI) remained confined to the area surrounding the inoculation site. *Cmm*382 formed biofilm-like structures composed of large bacterial aggregates on the interior of xylem walls as observed by confocal laser-scanning microscopy and scanning electron microscopy. These findings suggest that virulence factors located on the *chp/tomA* PAI or the plasmids are required for the effective movement of the pathogen and for the formation of cellular aggregates in tomato.

Laird *et al.* (2020) assessed *Bacillus velezensis* strain 1B-23 and *Bacillus* sp. strain 1D-12 for their potential biocontrol abilities against *Cmm* strain 98-1. Both *Bacillus* strains interfered with the growth of *Cmm* 98-1 *in vitro*, as determined by agar plate assays. Inoculation of *Cmm* 98-1-infected tomato plants with *Bac. velezenis* 1B-23 or *Bacillus* sp. 1D-12 led to significant reduction in disease incidence in a greenhouse trial. Liquid chromatography coupled to the Mass Spectrometry (LC-MS) of 1B-23 and 1D-12 extracts identified [Leu7] surfactin C13 (often called surfactin A), [Leu7] surfactin C14 (often called surfactin B) and [Leu7] surfactin C15 (often called surfactin C) in fractions of extracts that inhibited growth of *Cmm* strain 98-1.

Valenzuela *et al.* (2019) found that four strains of *Cmm*, out of 25 strains tested, were highly sensitive to streptomycin, showing minimum inhibitory concentration values (MIC) of <2 μg mL^{-1}. The remaining 21 strains possessed MIC of ≥100 μg mL^{-1} of streptomycin. The *strB* gene, encoding an aminoglycoside 6-phosphotransferase, which inactivates streptomycin, was detected in all the Chilean strains, including sensitive and resistant strains. In the 21 resistant strains, a mutation in codon 43 of the *rpsL* gene was determined, conferring high streptomycin resistance. Interestingly, the four streptomycin-sensitive *Cmm* strains did not possess this mutation. The study shows that the continuous use of streptomycin leads to emergence of resistant *Cmm* strains, directing researchers to look for novel alternatives to control this plant pathogenic bacterium.

Wassermann *et al.* (2020) genotypically characterised a collection of *Cmm* isolates from the main greenhouse tomato-producing areas of Argentina by BOX-PCR fingerprinting and screened for the presence of genes and plasmids involved in pathogenicity by PCR. BOX-PCR fingerprinting clustered strains into four groups. Group II was dominant and included the most virulent strains, while Group III was the smallest and had the least virulent strains. All local strains exhibited similar cellulolytic activity. Most of the examined strains carried two plasmids of similar size to those of NCPPB382, although there were strains with one or three plasmids. By PCR amplification of *repA*, pCM1 was detected only in strains belonging to Group III, which includes local strains closely related to reference strain NCPPB382. All analysed pathogenicity genes were widespread among strains, and so in strains belonging to Groups I and II, *celA* found on pCM1 in NCPPB382 could be found in the chromosome or in plasmids other than pCM1. This study contributes to a better understanding of the diversity of *Cmm* genetic profiles and virulence of strains present in Argentina. Such information could be useful for the selection of strains for screening of host resistance and development of resistant tomato varieties.

Nadzir *et al.* (2019) developed a new inoculation protocol on *in vitro* tomato plants to facilitate a more economical and higher-throughput disease screening. A new method using the PathoScreen system was tested to localise green fluorescent protein-tagged (GFP) *Cmm in planta* and to quantify the pathogen based on the percentage of corrected GFP (cGFP%). The system was sensitive in detecting the GFP-tagged *Cmm* in the shoots; in the roots, however, a high autofluorescence masked the detection and thus sensitivity of the assay. *In vitro* protocol was tested on several wild relatives of tomato, which were previously screened in a greenhouse assay. The correlation between wilt symptoms *in vitro* and in the greenhouse was overall moderate (r = 0.6462). The protocol worked well in differentiating the two parents that were used in the mapping studies. The study shows that the *in vitro* protocol can be efficiently used for resistance breeding in many tomato genotypes.

Thapa *et al.* (2020) utilised sequences of *Cla. michiganensis*-specific loci to develop a multiplex PCR-based diagnostic platform using two *Cla. michiganensis* chromosomal genes (*rhuM* and *tomA*) and an internal control amplifying both bacterial and plant DNA (16s ribosomal RNA). The multiplex PCR assay specifically detected *Cla. michiganensis* strains from a panel of 110 additional bacteria, including other *Clavibacter* spp. and bacterial pathogens of tomato. The assay was adapted to detect the presence of *Cla. michiganensis* in seed and tomato plant materials with high sensitivity and specificity. In conclusion, the described method represents a robust, specific tool for the detection of *Cla. michiganensis* in tomato seed and infected plants.

The disease symptoms observed in plants infected systemically by *Cmm* are wilting and cankers on the stem, whereas blister-like spots develop in locally infected leaves. A wide repertoire of serine proteases and cell wall degrading enzymes has been implicated in the development of wilt and canker symptoms. However, virulence factors involved in the formation of blister-like spots, which play an important role in *Cmm* secondary spread in tomato nurseries, are largely unknown. Chalupowicz *et al.* (2017) demonstrated that *Cmm* virulence factors play different roles during blister formation relative to wilting. Inoculation with a green fluorescent protein (GFP)-labelled *Cmm*382 indicated that penetration occurs mainly through trichomes. When spray inoculated on tomato leaves, the wild-type *Cmm*382 and *Cmm*100 (lacking plasmids pCM1 and pCM2) strains form blister-like spots on leaves, whereas *Cmm*27 (lacking the *chp*/*tomA* pathogenicity island) is non-pathogenic, indicating that plasmid-borne genes, which have a crucial role in wilting, are not required for blister formation. Conversely, mutations in chromosomal genes encoding serine proteases (*chpC* and *sbtA*), cell wall degrading enzymes (*pgaA* and *endX/Y*), a transcriptional regulator (*vatr2*), a putative perforin (*perF*) and a putative sortase (*srtA*) significantly affect disease incidence and the severity of blister formation. The transcript levels of these genes, as measured by quantitative reverse transcription-polymerase chain reaction, showed that, during blister formation, they are expressed early at 8–16 h after inoculation, whereas, during wilting, these are either expressed after 24–72 h or expressed at low levels. Plant gene expression studies suggest that *chpC* is involved in the suppression of host defence.

The viable, but non-culturable (VBNC) state is a unique survival strategy of bacteria in response to stress conditions. Bai *et al.* (2022) used RNA-sequencing analysis to monitor the mechanisms of the VBNC state during $CuSO_4$ induction in *Cla. michiganensis*. The results identified that numerous genes involved in stringent response, copper resistance and stress resistance were upregulated, and some involved in cell division were downregulated significantly. The authors investigated the importance of Rel, which is an essential enzyme in the synthesis of the molecular alarmone ppGpp, via the generation of a Δ*rel* mutant and its complementation strain. Biological characterisation revealed that a deficiency of *rel* reduced the bacterial growth, the production of exopolysaccharides and pathogenicity as well as ppGpp production. The Δ*rel* mutant increased the sensitivity to environmental stress, exhibiting reduced growth on minimal media and a propensity to enter the VBNC state in response to $CuSO_4$. These findings have important implications for the understanding of the survival mechanism and management of *Cla. michiganensis* and other phytopathogenic bacteria.

Peritore-Galve *et al.* (2021), in their review article, have summarised the information on epidemiology, host colonisation, and the bacterial genetics underlying the virulence and management of bacterial canker of tomato. They have also highlighted important areas of research into this pathosystem which have the potential to generate new strategies for the prevention and mitigation of this disease.

9.24.5 Management

The management of bacterial canker of tomato poses a significant challenge due to the lack of effective chemicals and resistant varieties and to the systemic nature of the causal agent. The following multifaceted approach, which mainly prevents or reduces the primary sources of inoculum, will reduce the incidence and severity of the disease.

1. Use only certified and pathogen-free seed, obtained from canker free plants for planting. Seed should be extracted only by the standard acetic acid extraction method or by the fermentation process. The seed extracted by these methods is free from seed coat contamination, but the embryonic infection of the seed is not completely controlled. Therefore, the seed must be treated with hot water or disinfected with acid or chlorine. In hot water treatment, the seed is pre-warmed in water at 38 °C for 10 minutes and then kept in hot water at 50 °C for 25 minutes. Immediately after the treatment, put the seed in cold water for 5 minutes. The disinfection of seed with acid is done by dipping the seed in a 0.8% solution of acetic acid for 24 h at 21 °C and with chlorine by dipping the seed in 1% sodium hypochlorite (20% bleach) for 20–40 minutes. For proper seed disinfection, pre-soaking of seed in water, ranging from 1–18 h, is essential.
2. Plant only certified disease-free transplants grown under a vigorous inspection programme.
3. Ensure the destruction of infected plant debris by its thorough decomposition. Soil solarisation, achieved by covering the soil with transparent polythene sheets for six weeks, proved effective in reducing the disease in plastic tunnels (Antoniou *et al.*, 1995).
4. Practise crop rotation for at least three years.
5. Destroy the volunteer tomato plants and weeds belonging to the Solanaceae family.
6. Sterilise seedbeds and soil in the greenhouse to destroy bacteria. Steam sterilisation is preferred over methyl bromide.
7. Pruning knives and other tools used in the greenhouse should be disinfected regularly. Prefer quaternary ammonium compounds over sodium hypochlorite or laundry bleach as a greenhouse disinfectant (Zitter, 1985).
8. Fixed copper sprays can be helpful in protecting the crop, particularly in superficial or localised infections. The combination of copper and mancozeb is better than copper alone. Preventive sprays which delay the disease development/outbreak until after the main fruit set will minimise the yield loss because fruit infection, responsible for major yield loss, occurs only on green immature fruits (less than 3 cm in diam).

The application of acibenzolar-S-methyl (benzo-[1, 2, 3]-thiadiazole-7-carbothioic acid-S-methyl ester, ASM; Bion 50 WG) activates the defence response of the host against the bacterium. Baysal *et al.* (2003) reported reduction in disease severity up to 76.3%, along with reduction in bacterial growth up to 68.2% in ASM-treated plants. The enhanced resistance was due to the significant increase in peroxidase and chitinase activity.

9. Barda *et al.* (2015) reported that the extracellular metabolites secreted by an epiphytic fungus *Pseudozyma aphidis* inhibited *Xan. campestris* pv. *vesicatoria*, *Xan. campestris* pv. *campestris*, *Pseudomonas syringae* pv. *tomato*, *Erwinia amylovora*, *Cla. michiganensis* and *Agrobacterium tumefaciens in vitro*. The application of *P. aphidis* spores on tomato plants in the greenhouse reduced the incidence of bacterial canker by 60% and the severity by 35%. Moreover, infected plants treated with *P. aphidis* were 25% taller than control plants. It was found that *P. aphidis* activated *PR1a* and other pathogenesis related genes in tomato plants and could trigger an induced-resistance response against *Cla. michiganensis* which proceeded in a salicylic acid-independent manner, as shown using NahG-transgenic tomato plants.
10. Wittmann *et al.* (2016) developed transgenic tomato plants by transferring the endolysin gene (*lys*) of bacteriophage CMP1. Endolysin gene encodes a peptidase that effectively reduces the *Cla. michiganensis* population by specifically hydrolysing its murein. The gene was transferred into tomato plants by *Agrobacterium*-mediated transformation. The presence of the gene was verified by PCR and the gene product was confirmed in immunoblots; it was expressed stably over three generations. Transgenic tomato plants did not show disease symptoms after infection with *Cmm*, but the significantly reduced populations of bacterial cells were identified in xylem sap and leaf extracts.
11. Tremblay *et al.* (2024) reported that ethanolic sugar maple autumn-shed leaf (SMASL) extract at 25 mg/mL completely eliminated the pathogen from tomato seeds without negatively impacting on germination. SMASL extract foliar spray applications using concentrations of 6.25 and 12.5 mg/mL significantly checked disease development under greenhouse and field conditions, showing better efficacy than copper octanoate.

REFERENCES

Antoniou, P. P., Tjamos, E. C., Andreou, M. T. and Panagopoulos, C. G. (1995). Effectiveness, modes of action and commercial application of soil solarization for control of *Clavibacter michiganensis* subsp. *michiganensis* of tomatoes. *Acta Hort.* 382: 119–28.

Bach, H. J., Jessen, I., Schloter, M. and Munch, J. C. 2003. A TaqMan-PCR protocol for quantification and differentiation of the phytopathogenic *Clavibacter michiganensis* subspecies. *J. Microbiol, Methods* 52: 85–91.

Bai, K., Jiang, N., Chen, X., Xu. X., Li. J. *et al.* (2022). RNA-seq analysis discovers the critical role of Rel in ppGpp synthesis, pathogenicity, and the VBNC state of *Clavibacter michiganensis*. *Phytopathology* 112: 1844–58.

Barda, O., Shalev, O., Alster, S., Buxdorf, K., Gafni, A. *et al.* (2015). *Pseudozyma aphidis* induces salicylic-acid-independent resistance to *Clavibacter michiganensis* in tomato plants. *Plant Dis.* 99: 621–26.

Baysal, O., Soylu, E. M. and Soylu, S. (2003). Induction of defence-related enzymes and resistance by the plant activator acibenzolar-S-methyl in tomato seedlings against bacterial canker caused by *Clavibacter michiganensis* subsp. *michiganensis*. *Plant Pathol.* 52: 747–53.

Braun, E. (2001). Bacterial canker of tomato. In: O. C. Maloy and T. D. Murray (eds.). *Encyclopedia of Plant Pathology*. John Wiley and Sons, Inc., New York, pp. 76–77.

Brochu, A.-S., Dumonceaux, T. J., Valenzuela, M., Bélanger, R. and Pérez-López, E. (2024). A new multiplex TaqMan qPCR for precise detection and quantification of *Clavibacter michiganensis* in seeds and plant tissue. *Plant Dis.* 108: 2272–82. https://doi.org/10.1094/PDIS-06-23-1194-SR

Chalupowicz, L., Barash, I., Reuven, M., Dror, O., Sharabani, G. *et al.* (2017). Differential contribution of *Clavibacter michiganensis* ssp. *michiganensis* virulence factors to systemic and local infection in tomato. *Mol. Plant Pathol.* 18: 336–46.

Chalupowicz, L., Zellermann, E.-M., Fluegel, M., Dror, O., Eichenlaub, R. *et al.* (2012). Colonization and movement of GFP-labeled *Clavibacter michiganensis* subsp. *michiganensis* during tomato infection. *Phytopatholgy* 102: 23–31.

Davis, M. J., Gillaspie, A. G. Jr., Vidaver, A. K. and Harris, R. W. (1984). *Clavibacter*: A new genus containing some phytopathogenic coryneform bacteria, including *Clavibacter xyli* subsp. *xyli* sp. nov., subsp. nov. and *Clavibacter xyli* subsp. *cynodontis* subsp. nov., pathogens that cause ratoon stunting disease of sugarcane and Bermudagrass stunting disease. *Int. J. Syst. Bacteriol.* 34: 107–17.

Frenkel, O., Bornestein, M., Shulhani, R., Sharabani, G., Sofer, M., *et al.* (2016). Secondary spread of *Clavibacter michiganensis* subsp. *michiganensis* in nurseries and the conditions leading to infection of tomato seedlings. *Eur. J. Plant Pathol.* 144: 569–79.

Ftayeh, R. M., von Tiedemann, A. and Rudolph, K. W. E. (2011). A new selective medium for isolation of *Clavibacter michiganensis* subsp. *michiganensis* from tomato plants and seed. *Phytopathology* 101: 1355–64.

Janse, J. D. (2005). *Phytobacteriology: Principles and Practice*. CABI Publishing, Wallingford, 360 pp.

Laird, M., Piccoli, D., Weselowski, B. McDowell, T., Renaud, J. *et al.* (2020). Surfactin-producing *Bacillus velezensis* 1B-23 and *Bacillus* sp. 1D-12 protect tomato against bacterial canker caused by *Clavibacter michiganensis* subsp. *michiganensis*. *J. Plant Pathol.* 102: 451–58. https://doi.org/10.1007/s42161-019-00461-w

Larrea-Sarmiento, A., Dobhal, S., Alvarez, A. M. and Arif, M. (2019). Insights into pathogenicity determinants of *Clavibacter michiganensis* subsp. *michiganensis* and their effects on disease expression. *Phytopathology* 109: S2.188.

Lelis, F. M. V., Czajkowski, R., de Souza, R. M., Ribeiro, D. H. and van der Wolf, J. M. (2014). Studies on the colonization of axenically grown tomato plants by a GFP-tagged strain of *Clavibacter michiganensis* subsp. *michiganensis*. *Eur. J. Plant Pathol.* 139: 53–66.

Li, X., Tambong, J., Yuan, K. X., Chen, W., Xu, H. *et al.* (2018). Re-classification of *Clavibacter michiganensis* subspecies on the basis of whole-genome and multilocus sequence analyses. *Int. J. Syst. Evol. Microbiol.* 68: 234–40.

Luo, L. X., Walters, C., Bolkan, H., Liu, X. L. and Li, J. Q. (2008). Quantification of viable cells of *Clavibacter michiganensis* subsp. *michiganensis* using a DNA binding dye and a real-time PCR assay. *Plant Pathol.* 57: 332–37.

Malliarakis, D., Pagoulatou, M. G., Mpalantinaki, E., Trantas, E., Ververidis, F. *et al.* (2023). Phylogenetic diversity of *Clavibacter michiganensis* subsp. *michiganensis* isolates causing bacterial canker of tomato in Greece. *J. Plant Pathol.* 105: 1403–19. https://doi.org/10.1007/s42161-023-01375-4

Nadzir, M. M. M., Lelis, F. M. V., Thapa, B., Ali, A. and Visser, R. G. F. (2019). Development of an *in vitro* protocol to screen *Clavibacter michiganensis* subsp. *michiganensis* pathogenicity in different *Solanum* species. *Plant Pathol.* 68: 42–48.

Nandi, M., Macdonald, J., Liu, P., Weselowski, B. and Yuan, Z. (2018). *Clavibacter michiganensis* ssp. *michiganensis*: Bacterial canker of tomato, molecular interactions and disease management. *Mol. Plant Pathol.* 19: 2036–50.

Peritore-Galve, F. C., Tancos, M. A. and Smart, C. D. (2021). Bacterial canker of tomato: Revisiting a global and economically damaging seedborne pathogen. *Plant Dis.* 105: 1581–95.

Sen, Y., van der Wolf, J., Visser, R. G. F. and van Heusden, S. (2015). Bacterial canker of tomato: Current knowledge of detection, management, resistance, and interactions. *Plant Dis.* 99: 4–13.

Smith, E. F. (1910). A new tomato disease of economic importance. *Science* 31: 794–96.

Thapa, S. P., O'Leary, M., Jacques, M.-A., Gilbertson, R. L. and Coaker, G. (2020). Comparative genomics to develop a specific multiplex PCR assay for detection of *Clavibacter michiganensis*. *Phytopathology* 110: 556–66.

Tremblay, V., Delisle-Houde, M., Demers, F., D'Amours, C., Filion, M. *et al.* (2024). Sugar maple leaf extracts: A new tool to control bacterial canker of tomato caused by *Clavibacter michiganensis* subsp. *michiganensis*. *Plant Pathol.* 73: 2123–31. https://doi.org/10.1111/ppa.13959

Valenzuela, M., Méndez, V., Montenegro, I., Besoain, X. and Seeger, M. (2019). Streptomycin resistance in *Clavibacter michiganensis* subsp. *michiganensis* strains from Chile is related to an *rpsL* gene mutation. *Plant Pathol.* 68: 426–33.

Vega, D. and Romero, A. M. (2016). Survival of *Clavibacter michiganensis* subsp. *michiganensis* in tomato debris under greenhouse conditions. *Plant Pathol.* 65: 545–50.

Wassermann, E., Montecchia, M. S., Garaventa, V. S., Correa, O. S. and Romero, A. M. (2020). Virulence and pCM1 plasmid carriage are related to BOX-PCR fingerprint type in strains of *Clavibacter michiganensis* subsp. *michiganensis* that cause bacterial wilt and canker of tomato in Argentina. *Plant Pathol.* 69: 723–32.

Wittmann, J., Brancato, C., Berendzen, K. W. and Dreiseikelmann, B. (2016). Development of a tomato plant resistant to *Clavibacter michiganensis* using the endolysin gene of bacteriophage CMP1 as a transgene. *Plant Pathol.* 65: 496–502.

Xu, X., Rajashekara, G., Paul, P. A. and Miller, S. A. (2012). Colonization of tomato seedlings by bioluminescent *Clavibacter michiganensis* subsp. *michiganensis* under different humidity regimes. *Phytopathology* 102: 177–84.

Yasuhara-Bell, J., Marrero, G. and Alvarez, A. M. (2014). Genes *clvA*, *clvF* and *clvG* are unique to *Clavibacter michiganensis* subsp. *michiganensis* and highly conserved. *Eur. J. Plant Pathol.* 140: 655–64.

Zitter, T. A. (1985). *Bacterial Diseases of Tomato*. Department of Plant Pathology, Cornell University, Ithaca, NY.

9.25 BACTERIAL SPECK OF TOMATO

Bacterial speck of tomato caused by *Pseudomonas syringae* pv. *tomato* was first described by Okabe in 1933 from Taiwan. The economic importance of the disease was realised in the late 1970s when it caused serious losses to winter tomato crop of 1977–1978 in southern Florida and in 1978 in southern transplant fields and tomato production areas of the northern United States. Cool and moist conditions favour the development of the disease

9.25.1 Host Range

Tomato is the only economically important host of the pathogen. The bacterium has also been isolated from a wide range of crop and weed species, such as *Arabidopsis thaliana*, *Lamium amplexicaule*, *Stellaria media* and *Gnaphalium* spp.

9.25.2 Symptoms

The symptoms appear on all the above-ground plant parts. Small, dark brown specks, 1–2 mm in diameter, often surrounded by discrete yellow haloes, appear on the leaves. Speck lesions coalesce to form blighted areas, leading to considerable leaf distortion. On field plants, the lesions are more concentrated at the leaf margins (Figure 9.24a). Leaf lesions of bacterial speck are difficult to distinguish from those of bacterial spot and young lesions of early blight. Unlike bacterial leaf spot, the leaf lesions of bacterial speck are not greasy; in bacterial speck, however, the leaf curling is more severe than that of bacterial spot.

In severe infection, irregular, dark brown lesions also appear on stem and petioles and blossoms are also blighted. Lesions on stem and petioles also cannot be distinguished from those caused by bacterial spot. The infection of 3- to 5-leaf stage seedlings results in the stunting of seedlings and a considerable loss in yield.

On immature green fruits (less than 3 cm in diameter), dark brown to black, slightly raised lesions but smaller (1.5 mm in diameter) than those of bacterial spot appear, which may be surrounded by dark green haloes. Lesions are very superficial and do not crack or become scabby like those of bacterial spot. Ripe fruits are not susceptible to infection but the black lesions formed on unripe fruits exist even after ripening (Figure 9.24b).

9.25.3 Causal Organism

Pseudomonas syringae pv. *tomato* (Okabe 1933) Young *et al.* 1978

Synonym: *Psm. tomato*

Bacterial cells are straight, Gram-stain-negative rods and measure 0.7–1.0 x 1.8–2.8 μm. The bacterium produces

(a) (b)

FIGURE 9.24 (a) Bacterial speck of tomato–Symptoms on leaf. (Source: www.aggie-horticulture.tamu.edu.) (b) Bacterial speck of tomato: Symptoms on fruits. (Courtesy: M. Babadoost, *Illinois Fruit and Vegetable News*, Vol. 13, No. 8.)

cream-coloured colonies with a diffusible green fluorescent pigment on King's medium B. It is strictly aerobic, motile having 1–2 polar flagella, produces levan on nutrient agar containing 5% sucrose but oxidase and arginine dihydrolase tests are negative. It differs from a closely related tomato pathogen, *Psm. syringae* pv. *syringae* by its fatty acid profile, ability to cause pits in pectate gel at pH 4.6 and inability to utilise erythritol or DL-lactate as a sole source of carbon (Cuppels, 2001). G + C content of DNA is 58–60 mol%. The bacterium elicits a hypersensitive reaction on tobacco but does not cause rotting of potato tubers. Two races of the bacterium, namely, race 0 and race 1, are known. Yan *et al.* (2008) have proposed the division of *Psm. syringae* pv. *tomato* into two pathovars.

The bacterium has been the focus of intensive research and hence extensively used to study the molecular mechanisms of host responses to infection, because *Psm. syringae* pv. *tomato* strain DC 3000 is the pathogen of a model plant *Arabidopsis thaliana*. A major part of information about the plant immune system has been obtained by studying the interaction of *Psm. syringae* pv. *tomato* DC 3000 with its two hosts, *A. thaliana* and tomato. Molecular basis of pathogenesis in *Psm syringae* pv. *tomato* is based on the coordinated expression of multiple pathogenicity and virulence factors such as toxins, extracellular proteins and polysaccharides; and translocation of proteins into plant cells by the type III (hrp) secretion system. The contribution of individual virulence factors to parasitism and disease development varies significantly between strains (Preston, 2000). The complete genome sequence of *Psm. syringae* pv. *tomato* strain DC 3000 determined by Feil *et al.* (2005) revealed that the genome (6.5 mbp) is composed of a circular chromosome and two plasmids.

Tambong *et al.* (2022) reported the draft whole-genome sequences of 33 *Psm. syringae* pv. *tomato* strains. Libraries were constructed using a Nextera DNA Flex prep kit (Illumina) following the manufacturer's instructions. The draft genome sequences were determined by paired-end sequencing using Illumina NextSeq technology at the Molecular Technologies Laboratory (Ottawa Research and Development Centre, Ottawa, Canada). The average size of the draft genome sequences is 6.305 Mbp with a range of 6.195 to 6.390 Mbp (average maximum contig = 381,352 bp and N_{50} = 139,729 bp). The average G + C content of the draft genome sequences is 58.59% (58.55 to 58.65%), with an average estimated coverage of 46.8× (36× to 67×).

9.25.4 Disease Cycle

The bacterium survives in diseased plant debris, on the roots and foliage of non-host plants, including weeds. Bacterium can survive in soil for more than one year. Bacterium can also survive in the crevices and cavities of seed coat. The survival in/on seed can be for a long time (over 20 years). Dissemination of the bacterium occurs through wind-driven rain, overhead irrigation and farm equipment. It may move considerable distances in the water droplets or aerosols generated by rain storms. The pathogen also spreads through contaminated seed and diseased transplants.

Penetration of the bacterium into the host is mainly through wounds, stomata and hydathodes. The studies have shown that stomata can play an active role in limiting bacterial invasions as part of plant innate immune system (Melotto *et al.*, 2008). As a counter-defence, *Psm. syringae* pv. *tomato* strain DC 3000 uses the virulence factor coronatine to actively open stomata. Hail and windstorms are often associated with increased disease severity.

Santamaría-Hernando *et al.* (2022) showed that the perception of GABA and L-Pro, two abundant components of the tomato apoplast, through the PsPto-PscC chemoreceptor drives the entry of *Psm. syringae* pv. *tomato* into the tomato apoplast. Recognition of both compounds by PsPto-PscC caused chemoattraction to both amino acids and participated in the regulation of GABA catabolism. Mutation of the PsPto-PscC chemoreceptor caused a reduced chemotactic response towards these compounds which, in turn, impaired entry and reduced virulence in tomato plants. Interestingly, GABA and L-Pro levels significantly increase in tomato plants upon pathogen infection and are involved in the regulation of the plant defence response. This finding illustrates how bacteria respond to plant signals produced during the interaction as cues to access the plant apoplast and to ensure efficient infection.

Disease is favoured by cool (21 °C) and moist (high R. H. and prolonged period of free moisture) weather. The resident population of the bacterium present on tomato plants can cause symptoms in 3–5 days under favourable weather conditions and under field conditions the symptoms can develop within 6–10 days. Fruits that develop during cool and moist weather can be severely infected. Fruits are most susceptible to infection from anthesis till they reach a diameter of 3 cm. Glandular and non-glandular leaf hairs present on ovaries are gradually lost after anthesis, leaving openings in the epidermis of young fruit. Bacterial cells associated with leaf hairs enter through these openings and cause fruit infection.

Psm. syringae pv. *tomato* produces a toxin called coronatine, which acts as a virulence factor and is responsible for the yellow haloes surrounding leaf lesions and the stunting of young seedlings. Bacterial populations that do not produce this toxin, grow poorly on host plants and cause little damage. A fragment of gene cluster-controlling coronatine production has been used as a DNA probe for disease diagnosis and epidemiologically studies of the pathogen.

Zaccardelli *et al.* (2005) developed a rapid method for the detection and identification of the bacterium from tomato plants and seeds based on the PCR amplification of *Psm. syringae* pv. *tomato* strain DC 3000 *hrpZpst* gene. The results confirmed the high specificity observed using pure cultures. Cuppels *et al.* (2006) developed PCR-based lesion assays using crude DNA extracts and primer sets (COR 1/2 for bacterial speck and BSX 1/2 for bacterial spot) for the detection of bacterial speck and bacterial spot pathogens.

Xin and He (2013) have reviewed the work to characterise the molecular mechanisms by which *Psm. syringae* pv. *tomato* DC 3000 causes disease in plants. Genomic analysis shows that the bacterium carries a large repertoire of potential virulence factors, including proteinaceous effectors that are secreted through the type III secretion system and a polyketide phytotoxin called coronatine, which structurally mimics the plant hormone jasmonate. Findings concerning the pathogenesis of *Psm. syringae* pv. *tomato* DC 3000 has not only provided several conceptual advances in understanding how a bacterial pathogen employs type III effectors to suppress plant immune responses and promote disease susceptibility, but also facilitated the discovery of the immune function of stomata and key components of jasmonate signalling in plants. The concepts derived from the study of pathogenesis of this bacterium may prove useful in understanding pathogenesis mechanisms of other plant pathogens.

Moyano *et al.* (2020) demonstrated the participation of *Psm. syringae* pv. *tomato* DC3000 (*Pto*) phytochromes and light in the bacterial physiology during the interaction with tomato plants. They found that *Pto* phytochromes were implicated in the control of some features, such as motility, biofilm formation, adhesion and emulsification capability related to the bacterial capability to enter into the plant apoplast and cause bacterial speck disease. Red light and bacteriophytochromes are important during the early colonisation stage of tomato phyllosphere, affecting *Pto* virulence. In addition, the development of disease symptoms in infiltrated leaflets is affected by light, which may be the consequence of type-two secretion system regulation.

Zhang *et al.* (2020) found that tomato seedlings treated with 0.5 mM Pip showed an increased resistance to *Psm. syringae* pv. *tomato* DC3000 compared to control. After pathogen infection, the expression of defence-related genes increased in plants pre-treated with Pip, while the accumulation of reactive oxygen species decreased. The data revealed that the exogenous application of Pip induced resistance against *Psm. syringae* pv. *tomato* DC3000 in tomatoes, possibly through the regulation of reactive oxygen species accumulation and defence-related gene expression.

Chai *et al.* (2020) used propidium monoazide (PMA) pre-treatment followed by a qPCR (PMA-qPCR) for quantifying viable cells of *Psm. syringae* pv. *tomato* in contaminated tomato seed. PMA could selectively bind to the chromosomal DNA of dead bacterial cells and, therefore, blocked the DNA amplification of qPCR. The primer pair Pst3F/Pst3R was designed based on gene *hrpZ* to specifically amplify and quantify *Psm. syringae* pv. *tomato* by qPCR. The PMA pre-treatment protocol was optimised for selectively detecting viable *Psm. syringae* pv. *tomato* cells, and the optimal PMA concentration and light exposure time were 10 μmol liter^{-1} and 10 minutes, respectively. In the sensitivity test, the detection limit of PMA-qPCR for detecting viable cells in bacterial suspension and artificially contaminated tomato seed was 10^2 CFU ml^{-1} and 11.86 CFU g^{-1}, respectively. For naturally contaminated tomato seed, viable *Psm. syringae* pv. *tomato* cells were quantified in 6 of the 19 samples tested, with infestation levels of approximately 10^2 to 10^4 CFU g^{-1}. The results show that the PMA-qPCR assay is a suitable tool for quantifying viable cells of *Psm. syringae* pv. *tomato* in tomato seed and can be used to avoid the potential risks of primary inoculum sources from contaminated seed.

O'Malley *et al.* (2020) re-evaluated the function of GacA in *Psm. syringae* pv. *tomato* DC3000. Contrary to previous reports, the authors discovered that GacA negatively regulates the expression of TIIISS genes in DC3000, and that GacA is not required for DC3000 virulence inside *Arabidopsis* leaf tissue. However, their findings show that GacA is required for the full virulence of leaf surface-inoculated bacteria. These data significantly revise the current understanding of GacS/GacA in regulating *Psm. syringae* virulence.

To improve the sensitivity and specificity of the tomato seed testing method for bacterial spot pathogens, Wang *et al.* (2022) developed a long-amplicon quantitative PCR (qPCR) assay coupled with propidium monoazide (PMA-qPCR) to quantify selectively four pathogenic *Xanthomonas* species in tomato seed. The optimised PMA-qPCR procedure was evaluated on pure bacterial suspensions, bacteria-spiked seed extracts and seed extracts of inoculated and naturally infected seed. A crude DNA extraction protocol was also developed. PMA-qPCR with crude bacterial DNA extracts resulted in the accurate quantification of 10^4 to 10^8 CFU/ml of viable bacteria when mixed with dead cells at concentrations as high as 10^7 CFU/ml in the seed extracts. With DNA purified from concentrated seed extracts, the PMA-qPCR assay was able to detect the DNA of the target pathogens in seed samples spiked with ≥75 CFU/ml (about 0.5 CFU/seed) of the viable pathogens. Latent class analysis of the inoculated and naturally infected seed samples showed that the PMA-qPCR assay had greater sensitivity than plating the seed extracts on the semi-selective modified Tween Medium B and CKTM media for all the four target species. Being much faster and more sensitive than the dilution plate method, the PMA-qPCR assay has the potential to be used as a standalone tool or in combination with the plating method to improve tomato seed testing and advance the production of clean seed.

Santamaría-Hernando *et al.* (2020) found that blue-light perception by *Psm. syringae* pv. *tomato* (*Pst*) DC3000 on leaf surfaces is required for optimal colonisation. Blue light triggers the activation of metabolic activity and increases the transcript levels of five chemoreceptors through the function of light oxygen voltage and BphP1 photoreceptors. The inactivation of *Pst*_1008 and *Pst*_2526 chemoreceptors causes a reduction in virulence. These findings indicate that during *Pst* DC3000 interaction with tomato plants, light perception, chemotaxis and virulence are highly interwoven processes.

9.25.5 Management

1. Use of resistant varieties is the most effective management of the disease. A bacterial speck resistance gene, *Pto* (single dominant), discovered in an Ontario breeding line ONT 7710 of *S. pimpinellifolium*, confers resistance to race 0 but not to race 1 of the pathogen. This was the first plant resistance gene to be isolated and characterised (Martin *et al.*, 1993).
2. Use pathogen-free seed or treat the seed with hot water (pre-warm at 38 °C for 10 minutes followed by 50 °C for 25 minutes) or 0.8% acetic acid or 1% sodium hypochlorite (20% bleach).
3. Use of disease-free transplants.
4. Destroy diseased plant debris, weeds and volunteer plants.
5. Use of drip irrigation to reduce leaf wetness.
6. Practise crop rotation for three years.
7. Protective sprays of fixed copper compounds are effective. As copper resistance is widespread among the *Psm. syringae* pv. *tomato* populations in the main tomato-growing areas, the copper compounds should be combined with mancozeb or chlorothalonil. Sprays should be given at regular intervals starting from anthesis until the first-formed fruits attain the size of 3 cm in diameter.
8. Treatment of tomato seeds, artificially inoculated with *Psm. syringae* pv. *tomato*, with a combination of mild chemo-thermal treatment, the inoculation of seed with *Azospirillum brasilense*, a plant growth-promoting bacterium and, later, one foliar application of a copper bactericide, nearly eliminated the disease even when the plants were grown under mist for six weeks. However, these treatments were ineffective when tried individually (Bashan and de-Bashan, 2002).
9. For the use of acibenzolar-S-methyl in disease management, refer to bacterial spot of tomato and pepper.

REFERENCES

Bashan, Y. and de-Bashan, L. E. (2002). Reduction of bacterial speck (*Pseudo-monas syringae* pv. *tomato*) of tomato by combined treatments of plant growth-promoting bacterium, *Azospirillum brasilense*, streptomycin sulphate, and chemo-thermal seed treatment. *Eur. J. Plant Pathol.* 108: 821–29.

Chai, A.-L., Ben, H.-Y., Guo, W.-T., Shi, Y.-X., Xie, X.-W. *et al.* (2020). Quantification of viable cells of *Pseudomonas syringae* pv. *tomato* in tomato seed using propidium monoazide and a Real-Time PCR assay. *Plant Dis.* 104: 2225–32.

Cuppels, D. A. (2001). Bacterial speck of tomato. In: O. C. Maloy and T. D. Murray (eds.). *Encyclopedia of Plant Pathology*. Vol. 1: John Wiley and Sons, Inc. New York, pp. 82–83.

Cuppels, D. A., Louws, F. J. and Ainsworth, T. (2006). Development and evaluation of PCR-based diagnostic assays for the bacterial speck and bacterial spot pathogens of tomato. *Plant Dis.* 90: 451–58.

Feil, H., Feil, W. S., Chain, P. Larimer, F., DiBartolo, G. *et al.* (2005). Comparison of the complete genome sequences of *Pseudomonas syringae* pv. *syringae* B728a and pv. *tomato* DC3000. *Proc. Natl. Acad. Sci.* USA 102 (31): 11064–69.

Martin, G. B., Brommonschenkel, S. H., Chunwongse, J., Frary, A., Ganal, M. W. *et al.* (1993). Map-based cloning of a protein kinase gene conferring disease resistance in tomato. *Science* 262: 1432–36.

Melotto, M., Underwood, W. and He, S. Y. (2008). Role of stomata in plant innate immunity and foliar bacterial diseases. *Annu. Rev. Phytopathol.* 46: 101–22.

Moyano, L., Carrau, A., Petrocelli, S., Kraiselburd, I., Gärtner, W. *et al.* (2020). Bacteriophytochromes from *Pseudomonas syringae* pv. *tomato* DC3000 modulate the early stages of plant colonization during bacterial speck disease. *Eur. J. Plant Pathol.* 156: 695–712. https://doi.org/10.1007/s10658-019-01918-5

O'Malley, M. R., Chien, C.-F., Peck, S. C., Lin, N.-C. and Anderson. J. C. (2020). A revised model for the role of GacS/GacA in regulating type III secretion by *Pseudomonas syringae* pv. *tomato* DC3000. *Mol. Plant Pathol.* 21: 139–44.

Okabe, N. (1933). Bacterial diseases of plants occurring in Formosa. II. Bacterial leaf spot of tomato. *J. Soc. Trop. Agri.* Taiwan 5: 26–36.

Preston, G. M. (2000). *Pseudomonas syringae* pv. *tomato*: The right pathogen, of the right plant, at the right time. *Mol. Plant Pathol.* 1: 263–75.

Santamaría-Hernando, S., Cerna-Vargas, J. P., Martínez-García, P. M., Polanco, S. F., Nebreda, S. *et al.* (2020). Blue-light perception by epiphytic *Pseudomonas syringae* drives chemoreceptor expression, enabling efficient plant infection. *Mol. Plant Pathol.* 21: 1606–19.

Santamaría-Hernando, S., López-Maroto, Á., Galvez-Roldán, C., Munar-Palmer, M., Monteagudo-Cascales, E. *et al.* (2022).

Pseudomonas syringae pv. *tomato* infection of tomato plants is mediated by GABA and L-Pro chemoperception. *Mol. Plant Pathol.* 23: 1433–45.

Tambong, J. T., Xu, R., Cuppels, D., Chapados, J., Gerdis, S. *et al.* (2022). Analysis of draft genome resources of thirty-three Canadian strains of *Pseudomonas syringae* pv. *tomato* isolated between 1992 and 2008 reveals achromobactin virulence cluster that is absent in the reference strain DC3000. *Phytopathology* 112: 968–72.

Wang, H., Wagnon, R., Moreno, D., Timilsina, S., Jones J. *et al.* (2022). A long-amplicon viability-qPCR test for quantifying living pathogens that cause bacterial spot in tomato seed. *Plant Dis.* 106: 1474–85.

Xin, X.-F. and He, S. Y. (2013). *Pseudomonas syringae* pv. *tomato* DC 3000: A model pathogen for probing disease susceptibility and hormone signaling in plants. *Annu. Rev. Phytopathol.* 51: 473–98.

Yan, S., Liu, H., Mohr, T. J. Jenrette, J. Chiodini, R. *et al.* (2008). Role of recombination in the evolution of Model plant pathogen *Pseudomonas syringae* pv. *tomato* DC3000, a very atypical tomato strain. *Appl. Environ. Microbiol.* 74: 3171–81.

Young, J. M., Dye, D. W. and Wilkie, J. P. (1978). Genus *Pseudomonas* Migula 1984. In: Young, J. M., Dye, D. W., Bradbury, J. F., Panagopoulos, C. G. and Robbs, C. F. (1978). A proposed nomenclature and classification for plant pathogenic bacteria. *N. Z. J. agric. Res.* 21: 153–77.

Zaccardelli, M., Spasiano, A., Bazzi, C. and Merighi, M. (2005). Identification and *in planta* detection of *Pseudomonas syringae* pv. *tomato* using PCR amplification of *hrpZ*$_{pst}$. *Eur. J. Plant Pathol.* 111: 85–90.

Zhang, H., Qiu, Y., Li, M., Song, F. and Jiang, M. (2020). Functions of pipecolic acid on induced resistance against *Botrytis cinerea* and *Pseudomonas syringae* pv. *tomato* DC3000 in tomato plants. *J. Phytopathol.* 168: 591–600.

9.26 BACTERIAL BLIGHT OF PEA

Pseudomonas syringae pv. *pisi* causes bacterial blight of pea (*Pisum sativum* L.); it was first reported by Sackett (1916) from Colorado, USA. It is a serious disease of pea (*Pisum sativum*) in cool and wet weather. In 2004, under favourable weather conditions, the disease caused close to 90% plant mortality in some fields in the Castilla y León region, the main pea-growing area of Spain. Although the epiphytotics of the disease in Australia are infrequent and may occur only once a decade, the field pea crop suffered a loss of 10% due to an epiphytotic in 1992. In the Wimmera region of Australia, disease incidence exceeding 30% has also been reported. Yield losses as high as 70% have been reported on an early sown pea cv. Sugar Snap from Horsham, Northwest Victoria, Australia. The disease also occurs in India and Armenia.

Psm. syringae pv. *syringae* has also been reported as a causal agent of bacterial blight of pea (Martin-Sanz *et al.*, 2011, 2012a). Martin-Sanz *et al.* (2011) have also reported that *Psm. syringae* pv. *syringae* has been observed to be as frequent as *Psm. syringae* pv. *pisi* as the cause of bacterial diseases in pea.

9.26.1 Host Range

In addition to *Pisum sativum*, *Psm. syringae* pv. *pisi* infects *P. sativum* var. *arvense, Lathyrus odoratus, L. latifolius, Vicia bengalensis, Lablab niger* and *Vigna* sp. (Bradbury, 1998).

9.26.2 Symptoms

The symptoms appear on all the above-ground plant parts. Initial symptoms are discrete, dark green, water-soaked spots on leaves, pods and stems. In cool and wet weather, the spots enlarge and coalesce to cover large areas. In warm and dry weather, lesion development is checked, resulting in smaller lesions, and water-soaking also becomes less obvious. With age, the lesions become darker and, finally, necrotic (Figure 9.25a and b). Older lesions on pods are brown and papery. In severely infected pods, the seeds may be covered with bacterial slime and the pathogen may penetrate the seeds through the funicle and micropyle. When the infection takes place at the seedling stage, the entire crop may be destroyed.

9.26.3 Causal Organism

Pseudomonas syringae pv. *pisi* (Sackett 1916) Young *et al.* 1978

Synonym: *Psm. pisi*, *Psm. syringae* pv. *syringae* van Hall 1902

The below-given description of the bacterium is taken mainly from Martin Sanz *et al.* (2005).

Bacterial cells are Gram-stain-negative, non-sporing rods, measuring 0.7 × 2.0–3.0 µm and motile by one to five polar flagella. Most strains produce a yellowish-green diffusible pigment on King's medium B, which fluoresces blue under ultraviolet light. Moderate growth occurs in 24 h on agar slants, which is filiform, glistening and greyish-white. The tests for nitrate reduction, starch hydrolysis, indole production and H_2S production are negative. An alkaline reaction is produced in litmus milk and soft curd formed is cleared. Gelatin is liquefied, but no lipolytic activity occurs. Acid but no gas is produced from glucose, galactose and sucrose. The bacterium is chemoorganotroph and obligate aerobe. Optimum growth occurs at 26–28 °C and at pH 6.5–7.5. The bacterium has the typical characteristics of a LOPAT group 1[a] green fluorescent pseudomonad. It can be detected from the infected seedlings with the help of ELISA. An avirulence gene designated as *avrPpiB* from race 3 has been cloned and sequenced.

Baltrus and Clark (2019) reported the sequencing and assembly of a complete genome sequence for *Psm. syringae* pv. *pisi* strain PP1. *Psm. syringae* phylogroup 2 strains usually contain a relatively low number of type III effectors in

(a) (b)

FIGURE 9.25 Bacterial blight of pea: (a) Vegetable pea (Courtesy of Julien Lanson, www.inra.frhyp3.) (b) Field pea. (Courtesy of Philip Wright, Department of Primary Industries, Orange, Australia.).

addition to multiple phytotoxins. Strain PP1 lacks (and likely has lost) these phytotoxins, and has regained numerous effectors compared to close relatives. This sequence extends a previously published analysis of virulence factors within this strain and enables the investigation of the mechanisms of acquisition for new effector families. The authors found that these effectors were acquired through horizontal transfer of multiple plasmids, an integrative and conjugative element (ICE), a transposon and also likely through the recombination of chromosomal regions.

Martin-Sanz *et al.* (2013) characterised a collection of 88 putative *Psm. syringae* pv. *syringae* strains, including 39 strains isolated from pea, by rep-PCR, MLST and *syrB* amplification, and evaluated these strains for pathogenicity and virulence. rep-PCR data grouped the strains from pea into two groups (1B and 1C), together with strains from other hosts; a third group (1A) was formed exclusively with strains isolated from non-legume species. The inoculations of these strains on two pea cultivars showed that the strains from groups 1A and 1C were less virulent than the strains from group 1B, suggesting a possible pathogenic specialisation in this group. These findings show the existence of genetically and pathogenically distinct *Psm. syringae* pv. *syringae* strain groups from pea, which can be useful for the diagnosis and epidemiology of the pathogen and also for resistance breeding.

Of 298 bacterial isolates collected from pea cultivars, landraces and breeding lines in North-Central Spain over several years, 225 of the isolates were identified as *Psm. syringae*, either pv. *pisi* (110 isolates) or pv. *syringae* (112 isolates) on the basis of biochemical and physiological characteristics and molecular markers. It is evident that *Psm. syringae* pv. *syringae* is as frequent as *Psm. syringae* pv. *pisi* as the cause of bacterial diseases in pea. Of 225 isolates, 222 were pathogenic on pea. Simultaneous infections of *Psm. syringae* pv. *pisi* and pv. *syringae* in the same fields were observed, suggesting the importance of resistance for both the pathovars in future commercial cultivars. A search for resistance among pea genotypes suitable for production in this region of Spain, or as breeding material, identified the presence of resistance genes for all the *Psm. syringae* pv. *pisi* races, with the exception of race 6. Pea cultivars Kelvendon Wonder, Cherokee, Isard, Iceberg, Messire and Attika were found to be suitable sources of resistance to *Psm. syringae* pv. *syringae* (Martin-Sanz *et al.* 2011).

Martin-Sanz *et al.* (2012b) examined a collection of 91 strains of *Psm. syringae* pv. *pisi* which included 53 strains from recent outbreaks in Spain and from 14 other countries and characterised them on the basis of 55 nutritional tests, genetic analysis (rep-PCR, amplification of AN3 and AN7 specific markers, and MLST) and pathogenicity on differential pea cultivars to identify races. Principal component analysis and distance dendrograms confirmed the existence of two genetic lineages within this pathovar, which are clearly discriminated by the AN3/AN7 markers, rep-PCR and MLST. Strains from races 1 and 7 amplified the AN3 marker, those from races 2, 6 and 8 amplified AN7, while strains of races 3, 4 and 5 amplified either AN3 or AN7. The Spanish collection diversity reflects the variability found in the worldwide collection, suggesting multiple introductions of the bacteria into Spain by contaminated seed lots.

There are eight races of the pathogen, which have been identified on the basis of their reactions to eight differentials. However, there is a possibility of there being new races not yet identified (Martin Sanz *et al.*, 2005). Race 2 is the most frequently distributed worldwide, followed by races 6 and 4. The gene-for-gene relationship between pea cultivars and pathogenic races is given in Table 9.26.

The identification of seven races of *Psm. syringae* pv. *pisi*, based on a set of eight different pea cultivars, allowed a gene-for-gene relationship to be proposed based on six matching pairs of resistance genes in the host and avirulence genes in the pathogen (Taylor *et al.*, 1989; Bevan *et al.*, 1995). On the basis of segregation among F_2 populations of crosses between differential cultivars sequentially inoculated with races of *Psm. syringae* pv. *pisi*, Bevan *et al.* (1995) secured evidence for four (and possibly six) putative resistance (*R*)/avirulence (*A*) gene pairs. *R1*, *R2* and *R3* are dominant resistance alleles at single loci; *R4* is a dominant allele at a single locus, which exhibits variable expression possibly dependent on genetic background. There is evidence that *R3* and *R4* are at linked loci. Homology tests provided evidence that the alleles *R2*, *R3* and *R4* were present in more than one cultivar. Two other alleles, *R5* and *R6*, were postulated to explain the observed segregation in certain crosses. The authors further inferred that out of races 2, 3 and 4, each carries a different single avirulence gene, race 6 carries no apparent avirulence gene and race 7 carries at least *A2*, *A3* and *A4*. Race 1 carries *A1*, *A3*, *A4* and possibly *A6* and race 5 carries *A2*, *A4* and possibly *A5* and *A6*.

9.26.4 Disease Cycle

The seed-borne inoculum is the main source of the perpetuation of the bacterium. The bacterium is internally and externally seed-borne and can survive for at least three years. Infected plant debris, particularly on soil surface and volunteer plants, are other sources of inoculum. Survival time of the bacterium is reduced by burying the debris. In southern Australia, it was reduced from 2 years to 11 months when the stubbles were buried 10 cm deep in the soil.

The penetration of the bacterium occurs through stomata, but it is mainly through wounds caused by hail, frost, storms, sand blasting and cultural operations. The ice-nucleating activity of the epiphytic cells of the bacterium increases frost damage. Local spread of the bacterium occurs through wind-splashed rain and overhead irrigation. Long-distance spread is through the infected seed. Cool and wet or humid weather favours the disease while warm and dry weather reduces the severity of the disease.

The epiphytic phase is the main characteristic of the life cycle of the bacterium. Bacterium survives, multiplies and spreads epiphytically in the crop. The inoculation of race 2 of the bacterium on susceptible cultivars resulted in the reduction of its population during the first three days; however, it then increased 1000 times more than the initial count without producing any symptoms. When it was inoculated on resistant cultivars, the bacteria did not multiply and the population levels decreased slightly. This indicates that the development of the resident phase is race-specific (Grondeau *et al.*, 1996).

Soil water content and the level of seed-borne inoculum greatly influence the transmission of the bacterium from seed to the seedlings. The high inoculum level and the high soil moisture content increase the number of infected seedlings and the number of lesions per plant (Roberts *et al.*, 1996).

The stem inoculation technique is generally used to screen pea germplasm, assuming that the responses would be same in other plant parts. However, Elvira-Recuenco (2000) showed that a resistant response in the stem was not always associated with the resistance in the leaf and/or pod. This was found for both race-specific and non-race-specific resistance. Elvira-Recuenco and associates (2003) found that race-specific resistance genes conferred resistance in the stem in a consistent way, but the leaf and pod response varied depending on the matching resistance and avirulence gene combination. *R2* generally conferred resistance in all the plant parts. *R3* and *R4* singly did not confer

TABLE 9.26
Gene-for-gene Relationship between Pea Cultivars and Races of *Pseudomonas syringae* pv. *pisi*

							Avirulence Gene						
Resistance Gene (*R*)							**1**	**2**	**3**	**4**	**5**	**6**	**7**
							1	.	.	.	.	.	.
							.	2	.	.	2	.	2
							3	.	3	.	.	.	3
							4	.	.	4	4	.	4
							.	.	.	.	(5)	.	.
							(6)	.	.	.	(6)	.	.
Kelvendon Wonder	.	.	.	.	.	.	+	+	+	+	+	+	+
Early Onward	.	2	.	.	.	.	+	_	+	+	_	+	_
Belinda	.	.	3	.	.	.	_	+	_	+	+	+	_
Hurst's Greenshaft	.	.	.	4	.	(6)	_	+	+	_	_	+	_
Partridge	.	.	3	4	.	.	_	+	_	_	_	+	_
Sleaford Triumph	.	2	.	4	(5)	.	_	_	+	_	_	+	_
Vinco	1	2	3	.	(5)	.	_	_	_	+	_	+	_
Fortune	.	2	3	4	.	.	_	_	_	_	_	+	_

Source: Bevan *et al.* (1995).
Note: + = susceptible response, _ = hypersensitive reaction.
Genes in brackets are partially confirmed, . = genes absent.

complete resistance in leaf and pod, but *R3* in combination with *R2* or *R4* enhanced leaf and pod resistance. Race-nonspecific resistance conferred stem resistance to all the races, leaf and pod resistance to races 2, 5 and 7 and variable reactions in leaves and pods to races 1, 3, 4 and 6.

Martin-Sanz *et al.* (2012a) studied the effects of frost on disease severity caused by *Psm. syringae* pv. *pisi* and *Psm. syringae* pv. *syringae*, finding that frost tolerance was effective in lowering the disease effects caused under frost-stress conditions, even in the absence of disease resistance genes. However, the highest degree of this protection is reached when frost tolerance and disease resistance genes are combined in the same genetic background.

9.26.5 Management

1. The cheapest and most effective method is the use of resistant cultivars. Many cultivars contain race-specific resistance genes for all the known races, with the exception of race 6.
2. The next best approach is the use of the pathogen-free seed. The following measures help in producing the pathogen-free seed.
    a. Production of seed in dry arid areas.
    b. Avoidance or less use of sprinkler irrigation.
    c. Field inspection from mid-flowering to late pod-fill stage. The removal of statutory certification controls in the UK in 1993 greatly increased the incidence of seed-borne infection (Reeves *et al.*, 1996). The seed test is available to detect seed-borne infection in the seed. It can detect the seed-borne infection level of less than 0.05%.
3. The following measures reduce the disease incidence by checking the spread of the pathogen.
    a. Use clean farm equipment.
    b. Avoid sprinkler irrigation.
    c. Destroy infected plant debris and volunteer plants.
4. Crop rotation for two years.
5. In Australia, the disease is successfully managed by adopting an integrated disease management approach consisting of crop rotation, the use of pathogen-free seed, the avoidance of planting in areas prone to frequent frosts or extreme wet weather, crop hygiene and avoiding early sowing. Seed treatment and application of foliar bactericides have limited use in controlling the disease (Hollaway *et al.*, 2007).

REFERENCES

Baltrus, D. A. and Clark, M. (2019). A complete genome sequence for *Pseudomonas syringae* pv. *pisi* PP1 highlights the importance of multiple modes of horizontal gene transfer during phytopathogen evolution. *Mol. Plant Pathol.* 20: 1013–18.

Bevan, J. R., Taylor, J. D., Crute, I. R., Hunter, P. J. and Vivian, A. (1995). Genetics of specific resistance in pea (*Pisum sativum*) cultivars to seven races of *Pseudomonas syringae* pv. *pisi*. *Plant Pathol.* 44: 98–108.

Bradbury, J. F. (1998). *Pseudomonas pisi. IMI Description of Fungi and Bacteria*. Sheet 126, CAB International, Wallingford.

Elvira-Recuenco, M. (2000). Sustainable control of pea bacterial blight: Approaches for durable genetic resistance and biocontrol by endophytic bacteria. PhD Thesis, Wageningin University, The Netherlands (cited by Martin Sanz *et al.*, 2005).

Elvira-Recuenco, M., Bevan, J. R. and Taylor, J. D. (2003). Differential responses to pea bacterial blight in stems, leaves and pods under glasshouse and field conditions. *Eur. J. Plant Pathol.* 109: 555–64.

Grondeau, C., Mabiala, A., Ait-Oumeziane, R. and Samson, R. (1996). Epiphytic life is the main characteristic of the life cycle of *Pseudomonas syringae* pv. *pisi*, pea bacterial blight agent. *Eur. J. Plant Pathol.* 102: 353–63.

Hollaway, G. J., Bretag, T. W. and Price, T. V. (2007). The epidemiology and management of bacterial blight (*Pseudomonas syringae* pv. *pisi*) of field pea (*Pisum sativum*) in Australia: A review. *Aus. J. Agric. Res.* 58: 1086–99.

Martin Sanz, A., Garcia Vaquero, C. A. and Caminero Saidana, C. (2005). Pea bacterial blight: A problem in Castilla y Leon. Grain Legumes No. 41, pp. 8–9.

Martín-Sanz, A., de la Vega, M. P. and Caminero, C. (2012a). Resistance to *Pseudomonas syringae* in a collection of pea germplasm under field and controlled conditions. *Plant Pathol.* 61: 375–87.

Martín-Sanz, A., de la Vega, M. P., Murillo, J. and Caminero, C. (2012b). Genetic, biochemical and pathogenic diversity of *Pseudomonas syringae* pv. *pisi* strains. *Plant Pathol.* 61: 1063–72.

Martín-Sanz, A., de la Vega, M. P., Murillo, J. and Caminero, C. (2013). Strains of *Pseudomonas syringae* pv. *syringae* from pea are phylogenetically and pathogenically diverse. *Phytopathology*, 103: 673–81.

Martín-Sanz, A., Palomo, J. L., de la Vega, M. P. and Caminero, C. (2011). Identification of pathovars and races of *Pseudomonas syringae*, the main causal agent of bacterial disease in pea in North-Central Spain, and the search for disease resistance. *Eur. J. Plant Pathol.* 129: 57–69.

Reeves, J. C., Hutchins, J. D. and Simpkins, S. A. (1996). The incidence of races of *Pseudomonas syringae* pv. *pisi* in UK pea (*Pisum sativum*) seed stocks, 1987-1994. *Plant Var. Seeds* 9: 1–8.

Roberts, S. J., Ridout, M. S., Peach, L. and Brough, J. (1996). Transmission of pea bacterial blight (*Pseudomonas syringae* pv. *pisi*) from seed to seedling: Effects of inoculum dose, inoculation method, temperature and soil moisture. *J. Appl. Bacteriol.* 81: 65–72.

Sackett, W. G. (1916). A bacterial stem blight of field and garden peas. *Bull. Colo. Agr. Expt. Sta.* 218: 1–43.

Taylor, J. D., Bevan, J. R., Crute, I. R., and Reader, S. L. (1989). Genetic relationship between races of *Pseudomonas syringae* pv. *pisi* and cultivars of *Pisum sativum*. *Plant Pathol.* 38: 364–75.

van Hall, C. J. J. (1902). *Bijdragen tot de kennis der Bakterieele Plantenzeikten*. Coöperatieve Drukkerij-vereeniging "Plantijn". Inaugural dissertation. Amsterdam.

Young, J. M., Dye, D. W., Bradbury, J. F., Panagopoulos, C. G. and Robbs, C. F. (1978). A proposed nomenclature and classification for plant pathogenic bacteria. *N. Z. J. Agric. Res.* 21: 153–177. https://doi.org/10.1080/00288233.1978.10427397

9.27 BACTERIAL WILT OF CUCURBITS

Bacterial wilt of cucurbits affects many cucurbitaceous species, although cucumbers and muskmelons are more susceptible to the pathogen. Squash, pumpkins, gourds and wild cucurbits are also infected, whereas the watermelon is resistant or rarely attacked. The disease is present throughout the United States and in Canada, to the east of the Rocky Mountains. It also occurs in Europe, Japan and South Africa.

The disease has also been reported from New Mexico on pumpkin and watermelon by Sanogo *et al.* (2011). Rojas *et al.* (2015) reviewed the research work to understand the cucurbit bacterial wilt pathosystem by recounting the early findings and updating our understanding of the disease cycle, including pathogen and vector biology. The authors have also highlighted the research areas that could lead to more efficient and ecologically based management of the disease.

(a)

FIGURE 9.26A Bacterial wilt of cucumber: Wilted plants. (Courtesy: Gerald Holmes, California Polytechnic State University at San Luis Obispo, Bugwood.org.)

9.27.1 Symptoms

The earliest symptom of the disease is the drooping of one or more leaves, followed by drooping and wilting of all the leaves of the infected vine. Wilt progresses through lateral shoots involving the entire plant. Wilted leaves shrivel and dry up, with infected stems turning pale and becoming soft. Before shrivelling and drying, cucumber and muskmelon leaves may turn dull green while those of squash and pumpkin become dull greyish-green and later turn bronze. The whole plant may wilt within two weeks of infection. Later on, wilted plants dry up and die (Figure 9.26a). Squash and pumpkin plants do not show rapid wilting, but may be dwarfed, and produce excessive blossoms and branches.

One diagnostic feature of the disease is the white bacterial ooze, which emerges t from the cut ends when infected stems are cut and pressed between the fingers. This sticky bacterial mass can be drawn into fine threads by touching the cut ends of the stem with a finger or a knife blade and gently pulling these away from the cut end (Figure 9.26b). A positive test is confirmation of the disease; however, a negative test does not indicate that the disease is absent. Another diagnostic test is to put a freshly cut infected stem in a glass of clean water and observe it. If bacterial wilt is present, the water will begin to turn milky at the cut end of the stem, ultimately turning the whole of the water milky.

The infection of the squash fruits usually occurs through the infected vines. Occasionally, however, it may be caused by beetles feeding on the blossoms and the rind of fruits. In the stored squash fruits, slimy rot progresses internally, without any external symptoms. With the further progress of the internal rot, dark spots or blotches appear on the outer surface, which enlarge and coalesce. Later on, soft rotting microorganisms invade these fruits, completely destroy them (Agrios, 2004).

(b)

FIGURE 9.26B Bacterial wilt of cucurbits: Strands of viscous bacterial ooze drawn from cut stem ends. (Courtesy: Michael Hoffmann.)

9.27.2 Causal Organism

Erwinia tracheiphila (Smith 1895) Bergey *et al.* 1923.

Synonym: *Erw. uredovora*

The following description of the bacterium is taken primarily from Hauben *et al.* (1998). Cells are facultatively anaerobic rods, Gram-stain-negative and motile by peritrichous flagella. Growth factors are required for growth and metabolism is fermentative. Bacterium grows very poorly on nutrient agar, but moderately well on yeast extract glucose chalk agar and glucose nutrient agar. H_2S is produced but gelatin is not liquefied. Tests for oxidase, urease, starch hydrolase, arginine dihydrolase, casein hydrolysis, phenylalanine deaminase and nitrate reduction are also negative.

Acid is produced from sucrose, fructose, galactose, rhamnose, glucose, glycerol and N-acetylglucosamine but not from arabinose, cellobiose, mannose, melibiose, lactose, maltose, raffinose, salicin, dulcitol, mannitol, sorbitol, trehalose and xylose. Bacterium utilises arabinose, maltose, mannose, ribose, salicin, sorbitol and sucrose but not adonitol, dulcitol, ethanol, xylitol, dextrin, cellobiose, xylose, lactose and starch as carbon sources.

Bacterial strains grow on arginine, glucosamine, leucine, phenylalanine, citrulline and tyrosine but not on betaine, choline, hydroxypyroline, thymine, tryptamine and xanthin as nitrogen sources. Strains are sensitive to ampicillin, amoxycillin, carbenicillin, cephalexine and kanamycin but not to fusidinic acid, methicillin and sulfafurazol. G + C content of the DNA of members of the species ranges from 50.0–52.0 mol%.

Rojas *et al.* (2013) investigated genetic variability in 12 *Erw. tracheiphila* strains, obtained from muskmelon, cucumber, or summer squash, by rep-PCR and cross-inoculation studies on different hosts. They provided the first evidence of genetic diversity within the pathogen and suggested that strain specificity is associated with plant host.

9.27.3 Disease Cycle

As the bacterium dies within 1–2 months in the dead plants, its survival in the diseased plant debris is not possible. The bacterium overwinters (survives) in the digestive tract of hibernating adults of two cucumber beetles, namely, striped cucumber beetle (*Acalymma vittatum*) and spotted cucumber beetle (*Diabrotica undecimpunctata*). The former is more common and, in North Dakota, it is the main disseminating agent of the bacterium.

In the spring, infected beetles feed on the leaves, causing deep wounds and depositing the bacteria in the wounds along with their faeces. Bacteria enter the xylem vessels exposed by wounds, multiplying there before spreading to other parts of the plant. Not all the beetles carry the bacterium and not all the wounds act as avenues of infection. Only the deep feeding wounds, in which xylem vessels are exposed, serve as avenues of infection. After entering the xylem vessels, bacteria multiply rapidly, producing polysaccharides and gums. The accumulation of bacterial cells, polysaccharides and gums, along with formation of tyloses, results in the blockade of vascular system, hindering the normal flow of water and nutrients. The plugging of xylem vessels is so extensive that it allows less than one-fifth of normal water flow, hence making it the primary cause of wilting. The rate of multiplication, and the period in which the bacterium multiplies, differ in susceptible and resistant cucumber cultivars. The generation time in susceptible and resistant cucumbers is 5.4 and 11.2 h, respectively. The bacterial population starts to decline on the fourth day in the resistant cucumber cultivars, while in the susceptible cultivars the peak population reaches on the sixth day (Watterson *et al.*, 1972).

Vrisman *et al.* (2016) studied the host colonisation dynamics of *Er. tracheiphila* using a bioluminescent strain (TedCu10-BL#9). Inoculated *Cucumis melo* plants showed bioluminescent bacteria in the inoculated leaf and petiole beginning 1-day post-inoculation (DPI) and the bacteria spread to roots via the stem by 2 DPI, reached the plant extremities 4 DPI; finally, the plants wilted 6 DPI. However, *Cucurbita* plants inoculated with TedCu10-BL#9 did not wilt, even at 35 DPI. Bioluminescent bacteria were detected 6 DPI in the main stem of squash and pumpkin plants, which harboured approximately 10^4 and 10^1 cfu g^{-1}, respectively, without symptoms.

Liu *et al.* (2018) provided experimental evidence that host age impacts the rate of symptom development in cucurbit bacterial wilt and that the movement of *Er. tracheiphila* within muskmelon plants occurs primarily in the downward direction.

Rojas and Gleason (2012) monitored the survival of *Erw. tracheiphila* on muskmelon leaves under various temperature and moisture conditions using a rifampicin-resistant strain of the bacterium. The survival of *Erw. tracheiphila* on wet muskmelon leaves depended on temperature ($P < 0.01$), with the greatest survival at 10 °C and 15 °C, and the least at 30 °C and 35 °C. Leaf wetness also impacted survival; an initial 12-hour dry period resulted in a 1,000- to 10,000-fold reduction in population size, followed by the stabilisation of the surviving population. It is evident that *Erw. tracheiphila* can survive on muskmelon leaves under a wide range of environmental conditions, suggesting that epiphytic populations might serve as a reservoir of inoculum for infections.

Beetles spread the bacteria through their contaminated mouth parts when they feed on healthy plants after feeding on infected plants. Infected wild cucurbits can also serve as a source of primary inoculum and from these plants the spread occurs through beetles. Cucumber beetles prefer wilted plants for feeding, resulting in an increase in number of contaminated beetles. The spread of the pathogen is fast under normal moisture conditions, because the activity of beetles decreases in prolonged cool and rainy weather. In extremely dry weather, the bacteria cannot penetrate the exposed vascular system due to the absence of a film of water. Each contaminated beetle can inoculate several healthy plants after once feeding on an infected plant.

9.27.4 Management

1. There is no cure of the cucurbit wilt. Hence, control is aimed at the prevention of the disease by controlling cucumber beetles.
2. Control of cucumber beetles with insecticide should be begun as soon as the first beetle is seen in the crop and sprays continued till the end of season at weekly intervals. Use any of the fast-killing insecticides such as methoxychlor, diazinon and carbryl (Sevin). Carbryl should be sprayed late in the day in order to prevent honey bee mortality. The organic control options for cucumber beetles include the foliar application of kaolin clay

(Surround WP), insecticidal soap containing neem oil and rotenone 1% dust.

3. Cover the plants with row covers, cones or other barriers to prevent the feeding of cucumber beetles on vines. Row covers should be placed over the plants early in the season, and their edges sealed to prevent the entry of the beetles. Row covers should be removed to allow pollination by insects. Rojas *et al.* (2011) studied the feasibility of the delayed removal of row covers to decrease the incidence of bacterial wilt of muskmelon and found that, in the years in which bacterial wilt occurred, the delayed removal of row covers for ten days resulted in the highest returns. However, when bacterial wilt was absent, the delayed-removal row-cover treatments gave the lowest returns. Results of the sensitivity analysis indicated that the delayed removal of row covers for ten days could be a cost-effective component of an integrated bacterial wilt suppression strategy for muskmelon, where bacterial wilt occurs in ≥50% of production seasons.
4. Place yellow sticky cups every 20 feet along the rows to trap the beetles. Replace these traps when they become dusty and non-sticky.
5. Remove and destroy the diseased plants to prevent further spread of the disease.
6. In the fall, remove and destroy the diseased plant debris in order to reduce sites for beetle hibernation and to minimise the survival of other cucurbit pathogens.
7. The cultivars resistant to bacterial wilt have little or no bitterness in their fruits. Cucurbitacin B and C are the compounds responsible for bitterness in the fruits; these compounds also attract cucumber beetles. Therefore, cultivars having no bitterness in fruits are most likely to be least susceptible to bacterial wilt.
8. To prevent squash rot in storage, harvest the fruits only from healthy vines and store them in clean and fumigated stores.
9. Do not plant cucurbits near weedy woods or brush, where the cucumber beetles are likely to overwinter.

REFERENCES

Agrios, G. N. (2004). *Plant Pathology*. 5th Ed. Elsevier Academic Press, San Diego, CA, 922 pp.

Bergey, D. H., Harrison, F. C., Breed, R. S., Hammer, B. W. and Huntoon, F. M. (1923). *Bergey's Manual of Determinative Bacteriology*. 1st Ed. Williams & Wilkins Co., Baltimore, MD.

Hauben, L., Moore, E. R. B., Vauterin, L. Steenackers, M., Mergaert, J. *et al.* (1998). Phylogenetic position of phytopathogens within the *Enterobacteriaceae*. *Syst. Appl. Microbiol.* 21: 384–97.

Liu, Q., Beattie, G. A., Saalau Rojas, E. and Gleason, M. L. (2018). Bacterial wilt symptoms are impacted by host age and involve net downward movement of *Erwinia tracheiphila* in muskmelon. *Eur. J. Plant Pathol.* 151: 803–10.

Rojas, E. S., Batzer, J. C., Beattie, G. A., Fleischer, S. J., Shapiro, L. R., *et al.* (2015). Bacterial wilt of cucurbits: Resurrecting a classic pathosystem. *Plant Dis.* 99: 564–74.

Rojas, E. S., Dixon, P. M., Batzer, J. C. and Gleason, M. L. (2013). Genetic and virulence variability among *Erwinia tracheiphila* strains recovered from different cucurbit hosts. *Phytopathology* 103: 900–05.

Rojas, E. S. and Gleason, M. L. (2012). Epiphytic survival of *Erwinia tracheiphila* on muskmelon (*Cucumis melo* L.). *Plant Dis.* 96: 62–66.

Rojas, E. S., Gleason, M. L., Batzer, J. C. and Duffy, M. (2011). Feasibility of delaying removal of row covers to suppress bacterial wilt of muskmelon (*Cucumis melo*). *Plant Dis.* 95: 729–34.

Sanogo, S., Etarock, B. F. and Clary, M. (2011). First report of bacterial wilt caused by *Erwinia tracheiphila* on pumpkin and watermelon in New Mexico. *Plant Dis.* 95: 1583.

Smith, E. F. (1895). *Bacillus tracheiphilus* sp. nov., die Ursache des Verwelkens verschiedener Cucurbitaceen. *Centralblatt für Bakteriologie und Parasitenkunde* 21: 364–73.

Vrisman, C. M., Deblais, L., Rajashekara, G. and Miller, S. A. (2016). Differential colonization dynamics of cucurbit hosts by *Erwinia tracheiphila*. *Phytopathology* 106: 684–92.

Watterson, J. C., Williams, P. H. and Durbin, R. D. (1972). Multiplication and movement of *Erwinia tracheiphila* in resistant and susceptible cucurbits. *Plant Dis. Reptr.* 56: 949–53.

9.28 ANGULAR LEAF SPOT OF CUCURBITS

Angular leaf spot of cucurbits, also called angular leaf spot of cucumber, is a widespread disease. It affects many cucurbit species but it is most severe on cucumber, honeydew melon and Zucchini squash. The disease also occurs on muskmelon, watermelon, cantaloupe, vegetable marrow, squashes and pumpkins. Yield losses occur mainly due to a smaller number of fruits and a reduction in weight per fruit because of the reduced assimilation of infected leaves. Losses in processing cucumbers may be as high as 50% or more in wet seasons when control measures are not undertaken.

Angular leaf spot of cucumber was first reported from the United States in 1913, and the causal bacterium was identified in 1915 by Smith and Bryan. Since then, the disease has spread to many other countries. Janse (2006) has given a comprehensive list of countries and regions where the disease occurs. This includes Algeria, Argentina, Australia, Brazil, Canada, China, Columbia, Egypt, large parts of Europe, Gabon, Iran, India, Israel, Japan, Jordan, Kazakhstan, Kenya, Korea, Laos, Mexico, New Zealand, the Philippines, Singapore, South Africa, Tajikistan, Thailand, Turkey, the USA, Uzbekistan, Venezuela and Zimbabwe. In temperate regions of the world, the disease can cause between a 37% and 40% reduction in fruit number and weight, respectively, besides rendering some fruits as unsaleable. In wet conditions, the disease renders over 80% of leaves infected by the tenth day of the first infection.

In India, the disease has been found in Kashmir valley, where its incidence and intensity ranges between 23.35 to 74.45% and 10.50 to 26.02%, respectively (Bhat *et al.*, 2007)

Newberry *et al.* (2016) collected 27 *Psm. syringae* strains causing foliar lesions and blighting on watermelon, cantaloupe and squash in Florida, Georgia and California over the course of several years. On the basis of genetic profiles obtained through both MLSA (*gyrB*, *rpoD*, *gapA* and *gltA*) and BOXPCR (BOXA1R), they identified 26 of the *Psm. syringae* strains to be distributed among three clades within genomospecies 1, and phylogenetically distinct from genomospecies 2 member *Psm. syringae* pv. *lachrymans*. Considerable genetic diversity among *Psm. syringae* strains infecting cucurbits is associated with the same disease, and reflects the larger ecological diversity of *Psm. syringae* populations from genomospecies 1.

Dutta *et al.* (2013a) reported *Xanthomonas cucurbitae* causing angular, necrotic spots with chlorotic halos on watermelon leaves, with no observable symptoms on fruits, from Georgia, USA. Dutta *et al.* (2013b) reported an outbreak of *Xan. cucurbitae* on pumpkin from Georgia, which caused severe symptoms on leaves and fruits. Leaves displayed small (2–3 mm), angular, water-soaked, yellow lesions, while fruits had small (2–3 mm), sunken, circular, dry lesions.

Jarial *et al.* (2011) reported bacterial spot of bottle gourd (*Lagenaria siceraria* syn. *L. vulgaris*) caused by *Xan. cucurbitae* (Bryan) Dowson from Himachal Pradesh. This was the first record of the disease on bottle gourd in the area. The symptoms were found on leaves, stem, floral parts and fruits. The survey, conducted at 64 locations of five districts of the state, revealed that the disease severity ranged from 12.50 to 78.33%, causing around 10.07 to 70.61% yield losses at different locations.

Bacterial spot of cucurbits, caused by *Xan. cucurbitae*, is a major problem in cucurbit-growing areas around the world. In the Midwestern region of the United States, pumpkin and squash fields can have greater than 90% infected fruits, leading to high yield losses. After performing restriction-site associated DNA sequencing (RAD-seq) analysis of *Xan. cucurbitae* isolates collected from the Midwestern region, Rai *et al.* (2024) selected five representative isolates for further characterisation, which included whole-genome sequencing and *in vitro* enzyme and *in planta* virulence assays. They found that minimal genetic diversity exists between these isolates, and that the isolates show differential virulence on different cucurbit host plants. This study contributes to our understanding of *Xan. cucurbitae* population dynamics in the Midwestern region and may assist with developing additional management strategies for controlling bacterial spot disease of cucurbits.

9.28.1 Symptoms

On cucumber, the symptoms appear on leaves, petioles, stems and fruits. On cotyledons, soft, transparent, round to irregular lesions are formed. Initially small, round to irregular, water-soaked spots appear on cucumber leaves. The spots enlarge until their spread is limited by large veins; hence the spots become angular in shape. The spots on the upper leaf surface turn whitish-grey to brown, while on the lower surface these appear gummy and shiny. In wet weather, bacterial ooze comes out from spots in the form of droplets on the lower leaf surface. In dry weather, bacterial ooze dries to form a thin, papery, white layer. As the infected areas die and dry, they shrink and tear away from the healthy leaf tissue, forming large irregular holes in the leaves (Agrios, 2004). Young fully expanded leaves are more susceptible than older leaves (Figure 9.26a). On stem and petioles, light yellow to light brown lesions appear, which exude abundant bacterial ooze. The severe infection of stem may lead to a blightening and the death of the whole plant.

On squash leaves, brown spots of varying sizes appear. The spots are surrounded by yellow haloes and the leaf area next to the yellow halo on the lower leaf surface appears water-soaked following wet weather.

The symptoms on watermelon leaves start as small, circular, dark lesions surrounded by yellow haloes. As the spots enlarge, these become angular, having white centres and covering a large area of the leaf (Figure 9.27).

Water-soaked, circular, usually superficial spots, much smaller than those on leaves, appear on the cucumber fruits. After the death of infected tissue, the centres of the lesions become chalky white and may show star-like cracks. Soft rot bacteria and fungi enter through these cracks and cause the rotting of the fruit accompanied by a foul smell (Figure 9.26b). Fruit infection can also occur during transit, or in storage. The infection of young fruits can cause deformity and the bending of fruits. The infection of young fruits can also result into their detaching from the vines. During wet weather, droplets of milky bacterial ooze frequently appear on the infection sites. Running down of the slime along the fruits causes secondary infection of the fruits (Anonymous, 2002).

FIGURE 9.27 Angular leaf spot of cucurbits: Angular spots on leaf. (Courtesy: Gerald Holmes, California Polytechnic State University at San Luis Obispo, Bugwood.org.)

9.28.2 Causal Organism

Pseudomonas syringae pv. *lachrymans* (Smith and Bryan 1915) Young *et al.* 1978

Synonym: *Psm. lachrymans*

Bacteria cells are short rods, non-sporing, Gram-stain-negative, motile by more than one polar flagellum. Poly-β-hydroxybutyrate granules are not accumulated, arginine dihydroxylase and oxidase are not produced, nitrates are reduced, protocatechuate is cleaved and levan is formed on nutrient agar containing 5% sucrose. Gelatin is liquefied, arbutin and esculin are hydrolysed, β-glucosidase test is positive and pectate gel pitting occurs at pH 4.8. Growth occurs on erythritol, inositol, mannitol, sorbitol, L-tartrate, trigonelline and quinate but not on D-tartrate, L- lactate, anthranilate, betaine and homoserine (Goto, 1992).

Significant variability in virulence, aggressiveness and genome structure has been reported among the numerous strains of the bacterium that have been identified (Olczak-Woltman *et al.*, 2007). Earlier, Krivchenko and Medvedeva (1985) reported 16 races of the bacterium and the occurrence of resistance against these races in cucumber cultivars.

Słomnicka *et al.* (2015) determined the diversity in pathogenic *Psm. syringae* pv. *lachrymans* and *Psm. syringae* pv. *syringae* strains isolated from cucurbit plants and found that pathogenicity tests of 22 strains on cucumber revealed remarkable differences in their virulence level. Genetic characterisation using MLST loci, as well as ITS1, ERIC and REP fingerprinting, allowed the grouping of strains into four phylogenetic groups. After amplicon sequencing and BLAST alignment, the collected strains complied with *Psm. syringae* pathovars *lachrymans* and *syringae* and also with saprophytic *Psm. fluorescens*. The strains of *Psm. syringae* pv. *lachrymans* may be categorised into at least two types differing in both disease symptoms and molecular characteristics. The results confirm that MLST-loci-based grouping of strains corresponds well to grouping based on pathogenicity tests.

Fullem *et al.* (2024) reported the diversity of Pseudomonads associated with bacterial leaf spot of cucurbits in the southeastern United States. Thirty-four isolates were identified as members of the *Psm. syringae* species complex, including *Psm. syringae* sensu stricto (12), *Psm. alliivorans* (12), *Psm. capsici* (nine) and *Psm. viridiflava* (one).

Słomnicka *et al.* (2018) reported the genome sequence of *Psm. syringae* pv. *lachrymans* strain 814/98, highly virulent to cucumber. Genome size was estimated to be 6.58 Mb, with 57.97% G C content. In all, 6,024 genes encoding proteins and 92 genes encoding RNAs were identified in this genome. The comparisons with the available sequenced genomes of pathovar *lachrymans*, as well as with other *Psm. syringae* pathovars, revealed the presence of three unique plasmids and 24 type III effector proteins in strain 814/98. Phylogenetic analyses of MLST loci and type III effector proteins clearly showed the existence of two distinct clusters of strains within pathovar *lachrymans*, which were grouped into either phylogroup 1 or 3, supporting non-monophyly within this pathovar.

9.28.3 Disease Cycle

The main sources of perpetuation of the bacterium are infected and/or infested seeds. In infected seeds, the bacterium resides in the seed coat and can survive for more than two years. Bhat (2009) reported that the bacterium survived up to 21 months in the infected seed, but the inoculum load of the seeds and the number of diseased seedlings produced decreased with the storage period. Inoculum load and the number of diseased seedlings immediately after the crop harvest were 5.3×10^3 cfu g^{-1} of seed and 19.44%, respectively, while the respective figures after 21 month's storage were 2.03×10^3 cfu g^{-1} of seed and 7.48%.

The bacterium also survives in the infected plant debris. Bhat (2009) also reported that in Kashmir valley, the bacterium survived for eight months in plant debris lying on the soil surface and buried in the soil, but the survival was better on the soil surface than when buried in soil. The dispersal of the pathogen occurs through wind-splashed rain, sprinkler irrigation, insects (spotted cucumber beetle, southern corn root worm), the hands of workers and farm machinery. Wind-blown soil, infected plant residue and irrigation water contaminated with the bacterium also act as dispersing agents. The penetration of the bacterium occurs through stomata and wounds.

Optimum conditions for disease development are high relative humidity and a temperature range of 23–28 °C. High relative humidity is more important than temperature for the outbreak of disease, and severe damage occurs when relative humidity is above 90%. Lesions formed when relative humidity is below 85%, remain small and do not develop into typical angular spots. Excessive nitrogen fertilisers favour the disease development.

Olczak-Woltman *et al.* (2009) studied the inheritance of resistance to the disease based on disease severity (number and size of necrotic and chlorotic lesions on the infected leaves) and the presence or absence of a chlorotic halos around the necrotic spots. They found that the disease severity appears to be controlled by multiple genes and the heritability of the resistance was estimated to be 53%. The authors further reported that the presence or absence of the chlorotic halo was to be governed by a single gene, with the presence of the halo (the susceptible phenotype) being a dominant character, and also identified a RAPD marker linked to the gene conferring the chlorotic halo. Genetic distance between the marker, OP-AO07, a polymorphic 420 bp amplicon in the DNA of the susceptible plants, and the locus encoding the chlorotic halo was estimated to be 13 cm.

Babadoost *et al.* (2019) reported that in the North Central Region more than 85% of pumpkin and winter squash fields had fruits infected with *Xanthomonas cucurbitae*. In some of the fields, more than 90% of fruits were affected. Screening of more than 500 pumpkin and squash cultivars and accessions did not identify any line with a reliable resistance to *Xan. cucurbitae. Xan. cucurbitae* survived for more than 24 months in infected pumpkin leaves and fruits, buried 5 and 15 cm deep under field conditions. *Xan. cucurbitae* also survived in naturally infected and artificially inoculated seeds at 4 and 22 °C for more than 18 months. Hot water treatment at 55 °C for 15 minutes or HCl treatment at 0.5% concentration for 40 min eradicated *Xan. cucurbitae* in pumpkin seeds. The evaluation of 27 chemical compounds and biocontrol agents for the management of *Xan. cucurbitae* in pumpkin fields revealed that mancozeb (Manzate PRO Stick), copper hydroxide (Kocide 3000), copper octanoate (Cueva SC), and an extract from *Reynoutria sachalinensis* (Regalia) were the most effective compounds in reducing the severity of bacterial spot on pumpkin leaves and fruits. Some of the bacterial isolates collected from pumpkin plants showed antagonism to *Xan. cucurbitae* in *in vitro* assays.

Zhang *et al.* (2021) showed that *Psm. amygdali* pv. *lachrymans* can be transmitted from the bottom leaf to the upper leaves inside the plant and excreted from the upper leaves through guttation. After the inoculation of a third leaf of cucumber with a bacterial suspension of *Psm. amygdali* pv. *lachrymans*, the bacterium was detected on the fifth leaf, the petiole, and the stem and in guttation drops. Healthy cucumber seedlings were infected by *Psm. amygdali* pv. *lachrymans* from the guttation droplets, indicating that guttation fluids containing *Psm. amygdali* pv. *lachrymans* could be a potential source of secondary transmission under high relative humidity.

Maglione *et al.* (2024) reported that winter rye cover crop sheltered competent squash phyllosphere bacteria to reduce the growth of *Psm. syringae* pv. *lachrymans* and angular leaf spot symptoms caused by it.

9.28.4 Management

1. Use pathogen-free and certified seed. Never use the seed from an infected crop. In the case of a non-certified seed, treat the seed in hot water (at 50 °C for 20 min) containing calcium propionate @ 4.4 oz/gal of water or acidic cupric acetate @ 6.7 oz/gal of water. Hot water treatment alone is not effective, but this treatment also does not completely eliminate the pathogen.
2. Practise crop rotation with non-host crops for three years.
3. Destroy the infected plant residue by burning or by burying.
4. Limit the use of sprinkler irrigation.
5. Avoid excessive use of nitrogenous fertilisers.
6. Avoid cultural operations and handling of plants when they are wet.
7. Minimise wounds to the plants, particularly to the cucumber fruits. Use smooth and flexible containers for storing fruits after harvest.
8. Control cucumber beetles and other insects by following a regular spray schedule.
9. Give protected sprays of cooper hydroxide (0.2%) or mancozeb (0.2%) at weekly intervals.
10. Root application of *Trichoderma asperellum* (T-203) caused accumulation of phytoalexins, induced systemic resistance against the bacterium in cucumber plants and caused 80% reduction in disease severity (Yedidia *et al.*, 2003).

REFERENCES

Agrios, G. N. (2004). *Plant Pathology*. 5th Ed. Elsevier Academic Press, San Diego, 922 pp.

Anonymous (2002). Angular leaf spot of cucurbits. Report on Plant Diseases No. 919. Department of Crop Sciences, Univ. of Illinois, Urbana, 3 pp.

Babadoost, M., Sulley, S., Thapa, S. and Zhang, X. (2019). Updates on management of bacterial spot of cucurbits incited by *Xanthomonas cucurbitae* (Abstr.) *Phytopathology* 109: S2.8.

Bhat, N. A. (2009). Survival of *Pseudomonas syringae* pv. *lachrymans*- incitant of angular leaf spot of cucumber under temperate conditions of Kashmir valley. *Indian Phytopathol.* 62: 429–34.

Bhat, N. A., Masoodi, S. D., Ahmad, M. and Zargar, M. Y. (2007). Occurrence and severity of angular leaf spot of cucumber in Kashmir valley. *Ann. Plant Prot. Sci.* 15: 410–13.

Dutta, B., Gitaitis, R. D., Lewis, K. J. and Langston, D. B. (2013a). A new report of *Xanthomonas cucurbitae* causing bacterial leaf spot of watermelon in Georgia, USA. *Plant Dis.* 97: 556.

Dutta, B., Gitaitis, R. D., Sanders, F. H., Booth, C., Smith, S. *et al.* (2013b). First report of bacterial leaf spot of pumpkin caused by *Xanthomonas cucurbitae* in Georgia, United States. *Plant Dis.* 97: 1375.

Fullem, K. R., Pena, M. M., Potnis, N., Goss, E. M., Minsavage, G. V. *et al.* (2024). Unexpected diversity of pseudomonads associated with bacterial leaf spot of cucurbits in the southeastern United States. *Plant Dis.* 108: 592–98. https://doi.org/10.1094/PDIS-06-23-1081-SR

Goto, M. (1992). *Fundamentals of Bacterial Plant Pathology*. Academic Press, Inc., San Diego, CA, 342 pp.

Janse, J. D. (2006). *Phytobacteriology: Principles and Practice*. CABI Publishing, Wallingford, 360 pp.

Jarial, K., Dogra, B., Mandradia, R. K., Kumar, S., Sharma, D. *et al.* (2011). Investigations on a new bacterial disease of bottle gourd in sub-tropical zone of Himachal Pradesh. *Plant Dis. Res.* 26: 68–75.

Krivchenko, V. I. and Medvedeva, N. I. (1985). Intraspecific differentiation of the bacterial pathogen of cucumber *Pseudomonas lachrymans*. Sbornik Nauchnykh Trudov po Prikladnoi Botanike. *Genetike i Selektsii* 92: 92–97.

Maglione, R., Ciotola, M., Cadieux, M., Toussaint, V., Laforest, M. *et al.* (2024). Winter rye cover crops shelter competent squash phyllosphere bacteria to reduce *Pseudomonas syringae* pv. *lachrymans* growth and angular leaf spot symptoms. *Phytopathology* 114: 1759–69. https://doi.org/10.1094/PHYTO-08-22-0291-R

Newberry, E. A., Jardini, T. M., Rubio, I., Roberts, P. D., Babu, B., *et al.* (2016). Angular leaf spot of cucurbits is associated with genetically diverse *Pseudomonas syringae* strains. *Plant Dis.* 100: 1397–1404.

Olczak-Woltman, H., Bartoszewski, G., Madry, W. and Niemirowicz-Szczytt, K. (2009). Inheritance of resistance to angular leaf spot (*Pseudomonas syringae* pv. *lachrymans*) in cucumber and identification of molecular markers linked to resistance. *Plant Pathol.* 58: 145–51.

Olczak-Woltman, H., Masny, A., Bartoszewski, G. Plucienniczak, A. and Niemirowicz-Szczytt, K. (2007). Genetic diversity of *Pseudomonas syringae* pv. *lachrymans* strains isolated from cucumber leaves collected in Poland. *Plant Pathol.* 56: 373–82.

Rai, R., Vittore, K. M., Pasion, J., Malvino, M. L., Mason, J. D. *et al.* (2024). Comparative genomics of *Xanthomonas cucurbitae* isolates collected from Midwestern United States pumpkin fields. *Plant Pathol.* 73: 2211–23.

Słomnicka, R., Olczak-Woltman, H., Bartoszewski, G. and Niemirowicz-Szczytt, K. (2015). Genetic and pathogenic diversity of *Pseudomonas syringae* strains isolated from cucurbits. *Eur. J. Plant Pathol.* 141: 1–14.

Słomnicka, R., Olczak-Woltman, H., Oskiera, M., Schollenberger, M., Niemirowicz-Szczytt, K. *et al.* (2018). Genome analysis of *Pseudomonas syringae* pv. *lachrymans* strain 814/98 indicates diversity within the pathovar. *Eur. J. Plant Pathol.* 151: 663–76.

Smith, E. F. and Bryan, M. K. (1915). Angular leaf-spot of cucumbers. *J. Agric. Res.* 5: 465–76.

Yedidia, I., Shoresh, M., Kerem, Z. Benhamou, N., Kapulnik, Y. *et al.* (2003). Concomitant induction of systemic resistance to *Pseudomonas syringae* pv. *lachrymans* in cucumber by *Trichoderma asperellum* (T-203) and accumulation of phytoalexins. *Appl. Environ. Microbiol.* 69: 7343–53.

Young, J. M., Dye, D. W. and Wilkie, J. P. (1978). Genus *Pseudomonas* Migula 1894. In: Young, J. M., Dye, D. W., Bradbury, J. F., Panagopoulos, C. G. and Robbs, C. F. (1978). A proposed nomenclature and classification of plant pathogenic bacteria. *N. Z. J. Agric. Res.* 21: 153–77.

Zhang, S., Wang, Y., Hu, T., Wang, S., Che, H. *et al.* (2021). Guttation fluids containing *Pseudomonas amygdali* pv. *lachrymans* are a potential source of secondary infection in cucumber. *Phytopathology* 111: 2162–67.

9.29 BACTERIAL BLIGHT OF COTTON

The disease is also known by other names, such as angular leaf spot and black arm of cotton. It was first reported from Alabama, USA in 1891 by Atkinson, who named it black rust. During 1900–1905, E. F. Smith proved Koch's postulates, described the causal organism and also established a relationship between the black arm and angular leaf spot phases of the disease, which had earlier been considered as different diseases. The disease occurs worldwide and causes heavy losses in some countries. In India, the disease was first noticed in Madras in 1918; it remained a minor disease till the introduction of tetraploid cottons. It is argued that in India susceptible cultivars phased out under the selection pressure of the pathogen, and the pathogen and the cultivars reached a balance. On the basis of this co-evolution, Knight and Hutchinson (1950) concluded that the disease originated in India. However, Patel and Kulkarni (1948) are of the view that the disease was introduced in the middle of the 19th century through exotic material.

During the 1940s, a campaign was launched in India to replace the rain-fed indigenous diploid cottons with a high-yielding but susceptible tetraploid, *Gossypium hirsutum*, and, to some extent, *G. barbadense* cottons. After the launch of this campaign, the severe epiphytotics of the disease occurred during 1948–52 and the disease became established in the 1960s. Irrigation and high doses of nitrogenous fertilisers applied to achieve the full potential of the introduced cotton also increased the disease's severity. In India, the disease occurs in all the cotton-growing areas and the losses are in the range 5–25% (Verma and Singh, 1972).

Malliarakis *et al.* (2021) confirmed the presence of bacterial blight of cotton in five commercial fields in the region of Drama in Greece, with yield losses reaching 30%. This is the first time that the causal agent of the disease, *Xan. citri* pv. *malvacearum*, although earlier observed and isolated, was fully identified by biochemical, molecular and pathogenicity tests in Greece.

The comprehensive review by Saini *et al.* (2024) attempts to provide insight into the dynamic nature of bacterial blight threatening cotton cultivation in India during the last century and the underpinning multifaceted research efforts that paved the way for countering the disease. An attempt is made to discuss the qualitative and quantitative studies on the genetics of bacterial blight for facilitating the breeding programmes.

9.29.1 Symptoms

There are four different types or phases of symptoms of the disease. It is not necessary, however, that all four of the symptoms should appear under in all situations. Their occurrence and predominance vary under different environmental conditions.

9.29.1.1 Seedling Blight

Water-soaked circular spots appear on the lower surface of cotyledons. The spots spread, forming irregular patches, and causing the drying and withering of the seedlings. Many plants may die. This phase is more common in countries like the USA, where the temperature is low at the time that the cotton is sown.

9.29.1.2 Leaf Spots

Water-soaked spots first appear on the lower surface of leaves. These spots enlarge, forming dark brown to black, angular dead areas and becoming also visible on the upper leaf surface. As the spread of these spots is bounded by veins and veinlets, these become angular in shape. The spots enlarge, forming irregular large patches (Figure 9.28). The infection extends along the veins, turning the veins black; the infected veins have a **finger-like appearance**. This phase is called **'vein blight'** or **'black vein'** (Figure 9.28a). From veins, the infection reaches the petioles through the midrib and infects the stem. Petiole infection leads to the

(a) (b)

FIGURE 9.28 (a) Bacterial blight of cotton: Angular spots and vein blight symptoms on leaves. (Courtesy: Stephen Allen, www.padil.gov.au, CCAL.) (b) Bacterial blight of cotton: Boll rot. (Courtesy: Stephen Allen, www.csiro.au, CCAL.)

severe defoliation of plants. Under humid weather conditions, yellowish bacterial exudate is formed on leaf lesions.

9.29.1.3 Black Arm

Linear, sunken, dark brown to black lesions appear on the stem and branches of the plants, eventually girdling them. In severely affected stems, there is deep cracking and gummosis, which results in the breaking off of the stem by strong winds. This phase is more common in Sudan and Uganda.

9.29.1.4 Boll Rot

Water-soaked and circular spots appear on bolls, which, later, turn brown and become sunken (Figure 9.28b). Severely infected bolls drop; in the case of mildly infected bolls, however, the infection progresses internally. Secondary organisms also enter the infected bolls and cause extensive boll rotting. The infected bolls may fail to open.

Sometimes, atypical symptoms of the disease appear, consisting of dull green, flaccid areas extending from the periphery of cotyledons and leaves, and elongated, water-soaked, dark brown areas on leaf lamina. A new type of symptomatology, named **"bacterial leaf blight"**, has been described by Malavolta *et al.* (2008) from the Sao Paulo state of Brazil. The new type of symptomatology is characterised by leaf blight, generally exhibiting chlorotic haloes and sometimes showing V-shaped lesions beginning at the border of the leaves. The bacterium causing the new kind of symptoms was identical to the type strain of *Xanthomonas axonopodis* pv. *malvacearum*.

Earlier it was considered that the bacterium invades only the parenchyma of the host. Since 1932, information has become available that the bacterium also invades the vascular system. From infected cotyledons, the bacterium moves through the vascular system of the petioles and reaches the stem. In the stem, it moves upward within the vascular system from where it breaks outward, causing cortical rotting of young stem tissues and petioles, and the infection of the leaf lamina. The movement of the bacterium in the susceptible cultivars indicates the probable systemic, or at least spreading, nature of the pathogen. The isolation of the bacterium from embryos and seeds also shows the systemic nature of the pathogen.

9.29.2 Causal Organism

Xanthomonas citri subsp. *malvacearum* (hereafter referred to as *Xcm*) (Smith 1901) Schaad *et al.* 2007, subsp. nov., nom. rev.

Synonym: *Xan. campestris* pv. *malvacearum, Xan. axonopodis* pv. *malvacearum*

Cells $A = \pi r^2$ rod-shaped and measure 0.9 × 1.2 µm; occur singly or in pairs, but rarely in chains, monotrichous, non-sporing, not acid-fast, encapsulated, chemoorganotrophic, facultatively aerobic and Gram-stain-negative. Metabolism is respiratory, never fermentative, catalase test is positive but oxidase negative. Asparagine is not utilised as a sole source of carbon and nitrogen. The bacterium may utilise methionine, glutamic acid and nicotinic acid in various combinations as minimal growth requirements. Triphenyl tetrazolium chloride @ 0.1 to 0.2% in nutrient agar inhibits its growth.

The bacterium liquefies gelatin, hydrolyses starch, digests casein but not fat. Produces H_2S but not indole, nitrate is not reduced; methyl red and Voges–Proskauer tests are negative, and can tolerate 3% sodium chloride. Sucrose, dextrose, maltose, lactose, levulose, xylose, glycerol, galactose and raffinose are utilised with acid production but without gas. Arabinose, rhamnose, salicin, mannitol and dulcitol are not utilised. The optimum temperature for growth is 31–32 °C while the minimum and maximum are at 6 and 42 °C, respectively. Plasmids varying in number and size are present in the cells (Das and Verma, 2003). Schaad *et al.* (2005), on the basis of DNA–DNA relatedness determined by sequencing the 16S–23S intergenic spacer regions and AFLP analysis, reclassified the bacterium as *Xan. smithii* subsp. *smithii* sp. nov. comb. nov. nom. nov. but later amended the classification and renamed it as *Xan. citri* subsp. *malvacearum* (*Xcm*) (Schaad *et al.*, 2006, 2007). Schaad *et al.* (2006) also reported the following characteristics of *Xcm*. Colonies of *Xcm* appear on yeast dextrose carbonate agar (YDC) and Fieldhouse Sasser (FS) agars after 40–44 and 56–60 h of incubation, respectively, at 28–30 °C. The bacterium produces an alkaline reaction in litmus milk without hydrolysis, liquefies gelatin but does not hydrolyse pectate. *Xcm* does not utilise melizitose. *Xcm* can be differentiated from *Xan. campestris* pv. *campestris* and most other *Xanthomonas* pathovars, subspecies and species by DNA–DNA reassociation assays, rep-PCR profiles, intergenic spacer (ITS) sequencing, serological methods, SDS-PAGE patterns of membrane proteins and phenotypic characters.

Zhai *et al.* (2013) reported high-quality draft sequences of a highly virulent *Xan. axonopodis* pv. *malvacearum* strain (Gottinger Sammlung Phytopathogener Bakterien [GSPB2388]) from Sudan and of a strain of race 18 (GSPB1386) from Nicaragua, using Illumina/Solexa paired-end sequencing. Short sequence reads were assembled into 61 scaffolds for GSPB2388 (N50 of 164 kb) and 127 scaffolds for GSPB1386 (N50 of 100 kb), with draft maps of roughly 5 Mb, which contain 4,665 and 4,520 protein-coding genes, respectively. Based on gene annotation and comparisons with plant pathogen proteins, 181 and 178 potential virulence-related genes, including genes encoding a major group of type III effectors, were identified from GSPB2388 and GSPB1386, respectively. Differential effectors and sequence diversity between the highly virulent strains and race 18 may enable the identification of key factors that contribute to the high virulence of highly virulent strains.

Nagrale *et al.* (2020) reported the occurrence and association of *Pantoea dispersa*, as a potential and principal pathogenic agent causing inner cotton (*G. hirsutum* L.) boll rots in Maharashtra state, India.

9.29.2.1 Races

Pathogenic variation in the bacterium is well established and to date now 32 races have been reported from different parts of the world (Table 9.29a). In 1946, Knight was probably the first person to suggest the existence of biological specialisation in the bacterium, when Stoneville-20, a line of upland cotton, resistant to the pathogen, was severely infected in the United States. In 1958, Bird and Hadley identified two races of the pathogen on the basis of their reaction to three lines of upland cotton. Brinkerhoff (1963), on the basis of inoculation of isolates collected from all over the world, on eight differential lines, including *G. hirsutum* and *G. barbadense*, identified ten new races. Later on, Hunter *et al.* (1968) developed a set of differentials (8 lines of upland cotton having different bacterial blight resistance genes) and identified 15 races of *Xcm*. Total number of races was raised to 17 when two more races, i.e., I_1 and I_2, were described from India (Table 9.29b).

Out of two sets of differentials developed by Nayudu (1964) and Hunter *et al.* (1968) for the identification of races of *Xcm*, the set developed by Hunter *et al.* (1968) is more suitable and is widely used. It contains isogenic lines with different bacterial blight resistance genes, which give stable reaction under different environmental conditions. Moreover, the race identification is based scientifically on the number of genes attacked, and on which, to date, all the isolates have given variable reactions.

TABLE 9.29A
Nomenclature of Races of *Xanthomonas citri* subsp. *malvacearum*. Part A: Primary Differentials for Race Number

Disease Reaction on Differentials: Susceptible = +, Resistant = –							Race Number	
I	II	III	IV	V	VI	VII	USA	India
+	–	–	–	–	–	–	13	1
+	+	–	–	–	–	–	1, 11	2
+	–	+	–	–	–	–	–	3
+	–	–	+	–	–	–	–	4
+	–	–	–	+	–	–	–	5
+	–	–	–	–	+	–	–	6
	+	+	–	–	–	–	2, 12	7
+	+	–	+	–	–	–	–	8
+	+	–	–	+	–	–	3	9
+	+	–	–	–	+	–	4	10
+	–	+	+	–	–	–	–	11
+	–	+	–	+	–	–	–	12
+	–	+	–	–	+	–	9	13
+	–	–	+	+	–	–	–	14
	–	–	+	–	+	–	–	15
+	–	–	–	+	+	–	19	16
+	+	+	+	–	–	–	–	17
+	+	+	–	+	–	–	15	18
+	+	+	–	–	+	–	17	19
+	+	–	+	+	–	–	6	20
+	+	–	+	–	+	–	–	21
+	+	–	–	+	+	–	5	22
+	–	+	+	+	–	–	–	23
+	–	+	+	–	+	–	–	24
+	–	+	–	+	+	–	–	25
+	–	–	+	+	+	–	–	26
+	+	+	+	+	–	–	8	27
+	+	+	+	–	+	–	16	28
+	+	+	–	+	+	–	14	29
+	+	–	+	+	+	–	7	30
+	–	+	+	+	+	–	–	31
+	+	+	+	+	+	–	10, 18	32

Source: Verma, J. P. and Raj, S., Blight and wilt of cotton, in: *Plant Diseases of International Importance*, Vol. IV, Diseases of Sugar, Forests and Plantation Crops, Mukhopadhyay, A. N. *et al.*, eds., Prentice Hall, Englewood Cliffs, N J, pp. 138–160, 1992. I, Acala-44 (no genes); II, Stoneville 2B-S9 (polygenes); III, Stoneville-20 (B_7 + polygenes); IV, Mebane B-1 (B_2 + polygenes); V, 1-10.B (B_{In} + polygenes); VI, 20-3 (B_N + polygenes); VII, 101-102.B (B_2B_3 + unknown).

TABLE 9.29B
Nomenclature of Races of *Xanthomonas citri* subsp. *malvacearum*. Part B: Additional Differentials for Biotypes of Race 32

Disease Reaction on Differentials: Susceptible = +, Resistant = –

VII	VIII	IX	X	Biotype Number
–	–	–	–	32.0
+	+	–	–	32.1 (probably race 18 of USA)
+	–	+	–	32.2
+	+	+	–	32.3 (Follin's isolate can be placed here)

Source: VII, 101–102.B (B_2B_3 + unknown); VIII, Empire **B4** (**B4** + polygenes); IX, Tamcot CAMD-E (**B2B3B7** + polygenes); X, a line yet to be developed.
Note: Even when the line X is attacked by a new race of *Xan. citri* subsp. *malvacearum*, the old race number will remain the same and valid. Additional lines can always be added to designate new races/biotypes. Further, additional information can always be given; e.g., race 32.3 can also attack the gene combination of TxBonham ($B_2B_3B_4B_7B_{Sm}$).

To date, 26 races have been described from India and these are *Xcm* R-1, 2, 4, 5, 7–10, 13, 14, 16–25 and 27–32. Race 32 was most widely distributed and was present at more than 83% of the locations tested. Out of 133 isolates tested, 59.25% belonged to this race. It was the most virulent race as it attacked five major bacterial blight resistance genes, namely, B_7, B_4, B_2, B_{In} and B_N. Other prevalent races were *Xcm* R-27, 28 and 29, which attacked three genes. Less virulent races, *Xcm* R-1, 2, 9 and 10, were restricted in their distribution (Verma and Raj, 1992).

Race *Xcm* R-18H described from Pakistan is capable of attacking five genes. A race from western and central Africa has infected the line 101-102.B, i.e., the combined genes B_2 + B_3, as well as other combined gene resistance. A race called post-Barkat race (PBR) is capable of attacking all single and developed combinations of major *B* genes for resistance in Sudan. As some of the presently known races are highly virulent and capable of attacking combined gene resistance, there is an urgent need to develop another set of differentials to properly identify these races.

More than one race may be present on the same plant, on the same leaf or even in the same lesion. Thirteen races were isolated from four different lesions formed on the same leaf of Acala-44, which is the most susceptible line of differentials, having no blight resistance gene. New races possessing more virulent genotypes could be generated in nature from mixed infections of pathogenic races (in a susceptible cultivar) in a synergistic reaction (Verma *et al.*, 1979).

Delannoy *et al.* (2005) reviewed the genetic basis of cotton resistance to the pathogen, with reference to resistance genes, resistance gene analogs and bacterial avirulence genes, along with the physiological mechanism involved in the hypersensitive reaction, and suggested that this host–pathogen interaction represents the most complex resistance gene/*avr* gene system yet known.

Essenberg *et al.* (2014) developed a pyramid of lines with all possible combinations of two and three genes using near-isogenic lines of upland cotton (*G. hirsutum*) carrying single, race-specific genes *B4*, *BIn* and *b7* for resistance to bacterial blight to achieve broader and higher resistance approaching that of L. A. Brinkerhoff's exceptional line Im216. Under field conditions in north-central Oklahoma, pyramid lines exhibited broader resistance to individual races and, consequently, higher resistance to a race mixture. Although some enhancement in resistance was found, the pyramided lines carrying two or three *B* genes did not acquire the level of resistance of Im216. In a growth chamber, bacterial populations of race 1 in and on leaves of the pyramid lines decreased significantly, with an increasing number of *B* genes in only one of four experiments. Older lines, Im216 and AcHR, exhibited considerably lower bacterial populations than any of the one-, two-, or three-*B*-gene lines.

9.29.3 Disease Cycle

Infected seed is the main source of perpetuation of the bacterium and it may carry the pathogen both externally (as a slimy mass in the fuzz) as well as internally. Other sources of perpetuation are infected self-sown seeds and infected plant debris. The bacterium does not survive in the plant debris up to the next season of crop when it is buried deep in soil and the soil remains wet. However, it can survive for seven months in trash on the soil surface. Bacterium also does not survive when the wheat crop is taken between two cotton crops. However, the plant debris serves as an important source of infection where cotton is grown as a continuous crop. It seems that, under northern Indian conditions, infected seed, including the self-sown seed, is an important source of primary inoculum.

Bacterium enters the host mainly through stomata. Insect punctures also serve as entry points for boll infection. The disease is favoured by high humidity and moderate temperature. Primary infection occurs at a relatively low temperature of 30 °C at which cotton germinates and grows, but secondary infection occurs at a relatively higher temperature, i.e., 35 °C. Presence of moisture is essential for the secondary spread of the pathogen during the first 48 h of infection. Relative humidity of 85% or above is essential for abundant infection. Dry and hot weather retard the normal development of the disease.

Wind-blown rain is the main disseminating agent of the bacterium. Surface irrigation and overhead irrigation also spread the bacterium. In some areas, wind-blown detached leaves and wind-blown dust may serve as a means of dissemination. Transmission of the bacterium by two cotton pests, namely, *Dysdercus koenigii* and *Earias* spp., is known but their role in dissemination and disease development is not clear. Bacterium survives in the excreta of larvae of *Earias* spp. For up to six days.

Jatropha curcas, *Lochnera pusilla*, *Thurbaria thespesioides*, *Eriodendron anfractosum* and *Thespesis* spp. act as a collateral host of the pathogen but their role in the disease epidemiology is not clearly defined.

Mijatovic *et al.* (2020) found that seed-to-seedling transmission of *Xa. citri* pv. *malvacearum* involves systemic dissemination through the vascular tissue of cotton plants.

According to Wheeler *et al.* (2020), a more modern concept of pathogen characterisation in *Xa. citri* subsp. *malvacearum* (*Xcm*) can be based on virulence factors, i.e. in type III effectors associated with its isolates. A series of *Xcm* isolates from across the USA were collected and tested for different type III effectors.

Mijatović *et al.* (2021) employed a wild-type (WT) field isolate of *Xa. citri* pv. *malvacearum* from Georgia (USA) to generate a non-pathogenic *hrcV* mutant lacking a functional type III secretion system (TIIISS-). They tagged WT and TIIISS-strains with an auto-bioluminescent Tn*7* reporter and compared the colonisation patterns of bacterial blight susceptible and resistant cotton seedlings using macroscopic image analysis and bacterial load enumeration. WT and T3SS-*Xan. citri* pv. *malvacearum* strains colonised cotyledons of resistant and susceptible cotton cultivars. However, *Xan. citri* pv. *malvacearum* populations were significantly higher in susceptible seedlings inoculated with the WT strain. Additionally, WT and T3SS-*Xan. citri* pv. *malvacearum* strains systemically colonised true leaves, although at different rates. Finally, they observed that seed-to-seedling transmission of *Xan. citri* pv. *malvacearum* may involve systemic spread through the vascular tissue of cotton plants. These findings yield novel insights into potential *Xan. citri* pv. *malvacearum* reservoirs for bacterial blight outbreaks.

Essenberg *et al.* (2023) analysed four cadalene and lacinilene phytoalexins from inoculated leaves of the near-isogenic cotton lines (NILs) carrying B_2, B_4, B_{In}, B_5 and b_7, respectively, a NIL carrying both B_4 and B_{In}, the susceptible parent line Ac44E and three highly resistant lines carrying multiple genes for bacterial blight resistance in a polygenic background. The lines were inoculated with race 1 of *Xan. citri* subsp. *malvacearum*, to which all the tested cotton lines (except Ac44E) exhibit resistance. They developed a rapid, semi-quantitative method for the estimation of phytoalexin amounts based on thin-layer chromatography and image analysis of the yellow-green autofluorescence of the lacinilenes. The cadalenes were detected by autofluorescence following their autooxidation to lacinilenes as the chromatographic solvents evaporated. Results showed that all lines that possess at least one *B* gene for bacterial blight resistance produced lacinilene and cadalene phytoalexins in response to infection. AcB_5 was exceptional in this activity.

9.29.4 Management

The eradication of seed-borne inoculum and the destruction of infected plant debris and self-sown seeds are of utmost importance for the management of the disease.

1. Treat the acid- or machine-delinted seeds with carboxin, oxycarboxin or Agrimycin–100 @ of 2 g kg^{-1} of seed.
2. Hot water treatment of seed at 56 °C for 10 min. is also effective.
3. Destroy the diseased plant debris and self-sown seedlings. Plant debris should be buried deep in the soil, followed by flooding of the field for 24–48 h.
4. Give protective sprays of phylloplane bacteria. Two species of *Flavobacterium* and one species each of *Pseudomonas* and *Aeromonas* are effective for control. This schedule should be included in the integrated management of the disease.
5. To check the secondary spread of the disease, spray the crop with Blitox (500 g) + Agrimycin–100 (20 g) or Streptocycline (3 g) per acre per spray. The first spray should be given on a 4- to 6-week-old crop and subsequent sprays at 10–15-day intervals. Three or more sprays are needed depending upon the severity of the disease. As reported by Verma and Raj (1992), these chemicals also possess some curative effect.
6. Use of resistant varieties.

REFERENCES

Brinkerhoff, L. A. (1963). Variability of *Xanthomonas malvacearum*: The cotton bacterial blight pathogen. *Oklahoma State Univ. Tech. Bull.* 98: 96 pp.

Das, I. K. and Verma, J. P. (2003). Plasmids in bacteria associated with bacterial leaf blight of cotton. *Indian Phytopathol.* 56: 180–82.

Delannoy, E., Lyon, B. R., Marmey, P., Jalloul, J. F., Daniel, J. L. *et al.* (2005). Resistance of cotton towards *Xanthomonas campestris* pv. *malvacearum*. *Annu. Rev. Phytopathol.* 43: 63–82.

Essenberg, M., Bayles, M. B., Pierce, M. L. and Verhalen, L. M. (2014). Pyramiding *B* genes in cotton achieves broader but not always higher resistance to bacterial blight. *Phytopathology* 104: 1088–97.

Essenberg, M., McNally, K. L., Bayles, M. B., Pierce, M. L., Hall, J. A. *et al.* (2023). Gene *B5* in cotton confers high and broad resistance to bacterial blight and conditions high amounts of sesquiterpenoid phytoalexins. *Phytopathology* 113: 812–23.

Hunter, R. E., Brinkerhoff, L. A. and Bird, L. S. (1968). The development of a set of upland cotton lines for differentiating races of *Xanthomonas malvacearum*. *Phytopathology* 58: 830–32.

Knight, R. L. and Hutchinson, J. B. (1950). The evolution of black arm resistance in cotton. *J. Genet.* 50: 36–58.

Malavolta, V. A. Jr., Destefano, S. A. L., Beriam, L. O. S., Jalloul, J. F., Daniel, J. L. (2008). Cotton leaf blight, a new symptomatology caused by *Xanthomonas axonopodis* pv. *malvacearum*. *Summa Phytopathol.* 34: 168–71.

Malliarakis, D., Simoglou, K. B., Grypari, D., Mpalantinaki, E., Pagoulatou, M. G. *et al.* (2021). Phenotypic and molecular characterization of *Xanthomonas citri* pv. *malvacearum* isolates obtained from bacterial blight diseased cotton plants in Greece. *J. Plant Pathol.* 103: 1137–40.

Mijatovic, J., Kemerait, R., Walcott, R. R. and Kvitko, B. H. (2020). Genetic requirements for seedling colonization

by *Xanthomonas citri* pv. *malvacearum* after cotton seed infection (Abstr.). *Phytopathology* 110: S2.94.
Mijatović, J., Severns, P. M., Kemerait, R. C., Walcott, R. R. and Kvitko, B. H. (2021). Patterns of seed-to-seedling transmission of *Xanthomonas citri* pv. *malvacearum*, the causal agent of cotton bacterial blight. *Phytopathology*, 111: 2176–84.
Nagrale, D. T., Gawande, S. P., Gokte-Narkhedkar, N. and Waghmare, V. N. (2020). Association of phytopathogenic *Pantoea dispersa* with inner boll rot of cotton (*Gossypium hirsutum* L.) in Maharashtra state, India. *Eur. J. Plant Pathol.* 158: 251–60. https://doi.org/10.1007/s10658-020-02071-0
Nayudu, M. V. (1964). Variation in *Xanthomonas malvacearum*. *Indian Cott. Grow. Rev.* 18: 350–55.
Patel, M. K. and Kulkarni, Y. S. (1948). *Xanthomonas malvacearum* (Erw. F. Smith) Dowson on exotic cottons in India. *Curr. Sci.* 17: 243–44.
Saini, A. S., Saini, S., Raj, K., Beniwal, J., Garima, G. *et al.* (2024). Bacterial blight: Once a menacing disease of cotton in India, now tamed and fading from research spotlight. *Plant Pathol.* 73: 1675–90. https://doi.org/10.1111/ppa.13921
Schaad, N. W., Postnikova, E., Lacy, G., Sechler, A., Agarkova, I. *et al.* (2005). Reclassification of *Xanthomonas campestris* pv. *citri* (ex Hasse 1915) Dye 1978 forms A, B, C, D and E as *X. smithii* subsp. *citri* (ex Hasse) sp. nov. nom. rev. comb. nov.; *X. fuscans* subsp. *aurantifolii* (ex Gabriel 1989) sp. nov. nom. rev. comb. nov.; and *X. alfalfae* subsp. *citrumelo* (ex Riker and Jones) Gabriel *et al.*, 1989 sp. nov. nom. rev. comb. nov.; *X. campestris* pv. *malvacearum* (ex Smith 1901) Dye 1978 as *X. smithii* subsp. *smithii* sp. nov. comb. nov. nom. nov.; *X. campestris* pv. *alfalfae* (ex Riker and Jones, 1935) Dye 1978 as *X. alfalfae* subsp. *alfalfae* (ex Riker *et. al.*, 1935) sp. nov. nom. rev.; and "var. *fuscans*" of *X. campestris* pv. *phaseoli* (ex Smith, 1897) Dye 1978 as *X. fuscans* subsp. *fuscans* sp. nov. *Syst. Appl. Microbiol.* 28: 494–518.
Schaad, N. W., Postnikova, E., Lacy, G., Sechler, A., Agarkova, I. *et al.* (2007). Validation of the publication of new names and new combinations previously effectively, but not validly, published outside the IJSEM. Vol. 57: pp. 893–97. Effective Publication: Schaad, N. W., Postnikova, E., Lacy, G., Sechler, A., Agarkova, I. *et al.* (2006). Emended classification of xanthomonad pathogens on citrus. *Syst. Appl. Microbiol.* 29: 690–95.
Smith, E. F. (1901). The cultural characters of *Pseudomonas hyacinthi*, *Ps. campestris*, *Ps. phaseoli*, and *Ps. stewarti*, four one-flagellate yellow bacteria parasitic on plants. *Bull., Division of Vegetable Physiology and Pathology, U S Department of Agri.* 28: 1–153.
Verma, J. P., Chowdhury, H. D. and Singh, R. P. (1979). Interactions between different races of *Xanthomonas malvacearum* in leaves of *Gossypium hirsutum*. *Z. Pflankrankh. Pflschutz.* 86: 460.
Verma, J. P. and Raj, S. (1992). Blight and wilt of cotton. In: A. N. Mukhopadhyay, J. Kumar, H. S. Chaube and U. S. Singh (eds.). *Plant Diseases of International Importance*. Vol. IV, *Diseases of Sugar, Forests and Plantation Crops*. Prentice Hall, Englewood Cliffs, NJ, pp. 138–60.
Verma, J. P. and Singh, R. P. (1972). Bacterial blight of Cotton. In: P. N. Patel (ed.). *Plant Bacteriology*. Vol. I, *Bacterial Diseases of Plants in India*. Indian Agricultural Research Institute, New Delhi, pp. 159–68.
Wheeler, T. A., Allen, T. W. Jr., Bart, R., Isakeit, T. S., Kemerait, R. *et al.* (2020). Bacterial blight of cotton: Attributes of the causal agent (*Xanthomonas citri* subsp. *malvacearum*) and impact of management on disease (Abstr.). *Phytopathology* 110: S1.22.
Zhai, J., Xia, Z., Liu, W., Jiang, X. and Huang, X. (2013). Genomic sequencing globally identifies functional genes and potential virulence-related effectors of *Xanthomonas axonopodis* pv. *malvacearum*. *Eur. J. Plant Pathol.* 136: 657–63.

BACTERIAL DISEASES OF BEAN

Common bean or common dry bean (*Phaseolus vulgaris* L.) suffers from five important bacterial diseases, namely, common blight, fuscous blight, halo blight, bacterial brown spot and bacterial wilt. The first four diseases are described below.

9.30 COMMON AND FUSCOUS BLIGHTS OF BEAN

Common bacterial blight, more frequently referred to as common blight, is an important disease of common bean (*Phaseolus vulgaris* L.). The disease causes significant yield losses in temperate and tropical areas and is one of the major constraints in dry bean production. Common blight was first reported in 1892 by Beach from New York and Halsted from New Jersey. In 1897, E. F. Smith described the causal bacterium and named it *Bacillus phaseoli*. Since then, the disease has been reported from many states of the USA and Canada, France, Hungary, Japan, Australia, Bulgaria, Madagascar, Norway, the Philippines, Spain, Turkey, New Zealand, Paraguay, Uruguay, Venezuela, Guatemala, Nicaragua, Honduras, Mexico, Argentina, Burundi, Brazil, Colombia, Uganda, Zaire, Zambia, Rwanda and South Africa. It is not yet established in a number of bean-growing countries of the European and Mediterranean Plant Protection Organisation (EPPO) region. Common blight pathogen is an A2 quarantine pest for EPPO.

In India, Patel and Jindal were the first to observe the disease in 1971 at Pune, Maharashtra, where it has been occurring for a long time. Osdaghi and Zademohamad (2016) reported the disease on lima bean (*P. lunatus* cv. Christmas) plants from East Azerbaijan and West Azerbaijan provinces, northwestern Iran, while it has already been reported on common bean and *mung* bean from several regions of Iran, including the central plain and southwestern provinces.

Although the disease is named common blight, it is not necessarily the most common one in any given situation.

Singh and Miklas (2015), in their review article, have stated that considerable information is available on many aspects of the disease. However, Andean and Middle American common bean cultivars, with high levels of combined resistance to less aggressive and aggressive bacterial strains in all aerial plant parts, are unavailable. They have reviewed the progress achieved in breeding for resistance, briefly describe problems faced and discuss strategies for integrated genetic improvement to develop cultivars resistant to common blight.

The disease occurs in temperate, subtropical and tropical areas, but the maximum losses are caused in tropical areas because the most severe outbreaks of the disease occur in warm humid weather. The losses result from the reduction in the photosynthetic area of plants and loss in quality and number of pods and seeds. In 1918, 75% of the fields in New York state of the USA were affected with common blight and the disease caused heavy losses. In the subsequent years, losses were in the range 20–50%. In 1953, the disease was widespread in Nebraska (USA) and it caused an estimated loss of over US$1 million. In 1976, in the USA, common blight was economically the most important disease among the bacterial diseases of beans, and it caused an estimated loss of US$4 million. Losses due to this disease in field bean crop in Ontario, Canada were more than 1252 tons in 1970 and 218 tons in 1972. A model for the assessment of yield losses from bacterial blight has been developed in Ontario, using aerial infrared photographic surveys to assess disease levels. In 1967, there were reports of yield losses up to 38% in dry bean in Ontario (Canada), and 10–20% in navy bean in Michigan (USA) (Gilbertson and Maxwell, 1992). Yield losses ranging from 17% to 22% were also reported from Colombia. In addition to loss in yield, the disease causes a reduction in the quality of seed through the discolouration of infected seeds.

In 1930, Burkholder described two bacterial diseases of bean, namely, fuscous blight and bacterial brown spot. The causal bacterium of the fuscous blight, isolated by him from bean plants collected from Switzerland, was very close to the pathogen of common blight and it also produced symptoms identical to common blight with some minor differences. Fuscous blight was first observed in Switzerland in 1924. Subsequently, it has also been reported from Australia, Yugoslavia, Canada, New Zealand, South America, Hungary, Germany and many other countries.

9.30.1 Host Range

In addition to common bean, the host range of *Xanthomonas axonopodis* pv. *phaseoli* (hereafter referred to as *Xap*) includes field bean, snap bean, lima bean, hyacinth bean, scarlet runner bean, soybean, *mung* bean, *urad* bean, *moth* bean, tepary bean, white flowering lupine and fenugreek (Partridge, 2003). *Xan. fuscans* subsp. *fuscans* (hereafter referred to as *Xff*), the causal agent of fuscous blight, infects field bean, snap bean, civet bean and scarlet runner bean.

9.30.2 Symptoms

The symptoms of common and fuscous blights are similar and generally cannot be distinguished under field conditions. These diseases can be identified by the isolation and identification of their causal pathogens or using molecular tools.

Both the diseases are primarily foliar diseases, but symptoms also appear on other above-ground plant parts. Most characteristic and diagnostic symptoms appear on trifoliate leaves. Small, water-soaked lesions may develop on cotyledons and/or primary leaves and the stems of seedlings grown from infected seeds.

On leaves, the symptoms start as small, water-soaked spots on the lower surface. The spots develop on the leaf margin or between the veins and there can be numerous spots on a leaf or leaflet. As these spots enlarge, their centres become necrotic and dry. Spots are surrounded by narrow bands of lemon-yellow tissue, which are diagnostic features of the disease. The yellow halo may be small or absent if the temperature rises above 27 °C. As the disease progresses, the spots enlarge, and on susceptible varieties continue to enlarge till they merge with each other, forming large dead areas of various shapes. Dark discoloured lesions also develop along the main vein, indicating systemic infection. Dry centres of the spots may tear and fall out, giving the leaves a tattered and shredded appearance. Badly infected leaves fall and severely infected plants defoliate (Figure 9.29a).

On stems, water-soaked, occasionally sunken lesions appear, which enlarge longitudinally and turn brown. The lesions frequently break open at the surface, exuding bacterial ooze. The occurrence of these lesions is nearer the vicinity of the first or cotyledonary node. Lesions may girdle the stem at the cotyledonary node, and this phase is known as **joint -rot**. Generally, the infected plants break off at this point due to the weight of the plant or action of the wind. On petioles and pulvinus, initially the lesions are water-soaked, but they may turn brown later on (Patel and Jindal, 1972).

On pods, small, circular, water-soaked spots appear which enlarge to form sunken, brick-red lesions. Pod lesions ultimately turn brown and differ in size, depending upon the age of the pod when infection occurs. In humid weather, a slimy bacterial exudate may be formed on pod lesions, which eventually dries, giving the lesions a glazed appearance (Figure 9.29b). In severely infected pods, seeds may be shrivelled, discoloured or reduced in size. Butter-yellow discolouration of infected seeds is more prominent on white-seeded varieties in comparison to seeds of coloured varieties.

Though the most symptoms of common and fuscous blights are similar, the intensity of seed discolouration and

(a)

FIGURE 9.29A Common blight of bean: Symptoms on aerial plant parts. (Courtesy: Robert Harveson and Howard Schwartz, www.plantmanagement.network.org)

(b)

FIGURE 9.29B Common blight of bean: Symptoms on pods. (Courtesy: Robert Harveson and Howard Schwartz, www.plantmanagement.network.org)

brown discolouration at the stem nodes is greater in the latter disease. In fuscous blight, the lesions on the stem are also more inclined to split the stem (Walker, 1969).

Common blight often occurs in association with fuscous blight. Since both the diseases occur simultaneously in many cases, it becomes difficult to assess the yield losses caused by them individually.

9.30.3 Causal Organisms

Common blight: *Xanthomonas axonopodis* pv. *phaseoli* (*Xap*) (Smith 1897) Vauterin *et al.* 1995

Synonym: *Xan. campestris* pv. *phaseoli*, *Xan. phaseoli*

Fuscous blight: *Xan. fuscans* subsp. *fuscans* (*Xff*) Schaad *et al.* 2007, subsp. nov.

Synonym: *Xan. campestris* pv. *phaseoli* var. *fuscans*, *Xan. axonopodis* pv. *phaseoli* var. *fuscans*

According to Young *et al.* (2004), *Xan. axonopodis* pv. *phaseoli* var. *fuscans* is not a valid name because it does not confirm either to the code (Rule 28 b) or to the standards (Standard 18.1).

On sucrose peptone agar (SPA) and yeast extract glucose calcium carbonate agar (YGCA), the colonies of *Xap* are yellow, convex, slimy and mucoid. On media containing tyrosine or on SPA, *Xff* produces a brown diffusible pigment, melanin. *Xff* strains also produce a dark internal pigment and, thus, can be distinguished from *Xap* on MXP medium. Both the bacterial species are strict aerobes, non-sporing and Gram-stain-negative rods. Xanthomonadins, having absorption peaks between 443 and 446 nm, are produced. The metabolism of glucose is strictly oxidative. The maximum temperature for growth is 38 °C. Other properties include hydrolysis of starch, casein and gelatin, lipolysis of Tween 80, proteolysis of milk, lack of pectinase activity in culture and the inability to reduce nitrates, to produce lysine dicarboxylase and to induce hypersensitive reaction in tobacco.

Xap and *Xff* are sufficiently distinct genetically to be classified into different groups based on 16S–23S intergenic spacer regions, restriction fragment length polymorphism, analysis of genomic and plasmid DNA, pulsed-field gel electrophoresis, DNA–DNA hybridisation and amplified DNA polymorphism.

Lopes *et al.* (2008) developed a semi-selective medium to detect *Xan. axonopodis* pv. *phaseoli* from bean seeds. The recovery frequency of bacterial colonies in relation to nutrient agar varied between 30 and 112%, depending upon the isolate.

Pathogenic variation exists in both the pathogens but evidence for the existence of races, identifiable on common bean differential genotypes, has yet to be found. *Xff* strains are generally more virulent than *Xap* strains and strains of the same pathogen from different locations also vary in virulence. Mutlu *et al.* (2008), on the basis of evaluation of common bean genotypes against 69 strains of *Xap* and 15 strains of *Xff*, found that *Xff* strains were more virulent than *Xap* strains having the same origin and pathogenic variation was more within *Xap* strains compared to *Xff* strains. Opio and Male-Kigiwa (1995) reported eight races of *Xap* based on their reaction in tepary bean (*P. acutifolius*).

Despite causing identical symptoms, *Xap* and *Xff* are genetically and phenotypically distinct species. No race structure exists for common bean, although pathogenic variability has been observed, particularly for *Xap* (O'Leary *et al.*, 2016).

To assess the relationship between *Xap* and *Xff*, O'Leary *et al.* (2016) sequenced whole genomes of five *Xap* and three *Xff* isolates, representative of global diversity. Complete genomes of all eight isolates were ~5 Mb, with single chromosome and one to three plasmids. One plasmid was conserved among all the isolates, whereas another was conserved in the three globally predominant *Xap* and all *Xff* isolates. A brown diffusible pigment, produced in culture by *Xff* strains resulting from the secretion and oxidation of homogentisic acid, was associated with a frame shift in the gene encoding homogentisate dioxygenase (hmgA). The deletion of the hmgA from *Xap* isolates lacking this mutation resulted in the production of brown pigment, while the complementation of a *Xff* isolate with a functional hmgA reduced pigment production substantially. This provides biological evidence that brown pigment production in *Xff* is due to the lack of a functional hmgA gene.

A DNA probe has been developed from plasmid DNA of *Xap* for the rapid detection and identification of *Xap* and *Xff* from bean plants and host debris. A 3.4 kilo base EcoRI fragment of *Xap* plasmid DNA, having a sequence that is repeated throughout the genomes of both the pathogens, was more specific than total plasmid DNA for this purpose (Gilbertson *et al.*, 1989).

9.30.4 Disease Cycle

The most important source of perpetuation of both the bacteria is infected or contaminated seed. Common blight bacterium can survive in the infected seed for 30 years. Internally infected and externally contaminated, but apparently healthy seeds are a more important inoculum source than the severely infected seeds. As low as 0.5% of infected seeds are sufficient to cause epiphytotics. The minimum number of bacteria per seed required to colonise seedlings is $10^{3\text{-}4}$. Infection or contamination of the seed occurs in different ways. Seed infection occurs via the vascular system where bacteria move through the pod pedicel into the funiculus and also through the raphe into the seed coat. Local pod infection can also result in the contamination and/or infection of seed. The bacteria move through the pod wall and contaminate the surface of developing seeds or penetrate beneath the seed coat.

He and Munkvold (2012) found that the vacuum extraction and centrifugation of seed extracts increased sensitivity for the detection of *Xap* in common bean seed and the highest sensitivity was obtained with the 3 h vacuum extraction followed by centrifugation. Bacterium was detected in 48 of 70 samples using the 3 h vacuum extraction with centrifugation, whereas only 35 of 70 field samples tested positive using overnight soaking, a significant difference. These steps would be valuable modifications to the current methods approved by the International Seed Testing Association.

Bastas and Sahin (2017) tested 198 dry bean seed samples of six cultivars collected from 12 provinces of the Central Anatolia Region of Turkey to assess these seed lots as primary inoculum sources of the major bacterial plant pathogens. It was found that 22.72%, 13.63%, 11.11%, 1.51% and 0.5% of seed samples tested were contaminated with five seed-borne bacterial plant pathogens, namely, *Psm. savastanoi* pv. *phaseolicola* (*Psp*), *Psm. syringae* pv. *syringae* (*Pss*), *Xap*, *Xan. axonopodis* pv. *phaseoli* var. *fuscans* (*Xapf*) and *Curtobacterium flaccumfaciens* pv. *flaccumfaciens* (*Cff*), respectively. The results showed that *Psp* and *Pss* were found together on cv. Cali, *Psp* and *Xap* on cv. Dermason and cv. Sira, and *Pss* and *Xap* on cv. Seker, cv. Dermason and cv. Cali. It is evident that the selection of a pathogen-free seed is very important in preventing the spread and effective management of seed-borne bacterial pathogens of bean.

Infected plant residue is another source of survival, particularly when it is not buried in soil and remains undecomposed, and also in areas where more than one crop of beans is taken from the same plot in a year. Resident bacteria on weed hosts and non-host plants also act as a source of survival. Gilbertson *et al.* (1989) developed a DNA probe from plasmid DNA of *Xap* for the rapid detection and identification of *Xap* and *Xff* from bean plants and host debris. A 3.4 kilo base EcoRI fragment of *Xap* plasmid DNA, having a sequence that is repeated throughout the genomes of both the pathogens, was more specific than total plasmid DNA for this purpose.

The penetration of bacteria into the host occurs through stomata, hydathodes and wounds caused by insects, hailstorms and cultural operations. Dissemination of the bacteria occurs through the infected seed, wind-splashed rain, sprinkler irrigation and implements. Seed is an important means for the short- and long-distance spread of the bacteria. The occurrence of devastating blight epiphytotics after wind-driven hailstorms is the result of the extensive spread of the bacteria by rain splashes and their entry through wounds caused by hailstorms. Overhead irrigation is another effective agent for the secondary spread of the bacteria. *Xap* can also be disseminated on the bodies of insects, such as *Diaprepes abbreviatus* and *Cerotoma ruficornis*, which may also provide avenues for the pathogen entry by creating wounds while feeding (Kaiser and Vakili, 1978). White flies (*Bemisia tabaci*), grasshoppers (*Melanoplus* sp.), Mexican bean beetles (*Epilachna varivestris*) and leaf miners have also been reported to transmit the bacterium.

When infected seeds germinate, the bacteria multiply and move as vascular pathogens throughout the large xylem vessels, causing the blocking of the vascular system and eventually the wilting of the plants. In most cases, the vascular infection results from seed-borne inoculum, but it can also occur from the penetration of bacteria through wounds. Stem cankers and leaf lesions may also result from the systemic infection. Alternatively, the bacteria may be splashed from the infected plant debris onto the seedlings, where these cause lesions on cotyledons and primary and trifoliate leaves under favourable conditions. The secondary spread of the bacteria from these lesions occurs through disseminating agents.

After entry into the host tissue through stomata and wounds, bacteria invade the intercellular spaces of the parenchyma tissue, dissolving the middle lamella slightly in advance. Affected cells collapse, and are penetrated by, the bacteria, leading to their disintegration and formation of lysigenous cavities. Bacteria then invade the xylem vessels in the stem and advance systemically in the larger xylem elements. Bacteria multiply in the xylem cells and, when large cell masses are produced, these either rupture or dissolve the cell walls and re-enter the parenchyma tissues. From the parenchyma tissue, the cells come out on the stem surface through stomata in the form of ooze and, after dissemination, enter the same stem and other plant parts. Hence, the bacteria invade both parenchyma and vascular tissue.

Infection or contamination of the seed occurs in different ways. Seed infection occurs via the vascular system, where bacteria move through the pod pedicel into the funiculus and also through the raphe into the seed coat. Local pod infection can also result in the contamination and/or infection of seed. Bacteria move through the pod wall and contaminate the surface of developing seeds or penetrate beneath the seed coat.

Disease development is favoured by warm humid weather. The optimum temperature for disease development is 24–32 °C. Under favourable weather conditions, bacteria multiply on the host surface and reach a threshold level before the symptoms develop. This threshold population is around 5×10^6 bacteria per 20 cm^2 of leaf tissue. This requirement of a minimum bacterial population for symptom development is probably the reason why the symptoms are usually observed during the reproductive stage of plant development. When the weather conditions are unfavourable for disease development, the bacteria colonise the plants without causing symptoms and become widespread on the plants.

Akhavan *et al.* (2013) reported that the epiphytic population of *Xap* plays a critical role in the development of the disease and subsequent epiphytotics. Epiphytic population of the bacterium in the field has two major parts: solitary cells (potentially planktonic) and biofilms (which are sources for providing and refreshing the solitary cell components). The mode of irrigation significantly effects the epiphytic population of *Xap*. The mean epiphytic population size in the field having an overhead-irrigation system is significantly higher than that involving furrow irrigation. A significant positive correlation between the epiphytic population size of *Xap* and disease severity has been reported in both the overhead-irrigated ($r = 0.64$) and the furrow-irrigated ($r = 0.44$) fields.

The age of the host plant, air temperature and host nutrition influence common blight development. Plants are susceptible at all growth stages, but young leaves are more susceptible than old leaves. Polysaccharide produced by *Xap* acts as a pathogenicity factor and disease severity is directly correlated with the amount produced by the pathogen. Possibly, it interferes with the movement of water and/or nutrients, leading to the production of lesions and wilting. Da Silva *et al.* (2009) observed that the bacterisation of bean seeds with a fluorescent *Pseudomonas* isolate (DFs 842) for 5 h at 10 °C and challenge inoculation with *Xap* caused metabolic alterations in bean plants due to the increase in total soluble protein content and polyphenol oxidase activity. An increase in polyphenol oxidase and a decrease in peroxidase activity induced resistance in bean plants against *Xap*.

Extracellular ATP (eATP) can function as a signalling molecule to regulate a wide range of cellular processes. Jiao *et al.* (2016) investigated the role of eATP in mediating the change of photosystem II (PSII) photochemistry of the tissues of bean leaves infected with *Xan. campestris* pv. *phaseoli*. The infection of the leaves with *Xan. campestris* pv. *phaseoli* caused a significant decrease in PSII maximal photochemical efficiency (Fv/Fm), maximum quantum efficiency of PSII photochemistry at illumination (Fv'/Fm'), PSII operating efficiency (ΦPSII), rate of non-cyclic electron transport through PSII (ETR), photochemical quenching (qP) and eATP level in tissues of the infected leaves. Nevertheless, the levels of non-photochemical quenching (qN) and the quantum yield of regulated energy dissipation of PSII (Y(NPQ)) were significantly increased. The application of exogenous ATP at 0.2 mm to uninfected leaves had no significant effect on any of the chlorophyll fluorescence parameters measured. However, in the tissues of infected leaves, the application of exogenous ATP alleviated the decreases of the Fv/Fm, Fv'/Fm', Φ_{PSII}, ETR, qP and eATP level and also abolished the increases of qN and Y(NPQ). The findings suggest that the change in PSII photochemistry by pathogen infection could be mediated by eATP.

Xie *et al.* (2012) compared the effectiveness of image analysis (IA) with an ordinal visual scale for the quantitative measurement of disease severity, its application in quantitative genetic studies and its effect on the estimates of genetic parameters. Studies were performed using eight backcross-derived families of common bean ($n = 172$) segregating for the molecular marker SU91, known to be associated with a QTL for resistance to common bacterial blight (CBB) and fuscous blight. Even though both IA and visual assessments were highly repeatable, IA was more sensitive for detecting quantitative differences between bean genotypes. CBB phenotypic difference between the two SU91 genotypic groups was consistently more than 5-fold for IA assessments, but generally only 2- to 3-fold for visual assessments. The findings suggest that the visual assessment results in an overestimation of the effect of QTL in genetic studies. This may be due to the lack of additivity and uneven intervals of the visual scale. Although the visual assessment of disease severity is a useful tool for general selection in breeding programmes, assessments using IA may be more suitable for phenotypic evaluations in quantitative genetic studies involving CBB resistance as well as other foliar diseases.

Donmez *et al.* (2013) screened 36 bean genotypes and two commercial cultivars, commonly grown in Erzurum and Erzincan provinces located in the Eastern Anatolia region of Turkey, for resistance to *Xap* under greenhouse and field conditions during 2001–2002. Among the 38 genotypes tested, only 36K was found to be resistant. In field, disease severity reduced the quality and quantity of bean seed.

Bhat *et al.* (2017) screened 41 bean genotypes by artificial inoculation during 2009 to 2010 in Srinagar, India, to identify genotypes resistant to *Xan. campestris* pv. *phaseoli*. Of the 41 genotypes screened, 2 were categorised as resistant, 2 as moderately resistant, 21 as moderately susceptible, 13 as susceptible and 3 as highly susceptible.

The epidemiology of fuscous blight is similar to that of common blight. Darsonval *et al.* (2009b) have shown the involvement of type III secretion system of *Xff* in its multiplication in the phyllosphere and systemic colonisation of bean plants, leading to its transmission to seeds. *hrp* regulatory genes of the bacterium play a major role in host colonisation processes. Strains with mutations in *hrp* regulatory genes, *hrpG* and *hrpX*, were impaired in their phyllospheric growth. Unlike the wild-type strain, strains with mutations in *hrp* genes were also not transmitted to seeds by a vascular pathway. However, the transmission of the bacterium to seeds by floral structures occurred with both wild and mutant strains. Darsonval *et al.* (2009a) also evaluated the role of bacterial attachment and biofilm formation in host colonisation processes for *Xff* and found that the mutations in five adhesion genes, namely, *pilA*, *fhab*, *xadA1*, *xadA2* and *yapH*, identified in *Xff*, altered the ability of all the mutants to adhere to polypropylene or seed. *PilA* was involved in adhesin and transmission to seed, and the mutation of *PilA* led to lower pathogenicity on beans. *YapH* was required for adhesion to seed, leaves and abiotic surfaces, but not for *in planta* transmission to seed or aggressiveness on leaves. However, none of the known adhesins was required for transmission to seed through floral structures. All mutants tested, except in *yapH*, were altered in their vascular transmission to seed.

Tumsa *et al.* (2020) evaluated 110 diverse accessions of common bean for common and fuscous bean blights resistance at three hot spot sites during the 2017 and 2018 seasons in Ethiopia. Genotype × site interaction had significant ($p <$ .05) effect on all assessed traits. This indicated the presence of marked variation among tested genotypes in common and fuscous blights resistance across the testing sites. Genotypes, including SEC21, SEC23, SMC21, VAX6, SEC12, SEC25, SMC22, VAX5, SEC20, SEC22, SEC24, SEC26, SMC16, SMC24, VAX6, SEC25, SEC21, SEC23 and SMC21, exhibited lower values of mean disease severity on leaf and pods and the AUD progress curve, and hence are useful genetic resources for future blight resistance breeding programmes. Nasir cultivar provided a grain yield of 3.45 ton/ha followed by VAX1 (2.86 ton/ha) and Hawassa Dume (2.83 ton/ha). Moreover, bean blight-resistant and high-yielding genotypes had more pods per plant and seeds per plant, making them ideal candidates for common bean breeding in Ethiopia or similar agro-ecologies which emphasised bean blight resistance and enhanced agronomic traits.

De Paiva *et al.* (2020) developed a single loop-mediated isothermal amplification (LAMP) assay for the specific detection of *Xan. phaseoli* pv. *phaseoli* (*Xpp*) and *Xan. citri* pv. *fuscans* (*Xcf*) from common bean plants and seeds. The detection limits for both the pathogens were 10 CFU/ml for cell suspensions and 1 fg of DNA, whereas in conventional PCR, the primers detected up to 10^5 CFU/ml and 1 ng of DNA. Specificity was confirmed by testing DNA from bean leaves, other *Xanthomonas* species, common fungal and bacterial bean pathogens and bacteria from the leaf microbiota. The method was tested with bean leaves inoculated with *Xpp*, and the pathogen could be detected from 4 h up to 15 days post-inoculation, even before disease symptoms were visible. When the method was applied to bacterium detection (*Xpp* or *Xcf*) in seed lots from infected plants, the bacterium detection rate was 100% (24 of 24). The pathogens were detected in seeds incubated for just 1 h in saline solution (0.85%), reducing the time needed for bacterium detection. The LAMP assay could be useful as a tool in bean bacterial blight management. The rapid and sensitive detection of bacteria in bean seed lots would reduce the risks of planting highly contaminated seeds in environments favourable to blight development.

9.30.5 Management

The management of common and fuscous blights is difficult due to the lack of resistant cultivars, effective chemicals and the seed-borne nature of the pathogens. However, the satisfactory control of both the diseases can be achieved by adopting the following integrated control measures:

1. Use the certified pathogen-free seed for sowing. Treat the seed with streptomycin/Streptocycline because the seed dust may contain the bacteria. Hot water treatment at 52 °C for 20 min or hot air treatment (23–32 h in dry hot (60 °C) air at 45–55% RH) has been found effective. The latter treatment does not appear to affect seed viability (Gondreau and Samson, 1994).
2. Use of resistant varieties is the best method for the management of these diseases. However, the breeding for disease resistance is complicated as different genes control resistance in pods and leaves. Great Northern varieties, namely, Beryl, Harris, Ivory and Marquis, are resistant to common blight pathogens. In comparison to common bean, scarlet runner bean has a higher level of resistance while the tepary bean has the highest level of resistance to common blight pathogen.
3. Destruction of infected plant debris by deep ploughing and by burying it in the soil.
4. Follow crop rotation with non-host crops for three years. In Romania, rotation with maize reduced 85% disease incidence.
5. Escape the disease by choosing a suitable planting date.
6. Destruction of volunteer bean plants early in the following year and of weed hosts.
7. Avoid or limit the use of sprinkler irrigation.
8. Do not use the run-off contaminated water for irrigation.
9. Avoid cultural operations when plants are wet.
10. Preventive sprays of copper fungicides (Kocide etc.) and antibiotics are helpful in reducing the disease severity as they reduce the epiphytic population of the bacteria. These sprays should be started early in the season and repeated at 7–10-day intervals. Use of antibiotics is not encouraged as it can lead to the development of bacterial strains resistant to the antibiotics.

    Application of Bion (0.05%) and BioZell-2000B (50% etheric oil of thyme, 20% oil of maize, 20% oil of anise and 10% oil of sesame) caused 68 and 50% reduction in common bean blight incidence along with 50 and 45% reduction in bacterial population, respectively (Abo-Elyouser, 2006).
11. Giorgio *et al.* (2016) reported that 60 out of 162 bean rhizobacteria inhibited *in vitro* the growth of selected virulent strains of both *Xan. axonopodis*. pv. *phaseoli* and its variety *fuscans*; and when applied to seeds before sowing, six of them reduced disease symptoms on bean in *in vivo* and greenhouse experiments.

12. Belete *et al.* (2021) reported that three applications of *Bacillus subtilis* str. QST 713 cell suspension (10^8 CFU mL^{-1}) significantly reduced the disease severity and incidence, disease infection rate and AUDPC over the control treatment. The plants treated with str. QST 713 showed a reduction of the infection rate by 40.36% and the disease incidence by 52.25%.

REFERENCES

Abo-Elyouser, K. A. M. (2006). Induction of systemic acquired resistance against common blight of bean (*Phaseolus vulgaris*) caused by *Xanthomonas campestris* pv. *phaseoli*. *Egypt. J. Phytopathol.* 34: 41–50.

Akhavan, A., Bahar, M., Askarian, H., Lak, M. R., Nazemi, A. *et al.* (2013). Bean common bacterial blight: Pathogen epiphytic life and effect of irrigation practices. *Springerplus* 2: 41.

Bastas, K. K. and Sahin, F. (2017). Evaluation of seedborne bacterial pathogens on common bean cultivars grown in central Anatolia region, Turkey. *Eur. J. Plant Pathol.* 147: 239–53.

Belete, T., Bastas, K. K., Francesconi, S. and Balestra, G. M. (2021). Biological effectiveness of *Bacillus subtilis* on common bean bacterial blight. *J. Plant. Pathol.* 103: 249–58. https://doi.org/10.1007/s42161-020-00727-8

Bhat, H. A., Ahmad, K., Ahanger, R. A., Wani, S. H., Bhat, A. H., *et al.* (2017). Resistant resources and transmission of bacterial blight of common beans (*Phaseolus vulgaris* L.). *J. Pharmacogn. Phytochem.* 6: 380–83.

da Silva, E. G., Moura, A. B., Bacarin, M. A. and Deuner, C. C. (2009). Metabolic alterations on bean plants originated from microbiolization of seeds with *Pseudomonas* sp. and inoculated with *Xanthomonas axonopodis* pv. *phaseoli*. *Summa Phytopathol.* 35: 98–104.

Darsonval, A., Darrasse, A., Durrand, K., Bureau, C., Cesbron, S. *et al.* (2009a). Adhesion and fitness in the bean phyllosphere and transmission to seed of *Xanthomonas fuscans* subsp. *fuscans*. *Mol. Plant-Microbe Interact.* 22: 747–57.

Darsonval, A., Darrasse, A., Meyer, D. Demarty, M., Durand, K. *et al.* (2009b). The type III secretion system of *Xanthomonas fuscans* subsp. *fuscans* is involved in the phyllosphere colonization process and in transmission to seeds of susceptible beans. *Appl. Environ. Microbiol.* 74: 2669–78.

de Paiva, B. A. R., Wendland, A., Teixeira, N. C. and Ferreira, M. A. S. V. (2020). Rapid detection of *Xanthomonas citri* pv. *fuscans* and *Xanthomonas phaseoli* pv. *phaseoli* in common bean by loop-mediated isothermal amplification. *Plant Dis.* 104: 98–203.

Donmez, M. F., Sahin, F. and Elkoca, E. (2013). Identification of bean genotypes from turkey resistance to common bacterial blight and halo blight diseases. *Acta Sci. Pol.* 12: 139–51.

Gilbertson, R. L. and Maxwell D. P. (1992). Common blight of beans. In: H. S. Chaube, U. S. Singh, A. N. Mukhopadhyay and J. Kumar (eds.). *Plant Diseases of International Importance*. Vol. II, *Diseases of Vegetables* and *Oilseed Crops*. Prentice Hall, Englewood Cliffs, NJ, pp. 18–39.

Gilbertson, R. L., Maxwell, D. P., Hagedorn, D. J. and Leong, S. A. (1989). Development and application of a plasmid DNA probe for detection of bacteria causing common bacterial blight of bean. *Phytopathology* 79: 518–25.

Giorgio, A., Cantore, P. L., Shanmugaiah, V., Lamorte, D. and Iacobellis, N. S. (2016). Rhizobacteria isolated from common bean in southern Italy as potential biocontrol agents against common bacterial blight. *Eur. J. Plant Pathol.* 144: 297–309.

Gondreau, C. and Samson, R. (1994). A review of thermotherapy to free plant material from pathogens, especially seeds from bacteria. *Crit. Rev. Plant Sci.*, 13: 57–75.

He, Y. and Munkvold, G. P. (2012). Comparison of extraction procedures for detection of *Xanthomonas axonopodis* pv. *phaseoli* in common bean seed. *Plant Pathol.* 61: 837–43.

Jiao, Q.-S., Feng, H.-Q., Tian, W.-Y., Bai, J.-Y., Sun, K., *et al.* (2016). Extracellular ATP functions in alleviating the decrease of PSII photochemistry caused by the infection of *Xanthomonas campestris* pv. *phaseoli*. *Plant Pathol.* 65: 819–25.

Kaiser, W. J. and Vakili, N. G. (1978). Insect transmission of pathogenic xanthomonads to bean and cowpea in Puerto Rico. *Phytopathology* 68: 1057–63.

Lopes, L. P., Alves, P. F. R., Zandona, C., Nunes, M. P. and Mehta, Y. R. (2008). A semi-selective medium to detect *Xanthomonas axonopodis* pv. *phaseoli* in bean seeds and its eradication through seed treatment with tolyfluanid. *Summa Phytopathol.* 34: 287–88.

Mutlu, N., Vidaver, A. K., Coyne, D. P., Steadman, J. R., Lambrecht, P. A. *et al.* (2008). Differential pathogenicity of *Xanthomonas campestris* pv. phaseoli and *X. fuscans* subsp. *fuscans* strains on bean genotypes with common blight resistance. *Plant Dis.* 92: 546–54.

O'Leary, M., Coaker, G., and Gilbertson, R. (2016). Complete genome sequences of *Xanthomonas axonopodis* pv. *phaseoli* and *X. fuscans* subsp. *fuscans* generated by long-read sequence technology (Abstr.). *Phytopathology* 106: S4.173.

Opio, F. and Male-Kigiwa, S. (1995). The status of bean breeding in Uganda. Breeding for disease resistance with emphasis on durability. *Proceedings Regional Workshop for Eastern, Central and Southern Africa*, October 2–6, 1994, Njoro, Kenya, pp. 110–13.

Osdaghi, E., and Zademohamad, A. A. (2016). *Phaseolus lunatus*, a new host of *Xanthomonas axonopodis* pv. *phaseoli* in Iran. *J. Phytopathol.* 164: 56–60.

Partridge, J. E. (2003). Common blight of bean. University of Nebraska, Lincoln. http://nu-distance.unl.edu/homer/disease/agron/drybeans/BeCmnBlt.html.

Patel, P. N. and Jindal, J. K. (1972). Common and fuscous blights of bean. In: P. N. Patel (ed.). *Plant Bacteriology*. Vol. I, *Bacterial Diseases of Plants in India*. Indian Agricultural Research Institute, New Delhi, pp. 44–58.

Schaad, N. W., Postnikova, E., Lacy, G., Sechler, A., Agarkova, I., *et al.* (2007). List of new names and new combinations previously effectively, but not validly, published. *Int. J. Syst. Evol. Microbiol.* Vol. 57: pp. 893–97. Effective Publication: Schaad, N. W., Postnikova, E., Lacy, G., Sechler, A., Agarkova, I., *et al.* 2006. Emended classification of xanthomonad pathogens on citrus. *Syst. Appl. Microbiol.* 29: 690–95.

Schwartz, H. F. (2007). Bacterial diseases of beans. Colorado State University. http://www.ext.colostate.edu/PUBS/crops/02913.html.

Singh, S. P. and Miklas, P. N. (2015). Breeding common bean for resistance to common blight: A review. *Crop Sci.* 55: 971–84.

Smith, E. F. (1897). Description of *Bacillus phaseoli* n. sp. *Bot. Gaz.* 24: 192.

Tumsa, K., Shimelis, H., Laing, M., Mukankusi, C. and Mathew, I. (2020). Identification of sources of resistance to common bacterial blight in common bean in Ethiopia. *J. Phytopathol.* 168: 707–20.

Vauterin, L., Hoste, B. Kersters, K. and Swings, J. (1995). Reclassification of *Xanthomonas*. *Int. J. Syst. Bacteriol.* 45: 472–89.

Walker, J. C. (1969). *Plant Pathology*. 3rd Ed. McGraw Hill Book Co., New York, 819 pp.

Xie, W., Yu, K., Pauls, K. P. and Navabi, A. (2012). Application of image analysis in studies of quantitative disease resistance, exemplified using common bacterial blight-common bean pathosystem. *Phytopathology* 102: 434–42.

Young, J. M., Bull, C. T., De Boer, S. H., Firrao, G., Saddler, G. E. *et al.* (2004). Names of plant pathogenic bacteria, 1864–2004. http://www.isppweb.org/names_bacterial_revised.asp

9.31 HALO BLIGHT OF BEAN

Halo blight of bean was first described by Burkholder (1926) from the United States. It has many similarities with common blight of bean and causes serious economic losses under favourable weather conditions. The disease is worldwide in distribution, wherever beans are grown under cool and wet climatic conditions. Occurrence of the disease has not been reported from India. Under favourable weather conditions, epiphytotics resulting from a few infected seeds may cause substantial yield losses.

9.31.1 Host Range

In addition to common bean, the host range of the bacterium includes field and snap beans, especially cranberry, red kidney and yellow-eyed beans, lima bean and scarlet runner bean. Artificially, the bacterium can infect tepary bean, adzuki bean, *mung* bean, soybean, *Macroptilium atropurpureum*, *Phaseolus polyanthus*, *P. polystachyus* and *Pueraria lobata*.

Taylor *et al.* (1996) reported ten alternative natural hosts, which included ten legume species representing seven different genera, namely, *Cajanus cajan*, *Desmodium* sp., *Lablab purpureus*, *M. atropurpureum*, *Neonotonia wightii*, *P. acutifolius*, *P. coccineus*, *P. lunatus*, *Vigna angularis* and *V. radiata*, in addition to the main host *P. vulgaris*. Of these alternative natural hosts, *Desmodium* sp. constituted a new host record.

9.31.2 Symptoms

Most symptoms of halo blight are similar to those of common blight. Symptoms of halo blight appear on all the above-ground plant parts. Water-soaked, round to irregular lesions are produced on cotyledons of seedlings raised from infected and/or infested seeds. Yellowish-brown to reddish-brown, water-soaked lesions appear on primary leaves. Severely infected plants show stunting and die.

Small (about 1.5 mm in diam), angular, greasy, water-soaked spots first appear on the lower surface of trifoliate leaves. As these spots enlarge and turn reddish-brown, characteristic light greenish-yellow haloes, 10 mm or more in diameter, often, but not always, appear around the spots, giving the leaves a yellowish appearance. Haloes, more prominent on young expanding leaves, appear due to the production of a toxin, phaseolotoxin by the bacterium and are a diagnostic symptom of the disease (Figure 9.30a). At temperatures above 27 °C, haloes formed are very small or may be absent. In 2004, Oguiza and associates reported the bacterial strains that do not have the genetic information for toxin production. In severe infections, the infection may become systemic, causing yellowing and death of new foliage. On stems, greasy, reddish-brown, elongated streaks appear. On pods, symptoms begin as small, water-soaked spots or streaks along the pod sutures. The spots, later on, turn reddish-brown and become sunken in the centre (Figure 9.30b). Under humid conditions, a light cream- or silver-coloured bacterial exudate may appear on the lesions. Seeds in infected pods are usually symptomless. However, in severely infected pods, seeds may be shrivelled, discoloured or reduced in size.

Pod symptoms of halo and common blight are similar. Therefore, the identification of these diseases should be based on leaf symptoms and the identification of the causal bacteria. Leaf symptoms of halo blight and bacterial brown spot may be confused due to the variation in margin/width of halo influenced by the prevailing temperature. In the case of brown spot, the spots are brown, necrotic having a narrow yellow halo.

(a)

FIGURE 9.30A Halo blight of bean: Symptoms on aerial plant parts. (Courtesy: Robert Harveson and Howard Schwartz, www.plantmanagement.network.org)

(b)

FIGURE 9.30B Halo blight of bean: Symptoms on pod. (Courtesy: Robert Harveson and Howard Schwartz, www.plantmanagement.network.org)

9.31.3 Causal Organism

Pseudomonas savastanoi pv. *phaseolicola* (Burkholder 1926) Gardan *et al.* 1992

Synonym: *Psm. phaseolicola*, *Psm. syringae* pv. *phaseolicola*

The below-given description of the bacterium is taken mainly from Burkholder (1926), who classified it as *Phytomonas medicaginis* var. *phaseolicola* nov. var.

Cells are rods with rounded ends, occur singly or in pairs and measure 0.5–1.25 × 1.5–3.0 µm. Gram-stain-negative, non-sporing, not acid fast, strict aerobes and motile with a polar flagellum. Colonies on nutrient agar are white to colourless or, at times, creamy, concentrically ringed with edges undulate. Growth of the colonies is slow to moderate. The bacterium does not reduce nitrate, does not produce H_2S and indole, does not liquefy gelatin and does not digest starch. Litmus milk is turned alkaline without curd formation and peptonisation. Pectate gel pitting occurs at pH 4.0. Weak acid production occurs from dextrose and saccharose but no acid from lactose and maltose and also no gas production from these sugars. Growth occurs on quinate and trigonelline but not on mannitol, inositol, sorbitol, erythritol, L-tartrate, D-tartrate, L-lactate, anthranilate, betaine and homoserine. Its ability to use rhamnose and raffinose as a sole source of carbon and sodium nitrate as a sole source of nitrogen varies with different strains. Maximum growth and production of phaseolotoxin occurs at 21–22 °C. Thermal death point is around 49 °C.

Taylor *et al.* (1996) reported nine races of the bacterium and, among them, race 6 being the predominant. They also carried out detailed race determination investigations on 175 selected isolates of *Psm. syringae* pv. *phaseolicola*, representative of the different geographical regions and hosts in which the pathogen was found, and identified nine races. Races 1, 2, 5, 6 and 7 were distributed worldwide, and race 6 was predominant. Other races were found mainly in Africa; races 3 and 4 in East/Central Africa and races 8 and 9 in southern Africa. Most isolates were obtained from the main host, *P. vulgaris*, while the alternative natural hosts included ten legume species belonging to seven different genera.

9.31.4 Disease Cycle

Infected and/or infested seed is the most important source of survival of the bacterium. The bacterium can survive for more than four years in infected seed. One contaminated seed in 16,000 is sufficient to cause severe epiphytotic under favourable weather conditions (Dillard and Legard, 1991). Infection of bean seeds occurs through lesions formed on pods and bacteria penetrate the seed coats without causing any visible symptoms. Infestation of seeds occurs during threshing, resulting from the adherence of dust of diseased tissue to seed surface and to wounds. As a source of primary inoculum, both infected and infested seeds are equally important (Goto, 1992).

Popović *et al.* (2014) described a diagnostic method based on nested-PCR, followed by ELISA and conventional bacteriology tests, for the rapid and reliable detection of 18 strains of *Psm. savastanoi* pv. *phaseolicola* isolated from infected bean leaves and seeds. The bacterium formed white, small and flat colonies on nutrient agar; creamy white, flat and circular on Milk-Tween agar; and light yellow, convex and shiny on modified sucrose peptone agar. On the basis of results of DAS- and pre T-cell antigen (PTA)-ELISA with respect to reactivity to specific antibodies, all the analysed strains belonged to *Psm. savastanoi* pv. *phaseolicola*. On inoculation, the bacterium formed greasy spots on bean pods and cotyledonary leaves and did not cause hypersensitive reaction on the leaves of tobacco and geranium.

Infected plant debris also acts as a source of survival of the bacterium. Survival in crop residue is at its maximum when it is on the soil surface. Dissemination of the bacterium occurs through wind-splashed rain, overhead irrigation, windstorms and farm machinery. Severe outbreaks of disease often occur after heavy rainstorms.

The disease is favoured by cool to moderate temperature (16–23 °C) and humid moist conditions. Temperatures above 27 °C inhibit the development of yellow haloes and systemic chlorosis. When weather becomes warmer and drier, plants recover from the disease and new growth is healthy.

Hydrogen sulphide (H_2S) is an important signalling molecule in both animals and plants, despite its toxic nature. In plants, H_2S controls stomatal apertures, thereby altering the ability of bacteria to invade plant tissues. Bacteria are known to generate H_2S as well as being exposed to plant-generated H_2S. During their interaction with plants, pathogenic bacteria are known to undergo alterations in their genomic complement. *Psp* strain 1302A undergoes loss of a section of DNA known as a genomic island (PSPGI-1) when exposed to the resistance response of plants. The loss of PSPGI-1 from *Psp* 1302A enables the pathogen to overcome the plants' resistance response and cause disease.

Neale *et al.* (2017), with the use of H_2S donor molecules, investigated changes induced in *Psp* 1302A genome, as demonstrated by excision of PSPGI-1. *Psp* 1302A cells were found to be resistant to low concentrations of H_2S. However, at sublethal H_2S concentrations, an increase in the expression of the PSPGI-1 encoded integrase gene (*xerC*), which is responsible for island excision, and a subsequent increase in the presence of the circular form of PSPGI-1, were detected. It shows that H_2S is able to initiate excision of PSPGI-1 from the *Psp* genome. Therefore, H_2S which may emanate from the plant has an effect on the genome structure of invading bacteria and their ability to cause disease in plants. The modulation of such plant signals may be a way to increase plant defence responses for crops in the future.

Tock *et al.* (2017) constructed high-resolution linkage maps for three recombinant inbred populations to map resistance to *Psm. syringae* pv. *phaseolicola* race 6 derived from the two common bean lines. This was complemented with a genome-wide association study of race 6 resistance in an Andean Diversity Panel of common bean. Genome-wide association study detected one QTL (HB5.1) on chromosome Pv05 for resistance to race 6 with significant influence on seed yield. Race 6 resistance from PI 150414 maps to a single major-effect QTL (HB4.2) on chromosome Pv04 and confers broad-spectrum resistance to eight other races of the pathogen. The resistance segregating in a Rojo × CAL 143 population maps to five chromosome arms and includes HB4.2. The same HB5.1 QTL, found in both Canadian Wonder × PI 150414 and Rojo × CAL 143 populations, was effective against race 6, but lacks broad resistance. The study provides evidence for marker-assisted breeding for more durable halo blight control in common bean by combining alleles of race-non-specific resistance (HB4.2 from PI 150414) and race-specific resistance (HB5.1 from cv. Rojo).

Bozkurt and Soylu (2011) studied the reaction of different bean cultivars to nine races of *Psm. syringae* pv. *phaseolicola*. Virulent bacterial races produced water-soaked lesions in susceptible and moderately resistant cultivars, but in moderately resistant cultivars the symptoms were later associated with more tissue browning around the sites of inoculation. In contrast, the resistant response produced the characteristic HR, characterised as small discrete browning and tissue collapse at the site of inoculation. No local cultivar was found completely resistant to all the races tested. Cultivars Sehirali-90 and Goynuk-98 showed resistant or moderately resistant reaction to five races, while cultivar Karacaşehir-90 showed the same reaction to six races. Analysis of bacterial growth and of the accumulation of isoflavonoid bean phytoalexin, phaseollin, carried out in plant tissues expressing compatible and incompatible interactions, revealed that HR was clearly associated with the restricted multiplication of bacteria during incompatible interactions. Pods treated with different races of the bacterium showed that a strong correlation was observed between the timing and extent of cell death, including the accumulation of phaseollin, being rapid and extensive in incompatible interactions compared to compatible interaction.

Donmez *et al.* (2013) screened 36 bean genotypes and two commercial cultivars, commonly grown in Erzurum and Erzincan provinces located in the eastern Anatolia region of Turkey, for resistance to *Psm. savastanoi* pv. *phaseolicola* under greenhouse and field conditions during 2001–2002. Among the 38 genotypes tested, only 36K was found to be resistant. In field, disease severity reduced the quality and quantity of bean seed.

Cooper *et al.* (2022) inoculated beans with virulent and avirulent races of *Psm. savastanoi* pv. *phaseolicola* to examine the role of hormone and phytoalexin production during immunity. With the exception of salicylic acid (SA), no other examined hormone consistently increased at sites of infection to greater levels in resistant beans compared with susceptible beans four days after inoculation. Daidzein, genistein, kievitone, phaseollin, phaseollidin, coumestrol and resveratrol substantially increased alongside SA in resistant beans but not in susceptible beans. *In vitro* culture assays revealed that SA, daidzein, genistein, coumestrol and resveratrol inhibited cultural growth of *Psm. savastanoi* pv. *phaseolicola* race 5. These results demonstrate that these phytoalexins may be regulated by SA and work with SA during effector-triggered immunity to restrict bacterial replication. This is the first report of antibiotic activity of daidzein, genistein and resveratrol to *Ps. savastanoi* pv. *phaseolicola*. These findings improve our understanding of the mechanistic output of effector-triggered immunity toward this bacterial pathogen of beans.

Arvizu-Gómez *et al.* (2024), using *Psm. savastanoi pv. phaseolicola*, evaluated intracellular redox state under low-temperature (18 °C) conditions. Generation of an oxidative-stress biosensor (pKL1) and fluorometry analyses with the *Psm. savastanoi pv. phaseolicola*-roGFP2 bioreporter strain were performed. At low temperatures (18 °C), the response capacity or roGFP2 oxidation diminished in relation to 28 °C. Likewise, changes in the intracellular redox potential of *Psm. savastanoi pv. phaseolicola* during its growth at 28 °C and 18°C were observed, with greater oxidation degree in cells grown at 18 °C. These findings demonstrated that low temperatures induce intracellular oxidative stress in *Psm. savastanoi pv. phaseolicola*. This is the first report of the relation of oxidative stress with low-temperature conditions in phytopathogenic bacteria.

9.31.5 Management

Management practices for the disease are the same as those for common and fuscous blights of bean.

REFERENCES

Arvizu-Gómez, J. L., Hernández-Morales, A., Llanos-Vargas, K. D., Olmedo-Álvarez, G., Campos-Guillén, J. *et al.* (2024). Influence of the low temperatures (18 °C) in the generation of intracellular oxidative stress in the phytopathogen bacterium *Pseudomonas savastanoi* pv. *phaseolicola* NPS3121. *J. Phytopathol.* 172: e13367. https://doi.org/10.1111/jph.13367

Bozkurt, İ. A. and Soylu, S. (2011). Determination of responses of different bean cultivars against races of *Pseudomonas syringae* pv. *phaseolicola*, causal agent of halo blight of bean. *Euphytica*, 179: 417–25.

Burkholder, W. H. (1926). A new bacterial disease of the bean. *Phytopathology* 16: 915–28.

Cooper, B., Campbell, K. B. and Garrett, W. M. (2022). Salicylic acid and phytoalexin induction by a bacterium that causes halo blight in beans. *Phytopathology* 112: 1766–75.

Dillard, H.R. and Legard, D.E. (1991). Bacterial diseases of beans. Fact sheet, Cornell University, Ithaca, New York. http://vegetablemdonline.ppath/cornell.edu.factsheets/Beans_Bacterial.htm.

Donmez, M. F., Sahin, F. and Elkoca, E. (2013). Identification of bean genotypes from Turkey resistance to common bacterial blight and halo blight diseases. *Acta Sci. Pol.* 12: 139–51.

Gardan, L., Bollet, C., Abu Ghorrah, M., Grimont, F. and Grimont, P. A. D. (1992). DNA relatedness among the pathovar strains of *Pseudomonas syringae* subsp. *savastanoi* Janse

(1982) and proposal of *Pseudomonas savastanoi* sp. nov. *Int. J. Syst. Bacteriol.* 42: 606–12.
Goto, M. (1992). *Fundamentals of Bacterial Plant Pathology*. Academic Press, Inc., San Diego, CA, 342 pp.
Neale, H., Deshappriya, N., Arnold, D., Wood, M. E., Whiteman, M. et al. (2017). Hydrogen sulfide causes excision of a genomic island in *Pseudomonas syringae* pv. *phaseolicola*. *Eur. J. Plant Pathol.* 149: 911–21.
Popović, T., Balaž, J. and Stanković, S. (2014). A method for the rapid detection and identification of halo blight pathogen on common bean. *Arch. Biol. Sci.* 66: 1393–1400.
Taylor, J. D., Teverson, D. M., Allen, D. J. *et al.* (1996). Identification and origin of races of *Pseudomonas syringae* pv. *phaseolicola* from Africa and other bean growing areas. *Plant Pathol.* 45: 469–78.
Tock, A. J., Fourie, D., Walley, P. G., Holub, E. B., Soler, A. et al. (2017). Genome-wide linkage and association mapping of halo blight resistance in common bean to race 6 of the globally important bacterial pathogen. *Front. Plant Sci.* 8: 1–17.

9.32 BACTERIAL BROWN SPOT OF BEAN

Burkholder in 1930 described two bacterial diseases of bean, one of which was bacterial brown spot. This disease is favoured by cool to moderate temperatures and high levels of moisture. Severe outbreaks of the disease occur when temperature is below 27 °C and humidity above 95% prevails for periods of 24 h or longer. Yield losses up to 55% have been reported in South Africa (Muedi *et al.*, 2011).

9.32.1 Symptoms

The disease is more severe on young foliage, with symptoms rarely seen on older leaves. Initially, small, water-soaked spots appear on the lower surface of trifoliate leaves. A narrow, greenish-yellow diffuse border which is about 1.5 mm in diameter may appear around some of the lesions. The spots enlarge ranging from 3–8 mm in diameter; on maturity, these give a typical "**brown spot**" appearance (Schwartz, 2007). The dead centres of the spots may fall out, producing shot-hole symptoms (Figure 9.31a). On the stem, systemic, sunken, longitudinal lesions appear. Sunken, brown spots appear on the pods. If the infection occurs during the early development of pods, infected pods may be twisted and kinked at the site of infection. Pale brown haloes, about 6 mm wide, occasionally appear around the spots on the pods (Figure 9.31b).

Leaf symptoms of bacterial brown spot may be confused with those of halo blight due to the variation in the margin/width of halo, influenced by temperature in the latter case.

9.32.2 Causal Organism

Pseudomonas syringae pv. *syringae* van Hall 1902
Synonym: *Psm. syringae* pv. *japonica*

Muedi *et al.* (2011) assessed the biochemical and genetic variability in *Psm. syringae* pv. *syringae* isolates, collected from dry bean-producing areas in South Africa, using levan production, oxidase production, pectinolytic activity, arginine dihydrolase production and tobacco hypersensitivity (LOPAT) tests and *SyrB* gene assessment. Biolog GN Microplates were used to assess carbon substrate utilisation. . The *SyrB* gene was present in 42% of isolates. The Biolog GN Microplates showed biochemical variation among isolates. Variable genomic patterns were observed in 48.5% of isolates by the BOX A1R primer and in 37.1% of isolates by the enterobacterial repetitive intergenic consensus (ERIC) 2 primer.

9.32.3 Disease Cycle

The bacterium survives in infected plant debris and infected seeds, but the former is more important. Infected bean stems and pod pieces play a more important role in the survival of bacterium. Disease epiphytotics are often traced to the survival of the bacterium in infected crop residue. The bacterium survives on hairy vetch and other weeds, but these are not considered an important source of survival. Non-certified seed lots also carry the bacterium from one crop season to the other. These seed lots are also a potential means of dissemination of bacterium to short- and long-distance places, including new areas. The dissemination of the bacterium also occurs through wind-splashed rain, overhead irrigation and farm equipment. Severe infections usually occur after major rainstorms.

(a) (b)

FIGURE 9.31 (a) Bacterial brown spot of bean: Symptoms on aerial plant parts. (Courtesy: Robert Harveson and Howard Schwartz, www.plantmanagement.network.org.) (b) Bacterial brown spot of bean: Symptoms on pod. (Courtesy: Robert Harveson and Howard Schwartz, www.plantmanagement.network.org)

Different strains of *Psm. syringae* pv. *syringae* causing diseases on various plant species grow as epiphytes on various plant species. *Psm. syringae* pv. *syringae* also grows epiphytically on snap and dry beans without causing disease symptoms. Symptoms of bacterial brown spot on beans generally occur following the development of large epiphytic populations of the bacteria. Severe infections usually develop after major rainstorms; however, the absence of symptoms before rainstorms does not mean that the causal bacteria were not present.

Muedi *et al.* (2015) assessed 27 common bean accessions over three seasons and locations for their resistance to *Psm. syringae* pv. *syringae* using artificial inoculation technique. Plots were rated for disease reaction on a modified 1–9 CIAT (Centro Internacional de Agricultura Tropical) scale, with 1 being resistant and 9 being susceptible. Disease ratings were used to construct the area under the disease progress curve (AUDPC). Highly significant genotype × environment interactions were recorded. VAX 4 and VAX 2 showed the lowest disease ratings of 1.8 and 2.0, and AUDPC values of 38.6 and 41.6, respectively; hence, they are potential resistance sources.

Salegua *et al.* (2020) screened dry bean genotypes against *Psm. syringae* pv. *syringae* and identified that 17.2% of evaluated germplasm were resistant and 45.3% moderately resistant. Genotypes ADP-0592, ADP-0790, ADP-0120 and ADP-0008 were selected for both resistance and high seed yield across the three environments. Genotypes ADP-0546, ADP-0630, ADP-0120 and ADP-0279 were selected for both high yield and resistance at Warden, whereas ADP-0038, ADP-0721 and ADP-0790 were selected for both traits at Middelburg and, lastly, ADP-0120 and ADP-0079 were selected for both traits at Potchefstroom. The best genotypes selected for both high yield and bacterial brown spot resistance had grain yield >1.45 t ha^{-1} across sites and >1.85 t ha^{-1} at individual sites; these out-yielded the best-performing check cultivar (1.13 t ha^{-1}) and the grand mean yield (0.87 t ha^{-1}) across sites. The AUDPC had a strong negative correlation ($r = -0.55$, $P < 0.001$) with grain yield at Potchefstroom. Medium-seeded genotypes showed a lower AUDPC than the large-seeded genotypes and indeterminate genotypes showed a lower AUDPC than determinate genotypes. The genotypes selected for resistance and yield can be utilised in future dry bean improvement efforts for the South African bean market.

9.32.4 Management

Management practices given for common and fuscous blights are applicable to this disease also.

REFERENCES

Muedi, H. T. H., Fourie, D. and Mclaren, N. W. (2011). Characterisation of bacterial brown spot pathogen from dry bean production areas of South Africa. *African Crop Sci. J.* 19: 357–67.

Muedi, H. T. H., McLaren, N. W. and Fourie, D. (2015). Reaction of selected common bean germplasm accessions to bacterial brown spot in South Africa. *Crop Prot.* 77: 87–93.

Salegua, V., Melis, R., Fourie, D., Sibiya, J. and Musvosvi, C. (2020). Screening Andean Diversity Panel dry bean lines for resistance to bacterial brown spot disease under field conditions in South Africa. *Plant Dis.* 104: 2509–14.

Schwartz, H. F. (2007). Bacterial diseases of beans. Colorado State University. http://www.ext.colostate.edu/PUBS/crops/02913.html.

van Hall, C. J. J. (1902). *Bijdragen tot de kennis der Bakterieele Plantenzeikten.* Coöperatieve Drukkerij-vereeniging “Plantijn”. Inaugural dissertation. Amsterdam.

BACTERIAL DISEASES OF SOYBEAN

Soybean (*Glycine max* L.) suffers from four important bacterial diseases, namely, pustule (*Xanthomonas axonopodis* pv. *glycines*), blight (*Pseudomonas savastanoi* pv. *glycinea*), wilt (*Curtobacterium flaccumfaciens* pv. *flaccumfaciens*) and wild fire (*Psm. syringae* pv. *tabaci*).

9.33 SOYBEAN PUSTULE

Soybean [*Glycine max* (L.) Merril.] pustule was first described from Japan by Nakano (1919). Hedges in 1922 reported it from the USA and named the causal bacterium as *Bacterium phaseoli* var. *sojense*. The disease occurs in most soybean-growing areas of the world, where warm weather and frequent rains occur during the growing season. Janse (2005) has listed 54 countries, including the USA, where the disease occurs. Severe infection causes premature defoliation which decreases yield by reducing the size and number of seeds. In an investigation reporting 11% loss in yield, 63% of it was attributed to the reduction in seed size.

In India, the disease was first reported by Uppal *et al.* (1938) from Jalgaon, Maharashtra. Now it is present in most parts of the country (Patel and Jindal, 1972).

Krawczyk and Borodynko-Filas (2020) reported an outbreak of a new disease of soybean plants, with symptoms resembling those described for the bacterial blight caused by *Pseudomonas syringae* pv. *glycinae* and bacterial pustule caused by *Xanthomonas axonopodis pv. glycines* in Polish soybean fields. Identification based on biological (Biolog Gen III) and molecular methods (16S rDNA and gyrB sequencing, PCR species-specific testing) revealed that the soy pathogenic bacterium is *Kosakonia cowanii*.

9.33.1 Host Range

In addition to soybean, the bacterium infects common bean (*Phaseolus vulgaris*), horse gram (*Macrotyloma uniflorum* = *Dolichos biflorus*), cowpea (*Vigna unguiculata*), lima bean (*Phaseolus lunatus*) redvine, buckwheat-vine (*Brunnichia cirrhosa*) and many other *Phaseolus* and *Vigna* spp.

9.33.2 Symptoms

The symptoms are confined mainly to the leaves and are most prevalent during the later part of the crop's growing season. Symptoms start as small, pale, yellowish-green spots (lesions) having dark reddish-brown centres, formed mostly on the upper leaf surface. As the disease progresses, the central part of each lesion develops into a minute, raised, light-coloured pustule, usually located on the lower surface of the leaf. Sometimes, leaf spots are formed without developing pustules. Lesions merge with each other, forming large, irregular areas of dead leaf tissue surrounded by yellowish margins. Dead areas may fall out due to the action of the wind, giving the leaves a ragged appearance (Figure 9.32a). Severe infection often results in partial defoliation. On pods, small, reddish brown, slightly raised spots may develop on the susceptible cultivars.

Symptoms of soybean pustule can be confused with those of soybean rust. Mature rust pustules have a round opening on the upper leaf surface for the release of spores. In the case of soybean pustule, if any opening is present, it is typically a linear crack across the surface of pustule. These features can be seen only with a magnification of 20X. The absence of water-soaking of bacterial pustules and their raised centres distinguish this disease from bacterial blight of soybean (Figure 9.32b). Additionally, in bacterial blight, the spots are more angular in shape (Giesler, 2006).

FIGURE 9.32A Bacterial pustule of soybean: Pustules on upper leaf surface (a), pustules on lower leaf surface (b) and magnified (10X) pustules on lower leaf surface (c). (Courtesy: Loren Giesler, www.ianrpubs.unl.edu)

FIGURE 9.32B Bacterial pustule of soybean: Leaf pustules (Courtesy: Daren Mueller, Iowa State University, Bugwood.org)

9.33.3 Causal Organism

Xanthomonas axonopodis pv. *glycines* (Nakano 1919) Vauterin *et al.* 1995

Synonym: *Xan. glycines, Xan. campestris* pv. *glycines*

The bacterium, being a pathovar of the *Xan. axonopodis* species, has the same morphological, cultural and biochemical characters as those of *Xan. axonopodis.* It can be distinguished from *Xan. axonopodis* and its other pathovars on the basis of host range. Description of *Xan. axonopodis* is given under the section on bacterial leaf spot of green gram. Some additional characteristics of the bacterium are given below. The colonies on leaf infusion agar are small, circular, and smooth having entire margin; they are initially pale yellow, but become deep yellow with age. The bacterium grows slowly in culture. The optimum growth occurs at 30–33 °C. Maximum and minimum temperatures at which growth can take place are 38 and 10 °C, respectively.

The virulence of *Xan. axonopodis* pv. *glycines* is associated with different genetic determinants, including a pathogenicity island that includes nine *hrp*, nine *hrc* and eight *hpa* genes which are regulated by HrpX and HrpG. This cluster includes genes encoding a type III secretion system that is essential for pathogenesis and for the induction of hypersensitive reaction (HR) on some plant species. HrpG seems to be essential for the induction of a nonspecific HR on different non-host plants. Mutation of HrpG shows that it plays critical roles in HR on tobacco. Another gene that encodes a pectate lyase in *Xan. axonopodis* pv. *glycines* has also been shown to be necessary for the induction of HR on tobacco and pepper, but not in relation to tomato, sesame and cucumber (Kaewnum *et al.*, 2006). The above-given information shows that *Xan. axonopodis* pv. *glycines* contains several genes that are required for disease development and the induction of HR.

Athinuwat *et al.* (2009) identified three races of the bacterium on pustule disease-resistant and -susceptible soybean cultivars based on virulence phenotype. For race 3, an *avrBs3* homolog, *avrXg1*, was identified that conferred resistance expressed as an HR on resistant cultivar Williams 82. Mutations in two predicted functional domains of *avrXg1* led to an increase in virulence on Williams 82 and an increase in bacterial population on susceptible cultivars. Expression of *avrXg1* in race 1, which is predicted to confer an unspecific HR, led to virulence on susceptible cultivars Spencer and PI 520733. Expression of *avrXg1* in race 2 that is predicted of carrying *avrBs3*-like genes resulted in increased virulence and fitness of pathogen on both resistant and susceptible cultivars. These results demonstrate the multifunctionality of *avrXg1* depending on pathogen and plant genetic backgrounds.

Li *et al.* (2022) used *Xan. axonopodis* pv. *glycines* strain ICMP5732 to generate DNA for *de novo* Nanopore and Illumina sequencing. After assembly, a high-quality complete genome, comprising one chromosome and two plasmids, was available. The genome sequence consists of a total of 4,588 prediction coding genes, including those for four transcription activator-like effectors.

9.33.4 Disease Cycle

The bacterium survives in the infected crop residue, infected seed and in the rhizosphere of wheat roots. The bacterium can survive in the seed for up to 2½ years. The possibility of survival in the infected crop residue is greater when the residue remains on the soil surface. The entry of the bacterium occurs through stomata and wounds. After entry into the host, the bacterium moves into intercellular spaces and multiplies. In response to infection, leaf cells increase in size and also divide at a faster rate. Increase in host growth and in bacterial cell mass in the localised area causes the bulging of epidermal cells on both the leaf surfaces. These raised areas rupture and become pustules.

Dissemination of the bacterium occurs through wind-splashed rain, overhead irrigation and cultural operations during wet foliage. Disease outbreaks occur after several days of rainstorms or hailstorms. The disease is favoured by warm and wet weather. Optimum temperature for disease development is 30–33 °C.

Sain and Gour (2013) reported that a fraction of partially purified toxin(s) number 4 and 7, collected using sephadex gel chromatography, was found to be highly toxic to the host plants as well as other plants on detached leaf/twig, and seedlings (more than 50% wilting). Partially purified toxin caused chlorosis in leaves of turnip, radish, castor, soybean and mustard (>50%). Toxin treatment also reduced 100% seed germination of clusterbean, gram, coriander, turnip and *urad* bean.

Athinuwat *et al.* (2018) reported that targeted mutations in *flgK* and *pilD* genes in strain KU-P-SW005 of *Xan. axonopodis* pv. *glycines* led to altered motility phenotypes. The *flgK* mutants lacked a monopolar flagellum and lost swimming motility, whereas the *pilD* mutant lacked type IV pili and was unable to move via twitching. The *flgK* and *pilD* mutants were also altered in biofilm production and caused lower levels of disease in susceptible soybean cultivar Spencer when compared to KU-P-SW005.

Typical symptoms of the disease are pustules surrounded by small yellow haloes. However, infection by PI 96188 causes pustules without chlorotic haloes, which indicates a resistant response. Kim *et al.* (2011) studied the mode of inheritance of this novel symptom by PI 96188 and investigated whether or not the gene controlling disease resistance in PI 96188 is identical to the *rxp* gene. PI 96188 was crossed with the resistant cultivar SS2-2. All F1 plants showed the same phenotype as SS2-2, and the F2 population segregated into 75 typical symptoms (haloes present): 28 novel symptoms (haloes absent) indicating the presence of a single recessive gene. To map the novel symptom in PI 96188, a population of 88 F7 recombinant inbred lines was developed from a cross between PI 96188 and the susceptible cultivar Jinjool. Disease resistance gene from PI 96188 was mapped on chromosome 10 (LG O) rather than chromosome 17 (LG D2). This gene was linked with the simple sequence repeat marker, Sat_108 at the distal end of chromosome 10. Thus, the disease resistance gene from PI 96188 was determined to be a new gene.

Narvel *et al.* (2001) mapped the *Rxp* locus in soybean that conditions reaction to bacterial pustule by simple sequence repeat marker analysis using a population of 116 F_4-derived lines from a cross between the resistant parent Young and the susceptible parent PI 416937. *Rxp* locus was mapped 3.9 cM from Satt372 and 12.4 cM from Satt014 on linkage group D2. Linkage associations were confirmed by identifying a close association between the simple sequence repeat genotype at each locus identified as flanking *Rxp* and the bacterial pustule reaction of individual lines derived from a population different from the one used for mapping. An analysis of molecular pedigree revealed that bacterial pustule-resistant cultivars inherited the resistance gene *rxp* from the ancestral cultivar CNS based on their consistent genotypic pattern at flanking marker loci. The results show that marker-assisted selection for *rxp* would be very effective.

Guo *et al.* (2019) showed that the deletion of Crp-like protein (*clp*) led to a significant reduction in the virulence of *Xan. axonopodis* pv. *glycines* in soybean, which was highly correlated with the drastic reductions in carbohydrates utilisation, extracellular polysaccharide (EPS) production, biofilm formation, cell motility and synthesis of cell wall degrading enzymes (CWDEs). These significantly impaired properties in the *clp* mutant were completely rescued by a single-copy integration of the wild-type *clp* into the mutant chromosome via homologous recombination. Interestingly, over-expression of *clp* in the wild-type strain resulted in significant increases in cell motility and synthesis of the CWDEs. Surprisingly, significant reductions in carbohydrates utilisation, EPS production, biofilm formation and the protease activity were observed in the wild-type strain over expressing *clp*, suggesting that Clp also plays a negative role in these properties. Furthermore, quantitative reverse transcription PCR analysis suggested that *clp* was positively regulated by the diffusible signal factor-mediated quorum-sensing system and the HrpG/HrpX cascade. It is evident that Clp functions as both activator and repressor in multiple biological processes in *Xan. axonopodis* pv. *glycines* that are essential for its full virulence.

Michalski *et al.* (1983) developed a computer-based advisory system for diagnosing soybean diseases in Illinois.

9.33.5 Management

1. Use of resistant varieties is the most effective approach for the management of this disease. Disease has been relegated to minor importance in the USA, mainly through host resistance. *Rxp* gene, originally present in cultivar CNS, a variety immune to soybean pustule, has been used to develop resistant varieties. *Rxp* gene confers resistance by requiring an increased number of bacterial cells for disease development. The *avrRxv* gene inhibits disease development by inducing a hypersensitive reaction. *Xan. axonopodis*

pv. *glycines* does not require some other genes, required by other phytopathogenic bacteria, for the elicitation of hypersensitive reaction.
2. Follow crop rotation for 2–3 years.
3. Destruction of infected crop residue by deep ploughing.
4. Avoid or limit the use of overhead irrigation.
5. Avoid cultural operations when foliage is wet.
6. Give protective sprays of copper fungicides.

REFERENCES

Athinuwat, D., Brooks, S., Burr, T. J. and Prathuangwong, S. (2018). Flagella and pili of *Xanthomonas axonopodis* pv. *glycines* are associated with motility, biofilm formation and virulence on soybean. *J. Phytopathol.* 166: 590–600.

Athinuwat, D., Prathuangwong, S., Cursino, L. and Burr, T. (2009). *Xanthomonas axonopodis* pv. *glycines* soybean cultivar virulence specificity is determined by *avrBs3* homolog *avrXg1*. *Phytopathology* 99: 996–1004.

Giesler, L. J. (2006). Bacterial diseases of soybean. University of Nebraska, Lincoln. http://extensionpublications.unl.edu/assets/pdf/g2058.pdf. Accessed December 10, 2018.

Guo, W., Gao, J., Chen, Q., Ma, B., Fang, Y. *et al.* (2019). Crp-like protein plays both positive and negative roles in regulating the pathogenicity of bacterial pustule pathogen *Xanthomonas axonopodis* pv. *glycines*. *Phytopathology* 109: 1171–83.

Janse, J. D. (2005). *Phytobacteriology: Principles and Practice*. CABI Publishing, Wallingford, 360 pp.

Kaewnum, S., Prathuangwong, S. and Burr, T. J. (2006). A pectate lyase homolog, *xagP*, in *Xanthomonas axonopodis* pv. *glycines* is associated with hypersensitive response induction on tobacco. *Phytopathology* 96: 1230–36.

Kim, K. H., Park, J.-H., Kim, M. Y., Heu, S. and Lee, S.-H. (2011). Genetic mapping of novel symptom in response to soybean bacterial leaf pustule in PI 96188. *J. Crop Sci. Biotechnol.* 14: 119–123.

Krawczyk, K. and Borodynko-Filas, N. (2020). *Kosakonia cowanii* as the new bacterial pathogen affecting soybean (*Glycine max* Willd.). *Eur. J. Plant Pathol.* 157: 173–83. https://doi.org/10.1007/s10658-020-01998-8

Li, X., Zhang, H., Peng, Y., Liu, Z., Xu, Z. *et al.* (2022). The complete genome resource of *Xanthomonas axonopodis* pv. *glycines* ICMP5732 causing pustule disease on soybean. *Plant Dis.* 106: 1743–45.

Michalski, R. S., Davis, J. H., Bisht, V. S. and Sinclair, J. B. (1983). A computer-based advisory system for diagnosing soybean diseases in Illinois. *Plant Dis.* 67: 459–63.

Nakano, K. (1919). Soybean leaf spot. *J. Plant Prot. Tokyo* 6: 217–21.

Narvel, J. M., Jakkula, L. R., Phillips, D. V., Wang, T., Lee, S.-H. *et al.* (2001). Molecular mapping of *Rxp* conditioning reaction to bacterial pustule in soybean. *J. Heredity* 92: 267–70.

Patel, P. N. and Jindal, J. K. (1972). Bacterial pustule of soybean. In: P. N. Patel (ed.). *Plant Bacteriology*. Vol. I, *Bacterial Diseases of Plants in India*. Indian Agricultural Research Institute, New Delhi, pp. 59–65.

Sain, S. K. and Gour, H. N. (2013). Pathological and physiobiochemical characterization of *Xanthomonas axonopodis* pv. *glycines*, incitant of *Glycine max* leaf pustules. *Indian Phytopathol.* 66: 20–27.

Uppal, B. N., Patel M. K. and Kamat, M. N. (1938). Bacterial leaf spot of soybean in Bombay. *J. Univ.* Bombay 6: 16–18.

Vauterin, L., Hoste, B., Kersters, K. and Swings, J. (1995). Reclassification of *Xanthomonas*. *Int. J. Syst. Bacteriol.* 45: 472–89.

9.34 BACTERIAL BLIGHT OF SOYBEAN

Bacterial blight of soybean (*Glycine max* L.) was first noticed by Heald in Nebraska (USA) in 1905, who described the disease and suggested its bacterial nature. In 1917, Johnson and Coerper isolated the bacterium, proved its pathogenicity and suggested it to be a member of the genus *Pseudomonas*. It is the most common bacterial disease of soybean in areas where cool and wet weather prevails.

The disease is most conspicuous in fields during mid-season as the bacterium remains active under moist conditions, until checked by hot and dry weather. It occurs throughout the United States, where many cultivars are susceptible. Frequent stormy conditions and rains, accompanied with cooler weather, have made this disease quite common in Nebraska soybean fields. Williams and Nyvall (1980) reported a yield reduction of 17.9%. Bacterium also considerably inhibits the germination of infected seeds and inhibition up to 68% has been reported (Nicholson and Sinclair, 1971). However, the disease has not caused significant economic losses in most areas of the world due to host resistance or plant tolerance. A survey undertaken in eight districts of Marathwada region of Maharashtra, India, during the *Kharif* seasons of 2009 and 2010, for recording the severity and incidence of the disease, revealed the average disease incidence of 14.5% and maximum pod blight severity on variety JS-335 (Jagtap *et al.*, 2012a).

The disease is widespread and occurs in Europe, appearing in Austria, Bulgaria, Czechoslovakia (former), Denmark, France, Hungary, Germany, Italy, Mainland Italy, Poland, Romania; Moldova, the Russian Far East, southern Russia, Sweden, the UK, Ukraine and the former Yugoslavia (Fed. Rep.); in Asia, being found in Brunei Darussalam, China, Hebei, Heilongjiang, Henan, Jiangsu, Jilin, Liaoning, Shaanxi, Shandong, Zhejiang, India (Madhya Pradesh, Maharashtra), Indonesia, Japan, Kazakhstan, Korea Republic, Mongolia, Pakistan, Republic of the Philippines and Taiwan; in Africa, in Egypt, South Africa, Zambia and Zimbabwe; in North America and Canada (Manitoba, Ontario), Mexico, the USA (Alabama, Delaware, Florida, Hawaii, Iowa, Illinois, Minnesota, Missouri, Nebraska, North Carolina, North Dakota, Ohio and Wisconsin); in South America, in Argentina, Brazil, Bahia, Goias, Parana, Rio Grande do Sul, Santa Catarina, Colombia and Venezuela; and in Oceania, in Australia (New South Wales, Northern Territory, Queensland, South Australia, Victoria) and New Zealand (CABI, 1999).

Yield losses in the range 18–22% have been reported under favourable weather conditions.

Krawczyk and Borodynko-Filas (2020) reported an outbreak of a new disease of soybean plants in Poland, with symptoms resembling those described for the bacterial blight caused by *Pseudomonas syringae* pv. *glycinae* and bacterial pustule caused by *Xanthomonas axonopodis pv.*

glycines. Identification based on biological (Biolog Gen III) and molecular methods (16S rDNA and gyrB sequencing, PCR species-specific testing) revealed that the soybean pathogenic bacteria is *Kosakonia cowanii*.

9.34.1 Host Range

In addition to soybean, the bacterium infects common bean, lima bean, snap bean, tepary bean and cowpea.

9.34.2 Symptoms

The disease is most prevalent in early season in contrast to soybean pustule, which appears late. The symptoms appear on all the above-ground plant parts, but are most evident on leaves in the middle to upper canopy. Generally, the lesions first appear on margins of cotyledons as brown spots, which enlarge and turn dark brown as the tissue collapses. Young plants may show stunting, and they may die if the infection reaches the growing point.

The symptoms are most conspicuous on leaves, although these also appear on stems, petioles and pods. Young leaves are most susceptible. Symptoms on the leaves start as small, angular, yellow to brown lesions. The lesions are usually water-soaked in the centre and surrounded by yellowish-green haloes. Older lesions turn reddish-brown to black as the tissue dies. Angular lesions enlarge in cool and wet weather and may merge to produce large, irregular dead areas. Eventually, the dried lesions fall out, giving the leaves a ragged appearance. Often, the leaves are badly shredded after strong winds and stormy rains. Infected young leaves become stunted, distorted and chlorotic. Severe infection results in the defoliation of lower leaves. Large, black lesions may also develop on stem and petioles (Figure 9.33).

FIGURE 9.33 Bacterial blight of soybean. (Courtesy: Clemson University – USDA Cooperative Extension Slide Series, Bugwood.org)

The symptoms on seed appear rarely. However, when lesions appear on pods, developing seeds may show some kinds of symptoms. Small, water-soaked lesions appear on pods, which enlarge and merge, giving rise to dark brown to black areas. Slimy bacterial ooze may cover the developing seeds. During storage, infected/infested seeds may become shrivelled, discoloured and sunken or raised lesions develop on them. Infected seeds lack vigour and germinate poorly.

Chlorosis, also called **systemic toxemia**, develops throughout the infected plants. It results from the production of a toxin, coronatine, by the bacterium in the older infected leaves and translocated throughout the plant. Strains of the bacterium carrying the *avrB* gene elicit a hypersensitive reaction in cultivars carrying the resistant gene *Rpg1*.

The symptoms of bacterial blight differ from those of soybean pustule in that, in the former, the lesions are water-soaked and flat in contrast to the raised centres of lesions of the latter.

The symptoms of bacterial blight can be confused with those of brown spot/Septoria leaf spot caused by *Septoria glycines*. Presence of yellowish green halos around the bacterial blight lesions distinguishes it from the latter. Both the diseases can occur together on the same plant, but bacterial blight is more common on young upper leaves, while brown spot is generally found on older, lower leaves of the plants.

9.34.3 Causal Organism

Pseudomonas savastanoi pv. *glycinea* (Coerper 1919) Gardan *et al.* 1992

Synonym: *Psm. syringae* pv. *glycinea*, *Psm. glycinea*

Gardan *et al.* (1992) found that *Psm. syringae* subsp. *savastanoi* strains belong to a DNA relatedness group that includes strains of *Psm. syringae* pv. *glycinea* and *Psm. syringae* pv. *phaseolicola*. This DNA group was distinct from those of *Psm. syringae* pv. *syringae* (including the type strain of *Psm. syringae*). The results of a numerical analysis were in accordance with DNA hybridisation data. Therefore, they elevated *Psm. syringae* subsp. *savastanoi* (Janse 1982) to species level as *Pseudomonas savastanoi* sp. nov., which includes *Psm. savastanoi* pv. *savastanoi*, *Psm. savastanoi* pv. *glycinea* and *Psm. savastanoi* pv. *phaseolicola*.

Psm. savastanoi (*ex* Smith 1908) Gardan *et al.* 1992: The below-given description of the species is based mainly on the publication of Gardan *et al.* (1992). Cells are Gram-stain-negative rods, measure 0.4–0.8 ×1.0–3.0 µm and are motile by means of one to four polar flagella. Rather slow growing. Colonies are white or creamy, smooth, flat and glistening with entire or erose margins on Yeast extract Bacto peptone glucose agar (YBGA). Blue fluorescent pigment is produced under UV light on King's medium B. Optimal temperature for growth is 24–26 °C. Metabolism is respiratory. Oxidase and arginine (Thornley) tests are negative and nitrates are not reduced. Esculin, gelatin, and starch are not hydrolysed. Assimilates sucrose, l-arabinose, caprylate, fumarate, gluconate, dl-glycerate, l-malate, pyruvate, citrate, d-α-alanine and l-proline, but does not assimilate lactose, l-xylose, adonitol, 2-aminobutyrate, dl-lactate, dl-3-hydroxybutyrate, d-(-)-tartrate, l-cysteine, l-methionine, and l-valine. The strains isolated from members of various genera of the *Oleaceae* and *Nerium oleander* do not produce levan, whereas strains isolated from *Fraxinus excelsior* (21%), *Phaseolus vulgaris* and *G. max* produce levan. The G + C content of the DNA is 60 mol % (De Ley, 1968). Produces a hypersensitive reaction on tobacco leaves.

Psm. savastanoi pv. *savastanoi* causes knots, galls and cankers on members of the various genera of the *Oleaceae* and *N. oleander*. *Psm. savastanoi* pv. *glycinea* causes bacterial blight of soybean and *Psm. savastanoi* pv. *phaseolicola* causes halo blight of bean.

Qi *et al.* (2011) sequenced genomes of two *Psm. savastanoi* pv. *glycinea* (*Psg*) strains, i.e., race 4 and B076, using 454 pyrosequencing. Genomes of both *Psg* strains share more than 4,900 highly conserved genes, indicating very low genetic diversity between *Psg* genomes. Though conserved, genome rearrangements and recombination events occur commonly within the two *Psg* genomes. When compared to each other, 437 and 163 specific genes were identified in B076 and race 4, respectively. Most specific genes are plasmid-borne, indicating that acquisition and maintenance of plasmids may represent a major mechanism to change the genetic composition of the genome and even acquire new virulence factors. Type three secretion gene clusters of *Psg* strains are near identical with that of *Psm. savastanoi* pv. *phaseolicola* (*Psp*) strain 1448A, and these shared 20 common effector genes. Moreover, the coronatine biosynthetic cluster is present on a large plasmid in strain B076, but not in race 4. *In silico* subtractive hybridisation-based comparative genomic analyses with nine sequenced phytopathogenic pseudomonads; identified dozens of specific islands, and revealed that the genomes of *Psg* strains are more similar to those belonging to the same genomospecies such as *Psp* 1448A than to other phytopathogenic pseudomonads. Number of highly conserved genes (core genome) among these decreased dramatically when more genomes were included in the subtraction, suggesting the diversification of pseudomonads, and further indicating the genome heterogeneity among pseudomonads. However, the number of specific genes did not change significantly, suggesting these genes are indeed specific in *Psg* genomes. These results strengthen the idea of a species complex of *Psm. syringae* and support the reclassification of *Psm. syringae* into different species.

Phytotoxin coronatine, produced by the bacterium in a temperature-dependent manner, consists of a polyketide, coronafacic acid (CFA) and an amino acid derivative, coronamic acid. It is produced optimally at 18 °C, whereas no detectable synthesis occurs at 28 °C. In 1966, seven races of the bacterium were identified from 104 isolates collected from five states of the USA, namely, Illinois, Iowa, Missouri, Minnesota and North Carolina. These races were identified on the basis of their reactions (resistant, intermediate and susceptible) on seven soybean cultivars, namely, Lindarin, Acme, Flambeau, Harosoy, Merit, Norchief, and Chippewa. In 1980, an eighth race was found in Australia. To date, nine races of the bacterium have been identified.

9.34.4 Disease Cycle

The bacterium survives in infected/infested seed and the first evidence of its seed–plant transmission was provided by Kendrick and Gardner (1921).

Yao *et al.* (2007) reported that the bacterium could be transmitted through infected seeds and the average incidences of infected seeds and seed–plant transmission rate were 7.6% and 0.48%, respectively. Infection/infestation of seed occurs through the pods during the growing season or at the time of threshing and cleaning.

Plasencia-Marquez *et al.* (2017) designed generic primers and evaluated the sensitivity of a BIO-PCR system to detect *Pseudomonas* spp. in soybean seeds. *Pseudomonas* strain isolated from soybean was diluted in seed extracts and the sensitivity of the BIO-PCR system using generic primers was evaluated. The PCR using PgB1 and PgB2 generic primers was highly specific to detect members of *Pseudomonas* genus. Using BIO-PCR, it was possible to detect bacterial strain until 10^2 cfu mL^{-1}.

The bacterium also survives in infected crop residue but the survival in crop residue is variable and is influenced by several factors. Survival is better under cold and dry conditions compared to warm and wet conditions. In Brazil, the bacterium overwinters on volunteer bean plants in winter wheat fields.

The bacterium grows epiphytically on leaves, buds and stems of soybean plants without causing any symptoms. Epiphytic growth is influenced by the reaction of the host to the bacterium. On the leaves of susceptible host cultivars, the population increased 1000-fold while on resistant cultivars, it remained either unchanged or declined. This epiphytic phase may serve as a continuous source of inoculum.

It is typically an early season disease favoured by cool and wet weather, which prevails during the early crop season. Optimum temperature for disease development is 21–27 °C. Warm and dry weather after the mid-season of crop prevents the disease development. Disease epiphytotics are a common seen after windy rainstorms coupled with cool weather.

The outbreaks that occur later in the season often follow windy rainstorms. Outbreaks typically develop several days after a rainstorm or hailstorm. Symptoms are most evident on new growth that is expanding at the time of the rain event.

Primary infection usually occurs on the cotyledons and these lesions serve as a major source of secondary inoculum. The spread of the bacterium occurs through wind-splashed rain, overhead irrigation and cultural operations especially during wet foliage and rubbing of wet leaves with each other. After entering through stomata and wounds, bacteria multiply in the intercellular spaces of mesophyll cells. Typical water-soaked lesions develop 5–7 days after infection. During multiplication in the host, the pathogen produces a toxin called **coronatine**, which is translocated throughout the plant, mainly through the phloem. Toxin interferes with photosynthesis and chlorophyll production, leading to systemic chlorosis and stunting of plants. Fully expanded trifoliate leaves do not show systemic chlorosis but it is at its maximum on young trifoliate leaves. Infected plants recover after a period of time and perform normal photosynthetic activity and chlorophyll production, but stunting remains permanently. Toxin produced by the bacterium produces chlorotic lesions also on tobacco, bean and tomato, indicating it is non-host-specific.

Gnanamanickam *et al.* (1982) reported a correlation between coronatine production by strains of *Psm. glycinea* in culture and the induction of systemic symptoms (chlorosis and stunting) in inoculated soybean plants. The application of purified preparations of coronatine to unifoliate leaves of soybean plants resulted in localised chlorosis, the development of chlorosis in subsequently developing trifoliate leaves and stunting of plant growth, similar to symptoms induced by infection. Coronatine was demonstrated in soybean leaves infected with the bacterium, but was not detected in healthy leaves. However, some pathogenic strains do not produce coronatine, which indicates that it may not be essential for pathogenicity.

Budde and Ullrich (2000) investigated the impact of temperature on coronatine production by *Psm. syringae* (= *Psm. savastanoi*) pv. *glycinea* PG4180 by inoculating soybean and tobacco plants. After spray inoculation, PG4180 caused typical bacterial blight symptoms on soybean plants when the bacterial culture was grown at 18 °C prior to inoculation, but not when derived from cultures grown at 28 °C. Disease outcome was quantified by determining the bacterial populations *in planta*. Temperature effect was not observed when PG4180 was artificially infiltrated into soybean leaves, indicating that bacterial invasion via natural plant openings and pre-inoculation temperature were important for phytotoxin synthesis. Inoculation of PG4180 on tobacco plants (incompatible host) elicited the HR irrespective of the bacterial pre-inoculation temperature. However, the HR was significantly delayed when tobacco plants were treated with cells of the CFA-overproducing derivative, PG4180.N9, which were derived from cultures grown at 18 °C, compared with parallels incubated at 28 °C. CFA biosynthesis by PG4180.N9 was optimal at 18° C and negligible at 28 °C. Impact of CFA synthesis on the HR studied with different growth media, mutants and transconjugants of PG4180, indicated that the amount of synthesised CFA, but not that of coronatine influenced the outcome of the HR. Experiments involving the application of purified coronafacoyl compounds suggested that the observed delay of the HR was mediated by CFA, providing further information on CFA's putative role as a molecular mimic of the plant signalling molecule, jasmonic acid.

Four resistance genes, *Rpg1*, *Rpg2*, *Rpg3* and *Rpg4*, have been identified, using a cross between cultivars, Flambeau × Merit. Line LL 489-605 is resistant. Soybean cultivars resistant to *Psm. syringae* pv. *glycinea* (*Psg*) exhibit an HR (necrotic) to infection. *Psg* strains carrying the *avrB* gene elicit the HR in soybean cultivars carrying the resistance gene *Rpg1*. *Psg* expressing *avrB* at a high level and capable of eliciting the HR in the absence of *de novo* bacterial RNA synthesis have been obtained in *in vitro* culture. Nutritional signals and regions within the *Psg* hrp gene cluster, an approximately 20-kilobase genomic region, also necessary for pathogenicity, control *avrB* transcription (Huynh *et al.*, 1989).

Resistant varieties have been developed in various countries, but most cultivars have moderate levels of resistance. At least nine races of the pathogen have been identified, and the occurrence of these races has complicated the breeding of resistant cultivars. Demonstration of a single dominant resistant gene, *Rpg1* in the host, together with molecular cloning of avirulence genes in the pathogen, suggested a gene-for-gene system (McGee 1992).

9.34.5 Management

The following measures can be helpful in reducing the disease incidence and severity.

1. Use of resistant varieties is the best method to manage the disease. Highly susceptible cultivars should not be planted in areas where the disease is a potential problem.
2. Use of pathogen-free seed.
3. Follow crop rotation with non-host crops for 2–3 years.
4. Practise deep tillage in fields where the disease is frequently severe. Cover soybean residues after harvest by tillage, if possible, or shred residue for quick decomposition. Differences in disease ratings and infestation occurred in one study after no tillage and conventional tillage systems with 83.3% and 70.3% of plants infected, respectively (Anaele *et al.*, 1990)
5. Avoid or limit the cultural operations when the foliage is wet.
6. Avoid or limit the overhead irrigation.
7. Biological control: *Bdellovibrio bacteriovorus*, a small comma-shaped bacterium isolated from the rhizosphere of soybean roots, inhibited the development of local and systemic symptoms of bacterial blight when inoculated onto soybean with *Psm. glycinea* at ratios of 9:1 and 99:1, respectively (Scherff, 1973a). A bacterium, characteristic of a group of yellow bacteria, isolated from soybean leaves infected with *Psm. glycinea* and designated as YB-3, was inoculated onto soybean leaves after mixing with *Psm. glycinea* cells. The mixture of *Psm. glycinea* and YB-3 in the ratio of 1:9 inhibited development of bacterial blight, while at 1:4 ratio, there was ca.

50% reduction in symptoms, and at 1:1 ratio there was no reduction in lesion development. There was complete inhibition of bacterial blight symptoms when 1:4 mixture of *Psm. glycinea* and YB-3 were incubated for 24–48 h before inoculation (Scherff, 1973b). It has been shown that naturally occurring epiphytes isolated from soybean leaves have a high antagonistic efficacy against *Psm. syringae* pv. *glycinea* (Volksch *et al.*, 1993; May *et al.*, 1997). One *Erwinia herbicola* (= *Pantoea agglomerans*) and one *Psm. syringae* strain possessed the ability to suppress the growth of the pathogen and the ability to establish and maintain epiphytic populations on soybeans plants under field conditions (Volksch *et al.*, 1996). This suggests that the spread of disease can be curtailed by these bacterial strains through competitive exclusion.

8. An epiphyte *Psm. syringae* pv. *syringae* 22d/93 (*Pss*22d), isolated from soybean leaves, had been characterised as a promising and species-specific biocontrol strain *in vitro* and *in planta* against *Psm. syringae* pv. *glycinea* (*Psg*). *Pss*22d produces three toxins, namely, syringomycin, syringopeptin and 3-methylarginine. In contrast to syringopeptin and syringomycin, 3-methylarginine inhibited the growth of *Psg in vitro*. Braun *et al.* (2010) reported that the wild-type *Pss*22d and its toxin-deficient mutants prevented development of disease symptoms normally caused by *Psg*. It is evident that none of the syringopeptin, syringomycin and 3-methylarginine are required for *Pss*22d's antagonistic activity *in planta*. Consequently, factors other than the three toxins may contribute to the intra-species antagonism *in planta*.
9. Use of chemicals: Jagtap *et al.* (2012b) reported that spray applications of Streptocycline (100 µg mL^{-1}) + copper oxychloride (0.25%) gave the lowest disease incidence (12.74%) along with highest seed yield (2,605 kg/ha) and test weight (14.33 g).
10. Give preventive sprays of copper fungicides.

REFERENCES

Anaele, A. O., Pacumbaba, R. P. and Bishnoi, U. R. (1990). The effects of conventional and no-tillage systems on incidence of bacterial blight and with interaction of row spacings and weed control methods on the performance of other agronomic characters of soybean. *J. Agron. Crop Sci.* 165: 312–18.

Bozkurt, İ. A. and Soylu, S. (2011). Determination of responses of different bean cultivars against races of *Pseudomonas syringae* pv. *phaseolicola*, causal agent of halo blight of bean. *Euphytica*, 179: 417–25.

Braun, S. D., Hofmann, J., Wensing, A., Weingart, H. and Ullrich, M. S. (2010). *In vitro* antibiosis by *Pseudomonas syringae* Pss22d, acting against the bacterial blight pathogen of soybean plants, does not influence *in planta* biocontrol. *J. Phytopathol.* 158: 288–95.

Budde, I. P. and Ullrich, M. S. (2000). Interactions of *Pseudomonas syringae* pv. *glycinea* with host and non-host plants in relation to temperature and phytotoxin synthesis. *Mol. Plant-Microbe Interact.* 13: 951–61.

CABI; EPPO. Distribution maps of plant diseases 1999 No. April (Edition 1) pp. Map 782.

Coerper, F. M. (1919). Bacterial blight of soybean. *J. Agr. Res.* 18: 179–94.

De Ley, J. (1968). DNA base composition and hybridization in the taxonomy of phytopathogenic bacteria. *Annu. Rev. Phytopathol.* 6: 63–90.

Gardan, L., Bollet, C., Ghorrah, M. A., Grimont, F. and Grimont P. A. D. (1992). DNA relatedness among the pathovar strains of *Pseudomonas syringae* subsp. *savastanoi* Janse 1982 and proposal of *Pseudomonas savastanoi* sp. nov. *Int. J. Syst. Bacteriol.* 42: 606–12.

Gnanamanickam, S. S., Starratt, A. N. and Ward, E. W. B. (1982). Coronatine production *in vitro* and *in vivo* and its relation to symptom development in bacterial blight of soybean. *Can. J. Bot.* 60: 645–50.

Huynh, T. V., Dahlbeck, D. and Staskawicz, B. J. (1989). Bacterial blight of soybean: regulation of a pathogen gene determining host cultivar specificity. *Science* 245 (4924): 1374–77.

Jagtap, G. P., Dhopte, S. B. and Dey, U. (2012a). Survey, surveillance and cultural characteristics of bacterial blight of soybean. *Afr. J. Agric. Res.* 7: 4559–63.

Jagtap, G. P., Dhopte, S. B. and Dey, U. (2012b). Bio-efficacy of different antibacterial antibiotic, plant extracts and bioagents against bacterial blight of soybean caused by *Pseudomonas syringae* pv. *glycinea*. *Scient. J. Microbiol.* 1: 1–9.

Janse, J. D. (1982). *Pseudomonas syringae* subsp. *savastanoi* (ex-Smith) subsp. nov., nom. rev., the bacterium causing excrescences on *Oleaceae* and *Nerium oleander* L. *Int. J. Syst. Bacteriol.* 32: 166–69.

Kendrick, J. B. and Gardner, M. W. (1921). Seed transmission of soybean bacterial blight. *Phytopathology* 11: 339.

Krawczyk, K. and Borodynko-Filas, N. (2020). *Kosakonia cowanii* as the new bacterial pathogen affecting soybean (*Glycine max* Willd.). *Eur. J. Plant Pathol.* 157: 173–83.

May, R., Volksch, B. and Kampmann, G. (1997). Antagonistic activities of epiphytic bacteria from soybean leaves against *Pseudomonas syringae* pv. *glycinea in vitro* and *in planta*. *Microb. Ecol.* 34: 118–24.

McGee, D. C. (1992). *Soybean Diseases: A Reference Source for Seed Technologists*. American Phytopathological Society, St. Paul, MN, 151 pp.

Nicholson, J. F. and Sinclair, J. B. (1971). Amsoy soybean seed germination inhibited by *Pseudomonas glycinea*. *Phytopathology* 61: 1390–93.

Plasencia-Marquez, O., Corzo, M., Rivero, D. and Martinez-Zubiaur, Y. (2017). BIO-PCR for *Pseudomonas* spp. detection in soybean (*Glycine max*) seeds (Abstr.). *Phytopathology* 107: S4.12.

Qi, M., Wang, D., Bradley, C. A. and Zhao, Y. (2011). Genome sequence analyses of *Pseudomonas savastanoi* pv. *glycinea* and subtractive hybridization-based comparative genomics with nine pseudomonads. *PLoS One* 6(1): e16451.

Scherff, R. H. (1973a). Control of bacterial blight of soybean by *Bdellovibrio bacteriovorus*. *Phytopathology* 63: 400–02.

Scherff, R. H. (1973b). Bacterial blight of soybeans as influenced by populations of yellow bacteria on leaves and buds. *Phytopathology* 63: 752–55.

Smith, E. F. (1908). Recent studies of the olive-tubercle organism. *Bull. Bur. Plant Ind., USDA* 131: 25–43.

Volksch, B., Nuske, J. and May, R. (1996). Characterization of two epiphytic bacteria from soybean leaves with antagonistic activities against *Pseudomonas syringae* pv. *glycinea*. *J. Basic Microbiol.* 36: 453–62.

Volksch, B., Ullrich, M. S. and Fritsche, W. (1993). Occurrence of antimicrobial activities of bacteria from soybean leaf spots. *J. Basic Microbiol.* 33: 349–55.

Williams, D. J. and Nyvall, R. F. (1980). Leaf infection and yield losses caused by brown spot and bacterial blight diseases of soybean. *Phytopathology* 70: 900–02.

Yao, H., Xie, G. and Jin, Y. (2007). Study on the causal agent of bacterial leaf spot of soybean in Shanghai. *Acta Agric. Shanghai* 23: 41–45.

9.35 BACTERIAL BLIGHT OF COWPEA

Bacterial blight of cowpea (*Vigna unguiculata* (L.) Walp.) was originally described as canker of cowpea by Burkholder (1944) from the USA. In India, it was first reported from Pune, Maharashtra by Patel and Diwan in 1950 and subsequently from many areas like, Tamil Nadu, Delhi, Punjab, Haryana, Uttar Pradesh and Karnataka. The disease also occurs in China, Brazil, Nigeria, Tanzania, Thailand, Venezuela, Zimbabwe, Uganda, Senegal, Sudan and South Africa.

High mortality of the seedlings is responsible for the major loss due to the disease. Plant mortality as high as 60% has been reported in the California Black-Eye variety in Texas, USA. In India, 15–20% seedling mortality was observed in certified seed production plots of Pusa Dofasli cultivar of the National Seeds Corporation of India in 1971 (Patel *et al.*, 1972).

9.35.1 Host Range

Different host species have been reported to be infected artificially, but there is little agreement among different workers on this aspect. It appears that, under natural conditions, cowpea is the main economic host. However, Sikirou and Wydra (2004) have reported *Sphenostylis stenocarpa* (African yam bean) as a potential host of the bacterium, for the first time on artificial inoculation. Gena *et al.* (2009) have reported successful inoculation and symptom production on French bean (*Phaseolus vulgaris*) and hyacinth bean (*Dolichos lablab*).

The bacterium multiplies epiphytically on five weed hosts, namely, *Parthenium hysterophorus, Tephrosia purpurea, Tylophora indica, Lega-scea mollies* and *Leucas aspera*; among them, the maximum population of the bacterium was found on *P. hysterophorus*.

9.35.2 Symptoms

First symptoms appear on cotyledons or primary leaves of seedlings developing from infected seeds. Infected cotyledons turn red and shrivel. On primary leaves, necrotic spots generally appear at the margin from where the bacterium moves to stem and then systemically proceeds to other plant parts. Severe infection damages cotyledons, primary leaves and growing point, resulting in the death of seedlings. In the surviving seedlings, the cankers usually develop at the points where cotyledons and primary leaves were attached. These cankers enlarge and turn brown, and ultimately the cankered stems break at this point due to the weight of plants and the action of fast winds.

FIGURE 9.34 Bacterial blight of cowpea. (Courtesy: Oliver Jeffrey; www.infonet-biovision.orgpustule)

Infection of trifoliate leaves results in the formation of light yellow, irregular to round spots (4–10 mm) having necrotic brown centres and deep yellow borders. These spots enlarge and may involve the whole leaf. Small veins within blighted areas turn reddish brown, become shrivelled and extend into the green leaf area. In severe infection, infection moves along the veins turning the area around the veins yellow (Figure 9.34). Later on, these leaves turn straw coloured and defoliate prematurely.

On infected pods, dark green, water-soaked streaks appear along the sutures. Severely infected pods turn yellow, shrivel and dry. Seeds produced in the severely infected pods are usually small, shrivelled and discoloured.

9.35.3 Causal Organism

Xanthomonas axonopodis pv. *vignicola* (Burkholder 1944) Vauterin *et al.* 1995

Synonym: *Xan. vignicola, Xan. campestris* pv. *vignicola*

Burkholder (1944) described the bacterium for the first time and named it *Xanthomonas vignicola* sp. nov. With the introduction of the pathovar system, it was changed to *Xan. campestris* pv. *vignicola* and, later, reclassified as *Xan. axonopodis* pv. *vignicola*. The below-given description of the bacterium is taken primarily from Burkholder (1944).

Bacterial cells are Gram-stain-negative, non-sporing rods (0.5–0.9 × 1.0–2.8 μm) with rounded ends and motile with a single polar flagellum. The colonies on potato dextrose agar are large, mucoid and pale yellow. Optimum temperature for growth is between 27–30 °C, with the maximum at 37 °C and minimum between 6 and 9 °C. Tests for starch hydrolysis, gelatin liquefaction, hydrogen sulphide production and lipolytic activity are positive, while those for indole and ammonia production and nitrate reduction are negative. No growth occurs in sodium salts of lactic, formic, succinic, tartaric and hippuric acids. Bacterial growth in citric and malic acid salts produces an alkaline reaction. Tyrosine and asparagine are not utilised as a combined source of carbon and nitrogen. The bacterium is aerobic and liquefies pectate medium. Khatri-Chhetri *et al.* (2003) reported that 55 strains of the bacterium obtained from 11 countries differed in metabolisation of different carbon sources using the Biolog GN microplate system; the strains belonging to different origins varied greatly but the variation was not related to the virulence of the strains.

Wydra *et al.* (2004) developed the first semi-selective medium, called cefazoline cellobiose-methionine medium, for the isolation of the bacterium. The colonies of the bacterium on this medium were white, round, raised and 0.2–1.8 mm in diameter after four days of incubation. The medium was helpful for the isolation of the bacterium from cowpea plant and soil samples.

9.35.4 Disease Cycle

Infected seed is an important source of perpetuation of the bacterium. Volunteer plants also act as a source of inoculum. Seed may be either externally or internally infected, or sometimes both. Heavily infested seeds usually give rise to diseased seedlings even under dry conditions, and primary infection through seed readily causes stem canker and results in high plant mortality. Khan (1996) observed 73.50% disease incidence in a crop raised from naturally infected seed. In another study, Soni (1982) reported that inoculation of cowpea plants at all growth stages, starting from seeds before sowing to flat green pod stage, led the bacterium to become seed-borne. Methods are available to detect the bacterium in the infected seeds (Soni and Thind, 1987, 1991).

Claudius-Cole *et al.* (2014) used various modifications of the plating and soaking methods to detect *Xan. axonopodis* pv. *vignicola* from cowpea seeds, collected from five locations within two agro-ecological regions in Nigeria. Percentage seed infection was significantly higher in Zaria, Minjibir (northern guinea savannah) and Iseyin (forest transition zone) than in the other production locations, in both plating and soaking assay methods. The seed coat alone was the most ($P \leq 0.05$) colonised part of the seed by the pathogen followed by the cotyledons, while the embryo was the least colonised. Pod inoculation was the most effective pathogenicity test, followed by inoculation of young plants, while inoculation of the excised cotyledons, stems and petioles was not reliable for expressing symptoms. Detached pods were equally efficient and took less time compared to the inoculation of young plants to prove the pathogenicity of the bacterium. The detection of *Xan. axonopodis* pv. *vignicola* was better with methods that breach the seed coat or which use the seed coat alone.

De Lima-Primo *et al.* (2015) reported that cowpea plants grown in the greenhouse and the field were inoculated with 10^5, 10^6 and 10^7 cfu mL^{-1} of *Xan. axonopodis* pv. *vignicola*, and seeds harvested from field plots were used to determine the seed plant transmission of the bacterium. In the greenhouse, plants were subjected to leaf wetness duration periods of 0, 4, 8, 12, 16, 20 and 24 h after inoculation. In another experiment, soybean, common bean and wild poinsettia were inoculated with *Xan. axonopodis* pv. *vignicola*. The plants inoculated with a concentration of 10^7 cfu mL^{-1} and subjected to 12 h of leaf wetness showed highest disease severity. Positive seed-plant transmission of the bacterium occurred in all the inoculum concentrations used. Hypersensitivity reaction was observed in wild poinsettia less than 24 h after inoculation.

The bacterium did not survive for more than three weeks in diseased plant debris under field conditions, indicating that diseased plant debris does not play any role in its survival (Khan, 1996). Infected crop residue and some weeds may act as source of primary inoculum when two cowpea crops are sown consecutively (Sikirou and Wydra, 2004).

Secondary spread of the bacterium occurs mostly through wind-splashed rain. Other means of secondary spread are overhead irrigation and ground irrigation. When the planting is done in furrows running along the slope, irrigation/rain water spreads the bacterium while moving from the higher level to the lower level of the field. Long-distance spread, or spread to the new areas, occurs through seed. Zandjanakou-Tachin *et al.* (2007) reported the presence of the bacterium in the faeces, intestines and on the legs and mandibles of *Zonocerus variegatus* (grasshopper) and indicated the possibility of the transmission of the bacterium through this insect.

The disease is severe in humid and warm weather. Optimum temperature for disease development is 28–34 °C. The disease is more severe in low-lying areas where water frequently stagnates, even if temporarily.

Rachael *et al.* (2015) screened 12 cowpea genotypes against three isolates collected from three different cowpea-growing zones, namely, Makurdi, Guma and Gboko local government areas in Benue State of Nigeria, designated as MKD388-1, GUM391 and GBK205-8, respectively, in screen house and field trials. Three genotypes, BOSADP, IT98K-1092–2 and IT98K-573-2-1, were consistently resistant in both pot and field experiments. Isolates did not differ significantly ($P < 0.05$) in their virulence means; rating 4.69 (MKD388-1), 4.50 (GUM391) and 4.33 (GBK205-8).

Singh and Patel (1977) reported that the resistance was governed by a single dominant gene in cowpea lines, namely, P 309, P 426 and P 910, and by a single recessive gene in Iron. In the case of Iron and P 910, the segregating populations showed some gradations in susceptible and resistant reactions indicating that, in addition to major genes, some modifying factors(s) may be involved.

9.35.5 Management

1. As the infected seed is only the main source of inoculum of the bacterium, the production of pathogen-free seed or the elimination of the seed-borne inoculum will control the disease. Different methods of seed treatment have been worked out by different workers. Jindal *et al.* (1989) recommended hot water (50 °C for 30 min.) or solar heat (soaking of seed in tap water from 8.00 a.m. to 12.00 noon and then spreading the seed in a thin layer exposed directly to sun rays for 5 h on a bright sunny day in summer) or Streptocycline (100 μg mL^{-1}) + Agallol (0.2%) or Streptocycline (100 μg mL^{-1}) + captan (0.2%) treatment. Jindal and Thind (1990a) found that *Erwinia herbicola* (recent name, *Pantoea agglomerans*) and *Penicillium oxalicum* were as effective as the above-mentioned four treatments in controlling seed-borne infection of the bacterium. These biocontrol agents also gave higher seed germination than hot water and chemical treatments.
2. Secondary spread of the pathogen should be checked by applying chemicals or biocontrol agents. Sprays of copper and copper + maneb mixtures have been found effective in controlling the disease along with a significant increase in grain yield (Gitaitis *et al.*, 1986). Jindal and Thind (1990b) have advocated the use of seed treatment with Streptocycline (100 μg mL^{-1}) + Agallol (2000 μg mL^{-1}) followed by three sprays of Streptocycline (100 μg mL^{-1}) + Bavistin (500 μg mL^{-1}) or two sprays of Streptocycline (100 μg mL^{-1}) + Bavistin (500 μg mL^{-1}) followed by a third spraying of Streptocycline (100 μg mL^{-1}) + Blitox 50-WP (2000 μg mL^{-1}). Combination of seed treatment and foliar sprays provided significantly better disease control and also gave a higher grain yield in comparison to seed treatment or spraying schedules alone. Later on, Jindal and Thind (1993) reported that foliar applications of *Bacillus* sp., *Erw. herbicola* and *P. oxalicum* gave significantly better control than Streptocycline (200 μg mL^{-1}), a potent and widely used bactericide.
3. Grow disease-resistant varieties. Two cultivars, namely, Tuv 12349 and Tuv 15549, are resistant to the disease in the field and in the greenhouse in Nigeria (Okechukwu and Ekpo, 2004).

REFERENCES

Burkholder, W. H. (1944). *Xanthomonas vignicola* sp. nov. pathogenic on cowpea and beans. *Phytopathology* 34: 430–32.

Claudius-Cole, A. O., Ekpo, E. J. A. and Schilder, A. M. C. (2014). Evaluation of detection methods for cowpea bacterial blight caused by *Xanthomonas axonopodis* pv. *vignicola* in Nigeria. *Trop. Agric. Res. & Ext.* 17: 77–85.

de Lima-Primo, H. E., de L. Nechet, K., de A. Halfeld-Vieira, B., de Oliveira, J. R., Mizubuti, E. S. G. *et al.* (2015). Epidemiological aspects of cowpea bacterial blight. *Trop. Plant Pathol.* 40: 46–55.

Gena, M., Shah, R., Jadon, K. S. and Mali, B. L. (2009). Host range of *Xanthomonas axonopodis* pv. *vignicola*, the incitant of bacterial blight in cowpea. *Indian Phytopathol.* 62: 539–40.

Gitaitis, R. D., Bell, D. K. and Smittle, D. A. (1986). Epidemiology and control of bacterial blight and canker of cowpea. *Plant Dis.* 70: 187–90.

Jindal, K. K. and Thind, B. S. (1990a). Microflora of cowpea seeds and its significance in the biological control of seed borne infection of *Xanthomonas campestris* pv. *vignicola*. *Seed Sci. & Technol.* 18: 393–403.

Jindal, K. K. and Thind, B. S. (1990b). Control of bacterial blight of cowpea (*Vigna unguiculata* (L.) Walp.). *Trop. Pest Manag.* 36: 211–13.

Jindal, K. K. and Thind, B. S. (1993). Biological control of bacterial blight of cowpea (*Vigna unguiculata* (L.) Walp.). *Phytopathol. Medit.* 32: 193–200.

Jindal, K. K., Thind, B. S. and Soni, P. S. (1989). Physical and chemical agents for the control of *Xanthomonas campestris* pv. *vignicola* from cowpea seeds. *Seed Sci. & Technol.* 17: 371–82.

Khan, A. N. A. (1996). Incidence and perpetuation of bacterial blight of cowpea caused by *Xanthomonas campestris* pv. *vignicola*. *Proc. 9th Int. Conf. Plant Path. Bact.* Chennai, pp. 349–56.

Khatri-Chhetri, G. B., Wydra, K. and Rudolph, K. (2003). Metabolic diversity of *Xanthomonas axonopodis* pv. *vignicola*, causal agent of cowpea bacterial blight and pustule. *Eur. J. Plant Pathol.* 109: 851–60.

Okechukwu, R. U. and Ekpo, E. J. A. (2004). Sources of resistance to cowpea bacterial blight disease in Nigeria. *Phytopathol. Z.* 152: 345–51.

Patel, P. N., Jindal, J. K. and Shekhawat, G. S. (1972). Bacterial blight (canker) of cowpea. In: P. N. Patel (ed.). *Plant Bacteriology*. Vol. I, *Bacterial Diseases of Plants in India*. Indian Agricultural Research Institute, New Delhi, pp. 79–83.

Rachael, D. T., Iheukwumere, C. C., and Lucky, O. (2015). Evaluation of selected cowpea genotypes for resistance to bacterial blight. *Int. J. Curr. Microbiol. App. Sci.* 4: 257–70.

Sikirou, R. and Wydra, K. (2004). Persistence of *Xanthomonas axonopodis* pv. *vignicola* in weeds and crop debris and identification of *Sphenostylis stenocarpa* as a potential new host. *Eur. J. Plant Pathol.* 110: 939–47.

Singh, D. and Patel, P. N. (1977). Studies on resistance in crops to bacterial diseases in India. VII. Inheritance of resistance to bacterial blight disease in cowpea. *Indian Phytopathol.* 30: 99–102.

Soni, P. S. (1982). Detection of phytopathogenic bacteria from cowpea and mung seeds. Ph. D. Thesis. Punjab Agricultural University, Ludhiana, 180 pp.

Soni, P. S. and Thind, B. S. (1987). Detection *of Xanthomonas campestris* pv. *vignicola* (Burkh.) Dye from cowpea seeds and *X. campestris* pv. *vignaeradiatae* (Sabet *et al.*) Dye from mung bean seeds by growing on tests. *Phytopathol. Medit.* 26: 1–6.

Soni, P. S. and Thind, B. S. (1991). Detection of *Xanthomonas campestris* pv. *vignaeradiatae* (Sabet *et al.*) Dye from green gram seeds and *X. campestris* pv. *vignicola* (Burkh.) Dye from cowpea seeds with the help of bacteriophages. *Plant Dis. Res.* 6: 6–11.

Vauterin, L., Hoste, B., Kersters, K. and Swings, J. (1995). Reclassification of *Xanthomonas*. *Int. J. Syst. Bacteriol.* 45: 472–89.

Wydra, K., Khatri-Chhetri, G., Mavridis, A. Sikirou, R. and Rudolph, K. (2004). A diagnostic medium for the semi-selective isolation and enumeration of *Xanthomonas axonopodis* pv. *vignicola*. *Eur. J. Plant Pathol.* 110: 991–1001.

Zandjanakou-Tachin, M., Fanou, A., Le Gall, P. and Wydra, K. (2007). Detection, survival and transmission of *Xanthomonas axonopodis* pv. *manihotis* and *X. axonopodis* pv. *vignicola*, causal agents of cassava and cowpea bacterial blight, respectively, in/by insect vectors. *Phytopathol. Z.*, 155: 159–69.

9.36 BACTERIAL LEAF SPOT OF GREEN GRAM

Green gram (*Vigna radiata* (L.) Wilczek) suffers from two bacterial diseases, namely, bacterial leaf spot and halo blight. Of these two diseases, the former is economically more important and widespread. Bacterial leaf spot of green gram or *mung* bean (local name of green gram in India) was first reported from China by Fang and associates in 1964. In India, its occurrence has been mentioned in the past, but detailed information on certain aspects of the disease has been published for the first time by Patel and his associates (Patel and Jindal, 1972; Patel *et al.*, 1972). The disease is widespread in green gram-growing areas of India. In addition to China and India, it also occurs in Pakistan (Iqbal *et al.*, 2003).

9.36.1 Host Range

Patel and Jindal (1972) reported that, in addition to green gram, the bacterium is moderately virulent on bean, lentil and Indian bean (*Dolichos lablab*), but weakly virulent on lima bean (*Phaseolus lunatus*) and *P. bracteatus*. The progress of infection on bean was not as fast and destructive as with the common and fuscous blight pathogens, the actual pathogens of this host. On bean, the spots were water-soaked but not raised as in green gram; these coalesced and developed into large necrotic patches.

9.36.2 Symptoms

The symptoms on cotyledons of the seedlings, raised from infected seeds, start as pinhead, circular, water-soaked spots, which are lighter green in colour when compared with neighbouring healthy tissue. The spots increase in size, turn brown, become necrotic and raised. In the case of severe infection, the spots may coalesce to cover the whole leaf. Longitudinal lesions, slits or cankers extending from soil level to the growing tip develop on the stems of seedlings grown from severely infected seeds.

On trifoliate leaves, circular to irregular, brown and distinctly raised spots appear. Generally, the spots are dry and necrotic from the beginning. In the beginning, these appear on either surface as superficial eruptions, but they gradually develop through the entire thickness of the leaflet and become corky and rough. After attaining some height, the eruptions begin to collapse due to the disintegration of infected cells, resulting in the formation of rifts. Sometimes, the rifts are also caused as a result of the disintegration of leaf tissue by the action of wind. In older spots, the superficial eruptions collapse, leaving behind saucer-shaped depressions. In a few cases, the shot-hole symptoms are also observed. Generally, the spots are isolated; however, in severe infections, these coalesce to form necrotic patches but they never develop into a typical leaf blight phase as in common blight of bean (Figure 9.35a and b). Lesions on petioles, petiolules or pulvinus are brown, flat or raised and may be elongated in the form of long streaks. Normally, the infected plants do not defoliate.

On stems, dark brown and elongated lesions are produced. In severely infected plants, dark brown, longitudinal slits or cankers develop on the stems, which often extend from soil level to the growing tip.

The lesions on pods are brown and raised, but they are rarely seen in nature. In artificially inoculated pods, brown and raised lesions appear. Sometimes, a thin layer of shining gummy bacterial ooze can also be seen on the outer surface. On the inner side of green pods, brownish lesions are formed just beneath the bacterial ooze. The seeds touching these lesions also develop brown spots. Seeds produced in the infected pods are generally small, shrivelled and discoloured. Most of the discoloured seeds are brown and yellowish, but they are also occasionally black. The normal or proper disease development occurs only on the monsoon season crop. In the summer crop, the symptoms develop on the cotyledonary and first trifoliate leaves, but subsequent spread of the disease does not occur.

As the pathogen is largely parenchymatous, infected plants are not completely killed. However, severe infections result in the necrosis of considerable leaf area, leading to a reduction in photosynthesis and a general weakening of the plants.

9.36.3 Causal organism

Xanthomonas axonopodis pv. *vignaeradiatae* (Sabet *et al.* 1969) Vauterin *et al.* 1995

Synonym: *Xan. campestris* pv. *vignaeradiatae*

The description of *Xan. axonopodis* Starr and Garces 1950 emend. Vauterin *et al.* 1995 is given below. Cells are straight rods, Gram-stain-negative, non-sporing and measure 0.4–0.6 × 0.8–2.0 µm; occur mostly singly or in pairs, occasionally short chains and rarely in filaments. Motile by a single polar flagellum, obligately aerobic and metabolism is strictly respiratory type. Colonies are usually yellow, smooth and butyrous, viscid or mucoid. Yellow pigments, xanthomonadins, are produced. Viscous or mucoid growth of the culture is due to the production of an extracellular heteropolysaccharide called xanthan.

(a) (b)

FIGURE 9.35 Pertains to bacterial leaf spot of green gram, (a) shows symptoms on upper leaf surface and (b) shows symptoms on lower surface of leaf.

Catalase is positive and oxidase is negative or weak. Urease and indole are not produced and nitrate is not reduced to nitrite. Litmus milk is not acidified. Most of the strains utilise D-galactose, melibiose, acetic acid, DL-lactic acid, malonic acid, L-leucine, L-serine and glycerol.

The members of this species are slow-growing and metabolically less active than those of *Xan. campestris*. Both these species can be clearly differentiated on the basis of phenotypic characters. *Xan. axonopodis* is the most heterogeneous species currently recognised in genus *Xanthomonas* and includes several pathovars. In general, these pathovars are not distinguishable by their phenotypic characters and their identification is reliable on the basis of their host range.

Xan. axonopodis pv. *vignaeradiatae*, being a pathovar of *Xan. axonopodis*, has the same morphological, cultural and biochemical characters as those of the latter. In addition to above mentioned characteristics, *Xan. axonopodis* pv. *vignaeradiatae* produces H_2S, liquefies gelatin, hydrolyses starch and ferments glucose, sucrose, lactose, mannose, xylose and raffinose producing only acid (Patel and Jindal, 1972). Thakur *et al.* (1977) confirmed the existence of six races in the bacterium. They further reported that (1) resistance in a differential to a race is governed by a single gene, (2) a gene governing resistance in a differential is dominant to its recessive allele governing susceptibility in another differential and (3) that a gene or allele governing resistance or susceptibility in the differentials to different races does not overlap.

Patro and Jindal (2000) have reported the existence of plasmids in the bacterium and their correlation with virulence. The highly virulent isolate contained the maximum number of plasmids. The plasmid-borne genes were shown to govern virulence, exopolysaccharide and pigment production and antibiotic resistance.

The bacterium produces exopolysaccharide, proteolytic and cellulolytic enzymes, which play a role in pathogenesis. Higher proteolytic and cellulolytic activities were observed in susceptible green gram cv. Pusa Baisakhi after 120 h of inoculation with a virulent culture (Dutta and Jindal, 2002).

9.36.4 Disease Cycle

Earlier, the bacterium was assumed to be seed-borne, but Soni (1982) established the seed-borne nature of the bacterium and showed that the seed is an important source of survival for the bacterium. The author further reported that the bacterium survived for one year in the seeds and the inoculation of the bacterium only at inflorescence and at the flat green pod stage led the bacterium to become seed-borne. The deformity of seeds, such as shrivelling, discolouration and small size, was only an indication, but not a sure test of seed-borne infection.

Soni and Thind (1987) compared different growing on tests for the detection of *Xan. axonopodis* pv. *vignaeradiatae* from green gram seeds and showed that traditional growing on the test was more suitable in the monsoon season when the conditions were highly favourable for disease development while incubation growing on the test proved more effective in the summer crop season. They further reported that wounded seed growing on test was most effective for detecting the lowest level of seed-borne infection and it detected 0.1% infected seeds in a seed lot. Bacteriophages have also been found effective for the detection of bacterium from shrivelled and discoloured green gram seeds (Soni and Thind, 1991).

The bacterium enters the host through stomata and wounds, and largely remains confined to the parenchyma tissue. Soni (1982), on the basis of recovery of the bacterium only from roots, primary and trifoliate leaves, suggested the non-systemic nature of the bacterium. Secondary spread of the bacterium is mainly through wind-splashed rain. Dissemination to distant places and new areas is mainly through the infected seed.

Most of the commercially grown varieties in India, including Pusa Baisakhi, are susceptible to the bacterium while Jalgaon 781 is resistant.

9.36.5 Management

1. As the bacterium perpetuates through infected seed, the use of bacteria-free seed and the eradication of seed-borne inoculum are the most effective and viable methods of disease management.
2. Seed treatment with hot water at 52 °C for 30 min, solar heat (for details see bacterial blight of cowpea), Streptocycline (100 µg mL^{-1}) + captan (0.2%) or Streptocycline (100 µg mL^{-1}) + Agallol (0.2%) have been found to be very effective for the eradication of seed-borne inoculum (Jindal and Thind, 1987).
3. For the control of secondary spread of the bacterium, four sprays, i.e., first two sprays of Streptocycline (100 µg mL^{-1}) + Bavistin (0.05%) followed by two sprays of Streptocycline (100 µg mL^{-1}) + Blitox (0.2%), at 10 days intervals, have been found effective by Jindal and Thind (1987). The authors also reported that the combinations of these sprays with hot water or solar heat treatments of seed proved most effective in managing the disease and caused a 69.9 and a 72.5% reduction in disease intensity, respectively. Gour and Ahmed (1983) recommended three sprays of Agrimycin–100 (0.025%) or Plantomycin (0.025%) + Bavistin (0.1%) for the control of the disease. Borah and associates (2000) have advocated the use of a phylloplane bacteria, *Bacillus* sp., for the control of disease and have recommended its preventive application. Rathore (2010), on the basis of three years of pooled data, reported that streptocycline (0.01%) sprays gave the minimum disease intensity (6.1%) and highest grain and fodder yields (10.28 and 20.59 q/ha) compared to 24.9% disease intensity, 5.76 q/ha grain yield, and 12.18 q/ha fodder yield obtained in control. Choudhary *et al.* (2019), on the basis of evaluation of different fungicides and antibiotics in pot experiments, obtained 89.72% disease control with Labistryn followed by Streptocyline (85.12%).

REFERENCES

Borah, P. K., Jindal, J. K. and Verma, J. P. (2000). Biological management of bacterial leaf spot of mungbean caused by *Xanthomonas axonopodis* pv. *vignaeradiatae*. *Indian Phytopathol.* 53: 384–94.

Choudhary, M., Singh, K., Khokhar, M. K., Rawal. P. and Shah, R. (2019). Management of bacterial leaf spot of greengram caused by *Xanthomonas axonopodis* pv. *vignaeradiatae*. *J. Mycol. & Pl. Pathol.* 49: 67–72.

Dutta, S. and Jindal, J. K. (2002). Hydrolytic enzymes and exo-polysaccharide production of *Xanthomonas axonopodis* pv. *vignaeradiatae* inciting bacterial leaf spot of mungbean. *Annl. Plant Prot. Sci.* 10: 295–98.

Gour, R. B. and Ahmed, S. R. (1983). Studies on the chemical control of bacterial blight of moong incited by *Xanthomonas phaseoli* (Smith) Dowson. *Pesticides* 17: 22–23.

Iqbal, S. M., Zubair, M., Anwar, M. *et al.* (2003). Resistance in mungbean to bacterial leaf spot disease. *Mycopathologia* 1: 81–83.

Jindal K. K. and Thind, B. S. (1987). Control of Cercospora and bacterial leaf spot of green gram. *Indian J. Agric. Sci.* 57: 372–75.

Patel, P. N. and Jindal, J. K. (1972). Bacterial leaf spot and halo blight diseases of mung bean (*Phaseolus aureus*) and other legumes in India. *Indian Phytopathol.* 25: 517–25.

Patel, P. N., Jindal, J. K. and Singh, D. (1972). Studies on resistance in crops to bacterial diseases in India. IV. Resistance in mung bean (*Phaseolus aureus*) to *Xanthomonas phaseoli*. *Indian Phytopathol.* 25: 526–29.

Patro, T. S. S. K. and Jindal, J. K. (2000). Expression of plasmid-borne characters in *Xanthomonas axonopodis* pv. *vignaeradiatae*. *Trop. Agr. Res.* 12: 86–94.

Rathore, B. S. (2010). Efficacy of streptocycline and plant extracts against bacterial leaf spot disease caused by *Xanthomonas axonopodis* pv. *vignaradiatae* of greengram. *Indian Phytopathol.* 63: 384–86.

Sabet, K. A., Ishag, F. and Khalil, O. (1969). Studies on the bacterial diseases of Sudan crops VII. New records. *Ann. Appl. Biol.* 63: 357–69.

Soni, P. S. (1982). Detection of phytopathogenic bacteria from cowpea and mung seeds. Ph.D. Thesis. Punjab Agricultural University, Ludhiana, 180 pp.

Soni, P. S. and Thind, B. S. (1987). Detection *of Xanthomonas campestris* pv. *vignicola* (Burkh.) Dye from cowpea seeds and *X. campestris* pv. *vignaeradiatae* (Sabet *et al.*) Dye from mung bean seeds by growing on tests. *Phytopathol. Medit.* 26: 1–6.

Soni, P. S. and Thind, B. S. (1991). Detection of *Xanthomonas campestris* pv. *vignaeradiatae* (Sabet *et al.*) Dye from green gram seeds and *X. campestris* pv. *vignicola* (Burkh.) Dye from cowpea seeds with the help of bacteriophages. *Plant Dis. Res.* 6: 6–11.

Starr, M. P. and Garces, O. C. (1950). El agente causante de la gomosis bacterial del pasto imperial en Colombia. *Revista FacultadNacional de Agronomia (Medellin)* 11: 73–83.

Thakur, R. P., Patel, P. N. and Verma, J. P. (1977). Studies on resistance in crops to bacterial diseases in India. XI. Genetic make-up of mungbean differentials of the races of bacterial leaf spot pathogen, *Xanthomonas phaseoli*. *Indian Phytopathol.* 30: 217–21.

Vauterin, L., Hoste, B., Kersters, K. and Swings, J. (1995). Reclassification of *Xanthomonas*. *Int. J. Syst. Bacteriol.* 45: 472–89.

9.37 HALO BLIGHT OF GREEN GRAM

Green gram or *mung* bean (vernacular name of green gram in India) suffers from four bacterial diseases: bacterial leaf spot caused by *Xanthomonas axonopodis* pv. *vignaeradiatae*; tan spot caused by *Curtobacterium flaccumfaciens* pv. *flaccumfaciens*; halo blight caused by *Pseudomonas savastanoi* pv. *phaseolicola*; and bacterial blight caused by *Xan. axonopodis* pv. *phaseoli*. In India, halo blight is comparatively of less economic importance than bacterial leaf spot.

Patel and Jindal (1970) observed halo blight for the first time in 1969 at the Indian Agricultural Research Institute, New Delhi when they found it on some *mung* bean lines resistant to bacterial leaf spot. Subsequently, Schmitthenner *et al.* (1971) reported a severe epiphytotic of the disease from Ohio in the USA.

During 2009–2014, halo blight of *mung* bean caused by *Psm. syringae* pv. *phaseolicola* was found in three of the main Chinese geographic regions, which were the sites of more than 90% of the *mung* bean-growing areas in China. The identity of the bacterium was confirmed on the basis of phenotypic characteristics, the physiological and biochemical properties, pathogenicity tests and fatty acid composition, in combination with specific PCRs and 16S–23S ribosomal DNA sequence analyses. The results indicate that the bacterium is likely of epidemiological significance on *mung* bean in China (Sun *et al.*, 2017). The disease also occurs in Australia, but the bacterium causing the disease has been identified as *Psm. savastanoi* pv. *phaseolicola*. In Australia, it was first reported from Queensland in the early 1980s (Ryley, 2008). Akhtar (1988) reported the occurrence of the disease from Pakistan.

Noble (2020) investigated the population genetics of *Psm. savastanoi* pv. *phaseolicola*, causing halo blight of mung bean (*Vigna radiata*). The pathogen and host populations were investigated at a board scale using field and glasshouse studies and in detail using molecular biology techniques, including qPCR and next-generation sequencing. The study found both the bacterial pathogen and the mung bean to have highly conserved genetic backgrounds. This will make it easier for breeders to target critical resistance genes to prevent the infection of the halo blight pathogen in future cultivars.

Sun *et al.* (2017) investigated the geographic distribution of halo blight of mung bean in China using consecutive multiyear field surveys and to confirm the identity of the causal agents. Between 2009 and 2014, surveys were conducted in 15 provinces covering seven geographic regions. The survey results revealed that the emerging mung bean disease has rapidly spread and it is prevalent in three of the main Chinese geographic regions, which contain more than 90% of the mung bean-growing areas in China. Based on the phenotypic characteristics, physiological and biochemical properties, pathogenicity tests, and fatty acid composition, in combination with specific PCRs and 16S–23S ribosomal DNA sequence analyses, the bacterium was identified as *Psm. syringae* pv. *phaseolicola*. According to authors' knowledge, this is the first report of *Psm. syringae* pv. *phaseolicola* causing halo blight on the mung bean in China. The findings also indicate that *Psm. syringae* pv. *phaseolicola* is likely to be of epidemiological significance on the mung bean in China.

9.37.1 Host Range

Schmitthenner *et al.* (1971) reported susceptible reaction on red kidney bean (*Phaseolus vulgaris*), lima bean (var. Fordhook 242) and soybean (var. Acme), in addition to *mung* bean. Schroth *et al.* (1971) also found that *mung* strain of pathogen could infect few varieties of bean, lima bean and soybean. Patel and Jindal (1972a) found that the infection on *Phaseolus atropurpureus*, *P. bracteatus*, *Dolichos lablab* and *D. biflorus* was as severe as on *mung* bean. On *P. mungo*, *P. trilobus* and *Lathyrus sativus*, only a few spots with distinct haloes were produced. Among the non-legumes, *Ficus religiosa* was susceptible. Only *P. bracteatus* and *P. atropurpureus* are susceptible to both the pathogens (causal agents of common bean and *mung* bean halo blights). Bean plants showed hypersensitive reaction within 24 h when inoculated with the *mung* bean pathogen. The bean bacterium attacked 10 of the 15 *Phaseolus* spp. and 2 other legumes, whereas the *mung* bean pathogen infected only 5 of 13 *Phaseolus* spp., 4 other legumes and a non-legume (Patel and Jindal, 1972a). *Mung* bean strain of the bacterium has a distinct host range than those of bean strain and they do not cross-infect (Patel and Walker, 1965).

9.37.2 Symptoms

On the seedlings developing from infected seeds, water-soaked spots first appear on cotyledonary leaves and then spread to the trifoliate leaves. Severe infection at this stage may result in the death of seedlings. On trifoliate leaves, the spots may be small or large, round or irregular, flat and water-soaked. Spots are surrounded by halo-like zones of greenish-yellow tissue, which result from the toxin produced by the bacteria multiplying in the water-soaked lesions. Young developing leaves on severely infected plants show systemic chlorosis due to the translocation of the toxin. Formation of halo zones and the development of systemic chlorosis are favoured by low temperature (15–20 °C) at nights, as in halo blight of common bean. The spots on old leaves are sometimes angular and do not show a distinct halo. Frequently, the infection spreads along the veins and veinlets turning them brown and necrotic and causing the distortion of the developing leaflets.

On petioles, long streaks appear which are first water-soaked but later on turn red. Pulvinus and petiolules of the severely infected leaves appear brown and sunken. Generally, there is no defoliation of the plants. On the stem, first water-soaked spots appear which gradually enlarge and turn into reddish streaks extending longitudinally. Such stems often split. On pods, water-soaked lesions appear and turn brown with time.

9.37.3 Causal Organism

Pseudomonas savastanoi pv. *phaseolicola mung* bean strain (Burkholder 1926) Gardan *et al.* 1992

Synonym: *Psm. phaseolicola mung* bean strain, *Psm. syringae* pv. *phaseolicola mung* bean strain

The first isolation of the *mung* bean halo blight pathogen appears to have been made by D. C. Graham of Scotland in 1966 from *mung* bean seeds. He named the bacterium *Pseudomonas phaseolicola* but presumably did not work with it further. Schroth *et al.* (1971) compared this culture with *Psm. phaseolicola* from different hosts and found it to be distinct from bean isolates in the utilisation of several nutritional substrates and host range. Schmitthenner *et al.* (1971) found that the *mung* bean bacterium was similar to both *Psm. phaseolicola* and *Psm. glycinea* in cultural characteristics. As only the *mung* bean isolates were virulent to *mung* bean, they proposed that the *mung* bean bacterium be named the *mung* bean strain of *Pseudomonas phaseolicola*. Keeping in view the host range and the overall similarity in bacteriological properties of the *mung* halo blight pathogen to that of the bean halo blight pathogen, Patel and Jindal (1972a) also identified it as *mung* strain of *Psm. phaseolicola*. Young *et al.* (1978) reclassified it as *Psm. syringae* pv. *phaseolicola mung* bean strain. Later on, Gardan *et al.* (1992) reclassified it as *Psm. savastanoi* pv. *phaseolicola mung* bean strain.

The below-given description of the *mung* bean halo blight bacterium is based on the studies carried out by Patel and Jindal (1972b).

Colonies on nutrient agar are circular, white, glistening, flat, entire and produce green fluorescent pigment on King's medium B. The cells are Gram-stain-negative rods, motile with 1–4 polar flagella and strictly aerobic. The bacterium utilises glucose oxidatively, utilises asparagine and can grow in the presence of 4.0% sodium chloride. Hydrolysis of gelatin is slow and milk is peptonised without curd formation. The tests for production of ammonia, urease and catalase and starch hydrolysis are positive while those for the production of indole, acetoin and H_2S, and reduction of nitrate are negative. The bacterium utilises sodium propionate and succinate but not sodium oxalate and tartrate.

Ryley (2008) reported that there are at least nine strains (races, pathotypes) of *Psm. savastanoi* pv. *phaseolicola* (causal agent of halo blight of *mung* bean in Australia), which has serious implications for resistance breeding; there are varieties that are immune to some strains but may be susceptible to others.

9.37.4 Disease Cycle

The bacterium is seed-borne, as reported by Schmitthenner *et al.* (1971), and its spread to different places occurs through seed. There is little information available on this disease and it is presumed that the disease cycle is similar to that of bean halo blight. The disease assumed serious proportions in Ohio, USA and the crop raised from infected seed gave a 60% less yield than the crop raised from healthy seed (Schmitthenner *et al.*, 1971). To date, the disease has not appeared in a severe form in India. Hence, there is a need to investigate, if the environmental or other factors are responsible for its low incidence in India, particularly in *Kharif* crop.

Noble *et al.* (2020) analysed the phenotypic and genotypic variation among 511 *Psm. savastanoi* pv. *phaseolicola* (*Psp*) isolates, causing halo blight in mung bean throughout Australia. *In planta* screening against a set of four mung bean cultivars with known susceptible and resistant reactions revealed five distinctive pathotypes. Isolates belonging to pathotype 2 were the most prevalent at 84% and were found to be highly pathogenic towards all tested mungbean genotypes. Further genetic testing for two known avirulence genes, *avrPphE* and *avrPphF*, identified the *avrPphE* gene in all the tested isolates and *avrPphF* present in only two. To identify candidate avirulence genes unique to *Psp* isolates infecting mung bean in Australia, a comparative genomics study was undertaken on the whole-genome sequences of two epidemiologically important *Psp* isolates, T11544 and K4287. This information has the potential to dramatically improve mungbean disease resistance.

Noble *et al.* (2022) optimised quantitative PCR (qPCR), SYBR assay, hydrolysis probe and conventional PCR, using the same primers against a plate-truthed dilution series of *Psm. savastanoi* pv. *phaseolicola*. The detection limit of the conventional PCR assay was approximately 9,000 CFU μL^{-1}, while both qPCR assays could detect 9 CFU μL^{-1}. These tests were then used to screen DNA extracted from 200 g allotments of 38 seed lots comprising six mungbean cultivars representing the primary Australian production area, and two seed lots of known infection status. From these lots, the pathogen was detected in six seed lots by

stagnates temporarily. It is also severe in furrow-sown fields where water contaminated with the pathogen moves from one end to another end of the furrow in muddy fields.

Shah *et al.* (1996) found that a temperature range of 20–28 °C and a relative humidity between 70 and 80% favoured the rapid progress of the disease. On the basis of epidemiological studies carried out for five years, from 1991–1995, they also developed a model to forecast the disease based on these parameters of the preceding week. Yadav and Nath (2006a) reported that a temperature range of 29–34 °C and relative humidity above 68% favoured the disease. They further reported that the maximum disease incidence of 58.3% was observed in the crop sown in the last week of July.

Kalaskar *et al.* (2014) investigated the activity of isoenzymes, peroxidase (PO) and polyphenoloxidase (PPO) in healthy and bacterial blight infected samples of the 12 genotypes of cluster bean. Range of 0.42 to 1.07 and 0.08 to 0.36 in OD/min/g of fresh weight tissues in healthy plants and of 0.19 to 1.20 and 0.09 to 0.52 in OD/min/g of fresh weight tissues in bacterial blight-infected plants was observed for PO and PPO activity, respectively. A decrease in the PO and PPO activity in bacterial blight-infected susceptible plants compared to bacterial blight-infected-resistant plants was observed. A decrease in the activity of PO (from 0.75 to 0.19) and of PPO (from 0.22 to 0.09) was at its highest in the most susceptible genotype, PNB. Authors suggest that an increase or no change in PO activity after the infection can be considered as a marker for resistance to bacterial blight.

Most of the improved varieties of cluster bean in India, such as Pusa Naubahar and Pusa Sadabahar, are highly susceptible to the disease. Branching-type varieties are also susceptible, but they suffer less from the disease because the new shoots compensate, to some extent, for the dead shoots. In addition to *guar*, bean acts as a natural host of the bacterium (Patel and Jindal, 1972).

Kaur *et al.* (2004) evaluated ten genotypes under field conditions and found that cv. HG 75 was highly resistant and cv. PNB was most susceptible to the pathogen.

9.38.4 Management

1. Use of pathogen-free seed and the eradication of seed-borne inoculum are the most effective and practical means of disease management. Seeds from surface-irrigated summer-grown crops are free from infection (Srivastava and Rao, 1963b, 1970).
2. Hot water treatment at 56 °C for 10 min eradicates the seed- borne inoculum. Seed dip in Streptocycline solution (200 µg mL^{-1}) for 3 h proved highly effective for eradicating the seed-borne inoculum (Rathore, 2000). Seed treatment with Streptocycline (500 µg mL^{-1}), followed by its foliar applications, gave minimum disease incidence and highest grain yield (Yadav and Nath, 2006b). Amin *et al.* (2017), on the basis of field trials conducted during *kharif* 2013 to 2015, reported that the soaking of seed in Streptocycline (250 µg mL^{-1}) solution for 30 min. plus a first spray of Streptocycline at 250 µg mL^{-1}, just at the time of the appearance of the disease and a second spray on the 15th day after the 1st spray, gave the maximum yield and minimum disease intensity. This treatment was closely followed by seed soaking in Streptocycline at 250 µg mL^{-1} for 30 min. plus 1st spray of Streptocycline at 200 µg mL^{-1} just at the disease appearance and the 2nd spray on the 15th day after the 1st spray.
3. Godara and Singh (2021) reported that seed soaking in Streptocycline (500 µg mL^{-1}) solution for one hour, followed by two sprays of Streptocycline (250 µg mL^{-1}) + copper oxychloride (0.2%) at a 15-day interval gave the minimum disease intensity (14.83%) and the highest grain yield (11.18 q/ha), along with a net return of Rs 16,270/ha.
4. Joshi *et al.* (2023) evaluated five bioagents and six chemicals against *Xan. axonopodis* pv. *cyamopsidis in vitro*. Among the bio-agents, *Pseudomonas fluorescens* was found to be the most effective with a maximum inhibition zone of 49.50 mm. Of the six antibacterial chemicals, Streptocycline was found to be the most effective at all the concentrations tested, with a maximum inhibition zone of 64.33 mm at 500 µg mL^{-1}. In the pot experiment, hot water treatment followed by seed biopriming with *Psm. fluorescens* and then spraying with Streptocycline (250 µg mL^{-1}) was found to be the most effective with the lowest disease intensity (10.11%), the highest disease reduction over control (68.69%), and the highest seed germination (93.33%) as well as the maximum plant length, at 18 cm and 15.50 cm for the healthy and the diseased plants, respectively.
5. Fresh leaf extracts of *Emblica officinalis* and *Eucalyptus* sp. have been demonstrated to be effective in reducing the disease under field conditions (Kaur *et al.*, 2005, 2006).
6. Use of resistant varieties.

REFERENCES

Amin, A. M., Patel, N. R., Jaiman, R. K., Prajapati, D. B. and Prajapati, B. G. (2017). Management of bacterial blight in clusterbean. *Environ. Ecol.* 35(2C): 1223–27.

Bagri, R. K., Goyal, S. K., Singh, J. and Shekhawat, S. S. (2019). Management of bacterial leaf blight disease of cluster bean in semi-arid tropic of Rajasthan. *Plant Dis. Res.* 34: 58–60.

Godara, S. L. and Singh, N. (2021). Management of bacterial leaf blight (*Xanthomonas axonopodis* pv. *cymopsidis*) of clusterbean in Rajasthan. *Indian Phytopathol.* 74: 223–27. https://doi.org/10.1007/s42360-020-00299-8

Joshi, S. H., Chattopadhyay, A., Singh, N. K., Jaiman, R. S. and Purohit, J. (2023). Bio-intensive management of

Xanthomonas axonopodis pv. *cyamopsidis* causing bacterial blight of clusterbean. *J. Mycol. & Pl. Pathol*. 53: 412–20.

Kalaskar, S. R., Shinde, A. S., Dhembre, V. M., Sheikh, W. A., Patil, V. S., *et al*. (2014). The role of peroxidase and polyphenoloxidase isoenzymes in resistance to bacterial leaf blight (*Xanthomonas campestris* pv. *cyamopsidis*) in clusterbean (*Cyamopsis tetragonoloba* (L.) Taub.). *J. Cell Tissue Res*. 14: 4577–80.

Kaur, B., Purkayastha, S., Dilbaghi, N. and Chaudhury, A. (2004) Evaluation of clusterbean genotypes for resistance to bacterial blight. *Ann. Agri. Bio. Res*. 9: 213–15.

Kaur, K., Jalali, I. and Chhabra, M. L. (2005). Efficacy of botanicals for the management of bacterial blight of clusterbean under disease stress conditions. *Proc. Natl. Symp. Plant Disease Diagnosis, Epidemiology and Management*, Hisar, 47 pp.

Kaur, K., Jalali, I. and Pal, V. (2006). Factors affecting the bioefficacy of phyto-extracts against *Xanthomonas axonopodis* pv. *cyamopsidis*. *Pl. Dis. Res*. 21: 164–67.

Kaur, K., Jalali, I. and Pal, V. (2010). Toxicity of botanicals against clusterbean bacterial blight pathogen. *J. Mycol. Pl. Pathol*. 40: 124–27.

Patel, A. J. and Patel, M. K. (1958). A new bacterial blight of *Cyamopsis tetragonaloba* (L.) Taub. *Curr. Sci*. 27: 258–59.

Patel, M. K., Dhande, G. W. and Kulkarni, Y. S. (1953). Bacterial leaf spot of *Cyamopsis tetragonaloba* (L.) Taub. *Curr. Sci*. 22: 183.

Patel, P. N. and Jindal, J. K. (1972). Bacterial blight of guar. In: P. N. Patel (ed.). *Plant Bacteriology*. Vol. I, *Bacterial Diseases of Plants in India*. Indian Agricultural Research Institute, New Delhi, pp. 66–69.

Rathore, B. S. (2000). Management of bacterial blight of clusterbean through seed dressers. *Plant Dis. Res*. 15: 89–92.

Ren, Y. Z., Yue, Y. L., Jin, G. X. and Du, Q. (2014). First report of bacterial blight of guar caused by *Xanthomonas axonopodis* pv. *cyamopsidis* in China. *Plant Dis*. 98: 840.

Shah, R., Dashora, P. K., Bhatnagar, M. K. and Mali, B. L. (1996). Forecasting model for bacterial blight of clusterbean. *Proc. 9th Int. Conf. Plant Path. Bact*. Chennai, pp. 525–27.

Srivastava, D. N. and Rao, Y. P. (1963a). Bacterial blight of guar. *Indian Phyto-pathol*. 16: 69–73.

Srivastava, D. N. and Rao, Y. P. (1963b). Seed transmission and control of bacterial blight of guar (*Cyamopsis tetragonaloba* (L.) Taub.). *Indian Phytopathol*. 16: 390–91.

Srivastava, D. N. and Rao, Y. P. (1970). Epidemiology and control of bacterial blight of guar (*Cyamopsis tetragonaloba* (L.) Taub.). *Indian Phytopathol. Soc. Bull. No* 6: 1–4.

Vauterin, L., Hoste, B., Kersters, K. and Swings, J. (1995). Reclassification of *Xanthomonas*. *Int. J. Syst. Bacteriol*. 45: 472–89.

Yadav, S. and Nath, R. (2006a). Effect of weather parameters and different sowing times on the occurrence of clusterbean bacterial blight. *J. Mycopathol. Res*. 44: 267–69.

Yadav, S. and Nath, R. (2006b). Efficacy of chemicals against bacterial blight of clusterbean. *J. Mycopathol. Res*. 44: 271–73.

9.39 BACTERIAL LEAF SPOT OF HYACINTH BEAN

Lablab bean, hyacinth bean, Indian bean or Indian butter bean (*Lablab purpureus*, syn. *Dolichos lablab*, *Lablab niger*), is also given another name in India, i.e., *sem phali*. The first natural occurrence of bacterial leaf spot on hyacinth bean was reported by Rangaswami and Gowda (1963) from Tamil Nadu, India. Subsequently, Patel and Jindal (1970) found it at the Indian Agricultural Research Institute, New Delhi and Amravati in Maharashtra, India. The disease also occurs in Sudan (Sabet, 1959).

9.39.1 Host Range

Rangaswami and Gowda (1963) reported that the bacterium infected *D. biflorus*, *Vigna catjang*, *Phaseolus vulgaris*, *Tophrosia purpurea* and *Cyamopsis psoralioides*. However, Patel and Jindal (1973) did not obtain infection on the above-mentioned hosts, with the exception of *P. vulgaris*. Susceptibility of *D. lablab* to *Xan. fuscans* subsp. *fuscans* and *Xan. axonopodis* pv. *glycines* has been reported. Whether or not *D. lablab* serves as a natural host of these pathogens is not known.

9.39.2 Symptoms

The disease is more severe during the rainy season. The symptoms appear mainly on leaves, in the form of round or irregular, water-soaked spots. These spots eventually turn brown. Several such spots may coalesce. In relatively dry weather, tissues surrounding closely situated spots may turn yellow and form large blotches, but this never leads to extensive blight. Due to severe spotting, leaves turn yellow and fall off prematurely. On stems, oblong, reddish-brown lesions appear, sometimes with greyish centres. On pods, initially the spots are water-soaked; later, however, they turn brown.

9.39.3 Causal Organism

Xanthomonas axonopodis pv. *phaseoli* hyacinth bean strain (Smith 1897) Vauterin *et al*. (1995)

Synonym: *Xan. phaseoli* hyacinth bean strain

9.39.4 Disease Cycle

The bacterium is seed-borne and also survives on vines growing perennially in the kitchen gardens. Warm and wet weather prevailing during the monsoon season favour the disease development. In hyacinth bean or Indian bean, the bacterium is largely confined to parenchymatous tissue while in the common bean (*P. vulgaris*), it enters the vascular system. New growth on infected vines is generally free from infection under dry conditions. A considerable reduction in yield occurs when plants are badly infected or disease severity is high in monsoon season. Seedling mortality occurs very rarely (Patel and Jindal, 1972).

Patel and Jindal (1973) evaluated 334 germplasm lines against the pathogen and found none of them to be immune or highly resistant. However, PLS 63, PLS 101, D.L. 2434 and M.S. 9491 showed moderate resistance as the lesions remained small and turned necrotic and the infected leaves did not defoliate.

9.39.5 Management

1. As the bacterium is seed-borne, the use of pathogen-free seed or treatment of infected seed can prove useful. The seed treatment recommended for bacterial blight of cowpea can also prove effective for this disease.
2. As the bacterium is mainly confined to the parenchyma, foliar sprays of Streptocycline @ 100 µg mL^{-1} can also prove effective.

REFERENCES

Patel, P. N. and Jindal, J. K. (1970). Bacterial diseases in seed legumes in 1968 and 1969. *Paper presented at the Fourth Annual Workshop Conference on Pulse Crops held at Ludhiana (Punjab) in April, 1970.*

Patel, P. N. and Jindal, J. K. (1972). Bacterial leaf spot of *Dolichos lablab*. In: P. N. Patel (ed.). *Plant Bacteriology*. Vol. I, *Bacterial Diseases of Plants in India*. Indian Agricultural Research Institute, New Delhi, pp. 84–86.

Patel, P. N. and Jindal, J. K. (1973). Studies on resistance in crops to bacterial diseases in India. VI: Reaction of moth bean (*Phaseolus acontifolius*) and hyacinth bean (*Dolichos lablab*) lines to bacterial blight and leaf spot diseases incited by *Xanthomonas phaseoli* strains. *Indian Phytopath.* 26: 607–08.

Rangaswami, G. and Gowda, S. S. (1963). On some bacterial diseases of ornamentals and vegetables in Madras state. *Indian Phytopathol.* 16: 74–85.

Sabet, K. A. (1959). Studies in the bacterial diseases of Sudan crops. III. On the occurrence, host range and taxonomy of the bacteria causing leaf blight diseases of certain leguminous plants. *Ann. Appl. Biol.* 47: 318–31.

Smith, E. F. (1897). Description of *Bacillus phaseoli* n. sp. *Botanical Gazette* 24: 192.

Vauterin, L., Hoste, B., Kersters, K. and Swings, J. (1995). Reclassification of *Xanthomonas*. *Int. J. Syst. Bacteriol.* 45: 472–89.

9.40 BACTERIAL LEAF SPOT AND STEM CANKER OF PIGEON PEA

Bacterial leaf spot and stem canker of pigeon pea (*Cajanus cajan*), also called *arhar* in India, was first reported from Bombay state by Kulkarni *et al.* (1950). Later on, it was also reported from other states, namely, Tamil Nadu, Punjab, Uttar Pradesh, Madhya Pradesh, Delhi and Haryana. The disease also occurs in China and Sudan.

9.40.1 Symptoms

The symptoms first appear as small, round, light brown, water-soaked spots on the upper surface of leaves, which later on turn dark brown, become angular in shape and are surrounded by haloes. The upper surface of the spots becomes raised due to exudation and the drying of the bacterial ooze. The spots may coalesce to form large lesions of 2 mm in diameter. Due to excessive spotting, the leaves turn yellow and defoliate prematurely. In the case of severe infection, the spots appear all over the leaf surface, with the infection extending to main and lateral veins, and petioles; resulting in yellowing and ultimately shedding of the leaves. On the main stem and lateral branches, dark brown cankers measuring 12.5 mm long and 2.5 mm wide appear. When these cankers are numerous and close to each other, they cause the peeling of the bark. Generally, the cankers on the stem are not very deep and do not cause girdling of the stems. A gummy bacterial exudate emerges from these cankers. An early and severe infection may cause the death of the seedlings (Patel and Jindal, 1972).

9.40.2 Causal Organism

Xanthomonas axonopodis pv. *cajani* (Kulkarni *et al.* 1950) Vauterin *et al.* 1995

Synonym: *Xan. campestris* pv. *cajani*

Kulkarni *et al.* (1950) described the bacterium for the first time and named it *Xanthomonas cajani* sp. nov. Dye (1978) changed it to *Xan. campestris* pv. *cajani* and, later, Vauterin *et al.* (1995) reclassified it as *Xan. axonopodis* pv. *cajani*. The below-given description of the bacterium is based primarily on the publication of Kulkarni *et al.* (1950).

Cells are short rods (0.9–1.4 × 1.3–2.2 µm). Gram-stain-negative, encapsulated, not acid-fast, non-sporing and motile by a single polar flagellum. Colonies on potato dextrose agar are naphthaline yellow, shining, pulvinate, smooth having entire margins and are 1.5 cm in diameter after seven days of incubation. Colonies on nutrient agar are round, slightly raised, yellow, shining and attain a diameter of 0.7 cm after four days of incubation. Bacterium reduces litmus, peptonises milk, liquefies gelatin, hydrolyses starch and casein, produces hydrogen sulphide and ammonia but nitrate reduction, M.R. (Methyl Red) and V.P. (Voges-Proskauer) tests are negative. Bacterium produces acid but no gas from glucose, sucrose, dextrose, lactose and maltose and does not utilise salicin. Citrates are utilised. Optimum temperature for growth is 26–32 °C and the thermal death point is at 51 °C. Growth is aerobic.

9.40.3 Disease Cycle

The bacterium is seed-borne. There is no information on its survival in the infected plant debris. The bacterium enters the host through stomata and wounds. The pathogen does not infect any other host (Kulkarni *et al.*, 1952)

The disease is generally more severe during the monsoon season (July–September) when the relative humidity is 80–90%, air temperature is in the range 24–31 °C and the crop is succulent. New leaves formed after the rains are generally free from infection. When the pigeon pea is grown as a mixed crop with sorghum, the high humidity and low temperature resulting from shading by the sorghum plants also increase the severity of the disease.

9.40.4 Management

1. Use of resistant varieties is the most effective and economical method.
2. Use of pathogen-free seed will ensure a disease-free crop because the infected seed is the only source of perpetuation of the bacterium. To control seed-borne infection, seed treatments used for managing bacterial blight of cowpea and bacterial blight of clusterbean can be tried.

REFERENCES

Dye, D. W. (1978). Genus IX *Xanthomonas* Dowson 1939. In: Young, J. M., Dye, D. W., Bradbury, J. F. *et al*. A proposed nomenclature and classification for plant pathogenic bacteria. *N. Z. J. Agric Res*. 21: 162–66.

Kulkarni, Y. S., Patel, M. K. and Abhyankar, S. G. (1950). A new bacterial leaf-spot and stem canker of pigeon pea. *Curr. Sci*. 19: 384.

Kulkarni, Y. S., Patel, M. K. and Abhyankar, S. G. (1952). A new bacterial leaf-spot and stem canker of pigeon pea. *Indian Phytopath*. 5: 21–22.

Patel, P. N. and Jindal, J. K. (1972). Bacterial leaf spot and stem canker of pigeon pea. In: P. N. Patel (ed.). *Plant Bacteriology*. Vol. I, *Bacterial Diseases of Plants in India*. Indian Agricultural Research Institute, New Delhi, pp. 87–89.

Vauterin, L., Hoste, B., Kersters, K. and Swings, J. (1995). Reclassification of *Xanthomonas*. *Int. J. Syst. Bacteriol*. 45: 472–89.

9.41 BACTERIAL BLIGHT OF SESAME

Bacterial blight of sesame (*Sesamum indicum* L.) was first reported from Sudan by Boughey in 1942. Sabet and Dowson (1960) described the symptoms and identified the bacterium as *Xanthomonas sesami*. The disease occurs in most of the sesame-growing regions of the world and causes considerable losses under favourable weather conditions. The complete loss of crop, particularly under rain-fed conditions, has been reported from Sudan (Sabet and Dowson, 1960). Vijayat and Chakravarti (1977) reported a 60% loss in the capsules under field conditions from Rajasthan, India. Approximately 20% loss in yield have been reported from the Jabalpur area of Madhya Pradesh, India by Shukla *et al*. (1972). The disease has also been reported from Pakistan (Fardoos, 2009) and Venezuela. Lee *et al*. (2005) reported the disease from Suwon, Hongchunand Yeonchun, Korea.

9.41.1 Symptoms

The plants are susceptible at all the growth stages. Small, water-soaked, dark green, marginal spots develop on the lower surface of cotyledons of the seedlings developing from infected seeds. In severe infection, the lesions spread rapidly, covering the whole cotyledons, which eventually dry. However, in cases of mild infection, small, scattered, brown, necrotic spots appear on the cotyledons. In highly susceptible varieties, the infection of the growing point may cause the death of seedlings. In surviving seedlings, dark brown, water-soaked spots appear on the true leaves. These spots turn dark olive-green, increase in size (to 2–3 mm) and turn dark reddish-brown to black. Spots are initially round, but often attain an angular shape. These may remain small or may coalesce to form large patches/blotches. Later on, the dead tissue of the spots dries out, becomes brittle and, ultimately, the leaves fall. Infection may also start from the margin of the leaves and proceeds towards the midrib forming light brown spots. Infected leaves ultimately show severe blight and finally dry.

On stem, the infection starts as a light brown discolouration and spreads along the stem forming slightly raised dark reddish-brown lesions. Honey dew-like bacterial ooze comes out from the infected areas of the stem. On capsules, oval, slightly raised, dark reddish-brown lesions are produced. Infection also occurs on petioles and flowers.

Bashir *et al*. (2007) found that the combined infection of *Pseudomonas syringae* pv. *sesami* and *Xan. campestris* pv. *sesami*, causal agents of bacterial leaf spot and bacterial blight of sesame, respectively, resulted in a more severe disease than the individual infections of these pathogens. Percentages of stem and leaf infection, and defoliation, were higher in combined infections than is the case with individual infections. Under such conditions, it becomes difficult to distinguish the symptoms of these individual diseases.

9.41.2 Causal Organism

Xanthomonas campestris pv. *sesami* (Sabet and Dowson 1960) Dye 1978

Synonym: *Xan. sesami*

The bacterium, being a pathovar of *Xan. campestris* species, has the same morphological, cultural and biochemical characters as those of *Xan. campestris*. It can be distinguished from *Xan. campestris* and its other pathovars on the basis of host range. The description of *Xan. campestris* is given under the section on black rot of crucifers.

9.41.3 Disease Cycle

Infected seeds carry the bacterium and effectively transmit it to the seedlings. The bacterium survives in the infected seeds up to 16 months and in soil from 4 to 6 months (Habish and Hamad, 1971). As the bacterium does not survive in the soil up to the sowing of the next crop, infected seed is the only source of perpetuation of the bacterium. Bacterium enters the host mainly through stomata and quickly invades vascular tissue. Secondary spread is mainly through wind-splashed rain. The spread to the distant areas occurs through the infected seed (Chand and Singh, 1972).

High temperature and high humidity favour disease development. Disease mainly develops in the rainy season, or when the relative humidity is high at night. Maximum disease intensity was observed in September 1993 and 1994 in Kanpur, Uttar Pradesh, India when the mean temperature ranged between 29.0 and 29.4 °C, relative humidity was 88.0 to 90.5% and the rainfall was 8.90 to 9.97 mm. There was a considerable reduction in disease severity by the end of September due to the decrease in temperature, relative humidity and rainfall (Srivastava *et al.*, 1997). The severity of the disease is also related to soil moisture (30–40%) and relative humidity (75–85%).

The age of leaves influences the severity of the disease. The disease index on older leaves is much less than that on fully expanded young leaves. The multiplication of the pathogen in the artificially inoculated leaves reached its maximum on the third day in both young and old leaves, but the number of bacterial cells in young leaves was ten times more than those in old leaves (Chand *et al.*, 1970).

Nitrogen and photoperiod influence the susceptibility of the varieties. Application of nitrogen under long photoperiod increased the susceptibility of variety Venezuela 51 while its susceptibility decreased under the short photoperiod. However, nitrogen under short photoperiod increased the susceptibility of variety Early Russian with little effect under long photoperiod. Firdous *et al.* (2007) reported that tomato gave a better hypersensitive reaction than potato to the bacterium, and high relative humidity and $1x10^9$ cfu mL^{-1} were required for the hypersensitive reaction.

Rehman *et al.* (2013) studied the effect of different concentrations of culture filtrate of *Xan. campestris* pv. *sesame* in Murashige and Skoog medium on seed germination and the height and root length of sesame seedlings, and reported that culture filtrate greatly affected the root length and height of seedlings, and slowed down the process of seed germination. Minimum root length and the smallest seedling height were obtained with 4% culture filtrate compared to the normal growth of seedlings in the control.

Inam-ul-Haq *et al.* (2016) evaluated ELISA with polyclonal antibodies (PAbs) for the detection of *Xan. campestris* pv. *sesami*. PAbs were prepared in rabbits against a pure culture of *Xan. campestris* pv. *sesami* and direct antigen coated-ELISA was used for the analysis of pathogen in sesame germplasm obtained from the National Agricultural Research Centre (NARC), Islamabad, Pakistan. The pathogen from 29 sesame varieties, categorised from resistant to susceptible, was isolated and used as an antigen. After incubation, optical density was calculated at 405 nm. The results revealed that lowest reaction (0.068) was observed in highly resistant sesame genotypes, SG-22 and SG-55, while the genotypes SG-34, SG-33 and SG-72 were categorised as resistant based on reactivity (0.073). The authors concluded that ELISA with PAbs should be preferred for the detection of *Xan. campestris* pv. *sesami* as these are more stable, target multiple epitopes on single antigen and are easy to produce.

Naqvi *et al.* (2012) screened sesame germplasm against *Xan. campestris* pv. *sesami*. Among the 20 entries screened at Faisalabad, no line was found immune, highly resistant, or resistant, but only four lines were found as moderately resistant, 14 as moderately susceptible and two were highly susceptible. Out of 119 lines of germplasm screened at NARC, none was found immune to the pathogen, three lines, SG-34, SG-22 and SG-55, were highly resistant, two lines, SG-72 and SG-33, were resistant, and ten lines were found to be moderately resistant. Forty-nine lines were categorised as moderately susceptible, 40 as susceptible and 15 as highly susceptible.

9.41.4 Management

The following measures are helpful in reducing the disease incidence.

1. Grow resistant varieties like T 58.
2. Sow the crop early, i.e., immediately after the onset of monsoon.
3. Sow the pathogen-free seed. In the absence of pathogen-free seed, the seed should be treated with hot water at 52 °C for 10 minutes followed by soaking in streptomycin solution (100 μg mL^{-1}) for 6–8 h. Chand *et al.* (1970) found that seed treatment with Ceresan Wet (0.1%), streptomycin (100 μg mL^{-1}) and their combination for 80 minutes reduced seedling infection but did not completely eliminate it. The secondary spread of the pathogen was so rapid that there were no differences in disease incidence among the treated and untreated seed plots.
4. Foliar sprays of Streptocycline solution (100 μg mL^{-1}) are recommended for the control of the disease, but the best control is achieved with the combination of seed treatment and foliar sprays. Rai and Srivastava (2003) achieved an effective control of the disease along with an increase in yield by seed treatment with Streptocycline (100 μg mL^{-1}) followed by 3 sprays of Streptocycline (100 μg mL^{-1}) + Blue Copper 50 (copper oxychloride @ 500 μg mL^{-1}) at 10 days intervals.

REFERENCES

Bashir, S., Irfan ul-Haque, M., Mukhtar, T., Irshad, G. and Hussain, M. A. (2007). Pathogenic variation in *Pseudomonas syringae* and *Xanthomonas campestris* pv. *sesami* associated with blight of sesame. *Pak. J. Bot.* 39: 939–43.

Chand, J. N., Mandloi, S. C. and Kulkarni, S. N. (1970). Studies on bacterial blight of sesamum caused by *Xanthomonas sesami*. Symposium on approaches to control important bacterial diseases of plants in India. *Indian Phytopathol. Soc. Bull. No.* 6: 19–23.

Chand, J. N. and Singh, D. (1972). Bacterial blight of sesamum. In: P. N. Patel (ed.). *Plant Bacteriology*. Vol. I, Bacterial Diseases of Plants in India. Indian Agricultural Research Institute, New Delhi, pp. 169–72.

Dye, D. W. 1978. Genus IX. *Xanthomonas* Dowson 1939. In: Young, J. M., Dye, D. W., Bradbury, J. F., Panagopoulos, C. G. and Robbs, C. F. (1978). A proposed nomenclature and classification for plant pathogenic bacteria. *N. Z. J. Agric. Res.* 21: 162–66.

Fardoos, S. (2009). Virulence analysis of *Xanthomonas campestris* pv. *sesami* and *Pseudomonas syringae* pv. *sesami* The causal organisms of Sesame (*Sesamum indicum* L.) bacterial blight. Ph. D. thesis, University of Arid Agriculture, Rawalpindi.

Firdous, S. S., Asghar, R., Haque, H. I. and Afzal, S. N. (2007). Development of Hypersensitive response by *Xanthomonas campestris* pv. *sesami* on *Lycopersicon esculentum* L. and *Solanum tuberosum* L. leaves. *Pak. J. Bot.* 39: 2135–39.

Habish, H. A. and Hamad, A. H. (1971). Survival and chemical control of *Xanthomonas sesami. F.A.O. Plant Prot. Bull.* No. 19: 36–40.

Inam-ul-Haq, M., Naqvi, S. F., Tahir, M. I., Rehman, H. M. Ahmed, R. *et al.* (2016). Serodiagnosis of *Xanthomonas campestris* pv. *sesami* causing bacterial leaf blight disease in sesame. *Pak. J. Agri. Sci.* 53: 535–39.

Lee, S.-D., Lee, J.-H., Kim, Y.-K., Heu, S.-G. and Ra, D.-S. (2005). Bacterial blight of sesame caused by *Xanthomonas campestris* pv. *sesami*. *Res. Plant Dis.* 11: 146–51.

Naqvi, S. F., Inam-ul-Haq, M., Tahir, M. I. and Mughal, S. M. (2012). Screening of sesame germplasm for resistance against the bacterial blight caused by *Xanthomonas campestris* pv. *sesami*. *Pak. J. Agri. Sci.* 49: 131–34.

Rai, M. and Srivastava, S. S. L. (2003). Field evaluation of some agro-chemicals for management of bacterial blight of sesame. *Farm Sci. J.* 12: 81–82.

Rehman, I., Deeba, F., ul Haque, M. I. and Saqlan Naqvi, S. M. (2013). Inhibition of sesame seedling growth by *Xanthomonas campestris* pv. *sesami* culture secretions. *J. Animal Plant Sci.* 23: 1207–10.

Sabet, K. A. and Dowson, W. J. (1960). Bacterial leaf spot of sesame (*Sesamum orientale* L.). *Phytopathol. Z.* 37: 252–58.

Shukla, B. N., Chand, J. N. and Kulkarni, S. N. (1972). Changes in sugar contents of sesame leaves infected with *Xanthomonas sesami*. *Nature* (London) 213 (5078): 813.

Srivastava, S. S. L., Rai, M. and Singh, D. V. (1997). Epidemiological study on bacterial blight of sesame. *Indian Phytopathol.* 50: 289–90.

Vijayat, R. and Chakravarti, B. P. (1977). Yield losses due to bacterial leaf spot of *Sesamum orientale* in Rajasthan. *Indian J. Plant Pathol.* 7: 97.

9.42 BACTERIAL LEAF SPOT OF SESAME

The disease was first observed by Malkoff in Bulgaria in 1903; he gave it the name "**black rot**". Later on, in 1906, Malkoff attributed it to *Bacterium sesami* but the Koch's postulates of the bacterium were proved by Kovacevski in 1930. Since then, the disease has been reported from Bulgaria, the USA, Japan, Greece, the former Yugoslavia, Egypt and India. In India, the disease was first reported from Delhi by Durgapal and Rao (1967). Now, the disease has spread to most states of India due to the movement of infected seed.

Isakeit *et al.* (2012) reported leaf spot of sesame caused by a *Xanthomonas* sp. from Texas, USA. The reports of *Xanthomonas arboricola*, the causal agent of bacterial spot of sesame, have increased in the Americas over the past decade. The disease was first documented in 1960 in Sudan, a large-scale producer of sesame. Since then, *Xan. arboricola* has been reported on sesame in other parts of the world. In the Americas, *Xan. arboricola* was first reported on sesame in Texas. A second report of the pathogen on the same host followed in 2019 in Sinaloa, Mexico and there was a third report in 2022 in North Carolina. Although not confirmed at the time of writing, the disease has also been observed in South Carolina. These reports indicate that *Xan. arboricola* is an emerging pathogen in sesame-producing regions across the USA (Montalban *et al.*, 2023).

Duin *et al.* (2023) reported *Psm. amygdali* pv. *sesami* causing water-soaked, angular lesions on sesame from North Carolina, USA.

Noh *et al.* (2024) observed dark angular necrotic spots on leaves and stems (100% incidence), blights and severe defoliation in a four-acre rainfed sesame field located in the Colleton County of South Carolina, USA in August 2022 and September 2023. The isolated bacterium was confirmed to be *Xan. euvesicatoria* pv. *sesami* (=*Xan. campestris* pv. *sesami*) based on atpD and dnaK sequences. The disease was first reported from Sudan and, later, from the United States and Mexico.

9.42.1 Host Range

Sesame appears to be the only natural host, although some other hosts are mildly infected artificially (Sutic and Dowson, 1962). However, these hosts do not play any role in the disease cycle.

9.42.2 Symptoms

The symptoms develop on all the above-ground plant parts as blackish spots varying in size. On leaves, the spots appear oil-soaked and have a star-like external border or they are delimited by the veins, giving them a distinct angular shape. From the spots, infection moves along the veins, turning these veins sooty in appearance. In the case of severe infections, the spots may cover the entire leaf area. Finally, the infected leaves curl, die and defoliate. Infection on the stem appears in the form of lesions, which generally encircle the stem, forming rings. The lesions appear on the whole length of the stem and frequently coalesce, forming large elongated areas. Capsules, infected at an early stage, turn black and do not bear seeds. Infection on older capsules results in the formation of minute spots (about 2 mm in diameter). Severely infected plants may die.

9.42.3 Causal Organism

Pseudomonas syringae pv. *sesami* (Malkoff 1906) Young *et al.* 1978

Synonym: *Psm. sesami*

The major characteristics of the bacterium are given below.

Bacterial cells are Gram-stain-negative, non-sporing, non-encapsulated rods, measuring 0.6–0.8 × 1.2–3.8 µm and motile by 2–5 polar flagella. Cells may occur singly or in pairs. Colonies on beef agar are circular, flat, striate, smooth and white having entire margins. Green fluorescent pigment is produced in culture. The bacterium liquefies gelatin rapidly, and gives an alkaline reaction in litmus milk without coagulation. Tests for starch hydrolysis, nitrate reduction and the production of indole and hydrogen sulphide are negative. Acid but no gas is produced from glucose; however, in the case of sucrose, lactose and glycerol, even acid is not produced. Temperature for optimum growth is 30 °C. The bacterium is facultatively aerobic.

Pathogenic variability exists in the bacterium. Strains differing in virulence and races giving differential reactions on sesame varieties have been reported. Thomas and Orellana (1962) identified two races of the pathogen, namely, race 1 and race 2. Race 2 infects variety Margo while race 1 does not. Varieties Delco and Venezuela 51 are susceptible to both the races while variety Early Russian is resistant. The two races are morphologically similar and also react alike in several standard physiological tests, but they differ in glucose-asparagine ratio requirements. At high sugar levels, race 2 requires less asparagine than race 1. Race 2 also liquefies gelatin and peptonises milk faster than race 1.

Thomas and Orellana (1962, 1963) found that the reactions of Margo, Early Russian and Venezuela 51 varieties to two races of the pathogen were related to differences in the concentrations of aspartic acid, histidine, leucine, cysteine and certain other amino acids and were associated closely with apparent differences in the ratio of reducing sugars to these amino acids. Delco and Venezuela 51, the most susceptible varieties, had the highest concentration of amino acids. Their findings substantiate the notion that the two races differ in the glucose asparagine ratios at which they grow.

9.42.4 Disease Cycle

The bacterium is transmitted through infected seed and this is probably the only source of perpetuation of the bacterium. Fawzi *et al.* (1991) found that the bacterium is internally seed-borne. After germination, the bacteria from the infected seeds infect the cotyledonary leaves; from there, the bacteria spread to all the above-ground plant parts. The bacterium enters the host through stomata and wounds and the infection remains confined to the parenchymatous and vascular tissues. The infection of the capsules results in the infection of the seeds. Warm temperature and wind-splashed frequent rains are the most favourable factors for the development of epiphytotics (Chand and Singh, 1972).

Firdous *et al.* (2009) conducted histological studies to elucidate the mode of infection of *Psm. syringae* pv. *sesami* and reported that the periodical examination of cleared discs taken from inoculated leaves showed the presence and multiplication of bacterial cells in depressions and around trichome bases for 24 h, before penetration through stomata and trichome basal cells. The authors further reported that the appearance of pin-point-sized spots three days after inoculation coincided with chloroplast damage; and the early appearance of small necrotic spots four days after inoculation revealed the presence of the bacterium in parenchymatous tissues, and from there it moved to transverse vascular systems.

Firdous *et al.* (2014) conducted further histopathological studies, using toluidine blue O, to elucidate the process of infection of *Psm. syringae* pv. *sesami* in a susceptible sesame genotype and reported that the bacterium presented as dark blue masses in infected tissues. The bacterium colonised substomatal and intercellular spaces of the spongy parenchyma cells, when initial water-soaking symptoms developed 2–3 days after inoculation. As water-soaking progressed, the disruption of mesophyll cells occurred; mesophyll tissues were surrounded by bacterium, followed by the thinning and disruption of the cell walls. Later on, when bacterial cells increased into space previously occupied by mesophyll cells, there were empty spaces without any differentiation of host tissues. Nevertheless, the bacterium was not found in vascular bundles (tracheary elements) of leaf and stem, but some phloem tissues of stem sections were found to be infected. Damage to the chloroplasts might have occurred due to the production of chlorosis or necrosis inducing toxins, and these toxins played a crucial role in the pathogenesis of *Psm. syringae* pv. *sesami* in sesame plants.

The application of nitrogen and photoperiod considerably influence the reaction of sesame varieties to the pathogen. Thomas (1965) investigated the effect of 12 and 16 h photoperiods per day along with and without supplemental nitrogen on the reaction of resistant variety Early Russian and susceptible variety Venezuela 51. The most favourable combination for disease development was a short light period coupled with supplemental nitrogen; with this combination, both the varieties gave susceptible reaction. However, with either a light period or without supplemental nitrogen, the original reactions of the varieties remained unchanged.

9.42.5 Management

1. Grow resistant varieties. White-seeded early varieties perform better than other varieties.
2. Seed treatments and foliar sprays recommended for bacterial blight of sesame are also effective against this disease.

REFERENCES

Chand, J. N. and Singh, D. (1972). Bacterial leaf spot of sesame. In: P. N. Patel (ed.). *Plant Bacteriology*. Vol. I, Bacterial Diseases of Plants in India. Indian Agricultural Research Institute, New Delhi, pp. 173–77.

Duin, I. M., D'Amico-Willman, K., Joglekar, P., Ritchie, D. F., Fagen, J. (2023). *Pseudomonas amygdali* pv. *sesami* and a bacteriophage isolated from sesame (*Sesamum indicum* L.). *Phytopathology* 113: S3.138.

Durgapal, J. C. and Rao, Y. P. (1967). Bacterial leaf spot of sesamum (*Sesamum orientale* L.) in India. *Indian Phytopathol.* 20: 178–79.

Fawzi, F. G., El Wakil, A. A., El Deeb, A. *et al.* (1991). Association of *Pseudomonas syringae* pv. *sesami* with some sesame seed lots in Egypt. *Egyptian J. Agric. Res.* 69: 685–93.

Firdous, S. S., Asghar, R., and Murtaza, G. 2014. Histopathology of leaf spot of sesame (*Sesamum indicum* L.) caused by *Pseudomonas syringae* pv. *sesami*. *J. Animal & Plant Sci.* 24: 814–19.

Firdous, S. S., Asghar, R., Ul-Haque, M. I. *et al.* (2009). Pathogenesis of *Pseudo-monas syringae* pv. *sesami* associated with sesame (*Sesamum indicum* L.) bacterial leaf spot. *Pak. J. Bot.* 41: 927–34.

Isakeit, T., Hassett, B. T., and Ong, K. L. 2012. First report of leaf spot of sesame caused by *Xanthomonas* sp. in the United States. *Plant Dis.* 96: 1222.

Malkoff, K. 1906. Weitere Untersuchungen uber die Bakterienkrankheit auf *Sesamum orientale*. *Centralblatt für Bakteriologie, Parasitenkunde und Infektionskrankheiten* 2 (16): 664–66.

Montalban, K., Huerta, A. I., D'Amico-Willman, K., Fagen, J., Ritchie, D. F. *et al.* (2023). An emerging pathogen: *Xanthomonas arboricola* on sesame (*Sesamum indicum* L.) in the United States. *Phytopathology* 113: S3.186.

Noh, E., Panth, M., Yang, X., Barnes, J. and Wang, H. (2024). First report of *Xanthomonas euvesicatoria* pv. *sesami* causing leaf spot on sesame (*Sesamum indicum*) in South Carolina, U.S.A. *Plant Dis.* 108: 2916. https://doi.org/10.1094/PDIS-05-24-0958-PDN

Sutic, D. and Dowson, W. J. (1962). Bacterial leaf spot of sesamum in Yugoslavia. *Phytopathol. Z.* 45: 57–65.

Thomas, C. A. (1965). Effect of photoperiod and nitrogen on reaction of sesame to *Pseudomonas sesami* and *Xanthomonas sesami*. *Plant Dis. Rptr* 49: 119–20.

Thomas, C. A. and Orellana, R. G. (1962). Resistance of sesame varieties and pathogenicity of strains of *Pseudomonas sesami* in relation to amino acids and reduced sugars (Abstr.). *Phytopathology* 52: 30.

Thomas, C. A. and Orellana, R. G. (1963). Amino compounds and reducing sugar in relation to resistance of sesame to *Pseudomonas sesami*. *Phytopathol. Z.* 46: 101–04.

Young, J. M., Dye, D. W. and Wilkie, J. P. (1978). Genus *Pseudomonas* Migula 1894. In: Young, J. M., Dye, D. W., Bradbury, J. F., Panagopoulos, C. G. and Robbs, C. F. (1978). A proposed nomenclature and classification for plant pathogenic bacteria. *N. Z. J. Agric. Res.* 21: 153–77.

9.43 BACTERIAL LEAF BLIGHT OF CASTOR BEAN

Castor bean (*Ricinus communis* L., family Euphorbiaceae) is the most important non-edible oilseed crop of the arid and semi-arid regions of India. It possesses exceptionally valuable qualities for the chemical industry. Castor bean oil is important in a wide range of industries such as nylon, fibres, soaps, detergents, dyes, greases, ointments, varnishes, costumes, jet engine lubricants, varnishes, cosmetics, perfumes *etc.* (Pathak, 2003). The castor bean has become a potential source of biofuel due to the constant pressure for renewable fuels. In India, castor bean is grown in the states of Andhra Pradesh, Gujarat, Tamil Nadu and Orissa. Castor bean crop is attacked by several diseases caused by bacteria, fungi and viruses, causing huge losses in yields. Bacterial leaf spot caused by *Xanthomonas axonopodis* pv. *ricini* (*Xar*) is the only important disease of bacterial etiology reported to threaten castor bean production. No cultivars expressing resistance to the pathogen are known.

Bacterial leaf blight of castor bean is prevalent in Korea and Kyushu, Japan (Yoshii and Takimoto, 1928). Arsenijevi *et al.* (1992) reported it from Yugoslavia for the first time in 1900. The disease is also prevalent in Brazil, Sudan, South Africa, the USA and Uganda, besides India. There was 25% loss in yield due to bacterial leaf blight in Texas and Oklahoma in the USA. In India, the disease was first reported by Patel *et al.* (1951) from Gujrat and thereafter had been reported from Andhra Pradesh, Tamil Nadu, Maharashtra, Rajasthan and Delhi. However, reliable information regarding the losses due to this disease in India is not available.

9.43.1 Symptoms

The disease first appears on the cotyledonary leaves as water-soaked lesions, which gradually turn dark brown to black. Small, round, water-soaked spots appear on young leaves, which may enlarge up to 2–5 mm in diameter and turn somewhat angular. Later on, these spots turn brown to black, dry up and become brittle. Under favourable weather conditions, these lesions coalesce and the infected leaves turn yellow, dry and defoliate. Black lesions may also appear on petioles, stem and branches. Blasting of immature capsules and breaking over of girdled racemes and stems also occur in very severe cases. No wilting of the affected plants occurs either in the field or in artificially inoculated plants. Bacterial ooze, in the form of shining beads, may be observed on the infected leaves in the morning hours or during humid weather.

9.43.2 Causal Organism

Xanthomonas axonopodis pv. *ricini* (Yoshii & Takimoto 1928) Vauterin *et al.* 1995

Synonym: *Xan. campestris* pv. *ricini*

Cells are Gram-stain-negative, non-sporing rods with rounded ends, occurring singly, in pairs, or in short chains; 0.4–0.9 by 1.3–2.6 µ in diameter, strictly aerobic, formed in chains, capsulated, non-acid-fast and motile by a single polar flagellum. On King's medium B, the colonies are white and fluorescent pigment is produced. Reactions in LOPAT tests are +, -, -, - and +, respectively. Colonies on nutrient agar are lemon-yellow, then turn brown and gradually changing to medium brown. Gelatin is liquefied, casein peptonised without coagulation, litmus milk becoming slightly acid and then reduced. Acid is formed from lactose but no gas is produced, nitrates are not reduced and there is a strong diastatic action. Bacterium makes a feeble growth in Uschinsky's solution, but not in Cohn's solution. Its thermal death point is 50 to 51 °C, maximum 39 °C, optimum 29 to 30 °C and minimum around 2.5 °C.

9.43.3 Disease Cycle

The bacterium is seed-bone and also survives in the infected plant debris. It enters the host through stomata, hydathodes and wounds. Short-distance spread occurs through wind-splashed rains and long-distance spread occurs through contaminated seed. Disease development is favoured by high humidity. Maximum disease development occurs when there are frequent showers with relative humidity ranging between 84–85 % and temperature above 28 °C.

9.43.4 Management

The following measures are helpful in minimising the incidence of the disease:

1. Seed treatment: Treat the seed with hot water at 58–60 °C for 10 minutes. Singh *et al.* (1996) reported that seed treatment with Streptocyline and Sandoz 6334 was effective for the management of the disease.
2. Removal and destruction of the diseased plant debris is recommended to reduce the soil-borne inoculum. Deep ploughing after harvest buries the infected plant debris, leading to a reduction in the pathogen's survival in the soil.
3. Destroy the possible alternate and collateral hosts.
4. Prophylactic sprays of Paushamycin (0.025 %) + copper oxychloride (0.3%) have been found very effective in reducing the disease incidence with an increase in yield.

REFERENCES

Arsenijevi, M., Masirevi, S. and Goto, M. (1992). First report of bacterial leaf spot of castor bean in Yugoslavia. *Plant Dis.* 76: 1283.

Patel, M. K., Kulkarni, Y. S. and Dhande, W. G. (1951). Bacterial leaf blight of castor. *Curr. Sci.* 20: 20.

Pathak, H. C. (2003). Emerging trends in castor seed development. *Proceedings of national seminar on castor seed, castor oil and its value-added products, 22nd May, 2003. Solvent Extractors Association of India, Ahmedabad, India*, pp. 54–62.

Singh, S., Majumadar, D. K. and Rehan, H. M. S. (1996). Evaluation of anti-inflammatory potential of fixed oil of *Ocimum sanctum* (Holybasal) and its possible mechanism of action. *J. Ethnopharmacology* 54: 19–26.

Vauterin L., Hoste B., Kersters K. and Swings J. (1995). Reclassification of Xanthomonas. *Int. J. Syst. Bacteriol.* 45: 472–89.

Yoshii, H. and Takimoto, S. (1928). Bacterial leafspot of castor bean and its pathogen. *J. Plant Prot.* Tokyo 15: 12–18.

9.44 CROWN GALL OF PLANTS

Most of the bacterial plant pathogens cause necrotic or hypoplastic effect on the host tissue. However, crown gall bacterium *(Agrobacterium tumefaciens)* has the opposite effect, i.e., it causes hyperplasia and hypertrophy of the affected tissue. The disease is worldwide in occurrence.

The disease has been known to occur in the USA since 1870. In 1897, Cavara from Italy was the first to isolate the bacterium from stem galls of grapes and establish its pathogenicity. However, between 1892 and 1894, numerous investigators from Europe and the USA worked on the disease without isolating the bacterium in culture. Smith and Townsend (1907) described the bacterium and named it *Bacterium tumefaciens*, which was subsequently changed to *Agrobacterium tumefaciens* by Conn in 1942.

In India, the occurrence of the disease was first reported by Singh in 1943, but its authentic report was made by Durgapal (1971) who isolated the bacterium and proved its pathogenicity. The disease occurs on cherry, peach, plum and pecan nut (*Carya illinoensis*).

Borges *et al.* (2019) reported *Agr. tumefaciens* biovar 1 infecting teak (*Tectona grandis*) in Brazil. The symptoms of crown gall on teak were observed in a clonal mini-nursery (~30% incidence) in June 2017 in Brazil. The phylogenetic tree formed a clade with both RhizTeca isolates and the type strain of *Agr. tumefaciens* [ATU16SRDF (GenBank D14500)] with "1" of posterior probability. This is the first report of *Agr. tumefaciens* biovar 1 infecting teak in Brazil, which could emerge as a novel and significant nursery disease.

Bai *et al.* (2022) isolated a bacterium from crown gall of Chinese dwarf cherry (*Cerasus humilis* (Bge.) and, based on its morphological, biochemical, pathogenicity and genetic characteristics, identified it as *Agr. tumefaciens*/biovar 1.

Two other diseases having some similarities with crown gall are caused by two other *Agrobacterium* species: *Agr. rhizogenes* causes hairy root of apple and some other plants; and *Agr. rubi* causes galls on canes of *Rubus* spp. (brambles). Hairy root of apple is described separately.

9.44.1 Host Range

Agr. tumefaciens has the widest host range of any known plant pathogenic bacteria. The bacterium infects a wide variety of dicotyledonous and also some gymnospermous plants. However, it does not infect monocots (grasses). More than 643 plant species belonging to 331 genera and 93 families are known to be infected by the bacterium. The disease is more severe on all the stone fruits. Apple, pear, blueberry, grapes, and some ornamental plants such as rose, are also severely infected. Many hosts are only known through artificial inoculation. Tomato acts as a very good indicator host, but has never been found to be a natural host. The detailed host range can be found in De Cleene and De Ley (1976) and Bradbury (1986) and also on http://www.cabicompendium.org/cpc/home.asp.

9.44.2 Symptoms

The galls or overgrowths occur commonly on the subterranean roots and crowns of trees and shrubs (Figure 9.36). These may also be formed on stem, branches, leaves and leaf veins. Aerial galls appear generally on grapevines, raspberry and black berry. Aerial galls are not common on fruit trees, although these can occur. Galls are usually formed in late spring or early summer.

Young galls are more or less spherical, white or flesh-coloured and soft. As these grow, their surface becomes convoluted. In the advanced stages of infection, the outer tissue of the galls turns dark brown to black due to death and the decay of peripheral cells. The breakdown of peripheral tumour tissue releases crown gall bacteria into the soil. As the galls age, these become hard, rough, woody and more cleft. Relatively soft galls are disintegrated while those consisting of a corky exterior and a hard-woody interior persist on the host plants. The size of galls may vary from 0.5–10 cm in diameter but in extreme cases they may be 30 cm or more in diameter.

Occasionally, the disease causes the formation of complex tumours, called **teratomas** These tumours may give rise to roots and buds and the buds give rise to shoots carrying leaves. In some hosts, secondary tumours develop at points distant from the primary tumours. These tumours are sterile, i.e., these do not contain bacteria and their tissue can be grown on chemically defined media. When the tissues of such tumours are transplanted into a healthy plant, sterile tumours are produced. Crown galls or tumours caused by *Agr. tumefaciens* differ from those of other hyperplastic diseases, since the former result from the uncontrolled growth or proliferation of host cells due to the transformation of infected cells, caused by the pathogen.

The aerial plant parts of mildly infected plants may appear normal. However, in case of severe infection, the plants are stunted, unthrifty and bear small and chlorotic leaves. In general, infected plants are more susceptible to adverse environmental conditions, particularly to winter injury. The yield of severely infected plants is reduced and most severely infected plants and vines may die.

FIGURE 9.36 Crown gall of plants. (Courtesy: missouribotanicalgarden.org)

Crown gall of grapes is an important disease of wine grapes in cold climates. In grapes, the galls are formed mostly on the aerial parts of the vines. Galls on grape roots are not typical, but the bacteria can induce a localised necrosis of roots. Most grape cultivars are susceptible to the disease. The incidence of grape crown gall from year to year is generally related to the severity of the preceding winter and the age of the vines. For more details on the disease, the readers can refer to the review article by Burr and Otten (1999).

Cane gall: Small protuberances or elongate ridges appear on the fruiting canes of *Rubus* spp., somewhat before the middle of growing season. These increase in size and may completely cover sections of the surface, especially in the lower portion of the cane. The gall tissue is whitish at first, but turns brown with age. As the season advances, severely infected canes split open, the leaves wither and the berries dry up. The simultaneous occurrence of crown gall, hairy root and wound overgrowths is commonly found in nature. Wound overgrowths, non-parasitic in nature, are generally confused with crown gall since they may occur along with the latter. Such overgrowths often occur on apple trees at the margin of root and scion, particularly when an imperfect union has been made (Walker, 1969).

Besides crown gall, some plant pathogenic bacteria produce galls on twigs, limbs and trunks of plants. In the latter case, bacterial pockets are found in light green to light yellow soft parenchymatous tissues surrounded by white woody tissues. Hence, the section of a tumour or gall should be examined for identification/differentiation of the disease.

9.44.3 Causal Organism

Agrobacterium tumefaciens (Smith & Townsend 1907) Conn 1942

Synonym: *Bacterium tumefaciens*, *Rhizobium radiobacter*

Allorhizobium vitis (Ophel and Kerr 1990) Mousavi *et al.* 2016* causes crown gall of grapevines

Synonym: *Agr. vitis*, *Rhiz. vitis*

For detailed taxonomic discussion and the reclassification of *Agrobacterium* species into *Rhizobium*, see Young *et al.* (2001). Garrity *et al.* (2005) have recognised both *Agrobacterium* and *Rhizobium* as separate genera in the family *Rhizobiaceae*. Hence, *Agr. tumefaciens* is used as a valid name here.

Bacterial cells are Gram-stain-negative, non-sporing rods, measure 0.6 -1.0 × 1.5–3.0 μm and motile by 1- 6 peritrichous flagella. Aerobic, chemoorganotrophs, soil inhabitants and possess a respiratory type of metabolism with oxygen as the

* References marked with an * are not cited in the bibliography.

terminal electron acceptor. The colonies are generally circular, smooth, convex and colourless to light beige. The growth on media containing carbohydrates is usually copious and slimy. The species is divided into three biovars, i.e., biovar 1, 2 and 3. *Agr. tumefaciens* strongly absorbs Congo red and aniline blue in contrast to little or no absorption by *Agr. rhizogenes. Agr. tumefaciens* produces an abundant growth on sodium selenate agar and calcium glycerophosphate medium with mannitol in contrast to no or very little growth by *Agr. rhizogenes*. Mol% of G + C of the DNA is 57–63 (T_m).

The Genome of *Agr. tumefaciens* str. C 58 consists of a circular chromosome (2.8 Mb), a linear chromosome (2.0 Mb) and two plasmids, namely, tumour-inducing plasmid called Ti- plasmid (206.479 kbp) and At-plasmid called the cryptic plasmid. Pathogenic ability, i.e., the ability to cause galls/tumours on the host plants, is governed by the Ti-plasmid. The strains which lack Ti-plasmid are non-pathogenic. The plants infected with *Agr. tumefaciens* str. C 58 produce nopaline type of opine. If the bacterium is grown near its maximum temperature for growth, i.e. 30 °C, then the bacterium loses its Ti-plasmid as well as its pathogenicity. Plasmid At of strain C 58 has been shown to be involved in the metabolism of opines and conjugation. The genome of strain C 58 was simultaneously sequenced by Goodner *et al.* (2001) and Wood *et al.* (2001).

Al-Karablieh and Khlaif (2002) found that the detection of *Agr. tumefaciens* from infected and symptomless plants by detecting *tmr* gene was more efficient than the routine isolation of the bacterium because the former method was more sensitive.

Agr. tumefaciens is an excellent genetic engineering tool. The DNA transformation capabilities of the bacterium have been extensively exploited in biotechnology as a means of inserting foreign genes into plants. Nester (2008) has rightly called it a **natural genetic engineer** and also a **household name** in plant genetic engineering and plant molecular genetics.

Agr. vitis Ophel and Kerr (1990) causes crown gall of grapes. Earlier, the strains of agrobacteria causing crown gall of grapes were classified as biovar 3 of *Agr. tumefaciens*. The taxonomic evaluation of biovar 3 strains by Ophel and Kerr in 1990 resulted in the naming of a new species, *Agr. vitis*. The bacterium is insensitive to agrocin 84 produced by *Agr. radiobacter* strain K 84. In contrast to *Agr. tumefaciens*, it has a limited host range. According to Kawaguchi (2009), *Rhizobium vitis* (syn. *Agr. vitis*) strains are genetically heterogeneous and can be separated into several genetic groups. Although *Agr. vitis* is mainly responsible for causing crown gall of grapes, *Agr. rhizogenes* and *Agr. tumefaciens* also infect grapes. Out of 56 tumorigenic Spanish grapevine strains of *Agrobacterium* spp., three strains were identified as *Agr. tumefaciens* (biovar 1), three as *Agr. rhizogenes* (biovar 2) and the remaining 50 as *Agr. vitis* (biovar 3) (Palacio-Bielsa *et al*., 2009). The strains of these three species were clearly distinguished on the basis 16S rRNA gene sequencing. However, no universal primer pair was found for the detection of these three species.

Johnson *et al.* (2013) developed real-time PCR assays employing DNA extraction methods that included magnetic capture hybridisation, immunomagnetic separation and extraction with the Mo Bio Powerfood kit for the detection of *Agr. vitis*. The assays incorporating magnetic capture hybridisation or immunomagnetic separation followed by real-time PCR were 10,000 times more sensitive than direct real-time PCR when tested using boiled bacterial cell suspensions, having detection thresholds of 10^1 cfu mL^{-1} compared with 10^5 cfu mL^{-1}. DNA extraction with the Powerfood DNA extraction kit was ten times more sensitive than direct real-time PCR, with a detection threshold of 10^4 cfu mL^{-1}. All the three assays were able to detect *Agr. vitis* in healthy-appearing grapevine cuttings taken from infected vines.

Sule *et al.* (2012) examined 40 strains of tumorigenic and non-tumorigenic strains of *Agr. tumefaciens*, *Agr. rhizogenes*, *Agr. rubi* and *Agr. vitis* using capillary isoelectric focusing. Isoelectric points of the investigated strains clearly differentiated these strains according to their respective species. Isoelectric points values obtained for *Agr. tumefaciens*, *Agr. rhizogenes*, *Agr. rubi* and *Agr. vitis* were 2.2, 4.0, 2.15 and 2.6, respectively. The differentiation corresponded to the phenotypic, PCR and fatty acid characterisations. Strains of *Rhizobium* species differed from *Agrobacterium* strains in their isoelectric points values. The merits of capillary isoelectric focusing over the phenotypic determinative tests, PCR and fatty acid analysis are its fast speed (15 min), relative simplicity and very small amount of chemicals used. This rapid and simple method is a major improvement over the classical methods of separation of *Agrobacterium* species and should prove useful for rapid characterisation of *Agrobacterium*-like colonies isolated from plant tumours for epidemiological and generic diversity studies.

9.44.4 Disease Cycle

The bacterium is an inhabitant of the soil and remains viable for several years. It can also survive in infected plant debris, on and around root surfaces by utilising leaked nutrients and in the outer portion of primary galls. The bacterium is widely distributed on the living planting stock. The bacterium penetrates the host through wounds, with fresh and deep wounds being more conducive for infection. The wounding of the plants results from pruning, cultural operations, growth cracks, grafting and chewing insects.

The dissemination of the bacterium occurs through wind-splashed rain and chewing insects. Abundant bacterial cells present on the surface of galls are washed down by rain water and carried away by the flowing water. Long-distance spread of the pathogen occurs through infected nursery and planting material, with bacterial cells being present on the surface or inside the host tissue. Tumour formation occurs between 10 and 33 °C. In some hosts, such as chrysanthemum, the bacterium also causes systemic infection.

The infection process and the host–pathogen interaction in this disease have been elucidated thoroughly. Nester (2008) has given a comprehensive account of host–parasite interaction leading to the development of tumours. Under favourable conditions, a single bacterial cell can cause infection. The wounding of host cells results in the release of cell sap containing phenolic and sugar compounds. The bacterial cells are attracted by these chemicals (such as acetosyringone), swim through soil water with the help of flagella and attach to the wounded plant cells. The attachment of bacteria to plant cells is accomplished through two distinct steps. After an initial weak and reversible attachment, the bacteria synthesise cellulose fibrils that anchor them to the wounded plant cells. The four main genes involved in this process are *chvA*, *chvB*, *pscA* and *att*. The products of the first three genes appear to be involved in the actual synthesis of the cellulose fibrils. These fibers also anchor the bacteria to each other, resulting in the formation of a microcolony. After the production of cellulose fibrils, a Ca^{2+}-dependent outer membrane protein called rhicadhesin is produced which also helps the bacteria to stick to the cell wall.

A large fragment of Ti-plasmid, called transfer DNA (or T-DNA), is excised from the plasmid, which, after insertion into the host cell, becomes incorporated at a semi-random location into the plant genome. For the transfer of T-DNA, *Agr. tumefaciens* uses a type IV secretion system, involving the production of T-pilus. The two-component sensor system of Vir A/Vir G is able to detect phenolic signals, particularly acetosyringone and hydroxyacetosyringone released by wounded plant cells. This results in a signal transduction event, activating the expression of 11 genes within the Vir B operon, which are responsible for the formation of T-pilus. First of all, Vir B propilin is formed, which is a polypeptide of 121 amino acids and is processed by the removal of 47 residues to form a T-pilus subunit. The subunit becomes circular through the formation of a peptide bond between two ends of the polypeptide. The products of the other Vir B genes transfer the subunits across the plasma membrane. Evidence provided by various studies shows that all of the genes from Vir B6 to Vir B10 might encode components of the transporter. An ATP-ase is also required for the active transport of the subunits (Wikipedia, 2008).

Before transferring T-DNA, it is removed from the Ti-plasmid. A Vir D1/D2 complex cuts the DNA at the left and right border sequences. Vir D2 protein is covalently attached to the 5' end. Vir D2 contains a motif that leads to the nucleoprotein complex being targeted to the type IV secretion system. In the cytoplasm of the host, T-DNA complex becomes coated with Vir E2 proteins, which are exported through the type IV secretion system independently from the T-DNA complex. Nuclear localisation signals located on the Vir E2 and Vir D2 are recognised by the importin alpha protein, which then associates with importin beta and the nuclear pore complex to transfer the T-DNA into the nucleus. Another protein, VIP 1, also plays an important role in the process, possibly by acting as an adapter to bring the Vir E2 to the importin. After entering the nucleus, VIP 2 may target the T-DNA to the areas of chromatin that are being actively transcribed so that the T-DNA can integrate into the host genome (Wikipedia, 2008).

Loyter *et al.* (2005) advanced the hypothesis that *Arabidopsis thaliana* Vir E2 interacting protein 1 (AtVIP 1), which associates with the T-complexes by binding to their Vir E2 component, acts as a molecular link between the T-complex and the histone constituents of the host chromatin. This hypothesis was based on their observations of interaction of AtVIP 1 with the histones in *in vitro* and *in planta*. Gelvin (2010), in his review article, has highlighted the role of plant proteins involved in *Agrobacterium*-mediated genetic transformation. The integration of T-DNA into a plant genome establishes a permanent transformation event, permitting stable expression of T-DNA-encoded transgenes. The transformation process is complex and requires the participation of numerous plant proteins. The review deals with the current information on plant proteins that contributes to *Agrobacterium*-mediated transformation, the roles of these proteins in the transformation process and the modern technologies that have been used to elucidate the cell biology of transformation.

T-DNA region contains genes, which cause gall formation. Among them, genes *iaam* and *laah* encode for the synthesis of indole acetic acid via the IAM pathway. As this biosynthetic pathway is not used in many plants for the production of IAA, this indicates that the plant has no molecular means of regulating it and that the IAA will be produced constitutively. The *ipt* and other genes encode the synthesis of cytokinins. The production of IAA and cytokinins stimulates cell division and gall formation and explains the hormonal independence of crown gall tissue. T-DNA region also contains genes which encode the synthesis of opines. Opines, namely, nopaline, octopine, agropine and agrocinopine, are present in the crown gall tissue and not in the healthy plant tissue. Infecting bacterium utilises these opines for its nourishment as a source of energy, carbon and usually nitrogen. Strains of *Agr. tumefaciens* are frequently classified according to the opines synthesised. Although the presence of the bacterium is necessary for the induction of crown gall, once the gall is initiated, the bacterium can be withdrawn without affecting the growth of galls.

Kawaguchi (2015) reported that a non-pathogenic strain of *Agr. vitis*, ARK-1, limits the development of crown gall of grapevine. The co-inoculation of grapevine shoots with ARK-1 and the tumorigenic (Ti) strain VAT03-9 with cell suspension having a 1:1 cell ratio resulted in a significantly lower expression of the virulence genes *virD2* and *virE2* of VAT03-9, one day after inoculation compared with expression levels when shoots were inoculated only with VAT03-9. ARK-1 began to suppress the VAT03-9 population seven days after inoculation. The biological control activity of ARK-1 seems to be based on the suppression of essential virulence genes.

Li *et al.* (2015) designed the primer pair ipt 3F/ipt 3R on the basis of the *ipt* gene, and this primer pair effectively amplified a 247 bp DNA fragment from the transfer DNA sequences of tumorigenic *Agrobacterium*. Using this primer pair, a sensitive real-time PCR assay was developed for quantifying tumourigenic *Agrobacterium* in soil as low as 10^2 cfu g^{-1}. The detection of tumorigenic *Agrobacterium* in 19 soil samples with this assay revealed $\leq 10^2$ cfu g^{-1} in maize or uncultivated soil and 10^3 to 10^6 cfu g^{-1} in soil planted with peach or cherry. Relationship between the disease severity and the tumourigenic *Agrobacterium* density in soil was verified by an artificial inoculation test. Disease incidence of 30–50% and a disease index of 20–30 were found in soil inoculated with *Agr. tumefaciens* at 10^3 cfu g^{-1}. The disease incidence increased to >80% when the inoculation density of *Agr. tumefaciens* was increased to $>10^5$ cfu g^{-1} of soil. These results indicated that tumorigenic *Agrobacterium* density in soil was positively correlated with the incidence of crown gall on peach seedlings, and the presence of 10^3 bacterial cells per gram of soil were unsuitable for peach planting. Determination of tumourigenic *Agrobacterium* density can improve the prediction of crown gall disease outbreaks as well as aid in the management of the disease.

Fuller *et al.* (2017) designed oligonucleotide primers and probes to target *virD2* for use in a molecular diagnostic tool, which relies on isothermal amplification and lateral-flow based detection. Oligonucleotide tools were tested in the assay and evaluated for detection limit and specificity in detecting alleles of *virD2*. One set of primers that successfully amplified *virD2* when used with an isothermal recombinase was selected. Both the tested probes had detection limits in picogram amounts of DNA. Probe 1 could detect all tested pathogenic isolates that represented the maximum diversity of *virD2*. Finally, the coupling of lateral-flow detection to the use of these oligonucleotide primers in isothermal amplification made the process simple and easy, and alleviated reliance on specialised tools necessary for molecular diagnostics. The assay is advancement for the rapid molecular detection of pathogenic *Agrobacterium* spp.

Yakabe *et al.* (2014) examined the possibility of Paradox (*Juglans hindsii* × *J. regia*) walnut seed acquiring *Agr. tumefaciens* based on harvest method used, prior to planting. Paradox seeds were collected directly from the mother tree without contacting the soil or picked up after lying on the orchard floor for up to 28 days, for a two-year period from two commercial nurseries. *Agr. tumefaciens* was never detected in or on the 2,650 seeds collected directly from the mother tree. Both virulent and avirulent *Agr. tumefaciens* strains were detected in and on the husk of nuts lying on the orchard floor at a frequency directly proportional to the time spent on the orchard floor. Irrespective of *Agr. tumefaciens* contamination in or on the husk, *Agr. tumefaciens* was never detected in the seed interior. Avoiding or reducing soil-borne populations of *Agr. tumefaciens* at the time of seed collection can play an important role in managing crown gall.

Johnson *et al.* (2016), using a highly sensitive magnetic capture hybridisation procedure along with real-time PCR, measured the distribution of tumorigenic *Agr. vitis* in dormant canes and green shoots of grapevines. Tumorigenic *Agr. vitis* was distributed from the basal to apical nodal and internodal tissues of canes as well as in non-lignified green shoots. In 2013, *Agr. vitis* was detected in up to 17% of shoot tips and 52% of meristems of greenhouse-grown plants initiated from known *Agr. vitis* contaminated cuttings. A lower frequency of detection was observed from surface-disinfected shoot tips (7%) in comparison to non-disinfected shoot tips (37%), suggesting an epiphytic survival on green tissues. In 2014, vines propagated from cuttings collected from crown gall-infected vines from a different vineyard yielded lower incidences of *Agr. vitis* from shoot tips, and the bacterium was not detected in meristems. Tumorigenic *Agr. vitis* was also detected in cuttings of wild grapevines that were collected both adjacent to and far removed from commercial vineyards.

Hao *et al.* (2017) investigated the effect of tumorigenic and non-tumorigenic *Agr. vitis* strains on grape graft take and found that tumorigenic *Agr. vitis* negatively affects graft strength and subsequent plant growth.

Yang *et al.* (2020) reported that ferritin plays an important role in the pathogenesis of *Agr. tumefaciens* through regulating iron homeostasis and oxidative stress survival.

Recently, Hooykaas (2023) have summarised some of the major discoveries concerning *Agr. tumefaciens*, including its biology, the diversity among agrobacteria, their taxonomic position, the diversity in Ti plasmids, the molecular mechanism used by the bacterium to transform plants, and the discovery of protein translocation from the bacteria to host cells as an essential feature of *Agrobacterium*-mediated transformation.

9.44.5 Management

As there is no cure for infected plants; the prevention of infection is essential. Therefore, the main emphasis should be laid on the production of clean nursery stock.

1. Promote a disease-free nursery by adopting the following measures.
    a. Nursery should be raised in the soil, free from the pathogen and chewing insects.
    b. Destroy/dispose of the infected stock.
    c. Budding should be preferred over grafting. If grafting is done, well-fitted grafts wrapped with antiseptic tape should be used.
    d. Inspection and certification of the stock should be mandatory.
    e. Avoid or minimise the wounding of nursery stock.
    f. Budding and grafting tools should be properly disinfected.
2. Application of antibiotics and other bactericides has generally not been successful, partially due to their short-lived activity in soil and partially due to the occurrence of the bacterium in the host tissue.

3. Soil fumigation tried in nurseries has also met with poor success, mainly due to the occurrence of the bacterium in the plant tissue. However, in some cases, the incidence of crown gall has increased after fumigation, resulting from the elimination of microorganisms which compete with the crown gall bacterium (Burr, 2001).
4. *Agr. radiobacter* strain K 84 is very effective in controlling crown gall. Seeds, cuttings and roots of young seedlings are dipped in cell suspension of strain K 84 before sowing. Since its first release in 1973 in Australia, it has been adopted in Africa, Europe, North and South America, and is used on commercial scale for the control of crown gall on stone fruits and roses. It was the first case of a microorganism to be used commercially for the control of a plant disease. However, in 1973, some reports from Greece indicated its non-effectiveness for controlling crown gall disease (Kerr, 1980).

   Agr. radiobacter str. K 84 produces a bacteriocin called agrocin 84, which is inhibitory to many pathogenic strains of *Agrobacterium*. The biosynthesis of agrocin 84 is encoded by a 48 kb plasmid, called pAgK84. It is a conjugative plasmid and can be transferred to the agrobacteria including pathogens. The transfer of this plasmid makes the recipient bacterium resistant to agrosin 84, thereby resulting in non-control of the recipient bacterium with str. K 84. This might have been one of the reasons for the failure of control of crown gall in Greece. Another failure of biological control in one peach nursery in Italy, due to transfer of pAgK84 from *Agr. rhizogenes* K 84 to natural pathogenic agrobacteria has also occurred (Raio *et al.*, 2009).

The transfer of pAgK84 to plant pathogenic agrobacteria posed a potential threat to the continued success of the biological control of crown gall. Prof. Allen Kerr of Waite Agricultural Research Institute, Adelaide, Australia, in collaboration with Dr Farrand of University of Illinois, identified the region of DNA that mediates the transfer of the plasmid pAgK84 to other bacteria. Prof. Kerr's team deleted this region and constructed a deletion mutant (transfer-deficient, Tra) and named it strain K 1026. In June 1987, strain K 1026 was released for experimental field trials at the Waite Agricultural Research Institute, Adelaide, Australia. This was only the third recombinant DNA organism to be field-tested anywhere in the world. In 1988, the Department of Agriculture, New South Wales, released strain K 1026 for commercial use as a pesticide, under the trade name
'No Gall'. The strain K 1026 was the first recombinant DNA organism to be released for commercial use anywhere in the world. Strain K 1026 has not been approved for release in any other country, with the exception of Australia, where strain K 84 is still used for biocontrol. Strain K 1026 is not intended to be used on fruit-bearing crops. Out of two biovars of *Agr. tumefaciens*, biovar 2 is more sensitive to biocontrol by strain K 84 or K 1026 of *Agr. radiobacter.*

Agr. radiobacter strain K84 is widely used for the control of crown gall disease. Sharma and Gupta (2017) have shown the possibility of using *Agr. radiobacter* strain UHFBA-218, a native strain isolated from rhizosphere soil of cherry root stock Colt (Solan, Himachal Pradesh, India), against *Agr. tumefaciens* isolate C 58. The frequency of conjugal transfer of agrocin genes and the acquisition of virulence genes (*VirD2* and *ipt*) was found to be minimal in strain UHFBA-218 compared to standard strain K84. Transconjugants obtained from the co-inoculation of *Agr. radiobacter* strain UHFBA-218 + *Agr. tumefaciens* isolates (C 58, I_1, and I_2) showed partial acquisition of virulence genes.

Gupta (2022) tested *Agr. radiobacter* strain UHFBA-218 over five years in Himachal Pradesh and obtained 92.14% disease control on peach, compared with 74.19% control provided by strain K-84 as seed treatment; it also provided 89% control as compared to 84% by strain K-84 as root dip on peach. Remarkable populations, i.e., 134.25×10^6 CFU/g and 154.7525×10^6 CFU/g of strain UHFBA-218, persisted over a year in rhizosphere soil of stone fruit plants when applied as seed treatment and root dip, respectively.

Sapna *et al.* (2011) reported that *in vitro* evaluation, *Agr. radiobacter* isolate Jd_2-(15) showed a maximum zone of inhibition of 4.16 cm against *Agr. tumefaciens*. In field evaluation, soil drench with Jd_2-(15) and soil solarisation (75 days) provided complete and 90.89% control of crown gall on peach seedlings.

Sharma *et al.* (2016) prepared a white stone powder-based formulation (Sol Gall) by mixing *Brevibacillus brevis* and *B. choshinensis* in a 1:1 ratio. Its soil drench application gave complete control of crown gall disease in peach nursery plants.

Asghari *et al.* (2020) demonstrated that two endophytic bacteria, namely, *Pseudomonas* sp. Sn48 and *Pantoea* sp. Sa14, isolated from domesticated and wild grapevines, can colonise roots and/or shoots of grapes and induce resistance in both above and below ground tissues of grapevines against *Agr. tumefaciens. Pseudomonas* sp. Sn48 and *Pantoea* sp. Sa14 are capable of inducing stilbenic phytoalexin production in grapevine tissues and prime plantlets for enhanced phytoalexin production after *Agr. tumefaciens* inoculation. It was also found that, in the majority of treatments, polyamine accumulation remained unchanged or slightly increased in plantlets treated with *Pseudomonas* sp. Sn48 and *Pantoea* sp. Sa14 compared with control. These findings reveal the efficacy of the selected endophytic bacteria in triggering grapevine resistance against *Agr. tumefaciens* and the possible use of these strains as an alternative to chemical control methods for grapevine crown gall disease management.

Yan *et al.* (2020) isolated *Bacillus albus* and *Bac. velezensis* from the rhizospheric soil of flowering cherry saplings and identified these based on morphological characteristics, physiological analysis, biochemical identification and multilocus sequence analysis. These strains showed distinct antagonistic activities against *Agr. tumefaciens.* In a greenhouse trial, the incidence of crown gall in flowering cherry seedlings without *Bac. albus* and *Bac. velezensis* treatments was 100%, and the mean diameter of the root nodules was 18.8 mm. The disease incidence in seedlings treated with *Bac. albus* and *Bac. velezensis* strains was 16.7% and 22.3%, respectively, and the mean diameter of the root nodules was 6.5 mm and 8.4 mm, respectively. In the field trials, the incidence of crown gall in untreated cherry saplings was 86%, while the incidence of crown gall in saplings treated with the *Bac. albus* and *Bac. velezensis* strains was 32% and 38%, respectively. These results indicate that *Bac. albus* and *Bac. velezensis* strains have a high potential as biocontrol agents for controlling crown gall of flowering cherry saplings.

The following precautions should be taken to ensure effective control with biocontrol agents.

1. Do not mix the biocontrol agent with chlorinated water.
2. Do not expose the biocontrol agent to direct sunlight.
3. Do not expose the biocontrol agent to either very high or freezing temperatures.

Strains K 84 and K 1026 are effective in controlling crown gall caused only by nopaline-type strains of *Agr. tumefaciens* and *Agr. rhizogenes* and not effective for crown gall of grapes caused by *Agr. vitis*. Dipping of basal ends of grape cuttings in cell suspension of *Rahnella aquatilis* HX2 reduced the incidence of grape crown gall to 30.8% compared to 93.5% in control in a three-year field trial (Chen *et al.*, 2007). F 2/5, a non-tumorigenic strain of *Agr. vitis* was also found effective for controlling crown gall of grapes. F 2/5 produces an antibiotic that inhibits the growth of many tumorigenic *Agr. vitis* strains *in vitro*, but antibiotic-minus mutants of F 2/5 were found to be as effective as the wild-type strain for the control of crown gall (Burr *et al.*, 1997). F 2/5 gives better disease control when applied 24 h prior to pathogen inoculation. Kawaguchi (2009) obtained an effective control of grapevine crown gall with a non-pathogenic *Rhib. vitis* (syn. *Agr. vitis*) strain VAR03–1. Soaking of roots of grapevine seedlings in a cell suspension of strain VAR03–1 before planting in infested soil caused significant reduction in tumour formation in a greenhouse test and also gave an effective control in the field. The established population of VAR03–1 strain was 10^6 cfu g^{-1} in the grapevine rhizosphere soil and it persisted for two years.

Agr. vitis strains ARK-1, ARK-2 and ARK-3 isolated from graft unions of nursery stock of grapevine were compared with *Agr. vitis* strain VAR03-1, a biocontrol agent effective against crown gall mentioned above. Among these three strains, strain ARK-1 inhibited maximum tumour formation, established its populations on roots of grapevine tree rootstock and persisted on roots for a year. None of the three strains produced a zone of inhibition against *Agr. vitis* (Ti) strain on yeast mannitol agar (YMA) medium. Strain ARK-1 did not reduce tumour incidence on the stems of grapevine when the strain was dead, or when only culture filtrate was used. These findings show the possibility that these three strains inhibit grapevine crown gall *in planta* by a different mechanism than that of VAR03-1. As strain ARK-1 was most effective in inhibiting tumour formation on grapevine, it appears to be a promising new agent for the control of grapevine crown gall (Kawaguchi and Inoue, 2012).

Kawaguchi *et al.* (2015) reported that soaking the roots of apple, Japanese pear, peach, rose and tomato in a cell suspension of *Agr.* (=*Rhizobium*) *vitis* ARK-1, before planting them into soil infected with tumorigenic *Agrobacterium* spp., reduced the number of plants developing crown gall tumours. Meta-analyses of the results from field trials of apple (six trials), Japanese pear (four trials) and peach (four trials), conducted during 2010–2013, showed integrated risk ratio (IRR) after treatment with ARK-1 to be 0.38 for apple crown gall, 0.16 for Japanese pear crown gall and 0.20 for peach crown gall, indicating that the disease incidence was significantly reduced by ARK-1 treatment. Meta-analyses of the results from three greenhouse trials of both rose and tomato showed IRR after treatment with ARK-1 to be 0.29 for rose crown gall and 0.16 for tomato crown gall, indicating that the disease incidence was significantly reduced by ARK-1 treatment.

The submersion of dormant grape cuttings in water bath at 55 °C for 30 min eradicates the inoculum from the cuttings except the survival of low levels of inoculum in tissues near the galls. The most effective way of producing *Agr. vitis* free grapevines is by propagating vines from shoot tips *in vitro* because the bacterium does not systemically invade grape shoots. Establishment of such a planting requires the selection of an area where soil is free from *Agr. vitis*.

Kawaguchi (2023) found that *Allorhizobium vitis* strain ARK-1 was the most effective antagonist and recommend its use for controlling grapevine crown gall. Kawaguchi *et al.* (2024) reported that while the control efficacy, of soaking of grape roots in cell suspension of a non-pathogenic *Allorhizobium vitis* strain ARK-1, exceeded that of soil injection, the latter method also proved effective in managing crown gall of grapevines. Therefore, soil injection is recommended as a viable approach for grapevine crown gall control.

REFERENCES

Al-Karablieh, N. and Khlaif, H. (2002). Occurrence and distribution of crown gall disease in Jordan. *Phytopathol. Medit.* 41: 226–34.

Asghari, S., Harighi, B., Ashengroph, M., Clement, C., Aziz, A. *et al.* (2020). Induction of systemic resistance to *Agrobacterium tumefaciens* by endophytic bacteria in grapevine. *Plant Pathol.* 69: 827–37.

Bai, J., Yao, T., Lan, X., Yang, Y. and Wang, Z. *et al.* (2022). Isolation of *Agrobacterium tumefaciens*/biovar 1 from the crown gall of *Cerasus humilis* in China. *J. Plant Pathol.* 104: 141–48. https://doi.org/10.1007/s42161-021-00947-6

Borges, R. C. F., Rossato, M., Albuquerque, G. M. R., Ferreira, M. A., Brasileiro, A. C. M. *et al.* (2019). Crown gall caused by *Agrobacterium tumefaciens* species complex: a novel nursery disease of *Tectona grandis* in Brazil. *J. Plant. Pathol.* 101: 445. https://doi.org/10.1007/s42161-018-00211-4

Bradbury, J. F. (1986). *Guide to the Plant Pathogenic Bacteria.* CAB Inter-national, Wallingford.

Burr, T. (2001). Crown gall. In: O. C. Maloy and T. D. Murray (eds.). *Encyclopedia of Plant Pathology*. Vol. 1. John Wiley and Sons, Inc., New York, pp. 263–65.

Burr, T. J. and Otten, L. (1999). Crown gall of grape: Biology and disease management. *Annu. Rev. Phytopathol.* 37: 53–80.

Burr, T. J., Reid, C. L., Tagliati, E., Bazzi, C. and Sule, S. (1997). Biological control of grape crown gall by strain F2/5 is not associated with agrocin production or competition for attachment sites on grape cells. *Phytopathology* 87: 705–11.

Chen, F., Guo, Y. B., Wang. J. H., Li, J. Y. and Wang, H. M. (2007). Biological control of grape crown gall by *Rahnella aquatilis* HX2. *Plant Dis.* 91: 957–63.

Conn, H. J. (1942). Validity of the genus *Acaligenes*. *J. Bacteriol.* 44: 353–60.

De Cleene, M. and De Ley, J. (1976). The host range of crown gall. *Bot. Rev.* 42: 389–466.

Durgapal, J. C. (1971). A preliminary note on bacterial diseases of temperate plants in India. *Indian Phytopathol.* 24: 379–82.

Fuller, S. L., Savory, E. A., Weisberg, A. J., Buser, J. Z., Gordon, M. I., *et al.* (2017). Isothermal amplification and lateral-flow assay for detecting crown-gall-causing *Agrobacterium* spp. *Phytopathology* 107: 1062–68.

Garrity, G. M., Bell, J. A. and Lilburn, T. (2005). The revised road map to the manual. In: D. J. Brenner, N. R. Krieg, J.T. Staley and G. M. Garrity (eds.). *Bergey's Manual of Systematic Bacteriology*. 2nd Ed., Vol. 2, The *Proteobacteria*, Part A, *Introductory Essays*. Springer-Verlag, New York, pp. 159–200.

Gelvin, S. B. (2010). Plant proteins involved in *Agrobacterium*-mediated genetic transformation. *Annu. Rev. Phytopathol.* 48: 45–68.

Goodner, B., Hinkle, G., Gattung, S., Miller, N., Blanchard, M. *et al.* (2001) Genome sequence of the plant pathogen and biotechnology agent *Agrobacterium tumefaciens* C58. *Science* 294: 2323–28.

Gupta. A. K. (2022). Crown gall in stone fruits: biological control by using *Agrobacterium radiobacter* strain UHFBA-218 and its suitability for commercial exploitation. *Indian Phytopathol.* 75: 831–44.

Hao, L., Kemmenoe, D., Canik Orel, D. and Burr, T. (2017). Investigation of tumorigenic and non-tumorigenic *Agrobacterium vitis* strains on grape graft take (Abstr.). *Phytopathology* 107: S2.8.

Hooykaas, P. J. J. (2023). The Ti plasmid, driver of *Agrobacterium* pathogenesis. *Phytopathology* 113: 594–604.

Johnson, K. L., Cronin, H., Reid, C. L. and Burr, T. J. (2016). Distribution of *Agrobacterium vitis* in grapevines and its relevance to pathogen elimination. *Plant Dis.* 100: 791–96.

Johnson, K. L., Zheng, D., Kaewnum, S., Reid, C. L. and Burr, T. (2013). Development of a magnetic capture hybridization Realtime PCR assay for detection of tumorigenic *Agrobacterium vitis* in grapevines. *Phytopathology* 103: 633–40.

Kawaguchi, A. (2009). Studies on the diagnosis and biological control of grapevine crown gall and phylogenetic analysis of tumorigenic *Rhizobium vitis*. *J. Gen. Plant Pathol.* 75: 462–63.

Kawaguchi, A. (2015). Biological control agent *Agrobacterium vitis* strain ARK-1 suppresses expression of the *virD2* and *virE2* genes in tumorigenic *A. vitis*. *Eur. J. Plant Pathol.* 143: 789–99.

Kawaguchi, A. (2023). Evaluation of biological control methods for grapevine crown gall through network meta-analysis. *Phytopathology* 113:S3.139.

Kawaguchi, A. and Inoue, K. (2012). New antagonistic strains of nonpathogenic *Agrobacterium vitis* to control grapevine crown gall. *J. Phytopathol.* 160: 509–18.

Kawaguchi, A., Inoue, K. and Tanina, K. (2015). Evaluation of the nonpathogenic *Agrobacterium vitis* strain ARK-1 for crown gall control in diverse plant species. *Plant Dis.* 99: 409–14.

Kawaguchi, A., Kirino, N., Inoue, K. and Noutoshi, Y. (2024). Biological control for grapevine crown gall through soil injection with *Allorhizobium vitis* strain ARK-1. *Eur. J. Plant Pathol.* https://doi.org/10.1007/s10658-024-02913-1

Kerr, A. (1980). Biological control of crown gall through production of agrocin 84. *Plant Dis.* 64: 25–30.

Li, Q., Guo, R.-J., Li, S.-D., Li, S.-F. and Wang, H.-Q. (2015). Determination of tumorigenic *Agrobacterium* density in soil by real-time PCR assay and its effect on crown gall disease severity. *Eur. J. Plant Pathol.* 142: 25–36.

Loyter, A., Rosenbluh, J., Zakai, N., Li, J., Kozlovsky, S. V., *et al.* (2005). The plant VirE2 interacting protein 1. A molecular link between the *Agrobacterium* T-complex and the host cell chromatin. *Plant Physiol.* 138: 1318–21.

Nester, E. (2008). *Agrobacterium*: The natural genetic engineer: 100 years later. APSnet Feature Story. http://www.apsnet.org/online/feature/Agrobacterium/.

Ophel, K. and Kerr, A. (1990). *Agrobacterium vitis* sp. nov. for strains of *Agrobacterium* biovar 3 from grapevines. *Int. J. Syst. Bacteriol.* 40: 236–41.

Palacio-Bielsa, A., Gonzalez-Abolafio, R., Alvarez, B., Lastra, B., Cambra, M. A. *et al.* (2009). Chromosomal and Ti plasmid characterization of tumorigenic strains of three *Agrobacterium* species isolated from grapevine tumours. *Plant Pathol.* 58: 584–93.

Raio, A., Peluso, R., Puopolo, G. and Zoina, A. (2009). Evidence of pAgK84 transfer from *Agrobacterium rhizogenes* K84 to natural pathogenic *Agrobacterium* spp. in an Italian peach nursery. *Plant Pathol.* 58: 745–53.

Sapna, Gupta A. K. and Khosla K. (2011). Bio-intensive management of crown gall in peach nursery. *Plant Dis. Research* 26: 183.

Sharma, A. and Gupta, A. K. (2017). New insights in the biological control of crown gall through native *Agrobacterium radiobacter* strain UHFBA- 218. *Plant Dis. Res.* 32: 137–52.

Sharma, A., Gupta, A. K., Singh, D., Khosla, K. and Mahajan, R. (2016). Biological control of crown gall disease in peach and cherry nursery plants by white stone powder based formulation (Sol Gall) of *Brevibacillus* spp. *Indian Phytopathol.* 69: 231–36.

Smith, E. F. and Townsend, C. O. (1907). A plant tumour of bacterial origin. *Science*, N. S. 26: 671–73.

Sule, S., Horka, M., Matouškova, H., Kubesova, A., Salplachta, J., *et al.* (2012). Characterization of *Agrobacterium* species by capillary isoelectric focusing. *Eur. J. Plant Pathol.* 132: 81–89.

Walker, J. C. (1969). *Plant Pathology*. 3rd Ed. McGraw-Hill Book Co. New York, 819 pp.

Wikipedia (2008). *Agrobacterium tumefaciens*- Wikipedia, the free encyclopedia. http://en.wikipedia.org/wiki/Agrobacterium_tumefaciens]

Wood, D. W., Setubal, J. C., Kaul, R. Monks, D. E., Kitajima, J. P. *et al.* (2001). The genome of the natural genetic engineer *Agrobacterium tumefaciens* C58. *Science* 294: 2317–23.

Yakabe, L. E., Parker, S. R. and Kluepfel, D. A. (2014). Incidence of *Agrobacterium tumefaciens* biovar 1 in and on "Paradox" (*Juglans hindsii* × *Juglans regia*) walnut seed collected from commercial nurseries. *Plant Dis.* 98: 766–70.

Yan, P., Zhang, X. T., Hu, L. J., Wang, Y. H., Zhu. M. L. *et al.* (2020). Two novel strains, *Bacillus albus* JK-XZ3 and *B. velezensis* JK-XZ8, with activity against *Cerasus* crown gall disease in Xuzhou, China. *Australasian Plant Pathol.* 49: 127–36. https://doi.org/10.1007/s13313-020-00682-z

Yang, J., Pan, X., Xu, Y., Li, Y., Xu, N. *et al.* (2020). *Agrobacterium tumefaciens* ferritins play an important role in full virulence through regulating iron homeostasis and oxidative stress survival. *Mol. Plant Pathol.* 21: 1167–78.

Young, J. M., Kuykendall, L. D., Martinez-Romero, E., Kerr, A. and Sawada, H. (2001). A revision of *Rhizobium* Frank 1889, with an emended description of the genus, and the inclusion of all species of *Agrobacterium* Conn 1942 and *Allorhizobium undicola* de Lajudie *et al.* 1998 as new combinations: *Rhizobium radiobacter*, *R. rhizogenes*, *R. rubi*, *R. undicola* and *R. vitis*. *Int. J. Syst. Evol. Microbiol.* 51: 89–103.

9.45 HAIRY ROOT OF APPLE

Hairy root is another important disease of apple, though it is less destructive than crown gall. It is also known by other names, such as '**woolly root**' and '**woolly knot**', and was long considered merely a form of crown gall. The bacterium also infects *Malus pumila*, *Rosa* spp., *Rubus* spp., rose, pear, raspberry, cotoneaster, hollyhock, honey locust, honey suckle, mulberry, peach, pea vine, quince, Russian olive and spirea. Artificially inoculated host range extends to more than 37 plant species, belonging to 30 genera and 15 families (De Cleene and De Ley, 1981). The disease is widely spread in apple nurseries and orchards in the USA. It is also found on rose in both California and Texas. The disease also occurs in Japan, Bulgaria and France. Pulawska and Kałużna (2012) also reported the occurrence of the disease on different host plants from Poland, and Kuzmanović *et al.* (2013) reported its occurrence on apricot trees from central Serbia.

9.45.1 Symptoms

The disease commonly occurs on grafted apple nursery trees, which are usually one, two or three years old. The production of a large number of small roots either directly from stems or from roots or from localised hard swellings, that often occur at graft unions of root and scion, is the characteristic feature of the disease (Walker, 1969). The production of roots may be as profuse as that of witches' broom. It can appear simultaneously with crown gall on the same plant and the symptoms can be confused in the early stage. On rose, the hairy roots originate at the ends of cuttings or at disbud scars. These malformations may be fleshy at first but eventually become fibrous. Infected apple trees become stunted.

9.45.2 Causal Organism

Agrobacterium rhizogenes (Riker *et al.* 1930) Conn 1942 emend. Sawada *et al.* 1993

Synonym: *Agr. radiobacter* biovar *rhizogenes*, *Agr. radiobacter* pv. *rhizogenes*

The below-given description of the bacterium is taken from Sawada *et al.* (1993) and Hayward and Waterston (1998). Cells are Gram-stain-negative, non-sporing rods, measure 0.6–1.0 × 1.5–3.0 µm, motile by 1–4 peritrichous flagella. Aerobic, possessing a strictly respiratory type of metabolism. The colonies are circular, convex, smooth and non-pigmented to light beige. Growth on carbohydrate-containing media is usually accompanied by copious, extracellular polysaccharide slime. Optimum temperature for growth is between 25 and 28 °C and no growth occurs at 35 °C. Strains require growth factors like biotin and do not grow in simple mineral-salts medium. Citrate and L-tyrosine are utilised. Acid is produced from dulcitol and -methyl-D-glucoside in the Ayers–Rupp–Johnson's synthetic medium containing 0.02% yeast extract. 3-Ketolactose is not produced from lactose and the arginine dihydrolase test is also negative. A large tumour-inducing plasmid (Ti) or a hairy root-inducing plasmid (Ri) is detected in many strains. Mol% G + C of the DNA is 59.0–63.0 (*Tm*). Like *Agr. tumefaciens*, it is also used as a tool in genetic engineering, though less frequently.

Chagas De Freitas and Taylor (2019) screened *Pseudomonas* collection both *in vitro* (clear zone assays) and *in planta* (reduction in disease severity) assays against *Agr. rhizogenes*. *In vitro* studies, they identified 14 *Pseudomonas* strains that exhibited zone-clearing activity. Further testing of these strains in planta revealed that two strains were also effective in reducing the disease incidence in tomato under soil and hydroponic conditions. The strains, not found effective *in vitro* testing, were also screened on tomato plants aiming to identify strains that may have been missed due to their non-performance in the *in vitro* study. From this testing, five additional strains reduced the disease incidence on tomato plants. No doubt, *in vitro* screening is helpful to identify potential strains, but strains with other potential modes of action could be lost in this method of testing.

9.45.3 Disease Cycle

The epidemiology and other aspects of the disease are comparatively much less worked out than that of crown gall. Like *Agr. tumefaciens*, *Agr. rhizogenes* is also an exclusively wound pathogen. It is readily disseminated on nursery stock, which may become infected from soil or by tools contaminated with the pathogen when cuttings and graftings are made. On rose, the bacteria may penetrate roots through injuries caused by the root-lesion nematode, *Pratylenchus vulnus*. The bacterium survives in steamed or untreated soil for more than one year. The virulence of the bacterium has been maintained in the culture for over 19 years. Other fundamental aspects of the disease cycle are similar to those of crown gall.

Bosmans *et al.* (2016) developed a SYBR Green-based quantitative real-time PCR assay based on the *rol* locus of the root-inducing-plasmid for detection and the quantification of *Agrobacterium* biovar 1 strains in hydroponic systems. The assay designed based on all *rolB* sequences currently available in GenBank, and developed using a collection of both target and non-target strains, was specific for rhizogenic *Agrobacterium* biovar 1 strains. Based on a calibration with artificially contaminated water samples mimicking hydroponic conditions, unknown bacterial concentrations could be accurately quantified in water samples from surveys carried out in different greenhouses. The detection limit obtained was of less than one cell mL^{-1} of water, following filtration of a volume of 500 mL. The qPCR assay is suitable for the routine detection of rhizogenic *Agrobacterium* biovar 1 strains in aqueous samples. Moreover, the assay can be used for pathogen assessment at the pre-symptomatic stage, allowing sufficient time to undertake the necessary preventive measures to control the disease in hydroponics before the infection of plants.

9.45.4 Management

Management approaches are the same as those for crown gall.

Chagas De Freitas *et al.* (2023) determined the potential of two *Pseudomonas* spp. (strain 1B1 and 93G8) to control hairy root disease caused by biovar 1 *Agr. rhizogenes*. There was a significant negative relationship between yield and disease severity index (DSI). Fruit number and fruit weight per plant were significantly higher in plants treated with *Pseudomonas* compared to untreated plants, with a DSI 42% lower for strain 1B1 and 38% lower for strain 93G8. The estimated net cash income (NCI) of *Pseudomonas* treatment was consistently higher relative to the NCI of untreated plants.

REFERENCES

Bosmans, L., Paeleman, A., Moerkens, R., Wittemans, L., Van Calenberge, B., *et al.* (2016). Development of a qPCR assay for detection and quantification of rhizogenic *Agrobacterium* biovar 1 strains. *Eur. J. Plant Pathol.* 145: 719–30.

Chagas De Freitas, C., Moraes, W. B. and Taylor, C. G. (2023). Examining the economic impact of beneficial *Pseudomonas* application for hairy root disease control on hydroponically grown tomatoes. *Phytopathology* 113: S3.165.

Chagas De Freitas, C. and Taylor, C. G. (2019). Examining the effectiveness of in vitro and in planta assay in assessing biocontrol activity of *Pseudomonas* strains against *Agrobacterium rhizogenes* (Abstr.). *Phytopathology* 109: S2.24.

Conn, H. J. (1942). Validity of the genus *Acaligenes*. *J. Bacteriol.* 44: 353–60.

De Cleene, M. and De Ley, J. (1981). The host range of infectious hairy-root. *Bot. Rev.* 47: 147–94.

Hayward, A. C. and Waterston, J. M. (1998). *Agrobacterium rhizogenes. IMI Description of Fungi and Bacteria*. Sheet 41, CAB International, Wallingford.

Kuzmanović, N., Ivanović, M., Prokić, A., Gasić, K., Blagojević, N., *et al.* (2013). Identification and characterization of *Agrobacterium* spp. isolated from apricot in Serbia. *Eur. J. Plant Pathol.* 137: 11–16.

Pulawska, J., and Kałużna, M. (2012). Phylogenetic relationship and genetic diversity of *Agrobacterium* spp. isolated in Poland based on *gyrB* gene sequence analysis and RAPD. *Eur. J. Plant Pathol.* 133: 379–90.

Riker, A. J., Banfield, W. M., Wright, W. H., Keitt, G. W. and Sagen, H. E. (1930). Studies on infectious hairy root of nursey apple trees. *J. Agric. Res.* 41: 507–40.

Sawada, H., Ieki, H., Oyaizu, H. and Matsumoto, S. (1993). Proposal for rejection of *Agrobacterium tumefaciens* and revised descriptions for the genus *Agrobacterium* and for *Agrobacterium radiobacter* and *Agrobacterium rhizogenes*. *Int. J. Syst. Bacteriol.* 43: 694–702.

Walker, J. C. (1969). *Plant Pathology*. 3rd Ed. McGraw-Hill Book Co. New York, 819 pp.

9.46 FIRE BLIGHT OF POME FRUIT TREES

Fire blight is probably the most devastating bacterial disease of some pome fruits and also the most feared by growers (Thomson, 2001). It is able to kill complete pome fruit orchards within a single growing season (Vanneste, 2000). It is most destructive on pears, but certain apple and quince varieties are also highly susceptible. Crabapple (*Malus* spp.) firethorn (*Pyracantha* spp.) and hawthorn (*Crataegus* spp.) are also frequently attacked. The bacterium also occasionally attacks rock spray (*Cotoneaster* spp.) The bacterium infects more than 180 plant species, including fruit trees, wild and ornamental plants, most of them belonging to the family Rosaceae. Kim *et al.* (2001) have reported a similar disease on Asian pear, *Pyrus pyrifolia* from Korea and Japan caused by *Erwinia pyrifoliae*. Lopez *et al.* (2011) reported *Erw. piriflorinigrans* sp. nov. as the cause of pear blossom necrosis in Valencia, Spain. The bacterium was identified using a polyphasic approach. Myung *et al.* (2016) reported fire blight of apple caused by *Erw. amylovora* from Korea.

Fire blight is of historical importance because this was the first plant disease shown to be caused by a bacterium. Professor T. J. Burrill from the University of Illinois, USA, through a series of experiments starting from 1878 to 1884 proved that the disease is caused by a bacterium. It is considered that fire blight originated from America and was known to be there for many years before it was discovered in the Hudson Valley of New York in 1794. The disease spread to most areas of the USA by the 1900s. Since then, it has spread to the whole of Europe (except Latvia and

Ukraine), Jordan, Iran, Israel, Lebanon, North America, New Zealand and Turkey. The disease is known to occur in more than 40 countries. To date, there is no confirmed report of its occurrence from India.

A survey in California in 1976 estimated a loss of $4.7 million due to fire blight while in Southwest Michigan alone the loss in 1991 was $3.8 million. Again, in 2000, a severe outbreak of fire blight of apples occurred in Michigan, which resulted into a loss of $42 million due to the removal of 350,000–450,000 apple trees covering approximately 1550–2300 acres (Longstroth, 2001). In 1998, apple and pear growers in Washington and northern Oregon suffered a loss of $ 68 million due to fire blight damage.

Shtienberg *et al.* (2015) have given an account of the existence of fire blight of pear in Israel over the last 30 years, highlighting unique characteristics that play a crucial role in the epidemiology of the disease under Israeli conditions, summarising the conclusions derived from the local experience and presenting future perspectives regarding fire blight management. The '**take-home**' message of the Israeli fire blight story is that the battle against this hazardous disease is never ending.

In their review article, Doukkali *et al.* (2022) emphasised on the aspects of the fire blight, namely, symptoms, disease cycle and pathogen development, as well as control methods and measures undertaken against the fire blight outbreak. The detection and control methods imposed by the government and employed by farmers drew much attention for the disease management. Integrated control strategies using tolerant cultivars in combination with sanitation practices, as well as cultural, physical, chemical and biological controls, have been discussed in this review.

Fire blight was first observed in Yili, Xinjiang Province, in Northwestern China, in 2016. Since then, it has spread to most of the pear-producing regions in Xinjiang Province and parts of Gansu Province. The disease caused severe damage to China's pear and apple industries, including the 2017 disease epidemic in Korla, Xinjiang, which caused an overall yield reduction of 30% to 50% in Korla and the destruction of over one million pear trees. Over the past few years, a combined effort of research, extension and education by the Chinese government, scientists, and fruit growers has greatly alleviated outbreaks and epiphytotics in affected regions, while successfully limiting the further spread of the disease to new geographical regions. Sun *et al.* (2023) have reviewed the occurrence, spread, and damage of fire blight to the Chinese fruit industry, as well as the management options used in China and their outcomes. They also discuss future perspectives for restraining the spread and alleviating the damage of the disease in China.

The fire blight research community has been meeting every three years since 1977 to exchange and discuss progress and new findings in the areas of resistance breeding and genetics, host–pathogen interaction, disease management, the spread of the pathogen and the genetics and population structure of *Erw. amylovora*. While until 2013 the meetings took place under the auspices of the International Society for Horticultural Science, since 2016 these have been held independently under the name of the International Symposium on Fire Blight (ISFB). The 1st ISFB was held in Girona, Spain, in 2016; the 2nd ISFB was held in Traverse City, MI, USA, in 2019, and the 3rd ISFB, in 2022, was hosted by the Institute for Breeding Research on Fruit Crops in Dresden-Pillnitz, which belongs to the Julius Kühn Institute (JKI), which is the Federal Research Centre for Cultivated Plants in Germany.

Erw. pyrifoliae, a causal agent of black shoot blight in apple and pear trees, is a plant pathogenic bacterium first reported in South Korea. The symptoms of black shoot blight are very similar to those of the fire blight disease in apple and pear trees caused by *Erw. amylovora*, as *Erw. pyrifoliae* has a genetically very close relationship with *Erw. amylovora*.

9.46.1 Symptoms

The symptoms are found on all the above-ground plant parts as well as on roots. Infected blossoms, leaves, twigs and branches give a burnt or scorched appearance, hence the name **"fire blight"**. The symptoms appear in the spring when the trees start a new growth. The first sign is the exudation of watery, light tan bacterial ooze from the twig, branch or trunk cankers (holdover cankers) formed in the previous season(s). The ooze turns dark after exposure to air, spreads to form a thin glistening layer, or dries down to form hard beads or films. The following different types or phases of symptoms appear. However, not all the types or phases appear in all the orchards or years or areas.

9.46.1.1 Blossom Blight

This is the most devastating phase of the disease. It is usually the first symptom to appear, within one to two weeks after bloom. Infected blossoms first appear water-soaked and then these begin to shrivel, wilt and eventually turn from brown to black. Infected blossoms may fall, but they often remain hanging on the trees. Infection of a single flower in the cluster usually results in the death of the whole spur. During the warm and humid weather, bacteria come out on the surface of infected blossoms in the form of ooze and are disseminated by pollinating insects (Figure 9.37a). This phase is critical for the outbreak of the fire blight epiphytotics because the bacteria can be readily disseminated throughout the orchard by pollinating insects.

9.46.1.2 Fruit Blight

Usually immature and unripened fruits are infected. Infection occurs either through an infected spur or directly through wounds or lenticels of the fruits. The infected fruits become water-soaked, turn brown, shrivel and finally turn black. Infected fruits usually remain clinging to the trees for several months. Bacterial ooze comes out on the surface of infected fruits in warm and humid weather (Figure 9.37b).

(a) (b) (c)

FIGURE 9.37 (a) Fire blight of apple: Blossom blight. (Source: Illinois Fruit and Vegetable News, Vol. 14 No. 7.) (b) Fire blight: Symptoms on Bartlett pear fruit. (Courtesy: Sugar David, www.plant-disease.ippc.orst.edu.) (c) Fire blight of apple: Shoot blight. (Courtesy: Florida Division of Plant Industry, Florida, Department of Agriculture and Consumer Services, Bugwood.org)

9.46.1.3 Shoot or Twig Blight

Also called **"blight strikes"**, these symptoms develop on actively growing vegetative shoots. Twigs or shoots may be infected directly, or infection from blossoms may move into these. Infected shoots wilt from the tip downward and form a diagnostic **"shepherd's crook"** (hooked) at the end. The bark of infected twigs turns brownish-black and is soft in the beginning but shrinks and hardens later on. The leaves on the infected twigs turn black and remain clinging to the twig (Figure 9.37c). This phase is more destructive to young and vigorously growing trees.

9.46.1.4 Leaf Blight

In addition to blightening of leaves on infected twigs, leaf blight occurs from the independent infection of leaves through wounds or through stomata and hydathodes. Necrotic, brown to black lesions appear along the midrib or main veins, between the veins and along the margins. As the blackening spreads, the leaves curl and shrivel, hang downwards and generally remain attached to the hooked and blighted twigs.

9.46.1.5 Trauma Blight

These symptoms are similar to shoot blight but are usually more random and widespread throughout the orchard. Symptoms result on account of the damage caused to the host by traumatic events like hail, frost or storm. The wounds caused by these traumatic events allow the bacteria to penetrate the host, bypassing the normal defences even in the resistant hosts.

9.46.1.6 Canker Blight

It is also called by other names like **"limb"**, **"trunk"** or **"body blight"**, depending on which part of the tree, the infection occurs. Cankers are formed as a result of bacteria travelling systemically into the wood tissue of the tree. From infected twigs and fruit spurs, infection moves down to the branches, scaffold limbs and trunks, forming cankers. In the beginning, the bark where canker is formed appears water-soaked and, later on, becoming discoloured, sunken and dry. Cankers formed on large limbs and trunk appear slightly to deeply depressed and occasionally develop cracks at the margins. When a canker enlarges and girdles the branch, the portion of the branch above the cankers dies. Many small cankers, less than 2.5 centimetres in diameter, usually not sunken and rarely cracked, often appear on small twigs. These usually go unnoticed due to their small size, but are a major source of inoculum for infection in the next spring. One active, overwintering canker contains sufficient bacteria to severely contaminate flower blossoms of trees present in ¼ to ½ acre (Smith, 2010). Cankers without cracks are termed indefinite and are more viable source of inoculum.

9.46.1.7 Collar and Root Blight

When the cankers develop at the base of the trunk, this is termed **collar blight**. The formation of cankers at collar or ground level results in the early death of plants.

The infection of the root by the bacterium results in **root blight** or **rootstock blight**. The primary route of entry of the bacterium into the rootstock is internally, often through apparently healthy limbs and trunks of trees that had only a few infected blossoms or shoots of the scion. After reaching the susceptible rootstock, the bacteria initiate the formation of new cankers that completely girdle and kill the tree in one to a few months. Some of the rootstocks, namely, M-26 and M-9, generally preferred in tree size controlling orchards, are highly susceptible to fire blight pathogen.

9.46.2 Causal Organism

Erwinia amylovora (Burrill 1882) Winslow *et al.* 1920

Synonym: *Bacterium amylovorum*

The bacterium is rod-shaped, Gram-Stain-negative, motile with peritrichous flagella and requires nicotinic acid as a growth factor. Colonies on nutrient agar or King's agar B are smooth, creamy-white and round with a glistening shine. Miller and Schroth's selective (MSS) medium, which utilises sorbitol as the carbon source, is suitable for the isolation of the bacterium due to the elimination of common bacterial species. The modification of MSS medium by Ishimaru and Klos (1984) is helpful for the isolation of the bacterium in the presence of a high population of *Pantoea agglomerans*. The optimum temperature for growth is around 27 °C. Gelatin is liquefied, levan is produced from sucrose but nitrate reduction and H_2S production are negative. Lactate, citrate, formate and trehalose are utilised as a sole source of carbon. The G + C content of the DNA is 53.6–54.1 mol%.

The bacterium contains a 40 kb segment of DNA that contains two large regions, termed *hrp* and *dsp*. Genes present in these regions code for harpin and probably for other proteins that are responsible for pathogenicity, hypersensitivity and perhaps host range. An extracellular polysaccharide (EPS) called amylovoran and the type III secretion system are two major, but separate virulence factors of the bacterium. The other virulence factors include lipopolysaccharides like levan, glucan and levansucrase. However, Billing (2011) has reported that neither the extracellular polysaccharide produced by *Erw. amylovora* is a toxin, nor responsible for the wilting symptom seen in the early stages of the disease.

Zeng *et al.* (2023), in their editorial article, have reviewed the information on *Erw. amylovora*. Tian *et al.* (2017) reported that the *Erw. amylovora* type VI secretion system (TVISS) plays a role in bacterial competition and virulence, possibly by influencing exopolysaccharide production. Of the 33 deletion mutants generated within three T6SS clusters of *Erw. amylovora*, 19 exhibited reduced virulence on immature pear fruit as compared with that of the wild-type strain. Among them, 6, 1 and 12 genes belonged to T6SS-1, T6SS-2 and T6SS-3 clusters, respectively. Interestingly, these 19 mutants also produced less amylovoran or levan or both.

Most strains of the bacterium contain a non-conjugative plasmid called the ubiquitous plasmid pEA29. Loss of this plasmid results in the reduction of virulence of the bacterium and the strains cured of this plasmid show thiamine auxotrophy. Presence of pEA29 influences the expression of chromosomal amylovoran biosynthetic operon during infection. Amylovoran is a pathogenicity determinant of the fire blight pathogen and thiamine biosynthesis is required for the high-level expression of the amylovoran biosynthetic operon during infection. Among the other plasmids found in *Erw. amylovora*, plasmid pEL60 is of special interest because it is an IncL/M plasmid with a high degree of similarity to the backbone of pCTX-M3, a multiple antibiotic resistance plasmid identified in the opportunistic human pathogen *Citrobacter freundii*. pEL60 is a cryptic plasmid, encoding only genes involved in replication, maintenance and conjugative transfer, while pCTX-M3 contains two additional insertions totalling 29.6 kb that encode a number of antibiotic resistance genes.

Koczan *et al.* (2009) also stressed the importance of amylovoran and levan in the pathogenesis of *Erw. amylovora*. They showed that biofilm formation played a critical role in the pathogenesis; amylovoran was required for biofilm formation and levan contributed to biofilm formation. Scanning electron microscopic analysis revealed an extensive biofilm formation in xylem vessels after inoculation with *Erw. amylovora* wild-type cells. On the other hand, Δams mutant cells were non-pathogenic and died rapidly following inoculation.

Castiblanco and Sundin (2018) investigated cellulose biosynthesis in *Erw. amylovora* Ea1189 and established that, in addition to amylovoran and levan, cellulose is a critical component of biofilms formed *in vitro* and in the apple host. They also demonstrated that, as reported for other bacterial species, this cellular process is also regulated at the post-translational level by the secondary messenger c-di-GMP through the cellulose synthase subunit BcsA. A role for the endoglucanase BcsZ in the c-di-GMP-dependent activation of cellulose biosynthesis in *Erw. amylovora* was also reported.

Lee *et al.* (2010) reported that virulence factor expression levels influenced the disease severity caused by wild isolates of *Erw. amylovora* on apple trees. On apple seedlings, three virulence characteristics, namely, amylovoran production, biofilm formation and growth in immature apple fruit, accounted for >70% of the variation in disease severity while in the case of greenhouse-grown trees, amylovoran production, biofilm formation, growth in immature apple fruit, hypersensitive cell death elicitation and sorbitol utilisation accounted for >75% of the variation in disease severity

Águila-Clares *et al.* (2018) investigated the response of *Erw. amylovora* to copper shock by a whole-genome microarray approach. Transcriptional analyses showed that, in the presence of copper, 23 genes showed increased expression and these genes were classified mainly into the transport and stress functional categories. Among these, the *copA* gene was strongly induced and regulated in a finely tuned manner by copper. Mutation of *copA*, *soxS*, *arcB*, *yjcE*, *ygcF*, *yhhQ*, *galF* and *EAM_3469* genes revealed that tolerance to copper in *Erw. amylovora* can be achieved by complex physiological mechanisms, including: (i) the control of copper homeostasis through, at least, the extrusion of Cu(I) by a P-type ATPase efflux pump CopA; and (ii) the overcoming of copper toxicity caused by oxidative stress by the expression of several reactive oxygen species-related genes, including the two major transcriptional factors SoxS and ArcB. Furthermore, complementation analyses demonstrated the important role of *copA* for copper tolerance in *Erw. amylovora*, not only *in vitro*, but also in inoculated pear shoots.

The first genome of *Erw. amylovora* to be published was the genome of the French isolate CFBP 1430 (Smits *et al.* 2010). Currently, 15 genome sequences of *Erw. amylovora* strains are publicly available, of which three isolates are reported to be specifically pathogenic on *Rubus* spp. (Mann *et al.* 2013; Smits *et al.* 2014).

Smits *et al.* (2017), in their review article, reported that the publication of the first *Erw. amylovora* genome has greatly accelerated and advanced our understanding of the fire blight pathogen. Further, the publications of multiple genomes made it clear that chromosomal diversity is relatively small and that most of the pan-genome variance is attributable to plasmids. In addition to gaining a more detailed view of the known virulence factors, genomics has enabled new breakthrough studies of virulence regulation mechanisms. Additionally, several niche adaptation and ecological fitness factors, though not directly influencing virulence, have been investigated in greater detail, providing novel insights into the physiology and ecology of the bacterium. Moreover, the application of genome data has yielded improved diagnostics and enabled population studies at different geographic scales.

In a study, via comparative approaches to the genome sequences of the strains of various *Erwinia* spp., Jin *et al.* (2023) designed specific primers targeting a hypothetical gene that is single copy and located in the chromosomal DNA of *Erw. amylovora*. This primer set specifically amplified the DNA of *Erw. amylovora* but no other bacteria, including *Erw. pyrifoliae*, *Pectobacterium* spp., *Pantoea* spp. and *Dickeya chrysanthemi*. Furthermore, this SYBR Green-based real-time PCR using the primer set allowed accurate estimation of the population of *Erw.* am*ylovora*.

9.46.3 Disease Cycle

The bacteria present in the dormant or holdover cankers, formed on the twigs, branches and stems in the previous year, are the main source of perpetuation. The bacteria present at the margins of the cankers are more active. The bacteria present in buds and apparently healthy wood also contribute to the survival, albeit to a lesser extent. Weishaupt *et al.* (2016) showed a potential role of fruit mummies (19.9% samples infected) and buds in overwintering, as well as a source of primary inoculum for dissemination of the pathogen early in the growing season. Blossoms of non-host plants growing close to infected trees were also shown to be colonised by *Erw. amylovora* and to enable epiphytic survival and propagation of bacteria. Sobiczewski *et al.* (2017) reported that *Erw. amylovora* has an ability to survive as a semi-necrotroph or necrotroph, which allows its overwintering in dead apple leaves.

In the next spring, when weather warms up, the bacteria present in the cankers begin to multiply and invade the healthy adjoining bark. During humid or wet weather, bacterial cells ooze out through the cracks and lenticels and this coincides with the opening of pear blossoms. Due to the sweet and sticky nature of bacterial ooze, various insects such as bees, flies and ants are attracted and get smeared with the ooze when they feed on it and spread it to all the flowers they visit subsequently. Splashing rain also spreads the bacteria from oozing cankers to flowers and other plant parts. Under dry conditions, the bacterial ooze dries to form long, narrow aerial strands, which are spread by wind currents.

Slack *et al.* (2017) made a microbiological examination of *Erw. amylovora* exopolysaccharide ooze and found an average of 10^8 cfu μL^{-1} in ooze droplets. Examination of apple host tissue at the site of emergence of ooze droplets, using scanning electron microscopy, revealed that ooze was not exuding through natural openings, but it was found on erumpent mounds and small (10 μm) tears in tissue. It is obvious that *Erw. amylovora* induced wounds in tissue provided the exit holes for ooze extrusion from the host. Analyses of *Erw. amylovora* populations in ooze droplets and within the stems from which ooze droplets emerged, indicated that approximately 9% of the total bacterial population from infected stems is diverted to ooze. Gene expression analyses revealed that *Erw. amylovora* cells in stem sections located above ooze droplets and in ooze droplets were actively expressing critical pathogenicity genes such as *hrpL*, *dspE* and *amsK*. The ooze that emerges from the host tissue acts as a source of large, concentrated populations of *Erw. amylovora*, and as the cells in ooze droplets are expressing genes required for pathogenesis, these are already primed for infection as and when they are dispersed from ooze to new infection courts.

Slack *et al.* (2022) tracked the populations of *Erw. amylovora* Ea110 on apple flower stigmas over the course of apple bloom in field studies conducted between 2016 and 2019. In 18 of 23 experiments, flower stigmas inoculated on the first day of opening were found to harbour large (10^6 to 10^7 cells per flower) populations of *Erw. amylovora* when assessed three to five days post-inoculation. However, populations inoculated on stigmas of flowers that were already open for three days did not reach 10^6 cells per flower, and populations inoculated on stigmas of flowers that were already open for five days never exceeded 10^4 cells per flower. During this study, ≥10-fold increases in *Erw. amylovora* stigma populations in a 24-h time period (termed population surges) were observed on 34.8, 20.0 and 4.0% of possible days on 1-, 3- and 5-day-open flowers, respectively. The population surges occurred on days with average temperatures as high as 24.5 and as low as 6.1 °C. Experiments incorporating more frequent sampling during days and overnight revealed that many population surges occurred between 10:00 p.m. and 2:00 a.m. A Pearson's correlation analysis of weather parameters occurring during surge events indicated that population surges were significantly associated with situations in which overnight temperatures increased or remained constant, in which wind speed decreased and in which relative humidity increased. These findings refine our knowledge of *Erw. amylovora* population dynamics and further indicate that *Erw. amylovora* is able to infect flowers during exposure to colder field temperatures than previously reported.

The bacteria present on the stigmas of flowers do not infect immediately, but grow epiphytically. Stigmas have a moist, nutrient-rich surface, which provides a favourable habitat for the multiplication of bacteria, increasing the number of cells up to $10^{6\text{-}7}$ per stigma. These bacteria are again spread to other flowers by rain or honeybees or other insects before the onset of environmental conditions favourable for infection. A mean temperature of 15.5 °C is required for the epiphytic growth of the bacteria on the stigmas.

With the help of rain or dew, the bacteria move from stigmas to the hypanthium or floral cup of the flower and enter the flowers through secretory cells (nectarthodes). After penetration into flowers, the bacteria multiply rapidly, producing necrotic symptoms.

The flowers can become infected within minutes after a rain or a heavy dew when the average daily temperature is 16 °C or higher. From the infected flower, infection spreads down the pedicel and then into the fruit spur. Infection of the spur causes the death of all the flowers, fruits and leaves borne on it. Shoot blight usually develops after blossom blight. The bacteria from infected flowers are disseminated by splashing rain and these enter leaves, twigs and branches. The penetration of leaves occurs mostly through wounds caused by insects, hailstorms, and strong winds but less frequently through stomata and hydathodes. From the leaf, infection spreads to the petiole and then to the twig or branch. Twigs and branches are penetrated directly through the lenticels and wounds and indirectly via flower or leaf infections. Once inside the host tissue, the bacteria move intercellularly through the parenchyma cells and in some cases through the xylem and phloem, leading to systemic infection. In systemic infection, surprisingly a single flower infection can cause the death of a tree in a few weeks. Unlike other wilt bacteria, *Erw. amylovora* moves rapidly from the vessels to the other tissues and causes the death of cells of these tissues. Blight and canker symptoms result from the death of these tissues.

Ideal conditions for infection and disease development include wet or humid weather coupled with daytime temperature ranging from 24–29 °C and a night temperature above 13 °C. High soil fertility, including excessive nitrogen which promotes succulent growth, also favours disease development. Vigorously growing shoots are most severely infected. Infection progresses very rapidly in the succulent tissue and may advance 10 to 30 cm in a day (Thomson, 1992).

Tancos *et al.* (2017) detected *Erw. amylovora* in asymptomatic budwood of trees more than 20 m away from the site of fire blight symptoms. In some seasons, there were significant ($P \leq 0.05$) differences in the incidence of *Erw. amylovora* in asymptomatic budwood collected from symptomatic trees and those up to 20 m away from them. The examination of individual bud dissections revealed that *Erw. amylovora* was within the tissue beneath the bud scales containing the meristem. There is an urgent need to better understand the factors that lead to the presence of *Erw. amylovora* in bud tissues in order to ensure the production of pathogen-free apple trees.

Billing (2011) has explained why views on host invasion by *Erw. amylovora* differ among workers. There are limited observations on orchard pear and apple trees; this is because in most experimental studies, young apple shoots on potted plants were used. Tissue maturity at the site of shoot inoculation is of prime importance. If xylem vessels are damaged, inoculum may be sucked into the vessels and the bacteria will multiply there. In younger tissue, there is less suction pressure. No one has suggested that migration in bark tissue cannot be a major route, and means of escape from mature xylem vessels have not been demonstrated and remain a matter for speculation.

Wöhner *et al.* (2018) confirmed the earlier finding that deletion of the type III secretion system abolishes the virulence of *Erw. amylovora.* The authors also established a new gene-for-gene relationship between the effector protein Eop1 and the fire blight-resistant ornamental apple cultivar Evereste and the wild species *Malus floribunda* 821.

Schachterle *et al.* (2022) reported that type III secretion system genes and flagellar genes are strongly co-expressed. Likewise, genes involved in biosynthesis of the exopolysaccharide amylovoran and sorbitol utilization had similar expression patterns. They further identified a group of 16 genes whose expression is increased and maintained at high levels throughout disease progression across time and tissues. They chose five of these genes for mutational analysis and observed that all the deletion mutants lacking these genes displayed a reduction in symptom development on apple shoots. Moreover, these induced genes were over-represented for genes involved in sulphur metabolism and cycling, suggesting the possibility of an important role for maintenance of oxidative homeostasis during apple flower infection.

Erfani-Moghadam and Zarei (2022) compared the activity of ascorbate peroxidase (APX), catalase (CAT), guaiacol peroxidase (POD) and phenylalanine ammonia-lyase (PAL), and the relevant biochemical compounds, including total phenol and flavonoid contents, in two pear cultivars in response to *Erw. amylovora* infection. All biochemical parameters were affected by the pathogen attack. However, the enzymatic responses of resistant and susceptible pear cultivars were completely different following *Erw. amylovora* infection. In cultivar Dargazi, the activity of CAT and APX was significantly suppressed in response to the pathogen infection during the early days after bacterial inoculation, while these were significantly activated in cultivar Duchess at the same stages. The POD showed a different pattern of activity from two other antioxidant enzymes. Pathogen attack induced the phenylpropanoid metabolism and caused an increase in PAL activity as well as total phenol and flavonoid content in both the pear cultivars, but a significantly higher rise was recorded in Dargazi than in Duchess. The data strongly suggest that the hypersensitivity response may contribute to the high resistance of Dargazi to *Erw. amylovora.* The results of the present study provide valuable information about the mechanisms underlying the fire blight resistance in Dargazi cultivar which would be highly beneficial for engineering fire blight resistance in *Pyrus* cultivars.

Mertoğlu *et al.* (2024) investigated the effects of mineral composition (N, P, K, Mg, Fe, Cu, and Zn) on the incidence of fire blight of pear. Minerals obtained from the leaves were not statistically correlated with resistance to fire blight; however the effect of all the minerals examined in the fruit, except for K, was found to be significant. Increased P and Mg concentrations were associated with disease resistance, while N, Zn, Fe, and Cu were associated with susceptibility. Nitrogen-causing susceptibility exerted 61% of this impact through itself directly and was sharply distinguished from other mineral compounds. The main effect of these minerals was in suppressing the negative effects of nitrogen on susceptibility by maintaining mineral balance.

Pirc *et al.* (2009) developed specific, sensitive and reliable TaqMan real-time PCR assays for the detection of *Erw. amylovora* targeting its chromosomal DNA (*ams*C gene and ITS region). Real-time PCR assays in combination with an automated DNA-extraction method based on magnetic beads (QuickPick TM) reliably detected at least 10^3 cells mL^{-1} (c. four cells per reaction) of the pathogen from blighted woody plant material. The assays were also able to detect the bacterium from symptomless plants, including cases where the subsequent isolation of the bacterium failed, giving an absolute quantification of *Erw. amylovora* before and after enrichment in liquid media, which provided proof for viability and the ability of the bacterium to multiply.

Gosch *et al.* (2012) described a LAMP of DNA assay for genomic DNA of *Erw. amylovora*, which relies on a highly specific primer design that does not allow amplification from typical DNA sources of the orchard biological system and from sample handling. The assay enabled the fast detection of down to approximately 20 cfu of pure *Erw. amylovora* or 100 fg genomic DNA (corresponding to approximately 25 *Erw. amyolovora* equivalents) per reaction in 45 min. The assay is suitable for field testing, as fast and reliable detection of *Erw. amylovora* in orchard samples is achieved with naked eye through a visual colour change with hydroxynaphthol blue.

Moradi *et al.* (2012) developed a reliable and rapid pathogen detection protocol that utilises LAMP for the detection of *Erw. amylovora*, using primers derived from the highly conserved fragment of the chromosomally *amsH* gene. The testing of 208 naturally infected samples revealed that the specificity value of LAMP was 84%, while conventional and nested PCR could detect only 59% and 73% of the whole collection, respectively. The specificity assay also showed that the LAMP protocol is species-specific for the detection of *Erw. Amylovora*, even in interspecies analysis.

Hinze *et al.* (2016) described an improved protocol to be used directly on whole bacteria in the field based on a real-time PCR approach previously published. The method allows for the early detection and quantification of the pathogen prior to the occurrence of first symptoms. There was a clear correlation between bacterial abundance and subsequent disease development; however, in some cases, no disease symptoms could be observed despite a pathogen load of up to 3.4×10^6 cells per blossom. Integration of the amount of pathogen detected into refined prediction algorithms may allow for the improvement of applied forecasting models, finally permitting a better abatement of fire blight.

DuPont *et al.* (2023) reported that the standard best management practice for reducing fire blight was branch removal at 30 cm below the proximal edge of visibly cankered tissue with sanitized loppers. This practice significantly reduced the number of new systemically-caused symptoms compared to the no-treatment control in seven of nine experiments with significant reductions in five experiments.

Emeriewen *et al.* (2023) reported that quantitative trait locus mapping for fire blight resistance in an F_2 population of *Malus fusca* MAL0045 uncovers novel resistance loci.

Certain orchard management practices, adopted in the USA more than two decades ago, have increased the frequency of more devastating outbreaks of fire blight. These are briefly summarised as below:

1. **High tree density**: Instead of 100–200 apple trees per acre, now 500–1000 trees per acre are planted.
2. **Tree size control**: In order to accomplish high tree densities, tree size-controlling rootstocks, namely, M–9 and M–26, are widely used. Both these rootstocks are highly susceptible to the disease (Steiner, 1998).
3. **Susceptible varieties**: Some apple varieties, namely, Gala, Fuji, Gingergold, Jonagold and Braeburn grown due to their demand in the fresh market, are highly susceptible to the disease. The combination of these varieties on highly susceptible rootstocks like M–9 and M–26 have further aggravated the problem.
4. **Training system**: New tree training systems adopted to make high-density plantings more productive, push the trees into early bearing. In this process, some natural physiological mechanisms that resist the progress of infection are reduced.

Dafny-Yelin *et al.* (2021) found that in fire blight-susceptible pear cultivars, a high-density planting system, associated with reduced tree vigour, presents a decreased risk of fire blight damage.

9.46.4 Management

No single control measure is effective. Therefore, an integrated and aggressive fire blight management programme should be followed. The following measures will help in minimising the disease severity.

1. **Sanitation**
   a. During the winter, all blighted twigs, branches, cankers and even the whole trees, if needed, should be cut and burnt. Infected twigs and branches should be cut at least 20–25 centimetres below the visible injury or canker.
   b. The orchard should be inspected in the early season to locate the cankers, which were not removed during the previous pruning because these become more prominent once the bacteria start oozing from them. These cankers should

be removed by cutting the twigs and branches at least 30 centimetres below the point of visible infection. Cutting tools should be disinfected after every cut with 10% commercial sodium hypochlorite solution (Wilcox, 2001).

2. **Horticultural practices**
   a. To reduce excessive succulency, trees should be grown in soil having balanced nutrition and a normal dose of nitrogen. Trees should also receive limited pruning.
   b. A proper insect control should be followed in the post-bloom period to control the insects responsible for the spread of bacteria.
   c. Moderately resistant and resistant varieties, if available, should be grown in areas where the disease is destructive.
   d. Growth-regulating acylcyclohexanedione compounds, namely, prohexadione-calcium (trade names Apogee® and Regalis®) and trinexapac-ethyl (trade names Moddus® and Primo®) offer a new, safe and promising method for the induction of resistance in pome fruit trees against fire blight and other pathogens and also for maintaining vegetative and reproductive equilibrium. Application of these compounds activates plant defence reactions, leading to the formation and storage of a biocidal compound, luteoforol in the treated tissue. The release of this compound as a result of pathogen infection causes the death of the pathogen and the adjacent plant cells.

Spinelli *et al.* (2004) reported that the combination of *Pan. agglomerans* and acylcyclohexanediones significantly reduced both blossom blight incidence and the percentage of shoot infection. Enhanced protection achieved with the combination resulted from the reduction in blossom blight incidence caused by the former and prevention of shoot infection due to the activation of plant defences by the latter. Spinelli *et al.* (2007) found that both prohexadione-calcium and trinexapac-ethyl caused significant reduction in blossom blight incidence, leading to significant reduction in shoot blight incidence. Both the compounds also caused 60–66% reduction in the migration of the bacterium in the plant tissue.

Johnson and Temple (2017) evaluated acibenzolar-S- methyl (ASM) for the control of fire blight on 3- to 14-year-old pear and apple trees. Suspensions of ASM (7.5–22.5 g of active ingredient per litre + silicone surfactant) were painted onto a 30 to 45 cm length of branch tissue immediately below primary pruning cuts or sprayed onto an 80 to 100 cm length of central trunk. In some experiments, a second ASM treatment was made in late June to early July in conjunction with secondary pruning of redeveloped cankers. Over a period of five years of field experiments, trees treated with ASM yielded 62% less diseased wood at the time of secondary and tertiary canker removal compared with non-ASM-treated trees. Tree mortality and the proportion of pruning cuts where fire blight redeveloped were also reduced by ASM treatment. Systemic acquired resistance induced by ASM application could prove an aid to pruning therapy in young fire blight-susceptible pear and apple trees where, after canker removal, disease symptoms frequently redevelop owing to residual cells of the pathogen distributed within symptomless portions of the tree.

3. **Chemical control**
Various chemicals, namely, copper oxychloride, Bordeaux mixture (0.5%) and streptomycin (100 μg $^{mL-1}$), have been used for the control of fire blight. Bordeaux mixture and copper oxychloride, as dormant sprays, have provided some, but not much, protection from the disease. Moreover, copper fungicides may cause a russeting or scarring of the fruit surface (Teviotdale, 2003).

Elkins *et al.* (2015) reported that epiphytic populations of *Erw. amylovora* on flowers at petal fall were significantly suppressed in three of four seasons in fixed copper + horticultural oil-treated plots compared with plots treated only with horticultural oil. Copper addition may add value to a fire blight management programme by delaying the increase of epiphytic populations of the bacterium in flowers.

Streptomycin is highly effective against blossom blight and also against trauma blight but the timings of its sprays are critical for effective management because this antibiotic is active for only three days after its application. For maximum benefit, streptomycin must be applied on the day when an infection event occurs, or at the latest a day after it. A delay in spraying for 24 h will result in a 10–20% reduction in control. For the prediction of infection event accurately, fire blight forecasting models, namely, **MARYBLYT** and **cougarBlight** can be effectively used.

The continuous use of streptomycin in many areas has resulted in the development of bacterial strains resistant to this antibiotic, hence making its use ineffective in these areas. Pathogen strains resistant to streptomycin were first observed in California pear orchards in 1971 and in Washington and Oregon in 1972 and then in the western United States, as well as in Michigan and Missouri. Russo *et al.* (2008) reported the occurrence of these resistant strains from New York. Streptomycin-resistant strains have also been reported from British Columbia (Canada), Lebanon, Israel, Mexico and New Zealand.

Tancos and Cox (2016) monitored the development of streptomycin resistance in *Erw. amylovora* in New York State from 2011 to 2014 using clustered regularly interspaced short palindromic repeat (CRISPR) spacer sequencing. The authors hypothesised that eradication efforts in 2002 failed and the resistant bacterial populations continued to spread throughout the state; resistance may also be developing at the same time and may also be due to the introduction of *Erw. amylovora* strains resistant to streptomycin from other regions of the United States. Increased understanding, as to how streptomycin-resistant *Erw. amylovora* isolates are introduced, evolve, or have become established, afforded by CRISPR profiling has been useful for disease management and restricting the movement of streptomycin resistance in New York.

In such areas, another broad-spectrum antibiotic, oxytetracycline, can be effectively used. For best results, it should be applied within 24 h before infection and also within 12–24 h after the infection.

Johnson *et al.* (2023) reported that, compared with the water-treated control, infection suppression after two bloom applications of an acidified commercial oxytetracycline formulation averaged 85.9 ± 0.4% compared with 72.2 ± 1.7% without an acidifier, but individual trials frequently had insufficient statistical power to separate among acidified and non-acidified antibiotic treatments.

McGhee and Sundin (2011) suggested Kasugamycin as an alternate antibiotic for fire blight management, but also cautioned about the possible role of non-target bacteria in the future transfer of a Kasugamycin resistance to *Erw. amylovora*.

Messenger, a biochemical pesticide, has shown some promise in controlling blossom blight phase when used in combination with streptomycin sprays. It is not bactericidal, but acts by activating the systemic acquired resistance genes of the host, which requires about 5–7 days. Therefore, its application should be made sufficiently in advance of blossom blight infection.

Slack *et al.* (2019) treated two young 'Gala' apple plantings with low rates of prohexadione-calcium (Pro-Ca) and acibenzolar-S-methyl (ASM). Rates of Pro-Ca ranged from 0.07–0.56 kg/ha and of ASM from 0.07–0.14 kg/ha; both as stand-alone and tandem treatments. Disease incidence with Pro-Ca at 0.14 kg/ha + ASM at 0.07 kg/ha was significantly lower than with either product alone. Lesion length was similar with a high rate of Pro-Ca alone, but with no significant impact on shoot length. The results show that low rates of Pro-Ca and ASM may work synergistically to reduce incidence of shoot blight on young apple trees.

4. **Biological control**

Johnson and Stockwell (1998) have highlighted the use of biocontrol agents for the management of fire blight. Blight-Ban A 506 (a product of *Pseudomonas fluorescens* strain A 506) is now available commercially for its control. Two applications of Blight-Ban caused 40–60% reduction in blossom blight. In another trial, its single application caused 50% reduction in fire blight. When this application was followed by weakly applications of streptomycin, however, the reduction in disease was 70%. Reduction in frost injury was also more in the combination treatment.

Stockwell *et al.* (2010) reported that *Pantoea. vagans* C9-1 suppressed fire blight in all the five trials in which the pathogen was applied as lyophilised cells, but in none of the trials when it was applied as freshly harvested cells. A mixture of *Psm. fluorescens* A506 and *Pan. vagans* C9-1 also suppressed fire blight, but the magnitude of disease suppression in all field trials (averaging 32%) was less than that attained with *Pan. vagans* C9-1 alone (42%).

Stockwell *et al.* (2011) hypothesised that strain *Psm. fluorescens* A506 decreases the biocontrol activity of *Pan. vagans* strain C9-1 and *Pan. agglomerans* strain Eh252 for fire blight, because A506 produces an extracellular protease in culture that degrades the peptide antibiotics produced by C9-1 and Eh252. This hypothesis was tested in five replicated field trials using strain A506 and A506 *aprX::Tn5*, an extracellular protease-deficient mutant, as individuals and in combination with C9-1 or Eh252 for the biological control of fire blight. On average, mixtures containing A506 *aprX::Tn5* were superior to those containing the wild-type strain, confirming that the extracellular protease of A506 decreased the biological control activity of C9-1 and Eh252 *in situ*. The mixtures of A506 *aprX::Tn5* and C9-1 or Eh252 were also superior to oxytetracycline or single biocontrol strains in suppressing pear fire blight. It is evident that certain biological control agents are mechanistically incompatible, in which one strain interferes with the mechanism by which a second strain suppresses plant disease. Mixtures composed of mechanistically compatible strains of biological control agents can suppress disease more effectively than individual biological control agents.

Mikiciński *et al.* (2016) evaluated *Psm. graminis* strain 49M isolated from the apple phyllosphere under laboratory and greenhouse conditions and found that it was equally, or even more, effective in protecting apple blossoms and apple terminal shoots than the reference strains A506 of *Psm. fluorescens* and C9-1 of *Pan. vagans*, as well as the bioproducts Blight-Ban A506, Blossom Protect and Hortocyna 18 SP (streptomycin), used for comparison. However, its effectiveness on pear fruitlets was slightly less, but significantly better (over 10%) than that of Blight-Ban A506. On its application onto apple blossoms in orchards, it effectively colonised blossoms during the entire bloom period in different weather conditions, including rain. The strain showed no pathogenicity toward plant tissue (pear fruitlets, apple blossoms, tobacco leaves) and is therefore a potential candidate for the development of a new biopesticide against fire blight.

Svircev *et al.* (2006) obtained an effective control of the disease, comparable to streptomycin, using *Pan. agglomerans* (a biocontrol agent of the pathogen) and phages specific to *Erw. amylovora.* In this case, *Pan. agglomerans* was also used to deliver and sustain phages on the blossom surface.

Nagy *et al.* (2015) reported that two phage strains, i.e., ΦEa104 and H5K, were able to translocate in apple seedlings in parts above ground level following their application to the roots and also in roots after spraying these onto the stem and leaves. A water suspension of phages effectively decreased severity of *Erw. amylovora* infection in apple seedlings following treatment of roots or aerial plant parts and application to the cotyledon, as judged by symptom bonitation. Demonstration of *in planta* translocation of *Erw. amylovora*-specific bacteriophages, and their role in reducing the fire blight severity may significantly contribute to a better control of *Erw. amylovora* and promote further investigations on penetration and translocation of phages into plants.

Gdanetz *et al.* (2024) assessed the efficacy of *Erwinia* phages and a commercial phage product for blossom blight control in the field across multiple locations in the eastern United States. In these field trials, disease control ranged from 0.0 to 82.7%, and the addition of an ultraviolet radiation protectant only resulted in significantly increased disease control in 2 of 12 tests. They also analysed microbial community population changes in response to phage application. Changes in bacterial community diversity metrics over time were not detected; however, relative abundances of target taxa were temporarily reduced after phage applications, indicating that these phage applications did not have deleterious effects on the flower microbiome. It was demonstrated that biological control of fire blight with phages is achievable, but a better understanding of phage-pathogen dynamics is required to optimize disease control efficacy.

Wallis and Cox (2020) evaluated pre-bloom prohexadione-calcium (PhCa) programmes for their effects on blossom and shoot blight management as well as their impact on shoot growth for three years in a mature 'Gala' orchard in New York. In all the three years of the study, all PhCa programmes resulted in less than 27% incidence (71% control) of blossom blight and less than 13% incidence (77% control) of shoot blight with minimal impact on tree growth. Inclusion of a biopesticide during bloom further reduced the incidence of blossom blight in one of the three years. Using light microscopy, it was found that cell walls in the cortical parenchyma of fruitlet pedicels on trees receiving pre-bloom PhCa applications were significantly thicker than those of untreated trees 40 days after full bloom and inoculation. These pre-bloom programmes can fit with standard production practices and may contribute toward the development of fire blight management programmes without the use of antibiotics.

Aćimović *et al.* (2021) evaluated eight fire blight post-infection spray schedules of prohexadione-calcium (PCA) alone or with acibenzolar-*S*-methyl (ASM) over two years. On mature trees of cultivar Royal Court, a single application of the PCA at the rate of 247 mg/litre at two to three days after inoculation resulted in 89.5 and 69.5% reductions, respectively, of shoot blight severity after inoculation. Two applications of PCA, at the rate of 247 mg/litre 12 or 14 days apart, with the first one applied two to three days after inoculation, resulted in 78.8 and 74.5% reduction of shoot blight severity in both years. A 100% control of canker incidence on perennial wood from infected shoots in both years was achieved with a single application of PCA (247 mg/litre) applied two or three days after inoculation. Three applications of PCA (125 mg/litre) + ASM (25 mg/litre) 12 to 16 days apart reduced canker incidence by 83.5 and 69% in the two years. The other programmes with lower PCA rates and frequencies of application, reduced shoot blight severity by 50.8 and 51.8% (PCA) and 62.6 to 72% and 59.3% (PCA + ASM) over two years, respectively. Reduction of canker incidence on wood by the other programmes was 66.5% and 69 to 90.4% in the two years, respectively. These treatments can reduce or prevent canker development and thus significantly abate tree losses in high-density apple orchards after the occurrence of fire blight epidemics.

Erw. amylovora relies on the type III secretion system (TIIISS) to cause fire blight. To evaluate the performance of TIIISS inhibitors in fire blight management, Yuan *et al.* (2023) conducted five field experiments in four states from 2015 to 2022 and observed reductions in blossom blight incidence as high as 96.7% compared with untreated trees. It is evident that TIIISS inhibitors exhibited good efficacy for the management of fire blight.

5. **Transgenic plants**
Transgenic apple lines, over-expressing the apple gene *MpNPR 1*, have shown increased resistance to fire blight pathogen (Malony *et al.*, 2006).

REFERENCES

Aćimović, S. G., Meredith, C. L., Santander, R. D. and Khodadadi, F. (2021). Proof of concept for shoot blight and fire blight canker management with post infection spray applications of prohexadione-calcium and acibenzolar-s-methyl in apple. *Plant Dis.* 105: 4095–4105.

Águila-Clares, B., Castiblanco, L. F., Quesada, J. M., Penyalver, R., Carbonell, J. *et al.* (2018). Transcriptional response of *Erwinia amylovora* to copper shock: *In vivo* role of the copA gene. *Mol. Plant Pathol.* 19: 169–79.

Billing, E. (2011). Fire blight. Why do views on host invasion by *Erwinia amylovora* differ? *Plant Pathol.* 60: 178–89.

Burrill, T. J. (1882). The bacteria: An account of their nature and effects, together with a systematic description of the species. *Rep. Bd. Tr. Ill. Ind. Univ.* 11: 93–157.

Castiblanco, L. F. and Sundin, G. W. (2018). Cellulose production, activated by cyclic di-GMP through BcsA and BcsZ, is a virulence factor and an essential determinant of the three-dimensional architectures of biofilms formed by *Erwinia amylovora* Ea1189. *Mol. Plant Pathol.* 19: 90–103.

Dafny-Yelin, M., Moy, J. C., Stern, R. A., Doron, I., Silberstein, M. *et al.* (2021). High-density 'Spadona' pear orchard shows reduced tree sensitivity to fire blight damage due to decreased tree vigour. *Phytopathol. Mediterr.* 60: 421–26.

Doukkali, L., Radouane, N., Ezrari, S., Tahiri, A., Tazi, B. *et al.* (2022). Lessons learnt from the fire blight epidemics: A mini review. *Indian Phytopathol.* 75: 611–25. https://doi.org/10.1007/s42360-022-00509-5

DuPont, S.T., Munir, M., Cox, K., Johnson, K., Peter, K. *et al.* (2023). Evaluation of pruning therapies in apple trees with fire blight. *J. Plant Pathol.* 105: 1695–1709. https://doi.org/10.1007/s42161-023-01447-5

Elkins, R. B., Temple, T. N., Shaffer, C. A., Ingels, C. A., Lindow, S. B. *et al.* (2015). Evaluation of dormant-stage inoculum sanitation as a component of a fire blight management program for fresh-market Bartlett pear. *Plant Dis.* 99: 1147–52.

Emeriewen, O. F., Richter, K., Flachowsky, H. and Peil, A. (2023). Quantitative trait locus mapping for fire blight resistance in an F_2 population of *Malus fusca* MAL0045 uncovers novel resistance loci. *Phytopathology* 113: 2222–29.

Erfani-Moghadam, J. and Zarei, A. (2022). Differential responses of pear cultivars to *Erwinia amylovora* infection; evidences of involvement the hypersensitivity response in pear resistance to fire blight. *Eur. J. Plant Pathol.* 162: 927–43.

Gdanetz, K., Dobbins, M. R., Villani, S. M., Outwater, C. A., Slack, S. M. *et al.* (2024). Multisite field evaluation of bacteriophages for fire blight management: Incorporation of ultraviolet radiation protectants and impact on the apple flower microbiome. *Phytopathology* 114: 1028–38. https://doi.org/10.1094/PHYTO-04-23-0145-KC

Gosch, C., Gottsberger, R. A., Stich, K. and Fischer, T. C. (2012). Blue EaLAMP-A specific and sensitive method for visual detection of genomic *Erwinia amylovora* DNA. *Eur. J. Plant Pathol.* 134: 835–45.

Hinze, M., Kohl, L., Kunz, S., Weishaupt, S., Ernst, M. *et al.* (2016). Real-time PCR detection of *Erwinia amylovora* on blossoms correlates with subsequent fire blight incidence. *Plant Pathol.* 65: 462–69.

Ishimaru, C. and Klos, E. J. (1984). New medium for detecting *Erwinia amylovora* and its use in epidemiological studies. *Phytopathology* 74: 1342–45.

Jin, Y. J., Lee, S. Y., Kong, H. G., Yang, S. I., Ham, H. *et al.* (2023). Novel detection and quantification approach of *Erwinia amylovora in vitro* and *in planta* using SYBR green-based Real-Time PCR assay. *Plant Dis.* 107: 624–27.

Johnson, K. B. and Stockwell, V. O. (1998). Management of fire blight: A case study in microbial ecology. *Annu. Rev. Phytopathol.* 36: 227–48.

Johnson, K. B. and Temple, T. N. (2017). Induction of systemic acquired resistance aids restoration of tree health in field grown pear and apple diseased with fire blight. *Plant Dis.* 101: 1263–68.

Johnson, K. B., Temple, T. N. and Achala, N. K. C. (2023). Acidifying spray suspensions of Oxytetracycline and Kasugamycin enhances their effectiveness for fire blight control in apple and pear. *Phytopathology* 113: 2205–14.

Kim, W. S., Hildebrand, M., Jock, S. and Geider, K. (2001). Molecular comparison of pathogenic bacteria from pear trees in Japan and the fire blight pathogen *Erwinia amylovora*. *Microbiology* 147: 2951–59.

Koczan, J. M., McGrath, M. J., Zhao, Y. and Sundin, G. W. (2009). Contribution of *Erwinia amylovora* exopolysaccharides amylovoran and levan to biofilm formation: Implications in pathogenicity. *Phytopathology* 99: 237–44.

Lee, S. A., Ngugi, H. K. and Halbrendt, N. O. (2010). Virulence characteristics accounting for fire blight disease severity in apple trees and seedlings. *Phytopathology* 100: 539–50.

Longstroth, M. (2001). The 2000 fire blight epidemic in Southwest Michigan apple orchards. *The Compact Fruit Tree.* 34: 16–19.

Lopez, M. M., Rosello, M., Llop, P., Ferrer, S, Christen, R., *et al.* (2011). *Erwinia piriflorinigrans* sp. nov., a novel pathogen that causes necrosis of pear blossoms. *Int. J. Syst. Evol. Microbiol.* 61: 561–67.

Malony, M., Borejsza-Wysocka, M. M., Aldwinckle, H. S., Jin, Q. L. and He, S. Y. (2006). Transgenic apple trees overexpressing the apple gene MpNPR 1 have increased resistance to fire blight. *Acta Hortic.* 704: 521–26.

Mann, R. A., Smits T. H. M., Buhlmann A., Blom J., Goesmann A. *et al.* (2013). Comparative genomics of 12 strains of *Erwinia amylovora* identifies a pan-genome with a large conserved core. *PLoS One* 8(2): e55644.

McGhee, G. C. and Sundin, G. W. (2011). Evaluation of Kasugamycin for fire blight management, effect on non-target bacteria, and assessment of kasugamycin resistance potential in *Erwinia amylovora*. *Phytopathology* 101: 192–204.

Mertoğlu, K., Evrenosoğlu, Y., Akkurt, E., Yeşilbaş, M. F. and Gülmezoğlu, N. (2024). Mineral composition modulates *Erwinia amylovora* resistance in pear based on path analysis. *Eur. J. Pl. Pathol.* 169: 593–99. https://doi.org/10.1007/s10658-024-02858-5

Mikiciński, A., Sobiczewski, P., Puławska, J. and Maciorowski, R. (2016). Control of fire blight (*Erwinia amylovora*) by a novel strain 49M of *Pseudomonas graminis* from the phyllosphere of apple (*Malus* spp.). *Eur. J. Plant Pathol.* 145: 265–76.

Moradi, A., Nasiri, J., Abdollahi, H. and Almasi, M. (2012). Development and evaluation of a loop-mediated isothermal amplification assay for detection of *Erwinia amylovora* based on chromosomal DNA. *Eur. J. Plant Pathol.* 133: 609–20.

Myung, I.-S., Lee, J.-Y., Yun, M.-J., Lee, Y.-H., Lee, Y.-K. *et al.* (2016). Fire blight of apple, caused by *Erwinia amylovora*, a new disease in Korea. *Plant Dis.* 100: 1774.

Nagy, J. K., Schwarzinger, I., Kunstler, A., Pogany, M. and Kiraly, L. (2015). Penetration and translocation of *Erwinia amylovora*-specific bacteriophages in apple- A possibility of enhanced control of fire blight. *Eur. J. Plant Pathol.* 142: 815–27.

Pirc, M., Ravnikar, M. Tomlinson, J. and Dreo, T. (2009). Improved fire blight diagnostics using quantitative real-time PCR detection of *Erwinia amylovora* chromosomal DNA. *Plant Pathol.* 58: 872–81.

Russo, N. L., Burr, T. J., Breth, D. I. and Aldwinckle, H. S. (2008). Isolation of streptomycin-resistant isolates of *Erwinia amylovora* in New York. *Plant Dis.* 92: 714–18.

Schachterle, J. K., Gdanetz, K., Pandya, I. and Sundin, G. W. (2022). Identification of novel virulence factors in *Erwinia amylovora* through temporal transcriptomic analysis of infected apple flowers under field conditions. *Mol. Plant Pathol.* 23: 855–69.

Shtienberg, D., Manulis-Sasson, S., Zilberstaine, M., Oppenheim, D. and Shwartz, H. (2015). The incessant battle against fire blight in pears: 30 years of challenges and successes in managing the disease in Israel. *Plant Dis.* 99: 1048–58.

Slack, S. M., Outwater, C. A. and Sundin, G. W. (2019). Possible synergistic effect of prohexadione-calcium and acibenzolar-S-methyl in managing shoot blight of apple (Abstr.). *Phytopathology* 109: S2.55.

Slack, S. M., Schachterle, J. K., Sweeney, E. M., Kharadi, R. R., Peng, J. *et al.* (2022). In-orchard population dynamics of *Erwinia amylovora* on apple flower stigmas. *Phytopathology* 112: 1214–25.

Smith, T. J. (2010). Fire blight management in the Pacific Northwest USA. http://county.wsu.edu/chelan-douglas/agriculture/treefruit/Pages/Fire_Blight.aspx.

Smits, T. H. M., Duffy, B., Sundin, G. W., Zhao, Y. F. and Rezzonico, F. (2017). *Erwinia amylovora* in the genomics era: From genomes to pathogen virulence, regulation, and disease control strategies. *J. Plant Pathol.* 99 (special issue): 7–23.

Smits, T. H. M., Guerrero-Prieto V. M., Hernandez-Escarcega G., Rezzonico F., Blom J. *et al.* (2014). Comparative genomics of *Erwinia amylovora* isolates from Mexico. *Acta Hortic.* (ISHS), pp. 173–77.

Smits, T. H. M., Rezzonico F., Kamber T., Blom J., Goesmann A. *et al.* (2010). Complete genome sequence of the fire blight pathogen *Erwinia amylovora* CFBP 1430 and comparison to other *Erwinia* spp. *Mol. Plant-Microbe Inter.* 23: 384–93.

Sobiczewski, P., Iakimova, E. T., Mikiciński, A., Węgrzynowicz-Lesiak, E. and Dyki, B. (2017). Necrotrophic behaviour of *Erwinia amylovora* in apple and tobacco leaf tissue. *Plant Pathol.* 66: 842–55.

Spinelli, F., Vanneste, J. L., Ciampolini, F., Cresti, M., Rademacher, W. *et al.* (2007). Potential and limits of acylcyclohexanediones for the control of blossom blight in apple and pear caused by *Erwinia amylovora. Plant Pathol.* 56: 702–10.

Spinelli, F., Vanneste, J. L., Cornish, D., Yu, J. and Costa, G. (2004). Growth-regulating acylcyclohexanediones, trinexapac-ethyl and prohexadione-calcium, decrease blossom blight incidence in pome fruits. *Acta Hort.* 704: 245–48.

Steiner, P. W. (1998). Problems in managing fire blight in high density orchards on M-9 and M-26 rootstocks. Presented in the *Annual Meeting of the Va./W.Va.* State Horticultural Societies, Roanoke, VA.

Stockwell, V. O., Johnson, K. B., Sugar, D. and Loper, J. E. (2010). Control of fire blight by *Pseudomonas fluorescens* A506 and *Pantoea vagans* C9-1 applied as single strains and mixed inocula. *Phytopathology* 100: 1330–39.

Stockwell, V. O., Johnson, K. B., Sugar, D. and Loper, J. E. (2011). Mechanistically compatible mixtures of bacterial antagonists improve biological control of fire blight of pear. *Phytopathology* 101: 113–23.

Sun, W., Gong, P., Zhao, Y., Ming, L., Zeng, Q. *et al.* (2023). Current situation of fire blight in China. *Phytopathology* 113: 2143–51.

Svircev, A. M., Lehman, S. M., Kim, W., Barszcz, E., Schneider, K. E. *et al.* (2006). Control of the fire blight pathogen with bacteriophages. *Proceedings of the 1st International Symposium on Biological Control of Bacterial Plant Diseases, Seeheim/Darmstadt, Germany.* Deutsche Bibliothek, Berlin.

Tancos, K. A., Borejsza-Wysocka, E., Kuehne, S., Breth, D. and Cox, K. D. (2017). Fire blight symptomatic shoots and the presence of *Erwinia amylovora* in asymptomatic apple budwood. *Plant Dis.* 101: 186–91.

Tancos, K. A. and Cox, K. D. (2016). Exploring diversity and origins of streptomycin-resistant *Erwinia amylovora* isolates in New York through CRISPR spacer arrays. *Plant Dis.* 100: 1307–13.

Teviotdale, B. L. (2003). Fire blight. IPM Education and Publications, Univ. Calif. Agric. Nat. Res. Publ. 7414.

Thomson, S. V. (1992). Fire blight of apple and pear. In: J. Kumar, H. S. Chaube, U. S. Singh and A. N. Mukhopadhyay (eds.). *Plant Diseases of International Importance.* Vol. III, *Diseases of Fruit Crops.* Prentice Hall, Englewood Cliffs, NJ, pp. 32–65.

Thomson, S. V. (2001). Fire blight. In: O. C. Maloy and T. D. Murray (eds.). *Encyclopedia of Plant Pathology.* John Wiley and Sons, Inc., New York. pp. 459–61.

Tian, Y., Zhao, Y., Shi, L., Cui, Z., Hu, B. *et al.* (2017). Type VI secretion systems of *Erwinia amylovora* contribute to bacterial competition, virulence, and exopolysaccharide production. *Phytopathology* 107: 654–61.

Vanneste, J. L. (2000). *Fire Blight: The Disease and Its Causative Agent, Erwinia Amylovora.* CABI, Wallingford.

Wallis, A. E. and Cox. K. D. (2020). Management of fire blight using pre-bloom application of prohexadione-calcium. *Plant Dis.* 104: 1048–54.

Weishaupt, S., Kohl, L., Kunz, S., Hinze, M., Ernst, M. *et al.* (2016). Alternative inoculum sources for fire blight: The potential role of fruit mummies and non-host plants. *Plant Pathol.* 65: 470–83.

Wilcox, W. (2001). Fire Blight. Tree fruit and berry pathology. http://www.nysaes.cornell.edu/pp/extension/tfabp/firepm.shtml.

Winslow, C. E. A., Broadhurst, J., Buchanan, R. E., Krumwiede, C. J., Rogers, L. A. *et al.* (1920). The families and genera of the bacteria. Final report of the Committee of the Society of American Bacteriologists on characterization and classification of bacterial types. *J. Bacteriol.* 5: 191–229.

Wohner, T. W., Richter, K., Sundin, G. W., Zhao, Y., Stockwell, V. O. *et al.* (2018). Inoculation of *Malus* genotypes with a set of *Erwinia amylovora* strains indicates a gene-for-gene relationship between the effector gene *eop1* and both *Malus floribunda* 821 and *Malus* "Evereste". *Plant Pathol.* 67: 938–47.

Yuan, X., Sundin, G. W., Zeng, Q., Johnson, K. B., Cox, K. D. *et al.* (2023). *Erwinia amylovora* type III secretion system inhibitors reduce fire blight infection under field conditions. *Phytopathology* 113: 2197–2204.

Zeng, Q., Slack, S. and Hassani, M. A. (2023). Pathogen spotlight on *Erwinia amylovora*- Recent advances in genomics, resistance breeding, and disease management. *Phytopathology* 113: 2140–42.

9.47 PEAR DECLINE

Pear (common pear, *Pyrus communis*) decline (PD) is an important disease affecting pear production around the world. '*Candidatus* Phytoplasma pyri' (subgroup 16SrX-C) is the etiological agent of this disease. In Argentina, Fernandez and Conci (2019) described the typical symptoms of the disease on pear trees located in Río Negro province. According to the authors, this was the first report of '*Can.* Phy. pyri' affecting pears in Argentina. Hunter *et al.* (2010) reported it from Ontario, Canada for the first time, but it appears that the disease has been present there for some time based on the number and distribution of positive samples obtained.

Pear trees infected by this pathogen exhibit symptoms of yellowing, reddening, loss of vigour, early blooming and, in some cases, a general collapse and death (Sabaté *et al.*, 2018). Apparently, disease development and symptoms expressed are highly influenced by the rootstock as well as by the stage of the disease (Seemüller *et al.*, 2011). Decline is much more prevalent on trees with root stocks of *P. ussuriensis* or *P. pyrifolia* or *P. pyricola* than on trees propagated on domestic *P. communis* roots.

Graft incompatibility, root rots and winter injury can be mistaken for pear decline. An abnormal, early red leaf colouration has been observed frequently on affected trees due to root damage.

9.47.1 Geographical Distribution

The disease is widespread in Germany, Italy and Switzerland, but locally established in Austria, the Czech Republic, France, Moldova, Netherlands, Slovakia, Slovenia, Spain and Yugoslavia; unconfirmed records have also come from Belgium and Russia. Parry's disease, a slightly different disease occurring in the UK, is now accepted to be caused by the pear decline phytoplasma. This disease also occurs in Libya, Canada, Africa and Australia. In the USA, the disease has been known since 1946 in the Pacific coast states (California, Oregon, Washington), and it is now present in parts of the northeastern states (Connecticut) and probably throughout commercial orchards in California.

9.47.2 Symptoms

Pear decline symptomatology comprises two types of symptoms: the slow decline phase and the quick decline phase.

9.47.2.1 Slow Decline Phase

In this phase, trees lose vigour over several seasons during which foliage becomes sparse with little or no terminal growth and a reduction in leaf size. Fruit set and size also decline. Slow decline occurs on more common rootstocks such as *P. communis*, *P. calleryana*, *P. pyrifolia, P. betulaefolia*, and *Cydonia oblonga*.

9.47.2.2 Quick Decline Phase

In this phase trees wilt, scorch and die in just a few weeks. It is often preceded by slow decline and reddening. This quick decline occurs on very susceptible rootstocks such as *P. ussuriensis* and *P. pyricola*.

Quick decline symptoms result from additional stress acting directly on roots (usually Phytophthora or Pythium root rot). The leaves turn red early due to starch accumulation in the upper parts of the tree.

Reddening may or may not accompany either slow or quick decline phases. Reddened leaves may curl downward with their margins rolled upwards. Leaf veins may be thickened with crinkled leaves that drop sooner than normal leaves. Reduced root systems may occur during all phases of the disease.

9.47.3 Causal Organism

Phytoplasma pyri Seemüller & Schneider, 2004

Synonym: '*Candidatus* Phytoplasma pyri' Seemüller & Schneider, 2004

Pear decline is caused by *Phy. pyri*, which belongs to the subgroup 16SrX-C of the apple proliferation group of phytoplasmas (Seemüller *et al.*, 1998). A brief account of this pathogen is given in OEPP/EPPO (2006).

9.47.4 Disease Cycle

Phy. pyri survives from season to season in the roots of infected trees. *Phytoplasma* population in the above-ground portions of the tree fluctuates through the year. The pathogen does not survive in above-ground phloem elements in late fall and winter, but it rebounds from root reservoirs once new phloem tissue is made in the spring.

Pear psylla (*Cacopsylla pyri)* transmits the pear decline phytoplasma from diseased to healthy plants. It acquires the phytoplasma after a few hours of feeding and remains infective for the rest of its life. Pear psyllid, *C. pyri*, also causes damage *per se* by injecting phytotoxins present in its saliva into leaves as it feeds.

Garcia-Chapa *et al.* (2005) studied the frequency of pear decline-positive insects and the transmission of pear

FIGURE 9.38 Pear decline (*Phytoplasma pyri*) plantdiseases.org

decline (PD) phytoplasma by *C. pyri* in Spain. Psyllids were used for experiments on phytoplasma transmission both to healthy *P. communis* trees and to an artificial feeding medium. Over a period of one year, about 100 psyllids were collected monthly from cv. Williams of pear trees, using the beating tray method, and were tested for the presence of PD phytoplasma. It was found that the frequency of PD positive psyllids changed through the year and that *C. pyri* transmitted the pear decline associated phytoplasma. Phytoplasma transmission was also effective under laboratory conditions with the use of a feeding medium. The relationship between PD positive *C. pyri*, pear decline phytoplasma transmission and the sex of the vector was also evaluated. Although the percentage of PD positive psyllids was similar in both genders, PD phytoplasma transmission by females was significantly higher than that by males. Since the sex ratio (male/female) was less than 1:1 for most of the year, these results should be taken into consideration for controlling pear decline in Mediterranean climates.

Carraro et al. (2001) *studied the* life cycle of pear decline phytoplasma in the vector *C. pyri*. For two consecutive years, groups of insects caught monthly from March to October were used for transmission trials. It was found that psylla was equally infective throughout the experimental period. In the second year, the testing of insects fed on dormant plants from November to March revealed that retention of infectivity of the pear decline phytoplasma in the vector lasted through the winter. The overwintering adults of *C. pyri* did not transmit the phytoplasmas to plants in dormancy, but they did, without any previous recharge feeding, after bud break in the spring. Control of the overwintering adults of *C. pyri* appears to be of fundamental importance for preventing the spread of pear decline. Pear decline phytoplasma can also be transmitted by grafting/budding infected stock onto healthy trees.

Dwarfing quince rootstocks have been highly tolerant to this disease. For example, Comice on quince BA-29 has been very successful. The cultivar William is incompatible with quince and will weaken and die just a few years after planting. The Farold series 40, 69 or 87 (also known as 'Old Home x Farmingdale') are also highly tolerant alternatives when grown on well-drained soils.

9.47.5 Management

The following measures can be helpful in reducing the incidence or checking the disease.

1. Use resistant or tolerant rootstocks.
2. Remove infected trees from nurseries and young orchards and destroy them.
3. Collect hardwood cuttings or scion stock in the winter or early spring and destroy them.
4. Use the best orchard management practices, including the best possible insect and disease control, irrigation, drainage, fertilisation, and pruning.
5. When grafting Asian pear trees over to European (*P. communis*) cultivars, graft below the union of the Asian pear with its rootstock to avoid creating a highly decline-susceptible tree.
6. Wild pear rootstocks (*P. communis*) should be preferred over the quince rootstocks for grafting.
7. Control pear psylla. In field trials, Vertimec 1.8 E C provides 4–5 weeks of protection under normal growing conditions. Its second application (10 ml in 10 litres of water) at the end of June just prior to a new population increase, strongly reduces egg numbers and nymph populations (Süle *et al.*, 2007).

REFERENCES

Carraro, L., Loi, N. and Ermacora, P. (2001). The life cycle of pear decline phytoplasma in the vector *Cacopsylla pyri*. *J. Plant Pathol.* 83: 87–90.

Fernandez, F. D. and Conci, L. R. (2019). First report of '*Candidatus* Phytoplasma pyri' related strain causing pear decline in Argentina. *Crop. Prot.* 121: 28–33.

Garcia-Chapa, M., Sabaté, J., Laviña, A. and Batlle, A. (2005). Role of *Cacopsylla pyri* in the epidemiology of pear decline in Spain. *Eur. J. Plant Pathol.* 111: 9–17.

Hunter, D. M., Svircev, A. M., Kaviani, M., Michelutti, R., Wang, L. *et al.* (2010). First report of pear decline caused by '*Candidatus* Phytoplasma pyri' in Ontario, Canada. *Plant Dis.* 94: 634.

OEPP/EPPO (2006). *Candidatus* Phytoplasma pyri. *Bulletin OEPP/EPPO Bulletin* 36: 127–28.

Sabaté, J., Rodon, J., Artigues, M., Laviña, A. and Batlle, A. (2018). Transmission of '*Candidatus* Phytoplasma pyri' by naturally infected *Cacopsylla pyri* to peach, an approach to the epidemiology of peach yellow leaf roll (PYLR) in Spain. *Plant Pathol.* 67: 978–86.

Seemüller, E., Marcone, C., Lauer, U., Ragozzino, A. and Göschl, M. (1998). Current status of molecular classification of the phytoplasmas. *J. Plant Pathol.* 80 (1): 3–26.
Seemüller, E. and Schneider, B (2004). '*Candidatus* Phytoplasma mali', '*Candidatus* Phytoplasma pyri' and '*Candidatus* Phytoplasma prunorum', the causal agents of apple proliferation, pear decline and European stone fruit yellows, respectively. *Int. J. Sys. Evol. Microb.* 54: 1217–26.
Seemüller, E., Schneider, B. and Jarausch, B. (2011). Pear decline phytoplasma. In: A. Hadidi, M. Barba, T. Candresse and W. Jelkmann (eds.). *Virus and Virus-Like Diseases of Pome and Stone Fruits*. APS Press, St Paul, MN, pp. 77–84.
Süle, S., Jenser, G. and Szita, É. (2007). Management of pear decline caused by '*Candidatus* Phytoplasma pyri' in Hungary. *Bull. Insectol.* 60 (2): 319–20.

9.48 CITRUS CANKER

Citrus canker is one of the most devastating diseases of citrus (*Citrus* spp.) trees. It is at present also one of the most feared diseases in the United States and some other countries. The losses occur due to the reduction in yield and market value of the infected fruits and also due to the premature fruit drop. Sometimes badly infected fruits may become unsalable. Fruits produced in canker-endemic regions suffer from a restricted market due to the enforcement of international plant quarantines.

The disease was first observed in 1913 in Florida. In 1915, Hasse identified the causal bacterium as *Pseudomonas citri*. The presence of the disease on citrus specimens collected from India (*C. medica* from Dehra Dun, Uttarakhand, India in 1827–31) and Indonesia (*C. aurantifolia* in 1842–44) and deposited in Kew Herbarium suggests that the disease probably originated in India or Indonesia. The reference to its Chinese origin is also found in the literature (Goto, 1992). The disease is endemic in Southeast Asia and Japan, from where it has spread to many citrus-growing countries. The disease occurs in severe form in seasons and/or regions characterised by warm and humid weather conditions. According to Goto (1992), the disease is particularly epiphytotic and damaging on seedlings and young trees, especially following storms (hurricanes) accompanied with rain under warm conditions. Full-grown trees show much less disease severity and damage.

Citrus canker has a long history in Florida, where it was repeatedly introduced and eradicated. It was introduced into the state in 1912, along with infected nursery plants from Japan; from there, it spread to all the Gulf States of the U S and beyond. It took about 20 years to eradicate the disease from Florida by destroying and burning of more than three million nursery and a quarter million fruit-bearing trees (Agrios, 2004). It then took a further 17 years (until 1949) to eliminate the disease from the entire United States. The appearance of a bacterial leaf spot (Xanthomonas leaf spot or citrus bacterial spot) resembling citrus canker in symptoms in 1984, and the appearance of actual citrus canker in 1986 in Manatee county of Florida, forced the government to begin a new process of eradication, which continued till 1992 and resulted in the destruction of more than 20 million nursery and orchard trees. Florida was declared free of citrus canker in early 1994. Unfortunately, however, the disease appeared again in urban Miami in October 1995 in commercial citrus in Manatee county in June 1997 and, subsequently, in many other counties of the state. In spite of the removal and destruction of 1.56 million commercial trees and 600,000 dooryard citrus trees state-wide, the disease-infected area increased to 1701 km^2 by March 2002.

Citrus canker is also present in Oceania, Argentina, Brazil, The Democratic Republic of Congo, Madagascar, Mauritius, Gabon, the Netherlands, Paraguay and Uruguay. The eradication programmes have succeeded in eradicating the disease from Australia, New Zealand, South Africa and Mozambique. The latest outbreak and eradication of citrus canker in Australia occurred in 1991. In Brazil, the disease was first noticed in 1957; from there it subsequently spread to Uruguay, Paraguay and Argentina. In spite of attempts to eradicate it, the disease has become established in these countries. However, eradication efforts in Brazil have kept the disease under good control (incidence below 0.22%) in Sao Paulo, the country's major citrus-growing state.

Several protocols for the eradication of the disease have been adopted in Sao Paulo state since its first detection in 1957. Among these, the one carried out from 1999 to 2009 was the most effective in terms of maintaining the disease at a very low level throughout the state. That protocol mandated periodic inspections and the removal of symptomatic and asymptomatic trees based on disease incidence in the block. After 2009, the eradication protocols were less stringent and, hence, the disease incidence continued to increase (Behlau *et al.* 2014).

In addition to the above-mentioned countries, the disease is also present in Bolivia, Senegal, Mali, Burkina Faso, Tanzania, Iran, Saudi Arabia, Yemen, Bangladesh, Comoros, Malaysia and Vietnam, but absent from Europe (Ference *et al.* 2018). It is one of the most important diseases of the international quarantine and all citrus-producing countries free of citrus canker strictly prohibit the import of citrus plants and fruits from disease-affected countries.

9.48.1 Host Range

Xan. citri subsp. *citri* infects a large number of citrus species. Among the most susceptible hosts are grapefruit, Mexican (Key) lime and lemons, trifoliate oranges, some sweet oranges and limes. In India, *kagzi* lime is the most susceptible host. In moist tropical and subtropical climates, these hosts suffer badly due to the disease. Mandarin oranges such as satsuma, cleopatra and ponkan are resistant. Other natural hosts include *Swinglea glutinosa*, *Fortunella margarita*, *F. japonica* and *Severinia buxifola*.

9.48.2 Symptoms

The symptoms appear on all the above-ground plant parts. Young citrus tissues during the second half of the expansion phase of the growth are at maximum susceptibility. Water-soaked, translucent, tiny spots, usually dark green with a raised convex surface, first appear on lower surface and then spread to the upper surface of the leaves. Initially, the spots are corky, having an oily appearance but they later turn brown and are surrounded by glossy, dark brown margins. As the disease progresses, the spots turn greyish or whitish, and the centre of the spot ruptures, exposing a light brown spongy mass and forming a crater-like depression. Yellow haloes are usually formed around such mature lesions. Centres of the spots may fall down, giving a shot-hole effect. The lesions are round to irregular in shape and have a diameter of 5–10 mm. As the disease severity increases, the spots coalesce, giving a scabby appearance (Figure 9.39a). The formation of lesions on the midrib or petioles of leaves results in more defoliation than the scattered lesions found only on the leaf blade. The feeding activities of Asian citrus leaf miner, *Phyllocnistis citrella*, increase the disease severity on leaves by exposing the leaf mesophyll tissues to splashed inoculum and thereby increasing the leaf infection.

As the disease progresses, the symptoms also appear on twigs, branches, fruits and thorns. The lesions on twigs are observed generally on more susceptible plant species like grapefruit, lime, sweet orange and trifoliate orange, etc. Lesions on twigs and branches are raised, corky and surrounded by an oily or water-soaked margin but no chlorotic haloes. The lesions may vary in length from one to several centimetres (Figure 9.39b). Severely infected twigs start to dry from the tip backward and severe defoliation occurs.

The symptoms on fruits are similar to those found on leaves with the exception that yellow haloes are not always present. The lesions are rough and corky, having more prominent crater-like depressions and remaining confined to the rind. Severely infected fruits become highly blemished and fetch very low prices, or may become unsalable (Figure 9.39c).

Vigorously growing citrus trees are highly susceptible to infection as each growth flush of the tree during a growing season is accompanied by a period of increased susceptibility. When leaves, twigs and fruits reach maturity, these become more resistant to infection due to the thickening of the cuticle and the hardening of the tissue.

To date, five types of citrus cankers or related forms have been reported affecting *Citrus* species. **Asiatic** or **true citrus canker** or **canker/cancrosis A**, originally found in Asia, is by far the most widespread and severe form of the disease (Gottwald *et al.*, 2002). It also affects a large number of *Citrus* species. The symptoms described above are of a typical true citrus canker.

False canker or **cancrosis B**, first found in South America, occurs mainly on lemons; it also appears to a lesser extent on lime, sour orange, sweet orange and citron, but is not found on grapefruit. It is found in Argentina, Paraguay and Uruguay.

Cancrosis/canker C or **Mexican lime canker**, originally found in the Sao Paulo state of Brazil, primarily affects Mexican lime and, to a lesser extent, lemon. In Mexican lime, it is most severe on Galego lemon but causes only slight infection on Tahiti lemon.

Rodriguez *et al.* (1985) reported disease-causing pustules on leaves and young shoots of Mexican lime but not infecting fruits, and on other citrus species along the Pacific coast of Mexico in 1981. It is called **canker D**.

In 1984, another bacterial disease of citrus, called nursery-form canker, was found in a nursery in Florida and subsequently in many other nurseries of the state (Schoulties *et al.*, 1987). The symptoms appeared on leaves and shoots of nursery plants but not on trees. The lesions formed were flat but occasionally sunken, necrotic having an extremely water-soaked appearance and chlorosis but never resembled erupted canker lesions. The host range of the bacterium included grapefruit, sweet and trifoliate oranges, tangor, tangerine, tangalo, mandarin, citrumelo and citrange. The disease was called Xanthomonas leaf spot or citrus bacterial spot and the pathogen was named *Xan. campestris* pv. *citrumelo*. The disease is now called **canker E**. The strains of pathogen causing cankers B, C and D were designated as *Xan. campestris* pv. *aurantifolia*.

The symptoms of citrus canker can be confused with those of other diseases such as scab, anthracnose, leprosis

(a) (b) (c)

FIGURE 9.39 (a) Citrus canker: Leaf spots. (b) Citrus canker: Twig cankers. (c) Citrus canker: Lesions on fruit.

and melanose, but the following three characteristic symptoms of citrus canker distinguish it from those of other diseases (Patel and Desai, 1972).

1. The lesions are raised on both surfaces of the leaves.
2. The lesions have a distinguished margin.
3. The lesions have crater-like depressions in the centre, which can be seen more clearly with the help of a hand lens.

Niculau *et al.* (2022) reported that MALDI-TOF MS combined with chemometric analysis was effective for differentiating the macromolecule profile of orange leaves with canker lesions, healthy leaves and leaves with phytotoxicity symptoms, proving that this technique may be used for the rapid diagnosis of citrus canker.

9.48.3 Causal Organisms

Xanthomonas citri subsp. *citri* (ex Hasse 1915) Gabriel *et al.* 1989, subsp.
nov. causes Asiatic or true citrus canker or canker A.
Synonym: *Xan. citri*, *Xan. campestris* pv. *citri*, *Xan. axonopodis* pv. *citri*
Xan. fuscans subsp. *aurantifolii* Schaad *et al.* 2007, subsp. nov. causes
cankers B, C and D.
Synonym: *Xan. campestris* pv. *aurantifolii*
Xan. alfalfae subsp. *citrumelonis* Schaad *et al.* 2007, subsp. nov. causes
canker E.
Synonym: *Xan. campestris* pv. *citrumelo*

The bacterium causing Asiatic or true citrus canker was first identified as *Pseudomonas citri* by Hasse in 1915. Following the establishment of genus *Xanthomonas*, it was transferred to this genus as *Xan. citri*. With the introduction of pathovar system, it was changed to *Xan. campestris* pv. *citri* and, in 1995, Vauterin *et al.* reclassified it as *Xan. axonopodis* pv. *citri*. Important morphological, physiological and biochemical characters of this bacterium described by Goto (1992) are given below.

Cells are Gram-stain-negative, non-sporing, straight rods and measure 0.5–0.75 × 1.5–2.0 µm. These are predominantly single, aerobic, non-encapsulated, having a polar flagellum and chemoorganotrophic. Cells do not accumulate poly-β-hydroxybutyrate granules and the metabolism of glucose is oxidative. The colonies on solid media are circular, convex, having an entire margin. Colonies are creamy-yellow and straw yellow on potato glucose agar and on nutrient agar, respectively, due to the production of xanthomonadins. Catalase test is positive while oxidase is negative or weak. Methionine or cysteine is essential for growth but serine inhibits the growth. Asparagine is not used as a sole source of carbon and nitrogen. The bacterium hydrolyses starch, esculin, casein, and Tween 80, liquefies gelatin and pectate gel and produces tyrosinase. Tests for nitrate reduction, arginine dihydrolase, urease and amino acid decarboxylases are negative. Glucose, sucrose, fructose, maltose lactose, galactose, xylose, mannose, glycerol, dextrin, starch and citrate are utilised as the sole sources of carbon. Optimum temperature for growth is 28° C.

To date, five different types of citrus cankers or related forms have been reported and these have been named as cankers A, B, C, D and E. Schaad *et al.* (2005) studied 44 strains of bacteria representing causal agents of all these five citrus canker types and on the basis of DNA–DNA relatedness assays, sequence of 16S–23S intergenic spacer regions and AFLP analysis grouped them into three distinct genotypes: taxon I included all "A" strains, taxon II contained all "B", "C" and "D" strains and taxon III contained all "E" strains. They named taxon I strains as *Xanthomonas smithii* subsp. *citri* (ex Hasse 1915) sp. nov. nom. rev. comb. nov., taxon II strains as *Xanthomonas fuscans* subsp. *aurantifolii* (ex Gabriel *et al.* 1989) sp. nov. nom. rev. comb. nov. and taxon III strains as *Xanthomonas alfalfae* subsp. *citrumelo* (ex Riker and Jones) Gabriel *et al.* 1989 sp. nov. nom. rev. comb. nov.

Schaad *et al.* (2006) emended the classification of xanthomonad pathogens on citrus as their earlier publication (Schaad *et al.*, 2005) contained nomenclature errors. *Xan. smithii* subsp. *citri* and *Xan. alfalfae* subsp. *citrumelo* were reclassified as *Xan. citri* subsp. *citri* and *Xan. alfalfae* subsp. *citrumelonis*, respectively. *Xan. fuscans* subsp. *aurantifolii* and *Xan. alfalfae* subsp. *citrumelonis* share many common characteristics. The below-given description of these three pathogens is taken from Schaad *et al.* (2006).

9.48.3.1 *Xan. citri* subsp. *citri*

Colonies of *Xan. citri* subsp. *citri* appear on YDC and FS agars after 40–44 and 56–60 h of incubation, respectively at 28–30° C. The bacterium utilises arabinose, maltose, lactose, cellobiose, mannitol and aspartic acid; hydrolyses pectate, liquefies gelatin, and causes an alkaline hydrolysis of litmus milk. *Xan. citri* subsp. *citri* produces large amounts of extracellular polysaccharides (EPS) both in culture media and in host tissue. Cells of the bacterium are rapidly killed when diluted in water, i.e., $< 10^6$ cells mL^{-1}. Therefore, it is evident that EPS protects the bacterial cells in water and from desiccation in air (in the bacterial ooze). Strains of *Xan. citri* subsp. *citri* are susceptible to bacteriophages CP1 and CP2, whereas those of *Xan. fuscans* subsp. *aurantifolii* are not. The bacterium may be distinguished from *Xan. campestris* and most other *Xanthomonas* species, subspecies and pathovars by DNA–DNA reassociation assays, ITS sequencing, rep-PCR profiles, and phenotypic traits.

Li *et al.* (2022) presented the genome sequence of *Xan. citri* pv. *citri* strain GD82 from a formalin-fixed citrus leaf specimen showing canker lesions collected in 1982 in Guangdong Province, China. GD82 genome consisted of

5,197,217 bp with G + C content of 64.8%, along with four circular plasmids: pXAC33 (32,377 bp), pXAC64 (63,972 bp), pXAC47 (47,810 bp) and pGD82.1 (219,560 bp). This is the oldest *Xan. citri* pv. *citri* genome from historical citrus canker specimens in China, which will enrich the current *Xan. citri* pv. *citri* genome database and facilitate genomic evolution research of *Xan. citri* pv. *citri*.

9.48.3.2 Xan. fuscans subsp. aurantifolii

The colonies of *Xan. fuscans* subsp. *aurantifolii* appear on YDC and FS agars after 50–60 and 70–76 h of incubation, respectively, at 28–30 °C. The strains of the bacterium produce a water-soluble brown pigment on several common agar media, including YDC. The bacterium utilises lactose, cellobiose and mannitol, and precipitates litmus milk. *Xan. fuscans* subsp. *aurantifolii* is distinguished from *Xan. citri* subsp. *citri* and *Xan. alfalfae* subsp. *citrumelonis* as it precipitates litmus milk and hydrolyses gelatin. *Xan. fuscans* subsp. *aurantifolii* does not utilise maltose or hydrolyse pectate whereas *Xan. citri* subsp. *citri* does. *Xan. fuscans* subsp. *aurantifolii* is differentiated from most other *Xanthomonas* species, subspecies and pathovars by DNA–DNA reassociation assays, rep-PCR profiles, ITS sequences, and phenotypic traits.

9.48.3.3 Xan. alfalfae subsp. citrumelonis

The colonies of the bacterium on YDC and FS agars appear after 30–34 and 40–44 h of incubation, respectively, at 28–30 °C. *Xan. alfalfae* subsp. *citrumelonis* utilises raffinose whereas *Xan. citri* subsp. *citri* does not. *Xa. alfalfae* subsp. *citrumelonis* can be differentiated from *Xan. fuscans* subsp. *aurantifolii* due to its more rapid growth on agar media, the liquefaction of gelatin and the utilisation of maltose. It is distinguished from *Xan. citri* subsp. *citri* by its ability to utilise raffinose, to produce acid from cellobiose and mannitol and also its faster growth on YDC and FS agars. *Xan. alfalfae* subsp. *citrumelonis* is distinguished from *Xan. campestris* pv. *campestris* and other *Xanthomonas* pathovars, subspecies and species by DNA–DNA reassociation assays, rep-PCR profiles, ITS sequences and phenotypic traits. These three subspecies can be distinguished from each other by serological assays.

Kositcharoenkul *et al.* (2011) developed a single-tube nested PCR for the detection of *Xan. citri* subsp. *citri*. The assay targets the *pthA* gene of *Xan. citri* subsp. *citri* and utilises different annealing temperatures for the two primer pairs. The assay reliably detected as few as 10^2 cells, was unaffected by the presence of PCR inhibitors and was 10-fold and 8,500-fold more sensitive than standard PCR and ELISA, respectively. The use of a washing method for DNA extraction increased the sensitivity compared to direct extraction from leaf tissue. The single-tube nested PCR detected *Xan. citri* subsp. *citri* from 67 samples out of 90 samples collected from affected pomelo orchards compared to 54 samples detected by standard PCR. It was also able to detect *Xan. citri* subsp. *citri* from samples both with and without symptoms. This assay can be used as a rapid and sensitive technique for routine *Xan. citri* subsp. *citri* detection in field samples for surveillance of citrus canker.

Yu *et al.* (2017) reported that the primer sets XcAsF/XcAswR, XaBF/XaBR and XaCF1/XaCR1 specifically differentiated *Xan. citri* subsp. *citri* A* and AW, *Xan. fuscans* subsp. *aurantifolii* B and *Xan. fuscans* subsp. *aurantifolii* C strains in PCR, respectively. With quantitative real-time PCR analysis, it was possible to detect a range of 1 to 10^7 bacterial cell(s), which will be helpful in detecting the species at the pre-symptom developmental stage of the quarantine period of the plant. PCR/primer efficiency was also demonstrated by using raw bacterial cells, which will reduce the considerable time and cost needed for genomic DNA isolation. The designed molecular markers and qPCR method will be useful for the routine diagnosis of *Xanthomonas* species, which will improve the management of citrus canker disease.

Marin *et al.* (2019) reported that the study of copper sensitivity of *Xan. citri* subsp. *citri* (*Xan. citri*) strains from Paraná state, Brazil, revealed the existence of copper-tolerant (Cu^T) but not copper-resistant (Cu^R) strains. Cu^T strains supported intermediate concentrations of copper in comparison to Cu^S (copper-susceptible) and Cu^R strains. Cu^T strains lack the gene clusters *copLAB* or *copABCD* responsible for copper resistance in Cu^R strains and the large plasmids (c. ≥200 kb) that normally carry these genes. The nucleotide sequences of chromosomal homologous genes *cohLAB*, involved in copper homeostasis, were 100% similar in strains of all phenotypes. Cu^T strains differed from Cu^S strains by the higher expression of the homologous chromosomal genes *cohA* and *cohB* in the presence of copper. Cu^T *Xan. citri* strains are not precursors of Cu^R strains and do not pose a threat to the efficient use of copper-based bactericides for the management of citrus canker in citrus orchards. Copper resistance and tolerance are distinct phenotypes and should not be used as synonyms. The proper characterisation of the sensitivity to copper leads to a more confident monitoring of the distribution of copper-resistant populations of *Xan. citri* and the adoption of containment measures only when necessary.

Using a genotyping method that did not require cultivating the bacterium or purifying DNA, Pruvost *et al.* (2019) deciphered the pathogen's spatial genetic structure at several microgeographic scales, down to a single lesion, in a situation of Asiatic citrus canker endemicity. In a grove where copper was recurrently applied for Asiatic citrus canker management, copper-susceptible and copper-resistant *Xan. citri* pv. *citri* coexisted and the bacterial population was structured as three genetic clusters, suggesting a polyclonal contamination. Canker lesions, which are of major biological significance as an inoculum source, may also represent a preferred niche for horizontal gene transfer. The study points out the potential of genotyping data for estimating the range of spatial dependency of plant bacterial pathogens, an important parameter for guiding disease management strategies.

Sidireddi *et al.* (2023) developed two rapid, cost-effective, and field-deployable LAMP assays, LAMP-77 for the detection of *Xan. citri* subsp. *citri* (*Xcc*) and LAMP-AvrGF1 to differentiate *Xcc* A^w strain from *Xcc* A. LAMP-77 primers were designed based on a portion of the pthA gene, a pathogenicity determinant gene of *Xcc*. AvrGF-1 primers were based on the *AvrGF1* gene, an avirulence gene of *Xcc* A^w that induces hypersensitivity response in grapefruit. This study optimized both LAMP assays for *Xcc* detection. The detection limit of the LAMP-77 primers was ca. $3.5{\times}10^3$ copies of the target molecule in the reaction while comparable detection limit (ca. $3.44{\times}10^3$ copies) with AvrGF-1 primers was obtained by increasing five times the amount of the input DNA. For the field applicability of the LAMP assays, the authors modified loop primers of both LAMP assays to detect the LAMP products on commercial nucleic acid lateral flow immunoassay cassette, enabling the detection of *Xcc* in the field by simple visual observation.

9.48.4 Disease Cycle

Due to the perennial nature of citrus trees, the bacterium survives on the infected trees, mainly in the lesions formed on branches and stems. The bacterial cells in stem lesions can remain viable for a long period (5–7 years). The bacterium is frequently found in soil, on citrus roots, or on weeds growing in orchards. Its population on weed roots does not decline in winter. The bacterium can survive for 2–3 months in leaf lesions, if the infected leaves are kept under dry conditions. The soil heavily infected with *Xan. citri* subsp. *citri* supports the inoculum level sufficient for infection only for a few weeks during the citrus growing season. However, to date, there is no direct evidence that the bacteria surviving in low numbers on weed hosts or in soil can lead to the development of epiphytotics.

The bacteria multiply in lesions formed on leaves, twigs and fruits. During wet weather, the bacteria ooze out from these lesions and are disseminated to new growth and other plants. Wind-driven rain is the main natural dispersal agent and a wind speed of >18 mph helps in the stomatal penetration of bacteria. Rainstorms also increase the water congestion of leaf tissue and citrus foliage can hold 7 microlitres of water per centimetre square of leaf area. Even one cell forced through stomatal openings in water-congested tissues can lead to successful infection. Pria *et al.* (2006) found that temperature had a greater effect than leaf wetness duration on the incidence of Asiatic citrus canker in Brazil. As 100% disease incidence occurred during a 4 h leaf wetness period, they suggested that minimum leaf wetness period is less than 4 h. Wind-driven rain is the main agent for short distance or local spread of the bacteria; in some cases, however, it may also cover a considerable distance. In Argentina, the wind-blown inoculum was detected up to 32 meters from the infected trees. Greater wind speeds consistently disperse more bacterial cells. In Florida, severe rainstorms and tropical storms carried the inoculum up to a distance of 7 miles.

Dos Santos *et al.* (2024) assessed the effect of the moisture chamber for 24 or 48 h before and/or after inoculation on the development of citrus canker lesions. The minimum concentrations required to infect 100% of the inoculated leaves using infiltration and spray inoculation were 10^4 and 10^7 cfu mL^{-1}, respectively. Intermediate severity levels, e.g., 3.2 to 11.4 and 2.5 to 6.7 lesions/cm^2, were observed using 10^3–10^4 cfu mL^{-1} for infiltration and 10^6–10^7 cfu mL^{-1} for spray inoculation. This severity level is suitable when the number of lesions in a given treatment needs to be determined.

Long-distance spread occurs generally through infected nursery plants, rainwater or floods. Strong winds and rainstorms increase the disease severity, leading to disease outbreaks by causing injuries to the host tissue, increasing humidity and by aiding in the dispersal of the inoculum. However, the size of injuries does not affect the disease severity because even a single bacterium can cause infection under the most favourable conditions (Goto, 1992). Prolonged humidity for about 12 h after artificial inoculation significantly increases the infection rate, whereas in the case of stomatal infection it is essential only for several minutes. The healing of wounds affects the chances of infection and this is much faster at higher temperatures around 25 °C.

The penetration of the bacterium into the host tissue occurs through stomata and more frequently through wounds caused by pruning, thorns and insects. Penetration through stomata generally occurs in young tissue. The number and size of stomata does not affect the disease's severity because the number of lesions does not differ significantly on leaves having more and wider stomata than on those having fewer and narrower stomata. Instead, the size of the lesion increases in susceptible varieties compared to resistant ones. The wounds caused by citrus leaf miners also serve as avenues for the entry of the bacterium and even the penetration of the resistant hosts also occurs due to the overcoming of the natural barriers of the host. Contaminated larvae of leaf miners also transport the bacteria through feeding galleries. The leaf miner wounds differ from natural wounds as these take more time to heal, i.e., 10–14 days in comparison to 1 day for natural wounds, and also require lower inoculum dose (1/1000 of the dose required for infection through natural openings) for infection. On some trees, citrus canker infections are confined entirely to leaf miner wounds. A study conducted in South Broward county of Florida in 2002 showed that approximately 60% of the diseased citrus canker leaves were injured by leaf miners and that the number of canker lesions increased significantly on the leaf miner infested leaves in comparison to leaves infected through stomata. Christiano *et al.* (2007) showed that wounds caused by Asian citrus leaf miner (*P. citrella*) on the leaves of Tahiti lime led to more disease severity than mechanical wounds and intact leaves. Maximum disease severity of 87% on leaves having pupal stage injury and of 63% on those having third instar stage injury, which was five times higher than that on intact leaves, was due to the greater amount of damage that exposes mesophyll cells to infection.

The optimum temperature for the infection of hosts lies between 20 and 30 °C. Excessive nitrogen fertilisation leads to more canker development due to the flushing of more shoots and succulent host tissue. The disease is most severe on young trees that flush many angular shoots than on mature or old trees. The angular shoots that develop from summer to autumn usually have a prolonged disease cycle due to the prolonged presence of young and susceptible shoots. The fruits also show infection for several months because their continuous growth provides suitable stomata for infection.

Francis *et al.* (2010) developed a detached leaf protocol for the rapid screening of germplasm for resistance to citrus canker (*Xan. citri* subsp. *citri*) and citrus bacterial spot (*Xan. alfalfae* subsp. *citrumelonis*) to evaluate limited quantities of leaf material. Detached leaf assay was found useful for the characterisation and differentiation of lesion phenotype for both the pathogens, permitting rapid screening of germplasm resistance based on the quantification of number of lesions and bacterial concentration.

Braido *et al.* (2014) developed two surface area diseased (SAD) sets i.e., a five-diagram set depicting five disease severities (0.5%, 2.0%, 8.0%, 27.0% and 40.0%) and a six-diagram set depicting six disease severities (0.5%, 1.0%, 3.0%, 9.0%, 20.0% and 40.0%) to improve the accuracy and reliability of visual estimates of Asiatic citrus canker on unripe (green) fruits of sweet orange. Both these sets significantly improved the accuracy and reliability of estimates, but the five-diagram set was significantly more accurate and reliable than the six-diagram set. Agreement, measured by Lin's concordance correlation coefficient, was 0.220 to 0.913 when SADs were not used, 0.814 to 0.955 when the five-diagram set was used and 0.863 to 0.925 when the six-diagram set was used. Although the SAD set was developed for sweet orange, doubtless it has the applicability to other citrus trees, including grapefruit.

Braido *et al.* (2015) also devised two SAD sets, i.e., five-diagram set depicting five disease severities (0.7%, 2.0%, 7.0%, 21.0% and 39.0%) and six-diagram set depicting six disease severities (0.7%, 2.0%, 4.0%, 10.0%, 21.0% and 39.0%) as aids for raters to improve the accuracy and reliability of visual estimates of canker severity on ripe fruits of sweet orange. Both five- and six-diagram sets improved the accuracy and reliability of raters' estimates, but the five-diagram set resulted in significantly more accurate and reliable estimates than the six-diagram set, and is thus recommended for use. Agreement, measured by Lin's concordance correlation coefficient (ρc), was 0.735 to 0.906 when SAD sets were not used, 0.931 to 0.985 when the five-diagram set was used, and 0.879 to 0.979 when the six-diagram set was used. Mean inter-rater reliability (R2) was 0.638 when SAD sets were not used, 0.889 when the five-diagram set was used and 0.834 when the six-diagram set was used.

Hu *et al.* (2016) compared the transcriptional profiles between citrus leaf tissue inoculated with *Xan. citri* subsp. *citri* 306 and a *pthA4*-deletion mutant strain (*Xan. citri* subsp. *citri* 306Δ*pthA4*) at 6, 48 and 120 h post-inoculation (hpi). PthA4-mediated infection significantly upregulated the expression of genes involved in cell wall degradation and modification, DNA packaging, G-protein, protein synthesis, sucrose metabolism and cell division functions at both 48 and 120 hpi, while the down-regulated genes were mainly enriched in photosynthesis, transport, secondary metabolism, cytochrome P450 and various plant defence-associated mechanisms. To validate microarray results, the gene expression of 26 genes representing genes associated with cell wall, immunity system and carbohydrate metabolism was confirmed using quantitative reverse-transcription polymerase chain reaction (PCR). Expression patterns of these genes at 48 and 120 hpi were consistent with the microarray results. The putative effector binding element for PthA4 (EBEPthA4) in the promoter regions of multiple genes upregulated by PthA4 was also identified, to which PthA4 might bind directly to control their gene expression. The investigation provides a dynamic picture of citrus genes regulated by PthA4 during the *Xan. citri* subsp. *citri* infection of citrus leaves at different stages and will be useful in further understanding the virulence mechanism of *Xan. citri* subsp. *citri* and identifying potential targets of PthA4.

Zhang *et al.* (2010) generated transgenic "Duncan" grapefruit (DG) and "Hamlin" sweet orange (Ham) expressing the Arabidopsis *NPR1* gene (*AtNPR1*), which is a key positive regulator of the long-lasting broad-spectrum resistance known as systemic acquired resistance (SAR). Overexpression of *AtNPR1* in citrus increased resistance to citrus canker, and the resistance was related with the expression levels of *AtNPR1* in the transgenic plants. The line DG 42-2 with the highest expression level of *AtNPR1* was also the most resistant, and it developed significantly fewer lesions, accompanied by a 10-fold reduction in population of *Xan. citri* subsp. *citri*. The lesions developed on DG 42-2 were also smaller and darker than those on the control plants and lacked callus formation. These lesion phenotypes resembled those on canker resistant kumquats and canker-susceptible citrus trees treated with SAR-inducing compounds. It is evident that the overexpression of *AtNPR1* in citrus is a promising approach for the development of more resistant cultivars to citrus canker.

Kobayashi *et al.* (2017) obtained transgenic "Pera" sweet orange [*C. sinensis* (L.) Osbeck] plants, constitutively expressing the sarcotoxin IA peptide fused to the PR1a signal peptide from *Nicotiana tabacum* for secretion in the intercellular space, by *Agrobacterium*-mediated transformation using thin sections of mature explants. Evaluation of citrus canker resistance in leaves of transgenic and non-transgenic plants was performed by inoculations of *Xan. citri* subsp. *citri* by infiltration and spraying. *Xan. citri* subsp. *citri* population was up to 2 log units lower in leaves of the transgenic plants compared to those of non-transgenic controls. The incidence of canker lesions was significantly higher in non-transformed controls (>10 lesions/cm^2)

than in the transgenic plants (<5 lesions/cm^2) after injection infiltration or spraying with *Xan. citri* subsp. *citri* inoculum. Accumulation of sarcotoxin IA peptide in sweet orange tissue did not cause any deleterious effects on the growth and development of the transgenic plants, indicating that this approach is suitable to provide resistance to citrus canker.

Mirzaei-Najafgholi *et al.* (2021) evaluated swarming motility, biofilm formation, resistance to H_2O_2 and xanthan production of 44 local strains of *Xan. citri* subsp. *citri* (*Xcc*) causing citrus canker in Iran. All the strains showed differential swarming motilities, biofilm formation abilities and xanthan production. The most virulent strain of *Xcc*, KVXCC1 exhibited the greatest capability to form biofilm on solid surfaces, xanthan production and pathogenicity on detached leaves of *Citrus aurantifolia* as well as positive chemotaxis toward *C. aurantifolia* extract. Fourteen common and commercial citrus genotypes evaluated for resistance to *Xcc* strain KVXCC1 were categorised into susceptible and resistant groups based on lesion number per inoculation site. *C. aurantifolia* was found to be the most susceptible cultivar while *C. aurantium, C. jumbhori* and *C. reticulata* cv. *ponkan* showed high and moderately high resistance against the bacterium.

Favaro *et al.* (2020) reported that the leaf surface of the 'Okitsu' cultivar contributes to resistance against *Xan. citri* subsp. *citri*. Ultrastructural analysis showed a thicker cuticle covering epidermal surface and guard cells in young 'Okitsu' leaves than in 'Clemenules' leaves. This thicker cuticle was associated with a smaller stomatal aperture and reduced cuticle permeability. Ultrastructural analysis showed a thicker cuticle covering epidermal surface and guard cells in young 'Okitsu' leaves than in 'Clemenules' leaves. This thicker cuticle was associated with a smaller stomatal aperture and reduced cuticle permeability. Furthermore, cuticular waxes extracted from young 'Okitsu' leaves have higher antibacterial activity against *Xan. citri* subsp. *citri* than 'Clemenules'. These data suggest that a faster development of epicuticular waxes in 'Okitsu' leaves play a central role in its resistance to *Xan. citri* subsp. *citri*.

Shi *et al.* (2015) compared the defence responses of citrus canker-resistant and citrus canker-susceptible genotypes to the *Xan. citri* subsp. *citri* (*Xcc*)-derived pathogen-associated molecular pattern (PAMP) flg22 (Xflg22) by analysing the expression of 20 citrus defence-associated genes. In the most resistant genotype, 'Nagami' kumquat, there was significant induction of several defence genes (*EDS1, NDR1, PBS1, RAR1, SGT1, PAL1, NPR2* and *NPR3*) as early as 6 h and up to 72 h after Xflg22 treatment. Contrary to this finding, the highly susceptible 'Duncan' grapefruit showed no induction of the same defence genes, even 120 h after treatment. Citrus genotypes with partial levels of resistance showed intermediate levels of transcriptional reprogramming that correlated with their resistance level. Xflg22 also triggered a rapid oxidative burst in all genotypes, which was higher and accompanied by the induction of PAMP-triggered immunity (PTI) marker genes (*WRKY22* and *GST1*) only in the more resistant genotypes. Pre-treatment with Xflg22 prior to *Xcc* inoculation inhibited bacterial growth in kumquat, but not in grapefruit. A flagellin-deficient *Xcc* strain (*XccΔfliC*) showed greater growth increase relative to wild-type *Xcc* in kumquat than in grapefruit. These findings show that Xflg22 initiates strong PTI in canker-resistant genotypes, but not in susceptible ones, and that a robust induction of PTI is an important component of citrus resistance to canker.

Jia *et al.* (2022) generated 19 genome-edited grapefruit plants with one biallelic mutant line (#Dun_{Yao}7), using 'Duncan' grapefruit epicotyls as explants. *Xan. citri* subsp. *citri* (*Xcc*) caused canker symptoms on wild-type and non-biallelic mutant plants but not on #Dun_{Yao}7. A *Xcc*PthA4 mutant containing the designer TAL effector dLOB1.5, which recognises a conserved sequence in both wild-type and #Dun_{Yao}7, caused canker symptoms on both wild-type and #Dun_{Yao}7. No off-target mutations were detected in #Dun_{Yao}7. This is the first study, which shows that CRISPR-mediated genome editing has been successfully used to generate disease-resistant plants for 'Duncan' grapefruit, paving the way for using disease-resistant varieties to control canker.

Duan *et al.* (2022) developed a rapid evaluation method based on enhanced yellow fluorescent protein (eYFP)-labeled *Xan. citri* subsp. *citri* (*Xcc*). The results demonstrated that eYFP does not affect the growth and virulence of *Xcc*. *Xcc*-eYFP allows measurement of bacterial titres, but is more efficient and rapid than the plate colony counting method. The evaluation of citrus germplasms from China, based on symptoms and bacterial titres, revealed that citrus germplasms 'Ichang' papeda and 'Huapi' kumquat were resistant, while 'Chongyi' wild mandarin, 'Mangshan' wild mandarin, 'Ledong' kumquat, 'Dali' citron, 'Yiliang' citron, 'Longyan' kumquat, 'Bawang' kumquat and 'Daoxian' wild mandarin were tolerant. These results can be leveraged in traditional breeding contexts or used to identify canker resistance genes to increase the disease resistance of commercial citrus varieties via biotechnological approaches.

de Oliveira Andrade *et al.* (2023) presented evidence that *Xan. aurantifolii* flagellin C protein (*XaFliC*) is sufficient to trigger a robust defence response that completely halts the development of canker lesions caused by virulent *Xan. citri* in sweet oranges but not in Mexican lime plants. The authors also showed that two polymorphic peptides of *XaFliC*, denoted here as flgIII-20 and flgIII-27, which are not related to the well-known flg-22 and flgII-28 peptides, also protected sweet orange leaves from *Xan. citri* infection. To the best of the author's knowledge, this is the first description of *FliC* as a pathogen-associated molecular pattern or molecular determinant of an immune response-induced by a citrus canker pathogen.

Franco de Lima *et al.* (2024) explored the use of antisense oligonucleotides (ASOs) with phosphorothioate modifications to transiently inhibit the gene expression of

CsLOB (*Lateral Organ Boundaries*) *1* in *Citrus sinensis*. They designed and validated three potential ASO sequences, which led to a significant reduction in disease symptoms compared with the control. The selected ASO3-*CsLOB1* significantly decreased the expression level of *CsLOB1* when delivered through two distinct delivery methods, and the reduction of the symptoms ranged from approximately 15 to 83%. Notably, plants treated with ASO3 did not exhibit an increase in symptom development over the evaluation period. This study highlights the efficacy of ASO technology, based on short oligonucleotide chemically modified sequences, as a promising tool for controlling phytopathogens without the need for genetic transformation or plant regeneration. The results demonstrate the potential of ASOs as a biotechnological tool for the management of citrus canker disease.

9.48.5 Management

1. The enforcement of strict quarantine is the best method of control in disease-free areas. *Xan. citri* subsp. *citri* is EPPO A 1 quarantine pest. Every effort should be made to eradicate the bacterium from such areas when it is introduced or found in these areas.
2. In areas where the disease is endemic, the following integrated management measures should be adopted to control it.

   a. Use of disease-free nursery and the production of pathogen-free budwood are the foundations of an integrated management programme.
   b. Use of resistant or field-resistant cultivars is the most important feature of the integrated disease management programme. Some 85% of citrus production in Japan is from the cultivation of a moderately resistant cultivar, Unshu (satsuma) orange.
   c. Cultural practices like the use of windbreaks and the pruning of diseased summer and autumn shoots are very effective. Pruning should be done during the dry season to prevent the dissemination of the pathogen. Windbreaks are most effective for the control of disease on susceptible varieties.
   d. On susceptible or highly susceptible cultivars and in areas prone to disease epiphytotics, the prevention of primary infection on spring shoots is very essential for the management of the disease.
   e. Periodical sprays of insecticides should be given to control the citrus leaf miner.
   f. Goto (1992) has given the criteria on the basis of which an epiphytotic can be predicted 1–2 months in advance. Based on that, the citrus orchardists can be advised to take timely steps for an effective chemical control.
   g. Preventive sprays of copper-based bactericides are used throughout the world to manage citrus canker. Copper sprays alone may control the disease on resistant or moderately resistant cultivars while the control on susceptible or highly susceptible cultivars will require the integration of several control measures. Timing and number of copper sprays will vary with weather conditions, age of trees and reaction of host cultivars. In general, however, 3–5 copper sprays are sufficient on cultivars having an intermediate level of resistance. A NASBA (nucleic acid sequence-based amplification) protocol using primers for *gum*D mRNA has been developed to assess the viability of this pathogen to a bactericidal treatment. The method is rapid, specific and sensitive and is able to detect viable bacterial cells using a hybridisation device, which allows the visualisation of the results in only 30 minutes (Scuderi *et al.*, 2010).

In India, Patel and Desai (1970) reported that the disease was effectively controlled within three years by removing the infected twigs every year and spraying the trees with 1% Bordeaux mixture on four occasions during the year. The first spray should be given in November–December, the second in February–March, the third before monsoon or the first fortnight of June and the fourth in July–August or when there is a break in the rains.

Machado *et al.* (2021) reported that copper sprayed before wounding and/or inoculation gave the greatest reduction (>90%) in the incidence of citrus canker. The reduction in cankers associated with wounds ranged from 91.8 to 96.1% for copper sprays applied from 24 h before to 0.5 h after wounding and inoculation, and was 67.8% when copper was applied up to 8 h after wounding and inoculation. As *Xan. citri* is splash-dispersed, the authors concluded that wounds need to be protected with copper sprays if rain is forecasted within 7 days of a wounding event. In cases when no rain is expected, there is no need to protect the trees with copper.

Continuous use of copper bactericides has resulted in the appearance of copper-tolerant strains of the bacteria in Argentina. Copper-resistant strains of *Xan. citri* pv. *citri* have also been reported from Réunion, France (Richard *et al.* 2017). In such areas, copper-mancozeb, a mixture of copper and mancozeb, should be used instead of copper alone.

Behlau *et al.* (2012) reported that there is a low, but significant likelihood of horizontal gene transfer of copper resistance genes from other xanthomonads or epiphytic bacteria to *Xan. citri* subsp. *citri* in nature.

Graham *et al.* (2016) evaluated the antimicrobial activity of two structurally different nano-zinc oxide materials, platelike Zinkicide SG4 and particulate Zinkicide SG6, against *Xan. citri* subsp. *citri*. In field trials conducted in southeast Florida in 2014 and 2015, the reduction of grapefruit canker incidence by Zinkicide SG4 and SG6 exceeded that of cuprous oxide and cuprous oxide/zinc oxide bactericides. No sign of phytotoxicity to the fruit rind was observed during either season. Antimicrobial activity of Zinkicide for the protection of leaves and fruits against *Xan. citri* subsp. *citri* was comparable or exceeded that for commercial copper and zinc oxide formulations, which may be attributed to the translaminar movement of Zinkicide.

Yu *et al.* (2016) demonstrated that bismerthiazol can effectively control citrus canker by both inhibiting the growth of *Xan. citri* subsp. *citri* and triggering the plant's host defence response through the expression of several pathogenesis-related genes (*PR1*, *PR2*, *CHI* and *RpRd1*) and the non-expression of PR genes (*NPR1*, *NPR2* and *NPR3*) in "Duncan" grapefruit, especially at early treatment times. Additionally, bismerthiazol induced the expression of the marker genes *CitCHS* and *CitCHI* in the flavonoid pathway and of the *PAL1* gene in the salicylic acid biosynthesis pathway at different times. Bismerthiazol also induced the expression of the priming defence-associated gene *AZI1*. The results show that the induction of the defence response in "Duncan" grapefruit by bismerthiazol may involve the SA signalling pathway and the priming defence, and that bismerthiazol may serve as an alternative to copper bactericides for the control of citrus canker.

Balogh *et al.* (2008) obtained an effective and consistent control of the disease with a bacteriophage. The phage application caused a 59% reduction in disease severity. However, phage treatment was ineffective, when applied with skimmed milk.

Ibrahim *et al.* (2017) reported that applications of phages formulated with skimmed milk and sucrose, in combination with ASM as soil drench, under field conditions significantly decreased the incidence of Asiatic citrus canker by 14.8% (trial 1) and 16.8% (trial 2) compared with untreated control plants. Overall, the inoculated plants treated with bacteriophages + ASM combination showed significant disease reduction under greenhouse and field conditions. The bacteriophages + ASM combination tested in these trials can be an effective tool in the integrated management programmes of Asiatic citrus canker.

Caicedo *et al.* (2016) reported that bacteria belonging to *Pseudomonas* and *Bacillus* genera have a strong ability to reduce the severity of citrus canker in sweet orange. Reduction in disease severity was associated with alteration in bacterial attachment and biofilm formation, the factors known to contribute to *Xan. citri* subsp. *citri* virulence. These quorum-quenching (interference with cell to cell signalling) bacteria may represent a highly valuable tool in biological control and offer an alternative to the traditional copper treatment currently used to treat citrus canker.

Ferreira *et al.* (2022) reported that the sprays of 32 mg metallic copper/m^3 given at 14-day intervals or of 42 mg/m^3 at 21-day intervals were the most cost-effective for the combined management of citrus canker and black spot (caused by *Phyllosticta citricarpa*).

Behlau *et al.* (2021) evaluated the efficacy of windbreak, copper sprays and leaf miner control in eight different combinations in a split-split plot design. *Casuarina cunninghamiana* trees were used as a wind break. Copper and leaf miner control sprays were performed every 21 days throughout the year. Individually, copper sprays gave the highest disease control followed by windbreak, while leaf miner control had no effect on the disease. The combination of windbreak and copper sprays gave the highest disease control. This combination reduced the incidence of diseased trees by approximately 60%, and the incidence of diseased leaves and fruit by ≥90%, increasing the yield by 2.0- to 2.6-fold in comparison to control plots. Copper sprays were important for reducing disease incidence and crop losses, whereas the windbreak had an additional contribution in minimising the incidence of cankered, non-marketable fruits.

REFERENCES

Agrios, G. N. (2004). *Plant Pathology*. 5th Ed. Elsevier Academic Press, San Diego, CA, 922 pp.

Balogh, B., Canteros, B. I., Stall, R. E. and Jones, J. B. (2008). Control of citrus canker and citrus bacterial spot with bacteriophages. *Plant Dis*. 92: 1048–52.

Behlau, F., Barelli, N. L. and Belasque, J. Jr. (2014). Lessons from a case of successful eradication of citrus canker in a citrus producing farm in Sao Paulo State. *Brazil. J. Plant Pathol*. 96: 561–68.

Behlau, F., Belasque, J. Jr., Leite, R. P. Jr., Filho, A. B., Gottwald, T. R. *et al*. (2021). Relative contribution of windbreak, copper sprays, and leaf miner control for citrus canker management and prevention of crop loss in sweet orange trees. *Plant Dis*. 105: 2097–2105.

Behlau, F., Canteros, B. I., Jones, J. B. and Graham, J. H. (2012). Copper resistance genes from different xanthomonads and citrus epiphytic bacteria confer resistance to *Xanthomonas citri* subsp. *citri*. *Eur. J. Plant Pathol*. 133: 949–63.

Braido, R., Goncalves-Zuliani, A. M. O., Janeiro, V., Carvalho, S. A., Belasque, J. Jr. *et al*. (2014). Development and validation of standard area diagrams as assessment aids for estimating the severity of citrus canker on unripe oranges. *Plant Dis*. 98: 1543–50.

Braido, R., Goncalves-Zuliani, A. M. O., Nocchi, P. T. R., Belasque, J. Jr., Janeiro, V. *et al*. (2015). A standard area diagram set to aid estimation of the severity of Asiatic citrus canker on ripe sweet orange fruit. *Eur. J. Plant Pathol*. 141: 327–37.

Caicedo, J. C., Villamizar, S., Ferro, M. I. T., Kupper, K. C. and Ferro, J. A. (2016). Bacteria from the citrus phylloplane can disrupt cell-cell signalling in *Xanthomonas citri* and reduce citrus canker disease severity. *Plant Pathol*. 65: 782–91.

Christiano, R. S. C., Pria, M. D., Jesus, W. C. Jr., Parra, J. R. P., Amorim, L. *et al*. (2007). Effect of citrus leaf-miner damage, mechanical damage and inoculum concentration on severity of symptoms of Asiatic citrus canker in Tahiti lime. *Crop Prot*. 26: 59–65.

de Oliveira Andrade, M., Cristina da Silva, J., Soprano, A. S., Shimo, H. M., Leme, A. F. P. et al. (2023). Suppression of citrus canker disease mediated by flagellin perception. *Mol. Plant Pathol*. 24(4): 331–45. doi:10.1111/mpp.13300 dos Santos, T. A., Marin, T. G. D., Gorayeb, E. S. and Behlau, F. (2024). Inoculum concentration and moisture chamber incubation period for inoculation of *Xanthomonas citri* subsp. *citri*, the causal agent of citrus canker. *Eur. J. Pl. Pathol.* 169: 461–71. https://doi.org/10.1007/s10658-024-02841-0

Duan, S., Long, Y., Cheng, S., Li, J., Ouyang, Z. *et al.* (2022). Rapid evaluation of the resistance of citrus germplasms against *Xanthomonas citri* subsp. *citri*. *Phytopathology* 112: 765–74.

Favaro, M. A., Molina, M. C., Roeschlin, R. A., Gadea, J., Gariglio, N., *et al.* (2020). Different responses in mandarin cultivars uncover a role of cuticular waxes in the resistance to citrus canker. *Phytopathology* 110: 1791–1801.

Ference, C. M., Gochez, A. M., Behalu, F., Wang, N., Graham, J. H. *et al.* (2018). Recent advances in the understanding of *Xanthomonas citri* ssp. *citri* pathogenesis and citrus canker disease management. *Mol. Plant Pathol.* 19: 1302–18.

Ferreira, D. H., Moreira, R. R., Silva, G. J. Jr. and Behlau, F. (2022). Copper rate and spray interval for joint management of citrus canker and citrus black spot in orange orchards. *Eur. J. Plant Pathol.* 163: 891–906. https://doi.org/10.1007/s10658-022-02527-5

Francis, M. I., Pena, A. and Graham, J. H. (2010). Detached leaf inoculation of germplasm for rapid screening of resistance to citrus canker and citrus bacterial spot. *Eur. J. Plant Pathol.* 127: 571–78.

Franco de Lima, L. F., Carvalho, I. G. B., de Souza-Neto, R. R., da Silva dos Santos, L. and Nascimento, C. A. (2024). Antisense oligonucleotide as a new technology application for *CsLOB1* gene silencing aiming at citrus canker resistance. *Phytopathology* 114: 1802–09. https://doi.org/10.1094/PHYTO-02-24-0058-KC

Gabriel, D. W., Kingsley, M. T., Hunter, J. E. and Gottwald, T. (1989). Reinstatement of *Xanthomonas citri* (ex Hasse) and *X. phaseoli* (ex Smith) to species and reclassification of all *X. campestris* pv. *citri* strains. *Int. J. Syst. Bacteriol.* 39: 14–22.

Goto, M. (1992). Citrus canker. In: J. Kumar, H. S. Chaube, U. S. Singh and A. N. Mukhopadhyay (eds). *Plant Diseases of International Importance*. Vol. III, *Diseases of Fruit Crops*. Prentice Hall, Englewood Cliffs, NJ, pp. 170–208.

Gottwald, T. R., Graham, J. H. and Schubert, T. S. (2002). Citrus canker: The pathogen and its impact. APSnet Feature Story. http://www.plantmana-gementnetwork.org/php/review/citruscanker.

Graham, J. H., Johnson, E. G., Myers, M. E., Young, M., Rajasekaran, P. *et al.* (2016). Potential of nano-formulated zinc oxide for control of citrus canker on grapefruit trees. *Plant Dis.* 100: 2442–47.

Hasse, C. H. (1915). *Pseudomonas citri*, the cause of citrus canker. *J. Agric. Res.* 4: 97–100.

Hu, Y., Duan, S., Zhang, Y., Shantharaj, D., Jones, J. B. *et al.* (2016). Temporal transcription profiling of sweet orange in response to PthA4-mediated *Xanthomonas citri* subsp. *citri* infection. *Phytopathology*, 106: 442–51

Ibrahim, Y. E., Saleh, A. A. and Al-Saleh, M. A. (2017). Management of Asiatic citrus canker under field conditions in Saudi Arabia using bacteriophages and acibenzolar-*S*-methyl. *Plant Dis.* 101: 761–65.

Jia, H., Omar, A. A., Orbović, V. and Wang, N. (2022). Biallelic editing of the *LOB1* promoter via CRISPR/Cas9 creates canker-resistant 'Duncan' grapefruit. *Phytopathology*, 112: 308–14.

Kobayashi, A. K., Vieira, L. G. E., Bespalhok Filho, J. C., Leite, R. P. Jr., Pereira, L. F. P. *et al.* 2017. Enhanced resistance to citrus canker in transgenic sweet orange expressing the sarcotoxin IA gene. *Eur. J. Plant Pathol.* 149: 865–73.

Kositcharoenkul, N., Chatchawankanphanich, O, Bhunchoth, A. and Kositratana, W. (2011). Detection of *Xanthomonas citri* subsp. *citri* from field samples using single-tube nested PCR. *Plant Pathol.* 60: 436–42.

Li, W., Li, Z., Zhang, L., Deng, X. and Zheng, Z. (2022). Genome sequence resource of *Xanthomonas citri* pv. *citri* from formalin-fixed citrus leaves specimen showing canker lesions collected in 1982. *Plant Dis.* 106: 304–06.

Machado, F. J., Marin, T. G. d. S., Canôas, F., da Silva Jr., G. J. and Behalu, F. (2021). Timing of copper sprays to protect mechanical wounds against infection by *Xanthomonas citri* subsp. *citri*, causal agent of citrus canker. *Eur. J. Plant Pathol.* 160: 683–92. https://doi.org/10.1007/s10658-021-02276-x

Marin, T. G. S., Galvanin, A. L., Lanza, F. E. and Behalu, F. (2019). Description of copper tolerant *Xanthomonas citri* subsp. *citri* and genotypic comparison with sensitive and resistant strains. *Plant Pathol.* 68: 1088–98.

Mirzaei-Najafgholi, H., Aeini, M., Tarighi, S. and Golmohammadi, M. (2021). Comparing bacterial properties in relation to the virulence factors of *Xanthomonas citri* subsp. *citri* strains and evaluating resistance of subtribe Citrinae cultivars to the most virulent strain. *J. Plant Pathol.* 103: 449–60. https://doi.org/10.1007/s42161-021-00760-1

Niculau, E. D. S., Ferreira, D., Behlau, E. R-F. F., Magnani, R. F., Toffano, L. et al. (2022). MALDI-TOF mass spectrometry for the diagnosis of citrus canker caused by *Xanthomonas citri* subsp. *citri*. *Molecules*, 27(24): 8947

Patel, R. S. and Desai, M. V. (1970). Control of citrus canker. *Indian J. Hort.* 27: 93–98.

Patel, R. S. and Desai, M. V. (1972). Citrus canker. In: P. N. Patel (ed.). *Plant Bacteriology*. Vol. I, *Bacterial Diseases of Plants in India*. Indian Agricultural Research Institute, New Delhi, pp. 184–91.

Pria, M. D., Christiano, R. C. S., Furtado, E. L. Amorim, L. and Filho, A. B. (2006). Effect of temperature and leaf wetness duration on infection of sweet oranges by Asiatic citrus canker. *Plant Pathol.* 55: 657–63.

Pruvost, O., Boyer, K., Ravigné, V., Richard, D. and Vernière, C. (2019). Deciphering how plant pathogenic bacteria disperse and meet: Molecular epidemiology of *Xanthomonas citri* pv. *citri* at microgeographic scales in a tropical area of Asiatic citrus canker endemicity. *Evol. Appl.* 12: 1523–38. PMID: 31462912; PMCID: PMC6708428.

Richard, D., Tribot, N., Boyer, C., Terville, M., Boyer, K. *et al.* (2017). First report of copper-resistant *Xanthomonas citri* pv. *citri* pathotype A causing Asiatic citrus canker in Reunion, France. *Plant Dis.* 101: 503.

Rodoriguez, S. G., Stapleton, J. J. and Civerolo, E. L. (1985). *Xanthomonas campestris* involved in Mexican lime bacteriosis in Colima, Mexico. *Proc. 6th Int. Conf. Plant Path. Bact.*, Maryland, pp. 658–62.

Schaad, N. W., Postnikova, E., Lacy, G., Sechler, A., Agarkova, I. *et al.* (2005). Reclassification of *Xanthomonas campestris* pv. *citri* (ex Hasse 1915) Dye 1978 forms A, B, C, D and E as *X. smithii* subsp. *citri* (ex Hasse) sp. nov. nom. rev. comb. nov.; *X. fuscans* subsp. *aurantifolii* (ex Gabriel 1989) sp. nov. nom. rev. comb. nov., and *X. alfalfae* subsp. *citrumelo* (ex Riker and Jones) Gabriel *et al.*, 1989 sp. nov. nom. rev. comb. nov.; *X. campestris* pv. *malvacearum* (ex Smith 1901) Dye 1978 as *X. smithii* subsp. *smithii* sp. nov. comb. nov. nom. nov.; *X. campestris* pv. *alfalfae* (ex Riker and Jones, 1935) Dye 1978 as *X. alfalfae* subsp. *alfalfae* (ex Riker *et. al.*, 1935) sp. nov. nom. rev.; and "var. *fuscans*" of *X. campestris* pv. *phaseoli* (ex Smith, 1897) Dye 1978 as *X. fuscans* subsp. *fuscans* sp. nov. *Syst. Appl. Microbiol.* 28: 494–518.

Schaad, N. W., Postnikova, E., Lacy, G., Sechler, A., Agarkova, I. *et al.* (2007). Validation of the publication of new names and new combinations previously effectively, but not validly, published outside the *Int. J. Syst. Evol. Microbiol.* 57: 893–97. Effective Publication: Schaad, N. W., Postnikova, E., Lacy, G., Sechler, A., Agarkova, I. *et al.* (2006). Emended classification of xanthomonad pathogens on citrus. *Syst. Appl. Microbiol.* 29: 690–95.

Schoulties, C. L., Civerolo, E. L., Miller, J. W., Stall, R. E., Krass, C. J. *et al.* (1987). Citrus canker in Florida. *Plant Dis.* 71: 388–95.

Scuderi, G., Golmohammadi, M., Cubero, J., Loópez, M. M., Civilleri, G. *et al.* (2010). Development of a simplified NASBA protocol for detecting viable cells of the citrus pathogen *Xanthomonas citri* subsp. *citri* under different treatments. *Plant Pathol.* 59: 764–72.

Shi, Q., Febres, V. J. and Jeffr (2015). Responsiveness of different citrus genotypes to the *Xanthomonas citri* ssp. *citri*-derived pathogen-associated molecular pattern (PAMP) flg22 correlates with resistance to citrus canker. *Mol. Plant Pathol.* 16: 507–20.

Sidireddi, S. H., Park, J. W., Setamou, M. and Kunta, M. (2023). A loop-mediated isothermal amplification (LAMP) assay for rapid field detection of citrus canker. *Phytopathology* 113: S3.75.

Yu, S. M., Ramkumar, G. and Lee, Y. H. (2017). Detection of *Xanthomonas citri* subsp. *citri* A*, AW and *X. fuscans* subsp. *aurantifolii* B, C using PCR and real-time PCR. *J. Plant Pathol.* 99: 461–67.

Yu, X., Armstrong, C. M., Zhou, M. and Duan, Y. (2016). Bismerthiazol inhibits *Xanthomonas citri* subsp. *citri* growth and induces differential expression of citrus defense-related genes. *Phytopathology* 106: 693–701.

Zhang, X., Francis, M. I., Dawson, W. O., Graham, J. H., Orbović, V. *et al.* (2010). Over-expression of the Arabidopsis *NPR1* gene in citrus increases resistance to citrus canker. *Eur. J. Plant Pathol.* 128: 91–100.

9.49 CITRUS HUANGLONGBING

This is the most damaging and devastating disease of citrus. The disease is also known by various other names such as **citrus greening disease, yellow shoot disease** or **yellow dragon disease**. Yellow dragon disease and yellow shoot disease are both English translations of the Chinese word **huanglongbing**. The International Organisation of Citrus Virologists, at its congress held in China in 1995, decided to adopt the original Chinese name "**huanglongbing**" (hereafter abbreviated as HLB) as the official name for the disease.

Southern China has long been considered as the area of origin for the Asian form of the disease. However, Gottwald *et al.* (2007) are of the view that earlier descriptions of citrus die-back in India suggest that the disease was present there at an earlier date. Capoor (1963) attributed citrus die-back to citrus tristeza virus after biological indexing studies but (Raychaudhuri *et al.* 1969, 1974) later showed that huanglongbing was a major component of the problem. Beattie *et al.* (2006) have hypothesised that the disease might have originated in Africa, possibly in a symptomless host such as *Verpris lanceolata*.

The disease is widespread in Asia (eastern, central and southern), Africa, the Arabian Peninsula and Brazil. The disease was noticed in Florida State on 24 August, 2005 on a pummelo tree in a commercial tropical fruit nursery in Florida city. Since then, it has spread to most of southern Florida. In Brazil, the disease was first reported in 2004 from São Paulo state and both Asian and American forms of the disease occur there. According to Pietersen *et al.* (2010), to date only '*Candidatus* Liberibacter africanus' is associated with the disease in commercial citrus in South Africa in addition to '*Can.* Lib. africanus subsp. capensis', which has been found only in South Africa infecting an indigenous rutaceous species, *Calodendrum capense* (cape chestnut). The disease has not been reported from Australia, Mexico, Central America and the Mediterranean citrus-growing areas. Citrus-growing countries, where the disease is absent, are actively guarding against the entry of the pathogen.

Bove (2006) has correctly stated that HLB is the most important, destructive and devastating disease of citrus in the world. There are no known sources of resistance and the management of the disease is difficult and expensive. Reduction in yield can reach to the extent of 30 to 100%, depending on the proportion of the affected canopy. Affected orchards become economically unviable in 7–10 years after planting. About 100 million trees were affected and have been destroyed in many countries of South and Southeast Asia, India, Indonesia, the Philippines, the Arabian Peninsula and South Africa. Since 2004, more than 500,000 trees have been officially destroyed in Brazil due to HLB and around 300,000–400,000 trees unofficially destroyed by commercial citrus growers (Gottwald *et al.*, 2007). Catastrophic losses due to the disease from India during the 1960s were reported by Fraser *et al.* (1966). In the past three decades or so, 12 reviews have been published on the disease (Bove, 2006; da Graca, 1991; da Graca and Korsten, 2004; da Graca *et al.*, 2022: Garnier and Bove, 1993; Halbert and Manjunath, 2004; Gottwald *et al.*, 2007; Gottwald, 2010: Killiny, 2022; Pandey *et al.*, 2022; Wang and Trivedi, 2013; Wang *et al.*, 2017).

The review published by Gottwald *et al.* (2007) deals with the epidemiology of HLB and the implications of the temporal increase and spatial spread of the disease on prospects for disease control and mitigation. This review also contains concise information on the main features and symptoms of the disease, the causal organisms and their vectors. Last review published by Gottwald (2010) deals with epidemiological information of the disease.

Wang and Trivedi (2013), in their review article, have presented information on the research on citrus, '*Can.* Lib. asiaticus' and psyllid interactions with specific emphasis on evolutionary relationships among '*Can.* Liberibacter spp.', genetic diversity, host range, genome analysis, transmission, virulence mechanisms and the ecological importance of HLB. Review by Wang *et al.* (2017) deals with the progress made in understanding the HLB disease pyramid, and how the microbiome affects the HLB disease pyramid, including the interaction between HLB and the citrus microbiome, the interaction between liberibacters and psyllids, the interaction between liberibacters and gut microbiota in psyllids and the effect of HLB on selected above and below ground citrus pathogens. Their implications for disease management are also discussed.

Merfa *et al.* (2019) reported that the inability to grow '*Can.* Lib. asiaticus' (*C*Las) in culture (except transiently) remains a significant scientific challenge toward effective HLB management. Lack of axenic cultures has restricted functional genomic analyses, transfer of *C*Las to either insect or plant hosts for fulfilment of Koch's postulates, characterisation of host–pathogen interactions and effective screening of antibacterial compounds. In the past decade or so, substantial progress has been made toward *C*Las culturing: (i) three reports of transient *C*Las cultures were published, (ii) a new species of *Liberibacter* (*Lib. crescens* strain BT-1) was identified and axenically cultured from diseased mountain papaya, (iii) psyllid haemolymph and citrus phloem sap were biochemically characterised, (iv) *C*Las phages were identified and lytic genes possibly affecting *C*Las growth were described and (v) genomic sequences of 15 *C*Las strains were made available. In addition, the development of *Lib. crescens* as a surrogate host for the functional analyses of *C*Las genes, has provided valuable insights into *C*Las pathogenesis and its physiological dependence on the host cell. In this review, the authors have summarised the conclusions from these important studies.

Saha *et al.* (2017) have designed a web portal with information for consumers and growers as well as genomics resources for citrus, Asian citrus psyllid and '*Can.* Lib. asiaticus'. This is an open source software project with the goal of developing a system for storing, integrating and analysing all-omics data related to huanglongbing. The portal can be accessed at https://citrusgreening.org/.

Killiny (2022) has reviewed the findings concerning the relationship of '*Can.* Lib. asiaticus' with its insect vector, *Diaphorina citri* and molecular features of the acquisition of '*Can.* Lib. asiaticus' from the plant host and its growth and circulation within *D. citri*, as well as its transmission to plants, are presented. In addition, the effects of '*Can.* Lib. asiaticus' on the energy metabolism (respiration, tricarboxylic acid cycle and adenosine triphosphate production), metabolic pathways, immune system, endosymbionts and detoxification enzymes of *D. citri* are discussed, together with other impacts such as shorter lifespan, altered feeding behaviour and higher fecundity. Although '*Can.* Lib. asiaticus' has significant negative effects on its insect vector, it increases its vector fitness, indicating that it develops a mutualistic relationship with its vector. This review will help in understanding the specific interactions between '*Can.* Lib. asiaticus' and its psyllid vector in order to design innovative management strategies.

Da Graça *et al.* (2022) have presented a review on various aspects of the African form of HLB, including its specific bacterial species and subspecies, its main insect vector, its geographic distribution and current management strategies.

Pandey *et al.* (2022) have reviewed recent studies on the genetic information of flagellar and type IV pili biosynthesis, their expression profiles and the movement of *Can.* Liberibacter spp. inside the plant and insect hosts. The transcriptomic, proteomic and metabolomic studies of susceptible and tolerant plant genotypes to *Can.* Liberibacter spp. infection and how *Can.* Liberibacter spp. adapt in plants are also reviewed. Analyses of the interactions between plants and *Can.* Liberibacter spp. imply the involvement of immune response in the *Can.* Liberibacter pathosystems. The movement of *Can.* Liberibacter spp. inside the host and its interactions with plants leading to symptom development are also discussed.

9.49.1 Host Range

HLB bacteria infect most citrus species, hybrids and many citrus relatives. The most severely infected hosts are sweet oranges, mandarins, mandarin hybrids and tangelo (Halbert and Manjunath, 2004). HLB in India and Africa is more severe on mandarins and sweet oranges while limes, lemons, grapefruit and pummelos show tolerant reaction. However, Asian forms of HLB found in Taiwan and Florida and the American form reported from Brazil affect severely all commercial citrus species. It is possible that strains of HLB pathogen adapt to citrus species over a period of time, which is evident from the fact that grapefruit and pummelo, once considered resistant to the pathogen in Taiwan, are now quite susceptible.

Some citrus relatives, namely, *Atalantia missionis*, *Balsamocitrus dawei*, *Clausena indica*, *Microcitrus australasica*, *Swinglea glutinosa* and *Severinia buxifolia*, are also severely infected. The other citrus relatives infected are *Limonia acidissima*, *Murraya paniculata* and *Verpris lanceolata*. There is disagreement in the literature about the status of *M. paniculata* as a host of HLB pathogens. It is a host in Brazil, but not in Taiwan. *Catharanthus roseus* (pink periwinkle) and *Nicotiana xanthii* (tobacco) are the non-rutaceous reported hosts. Tomato has also been shown to be a host for '*Can.* Lib. asiaticus' and the bacterium was transmitted through dodder (Duan *et al.*, 2008).

9.49.2 Symptoms

The symptoms vary with host species/cultivar, age of host, time of infection and biotic and abiotic factors that influence the growth of trees. The most characteristic foliage symptoms are the blotchy mottling and yellowing of leaves (Figure 9.40a). In blotchy mottling, the infected leaves exhibit a pattern of yellow and dark green areas, lacking clear limits between the colours. The patterns of blotchy mottling are asymmetrical on two halves of the leaf. Small yellow leaves appear on a single shoot/branch or on a section of the tree canopy and yellowing gradually spreads to the entire tree (Figure 9.49b). The appearance of a yellow shoot or shoots has led to the name **"yellow shoot disease"**. The newest leaves may show symptoms, similar to zinc deficiency with yellow venation, while the older leaves have the characteristic greening mottle. Other symptoms are reduced foliage, severe die-back of twigs or limbs, poor flowering and stunting of trees.

The fruits on diseased trees are small, poorly coloured and often misshaped. Seeds in symptomatic fruits are often small, dark and aborted and the fruits remain green on stylar end even after maturity (Figure 9.40c), hence the name **greening**. The fruit development can be asymmetrical and lopsided (Figure 9.40d). The diseased fruits have a sour bitter taste and drop prematurely.

The following symptoms of huanglongbing help in differentiating it from chronic citrus decline caused by citrus tristeza virus, blight and citrus variegated chlorosis. Yellowing of leaf midrib and veins and slight swelling of veins in early stage of the disease. The seeds are aborted in small and misshaped fruits. The fruits, especially small, have a bitter taste compared to sweet taste found in other diseases.

The above-given symptoms are sufficient to diagnose the disease under field conditions, but final confirmation by other methods is required for regulatory measures. Before the development of molecular techniques, the diagnosis of the disease was done by biological indexing, thin layer chromatography for determining the presence of gentisoyl-β-glucoside, although not all HLB-infected citrus species produce it, the use of monoclonal antibodies and the use of indicator plants like sweet orange, Duncan grapefruit and sweet tangor seedlings. A staining technique employing gentisic acid has been found to be very useful in the survey work. However, recently developed techniques such as DNA hybridisation and PCR are rapid, sensitive and specific for the diagnosis of HLB in the laboratory.

PCR is now the main confirmatory test and is routinely used in many laboratories, including those in Florida and Brazil. Das *et al.* (2007) confirmed the presence of the disease in the northeastern region of India by PCR technique. Manjunath *et al.* (2008) developed a TaqMan-based real-time quantitative PCR method to detect '*Can.* Lib. asiaticus' from the nymphs and adults of *Diaphorina citri* collected from apparently healthy and diseased citrus trees. This finding is highly significant as the presence of the bacterium in an area can be detected one to several years before the development of disease symptoms and hence the strategies for disease management can be initiated in advance. Li *et al.* (2008), using TaqMan real-time PCR, established a universal standard curve (based on DNA extracted from naturally infected plants) for the quantification of the pathogen in various tissues of different citrus species planted in different geographical regions for studying the host–pathogen interactions and epidemiology and for devising disease management strategies.

Three types of HLB, namely, Asian, African and American, occur in nature. The symptoms of Asian and African forms are generally the same. However, the Asian form is considered more severe because it causes more extensive and severe die-back, eventually killing the trees. The bacterium causing the Asian form is also more

(a) (b) (c) (d)

FIGURE 9.40 (a) Citrus huanglongbing: Leaf symptoms. (b) Citrus huanglongbing: Yellow shoot (Courtesy: Tim Gottwald). (c) Citrus huanglongbing: Greening on stylar end of fruits (Courtesy: Tim Gottwald). (d) Citrus huanglongbing: Asymmetrical lopsided fruit development.

heat-tolerant. The symptoms of the African form are suppressed by long exposure to temperature above 30 °C. Hence, it is usually found in citrus at elevations above 700 m, while the Asian form occurs in lower-lying, hotter areas. The American form appears to resemble closely the Asian form in symptom expression and severity, but its causal agent is also intolerant to heat, like that of the African type.

Chen *et al.* (2009) have shown the presence of a phytoplasma related to '*Can.* Phytoplasma asteris' in citrus infected with '*Can.* Lib. asiaticus'. Out of 141 samples collected from Guangdong Province, PR China, 69 (or 48.9%) of the samples were positive for both '*Can.* Phy. asteris' and '*Can.* Lib. asiaticus'. After the transmission from symptomatic citrus to periwinkle (*Catharanthus roseus*) via dodder (*Cuscuta campestris*), both phytoplasma and '*Can.* Lib. asiaticus' were detected from the infected periwinkle. In addition to yellowing/mottling, the infected periwinkle plants showed symptoms of virescence and phyllody which are commonly associated with phytoplasma diseases.

9.49.3 Causal Organisms

Liberibacter asiaticus Jagoueix-Eveillard *et al.* 1994 causes Asian form of HLB of citrus and citrus relatives

Liberibacter africanus Jagoueix-Eveillard *et al.* 1994 causes African form of HLB of citrus and citrus relatives

Liberibacter africanus subsp. *capensis* Garnier *et al.* 2000 causes HLB of cape chestnut (*Calodendrum capense*).

Liberibacter africanus subsp. *clausenae* Roberts *et al.* 2015 causes HLB of Horse wood (*Clausena*)

Liberibacter africanus subsp. *tecleae* Roberts and Pietersen 2017 causes HLB of flaky cherry-orange (*Teclea gerrardii*)

Liberibacter africanus subsp. *vepridis* Roberts *et al.* 2015 causes HLB of white ironwood (Vepris)

Liberibacter africanus subsp. *zanthoxyli* Roberts *et al.* 2015 causes HLB of small forest knob wood (Zanthoxylum)

Liberibacter americanus Teixeira *et al.* 2005 causes American form of HLB of citrus and citrus relatives

For many years, HLB was considered to be caused by mineral deficiencies/toxicities or water logging. Later on, the transmission of the causal agent through grafting and insects led to the conclusion that it was a virus. The electronic microscopic studies revealed the presence of mycoplasma-like organisms (MLOs) in the phloem sieve elements of infected plants but not in healthy plants. However, Garnier and Bove in 1978, after close examination of these organisms, found that they have thicker envelops than MLOs and hence they are true bacteria

The above-mentioned eight pathogens are fastidious, phloem-limited and Gram-stain-negative bacteria. '*Can.* Lib. asiaticus' requires high temperature than '*Can.* Lib. africanus' for optimum expression of the disease.

Roberts *et al.* (2015) extracted total DNA from *Clausena anisata*, *Vepris lanceolata* and *Zanthoxylum capense* trees and tested for the presence of liberibacters by a generic Liberibacter TaqMan real-time PCR assay. Liberibacters present in positive samples were characterised by amplifying and sequencing *rplj*, *omp* and 16S rRNA gene regions. Phylogenetic analysis of the *rplj* and *omp* gene regions revealed unique clusters for liberibacters associated with each of the tree species. The following names were proposed for these novel liberibacters: '*Candidatus* Liberibacter africanus subsp. clausenae', '*Candidatus* Liberibacter africanus subsp. vepridis', and '*Candidatus* Liberibacter africanus subsp. zanthoxyli' which infect *C. anisata*, *V. lanceolata* and *Z. capense*, respectively.

Sechler *et al.* (2009) designed a medium called Liber A to culture all the three Liberibacter spp. that cause HLB. Medium containing citrus vein extract and a growth factor supported the growth of *Liberibacter* spp. for four or five single colony transfers before the decline of viability. The colonies of '*Can.* Lib. asiaticus' were irregularly-shaped, convex and 0.1–0.3 mm after 3–4 days of incubation. Two strains of '*Can.* Lib. asiaticus' and one of '*Can.* Lib. americanus' grown on this medium proved pathogenic to citrus and were isolated from non-inoculated tissues of inoculated trees and seedlings some 9 and 2 months later, respectively. Identity of the bacteria was confirmed by real-time PCR and 16S rDNA sequencing. This is the first report of the culturing and pathogenicity of '*Can.* Lib. asiaticus' and '*Can.* Lib. americanus' associated with HLB.

Duan *et al.* (2020) developed a semi-selective media based on Biolog assays and antibiotic screening and optimised the growth conditions for the cultivation of *Lib. asiaticus* (*Las*) *in vitro*. Under optimised conditions, they were able to culture *Las* bacteria to a growth peak in 2–3 weeks in the liquid medium, which was estimated as 10^6–10^7 cells per/ml with an approximately 500-fold increase based on qPCR. The cultured *Las* bacteria remained in a dynamic growth for more than 12 months and displayed limited growth in subcultures. The presence and growth of the *Las* bacterial cells were also confirmed by fluorescence *in situ* hybridisation with *Las*-specific probes and gene expression profiling, which targeted ca. 500 *Las* metabolism-related genes. It is important to note that *Las* bacterial growth in the medium relied on the presence of a helper bacterium. Cultured *Las* was inoculated back to citrus seedlings via psyllid feeding and inoculation. Although less than 5% of the fed psyllids and inoculated plants became positive, this is the first demonstration of significant growth of *Las* bacteria *in vitro* and the successful inoculation of cultured *Las* back to psyllids and citrus plants.

Duan *et al.* (2009) reported the complete genome sequence of '*Can.* Lib. asiaticus', one of the causal agents of HLB. A complete circular genome of this bacterium was obtained by metagenomics, using the DNA extracted from a single '*Can.* Lib. asiaticus'-infected psyllid. The 1.23 mb genome has an average G + C content of 36.5%. A high percentage of genes involved in both cell motility (4.5%) and active transport in general (8.0%) may be contributing to its virulence. The bacterium appears to have a limited ability for aerobic respiration and is likely auxotrophic for at least five amino acids. As the bacterium is intracellular in nature, it lacks type III and type IV secretion systems as well as typical free-living or plant-colonising extracellular degradative enzymes. However, it seems that it has all the type I secretion system genes needed for both multidrug efflux and toxin effector secretion. This is the first genome sequence of a fastidious alpha proteobacteria that is both an intracellular plant pathogen and insect symbiont.

Das *et al.* (2023) reported the whole-genome sequence of a '*Can.* Liberibacter asiaticus' (*C*Las) strain obtained from the leaf midrib of an HLB-infected Khasi mandarin tree in Assam, India. The draft genome of strain AS-TNSK3 is 1236955 bp with G + C content of 36.5% and harbours both Type 1- and Type 2-like prophages. The annotation of AS-TNSK3 genome identifies 1084 open reading frames (ORFs) and 45 RNA genes. This *C*Las whole-genome sequence is the first report from Northeast India, considered to be the centre of origin of the genus *Citrus*.

Wulff *et al.* (2009) have reported rRNA operons and genome size of another liberibacter, '*Can.* Lib. americanus'. The sequences of the entire 23S rRNA and 5S rRNA genes from '*Can.* Lib. americanus' have been obtained, using a consensus primer designed on known $tRNA^{Met}$ sequences of rhizobia. The size of the '*Can.* Lib. americanus' genome was determined by PFGE, using '*Can.* Lib. americanus' -infected periwinkle plants for bacterial enrichment, and was found to be close to 1.31 mbp. In order to determine the number of ribosomal operons on the Lam genome, probes designed to detect the 16S rRNA gene and the 3' end of the 23S rRNA gene were developed and used for Southern hybridisation with I-*Ceu*I-treated genomic DNA. These results suggest that there are three ribosomal operons in a circular genome. '*Can.* Lib. americanus' is the first liberibacter species for which such data are available.

Tomimura *et al.* (2009) investigated the genetic diversity and relationship among '*Can.* Lib. asiaticus' isolates from different hosts and distinct geographical areas in Southeast Asia. Genetic diversity was estimated by sequencing four well-characterised DNA fragments, namely, the 16S ribosomal DNA (rDNA) and 16S/23S intergenic spacer regions, the outer membrane protein (*omp*) gene region, the *trm*U-*tuf*B-*sec*E-*nus*G-*rp*/KAJL-*rpo*B region (gene cluster region), and the bacteriophage-type DNA polymerase region. The sequences of the 16S rDNA and 16S/23S intergenic spacer regions were identical in all the isolates, but nucleotide substitutions were observed in both the *omp* gene and the gene cluster regions. Extended bacteriophage-type DNA polymerase sequences acquired by thermal asymmetric interlaced PCR provided the most sequence diversity among the isolates. Phylogenetic analysis of the bacteriophage-type DNA polymerase sequences revealed three clusters in the southeast Asian population of the pathogen and all the Indonesian isolates were clustered in one group. The remaining two clusters were not correlated with geographic distribution and the differences in genetic sequences also did not reflect differences in the hosts, i.e., mandarin or pummelo, from which they were isolated. The bacteriophage-type DNA polymerase region would be useful for molecular differentiation among southeast Asian isolates of '*Can.* Lib. asiaticus'.

Ghosh *et al.* (2015) characterised the variability in the Indian '*Can.* Lib. asiaticus' based on the tandem repeats at the genomic locus CLIBASIA_01645 and categorised it into four classes based on the tandem repeat number (TRN); Class I (TRN ≤ 5), Class II (TRN $> 5 \leq 10$), Class III (TRN $> 10 \leq 15$) and Class IV (TRN > 15). It was found that the Indian population of '*Can.* Lib. asiaticus' is more diverse than compared to Florida and Guangdong populations, which showed less diversity. While TRN5 and TRN7 genotypes dominated Florida and Guangdong populations, respectively, the Indian '*Can.* Lib. asiaticus' populations with TRN copy numbers 9, 10, 11, 12 and 13 were widely distributed throughout the country. Additionally, TRN2 and TRN17 genotypes were also observed among the Indian '*Can.* Lib. asiaticus' populations. Predominant '*Can.* Lib. asiaticus' genotypes from the northeastern region of India were TRN6 and TRN7 (53.12%) and surprisingly similar to neighbouring South China populations. Citrus cultivars showed no preference for any specific '*Can.* Li. asiaticus' genotype.

Pandey *et al.* (2024) have shed light on the pathogenicity of '*Can.* Liberibacter asiaticus' by providing useful information on the temporal dynamics of reactive oxygen species production, phloem cell death, and '*Can.* Liberibacter asiaticus' growth and have also provided useful information in determining the timing for application of antioxidants and antimicrobial agents to control HLB.

Motghare *et al.* (2018) sequenced and characterised four crucial protein genes from 18 HLB samples collected across India. The four protein genes include two component response regulator protein, periplasmic solute binding protein, putative protease IV transmembrane protein and serine protease do-like protein of the causal bacterium. Secondary and tertiary structures of the proteins were also predicted using PSIPRED and TMHMM software, respectively. Phylogenetic analyses and secondary protein structure prediction reveals that these four proteins are unique, essential and conserved in '*Can.* Lib. asiaticus' populations infecting different citrus cultivars prevalent in India. Further, these structures gave an idea about active sites which can act as perfect target sites for blocking. These bacterial encoded proteins are potential targets of antimicrobial inhibitors.

Higgins *et al.* (2024) synthesized a subset of computationally derived 20-mer peptide fragments from 182 nodule-specific cysteine-rich (NCR) peptides to identify those with activity against the unculturable vascular pathogen,

'*Can*. Lib. asiaticus' (*C*Las), associated with citrus greening disease. Grounded in a design of experiments framework, the authors evaluated the peptides in a screening pipeline involving three distinct assays: a bacterial culture assay with *Lib. crescens*; a *C*Las-infected excised citrus leaf assay; and an assay to evaluate the effects on bacterial acquisition by the nymphal stage of hemipteran vector, *Diaphorina citri*. A subset of the 20-mer NCR peptide fragments inhibits both *C*Las growth in citrus leaves and *C*Las acquisition by *D. citri*. Two peptides induced higher levels of *D. citri* mortality. These findings reveal 20-mer NCR peptides as a new class of plant-derived biopesticide molecules to control citrus greening disease.

9.49.4 Disease Cycle

As the host trees infected by the bacteria are perennial in nature, the bacteria survive in/on the infected hosts. The pathogen is spread by infected nursery plants, by grafting diseased budwood, dodder and two species of psyllid insects, namely, Asian citrus psylla, *Diaphornia citri* and African citrus psylla, *Trioza erytreae*. Natural transmission of Asian and American HLB is by *D. citri* while that of African HLB is by *T. erytreae*. Temperature tolerance and sensitivities of these two vector species match with those of HLB type, these are naturally associated with; i.e., *D. citri* is heat-tolerant, whereas *T. erytreae* is sensitive to high temperature. However, under experimental conditions, both the insect species can transmit both the pathogens causing Asian and African HLB. Adults of both the insect species can acquire and transmit the bacteria, but the acquisition periods reported by different workers vary greatly. An acquisition period of 15–30 min and 5 h minimum for *D. citri* and 24 h for *T. erytreae* has been reported. Nymphs of both the species acquire the bacteria, which are transmitted by the adults in a persistent manner. There is also evidence for transovarial transmission by *T. erytreae*. Other species of psyllid have been found infecting citrus, but none of these have been found to act as a vector.

Hartung *et al.* (2010) could not prove the seed–plant transmission of *Lib. asiaticus* and were also unable to confirm an earlier report of transmission of the bacterium through seed.

Coletta-Filho *et al.* (2014) reported that pathogen acquisition by *D. citri* was positively associated with plant infection level and time since inoculation, with acquisition occurring as early 60 days after inoculation, while the trees showed the first disease symptoms around 200 days after inoculation. These results suggest that there is ample potential for psyllids to acquire the pathogen from trees during the asymptomatic phase of infection. If so, this could limit the effectiveness of tree rouging as a disease management tool and would likely explain the rapid spread observed for the disease in the field.

Lib. asiaticus (*Las*) is transmitted in a persistent, circulative and propagative manner by *D. citri*. Killiny *et al.* (2017) showed that the levels of ATP and many other nucleotides were significantly higher in *Las* infected psyllids than in healthy psyllids. These results indicated that *Las* stimulated *D. citri* to produce more ATP and many other energetic nucleotides, while it may inhibit their consumption by *D. citri*. As a result of ATP accumulation, the adenylated energy charge (AEC) increased and the AMP/ATP and ADP/ATP ratios decreased in *Las*-infected *D. citri* psyllids. Survival analysis confirmed a shorter life span for *Las*-infected *D. citri* psyllids. In conclusion, *Las* alters the energy metabolism of its psyllid vector, *D. citri*, in order to secure its need for energetic nucleotides.

Beloti *et al.* (2018) did not obtain the successful transmission of *Lib. asiaticus* by *D. citri* to *Murraya koenigii* (curry leaf) seedlings and rated it as immune. As *M. koenigii* is attractive for *D. citri* and is immune to infection by *Lib. asiaticus*, they postulated that *M. koenigii* is an excellent candidate for use as a trap crop to improve the management of the insect vector and consequently of huanglongbing.

Ammar *et al.* (2011) detected *Lib. asiaticus* by fluorescence *in situ* hybridisation (FISH) and confocal laser scanning microscopy in the filter chamber, midgut, Malpighian tubules, haemolymph, salivary glands, ovaries and in muscle and fat tissues of *D. citri* that acquired *Lib. asiaticus* from infected plants, as well as in the phloem of infected citrus leaves. *Lib. asiaticus* appeared as pleomorphic bodies or short thin rods that were much more dispersed and individually distinct in citrus leaf phloem and in *D. citri* haemolymph, but more densely aggregated in cells of the salivary glands and other *D. citri* organs and tissues. It is the first *in situ* demonstration of *Lib. asiaticus* infection in various psyllid organs and tissues and shows the near-systemic infection of *D. citri* by *Lib. asiaticus*. Canale *et al.* (2017) reported that the long latent period and the persistence of transmission are indirect evidences of circulative propagation of *Lib. asiaticus* in *D. citri*.

Johnson *et al.* (2014) emphasised the significance of early root infection to tree health and suggested a model for phloem-limited bacterial movement from the initial insect feeding site to the roots where it replicates, damages the host root system and then spreads to the rest of the canopy during subsequent leaf flushes. Hilf and Lewis (2016) reported that grafting with intact asymptomatic and HLB-symptomatic leaves resulted in the infection of 61 of 78 (78%) and 35 of 41 (85%) plants with '*Can*. Lib. asiaticus', respectively. In addition, inoculum consisting of the leaf petiole only or only an inoculum tissue remnant under the bark of the receptor tree resulted in 6 of 12 (50%) and 7 of 31 (23%) infected trees, respectively. The majority of grafted plants developed the foliar blotchy mottle symptom considered diagnostic for HLB, while some plants also displayed the stunted, chlorotic shoots for which the disease is named. The qPCR data, together with the symptoms, displayed/demonstrated that individual leaves from infected trees can serve as effective inoculum sources for the transmission and propagation of *Lib. asiaticus* via grafting. Although, the HLB pathogens are bacteria, these do not spread by casual contamination of personnel and tools or by wind and rain.

Davis *et al.* (2016) determined the distribution of *Lib. asiaticus* within bark tissue of mature infected citrus trees using the bark samples collected at the graft union and at standard intervals in the canopy above and the root system below. The highest populations of *Lib. asiaticus* resided in the small diameter branches and canopy tissue of the sampled trees, while populations decreased by 2–3 logs in the larger diameter trunk, graft union and upper root zone. There are several possible explanations for the distinct concentration pattern found in the infected trees that could assist in identifying and improving methods of control for HLB. Lopes *et al.* (2013) reported that *Lib. asiaticus* titres and the percentages of *Lib. asiaticus*-positive psyllids were lower in plants growing in the warmer temperature regime (24 °C–38 °C) than in plants maintained in the cooler regimes. It shows that the lower incidence and slower spread of *Lib. asiaticus* in warmer regions of Sao Paulo state of Brazil are related to the influence of ambient temperatures on titres of *Lib. asiaticus* in leaves.

Flores-Sanchez *et al.* (2017) evaluated the fit of diffusion and classic disease gradient models to large-scale HLB spatial data originated from initial foci to improve sampling, monitoring and control strategies for *D. citri*, vector of *Lib. asiaticus*. The diffusion model provided the best fit among the family of time-gradient curves ($r2$ = 0.90–0.99) due to the flexibility of a three-parameter model. The gradients were well conformed to the model in a 25–82.6 km range, having the east–west direction the longest effect. Out of four transect routes, namely, Yuc-1, Yuc-2, QRoo-1 and QRoo-2, selected based on the directionality of the prevailing winds and foci location of HLB-infected plants, Yuc-2 and QRoo-2 transects showed 82.6 and 43.9 km gradients with diffusion coefficients of 0.15 and 0.09, respectively. The study constitutes the first quantitative evidence of the regional spread of *Lib. asiaticus* from a single focus and the application of a flexible model that improved the fit and allowed to better compare different gradients. These results are useful to determine the size of regional areas of *D. citri* control, a management programme currently implemented in Mexico to combat HLB.

Hilf *et al.* (2013) observed bacterial cells with the morphology and physical dimensions appropriate for *Lib. asiaticus* in phloem sieve elements in the vascular bundle of grapefruit seed coats using transmission electron microscopy (TEM). Fluorescence *in situ* hybridisation (FISH) analyses using probes complementary to the *Lib. asiaticus* 16S rRNA gene revealed bacterial cells in the vascular tissue of intact seed coats of grapefruit and pummelo, and in fragmented vascular bundles excised from grapefruit seed coats. Physical measurements and the morphology of individual bacterial cells were consistent with those ascribed in the literature to *Lib. asiaticus*. However, no bacterial cells were observed in seed preparations from the fruits of non-infected trees. A small library of clones amplified from seed coats from a non-infected tree using degenerate primers targeting procaryote 16S rRNA gene sequences contained no *Lib. asiaticus* sequences, while 95% of the sequences in a similar library from DNA from seed coats from an infected tree were identified as *Lib. asiaticus*, providing molecular genetic corroboration that the bacterial cells observed by TEM and FISH in seed coats from infected trees were of *Lib. asiaticus*.

Kim and associates (2009) provided microscopic and microarray analyses of response of sweet orange (*Citrus sinensis*) to *Lib. asiaticus* infection. The microarray analysis indicated that the pathogen infection affected expression of 624 genes, and the proteins encoded by some of these genes were associated with sugar metabolism, plant defence, phytohormone and cell wall metabolism. The anatomical analyses indicated that HLB bacterium infection caused phloem disruption, sucrose accumulation and plugged sieve pores. Phloem blockade occurred due to the plugging of sieve pores rather than the formation of bacterial cell aggregates because *Lib. asiaticus* does not form aggregates in citrus. The up-regulation of three key starch genes, ADP-glucose pyrophosphorylase, starch synthase, and glucose-bound starch synthase, and also the starch debranching enzyme likely contributed to the accumulation of starch in HLB-affected leaves. Up-regulation of *pp2* gene is related to callose deposition to plug sieve pores in the infected plants.

Ramadugu *et al.* (2016) screened 65 *Citrus* accessions and 33 accessions belonging to 20 other genera closely related to *Citrus* for resistance to '*Can.* Lib. asiaticus' and found 2 immune, 6 resistant and 14 tolerant accessions. Most citrus cultivars were rated as susceptible: 15 citrons, lemons and limes retained leaves in spite of the disease status. Resistance and high levels of field tolerance were observed in plants of many genera other than *Citrus*. Disease resistance/tolerance was observed in Australian citrus relative genera *Eremocitrus* and *Microcitrus*, which are sexually compatible with citrus and may be useful in future breeding trials to impart resistance against '*Can.* Lib. Asiaticus' to cultivated citrus.

HLB pathogens and their insect vectors, Asian citrus psyllids (ACP) and African citrus psyllids (AfCP), are invading citrus-producing regions where HLB was absent previously, including the Mediterranean basin. Importantly, the Mediterranean region is one of the two major citrus-producing areas which are not impacted by HLB. Wang (2020) has provided a short perspective regarding: a) information related to the distribution of the HLB pathogens and psyllid vectors in this region and neighbouring countries, b) predicted distribution of the HLB for this region, c) the possible evolution of Liberibacters and how these could have established their relationship with different hosts and d) approaches to protect from HLB. The author has specifically emphasised quarantine measures against AfCP, ACP, *Lib. asiaticus*, *Lib. africanus* and *Lib. americanum*, the early detection and diagnosis of HLB and removal of inoculum, the surveillance of AfCP and ACP and eradication of citrus psyllids once identified.

Periwinkle (*Catharanthus roseus*) is known to carry a greater number of cells of the bacterium causing HLB and, hence, develops HLB symptoms in a short period of time. Therefore, it is an ideal host plant for studying the HLB bacterium (Liu *et al*., 2019).

Fu *et al*. (2019) tested a set of stem and petiole tissue samples by TPE-qPCR and immunoassay. Although the detection rate by TPE-qPCR was higher than by immunoassay, about 6% of tissue prints were positive by immunoassay but not by TPE-qPCR. Thus, a higher detection rate would be achieved by combining TPE-qPCR with immunoassay. Significant differences were observed in the performance of nitrocellulose membranes from different manufacturers in these assays. Immuno tissue prints showed that the spatial distribution pattern of '*Can*. Lib. asiaticus' (CLas) infection varied widely from one sample to another, but the patterns were highly correlated among serial sections from the same sample, suggesting that CLas preferentially colonises adjacent phloem cells in a vertical rather than horizontal direction.

Huang *et al*. (2024) proposed a mutation-enhanced single-tube duplex PCR (mSTD-PCR) containing Internal standard (IS) with mutant-type primers. By introducing the 3′-terminal mutation in the primer of IS to weaken its amplification reaction and its inhibition of '*Can*. Liberibacter asiaticus' (CLas) detection, the sensitivity and quantitative accuracy of CLas detection will not be affected by IS. In evaluating the sensitivity of CLas detection using simulation samples, the mSTD-PCR showed consistent sensitivity at 25 copies per test compared with the single-plex CLas assay. The detection result of 30 leaves and 30 root samples showed that the mSTD-PCR could recognize false-negative results caused by the PCR inhibitors and reduce workload by 48% compared with the single-plex CLas assay. Generally, the proposed mSTD-PCR provides a reliable, efficient, inhibitor-monitorable, quantitative screening method for accurately controlling HLB and a universal method for establishing a PCR assay for various pathogens.

Yang *et al*. (2020) treated hydroponic cultures of *Lib. asiaticus* (*Las*)-infected citrus roots with sulfadimethoxine sodium (SDX), indole butyric acid (IBA), or water for 60 days to evaluate root metabolism and resistance against *Las* via transcriptomic analysis. It was found that SDX and IBA treatments increased the active root surface area in HLB-affected citrus, which may be due to root hair and lateral root growth. *Las* titres in HLB-affected citrus roots treated with SDX were lower than those of IBA and control treatments. Function categorisation indicated that plant hormone signal transduction and plant pathogen interaction were the most markedly enriched pathways in the HLB-affected citrus root after SDX treatment. Expression of genes involved in the biosynthesis of auxin and ethylene, which are related to root hair and lateral root growth, were up-regulated by SDX. Moreover, SDX also induced genes related to the metabolism of jasmonates, brassinosteroids, reactive oxygen species and secondary metabolites, which are beneficial for resistance against HLB. The authors proposed a model for SDX in regulating metabolic pathways in the roots and resistance against *Las* in HLB-affected citrus roots, which is beneficial for recovering an HLB-affected citrus root system and combating *Las*.

Fu *et al*. (2020) surveyed the *Las* populations from nine provinces for the presence of prophage. Two major prophage typing groups (PTGs) were found to be associated with two different altitude regions: strains of *Las* in PTG1 from high-altitude regions (HAR) principally contained prophage Type 1 only or Types 1 and 3, whereas strains of *Las* in PTG2 from low-altitude regions (LAR) mainly contained prophage Type 2. The discovery of these *Las* population patterns provides evidence for the independent origins of HLB in HAR and LAR. Guangdong province is the generally recognised domestic region of origin for HLB and is primarily responsible for the dissemination of HLB in LAR through the transporting of seedlings. Both Yunnan and Sichuan provinces are the probable regions of origin for HLB in HAR. PTG2 was further divided into two subgroups: PTG2-1, found in Guangdong, Fujian and Guangxi; and PTG2-2, found in Jiangxi, Zhejiang and Hunan. These regions and prophage types are correlated with early and late introductions of HLB in LAR. These molecular analyses were supported by studying the history of the dissemination of HLB in historical documents.

Raiol *et al*. (2021a) sampled non-girdled and fully or partially girdled stems of potted plants of 'Pera' sweet orange, graft-inoculated with *Lib. asiaticus* (*Las*) above or below girdling, in distinct regions and assessed for the presence of *Las* by qPCR, 6 months post-inoculation. *Las* invaded all regions of partially and non-girdled plants but remained restricted to the inoculated regions of fully girdled plants, evidencing that *in planta* bacterium movement is limited to the phloem. In fully girdled plants, starch accumulated above the girdling site, probably because of changes in flow of phloem sap. To study the influence of actively growing tissues, inoculated 'Valencia' sweet orange plants were kept intact or were top- or root-pruned to force the production of new tissues, and sampled at 15-day intervals. *Las* migrated rapidly, most predominantly toward newly developing root and leaf tissues. The rapid and predominant movement of *Las* to newly developed shoots and roots would explain the failures of canopy heat treatments and pruning to cure HLB-affected trees, and also reinforces the need to protect rapidly growing new shoots from feeding by *Diaphorina citri* in order to minimise the transmission and spread of the pathogen by the vector within and between orchards.

Healthy and HLB-affected 'Midsweet' sweet orange grafted on Kuharske citrange and irrigated every 4–5 days with phosphate buffers at pH 5.8, 6.4, 7.0 and 8.0 were monitored for 6 weeks (Ghimire *et al*., 2019). HLB-affected plants at pH 5.8 showed a greater increase in plant height with 100% survival rate compared to the highest death rate at pH 8.0. Soil media of HLB-affected plants had higher electrical conductivity than healthy plants at pH 8.0, indicating poor nutrient uptake by roots, not just nutrient unavailability. At pH 8.0, HLB-affected plants had more leaf drop, leaving just 0.25 g leaf fresh weight on the plant compared to 15 g on healthy plants. Among all treatments, HLB-affected plants at pH 8.0 had the most limited root growth and development pattern, leading to lowest root dry weight. In contrast, at pH 5.8, HLB-affected plants had the same dry root weight as healthy plants. HLB symptoms were more prominent at pH 8.0, suggesting management of irrigation water pH as a potential option for minimising HLB effects.

Park *et al.* (2020) reported that isothermal amplification technology such as recombinase polymerase amplification (RPA) offers an alternative to PCR for the development of rapid, sensitive and field-deployable diagnostic detection assays. The authors developed an RPA assay targeting the nrdB gene, of which there are five copies in the *Lib. asiaticus* (*Las*) genome and evaluated its applicability for CLas detection using DNA extracts prepared from citrus leaf and fibrous root tissue.

Pandey and Wang (2019) reported a new method for the targeted early detection of *Lib. asiaticus* in cultivar Valencia sweet orange (*Citrus sinensis*) before HLB symptom expression, taking the advantage of the fact that *Lib. asiaticus* remains around the Asian citrus psyllid (ACP) feeding site immediately after transmission into the young flush and before flush maturation. ACPs secrete salivary sheaths at their feeding sites, which can be visualised using Coomassie brilliant blue staining owing to the presence of salivary sheaths secreted by ACP. Epifluorescence and confocal microscopy indicate the presence of salivary sheaths beneath the blue spots on ACP-fed leaves. Quantitative real-time PCR and conventional PCR assays are able to detect *Lib. asiaticus* in the ACP feeding surrounding areas as early as 2 to 20 days after ACP feeding. This finding lays a foundation to develop much-needed tools for the early diagnosis of HLB before symptom expression, thus assisting *Lib. asiaticus* inoculum removal and preventing HLB from spreading.

Pandey *et al.* (2021) showed that Asian citrus psyllids (ACPs) were unable to transmit '*Can.* Lib. asiaticus' (*C*Las) via feeding on mature leaves. It is estimated that it takes three years at most for *C*Las to infect the whole tree. Overall, the spatiotemporal detection of *C*Las in different tissues after ACP transmission helps visualise the infection process of *C*Las in planta and subsequent HLB symptom development, and provides evidence showing that young leaves should be the focus of HLB management.

Raiol *et al.* (2021b) reported that a temperature of 25.7 °C was the optimum condition for plant colonisation by '*Can.* Lib. asiaticus'. This explains the higher impact and incidence of huanglongbing disease during the winter months or regions of milder climates in Brazil.

Alquézar *et al.* (2022), in their review article, have advocated that HLB management can be approached in terms of prevention, limiting or avoiding pathogens and associated vectors to reach an area, or, in terms of control, trying to reduce the impact of the disease by adopting different cultural strategies depending on infestation/infection levels. In both cases, control of psyllid populations is currently the best way to stop HLB spread. Best cultural actions (CHMAs, TPS system) to attain this goal and, thus, able to limit HLB spread, and ongoing research in this regard is summarised in this review.

Pulici *et al.* (2022) studied the effects of local infection on systemic disease development using split-root rhizoboxes and late summer below-the-split bud inoculation. Upward bacterial movement from roots is linked to seasonal flushes and '*Can.* Lib. asiaticus' (*C*Las) population in roots. *C*Las stayed isolated to one side of the roots for at least eight months, until the spring flush. HLB caused differential root responses depending on tree age at the time of infection. Systemic effects, independent of *C*Las movement, occur very early after infection. Stimulation of root growth occurred on non-infected roots prior to *C*Las detection in 1.5-year-old trees but decreased in 2.5-year-old trees. Independent of tree age, root growth was stimulated during spring root flushes after the stabilisation of the *C*Las population. Root die-back began simultaneously with the detection of *C*Las in roots (6 weeks post-inoculation). Infection and tree age altered root lifespan. In total, 1.5-year-old *C*Las-infected roots from summer and fall flushes had 3 and 6 weeks reduced lifespan. In contrast, lifespan of 2.5-year-old *C*Las-infected plants was unaffected. The season also affected root lifespan with late summer root flush lifespan being three times shorter than fall or spring root flushes. Split-root inoculation allowed the study of local and systemic effects of *C*Las infection in roots, information crucial to prolonging the productivity of HLB-affected trees.

Rattner *et al.* (2022) developed recombinase polymerase amplification (RPA) assay to detect '*Can.* Lib. asiaticus' (*C*Las) in crude citrus extracts utilising isothermal amplification, without the need for DNA purification. Primers were designed to amplify a region of the *C*Las RNR gene, and a fluorescent labelled probe allowed for the detection of the amplicon in real time within 8 min. at 39° C. The assay was specific to *C*Las, and the sensitivity was comparable to qPCR, with a detection limit cycle threshold of 34. Additionally, the RPA assay was combined with a lateral flow device for a point-of-use assay that is field-deployable. Both assays were 100% accurate in detecting *C*Las in fresh citrus crude extracts from leaf midribs and roots for five California strains of *C*Las tested in the Contained Research Facility in Davis, California. This assay will be important for distinguishing *C*Las-infected trees in California from those infected by other pathogens that cause similar disease symptoms and can help control Huanglongbing spread.

Ribeiro *et al.* (2023) conducted RNA-seq analyses on citrus huanglongbing (HLB)-susceptible Valencia sweet orange and HLB-tolerant mandarin 'LB8-9' in winter, spring, summer and fall. Significant variations in differentially expressed genes (DEGs) related to HLB were observed among the four seasons. For both cultivars, the highest number of DEGs were found in the spring. CLas infection stimulates the expression of immune-related genes such as NBS-LRR, RLK, RLCK, CDPK, MAPK pathway, ROS, and PR genes in both cultivars, consistent with the model that HLB is a pathogen-triggered immune disease. HLB-positive mandarin 'LB8-9' trees contained higher concentrations of maltose and sucrose, which are known to scavenge ROS. In addition, mandarin 'LB8-9' showed higher expression of genes involved in phloem regeneration, which might contribute to its HLB tolerance. The investigation sheds light on the pathogenicity mechanism of the HLB pathosystem and the tolerance mechanism against HLB, providing valuable insights into HLB management.

Successful primary spread of huanglongbing (HLB) pathogen within the citrus orchard occurs through the Asian citrus psyllid (ACP; *Diaphorina citri*) carrying the HLB bacterium '*Can.* Lib. asiaticus' (*C*Las) from external areas to blocks located in the orchard periphery. Primiano *et al.* (2023) evaluated the effect of two citrus planting directions, either parallel or perpendicular to the orchard's edge, by comparing the HLB gradient and incidence in blocks located at the orchard's edge by the analysis of satellite images. Six areas with two neighbouring blocks (one planted in parallel and the other in a perpendicular direction) were evaluated in two periods. Total HLB incidence in blocks with a perpendicular planting direction was higher (±30.3%) than in blocks with a parallel planting direction (±22.8%). Overall, planting parallel to the orchard's edge is beneficial in reducing HLB incidence in orchards. This strategy should be integrated into HLB management programmes.

Archer *et al.* (2023) injected mature 'Valencia' and 'Hamlin' sweet orange (*Citrus sinensis*) and 'Duncan' grapefruit (*C. paradisi*) trees with oxytetracycline (OTC) in the spring and/or fall. Pre-harvest fruit drop, yield and fruit quality were measured at harvest. OTC injections caused a reduction in fruit drop, an increase in fruit yield and fruit size, and improved juice quality. However, results varied due to the timing of injection and were not consistent across all three varieties. Residue analysis at different time points after injection suggests that trunk injection effectively delivers therapeutics to mature citrus trees. The study provides fundamental information on the short-term benefits associated with trunk injection of OTC for HLB management in citrus groves. The potential for use of trunk injection at the commercial scale and the possible risks are discussed.

Pandey *et al.* (2023), using transmission electron microscopy, found that the mean diameter of the sieve pores decreased in young leaves of Huanglongbing (HLB)-positive trees after '*Can.* Lib. asiaticus' (*C*Las) infection, consistent with *C*Las-triggered callose deposition. RNA-seq-based global expression analysis of young leaves of HLB-positive sweet orange with (*C*Las-Pos) and without (*C*Las-Neg) detectable *C*Las demonstrated a significant impact on gene expression in young leaves, including on the expression of genes involved in host immunity, stress response and plant hormone biosynthesis and signalling. *C*Las-Pos and *C*Las-Neg expression data displayed distinct patterns. The number of upregulated genes was higher than that of the downregulated genes in *C*Las-Pos for plant–pathogen interactions, glutathione metabolism, peroxisome and calcium signalling, which are commonly associated with pathogen infections, compared with the healthy control. On the contrary, the number of upregulated genes was lower than that of the downregulated genes in *C*Las-Neg for genes involved in plant–pathogen interactions and peroxisome biogenesis/metabolism. Additionally, a time-course quantitative reverse transcription-PCR-based expression analysis visualised the induced expression of companion cell-specific genes, phloem protein 2 genes, and sucrose transport genes in young flushes triggered by *C*Las. This study advances our understanding of early events during *C*Las infection of citrus young flushes.

Du *et al.* (2023) showed that AGH17488, a secreted protein encoded by the prophage region of the *C*Las genome, suppresses plant immunity via targeting the host ascorbate peroxidase 6 (APX6) protein in *Nicotiana benthamiana* and *Citrus sinensis*. The transient expression of *AGH17488* reduced the chloroplast localisation of APX6 and its enzymatic activity, inhibited the accumulation of reactive oxygen species (H_2O_2 and O_2^-) and the lipid oxidation end product, malondialdehyde in plants, and promoted the proliferation of *Psm. syringae* pv. *tomato* DC3000 and *Xan. citri* subsp. *citri*. The study reveals a novel mechanism underlying how *C*Las uses a prophage-encoded effector, AGH17488, to target a reactive oxygen species accumulation-related gene, *APX6*, in the host to facilitate its infection.

Zhang *et al.* (2024) reported that effector *C*Las0185 targets methionine sulphoxide reductase B1 of *Citrus sinensis* to promote the multiplication of *Lib. asiaticus* via enhancing enzymatic activity of ascorbate peroxidase 1.

9.49.5 Management

1. Enforcement of quarantines to prevent the entry of the pathogen into pathogen-free areas.
2. Use of pathogen-free propagating plant material.
3. Removal and destruction of infected trees as soon as these are noticed. In South Africa, a management strategy consisting of the production of disease-free budwood and nursery trees, psyllid control and the removal of infected trees was implemented. Le Roux *et al.* (2006) found that, as a result of this strategy, the disease incidence of 38% found in the 1960s dropped to less than 1% in 2006. When HLB was reported in Brazil, the citrus industry in Sao Paulo state of Brazil had already moved to budwood isolation, a nursery disease

certification programme, nursery industry isolation and requiring all nursery production to be in secure insect-proof screen houses.

The Florida Department of Agriculture and Consumer Services (FDACS) has strengthened its existing bud-wood nursery certification programme by expanding the testing for HLB and the movement of certified bud-wood sources to an isolated greenhouse location in northern Florida outside the commercial production area. The FDACS has desired all the field-grown nursery production to cease and that no field-grown citrus nursery trees be sold after January 2008. Further, all subsequent nursery production must be in insect-proof greenhouses or screen houses with a requisite one-mile buffer where no HLB-infected commercial or residential citrus has been detected (Gottwald *et al.*, 2007).

4. da Graca (1991) reported that tetracycline injections or sprays reduced foliar and fruit incidence ranging 10–97%. However, Buitendag and von Broembsen (2003) have discouraged the use of antibiotics for the control of the disease because only partial success was achieved. Zhang *et al.* (2010) devised an optimised system using *Lib. asiaticus*-infected periwinkle cuttings to screen chemicals effective for the control of this bacterium. The vermiculite was used as a growth medium and a fertilisation routine consisted of half-strength Murashige and Tucker medium supplemented with naphthalene acetic acid (4 μg mL^{-1}) and indole-3-butyric acid (4 μg mL^{-1}). The presence of *Lib. asiaticus* in the infected cuttings was detected using both nested PCR and quantitative real-time PCR. This system allowed a plant regeneration rate of 60.6% for *Lib. asiaticus*-infected cuttings in contrast to <1% regeneration rate with water alone. When treated with penicillin G sodium at 50 μg mL^{-1}, all plants regenerated from *Lib. asiaticus*-infected cuttings were free from infection of *Lib. asiaticus*. Treatment with 2, 2-dibromo-3-nitrilopropionamide at 200 μL^{-1} also significantly reduced the percentage of infected plants and the titre of the bacterium. Zhang *et al.* (2011) reported that the combined application of penicillin and streptomycin via trunk injection or root soaking eliminated or suppressed the *Lib. asiaticus* bacterium in the HLB-affected citrus plants.

Li *et al.* (2016) reported that three or four applications of ascorbic acid (AA) (60–600 μM), β-aminobutyric acid (BABA) (0.2–1.0 mM), 2,1,3-benzothiadiazole (BTH) (1.0 mM), 2,6-dichloroisonicotinic acid (0.1 mM), non-metabolisable glucose analog 2-deoxy-d-glucose (2-DDG) (100 μM), BABA (1.0 mM) + BTH (1.0 mM), BTH (1.0 mM) + AA (600 μM) and BTH (1.0 mM) + 2-DDG (100 μM) during each season, slowed down the growth of '*Can.* Lib. asiaticus' in the infected plants and reduced HLB disease severity by approximately 15% to 30% compared to the non-treated control, depending on the age and initial HLB severity of infected trees. These treatments also conferred positive effect on fruit yield and quality. These findings indicate that plant defence inducers may be part of a useful strategy for the management of citrus HLB.

Hu and Wang (2016) evaluated the spatiotemporal dynamics of oxytetracycline hydrochloride (OTC) and its effectiveness for controlling HLB via trunk injection. Uniform distribution of OTC throughout tree canopies and root system was achieved two days after trunk injection. High concentrations of OTC (>850 μg kg^{-1}) remained in leaf and root at least for one month and moderate concentrations (>500 μg kg^{-1}) persisted for more than nine months. The reduction of '*Can.* Lib. asiaticus' populations in root system and leaves of OTC treated trees were over 95% and 99% (i.e., 1.76 and 2.19 log reduction) between 2- and 28-days post-injection. The health of trees receiving OTC treatment improved, fruit yield increased and juice acidity reduced compared to water-injected control, even though the differences were not statistically significant during the test period. It is evident that trunk injection of OTC can be used as an effective measure for the integrated management of citrus HLB.

Hu *et al.* (2018) conducted three field trials in which they evaluated eight plant defence activators and three antibiotics to control HLB by trunk injection of young and mature sweet orange trees. Four trunk injections, of salicylic acid, oxalic acid, acibenzolar-*S*-methyl and potassium phosphate, provided significant control of the disease by suppressing '*Can.* Lib. asiaticus' titre and disease progress. Trunk injection of penicillin, streptomycin and oxytetracycline hydrochloride resulted in excellent control of the disease. In general, antibiotics were more effective in the reduction of *Lib. asiaticus* titre and symptom development than plant defence activators. These treatments also increased yield and improved fruit quality

5. Thermotherapy of bud-wood at 47 °C for 4 h also gave 100% control (da Graca, 1991). However, this method is considered impracticable for large-scale use. Hoffman *et al.* (2013) reported that continuous thermal exposure to 40 to 42 °C for a minimum period of 48 h was sufficient to significantly reduce titre or eliminate *Lib. asiaticus* bacteria entirely from HLB-affected citrus seedlings. The results of qPCR after treatment illustrated significant decreases in the *Lib. asiaticus* bacterial titre, combined with healthy vigorous growth by all surviving trees. Repeated qPCR testing confirmed that previously infected, heat-treated plants showed no detectable levels of *Lib. asiaticus*, while untreated control plants remained highly infected. The treatment may prove useful for the control of Liberibacter-infected nursery and greenhouse plants

6. Chemical control of psyllid vectors is probably the easiest strategy to control HLB. Biological control of psyllid vectors have also been tried and found successful in some cases. Two hymenopterous ectoparasites, namely, *Tamarixia dryl* and *T. radiatus*, parasitise nymphs of both the psyllid vectors. These parasites significantly reduced the psyllid populations and lessened the damage of HLB on Reunion Island. However, the biological control with these parasites in Taiwan was not effective due to the presence of indigenous populations of hyperparasites, which attack these hymenopterous biocontrol agents. These hyperparasites are not present on Reunion Island. In Florida, the introduction of *T. radiatus* caused 4–70% reduction in psyllid populations. Another internal parasitoid, *Diaphorencyrtus aligarhensis*, has also been found to attack *D. citri*. Unfortunately, the use of insecticides to control psyllid vectors of HLB also reduces the population of biocontrol insects (Gottwald *et al*., 2007).
7. In an international workshop on HLB held in Ishigaki, Japan, a novel strategy to control HLB was approved. This strategy is to interplant citrus with guava, which apparently inhibits the psyllid vector. The inhibition can be by attracting the vector to guava and killing it when it feeds, presumably by some toxin or by repelling or confusing the vector through the release of volatile compounds. The exact effect is not known but preliminary data obtained are quite encouraging (Beattie *et al*., 2006).

Hopkins and Benyamini (2020) evaluated EB92-1 for the biocontrol of citrus greening on two-year-old sweet orange trees in Florida in 2014/2015 and found that only 8% of the treated trees had severe symptoms after six years compared with 40% of the untreated trees. After six years, yield was reduced by 40% in the untreated control compared with the EB92-1 treated. While a single treatment with EB92-1 at planting can provide control of Pierce's Disease for several years, management of citrus greening may require treatment after every 18–24 months to suppress the disease.

Yuan *et al*. (2021) suggested a three-pronged approach to control HLB, namely removal of HLB-symptomatic trees and their replacement with HLB-free trees and psyllid control. They investigated the efficacy of region-wide comprehensive implementation of this scheme to control HLB in Gannan region, China, with a total planted citrus acreage of over 110,000 ha from 2013 to 2019. With the region-wide implementation of comprehensive HLB management, the overall HLB incidence in Gannan decreased from 19.71% in 2014 to 3.86% in 2019. A partial implementation of such a programme (without a comprehensive inoculum removal) at the regional level in Brazil resulted in HLB incidence increasing from 1.89% in 2010 to 19.02% in 2019.

Li *et al*. (2021) studied the residue dynamics of Streptomycin (STR) and its effect on '*Can*. Lib. asiaticus' (*C*Las) titres *in planta*, applied by foliar spray and trunk injection on three-year-old citrus trees that were naturally infected by *C*Las in the field. The spray application did not significantly affect *C*Las titres in the leaves of treated plants as determined by quantitative PCR. After trunk injection, peak levels of STR were observed 7 to 14 days post-application (dpa) in the leaf and root tissues, and near-peak levels were sustained for another 14 days before a significant decline. *C*Las titres in leaves were significantly reduced by trunk injection of STR at 1.0 or 2.0 g/tree, starting from 7 dpa and throughout the experimental period. The reduction of *C*Las titres was positively correlated with STR residue levels in leaves. The *in planta* minimum effective concentration of STR needed to suppress the *C*Las titre to an undetectable level (cycle threshold ≥36.0) was 1.92 μg/g fresh weight. The determination of the *in planta* minimum effective concentration of STR against *C*Las and its spatiotemporal residue levels in planta provides the guidance for the use of STR for HLB management.

REFERENCES

Alquézar, B., Carmona, L., Bennici, S., Miranda, M. P., Bassanezi, R. B. *et al*. (2022). Cultural management of huanglongbing: current status and ongoing research. *Phytopathology*, 112: 11–25.

Ammar, E.-D., Shatters, R. G. Jr. and Hall, D. G. (2011). Localization of '*Candidatus* Liberibacter asiaticus', associated with citrus huanglongbing disease, in its psyllid vector using fluorescence *in situ* hybridization. *J. Phytopathol*. 159: 726–34.

Archer, L., Kunwar, S., Alferez, F., Batuman, O. and Albrecht, U. (2023). Trunk injection of oxytetracycline for huanglongbing management in mature grapefruit and sweet orange trees. *Phytopathology*, 113: 1010–21.

Beattie, G. A. C., Holford, P., Mabberley, D. J., Haigh, A. M., Bayer, R. *et al*. (2006). Aspects and insights of Australia-Asia collaborative research on Huanglongbing. *Proc. Int. Workshop for the Prevention of Citrus Greening Disease in Severely Infected Areas*. Int. Res. Div., Agric. Forestry Fisheries Res. Counc. Secretariat, Ministry of Agric., Forestry and Fisheries, Tokyo, Japan, pp. 47–64.

Beloti, V. H., Alves, G. R., Coletta-Filho, H. D. and Yamamoto, P. T. (2018). The Asian citrus psyllid host *Murraya koenigii* is immune to citrus huanglongbing pathogen '*Candidatus* Liberibacter asiaticus'. *Phytopathology*, 108: 1089–94.

Bove, J. M. (2006). Huanglongbing: A destructive, newly-emerging, century-old disease of citrus. *J. Plant Pathol*. 88: 7–37.

Buitendag, C. H. and von Broembsen, L. A. (2003). Living with citrus greening disease. In: P. Moreno, J. V. da Graca and L. W. Timmer (eds). *Proc. 12th Conf. Int. Organ. Citrus Virol*. IOCV, Riverside, CA, pp. 269–71.

Canale, M. C., Tomaseto, A. F., Haddad, M. d. L. Coletta-Filho, H. D. and Lopes, J. R. S. (2017). Latency and persistence of '*Candidatus* Liberibacter asiaticus' in its psyllid vector, *Diaphorina citri* (Hemiptera: Liviidae). *Phytopathology*, 107: 264–72.

Capoor, S. P. (1963). Decline of citrus trees in India. *Bull. Nat. Inst. Sci. India*. 24: 48–64.

Chen, J., Pu, X., Deng, X., Liu, S., Li, H. *et al.* (2009). A phytoplasma related to '*Candidatus* Phytoplasma asteris' detected in citrus showing huanglongbing (yellow shoot disease) symptoms in Guangdong, P. R. China. *Phytopathology* 99: 236–42.

Coletta-Filho, H. D., Daugherty, M. P., Ferreira, C. and Lopes, J. R. S. (2014). Temporal progression of '*Candidatus* Liberibacter asiaticus' infection in citrus and acquisition efficiency by *Diaphorina citri*. *Phytopathology* 104: 416–21.

da Graca, J. V. (1991). Citrus greening disease. *Annu. Rev. Phytopathol*. 29: 109–36.

da Graça, J. V., Cook, G., Ajene, I. J., Grout, T. G., Pietersen, G. *et al.* (2022). A review of the '*Candidatus* Liberibacter africanus' citrus pathosystem in Africa. *Phytopathology* 112: 44–54.

da Graca, J. V. and Korsten, L. (2004). Citrus huanglongbing: Review, present status and future strategies. In: S. A. M. H. Naqvi (ed.). *Diseases of Fruits and Vegetables*. Vol. I. Kluwer Academic Press, Dordrecht, pp. 229–45.

Das, A. K., Jagannadham, P. T. K., Bhate, R., Srivastava, A. K., Bora, P. *et al.* (2023). Draft whole-genome sequence of '*Candidatus* Liberibacter asiaticus' strain AS-TNSK3 from a Khasi mandarin (*Citrus reticulata*) tree in Northeast India. *J. Plant Pathol.* 105: 1683–86. https://doi.org/10.1007/s42161-023-01486-y

Das, A. K., Rao, C. N. and Singh Shyam (2007). Presence of citrus greening (huanglongbing) disease and its psyllid vector in North-eastern region of India confirmed by PCR technique. *Curr. Sci*. 92: 1759–63.

Davis, H., Orrock, J. and Johnson, E. (2016). Distribution of '*Candidatus* Liberibacter asiaticus' within bark tissue of mature infected citrus trees (Abstr.). *Phytopathology* 106: S2.8.

Du, J., Wang, Q., Shi, H., Zhou, C., He, J. *et al.* (2023). A prophage-encoded effector from '*Candidatus* Liberibacter asiaticus' targets ascorbate peroxidase 6 in citrus to facilitate bacterial infection. *Mol. Plant Pathol*. 24: 302–16.

Duan, Y., Armstrong, C., Zheng, D., Yao, W., Wu, B. *et al.* (2020). *In vitro* growth dynamics of *Liberibacter asiaticus*, the causal agent of citrus huanglongbing (Abstr.). *Phytopathology* 110: S2.76–2.77.

Duan, Y., Zhou, L., Hall, D. G., Li, W., Doddapaneni, H. *et al.* (2009). Complete genome sequence of citrus huanglongbing bacterium, '*Candidatus* Liberibacter asiaticus' obtained through metagenomics. *Mol. Plant–Microbe Interact*. 22: 1011–20.

Duan, Y. P., Gottwald, T., Zhou, L. J. and Gabriel D. W. (2008). First report of dodder transmission of '*Candidatus* Liberibacter asiaticus' to tomato (*Lycopersicon esculentum*). *Plant Dis*. 92: 831.

Flores-Sanchez, J. L., Mora-Aguilera, G., Loeza-Kuk, E., Lopez-Arroyo, J. I., Gutierrez-Espinosa, M. A. *et al.* 2017. Diffusion model for describing the regional spread of huanglongbing from first-reported outbreaks and basing an area wide disease management strategy. *Plant Dis*. 101: 1119–27.

Fraser, L. R., Singh, D., Capoor, S. P. and Nariani, T. K. (1966). Greening virus, the likely cause of citrus dieback in India. *FAO Plant Prot. Bull*. 14: 127–30.

Fu, S., Bai, Z., Su, H., Liu, J., Hartung, J. S. *et al.* (2020). Occurrence of prophage and historical perspectives associated with the dissemination of huanglongbing in mainland China. *Plant Pathol*. 69: 132–38.

Fu, S. M., Liu, H. W., Liu, Q. H., Zhou, C. Y. and Hartung J. S. (2019). Detection of '*Candidatus* Liberibacter asiaticus' in citrus by concurrent tissue print-based qPCR and immunoassay. *Plant Pathol*. 68: 796–803.

Garnier, M. and Bove, J. M. (1993). Citrus greening disease and the greening bacterium. In: P. Moreno, J. V. da Graca and L. W. Timmer (eds.). *Proc. 12th Conf. Int. Organ. Citrus* Virol. IOCV, Riverside, CA, pp. 212–19.

Garnier, M., Jagoueix-Eveillard, S., Cronje, P. R., Le Roux, H. F. and Bove, J. M. (2000). Genomic characterization of a Liberibacter present in an ornamental rutaceous tree, *Calodendrum capense*, in the Western Cape province of South Africa. Proposal of '*Candidatus* Liberibacter africanus subsp. capensis'. *Int. J. Syst. Evol. Micrbiol*. 50: 2119–25.

Ghimire, L., Sharma, S., Kadyampakeni, D., Johnson, E. G. and Vashisth, T. (2019). Huanglongbing-irrigation pH interaction affects disease progression (Abstr.). *Phytopathology* 109: S2.31.

Ghosh, D. K., Bhose, S., Motghare, M., Warghane, A., Mukherjee, K. *et al.* (2015). Genetic diversity of the Indian populations of '*Candidatus* Liberibacter asiaticus' based on the tandem repeat variability in a genomic locus. *Phytopathology* 105: 1043–49.

Gottwald, T. R. (2010). Current epidemiological understanding of citrus huanglongbing. *Annu. Rev. Phytopathol*. 48: 119–39.

Gottwald, T. R., da Graca, J. V. and Bassanezi, R. B. (2007). Citrus huanglongbing: The pathogen and its impact. APSnet Feature Story. http://www.apsnet.org/huanglongbing.

Halbert, S. E. and Manjunath, K. L. (2004). Asian citrus psyllids (Sternorrhyncha: Psyllidae) and greening disease in citrus: A literature review and assessment of risk in Florida. *Fla. Entomol*. 87: 330–54.

Hartung, J. S., Halbert, S. E., Pelz-Stelinski, K., Brlansky, R. H., Chen, C. *et al.* (2010). Lack of evidence for transmission of '*Candidatus* Liberibacter asiaticus' through citrus seed taken from affected fruit. *Plant Dis*. 94: 1200–05.

Higgins, S. A., Igwe, D. O., Coradetti, S., Ramsey, J. S., DeBlasio, S. L. *et al.* (2024). Plant-derived, nodule-specific cysteine-rich peptides as a novel source of biopesticides for controlling citrus greening disease. *Bacteriology* 114: 971–81. https://doi.org/10.1094/PHYTO-09-23-0322-KC

Hilf, M. E. and Lewis, R. S. (2016). Transmission and propagation of '*Candidatus* Liberibacter asiaticus' by grafting with individual citrus leaves. *Phytopathology* 106: 452–58.

Hilf, M. E., Sims, K. R., Folimonova, S. Y. and Achor, D. S. (2013). Visualization of '*Candidatus* Liberibacter asiaticus' cells in the vascular bundle of citrus seed coats with fluorescence in situ hybridization and transmission electron microscopy. *Phytopathology* 103: 545–54.

Hoffman, M. T., Doud, M. S., Williams, L., Zhang, M.-Q., Ding, F. *et al.* (2013). Heat treatment eliminates '*Candidatus* Liberibacter asiaticus' from infected citrus trees under controlled conditions. *Phytopathology* 103: 15–22.

Hopkins D. L. and Benyamini Y. (2020). Field trials demonstrate biological control of Pierce's disease in Florida, Georgia, and California and citrus greening in Florida (Abstr.). *Phytopathology* 110: S2.126.

Hu, J., Jiang, J. and Wang, N. (2018). Control of citrus huanglongbing via trunk injection of plant defense activators and antibiotics. *Phytopathology* 108: 186–95.

Hu, J. and Wang, N. (2016). Evaluation of the spatiotemporal dynamics of oxytetracycline and its control effect against citrus huanglongbing via trunk injection. *Phytopathology* 106: 1495–1503.

Huang, W., Zhong, Z., Lin, Z., Zhang, J., Liu, J. *et al.* (2024). An inhibitor-monitorable single-tube duplex quantitative Real-time PCR assay for the detection of '*Candidatus* Liberibacter asiaticus'. *Plant Dis.* 108: 1157–64. https://doi.org/10.1094/PDIS-09-22-2168-SR

Jagoueix, S., Bove, J. M. and Garnier, M. (1994). The phloem limited bacterium of greening disease of citrus is a member of the α subdivision of the Proteobacteria. *Int. J. Syst. Bacteriol.* 44: 379–86.

Johnson, E. G., Wu, J., Bright, D. B. and Graham, J. H. (2014). Association of '*Candidatus* Liberibacter asiaticus' root infection, but not phloem plugging with root loss on huanglongbing-affected trees prior to appearance of foliar symptoms. *Plant Pathol.* 63: 290–98.

Killiny, N. (2022). Made for each other: Vector-pathogen interfaces in the huanglongbing pathosystem. *Phytopathology* 112: 26–43.

Killiny, N., Hijaz, F., Ebert, T. A., and Rogers, M. E. (2017). A plant bacterial pathogen manipulates its insect vector's energy metabolism. *Appl. Environ. Microbiol.* 83: e03005–e03016

Kim, J. S., Sagaram, U. S., Burns, J. K., Li, J.-L. and Wang, N. (2009). Response of sweet orange (*Citrus sinensis*) to '*Candidatus* Liberibacter asiaticus' infection: Microscopy and microarray analyses. *Phytopathology* 99: 50–57.

Le Roux, H. F., van Vuuren, S. P. and Manicom, B. Q. (2006). Huanglongbing in South Africa. *Proc. Huanglongbing-Greening Int. Workshop*, Ribeirao Preto, Brazil, pp. 5–9.

Li, J., Kolbasov, V. G., Lee, D., Pang, Z., Huang, Y. *et al.* (2021). Residue dynamics of streptomycin in citrus delivered by foliar spray and trunk injection and effect on '*Candidatus* Liberibacter asiaticus' titer. *Phytopathology* 111: 1095–1103.

Li, J., Trivedi, P. and Wang, N. (2016). Field evaluation of plant defense inducers for the control of citrus huanglongbing. *Phytopathology*, Vol. 106: 37–46. https://doi.org/10.1094/PHYTO-08-15-0196-R

Li, W., Li, D., Twieg, E. Hartung, J. S. and Levy, L. (2008). Optimized quantification of unculturable '*Candidatus* Liberibacter spp.' in host plants using real-time PCR. *Plant Dis.* 92: 854–61.

Liu, X., Zheng, Y., Wang-Pruski, G. Gan, Y., Zhang, B. (2019). Transcriptome profiling of periwinkle infected with Huanglongbing ('*Candidatus* Liberibacter asiaticus'). *Eur. J. Plant Pathol.* 153: 891–906. https://doi.org/10.1007/s10658-018-01607-9

Lopes, S. A., Luiz, F. Q. B. F., Martins, E. C., Fassini, C. G., Sousa, M. C. *et al.* (2013). '*Candidatus* Liberibacter asiaticus' titers in citrus and acquisition rates by *Diaphorina citri* are decreased by higher temperature. *Plant Dis.* 97: 1563–70.

Manjunath, K. L., Halbert, S. E., Ramadugu, C., Webb, S. and Lee, R. F. (2008). Detection of '*Candidatus* Liberibacter asiaticus' in *Diaphorina citri* and its importance in the management of citrus huanglongbing in Florida. *Phytopathology* 98: 387–96.

Merfa, M. V., Pérez-López, E., Naranjo, E., Jain, M., Gabriel, D. W. *et al.* (2019). Progress and obstacles in culturing '*Candidatus* Liberibacter asiaticus', the bacterium associated with huanglongbing. *Phytopathology* 109: 1092–1101.

Motghare, M., Shukla, P. K., Kumar, P., Sharma, A. K. and Ghosh, D. K. (2018). Characterization of four vital protein encoding genes of *Candidatus* Liberibacter asiaticus, the causal agent of citrus greening disease. *Indian Phytopathol.* 71: 197–205. https://doi.org/10.1007/s42360-018-0024-0

Pandey, S. S., da Costa Vasconcelos, F. N. and Wang, N. (2021). Spatiotemporal dynamics of '*Candidatus* Liberibacter asiaticus' colonization inside citrus plant and huanglongbing disease development. *Phytopathology* 111: 921–28.

Pandey, S. S., Hendrich, C., Andrade, M. O. and Wang, N. (2022). '*Candidatus* Liberibacter: From movement, host responses, to symptom development of citrus huanglongbing. *Phytopathology* 112: 55–68.

Pandey, S. S., Li, J., Oswalt, C. and Wang, N. (2024). Dynamics of '*Candidatus* Liberibacter asiaticus' growth, concentrations of reactive oxygen species, and ion leakage in huanglongbing-positive sweet orange. *Phytopathology* 114: 961–70. https://doi.org/10.1094/PHYTO-08-23-0294-KC

Pandey, S. S. and Wang, N. (2019). Targeted early detection of citrus Huanglongbing causal agent '*Candidatus* Liberibacter asiaticus' before symptom expression. *Phytopathology* 109: 952–59.

Pandey, S. S., Xu, J., Achor, D. S., Li, J. and Wang, N. (2023). Microscopic and transcriptomic analyses of early events triggered by '*Candidatus* Liberibacter asiaticus' in young flushes of huanglongbing-positive citrus trees. *Phytopathology* 113: 985–97.

Park, J. W., Kunta, M., Da Graça, J. V. and Braswell, W. E. (2020). Detection of '*Candidatus* Liberibacter asiaticus' by recombinase polymerase amplification (RPA) assay (Abstr.). *Phytopathology* 110: S2.12.

Pietersen, G., Arrebola, E., Breytenbach, J. H. J., Korsten, L., le Roux, H. F. *et al.* (2010). A surrey for '*Candidatus* Liberibacter' species in South Africa confirms the presence of only '*Ca.* L. africanus' in commercial citrus. *Plant Dis.* 94: 244–49.

Primiano, I. V., Eduardo, W. I., Amorim, L., Filho, A. B., Montesino, L. H. *et al.* (2023). The effect of citrus planting orientation on the gradient and incidence of huanglongbing. *Plant Pathol.* 72: 686–95.

Pulici, J. D. V. S., Murata, M. M. and Johnson, E. G. (2022). Early physiological plant response and systemic effects of huanglongbing infection in split-root plants. *Phytopathology* 112: 1833–43.

Raiol, L. L. Jr., Cifuentes-Arenas, J. C., Cunniffe, N. J., Turgeon, R. and Lopes, S. A. (2021b). Modeling '*Candidatus* Liberibacter asiaticus' movement within citrus plants. *Phytopathology* 111: 1711–19.

Raiol, L. L. Jr., Cifuentes-Arenas, J. C., de Carvalho, E. V., Girardi, E. A. and Lopes, S. A. (2021a). Evidence that '*Candidatus* Liberibacter asiaticus' moves predominantly toward new tissue growth in citrus plants. *Plant Dis.* 105: 34–42.

Ramadugu, C., Keremane, M. L., Halbert, S. E., Duan, Y. P., Roose, M. L. *et al.* (2016). Long-term field evaluation reveals huanglongbing resistance in *Citrus* relatives. *Plant Dis.* 100: 1858–69.

Rattner, R. J., Godfrey, K. E., Hajeri, S. and Yokomi, R. K. (2022). An improved recombinase polymerase amplification coupled with lateral flow assay for rapid field detection of '*Candidatus* Liberibacter asiaticus'. *Plant Dis.* 106: 3091–99

Raychaudhuri, S. P., Nariani, T. K., Gosh, S., Viswanath, S. M. and Kumar, D. (1974). Recent studies on citrus greening in India. In: L. G. Weathers and M. Cohen (eds.). *Proc. 6th Conf. Int. Organ. Citrus Virol. Univ. California, Div. Agric. Sci.*, pp. 53–57.

Raychaudhuri, S. P., Nariani, T. K. and Lele, V. C. (1969). Citrus die-back problem in India. *Proc. 1st Int. Citrus Symp.* 3: 1433–37.

Ribeiro, C., Xu, J., Hendrich, C., Pandey, S. S., Yu, Q. *et al.* (2023). Seasonal transcriptome profiling of susceptible and tolerant citrus cultivars to citrus huanglongbing. *Phytopathology* 113: 286–98.

Roberts, R. and Pietersen, G. (2017). A novel subspecies of '*Candidatus* Liberibacter africanus' found on native *Teclea gerrardii* (Family: Rutaceae) from South Africa. *Antonie Van Leeuwenhoek*. 110: 437–44.

Roberts, R., Steenkamp, E. T. and Pietersen, G. (2015). Three novel lineages of '*Candidatus* Liberibacter africanus' associated with native rutaceous hosts of *Trioza erytreae* in South Africa. *Int. J. Syst. Evol. Microbiol*. 65: 723–31.

Saha, S., Flores, M., Hosmani, P., Fernandez-Pozo, N., Brown, S. *et al.* (2017). Citrus greening.org—A systems biology resource for vector biologists (Abstr.). *Phytopathology* 107: S2.11–12.

Sechler, A., Schuenzel, E. L., Cooke, P. Donnua, S., Thaveechai, N. *et al.* (2009). Cultivation of '*Candidatus* Liberibacter asiaticus', '*Ca*. L. africanus' and '*Ca*. L. americanus' associated with huanglongbing. *Phytopathology* 99: 480–86.

Teixeira, D. C., Saillard, C., Eveillard, S., Danet, J. L., da Costa, P. I. *et al.* (2005). '*Candidatus* Liberibacter americanus', associated with citrus huanglongbing (greening disease) in Sao Paulo State, Brazil. *Int. J. Syst. Evol. Microbiol*. 55: 1857–62.

Tomimura, K., Miyata, S., Furuya, N., Kubota, K., Okuda, M. *et al.* (2009). Evaluation of genetic diversity among '*Candidatus* Liberibacter asiaticus' isolates collected in Southeast Asia. *Phytopathology* 99: 1062–69.

Wang, N. (2020). A perspective of citrus Huanglongbing in the context of the Mediterranean Basin. *J. Plant Pathol*. 102: 635–40. https://doi.org/10.1007/s42161-020-00555-w

Wang, N., Stelinski, L. L., Pelz-Stelinski, K. S., Graham, J. H. and Zhang, Y. (2017). Tale of the huanglongbing disease pyramid in the context of the citrus microbiome. *Phytopathology* 107: 380–87.

Wang, N. and Trivedi, P. (2013). Citrus huanglongbing: A newly relevant disease presents unprecedented challenges. *Phytopathology* 103: 652–65.

Wulff, N. A., Eveillard, S., Foissac, X., Ayres, A. J. and Bove, J.-M. (2009). rRNA operons and genome size of '*Candidatus* Liberibacter americanus', a bacterium associated with citrus huanglongbing in Brazil. *Int. J. Syst. Evol. Microbiol*. **59**: 1984–91.

Yang, C., Powell, C. A., Duan, Y., Ancona, V., Huang, J. *et al.* (2020). Transcriptomic analysis reveals root metabolic alteration and induction of huanglongbing resistance by sulphonamide antibiotics in huanglongbing-affected citrus plants. *Plant Pathol*. 69: 733–43.

Yuan, X., Chen, C., Bassanezi, R. B., Wu, F., Feng, Z. *et al.* (2021). Region-wide comprehensive implementation of rogueing infected trees, tree replacement, and insecticide applications successfully controls citrus huanglongbing. *Phytopathology* 111: 1361–68.

Zhang, M., Duan, Y., Zhou, L., Turechek, W. W., Stover, E. *et al.* (2010). Screening molecules for control of citrus huanglongbing using an optimized regeneration system for '*Candidatus* Liberibacter asiaticus'-infected periwinkle (*Catharanthus roseus*) cuttings. *Phytopathology* 100: 239–45.

Zhang, M., Powell, C. A., Zhou, L., He, Z., Stover, E. *et al.* (2011). Chemical compounds effective against the citrus huanglongbing bacterium '*Candidatus* Liberibacter asiaticus' in planta. *Phytopathology* 101: 1097–1103.

Zhang, S., Wang, X., Zhao, T. and Zhou, C. (2024). Effector *C*Las0185 targets methionine sulphoxide reductase B1 of *Citrus sinensis* to promote multiplication of '*Candidatus* Liberibacter asiaticus' via enhancing enzymatic activity of ascorbate peroxidase 1. *Mol. Pl. Pathol*. 25: e70002. https://doi.org/10.1111/mpp.70002

9.50 CITRUS STUBBORN DISEASE

The symptoms of the disease reflect the name of the disease. Diseased trees remain stunted and fail to grow despite the best efforts of the orchardists. It is a serious disease of certain citrus species and occurs in hot, dry areas. The disease is considered the greatest threat to the production of grapefruit and sweet oranges in California and also in some Mediterranean countries. Due to the slow progress of the symptoms, the spread of the disease often goes unnoticed. The loss occurs due to a reduction in the yield and the low market value of the produce.

The disease was first reported from California in 1915, affecting navel oranges. It affected most commercial citrus varieties in California, but grapefruit, tangelos and navel and Valencia oranges were most severely infected. During the mid-1950s and early 1960s, the disease appeared on more than two million trees, with the incidence ranging from 50–100% in some orchards in California. This large-scale infestation of trees caused an alarm for the research scientists and the investigations for the management of the disease were intensified. Since its first appearance, the disease has spread to many countries, including western United States, the eastern Mediterranean Basin, the Middle East, northern Africa and Brazil. The disease is not present in Florida, but the strict quarantines are enforced to regulate the entry of citrus material into the state. The presence of the disease in Australia is also in question.

Etiological studies during the 1930s showed that the causal agent could be transmitted by grafting. For several years, the disease was thought to be caused by a virus due to its graft transmissibility and the non-isolation of any specific pathogen. Suppression of the disease by tetracycline and not by penicillin demonstrated by Igwegbe and Calavan (1970) and later on electron microscopic observations indicating the presence of mycoplasma-like organisms in the phloem element of diseased but not healthy citrus trees, indicated that the causal agent is a member of the class Mollicutes. Fudl-Allah *et al.* (1972) cultured an organism from infected citrus trees, followed by its characterisation and naming (as *Spiroplasma citri)* by Saglio *et al.* (1973).

9.50.1 Host Range

Citrus is the main economic host of *Spr. citri*. Most citrus species and cultivars are infected, some of them without symptoms. Sweet oranges, tangelos, grapefruit, mandarins and their hybrids show maximum disease severity. Trifoliate oranges and their hybrids, acid limes and lemons show mild symptoms on artificial inoculation. Many dicots and some monocots, including weeds (several species of family Brassicaceae), are also the hosts of *Spr. citri*, but many of them show no symptoms. Non-citrus host species include pink periwinkle (*Catharanthus roseus*), London rocket (*Sisymbrium irio*), horseradish (*Armoracia rusticana*), wild radish (*Raphanus raphanistrum*), turnip, Bermuda grass, Johnson grass, fox-glove, rice, peach, cherry, pear, sesame, zinnia, aster and marigold. The bacterium also causes brittle root in cultivated horseradish (Anonymous, 2004).

9.50.2 Symptoms

The symptoms of the disease are most pronounced under warm conditions. In cool weather, even highly susceptible cultivars may remain symptomless. The symptoms vary greatly and often some of them are expressed only once. Leaves on infected trees are small, cupped, abnormally upright, often mottled and chlorotic (Figure 9.41a).
Excessive defoliation of affected trees occurs in the winter. Mottled appearance and the veinal chlorosis of leaves can be confused with the symptoms caused by citrus tristeza virus and/or the cachexia viroid (Brown, 1992). Twigs and branches on the affected trees generally show upright growth. Shortening of internodes and the production of more shoots may result in the so-called 'witches' broom' appearance. Some of the affected twigs suffer from die-back and the bark may be abnormally thickened. In the advanced stage, the wilting of affected branches is a characteristic symptom of the disease. The disease is rarely lethal, but the trees may show severe stunting. Affected trees may bloom prolifically more than once during the season, many fruits drop prematurely leading to reduction in yield.

Affected trees generally produce smaller and fewer fruits, which are generally lopsided and resemble acorns (Figure 9.41b). The rind of such fruits is abnormally thick and green at the base and thin and coloured at the tip. Fruits from affected trees are usually sour, bitter or insipid and the seeds in such fruits are poorly developed and frequently aborted (Agrios, 2004). Symptoms on mature trees are generally less conspicuous. The symptoms of the disease are persistent, remaining evident even when the tree is top-worked with healthy budwood, hence the name '**stubborn**'. Some hosts show no symptoms, while others show only mild stunting and the deformation of leaves. Non-citrus hosts may show symptoms of stunting and other deformities. Titre of *Spr. citri* but not the bacterial genotype is, at least in part, responsible for the severity of the disease in

(a)

FIGURE 9.41A Citrus stubborn: Symptoms on aerial plant parts. (Courtesy: J.M. Bové, INRA Centre de Recherches de Bordeaux, Bugwood.org)

(b)

FIGURE 9.41B Citrus stubborn: Lopsided development of sweet orange fruit. (Courtesy: J.M. Bové, INRA Centre de Recherches de Bordeaux, Bugwood.org)

diseased citrus trees. Quantitative PCR showed that the spiroplasma titre was more than 6000 times higher in fruits from severely symptomatic than from mildly symptomatic trees (Mello *et al.*, 2010).

As the symptoms of citrus stubborn are readily confused with other diseases/maladies, the proper identification of disease and its confirmation by greenhouse tests is essential for the proper diagnosis of the disease. The diagnosis of the disease in the field should be carried out in warm weather, when the day temperature continuously exceeds 27 °C, because the symptoms are most prominent at this temperature. Its confirmation should be done by grafting on indicator plants and culturing the causal organism. Marsh or Duncan grapefruit and Madam Vinous sweet orange are very good indicator hosts and the grafting should be done on the seedlings of these hosts in a warm greenhouse (day temperature 27–35 °C and night temperature 21–24 °C). Three growth flushes should be observed for leaf symptoms (misshapen and mottling) and the stunting of seedlings. The confirmation of the disease can also be carried out by the identification of the pathogen on the basis of helical morphology and motility by dark field microscopy and/or serological techniques like Elisa. Nowadays, DNA probes and PCR techniques are routinely used in many laboratories.

9.50.3 Causal Organism

Spiroplasma citri Saglio *et al.* 1973.

The bacterium is motile, helical in shape, filamentous and lacks a true cell wall. Helical filaments are usually 100–200 nm in diameter and 2–4 um in length. Colonies on solid media containing 20% horse serum and 0.8% Noble agar (Difco) are umbonate and 60–150 μm in diam. Moderate turbidity is produced in liquid cultures. Acid is produced from glucose and fructose, but delayed from mannose. Arginine hydrolysis and phosphatase activity are positive and tetrazolium chloride is anaerobically reduced. The optimum temperature for growth is 32 °C. Mol% G + C of the DNA is 25–27 (Tm, BD). DNA hybridisation tests between *Spr. citri* and other established species of the genus show no significant genetic relatedness (<1%). Its viability is increased by passage through insect vectors (leafhoppers).

The detection of *Spr. citri* by PCR has been done using primers based on gene sequences for spiralin, the most abundant membrane protein of the bacterium. However, these primers lack the sensitivity needed for consistent pathogen detection from field citrus trees. Because the spiralin gene is present as a single copy in the chromosome of all *Spr. citri* strains examined, PCR detection based on multiple-copy membrane protein genes like P89 or P58 will certainly be more efficient. Yokomi *et al.* (2008) developed a PCR technique for the detection of *Spr. citri* using primers based on sequences of the P89 putative adhesion gene and P58 putative adhesion multigene of the bacterium. PCR results matched with those of pathogen, culturing 85–100% of the time depending on the primers used. In addition, the method also detected the bacterium from 5–10% of trees giving negative results by culturing of the pathogen, which showed that PCR performed as well as, or better than, pathogen culturing. However, real-time PCR developed with detection limits estimated to be between 10^{-4} and 10^{-6} ng by the serial dilution of a recombinant *Spr. citri* plasmid into DNA extracts from healthy sweet orange proved to be the best method for detection.

Shi *et al.* (2014) developed a novel diagnostic method for the detection of *Spr. citri*, using a *Spr. citri*-secreted protein as the detection marker. Using mass spectrometry analysis, they identified a unique secreted protein from *Spr. citri* that is highly expressed in the presence of citrus phloem extract. ScCCPP1, an antibody generated against this protein, was able to distinguish *Spr. citri*-infected citrus and periwinkle from healthy plants. Besides, the antiserum can be used to detect the pathogen using a simple direct tissue print assay without the need for sample processing or specialised laboratory equipment and may be suitable for field surveys. The results provide a novel concept of using pathogen-secreted protein as a marker for the diagnosis of a citrus bacterial disease and can probably be applied to other plant diseases.

Wang *et al.* (2015) developed two primer sets, Php-orf1 and Php-orf3, from conserved prophage sequences in *Spr. citri* genome to improve sensitivity for qPCR detection of *Spr. citri* from infected citrus trees. These sequences improved at least 10-fold sensitivity for the detection of *Spr. citri*-infected trees and no false-positive samples were detected with any of the primer sets. The enhanced sensitivity resulted from the higher copy number of prophage genes in *Spr. citri* genome and, thus, improved citrus stubborn disease diagnosis from field samples.

Drais *et al.* (2019) developed a Loop-Mediated Isothermal Amplification technique (LAMP) to detect *Spr. citri*, targeting the spiralin gene. The protocol was optimised for crude plant extracts from infected trees to allow on-site field testing. The LAMP assay showed high specificity to *Spr. citri* and detected DNA to a level of 100 fg/μl with no inhibition by crude plant extracts. Although the LAMP assay was nine times less sensitive than qPCR with purified DNA templates, it performed well in field validations using a portable BioRanger device with citrus crude extracts. LAMP assay showed detection efficiency and percentage yes/no calls similar to those obtained by real-time PCR conducted with DNA extracted and purified from the same sample. The LAMP procedure allows growers, pest control or diagnostic services to rapidly test for *Spr. citri* in the field without a laboratory or DNA purification.

9.50.4 Disease Cycle

The bacterium survives in infected hosts, where it is found in the phloem. The pathogen is spread naturally in citrus orchards by several leafhoppers, such as *Scaphytopius nitridus*, *Circulifer tenellus* (beet leafhopper) and *Neoaliturus haemoceps*. Transmission (spread) of the pathogen also occurs through budding and grafting but neither through seed, nor mechanically. The natural spread of the pathogen is most rapid in young citrus orchards.

C. tenellus is the most efficient vector in nature. The bacterium has a wide range of hosts, including numerous weeds of the family Brassicaceae. *Spr. citri* multiplies in *C. tenellus* before its transmission into the host. After ingestion by the leafhopper, the pathogen traverses the gut wall, enters the haemocoel and multiplies during the latent period, which ranges from days to weeks. Then the cells move to salivary glands, where they continue to multiply in the cell cytoplasm. Finally, the pathogen cells are deposited with salvia into the phloem cells of host plants. In some insects, effects such as shortened life span, a reduction in egg numbers and the weakening of flight muscles and other organs are produced. This indicates that the insects also act as hosts for the pathogen. However, there is no evidence of transovarial transmission (Fletcher, 2001).

The colonisation of phloem sieve tube cells by the pathogen results in the occlusion and necrosis of these cells. However, the sieve cells are not uniformly invaded; some sieve tubes have either no or just a few cells while others are completely occluded. Occlusion of tubes interferes with the movement of nutrients, leading to manifestation of some of the disease symptoms. Loss of chlorophyll may result due to the breakdown of anthocyanin pigments. Stunting can also occur either due to a reduction in the availability of nutrients or variation in the normal balance of plant growth regulators. *Spr. citri* has been reported to cause changes in the levels of both auxins and cytokinins.

9.50.5 Management

The following measures should be adopted to manage the disease.

1. Enforcement of strict quarantines to prevent the entry of the pathogen in disease-free areas like Australia and Florida in the USA.

2. Use of pathogen-free budwood and rootstocks (Anonymous, 2003).
3. Early detection, removal and destruction of diseased trees.
4. Remove and destroy all the Brassica weed hosts in field borders and within groves to limit the spread of the pathogen to healthy trees. Potential insect vectors on the weeds should be killed before their removal.
5. As tetracycline treatment of young citrus trees has shown encouraging results, it should be tried at a commercial scale to test its efficacy.

REFERENCES

Agrios, G. N. (2004). *Plant Pathology*. 5th Ed. Elsevier Academic Press, San Diego, CA, 922 pp.

Anonymous (2003). Citrus stubborn disease. University of California, IPM. http://www.ipm.ucdavis.edu/PMG/r107101211.html.

Anonymous (2004). Exotic threats of citrus: Stubborn disease of citrus. Fact sheet, Plant Health Australia. www.planthealthaustralia.com.au.

Brown, L. G. (1992). Stubborn disease of citrus. Plant Pathology Circular No. 351. Fla. Dept. Agric. and Consumer Services, Division of Plant Industry, Gainesville, FL.

Drais, M. I., Maheshwari, Y., Selvaraj, V., Varvaro, L., Yokomi, R. *et al.* (2019). Development and validation of a loop-mediated isothermal amplification technique (LAMP) for the detection of *Spiroplasma citri*, the causal agent of citrus stubborn disease. *Eur. J. Plant Pathol.* 155: 125–34.

Fletcher, J. (2001). Citrus stubborn disease. In: O. C. Maloy and T. D. Murray (eds.). *Encyclopedia of Plant Pathology*. Vol. 1. John Wiley and Sons, Inc., New York, pp. 222–24.

Fudl-Allah, A. E. -S. A., Calavan, E. C. and Igwegbe, E. C. K. (1972). Culture of a mycoplasmalike organism associated with stubborn disease of citrus. *Phytopathology* 62: 729–31.

Igwegbe, E. C. K. and Calavan, E. C. (1970). Occurrence of mycoplasma like bodies in phloem of stubborn-infected citrus seedlings. *Phytopathology* 60: 1525–26.

Mello, A. F. S., Yokomi, R. K., Melcher, U., Chen, J. C. and Fletcher, J. (2010). Citrus stubborn severity is associated with *Spiroplasma citri* titre but not with bacterial genotype. *Plant Dis.* 94: 75–82.

Saglio, P., Lhospital, M., Lafleche, D., Dupont, G., Bove, J. M. *et al.* (1973). *Spiroplasma citri* gen. and sp. n.: A mycoplasma-like organism associated with "stubborn" disease of citrus. *Int. J. Syst. Bacteriol.* 23: 191–204.

Shi, J., Pagliaccia, D., Morgan, R., Qiao, Y., Pan, S. *et al.* (2014). Novel diagnosis for citrus stubborn disease by detection of a *Spiroplasma citri*-secreted protein. *Phytopathology* 104: 188–95.

Wang, X., Doddapaneni, H., Chen, J. and Yokomi, R. K. (2015). Improved real-time PCR diagnosis of citrus stubborn disease by targeting prophage genes of *Spiroplasma citri*. *Plant Dis.* 99: 149–54.

Yokomi, R. K., Mello, A. F. S., Saponari, M. and Fletcher, J. (2008). Polymerase chain reaction-based detection of *Spiroplasma citri* associated with citrus stubborn disease. *Plant Dis.* 92: 253–60.

9.51 BACTERIAL CANKER AND GUMMOSIS OF STONE FRUIT TREES

Bacterial canker and gummosis principally affects stone fruit trees. The disease is most severe on sweet cherry (*Prunus avium*), particularly on young trees, and the trees under stress are much more susceptible than healthy trees. The disease is also a major limiting factor for timber production from wild cherry. Pome fruits like apple and pear, rootstocks and some ornamental trees are also infected, but to a lesser extent. Out of *Pseudomonas syringae* pv. *syringae* and *Psm. syringae* pv. *morsprunorum*, the causal agents of the disease, the latter has a limited host range and is confined mainly to tart cherry and plum. The disease also occurs on sour cherry (*P. cerasus*), apricot (*P. armeniaca*), almond (*P. amygdalus*) and peach (*P. persica*). *Psm. syringae* pv. *syringae* has a wide host range and is more virulent on sweet cherry and other stone fruits. The disease is also known by other names like **bud blast, blossom blast, spur and twig blight, sour sap** and **die-back**.

Disease outbreaks are usually associated with prolonged cold wet weather accompanied with late spring frost. *Psm. syringae* pv. *morsprunorum* is widespread in Europe and is present in eastern North America and South Africa. In South Africa, bacterial canker caused by *Psm. syringae* pv. *syringae* is one of the most important diseases of stone fruit trees, and annual losses probably exceed US$10 million (Hattingh *et al.*, 1989).

A third pathogen, namely, *Psm. syringae* pv. *persicae* which causes leaf spots, cankers and gummosis of fruits on peach trees, has been reported from France, where it killed more than one million trees. One or more of these pathogens have also been reported from Australia, Canada, India, Japan, Lebanon, the UK and the USA.

In their review, Marroni *et al.* (2024) reported 20–50% losses in young orchards of sweet cherry in New Zealand; sometimes the removal of entire cherry blocks was observed. They also reported on the current knowledge of *Psm. syringae* pathovars causing bacterial canker of *Prunus* species, with specific focus on sweet cherry in New Zealand, and covering their identification, the infection process, virulence-associated factors, epidemiology, symptoms and management strategies.

9.51.1 Symptoms

The symptoms appear on all the above-ground plant parts. The most characteristic symptom is the formation of cankers accompanied by the exudation of gum. This exudation occurs in most cankers early in the growing season. The gum comes out through injuries or cracks in the bark and runs down the surface of limbs. New cankers are generally formed in late winter and early spring, and these generally develop on trunk, limbs and twigs. Cankers usually develop at the bud union, in pruning wounds or when the pathogen invades woody tissue following the infection of blossoms or dormant buds. The spread of the cankers is mostly upward and slightly downwards and to the sides. The cankered areas are slightly sunken and often darker than the surrounding healthy bark. On cherry, the cankers are typically elongated in shape. Cankers expand rapidly in spring and may girdle trunks, branches and twigs. The leaves on the terminal portions of cankered limbs and branches become curled, droop, first turn light green, then yellow and finally die. Within a few weeks, the branch or the plant parts above the cankers also die. The removal of cankered bark exposes the underlying bright orange to brown tissue, often emitting a strong sour smell. Cankers on the shoots also develop from infection occurring through leaf scars. Infected young shoots wilt and die back from the tip.

On leaves, water-soaked spots appear, which are small, circular to angular in shape, and often surrounded by chlorotic haloes in the early stage. Later on, spots become brown, dry and brittle and eventually fall out forming shot-holes or giving a tattered appearance to the leaves. The spots may also coalesce, especially at the margins of the leaves, to form large patches of dead tissue. In sweet cherry, leaf spots typically occur at the margins of the leaves. However, symptoms on leaves are apparently limited to some fruit cultivars growing in regions where summers are wet and humid (Hattingh and Roos, 1992).

Infected dormant buds turn brown and fail to open or break in spring; this phase is referred to as **"dormant bud blast"** or **"dead bud condition"**. Small cankers often develop at the base of these buds. Cherry, apricot and pear generally suffer heavily from dormant bud blast in certain areas. Some infected dormant buds open in spring, but these collapse in early summer, exhibiting wilted leaves and dried-up fruits. Both leaf and flower buds are equally damaged. In severe cases, 90% or more buds on a tree may die.

Infection of the blossoms occurs in spring; the infected blossoms turn from brown to black and die. Blossom infection usually spreads into the fruit-bearing spur, causing the collapse of the whole blossom cluster. During full bloom, the poor blooming of the infected trees is quite apparent.

On green cherry fruits, the lesions are chocolate-brown, surrounded by water-soaked tissue. As the lesions and fruit age, the affected tissues collapse, forming deep black depressions in the flesh with margins turning yellow to red. Elliptical and brown lesions having water-soaked margins appear on fruit stalks. Fruit infections occur sporadically, but these can be severe when prolonged wet and cold weather prevails during or shortly after bloom (Jones, 2001).

The symptoms of bacterial canker can be confused with those of Cytospora canker, but these can be distinguished on the basis of differences in the canker margin. There is a distinct margin at the top and bottom of Cytospora cankers, while the margins of bacterial cankers are irregular and streaks of discoloured strands frequently extend into healthy tissue.

9.51.2 Causal Organisms

1. *Pseudomonas syringae* pv. *syringae* van Hall 1902

Synonym: *Psm. syringae* pv. *japonica*

2. *Psm. syringae* pv. *morsprunorum* (Wormald 1931) Young *et al.* 1978

Synonym: *Psm. amygdali* pv. *morsprunorum*

3. *Psm. syringae* pv. *persicae* (Prunier *et al.* 1970) Young *et al.* 1978

Synonym: *Psm. morsprunorum* f. sp. *persicae*

The cells are straight rods, measure 0.7 – 1.2 × 1.5 µm, Gram-stain-negative, aerobic, motile having polar flagella and do not accumulate poly-β-hydroxybutyrate as a carbon reserve. The bacterium gives negative reaction for oxidase and arginine dihydrolase. Colonies on King's medium B are circular, smooth and glistening with entire or lobate margins, and produce light green fluorescent pigment. Pathovars pv.
syringae and pv. *morsprunorum* can be differentiated from each other on the basis of the GATTa (gelatin liquefaction, esculin hydrolysis, tyrosinase activity and tartrate utilisation) scheme. *Psm. syringae* pv. *syringae* is positive for the first two tests and negative for the last two tests while the reverse is true for *Psm. syringae* pv. *morsprunorum*. The LOPAT scheme can also be used for the differentiation of these pathovars. Both the pathovars elicit hypersensitive reaction on tobacco leaves.

Most strains of *Psm. syringae* pv. *syringae* produce the phytotoxin syringomycin, which plays a role in the virulence of the pathogen. Two genes, namely, *syrA* and *syrB*, involved in the production of this toxin have been identified and analysed in strain B 301D of the bacterium (Xu and Gross, 1988). Further study revealed that *syrA* is a regulatory gene and is required for both syringomycin production and pathogenicity. Both the genes are required for the formation of proteins Sr4 and Sr5, which are believed to be components of the syringomycin synthetase complex. Probes and PCR primers for genes required for syringomycin toxin are highly reliable for the detection of pathogenic strains of this pathovar.

Many strains of *Psm. syringae* pv. *syringae* act as ice-nuclei. *Psm. syringae* pv. *syringae* contains plasmid-mediated genes for copper resistance. Many strains of *Psm. syringae* pv. *morsprunorum* contain genes for coronatine production. Complete genome sequencing of *Psm. syringae* pv. *syringae* strain B 728a by Feil *et al.* (2005) revealed that the genome (6.1 mb) contains a circular chromosome and no plasmid. Among the 976 protein-encoding genes, there are genes that encode for syringopeptin, syringomycin, indole acetic acid biosynthesis, arginine degradation and production of ice-nuclei. Certain unique genes, such as for ectoine synthase, DNA repair and antibiotic production, may contribute to the epiphytic fitness and stress tolerance of the bacterium.

Abbasi *et al.* (2013) studied genetic diversity among 78 Iranian strains of *Psm. syringae* pv. *syringae* by extracting genomic DNA from strains and using it in rep-PCR and IS50-PCR analysis. Cluster analysis was performed using UPGMA. The strains of the bacterium were separated into nine distinguishable genotypic groups by the combination dataset of both rep-PCR and IS50-PCR at 73% similarity level. There was no significant correlation between the genetic diversity and geographical origin of the isolates. These results show that a combination of rep-PCR and IS50-PCR finger printing can be used as a high-resolution genomic finger-printing method for elucidating diversity among the strains of this bacterium. The existence of a considerable genetic diversity among the strains of bacterium causing canker of stone fruit trees in Iran, reported here, will be of immense use in the development of resistant genotypes against this bacterial pathogen.

A third pathogen, *Psm. syringae* pv. *persicae*, is more closely related to *Psm. syringae* pv. *morsprunorum*. It does not produce fluorescent pigment on King's medium B, but it does do so on casamino acid sucrose agar.

9.51.3 Disease Cycle

Three above-mentioned bacterial pathogens perpetuate in the margins of active cankers, infected buds and systemically in the vascular systems of certain hosts. The bacteria also survive epiphytically on blossoms, buds and leaves of healthy and infected plants and also on non-host plants, including weeds. Bacterial population on blossoms can build up to 10^{4-6} cfu/blossom (Kennelly *et al.*, 2007). In fact, the concept of a plant pathogen surviving as an epiphyte on healthy plant tissue was first proposed in studies of *Psm. syringae* pv. *morsprunorum*. Epiphytic populations of these bacteria also establish on the surface of leaves. The bacteria present in the buds colonise the new leaves as these emerge in spring. Environmental conditions influence the multiplication of bacteria, causing fluctuations in population levels. During adverse atmospheric conditions like warm, dry spells, epiphytic bacteria enter through stomata and colonise the intercellular spaces of spongy parenchyma. On reaching the uninvaded stomatal chambers, these multiply profusely and the cell masses extrude through stomata. It appears that epiphytic populations are constantly replenished in this way (Hattingh and Roos, 1992). In plum, *Psm. syringae* pv. *syringae* spreads from colonised blossoms, which are not killed, to the developing seeds and survives in the seed.

The dissemination of the bacteria occurs through wind-splashed rain, water splashes, pruning tools and infected bud-woods and nursery plants. The penetration of the bacteria into the host occurs through wounds, stomata and injuries caused by frost and severe storms on blossoms, buds and leaves. Spotts *et al.* (2010) showed that all seven types of injuries and wounds, whether natural or operational, facilitated the infection of sweet cherry by *Psm. syringae* pv. *syringae*. The infection of inoculated wounds made in spring and summer (heading cuts when trees were planted, scoring cuts and summer pruning) resulted in the maximum canker incidence and severity. Inoculation of heading cuts resulted in the highest tree mortality of 86%. Bacterial cankers did not develop in summer or winter when active cankers were cut and the same pruning tool was used immediately to make heading cuts on healthy trees. Heading cuts became resistant to infection after one week in summer and three weeks in winter.

Infection and the progress of the disease is greatly influenced by environmental factors. Free water on leaf surface and high relative humidity are essential for at least 24 h for a significant leaf infection to occur. Symptoms appear after about five days at 21–27° C. The disease is usually severe after cold winters and prolonged spring rains. A warning system based on temperature and free moisture conditions for the prediction of blossom blight infection in the field has been developed (Latorre *et al.*, 2002). The development of the cankers is governed largely by the weather and the growth of the host. The cankers expand in the fall and winter, but the maximum development occurs in the period between the end of cold weather and the beginning of the rapid tree growth in the spring. Active growth of the host and high temperature in the spring result in the formation of callus tissue around the cankers, which checks the further advance of the cankers. Some cankers are inactivated permanently while others restart the growth when favourable conditions for the growth of bacteria return. Fresh infections during the active growing season cause hardly any significant loss as their development is quickly arrested by the callus tissue.

Frost injury predisposes the host tissue to infection, but the infection is governed by the extent of wet weather during the thawing period. Many strains of *Psm. syringae* pv. *syringae* act as ice-nuclei and hence cause frost injury to plants by forming ice at relatively higher freezing temperatures. These bacteria also produce bacteriocins, which inhibit the growth of non-ice nucleation-active bacteria, thereby making their own growth less competitive. Frost injury predisposes the buds, flowers and young leaves to infection. The infection of buds starts at the base of outer bud scales and advances, causing the death of buds. Infection may spread downwards, killing the stem tissues around the base of the bud. The bacteria penetrate flowers through natural openings and wounds, and spread quickly through floral parts under humid conditions. Infection may advance into the spur and further to the twig. Young and succulent leaves are frequently infected during cool and wet springs and the bacteria enter through stomata. Bacteria spread intercellularly and cause the collapse and death of cells, leading to the formation of necrotic and angular spots. Infections on mature leaves are rare. Under humid conditions, the bacteria ooze out from infected plant parts through stomata, lenticels and wounds formed in the infected tissue and are spread to other plant parts by various disseminating agencies.

The limbs generally become infected during the fall and early winter. The entry of bacteria occurs through natural and pruning wounds, leaf scars and also through the bases of infected buds and spurs. After entry into the host, bacteria multiply and move intercellularly into the bark and then further into the ray parenchyma of the phloem and xylem. In advanced stages of infection, bacteria cause the breakdown of parenchyma cells, forming cavities full of bacterial cells. Occasionally, the bacteria invade the xylem vessels but these do not move for long distances in these vessels (Agrios, 2004).

Donmez *et al.* (2010), on the basis of artificial inoculation screening, reported that apricot cultivars Hasanbey, Cöloğlu, Soğanci and Şekerpare were resistant, while Şam, Tokaloğlu (Erzincan) and Erken Ağerik were susceptible to *Psm. syringae* pv. *syringae*.

9.51.4 Management

Presently, successful management of the disease is not possible due to the lack of effective chemicals and biocontrol agents, the lack of host resistance and the endophytic nature of the pathogen, during some phases of the diseases cycle. However, the following cultural practices and preventive measures will certainly reduce the incidence and severity of the disease.

1. Most if not all, factors that reduce the vigour of trees probably favour disease development. Therefore, avoiding or reducing the effect of these stress factors will certainly minimise the severity of the disease.
2. Plant only healthy nursery trees in the orchards.
3. Use only healthy bud-wood for propagation.
4. Susceptible varieties should be propagated on rootstocks resistant to the disease and grafting should be as high as possible. Mazzard seedlings, particularly the F12-1, are the most resistant rootstocks.
5. Prune the cankered limbs at least 15 cm below the visible cankers. Pruning should be avoided in early spring and fall, as the causal bacteria are active during this period. Pruning tools should be disinfected frequently with 10% clorox (household bleach) or shellack thinner (70% ethyl alcohol). Cankers on trunks and large branches can be removed by cauterisation with a hand-held propane burner. The flame is directed particularly at the margins of the canker till the underlying tissue crackles and chars. This has been used with some success in New Zealand.
6. Destroy badly infected young plantings.
7. Protective sprays of fixed copper compounds (copper oxychloride @ 8–12 lb A^{-1}) or Bordeaux mixture (12:12:100) have given a partial control of the disease in nursery and orchards. Nevertheless, pathogen strains resistant to copper are common in orchards in Pacific Northwest and other areas with a history of copper use. Moreover, copper is phytotoxic to most stone fruits, particularly to sweet cherry, and its repeated applications aggravate it further. Cherry nursery plants should be sprayed before digging for planting, and in the orchard; sprays should be given in the fall and in the early spring before blossoming. An airblast sprayer should be preferred over other sprayers as it would blow off all except the terminal leaves from the trees and thereby reduce the chances of bacterial penetration through leaf scars, further leading to leaf infection.
8. Acibenzolar-S-methyl (ASM, sold under the trade name of Actigard in the United States and Bion in Europe) is a compound that enhances the natural defences of plants by inducing systemic acquired resistance. Monthly applications of ASM moderately reduced bacterial decline of hazelnut caused by *Psm. syringae* pv. *syringae* in Italy.
9. Avoid ring nematode-infested soils for planting orchards, as the high population of these nematodes is generally associated with high disease incidence.

REFERENCES

Abbasi, V., Rahimian, H. and Tajick-Ghanbari, M. A. (2013). Genetic variability of Iranian strains of *Pseudomonas syringae* pv. *syringae* causing bacterial canker disease of stone fruits. *Eur. J. Plant Pathol*. 135: 225–35.

Agrios, G. N. (2004). *Plant Pathology*. 5th Ed. Elsevier Academic Press, San Diego, CA, 922 pp.

Donmez, M. F., Karlidag, H. and Esitken, A. (2010). Identification of resistance to bacterial canker (*Pseudomonas syringae* pv. *syringae*) disease on apricot genotypes grown in Turkey. *Eur. J. Plant Pathol*. 126: 241–47.

Feil, H., Feil, W. S., Chain, P., Larimer, F., DiBartolo, G. *et al.* (2005). Comparison of the complete genome sequences of *Pseudomonas syringae* pv. *syringae* B728a and pv. *tomato* DC 3000. *Proc. Natl. Acad. Sci*. USA, 102: 11064–69.

Hattingh, M. J. and Roos, I. M. M. (1992). Bacterial canker of stone fruits. In: J. Kumar, H. S. Chaube, U. S. Singh and A. N. Mukhopadhyay (eds.). *Plant Diseases of International Importance*. Vol. III, *Diseases of Fruit Crops*. Prentice Hall, Englewood Cliffs, NJ, pp. 394–404.

Hattingh, M. J., Roos, I. M. M. and Mansvelt, E. L. (1989). Infection and systemic invasion of deciduous fruit trees by *Pseudomonas syringae* in South Africa. *Plant Dis*. 73: 784–89.

Jones, A. L. (2001). Bacterial canker of stone fruit trees. In: O. C. Maloy and T. D. Murray (eds.). *Encyclopedia of Plant Pathology*. Vol. 1. John Wiley and Sons, Inc., New York, pp. 74–76.

Kennelly, M. M., Cazorla, F. M., de Vicente, A., Ramos, C. and Sundin, G. W. (2007). *Pseudomonas syringae* diseases of fruit trees: Progress toward understanding and control. *Plant Dis*. 91: 4–17.

Latorre, B. A., Rioja, M. E. and Lillo, C. (2002). The effect of temperature on infection and a warning system for pear blossom blast caused by *Pseudomonas syringae* pv. *syringae*. *Crop. Prot*. 21: 33–39.

Marroni, M. V., Casonato, S., Pitman, A. R., Visnovsky, S. B., Beresford, R. B. *et al.* (2024). Review of *Pseudomonas* species causing bacterial canker of *Prunus* species with emphasis on sweet cherry (*Prunus avium*) in New Zealand. *Eur. J. Plant Pathol.* 168: 297–314. https://doi.org/10.1007/s10658-023-02755-3

Prunier, J., Luisetti, J. and Gardan, L. (1970). Etudes sur les bacterioses des arbres fruitiers. II. Caracterisation d'un *Pseudomonas* non-fluorescent agent d'une bacterioses nouvelle du pecher. *Ann. Phytopathol.* 2: 181–97.

Spotts, R. A., Wallis, K. M., Serdani, M. and Azarenko, A. N. (2010). Bacterial canker of sweet cherry in Oregon–infection of horticultural and natural wounds, and resistance of cultivar and rootstock combinations. *Plant Dis.* 94: 345–50.

van Hall, C. J. J. (1902). Bijdragen tot de kennis der Bakterieele Plantenzeikten. Cooperatieve Drukkerij-vereeniging "Plantijn," Inaugural dissertation, Amsterdam.

Wormald, H. (1931). Bacterial diseases of stone fruit trees in Britain. III. The symptoms of bacterial canker in plum trees. *J. Pomol. Hort. Sci.* 9: 239–56.

Xu, G. W. and Gross, D. C. (1988). Physical and functional analyses of the *syrA* and *syrB* genes involved in syringomycin production by *Pseudomonas syringae* pv. *syringae*. *J. Bacteriol.* 170: 5680–88.

Young, J. M., Dye, D. W. and Wilkie, J. P. (1978). Genus *Pseudomonas* Migula 1894. In: Young, J. M., Dye, D. W., Bradbury, J. F., Panagopoulos, C. G. and Robbs, C. F. (1978). A proposed nomenclature and classification for plant pathogenic bacteria. *N. Z. J. Agric. Res.* 21: 153–77.

9.52 BACTERIAL CANKER OF MANGO

The disease is also called **bacterial black spot** or **angular leaf spot of mango** (*Mangifera indica* L.). Doidge in 1909 reported the disease for the first time from South Africa. Since then, it has been reported from Australia, the Comoro Islands, Japan, Kenya, Mauritius, Malaysia, New California, Pakistan, the Philippines, Reunion Island, Taiwan, Thailand and the United Arab Emirates. The disease probably occurs throughout southern and eastern Africa and in most parts of Asia. It also occurs in Brazil (Gagnevin and Pruvost, 2001). In India, it was first reported by Patel *et al.* (1948) as bacterial leaf spot and is prevalent in Uttar Pradesh, Bihar, Maharashtra, Tamil Nadu, Delhi, Haryana and Punjab. Most recently, it has been reported from Ghana (Honger *et al.*, 2024).

Yasuhara-Bell *et al.* (2013) reported the occurrence of bacterial black spot of mango for the first time from Hawai'i or anywhere in the United States. Sanahuja *et al.* (2016) confirmed the occurrence of bacterial canker of mango for the first time in the Americas. Zombre *et al.* (2017) reported bacterial canker of mango from Togo. Uribe-Lorio and Wang (2017) reported *Erwinia billingiae* as a causal agent of bacterial canker on mango trees in Costa Rica. Phylogenetic analysis of 16S rRNA and concatenated MLST genes showed the Costa Rican strain closely related to *Erw. billingiae*.

The disease is a limiting factor in mango production because it is very difficult to control the disease, particularly on susceptible varieties. It can be very destructive in areas where high temperature and rainfall occur concomitantly. Most of the commercial varieties are highly susceptible to the pathogen and the infection results in drastic yield losses. These losses result from premature fruit drop, a reduction in fruit quality, the severe defoliation of plants and fruit rot in storage. Insect and fungal invasions through fruit cankers further aggravate the losses. Fruit infections ranging from 50–80% are common on highly susceptible cultivars. In 1966 and 1967, severe epiphytotics of the disease caused almost 100% fruit loss on most susceptible cultivars in most of the mango-growing areas of South Africa. Fruit loss was mainly due to the fruit drop and the non-marketability of the fruit. The estimated loss alone in 1966 was approximately US$1 million (Gagnevin and Pruvost, 2001).

In India, the pathogen has been observed to cause 20–50% fruit infection in Langra cultivar and 10–85% in seedling mangoes under natural conditions. In hybrids and cv. Bangalora, 37–40% fruits showed infection, of which 20–25% dropped. In many hybrids, even 100% fruit infection was observed. Infected fruits, including those invaded by insects and fungi rot in the orchard during ripening (Shekhawat and Patel, 1972). Kishun (1982) observed 20–80, 10–70, 10–40 and 10–55% fruit drop in Alphanso, Pairi, Totapuri and local cultivars, respectively and the maximum fruit drop in all the cultivars occurred 30–60 days after fruit set. The author further reported that the fruit rot in storage was 5–80, 11–67, 10–100 and 5–80% in Alphanso, Pairi, Totapuri and local varieties, respectively.

9.52.1 Host Range

Xan. citri pv. *mangiferaeindicae* have been isolated from cashew (*Anacardium occidentale* L.) in India but these strains were lost. Non-pigmented strains were also isolated from ambarella in the French West Indies and from Brazilian pepper (*Schinus terebinthifolius* Raddi) in Reunion. Ambarella strains are not only taxonomically distinct from mango isolates but are also much less aggressive on mango. In contrast, isolates from Brazilian pepper cannot be distinguished from mango isolates by most phenotypic and genetic analyses, but these are also less aggressive on mango compared to mango isolates. Hence, neither the ambarella nor Brazilian pepper strains pose any serious threat to mango.

Zombre *et al.* (2016) reported the first outbreak of a cashew bacterial disease in Burkina Faso (Western Africa), where *Xan. citri* pv. *mangiferaeindicae* recently emerged on mango. A comprehensive molecular characterisation, based on multilocus sequence analysis (MLSA), supplemented with pathogenicity assays of isolates obtained during the outbreak, indicated that the causal agent on cashew in Burkina Faso was *Xan. citri* pv. *mangiferaeindicae* and not *Xan. citri* pv. *anacardii*. Pathogenicity data supported by population biology in Burkina Faso suggested a lack of host specialisation. Therefore, the inoculum from each crop can be potentially harmful to both the host species. Symptoms induced on cashew leaves and fruit by *Xan. citri* pv. *mangiferaeindicae* and non-pigmented strains of *Xan.*

citri pv. *anacardii* are similar, although the causative bacteria are genetically different.

9.52.2 Symptoms

The symptoms appear on the above-ground parts of the mango trees. Leaf and fruit symptoms are most common, but the symptoms on twigs and branches occur on highly susceptible cultivars or when the infection is very severe. On leaves, the symptoms appear as minute, water-soaked, irregular spots, which enlarge, turn brown or dark brown, become angular due to delineation by veins and finally appear cankerous and raised. Individual spots are usually smaller than 0.5 cm^2. Most of the lesions are surrounded by chlorotic haloes, which are more pronounced on young leaves and may extend 5 mm beyond the necrotic areas. On old leaves, the haloes are narrow and can be best seen when the leaf is held against light. Several lesions may coalesce to form larger necrotic cankerous patches. In old lesions, the upper epidermis becomes loose, greyish white and sometimes peels-off. Severe leaf infection may result in yellowing of the leaves and defoliation (Figure 9.42).

On fruits, the symptoms appear as small, water-soaked spots on lenticels. These spots later become star-shaped, develop into cankers, which are much more raised and darker than those on leaves, and often burst open, releasing a highly contagious gummy substance containing bacterial cells. The insect larvae and fungi enter fruits through the openings of these lesions and cause rotting and fruit drop. Often, a "**tear stain**" infection pattern is observed on the fruit. Many times, fruits in bunches appear healthy but their close examination reveals the presence of lesions at the points where these touch each other.

FIGURE 9.42 Bacterial canker of mango: leaf symptoms.

On twigs and branches, the lesions are raised having longitudinal rifts, predominantly formed on the under surface but without any gummy ooze. Several lesions may run together forming longitudinal scars. The disease is not systemic in nature. Bud necrosis and inflorescence cankers are rarely observed and only on highly susceptible cultivars.

Susceptibility of the plant organs varies with time for a given cultivar. Young leaves are resistant, likely because these do not have functional stomata. However, these become very susceptible when these enlarge, with lesions appearing just after leaf hardening. Leaf susceptibility then decreases with time. Fruit susceptibility increases over time and is highest during the period preceding harvest. This correlates with maximum receptivity of the lenticels.

9.52.3 Causal Organism

Xanthomonas axonopodis pv. *mangiferaeindicae* (Patel *et al.* 1948) Ah-You *et al.* 2007 comb. nov.

Synonym: *Xan. campestris* pv. *mangiferaeindicae*, *Xan. citri* pv. *mangiferaeindicae*

The causal agent of mango bacterial black spot (MBBS), *Xan. citri* pv. *mangiferaeindicae*, was first isolated in South Africa (Doidge, 1915), then reported from tropical and subtropical mango-producing areas of India, Pakistan, Brazil, the Philippines, China and the United States (Gagnevin and Pruvost, 2001; Gagnevin *et al.*, 1997; Patel *et al.*, 1948).

Patel *et al.* (1948) first described the bacterium and named it *Pseudomonas mangiferae-indicae* sp. nov. Robbs *et al.* (1974) reclassified the bacterium as *Xan. campestris* subsp. *mangiferae-indicae* sp. nov. With the introduction of the pathovar system for the classification of plant pathogenic bacteria, it was changed to *Xan. campestris* pv. *mangiferaeindicae*. Later on, Ah-You *et al.* (2007) changed it to *Xan. axonopodis* pv. *mangiferaeindicae* and then, finally, to *Xan. citri* pv. *mangiferaeindicae* (Ah-You *et al.*, 2009). The name *Xan. citri* pv. *mangiferaeindicae* given by Ah-You *et al.* (2009) does not confirm to Standards 17 and 21.

The bacterium has all the characteristics of genus *Xanthomonas* except for the production of yellow pigment. It produces non-pigmented (creamy-white) colonies on agar media. However, a few yellow-pigmented strains have been isolated from mango in Brazil, Florida, South Africa and Reunion Island. Atypical strains, including yellow-pigmented strains collected from various countries and the non-pigmented strains isolated from mango in Brazil and from ambarella (*Spondias cytherea* Sonn.) in the French West Indies, are phenotypically and genetically distinct from *Xan. citri* pv. *mangiferaeindicae*. The below-given description of the bacterium is taken primarily from Manicom and Wallis (1984).

Bacterial cells are Gram-stain-negative, aerobic, nonsporulating rods (0.4–0.5 × 1.0–1.5 µm) and motile by a single polar flagellum. Colonies on nutrient agar or King's medium B are round, shallow and convex, having entire margins. Colony colour is initially smoke grey, but soon changes to creamy-white. On potato dextrose agar, the colonies are circular, smooth, glistening, pulvinate having entire margins and measure 1.0–1.5 cm in diameter after seven days of incubation. The bacterium is strictly oxidative. The optimum temperature for growth is 27 °C, but the growth can occur up to 37 °C. The thermal death point is around 55 °C.

Tests are positive for esculin hydrolysis, caseinase, catalase, cellulase, gelatinase, hydrogen sulphide, levan and lipase production and starch hydrolysis, but negative for acetoin production, arginine dihydrolase activity, indole and ammonia production, 2-ketogluconate production, nitrate reduction, cytochrome oxidase, poly-β-hdroxybutyrate synthesis, urease production, growth on asparagine as sole source of C and N, soft rot of potatoes and hypersensitive reaction on tobacco. The bacterium produces acid from glucose, sucrose, xylose, galactose, mannose, D-ribose, trehalose, cellobiose, melibiose, melezitose, esculin, dextrin, glycerol and sorbitol when growth occurs, utilising these compounds. SXTPA, a semi-selective medium, was developed by Chand and Kishun (1995) for the isolation of the bacterium. Two semi-selective media, namely, KC and NCMT3, having plating efficiencies ranging from 76–104% and from 78–132%, respectively have also been developed for the isolation of the bacterium (Pruvost *et al*., 2005). In the latter case, the repeated isolations of the pathogen were possible even from asymptomatic leaves.

Most of the strains produce heat stable toxin (Kishun and Shukla, 2000). Gagnevin *et al*. (1997) assessed the genetic diversity among 138 strains of the bacterium isolated from mango, ambarella and pepper tree in 14 countries and found that 11 strains did not belong to *Xan. campestris* pv. *mangiferaeindicae*. The authors further revealed that the remaining 127 strains formed four groups, of which the group with the greatest diversity consisted of strains from Southeast Asia where mango originated. Another group contained only strains isolated from pepper trees in Reunion, indicating that pepper tree may not be a collateral host for bacterial leaf spot pathogen.

Ah-You *et al*. (2007), on the basis of pathogenic variation and host range, divided and reclassified *Xan. campestris* pv. *mangiferaeindicae* into three pathovars, namely, *Xan. axonopodis* pv. *mangiferaeindicae* pathogenic on mango and cashew, *Xan. axonopodis* pv. *anacardii* pathogenic on cashew and *Xan. axonopodis* pv. *spondiae* pathogenic on ambarella and mombin. These three pathovars also correspond to three genetically divergent groups of *Xan. axonopodis* identified by Rademaker on the basis of AFLP data. Ah-You *et al*. (2009), using amplified fragment length polymorphism, MLSA and DNA–DNA hybridisation for the genotypic classification of *Xanthomonas* pathovars associated with the plant family Ancardiaceae, identified causal agents of mango bacterial canker and cashew bacterial spot as *Xan. citri* pv. *mangiferaeindicae* and *Xan. citri* pv. *anacardii*, respectively.

To date, the genomes of three strains of *Xan. citri* pv. *mangiferaeindicae* have been sequenced. These are LMG941 strain from India (Midha *et al*., 2012); XC01 strain from Guangxi, China (Liu *et al*., 2018), and L242 from the United States (Dommel *et al*., 2019). However, there are obvious contradictions and inconsistency among the three genomes, because the genome sizes of strains LMG941, XC01, and L242 are 5.11 Mb (Midha *et al*., 2012), 3.86 Mb (Liu *et al*., 2018) and 5.30 Mb (Dommel *et al*., 2019), respectively. Strikingly, in the first two genomes, there is no information about the transcription activator-like effector (TALE) genes, the important virulence and avirulence genes in most pathovars of *Xan. citri* (An *et al*., 2020). For strain LMG941, the highly repetitive TALE gene sequences were not correctly assembled because the genome was sequenced on the Roche 454 GS (FLX Titanium) pyrosequencing platform. However, in the case of strain XC01, the correctness of the extremely small genome in *Xan. citri* pv. *mangiferaeindicae* deserves further consideration. Bie *et al*. (2022) reported the complete genome sequence for strain GXG07 generated by long-read sequencing.

9.52.4 Disease Cycle

The bacterium survives in leaf lesions and in cankers formed on the twigs and branches, but survival is more effective in leaf lesions. Three-month-old leaf lesions on a susceptible cultivar may contain 10^7 cfu per lesion while the population in 18-month-old lesions was 10^5 cfu per lesion. The population of the bacterium in leaf lesions decreases relatively slowly under controlled conditions. The role of twig cankers is also important in the survival of the bacterium as these persist on the trees while the infected leaves defoliate prematurely.

The bacterium enters the host through stomata, lenticels and wounds. Under North Indian conditions, the infection on fruits starts by the end of April and the disease progresses considerably even during summer months. On leaves and twigs, fresh infections develop only after the rains and the severity of the disease is at its height in July and August. During winter, the disease remains practically dormant.

The population of the bacterium on mango tree surfaces can average as high as 10^6 cfu per leaf or fruit. Surface population varies with climatic conditions, but is correlated with the number of lesions per leaf or fruit. Humid conditions favour the epiphytic growth. Presence of free water allows the release and redistribution of bacteria from the ruptured epidermis. The population quickly decreases to undetectable levels during dry conditions. Populations of $10^{4\text{-}6}$ cfu per fruit can be found on young fruits under natural conditions and these probably constitute inoculum associated with latent infection, which will be responsible for symptoms when the fruit receptivity increases. Kishun and Chand (1993) have reported the epiphytic survival of the bacterium on 18 weeds found in and around the mango orchards.

Long-distance dissemination or dissemination to the pathogen-free areas occurs through contaminated or infected propagating plant material. Seed–plant transmission of the bacterium has not been proved. Wind-splashed rains and overhead irrigation are important for local or short-distance dissemination of the bacterium. In India, Kishun (1986) reported that *Myllocerus discolor* var. *variegata* (ashy weevil), *Orthaga evadrusalis* (leaf webber) and *Cantheconidia furcellata* (bug) transmitted the pathogen from the diseased to healthy plants. Later on, Kishun and Chand (1989) extended the list of such insects to 22, including the first two insects mentioned above.

9.52.5 Management

1. Use of a disease-free nursery is the most effective approach for the management of the disease. Both the root stocks and scion wood used should be free from the pathogen. However, it is very difficult to determine whether or not the planting material is free from the pathogen, due to the latent infection of the pathogen or the presence of its epiphytic population.
2. Good nursery management practices such as disinfection of budwood using bactericides, protection of plants from wind-driven rains by the use of windbreaks, destruction of source of inoculum and reduction of epiphytic bacterial populations by copper sprays should be followed.
3. Chemical control alone is generally ineffective, but better control of the disease has been achieved by following an integrated approach that includes destroying inoculum sources by pruning, the use of windbreaks and drip irrigation and regularly spraying copper fungicides. In South Africa, more than 20 annual sprays of copper are given on susceptible cultivars. There is every likelihood that it will result in the development of pathogen population resistant to the copper. Three sprays of Streptocycline (100 µg mL^{-1}) at 15- day intervals, starting when the fruits are of pea seed size, have been tried in India and found effective in some cases. Fruit stones used for seedlings should be dipped in Streptocycline solution (100 µg mL^{-1}) for four hours before sowing (personal communication from Ram Kishun).
4. Attempts have been made to manage the disease by the use of antagonistic bacteria (Kishun *et al.*, 2006), where the single application of these bacteria reduced the disease.

Use of resistant varieties is another very effective method of disease management; unfortunately, however, there is hardly any cultivar with complete resistance to the pathogen. Heidi and Sensation cultivars are partially resistant to the pathogen. In India, Bombay Green has been found to be resistant to the pathogen under both natural and artificial inoculation conditions.

REFERENCES

Ah-You, N., Gagnevin, L., Chiroleu, F., Jouen, E., Neto, J. R. *et al.* (2007). Pathological variations within *Xanthomonas campestris* pv. *mangiferaeindicae* support its separation into three distinct pathovars that can be distinguished by amplified fragment length polymorphism. *Phytopathology* 97: 1568–77.

Ah-You, N., Gagnevin, L., Grimont, P. A. D., Brisse, S., Nesme, X. *et al.* (2009). Polyphasic characterization of xanthomonads pathogenic to members of the Anacardiaceae and their relatedness to species of *Xanthomonas*. *Int. J. Syst. Evol. Microbiol.* 59: 306–18.

An, S. Q., Potnis, N., Dow, M., Vorhölter, F. J., He, Y. Q. *et al.* (2020). Mechanistic insights into host adaptation, virulence and epidemiology of the phytopathogen *Xanthomonas*. *FEMS Microbiol. Rev.* 44: 1–32.

Bie, F., Li, Y., Liu, Z., Qin, M., Li, S. *et al.* (2022). High-quality genome resource of mango bacterial black spot pathogen *Xanthomonas citri* pv. *mangiferaeindicae* GXG07 isolated from Guangxi, China. *Plant Dis.* 106: 1027–30.

Chand, R. and Kishun, R. (1995). A semi selective medium for the isolation of bacterial pathogen *Xanthomonas campestris* pv. *mangiferaeindicae*. *African Plant Prot.* 1: 45–48.

Doidge, E. M. (1915). A bacterial disease of the mango, *Bacillus mangiferae* n. sp. *Ann. Appl. Biol.* 2: 1–45.

Dommel, M. R., Huguet-Tapia, J. C. and White, F. F. (2019). TAL effector composition and characterization of *Xanthomonas citri* pv. *mangiferaeindicae*. Abstr. 18th Congress IS-MPMI. *Mol. Plant-Microbe Interact.* 32: S186.

Gagnevin, L., Leach, J. E. and Pruvost, O. (1997). Genomic variability of *Xanthomonas* pathovar *mangiferaeindicae*, agent of mango bacterial black spot. *Appl. Environ. Microbiol.* 63: 246–53.

Gagnevin, L. and Pruvost, O. (2001). Epidemiology and control of mango bacterial black spot. *Plant Dis.* 85: 928–35.

Honger, J. O., Ofoe, R., Osabutey, S. and Abbey, A. A. (2024). Importance, etiology and management of the bacterial black spot disease of mango. *Indian Phytopath.* 77: 681–91. https://doi.org/10.1007/s42360-024-00764-8

Kishun, R. (1982). Loss in mango fruit due to bacterial canker *Xanthomonas mangiferaeindicae*. *Proc. 5th Int. Conf. Plant Path. Bact.* Cali., Colombia, pp. 181–84.

Kishun, R. (1986). Role of insects in transmission and survival of *Xanthomonas campestris* pv. *mangiferaeindicae*. *Indian Phytopathol.* 39: 509–11.

Kishun, R. and Chand, R. (1989). Mechanical transmission of *Xanthomonas campestris* pv. *mangiferaeindicae* through insects. *Indian J. Plant Pathol.* 7: 112–14.

Kishun, R. and Chand, R. (1993). Epiphytic survival of *Xanthomonas campestris* pv. *mangiferaeindicae* on weeds and its role in MBCD. *Plant Dis. Res.* 9: 35–40.

Kishun, R., Mishra, D., Ram, R. A. and Verma, A. K. (2006). Management of mango bacterial canker through antagonists. *J. Eco-friendly Agric.* 1: 54–56.

Kishun, R. and Shukla, R. S. (2000). Evaluation of *Xanthomonas campestris* pv. *mangiferaeindicae* for toxins and their phytotoxicity. *Proc. Int. Conf. Integrated Plant Dis. Mang. for Sustainable Agriculture*, pp. 743–44.

Liu, F., Ou, X. and Zhan, R. (2018). Whole-genome sequence and genome annotation of *Xanthomonas citri* pv. *mangiferaeindicae*, causal agent of bacterial black spot on *Mangifera indica*. *Arch. Microbiol.* 200: 835–40.

Manicom, B. Q. and Wallis, F. M. (1984). Further characterization of *Xantho-monas campestris* pv. *mangiferaeindicae*. *Int. J. Syst. Bacteriol.* 34: 77–79.

Midha, S., Ranjan, M., Sharma, V., Pinnaka, A. K. and Patil, P. B. (2012). Genome sequence of *Xanthomonas citri* pv. *mangiferaeindicae* strain LMG941. *J. Bacteriol.* 194: 3031.

Patel, M. K., Moniz, L. and Kulkarni, Y. S. (1948). A new bacterial disease of *Mangifera indica* L. *Curr. Sci.* 17: 189–90.

Pruvost, O., Roumagnac, P., Gaube, C., Chiroleu, F. and Gagnevin, L. (2005). New media for the semiselective isolation and enumeration of *Xanthomonas campestris* pv. *mangiferaeindicae*, the causal agent of mango bacterial black spot. *J. Appl. Microbiol.* 99: 803–15.

Robbs, C. F., Ribeiro, R. de L. D. and Kimura, O. (1974). Sobre el posicao taxonomica de *Pseudomonas mangiferaeindicae* Patel *et al.* 1948, agente causal da "Mancha bacteriana" das folhas da Mangira (*Mangifera indica* L.). Arq. *Univ. Fed. Rural Rio de Janeiro* 4: 11–14.

Sanahuja, G., Ploetz, R. C., Lopez, P., Konkol, J. L., Palmateer, A. J. *et al.* (2016). Bacterial canker of mango, *Mangifera indica*, caused by *Xanthomonas citri* pv. *mangiferaeindicae*, confirmed for the first time in the Americas. *Plant Dis.* 100: 2520.

Shekhawat, G. S. and Patel, P. N. (1972). Bacterial canker of mango. In: P. N. Patel (ed.). *Plant Bacteriology*. Vol. I, *Bacterial Diseases of Plants in India*. Indian Agricultural Research Institute, New Delhi, pp. 192–95.

Uribe-Lorio, L. and Wang, A. W. (2017). Identification of *Erwinia billingiae* as a causal agent of bacterial canker on mango (*Mangifera indica*) trees in Costa Rica (Abstr.). *Phytopathology* 107: S4.12–13.

Yasuhara-Bell, J., de Silva, A. S., Alvarez, A. M., Shimabuku, R. and Ko, M. (2013). First report in Hawai'i of *Xanthomonas citri* pv. *mangiferaeindicae* causing bacterial black spot on *Mangifera indica*. *Plant Dis.* 97: 1244.

Zombre, C., Sankara, P., Ouedraogo, S. L., Wonni, I., Boyer, K. *et al.* (2016). Natural infection of cashew (*Anacardium occidentale*) by *Xanthomonas citri* pv. *mangiferaeindicae* in Burkina Faso. *Plant Dis.* 100: 718–23.

Zombre, C., Wonni, I., Ouedraogo, S. L., Kpemoua, K. E., Assignon, K. *et al.* (2017). First report of *Xanthomonas citri* pv. *mangiferaeindicae* causing mango bacterial canker on *Mangifera indica* in Togo. *Plant Dis.* 101: 503.

9.53 PIERCE'S DISEASE OF GRAPEVINE

Pierce's disease (hereafter referred to as PD), named after N. B. Pierce, is a lethal disease of grapevines. It was first observed in 1884 around Anaheim, when a mysterious epiphytotic of the disease affected vines in southern California. Earlier, the disease was believed to be caused by a virus. In 1973, a xylem-limited rickettsia-like bacterium was implicated with the disease on the basis of an electron-microscopic examination of the infected plant tissue. Later on, in 1978, Davis and associates isolated the bacterium and proved its pathogenicity.

The disease occurs in the United States, Mexico, Brazil, Argentina, Paraguay, Costa Rica, Slovenia and Venezuela. In North America, PD is mainly confined to areas which experience mild winters. Disease is less prevalent where winter temperatures are cool. It is present in the southern United States of Florida, Texas and California, and in the eastern states it extends up to Virginia. In Florida and other southeastern states, the disease has prohibited the commercial cultivation of European grape varieties. With the introduction and establishment of glassy-winged sharpshooter in California after 1990, the status of the disease changed from minor to severe with an enormous increase in losses (Anonymous, 2000). Su *et al.* (2013) reported Pierce's disease of grapevines caused by *Xylella fastidiosa* for the first time from Taiwan, confirming this as the first report from the Asian continent.

Moralejo *et al.* (2024) reported an outbreak of *Xyl. fastidiosa* subsp. *pauca* ST53 affecting wild and cultivated olive trees on the island of Mallorca, Spain. Chatterjee *et al.* (2008), in their review article, have highlighted the progress made in understanding the process by which *Xyl. fastidiosa* spreads within the xylem vessels of susceptible plants as well as the traits that contribute to its acquisition and transmission by sharpshooter vectors. Its complex lifestyle, as both a plant and insect colonist, involves traits that are in conflict with these stages, thus apparently necessitating the use of a gene regulatory scheme that allows cells expressing different traits to co-occur in the plant.

Janse and Obradovic (2010), in their mini-review, have given an elaborate account of host species attacked by *Xyl. fastidiosa* and its subspecies. The authors also concluded that *Xyl. fastidiosa* is a real and emerging threat for Europe, not only for *Vitis* and *Citrus*, but also for stone fruits (almond, peach and plum) and oleander. It is difficult to prevent its entry and difficult to control once established; hence, they deserve more attention than they have done to date. Resistance in European grapes is scarce or even absent. Vector control in the USA has not proved to be very effective. Cultural practices to keep plants in optimum condition are of importance, but not sufficient, and the use of avirulent strains for cross-protection is still in its infancy.

Rapicavoli *et al.* (2018), in their review, focus on the status of *Xyl. fastidiosa* pathogenic associations in plant hosts in which the bacterium is either endemic or has been recently introduced. Plant diseases associated with *Xyl. fastidiosa* have been documented for over a century, and much about what is known in the context of host–pathogen interactions are based on hosts, such as grape and citrus, in which this pathogen has been well described. Recent attention has focused on newly emerging *Xyl. fastidiosa* diseases, such as in olives.

9.53.1 Host Range

Strains of *Xyl. fastidiosa* subsp. *fastidiosa* causing PD have a very wide host range and this has been further expanded by including the hosts showing positive insect transmission of the bacterium. Hopkins (1989) reported that many grasses, legumes and perennial trees belonging to at least 28 families are infected by the bacterium, but most of them do not show any symptoms. Hill and Purcell (1997) found that 94 out of 155 tested species were the host of the bacterium. Wistrom and Purcell (2005), on the basis of greenhouse studies, showed that out of 29 plant species tested, 27 were the host of the bacterium. Common sunflower, coclebur, annual bursage, morning glory, sacred datura, horseweed, poison hemlock and faba bean were the most frequently infected hosts. Purcell (2006) has also compiled a comprehensive list of host species of PD strains.

There are strains of the bacterium other than the PD strain. These strains cause different diseases in different hosts in different regions. These diseases include phony peach disease, alfalfa dwarf disease, citrus variegated chlorosis, plum leaf scald and leaf scorch diseases of coffee, oleander, oak, elm, maple, mulberry, plum and sycamore. Some of these strains also cross-infect hosts for *Xyl. fastidiosa*.

Casarin *et al.* (2023). demonstrated that indigenous European Salicaceae such as *Salix alba* and *Populus tremula* are new potential hosts for *Xyl. fastidiosa*.

9.53.2 Symptoms

The symptoms of PD observed in the summer and the fall are more reliable for its identification than spring symptoms. Vines infected in spring show the first symptom of water stress in midsummer. The margins of leaves become slightly yellow or red, in white and red varieties, respectively. As the infection progresses, leaf margins dry and turn brown in concentric zones (Figure 9.43). Dried leaf laminas fall, leaving behind dried petioles having burnt-appearing tips, called **matchsticks**, attached to the vines, even after normal leaf fall. The wood of infected canes matures irregularly, forming patches of immature green bark surrounded by mature brown bark, and vice versa. Grape clusters on infected vines stop growing, shrivel and dry up. Leaf symptoms may vary on different grape cultivars. In Chenin Blanc and Thompson Seedless cultivars, discolouration and scorching of leaves may appear in sectors instead of along the margins (Anonymous, 2006).

FIGURE 9.43 Pierce's disease of grapevine. (Courtesy: California Dept. of Food and Agriculture, PD Control Programme, www.pdgwss.net)

Infection gradually spreads from the point of infection to the upper side and, more slowly, towards the base of the vines. Die-back of shoots and roots also occurs. Climatic variation in different regions can influence the severity and timing of the symptoms, but not the type. In hot climates, the disease progresses rapidly even under adequate soil moisture conditions due to higher moisture stress. Shoots of infected vines grow slowly and vines deteriorate rapidly.

Infected plants show a delayed and stunted growth in the next spring season. Some canes and spurs may fail to bud out. Newly developed leaves show chlorotic areas between the veins. Late in the season, leaves become scorched and scorching occurs first on the oldest leaves. Yellow to brown streaks are seen in longitudinal and cross-sections of the current season wood of the infected canes. Gum formation in vessels and other cells, and formation of tyloses also occur in this wood (Agrios, 2004).

Due to the rapid movement of the bacteria, young vines are more susceptible to the pathogen than the old vines. The length of time for which an infected vine can survive depends on climate, age and the variety of vines. A 1-year-old wine of a susceptible variety like Pinot Noir can die the same year after infection, while chronically infected 10-year-old vines of Chenin Blanc or Ruby Cabernet can live for more than 5 years. Krell *et al.* (2006) reported that diagnosis of the disease based on foliar symptoms was the only consistent indicator of infection in *Xylella fastidiosa*-positive grapevines and samples from the basal portion of a cane increased the probability of its detection. Schaad *et al.* (2002) has highlighted the use of a real-time PCR requiring only one hour for the on-site diagnosis of PD from asymptomatic vines.

9.53.3 Causal Organism

Xylella fastidiosa subsp. *fastidiosa* Wells *et al.* 1987
Synonym: *Xyl. fastidiosa*

In 1973, the pathogen reported to be associated with this disease was referred to as "**rickettsia-like**" because of its similarities in morphology and ultra-structure with members of the Rickettsiaceae family. Wells *et al.* (1987) proposed a new genus *Xylella* and included all the strains of fastidious, Gram-stain-negative, xylem-limited bacteria under a single species called *Xyl. fastidiosa*.

The species epithet *fastidiosa* refers to nutritional fastidiousness, particularly with regard to the primary isolation of the bacterium. It does not grow on most common bacteriological media used. However, it can be cultured on specialised media like buffered cysteine-yeast extract (BCYE) agar containing charcoal. Periwinkle wilt (PW) medium has gained the widest acceptance, and supports the growth of all known strains of *Xyl. fastidiosa*. Almeida *et al.* (2004) has proposed a simpler solid medium for supporting the growth of PD strains of *Xyl. fastidiosa*. The following description of the bacterium is taken mainly from Wells *et al.* (1987).

Cells are Gram-stain-negative, non-flagellate, straight rods measuring 0.25–0.35 × 0.9–3.5 µm. Cells are predominantly single, but they sometimes form filamentous strands under certain conditions. It is an obligate aerobe, inhibited by 2.5% carbon dioxide. It is non-pigmented, non-halophilic, Kovac's oxidase negative and catalase positive. The bacterium hydrolyses gelatin, utilises hippurate and most strains produce beta-lactamase. It does not ferment D-glucose, does not produce indole and H_2S, and is also negative for beta-galactosidase, lipase, amylase, coagulase and phosphatase tests. Optimum temperature for growth is 26–28 °C and optimum pH is 6.5–6.9. G + C content of the DNA is 51.0–52.4 mol%.

Although Wells *et al.* (1987) recognised only one species to include all the strains, there are definitely different pathotypes, most of which have not been characterised and compared with each other (Hopkins and Purcell, 2002). Schaad *et al.* (2004) proposed three new subspecies, namely, *Xyl. fastidiosa* subsp. *piercei*, *Xyl. fastidiosa* subsp. *multiplex* and *Xyl. fastidiosa* subsp. *pauca*. *Xyl. fastidiosa* subsp. *pauca* has not been validated due to the lack of deposition of a culture of the type strain. As the type strain designated for subsp. *piercei* is the type strain of the species, the name *Xyl. fastidiosa* subsp. *fastidiosa* takes priority over *Xyl. fastidiosa* subsp. *piercei*. The study of isolates on the basis of which Schaad *et al.* (2004) proposed three new subspecies of *Xyl. fastidiosa* were: strains from grape, alfalfa, almond and maple classified as *Xyl. fastidiosa* subsp. *piercei*, those from peach, plum, elm, pigeon grape, sycamore and almond as *Xyl. fastidiosa* subsp. *multiplex*, and those only from citrus as *Xyl. fastidiosa* subsp. *pauca*.

A rapid isothermal assay for specific detection of *Xyl. fastidiosa* using the advanced recombinase-polymerase amplification technology has been developed by Agdia. The assay performs both as a real-time and an endpoint test from a single reaction tube at 39° C for 20 min. Reaction template is simply prepared by soaking 50 mg of petiole cross-sections in 0.5 mL AMP1 extraction buffer for 10 min. The assay reacted to over 27 isolates from a dozen hosts, including grapevine and citrus, while consistently detecting 22 copies of spiked *Xyl. fastidiosa* genome in petiole extract (1:10, W/V). No reaction background was observed in host tissue such as grapevine, citrus, olive, almond, coffee, blueberry and blackberry. No cross-reaction was observed for *Escherichia coli*, *Xanthomonas*, *Erwinia* and *Pseudomonas* species. This assay provides users with a fast and reliable tool to monitor the regional and international movements of *Xyl. fastidiosa* (Li *et al.*, 2017).

Simpson *et al.* (2000) determined the complete genome sequence of clone 9a5C, the causal agent of citrus variegated chlorosis. This was the first plant pathogenic bacterium for which the complete genome was sequenced.

Van Sluys *et al.* (2003) reported the genome sequence of *Xyl. fastidiosa* (Temecula strain), isolated from a naturally infected grapevine with PD in a wine-grape-growing region of California. Comparative analyses with a previously sequenced *Xyl. fastidiosa* strain 9a5c causing CVC (Simpson *et al.* 2000) revealed that 98% of the PD *Xyl. fastidiosa* Temecula genes are shared with those of CVC *Xyl. fastidiosa*. Genomic differences are limited to phage-associated chromosomal rearrangements and deletions that also account for the strain-specific genes present in each genome. Genomic islands, one in each genome, were identified, and their presence in other *Xyl. fastidiosa* strains was analysed. They concluded that these two organisms have identical metabolic functions and are likely to use a common set of genes in plant colonisation and pathogenesis, permitting the convergence of functional genomic strategies. Comparative analyses of the complete sequences and annotations of PD *Xyl. fastidiosa* and a *Xyl. fastidiosa* representative of the CVC group isolated from Brazil revealed that these strains exhibit remarkably limited genomic variability and share 95.7% amino acid identity in equivalent regions. There are only three genomic rearrangements, two identified genomic islands, 41 PD *Xyl. fastidiosa* and 152 CVC *Xyl. fastidiosa* strain-specific genes and some genes harbouring frame shifts.

Gomila *et al.* (2019) reported a *de novo*-assembled draft genome sequence of the first isolates causing Pierce's disease in Europe: *Xyl. fastidiosa* subsp. *fastidiosa* strains XYL1732/17 and XYL2055/17. Both strains were isolated from grapevines (*Vitis vinifera*), showing Pierce's disease symptoms at two different locations in Mallorca, Spain. XYL1732/17 genome is 2,444,109 bp long, with a G + C content of 51.5%; it contains 2,359 open reading frames and 48 tRNA genes. XYL2055/17 genome is 2,456,780 bp long, with a G + C content of 51.5%; it contains 2,384 open reading frames and 48 tRNA genes.

There are currently >140 *Xyl. fastidiosa* whole-genome sequences deposited in the GenBank bacterial genome database, but none are from *Xyl. fastidiosa* subsp. *multiplex* strains associated with phony peach disease. Johnson *et al.* (2022) reported a genome resource for *Xyl. fastidiosa* subsp. *multiplex* strain P5A2 causing phony peach disease in the southeastern United States.

Giampetruzzi *et al.* (2023) described the draft genome sequences of *Xyl. fastidiosa* strain Alm_Lz_1, obtained after its successful isolation in pure culture from a symptomatic tree of *Prunus dulcis*. Genomic DNA of strain Alm_Lz_1 was extracted from pure culture grown on buffered charcoal yeast extract agar medium (Wells *et al.*, 1981) using a commercial DNA purification kit. The whole-genome sequencing library was paired-end sequenced with a NovaSeq 6000 Illumina platform.

Illumina sequencing yielded a total of 14,027,398 high-quality paired reads (2 × 150 bp) (Q30% of 92.23%). *De novo* genome assembly was done using SPAdes 3.13.0 (Bankevich *et al.*, 2012) using multiple kmer (31,51,81,101,121) and trying to reduce the number of mismatches and short indels (careful option). Presence of contigs annotated as plasmid sequence was searched running a plasmid SPAdes tool (Antipov *et al.*, 2019). Final assemblies of the bacterial chromosomes resulted in 131 contigs, with an equal GC content of 51.73% (N50s was for 193,575). Average nucleotide coverage of the chromosomal genome was 850×. Whole-genome comparison by Roary tool (v3.11.2) (Page *et al.*, 2015) was used to generate alignments of accessory and core genes between subsp. *multiplex* strains and the newly sequenced Alm_Lz_1 strains (1,378 core genes, 3,385 accessory genes) re-annotated by Prokka tool (Seemann, 2014).

Kandel (2015) described natural competence (NC), a mode of horizontal gene transfer in which bacterial cells take up free DNA from the environment and recombine it, in *Xyl. fastidiosa*. To more closely approximate *in planta* conditions, the author used a microfluidic chamber setting (MC) that resembles the plant xylem vessels, which *Xyl. fastidiosa* naturally inhabits and culture medium supplemented with different concentrations of grapevine sap to create a natural environment mimic system (NEMS). Using combinations of live and dead antibiotic marker-tagged *Xyl. fastidiosa* mutant cells, it was demonstrated that NC occurs in the NEMS and that the recombination frequency is comparable to that obtained from experiments conducted on agar plates and in MC containing only artificial media. These results suggest that NC can occur in nature and might be responsible for the recently described host and geographical shifts in *Xyl. fastidiosa*.

9.53.4 Disease Cycle

The mechanisms associated with pathogenicity and virulence involve toxins, antibiotics, ion sequestration, and bacterium–host interactions mediated by a range of proteins.

After penetration into grapevines, the bacteria multiply for several months in a water-conducting system of plants and spread through the xylem system. The multiplication of the bacteria causes blocking of xylem vessels and this blockade is responsible for most of the external symptoms of the disease. Roper *et al.* (2007) hypothesised on the basis of presence of genes in the genome of *Xyl. fastidiosa* that several cell wall-degrading enzymes, such as beta-1,4 endoglucanases, xylanases, xylosidases and polygalacturonase (PG) produced by *Xyl. fastidiosa* breakdown pit membranes. They further demonstrated that among them, PG encoded by *pglA* gene is a critical virulence factor and is required by *Xyl. fastidiosa* for the successful infection of grapevines because a mutant in *pglA* lost the pathogenicity.

Wallis and Chen (2012) examined induction of phenolic compounds in grapevines of cv. Thompson Seedless infected with *Xyl. fastidiosa* over a six-month period. Two months post-inoculation with the bacterium, catechin, astringin and digalloylquinic acid were found at greater levels in xylem sap; multiple catechins, procyanidins and stilbenoids were found at greater levels in xylem tissues; and precursors to lignin and condensed tannins were found at greater levels in xylem cell walls. However, such large-scale inductions of phenolic compounds were not observed four months after inoculation. Six months after inoculation, infected plants showed significantly reduced phenolic levels in xylem sap and tissues, including lowered levels of lignin and condensed tannins compared to control plants. At six months, infected plants showed severe disease symptoms and most photosynthetic tissue was abscised. It appears that even though grapevine plants may initially respond to *Xyl. fastidiosa* infection with the increased production of phenolic compounds, ultimately, the bacterium brings the grapevine plants into a state of decline, whereby diseased plants no longer have the ability to support secondary metabolite production, including defence-associated phenolic compounds.

Burbank and Stenger (2017) characterised the role of the DinJ/RelE toxin-antitoxin system in controlling bacterial proliferation and population size during plant colonisation. The ratio of antitoxin/toxin expressed is dependent on bacterial growth conditions, with lower amounts of antitoxin present under conditions designed to mimic grapevine xylem sap. A knockout mutant of DinJ/RelE exhibits a hypervirulent phenotype, with higher bacterial populations and increased symptom development and plant decline. It is probable that DinJ/RelE acts to prevent excessive population growth, thereby contributing to the ability of the pathogen to spread systemically without completely blocking the xylem vessels and increasing the probability of acquisition by the insect vector.

Shi *et al.* (2013) reported that when *Xyl. fastidiosa* is grown in xylem fluid from the susceptible species, bacterial cells formed a heavier biofilm compared to those in xylem fluid from the resistant species. Compared with xylem fluid of *V. smalliana* (resistant species), xylem fluid of *V. vinifera* (susceptible species) stimulated the expression of *Xyl. fastidiosa* genes *rpfC*, *gacA*, *xrvA*, *gcvR* and *cysB*, involved in virulence regulation, and genes *pilI*, *pilU*, *pilE* and *pilG* involved in biogenesis of pili and twitching motility. The increased expression of virulence genes likely contributes to the expression of PD symptoms in the susceptible grapevines, whereas reduced expression of these genes in resistant grapevines may lead to the suppression of symptoms.

Shi and Lin (2018) elucidate the roles of *pilG* in the pathogenicity of *Xyl. fastidiosa* by generating the *pilG*-deletion mutant *Xf*Δ*pilG* and complemented strain *Xf*Δ*pilG*-C. While all the strains showed similar growth curves *in vitro*, *Xf*Δ*pliG* showed a significant reduction in cell-matrix adherence and biofilm production compared to wild-type *Xyl. fastidiosa* and *Xf*Δ*pilG*-C. The genes *pilE*, *pilU*, *pilT* and *pilS* were downregulated in *Xf*Δ*pliG* compared to its complemented strain and wild-type *Xyl. fastidiosa*. No symptoms of Pierce's disease appeared in grapevines inoculated with *Xf*Δ*pilG*, while the grapevines inoculated with the wild-type *Xyl. fastidiosa* and complemented strain *Xf*Δ*pilG*-C produced typical disease symptoms. The results indicate that *pilG* plays a role in *Xyl. fastidiosa* virulence in grapevines.

The multiplication and movement of the bacterium varies in different plant species. In *V. vinifera* and wild grapes, the bacterium multiplies and reaches a concentration of 10

d'Azur region; (3) positive adults were found during winter; and (4) the bacterial load in *P. spumarius* (droplet digital PCR) usually ranged from 10^3 to 10^4 cells per insect, but could be as high as 10^5 or 10^6 cells per insect for some individuals (13%). Subspecies and sequence types detected in *P. spumarius* corresponded to the situation officially reported for plants in the same areas (Cunty *et al.*, 2020).

To improve the reliability of the TaqMan assay, currently widely used in surveys, Bonants *et al.* (2019) developed a triplex TaqMan assay, in which two specific and sensitive TaqMan assays, previously designed for *Xyl. fastidiosa*, were combined with an internal control. The triplex assay exhibited the same diagnostic sensitivity as the simplex assays. DNA extracts from plant material free of *Xyl. fastidiosa*, from artificially inoculated hosts plants or from naturally infected plants sampled in France, Spain and Italy, or intercepted in Austria and the Netherlands, were analysed for the presence of *Xyl. fastidiosa* using the metagenomic approach. In all samples, even in samples with a low infection level, but not in the pathogen-free samples, DNA reads, specific for *Xyl. fastidiosa*, were detected. In most cases, the pathogen could be identified to the subspecies level, and for one sample even the whole genome could be assembled and the sequence type could be determined. All results of next-generation sequencing analysed samples were confirmed with the triplex TaqMan PCR and loop-mediated isothermal amplification.

Hao *et al.* (2017) examined the PD1311 gene, which encodes a putative acyl-coenzyme A (acyl-CoA) synthetase, and is highly conserved across *Xylella* species. It was determined that PD1311 is required for virulence, as the deletion mutant, ΔPD1311, was unable to cause disease on grapevines. ΔPD1311 strain was impaired in actions known to be associated with PD development, including motility, aggregation and biofilm formation. ΔPD1311 also expressed enhanced sensitivity to H_2O_2 and polymyxin B, and showed reduced survival in grapevine sap, when compared with wild-type *Xyl. fastidiosa* Temecula 1 (TM1). Following inoculation, ΔPD1311 could not be detected in grape shoots, which may be related to its altered growth and sensitivity phenotypes. Inoculation with ΔPD1311, two weeks prior to TM1 prevented the development of PD in a significant fraction of vines and eliminated detectable levels of TM1. In contrast, vines inoculated simultaneously with TM1 and ΔPD1311 developed disease at the same level as TM1 alone. In these vines, TM1 populations were distributed similarly to populations in TM1-only inoculated plants. These findings suggest that, through an indirect mechanism, the pretreatment of vines with ΔPD1311 suppresses pathogen population and disease development.

Müller *et al.* (2022) observed the natural incidence of *Xyl. fastidiosa* on 12 weed species prevalent in plum orchards, having a high incidence of plum leaf scald in Brazil; and also determined the colonisation capacity of three sequence types (STs 26, 67 and 71) of *Xyl. fastidiosa* isolated from plum trees. Nine species, namely, *Bidens pilosa*, *Lepidium ruderale*, *Lolium multiflorum*, *Plantago major*, *Parthenium hysterosphorus*, *Raphanus sativus*, *Rumex* sp., *Solanum americanum* and *Vernonia* sp., tested positive by PCR for *Xyl. fastidiosa* infection. *L. ruderale* and *L. multiflorum* had the highest infection rates of 70% and 90% in root and leaf samples, respectively, and 100% in leaf samples of *L. multiflorum*. Mechanical inoculation allowed the systemic infection of the bacteria in *B. pilosa*, *L. ruderale*, *R. sativus* and *S. americanum*. These observations show that several weed species in plum orchards are hosts of representative STs of *Xyl. fastidiosa* that infect plum, and might play a role in the spread of the pathogen to plum trees.

Combining Oxford Nanopore Technologies and Illumina sequencing data, Velasco-Amo *et al.* (2023) assembled high-quality complete genomes of seven *Xyl. fastidiosa* subsp. *fastidiosa* strains isolated from different plants in Spain, the USA and Mexico. Comparative genomic analyses discovered differences in plasmid content among strains, including plasmids that had been overlooked previously when using the Illumina sequencing platform alone. Interestingly, in strain CFBP8073, intercepted in France from plants imported from Mexico, three plasmids were identified, including two (plasmids pXF-P1.CFBP8073 and pXF-P2.CFBP8073) not previously described in *Xyl. fastidiosa* and one (pXF5823.CFBP8073) almost identical to a plasmid described in a *Xyl. fastidiosa* strain from citrus. Plasmids found in the Spanish strains here were similar to those described previously in other strains from the same subspecies and ST1 isolated in the Balearic Islands and the USA.

De La Fuente *et al.* (2024) reviewed intrinsic problems that arise from the fastidious growth of *Xyl. fastidiosa*, the lack of specificity for insect transmission, and the economic and social importance of perennial mature woody plant hosts. Epidemiological models and predictions of pathogen establishment and disease expansion, vital for preparedness, are based on very limited data. Most of the current knowledge has been gathered from a few pathosystems, whereas several hundred remain to be studied, probably including those that will become the centre of the next epiphytotic. Unfortunately, aspects of a particular pathosystem are not always transferable to others. The authors recommend the diversification of research topics of both fundamental and applied nature addressing multiple pathosystems. Increasing preparedness through knowledge acquisition is the best strategy to anticipate and manage diseases caused by this pathogen, described as "the most dangerous plant pathogenic bacterium known worldwide".

9.53.5 Management

The management of the disease is difficult due to the systemic nature of the pathogen and wide habitats of the vectoring insects. However, the following measures will certainly help in reducing the disease incidence.

1. Insecticidal control of insect vectors in habitats adjacent to the vineyards helps in reducing the disease incidence but it is not viable in highly

susceptible cultivars or in vineyards having less than three-year-old vines.

2. Plant less susceptible or tolerant cultivars in vineyards that are near a known or suspected hotspot of the disease.
3. Use systemic insecticides to control insect vectors on vines and also on a 50–100 feet-wide strip of natural vegetation along the vineyard edge. Tubajika *et al.* (2007) achieved 88%, 78% and 78% disease control with the application of kaolin, harpin and imidacloprid insecticides, respectively, in Kern county of California.
4. Breeding of insect vectors on irrigated grasses should be controlled by weed control instead of insecticide control.
5. Remove and destroy chronically infected vines, i.e., vines showing disease symptoms for more than one year.
6. Also remove vines showing extensive foliar symptoms and severe die-back even if it is their first year of infection.
7. Early disease detection and vine removal is recommended in vineyards that experience influxes of glassy-winged sharpshooters.
8. Hot water treatment of cuttings and propagation buds at 45 °C for 3 h.
9. *Xyl. fastidiosa* subsp. *fastidiosa* is a quarantined pest (EPPO A1) in many countries. The enforcement of quarantine regulations will prevent the entry of the pathogen in disease-free areas.
10. Tetracycline is effective in reducing the disease incidence and severity, but its efficacy and viability needs to be tested at the field level.
11. Cross-protection studies carried out by Donald Hopkins at the University of Florida, USA, using a benign strain of PD bacterium, reduced the death of Cabernet Sauvignon vines to less than 20%. Although 20% vine death may not sound that encouraging, it is significant in the sense that the cultivar like Cabernet Sauvignon cannot normally be grown in Florida, the hottest PD spot in the U S.
12. Meyer and Kirkpatrick (2011) reported that foliar and drench applications of the plant growth regulator, abscisic acid to 1-year-old potted *V. vinifera* Pinot Noir and Cabernet Sauvignon vines infected with *Xyl. fastidiosa* showed some effectiveness in the cure of affected grapevines. Pinot Noir grapevines treated with drench applications of abscisic acid showed significantly greater disease cure than the unsprayed control plants. It has been shown that plant phenolics have antimicrobial properties, and the authors found a positive correlation between effective abscisic acid treatments and the total phenolic compound content of xylem sap extracted from Pinot Noir vines.
13. Fletcher and Wayadande (2002) have suggested a novel approach of paratransgenesis to control plant diseases caused by *Xyl. fastidiosa*. In this approach, non-*Xylella* enteric bacteria, i.e., normally gut bacteria of leafhoppers will be transformed to secrete toxin(s) lethal to *Xyl. fastidiosa* subsp. *fastidiosa*. These transformed bacteria will pass from insect to insect via normal coprophagus activity (ingestion of faecal matter) and these paratransgenic insects will harm the *Xyl. fastidiosa* subsp. *fastidiosa* cells acquired by them.
14. Hopkins (2005) reported that biological control by inoculation of susceptible grapevines with benign strains of *Xyl. fastidiosa*, especially strain EB92-1, have the potential to control PD in commercial vineyards in Florida and other areas where the disease is endemic.

Hopkins and Benyamini (2020) reported that injection of a benign strain of *Xyl. fastidiosa* (EB92-1) into the xylem vessels of *V. vinifera* 'Cabernet Sauvignon' vines controlled PD of grapevine for 14 years in the University of Florida vineyard. In eight-year field trials in central Florida, strain EB92-1 provided significant control of PD in *V. vinifera* 'Cabernet Franc', 'Chardonnay', 'Chenin Blanc' and 'Ruby Cabernet', as well as *V. aestivalis* 'Cynthiana' and a French American interspecific hybrid 'Chambourcin'. EB92-1 provided biological control of PD in a new planting of 'Malbec' for seven years and in a mature planting of 'Mourvedra' for five years in Dahlonega Georgia. In the UC Riverside vineyard, EB92-1 provided significant control of PD for six years in 'Cabernet Sauvignon' and 'Pinot Noir'.

Baccari *et al.* (2019) reported that while most endophytic bacteria found within grape did not grow or multiply when inoculated into mature grape vines, *Paraburkholderia phytofirmans* strain PsJN achieved population sizes as large as 10^6 cells/g and moved one metre or more within four weeks after inoculation into vines. *Xyl. fastidiosa*, when inoculated alone, achieved large population sizes and moved extensively in grape, but few viable cells were recovered from plants in which it was co-inoculated with strain PsJN and the incidence of leaves exhibiting scorching symptoms typical of Pierce's disease was consistently greatly reduced compared to control plants. The suppression of disease symptoms occurred not only when strain PsJN was co-inoculated with the pathogen by puncturing stems in the same site in plants, but also when inoculated at the same time but at different sites in the plant. Large population sizes of strain PsJN could be established in both leaf lamina and petioles by the topical application of cell suspensions in 0.2% of an organo-silicon surfactant, and such treatments were as effective as direct puncture inoculations of this biocontrol strain in reducing disease severity. While inoculation of strain PsJN into plants by either method at the same time as or even four weeks after that of the pathogen resulted in large reductions in disease severity, much less disease control was achieved by inoculation of PsJN four weeks prior to that of the pathogen. The expression of grapevine PR1 and ETR1 within three weeks of inoculation was substantially higher in plants inoculated with both *Xyl. fastidiosa* and strain PsJN compared with that in plants inoculated only with the pathogen or strain PsJN, suggesting that this biological control agent reduces disease by priming the expression of innate disease resistance pathways in plants that otherwise would have exhibited minimal responses to the pathogen. Strain PsJN thus appears highly efficacious for the control of Pierce's disease when used as an eradicant treatment that can be easily made even by spray application.

REFERENCES

Agrios, G. N. (2004). *Plant Pathology*. 5th Ed. Elsevier Academic Press, San Diego, CA, 922 pp.

Almeida, R. P. P., Mann, R. and Purcell, A. H. (2004). *Xylella fastidiosa* cultivation on a minimal solid defined medium. *Curr. Microbiol*. 48: 368–72.

Anonymous (2000). Pierce's disease. Hortguard threat data sheet. Department of Agriculture, Government of Western Australia, pp. 10.

Anonymous (2006). Grape Pierce's disease. IPM Programme, University of California. http://www.ipm.ucdavis.edu/PMG/r302101211.html.

Antipov, D., Raiko, M., Lapidus, A. and Pevzner, P. A. (2019). Plasmid detection and assembly in genomic and metagenomic data sets. *Genome Res*. 29: 961–68. https://doi.org/10.1101/gr.241299.118

Baccari, C., Antonova, E. and Lindow, S. (2019). Biological control of Pierce's disease of grape by an endophytic bacterium. *Phytopathology* 109: 248–56.

Baccari, C. and Lindow, S. E. (2011). Assessment of the process of movement of *Xylella fastidiosa* within susceptible and resistant grape cultivars. *Phytopathology* 101: 77–84.

Backus, E. A. and Morgan, D. J. W. (2011). Spatiotemporal colonization of *Xylella fastidiosa* in its vector supports the role of egestion in the inoculation mechanism of foregut-borne plant pathogens. *Phytopathology* 101: 912–22.

Backus, E. A., Shugart, H. J., Rogers, E. E., Morgan, J. K. and Shatters, R. (2015). Direct evidence of egestion and salivation of *Xylella fastidiosa* suggests sharpshooters can be "flying syringes". *Phytopathology* 105: 608–20.

Bankevich, A., Nurk, S., Antipov, D., Gurevich, A. A., Dvorkin, M. *et al*. (2012). SPAdes: A new genome assembly algorithm and its applications to single-cell sequencing. *J. Comput. Biol*. 19: 455–77. https://doi.org/10.1089/cmb.2012.0021

Bonants, P., Griekspoor, Y., Houwers, I., Krijger, M., van der Zouwen, P. *et al*. (2019). Development and evaluation of a triplex TaqMan assay and next-generation sequence analysis for improved detection of *Xylella* in plant material. *Plant Dis*. 103: 645–55.

Burbank, L. P. and Stenger, D. C. (2017). The DinJ/RelE toxin-antitoxin system suppresses bacterial proliferation and virulence of *Xylella fastidiosa* in grapevine. *Phytopathology* 107: 388–94.

Casarin, N., Hasbroucq, S., Pesenti, L. *et al*. (2023). Salicaceae as potential host plants of *Xylella fastidiosa* in European temperate regions. *Eur. J. Plant Pathol*. 165: 489–507. https://doi.org/10.1007/s10658-022-02622-7

Chatterjee, S., Almeida, R. P. P. and Lindow, S. (2008). Living in two worlds: The plant and insect lifestyles of *Xylella fastidiosa. Annu. Rev. Phytopathol*. 46: 243–71.

Cornara, D., Sicard, A., Zeilinger, A. R., Porcelli, F., Purcell, A. H. *et al*. (2016). Transmission of *Xylella fastidiosa* to grapevine by the meadow spittlebug. *Phytopathology* 106: 1285–90.

Cunty, A., Legendre, B., de Jerphanion, P., Juteau, V., Forveille, A. *et al*. (2020). *Xylella fastidiosa* subspecies and sequence types detected in *Philaenus spumarius* and in infected plants in France share the same locations. *Plant Pathol*. 69: 1798–1811.

De La Fuente, L., Navas-Cortés, J. A. and Landa, B. B. (2024). Ten challenges to understanding and managing the insect-transmitted, xylem-limited bacterial pathogen *Xylella fastidiosa. Phytopathology* 114: 869–84. https://doi.org/10.1094/PHYTO-12-23-0476-KC

Fletcher, J. and Wayadande, A. (2002). Fastidious vascular-colonizing bacteria. APS Education Center. http://www.apsnet.org/education/introplantpath/pathogengroups/fastidious.

Francis, M., Civerolo, E. L. and Bruening, G. (2008). Improved bioassay of *Xylella fastidiosa* using *Nicotiana tabacum* cultivar SR1. *Plant Dis*. 92: 14–20.

Giampetruzzi, A., Loconsole, G., Zicca, S., Boscia, D., Balestra, G. M. *et al*. (2023). Draft genome sequence resource of *Xylella fastidiosa* strain Alm_Lz_1 associated with a new outbreak in Lazio, Italy. *Phytopathology* 113: 108–11.

Gomila, M., Moralejo, E., Busquets, A., Segui, G., Olmo, D. *et al*. (2019). Draft genome resources of two strains of *Xylella fastidiosa* XYL1732/17 and XYL2055/17 Isolated from Mallorca vineyards. *Phytopathology* 109: 222–24.

Hao, L., Johnson, K., Cursino, L., Mowery, P. and Burr, T. J. (2017). Characterization of the *Xylella fastidiosa* PD1311 gene mutant and its suppression of Pierce's disease on grapevines. *Mol. Plant Pathol*. 18: 684–94.

Hill, B. L. and Purcell, A. H. (1997). Populations of *Xylella fastidiosa* in plants required for transmission by an efficient vector. *Phytopathology* 87: 1197–1201.

Hopkins, D. L. (1989). *Xylella fastidiosa*: Xylem-limited bacterial pathogen of plants. *Annu. Rev. Phytopathol*. 27: 271–90.

Hopkins, D. L. (2005). Biological control of Pierce's disease in the vineyard with strains of *Xylella fastidiosa* benign to grapevine. *Plant Dis*. 89: 1348–52.

Hopkins D. L. and Benyamini Y. (2020). Field trials demonstrate biological control of Pierce's disease in Florida, Georgia, and California and citrus greening in Florida (Abstr.). *Phytopathology* 110: S2.126.

Hopkins, D. L. and Purcell, A. H. (2002). *Xylella fastidiosa*: Cause of Pierce's disease of grapevine and other emergent diseases. *Plant Dis*. 86: 1056–66.

Janse, J. D. and Obradovic, A. (2010). *Xylella fastidiosa*: Its biology, diagnosis, control and risks. *J. Plant Pathol*. 92: S1.35–48.

Johnson, K. A., Bock, C. H., Brannen, P. M. and Chen. J. (2022). A genome resource for *Xylella fastidiosa* subsp. *multiplex* strain P5A2 causing phony peach disease in the Southeastern United States. *Phytopathology* 112: 2466–70.

Kandel, P. P. (2015). Genetic recombination of *Xylella fastidiosa* cultured in microfluidic chambers with grapevine sap. *Phytopathology* 105: S2.6.

Krell, R. K., Perring, T. M., Farrar, C. A., Park, Y.-L. and Gispert, C. (2006). Intraplant sampling of grapevines for pierce's disease diagnosis. *Plant Dis*. 90: 351–57.

Li, R., Russell, P., Zhang, S., Davenport, B. and Eads, A. (2017). Rapid isothermal detection of *Xylella fastidiosa*

through recombinase polymerase amplification (Abstr.). *Phytopathology* 107: S4.12.
Lieth, J. H., Meyer, M. M., Yeo, K.-H. and Kirkpatrick, B. C. (2011). Modelling cold curing of Pierce's disease in *Vitis vinifera* "Pinot Noir" and "Cabernet Sauvignon" grapevines in California. *Phytopathology* 101: 1492–1500.
Merfa, M. V. and De La Fuente, L. (2019). Role of pil genes in movement and natural competence of the plant pathogen *Xylella fastidiosa* (Abstr.). *Phytopathology* 109: S2.98.
Meyer, M. M. and Kirkpatrick, B. C. (2011). Exogenous applications of abscisic acid increase curing of Pierce's disease-affected grapevines growing in pots. *Plant Dis.* 95: 173–77.
Moralejo, E., Quetglas, B., Montesinos, M., Adrover, F., Olmo, D. *et al.* (2024). Outbreak of *Xylella fastidiosa* subsp. *pauca* ST53 affecting wild and cultivated olive trees on the island of Mallorca, Spain. *Phytopath. Mediterr.* 63: 475–80.
Müller, C., Esteves, M. B., Kleina, H. T., de Melo Sales, T., Liva, K. B. *et al.* (2022). Weeds as alternative hosts of *Xylella fastidiosa* in Brazilian plum orchards. *J. Plant Pathol.* 104: 487–93.
Page, A. J., Cummins, C. A., Hunt, M., Wong, V. K., Reuter, S. *et al.* (2015). Roary: Rapid large-scale prokaryote pan genome analysis. *Bioinformatics* 31: 3691–93. https://doi.org/10.1093/bioinformatics/btv421
Purcell, A. H. (2006). *Xylella fastidiosa*. A scientific and community internet resource on plant diseases caused by the bacterium *Xylella fastidiosa*. http://nature.berkeley.edu/xylella/central.hosts.htm.
Rapicavoli, J., Ingel, B., Blanco-Ulate, B., Cantu, D. and Roper, C. (2018). *Xylella fastidiosa*: an examination of a re-emerging plant pathogen. *Mol. Plant Pathol.* 19: 786–800.
Roper, M. C., Greve, L. C., Warren, J. G., Labavitch, J. M. and Kirkpatrick, B. C. (2007). *Xylella fastidiosa* requires polygalacturonase for colonization and pathogenicity in *Vitis vinifera* grapevines. *Mol. Plant-Microbe Interact.* 20: 411–19.
Schaad, N. W., Opgenorth, D. and Gaush, P. (2002). Real-time polymerase chain reaction for one-hr on-site diagnosis of Pierce's disease of grape in early season asymptomatic vines. *Phytopathology* 92: 721–28.
Schaad, N. W., Postnikova, E., Lacy, G., Fatmi, M. and Chang, C.-J. (2004). *Xylella fastidiosa* subspecies: *X. fastidiosa* subsp. *fastidiosa*, subsp. nov., *X. fastidiosa* subsp. *multiplex* subsp. nov. and *X. fastidiosa* subsp. *pauca* subsp. nov. *Syst. Appl. Microbiol.* 27: 290–300.
Seemann, T. (2014). Prokka: Rapid prokaryotic genome annotation. *Bioinformatics* 30: 2068–69. https://doi.org/10.1093/bioinformatics/btu153
Shi, X., Bi, J., Morse, J. G., Toscano, N. C. and Cooksey, D. A. (2013). Effect of xylem fluid from susceptible and resistant grapevines on developmental biology of *Xylella fastidiosa*. *Eur. J. Plant Pathol.* 135: 127–35.
Shi, X. and Lin, H. (2018). The chemotaxis regulator *pilG* of *Xylella fastidiosa* is required for virulence in *Vitis vinifera* grapevines. *Eur. J. Plant Pathol.* 150: 351–62.
Simpson, A. J. G., Reinach, F. C., Arruda, P., Abreu, F. A., Acencio, M. *et al.* (2000). The genome sequence of the plant pathogen, *Xylella fastidiosa*. *Nature* 406: 151–57.
Stenger, D. C., Burbank, L. P., Krugner, R. and Sisterson, M. S. (2019). Individual field-collected glassy-winged sharpshooter vectors harbour sequences from two *Xylella fastidiosa* subspecies. *Eur. J. Plant Pathol.* 155: 329–38.
Su, C.-C., Chang, C. J., Chang, C.-M., Shih, H.-T., Tzeng, K.-C. *et al.* (2013). Pierce's disease of grapevines in Taiwan: Isolation, cultivation and pathogenicity of *Xylella fastidiosa*. *J. Phytopathol.* 161: 389–96.
Tubajika K. M., Civerolo, E. L., Puterka, G. J., Hashim, J. M., Luvisi, D. A. *et al.* (2007). The effects of kaolin, harpin and imidacloprid on development of Pierce's disease in grape. *Crop Prot.* 26: 92–99.
Van Sluys, M. A., de Oliveira, M. C., Monteiro-Vitorello, C. B., Miyaki, C. Y. and Furlan, L. R. (2003). Comparative analyses of the complete genome sequences of pierce's disease and citrus variegated chlorosis strains of *Xylella fastidiosa*. *J. Bacteriol.* 185: 1018–26.
Varela, L. G. Pierce's disease in the North coast. University of California. Cooperative Extension and Statewide IPM Project. http://www.cnr.berkeley.edu/xylella/pd97.html.
Velasco-Amo, M. P., Arias-Giraldo, L. F., Román-Écija, M., De La Fuente, L., Marco-Noales, E. *et al.* (2023). Complete circularized genome resources of seven strains of *Xylella fastidiosa* subsp. *fastidiosa* Using hybrid assembly reveals unknown plasmids. *Phytopathology* 113: 1128–32.
Wallis, C. M. and Chen, J. (2012). Grapevine phenolic compounds in xylem sap and tissues are significantly altered during infection by *Xylella fastidiosa*. *Phytopathology* 102: 816–26.
Wells, J. M., Raju, B. C., Hung, H. Y., Weisburg, W. G., Mandelco-Paul, L. *et al.* (1987). *Xylella fastidiosa* gen. nov., sp. nov: Gram-negative, xylem-limited, fastidious plant bacteria related to *Xanthomonas* spp. *Int. J. Syst. Bacteriol.* 37: 136–43.
Wells, J. M., Raju, B. C., Nyland, G. and Lowe, S. K. (1981). Medium for isolation and growth of bacteria associated with plum leaf scald and phony peach diseases. *Appl. Environ. Microbiol.* 42: 357–63.
Wistrom, C. and Purcell, A. H. (2005). The fate of *Xylella fastidiosa* in vineyard weeds and other alternate hosts in California. *Plant Dis.* 89: 994–99.

9.54 BACTERIAL CANKER OF GRAPEVINE

Grapevine (*Vitis vinifera* L.) suffers from several bacterial diseases, namely, bacterial blight (*Xylophilus ampelinus*), bacterial necrosis (*Pseudomonas syringae* pv. *syringae*), crown gall (*Agrobacterium vitis*), Pierce's disease (*Xylella fastidiosa* subsp. *fastidiosa*) and bacterial canker (*Xanthomonas campestris* pv. *viticola*). Pierce's disease and crown gall are described elsewhere in the book and bacterial canker is described here. It is also known by another name, i.e., **bacterial leaf spot of grapevine**.

The disease was first observed by Nayudu (1972) from Andhra Pradesh, India. Subsequently, the disease has been reported from other states of India, namely, Punjab, Maharashtra and Karnataka, and it also occurs in other grape-growing areas of the country. In Punjab, the disease is severe on commercial grown cultivars like Anabe-e-shahi, Perlette and Thompson seedless. In certain areas of Punjab, the disease incidence as high as 93% and disease intensity up to 29% have been reported. Chand and Kishun (1990) reported an outbreak of the disease from India and concluded that primarily the more extensive cultivation of the susceptible seedless varieties was the cause of the outbreak.

The disease also occurs in Brazil, where it was first observed in 1998 in vineyards in the Sumedio of the Sao Francisco river basin area of Pernambuco state. The disease incidence was as high as 100% and, in some cases, the yield losses were nearly total (Lima *et al*., 1999). The disease also occurs in other states of Brazil, namely, Piaui, Bahia, Ceara and Roraima. In Brazil, it is a quarantined disease, officially restricted to these states, including Pernambuco (Halfeld-Vieira and Nechet, 2006).

9.54.1 Host Range

Chand and Kishun (1990) tested 16 plant species and found that only mango was infected. Although mango does not show infection under natural conditions, its infection under artificial inoculated conditions indicates that it may serve as an additional source of inoculum of the bacterium. Neem (*Azadirachta indica* A. Juss.), reported as a colateral host of the bacterium by Nayudu (1972), also did not show any infection.

9.54.2 Symptoms

The symptoms are found on all the above-ground plant parts. The symptoms first appear on young leaves as small, water-soaked lesions on the lower surface surrounded by yellow haloes. The lesions enlarge and become angular, cankerous and dark brown. Lesions then also appear on the upper leaf surface. Sometimes, the lesions coalesce, causing the death of extensive leaf areas. Spots are rough to touch due to drying of gummy bacterial ooze. Cankers on the veins are also common. Vein infection may range from 1 mm long to entire length of the vein. On petioles, the lesions are brown, elongated and cankerous, and measure 1–5 mm in length (Figure 9.44).

On stems, brown to black, elongated and raised cankers appear. In the advanced stage, cankers split longitudinally in the centre, exposing the vascular discolouration. Severe infection on young twigs and branches results in the drying and killing of twigs. On the berries, brown to black, irregular to circular, raised and skin-deep cankers appear. The cankers are formed mainly on the inner berries of the bunch. The symptoms on berries appear, when these are nearing maturity.

FIGURE 9.44 Bacterial canker of grapevine: Leaf symptoms.

In Punjab, India, the disease appears in the third or fourth week of March and the progress of the disease continues on the new plant growth up to the middle of May. From mid-May to mid-July, fresh infections do not occur on young plant parts. With the onset of rains, the fresh infections start occurring from mid-July and the progress of the disease continues up to September.

9.54.3 Causal Organism

Xanthomonas campestris pv. *viticola* (Nayudu 1972) Dye 1978.

Synonym: *Pseudomonas viticola*

The colonies are white, convex, and mucoid and no fluorescent pigment is produced in culture media. Cells are Gram-stain-negative rods, non-sporing and grow aerobically. The bacterium hydrolyses esculin and starch, and produces acid from glucose, cellobiose, trehalose, melibiose and arabinose; it is weakly positive for glycerol. Acid is not produced from dulcitol and the tests for oxidase and urease are also negative. Asparagine is not utilised as a sole source of carbon and nitrogen. The bacterium produces hypersensitive reaction on tomato and tobacco in 48 hours.

On the basis of combined analysis of PCR patterns obtained with primers REP, ERIC and BOX, Trindade *et al*. (2005) found a high degree of similarity between Brazilian strains and Indian-type strain NCPPB 2475. Trindade *et al*. (2007) designed primers Xcv1F/Xcv3R and RST2/Xcv3R based on the partial sequence of *hrpB* gene for the detection and identification of *Xan. campestris* pv. *viticola* (*Xcv*). The detection limit of primer RST2/Xcv3R was 10^4 cfu mL^{-1} and the sensitivity of Xcv1F/Xcv3R and RST2/Xcv3R was 10 pg and 1 pg of purified bacterial DNA, respectively. Presence of the bacterium in tissues of grapevine petioles, previously inoculated with it, could not be detected by PCR using macerated extract added directly in the reaction. However, the amplification was positive with the introduction of an agar-plating step prior to PCR.

Keeping in view the doubtful taxonomic position of *Xan. campestris* pv. *viticola* and pigmented isolates pathogenic to cashew plant, da Gama *et al*. (2018) used the multi-locus sequence analysis (MLSA) technique, average nucleotide identity values and tetranucleotide frequency correlation coefficients to analyse their phylogenetic relationship in relation to other *Xanthomonas* species. On the basis of results obtained, the authors suggested the taxonomic repositioning of *Xan. campestris* pv. *viticola* (Nayudu 1972) Dye 1978 as *Xan. citri* pv. *viticola* (Nayudu 1972) Dye 1978 comb. nov. and emendation of the description of *Xan. citri* pv. *anacardii* to include pigmented isolates pathogenic to cashew plant.

Kamble *et al*. (2017) evaluated *in vitro* 11 chemicals along with three biocontrol agents against *Xan. campestris* pv. *viticola*. Among the chemicals, Streptocycline, mancozeb and bronopol showed significant inhibition of the pathogen, while kasugamycin, copper oxychloride and copper hydroxide exhibited only low efficacy. Copper sulphate, validamycin, difenoconazole, carbendazim and potassium phosphate showed no inhibitory effect at 50, 100, 500, 1,000, 2,000 and 3,000 μg mL^{-1}. Among the biocontrol agents, *Bacillus subtilis* and *Trichoderma asperolloides* showed the inhibitory effect against the bacterium.

Da Silva *et al.* (2020) evaluated the effects of crude aqueous extracts of 12 plant species, including Caatinga plants, on planktonic growth and biofilm formation of *Xan. citri* pv. *viticola* (*Xcv*). Crude aqueous extracts of Caatinga plants, collected from Catimbau Valley, Brazil, were evaluated *in vitro* against four strains of *Xcv* to determine the antibacterial and antibiofilm (formation) activities of these plants. Only the extracts obtained from leaves of Caatinga plants (*Homalolepis ferruginea*) showed antibiofilm and antibacterial activity against all the tested strains. The extract reduced biofilm formation and planktonic growth by up to 54.6% and 60.2%, respectively, thus making *H. ferruginea* a source of promising biomolecules for the management of grapevine bacterial canker.

Villela *et al.* (2019) tested symptomatic and asymptomatic grapevines by qPCR (real-time PCR) with *hrpB* primers and cPCR (conventional PCR) with xanthomonadin primers under natural infection conditions. In both cases, plant extracts were enriched for 36–72 h. *Xanthomonas citri* pv. *viticola* (*Xcv*) was detected in all symptomatic samples by qPCR and the result was confirmed by cPCR. For the asymptomatic samples, *Xcv* was detected in 93.4% with qPCR and in 89.5% with cPCR. These two methods offer advantages in terms of sensitivity and specificity, and these could be useful in quarantine programs, the certification of grapevine-propagating material and the detection of inoculum sources in alternative hosts, contributing to the prevention of pathogen spread to disease-free areas.

9.54.4 Disease Cycle

The bacterium survives mainly in the stem cankers. The survival also occurs in the infected crop residue. The local dissemination of the bacterium occurs through wind-splashed rain, harvesting and pruning tools. The long-distance dissemination of the pathogen occurs through infected propagating material.

Naue *et al.* (2014) evaluated the efficacy of seven sanitising agents to disinfect grapevine cutting tools and water contaminated with *Xan. campestris* pv. *viticola* 2Rif. Sodium hypochlorite (20,000 μg mL^{-1}) and dodecyl dimethyl ammonium chloride (1,140 μg mL^{-1}) produced the largest inhibition zones against four isolates of the bacterium and were 100% effective in disinfecting contaminated harvest shears. These two sanitising agents maintained their efficacy for 8 h.

The bacterium survived for 24 h on harvest shears and was spread by contaminated harvest shears until the 24th cut, on average. Water disinfection was successfully achieved with dodecyl dimethyl ammonium chloride (570 μg mL^{-1}), sodium hypochlorite (5,000 μg mL^{-1}) and benzalkonium chloride (122.5 μg mL^{-1}).

9.54.5 Management

1. Use disease-free propagating material. Hot water treatment of cuttings at 52 °C for 30 min reduces the infection.
2. Practise rouging and drastic pruning of the infected vines. The time of pruning also affects disease management. Chand *et al.* (1991) found that disease intensity was 76.8% at three locations in Karnataka, India when vines were pruned between 1 and 15 September while it was 5.6% in vineyards pruned after 10 October. Pruning the vines from the second week of October and onwards gave the best results in the grape-producing areas of Maharashtra and North Karnataka.
3. Disinfect harvesting and pruning tools.
4. Use windbreaks to reduce pathogen dissemination.
5. Give protective sprays of copper and thiocarbamate fungicides (Nascimento and Mariano, 2004). Under Punjab (India) conditions, 4 sprays at 15 days intervals starting from 15 July gave an effective control of the disease. Bordeaux mixture (0.8%) proved most effective, followed by Streptocycline (200 μg mL^{-1}).

Shivananda *et al.* (2011) reported that the application of Streptocycline (500 μg mL^{-1}) + copper oxychloride (2,000 μg mL^{-1}) thrice at a 20-day interval was most effective, as it gave a minimum percent disease index (PDI) (29.86%), a maximum yield (26.95 t ha^{-1}), more number of bunches produced per plant (20.03), the lowest infected bunches per plant (3.40) and the maximum single bunch weight (925 g). This was followed by Streptocycline (500 μg mL^{-1}) alone, which gave PDI (35.35%), yield (23.50 t ha^{-1}), bunches produced per plant (17.13), number of infected bunches per plant (5.05) and single bunch weight (862 g). Among the biocontrol agents, *Bacillus subtilis* at the rate of 5,000 μg mL^{-1} proved most effective as it gave PDI of 45.48%, yield 14.23 t ha^{-1}, 9.35 bunches produced per plant, 7.10 infected bunches per plant and single bunch weight of 413 g. This was followed by *Pseudomonas fluorescens* at 5,000 μg mL^{-1}, which gave PDI of 47.09%, yield of 11.35 t ha^{-1}, 7.43 bunches produced per plant, 7.08 infected bunches per plant and single bunch weight of 318 g.

REFERENCES

Chand, R. and Kishun, R. (1990). Outbreak of grapevine bacterial canker disease in India. *Vitis* 29: 183–88.

Chand, R., Patil, B. P. and Kishun, R. (1991). Management of bacterial canker disease (*Xanthomonas campestris* pv. *viticola*) of grapevine (*Vitis vinifera*) by pruning. *Indian J. Agri. Sci.* 61: 220–22.

da Gama, M. A. S., Mariano, R., de Lima R., da Silva Jr. W. J., de Farias, A. R. G. *et al.* 2018. Taxonomic repositioning of *Xanthomonas campestris* pv. *viticola* (Nayudu 1972) Dye 1978 as *Xanthomonas citri* pv. *viticola* (Nayudu 1972) Dye 1978 comb. nov. and emendation of the description of *Xanthomonas citri* pv. *anacardii* to include pigmented isolates pathogenic to cashew plant. *Phytopathology* 108: 1143–53.

da Silva, P. H. R., da Conceição, C. S., Leal, C. M., da Silva, W. J., da Gama, M. A. S. (2020). Caatinga biome plant extracts affect the planktonic growth and biofilm formation of *Xanthomonas citri* pv. *viticola*. *J. Plant Pathol.* 102: 1245–50. https://doi.org/10.1007/s42161-020-00585-4

Dye, D. W. (1978). Genus IX. *Xanthomonas* Dowson 1939. In Young, J. M., Dye, D. W., Bradbury, J. F., Panagopoulos, C. G. and Robbs, C. F. (1978). A proposed nomenclature and classification for plant pathogenic bacteria. *N. Z. J. Agric. Res.* 21: 162–66.

Halfeld-Vieira, B. A. and Nechet, K. de L. (2006). Bacterial canker of grapevine in Roraima, Brazil. *Fitopatologia Brasileira* 31: 604.

Kamble, A. K., Sawant, S. D., Saha, S. and Sawant, I. S. (2017). *In vitro* efficacy of different chemicals and biological agents against *Xanthomonas campestris* pv. *viticola* causing bacterial leaf spot of grapes. *Int. J. Agric. Sci.* 9: 4427–30.

Lima, M. F., Ferreira, M. A. S. V., Moreira, W. A. and Dianese, J. C. (1999). Bacterial canker of grapevine in Brazil. *Fitopatologia Brasileira* 24: 440–43.

Nascimento, A. R. P. and Mariano, R. de L. R. (2004). Bacterial canker of grape-vine: Etiology, epidemiology and control strategies. *Ciencia Rural* 34: 301–07.

Naue, C. R., Costa, V. S. O., Barbosa, M. A. G., Batista, D. C., Souza, E. B. *et al.* (2014). *Xanthomonas campestris* pv. *viticola* on grapevine cutting tools and water: Survival and disinfection. *J. Plant Pathol.* 96: 451–58.

Nayudu, M. V. (1972). *Pseudomonas viticola* sp. nov., incitant of a new bacterial disease of grapevine. *Phytopathol. Z.* 73: 183.

Shivananda, J., Ravikumar, M. R. and Hiremani, N. (2011). Evaluation of different chemicals and bioagents against bacterial leaf spot of grapevine and their effect on yield and yield parameters. *Int. J. Plant Prot.* 4: 377–80.

Trindade, L. C. da, Marques, E., Lopes, D. B. and Ferreira, M. A. da S. V. (2007). Development of a molecular method for detection and identification of *Xanthomonas campestris* pv. *viticola*. *Summa Phytopathol.* 33: 16–23.

Trindade, L. C., Lima, M. F. and Ferreira, M. A. S. V. (2005). Molecular characterization of Brazilian strains of *Xanthomonas campestris* pv. *viticola* by rep-PCR fingerprinting. *Fitopatologia Brasileira* 30: 46–54.

Villela, J. G. A., Ritschel, P., Barbosa, M. A. G., Baccin, K. M. S., Rossato, M. *et al.* (2019). Detection of *Xanthomonas citri* pv. *viticola* on grapevine by real-time PCR and BIO-PCR using primers designed from pathogenicity and xanthomonadin gene sequences. *Eur. J. Plant Pathol.* 155: 445–59. https://doi.org/10.1007/s10658-019-01779-y

9.55 PAPAYA BUNCHY TOP DISEASE

The papaya plant (*Carica papaya* L.), also called papaw or pawpaw, is one of the 22 accepted species in the genus *Carica* of the family Caricaceae. It was first domesticated in Mesoamerica (modern-day southern Mexico and central America). Papaya is native to the American tropics, but is now cultivated widely throughout the tropical and subtropical areas of the world for both local consumption and export of fruit and, to a lesser extent, for the production of papain. Papaya fruit has become the fourth most traded tropical fruit in the world after banana, mango and pineapple. In 2020, India was responsible for 43% of the world's papaya fruit production.

Papaya bunchy top (abbreviated as PBT) disease is one of the most economically important diseases of the fruit in the American tropics (Davis, 1994). PBT was first reported from Puerto Rico in 1931 (Cook, 1931). Papaya diseases of unknown etiology, but similar in symptomology to PBT, were also reported at about the same type from Jamaica (Smith, 1929) and the Dominican Republic (Ciferri, 1930).

Due to the inability to isolate the pathogen of PBT disease, Koch's postulates could not be fulfilled. Therefore, the earlier scientists believed that papaya bunchy top disease was caused by a virus or a mycoplasma-like organism (MLO) or a phytoplasma. As the identity of the pathogen was unknown, the identification of all the possible causal agents was surmised solely based on commonly associated symptoms.

Wei *et al.* (2017) reported PBT disease due to '*Candidatus* Phytoplasma brasiliense' for the first time from northern Peru. In 2018, a new papaya disease occurred in Ibadan, Nigeria. The affected papaya trees exhibited symptoms including excessive shoot proliferation at the top of the crown, leaf mosaic and crinkling and die-back. This disease was designated as Nigerian papaya bunchy top (abbreviated as NGPBT) disease. Phytoplasmas were detected in diseased papaya samples by semi-nested PCR using phytoplasma 16 S rRNA gene-specific primers. The subsequent determination and analysis of the amplified 16 S rRNA gene sequences and phylogenetic analysis revealed that NGPBT phytoplasmas were most closely related to '*Can.* Phy. convolvuli' in the Stolbur (16SrXII) group. Virtual restriction fragment length polymorphism (RFLP) analysis, based on the 16 S rRNA gene F2nR2 fragment, showed that NGPBT strains represented a novel subgroup in the Stolbur phytoplasma group, designated as 16SrXII-O. This is the first report of a phytoplasma disease occurring in papaya in Nigeria. Identification of the NGPBT strains not only broadened the genetic diversity of the Stolbur phytoplasma group but also signalled an expansion of the phytoplasma group's geographic distribution. Notably, two previously reported phytoplasmas associated with papaya diseases in Australia and Taiwan also belong to the 16SrXII group and exhibit symptoms similar to NGPBT disease (Kazeem *et al.*, 2021). PBT disease is found in tropical areas within the Americas.

The disease named bunchy top symptom was first observed from Villa Clara province, Cuba in 2003, but it is now widespread throughout the country (Arocha *et al.*, 2007). The presence of phytoplasma causing this disease, in weed species and its putative vector *Empoasca papayae*, suggests that these plants and the leafhopper may play a role in the spread of this disease.

9.55.1 Host Range

Papaya is the only known host of '*Can.* Phy. caricae' (Davis, 1993).

9.55.2 Symptoms

Typical symptoms appear around 30 to 45 days after infection. Early symptoms are chlorosis and stunting of young leaves, accompanied by internode and petiole elongation. As the disease progresses, petioles and leaves become rigid and thickened. Chlorosis spreads to older leaves followed by necrosis, and water-soaked spots appear on the petioles and stems. During advanced stages of the disease development, infected plants become denuded, with the exception of a few stunted leaves that remain at the apex. Die-back can also occur in some varieties of papaya when infected (Figure 9.45).

The disease development rate and severity of symptoms vary greatly with host genotype and region. In diseased plants, flowers and fruits are rarely produced. If infected plants produce any fruit, this tends to be bitter and hence unmarketable. A lack of latex flow from the leaves, stems, petioles and fruits, which used to be regarded as a very important symptom for disease diagnosis, is now found to be unreliable (Webb and Davis, 1987). Hence, it is necessary to look for other important symptoms before making a final diagnosis.

FIGURE 9.45 Papaya bunchi top disease: Yellowing, wilting and drying of leaves.

9.55.3 Causal Organism

'*Candidatus* Phytoplasma caricae' Arocha *et al.* 2005

It is a Gram-stain-negative, rod shaped and laticifer-inhibiting bacterium. Cells are 0.8 to 1.6 μm long and 0.25 to 0.35 μm wide.

The reference strain PD1 of '*Can.* Phytoplasma pyri' (GenBank accession number AJ542543) was described along with '*Can.* Phytoplasma mali' and '*Can.* Phytoplasma prunorum' and sharing a 16S rRNA sequence identity >97.50% compared with both. A total of 34 strains having >95% of sequence available showing 99.14% to 100% of sequence identity, compared to the reference strain. Among strains classified within '*Can.* Phytoplasma mali' and '*Can.* Phytoplasma prunorum', 64 show 98.68% to 99.27% and seven 97.50% to 98.64% of sequence identity in comparison with the reference strain. The closely related '*Can.* Phytoplasma pyri' PD1 and '*Can.* Phytoplasma mali' AT strains can be separated by sequences of additional genetic markers as mentioned for '*Can.* Phytoplasma mali'. Differentiation is also possible based on differential insect vector transmission (Seemüller and Schneider, 2004).

Davis *et al.* (1996) found rod-shaped bacteria measuring 0.25–0.35 μm in width and 0.8 to 1.6 μm in length associated with the disease. Their cell wall ultrastructure resembled that of Gram-stain-negative bacteria, except that a peptidoglycan layer was not evident in the periplasmic space. The bacteria appeared to colonise laticifers.

The bacteria are found within the sap of the plant. The bacteria infect the plant cells, destroy the cytoplasm along with its organelles, causing the eventual collapse of the cells (Davis, 1993).

9.55.4 Disease Cycle

PBT phytoplasma survives in infected papaya plants. Two leafhopper species, namely, *Empoasca papayae* Oman (Sein) (Adsuar, 1946) and *E. stevensi* Young (Haque and Parasram, 1973), have been identified as vectors of PBT phytoplasma. Agrios (1969) also reported vectoring of PBT phytoplasma by two species of leafhopper, namely, *E. papayae* Oman (Sein) and *E. stevensi* Young.

The insect species that vector the pathogen vary with the geographical location (Davis *et al.*, 1996). *E. papayae* is more commonly associated with vectoring the pathogen because of its widespread distribution throughout the American tropics. *E. stevensi* has a smaller range of distribution and is mostly attributed to cases in Trinidad (Ploetz, 1994).

9.55.5 Management

The following measures can be helpful to reduce the disease incidence or to control it.

1. Papaya cultivars tolerant to the PBT pathogen have been found and bred to check the spread of the disease. This is the most practical method of disease management, but the use of these cultivars is limited to areas having a low disease pressure.
2. Use of clean propagation material and quick removal of diseased trees are also necessary to prevent the spread of the pathogen (Agrios, 1969).
3. Use of insecticides has also proved effective in limiting the population of leafhopper vectors (Davis, 1993).
4. Soil application of chlortetracycline or tetracycline hydrochloride, around infected plants, has shown some success in managing the disease, but has not yet become commercially viable.

REFERENCES

Adsuar, J. (1946). Transmission of papaya bunchy top by leafhoppers of genus *Empoasca*. *Science* 103: 316.

Agrios, G. N. (1969). *Plant Pathology*. Elsevier Academic Press, New York. ISBN 978-0120445653.

Arocha, Y., Lopez, M., Pinol, B., Fernandez, M., Picornell, B. *et al.* (2005). '*Candidatus* Phytoplasma graminis' and '*Candidatus* Phytoplasma caricae', two novel phytoplasmas associated with diseases of sugarcane, weeds and papaya in Cuba. *Int. J. Syst. Evol. Microbiol.* 55: 2451–63.

Arocha, Y., Piñol, B., López, M., Miranda, I., Almeida, R. *et al.* (2007). 'Bunchy top symptom' of papaya in Cuba: new insights. *Bull. Insectology* 60 (2): 393–94.

Ciferri, R. (1930). Phytopathological survey of Santo Domingo 1925–1929. *J. Agric.* Univ. P. R. 14: 5–44.

Cook, M. T. (1931). New virus diseases of plants in Puerto Rico. *J. Agric.* Univ. P. R. 15: 193–95.

Davis, M. J. (1993). Papaya Bunchy Top, MLO. Crop Knowledge Master.

Davis, M. J. (1994). Bunchy top. In: R. C. Ploetz, G. A. Zentmeyer, W. T. Nishijima, K. G. Rohrback and H. D. Ohr (eds.). *Compendium of Tropical Fruit Diseases*. American Phytopathological Society, St. Paul, MN, pp. 69–70.

Davis, M. J., Kramer, J. B., Ferwerda, F. H. and Brunner, B. R. (1996). Association of a bacterium and not a phytoplasma with papaya bunchy top disease. *Phytopathology* 86: 102–09.

Haque, S. Q. and Parasram, S. (1973). *Empoaska stevensi*, a new vector of bunchy top disease of papaya. *Plant Dis. Rep.* 57: 412–13.

Kazeem, S. A., Inaba, J., Zhao, Y., Zwolinska, A., Ogunfunmilayo, A. O. *et al.* (2021). Molecular identification and characterization of '*Candidatus* Phytoplasma convolvuli'-related strains (representing a new 16SrXII-O subgroup) associated with papaya bunchy top disease in Nigeria. *Crop Prot.* 148, October 2021, 105731

Ploetz, R. C. (1994). In: R. C. Ploetz, G. A. Zentmyer, W. T. Nishijima, K. G. Rohrbach, and H. D. Ohr (eds.). *Compendium of Tropical Fruit Diseases*. American Phytophathological Society, St. Paul, MN.

Seemüller, E. and Schneider, B. (2004). '*Candidatus* Phytoplasma mali', '*Candidatus* Phytoplasma pyri' and '*Candidatus* Phytoplasma prunorum', the causal agents of apple proliferation, pear decline and European stone fruit yellows, respectively. *Int. J. Syst. Evol. Microbiol.* 54: 1217–26.

Smith, F. E. V. (1929). Plant diseases in Jamaica in 1928. Page 19 in: Annual Report of the Department of Science and Agriculture of Jamaica for 1928. Government Printing Office, Kingston, Jamaica.

Webb, R. R. and Davis, M. J. (1987). Unreliability of latex-flow test for diagnosis of bunchy-top of papaya caused by a mycoplasma-like organism. *Plant Dis.* 71: 192.

Wei, W., Perez-Lopez, E., Davis, R. E., Bermudez-Diaz, L., Granda-Wong, C. *et al.* (2017). '*Candidatus* Phytoplasma brasiliense'-related strains associated with papaya bunchy top disease in northern Peru represent a distinct geographic lineage. *Crop Prot.* 92: 99–106.

9.56 BACTERIAL CANKER OF KIWI

Bacterial canker of kiwi (staminate kiwi, *Actinidia deliciosa*) was first reported from California, USA in 1980. Later on, in 1983, the causal bacterium was identified as *Pseudomonas syringae*. Serizawa *et al.* (1989) reported the disease on Chinese gooseberry (*A. chinensis*) from Japan in the early 1980s, where its severity is one of the limiting factors for kiwifruit production. According to Goto (1992), the disease found in Japan is more destructive and such a form of the disease has not been found in other areas of the world. In the EPPO region, the disease was first noticed in northern Italy in 1992, where it remained sporadic with a low incidence for 15 years. However, in 2007/2008, economic losses started occurring particularly in the Lazio region, and the spread of the disease to other kiwifruit-producing regions in Italy began to raise concerns. Due to the emerging problem of *Psm. syringae* pv. *actinidiae* (hereafter referred to as *Psa*) in the Mediterranean region, the EPPO Secretariat decided to add it to the EPPO Alert List. Ferrante and Scortichini (2010) reported a severe epiphytotic of the disease during 2008–09 from central Italy on *A. chinensis* (yellow kiwifruit). In Italy, it is estimated that the economic losses, including the impact on trade, resulting from *Psm. syringae* pv. *actinidiae* have reached two million euros (Balestra *et al.*, 2009). The disease has also been reported from China, New Zealand and Korea. The disease also poses a serious problem in Korea, where it was first noticed in 1988. The disease has also been reported from Australia, Spain, and Chile.

Kim *et al.* (2017), in their review article, reported the appearance of biovar *Psa*3 in Korea in 2011 (the biovar *Psa*2 was present since 1988) that caused tremendous economic losses by destroying many vines or orchards of yellow-fleshed kiwifruit cultivars in one or several growing seasons. Bacterial canker epiphytotics caused by both *Psa*2 and *Psa*3 biovars were prevalent in Korea in recent years. In this review, the authors also summarised the symptomatology, etiology, disease cycle, diagnosis and epidemiology of kiwifruit bacterial canker in Korea.

Koh *et al.* (2012) reported *Pectobacterium carotovorum* subsp. *actinidiae* as a new bacterial pathogen causing kiwifruit canker-like symptoms in Korea. Wu et al. (2017) reported this bacterium causing kiwifruit bacterial canker-like symptoms in China, this being the first report of this disease from China and also outside of Korea.

Lu *et al.* (2021) identified the key determinants that control the virulence of *Pec. actinidiae*, the causal agent of summer canker of kiwi, in kiwifruit by genomic and functional analyses. The analysis of four *Pec. actinidiae* isolates indicated low genetic variability, with an average of 98.7% genome-level sequence similarity and 82% shared protein-coding gene content. Phylogenetic analysis, based on both bulk single nucleotide polymorphisms (SNPs) and single-copy genes, revealed that *Pec. actinidiae* strains cluster into a single clade, which is closely related to the clades of *Pec. odoriferum* (species with a completely different host range). Through comparison between these two clades of strains, 746 unique core orthologs/genes were clustered in the clades of *Pec. actinidiae*, especially key virulence determinants involved in the biosynthesis of secretion systems (type III, IV and VI), iron, flagellar structure and the quorum-sensing system. These findings provide insights into the pathogenomics underlying the genetic diversification and evolution of pathogenicity in *Pec. actinidiae* species.

The review by Santos *et al.* (2023) provides an in-depth and critical analysis on the scientific progress made for developing alternative tools for sustainable kiwifruit bacterial canker (KBC) management. Their status in terms of technological maturity is discussed and a set of opportunities and threats are also presented. The gradual replacement of susceptible kiwifruit cultivars with more tolerant ones significantly reduced KBC incidence and was a major milestone for *Pseudomonas syringae* pv. *actinidiae* (*Psa*) containment, which highlights the importance of plant breeding. Nonetheless, this is a very laborious process. Moreover, the potential threat of *Psa* evolving to more virulent biovars, or resistant lineages to existing control methods, strengthens the need to keep on exploring effective and more environmentally friendly tools for KBC management. Currently, plant elicitors and beneficial fungi and bacteria are already being used in the field with some degree of success. Precision agriculture technologies, for improving early disease detection and preventing pathogen dispersal, are also being developed and optimised. These include hyperspectral technologies and forecast models for *Psa* risk assessment, with the latter being slightly more advanced in terms of technological maturity. Additionally, plant protection products based on innovative formulations with molecules with antibacterial activity against *Psa* (*e.g.*, essential oils, phages and antimicrobial peptides) have been validated primarily in laboratory trials and with few compounds already reaching field application. The lessons learned with this pandemic disease, and the acquired scientific and technological knowledge, can be of importance in sustainably managing other plant diseases and handling future pandemic outbreaks.

Since 2012, a new pathogenic syndrome has frequently been observed in many areas of kiwifruit cultivation in Italy. The main symptoms include an initial withering of the leaves followed by a total and sudden collapse of plants, mainly occurring during summer. The withered leaves fall and the main and secondary feeder roots appear rotten, sometimes showing a reddish-brown discolouration. The disease, which affects both the green and the yellow-fleshed cultivars, has been called **kiwifruit vine decline** and is locally known as ***moria***. The syndrome has been found consistently associated with the water-logging of soil, which frequently occurs, either after the traditional agronomical practice of irrigating orchards through surface irrigation or following very heavy rainfall. The role played by bacteria in this syndrome has not been investigated to date. Spigaglia *et al.* (2020) isolated *Clostridium* spp. from both rotten roots and soils obtained from Italian kiwifruit orchards affected by this syndrome and identified these species as *Clostridium bifermentans* and *Clostridium subterminale*. Both the species incited symptoms in kiwifruit, in *in vivo* and *in vitro* pathogenicity tests, indicating for the first time that anaerobic bacteria are able to cause damage to woody crops. Soil water-logging seems to potentially favour the colonisation of kiwifruit roots by anaerobic bacteria, probably because the saturation of the soil can facilitate the proliferation and persistence of these bacteria during long periods of the vegetative growth of the crop. The occurrence of these anaerobic bacteria does not exclude the possibility that other microorganisms can play additional/synergic role(s) in causing the kiwifruit vine decline.

9.56.1 Host Range

Psm. syringae pv. *actinidiae* naturally infects only kiwi plants, both cultivated and wild. It can infect peach and Mume on artificial inoculation. *Psm. syringae* pv. *syringae* is pathogenic to cowpea and carrots in addition to kiwi.

9.56.2 Symptoms

The symptoms appear on almost all the above-ground plant parts. Symptoms appear on trunks and limbs from late winter to early spring, after the break in the dormancy of vines. The most conspicuous symptom is the red-rusty exudation that appears on buds, the joints of branches and leaves, and pruning scars. The bark of affected parts deeply shrivels and often dries, leading to the formation of cracks.

On leaves, the infection is generally limited to young expanding leaves. Initially, small, water-soaked, light green, scattered spots appear on leaf lamina. The spots increase in size (2–3 mm in diameter), turn brown to dark brown and become angular in shape due to the restrictions imposed by veins and veinlets. The spots are surrounded by bright yellow haloes. Bacterial ooze is generally formed on the lower surface of leaves. In the case of severe infection, the entire leaf may be blighted and shrivelled (Figure 9.46a).

Water-soaked, dark green, elongated lesions appear on the infected canes. As the infection advances, longitudinal cracks often appear, leading to the wilting and blightening of the whole shoot and the exudation of bacterial ooze (Figure 9.46b). The removal of the bark usually reveals a brown discolouration of the external vascular tissues and a reddening of the tissues beneath lenticels. Infected flowers turn brown and often wither before opening. Infected flowers that open generally have under-developed petals. Necrotic lesions are also formed on the sepals (Goto, 1992).

Another disease of kiwi, called **bacterial blossom blight** or **blossom rot**, has some similarities with bacterial canker. Blossom blight occurs only on young flowers and is characterised by the browning of petals and the shedding of premature flowers. Severity of the disease increases with rainfall during the flowering period. *Psm. syringae*, *Psm. viridiflava* and *Psm. syringae* pv. *actinidiae* form the causal complex, which causes blossom blight.

FIGURE 9.46A Bacterial canker of kiwi: Symptoms on leaf. (Courtesy: Riccardo Bugiani, Plant Protection Service of Emilia-Romagna region, Italy.)

FIGURE 9.46B Bacterial canker of kiwi: Stem canker. (Courtesy: Riccardo Bugiani, Plant Protection Service of Emilia-Romagna region, Italy.)

9.56.3 Causal Organisms

1. *Pseudomonas syringae* pv. *actinidiae* Takikawa *et al.* 1989
   Synonym: *Psm. syringae* pv. *avellanae*
2. *Psm. syringae* pv. *syringae* van Hall 1902
   Synonym: *Psm. syringae* pv. *japonica*
3. *Psm. syringae* pv. *actinidifoliorum* pv. nov. Cunty *et al.* 2015

9.56.3.1 Psm. syringae pv. actinidiae

This was first identified by Takikawa *et al.* (1989) as a new pathovar of *Psm. syringae*. Ushiyama *et al.* (1992) are of the view that *Psm. syringae* pv. *actinidiae* (abbreviated as *Psa*) might have originated from *Actinidia* wild species, namely, *A. arguta* and *A. kolomikta*, which are distributed in northern areas of Japan and show natural infection.

Bacterial cells are Gram-stain-negative, non-sporing rods, motile by polar flagella. The bacterium does not produce fluorescent pigment on King's medium B. Neither β-galactosidase or urease are produced, and nor is gelatin liquefied. The bacterium does not utilise DL-lactate, caprate, erythritol, xylose, L-histidine and trigonelline (Goto, 1992). Some strains of the bacterium produce phaseolotoxin, as does *Psm. syringae* pv. *phaseolicola*, the causal agent of bean (*Phaseolus vulgaris*) halo blight. However, the Korean strains of the bacterium were found to produce coronatine but not phaseolotoxin (Han *et al.*, 2003). Ferrante and Scortichini (2010) reported that none of the strains isolated from central Italy during 2008–09 epiphytotic possessed genes coding for phaseolotoxin or coronatine, but all had an effector protein, namely, *hopA1*, differentiating these from the strains causing past outbreaks in Japan and Italy. They further reported that the strains which caused the outbreak in central Italy appears to be distinct from those found previously in Japan, South Korea and Italy.

Four populations of *Psm. syringae pv. actinidifoliorum. (Psa)*, namely, *Psa*1, *Psa*2, *Psa*3 and *Psa*4, have been described and *Psa*4 is less virulent than the other three populations. Also, multi-locus sequence typing of housekeeping genes or broad *Psa* strain genome comparisons revealed that *Psa* 4 strains cluster separately from other three *Psa* populations. Furthermore, repetitive sequence PCR fingerprinting, type III secretion system effector protein genes detection and colony morphology clearly indicate the distinctiveness of *Psa* 4 strains from the other three *Psa* populations. Rep-PCR molecular typing revealed a high similarity of the *Psa* 4 strains with members of the *Psm. avellanae* species. It was concluded that only *Psa*1, *Psa*2 and *Psa*3 populations, pathogenic to *Actinidia* species should be included in *Psa*, and the *Psa*4 strains isolated in New Zealand do not belong to the pathovar *actinidiae*; most probably, they belong to a new still unnamed pathovar (Ferrante and Scortichini, 2015). Cunty *et al.* (2015) performed an MLSA based on four housekeeping genes (*gapA*, *gltA*, *gyrB* and *rpoD*) on 72 strains representative of the French outbreak that has occurred since the beginning of 2010, and found that all the strains fell into two phylogenetic groups; one clonal corresponding to biovar 3 and the other corresponding to biovar 4. Strains of biovar 4 were found to be substantially different from those of the other biovars as they were less aggressive and caused only leaf spots, whereas *Psa* biovars 1, 2 and 3 also caused canker and shoot die-back symptoms. Based on these pathogenic differences, supported by phenotypic, genetic and phylogenetic differences, the authors proposed that *Psa* biovar 4 should be renamed as *Psm. syringae* pv. *actinidifoliorum* pv. nov.

Psa has an extensive global distribution, yet the isolates causing widespread losses to the kiwifruit industry can all be traced to a single MLSA group, *Psa*3 (Chapman *et al.*, 2012).

Patel *et al.* (2017) performed a genetic screen in order to identify transposon mutants altered in the lipolytic activity and aimed to identify the set of secretion and global regulatory loci that control lipolytic activity and also play important roles in *in planta* fitness. The screen for altered lipolytic activity phenotype identified a total of 58 Tn*5* transposon mutants. The mapping of all these Tn*5* mutants revealed that the transposons were inserted in genes that play roles in cell division, chemotaxis, metabolism, movement, recombination, regulation, signal transduction and transport as well as a few unknown functions. Several of these identified *Psa* Tn*5* mutants, notably the functions affected in phosphomannomutase AlgC, lipid A biosynthesis acyltransferase, glutamate-cysteine ligase and the type IV pilus protein PilI, were also found affected in *in planta* survival and/or growth in kiwifruit plants.

Rees-George *et al.* (2010) found that several PCR primers designed earlier to identify *Psm. syringae* pv. *actinidiae* were not specific. Therefore, the authors designed two new sets of PCR primers, PsaF1/R2 and PsaF3/R4, to be complementary to a portion of the 16S–23S rDNA intertranscribed spacer regions. These primers amplified a DNA fragment from strains of *Psm. syringae* pv. *actinidiae* and a tea pathogen strain, *Psm. syringae* pv. *theae*, but not from 56 strains of bacteria belonging to six genera and 17 species. It was not possible to distinguish *Psm. syringae* pv. *actinidiae* from the phylogenetically similar *Psm. syringae* pv. *theae* using the ITS, *gyrB, acnB, rpoD, pgi* or *cts* gene regions to design PCR primers. As *Psm. syringae* pv. *theae* is unlikely to be present in kiwifruit, primers PsaF1/R2 and PsaF3/R4 can be used to detect *Psm. syringae* pv. *actinidiae* from kiwifruit tissue.

9.56.3.2 Psm. syringae pv. syringae

Psm. syringae pv. *syringae*, the bacterium first isolated from kiwi canker in 1983, differs from *Psm. syringae* pv. *actinidiae* in liquefying gelatin and producing fluorescent pigment. Strains of both the above-given pathogens resistant to streptomycin occur in Korea and Japan (Han *et al.*, 2004). Strains of *Psm. syringae* pv. *actinidiae* resistant to copper sulphate occur in Japan and the resistance to copper sulphate operates through the copper-trapping system (Masami *et al.*, 2004). Strains of *Psm. syringae* resistant to copper have also been reported from New Zealand (Vanneste and Voyl, 2003).

9.56.3.3 Psm. syringae pv. actinidifoliorum pv. nov.

The strains of this bacterium, i.e., biovar 4, are substantially different from those of *Psa* biovars 1, 2 and 3. Strains of biovar 4 are less aggressive and cause only leaf spots, whereas strains of *Psa* biovars 1, 2 and 3 also cause canker and shoot die-back.

Turco *et al.* (2024) have provided the first comparative genome analysis of 153 *Psm. syringae* pv. *actinidiae* genome sequences available in the NBCI public database. The results of this pangenome analysis, beside supporting previous phylogenetic characterisations, provide new details about the pathogenicity features of the different biovars.

Cesari *et al.* (2023) isolated 23 isolates of *Psm. syringae* pv. *actinidiae* from *Actinidia chinensis* orchards of Central Italy planted with cultivars Hayward and Zespri G3®. All the isolates showed resistance to 1.2 mM copper sulphate added to mannitol-glutamate-yeast extract (MGY) medium. The *Psm. syringae* pv. *actinidiae* strains that showed resistance to copper have the *copA* and *copD* genes that regulate the efflux of copper through the inner membrane and the copper sensor *copS* gene. They do not possess the *cusABC* complex that confers resistance to the high concentration of copper. Representative copper-resistant strains were proven pathogenic to kiwifruit plants upon artificial inoculation. This is the first record of copper-resistant *Psm. syringae* pv. *actinidiae* strains in Italy and, as far as we know, in Europe.

9.56.4 Disease Cycle

The bacteria survive in cankers formed on the infected vines. In Japan, infected *Actinidia* wild spp. (*A. arguta* and *A. kolomikta*) also help in the perpetuation of the bacterium and act as a source of infection for cultivated vines. Bacterial cells ooze out from cankers formed on trunks, limbs and canes, and are distributed by wind-splashed rain and overhead irrigation. Long-distance spread occurs through infected planting material. The bacteria enter the host through stomata, lenticels and wounds caused by pruning and cultural operations. Mature leaves and canes are not infected in summer, and even the penetration through wounds does not result in infection. The disease is severe when the mean temperature in early spring remains below 15 °C and is coupled with frequent rains. Winter cankers cause more yield loss as these reduce the size of productive vines by damaging the main vine structure. Spring cankers have less direct effect on the yield, but are important for the formation of winter cankers in the next year. Four species of *Actinidia*, namely, *A. deliciosa, A. chinensis, A. arguta* and *A. kolomikta*, are susceptible to the pathogen. There is no information on the susceptibility of other *Actinidia* species. Observations made in Italy suggest that the damage is more severe on yellow-fleshed kiwifruit cultivars, namely, *A. chinensis* cvs. Hort 16A and Jin Tao, than on the more widely grown green-fleshed cv. Hayward of *A. deliciosa*.

Xie *et al.* (2023) examined the virulence of 377 isolates of *Psm. syringae* pv. *actinidiae* biovar 3 (*Psa*3), isolated from 76 kiwifruit orchards suffering from bacterial canker disease. Eighty-four non-pathogenic isolates were identified in 40 orchards. A non-pathogenic isolate G166 was found to be defective in *hrpL* transcription and the downstream type III secretion system (TIIISS)-dependent phenotypes. Comparative genomics and complementary expression assay revealed that a single-base "G" insertion in the *hrpL* promoter blocks gene transcription by reducing promoter activity. The electrophoretic mobility shift assay showed that the genetic variation impairs σ^{54}/promoter binding during gene transcription under *hrp*-inducing conditions, resulting in lower expression of *hrpL*. A PCR-restriction fragment length polymorphism assay, performed to trace the evolutionary history of this mutation, revealed the independent onset of genetic variations in natural *Psa*3 populations. It was also found that non-pathogenic variants outperformed virulent *Psa*3 bacteria for both epiphytic and apoplast colonisation of kiwifruit leaves in mixed inoculations. This study highlights a novel mechanism for the loss of virulence in *Psa*3 and provides insight into bacterial adaptive evolution under natural settings.

Ferrante and Scortichini (2014) reported that autumn and winter frosts promote *Psa* multiplication in the inoculated twigs of both *A. chinensis* and *A. deliciosa*. The damage from frost, freeze thawing and the accumulation of *Psa* in *Actinidia* twigs promotes the migration of the pathogen within and between the orchards. The results obtained in this study confirmed that *A. deliciosa* is more frost-tolerant than *A. chinensis*, autumn frosts are more dangerous to these hosts than winter frosts, and, in the absence of *Psa*, young kiwifruit plants remain sensitive to frost.

Li *et al.* (2005) reported that the resistance of cultivars was related to the contents of lignin and soluble sugars in annual twigs and foliage. Contents of both the compounds were higher in resistant cultivars than those of susceptible cultivars. After infection, soluble sugar decreased in both the cultivars but the decrease was much greater in susceptible cultivar than in resistant cultivar. In contrast, the lignin content increased in both the cultivars and the increase in resistant cultivar was much higher in comparison to susceptible cultivar. Balestra *et al.* (2001) reported that staminate kiwi cv. GTH, highly resistant to *Psm. syringae* pv. *syringae*, acquired susceptibility when transformed with *rolABC* of *Agrobacterium rhizogenes*. Susceptibility was inherited by the T1 offspring containing *rolABC* genes, derived by crossing pollens of transgenic *rolABC* GTH with the pistilate cv. Hayward (highly susceptible). Susceptibility was probably correlated to high nitrogen content in the leaves of *rolABC* plants. The high resistance of the staminate GTH was inherited by all the offsprings lacking *rolABC* gene.

Reglinski *et al.* (2024) reported that application of saccharin at 0.25, 0.5, 1.0 and 2.0 g/L, one week before inoculation with *Psm. syringae* pv. *actinidiae* biovar 3 (*Psa*), induced a dose-dependent reduction in leaf necrosis in two *Actinidia chinensis* cultivars, Hayward and Zesy002, in the glasshouse plants. Saccharin at the rate of 2.0 g^{L-1} reduced leaf necrosis in Hayward by 77% and in Zesy002 by over 90%. The development of induced resistance (IR) in both cultivars was concomitant with the accumulation of salicylic acid (SA) and salicylate glycoside (SAG), and the upregulation of SA-pathway genes (*PR1* and *PR2*) in treated leaves. In orchard-grown Hayward vines, saccharin (1.0 g^{L-1}) induced the upregulation of SA-pathway genes and reduced *Psa* leaf necrosis and flower bud rot by 50% and 25%, respectively, compared with controls. However, saccharin residues were detected in fruit collected from vines that were sprayed with saccharin (1.0 g^{L-1}) before flowering. Residue level correlated with application frequency and timing and was highest (0.051 mg/kg) in vines that received three pre-flowering sprays. This exceeds the default maximum residue limit (MRL; 0.01 mg/kg) for some export markets and could limit the use of saccharin for disease management.

Beatrice *et al.* (2017) evaluated the chitosan's elicitation effect to induce systemic acquired resistance (SAR) in kiwi plants for the control of *Psa* infection. Chitosan elicited a systemic response in kiwi plants with intensity comparable to other well-known signalling compounds such as salicylic acid, methyl jasmonate, or ethylene. The data obtained from chitosan treatments in *in vitro* cultures were confirmed in plants grown in a greenhouse, in which the combination of chitosan treatment and the bacterial inoculum had the greatest effect on pathogenesis related proteins. The study also proved that chitosan, leading to an increased expression of both pathogenesis-related proteins, has a role in kiwifruit defence reactions.

Balestra *et al.* (2013) developed and validated a new multiplex PCR assay using 32 *Psa* isolates and 15 non-*Psa* strains and correctly assigned *Psa* strains to three different haplotypes: a Japanese/Korean group, a European group and a Chinese group. The described PCR assays has a limit of detection of approximately 5–50 pg of purified DNA or of 5×10^2 bacteria/PCR and were shown to work with both artificially and naturally infected plant tissues. Hence, the described method offers a suitable tool for the detection of *Psa* and haplotype attribution, in particular, when testing a high number of samples during surveillance and prevention activities.

Ruinelli *et al.* (2017) developed a rapid, sensitive and reliable field-based assay for the detection of *Psa*, using a comparative genomic approach on the publicly available *Psa* genomic data to select unique target regions for the development of two LAMP assays able to detect *Psa* and to discriminate strains belonging to the highly virulent and globally spreading *Psa* biovar 3. Both the LAMP assays showed specificity in accordance with their target and were able to detect reliably 125 cfu per reaction in less than 30 minutes. Developed assays were also able to detect the presence of *Psa* in naturally infected kiwifruit material with and without symptoms, thereby increasing the potential of the LAMP assays for phytosanitary use.

Chen *et al.* (2018) raised a polyclonal antiserum to the bacterial effector-hopz5 (PAb:hopz5) and tested its specificity. PAb:hopz5 was able to detect *Psa* from culture and infected plant samples, thus representing a suitable tool for the immunodetection of the causal agent of kiwifruit bacterial canker in field samples. Detection sensitivity of a colloidal gold immunochromatographic strip was high, up to 2.2×10^3 cfu mL^{-1}. Therefore, PAb:hopz5 constitutes a suitable and accurate tool for detection of *Psa* and haplotype distribution.

Kim *et al.* (2018) standardised an infection risk model used in New Zealand by modifying the temperature and rainfall functions of the existing model for use in Korea. Analyses using statistical correlation and prediction–realisation tables revealed that the modified model is valid with high agreement (a correlation coefficient of 0.85 and an accuracy of 85.7%, respectively) between the observed disease incidence and simulated disease risk from the model. The model was also found to be highly sensitive to the presence or absence of rainfall than any other weather-variable inputs. The modified *Psa* risk model can be used to provide practical and applicable information for timely disease control to the kiwifruit farmers in Korea.

Ye *et al.* (2020) proposed that kiwifruit cultivars, namely, 'Wanjin', 'Donghong', 'Huate', 'Fujianmaohua', 'Ganmi No. 6', 'Hongguoruanzao', 'Ruanzao No. 2' and 'Jialv', should be used as parental materials to breed new cultivars tolerant to bacterial canker for future kiwifruit production.

Wang *et al.* (2021) used genotype Gui-1 in the *Actinidia guilinensis* species as a rootstock to evaluate its effect on the resistance of the highly susceptible cultivar 'Hongyang' (*A. chinensis*) with an *in vitro* assay. The results show that Gui-1 significantly improved the resistance of 'Hongyang' scion to *Psm. syringae* pv. *actinidiae* (Psa) without affecting the fruit quality, and also showed that the *in vitro* cane could still maintain the rootstock-induced resistance, but the resistance imparted to the scion by the rootstock might be spatially and temporally uneven. The study provides the first laboratory evidence of grafting-induced resistance to Psa, which provides a foundation for the use of resistant rootstock in kiwifruit production. The investigation also highlights implications for the study of the mechanism of rootstock–scion interaction, especially those related to the resistance to non-soil-borne disease.

Wang *et al.* (2020) tested the response of 104 wild genotypes of 30 *Actinidia* species (including 37 taxa) to *Psm. syringae* pv. *actinidiae* (*Psa*) *in vitro* bioassay, and identified a considerable number of individuals from different species with tolerance or high resistance to *Psa*. The results showed high consistency between years of trial. This is the first large-scale evaluation of diverse *Actinidia* species with resistance to *Psa* through an *in vitro* bioassay. The resistant genotypes of *A. chinensis* identified can be used in future kiwifruit improvement programmes. These findings should also help provide an understanding of the resistance of hosts to *Psa*.

Barrett-Manako *et al.* (2021) evaluated three alternative *Psm. syringae* pv. *actinidiae* quantification methods, namely, start of growth time, quantitative PCR (qPCR), and droplet digital PCR (ddPCR), in two infected kiwifruit cultivars. Method performance in each case was compared with the colony counting method. Methods were validated using calibration curves obtained with serial dilutions of *Psm. syringae* pv. *actinidiae* biovar 3 (*Psa*3) inoculum and standard growth curves obtained from kiwifruit samples infected with *Psa*3 inoculum. All the three alternative methods showed high correlation ($r > 0.85$) with the colony counting method. qPCR and ddPCR were very specific, sensitive (5×10^2 CFU/cm^2), highly correlated to each other ($r = 0.955$), and flexible, allowing for sample storage. The inclusion of a kiwifruit biomass marker increased the methods' accuracy. The qPCR method was efficient and allowed for high-throughput processing and the ddPCR method showed highly accurate results but was more expensive and time-consuming. While not ideal for high-throughput processing, ddPCR was useful in developing accurate standard curves for the qPCR method. The combination of the two methods is high-throughput, specific for *Psa*3 quantification, and useful for research studies like disease phenotyping and host–pathogen interactions.

The bio-bactericide $YBCA_5$ is a strain of *Aureobasidium pullulans*, which mitigates the leaf spot symptoms of bacterial canker in kiwifruit caused by *Psm. syringae* pv. *actinidiae* (*Psa*) biovar 3. Using scanning electron microscopy (SEM), Rheinländer *et al.* (2021) studied the mode of action of $YBCA_5$ against *Psa* by investigating the interactions between these two microorganisms *in planta*. The investigation was carried out by applying either $YBCA_5$ or *Psa* separately and the combination of both on the abaxial surface of kiwifruit leaves (*Actinidia chinensis* var. *deliciosa* 'Hayward'). SEM observations revealed that blastospores of $YBCA_5$ occasionally lodged inside the outer cavity of the stomata and commenced multiplying (budding) within seven days after application. Within 10–14 days, mycelium was observed extending across the phylloplane. On leaves treated with $YBCA_5$ and *Psa*, both the microorganisms could be differentiated based on differences in their size; *Psa* cells measuring approximately a quarter of the length of the $YBCA_5$ blastospores (1.6 ± 0.02 µm and 6.1 ± 0.2 µm, respectively). The *Psa* cells adhered to the $YBCA_5$ blastospores and hyphae. Seven days after inoculation, *Psa* cells adjacent to $YBCA_5$ were deflated; their typical rod-shape was distorted and they appeared unable to infect the stomata. On average, distorted *Psa* cells were 20% shorter and 30% wider compared with those on leaves without $YBCA_5$. Ten days after inoculation, the *Psa* cells were appressed against the phylloplane and appeared to have collapsed.

The authors hypothesised that this phenomenon is associated with extracellular substances exuded by $YBCA_5$.

Campigli *et al.* (2023) demonstrated that by proving qPCR's capacity to detect latent infections of *Psm. syringae* pv. *actinidia*e biovar 3 in artificially inoculated leaves, their (qPCRPsa3 and qPCRPv7) potential usefulness in surveillance programmes and for epidemiological studies of *Psm. syringae* pv. *actinidia*e biovar 3 was verified.

9.56.5 Management

The following measures are effective in minimising the disease incidence.

1. Use disease-free nursery plants.
2. Use windbreaks to prevent injuries to the vines from strong winds.
3. During pruning, if infected vines show bacterial ooze at the cut surface, these should be cut back to the healthy portion.
4. During pruning, disinfect the scissors regularly with disinfectants like sodium hypochlorite solution.
5. Avoid or minimise the use of overhead irrigation.
6. Streptomycin and copper sulphate are generally used to control the disease in orchards. However, their continuous use has resulted in the development of pathogen strains resistant to both the chemicals and this has created a serious problem in the management of the disease. Strains of *Psm. syringae* pv. *actinidiae* resistant to streptomycin and copper sulphate occur in Japan (Goto *et al.*, 1994; Lee *et al.*, 2005) and those of *Psm. syringae* in New Zealand (Vanneste and Voyl, 2003). Han *et al.* (2004) also showed that Japanese and Korean isolates of *Psm. syringae* pv. *actinidiae* and *Psm syringae* pv. *syringae* harboured *strA-strB* streptomycin resistance genes for a streptomycin-resistance determinant.

In areas having the prevalence of copper/antibiotic-resistant strains of the pathogen, Scortichini (2014) compared the field efficacy of a chitosan-based compound with a copper compound in a three-year trial. Chitosan caused no phytotoxic effect on plants and fruits and showed an overall higher level of performance than the copper compound in reducing the disease severity throughout the trial. Chitosan also significantly reduced the presence of exudates on trunk and leaders recorded at the end of winter.

Scortichini (2016) tested the field efficacy of a zinc (4.7%)–copper (2.6%)–hydracid of citric acid (21.4%) biocomplex compound (applied three times, at a dose of 1.0%) to reduce the occurrence of bacterial exudates oozing out from the main trunk and leaders in early spring, during the winters of 2013–14 and 2014–15 at four sites in Italy. The biocomplex compound was more effective than 1% Bordeaux mixture used under the same conditions. This biocomplex formulation could be effectively used to reduce the amount of exudates produced by the bacterium in areas where winter frost occurs regularly.

Fiorillo *et al.* (2023) demonstrated that the phage φPSA2 (previously characterised) contains the spread of *Psm. syringae* pv. *actinidiae* inside plant tissue and reduces the symptoms of the disease. The data presented are a strong indication for the efficiency of this phage and open the possibility of developing a phage therapy based on φPSA2 to control the bacterial canker of kiwifruit.

REFERENCES

Balestra, G. M., Mazaglia, A., Quattrucci, A. Renzi, M. and Rossetti, A. (2009). Occurrence of *Pseudomonas syringae* pv. *actinidiae* in Jin Tao kiwi plants in Italy. *Phytopathol. Medit.* 48: 299–301.

Balestra, G. M., Rugni, E. and Varvaro, L. (2001). Increased susceptibility to *Pseudomonas syringae* pv. *syringae* and *Pseudomonas viridiflava* of kiwi plants having transgenic *rolABC* genes and its inheritance in the T1 offspring. *Phytopathol. Z.* 149: 189–94.

Balestra, G. M., Taratufolo, M. C., Vinatzer, B. A. and Mazzaglia, A. (2013). A multiplex PCR assay for detection of *Pseudomonas syringae* pv. *actinidiae* and differentiation of populations with different geographic origin. *Plant Dis.* 97: 472–78.

Barrett-Manako, K., Andersen, M., Martínez-Sánchez, M., Jenkins, H., Hunter, S. *et al.* (2021). Real-time PCR and droplet digital PCR are accurate and reliable methods to quantify *Pseudomonas syringae* pv. *actinidiae* biovar 3 in kiwifruit infected plantlets. *Plant Dis.* 105: 1748–57.

Beatrice, C., Linthorst, J. M. H., Cinzia, F. and Luca, R. (2017). Enhancement of *PR1* and *PR5* gene expressions by chitosan treatment in kiwifruit plants inoculated with *Pseudomonas syringae* pv. *actinidiae*. *Eur. J. Plant Pathol.* 148: 163–79.

Campigli, S., Luti, S., Martellini, T., Rizzo, D., Bartolini, L. *et al.* (2023). TaqMan QPCR assays improve *Pseudomonas syringae* pv. *actinidia*e biovar 3 and *P. viridiflava* (PG07) detection within the *Pseudomonas* sp. community of kiwifruit. *Phytopathol. Mediterr.* 62: 95–114. https://doi.org/10.36253/phyto-14400

Cesari, E., Marocchi, F., L'Aurora, A., Pucci, N., Scala, V. *et al.* (2023). Occurrence of copper-resistant *Pseudomonas syringae* pv. *actinidiae* strains in kiwifruit orchards of Central Italy. *J. Phytopath.* 171: 768–74.

Chapman, J. R., Taylor, R. K., Weir, B. S., Romberg, M. K., Vanneste, J. L. *et al.* (2012). Phylogenetic relationships among global populations of *Pseudomonas syringae* pv. *actinidiae*. *Phytopathology* 102: 1034–44.

Chen, H., Hu, Y., Qin, K., Yang, X., Jia, Z. *et al.* (2018). A serological approach for the identification of the effector *hopz5* of *Pseudomonas syringae* pv. *actinidiae*: A tool for the rapid immunodetection of kiwifruit bacterial canker. *J. Plant Pathol.* 100: 171–77.

Cunty, A., Poliakoff, F., Rivoal, C., Cesbron, S., Fischer-Le Saux, M. *et al.* (2015). Characterization of *Pseudomonas syringae* pv. *actinidiae* (Psa) isolated from France and assignment of Psa biovar 4 to a *de novo* pathovar: *Pseudomonas syringae* pv. *actinidifoliorum* pv. nov. *Plant Pathol.* 64: 582–96.

Ferrante, P. and Scortichini, M. (2010). Molecular and phenotypic features of *Pseudomonas syringae* pv. *actinidiae* isolated during recent epidemics of bacterial canker on yellow

kiwifruit (*Actinidia chinensis*) in central Italy. *Plant Pathol.* 59: 954–62.
Ferrante, P. and Scortichini, M. (2014). Frost promotes the pathogenicity of *Pseudomonas syringae* pv. *actinidiae* in *Actinidia chinensis* and *A. deliciosa* plants. *Plant Pathol.* 63: 12–19.
Ferrante, P. and Scortichini, M. (2015). Redefining the global populations of *Pseudomonas syringae* pv. *actinidiae* based on pathogenic, molecular and phenotypic characteristics. *Plant Pathol.* 64: 51–62.
Fiorillo, A., Frezza, D., Di Lallo, G. and Visconti, S. (2023). A phage therapy model for the prevention of *Pseudomonas syringae* pv. *actinidiae* infection of kiwifruit plants. *Plant Dis.* 107: 267–71.
Goto, M. (1992). *Fundamentals of Bacterial Plant Pathology*. Academic Press, Inc, San Diego, CA, 342 pp.
Goto, M., Hikota, T., Nakajima, M. Takikawa, Y. and Tsuyumu, S. (1994). Occurrence and properties of copper-resistance in plant pathogenic bacteria. *Ann. Phytopathol. Soc. Japan* 60: 147–53.
Han, H. S., Koh, Y. J., Hur, J.-S. and Jung, J. S. (2003). Identification and characterization of coronatine-producing *Pseudomonas syringae* pv. *actinidiae*. *J. Microbiol. Biotechnol.* 13: 110–18.
Han, H. S., Koh, Y. J., Hur, J.-S. and Jung, S. (2004). Occurrence of the *strA-strB* streptomycin resistance genes in *Pseudomonas* species isolated from kiwifruit plants. *J. Microbiol.* 42: 365–68.
Kim, G. H., Jung, J. S. and Koh, Y. J. (2017). Occurrence and epidemics of bacterial canker of kiwifruit in Korea. *The Plant Pathol. J.* 33: 351–61.
Kim, K.-H., Son, K. I. and Koh, Y. J. (2018). Adaptation of the New Zealand Psa risk model for forecasting kiwifruit bacterial canker in Korea. *Plant Pathol.* 67: 1208–19.
Koh, Y. J., Kim, G. H., Lee, Y. S., Sohn, S. H., Koh, H. S. *et al.* (2012). *Pectobacterium carotovorum* subsp. *actinidiae* subsp. nov., a new bacterial pathogen causing canker-like symptoms in yellow kiwifruit, *Actinidia chinensis*. *N. Z. J. Crop Hortic. Sci.* 40: 269–79.
Lee, J. H., Kim, J. H., Kim, G. H., Jung, J. S., Hur, J.-S. *et al.* (2005). Comparative analysis of Korean and Japanese strains of *Pseudomonas syringae* pv. *actinidiae* causing bacterial canker of kiwifruit. *Plant Pathol. J.* 21: 119–26.
Li, M., Tan, G., Li, Y. and Zhou, Z. (2005). Relationship between contents of lignin and soluble sugar in plants of kiwifruit cultivars and their resistance to kiwifruit bacterial canker infected by *Pseudomonas syringae* pv. *actinidiae*. *Acta Phytophylacica Sinica* 32: 138–42.
Lu, Q., Yan, F., Liu, Y., Li, Q., Yang, M. *et al.* (2021). Comparative genomic analyses reveal functional insights into key determinants of the pathogenesis of *Pectobacterium actinidiae* in kiwifruit. *Phytopathology* 111: 789–98.
Masami, N., Masao, G., Katsumi, A. and Tadaaki, H. (2004). Nucleotide sequence and organization of copper resistance genes from *Pseudomonas syringae* pv. *actinidiae*. *Eur. J. Plant Pathol.* 110: 223–26.
Patel, H. K., Ferrante, P., Xianfa, M., Javvadi, S. G., Subramoni, S. *et al.* (2017). Identification of loci of *Pseudomonas syringae* pv. *actinidiae* involved in lipolytic activity and their role in colonization of kiwifruit leaves. *Phytopathology* 107: 645–53.
Rees-George, J., Vanneste, J. L., Cornish, D. A. Pushparajah, I. P. S., Yu. J. *et al.* (2010). Detection of *Pseudomonas syringae* pv. *actinidiae* using polymerase chain reaction (PCR) primers based on the 16S-23S rDNA intertranscribed spacer region and comparison with PCR primers based on other gene regions. *Plant Pathol.* 59: 453–64.
Reglinski, T., Wurms, K., Northcott, G., Taylor, J., Chee, A. A. *et al.* (2024). Saccharin induces resistance against *Pseudomonas syringae* pv. *actinidiae* (*Psa* biovar 3) in glasshouse kiwifruit and orchard vines. *Plant Pathol.* 73: 2440–52.
Rheinländer, P. A., Sutherland, P. W. and Elmer, P. A. G. (2021). Visualisation of the mode of action of a biological control agent, *Aureobasidium pullulans* (strain $YBCA_5$) against *Pseudomonas syringae* pv. *actinidiae* biovar 3 on the kiwifruit phylloplane. *Australasian Plant Pathol.* 50: 379–88. https://doi.org/10.1007/s13313-021-00783-3
Ruinelli, M., Schneeberger, P. H. H., Ferrante, P., Buhlmann, A., Scortichini, M. *et al.* (2017). Comparative genomics-informed design of two LAMP assays for detection of the kiwifruit pathogen *Pseudomonas syringae* pv. *actinidiae* and discrimination of isolates belonging to the pandemic biovar 3. *Plant Pathol.* 66: 140–49.
Santos, M. G., da Silva, M. N., Vasconcelos, M. W. and Carvalho, S. M. P. (2023). Scientific and technological advances in the development of sustainable disease management tools: A case study on kiwifruit bacterial canker. *Front. Plant Sci.* 14: 1306420.
Scortichini, M. (2014). Field efficacy of chitosan to control *Pseudomonas syringae* pv. *actinidiae*, the causal agent of kiwifruit bacterial canker. *Eur. J. Plant Pathol.* 140: 887–92.
Scortichini, M. (2016). Field efficacy of a zinc-copper-hydracid of citric acid biocomplex compound to reduce oozing from winter cankers caused by *Pseudomonas syringae* pv. *actinidiae* to *Actinidia* spp. *J. Plant Pathol.* 98: 651–55.
Serizawa, S., Ichikawa, T., Takikawa, Y., Tsuyumu, S. and Goto, M. (1989). Occurrence of bacterial canker of kiwi-fruit in Japan: Description of symptoms, isolation of the pathogen and screening of bactericides. *Ann. Phytopathol. Soc. Japan* 55: 427–36.
Spigaglia, P., Barbanti, F., Marocchi, F., Mastroleo, M., Baretta, M. *et al.* (2020). *Clostridium bifermentans* and *C. subterminale* are associated with kiwifruit vine decline, known as *moria*, in Italy. *Plant Pathol.* 69: 765–74.
Takikawa, Y., Serizawa, S., Ichikawa, T., Tsuyumu, S. and Goto, M. (1989). *Pseudomonas syringae* pv. *actinidiae* pv. nov.: The causal bacterium of canker of kiwifruit in Japan. *Ann. Phytopathol. Soc. Japan* 55: 437–44.
Turco, S., Ferrucci, A., Drais, M. I. Mazzaglia, A. (2024). A Genome-wide comparative analysis to characterise the phylogenetic relationship and the pathogenicity of *Pseudomonas syringae* pv. *actinidiae* strains. *Eur. J. Plant Pathol.* 168: 97–107. https://doi.org/10.1007/s10658-023-02731-x
Ushiyama, K., Kita, N., Aono, N. Ogawa, J. and Fujii, H. (1992). Bacterial canker disease of wild actinidia plants as the infection source outbreak of bacterial canker of kiwifruit caused by *Pseudomonas syringae* pv. *actinidae*. *Ann. Phytopathol. Soc. Japan* 58: 426–30.
van Hall, C. J. J. (1902). Bijdragen tot de kennis der Bakterieele Plantenzeikten. Cooperatieve Drukkerij-vereeniging "Plantijn," Inaugural dissertation, Amsterdam.
Vanneste, J. L. and Voyl, M. D. (2003). Genetic basis of copper resistance in New Zealand strains of *Pseudomonas syringae*. *New Zealand Plant Prot.* 56: 109–12.
Wang, F-m., Li, J-w., Ye, K-y., Gong, H-j., Liu, P-p. *et al.* (2021). Preliminary report on the improved resistance towards *Pseudomonas syringae* pv. *actinidiae* of cultivated kiwifruit (*Actinidia chinensis*) when grafted onto wild *Actinidia guilinensis* rootstock *in vitro*. *J. Plant Pathol.* 103: 51–54. https://doi.org/10.1007/s42161-020-00719-8

Wu, W. X., Liu, Y., Huang, X. Q. and Zhang, L. (2017). First report of summer canker caused by *Pectobacterium carotovorum* subsp. *actinidiae* on kiwifruit in China's Sichuan Province. *Plant Dis.* 101: 1540.

Wang, F.-M., Mo, Q.-H., Ye, K.-Y., Gong, H.-J., Qi, B.-B. *et al.* (2020). Evaluation of the wild *Actinidia* germplasm for resistance to *Pseudomonas syringae* pv. *actinidiae*. *Plant Pathol.* 69: 979–89.

Xie, T., Wu, X., Luo, L., Qu, Y., Fan, R. *et al.* (2023). Natural variation in the hrpL promoter renders the phytopathogen *Pseudomonas syringae* pv. *actinidiae* non-pathogenic. *Mol. Plant Pathol.* 24: 262–71.

Ye, Zf., Wei, Pf., Xue, Sz., Liu, P., Jia, B. *et al.* (2020). Characterization of *Pseudomonas syringae* pv. *actinidiae* Biovar 3 strain and cultivar tolerance to kiwifruit bacterial canker in China. *J. Plant Pathol.* 102: 443–50. https://doi.org/10.1007/s42161-019-00459-4

9.57 XANTHOMONAS WILT OF BANANA

Bananas (*Musa* spp.) are the world's most important fruit crop in terms of production volume and trade (Faostat, 2014). Although a major staple in Africa, Asia and Latin America, only 13% of bananas produced are internationally traded (Lescot, 2013), indicating the fruit's importance for domestic markets and food security. In East and Central Africa, it is a significant dietary component, ranging from about 20% of daily total food intake in Uganda up to 80% in parts of Rwanda (Abele *et al.*, 2007).

Until recently, bacterial diseases of bananas have not received an equal amount of attention compared to other major threats to banana production, such as the fungal diseases, namely black leaf streak (*Mycosphaerella fijiensis*) and Fusarium wilt (*Fusarium oxysporum* f. sp. *cubense*). However, bacterial pathogens cause significant impacts on bananas globally and management practices are not always well known or widely adopted by farmers. Bacterial diseases of bananas can be divided into three groups: (1) Ralstonia-associated diseases (Moko/Bugtok disease caused by *Ralstonia solanacearum* and banana blood disease caused by *Ral. syzygii* subsp. *celebesensis*); (2) Xanthomonas wilt of banana caused by *Xanthomonas vasicola* pv. *musacearum*; and (3) Erwinia-associated diseases (bacterial head rot or tip-over disease caused by *Erwinia carotovora* ssp. *carotovora* and *Erw. chrysanthemi*, bacterial rhizome and pseudostem wet rot caused by *Dickeya paradisiaca*, formerly named as *Erw. chrysanthemi* pv. *paradisiaca*).

Castellani (1939) observed the symptoms of Xanthomonas wilt for the first time on enset in Ethiopia in the 1930s. The disease was, however, first identified in Ethiopia as Xanthomonas wilt in 1968 on enset (Yirgou and Bradbury, 1968) and subsequently on banana in 1974 (Yirgou and Bradbury, 1974). Since 2001, it has been reported from Uganda (Tushemereirwe *et al.*, 2003), the eastern Democratic Republic of Congo (hereafter abbreviated as DR Congo) (Ndungo *et al.*, 2004), Rwanda, Tanzania (Carter *et al.*, 2010), Kenya (Mbaka *et al.*, 2009; Carter *et al.*, 2010) and Burundi (*Carter et al.*, 2010). In DR Congo, Xanthomonas wilt has also been reported by the provincial agricultural research and extension services, in Uvira and Fizi in South Kivu province, in the Kalemie territory of northern Katanga province, and in Tshopo district in Oriental province (Anonymous, 2012). Since 2001, the disease has become the most important and widespread disease of *Musa* in East and Central Africa. Its non-discriminate infection of all *Musa* cultivars in this region (Ssekiwoko *et al.*, 2006c) and ability to cause up to 100% yield loss, especially in ABB type bananas, severely compromises food security and the livelihoods of banana-based farming households (Tushemereirwe *et al.*, 2003, 2004; Kagezi *et al.*, 2006; Ssekiwoko *et al.*, 2006a, 2006b). In fact, low soil fertility and Xanthomonas wilt are currently considered as the two major threats to banana productivity in the East African Great Lakes region (Kalyebara *et al.*, 2007).

Economic losses in the range of US$2–8 billion due to banana Xanthomonas wilt (here after abbreviated as BXW) have been reported over a decade within East Africa (Nkuba *et al.*, 2015; Tripathi *et al.*, 2009). As a result of the inconsistent adoption of recommended management options (Shimwela *et al.*, 2016) and the lack of a resistant source in *Musa* germplasm (Tripathi *et al.*, 2017), *Xan. vasicola* pv. *musacearum* is likely to remain a continuing threat within East and Central Africa.

9.57.1 Host Range

Banana, plantain and enset are the only known natural hosts of *Xan. vasicola* pv. *musacearum* (hereafter abbreviated as *Xvm*) (Thwaites *et al.*, 2000). All banana cultivars, with the exception of *Musa balbisiana*, are susceptible to the pathogen (Ssekiwoko *et al.*, 2006a; Tripathi *et al.*, 2009). Cultivated enset in Ethiopia is also very susceptible. However, to date *Xvm* has not been isolated from wild enset growing in DRC and Rwanda, and no plants have been observed with natural symptoms.

9.57.2 Symptoms

Visible symptoms after an insect-mediated infection on the male inflorescence part include the wilting of male bud bracts, followed by the drying of the rachis coupled with bacterial exudation, often followed by the premature ripening of some or all the fruits, and, eventually, the wilting and death of the entire plant. An internal cross-section of an infected floral stalk shows yellowish bacterial ooze from the vascular bundles, while a cross-section of a fruit shows rusty brown stains in the fruit pulp. Late floral symptoms (when the banana bunch is physiologically mature) have also been reported due to tool-mediated infections. Xanthomonas wilt bacteria entering the corm, roots, pseudostem and leaves of banana plants, through the use of garden tools, will first cause a progressive yellowing and wilting of the leaves. In addition, a yellow- or cream-coloured ooze, typical of many bacterial infections, exudes within a couple of minutes of cut tissue of an infected plant, and copious quantities of ooze may be produced over a period of several hours. A cross-section of a diseased pseudostem reveals brown or yellow streaks in the vascular tissue and yellowish bacterial ooze. The affected pseudostems most often wilt and die (Figure 9.47)

The most distinctive symptom is the premature ripening of the fruit. Single bananas start to ripen while the rest of the bunch remains green and the inside of the fruits has a brown/black discolouration. The male flower bud turns black, shrivels and eventually dies. Young emerging leaves turn yellow, wilt, fold over and collapse. If the fruit and leaf stalk or pseudostem of an infected plant is cut, cream/yellow bacterial ooze starts to flow from the cut end. Fusarium wilt (or Panama disease) can cause symptoms similar to Xanthomonas banana wilt, but this also causes the yellowing of older leaves first, does not affect the fruit and causes a dark staining inside the pseudostem which is absent in the latter.

The symptoms on enset are slightly different. The inner folds of the drooping heart leaf are covered with greyish-brown patches and when the leaf eventually emerges at the petiole, yellowish bacterial slime oozes from the vascular bundles (Yirgou and Bradbury, 1968). All the leaves wilt, bend over and wither, causing death of the entire enset plant (Yirgou and Bradbury, 1968).

The symptoms of Xanthomonas wilt do not differ markedly from those of caused by *Ral. solanacearum*. The incubation period for Xanthomonas wilt is about three weeks, but for *Ral. solanacearum* it depends on the cultivar, the plant growth stage, the mode of pathogen transmission (i.e., through an infection court on the male inflorescence part or on the pseudostem or leaves) and environmental conditions.

Xvm can be isolated on different culture media, such as cellobiose cephalexin agar (Tripathi *et al.*, 2007), yeast extract peptone glucose agar (YPGA) (Mwangi *et al.*, 2007) and yeast extract tryptone sucrose agar (YTSA)–cephalexin–cycloheximide (Tripathi *et al.*, 2007). A semi-selective culture medium can be made by adding cephalexin and 5-fluorouracil. 5-fluorouracil eliminates fluorescent pseudomonads (Sijam *et al.*, 1991), whereas cephalexin suppresses most Gram-staining-positive bacteria and is reported to inhibit *Erwinia* spp. (Schaad *et al.*, 2001). *Xvm* can be isolated onto YPGA from all plant parts, insects and soil.

FIGURE 9.47 Premature fruit ripening and fruit discolouration, yellowing and wilting of leaves and exudation of bacterial ooze from pseudostem.

9.57.3 Causal Organism

Xanthomonas vasicola pv. *musacearum* (Yirgou and Bradbury 1968) Studholme *et al.* 2020

Synonym: *Xan. campestris* pv. *musacearum* (Yirgou and Bradbury 1968) Dye 1978

and *Xan. musacearum* Yirgou and Bradbury 1968

The causal agent of Xanthomonas wilt of banana was originally described as *Xan. musacearum* (Yirgou and Bradbury, 1968), before being subsequently classified as *Xan. campestris* pv. *musacearum* (abbreviated as *Xcm*) by Young *et al.* (1978). Aritua *et al.* (2008) showed that strains of *Xcm* are homogeneous and very similar to *Xan. vasicola*. *Xan. vasicola* includes causal agents of several economically important diseases, including the pathovars *Xan. vasicola* pv. *holcicola* (abbreviated as *Xvh*) pathogenic to sorghum, and *Xan. vasicola* pv. *vasculorum* (abbreviated as *Xvv*) pathogenic to sugarcane (*Saccharum officinarum*) and maize (*Zea mays*) (Qhobela and Claflin, 1988; Vauterin *et al.*, 1992, 1995). Aritua *et al.* (2008) demonstrated that strains of *Xvh* and *Xvv* were non-pathogenic on banana but were pathogenic on maize, whereas *Xcm* strains were pathogenic on both banana and maize. These pathogenicity data suggest a host-jump by a strain of *Xvh* or *Xvv* onto a *Musa* species, because the *Xcm* strains retained pathogenicity to maize (Aritua *et al.*, 2008). Pathovar *musacearum* is distinguished on the basis of its phytopathogenic specialisation and is distinguished from other pathovars by its *gyrB* gene sequence (Parkinson *et al.*, 2009) and genome-wide sequence analysis.

Xvm is a Gram-stain-negative, aerobic, rod-shaped bacterium, measures 0.7–0.9 × 1.8–2.0 µm, motile by a single polar flagellum and belongs to the genus *Xanthomonas* in the class "*Gammaproteobacteria*" of phylum *Proteobacteria*. The optimum growth on nutrient agar is obtained at 25–28 °C. It produces distinct yellow, mucoid, circular and convex colonies after three days of incubation. Yellow colouration is characteristic of xanthomonads and is a result of the presence of brominated aryl polyenes, called xanthomonadins. *Xvm* does not accumulate poly-β-hydroxybutyric acid, is negative for oxidase, tyrosinase, nitrate reduction, starch hydrolysis and gelatinase tests, and induces a hypersensitive response in tobacco leaves (although this response may be weak). Biochemical characteristics include urease production, hydrolysis of aesculin, production of hydrogen sulphide from peptone, catalase production and utilisation of sorbitol, dulcitol and salicin. Gelatin is slowly liquefied and starch is not hydrolysed. Growth is quite rapid and very mucoid when cultured on yeast peptone sucrose agar media for 48 h at 28 °C.

Further characterisation of *Xvm* using fatty acid methyl ester analysis, *gyrB* gene sequencing and repetitive element sequence-based PCR has revealed that *Xvm* is more closely related to *Xan. vasicola* pv. *vasculorum* and *Xan. vasicola* pv. *holcicola* than to *Xan. campestris* (Aritua *et al.*, 2008; Parkinson *et al.*, 2009).

According to Bradbury (1986), the natural hosts of *Xvm* include *Ensete ventricosum* (enset) and *Musa* spp. (banana). Additional hosts by inoculation include *Saccharum* sp. (sugarcane) and *Zea mays* (maize) and the disease is exhibited as a bacterial wilt where leaves wilt and wither; yellowish bacterial masses are found in vascular tissue and parenchyma. It is not known if the strains being transferred to this taxon conform to the species description for metabolic activity (Studholme *et al.*, 2020).

Xvm is a genetically highly monomorphic pathogen on perennial banana and enset crops, with no known diversity between currently studied isolates. Studies by Aritua *et al.* (2008), using Rep-PCR, revealed that *Xvm* strains have identical profiles and are thus homogeneous in Uganda, DRC, Rwanda and Ethiopia, and hence clonal. However, whole-genome sequencing revealed 272 SNPs amongst a small collection of 13 *Xvm* isolates from seven East and Central African countries (Wasukira *et al.*, 2012). According to Wasukira *et al.* (2012), *Xcm* from enset and banana differ at 67 SNP positions that include non-silent polymorphisms in several potential virulence genes. Wasukira *et al.* (2012) also separated *Xvm*, at 86 SNP positions, into two distinct sublineages, with isolates from Ethiopia, DRC and Rwanda in sublineage I and isolates from Burundi, Kenya, Tanzania and Uganda in sublineage II. As *Xvm* is able to infect both banana and enset, it is unclear whether these differences have any biological significance.

Adikini *et al.* (2011) developed a specific and rapid diagnostic tool to detect *Xcm*. PCR primers were developed from intergenic regions of *Xcm* following its partial sequence and were tested for specificity to *Xcm* strains collected from various regions in Uganda. These were also tested for specificity against related *Xanthomonas* species from the vasicola group, *Xanthomonas* species pathogenic to other crops and against those existing saprophytically on banana plants. Seven primer sets (*Xcm*12, *Xcm*35, *Xcm*36, *Xcm*38, *Xcm*44, *Xcm*47 and *Xcm*48) were found to be very specific to *Xcm*. These primer sets directed the amplification of the expected product for all 52 strains of *Xcm* collected from different locations in Uganda. No amplification products were obtained with unrelated phytopathogenic bacteria or endophytic/epiphytic bacteria from banana. A detection limit of 10^3 CFU mL^{-1} corresponding to about four cells per PCR reaction was observed when *Xcm* cells were used for all the seven primer sets. The DNA samples from symptomless plant tissues also tested positive with primer set *Xcm*38. The specific PCR method described here is a valuable diagnostic tool that can be used to detect the pathogen at early stages of infection and monitor disease.

Adriko *et al.* (2012) developed a pathovar-specific PCR for the detection of *Xcm* by the amplification of a 265-bp region of the gene encoding the general secretion pathway protein D (GspD). A distinct DNA fragment of the expected size was amplified from genomic DNA from all the 12 *Xcm* isolates tested and no amplification of DNA was observed from other xanthomonads or plant-associated bacteria, including the two closely related species *Xvh* and *Xan.*

axonopodis pv. *vasculorum*. The *Xcm*-specific PCR was successfully multiplexed with internal control primers targeting 16S rDNA for application on DNA from bacterial cultures and with primers targeting plant mitochondrial 26S rDNA for application on DNA extracted from plant material. The diagnostic discrimination of healthy and infected plants was subsequently demonstrated in tests on artificially inoculated screen house cultivars of banana and field bananas with and without symptoms sampled from different parts of Uganda. The finding demonstrated a robust and specific *Xcm* diagnostic tool with the added advantage of applying internal PCR controls for direct quality assessment of results.

In addition, and based on comparisons of genome-wide sequence data from multiple isolates of *Xcm* and multiple strains of *Xvv*, Wasukira *et al.* (2012) identified genes specific to *Xcm* that could be used to specifically detect *Xcm* by PCR-based methods.

Karamura *et al.* (2015) carried out large-scale comparative pathogenicity studies using *Xan. vasicola* strains and *Xcm* strains on banana, maize and sugarcane. Six strains of *Xcm* used in the experiments caused disease in sugarcane and banana but not on maize. Two and four strains of *Xvv* and *Xvh*, respectively, were not only pathogenic on maize and sugarcane; each also caused distinct symptoms on maize. *Xvv* caused deformation of the maize plants while *Xvh* caused stunted growth.

Nakato *et al.* (2021) reported that the genotyping of 97 isolates and the whole-genome sequencing of 15 isolates of *Xan. vasicola* pv. *musacearum* (*Xvm*) revealed not only the presence of SL 2 in Ethiopia, but additional diversity beyond SL 1 and SL 2 in four new clades. Moreover, SL 2 was detected in the Democratic Republic of Congo, where previously SL 1 was the only clade reported. These results demonstrate a greater range of genetic diversity among *Xvm* isolates than previously reported, especially in Ethiopia, and they further support the hypothesis that the East/Central African Xanthomonas wilt epidemic has been caused by a restricted set of genotypes drawn from a highly diverse pathogen pool in Ethiopia.

9.57.4 Disease Cycle

Ocimati *et al.* (2015) investigated *Xcm* survival and latent infections in subsequent generations, the survival of latently infected planting materials (suckers), the incidence of latent infections in symptomless plants in mats having diseased plants, and Xanthomonas wilt status across farms and markets in districts previously devastated but currently endemic. Ongoing experiments were protected from new infections. Latent bacteria at low levels were detected in up to 20% of the third-generation suckers, with a significant ($P < 0{\cdot}05$) reduction (43–20%) in subsequent generations. Only 3–6% of latently infected suckers succumbed to the wilt. The incidence of *Xcm* in symptomless suckers from farmers' fields (with up to 70% incidence) was low (3%) while it increased (to 8–25%) with disease severity in mats in controlled experiments. In the surveyed districts, the incidence had significantly declined with yields observed to have recovered relative to earlier reports, although latent infections remained high. This investigation provides evidence that if new infections are prevented, fields with high wilt incidence can be rejuvenated. It showed incomplete systemic movement of the bacterium in mats coupled to a gradual decline of bacterial load in subsequent generations to levels that cannot initiate infection. These findings explain the current successes of farmers practising single diseased plant removal instead of whole mat rouging, and gives hope to farmers lacking access to clean planting material.

Ocimati *et al.* (2013a) examined the movement of *Xcm* into plants and mats, in relation to the disease incubation period. Mature Pisang Awak and East African highland (AAA-EA) plants were inoculated with *Xcm* through abscission wounds of female bracts, male bud bracts, male flowers, a combination of male bud bracts and flowers, and by cutting male buds with a contaminated machete. Thirty plants per genotype and treatment were monitored for 24 months for disease symptoms. All floral entry points resulted in disease development, with the highest incidence in combined male bract and male flower abscission wound inoculations. The study confirmed the systemic nature of *Xcm* and the ability of the pathogen to survive within the mat for a period of 5–16 months without causing any disease. Therefore, the reliance on the expression of disease symptoms to manage the wilt is incorrect. The long incubation period in lateral shoots may explain the current resurgence of the disease in areas where the disease was thought to have been successfully eradicated.

In the absence of natural host plant resistance among banana cultivars, Tripathi *et al.* (2010) evaluated the constitutive expression of the hypersensitivity response-assisting protein (*Hrap*) gene from sweet pepper (*Capsicum annuum*) for its ability to confer resistance to BXW. Transgenic lines expressing the *Hrap* gene under the regulation of the constitutive CaMV35S promoter were generated using embryogenic cell suspensions of two banana cultivars, Sukali Ndiizi and Mpologoma. These lines were characterised by molecular analysis, and were challenged with *Xcm* to analyse the efficacy of the *Hrap* gene against BXW. Six of the eight transgenic lines expressing *Hrap* showed no symptoms of infection after the artificial inoculation of potted plants in the screen house, while control non-transgenic plants showed severe symptoms resulting in complete wilting. It is evident that the constitutive expression of the sweet pepper *Hrap* gene in banana results in enhanced resistance to BXW. This development of transgenic banana varieties resistant to BXW will boost the arsenal available to fight this epidemic disease and save livelihoods in the Great Lakes region of East and Central Africa.

All banana varieties, with the exception of *M. balbisiana*, are susceptible to *Xcm*. Despite being resistant to *Xcm* infection, *M. balbisiana* is not preferred for breeding because it belongs to the BB genome subgroup, while most edible bananas are of the A genome. To identify potential sources of resistance to *Xcm*, Nakato *et al.* (2019) evaluated 72 banana accessions representing the *Musa* genetic diversity in an outdoor confined potted trial. The midribs of the youngest leaf of 3-month-old banana plants were inoculated with 10^8 CFU mL^{-1} of *Xcm* isolate USY13P, and symptom development recorded weekly for four months. The results confirmed that *M. balbisiana* genotypes are indeed resistant to *Xcm*. Varieties within the *M. acuminata* subsp. *zebrina* (AA) set were further identified as potentially useful sources of *Xcm* resistance. These findings reveal the possibility to develop banana and plantain varieties having tolerance to *Xcm*.

Ten banana cultivars were evaluated for their relative susceptibility to *Xcm* by injecting bacterial inoculum into the pseudostem of *in vitro* plantlets as well as potted plants. Various banana cultivars showed significant differences in susceptibility to *Xcm*. The beer banana cultivar Pisang Awak and the dessert cultivars Dwarf Cavendish, Giant Cavendish and FHIA-17 were found to be highly susceptible, whereas East African Highland banana cultivar Nakitembe was found to be the least susceptible. The other cultivars, Mpologoma, Mbwazirume, Sukali Ndiizi and FHIA-25, were found to be susceptible. Diploid parent *M. balbisiana* was found to be resistant. As all the commercial cultivars are susceptible, there is an urgent need for development of new resistant cultivars for the control of the disease (Tripathi and Tripathi, 2009).

Nakato *et al.* (2018) have reviewed the information on the impact of *Xcm* as a major constraint to banana, plantain and enset production in Central and East Africa over the past decade. The information is provided on host range, pathogen diversity, symptoms, pathogen distribution, ecology and epidemiology and disease management.

Tripathi *et al.* (2009), in their article entitled "Xanthomonas wilt, a threat to banana production in East and Central Africa", has given comprehensive information on various aspects of the disease. The aspects dealt with include economic and social impact, causal organism, symptoms, disease diagnosis, disease cycle/epidemiology, the susceptibility of commonly grown cultivars, and control and management strategies, including cultural practices, wilt-escaping, tolerant and resistant cultivars. Although much of the emphasis on management of BXW has been based on the information on other bacterial wilt diseases, more basic and strategic information on this pathogen is required. The development of disease-resistant banana cultivars remains a high priority because farmers are reluctant to employ labour-intensive disease control measures. Nevertheless, some countries such as Uganda and Tanzania have achieved remarkable success in containing the impact of the disease through a combination of management approaches.

In the field, *Xvm* is mainly transmitted through the use of contaminated farming implements (Addis *et al.*, 2010; Biruma *et al.*, 2007; Eden-Green, 2004; Kagezi *et al.*, 2006; Tinzaara *et al.*, 2006). Experimentally, it has been shown that *Xvm* can survive on stainless steel for up to 20 days and on non-stainless steel for up to 6 days (Buregyeya *et al.*, 2008). Farming tools have been shown to be responsible for the long-distance spread of the pathogen and sporadic outbreaks of the disease (Tushemereirwe *et al.*, 2006). In Tanzania, Shimwela *et al.* (2016) have observed continued transmission of *Xvm* resulting from inconsistent tool sterilisation and the exposure of *Xvm* to rain. Poor farmers' fields are at a high risk of infection because they tend to borrow tools from their neighbours and the recommended management practices are difficult to enforce. It is therefore recommended that there should be limited cutting of BXW-affected plants in dry periods and that farm tools should be sterilised by placing them in a fire. Intercropping banana with annual crops, such as groundnuts, beans and maize, which demand regular weeding, increases the risk of *Xvm* transmission through injuries to the pseudostem, corm and roots of the banana plant (Ocimati *et al.*, 2013b). The ability of *Xvm* to spread within the corm has also been reported by Ssekiwoko *et al.* (2006a).

Vectors such as insects, birds, bats and ruminants are all important means for the spreading of *Xvm* within and across plantations (Buddenhagen, 2006; Karamura *et al.*, 2008; Tinzaara *et al.*, 2006). *Xvm* vector transmission is, however, dependent on the size and population of the vector (Nakato *et al.*, 2014). For example, bees, wasps and birds may be present in small numbers, but are capable of visiting larger areas and hence can transmit the pathogen more readily, compared with drosophilids, which are present in large numbers, but spend most of their lives on fewer banana plants (Smith *et al.*, 2008). However, vector population thresholds that favour *Xvm* spread have not been documented. Insect vectors pick up the exudate from the male bud scars of diseased plants and inoculate the cushions (to which the male flowers were attached) of healthy plants. The distances over which insect vectors transmit *Xcm* have not been determined, but could be relatively close based on the lifestyles of these insects. For example, drosophilids could be responsible for within-farm transmission at distances not exceeding 3 m (the recommended spacing between banana mats), whereas bees and wasps are capable of moving longer distances, and hence could be responsible for transmission at distances exceeding this amount.

It has been observed that insect transmission of *Xvm* only results in floral symptoms, thereby affirming the importance of male bud bract wounds as entry points (Nakato *et al.*, 2014). Insects rarely transmit *Xvm* through the female inflorescence; however, when they do, the bacterial numbers are too few to induce disease symptoms (Nakato *et al.*, 2014). This is because fewer female inflorescences open daily and the scars dehydrate more rapidly compared with the larger number of male inflorescences produced that open daily, thus increasing the risk of *Xvm* spread by insects. Insect transmission of *Xvm* is more prevalent in cultivars that shed male flowers and bracts, e.g. 'Pisang Awak' (ABB), 'Ney Poovan' (AB), 'Bluggoe' (ABB) and 'Gros Michel' (AAA) (Addis *et al.*, 2004; Blomme *et al.*, 2005; Karamura *et al.*, 1998). Cultivars with persistent male flowers escape the insect transmission of *Xvm*, but are not grown as pure stands and thus have no impact as barriers to *Xvm* spread. Similarly, insect *Xvm* transmission is also dependent on altitude. In areas dominated by cultivars with dehiscent bracts, such as 'Pisang Awak' (ABB), and at an altitude below 1700 m above sea level (masl), insect transmission of *Xvm* is more prevalent. According to Addis *et al.* (2004) and Blomme *et al.* (2005), there is limited insect activity above 1700 metres above sea level, and hence fewer floral infections. Similar observations were also made by Shimelash *et al.* (2008) and Tripathi *et al.* (2009). Were *et al.* (2015) reported that the banana weevil, *Cosmopolites sordidus*, was a potential vector of *Xvm* as a result of the ability of the pathogen to survive on and within the insect.

Between distant farms, infected planting material is the main mode of transmission of *Xvm* (Karamura *et al.*, 2008). *Xvm* may remain latent for 7–14 days within the plant and, because of its systemic nature, it spreads through the planting of material (Ssekiwoko *et al.*, 2006a). *Xvm* may also spread latently through the banana bunch, but this depends entirely on whether the bunch residues are discarded or used as mulch in banana plantations (Nakato *et al.*, 2013). *Xvm* survival in soil is affected by the soil moisture content. *Xvm* has poor competitive ability and therefore succumbs to competition from other microbes. In the absence of competition, *Xvm* cannot persist in moist soils for a period longer than 90 days, or 30 days in dry soils (Mwebaze *et al.*, 2006). Rainfall plays an important role in the development and dissemination of *Xvm*. Manyong *et al.* (2008) and Tripathi *et al.* (2009) reported a positive correlation between BXW and rainfall. Similarly, studies on the spatiotemporal distribution of *Xvm* in the Kagera region of Tanzania further emphasized the role of rainfall in the short-distance spread of BXW (Shimwela *et al.*, 2016).

Ocimati *et al.* (2022) reported that *Xan. vasicola* pv. *musacearum* (*Xvm*) spread within mats is influenced by the amount of inoculum, and the physiological stage of the mother plant and attached suckers. The position of suckers, predominantly at the bottom of corms, also protects them from infection. Measures that reduce *Xvm* inoculum build-up in mats will thus minimise *Xvm* spread within mats.

Ochola *et al.* (2015) investigated the effects of water stress on the development of BXW. In the absence of host resistance, all inoculated banana plants succumbed to *Xcm* infection and symptoms were expressed, on average, 14 days post-inoculation (dpi). Water stress effects were significant ($P < 0{\cdot}05$) for the disease incubation period, and significant ($P < 0{\cdot}001$) for incidence and severity. Data revealed that BXW development is hastened by the combined effect of water stress before and after inoculation compared to when plants are stressed only before inoculation or maintained stress-free. A two-fold increase in evaluated disease parameters suggests that both the timing of infection and the duration of exposure to water stress are important in determining the shortened resident phase of *Xcm* to multiply and rapidly spread in the vascular tissues of the banana. This study has shown that water-stressed banana plants are physiologically weak and hence more vulnerable to *Xcm* infections.

Mburu *et al.* (2016) evaluated the effect of silicon supplementation on resistance to BXW using various banana cultivars. Silicon application at a concentration of 200 mg per plant per week was found to be optimal in enhancing resistance to BXW since there was no detrimental effect on plant growth. The continuous supply of silicon before and after pathogen inoculation led to a significantly higher level of resistance to BXW in all the banana cultivars tested in comparison to inoculated plants to which silicon had not been applied. Banana plants treated with silicon before pathogen inoculation only, also exhibited high protection against BXW, similar to the plants treated continuously with silicon. The total silicon content in leaves increased significantly in silicon-treated plants in comparison with non-silicon-treated control plants. The amount of silicon accumulation was directly correlated to the duration of application; plants treated with silicon continuously showed significantly higher amounts of silicon accumulation in leaves than plants where silicon application was terminated following bacterial inoculation, or when silicon application started immediately after pathogen inoculation. The silicon-applied plants also showed higher activity of the peroxidase enzyme in comparison to non-silicon-treated control plants. These findings reveal that the application of silicon enhances resistance to BXW and may provide an alternative to disease management strategy.

The role of *R* genes, pathogenesis-related (*PR*) genes and nonexpressor of pathogenesis-related 1 protein (NPR1) in BXW development and cultivar resistance to *Xvm* has been investigated by Ssekiwoko *et al.* (2015). The results showed that *Xvm* deactivates the plant detection system and related downstream reactions, thereby preventing the plants from defending themselves against pathogen attack, leading to disease development. Although *Xvm* successfully established itself in the resistant cultivar, leading to the death of a single leaf, migration within the resistant cultivar did not proceed to other plant tissues. Similar responses have been observed in other *Musa* species artificially inoculated with *Xvm*, resulting in slowed disease progression. Next-generation sequencing approaches exploring the *in planta* transcriptomics of both the host and the pathogen may provide clues about the specific genes involved in pathogenicity and defence responses, thereby assisting in the development of resistant varieties.

Resistance to *Xvm* in commonly grown banana cultivars has not been identified; however, *M. balbisiana*, the wild relative of the banana, has shown some resistance (Ssekiwoko *et al.*, 2006a, 2015; Tripathi *et al.*, 2008, 2009). For example, *in vitro* screening studies identified 'Pisang Awak' as highly susceptible, 'Nakitembe' as moderately resistant and *M. balbisiana* as resistant (Tripathi *et al.*, 2008). Subsequent studies evaluating banana cultivars for resistance to *Xvm* consistently identified *M. balbisiana* as resistant. The resistance observed in this *Musa* species is probably quantitative and associated with the limited spread of *Xvm* into the plant tissue. The HR-like symptoms observed in *M. balbisiana* are probably an indication of a reaction by the plant to the pathogen, stimulated by specific elicitors, as *Xvm* multiplication is partial, only occurring in a single leaf and not spreading into other tissues of the plant (Ssekiwoko *et al.*, 2015). Although the cultivar 'Nakitembe' may not possess cell-mediated resistance, it is likely to escape insect-mediated infections in the field because of its persistent male flowers and bracts that prevent the exposure of cushions on which insects can land (Tripathi and Tripathi, 2009). These results have prompted studies to explore the mechanisms used by *M. balbisiana* to limit disease expression.

Attempts to understand the mechanisms that play a role in conferring resistance to *Xvm* within *M. balbisiana* have revealed that resistance is not conferred by commonly known pathways, such as HR, systemic acquired resistance (SAR) or induced systemic resistance (ISR) (Ssekiwoko *et al.*, 2015). Defence-related gene expression in *M. balbisiana*, such as proteinase 3 antigen (PR3), NPR1 and MbNBS, declined within the first 6 h post-inoculation, but recovered much later in the infection process. This may explain why BXW symptoms and death occurred only in the inoculated leaf of *M. balbisiana* and not in the rest of the plant (Ssekiwoko *et al.*, 2015). There is thus a need to further explore the role of the entire gene profile expressed at the point of infection and during the infection process. Similarly, no resistance has been identified in enset. However, some clones have shown a high level of tolerance to *Xvm* infection. For example, the evaluation of 20 enset clones has shown that 'Mezya', 'Hiniba', 'Bedadet', 'Sorpie' and 'Sigezasarum' are tolerant because of the long disease incubation period and low levels of disease incidence (below 30%), whereas 'Warke Bidu', 'Awenyi', 'Kekar', 'Astara', 'Bufare', 'Geziwet 2', 'Gulumo' and 'Kulo' are susceptible because of the short disease incubation time and 100% incidence (Hunduma *et al.*, 2015; Welde-Michael *et al.*, 2008).

The use of genetic engineering to develop resistant bananas to *Xvm*, employing the plant ferrodoxin-like gene (*Pflp*) and hypersensitive resistance-assisting gene (*Hrap*) from sweet pepper (*Capsicum annum*), have yielded promising results (Chen *et al.*, 2000; Lin *et al.*, 1997). Banana lines carrying these two genes were evaluated under *in vitro* and *in vivo* conditions and later in confined field trials. The results demonstrated that the transgenes can provide resistance to *Xvm* without altering plant physiology (Namukwaya *et al.*, 2012; Nordling, 2010; Tripathi *et al.*, 2010). Similarly, the rice pattern recognition receptor (PRR), the *Xa21* gene, was considered a good candidate for the engineering of transgenic resistance to *Xvm* because of its broad-spectrum resistance against the rice pathogen, *Xan. oryzae* pv. *oryzae* (*Xoo*) (Song *et al.*, 1995; Wang *et al.*, 1996). According to Tripathi *et al.* (2014), the *Xa21* gene confers resistance to *Xvm* because the *Xoo raxSTAB* operon, required for the activation of *Xa21*-mediated immunity, is highly conserved in diverse *Xanthomonas* species, including *Xvm*. The analysis of transgenes with the *Xa21* gene resulted in enhanced resistance to *Xvm* with no observable visible physiological effects and no measurable morphological differences between the transgenes and the control plants (Tripathi *et al.*, 2014). However, as PRR-mediated immunity can be overcome through the suppression of the immune response by bacterial effectors (Dodds and Rathjen, 2010), it was suggested that *Xa21* could be used in combination with *Hrap* and/or *Pflp*, which showed enhanced resistance against *Xvm* in previous reports (Namukwaya *et al.*, 2012; Tripathi *et al.*, 2010, 2013). Nimusiima *et al.* (2015) observed no detectable effect of the transgenes on the inhabiting bacterial community resulting from genetic modification. Such transgenic banana material will greatly contribute to the control and containment of *Xvm* epidemics.

Hodgetts *et al.* (2015) developed a polyclonal antibody (pAb) and deployed in a lateral flow device (LFD) format to allow rapid in-field detection of *Xcm*. Pure cultures of *Xcm* were used to immunise a rabbit, the IgG antibodies purified from the serum and the resulting polyclonal antibodies tested using ELISA and LFD. Testing against a wide range of bacterial species showed the pAb detected all strains of *Xcm*, representing isolates from seven countries and the known genetic diversity of *Xcm*. The pAb also detected the closely related *Xan. axonopodis* pv. *vasculorum* (*Xav*), primarily a sugarcane pathogen. Detection was successful in both naturally and experimentally infected banana plants, and the LFD limit of detection was 10^5 cells mL^{-1}. Whilst the pAb is not fully specific for *Xcm*, *Xav* has never been found in banana. Therefore, the LFD can be used as a first-line screening tool to detect *Xcm* in the field. Testing by LFD requires no equipment, can be performed by non-scientists and is cost-effective. Therefore, this LFD provides a vital tool to help in the management and control of *Xcm*.

Tripathi *et al.* (2019) reported the first comparative transcriptome profiles of the susceptible and resistant genotypes of banana during early infection with *Xcm* and provided insights into the defence mechanism in *Musa balbisiana*, which can be used for the genetic improvement of commonly cultivated banana varieties.

The single diseased stem removal (SDSR) technology (i.e. the timely cutting of diseased banana plants) is an important Xanthomonas wilt (XW) management method being promoted in XW-endemic zones of East/Central Africa. Proper SDSR application leads to rapid field recovery. Nitamwira *et al.* (2019) reported that XW incidence in suckers increased with a delay in the cutting of symptomatic mother plants, with the lowest incidence (0.27%) observed when symptomatic plants were immediately cut upon symptom expression and the highest (1.61%) when plants decayed *in situ*. Hence, immediate cutting of plants after first visible symptoms is recommended to farmers. SDSR was also found to be as effective at the ABB banana-dominated low-altitude sites, as previously reported for highland bananas and environments.

Rutikanga *et al.* (2016) has advocated the cultivation of non-host crops after uprooting *Xan. campestris* pv. *musacearum* infected banana plants for breaking the Xanthomonas wilt disease cycle in fields. Lengthening the time under break crops to two and four seasons, respectively, reduced the incidence to 3% and zero in Rwanda and 0–8% in the two-season experiments in DR Congo. Two seasons with a break crop plus collective XW control are recommended in these agro-ecosystems.

Kilwenge *et al.* (2023) reported that Xanthomonas wilt of banana is a major threat to banana production in Rwanda, causing up to 100% yield loss. They found that precipitation during the coldest quarter, average maximum monthly temperature, annual precipitation, and elevation were the strongest predictors, which were responsible for 22.1, 13.0, 12.6, and 9.4% of the observed incidence variability, respectively, while the mean temperature of the coldest quarter had the highest gain in isolation.

9.57.5 Management

1. Once the plant is infected, it will die. Remove infected plants as soon as they are encountered. Cut the infected plants into pieces and bury deep outside the banana farm.
2. Remove the male bud before it opens, when the last hand is exposed. The practice of de-budding by means of a forked wooden stick just after the formation of the last hand is an effective control measure for all bacterial wilts of *Musa* spp., incidentally also resulting in larger/bigger and more evenly filled fruits. The risk of the spread of the pathogen is reduced, if de-budding is done with a forked stick rather than with a cutlass.
3. Cleaning of garden/field tools during routine plantation and sanitation practices, in fields infected with *Xvm* or *Ralstonia* sp., can be done using a 20% solution of household bleach (sodium hypochlorite 3.5%) or sterilised using heat from a hot fire.
4. Do not use suckers (planting materials) from an infected farm to start a new plantation.
5. Herbicides can be used successfully to destroy infected plants. Mature plants can be injected with 2, 4- dichlorophenoxy acetic acid (2, 4-D). At least 1.6 ml of the original concentration of 2, 4-D (irrespective of the dilution) needs to be applied per mature plant. The effect of 2, 4-D is visible two weeks after injection when the pseudostems of injected plants break at the base and fall off. By this time the 2, 4-D has reached the underground corm, which begins to rot. This method is much easier and cheaper than manually rouging and burying infected plants.
6. Although no banana cultivar resistant to the disease has been identified to date, it appears that certain cultivars have mechanisms to 'escape' the disease. The management and control of the disease involves methods that reduce the inoculum's density and the spread of the pathogen. Removal of the male bud (de-budding) has proven to be very effective in preventing the disease incidence since the male bud appears to be the primary infection site. While germplasm screening for disease resistance is in progress, efforts to genetically engineer resistance in some banana cultivars are also making good progress.
7. Nakakawa *et al.* (2017) designed an optimal control framework in which the use of clean planting materials, debudding, disinfection of tools and rouging were considered as control measures of BXW within a plantation of multiple cultivars. A combination of both prevention and containment controls yielded the greatest decline in infection levels of cultivars. They further emphasised that for effective disease management, it is important to assess the endemic level of the plantation before the application of control measures, and, once implemented, this should be maintained even when the disease is undetectable to eliminate possible resurgence.
8. During the initial epiphytotic phase of *Xvm* outbreaks, integrated management approaches are recommended. These include uprooting, chopping and burying of diseased mats or stools (i.e. clumps of plants formed from the same parent plant), the removal of the male bud with a forked stick on the formation of the last cluster of fruit to limit insect transmission and the cleaning of farming implements with sodium hypochlorite. This integrated approach is time-consuming and labour-intensive for resource-constrained farmers (Tushemereirwe *et al.*, 2006). These recommendations are founded on the management principles used to deal with other banana bacterial wilt diseases, such as Moko disease (Thwaites *et al.*, 2000).
9. Mburu *et al.* (2016) reported that silicon concentrations above 200 mg/week induce resistance to BXW and thus may augment existing control measures through integrated *Xvm* management options.
10. Practices, such as the continuous monitoring of plantations and removal of symptomatic plants, referred to as single diseased stem removal (abbreviated as SDSR), are encouraged. SDSR lowers Xvm inoculum, preventing it from spreading to neighbouring suckers and thus reducing disease incidence (Kubiriba et al., 2012). However, the impact of SDSR may be more effective during the dry season, when pathogen transmission is less, than during the rainy season.
11. According to Blomme *et al.* (2017), SDSR is a complementary practice to complete mat uprooting, applicable when latent infections and

incomplete systemic infections have occurred. They further noted that the choice of which method to use, either complete mat uprooting or SDSR, will depend on the farming objectives and disease occurrence. Complete mat uprooting is recommended where the disease is reported for the first time, with the aim of removing a large portion of the inoculum. This method appeals to intensive, market-oriented farmers, whereas SDSR is the more favoured method of subsistence farmers.

12. Uwamahoro *et al.* (2019) have advocated the proper implementation of management practices and cultivation of a mixture of indigenous and improved cultivars.

REFERENCES

Abele, S., Twine, E. and Legg, C. (2007). *Food security in Rwanda. C3P Food Security Briefs No.5, Ibadan, Nigeria (IITA).* http://c3project.iita.org/

Addis, T., Handoro, F. and Blomme, G. (2004). Bacterial wilt (*Xanthomonas campestris* pv. musacearum) on Enset and banana in Ethiopia. *InfoMusa* 13: 44–45.

Addis, T., Turyagyenda, L. F., Alemu, T., Karamura, E. and Blomme, G. (2010). Garden tool transmission of *Xanthomonas campestris* pv. *musacearum* on banana (*Musa* spp.) and enset in Ethiopia. *Acta Hort.* 879: 367–72. https://doi.org/10.17660/ActaHortic.2010.879.39

Adikini, S., Tripathi, L., Beed, F., Tusiime, G., Magembe, E. M. *et al.* (2011). Development of a specific molecular tool for detecting *Xanthomonas campestris* pv. *musacearum*. *Plant Pathol.* 60: 443–52.

Adriko, J., Aritua, V., Mortensena, C. N., Tushemereirwe, W. K., Kubiriba, J. *et al.*, (2012). Multiplex PCR for specific and robust detection of *Xanthomonas campestris* pv. *musacearum* in pure culture and infected plant material. *Plant Pathol.* 61: 489–97.

Anonymous (2012). *Guide de Terrain sur le Contrôle et la Gestion du BXW (Wilt Bactérien du bananier) en Province du Nord et Sud Kivu, RDC.* North Kivu: Commission Provinciale de lutte contre les maladies du bananier au Nord Kivu. 27.

Aritua, V., Parkinson, N., Thwaites, R., Heeney, J. V., Jones, D. R. *et al.* (2008). Characterization of the *Xanthomonas* sp. causing wilt of enset and banana and its proposed reclassification as a strain of *X. vasicola*. *Plant Pathol.* 57: 170–77. https://doi.org/10.1111/j.1365-3059.2007.01687.x

Biruma, M., Pillay, M., Tripathi, L., Blomme, G., Abele, S. *et al.* (2007). Banana Xanthomonas wilt: A review of the disease, management strategies and future research directions. *Afric. J. Biotechnol.* 6: 953–62.

Blomme, G., Mukasa, H., Ssekiwoko, F. and Eden-Green, S. (2005). On-farm assessment of banana bacterial wilt control options. *Afric. Crop Sci. Conf. Proc.* 7: 317–20.

Blomme, G., Ocimati, W., Sivirihauma, C., Vutseme, L., Mariamu, B. *et al.* (2017). A control package revolving around the removal of single diseased banana stems is effective for the restoration of Xanthomonas wilt infected fields. *Eur. J. Plant Pathol.* 149(2): 385–400. https://doi.org/10.1007/s10658-017-1189-6

Bradbury, J. F. (1986). *Guide to Plant Pathogenic Bacteria*. Kew: CAB International Mycological Institute.

Buddenhagen, I. (2006). Managing banana bacterial wilts in latin America. In: E. Karamura, M. Osiru, G. Blomme, C. Lusty, and C. Pic (eds.). *Developing a Regional Strategy to Address the Outbreak of Xanthomonas Wilt in East and Central Africa. Proceedings of the Banana Xanthomonas Wilt Regional Preparedness and Strategy Development Workshop*. Kampala, Uganda, 14–18 February 2005. International Network for the Improvement of Banana and Plantain (INIBAP), Montpellier, pp. 23–25.

Buregyeya, H. G., Tusiime, J., Kubiriba, J. and Tushmereirwe, W. K. (2008). Evaluation of distant transmission of banana bacterial wilt in Uganda. *Paper presented atthe conference on Banana and Plantain in Africa: Harnessing International Partnerships to Increase Research Impact, October* 5–9, 2008, Mombasa.

Carter, B. A., Reeder, R., Mgenzi, S. R., Kinyua, Z. M., Mbaka, J. N. *et al.* (2010). Identification of *Xanthomonas vasicola* (formerly *X. campestris* pv. *musacearum*), causative organism of banana Xanthomonas wilt, in Tanzania, Kenya and Burundi. *Plant Pathol.* 59: 403. https://doi.org/10.1111/j.1365-3059.2009.02124.x

Castellani, E. (1939). Su un marciume dell' ensete. *L'Agric. Coloniale Firenze* 33: 297–300.

Chen, C. H., Lin, H. J., Ger, M. J., Chow, D. and Feng, T. Y. (2000). The cloning and characterization of a hypersensitive response assisting protein that may be associated with the harpin-mediated hypersensitive response. *Plant Mol. Biol.* 43: 429–38. https://doi.org/10.1023/A:1006448611432

Dodds, P. N. and Rathjen, J. P. (2010). Plant immunity: Towards an integrated view of plant–pathogen interactions. *Nat. Rev. Genet.* 11: 539–48.

Dye, D. W. (1978). In: Young, J. M., Dye, D. W., Bradbury, J. F., Panagopoulos, C. G. and Robbs, C. F. (1978). A proposed nomenclature and classification for plant pathogenic bacteria. *N. Zeal. J. Agric. Res.* 21: 153–77. https://doi.org/10.1080/00288233.1978.10427397

Eden-Green, S. (2004). Focus on bacterial wilt. How can the advance of banana Xanthomonas wilt be halted? *InfoMusa* 13: 38–41.

FAOSTAT (2014). *Banana and Plantain Surface and Production in 2013*. http://faostat.fao.org/site/339/default.aspx

Hodgetts, J., Karamura, G., Johnson, G., Hall, J., Perkins, K. *et al.* (2015). Development of a lateral flow device for in-field detection and evaluation of PCR-based diagnostic methods for *Xanthomonas campestris* pv. *musacearum*, the causal agent of banana Xanthomonas wilt. *Plant Pathol.* 64: 559–67.

Hunduma, T., Sadessa, K., Hilu, E. and Oli, M. (2015). Evaluation of enset clones resistance against enset bacterial wilt disease (*Xanthomonas campestris* pv. *musacearum*). *J. Vet. Sci. Technol.* 6: 232.

Kagezi, G. H., Kangire, A., Tushmereirwe, W., Bagamba, F., Kikulwe, E. *et al.* (2006). Banana bacterial wilt incidence in Uganda. *Afric. Crop Sci. J.* 14: 83–91.

Kalyebara, R., Wood, S. and Abodi, P. M. (2007). Assessing the potential impact of selected technologies on the banana industry in Uganda. In: M. Smale and W. K. Tushemereirwe (eds.). *An Economic Assessment of Banana Genetic Improvement and Innovation in the Lake Victoria Region of Uganda and Tanzania. IFPRI Research Report 155.* IFPRI, Washington, DC, pp. 141–56.

Karamura, E., Frison, E., Karamura, D. A. and Sharrock, S. (1998). *Banana Production Systems in Eastern and Southern Africa.* Bananas and Food Security, INIBAP, Montpellier, pp. 401–12.

Karamura, E. B., Turyagyenda, F. L., Tinzaara, W., Blomme, G., Molina, A. *et al.* (2008). Xanthomonas wilt of bananas in East and Central Africa. *Diagnostic and Management Guide. Bioversity International, Uganda. INIBAP.*

Karamura, G., Smith, J., Studholme, D., Kubiriba, J. and Karamura, E. (2015). Comparative pathogenicity studies of the *Xanthomonas vasicola* species on maize, sugarcane and banana. *Afr. J. Plant Sci.* 9: 385–400. https://doi.org/10.5897/AJPS2015.1327

Kilwenge, R., Adewopo, J., Manners, R., Mwizerwa, C., Kabirigi, M. *et al.* (2023). Climate-related risk modeling of banana Xanthomonas wilt disease incidence in the cropland area of Rwanda. *Plant Dis.* 107: 2017–26.

Kubiriba, J., Karamura, E. B., Jogo, W., Tushemereirwe, W. K. and Tinzaara, W. (2012). Community mobilization: A key to effective control of banana Xanthomonas wilt. *J. Dev. Agric. Econ.* 4: 125–131.

Lescot, T. (2013). World plantain and banana production systems. *Proceedings XX International Meeting ACORBAT*, September 9–13, Fortaleza, pp. 26–34.

Lin, H. J., Cheng, H. Y., Chen, C. H., Huang, H. C. and Feng, T. Y. (1997). Plant amphipathic proteins delay the hypersensitive response caused by harpin Pss and *Pseudomonas syringae* pv. syringae. *Physiol. Mol. Plant Pathol.* 51: 367–76. https://doi.org/10.1006/pmpp.1997.0121

Manyong, V. M., Legg, C., Mwangi, M., Nakato, V., Coyne, D. *et al.* (2008). The potential benefits of GIS techniques in disease and pest control: An example based on a regional project in Central Africa. In: T. Dubois, C. Staver, S. Hauser, and D. Coyne (eds.). *IV International Symposium on Banana: International Conference on Banana and Plantain in Africa*, Vol. 879. Harnessing International: Acta Horticulturae, pp. 333–40.

Mbaka, J. N., Nakato, V. G., Auma, J. and Odero, B. (2009). Status of banana Xanthomonas wilt in Western Kenya and factors enhancing its spread. *Proc. Afr. Crop Sci. Conf.* 9, 673–76.

Mburu, K., Oduor, R., Mgutu, A. and Tripathi, L. (2016) Silicon application enhances resistance to *Xanthomonas* wilt disease in banana. *Plant Pathol.* 65: 807–15. Mgenzi, S. R. B., Muchunguzi, D., Mutagwaba, T., Mkondo, F. and Mohamed, R. (2006). An Outbreak of Banana Bacterial wilt in Muleba district, Kagera region, Tanzania. Disease report, Bukoba: Maruku Agriculture Research and Development Institute, Tanzania.

Mwangi, M., Mwebaze, M., Bandyopadhyay, R., Aritua, V., Eden-Green, S. *et al.*, (2007). Development of a semiselective medium for isolating *Xanthomonas campestris* pv. *musacearum* from insect vectors, infected plant material and soil. *Plant Pathol.* 56: 383–90.

Mwebaze, J. M., Tusime, G., Tushmereirwe, W. K. and Kubiriba, J. (2006). The survival of *Xanthomonas campestris* pv. *musacearum* in soil and plant debris. *Afr. Crop Sci. J.* 14: 121–27.

Nakakawa, J., Mugisha, J. Y. T., Shaw, M. W., Tinzaara, W. and Karamura, E. (2017). Banana Xanthomonas wilt infection: The role of debudding and roguing as control options within a mixed cultivar plantation. *Int. J. Math. Math. Sci.* Article ID 4865015, 13 pp. http://hdl.handle.net/10568/89844

Nakato, G. V., Beed, F., Ramathani, I., Rwomushana, I. and Opio, F. (2013). Risk of banana Xanthomonas wilt spread through trade. *J. Crop Prot.* 2: 151–61.

Nakato, G. V., Christelova, P., Were, E., Nyine, M., Coutinho, T. A. *et al.* (2019). Sources of resistance in *Musa* to *Xanthomonas campestris* pv. *musacearum*, the causal agent of banana Xanthomonas wilt. *Plant Pathol.* 68: 49–59.

Nakato, G. V., Ocimati, W., Blomme, G., Fiaboe, K. K. M. and Beed, F. (2014). Comparative importance of infection routes for banana Xanthomonas wilt and implications on disease epidemiology and management. *Can. J. Plant Pathol.* 36: 418–27.

Nakato, G. V., Studholme, D. J., Blomme, G., Grant, M., Coutinho, T. A. *et al.* (2021). SNP-based genotyping and whole-genome sequencing reveal previously unknown genetic diversity in *Xanthomonas vasicola* pv. *musacearum*, causal agent of banana Xanthomonas wilt, in its presumed Ethiopian origin. *Plant Pathol.* 70: 534–43.

Nakato, V., Mahuku, G. and Coutinho, T. (2018). *Xanthomonas campestris* pv. *musacearum*: A major constraint to banana, plantain and enset production in central and east Africa over the past decade. *Mol. Plant Pathol.* 19: 525–36.

Namukwaya, B., Tripathi, L., Tripathi, J. N., Arinaitwe, G., Mukasa, S. B. *et al.* (2012). Transgenic banana expressing Pflp gene confers enhanced resistance to Xanthomonas wilt disease. *Transgenic Res.* 12, 855–65. https://doi.org/10.1007/s11248-011-9574-y

Ndungo, V., Bakelana, K., Eden-Green, S. and Blomme, G. (2004). An outbreak of banana Xanthomonas wilt (*Xanthomonas campestris* pv. *musacearum*) in the democratic republic of Congo. *InfoMusa* 13: 43–44. https://doi.org/10.1111/mpp.12578

Nimusiima, J., Köberl, M., Tumuhairwe, J. B., Kubiriba, J., Staver, C. *et al.* (2015). Transgenic banana plants expressing Xanthomonas wilt resistance genes revealed a stable non-target bacterial colonization structure. *Sci. Rep.* 5, 18 078.

Nkuba, J., Tinzaara, W., Night, G., Niko, N., Jogo, W. *et al.* (2015). Adverse impact of Banana Xanthomonas wilt on farmers livelihoods in eastern and Central Africa. *Afr. J. Plant Sci.* 9: 279–86. https://doi.org/10.5897/AJPS2015.1292

Nordling, L. (2010). Uganda prepares to plant transgenic bananas. *Nature.* ISSN: 0028-0836. http://www.nature.com/news/2010/101001/full/news.2010.509.html

Ntamwira, J., Blomme, G., Bahati, L. and Ocimati, W. (2019). Effect of timing of diseased plant cutting, altitude and banana cultivar on efficacy of singly removing Xanthomonas wilt infected banana plants. *Eur. J. Plant Pathol.* 154: 477–89. https://doi.org/10.1007/s10658-019-01671-9

Ochola, D., Ocimati, W., Tinzaara, W., Blomme, G. and Karamura, E. B. (2015). Effect of water stress on the development of banana Xanthomonas wilt disease. *Plant Pathol.* 64: 552–58.

Ocimati, W., Nakato, G. V., Fiaboe, K. K. M., Beed, F. and Blomme, G. (2015). Incomplete systemic movement of *Xanthomonas campestris* pv. *musacearum* and the occurrence of latent infections in Xanthomonas wilt-infected banana mats. *Plant Pathol.* 64: 81–90. https://doi.org/10.1111/ppa.12233

Ocimati, W., Ssekiwoko, F., Karamura, E., Tinzaara, W., Eden-Green, S. *et al.* (2013a). Systemicity of *Xanthomonas campestris* pv. *musacearum* and time to disease expression after inflorescence infection in East African highland and Pisang Awak bananas in Uganda. *Plant Pathol.* 62: 777–85.

Ocimati, W., Ssekiwoko, F., Karamura, E. B., Tinzaara, W. and Blomme, G. (2013b). Does *Xanthomonas campestris* pv. *musacearum* colonize banana cord root tissue? *Acta Hort.* 986: 103–09. https://doi.org/10.17660/ActaHortic.2013.986.8

Ocimati, W., Tazuba, A. F. and Blomme, G. (2022). Spread of *Xanthomonas vasicola* pv. *musacearum* within banana mats: Implications for Xanthomonas wilt management. *J. Plant Pathol.* 104: 1111–22. https://doi.org/10.1007/s42161-022-01165-4

Parkinson, N., Cowie, C., Heeney, J. and Stead, D. (2009). Phylogenetic structure of *Xanthomonas* determined by comparison of *gyrB* sequences. *Int. J. Syst. Evol. Microbiol.* 59: 264–74.

Qhobela, M. and Claflin, L. E. (1988). The taxonomic position of *Xanthomonas campestris* pv. *holcicola* (Elliott 1930) Dye

1978 and *Xanthomonas campestris* pv. *vasculorum* (Cobb 1893) Dye 1978. *Phytopathology* 77: 1766.

Rutikanga, A., Sivirihauma, C., Ocimati, W., Night, G., Murekezi, C. *et al.* (2016). Breaking the cycle of *Xanthomonas campestris* pv. *musacearum* in infected fields through the cultivation of annual crops and disease control in adjacent fields. *J. Phytopath.* 164: 659–70.

Schaad, N. W., Jones, J. B. and Chun, W. (2001). *Laboratory Guide for Identification of Plant Pathogenic Bacteria.* 3rd Ed. APS Press, St. Paul, MN.

Shimelash, D., Alemu, T., Addis, T., Turyagyenda, F. L. and Blomme, G. (2008). Banana Xanthomonas wilt in Ethiopia: Occurrence and insect vector transmission. *Afr. Crop Sci. J.* 16: 75–87.

Shimwela, M. M., Ploetz, R. C., Beed, F. D., Jones, J. B., Blackburn, J. K. *et al.* (2016). Banana Xanthomonas wilt continues to spread in Tanzania despite an intensive symptomatic plant removal campaign: An impending socio-economic and ecological disaster. *Food Security* 8: 939–51. https://doi.org/10.1007/s12571-016-0609-3

Sijam, K., Chang, C. J. and Gitaitis, R. D. (1991). An agar medium for the isolation and identification of *Xanthomonas campestris* pv. *vesicatoria* from seed. *Phytopathology* 81: 831–34.

Smith, J. J., Jones, D. R., Karamura, E., Blomme, G. and Turyagyenda, F. L. (2008). An analysis of the risk from *Xanthomonas campestris* pv. *musacearum* to banana cultivation in eastern, central and southern Africa. Bioversity International, Montpellier, n. 35 p.

Song, W. Y., Wang, G. L., Chen, L. L., Kim, H. S., Pi, L. Y. *et al.* (1995). A receptor kinase-like protein encoded by the rice disease resistance gene, *Xa21*. *Science* 270: 1804–06.

Ssekiwoko, F., Kiggundu, A., Tushemereirwe, W. K., Karamura, E. and Kunert, K. (2015). *Xanthomonas vasicola* pv. *musacearum* down-regulates selected defense genes during its interaction with both resistant and susceptible banana. *Physiol. Mol. Plant Pathol.* 90: 21–6.

Ssekiwoko, F., Taligoola, H. K. and Tushemereirwe, W. K. (2006a). *Xanthomonas campestris* pv. *musacearum* host range in Uganda. *Afr. Crop Sci. J.* 14: 111–20.

Ssekiwoko, F., Turyagyenda, L. F., Mukasa, H., Eden-Green, S. and Blomme, G. (2006b). Systemicity of *Xanthomonas campestris* pv. *musacearum* in flower-infected banana plants. In: *Proceedings of the XVII ACORBAT International Meeting: Banana: A Sustainable Business*. Joinville, Santa Catarina.

Ssekiwoko, F., Tushemereirwe, W., Batte, M., Ragama, P. E. and Kumakech, A. (2006c). Reaction of banana germplasm to inoculation with *Xanthomonas campestris* pv. *musacearum*. *Afr. Crop. Sci. J.* 14: 151–55.

Studholme, D. J., Wicker, E., Abrare, S. M., Aspin, A. and Bogdanove, A. (2020). Transfer of *Xanthomonas campestris* pv. *arecae* and *X. campestris* pv. *musacearum* to *X. vasicola* (Vauterin) as *X. vasicola* pv. *arecae* comb. nov. and *X. vasicola* pv. *musacearum* comb. nov. and description of *X. vasicola* pv. *vasculorum* pv. nov. *Phytopathology* 110: 1153–60.

Thwaites, R., Eden-Green, S. J. and Black, R. (2000). Diseases caused by bacteria. In: D. R. Jones (ed.). *Diseases of Banana, Abacá and Enset*. CAB International, Wallingford, pp. 213–39.

Tinzaara, W., Gold, C. S., Ssekiwoko, F., Bandyopadhyay, R., Abera, A. *et al.* (2006). Role of insects in the transmission of banana bacterial wilt. *Afr. Crop Sci. J.* 14: 105–11.

Tripathi, L., Atkinson, H., Roderick, H., Kubiriba, J. and Tripathi, J. N. (2017). Genetically engineered bananas resistant to Xanthomonas wilt disease and nematodes. *Food Energy Secur.* 6(2): 37–47.

Tripathi, L., Mwaka, H., Tripathi, J. N. and Tushemereirwe, W. K. (2010). Expression of sweet pepper *Hrap* gene in banana enhances resistance to *Xanthomonas campestris* pv. *musacearum*. *Mol. Plant Pathol.* 11: 721–31. https://doi.org/10.1111/j.1364-3703.2010.00639.x

Tripathi, L., Mwangi, M., Abele, S., Aritua, V., Tushemereirwe, W. K. *et al.* (2009). Xanthomonas wilt: A threat to banana production in East and Central Africa. *Plant Dis.* 93: 440–51. https://doi.org/10.1094/PDIS-93-5-0440

Tripathi, L., Odipio, J., Tripathi, J. N. and Tusiime, G. (2008). A rapid technique for screening banana cultivars for resistance to Xanthomonas wilt. *Eur. J. Plant Pathol.* 121: 9–19.

Tripathi, L. and Tripathi, J. N. (2009). Relative susceptibility of banana cultivars to *Xanthomonas campestris* pv. *musacearum*. *Afr. J. Biotechnol.* 8(20): 5343–50.

Tripathi, L., Tripathi, J. N., Kiggundu, A., Korie, S., Shotkoski, F. *et al.* (2014). Field trial of Xanthomonas wilt disease-resistant bananas in East Africa. *Nat. Biotechnol.* 32: 868–70. https://doi.org/10.1038/nbt.3007

Tripathi, L., Tripathi, J. N., Shah, T., Muiruri, K. S. and Katari, M. (2019). Molecular basis of disease resistance in banana progenitor *Musa balbisiana* against *Xanthomonas campestris* pv. *musacearum*. *Sci. Rep.* 9(1): 7007. https://doi.org/10.1038/s41598-019-43421-1

Tripathi, L., Tripathi, J. N., Tushemereirwe, W. K., Arinaitwe, G. and Kiggundu, A. (2013). Transgenic bananas with enhanced resistance against Xanthomonas wilt disease. *Acta Hortic.* 974: 81–90.

Tripathi, L., Tripathi, J. N., Tushemereirwe, W. K. and Bandyopadhyay, R. (2007). Development of a semi-selective medium for isolation of *Xanthomonas campestris* pv. *musacearum* from banana plants. *Eur. J. Plant Pathol.* 117: 177–86.

Tushemereirwe, W., Kangire, A., Smith, J., Ssekiwoko, F., Nakyanzi, M. *et al.* (2003). An outbreak of bacterial wilt on banana in Uganda. *InfoMusa* 12: 6–8.

Tushemereirwe, W., Kangire, A., Ssekiwoko, F., Offord, L. C., Crozier, J. *et al.* (2004). First report of *Xanthomonas campestris* pv. *musacearum* on banana in Uganda. *Plant Pathol.* 53: 802. https://doi.org/10.1111/j.1365-3059.2004.01090.x

Tushemereirwe, W. K., Okaasai, O., Kubiriba, J., Nankinga, C., Muhangi, J. *et al.* (2006). Status of banana bacterial wilt in Uganda. Special issue. Banana Bacterial Wilt in Uganda: A disease that threatens livelihoods. *Afr. Crop Sci. J.* 14: 73–82.

Uwamahoro, F., Berlin, A., Bylund, H., Bucagu, C. and Yuen, J. (2019). Management strategies for banana *Xanthomonas* wilt in Rwanda include mixing indigenous and improved cultivars. *Agron. Sustain. Dev.* 39: Article number 22. https://doi.org/10.1007/s13593-019-0569-z

Vauterin, L., Hoste, B., Kersters, K. and Swings, J. (1995). Reclassification of *Xanthomonas*. *Int. J. Syst. Bacteriol.* 45: 472–89. https://doi.org/10.1099/00207713-45-3-472

Vauterin, L., Yang, P., Hoste, B., Pot, B., Swings, J. *et al.* (1992). Taxonomy of xanthomonads from cereals and grasses based on SDS-PAGE of proteins, fatty-acid analysis and DNA hybridization. *J. Gen. Microbiol.* 138: 1467–77. https://doi.org/10.1099/00221287-138-7-1467

Wang, G. L., Song, W. Y., Ruan, D. L., Sideris, S. and Ronald, P. C. (1996). The cloned gene *Xa21* confers resistance to multiple *Xanthomonas oryzae* pv. *oryzae* isolates in transgenic plants. *Mol. Plant–Microbe Interact.* 9: 850–55.

Wasukira, A., Tayebwa, J., Thwaites, R., Paszkiewicz, K., Aritua, V. *et al.* (2012). Genome-wide sequencing reveals two major sub-lineages in the genetically monomorphic pathogen

Xanthomonas campestris pv. *musacearum*. *Genes* 3: 361–77. https://doi.org/10.3390/genes3030361

Welde-Michael, G., Bobosha, K., Addis, T., Blomme, G., Mekonnen, S. *et al.* (2008). Mechanical transmission and survival of bacterial wilt on enset. *Afr. Crop Sci. J.* 16: 97–102.

Were, E., Nakato, G. V., Ocimati, W., Ramathani, I., Olal, S. *et al.* (2015). The banana weevil, *Cosmopolites sordidus* (Germar), is a potential vector of *Xanthomonas campestris* pv. *musacearum* in bananas. *Can. J. Plant Pathol.* 37: 427–34.

Yirgou, D. and Bradbury, J. F. (1968). Bacterial wilt of enset (*Ensete ventricosum*) incited by *Xanthomonas musacearum* sp. n. *Phytopathology* 58: 111–12.

Yirgou, D. and Bradbury, J. F. (1974). A note on wilt of banana caused by the enset wilt organism *Xanthomonas musacearum*. *East Afr. Agric. For. J.* 40: 111–14.

Young, J. M., Dye, D. W., Bradbury, J. F., Panagopoulos, C. G. and Robbs, C. F. (1978). A proposed nomenclature and classification for plant pathogenic bacteria. *N. Zeal. J. Agric. Res.* 21: 153–77. https://doi.org/10.1080/00288233.1978.10427397

9.58 YELLOW PUSTULE OF TULIP

Tulips belongs to the genus *Tulipa*. This is a genus of about 100 species of bulbous herbs in the lily family (Liliaceae), native to Central Asia and Turkey. These are spring-blooming perennial herbaceous geophytes having bulbs as storage organs. The tulips are among the most popular of all garden flowers, and numerous cultivars and varieties have been developed. Tulip suffers from many diseases caused by plant pathogenic fungi, bacteria and viruses, including yellow pustule. Garden tulip (*Tulipa gesneriana*) is a commonly affected host of this disease.

In 1969, Saaltink and Maas Geesteranus described a new bacterial disease of tulip caused by *Corynebacterium oortii* nov. sp. from the Netherlands. This disease had been known there by the name of ‘geelpok’ (yellow spot) of the bulb or ‘helsvuur’ (hell fire) of the foliage, but its cause remained obscure for many years. The symptoms consisted of yellow spots on the bulbs and silvery spots and streaks on the leaves.

A strikingly similar disorder of tulip has been seen in Japan since around 1960. Because of the peculiar appearance of affected plants, which has no resemblance to any other established diseases of tulip in Japan, Japanese growers called it by various names: ‘Bakuretsu-byo’ (bursting disease) in Niigata prefecture; ‘Kusha-kusha’ (crumpling) in Toyama prefecture; and ‘Boroboro’ (crumbling) in Fukui prefecture. Its outbreak and severity varied greatly from year to year, field to field and/or cultivar to cultivar and seemed strongly influenced by climatic conditions. The year 1972 saw an unusual outbreak of this disorder throughout the tulip areas, causing considerable losses to the bulb crops. Kobayashi and Obata (1974) critically examined diseased tulip plants in the light of the finding of Saaltink and Maas Geesteranus (1969) and demonstrated that this disorder is identical with the Dutch disease in terms of both symptomatology and etiology. This disease is now widespread in tulip-growing areas throughout Japan and is a cause of concern for the production of bulbs and cut flowers. The name ‘Kaiyo-byo’ (bacterial canker) was recommended as a standard name for this disease in the country.

9.58.1 Geographical Distribution

The disease is present in Denmark, Germany, Japan, the U K and The Netherlands. The disease is not of much economic importance (Janse, 2005).

9.58.2 Symptoms

There are two phases of the symptoms i.e., hell-fire (‘helsvuur’) of the leaves and yellow pustule (‘Geel-pok’) of tulip bulbs.

9.58.2.1 Hell-fire

On the leaves, silver-grey spots up to 5 mm in diameter are the first symptoms. In later stages after some frosty nights, upper and lower epidermises may crack badly and become loosened from the other tissue, usually along the veins, hence the name **‘hell-fire’**. The vascular tissue of leaves and bulbs shows some yellowish discolouration. Severely infected plants are stunted and the flower buds wither easily. Infected plants may also have leaves with areas in which the upper and lower epidermises are badly cracked, giving a characteristic rough appearance to the leaves. These plants show yellowish stem interiors that run down to thc young growing bulbs (Bradbury, 1973). Rupturing of the epidermal tissue also appears on the stem, principally close to the flower bud. The flower petals may exhibit discoloured fleckings and blistering cankers. Affected parts of the plants, whether leaf, stem or inflorescence, show various degrees of distortion depending upon the severity of the disease. Plants which are heavily infected in early growth stage frequently topple and wither away. In some plants, a heavy cracking of the upper or lower epidermis occurs without any sign of silvery symptoms.

9.58.2.2 Yellow Pustule

On bulbs, tiny slightly swollen white to grey spots occur on the outermost white scales which later become raised and turn yellow, leading to the name **‘yellow pustule’**. Initial symptoms develop mainly at the tip or middle portion of the first and second sprouting leaves, but less often on upper leaves. The symptom development enhances rapidly as the growth of tulip accelerates, but it tends to be arrested toward the blooming stage. Less severely affected bulbs develop into stunted plants with leaves showing silvery streaks along their veins (Bradbury, 1973).

FIGURE 9.48 Yellowing and wrinkling of leaf margins and tips, and yellow pustules on bulbs.

9.58.3 Causal Organism

Curtobacterium flaccumfaciens pv. *oortii* (Saaltink and Maas Geesteranus 1969) Collins & Jones 1983

Synonyms:

1. *Corynebacterium flaccumfaciens* pv. *oortii* Saaltink & Maas Geesteranus 1969
2. *Cor. oortii* Saaltink & Mass Geesteranus 1969

The bacteria are Gram-stain-positive, non-sporing rods with rounded ends, motile with a single polar flagellum and measure 0.8 (0.5–1.3) × 1 .6 (0.9–2.6) µ. On beef-extract peptone agar, the colonies are moderate, filiform, slightly raised and measure 1.0–2.0 mm in diameter after 3 days of incubation at 25 °C. On potato dextrose agar, colonies are white having slightly creamy tinge and measure 2.0 mm in diameter after 4 days of incubation.

Growth is aerobic and gelatin liquefaction is slow. The Voges–Proskauer reaction is negative in glucose phosphate peptone medium. Starch is not hydrolysed, citrate is not utilised, methyl red test and hydrolysis of urea are negative. Glucose decomposition is positive under aerobic conditions but negative under anaerobic conditions. Oxidase reaction is negative but catalase reaction is positive. Temperature requirements: optimum 25–30 °C, minimum below 5 °C and maximum 37 °C. Thermal death point is 54 °C.

9.58.4 Disease Cycle

Cur. flaccumfaciens pv. *oortii* is a systemic vascular pathogen and a wound parasite. The bacterium survives in infected bulbs and diseased plant refuse left in the field. The transmission of the bacterium to the newly growing plants is from infected bulbs, and transmission between plants in the field is by wind-blown rain and sand and by contaminated implements used to cut the flowers.

Cold weather and wounds caused by the rubbing of leaves, hail and cultivation practices in spring favour disease development.

Though the disease can be observed on most cultivars, it is more common on those cultivars belonging to the breed of Darwin, such as William Pitt, Red Pitt, Paul Richter etc. Cottage and Darwin Hybrid groups are also severely infected. Diseased plants appear in patches, scattered around in a field. There are no apparent relations between the rate of disease occurrence and the soil conditions.

9.58.5 Management

The following measures can help to reduce or check the disease incidence.

1. Preventive sprays or dipping in Captan fungicide containing pesticides can provide some protection.
2. Removal and destruction of infected plants and bulbs.
3. Avoiding working in the fields during humid weather conditions (especially when removing the flowers) reduces the risk of infection.
4. Late planting that avoids critical night frost may be beneficial.

REFERENCES

Bradbury, J. F. (1973). Descriptions of Fungi and Bacteria. *Corynebacterium oortii*. IMI Descriptions of Fungi and Bacteria, 38, Sheet 375.

Collins, M. D. and Jones, D. (1983). Reclassification of *Corynebacterium flaccumfaciens*, *Corynebacterium betae*, *Corynebacterium oortii* and *Corynebacterium poinsettiae* in the genus *Curtobacterium*, as *Curtobacterium flaccumfaciens* comb. nov. *Microbiology* 129: 3545–48.

Janse, J. D. (2005). Bacteria pathogenic to monocotyledonous plants, bulbaceous plants. In: *Phytobacteriology: Principles and Practice*. CABI Publishing, Wallingford, pp. 174–75.

Kobayashi, T. and Obata, T. (1974). Occurrence of *Corynebacterium oortii* Saaltink et Maas Geesteranus on tulip in Japan. *Res. Bull. Pl. Prot. Japan*, No. 12: 7–13.

Saaltink, G. J. and Maas Geesteranus, H. P. (1969). A new disease in tulip caused by *Corynebacterium oortii* nov. spec. *Netherlands J. Plant Pathol.* 75: 123–28. https://doi.org/10.1007/BF02137204

9.59 BACTERIAL BLIGHT OF GLADIOLUS

Gladiolus is a genus of perennial cormous-flowering plants in the iris family (Iridaceae). It is sometimes called the 'sword lily', but is usually called by its generic name (plural gladioli). The members of the genus grow in Asia, the Mediterranean area of Europe, South Africa and tropical Africa. The centre of diversity is in the Cape Floristic Region. The genus *Gladiolus* contains about 300 species. The World Checklist of Selected Plant Families had over 276 species in 1988, as of February 2017, the number stood at 300 (Wikipedia reference No. 16 n.d.).

In the summer of 1922, a disease-causing serious injury to the leaves of gladiolus (*Gladiolus* × *hortulanus*) was reported by a large number of growers from the Midwest of the USA. In 1923, in the same fields, and especially in poorly drained areas, the disease appeared on some young plants, resulting in a large number being destroyed. Young stock appears to be much more susceptible to this disease than mature plants. Of 16 varieties studied in the field, the young stock of 15 varieties was severely attacked, while mature plants of the same varieties showed only slight spotting or no infection. The one exception was in the variety Schwaben where mature plants were nearly as badly infected as the young stock. In contrasty, no trace of this disease was found on either young or mature plants of the variety Mrs. Frank Pendleton, although these were growing in the same fields and under the same environmental conditions. Microscopic examination of the discoloured leaf areas revealed bacterial mass in the tissues. A review of the literature revealed that it seems to be the first report of this blight on the gladiolus (McCulloch, 1924). The disease is also known by the name of '**leaf spot and scorch of gladiolus**' (Janse, 2005).

9.59.1 Geographical Distribution

McCulloch (1924) reported it from the USA for the first time. The disease also occurs in Australia, Canada, Finland, the Netherlands, South Africa and around the rest of the USA.

9.59.2 Symptoms

Dark green, water-soaked, leaf spots that are more or less rectangular, are the first symptoms of the disease. These leaf spots later turn brown and necrotic. Bacterial ooze is present on the lesions under humid conditions. The presence of many leaf spots and their collision leads to reduction in growth of plants and corms.

Lesions are first visible as narrow, horizontal and water-soaked spots, but the later progress of the infection is mostly in a longitudinal direction so that the spots appear more or less as squares or rectangles. Adjoining lesions unite until often the entire leaf from the tip to the base is involved. This widespread infection of the leaf is most noticeable in the young stock. All parts of the leaves of blooming plants are subjected to attack, but the infection is quite often found only on the middle area of the leaf. As seen in fresh specimens, these spots are translucent and water-soaked. In reflected light, the colour is very dark green; in transmitted light, it is bright green. The translucency of the spots is permanent, but when old and dry, the colours are of various shades of brown.

A considerable amount of bacterial exudate flows from the affected tissues. Under dry conditions, this exudate forms slender, twisted, white columns of between 2 and 10 mm in length. On drying, it may form a thin brittle layer or it may occur in the form of numerous small beads. Under moist conditions, it forms a viscid film over the leaf surface. Wind-blown soil, tiny insects and various small particles become embedded in this exudate, often in sufficient quantities to cover the leaf surface. The exudate seems to be composed entirely of bacteria; it dissolves readily in water and undoubtedly is a factor in the dissemination of the pathogen. Based on the presence of copious exudate in the lesions, McCulloch (1924) proposed the name *Bacterium gummisudans* for the causal bacterium of this disease.

9.59.3 Causal Organism

Xanthomonas campestris pv. *gummisudans* (McCulloch 1924) Dye 1978

Synonym: *Bacterium gummisudans* n. sp.

Bacterial cells are Gram-stain-negative, aerobic, capsulated, non-sporing, short rods, rounded at the ends, measure 0.6 to 0.8 μ by 1.0 to 2.8 μ, and are motile by means of one polar flagellum. A yellow, viscid growth is formed on culture media, nitrates are not reduced, casein in milk is digested and gelatin is liquefied. Acid is produced from dextrose and saccharose; ammonia and hydrogen sulphide are also produced. Optimum temperature for growth is around 30 °C, maximum 36 °C, minimum below 2 °C, and the thermal death point is around 50 °C. The bacterium is sensitive to sodium chloride and acids, readily killed by drying and by exposure to sunlight, and not acid-fast.

9.59.4 Disease Cycle

Microscopic examination of the lesions reveals that the organism attacks the parenchyma. The intercellular spaces and the cavities formed by the disintegration of cell walls become packed with the bacteria.

Bacterial cells enter the host through stomata. In the earliest stages, the bacteria are found only in the stomatal chamber, from which they spread into and fill adjoining intercellular spaces. Some chemical change occurs at this stage in the cell walls of the host, adjoining the masses of bacteria, as evidenced by their increased affinity for stains. Whether they are later dissolved or only crushed by pressure and thereby changed chemically, as to be vague and doubtful, is not yet determined, but cavities occur and become entirely filled with the pathogen. In bigger lesions, no definite cell wall structure remains in the central space. In some large cavities, the bacteria are arranged in layers (McCulloch, 1924).

9.59.5 Management

The following measures can help to reduce or check disease incidence.

1. Use pathogen-free or treated corms for sowing.
2. Removal and destruction of infected plants and bulbs.
3. Avoiding working in the fields during humid weather conditions reduces the risk of infection.
4. If possible, plant foliage should be kept dry.
5. Grow the resistant varieties.
6. Chemical control of the disease has not yet been worked out, but there is a possibility for keeping in mind, the efficacy of available chemicals.

REFERENCES

Dye, D. W. (1978). Genus IX. *Xanthomonas* Dowson 1939. In Young, J. M., Dye, D. W., Bradbury, J. F., Panagopoulos, C. G. and Robbs, C. F. (1978). A proposed nomenclature and classification for plant pathogenic bacteria. *N. Z. J. Agric. Res*. 21: 162–66.

Janse, J. D. (2005). Bacteria pathogenic to monocotyledonous plants, bulbaceous plants. In: *Phytobacteriology: Principles and Practice*. CABI Publishing, Wallingford, pp. 180–81.

McCulloch, L. (1924). A bacterial blight of gladioli. *J. Agric. Res*. 27: 225–34.

Wikipedia reference No. 16. (n.d.) "Search for *Gladiolus*". *World Checklist of Selected Plant Families*. Royal Botanic Gardens, Kew. Retrieved 2 February 2017.

9.60 SANDAL SPIKE DISEASE

Sandalwood tree (*Santalum album* Linn., family Santalaceae) is a highly valued tree in India. The heartwood of this tree is extensively used for carving, a traditional art in Karnataka, Tamil Nadu and Kerala. East Indian sandalwood oil is used in the perfume industry (particularly in high-priced perfumes) and also in the pharmaceutical industries. The use of sandalwood in religious ceremonies and rituals dates back to several centuries. The tree is predominantly distributed in the southern part of Karnataka and the northern part of Tamil Nadu. Due to stringent monopolistic rules, sandalwood tree has been declared a government property in these two states.

For a long time, sandalwood was not cultivated and the natural population also dwindled due to extensive illegal felling and smuggling. Therefore, in 1997 the sandalwood tree was placed under the 'Vulnerable' category by the International Union for Conservation of Nature. Consequently, the governments of Karnataka and Tamil Nadu retracted the monopolistic policy, in 2001 and 2002 respectively, and liberalised the policy for sandalwood cultivation and harvest by individual entrepreneurs and corporate bodies. The sandalwood tree is highly susceptible to sandal spike disease caused by a non-culturable phytoplasma seen exclusively in the phloem of diseased tissues. Infected trees die within one to two years of the appearance of disease symptoms. As early as 1792, Tippu Sultan declared it a 'Royal Tree' of Mysuru. It is currently rated to be the most precious and valuable tree in India.

India's sandalwood trees, particularly in Karnataka, are facing a serious threat with the return of the destructive **sandal spike disease**. The infection has resurfaced in the aromatic tree's natural habitats in Karnataka and Kerala (Kumar, 2020). According to a study undertaken by R. Sundararaj and R. Raja Rishi of the Bengaluru-based Institute of Wood Science and Technology (IWST), the natural populations of sandal trees in Marayoor of Kerala and various reserve forests in Karnataka, including MM Hills, are heavily infected with the disease. Between 1% and 5% of sandalwood trees are lost annually due to this disease. Accordingly, scientists warn that it could wipe out the entire natural population, if measures are not taken to prevent its spread. They fear that any delay in arresting the trend may result in the disease spreading to cultivated sandalwood trees.

The disease was first reported in Kodagu, Karnataka in 1899. More than a million sandalwood trees were removed from the Kodagu and Mysuru region between 1903 and 1916, prompting the Maharaja of Mysuru to announce a reward of Rs. 10,000 in 1907 for anyone who was able to find a remedy. In the period 1917–1925, some 98,734 trees were extracted in Salem due to this disease. Such was the impact of this disease in Karnataka that the growing stock was reduced to around 25% of its initial level between 1980 and 2000.

The rapid spread of the infection was largely due to restrictions on green felling in forests, which allowed vectors to spread the disease to healthy trees, according to the IWST study. It is very difficult to identify the symptoms of sandal spike disease in the initial stage. It can only be detected when the tree becomes completely affected. It is significant to note that the price of Indian sandalwood and its oil has risen significantly since 1995 (at a rate of 20% annually), mainly due to the depletion of its production. The production of heartwood has decreased from 4000 tons in the 1930s to a mere 300 tons at present, with the prices correspondingly rising from ₹12,000 to ₹29,500 per kg.

Parthasarathi and Venkatesan (1982) reviewed the work done on sandal spike disease under various heads: symptoms of the disease, causal organism, collateral hosts of the pathogen, rank vegetation and disease incidence, vectors of the pathogen, culture and serology, physiology of the spiked sandal tree, and attempts to control the disease.

During the survey of a one-year-old sandalwood plantation in Hagalwadi village, Gubbi taluk, Tumkur district of Karnataka, Kumar and Joshi (2012) observed young plants showing symptoms of a sandal spike-like disease. The infected sandalwood plants had all the typical symptoms of sandal spike disease, such as small leaves, stunted height, the malformation of laminae and the affected leaves having a spike-like appearance when compared with the normal healthy sandalwood plant. The forest area showing disease symptoms was also invaded by a common weed *Stachytarpheta indica*, which also displayed similar disease symptoms. This was the first case of the manifestation of spike disease symptoms being observed in a plantation.

The fluorescent spots were detected exclusively in the phloem of diseased sandal tissues and not in healthy tissues. Ultrastructural studies using scanning electron microscopy confirmed the pleomorphic nature of sandal spike phytoplasma. Indirect enzyme linked immunosorbent assay (ELISA) was found to be very sensitive for detecting the pathogen in comparison with direct ELISA. In the case of indirect ELISA, one hour was found to be optimum for both antigen coating and incubation of antibody. The techniques could not detect phytoplasma in the healthy sandal plants and in witches' broom-affected *Zizyphus oenoplia*. The techniques confirmed that sandal spike phytoplasma is specific to sandal, and that it is not transmitted through the host plants. Sandal spike phytoplasma was not detected using ELISA tests in the salivary gland and intestine of the insect vector, *Redarator bimaculatus*, collected from diseased trees. Indirect ELISA tests detected infection in trees six months prior to the external expression of disease symptoms, confirming its potential for the early detection of disease through these tests (Balasundaran and Muralidharan, 2004).

Sundararaj *et al.* (2022) have published a review on the pathogenic diseases of Indian sandalwood, including sandal spike disease. The purpose of this review is mainly to inform the scientific community about reported deleterious pathogenic diseases of sandalwood in India, China, Australia, Bangladesh and their impact on the quality and yield of sandalwood.

FIGURE 9.49 Sandalwood branches and twigs showing typical symptoms of shoot spike, little leaf and discolouration of leaves.

9.60.1 Symptoms

Spike disease is characterised by an extreme reduction in leaf size accompanied by the stiffening and reduction of internode length. In the advanced stage, the entire shoot gives the appearance of a spike inflorescence. Spiked trees die within one or two years of the appearance of visible symptoms.

9.60.2 Causal Organism

'*Candidatus* Phytoplasma asteris' strains of 16SrI-B and 16SrXI-B phytoplasma subgroups

Sandal spike phytoplasma was first detected by electron microscopy in 1969 by three independent groups (Dijkstra and Ie, 1969; Hull *et al.*, 1969, Verma *et al.*, 1969). Khan *et al.* (2006) identified a subgroup 16SrI-B phytoplasma associated with sandal spike disease, marking the first definite identification of this subgroup.

Sandal spike phytoplasma is a pleomorphic microorganism and is the smallest organism capable of independent replication (i.e. it does not need a host). The pathogen is around 0.4 to 1.0 micrometres in diameter, and has a cell membrane, ribosome and DNA. The amino acids cysteine, methionine and tryptophan are absent in sandal spike phytoplasma.

Thomas and Balasundaran (1998) stained free-hand sections of spiked sandal and healthy sandal tissue with DNA-binding fluorochrome, 4, 6 diamidino-2-phenyl indole (DAPI), in order to detect the presence of sandal spike phytoplasma. Yellow-green fluorescence was detected in the phloem of the diseased sandal tissue while the phloem tissue of healthy sandal and of host plants growing close to the spike-affected sandal in the field, as well as in the glasshouse, did not show fluorescence. Intensity of fluorescence was higher in the young stem and inner bark tissue than in the root, petiole and leaf tissue. When the efficiency of DAPI stain was compared with that of Dienes' stain to detect phytoplasma, the latter was found to be less efficient.

Thomas and Balasundaran (1999) screened spike-affected sandal tissue to detect the presence of phytoplasma using the PCR technique. Oligonucleotide primers specific to the conserved region of 16S rRNA gene were used to amplify a 558 bp sequence of the phytoplasma. Four DNA fragments were obtained when the PCR products, after 20 cycles of amplification, were subjected to restriction fragment length polymorphism analysis with *Alu* I restriction endonuclease. The technique confirmed that sandal spike phytoplasma belongs to group I of the 11 major phytoplasma groups.

Selvaraj *et al.* (2023) reported that there are various biological, histopathological, electron microscopic and molecular methods available for the detection and identification of sandalwood spike phytoplasma. Biological and histopathological tests are time-consuming and ambiguous. Serological tests like ELISA and molecular tests like PCR are more versatile and accurate for the detection of phytoplasma. They reported that the application of various molecular techniques such as nested PCR, RFLP and the sequence analysis of 16S ribosomal RNA led to identify the sandalwood spike disease (SSD) phytoplasma isolates as a member of 16SrI-B and 16SrXI-B phytoplasma subgroups. In this publication, various diagnostic methods used for the detection and identification of SSD phytoplasma are summarised.

Thomas and Balasundaran (2001) reported a differential filtration technique for the purification of sandal spike phytoplasma. The purity was assessed using scanning and transmission electron microscopy. Polyclonal antibodies raised against the purified phytoplasma were used to detect this pathogen using a double diffusion test, an enzyme-linked immunosorbent assay (ELISA) and a dot immunobinding assay (DIBA). The size of the antigenic protein was revealed by immunoblotting (Western blotting). Inner bark and stem tissues from diseased branches were utilised for the purification of phytoplasma.

In the case of transmission electron microscopy, the technique of Jiang and Chen (1987) was used and so it was for DIBA. Electrophoresis was done following the protocol of Saeed *et al.* (1992). The immunoblotted membrane was probed with phytoplasma-specific antibody as in DIBA and the position of the band was calculated with respect to the marker proteins.

Scanning electron micrographs of partially purified pellets of phytoplasma prior to cross-absorption with antiserum against healthy plant extract showed plant debris among phytoplasma cells, whereas impurities were sparse when the pellets were subjected to cross absorption.

DIBA was highly specific and rapid since in this technique, red-coloured dots were visualised on the membrane within three hours, but the sensitivity of the test was very low compared to ELISA. The use of immunogenic probes at 1:500, 1:1000 and 1:2000 dilutions could help to detect the presence of the antigen, but 1:500 dilution was found to be more sensitive when detecting the antigen. One advantage of DIBA is that a large number of samples can be processed within two to three hours rather than a day required for conventional ELISA procedures; however, the test was found to be less sensitive when compared to ELISA. Washing the membrane with deionized water rather than using PBS or TBS prior to the application of the sample was found to be superior. The technique could detect > 1:100 dilution of purified phytoplasma, the lower sensitivity of the technique to detect phytoplasma in crude extract may be attributed to the contaminating plant proteins.

Sandal spike phytoplasma antigenic protein has a molecular weight of 14,000 Daltons and could be detected in both purified and semi-purified phytoplasma, but the same was absent in healthy sandal. The silver binding method of protein estimation is more sensitive than the conventional Lowry method, probably because of the low concentration of aromatic amino acids.

9.60.3 Disease Cycle

Phytoplasma-causing sandal spike exhibits two types of spread: (1) The sporadic outbreaks affecting isolated patches of sandal far removed from any apparent sources of infection; and(2) Secondary attacks responsible for the radial spread of the disease around the centre of primary infection (Sreenivasaya and Rangaswami, 1934).

Spike disease phytoplasma was shown to be insect-transmitted as early as 1934. Insect vectors found responsible for it are *Moonia albimaculata*, *Coelidia indica* and *Nephotettix virescens*. Sen-Sarma (1978) worked out biology, host plants and transmission by these vectors. The population of *N. virescens* is the highest in October–November. It primarily breeds on rice and gramineous hosts.

Ghosh *et al.* (1985) suspected *Zizyphus oenoplia, Stachytarpheta* sp. and *Jasminum rigidum*, showing witches' broom symptoms at several places, to be the collateral hosts of the pathogen. *Redarator bimaculatus* was identified as an insect vector of the pathogen through transmission studies, and by the presence of MLO in the tissues of the intestine and salivary glands of the insects fed on spiked sandal, and in the phloem tissues of plants infected through the vector.

Shafiullah and Naik (2003) found that there is a positive correlation between the population of *Coelidia indica* and the incidence of spike disease. Regression equation (Y= 3.232 + 1.62 X) fit for this relation suggested that every 1% increase in the population of *C. indica* resulted in a 1.62% increase in spike disease incidence.

9.60.4 Management

1. Rao *et al.* (1975) found that spiked plants of sandal showed recovery in disease symptoms 25–30 days after treatment with benlate (1 g per tree) + terramycin (300 mg per tree) or terramycin alone; and 30–35 days with benlate + ledermycin (300 mg per tree) or ledermycin alone. However, it took longer time with treatments with achromycin (300 mg per tree), aureomycin (300 mg per tree) or benlate. The recovery lasted for 120 to 150 days in mixed treatments whereas when treated with ledermycin or terramycin alone it lasted 50–60 days, with benlate 80–90 days and with achromycin or aureomycin 30–40 days. Treatment with Bavistin, Plantvax and Vitavax had no effect on disease control. The treatment with benlate + terramycin was the best of all the treatments.
2. Ghosh *et al.* (1985) observed that infusion of aqueous solution of tetracycline antibiotics, 1 g per 1 L of water, in spiked trees gave remission

of disease symptoms lasting for three to five months. Higher doses of tetracyclines i.e., 2 to 8 g/tree, prolonged the remission period up to seven to eight months.
3. Raychaudhuri *et al.* (1972) reported that the application of dimethyl chlortetracycline hydrochloride by the girdling method arrested the disease and led to the complete recovery of spiked shoots after around 60 days.

REFERENCES

Balasundaran, M. and Muralidharan, E. M. (2004). Development of spike disease resistant sandal seedlings through biotechnology involving ELISA technique and tissue culture. Kerala Forest Research Institute, Trichur, Report, 64 pp.

Dijkstra, J. and Ie, T. S. (1969). Presence of mycoplasma-like bodies in the phloem of sandal affected with spike disease. *Neth. J. Plant Pathol.* 75 (6): 374–78.

Ghosh, S. K., Balasundaran, M. and Mohamed Ali, M. I. (1985). Studies on the spike disease of sandal. Kerala Forest Research Institute Peechi, Thrissur, Report, 56 pp.

Hull, R., Horne, T. W. and Nayar, R. M. (1969). Mycoplasma-like bodies associated with sandal spike disease. *Nature* 224: 1121–22.

Jiang, Y. P. and Chen, T. A. (1987). Purification of mycoplasma-like organisms from lettuce with aster yellows disease. *Phytopathology* 77: 949–53.

Khan, J. A., Srivastava, P. and Singh, S. K. (2006). Identification of a '*Candidatus* Phytoplasma asteris'-related strain associated with spike disease of sandal (*Santalum album*) in India. *New Dis. Reports* 12: 44, (BSPP).

Kumar, A. N. A. and Joshi, G. (2012). Incidence of Sandal spike symptoms in a one-year-old plantation in Karnataka. *Curr. Sci.* 103: 613–14.

Kumar, B. S. S. (2020). Invisible killer threatens India's sandal wood forests. *The Hindu, Bengaluru*, September 28, 2020.

Parthasarathi, K. and Venkatesan, K. R. (1982). Sandal spike disease. *Curr. Sci.* 51 (5): 225–30.

Rao, P. S., Srimathi, R. A., Nag, K. C., Raychaudhuri, S. P., Gosh, S. K. *et al.* (1975). Response of spike disease of sandal to mixed treatment with antibiotics and fungicides. *Proc. Indian Natl. Sci. Acad.* 41, B, (4): 340–42.

Raychaudhuri, S. P., Chenulu, V. V., Ghosh, S. K., Varma, A., Rao, P. S. *et al.* (1972). Chemical control of spike disease of Sandal. *Curr. Sci.* 41 (2): 72–73.

Saeed, E., Rage, P. and Cousin, M. T. (1992). Determination of the antigenic protein size associated with faba bean phyllody MLO using SDS-PAGE electrophoresis and immunotransfer. *J. Phytopath.* 136: 1–8.

Selvaraj, V., Suryanarayana, V., Venkataravanappa, V., Rao, G. P., & Mandal, B. (2023). Diagnosis of Phytoplasma Associated with the Sandalwood Spike Disease. *Indian J. Entomol.* 85 (4): 1115–1126. https://doi.org/10.55446/IJE.2022.456

Sen-Sarma, P. K. (1978). Sandal spike disease – A vector-transmitted disease. *Proc. Indian Natl. Sci. Acad.* B 46 (6): 817–21.

Shafiullah, B. and Naik, S. T. (2003). Association of some host plants and insects with sandal spike disease. *Indian Forester* 129 (3): 393–400.

Sreenivasaya, M. and Rangaswami, S. (1934). Field studies in the spike-disease of sandal (*Santalum album* Linn.). I. Observations on the natural dissemination of spike. *Proc. Indian Acad. Sci.* Sec. B, 1 (4): 143–54.

Sundararaj, R., Swetha, P. and Rishi, R. R. (2022). Pathogenic diseases of Indian sandalwood (*Santalum album* L.), a review. *J. Plant Pathol.* 104: 1321–36. https://doi.org/10.1007/s42161-022-01208-w

Thomas, S. and Balasundaran, M. (1998). *In situ* detection of phytoplasma in spike-disease-affected sandal using DAPI stain. *Curr. Sci.* 74 (11): 989–93.

Thomas, S. and Balasundaran, M. (1999). Detection of sandal spike phytoplasma by polymerase chain reaction. *Curr. Sci.* 76 (12): 1574–76.

Thomas, S. and Balasundaran, M. (2001). Purification of sandal spike phytoplasma for the production of polyclonal antibody. *Curr. Sci.* 80 (12): 1489–94.

Verma, A., Chenulu, V. V., Raychaudhuri, S. P., Prakash, N. and Rao, P. S. (1969). Mycoplasma-like bodies in tissue infected with sandal spike and brinjal little leaf. *Indian Phytopath.* 22: 289–91.

9.61 BACTERIAL BLIGHT OF MULBERRY

Bacterial blight of mulberry (*Morus alba* L.) was first reported from the United States in 1983. Prior to this date, it had been incompletely described by Boyer and Lambert from France in 1893. This is the earliest record of a disease caused by a *Pseudomonas* species. The occurrence of the disease in Japan was confirmed in 1901. The disease has also been reported from the eastern Anatolia region of Turkey, where its incidence was nearly 100% in spring 1999 (Sahin *et al.*, 1999). The disease also occurs in Korea DPR, the Republic of Korea, China, Brazil, Bulgaria, Canada, Australia, Germany, Hungary, New Zealand, Romania, South Africa, Uganda, the UK, Iran, Italy, Pakistan and India. In India, the disease is widespread in Karnataka state, where mulberry is widely grown due to the presence of the sericulture industry. The disease incidence was 45.9% and the percentage of infested gardens ranged from 80–85 in Kollegal area of Karnataka state in the rainy season of 1998 (Sharma *et al.*, 2000).

Initially, the symptoms of the disease appear in the cool, wet weather of the spring season, but the severity of the disease increases in the rainy season during June to July. The severity of the disease decreases after the rainy season, but it appears again in autumn in a mild form.

Bacterial leaf spot incited by *Xanthomonas campestris* pv. *mori* is another devastating foliar disease of mulberry reported globally (Banerjee *et al.* 2012). Allahverdi *et al.* (2016) reported bacterial canker of white mulberry caused by *Citrobacter freundii* from northwestern Iran. The symptoms of disease were dark brown to black, elongated lesions on the trunk. Dead barks were easily removed from the underlying tissues and a dark brown gum exuded from the shoot and trunk cankers.

Krawczyk and Łochyńska (2020) observed mulberry trees having chlorotic angular leaf spots and blotches, in one of the Polish mulberry plantations. The disease intensity increased in spring under humid conditions and it spread to developing blossoms and leaves. Diseased leaves were collected during the growing seasons of 2017 and 2018 from 18-year-old trees cultivated at the Experimental

Farm of the Institute of Natural Fibres and Medicinal Plants in Petkowo, Poland (52°12′40″N 17°15′31″E). Leaf lesions exhibited bacterial streaming and bacterial colonies were readily isolated on tryptic soy agar. Biochemical (Biolog Gen III system v. 2.8.0), molecular (16S rDNA and *gyr*B genes) and pathogenicity testing in the greenhouse identified the causal agent as *Pseudomonas syringae*. To their knowledge, this is the first description and characterisation of *Psm. syringae* affecting mulberry trees not just in Poland but throughout Europe.

Hseu *et al.* (2018) noticed an unknown disease on the commercial orchards of fruit mulberry (*Morus alba*) in Yuan shan, Yi-lan, which had been occurring since 2013. The symptoms began with small water-soaked lesions with haloes on the edge of the leaves, before they gradually coalesced and caused leaf deformation. The infected fruits also showed black flecks initially, flecks merged gradually and fruits deformed and dropped. In severe cases, twig tips wilted and dried up. The causal agent was identified as *Pseudomonas* species using traditional physiological and biochemical tests. PCR tests indicated that tested isolates could generate a specific 144 bp fragment by using PsyF/PsyR primer pair specific to the *Psm. syringae* group. In addition, LOPAT and Biolog tests indicated that the bacterium is within *Psm. syringae* group, especially similar to *Psm. syringae* pv. *mori*. In advance, the bacterium has proved to have 99% sequence similarity when compared with *Psm. syringae* pv. *mori* (AB001446.1) in the gene bank by using 16S rDNA analysis. Furthermore, a preliminary chemical screening using ten commercial formulated agrochemicals found that streptomycin + tetracycline, thiophanate methyl + streptomycin and Kasugamycin + copper oxychloride showed better control efficacy *in vitro*.

Basavand *et al.* (2021) observed a wet wood disease on *Morus* trees (Black mulberry, White mulberry and Red mulberry), in the municipal lands and parks of Rafsanjan city (Kerman Province), Iran, in the summer and spring of 2016. The typical symptoms were wet wood symptoms characterised by canker, stem weeping and liquid exudates on the trunks of infected trees. The causal agent of the disease, based on its biochemical and phenotypic feature, was identified as *Klebsiella oxytoca* and this was confirmed using sequence analysis of *atpD* gene. Artificial inoculation of the predominant isolated bacterial strain on one-year-old mulberry seedlings, the necrotic symptoms were observed on the inner bark of inoculated seedlings. After fulfilling the Koch's postulates, the re-isolated bacterium was identified as *Klebsiella oxytoca* based on colony morphology, phenotypic and biochemical tests. The characteristics of *Kl. oxytoca* strains as the causal agent of wet wood disease of Red mulberry (*M. rubra*), White mulberry (*M. alba*) and Black mulberry (*M. nigra*) are described for the first time.

9.61.1 Host Range

Mulberry is the only naturally infected host. In addition to *M. alba*, *M. alba* var. *taratarica*, *M. bombycis* (Japanese mulberry), *M. kagayamae*, *M. latifolia*, *M. nigra* (black mulberry) and *M. ruba* (red mulberry) have been reported as natural hosts of *Psm. syringae* pv. *mori* (Janse, 2005). With artificial inoculation, the bacterium also infects kidney beans (Goto, 1992).

FIGURE 9.50 Bacterial blight of mulberry. (Courtesy: William M. Brown Jr., Bugwood.org)

9.61.2 Symptoms

The symptoms appear on leaves, twigs and buds. On leaves, small, water-soaked, irregular, dark green, sunken spots appear. The spots turn brown to black with yellow borders and translucent edges as these age. The spots enlarge, become angular in shape and often fall out. Elongated, dark brown and slightly sunken spots appear on large veins, midribs and petioles of the leaves. The occurrence of these spots causes the curling and distortion of the leaves due to the unbalanced growth of lamina. Diseased leaves at the tips of twigs wilt and, later die. Infected buds may become disfigured as these swell.

Dark brown to black, elongated, more or less sunken streaks appear on twigs. The streaks may coalesce to girdle the shoots, resulting in the die-back symptoms. In humid weather, creamy-yellowish bacterial ooze exudes from the lenticels of the infected parts. Severely infected young trees often appear stunted.

9.61.3 Causal Organism

Pseudomonas syringae pv. *mori* (Boyer & Lambert 1893) Young *et al.*, 1978

Synonym: *Psm. mori*

Cells are Gram-stain-negative, aerobic rods, measure 0.9–1.3 by 1.8–4.5 μm and motile with a polar flagellum. Green fluorescent pigment is produced on King's medium B. Colonies on agar media are white, slow-growing, smooth, flat with entire margins and becoming undulate. Tests for gelatin liquefaction, nitrate reduction, hydrogen sulphide and oxidase production are negative while those for catalase and levan production are positive. Milk becomes clear with alkaline reaction. Arbutin is hydrolysed and D-gluconate, 2-ketoglutarate, inositol, mannitol, sorbitol and trigonelline are used as a sole source of carbon. DL-lactate, L-tartrate, quinate, erythritol and DL-homoserine are not used as a sole source of carbon. On the basis of fatty acid methyl ester analysis by gas chromatography, the bacterium confirms to *Psm. syringae* identification, with a similarity index ranging from 0.82–0.94 (Sahin *et al.*, 1999).

Sharma *et al.* (2000) reported that both *Psm. syringae* pv. *mori* and *Xanthomonas campestris* pv. *mori* were responsible for causing bacterial blight of mulberry and they also reported the synergistic effect of both the bacteria on disease development. There is a need to study the diseases caused by these bacteria individually and to distinguish these diseases on the basis of symptoms. Another report by Nagaraj *et al.* (2001), reporting *Xan. campestris* pv. *moricola* as the cause of bacterial blight of mulberry from Karnataka, India has further added to the confusion regarding the causal agent(s) of the disease. *Xan. campestris* pvs. *mori* and *moricola* are not included in the validly published names of plant pathogenic bacteria enlisted by Young *et al.* (2004). However, Bull *et al.* (2010) have included *Xan. campestris* pv. *mori* in the comprehensive list of names of plant pathogenic bacteria published during 1980 to 2007.

9.61.4 Disease Cycle

The bacterium survives in the cankers formed on the twigs. Pruned shoots and fallen leaves on ground, if not collected and destroyed, also serve as a source of inoculum. The bacterium is not transmitted through seed. Bacterial cells oozing from twig lesions are disseminated by wind-splashed rain. Secondary spread of the bacterium occurs through rain from freshly formed lesions.

The increased severity of the disease during the rainy season is due to the high humidity and dissemination of the bacterium caused by wind-splashed rain. Gangwar and Thangavelu (1995) found a highly significant positive correlation between disease severity and rainfall and relative humidity, but a non-significant correlation with minimum and maximum temperatures. Close spacing of trees, high doses of nitrogenous fertilisers, poor field sanitation and the wounding of trees also increase the severity of the disease.

Among the indigenous mulberry varieties, grown in India, S 3, S 13, S 14 and S 1531 are highly resistant and K 2, MR 2, S 25, S 34, S 54, S 1, CSRS 1, Sujanpur and SLM 1 are resistant to the pathogen. Some of the exotic germplasms, namely, English Black and *M. australis*, are also resistant (Gangwar and Thangavelu, 1995).

Sato *et al.* (1982) reported that a study of 17 isolates did not reveal apparent correlations between plasmid pattern and toxigenicity, phage sensitivity or other phenotypes tested. All strains tested accepted freely the non-conjugative plasmid RSF1010 through mobilisation by naturally occurring conjugative plasmid pBPW1 of *Psm. syringae* pv. *tabaci* Br2, but re-transferred RSF1010 into *Escherichia coli* at very low frequency. Therefore, this indicates that these possess an indigenous fertility system which is either naturally repressed or mobilises RSF1010 with low efficiency.

9.61.5 Management

1. Use of resistant varieties and healthy planting material are the most effective management practices.
2. Minimise wounds to the limbs and new shoots.
3. Prune the affected shoots and branches in the late dormant season and destroy these. Pruning should be done back to healthy wood and the pruning tools should be disinfected with 70% alcohol or 10% bleach between cuts.
4. Do not apply excessive doses of nitrogenous fertilisers.
5. Plant the trees at proper spacing to provide good air circulation.
6. Spray Kocide 2000 T/N/O @ 0.75–3 lb/A as a dormant spray and @ 0.75–2 lb/A when new growth starts. The application of Kocide is not recommended on trees grown for edible fruit.

REFERENCES

Allahverdi, T., Rahimian, H. and Ravanlou, A. (2016). First report of bacterial canker in mulberry caused by *Citrobacter freundii* in Iran. *Plant Dis.* 100: 1774.

Banerjee, R., Das, N. K., Doss, S. G., Saha, A. K., Bajpai, A. K. *et al.* (2012). Narrow sense heritability estimates of bacterial leaf spot resistance in pseudo F2(F1) population of mulberry (*Morus* spp.). *Eur. J. Plant Pathol.* 133: 537–44.

Basavand, E., Khodaygan, P., Rahimian, H., Ganjeh, A. Molaei, S. *et al.* (2021). Identification and characterization of *Klebsiella oxytoca* strains associated with wetwood disease of *Morus* trees. *Indian Phytopathol.* 74: 1123–27. https://doi.org/10.1007/s42360-021-00360-0

Boyer, G. and Lambert, F. (1893). Sur deux nouvelles maladies du murier. *C. R. Hebd. Séanc. Aced. Sci.* 117: 342–43.

Bull, C. T., De Boer, S. H., Denny, T. P., Firrao, G., Fischer-Le Saux, M. *et al.* (2010). Comprehensive list of names of plant pathogenic bacteria, 1980–2007. *J. Plant Pathol.* 92: 551–92.

Gangwar, S. K. and Thangavelu, K. (1995). Varietal and seasonal incidence of bacterial leaf blight disease of mulberry caused by *Pseudomonas mori* (Boyer *et* Lambert) Stevens. *Indian J. Sericulture* 34: 110–13.

Goto, M. (1992). *Fundamentals of Bacterial Plant Pathology.* Academic Press, Inc. San Diego, CA, 342 pp.

Hseu, S. H., Sung, C. J., Lee, H. Y., Tsai, S. F., Hsu, S. Y. *et al.* (2018). Bacterial blight disease of mulberry caused by *Pseudomonas syringae* pv. *mori*. *J. Plant Med.* 60: 41–47.

Janse, J. D. (2005). *Phytobacteriology: Principles and Practice.* CABI Publishing, Wallingford, 360 pp.

Krawczyk, K. and Łochyńska, M. (2020). Identification and characterization of *Pseudomonas syringae* pv. *mori* affecting white mulberry (*Morus alba*) in Poland. *Eur. J. Plant Pathol.* 158: 281–91. https://doi.org/10.1007/s10658-020-02074-x
Nagaraj, M. S., Srinivasachary and Khan, A. N. A. (2001). Studies on the management of *Xanthomonas campestris* pv. *moricola* of mulberry. *J. Mycopathol. Res.* 39: 87–90.
Sahin, F., Kotan, R. and Donmez, M. F. (1999). First report of bacterial blight of mulberries caused by *Pseudomonas syringae* pv. *mori* in the eastern Anatolia region of Turkey. *Plant Dis.* 83: 1176.
Sato, M., Staskawicz, B. J. and Panopoulos, N. J. (1982). Indigenous plasmids of *Pseudomonas syringae* pv. *mori*, the causal agent of bacterial blight of mulberry. *Ann. Phytopath. Soc.* Japan 48: 27–33.
Sharma, D. D., Baqual, M. F., Gupta, V. P. and Chandrashekar, D. S. (2000). A survey on the occurrence of bacterial blight disease complex in mulberry. *Indian J. Sericulture* 39: 113–16.
Young, J. M., Bull, C. T., De Boer, S. H., Firrao, G., Saddler, G. E. *et al.* (2004). Names of plant pathogenic bacteria, 1864–2004. http://www.isppweb.org/about_tppb.asp.
Young, J. M., Dye, D. W. and Wilkie, J. P. 1978. Genus *Pseudomonas* Migula 1894. In: Young, J. M., Dye, D. W., Bradbury, J. F., Panagopoulos, C. G., and Robbs, C. F. 1978. A proposed nomenclature and classification for plant pathogenic bacteria. *N. Z. J. Agric. Res.* 21: 153–77.

9.62 ANGULAR LEAF SPOT OF *PEEPAL*

Peepal or *Pipal* (*Ficus religiosa* L.) is a tree, which is generally grown for shade in India. It is also a religious and sacred tree and is worshiped in the country. An angular leaf spot of bacterial nature was first observed on this tree in 1968 at the Indian Agricultural Research Institute, New Delhi, India. Later on, it was also observed in Uttar Pradesh, India (Jindal and Patel, 1972a) and now, it also occurs in other states such as Punjab, Haryana etc. According to Bradbury (1986), this disease has been reported only from India (Delhi and nearby places), Italy (Cavara's original description) and possibly from Turkey.

Duff (1991) observed a severe leaf drop on *F. microcarpa* Lf. var. *hillii* Corner in a Darwin nursery. Yellow to brown angular lesions of varying sizes were observed on fallen leaves and also on leaves that were still attached to the trees. Other plant species found to exhibit the same symptoms included *F. microcarpa* var. *hillii* cv. Variegated and *Ficus* sp. Severely infected plants shed their leaves, with subsequent new growth remaining free from any lesions for only a short period of time. Symptoms were more pronounced during the spring and summer due to the wet season, while they declined in winter due to the dry season.

9.62.1 Host Range

Jindal and Patel (1972a) found that on artificial inoculation, *F. infectoria* and *F. glomerata* were as susceptible as the natural host, *F. religiosa*; whereas *F. carica, F. bengalensis* and *Morus alba* showed small, water-soaked spots, which quickly turned necrotic and did not expand.

FIGURE 9.51 Angular leaf spot of peeple: Symptoms on upper leaf surface (left) and on lower leaf surface (right).

Hence, these are not considered susceptible hosts. Duff (1991) reported *F. microcarpa* var. *hillii, F. microcarpa* var. *hillii* cv. Variegated and *Ficus sp.* as hosts of this bacterium.

9.62.2 Symptoms

Numerous, water-soaked, translucent, angular spots appear on the infected leaves. Later on, these spots turn brown, become necrotic and are surrounded by chlorotic haloes. Several spots coalesce, covering large areas on the leaf lamina and may even cover the entire leaf (Figure 9.51). Such leaves soon become dry and wilt. In severely infected trees, all of the leaves may be generally infected. Occasionally, the infection spreads to leaf veins and petioles. Light brown to dark brown, small lesions may also appear on the twigs.

9.62.3 Causal Organism

Xanthomonas campestris pv. *fici* (Cavara 1905) Dye 1978

Synonym: *Xan. fici*

Jindal and Patel (1972a) reported the bacterium for the first time from *F. religiosa* and named it as *Xanthomonas fici* sp. nov. However, from the recent name of the bacterium given above, it appears that Cavara (1905) had described or named this bacterium at an earlier date. The description of the bacterium given below is taken from Jindal and Patel (1972b).

Cells are Gram-stain-negative rods having rounded ends; usually occur singly but occasionally in pairs, non-encapsulated, non-sporing, not acid-fast and motile with a single polar flagellum. The bacterium is strictly aerobic.

The colonies on nutrient agar are circular, raised, glistening and yellow. The bacterium utilises glucose oxidatively, hydrolyses starch, liquefies gelatin, produces H_2S

and ammonia but not indole, acetoin and urease. Asparagine is not utilised as a sole source of carbon and nitrogen. Milk is turned alkaline and peptonised without coagulation, litmus is reduced and methyl red test is negative. Growth is retarded by 2.0% and suppressed by 3.0% sodium chloride. Catalase is produced. Acid but no gas is produced from mannose, sucrose, glucose, galactose, arabinose, cellobiose, lactose, trehalose, fructose, xylose, raffinose, melibiose, glycogen and dextrin. Neither acid nor gas is produced from *meso*-erythrit, ribose, sorbose, rhamnose, sorbitol, adonitol, dulcitol, inositol, mannitol, inulin and salicin. The bacterium utilises sodium propionate and succinate but not oxalate and tartrate. Unlike other *Xanthomonas* spp., the bacterium reduces nitrate to nitrite.

9.62.4 Disease Cycle

As all the leaves on *peepal* trees are shed in the winter, the bacterium probably survives in lesions formed on branches, growing buds or leaf scars.

Generally, the disease does not appear on new leaves developing in the summer months (April to June) until the onset of the monsoon. It indicates that high humidity and spattering rains are required for the multiplication and dissemination of the pathogen. The bacterium enters through stomata and remains confined largely to the parenchymatous tissue (Jindal and Patel, 1972a).

9.62.5 Management

No information is available on the management of the disease. As the disease is largely parenchymatous, it can be effectively controlled by 2–3 sprays of Streptocycline @ 100 μg mL^{-1}.

REFERENCES

Bradbury, J. F. (1986). *Guide to Plant Pathogenic Bacteria*. CAB International Mycological Institute, Kew, 220 pp.

Cavara, F. (1905). Bacteriosi del fico. Atti della Accademia Gioenia di Scienze Naturali, Catania 18: memo 14, pp. 1–17.

Duff, J. (1991). Angular leaf spot of ornamental *Ficus* spp. due to *Xanthomonas campestris* pv. *fici*. *Australasian Plant Pathol.* 20: 1–2.

Dye, D. W. (1978). Genus IX *Xanthomonas* Dowson 1939. In: Young, J. M., Dye, D. W., Bradbury, J. F., Panagopoulos, C. G. and Robbs, C. F. A proposed nomenclature and classification for plant pathogenic bacteria. *N. Z. J. Agric Res.* 21: 162–66.

Jindal, J. K. and Patel, P. N. (1972a). Angular leaf spot of pipal (*Ficus religiosa*) due to *Xanthomonas fici* sp. nov. *Indian Phytopathol.* 25: 164–66.

Jindal, J. K. and Patel, P. N. (1972b). Angular leaf spot of pipal. In: P. N. Patel (ed.). *Plant Bacteriology*. Vol. I, *Bacterial Diseases of Plants in India*. Indian Agricultural Research Institute, New Delhi, pp. 245–46.

Index

Y

For Product Safety Concerns and Information please contact our EU representative GPSR@taylorandfrancis.com Taylor & Francis Verlag GmbH, Kaufingerstraße 24, 80331 München, Germany